SOLUBILITIES OF
INORGANIC AND ORGANIC COMPOUNDS

SOLUBILITIES OF INORGANIC AND ORGANIC COMPOUNDS

VOLUME 3

Ternary and Multicomponent
Systems of Inorganic Substances

PART 3

Edited by
Dr. HOWARD L. SILCOCK

PERGAMON PRESS

OXFORD · NEW YORK · TORONTO · SYDNEY · PARIS · FRANKFURT

U.K.	Pergamon Press Ltd., Headington Hill Hall, Oxford OX3 0BW, England
U.S.A.	Pergamon Press Inc., Maxwell House, Fairview Park, Elmsford, New York 10523, U.S.A.
CANADA	Pergamon of Canada, Suite 104, 150 Consumers Road, Willowdale, Ontario, M2J 1P9, Canada
AUSTRALIA	Pergamon Press (Aust.) Pty. Ltd., P.O. Box 544, Potts Point, N.S.W. 2011, Australia
FRANCE	Pergamon Press SARL, 24 rue des Ecoles, 75240 Paris, Cedex 05, France
FEDERAL REPUBLIC OF GERMANY	Pergamon Press GmbH, 6242 Kronberg-Taunus, Pferdstrasse 1, Federal Republic of Germany

First edition 1979

British Library Cataloguing in Publication Data

Solubilities of inorganic and organic compounds.
Vol.3: Ternary and multicomponent systems of inorganic substances
1. Solubility - Tables
I. Silcock, H II. Akademii nauk SSSR
541'.342'0212 QD543 79-40319
ISBN 0-08-023570-0
ISBN 0-08-023599-9 Set of 3 Vols in 7

Printed in Great Britain by A. Wheaton & Co. Ltd., Exeter

PART 3

Ca_2CO_3—MeX—H_2O

$t = 25$

Salt		Concn. in aqueous soln., g/l	
Name	Formula	Salt	Calcium Carbonate
Sodium Chloride	NaCl	0.0365	0.0161
»	»	0.0456	0.0161
»	»	0.0608	0.0179
»	»	0.0729	0.0179
»	»	0.0912	0.0179
»	»	0.182	0.0241
»	»	0.243	0.0250
»	»	0.365	0.0286
»	»	0.729	0.0520
Sodium Bromide	NaBr	0.103	0.0116
»	»	0.129	0.0125
»	»	0.171	0.0152
»	»	0.206	0.0161
»	»	0.257	0.0171
»	»	0.514	0.0205
»	»	0.686	0.0234
»	»	1.029	0.0250
»	»	2.058	0.0409
Sodium Chlorate	$NaClO_3$	0.106	0.0170
»	»	0.133	0.0179
»	»	0.177	0.0161
»	»	0.213	0.0161
»	»	0.266	0.0161
»	»	0.532	0.0250
»	»	0.710	0.0277
»	»	1.064	0.0320
»	»	2.129	0.0634
Sodium Nitrate	$NaNO_3$	0.085	0.0161
»	»	0.106	0.0161
»	»	0.141	0.0179
»	»	0.170	0.0179

Table № 3739 (continuation)

Salt		Concn. in aqueous soln., g/l	
Name	Formula	Salt	Calcium Carbonate
Sodium Nitrate	$NaNO_3$	0.212	0.0193
»	»	0.425	0.0286
»	»	0.567	0.0304
»	»	0.850	0.0345
»	»	1.700	0.0678
Potassium Chloride	KCl	0.0746	0.0143
»	»	0.124	0.0150
»	»	0.186	0.0170
»	»	0.374	0.0241
»	»	0.497	0.0250
»	»	0.746	0.0268
»	»	1.491	0.0320
Potassium Bromide	KBr	0.119	0.0152
»	»	0.198	0.0161
»	»	0.595	0.0214
»	»	1.190	0.0241
»	»	2.380	0.0417

№ 3740 [2062]

SODIUM CHLORIDE–CALCIUM CARBONATE –WATER

$NaCl—CaCO_3—H_2O$

$t = 100$

Content in aqueous soln.		Comp. of solid phase
Sodium Chloride, g/l	Calcium Oxide CaO, mol/l	
17.5	0.00060	$CaCO_3$
2.5	0.00045	»
0.5	0.00040	»
0	0.000375	»

№ 3741 **SODIUM CHLORIDE–CALCIUM CARBONATE–WATER** [824]

$NaCl—CaCO_3—H_2O$

Content in aqueous soln.		t	pH	Comp. of solid phase
Sodium Chloride wt %	Calcium Carbonate, mg-eq/l			
25	1.64	16—18	8.0	$CaCO_3$
20	2.31	16—18	7.9	»
15	2.72	16—18	8.1	»
8	3.27	16—18	8.2	»
3.5	3.17	16—18	8.1	»
1	2.80	16—18	8.0	»

Table № 3741 (continuation)

Content in aqueous soln.		t	pH	Comp. of solid phase
Sodium Chloride wt %	Calcium Carbonate, mg–eq/l			
25	1.55	20—30	8.3	$CaCO_3$
20	1.95	20—30	8.1	»
15	2.34	20—30	8.3	»
8	2.85	20—30	8.4	»
3.5	2.80	20—30	8.4	»
1	2.40	20—30	8.3	»

№ 3742 SODIUM CHLORIDE–CALCIUM SULFATE–WATER [1315]

$NaCl—CaSO_4—H_2O$

Content in aqueous soln., g/l		t	Comp. of solid phase
Sodium Chloride	Calcium Sulfate		
139.9	7.4	15	$CaSO_4 \cdot 2H_2O$
51.4	5.6	15	»
31.1	4.8	15	»
10.6	3.7	15	»
5.1	3.1	15	»
1.1	2.5	15	»
0.6	2.3	15	»
315.55	5.37	23	$NaCl + CaSO_4 \cdot 2H_2O$
306.40	5.68	23	$CaSO_4 \cdot 2H_2O$
229.70	7.03	23	»
197.20	7.25	23	»
129.50	7.50	23	»
75.58	6.74	23	»

№ 3743 SODIUM CHLORIDE–CALCIUM SULFATE–WATER [657]

$NaCl—CaSO_4—H_2O$

$t = 20$

Content in aqueous soln., wt %		Comp. of solid phase	Content in aqueous soln., wt %		Comp. of solid phase
Sodium Chloride	Calcium Sulfate		Sodium Chloride	Calcium Sulfate	
0.5	0.2889	$CaSO_4 \cdot 2H_2O$	5	0.5574	$CaSO_4 \cdot 2H_2O$
1	0.3365	»	10	0.6538	»
2	0.4260	»	15	0.6422	»
3	0.4572	»	20	0.5899	»
4	0.5277	»			

№ 3744 SODIUM CHLORIDE–CALCIUM SULFATE–WATER [1477]

$NaCl—CaSO_4—H_2O$

$t = 14$

Content in aqueous soln., g/l		Comp. of solid phase	Content in aqueous soln., g/l		Comp. of solid phase
Sodium Chloride	Calcium Sulfate		Sodium Chloride	Calcium Sulfate	
292.6	5.30	$CaSO_4 \cdot 2H_2O$	102.3	6.90	$CaSO_4 \cdot 2H_2O$
263.2	5.50	»	87.75	6.58	»
234.0	5.90	»	58.50	5.72	»
204.7	6.30	»	29.25	4.40	»
175.6	6.80	»	14.62	3.68	»
160.8	7.00	»	11.70	3.41	»
146.2	7.10	»	5.850	2.79	»
131.6	7.20	»	2.925	2.32	»
117.0	7.10	»	0	1.70	»

№ 3745 SODIUM CHLORIDE–CALCIUM SULFATE–WATER [1404]

$NaCl—CaSO_4—H_2O$

$t = 14$

Content in aqueous soln., g/l		Comp. of solid phase	Content in aqueous soln., g/l		Comp. of solid phase
Sodium Chloride	Calcium Sulfate		Sodium Chloride	Calcium Sulfate	
31.3	15.83	$CaSO_4 \cdot 2H_2O$	47.7	8.26	$CaSO_4 \cdot 2H_2O$
231.5	12.75	»	24.4	6.35	»
142.2	11.93	»	0	2.0	»
95.0	10.56	»			

№ 3746 SODIUM CHLORIDE–CALCIUM SULFATE–WATER [1505]

$NaCl—CaSO_4—H_2O$

$t = 25$

Content in aqueous soln., mol/l		Comp. of solid phase	Content in aqueous soln., mol/l		Comp. of solid phase
NaCl	$CaSO_4$		NaCl	$CaSO_4$	
5.00	0.0389	$CaSO_4 \cdot 2H_2O$	1.00	0.0458	$CaSO_4 \cdot H_2O$
3.00	0.0500	»	0.10	0.0253	»
2.00	0.0549	»			

№ 3747 [2481]

SODIUM CHLORIDE–CALCIUM SULFATE–WATER

$NaCl—CaSO_4—H_2O$

Content of Sodium Chloride in aqueous soln., mol/l	Solubility of Calcium Sulfate in aqueous solution of Sodium Chloride wt %	t	Comp. of solid phase
5	0.51	25	$CaSO_4 \cdot 2H_2O$
4	0.58	25	»
2.5	0.644	25	»
1.5	0.605	25	»
1	0.575	25	»
0.1	0.295	25	»
5	0.49	40	»
2.5	0.66	40	»
2	0.64	40	»
1.5	0.605	40	»
1	0.58	40	»
0.2	0.35	40	»
0.1	0.295	40	»
4	0.61	25	$CaSO_4 \cdot 0.5H_2O$
2	0.71	25	»
1	0.61	25	»
0.65	0.54	25	»
0	0.26	25	»
5	0.54	40	»
4	0.615	40	»
1	0.60	40	»
0	0.23	40	»

№ 3748]334]

SODIUM CHLORIDE–CALCIUM SULFATE–WATER

$NaCl—CaSO_4—H_2O$

Content in aqueous soln., wt %		t	Comp. of solid phase
Sodium Chloride	Calcium Sulfate		
27.85	0.336	100	$CaSO_4$
26.16	0.368	100	»
20.52	0.514	100	$CaSO_4 \cdot 0.5H_2O$
14.63	0.585	100	»
9.83	0.663	100	$CaSO_4 \cdot 2H_2O$
5.13	0.548	100	»
24.48	0.295	125	$CaSO_4$
20.13	0.436	125	»
15.07	0.467	125	—
9.98	0.472	125	—
5.13	0.402	125	—
29.73	0.233	150	$CaSO_4$
20.00	0.247	150	»
15.21	0.242	150	»
10.01	0.210	150	»

Table № 3748 (continuation)

Content in aqueous soln., wt %		t	Comp. of solid phase
Sodium Chloride	Calcium Sulfate		
5.01	0.160	150	$CaSO_4$
30.15	0.233	175	»
20.35	0.213	175	»
15.01	0.210	175	»
9.77	0.170	175	»
5.12	0.117	175	»
30.80	0.220	200	»
20.33	0.182	200	»
15.39	0.166	200	»
10.02	0.130	200	»
5.22	0.094	200	»

№ 3749 SODIUM CHLORIDE–CALCIUM SULFATE–WATER [1653]

$NaCl—CaSO_4—H_2O$

$t = 19$

Content in aqueous soln., g/l		Comp. of solid phase	Content in aqueous soln., g/l		Comp. of solid phase
Sodium Chloride	Calcium Sulfate		Sodium Chloride	Calcium Sulfate	
302	5.65	$CaSO_4 \cdot 2H_2O$	95	6.80	$CaSO_4 \cdot 2H_2O$
205	6.92	»	55	5.87	»
139	7.20	»	51	5.69	»
127	7.18	»	10	3.57	»
108	6.95	»			

№ 3750 SODIUM CHLORIDE–CALCIUM SULFATE–WATER [993]

$NaCl—CaSO_4—H_2O$

$t = 25$

Content in aqueous soln., g/l		Content in aqueous soln., g/l		Content in aqueous soln., g/l	
Calcium Sulfate	Sodium Chloride	Calcium Sulfate	Sodium Chloride	Calcium Sulfate	Sodium Chloride
4.31	19.81	7.24	138.76	6.65	231.52
6.00	50.03	7.23	152.15	6.10	283.48
6.87	86.10	7.16	168.16	5.82	317.21
7.22	119.88	7.06	180.73		

№ 3751 [994]

SODIUM CHLORIDE–CALCIUM SULFATE–WATER

$NaCl—CaSO_4—H_2O$

$t = 25$

Content in aqueous soln., g/l		d_t	Comp. of solid phase
Sodium Chloride	Calcium Sulfate		
317.23	5.816	1.2014	$CaSO_4 \cdot 2H_2O$
283.50	6.102	1.1812	»
231.51	6.654	1.1501	»
180.75	7.056	1.1214	»
168.20	7.161	1.1140	»
152.13	7.230	1.1049	»
138.76	7.236	1.0958	»
119.87	7.216	1.0837	»
86.10	6.865	1.0612	»
50.035	6.003	1.0379	»
19.807	4.312	1.0153	»
14.870	3.850	1.0114	»
10.072	3.405	1.0077	»
4.972	2.899	1.0034	»
2.488	2.438	1.0013	»
0	2.083	0.9993	»

№ 3752 [1006]

SODIUM CHLORIDE–CALCIUM SULFATE–WATER

$NaCl—CaSO_4—H_2O$

$t = 25$

Content in aqueous soln., g/l		Content in aqueous soln., g/l		Content in aqueous soln., g/l	
Sodium Chloride	Calcium Sulfate	Sodium Chloride	Calcium Sulfate	Sodium Chloride	Calcium Sulfate
246.20	6.280	129.60	7.480	5.67	3.140
196.40	6.640	47.36	6.100	0.71	2.250
161.60	6.900	15.78	4.350	0	2.085

№ 3753 [2115]

SODIUM CHLORIDE–CALCIUM SULFATE–WATER

$NaCl—CaSO_4—H_2O$

$t = 25$

Content in saturated aqueous soln., g/100 g Water		d	Comp. of solid phase
Sodium Chloride	Calcium Sulfate		
0.00	0.262	0.9994	$CaSO_4$
4.07	0.601	1.0300	»
9.05	0.807	1.0636	»
13.99	0.816	1.0937	»
15.01	0.835	1.0995	»
18.18	0.793	1.1202	»
19.80	0.802	1.1247	»
25.03	0.700	1.1519	»
29.78	0.625	1.1741	»
35.50	0.536	1.2012	»
35.69	0.528	1.2013	»
36.64	0.395	1.2005	$CaSO_4+NaCl$
35.67	0.173	1.1993	NaCl
35.96	0.0	1.1984	»

№ 3754 [488]

SODIUM CHLORIDE–CALCIUM SULFATE–WATER

$NaCl—CaSO_4—H_2O$

$t = 35$

Comp. of saturated soln., wt %			Comp. of solid phase
Calcium Sulfate	Sodium Chloride	Water	
0.211	0.00	99.789	$CaSO_4 \cdot 2H_2O$
0.400	1.43	98.170	»
0.649	8.62	90.731	»
0.690	14.35	84.960	»
0.636	17.22	82.144	»
0.480	24.04	75.480	»
0.472	26.26	73.268	$CaSO_4 \cdot 2H_2O+NaCl$
0.000	26.76	73.24	NaCl

№ 3755 SODIUM CHLORIDE–CALCIUM SULFATE–WATER [1247]

$NaCl—CaSO_4—H_2O$

Concn. of Sodium Chloride in aqueous soln., mol/1000 g Water	Solubility of Calcium Sulfate in aqueous soln. of Sodium Chloride, g/100 g Water	t	Concn. of Sodium Chloride in aqueous soln., mol/1000 g Water	Solubility of Calcium Sulfate in aqueous soln. of Sodium Chloride, g/100 g Water	t
0.00	0.275	25	0.00	0.225	40
1.00	0.751	25	0.94	0.630	40
1.59	0.816	25	1.77	0.691	40
2.58	0.831	25	2.90	0.697	40
3.70	0.770	25	3.70	0.640	40
4.64	0.664	25	4.70	0.572	40
5.24	0.631	25	5.60	0.508	40
0.00	0.262	30	0.00	0.190	50
0.92	0.688	30	0.93	0.522	50
1.72	0.764	30	1.80	0.611	50
2.85	0.768	30	2.88	0.619	50
3.70	0.720	30	3.70	0.590	50
4.64	0.635	30	4.71	0.521	50
5.55	0.564	30	6.30	0.486	50

№ 3756 [2142]

SODIUM CHLORIDE–CACLIUM SULFATE–WATER

$NaCl—CaSO_4—H_2O$

Content of Sodium Chloride in aqueous soln., mol/1000 g Water	Solubility of Calcium Sulfate in aqueous soln. of Sodium Chloride, mol/1000 g Water	t	Comp. of solid phase
0.0	0.0151	25	$CaSO_4 \cdot 2H_2O$
0.01	0.0162	25	»
0.03	0.0179	25	»
0.06	0.0200	25	»
0.1	0.0224	25	»
0.3	0.0306	25	»
0.6	0.0383	25	»
1.0	0.0450	25	»
2.0	0.0545	25	»
3.0	0.0564	25	»
4.0	0.0542	25	»
6.0	0.0424	25	»
0.0	0.0156	40	»
0.01	0.0167	40	»
0.03	0.0184	40	»
0.06	0.0205	40	»
0.1	0.0228	40	»
0.3	0.0310	40	»
0.6	0.0390	40	»
1.0	0.0458	40	»
2.0	0.0540	40	»
3.0	0.0558	40	»
4.0	0.0536	40	»
6.0	0.0440	40	»

Table № 3756 (continuation)

Concn. of Sodium Chloride in aqueous soln., mol/1000 g Water	Solubility of Calcium Sulfate in aqueous soln. of Sodium Chloride, mol/1000 g Water	t	Comp. of solid phase
0.0	0.0149	60	$CaSO_4 \cdot 2H_2O$
0.01	0.0158	60	»
0.03	0.0174	60	»
0.06	0.0196	60	»
0.1	0.0220	60	»
0.3	0.0307	60	»
0.6	0.0390	60	»
1.0	0.0464	60	»
2.0	0.0553	60	»
3.0	0.0577	60	»
4.0	0.0566	60	»
6.0	0.0490	60	»
0.0	0.0113	100	»
0.01	0.0124	100	»
0.03	0.0141	100	»
0.06	0.0165	100	»
0.1	0.0195	100	»
0.3	0.0287	100	»
0.6	0.0376	100	»
1.0	0.0459	100	»
2.0	0.0510	100	»
3.0	0.0515	100	»
4.0	0.0505	100	»
6.0	0.045	100	»
0.0	0.00553	100	$CaSO_4$
0.01	0.00630	100	»
0.03	0.00760	100	»
0.06	0.00904	100	»
0.1	0.0107	100	»
0.3	0.0164	100	»
0.6	0.0220	100	»
1.0	0.0272	100	»
2.0	0.0322	100	»
3.0	0.0340	100	»
4.0	0.0345	100	»
6.0	0.033	100	»
0.0	0.00680	125	$CaSO_4 \cdot 0.5H_2O$
0.01	0.00780	125	»
0.03	0.00935	125	»
0.06	0.0112	125	»
0.1	0.0133	125	»
0.3	0.0207	125	»
0.6	0.0282	125	»
1.0	0.0353	125	»
2.0	0.0471	125	»
3.0	0.0550	125	»
4.0	0.0612	125	»
6.0	0.0705	125	»
10.0	0.0820	125	»
0.0	0.00247	125	$CaSO_4$
0.01	0.00311	125	»

Table № 3756 (continuation)

Concn. of Sodium Chloride in aqueous soln., mol/1000 g Water	Solubility of Calcium Sulfate in aqueous soln. of Sodium Chloride, mol/1000 g Water	t	Comp. of solid phase
0.03	0.00391	125	$CaSO_4$
0.06	0.00482	125	»
0.1	0.00586	125	»
0.3	0.00950	125	»
0.6	0.0130	125	»
1.0	0.0166	125	»
2.0	0.0225	125	»
3.0	0.0265	125	»
4.0	0.0294	125	»
6.0	0.0340	125	»
10.0	0.0394	125	»
0.0	0.00154	150	»
0.01	0.00201	150	»
0.03	0.00269	150	»
0.06	0.00333	150	»
0.1	0.00409	150	»
0.3	0.00692	150	»
0.6	0.00979	150	»
1.0	0.0127	150	»
2.0	0.0175	150	»
3.0	0.0207	150	»
4.0	0.0232	150	»
6.0	0.0270	150	»
10.0	0.0320	150	»
0.0	0.000820	175	»
0.01	0.00117	175	»
0.03	0.00161	175	»
0.06	0.00209	175	»
0.1	0.00267	175	»
0.3	0.00476	175	»
0.6	0.00700	175	»
1.0	0.00918	175	»
2.0	0.0131	175	»
3.0	0.0159	175	»
4.0	0.0180	175	»
6.0	0.0213	175	»
10.0	0.0256	175	»
0.0	0.000498	200	»
0.01	0.000718	200	»
0.03	0.00104	200	»
0.06	0.00141	200	»
0.1	0.00181	200	»
0.3	0.00344	200	»
0.6	0.00524	200	»
1.0	0.00712	200	»
2.0	0.0106	200	»
3.0	0.0131	200	»
4.0	0.0150	200	»
6.0	0.0182	200	»
10.0	0.0220	200	»

№ 3757 SODIUM CHLORIDE–GYPSUM–WATER [2115]

$NaCl—CaSO_4 \cdot 2H_2O—H_2O$

Content in saturated aqueous soln., g/100 g Water		t	d	Comp. of solid phase
Sodium Chloride	Gypsum			
0.0	0.206	25	0.9992	$CaSO_4 \cdot 2H_2O$
3.95	0.537	25	1.0287	»
6.13	0.625	25	1.0436	»
8.70	0.698	25	1.0598	»
13.55	0.766	25	1.0891	»
16.82	0.778	25	1.1079	»
20.57	0.765	25	1.1281	»
23.53	0.753	25	1.1434	»
28.47	0.705	25	1.670	»
33.99	0.632	25	1.1958	»
35.70	0.616	25	1.2021	$CaSO_4 \cdot 2H_2O+NaCl$
35.61	0.615	25	1.2020	»
35.67	0.419	25	1.2009	NaCl
35.89	0.261	25	1.1999	»
35.68	0.145	25	1.1992	»
35.86	0.061	25	1.1990	»
35.96	0.00	25	1.1984	»
0.00	0.209	35	0.9962	$CaSO_4 \cdot 2H_2O$
4.99	0.592	35	1.0326	»
11.00	0.743	35	1.0708	»
15.77	0.782	35	1.0980	»
19.96	0.783	35	1.1207	»
25.33	0.723	35	1.1489	»
29.98	0.670	35	1.1696	»
35.87	0.615	35	1.1969	$CaSO_4 \cdot 2H_2O+NaCl$
35.70	0.461	35	1.1970	NaCl
35.90	0.245	35	1.1956	»
36.02	0.00	35	1.1935	»

№ 3758 SODIUM CHLORIDE–GYPSUM–WATER [1246]

$NaCl—CaSO_4 \cdot 2H_2O—H_2O$

Concn. of Sodium Chloride in aqueous soln., mol/1000 g Water	Solubility of Gypsum, in conversion to $CaSO_4$, in aqueous soln. of Sodium Chloride, g/100 g Water	t	Concn. of Sodium Chloride in aqueous soln., mol/1000 g Water	Solubility of Gypsum, in conversion to $CaSO_4$, in aqueous soln. of Sodium Chloride, g/100 g Water	t
0.00	0.207	25	0.92	0.600	30
1.00	0.614	25	1.72	0.713	30
1.59	0.709	25	2.85	0.755	30
2.58	0.773	25	3.70	0.726	30
3.70	0.741	25	4.64	0.658	30
4.64	0.698	25	5.55	0.616	30
5.24	0.647	25	0.00	0.218	40
0.00	0.222	30	0.94	0.607	40

Table № 3758 (continuation)

Concn. of Sodium Chloride in aqueous soln., mol/1000 g Water	Solubility of Gypsum, in conversion to $CaSO_4$, in aqueous soln. of Sodium Chloride, g/100 g Water	t	Concn. of Sodium Chloride in aqueous soln., mol/1000 g Water	Solubility of Gypsum, in conversion to $CaSO_4$, in aqueous soln. of Sodium Chloride, g/100 g Water	t
1.77	0.720	40	0.93	0.559	50
2.90	0.760	40	1.80	0.722	50
3.70	0.736	40	2.88	0.775	50
4.70	0.690	40	3.70	0.752	50
5.60	0.632	40	4.71	0.700	50
0.00	0.212	50	5.87	0.638	50

№ 3759 [1102]

SODIUM CHLORIDE–CALCIUM SULFATE–CALCIUM CHLORIDE–WATER

$NaCl—CaSO_4—CaCl_2—H_2O$

Content in aqueous soln., wt %			t	Comp. of solid phase
CaO	SO_3	Cl^-		
0.1838	0.2548	15.253	0	$NaCl + CaSO_4 \cdot 2H_2O$
0.1845	0.2602	15.920	10	»
0.1899	0.2670	15.967	25	»
0.2035	0.2781	16.123	40	»
0.2100	0.2841	16.270	50	»
0.2185	0.2968	16.324	62.5	»
0.2240	0.2204	16.459	65	NaCl
0.2299	0.2135	16.486	71	»
0.2309	0.2109	16.524	75	»
0.2225	0.2069	16.670	85	»
0.2068	0.2007	17.128	90	»

№ 3760 [334]

SODIUM CHLORIDE–CALCIUM SULFATE–CALCIUM CHLORIDE–WATER

$NaCl—CaSO_4—CaCl_2—H_2O$

Content in aqueous soln., wt %									Comp. of solid phase
t = 100			t = 150			t = 175			
Sodium Chloride	Calcium Chloride	Calcium Sulfate	Sodium Chloride	Calcium Chloride	Calcium Sulfate	Sodium Chloride	Calcium Chloride	Calcium Sulfate	
Saturated soln.	0	0.336	Saturated soln.	0	0.233	Saturated soln.	0	0.229	$CaSO_4$
»	0.30	0.280	»	0.46	0.093	»	0.45	0.099	»
»	0.60	0.219	»	1.01	0.053	»	0.98	0.047	»
»	1.14	0.153	»	2.97	0.031	»	2.93	0.027	»
»	3.14	0.090	»	9.79	0.023	»	9.80	0.017	»

Content in aqueous soln., wt %									Comp. of solid phase
$t = 100$			$t = 150$			$t = 175$			
Sodium Chloride	Calcium Chloride	Calcium Sulfate	Sodium Chloride	Calcium Chloride	Calcium Sulfate	Sodium Chloride	Calcium Chloride	Calcium Sulfate	
Saturated soln.	10.45	0.047	—	—	—	—	—	—	$CaSO_4$
15	0	0.590	15	0	0.236	15	0	0.210	»
15	0.29	0.490	15	0.52	0.075	15	0.54	0.068	»
15	0.62	0.386	15	1.00	0.067	15	1.04	0.056	»
15	1.06	0.289	15	3.06	0.030	15	2.99	0.026	»
15	3.01	0.163	15	9.45	0.020	15	9.55	0.021	»
15	9.76	0.068	—	—	—	—	—	—	»
5	0	0.540	5	0	0.156	5	0	0.116	»
5	0.27	0.419	5	0.51	0.045	5	0.53	0.039	»
5	0.53	0.330	5	1.03	0.035	5	0.99	0.029	»
5	1.02	0.259	5	3.15	0.034	5	2.97	0.025	»
5	2.97	0.170	5	10.00	0.042	5	9.70	0.025	»
5	9.59	0.129	—	—	—	—	—	—	»

№ 3761 [766]

SODIUM CHLORIDE–CALCIUM SULFATE–CALCIUM CHLORIDE–WATER

$NaCl—CaSO_4—CaCl_2—H_2O$

$t = 25$

Content in aqueous soln., wt %			d	Comp. of solid phase
Sodium Chloride	Calcium Chloride	Calcium Sulfate		
29.77	3.1	0.0409	1.301	$NaCl + CaSO_4 \cdot 2H_2O$
18.92	9.53	0.0396	1.205	»
16.17	11.56	0.041	1.204	»
12.03	13.36	0.0438	1.203	»
8.51	18.22	0.050	1.201	»
6.66	19.14	0.060	—	»
2.9	22.4	0.102	1.214	»
1.58	24.39	0.136	1.203	»
0.738	25.37	0.224	1.206	»
0.111	26.3	0.482	1.17	»
42.7	1.2	Traces	—	$NaCl + CaSO_4 + CaCl_2$

SODIUM CHLORIDE–CALCIUM THIOSULFATE–CALCIUM CHLORIDE–WATER

$NaCl—CaS_2O_3—CaCl_2—H_2O$

$t = 25$

Content in aqueous soln., wt %				Comp. of solid phase
Ca^{2+}	Na^+	Cl^-	$S_2O_3^{2-}$	
5.14	Not determined	33.09	2.80	$Na_2S_2O_3 \cdot 5H_2O + CaS_2O_3 \cdot 6H_2O + NaCl$
5.32	»	21.81	3.29	$CaS_2O_3 \cdot 6H_2O + NaCl$
8.58	»	16.43	10.78	»
10.69	»	9.86	15.36	»
11.56	»	8.30	16.52	»
12.23	»	8.00	17.91	»
13.14	0.41	7.19	19.29	»
13.59	0.29	6.37	20.57	$CaS_2O_3 \cdot 6H_2O + NaCl + CaS_2O_3 \cdot H_2O$
—	—	5.88	20.88	$CaS_2O_3 \cdot 6H_2O + CaS_2O_3 \cdot H_2O$
13.10	0.32	5.05	20.36	$CaS_2O_3 \cdot H_2O + NaCl$
13.34	0.35	4.65	21.16	$CaS_2O_3 \cdot H_2O + CaCl_2 \cdot 6H_2O + NaCl$
—	—	0.76	28.70	$CaS_2O_3 \cdot 6H_2O + CaCl_2 \cdot 6H_2O$

№ 3763 [1328]

SODIUM CHLORIDE–CALCIUM CHLORIDE–WATER

$NaCl—CaCl_2—H_2O$

$t = 25$

Comp. of saturated soln., wt %			d_{25}	Comp. of solid phase
Sodium Chloride	Calcium Chloride	Water		
0.00	45.65	54.35	—	$CaCl_2 \cdot 6H_2O$
1.02	43.52	55.46	1.4441	$CaCl_2 \cdot 6H_2O + NaCl$
1.02	36.52	62.46	1.3651	$NaCl$
1.16	34.44	64.40	1.3463	»
5.37	25.45	69.18	1.2831	»
7.60	21.37	71.03	1.2653	»
13.62	14.11	72.27	1.2367	»
23.81	2.87	73.32	1.2080	»
26.36	0.00	73.64	1.2030	»

№ 3764 [2315]

SODIUM CHLORIDE–CALCIUM CHLORIDE–WATER

$NaCl—CaCl_2—H_2O$

Comp. of saturated soln., wt %			t	Comp. of solid phase
Sodium Chloride	Calcium Chloride	Water		
0.00	57.0	43.0	50	$CaCl_2 \cdot 2H_2O$
0.9	56.3	42.8	50	$CaCl_2 \cdot 2H_2O + NaCl$
3.6	30.9	65.5	50	$NaCl$

Comp. of saturated soln., wt %			t	Comp. of solid phase
Sodium Chloride	Calcium Chloride	Water		
13.2	15.1	71.7	50	NaCl
24.4	3.3	72.3	50	»
26.8	0.0	73.2	50	»
0.8	58.1	41.1	94.5	$CaCl_2 \cdot 2H_2O + NaCl$
1.1	57.4	41.5	94.5	NaCl
1.3	45.7	53.0	94.5	»
4.3	32.8	62.9	94.5	»
15.0	15.3	69.7	94.5	»
18.1	11.5	70.4	94.5	»

№ 3765 SODIUM CHLORIDE–CALCIUM CHLORIDE–WATER [2207]

$NaCl—CaCl_2—H_2O$

$t = 25$

Comp. of saturated soln., wt %			d_{25}	Comp. of saturated soln., wt %			d_{25}
Sodium Chloride	Calcium Chloride	Water		Sodium Chloride	Calcium Chloride	Water	
25.30	1.10	73.60	1.207	17.55	9.50	72.95	1.225
24.32	2.16	73.52	1.210	15.91	11.48	72.61	1.233
23.37	3.22	73.41	1.209	10.54	17.77	71.69	1.241
20.43	5.45	74.12	1.216	8.05	21.0	70.95	1.257
19.17	7.40	73.43	1.220	5.63	24.58	69.79	1.276

№ 3766 SODIUM CHLORIDE–CALCIUM CHLORIDE–WATER [2317]

$NaCl—CaCl_2—H_2O$

Content in aqueous soln., wt %		t	Comp. of solid phase
Sodium Chloride	Calcium Chloride		
24.4	3.3	50	NaCl
13.2	15.1	50	»
3.60	30.9	50	»
0.90	56.3	50	$NaCl + CaCl_2 \cdot 2H_2O$
18.1	11.5	94.5	NaCl
15.0	15.3	94.5	»
14.0	16.3	94.5	»
4.3	32.8	94.5	»
1.3	45.7	94.5	»
1.2	52.7	94.5	»
1.1	57.4	94.5	»
0.8	58.1	94.5	$NaCl + CaCl_2 \cdot 2H_2O$

№ 3767 SODIUM CHLORIDE–CALCIUM CHLORIDE–WATER [333]

$NaCl—CaCl_2—H_2O$

$t = 0$

Content in aqueous soln., wt %		Comp. of solid phase	Content in aqueous soln., wt %		Comp. of solid phase
Sodium Chloride	Calcium Chloride		Sodium Chloride	Calcium Chloride	
26.23	0	NaCl	1.67	30.33	NaCl
22.70	3.75	»	1.68	30.42	»
16.37	8.26	»	1.18	31.59	»
18.02	10.84	»	1.02	31.67	»
9.11	18.66	»	0.69	32.53	»
4.70	24.70	»	0.70	33.10	»
2.02	29.35	»	0.78	33.49	»
1.94	29.53	»	0.62	34.87	»

№ 3768 SODIUM CHLORIDE–CALCIUM CHLORIDE–WATER [1881]

$NaCl—CaCl_2—H_2O$

$t = 0$

Comp. of saturated soln., wt %			Comp. of solid phase
Sodium Chloride	Calcium Chloride	Water	
0.00	37.55	62.45	$CaCl_2 \cdot 6H_2O$
0.31	37.50	62.19	$CaCl_2 \cdot 6H_2O + NaCl$
0.93	35.31	63.76	NaCl
3.59	27.14	69.27	»
7.78	20.60	71.62	»
11.36	16.12	72.52	»
17.28	9.53	73.19	»
21.19	5.33	73.48	»
26.42	0.00	73.58	$NaCl \cdot 2H_2O$

№ 3769 SODIUM CHLORIDE–CALCIUM CHLORIDE–WATER [409]

$NaCl—CaCl_2—H_2O$

Content in aqueous soln., wt %		t	Comp. of solid phase
Sodium Chloride	Calcium Chloride		
16.02	7.47	−23	$NaCl \cdot 2H_2O +$ ice
19.75	3.47	−22	»
23.58	1.94	− 5	$NaCl \cdot 2H_2O$
21.73	4.05	− 5	»
16.43	9.85	− 5	»
11.28	16.57	− 5	$NaCl \cdot 2H_2O + NaCl$
9.40	18.78	− 5	NaCl
4.95	25.00	− 5	»
1.78	31.35	− 5	»
0.81	36.12	− 5	$NaCl + CaCl_2 \cdot 6H_2O$

№ 3770 SODIUM CHLORIDE–CALCIUM CHLORIDE–WATER [510]

$NaCl—CaCl_2—H_2O$

Content in aqueous soln., wt %		t	Comp. of solid phase
Sodium Chloride	Calcium Chloride		
8.56	19.25	−10	$NaCl \cdot 2H_2O$
1.93	31.20	−10	»
0.80	35.34	−10	$NaCl \cdot 2H_2O + CaCl_2 \cdot 6H_2O$
11.0	16.0	0	NaCl
1.80	32.06	0	»
0.68	36.62	0	$NaCl + CaCl_2 \cdot 6H_2O$
0	37.25	0	$CaCl_2 \cdot 6H_2O$
0.43	44.52	25	$NaCl + CaCl_2 \cdot 6H_2O$
0	44.70	25	$CaCl_2 \cdot 6H_2O$
17.09	10.69	50	NaCl
5.70	25.46	50	»
1.69	35.83	50	»
0.44	56.65	50	$NaCl + CaCl_2 \cdot 2H_2O$
0	56.85	50	$CaCl_2 \cdot 2H_2O$
15.10	15.56	100	NaCl
4.10	32.68	100	»
1.27	51.70	100	»
1.08	60.28	100	$NaCl + CaCl_2 \cdot 2H_2O$
0	61.05	100	$CaCl_2 \cdot 2H_2O$
2.16	48.20	130	NaCl
1.85	54.32	130	»
1.70	62.56	130	$NaCl + CaCl_2 \cdot 2H_2O$
0	63.70	130	$CaCl_2 \cdot 2H_2O$

№ 3771 SODIUM CHLORIDE–CALCIUM CHLORIDE–WATER [1020]

$NaCl—CaCl_2—H_2O$

Comp. of saturated soln., wt %			t	Comp. of solid phase
Calcium Chloride	Sodium Chloride	Water		
0.00	26.80	73.20	25.0	NaCl
0.00	26.23	73.77	0	»
14.10	14.84	71.06	24.0	»
13.80	14.60	71.60	0	»
13.80	14.80	71.40	− 5.0	»
13.21	13.83	72.96	− 1.0	»
13.64	14.36	72.00	− 5.7	$NaCl + NaCl \cdot 2H_2O$
0.00	25.60	74.40	− 5.0	$NaCl \cdot 2H_2O$
13.51	14.22	72.27	− 7.5	»
0.00	23.60	76.40	−20.0	»
13.16	13.86	72.98	−11.5	»
13.40	14.20	72.40	−10.0	»
0.00	25.00	75.00	−10.0	»
0.00	24.40	75.60	−15.0	»
13.00	13.80	73.20	−15.0	»

Table № 3771 (continuation)

Comp. of saturated soln., wt %			t	Comp. of solid phase
Calcium Chloride	Sodium Chloride	Water		
12.90	13.60	73.50	—20.0	$NaCl \cdot 2H_2O$
12.80	13.40	73.80	—25.0	»
12.73	13.41	73.86	—19.5	»
12.60	13.20	74.20	—26.7	$NaCl \cdot 2H_2O$+ ice
11.20	13.80	75.00	—25.0	»
0.00	13.50	86.50	—10.0	Ice
0.00	19.00	81.00	—15.0	»
0.00	22.40	77.60	—20.0	»
12.30	12.80	74.90	—25.0	»
12.27	12.92	74.81	—23.7	»
11.40	12.00	76.60	—20.0	Ice
11.38	11.93	76.69	—21.2	»
10.00	10.50	79.50	—15.0	»
9.64	10.15	80.21	—14.7	»
8.40	0.00	91.60	— 5.0	»
7.93	7.78	84.29	— 9.5	»
7.80	8.20	84.00	—10.0	»
7.60	1.20	91.20	— 5.0	»
5.34	5.62	89.04	— 6.8	»
4.40	4.60	91.00	— 5.0	»
0.00	7.40	92.60	— 5.0	»
27.30	4.70	68.00	25.0	NaCl
27.13	4.61	68.26	25.0	»
26.82	4.56	68.62	13.0	»
26.64	4.52	68.84	5.2	»
26.60	4.60	68.80	0	»
26.48	4.50	69.02	—11.2	»
26.40	4.60	69.00	— 5.0	»
26.32	4.47	69.21	—14.0	»
26.10	4.30	69.60	—10.0	»
26.00	4.20	69.80	—16.5	$NaCl + NaCl \cdot 2H_2O$
24.00	5.80	70.20	—10.0	»
4.80	21.20	74.00	— 5.0	»
25.80	4.80	69.40	—15.0	»
25.84	4.39	69.77	—17.0	$NaCl \cdot 2H_2O$
25.43	4.32	70.25	—21.5	»
25.40	4.40	70.20	—20.0	»
24.80	4.20	71.00	—25.0	»
24.70	4.20	71.10	—27.2	»
24.20	4.20	71.60	—35.0	»
26.20	3.10	70.70	—40.0	$NaCl \cdot 2H_2O$+ ice
24.00	4.00	72.00	—36.0	»
23.50	4.50	72.00	—35.0	»
19.60	7.00	73.40	—30.0	»
23.80	4.00	72.20	—35.0	Ice
23.72	4.00	72.28	—35.5	»
22.00	3.80	74.20	—30.0	»
20.40	3.60	76.00	—25.0	»
20.35	3.46	76.19	—25.0	»
18.40	3.20	78.40	—20.0	»
16.20	2.70	81.10	—15.0	»

Table № 3771 (continuation)

Comp. of saturated soln., wt %			t	Comp. of solid phase
Calcium Chloride	Sodium Chloride	Water		
15.75	2.68	81.57	—14.2	Ice
13.00	2.20	84.80	—10.0	»
12.46	2.12	85.42	— 9.5	»
10.03	1.71	88.26	— 6.7	»
35.60	1.40	63.00	25.0	NaC
35.48	2.88	62.54	40.0	»
32.40	2.80	64.80	25.0	»
31.68	2.64	65.68	17.0	»
31.40	2.60	66.00	0	»
31.29	2.61	60.10	— 5.0	»
30.80	2.42	66.78	— 5.0	»
30.70	2.40	66.90	—10.0	»
30.65	2.36	66.99	—15.0	»
30.60	2.30	67.10	—20.0	»
30.47	2.56	66.97	—20.8	$NaCl+NaCl \cdot 2H_2O$
29.00	3.00	68.00	—20.0	»
32.60	1.30	66.10	—25.0	$NaCl \cdot 2H_2O$
30.40	2.53	67.07	—22.7	»
30.16	2.51	67.33	—25.0	»
29.82	2.49	67.69	—25.8	$NaCl \cdot 2H_2O$
29.80	2.20	68.00	—25.0	»
29.53	2.46	68.01	—29.5	»
29.40	2.20	68.40	—30.0	»
29.14	2.43	68.43	—31.7	»
28.80	2.20	69.00	—35.0	»
28.60	2.38	69.02	—38.0	»
28.60	2.10	69.30	—40.0	»
28.30	2.20	69.50	—45.0	»
24.70	1.20	74.40	—30.0	»
28.16	2.35	69.49	—46.8	$NaCl \cdot 2H_2O+$ ice
27.40	2.50	70.10	—45.0	»
27.80	2.20	70.00	—45.0	Ice
27.64	2.30	70.06	—43.5	»
27.20	1.00	71.80	—40.0	»
26.80	2.00	72.20	—40.0	»
26.25	2.19	71.56	—38.0	»
25.80	1.00	73.20	—35.0	»
25.60	1.80	72.60	—35.0	»
23.80	1.80	74.40	—30.0	»
20.67	1.72	77.61	—22.0	»
16.38	1.36	82.26	—14.0	»
35.99	1.40	62.61	30.0	NaCl
35.05	1.36	63.59	20.0	»
34.51	1.34	64.15	10.0	»
34.40	1.40	64.20	0	»
33.80	1.40	64.80	— 5.0	»
33.60	1.40	65.00	—15.0	»
33.30	1.30	65.40	—20.0	»
43.50	1.00	55.50	25.0	$NaCl+CaCl_2 \cdot 6H_2O$
37.60	1.10	61.30	0	»
34.20	1.10	64.70	— 5.0	»

Table № 3771 (continuation)

Comp. of saturated soln., wt%			t	Comp. of solid phase
Calcium Chloride	Sodium Chloride	Water		
33.80	1.20	65.00	—10.0	$NaCl+CaCl_2\cdot 6H_2O$
33.70	1.30	65.00	—15.0	»
33.58	1.30	65.12	—19.0	»
33.40	1.35	65.25	—20.0	»
32.40	1.35	66.25	—22.4	$NaCl\cdot 2H_2O+NaCl+CaCl_2\cdot 6H_2O$
32.72	1.27	66.01	—26.2	$CaCl_2\cdot 6H_2O$
31.82	1.23	66.95	—33.5	»
31.14	1.21	67.65	—39.5	»
30.40	1.20	68.40	—45.0	»
30.40	0.00	69.60	—49.8	$CaCl_2\cdot 6H_2O+$ ice
29.76	1.15	69.09	—50.7	»
28.74	1.11	70.15	—45.5	Ice
28.60	1.20	70.20	—45.0	»
24.20	1.00	74.80	—30.0	»
22.85	0.89	76.26	—25.8	»
16.51	0.64	82.85	—12.2	»
11.17	0.43	88.40	— 6.5	»

№ 3772 SODIUM CHLORIDE–CALCIUM CHLORIDE–WATER [314]

$NaCl—CaCl_2—H_2O$

$t = 25$

Content in aqueous soln., wt %		Comp. of solid phase	Content in aqueous soln., wt %		Comp. of solid phase
Sodium Chloride	Calcium Chloride		Sodium Chloride	Calcium Chloride	
2.83	30.39	NaCl	0.31	44.0	NaCl
0.60	39.25	»	0.27	50.83	NaCl (metastable)

№ 3773 SODIUM CHLORIDE–CALCIUM CHLORIDE–WATER [1106]

$NaCl—CaCl_2—H_2O$

Content in aqueous soln., wt %		t	Comp. of solid phase
Sodium Chloride	Calcium Chloride		
26.4	0	18	NaCl
22.6	3.0	18	»
19.6	6.3	18	»
14.3	12.3	18	»
11.8	15.1	18	»
9.6	18.2	18	»
3.5	28.1	18	»
0.62	40.80	18	$NaCl+CaCl_2\cdot 6H_2O$
0	41.97	18	$CaCl_2\cdot 6H_2O$
27.0	0	50	NaCl

Table № 3773 (continuation)

Comp. of saturated soln., wt %		t	Comp. of solid phase
Sodium Chloride	Calcium Chloride		
23.6	3.0	50	NaCl
20.6	6.3	50	»
15.0	12.3	50	»
10.5	18.2	50	»
4.3	28.1	50	»
0.8	41.5	50	»
0.65	56.45	50	$NaCl + CaCl_2 \cdot 2H_2O$
0	56.54	50	$CaCl_2 \cdot 2H_2O$
28.3	0	95	NaCl
25.2	3.0	95	»
21.3	6.4	95	»
16.7	12.8	95	»
12.8	17.6	95	»
6.1	28.1	95	»
1.9	41.5	95	»
1.00	60.15	95	$NaCl + CaCl_2 \cdot 2H_2O$
0	60.28	95	$CaCl_2 \cdot 2H_2O$

№ 3774 SODIUM CHLORIDE–CALCIUM CHLORIDE–WATER [1107]

$NaCl—CaCl_2—H_2O$

Content in aqueous soln., wt %		t	Comp. of solid phase
Sodium Chloride	Calcium Chloride		
26.5	0	18	NaCl
0.42	41.60	18	$NaCl + CaCl_2 \cdot 6H_2O$
0	42.05	18	$CaCl_2 \cdot 6H_2O$
27.0	0	50	NaCl
0.62	56.43	50	$NaCl + CaCl_2 \cdot 2H_2O$
0	56.55	50	$CaCl_2 \cdot 2H_2O$
28.3	0	95	NaCl
0.98	60.10	95	$NaCl + CaCl_2 \cdot 2H_2O$
0	60.20	95	$CaCl_2 \cdot 2H_2O$

№ 3775 SODIUM CHLORIDE–CALCIUM CHLORIDE–WATER [270]

$NaCl—CaCl_2—H_2O$

$t = 25$

Comp. of saturated soln., wt %			d	Comp. of solid phase
Calcium Chloride	Sodium Chloride	Water		
48.39	0.53	51.08	—	α-$CaCl_2 \cdot 4H_2O + NaCl$
51.31	0.35	48.34	—	β-$CaCl_2 \cdot 4H_2O + NaCl$
52.49	0.12	47.39	—	γ-$CaCl_2 \cdot 6H_2O + NaCl$
45.57	0.00	54.43	1.4487	$CaCl_2 \cdot 6H_2O$

Table № 3775 (continuation)

Comp. of saturated soln., wt %			d	Comp. of solid phase
Calcium Chloride	Sodium Chloride	Water		
43.52	1.03	55.45	1.4441	$CaCl_2 \cdot 6H_2O + NaCl$
34.36	1.18	64.46	1.3463	NaCl
25.48	5.35	69.17	1.2838	»
14.14	13.62	72.24	1.2367	»
11.48	16.96	71.56	1.2315	»
7.46	25.59	66.95	1.2237	»
0.00	26.41	73.59	1.1998	»

№ 3776

SODIUM CHLORIDE–CALCIUM SULFATE–WATER [996]

$NaCl—CaSO_4—H_2O$

$t = 25$

Conc. of Sodium Chloride in aqueous soln., mol/l	Solubility of Calcium Sulfate in aqueous soln. of Sodium Chloride, mol/l	d_4	Conc. of Sodium Chloride in aqueous soln., mol/l	Solubility of Calcium Sulfate in aqueous soln. of Sodium Chloride, mol/l	d_4
0.0432	0.0180	1.0013	2.0130	0.0544	1.0815
0.0854	0.0214	1.0036	2.4990	0.0560	1.0958
0.1730	0.0252	1.0079	2.7540	0.0562	1.1028
0.3420	0.0320	1.0160	3.0650	0.0560	1.1091
0.8720	0.0449	1.0396	3.3120	0.0555	1.1136
1.5910	0.0521	1.0640			

№ 3777 [1796]

SODIUM CHLORIDE–CALCIUM CHLORIDE–ZINC CHLORIDE–WATER

$NaCl—CaCl_2—ZnCl_2—H_2O$

$t = 15$

Comp. of solvent, wt %			Solubility of Sodium Chloride wt %	d		Comp. of solid phase
Calcium Chloride	Zinc Chloride	Water		$t = 25$	$t = 50$	
41.1	0.0	58.9	0.46	1.407	1.391	$CaCl_2 \cdot 6H_2O$
38.8	6.0	55.2	0.46	1.452	1.436	»
36.9	11.5	51.6	0.38	1.496	1.479	»
33.7	23.6	42.7	0.31	1.605	—	»
34.5	23.3	42.2	0.31	1.609	1.592	$CaCl_2 \cdot 6H_2O + CaCl_2 \cdot 4H_2O$
33.2	26.6	40.2	0.29	1.649	—	$CaCl_2 \cdot 4H_2O$
33.2	28.5	38.3	0.29	1.682	—	$CaCl_2 \cdot 4H_2O + CaCl_2 \cdot ZnCl_2 \cdot (H_2O)_x$
31.6	30.6	37.8	0.36	1.679	—	$CaCl_2 \cdot ZnCl_2 \cdot (H_2O)_x$
28.9	32.9	38.2	0.83	1.673	—	»

Comp. of solvent, wt %			Solubility of Sodium Chloride wt %	d		Comp. of solid phase
Calcium Chloride	Zinc Chloride	Water		t = 25	t = 50	
25.1	39.3	35.6	0.00	—	—	—
24.8	42.1	33.1	2.02	1.780	1.762	$CaCl_2 \cdot ZnCl_2 \cdot 5H_2O$
21.7	47.6	30.7	2.13	1.824	—	»
17.6	54.5	27.9	2.45	1.913	1.891	$CaCl_2 \cdot ZnCl_2 \cdot 4H_2O$ or $CaCl_2 \cdot ZnCl_2 \cdot 5H_2O$
16.6	58.1	25.3	1.90	1.945	—	»
16.8	62.0	21.2	1.47	2.028	2.006	»
12.7	69.9	19.4	1.59	2.092	2.069	$ZnCl_2 \cdot (H_2O)_x$
9.2	70.5	20.3	1.83	—	—	$ZnCl_2$
4.2	75.0	20.8	2.11	—	—	»
4.0	76.4	19.6	2.07	—	—	»
1.9	78.6	19.5	2.24	2.166	—	»
0.0	80.8	19.2	0.00	2.194	—	»

№ 3778 SODIUM CHLORIDE–CALCIUM IODATE–WATER [1745]

$NaCl—Ca(IO_3)_2—H_2O$

$t = 25$

Concn. of Sodium Chloride in aqueous soln., g/l	Solubility of Calcium Iodate in aqueous soln. of Sodium Chloride, g/l	Concn. of Sodium Chloride in aqueous soln., g/l	Solubility of Calcium Iodate in aqueous soln. of Sodium Chloride, g/l
0.0	3.110	29.225	5.747
2.922	3.773	43.837	6.426
5.845	4.102	58.45	7.280
8.767	4.379	87.67	8.247
17.535	5.081	116.9	9.222

№ 3779 [1183]

SODIUM CHLORIDE–CALCIUM IODATE–WATER

$NaCl—Ca(IO_3)_2—H_2O$

$t = 0$

Concn. in aqueous soln., wt %	
Sodium Chloride	Calcium Chloride
0.00	$90.88 \cdot 10^{-5}$
$5.79 \cdot 10^{-5}$	$92.35 \cdot 10^{-5}$
$11.59 \cdot 10^{-5}$	$94.38 \cdot 10^{-5}$
$17.38 \cdot 10^{-5}$	$95.78 \cdot 10^{-5}$
$23.18 \cdot 10^{-5}$	$96.90 \cdot 10^{-5}$

№ 3780 [1786]

SODIUM CHLORIDE–MANGANESE–TETRAHYDROGEN ORTHOBORATE–WATER

$NaCl—MnH_4(BO_3)_2—H_2O$

Concn. in aqueous soln., g/l		t
Manganese Tetrahydrogen Orthoborate	Sodium Chloride	
1.31	20	18
0.60	20	60
0.29	20	80

№ 3781 SODIUM CHLORIDE–MANGANESE (II) CHLORIDE–WATER [278]

$NaCl—MnCl_2—H_2O$

$t = 25$

Comp. of saturated soln., wt %			d	Comp. of solid phase
Manganese Chloride	Sodium Chloride	Water		
0.0	26.42	73.58	1.199	NaCl
7.57	20.66	71.77	1.237	»
8.72	20.45	70.83	1.245	»
18.70	13.78	67.52	1.301	»
21.95	12.76	65.29	1.325	»
26.95	9.69	63.36	1.358	»
31.30	7.90	60.80	1.414	»
35.02	7.12	57.86	1.432	»
35.60	6.25	58.15	1.458	»
39.50	5.50	55.00	1.500	»
40.61	4.97	54.42	1.507	$NaCl + MnCl_2 \cdot 4H_2O$
40.79	4.70	54.51	—	»
40.82	4.54	54.64	—	$MnCl_2 \cdot 4H_2O$
42.27	2.04	55.69	1.498	»
43.51	0.0	56.49	1.494	»

№ 3782 [2735]

SODIUM CHLORIDE–IRON (II) HYDROXIDE–WATER

$NaCl—Fe(OH)_2—H_2O$

$t = 25$

Concn. of Sodium Chloride in aqueous soln., g/l	Solubility of Iron Hydroxide in aqueous soln. of Sodium Chloride, g/l	Concn. of Sodium Chloride in aqueous soln., g/l	Solubility of Iron Hydroxide in aqueous soln. of Sodium Chloride, g/l
0.0	0.00375	117.0	0.0042
29.25	0.0041	146.25	0.00415
58.5	0.0042	175.0	0.004
87.75	0.00425	234.0	0.00375

№ 3783 SODIUM CHLORIDE–IRON (III) CHLORIDE–WATER [1854]

$NaCl—FeCl_3—H_2O$

$t = 21$

Comp. of saturated soln., wt %			Comp. of solid phase
Iron Chloride	Sodium Chloride	Water	
0.00	36.10	63.90	NaCl
24.27	9.10	66.63	Mixed crystals
25.40	8.45	66.15	»
26.40	5.25	68.35	»
38.15	3.90	57.95	»
45.38	2.45	52.17	»
46.75	2.11	51.14	»
83.39	0.00	16.61	$FeCl_3$

№ 3784 SODIUM CHLORIDE–COBALT (II) CHLORIDE–WATER [2161]

$NaCl—CoCl_2—H_2O$

$t = 20$

Comp. of saturated soln., wt %			Comp. of solid phase
Cobalt Chloride	Sodium Chloride	Water	
34.98	0.0	65.02	$CoCl_2 \cdot 6H_2O$
33.38	3.35	63.27	»
33.05	3.65	63.30	$CoCl_2 \cdot 6H_2O + NaCl$
32.84	3.91	63.25	NaCl
28.26	6.07	65.67	»
27.73	6.63	65.64	»
21.45	10.32	68.23	»
13.63	15.71	70.66	»
8.20	19.40	72.40	»
0.00	26.09	73.91	»

№ 3785 SODIUM CHLORIDE–COBALT (II) CHLORIDE–WATER [1192]

$NaCl—CoCl_2—H_2O$

Comp. of saturated soln., wt %			t	Comp. of solid phase
Sodium Chloride	Cobalt Chloride	Water		
1.8	35.3	62.9	25	$CoCl_2 \cdot 6H_2O$
4.0	33.2	62.8	25	»
4.5	33.0	62.5	25	$CoCl_2 \cdot 6H_2O + NaCl$
9.15	23.8	67.05	25	NaCl
12.2	18.6	69.2	25	»
15.62	13.5	70.88	25	»
20.0	7.02	72.98	25	»
25.2	2.64	72.16	25	»
2.63	38.8	58.57	38	$CoCl_2 \cdot 6H_2O$
3.34	38.3	58.36	38	$CoCl_2 . 6H_2O + NaCl$
7.61	27.8	64.59	38	NaCl
12.7	17.35	69.95	38	»
18.1	9.15	72.75	38	»
2.64	42.9	54.46	45	$CoCl_2 \cdot 6H_2O + NaCl$
1.8	46.0	52.2	50	$CoCl_2 \cdot 4H_2O$
2.5	46.0	51.5	50	»
2.64	45.1	52.26	50	$CoCl_2 \cdot 4H_2O + NaCl$
17.3	13.0	69.7	50	NaCl
21.2	7.8	71.0	50	»
23.3	4.93	71.77	50	»
2.61	45.2	52.19	51	$CoCl_2 \cdot 4H_2O + NaCl$
0.27	49.1	50.63	60	$CoCl_2 \cdot 2H_2O$
2.02	48.3	49.68	60	»
2.42	47.5	50.08	60	»
0.00	41.01	58.99	98	—
1.70	39.91	58.39	98	—
3.91	39.09	57.00	98	—
5.34	38.08	56.58	98	—

Comp. of saturated soln., wt %			t	Comp. of solid phase
Sodium Chloride	Cobalt Chloride	Water		
5.69	35.70	58.61	98	—
7.90	23.14	68.96	98	—
11.90	12.98	75.12	98	—
14.61	7.26	78.13	98	—
15.85	4.61	79.54	98	—
17.53	1.32	81.15	98	—
18.66	0.00	81.34	98	—
2.30	50.3	47.40	100	$CoCl_2 \cdot 2H_2O$
3.63	50.1	46.27	100	»
4.99	49.4	45.61	100	»
5.30	48.6	46.10	100	»
5.76	50.0	44.24	100	$CoCl_2 \cdot 2H_2O + NaCl$
6.44	44.4	49.16	100	NaCl
9.04	37.2	53.76	100	»
14.16	24.9	60.94	100	»
21.7	9.9	68.4	100	»

№ 3786 SODIUM CHLORIDE–COBALT (II) CHLORIDE–WATER [1643]

$NaCl—CoCl_2—H_2O$

$t = 25$

Comp. of saturated soln., wt %			Comp. of solid phase
Cobalt Chloride	Sodium Chloride	Water	
35.67	0.00	64.33	$CoCl_2 \cdot 6H_2O$
34.59	2.01	63.40	»
33.22	4.23	64.55	»
32.92	4.69	62.39	$CoCl_2 \cdot 6H_2O + NaCl$
31.62	4.58	63.80	NaCl
25.82	7.82	66.36	»
16.00	14.11	69.89	»
7.17	20.73	72.10	»
0.00	26.50	73.50	»

№ 3787 SODIUM CHLORIDE–COBALT (II) CHLORIDE–WATER [1215]

$NaCl—CoCl_2—H_2O$

Comp. of saturated soln., wt %			t	Comp. of solid phase
Cobalt Chloride	Sodium Chloride	Water		
2.64	25.20	72.16	25	NaCl
6.42	20.60	72.98	25	»
7.02	20.00	72.98	25	»
10.20	18.25	71.55	25	»

Table № 3787 (continuation)

Comp. of saturated soln., wt %			t	Comp. of solid phase
Cobalt Chloride	Sodium Chloride	Water		
13.50	15.62	70.88	25	NaCl
14.50	15.10	70.40	25	»
18.60	12.20	69.20	25	»
23.80	9.15	67.05	25	»
33.00	4.50	62.50	25	$CoCl_2 \cdot 6H_2O + NaCl$
32.90	4.69	62.41	25	»
33.20	4.00	62.80	25	$CoCl_2 \cdot 6H_2O$
35.30	1.80	62.90	25	»
9.15	18.10	72.75	38	NaCl
17.35	12.70	69.95	38	»
27.80	7.61	64.59	38	»
34.30	4.81	60.89	38	»
38.30	3.34	58.36	38	$NaCl + CoCl_2 \cdot 6H_2O$
38.80	2.63	58.57	38	$CoCl_2 \cdot 6H_2O$
43.30	0.84	55.86	45	»
42.80	2.00	55.20	45	»
43.40	2.10	54.50	45	»
43.00	2.52	54.48	45	»
42.90	2.64	54.46	45	$CoCl_2 \cdot 6H_2O + NaCl$
4.93	23.30	71.77	50	NaCl
7.80	21.20	71.00	50	»
13.00	17.30	69.70	50	»
45.10	2.64	52.26	50	$CoCl_2 \cdot 4H_2O + NaCl$
46.00	2.50	51.50	50	$CoCl_2 \cdot 4H_2O$
46.10	2.00	51.90	50	»
46.00	1.80	52.20	50	»
45.20	2.61	52.19	51	$CoCl_2 \cdot 4H_2O + NaCl$
46.80	1.59	51.61	51	$CoCl_2 \cdot 4H_2O$
46.90	0.71	52.39	51	»
46.60	0.45	52.95	51	»
49.10	0.27	50.63	60	$CoCl_2 \cdot 2H_2O$
48.60	0.69	50.71	60	»
48.30	2.02	49.68	60	»
47.50	2.42	50.08	60	»
47.00	2.45	50.55	60	»
47.50	2.83	49.67	60	$CoCl_2 \cdot 2H_2O + NaCl$
50.30	2.30	47.40	100	$CoCl_2 \cdot 2H_2O$
50.10	3.63	46.27	100	»
49.40	4.99	45.61	100	»
48.60	5.30	46.10	100	»
50.00	5.76	44.24	100	$CoCl_2 + NaCl$
44.40	6.44	49.16	100	NaCl
37.20	9.04	53.76	100	»
24.90	14.16	60.94	100	»
9.90	21.70	68.40	100	»

№ 3788 [2669]

SODIUM CHLORIDE–COBALT (II) IODATE–WATER

$NaCl—Co(IO_3)_2—H_2O$

$t = 30$

Concn. in aqueous soln., wt %		d_4	Comp. of solid phase
Sodium Chloride	Cobalt Iodate		
0.0	0.476	1.0006	$Co(IO_3)_2 \cdot 2H_2O$
0.0170	0.486	1.0007	»
0.0304	0.491	1.0008	»
0.0403	0.495	1.0009	»
0.0649	0.507	1.0010	»
0.0818	0.511	1.0011	»
0.156	0.535	1.0016	»
0.207	0.548	1.0020	»
0.311	0.568	1.0029	»
0.550	0.609	1.0054	»
0.847	0.646	1.0085	»

№ 3789 [1812]

SODIUM CHLORIDE–COPPER (II) SULFATE–WATER

$NaCl—CuSO_4—H_2O$

$t = 25$

Concn. in aqueous soln., g/l	
Copper Sulfate	Sodium Chloride
224.1	21.0
227.6	77.2
240.5	147.9

№ 3790 [2014]

SODIUM CHLORIDE–COPPER (I) CHLORIDE–WATER

$NaCl—CuCl—H_2O$

$t = 26.5$

Comp. of saturated soln., wt %			Comp. of solid phase
Sodium Chloride	Copper Chloride	Water	
0.00	1.52	98.48	CuCl
9.50	2.77	87.73	»
16.19	5.70	78.11	»
16.10	24.20	59.70	»
19.66	26.50	53.84	»
22.0	28.59	49.41	»
27.62	22.10	50.28	NaCl
28.60	21.0	50.40	»
30.0	11.02	58.98	»

№ 3791 [1876]

SODIUM CHLORIDE–COPPER (II) CHLORIDE–WATER

$NaCl—CuCl_2—H_2O$

Comp. of solvent, wt %		Solubility of Copper Chloride, g/l	t
Sodium Chloride	Water		
Saturated solution		89	11
»		119	40
»		169	90
15	85	36	11
15	85	60	40
15	85	103	90
5	95	0.0	11
5	95	11	40
5	95	26	90

№ 3792 [2500]

SODIUM CHLORIDE–COPPER (II) CHLORIDE–WATER

$NaCl—CuCl_2—H_2O$

$t = 30$

Comp. of saturated soln., wt %			Comp. of solid phase
Sodium Chloride	Copper Chloride	Water	
0.00	43.95	56.05	$CuCl_2 \cdot 2H_2O$
3.10	41.14	55.76	»
4.28	41.06	54.66	»
6.41	39.40	54.19	»
10.25	36.86	52.89	$CuCl_2+NaCl$
12.02	32.38	55.60	NaCl
12.25	32.40	55.35	»
13.54	28.64	57.82	»
15.40	23.72	60.88	»
18.44	16.98	64.58	»
20.61	11.03	68.36	»
26.47	0.00	73.53	»

№ 3793 [1597]

SODIUM CHLORIDE–COPPER CHLORIDE–WATER

$NaCl—CuCl_2—H_2O$

$t = 19$

Concn. in aqueous soln., g/l		Comp. of solid phase	Concn. in aqueous soln., g/l		Comp. of solid phase
Sodium Chloride	Copper Chloride		Sodium Chloride	Copper Chloride	
9.3	1.2	$CuCl_2 \cdot 2H_2O$	243.2	129.0	$CuCl_2 \cdot 2H_2O$
47.2	5.4	»	369.8	298.3	$CuCl_2+NaCl$
80.2	12.3	»	361.5	245.8	NaCl
123.1	29.2	»	329.0	50.4	»
170.7	54.4	»	317.0	0.0	»

$NaCl—CuCl_2—H_2O$

Comp. of saturated soln., wt %			t	Comp. of solid phase
Sodium Chloride	Copper Chloride	Water		
26.45	0.00	73.55	25	NaCl
26.08	1.07	72.85	25	»
25.16	2.93	71.91	25	»
25.12	4.75	70.13	25	»
24.15	6.02	69.83	25	»
22.39	9.35	68.26	25	»
21.99	11.71	66.30	25	»
21.29	15.98	62.73	25	»
20.23	18.60	61.17	25	»
19.90	20.30	59.80	25	»
18.57	24.34	57.09	25	»
16.45	28.43	55.12	25	»
15.00	31.39	53.61	25	»
11.95	34.35	53.70	25	»
8.66	38.31	53.03	25	»
8.96	38.24	52.80	25	$NaCl + CuCl_2 \cdot NaCl \cdot 2H_2O$
8.86	38.10	53.04	25	»
8.65	38.32	53.03	25	»
8.23	38.26	53.51	25	»
8.83	38.75	52.42	25	»
8.14	37.42	54.44	25	$CuCl_2 \cdot NaCl \cdot 2H_2O$
7.86	38.81	53.33	25	»
6.20	40.20	53.60	25	»
4.53	41.68	53.79	25	»
4.40	41.46	54.14	25	$CuCl_2 \cdot 2H_2O$
3.53	41.20	55.27	25	»
2.15	42.00	55.85	25	»
0.81	42.15	57.04	25	»
0.00	43.32	56.68	25	»
26.60	0.00	73.40	35	NaCl
24.45	4.16	71.39	35	»
23.62	8.20	68.18	35	»
22.02	12.12	65.86	35	»
20.84	15.91	63.25	35	»
18.54	22.78	58.68	35	»
15.76	28.42	55.82	35	»
12.40	35.04	52.56	35	»
10.82	38.02	51.16	35	»
10.90	37.90	51.20	35	$NaCl + CuCl_2 \cdot NaCl \cdot 2H_2O$
10.80	37.84	51.36	35	»
10.82	37.78	51.40	35	»
10.78	37.92	51.30	35	$CuCl_2 \cdot NaCl \cdot 2H_2O$
9.40	39.13	51.47	35	»
8.53	39.44	52.03	35	»
6.49	40.52	52.99	35	»
5.12	42.10	53.78	35	»
4.98	41.96	53.06	35	$CuCl_2 \cdot 2H_2O$
3.12	42.80	54.08	35	»
1.71	43.60	54.69	35	»
0.00	44.74	55.26	35	»

№ 3795 SODIUM CHLORIDE–COPPER (II) CHLORIDE–WATER 279, 417]

$NaCl—CuCl_2—H_2O$

$t = 25$

Comp. of saturated soln., wt %			Comp. of solid phase
Sodium Chloride	Copper Chloride	Water	
8.96	38.24	52.80	$NaCl + CuCl_2 \cdot NaCl \cdot 2H_2O$
8.86	38.10	53.04	»
8.65	38.32	53.03	»
8.23	38.26	53.51	»
8.83	38.75	52.42	»
8.14	37.42	54.44	$CuCl_2 \cdot NaCl \cdot 2H_2O$
7.86	38.81	53.33	»
6.20	40.20	53.60	»
4.53	41.68	53.79	»
4.40	41.46	54.14	$CuCl_2 \cdot 2H_2O$
3.53	41.20	55.27	»
2.15	42.00	55.85	»
0.81	42.15	57.04	»
0.00	43.32	56.68	»

№ 3796 [279, 417]

SODIUM CHLORIDE–COPPER (II) CHLORIDE–WATER

$NaCl—CuCl_2—H_2O$

$t = 25$

Comp. of saturated soln., wt %			Comp. of solid phase
Copper Chloride	Sodium Chloride	Water	
0.00	26.45	73.55	NaCl
1.07	26.08	72.85	»
2.93	25.16	71.91	»
4.75	25.12	70.13	»
6.02	24.15	69.83	»
9.35	22.39	68.26	»
11.71	21.99	66.30	»
15.98	21.29	62.73	»
18.60	20.23	61.17	»
20.30	19.90	59.80	»
24.34	18.57	57.09	»
28.43	16.45	55.12	»
31.39	15.00	53.61	»
34.35	11.95	53.70	»
38.31	8.66	53.03	»

№ 3797 [418]

SODIUM CHLORIDE–COPPER (I) CHLORIDE–WATER

$NaCl—CuCl—H_2O$

$t = 25$

Comp. of saturated soln., wt %			Comp. of solid phase
Copper Chloride	Sodium Chloride	Water	
0.0	26.45	73.55	NaCl
1.65	26.78	71.57	»
4.06	28.40	67.54	»
8.36	29.75	61.89	»
10.90	30.03	59.07	»
14.20	29.88	55.92	»
18.40	29.62	51.98	»
21.70	28.90	49.40	»
24.80	27.10	48.10	»
28.60	24.15	52.75	»
28.75	23.90	47.35	NaCl+CuCl
28.73	23.85	47.42	»
25.24	20.15	54.61	CuCl
10.40	15.20	74.40	»
4.10	10.05	85.85	»
2.05	4.11	93.84	»
1.63	0.00	98.37	

№ 3798 [2596]

SODIUM CHLORIDE–ZINC HYDROXIDE–WATER

$NaCl—Zn(OH)_2—H_2O$

$t = 16—20$

Comp. of solvent, wt %		Solubility of Zinc Hydroxide, g/l
Sodium Chloride	Water	
1	99	0.051

№ 3799 [1353]

SODIUM CHLORIDE–ZINC CARBONATE–WATER

$NaCl—ZnCO_3—H_2O$

$t = 14$

Comp. of solvent, wt %		Solubility of Zinc Carbonate, g/l
Sodium Chloride	Water	
5.85	94.15	0.0586
7.45	92.55	0.0477

№ 3800 SODIUM CHLORIDE–STRONTIUM CARBONATE–WATER [2262]

$NaCl—SrCO_3—H_2O$

Comp. of solvent, wt %		Solubility of Strontium Carbonate, g/1000 g Water	t	Comp. of solvent, wt %		Solubility of Strontium Carbonate, g/1000 g Water	t
Sodium Chloride	Water			Sodium Chloride	Water		
0.0	100.0	0.0081	25	0.0	100.0	0.010	40
0.117	99.883	0.011	25	0.117	99.883	0.013	40
0.291	99.709	0.014	25	0.291	99.709	0.016	40
0.58	99.42	0.019	25	0.58	99.42	0.022	40
1.15	98.85	0.029	25	1.15	98.85	0.032	40
1.72	98.28	0.037	25	1.72	98.28	0.041	40
2.28	97.72	0.043	25	2.28	97.72	0.048	40
2.84	97.16	0.048	25	2.84	97.16	0.054	40
5.52	94.48	0.065	25	5.52	94.48	0.076	40
14.91	85.09	0.101	25	14.91	85.09	0.127	40

№ 3801 [2703]

SODIUM CHLORIDE–STRONTIUM SULFATE–WATER

$NaCl—SrSO_4—H_2O$

$t = 25$

Comp. of solvent, wt %		Solubility of Strontium Sulfate, wt %
Sodium Chloride	Water	
8.44	91.56	0.164
15.54	84.46	0.217
22.17	77.83	0.180

№ 3802 [1693]

SODIUM CHLORIDE–STRONTIUM SULFATE–WATER

$NaCl—SrSO_4—H_2O$

Comp. of solvent, wt %		Solubility of Strontium Sulfate, g/l	t
Sodium Chloride	Water		
10	90	0.630	20
10	90	0.691	40

$NaCl—SrCl_2—H_2O$

Comp. of saturated soln., wt %			t	Comp. of solid phase
Strontium Chloride	Sodium Chloride	Water		
0.0	26.0	74.0	18	NaCl
9.0	20.1	70.9	18	»
18.0	14.2	67.8	18	»
24.3	10.1	65.6	18	»
26.7	8.7	64.6	18	»
27.1	8.6	64.3	18	$SrCl_2 \cdot 6H_2O + NaCl$
28.0	7.2	64.8	18	$SrCl_2 \cdot 6H_2O$
29.6	5.5	64.9	18	»
34.3	0.0	65.7	18	»
0.0	27.0	73.0	60	NaCl
8.9	21.2	69.9	60	»
17.8	15.2	67.0	60	»
24.0	11.6	64.4	60	»
36.9	5.1	58.0	60	»
44.0	3.0	53.0	60	»
44.3	2.9	52.8	60	$SrCl_2 \cdot 2H_2O + NaCl$
44.8	2.2	53.0	60	$SrCl_2 \cdot 2H_2O$
45.8	0.9	53.3	60	»
46.1	0.5	53.4	60	$SrCl_2 \cdot 6H_2O$
46.2	0.0	53.8	60	»
0.0	28.5	71.5	100	NaCl
8.8	22.5	68.7	100	»
17.5	17.2	65.3	100	»
23.6	13.1	63.3	100	»
36.4	6.6	57.0	100	»
48.0	3.8	48.2	100	»
48.2	3.6	48.2	100	$NaCl + SrCl_2 \cdot 2H_2O$
48.7	3.5	47.8	100	$SrCl_2 \cdot 2H_2O$
50.8	0.0	49.2	100	»
27.1	8.6	64.3	18.0	$SrCl_2 \cdot 6H_2O + NaCl$
35.2	5.1	59.7	40.0	»
39.3	3.9	56.8	50.0	»
41.5	3.4	55.1	55.0	»
43.6	3.1	53.3	58.7	»
44.2	2.9	52.9	59.6	$SrCl_2 \cdot 2H_2O + NaCl$
44.3	2.9	52.8	60.0	»
44.9	3.1	52.0	70.0	»
45.9	3.3	50.8	80.0	»
46.6	3.4	50.0	90.0	»
48.2	3.6	48.2	100.0	»
49.8	3.7	46.5	110.0	»
50.1	3.8	46.1	114.0	»

№ 3804 SODIUM CHLORIDE–OXOZIRCONIUM CHLORIDE–WATER [93]

$NaCl—ZrOCl_2—H_2O$

$t = 25$

Comp. of saturated soln., wt %			Comp. of solid phase
Sodium Chloride	Oxozirconium Chloride	Water	
26.71	0.00	73.29	NaCl
20.63	7.82	71.45	»
16.51	14.14	69.35	»
13.06	19.26	67.68	»
12.50	20.02	67.48	»
8.08	27.66	64.26	»
7.51	28.53	63.96	»
7.08	28.70	64.22	$NaCl+ZnOCl_2 \cdot 8H_2O$
7.00	28.90	64.10	$ZrOCl_2 \cdot 8H_2O$
5.52	31.00	63.48	»
2.81	34.00	63.19	»
0.00	37.60	62.40	»

№ 3805 [2486]

SODIUM CHLORIDE–SILVER CHLORIDE–WATER

$NaCl—AgCl—H_2O$

$t = 100$

Comp. of saturated soln., wt %		
Sodium Chloride	Silver Chloride	Water
26.3	0.460	73.2

№ 3806 SODIUM CHLORIDE–SILVER CHLORIDE–WATER [2486]

$NaCl—AgCl—H_2O$

Concn. in aqueous soln., wt %		t	Concn. in aqueous soln., wt %		t
Sodium Chloride	Silver Chloride		Sodium Chloride	Silver Chloride	
10.00	0.0025	15	14.0	0.054	80
14.29	0.0071	15	14.0	0.069	90
18.18	0.0182	15	14.0	0.090	100
21.98	0.0439	15	14.0	0.107	104
23.53	0.0706	15	26.3	0.128	15
25.64	0.103	15	26.3	0.132	30
26.31	0.127	15	26.3	0.158	40
14.0	0.007	15	26.3	0.184	50
14.0	0.011	30	26.3	0.263	70
14.0	0.014	40	26.3	0.315	80
14.0	0.023	50	26.3	0.368	90
14.0	0.042	70			

№ 3807 SODIUM CHLORIDE–SILVER CHLORIDE–WATER [1135]

$NaCl—AgCl—H_2O$

Concn. in aqueous soln., g/l		t	Concn. in aqueous soln., g/l		t
Sodium Chloride	Silver Chloride		Sodium Chloride	Silver Chloride	
34.3	0.0018	20	76.7	0.236	90
34.3	0.016	50	115.0	0.031	20
34.3	0.067	90	115.0	0.124	50
46.0	0.0025	20	115.0	0.436	90
46.0	0.025	50	153.0	0.090	20
46.0	0.100	90	153.0	0.191	50
57.5	0.0047	20	153.0	0.732	90
57.5	0.034	50	230.0	0.313	20
57.5	0.135	90	230.0	0.889	50
76.7	0.0125	20	230.0	1.706	90
76.7	0.058	50			

№ 3808 SODIUM CHLORIDE–SILVER CHLORIDE–WATER [1650]

$NaCl—AgCl—H_2O$

$t = 25$

Concn. in aqueous soln., g/l		Concn. in aqueous soln., g/l		Concn. in aqueous soln., g/l	
Sodium Chloride	Silver Chloride	Sodium Chloride	Silver Chloride	Sodium Chloride	Silver Chloride
54.53	0.0123	132.8	0.0815	219.0	0.3521
69.55	0.0186	155.4	0.1213	232.5	0.4117
83.76	0.0263	166.1	0.1487	255.0	0.5448
94.51	0.0350	191.1	0.2264	265.1	0.6146
109.4	0.0498	202.9	0.2713	294.5	0.8635

№ 3809 SODIUM CHLORIDE–SILVER CHLORIDE–WATER [1651]

$NaCl—AgCl—H_2O$

$t = 25$

Concn. in aqueous soln., g/l		Concn. in aqueous soln., g/l	
Sodium Chloride	Silver Chloride	Sodium Chloride	Silver Chloride
0.2256	0.00010	5.169	0.00052
0.5401	0.00013	20.78	0.0025
1.027	0.00019	29.88	0.0040
2.136	0.00027	56.97	0.0115

№ 3810 SODIUM CHLORIDE–SILVER CHLORIDE–WATER [1965]

$NaCl—AgCl—H_2O$

$t = 25$

Concn. in aqueous soln., g/l		Concn. in aqueous soln., g/l		Concn. in aqueous soln., g/l	
Sodium Chloride	Silver Chloride	Sodium Chloride	Silver Chloride	Sodium Chloride	Silver Chloride
52.65	0.0141	71.37	0.297	112.9	0.630
52.65	0.412	81.32	0.0262	112.9	0.549
52.65	0.143	81.32	0.548	129.8	0.0777
64.35	0.0187	81.32	0.357	129.8	0.689
64.35	0.453	94.18	0.0362	129.8	0.732
64.35	0.196	94.18	0.589	138.6	0.0893
71.37	0.0199	94.18	0.448	138,6	0.717
71.37	0.487	112.9	0.0561	138.6	0.843

№ 3811 [2329]

SODIUM CHLORIDE–SILVER CHLORIDE–WATER

$NaCl—AgCl—H_2O$

$t = 25$

Concn. in aqueous soln., g/l	
Sodium Chloride	Silver Chloride
0.0584	0.000054
0.1461	0.000041
0.5845	0.000068
5.845	0.000427
14.613	0.00129
29.225	0.00358
43.838	0.00741
58.45	0.01248

№ 3812 [2331]

SODIUM CHLORIDE–SILVER CHLORIDE–WATER

$NaCl—AgCl—H_2O$

$t = 25$

Concn. of Sodium Chloride in aqueous soln., g/l	Solubility of Silver Chloride in aqueous soln. of Sodium Chloride g/l
0.0058	0.000396
0.0584	0.000067
0.1461	0.000053
0.5845	0.000067
5.845	0.000417
29.22	0.00345
58.45	0.01208

№ 3813 SODIUM CHLORIDE–SILVER CHLORIDE–WATER [1941]

$NaCl—AgCl—H_2O$

Concn. in aqueous soln., g/l		t	Concn. in aqueous soln., g/l		t
Sodium Chloride	Silver Chloride		Sodium Chloride	Silver Chloride	
0.00322	0.000272	15	3.22	0.0001434	15
0.00644	0.000129	15	6.44	0.0002581	15
0.0322	0.0000459	15	0.00292	0.000774	25
0.0644	0.0000373	15	0.00585	0.000287	25
0.322	0.0000373	15	0.0292	0.0000989	25
0.644	0.0000488	15	0.0585	0.0000746	25

Table № 3813 (continuation)

Concn. in aqueous soln., g/l		t	Concn. in aqueous soln., g/l		t
Sodium Chloride	Silver Chloride		Sodium Chloride	Silver Chloride	
0.292	0.0000832	25	0.0323	0.000244	35
0.585	0.0000946	25	0.0644	0.000172	35
2.92	0.000244	25	0.322	0.000158	35
5.85	0.000545	25	0.644	0.000201	35
0.00376	0.000133	35	3.22	0.000559	35
0.00673	0.000717	35	6.44	0.001018	35

№ 3814 SODIUM CHLORIDE–SILVER CHLORIDE–WATER [2077]

$NaCl—AgCl—H_2O$

$t = 18$

Concn. of Sodium Chloride in aqueous soln., g/l	Solubility of Silver Chloride in aqueous soln. of Sodium Chloride, g/l	Concn. of Sodium Chloride in aqueous soln., g/l	Solubility of Silver Chloride in aqueous soln. of Sodium Chloride, g/l
0.00585	$1.49 \cdot 10^{-4}$	1.170	$0.844 \cdot 10^{-4}$
0.01170	$0.949 \cdot 10^{-4}$	2.925	$1.65 \cdot 10^{-4}$
0.02925	$0.58 \cdot 10^{-4}$	5.845	$2.92 \cdot 10^{-4}$
0.05850	$0.46 \cdot 10^{-4}$	11.79	$5.7 \cdot 10^{-4}$
0.1170	$0.407 \cdot 10^{-4}$	29.25	$2.76 \cdot 10^{-3}$
0.2925	$0.476 \cdot 10^{-4}$	58.45	$1.008 \cdot 10^{-2}$
0.5850	$0.605 \cdot 10^{-4}$	117.9	$4.73 \cdot 10^{-2}$

№ 3815 [2486]

SODIUM CHLORIDE–SILVER IODIDE–WATER

$NaCl— AgI —H_2O$

$t = 15$

Concn. in aqueous soln., wt %	
Sodium Chloride	Silver Iodide
25.00	0.00072
26.31	0.0244

№ 3816 SODIUM CHLORIDE–CADMIUM CHLORIDE–WATER [2615]

$NaCl—CdCl_2—H_2O$

Comp. of saturated soln., wt %			t	Comp. of solid phase
Sodium Chloride	Cadmium Chloride	Water		
0.00	52.67	47.33	19.3	$CdCl_2 \cdot 2.5H_2O$
3.35	52.03	44.62	19.3	$CdCl_2 \cdot 2\frac{1}{2}H_2O + CdCl_2 \cdot 2NaCl \cdot 3H_2O$
6.18	43.15	50.67	19.3	$CdCl_2 \cdot 2NaCl \cdot 3H_2O$
15.49	24.15	60.36	19.3	»
25.75	4.16	70.09	19.3	$CdCl_2 \cdot 2NaCl \cdot 3H_2O + NaCl$
26.38	0.0	73.62	19.3	NaCl
3.97	54.76	41.27	29.7	$CdCl_2 \cdot 2\frac{1}{2}H_2O + CdCl_2 \cdot 2NaCl \cdot 3H_2O$
4.32	52.87	42.81	29.7	$CdCl_2 \cdot 2NaCl \cdot 3H_2O$
5.90	48.46	45.64	29.7	»
7.46	44.11	48.43	29.7	»
16.40	25.55	58.05	29.7	»
25.54	6.41	68.05	29.7	$CdCl_2 \cdot 2NaCl \cdot 3H_2O + NaCl$
6.00	54.34	39.66	40.1	$CdCl_2 \cdot H_2O + CdCl_2 \cdot 2NaCl \cdot 3H_2O$
16.60	27.4	56.29	40.1	$CaCl_2 \cdot 2NaCl \cdot 3H_2O$
25.19	8.78	66.03	40.1	$CdCl_2 \cdot 2NaCl \cdot 3H_2O + NaCl$
7.36	54.11	38.53	54.5	$CdCl_2 \cdot H_2O + CdCl_2 \cdot 2NaCl \cdot 3H_2O$
17.75	28.41	53.84	54.5	$CdCl_2 \cdot 2NaCl \cdot 3H_2O$
24.16	13.93	62.91	54.5	$CdCl_2 \cdot 2NaCl \cdot 3H_2O + NaCl$
26.91	0.00	73.09	54.5	NaCl

№ 3817 SODIUM CHLORIDE–CADMIUM CHLORIDE–WATER [1166]

$NaCl—CdCl_2—H_2O$

$t = 25$

Concn. in aqueous soln., wt %		Comp. of solid phase
Sodium Chloride	Cadmium Chloride	
25.90	2.60	NaCl
25.50	5.25	»
23.14	8.62	$NaCl + 2NaCl \cdot CdCl_2 \cdot 3H_2O$
19.95	14.42	$2NaCl \cdot CdCl_2 \cdot 3H_2O$
18.36	17.37	»
16.97	22.26	»
15.37	27.34	»
14.46	30.55	»
13.78	31.76	$2NaCl \cdot CdCl_2 \cdot 3H_2O + 3NaCl \cdot 4CdCl_2 \cdot 14H_2O$
12.82	33.32	$3NaCl \cdot 4CdCl_2 \cdot 14H_2O$
8.65	40.75	»
6.72	45.09	»
4.68	50.93	»
3.95	53.71	$3NaCl \cdot 4CdCl_2 \cdot 14H_2O + CdCl_2 \cdot 2.5H_2O$
2.61	53.55	$CdCl_2 \cdot 2.5H_2O$
0.00	54.62	»

SODIUM CHLORIDE–CADMIUM CHLORIDE–WATER

$NaCl—CdCl_2—H_2O$

Concn. in aqueous soln., wt %		d_4^t	t	Comp. of solid phase
Cadmium Chloride	Sodium Chloride			
46.45	2.5	—	0	$CdCl_2 \cdot 2.5H_2O + 4CdCl_2 \cdot 3NaCl \cdot 14H_2O$
42.3	3.6	1.587	0	$4CdCl_2 \cdot 3NaCl \cdot 14H_2O$
38.7	4.4	1.533	0	»
29.0	7.75	1.409	0	»
25.4	8.65	1.357	0	»
19.7	13.0	—	0	»
17.8	15.2	—	0	$4CdCl_2 \cdot 3NaCl \cdot 14H_2O + CdCl_2 \cdot 2NaCl \cdot 3H_2O$
6.8	20.5	1.236	0	$CdCl_2 \cdot 2NaCl \cdot 3H_2O$
5.55	21.35	—	0	»
3.15	24.0	—	0	»
4.95	21.95	—	0	$CdCl_2 \cdot 2NaCl \cdot 3H_2O + CdCl_2 \cdot 8NaCl \cdot 16H_2O$
5.05	21.9	—	0	$6CdCl_2 \cdot 5NaCl \cdot H_2O + CdCl_2 \cdot 8NaCl \cdot 16H_2O$
4.45	22.1	—	0	$CdCl_2 \cdot 8NaCl \cdot 16H_2O$
3.15	22.55	—	0	»
1.1	23.65	—	0	»
0.75	24.0	—	0	»
2.1	25.9	—	0	$CdCl_2 \cdot 2NaCl \cdot 3H_2O + NaCl$
0.07	26.35	—	0	$CdCl_2 \cdot 8NaCl \cdot 16H_2O + NaCl + NaCl \cdot 2H_2O$
51.6	3.6	1.770	19.3	$CdCl_2 \cdot 2.5 \cdot H_2O + 4CdCl_2 \cdot 3NaCl \cdot 13H_2O$
50.2	4.1	1.745	19.3	$4CdCl_2 \cdot 3NaCl \cdot 13H_2O$
41.0	7.3	1.600	19.3	»
36.3	9.3	1.537	19.3	»
28.4	14.5	1.464	19.3	$4CdCl_2 \cdot 3NaCl \cdot 13H_2O + CdCl_2 \cdot 2NaCl \cdot 3H_2O$
4.3	25.8	1.243	19.3	$CdCl_2 \cdot 2NaCl \cdot 3H_2O + NaCl$
54.2	6.2	1.860	39	$CdCl_2 \cdot H_2O + 4CdCl_2 \cdot 3NaCl \cdot 13H_2O$
50.4	7.8	1.792	39	$4CdCl_2 \cdot 3NaCl \cdot 13H_2O$
46.5	9.6	1.729	39	»
42.6	11.7	1.678	39	»
39.7	13.3	1.639	39	$CdCl_2 \cdot NaCl \cdot 2H_2O + CdCl_2 \cdot 2NaCl \cdot 3H_2O$
37.5	13.8	1.600	39	$CdCl_2 \cdot 2NaCl \cdot 3H_2O$
8.5	25.45	1.278	39	$CdCl_2 \cdot 2NaCl \cdot 3H_2O + NaCl$
53.4	9.0	1.866	60	$CdCl_2 \cdot H_2O + CdCl_2 \cdot NaCl \cdot 2H_2O$
49.9	10.6	1.803	60	$CdCl_2 \cdot NaCl \cdot 2H_2O$
47.4	11.6	1.764	60	»
42.6	14.0	1.686	60	»
41.1	14.7	1.662	60	$CdCl_2 \cdot NaCl \cdot 2H_2O + CdCl_2 \cdot 2NaCl \cdot 3H_2O$
36.5	16.3	1.588	60	$CdCl_2 \cdot 2NaCl \cdot 3H_2O$
15.1	24.8	1.388	60	$CdCl \cdot 2NaCl \cdot 3H_2O + NaCl$
57.8	2.6	—	100	$CdCl_2 \cdot H_2O$
56.0	5.9	—	100	»
54.9	11.2	—	100	»
54.7	12.6	—	100	»
53.6	13.0	—	100	$CdCl_2 \cdot NaCl \cdot 2H_2O$
48.6	15.6	—	100	»
47.6	16.4	—	100	$CdCl_2 \cdot NaCl \cdot 2H_2O + CdCl_2 \cdot 2NaCl \cdot 3H_2O$
39.1	19.7	—	100	$CdCl_2 \cdot 2NaCl \cdot 3H_2O$
28.9	24.0	—	100	$CdCl_2 \cdot 2NaCl \cdot 3H_2O + NaCl$
27.0	24.3	—	100	$NaCl$
18.0	26.3	—	100	»

$NaCl—CsCl—H_2O$

Comp. of saturated soln., wt %			t	d_4^t	Comp. of solid phase
Sodium Chloride	Cesium Chloride	Water			
26.58	0.00	73.42	25	1.119	NaCl
21.77	8.31	69.92	25	1.306	»
16.78	27.98	55.24	25	1.476	»
12.37	41.76	45.87	25	1.640	»
11.95	42.33	45.72	25	1.642	»
11.95	42.88	45.77	25	1.644	NaCl+ Solid soln.
8.37	49.26	43.37	25	1.710	Solid soln.
2.98	57.32	39.70	25	1.861	»
1.60	62.94	35.46	25	1.900	»
0.97	64.16	34.87	25	1.903	»
0.62	64.85	34.53	25	1.915	»
0.00	65.77	34.27	25	1.924	CsCl
26.78	0.00	73.22	50	1.188	NaCl
23.78	7.88	68.34	50	1.270	»
22.16	12.82	65.02	50	1.285	»
17.33	26.21	56.46	50	1.411	»
12.75	40.30	46.95	50	1.571	»
9.27	50.97	39.79	50	1.730	»
8.65	53.17	38.18	50	1.774	NaCl+ Solid soln.
6.22	57.00	36.78	50	—	solid soln.
5.06	59.53	35.00	50	1.855	»
3.51	62.49	34.00	50	1.900	»
1.95	65.66	32.35	50	1.945	»
1.30	66.40	32.30	50	—	solid soln.
0.00	68.61	31.39	50	1.994	CsCl
27.24	0.00	72.76	75	1.181	NaCl
23.01	11.60	65.30	75	1.266	»
20.24	17.20	62.56	75	1.330	»
17.66	26.85	55.49	75	1.379	»
13.30	36.10	50.60	75	1.512	»
10.86	45.55	43.59	75	1.688	»
7.20	57.99	34.81	75	1.844	»
6.75	59.62	33.63	75	1.889	NaCl+ solid soln.
5.14	62.45	32.43	75	1.919	solid soln.
2.90	66.19	30.91	75	1.942	»
0.43	69.47	30.10	75	—	»
0.00	70.62	29.38	75	2.039	CsCl

№ 3820 [1812]

SODIUM CHLORIDE–BARIUM HYDROXIDE–WATER

$NaCl—Ba(OH)_2—H_2O$

$t = 25$

Content in aqueous soln., g/l	
Sodium Chloride	Barium Hydroxide
0.0	47.6
42.7	54.0
83.7	59.9
165.1	69.1

№ 3821 [1352]

SODIUM CHLORIDE–BARIUM CARBONATE–WATER

$NaCl—BaCO_3—H_2O$

Comp. of solvent, wt %		Solubility of Barium Carbonate, g/l	t
Sodium Chloride	Water		
0.15	99.85	0.0587	b.p.
1.0	99.0	0.0787	»
3.0	97.0	0.1056	»
10.0	90.0	0.1573	»
30.0	70.0	0.2784	»
10.0	90.0	0.1085	10
10.0	90.0	0.1126	20
10.0	90.0	0.1231	40
10.0	90.0	0.1303	60
10.0	90.0	0.1418	80

№ 3822 [2662]

SODIUM CHLORIDE–BARIUM CARBONATE–WATER

$NaCl—BaCO_3—H_2O$

Comp. of solvent, wt %		Solubility of Barium Carbonate, g/1000 g Water	
Sodium Chloride	Water	$t = 25$	$t = 40$
0.0	100.0	0.0180	0.0241
0.117	99.883	0.0231	0.0336
0.291	99.709	0.0299	0.0438
0.58	99.42	0.0401	0.0554
1.15	98.85	0.0567	0.0726
1.72	98.28	0.0705	0.0848
2.28	97.72	0.0825	0.0918
2.84	97.16	0.093	0.101
5.52	94.48	0.136	0.122
14.91	85.09	0.196	0.170

№ 3823 SODIUM CHLORIDE–BARIUM METASILICATE–WATER [2020]

$NaCl—BaSiO_3 \cdot 6H_2O—H_2O$

$t = 20$

Concn. of Sodium Chloride in aqueous soln., g/l	Solubility of Barium Metasilicate Hexahydrate in aqueous soln. of Sodium Chloride, g/l	Concn. of Sodium Chloride in aqueous soln., g/l	Solubility of Barium Metasilicate Hexahydrate in aqueous soln. of Sodium Chloride, g/l
0.2922	1.221	1.1688	1.337
0.5844	1.231	1.7432	1.367

№ 3824 SODIUM CHLORIDE–BARIUM SULFATE–WATER [2639]

$NaCl—BaSO_4—H_2O$

Concn. of Sodium Chloride in aqueous soln., g-mol/1000 g Water	Solubility product of $BaSO_4$*	t	Concn. of Sodium Chloride in aqueous soln., g-mol/1000 g Water	Solubility product of $BaSO_4$*	t
0.000	$1.166 \cdot 10^{-5}$	25	4.500	$342.250 \cdot 10^{-5}$	35
0.001	$1.638 \cdot 10^{-5}$	25	5.000	$376.360 \cdot 10^{-5}$	35
0.005	$2.560 \cdot 10^{-5}$	25	0.000	$2.074 \cdot 10^{-5}$	50
0.010	$3.386 \cdot 10^{-5}$	25	0.001	$2.856 \cdot 10^{-5}$	50
0.050	$9.120 \cdot 10^{-5}$	25	0.005	$4.410 \cdot 10^{-5}$	50
0.100	$15.366 \cdot 10^{-5}$	25	0.010	$5.905 \cdot 10^{-5}$	50
0.200	$27.04 \cdot 10^{-5}$	25	0.050	$16.000 \cdot 10^{-5}$	50
0.400	$44.89 \cdot 10^{-5}$		0.100	$27.040 \cdot 10^{-5}$	50
0.600	$60.84 \cdot 10^{-5}$	25	0.200	$47.610 \cdot 10^{-5}$	50
0.800	$77.44 \cdot 10^{-5}$	25	0.400	$79.210 \cdot 10^{-5}$	50
1.000	$92.16 \cdot 10^{-5}$	25	0.600	$110.250 \cdot 10^{-5}$	50
1.500	$125.44 \cdot 10^{-5}$	25	0.800	$136.890 \cdot 10^{-5}$	50
2.000	$156.25 \cdot 10^{-5}$	25	1.000	$163.840 \cdot 10^{-5}$	50
2.500	$182.25 \cdot 10^{-5}$	25	1.500	$222.010 \cdot 10^{-5}$	50
3.000	$207.36 \cdot 10^{-5}$	25	2.000	$272.250 \cdot 10^{-5}$	50
3.500	$234.09 \cdot 10^{-5}$	25	2.500	$313.290 \cdot 10^{-5}$	50
4.000	$259.21 \cdot 10^{-5}$	25	3.000	$349.690 \cdot 10^{-5}$	50
4.500	$285.61 \cdot 10^{-5}$	25	3.500	$388.090 \cdot 10^{-5}$	50
5.0	$313.290 \cdot 10^{-5}$	25	4.000	$424.360 \cdot 10^{-5}$	50
0.000	$1.562 \cdot 10^{-5}$	35	4.500	$457.960 \cdot 10^{-5}$	50
0.001	$2.132 \cdot 10^{-5}$	35	5.000	$492.840 \cdot 10^{-5}$	50
0.005	$3.276 \cdot 10^{-5}$	35	0.000	$2.372 \cdot 10^{-5}$	65
0.010	$4.326 \cdot 10^{-5}$	35	0.001	$2.349 \cdot 10^{-5}$	65
0.050	$12.180 \cdot 10^{-5}$	35	0.005	$3.026 \cdot 10^{-5}$	65
0.100	$19.981 \cdot 10^{-5}$	35	0.010	$7.290 \cdot 10^{-5}$	65
0.200	$33.640 \cdot 10^{-5}$	35	0.050	$19.097 \cdot 10^{-5}$	65
0.400	$56.250 \cdot 10^{-5}$	35	0.100	$33.408 \cdot 10^{-5}$	65
0.600	$77.440 \cdot 10^{-5}$	35	0.200	$59.290 \cdot 10^{-5}$	65
0.800	$96.130 \cdot 10^{-5}$	35	0.400	$106.090 \cdot 10^{-5}$	65
1.000	$112.360 \cdot 10^{-5}$	35	0.600	$153.760 \cdot 10^{-5}$	65
1.500	$153.760 \cdot 10^{-5}$	35	0.800	$201.640 \cdot 10^{-5}$	65
2.000	$190.440 \cdot 10^{-5}$	35	1.000	$240.250 \cdot 10^{-5}$	65
2.500	$219.040 \cdot 10^{-5}$	35	1.500	$324.000 \cdot 10^{-5}$	65
3.000	$246.490 \cdot 10^{-5}$	35	2.000	$396.010 \cdot 10^{-5}$	65
3.500	$275.560 \cdot 10^{-5}$	35	2.500	$449.440 \cdot 10^{-5}$	65
4.000	$306.250 \cdot 10^{-5}$	35	3.000	$497.290 \cdot 10^{-5}$	65
3.500	$538.240 \cdot 10^{-5}$	65	5.000	$954.810 \cdot 10^{-5}$	80

Table № 3824 (continuation)

Concn. of Sodium Chloride in aqueous soln., g-mol/1000 g Water	Solubility product of $BaSO_4$*	t	Concn. of Sodium Chloride in aqueous soln., g-mol/1000 g Water	Solubility product of $BaSO_4$*	t
4.000	$580.810 \cdot 10^{-5}$	65	0.000	$2.822 \cdot 10^{-5}$	95
4.500	$630.010 \cdot 10^{-5}$	65	0.001	$3.881 \cdot 10^{-5}$	95
5.000	$676.000 \cdot 10^{-5}$	65	0.005	$6.502 \cdot 10^{-5}$	95
0.000	$2.624 \cdot 10^{-5}$	80	0.010	$9.302 \cdot 10^{-5}$	95
0.005	$6.101 \cdot 10^{-5}$	80	0.050	$23.522 \cdot 10^{-5}$	95
0.010	$8.410 \cdot 10^{-5}$	80	0.100	$39.690 \cdot 10^{-5}$	95
0.050	$21.809 \cdot 10^{-5}$	80	0.200	$77.440 \cdot 10^{-5}$	95
0.100	$37.577 \cdot 10^{-5}$	80	0.400	$161.290 \cdot 10^{-5}$	95
0.200	$70.640 \cdot 10^{-5}$	80	0.600	$249.640 \cdot 10^{-5}$	95
0.400	$136.890 \cdot 10^{-5}$	80	0.800	$334.890 \cdot 10^{-5}$	95
0.600	$204.490 \cdot 10^{-5}$	80	1.000	$420.250 \cdot 10^{-5}$	95
0.800	$265.690 \cdot 10^{-5}$	80	1.500	$620.010 \cdot 10^{-5}$	95
1.000	$327.61 \cdot 10^{-5}$	80	2.000	$789.610 \cdot 10^{-5}$	95
1.500	$449.44 \cdot 10^{-5}$	80	2.500	$936.360 \cdot 10^{-5}$	95
2.000	$561.690 \cdot 10^{-5}$	80	3.000	$1075.840 \cdot 10^{-5}$	95
2.500	$635.040 \cdot 10^{-5}$	80	3.500	$1204.090 \cdot 10^{-5}$	95
3.000	$702.250 \cdot 10^{-5}$	80	4.000	$1324.960 \cdot 10^{-5}$	95
3.500	$767.290 \cdot 10^{-5}$	80	4.500	$1447.000 \cdot 10^{-5}$	95
4.000	$829.440 \cdot 10^{-5}$	80	5.000	$1576.090 \cdot 10^{-5}$	95
4.500	$894.010 \cdot 10^{-5}$	80			

*The concns. of Ba^{2+} and SO_4^{2-} ions are expressed in g-mol/1000 g soln.

№ 3825 [1951]

SODIUM CHLORIDE–BARIUM CHLORIDE–WATER

$NaCl—BaCl_2—H_2O$

$t = 17$

Comp. of saturated soln., wt %		
Sodium Chloride	Barium Chloride	Water
24.89	2.96	72.15

№ 3826 [3251]

SODIUM CHLORIDE–BARIUM CHLORIDE–WATER

$NaCl—BaCl_2—H_2O$

Content in aqueous soln., wt %		t	Comp. of solid phase
Sodium Chloride	Barium Chloride		
24.80	2.5	10	$NaCl + BaCl_2 \cdot 2H_2O$
7.40	16.2	10	$BaCl_2 \cdot 2H_2O$
0	25.0	10	»
24.5	3.0	20	$NaCl + BaCl_2 \cdot 2H_2O$
7.33	17.4	20	$BaCl_2 \cdot 2H_2O$

Table № 3826 (continuation)

Content in aqueous soln., wt %		t	Comp. of solid phase
Sodium Chloride	Barium Chloride		
0	26.4	20	$BaCl_2 \cdot 2H_2O$
24.3	3.6	30	$NaCl+BaCl_2 \cdot 2H_2O$
7.26	18.7	30	$BaCl_2 \cdot 2H_2O$
0	27.7	30	»
24.0	4.5	40	$NaCl+BaCl_2 \cdot 2H_2O$
7.18	20.1	40	$BaCl_2 \cdot 2H_2O$
0	29.0	40	»
23.7	5.5	50	$NaCl+BaCl_2 \cdot 2H_2O$
7.10	21.15	50	$BaCl_2 \cdot 2H_2O$
0	30.3	50	»

№ 3827 SODIUM CHLORIDE–BARIUM CHLORIDE–WATER [2449]

$NaCl—BaCl_2—H_2O$

Comp. of saturated soln., wt %			t	Comp. of saturated soln., wt %			t
Barium Chloride	Sodium Chloride	Water		Barium Chloride	Sodium Chloride	Water	
2.9	25.0	72.1	20	9.4	22.8	67.2	80
4.5	23.0	72.5	40	11.8	22.2	66.0	100
6.8	23.4	69.8	60				

№ 3828 [2501]

SODIUM CHLORIDE–BARIUM CHLORIDE–WATER

$NaCl—BaCl_2—H_2O$

$t = 30$

Comp. of saturated soln., wt %			Comp. of solid phase
Barium Chloride	Sodium Chloride	Water	
0.00	26.47	73.53	NaCl
2.27	25.28	72.44	»
3.80	23.77	72.43	$NaCl+BaCl_2 \cdot 2H_2O$
5.76	20.25	73.99	$BaCl_2 \cdot 2H_2O$
8.19	17.89	73.92	»
12.25	13.39	74.36	»
15.83	10.06	74.11	»
20.93	5.39	73.68	»
24.24	2.76	73.00	»
27.60	0.00	72.40	»

№ 3829 [1606]

SODIUM CHLORIDE–BARIUM CHLORIDE–WATER

$NaCl—BaCl_2—H_2O$

$t = 20$

Comp. of saturated soln., wt %			Comp. of solid phase
Sodium Chloride	Barium Chloride	Water	
7.80	16.89	75.31	$BaCl_2 \cdot 2H_2O$
9.54	14.99	75.47	»
14.50	10.04	75.46	»
18.52	6.51	74.97	»
21.23	4.17	74.60	»
24.59	3.07	72.34	$BaCl_2 \cdot 2H_2O+NaCl$
25.28	1.65	73.07	NaCl

№ 3830 [1502]

SODIUM CHLORIDE–BARIUM CHLORIDE–WATER

$NaCl—BaCl_2—H_2O$

$t = 20$

Comp. of saturated soln., wt %			Comp. of solid phase
Sodium Chloride	Barium Chloride	Water	
0.00	26.20	73.80	$BaCl_2 \cdot 2H_2O$
4.03	25.00	70.97	»
8.00	24.50	77.50	»
12.00	22.12	65.88	»
14.56	21.51	63.93	»
16.50	21.29	62.21	»
19.53	18.52	61.95	NaCl
20.32	12.10	67.58	»
22.01	9.98	68.01	»
23.51	4.26	72.23	»
26.50	0.00	73.50	»

№ 3831 [62]

SODIUM CHLORIDE–BARIUM CHLORIDE–WATER

$NaCl—BaCl_2—H_2O$

$t = 20$

Concn. in aqueous soln., wt %		Comp. of solid phase	Concn. in aqueous soln., wt %		Comp. of solid phase
Barium Chloride	Sodium Chloride		Barium Chloride	Sodium Chloride	
16.89	7.38	$BaCl_2 \cdot 2H_2O$	4.17	21.23	$BaCl_2 \cdot 2H_2O$
10.04	14.5	»	3.07	24.59	$BaCl_2 \cdot 2H_2O+NaCl$
6.51	18.52	»	1.65	25.28	NaCl

№ 3832 [1229]

SODIUM CHLORIDE–BARIUM CHROMATE–WATER

$NaCl—BaCrO_4—H_2O$

$t = 25$

Comp. of saturated soln., wt %	
Sodium Chloride	Barium Chromate
3.65	0.00709
14.61	0.01133

№ 3833 [1001]

SODIUM CHLORIDE–LANTHANIUM CHLORIDE–WATER

$NaCl—LaCl_3—H_2O$

$t = 25$

Comp. of saturated soln., wt %			Comp. of solid phase
Lanthanum Chloride	Sodium Chloride	Water	
49.00	0.00	51.00	$LaCl_3 \cdot 7H_2O$
48.72	0.20	51.08	»
47.98	0.97	51.05	$LaCl_3 \cdot 7H_2O + NaCl$
47.92	0.99	51.09	»
48.00	0.98	51.02	»
47.96	0.95	51.09	»
47.93	0.99	51.08	»
47.56	1.20	51.24	NaCl
46.54	1.25	52.21	»
43.84	1.62	55.54	»
42.36	1.98	55.66	»
40.02	2.39	57.59	»
35.30	4.38	60.32	»
21.82	11.55	66.63	»
15.45	16.00	68.55	»
11.04	18.60	70.36	»
9.72	19.47	70.81	»
8.23	20.36	71.41	»
4.29	23.21	72.50	»
2.64	24.73	72.63	»
0.00	26.40	73.60	»

№ 3834 [1677]

SODIUM CHLORIDE–LANTHANUM IODATE–WATER

$NaCl—La(IO_3)_3—H_2O$

$t = 25$

Concn. of Sodium Chloride in aqueous soln., g/l	Solubility of Lanthanum Iodate in aqueous soln. of Sodium Chloride, g/l
5.85	1.0798

№ 3835 SODIUM CHLORIDE–CERIUM CHLORIDE–WATER [959]

$NaCl—CeCl_3—H_2O$

Comp. of saturated soln., wt %			t	Comp. of solid phase
Cerium Chloride	Sodium Chloride	Water		
0.0	26.5	73.5	30	NaCl
12.4	17.2	70.4	30	»
24.5	9.5	66.0	30	»
27.7	7.8	64.5	30	»
38.4	4.0	57.6	30	»
48.5	2.0	49.5	30	$NaCl+CeCl_3 \cdot 7H_2O$
49.0	1.3	49.7	30	$CeCl_3 \cdot 7H_2O$
49.2	0.0	50.8	30	»
0.0	27.0	73.0	50	NaCl
12.3	17.8	69.9	50	»
26.5	9.0	64.5	50	»
27.5	8.5	64.0	50	»
38.0	5.0	57.0	50	»
50.5	2.5	47.0	50	$NaCl+CeCl_3 \cdot 7H_2O$
51.2	1.1	47.7	50	$CeCl_3 \cdot 7H_2O$
51.7	0.0	48.3	50	»

№ 3836 [1001]

SODIUM CHLORIDE–NEODYMIUM CHLORIDE–WATER

$NaCl—NdCl_3—H_2O$

$t = 25$

Comp. of saturated soln., wt %			Comp. of solid phase
Neodymium Chloride	Sodium Chloride	Water	
49.94	0.00	50.06	$NdCl_3 \cdot 6H_2O$
48.73	0.61	50.66	$NdCl_3 \cdot 6H_2O+NaCl$
48.71	0.62	50.67	»
48.72	0.61	50.67	»
48.69	0.61	50.70	NaCl
46.28	1.82	51.80	»
42.14	2.80	55.04	»

Table № 3836 (continuation)

Comp. of saturated soln., wt %			Comp. of solid phase
Neodymium Chloride	Sodium Chloride	Water	
34.44	5.90	59.66	NaCl
30.02	7.75	62.23	»
23.12	11.68	65.20	»
14.34	16.74	68.92	»
7.85	20.28	71.85	»
4.50	23.56	71.94	»
2.26	24.74	73.00	»
1.78	25.64	72.58	»
0.00	26.36	73.64	»

№ 3837 SODIUM CHLORIDE–PLATINUM CHLORIDE–WATER [445]

$NaCl—PtCl_4—H_2O$

Concn. in aqueous soln., wt %		t	Comp. of solid phase
Sodium Chloride	Platinum Chloride		
0.0	62.38	40	$PtCl_4 \cdot 5H_2O$
0.11	61.47	40	$PtCl_4 \cdot 5H_2O + Na_2PtCl_6 \cdot 6H_2O$
2.91	60.43	40	$Na_2PtCl_6 \cdot 6H_2O$
4.35	59.05	40	»
5.28	56.33	40	»
6.19	53.60	40	»
7.43	50.03	40	»
16.93	42.79	40	»
11.11	40.31	40	»
13.14	37.89	40	»
14.66	33.60	40	»
20.69	30.57	40	»
20.18	29.26	40	»
23.80	25.54	40	»
23.33	24.36	40	»
29.71	20.37	40	$Na_2PtCl_6 \cdot 6H_2O + NaCl$
29.05	19.79	40	NaCl
28.83	18.76	40	»
28.50	17.48	40	»
24.44	19.75	40	»
28.14	1.78	40	»
27.04	0.0	40	»
0.0	74.75	60	$PtCl_4 \cdot 4H_2O$
1.71	72.35	60	»
3.84	69.25	60	»
4.75	68.11	60	$PtCl_4 \cdot 4H_2O + Na_2PtCl_6 \cdot 6H_2O$
5.15	66.54	60	$Na_2PtCl_6 \cdot 6H_2O$
5.98	65.41	60	»
6.55	61.96	60	»
7.75	59.67	60	»

Table № 3837 (continuation)

Concn. in aqueous soln., wt %		t	Comp. of solid phase
Sodium Chloride	Platinum Chloride		
8.91	56.09	60	$Na_2PtCl_6 \cdot 6H_2O$
10.64	51.80	60	»
11.95	48.30	60	»
15.23	43.90	60	»
16.19	40.98	60	»
18.64	38.04	60	»
21.30	34.56	60	»
23.69	35.13	60	»
25.49	33.87	60	»
29.04	32.48	60	»
30.97	31.93	60	»
31.87	31.73	60	»
33.44	31.13	60	$Na_2PtCl_6 \cdot 6H_2O + NaCl$
32.35	29.95	60	NaCl
31.30	25.37	60	»
29.27	20.79	60	»
29.02	11.73	60	»
26.95	5.89	60	»
27.15	0.0	60	»
0.0	78.60	80	$PtCl_4 \cdot 3H_2O$
1.89	75.88	80	»
3.55	73.25	80	»
5.19	71.01	80	$PtCl_4 \cdot 3H_2O + Na_2PtCl_6 \cdot 6H_2O$
7.73	63.69	80	$Na_2PtCl_6 \cdot 6H_2O$
16.46	47.47	80	»
12.68	53.18	80	»
18.03	44.66	80	»
21.98	40.18	80	»
28.88	34.19	80	»
34.26	29.94	80	$Na_2PtCl_6 \cdot 6H_2O + NaCl$
34.28	25.67	80	NaCl
33.09	20.92	80	»
33.06	17.10	80	»
31.99	13.92	80	»
30.48	7.03	80	»
29.26	3.57	80	»
27.55	0.0	80	»
0.0	82.50	90	$PtCl_4 \cdot 3H_2O$
1.85	76.55	90	»
3.58	76.34	90	»
4.83	75.66	90	»
6.36	72.75	90	$PtCl_4 \cdot 3H_2O + Na_2PtCl_6 \cdot 6H_2O$
7.50	70.21	90	$Na_2PtCl_6 \cdot 6H_2O$
8.54	66.18	90	»
10.44	63.48	90	»
13.25	57.66	90	»
17.52	50.58	90	»
14.12	46.22	90	»
24.43	41.35	90	»
36.48	32.11	90	$Na_2PtCl_6 \cdot 6H_2O + NaCl$
35.67	29.44	90	NaCl

Table № 3837 (continuation)

Concn. in aqueous soln., wt %		t	Comp. of solid phase
Sodium Chloride	Platinum Chloride		
35.61	25.69	90	NaCl
35.17	22.12	90	»
34.48	15.70	90	»
33.03	8.93	90	»
31.24	3.79	90	»
28.01	0.0	90	»
0.0	85.11	98	$PtCl_4 \cdot 3H_2O$
1.36	82.64	98	»
2.69	80.52	98	»
3.33	78.89	98	»
4.37	78.15	98	»
4.60	77.29	98	»
5.83	76.27	98	$PtCl_4 \cdot 3H_2O + Na_2PtCl_6 \cdot 6H_2O$
7.24	72.86	98	$Na_2PtCl_6 \cdot 6H_2O$
9.57	68.15	98	»
12.13	63.49	98	»
17.26	57.02	98	»
18.52	53.39	98	»
21.87	47.19	98	»
25.34	43.60	98	»
30.57	41.99	98	»
38.20	32.34	98	»
43.73	32.10	98	$Na_2PtCl_6 \cdot 6H_2O + NaCl$
42.65	27.42	98	NaCl
39.61	23.14	98	»
38.38	18.60	98	»
35.26	12.01	98	»
31.40	1.44	98	»
28.40	0.0	98	»

№ 3838 [1818]
SODIUM CHLORIDE–MERCURY (II) OXIDE–WATER
$NaCl—HgO—H_2O$

$t = 25$

Concn. in aqueous soln., g/l	
Mercury Oxide	Sodium Chloride
0.28158	61.4
0.38988	90.75
0.49818	124
0.80142	183
0.98553	246.6
1.71114	318.6

№ 3839 SODIUM CHLORIDE–MERCURY (I) CHLORIDE–WATER [2412]
$NaCl—HgCl—H_2O$

$t = 25$

Concn. in aqueous soln., g/l		d	Comp. of solid phase
Mercury Chloride	Sodium Chloride		
0.0041	5.85	—	HgCl+0.1 g Hg
0.041	58.50	1.040	»
0.129	119	1.078	»
0.194	148.25	1.093	»
0.380	222.3	1.142	»
0.643	292.5	1.188	»

№ 3840 [1862]

SODIUM CHLORIDE–MERCURY (II) CHLORIDE–WATER

$NaCl—HgCl_2—H_2O$

Comp. of solvent, wt %		Solubility of Mercury Chloride, g/l	t	Comp. of solvent, wt %		Solubility of Mercury Chloride, g/l	t
Sodium Chloride	Water			Sodium Chloride	Water		
0.5	99.5	100	15	10.0	90.0	580	15
0.5	99.5	130	65	10.0	90.0	680	65
0.5	99.5	440	100	10.0	90.0	1100	100
1.0	99.0	140	15	25	75	1200	15
1.0	99.0	180	65	25	75	1420	65
1.0	99.0	480	100	25	75	1960	100
5.0	95.0	300	15	26	74	1280	15
5.0	95.0	360	65	26	74	1520	65
5.0	95.0	640	100	26	74	2080	100

№ 3841 [1632]

SODIUM CHLORIDE–MERCURY (II) CHLORIDE–WATER

$NaCl—HgCl_2—H_2O$

Comp. of saturated soln., wt %			t	Comp. of solid phase
Sodium Chloride	Mercury Chloride	Water		
26.5	0.00	73.5	25	NaCl
18.66	51.35	29.99	25	$NaCl + NaCl \cdot HgCl_2 \cdot 2H_2O$
18.71	51.32	29.97	25	»
18.64	51.42	29.94	25	»
18.87	51.26	29.87	25	»
14.97	57.74	27.29	25	$NaCl \cdot HgCl_2 \cdot 2H_2O$
14.03	59.69	26.28	25	»
13.25	62.16	24.59	25	»
13.17	62.59	24.24	25	»
12.97	62.50	24.53	25	$NaCl \cdot HgCl_2 \cdot 2H_2O + HgCl_2$
13.14	62.48	24.38	25	»
13.15	62.55	24.30	25	»
19.46	46.49	34.05	10.3	—
19.48	46.50	34.02	10.3	—

№ 3842 [1815]

SODIUM CHLORIDE–MERCURY (II) CHLORIDE–WATER

$NaCl—HgCl_2—H_2O$

$t = 25$

Concn. in aqueous soln., g/l	
Mercury Chloride	Sodium Chloride
101	18.88
152.9	39.1
203.4	63.1
323.5	108.4
549	182.3
933	296.7

№ 3843 [406]

SODIUM CHLORIDE–THALLIUM HEXANITROCOBALTATE–WATER

$NaCl—Tl_3[Co(NO_2)_6]—H_2O$

$t = 20$

Content in aqueous soln.	
Sodium Chloride mol/l	Thallium Hexanitrocobaltate, $[mol/l]\cdot10^5$
0.000	8.5
0.001	9.3
0.01	11.1
0.1	12.8
0.25	16.8
0.5	22.4
1.0	30.5

№ 3844 [2264]

SODIUM CHLORIDE–THALLIUM (I) CHLORIDE–WATER

$NaCl—TlCl—H_2O$

$t = 25$

Concn. of Sodium Chloride in aqueous soln., g/l	Solubility of Thallium Chloride in aqueous soln., of Sodium Chloride, g/l
1.462	2.084
2.925	1.420
5.85	0.947
11.70	0.6501

№ 3845 [2039]

SODIUM CHLORIDE–THALLIUM (I) IODATE–WATER

$NaCl—TlIO_3—H_2O$

$t = 25$

Concn. of Sodium Chloride in aqueous soln., g/l	Solubility of Thallium Iodate in aqueous soln. of of Sodium Chloride, g/l
5.85	0.994

№ 3846 [1585]

SODIUM CHLORIDE–THALLIUM (I) IODATE–WATER

$NaCl—TlIO_3—H_2O$

$t = 25$

Concn. of Sodium Chloride in aqueous soln., g/l	Solubility of Thallium Iodate in aqueous soln. of Sodium Chloride, g/l
1.179	0.801
2.340	0.861
5.850	0.994

№ 3847 [1175]

SODIUM CHLORIDE–LEAD (II) SULFATE–WATER

$NaCl—PbSO_4—H_2O$

$t = 18$

Concn. of Sodium Chloride in aqueous soln., g/l	Solubility of Lead Sulfate in aqueous soln. of Sodium Chloride, g/l
5.85	0.1119
11.70	0.1873
17.55	0.2651
23.40	0.3376

№ 3848 SODIUM CHLORIDE–LEAD (II) SULFATE–WATER [291]

$NaCl—PbSO_4—H_2O$

$t = 25$

Concn. of Sodium Chloride in aqueous soln., mol/l	Solubility of Lead Sulfate in aqueous soln. of Sodium Chloride, mol/l	Concn. of Sodium Chloride in aqueous soln., mol/l	Solubility of Lead Sulfate in aqueous soln. of Sodium Chloride, mol/l
$1 \cdot 10^{-3}$	$1.58 \cdot 10^{-4}$	$5 \cdot 10^{-1}$	$2.12 \cdot 10^{-3}$
$5 \cdot 10^{-3}$	$2.01 \cdot 10^{-4}$	1.00	$5.36 \cdot 10^{-3}$
$1 \cdot 10^{-2}$	$2.37 \cdot 10^{-4}$	2.00	$7.69 \cdot 10^{-3}$
$5 \cdot 10^{-2}$	$3.50 \cdot 10^{-4}$	3.00	$1.37 \cdot 10^{-2}$
$1 \cdot 10^{-1}$	$5.59 \cdot 10^{-4}$		

№ 3849 SODIUM CHLORIDE–LEAD (II) CHLORIDE–WATER [1488]

$NaCl—PbCl_2—H_2O$

Comp. of saturated soln., wt %			t	Comp. of solid phase
Lead Chloride	Sodium Chloride	Water		
0.82	0.00	99.18	13	$PbCl_2$
0.18	1.0	98.82	13	»
0.09	4.97	94.94	13	»
0.19	13.39	86.42	13	»
0.31	16.46	83.23	13	»
1.10	23.46	75.44	13	»
1.88	26.17	71.95	13	$PbCl_2+NaCl$
0.75	26.29	72.96	13	NaCl
0.0	26.33	73.67	13	»
0.57	2.07	97.36	50	$PbCl_2$
0.38	5.36	94.26	50	»
0.58	9.57	89.85	50	»
1.20	17.22	81.58	50	»
2.41	22.94	74.65	50	»
4.04	25.51	70.45	50	»
4.94	26.46	68.60	50	$PbCl_2+NaCl$
3.88	26.48	69.64	50	NaCl
1.50	26.89	71.61	50	»
1.65	2.01	96.34	100	$PbCl_2$
1.53	5.00	93.47	100	»
2.54	13.12	84.34	100	»
4.95	19.55	75.50	100	»
9.00	24.01	66.99	100	»
10.47	25.21	64.32	100	»
11.92	26.27	61.81	100	$PbCl_2+NaCl$
6.86	27.05	66.09	100	NaCl
2.82	27.09	70.09	100	»

№ 3850 SODIUM CHLORIDE–LEAD (II) CHLORIDE–WATER [1817]

$NaCl—PbCl_2—H_2O$

$t = 25$

Concn. of Sodium Chloride in aqueous soln., g/l	Solubility of Lead Chloride in aqueous soln. of Sodium Chloride, g/l	Concn. of Sodium Chloride in aqueous soln., g/l	Solubility of Lead Chloride in aqueous soln. of Sodium Chloride, g/l
0.0	10.84	119.9	2.67
29.84	1.80	177.2	3.56
59.67	1.76	239.8	12.18

№ 3851 SODIUM CHLORIDE–LEAD (II) CHLORIDE–WATER [1965]

$NaCl—PbCl_2—H_2O$

$t = 25$

Concn. in aqueous soln., g/l			
Lead Chloride	Sodium Chloride	Lead Chloride	Sodium Chloride
2.23	14.63	3.29	143.91
1.90	23.40	3.63	155.61
1.81	29.25	4.09	169.65
1.66	40.95	4.44	179.01
1.67	50.31	4.93	192.46
1.75	66.69	5.52	207.09
1.95	85.41	6.13	221.13
2.04	91.26	6.96	236.34
2.40	108.22	7.55	244.53
2.51	114.66	8.76	256.8
2.92	131.62	9.75	263.25

Remark. Solid phase above 0.5 mol/l $NaCl—NaCl \cdot 2PbCl_2$.

№ 3852 [1484]

SODIUM CHLORIDE–LEAD (II) CHLORIDE–WATER

$NaCl—PbCl_2—H_2O$

$t = 25$

Comp. of saturated soln., wt %			Comp. of solid phase
Sodium Chloride	Lead Chloride	Water	
26.44	0.00	73.56	NaCl
26.47	0.60	72.93	»
26.45	1.45	72.10	»
26.33	2.48	71.19	$NaCl + PbCl_2$
23.85	1.60	74.55	$PbCl_2$
21.64	1.07	77.29	»
19.21	0.71	80.08	»
16.58	0.50	82.92	»
12.47	0.31	87.22	»

Table № 3852 (continuation)

Comp. of saturated soln., wt %			Comp. of solid phase
Sodium Chloride	Lead Chloride	Water	
9.06	0.19	90.75	$PbCl_2$
7.39	0.17	91.44	»
5.64	0.16	94.20	»
4.75	0.16	95.09	»
3.84	0.16	96.00	»
1.96	0.20	97.84	»
1.53	0.25	98.22	»
0.49	0.71	98.80	»
0.00	1.17	98.83	»

№ 3853 SODIUM CHLORIDE–LEAD (II) CHLORIDE–WATER [234]

$NaCl—PbCl_2—H_2O$

Comp. of saturated soln., wt %			t	Comp. of solid phase
Lead Chloride	Sodium Chloride	Water		
1.75	0.00	98.25	50	$PbCl_2$
0.57	2.07	97.36	50	»
0.38	5.36	94.25	50	»
0.58	9.57	89.85	50	»
1.20	17.22	11.58	50	»
1.52	20.21	78.27	50	»
2.41	22.94	74.65	50	»
4.04	25.51	70.45	50	»
4.92	26.46	68.62	50	$PbCl_2+NaCl$
4.24	26.38	69.38	50	NaCl
3.88	26.48	69.64	50	»
2.08	26.67	71.25	50	»
1.50	26.89	72.61	50	»
0.00	26.99	73.01	50	»
3.10	0.00	96.90	100	$PbCl_2$
1.65	2.01	96.34	100	»
1.60	2.05	96.35	100	»
1.50	4.10	94.40	100	»
1.53	5.00	93.47	100	»
1.67	6.71	91.62	100	»
2.54	13.12	84.34	100	»
4.95	19.55	75.50	100	»
9.00	24.01	66.99	100	»
9.12	24.30	66.58	100	»
9.58	24.47	65.85	100	»
10.47	25.21	64.32	100	»
11.90	26.05	62.05	100	»
11.92	26.27	61.81	100	$PbCl_2+NaCl$
6.86	27.05	66.09	100	NaCl
2.82	27.09	70.09	100	»
0.00	27.80	72.20	100	»

№ 3854 [2074]

SODIUM CHLORIDE–LEAD (II) CHLORIDE–WATER

$NaCl—PbCl_2—H_2O$

$t = 25$

Concn. of Sodium Chloride in aqueous soln., mol/1000 g Water	Solubility of Lead Chloride in aqueous soln. of Sodium Chloride, g/1000 g Water
0.4723	1.870
1.0125	1.784
2.02650	2.743
4.0216	8.367

№ 3855 [1817]

SODIUM CHLORIDE–LEAD (II) BROMIDE–WATER

$NaCl—PbBr_2—H_2O$

$t = 25$

Concn. of Sodium Chloride in aqueous soln., g/l	Solubility of Lead Bromide in aqueous soln.. of Sodium Chloride, g/l
0.00	9.63
27.26	1.67
55.10	2.07
109.6	3.10
164.7	5.38
219.8	11.20
274.9	22.65

№ 3856 **SODIUM CHLORIDE–LEAD (II) IODIDE–WATER** [1300]

$NaCl—PbI_2—H_2O$

$t = 25$

Concn. in aqueous soln., wt %		Comp. of solid phase
Lead Iodide	Sodium Chloride	
0.0758	0.0	PbI_2
0.0778	0.029	»
0.0859	0.059	»
0.0951	0.116	»
0.110	0.234	»
0.141	0.586	»
0.164	1.170	»
0.179	2.980	»

№ 3857 [2040]

SODIUM CHLORIDE–LEAD (II) IODATE–WATER

$NaCl—Pb(IO_3)_2—H_2O$

$t = 25$

Concn. of Sodium Chloride in aqueous soln., g/l	Solubility of Lead Iodate in aqueous soln. of Sodium Chloride, g/l
5.85	0.0462

№ 3858 [2398]

SODIUM CHLORATE–SODIUM CHROMATE–WATER

$NaClO_3—Na_2CrO_4—H_2O$

Concn. in aqueous soln., wt %		t	Comp. of solid phase
Sodium Chlorate	Sodium Chromate		
48.28	0.0	19	$NaClO_3$
41.91	6.43	19	»
33.59	14.56	19	»
21.57	27.00	19	»

Table № 3858 (continuation)

Concn. in aqueous soln., wt %		t	Comp. of solid phase
Sodium Chlorate	Sodium Chromate		
15.02	35.04	19	$NaClO_3+Na_2CrO_4 \cdot 6H_2O$
10.70	37.26	19	$Na_2CrO_4 \cdot 6H_2O$
5.14	40.60	19	»
2.31	42.26	19	»
0.0	43.63	19	».
50.06	0.0	25	$NaClO_3$
43.88	5.95	25	»
37.06	12.45	25	»
29.30	20.42	25	»
21.50	28.51	25	»
15.65	35.18	25	»
14.43	36.43	25	$NaClO_3+Na_2CrO_4 \cdot 4H_2O$
9.82	39.47	25	$Na_2CrO_4 \cdot 4H_2O$
7.34	41.04	25	»
0.0	45.59	25	$Na_2CrO_4 \cdot 6H_2O$
55.49	0.0	50	$NaClO_3$
48.49	6.36	50	»
36.71	18.37	50	»
23.55	31.45	50	»
15.81	40.80	50	»
13.86	43.14	50	$NaClO_3+Na_2CrO_4 \cdot 4H_2O$
11.54	44.21	50	$Na_2CrO_4 \cdot 4H_2O$
6.20	47.32	50	»
0.0	50.66	50	»

№ 3859 SODIUM CHLORATE–SODIUM BROMIDE–WATER [2400]

$NaClO_3—NaBr—H_2O$

$t = 25$

Concn. in aqueous soln., wt %		Comp. of solid phase	Concn. in aqueous soln., wt %		Comp. of solid phase
Sodium Chlorate	Sodium Bromide		Sodium Chlorate	Sodium Bromide	
50.10	0.0	$NaClO_3$	13.85	40.36	$NaClO_3+NaBr \cdot 2H_2O$
37.93	11.86	»	14.03	40.11	»
29.54	20.72	»	13.98	40.18	»
20.87	30.75	»	13.89	40.28	»
16.29	36.77	»	12.38	41.16	$NaBr \cdot 2H_2O$
15.37	38.10	»	8.07	43.74	»
13.87	40.32	$NaClO_3+NaBr \cdot 2H_2O$	7.22	44.28	»
13.97	40.18	»	0	48.49	»

№ 3860 [2620]

SODIUM CHLORIDE–SODIUM BROMATE–WATER

$NaClO_3—NaBrO_3—H_2O$

$t = 25$

Comp. of saturated soln., wt %			Comp. of solid phase
Sodium Bromate	Sodium Chlorate	Water	
28.29	0.00	71.71	$NaBrO_3$
16.46	18.91	64.63	»
12.20	28.03	59.77	»
8.68	36.75	54.57	»
7.14	40.98	51.88	»
7.00	41.47	51.53	S. S. I (2.02)
6.54	42.62	50.84	S. S. I (1.70)
5.99	43.66	50.35	S. S. I (6.84)
6.05	43.55	50.40	S. S. I+S. S. II (24.3)
5.33	44.64	50.03	S. S. II (40.4)
5.07	44.98	49.95	S. S. II (64.2)
4.49	45.66	49.85	S. S. II (71.0)
3.79	46.46	49.75	S. S. II (83.2)
3.69	46.56	49.75	S. S. II (85.0)
2.84	47.42	49.74	S. S. II (89.4)
1.89	48.36	49.75	S. S. II (91.0)
0.96	49.16	49.88	S. S. II (98.0)
0.79	49.36	49.75	S. S. II (98.5)
0.00	50.07	49.93	$NaClO_3$

Note. S.S.I and S.S.II are solid solns. having the indicated percentage wt of $NaClO_3$

№ 3861 [2403]

SODIUM CHLORATE–SODIUM MOLYBDATE–WATER

$NaClO_3—Na_2MoO_4—H_2O$

$t = 25$

Concn. in aqueous soln., wt %		d_4	Comp. of solid phase
Sodium Molybdate	Sodium Chlorate		
39.38	0.00	1.432	$Na_2MoO_4 \cdot 2H_2O$
36.11	4.23	1.441	»
36.42	9.04	1.441	»
28.53	14.12	1.440	»
22.83	21.94	1.442	»
17.95	29.14	1.453	»
14.59	34.39	1.466	»
13.04	37.05	1.472	»
11.77	39.21	1.478	$Na_2MoO_4 \cdot 2H_2O+NaClO_3$
11.75	39.25	1.479	»
11.81	39.17	1.481	»
11.77	39.21	1.479	»
11.74	39.29	1.476	$NaClO_3$
8.87	41.85	1.465	»
5.72	44.70	1.456	»
2.60	47.60	1.438	»
0.00	50.02	1.433	»

№ 3862 SODIUM CHLORATE–SODIUM IODIDE–WATER [2400]

$NaClO_3—NaI—H_2O$

$t = 25$

Concn. in aqueous soln., wt %		Comp. of solid phase	Concn. of aqueous soln., wt %		Comp. of solid phase
Sodium Chlorate	Sodium Iodide		Sodium Chlorate	Sodium Iodide	
50.10	0.0	$NaClO_3$	4.08	61.79	$NaClO_3 + NaI \cdot 2H_2O$
38.72	12.40	»	4.20	61.73	»
27.62	25.53	»	4.51	61.61	»
18.67	36.53	»	4.32	61.68	»
10.28	48.78	»	2.83	62.65	$NaI \cdot 2H_2O$
7.11	54.63	»	1.43	63.67	»
5.44	58.56	»	1.22	64.00	»
4.50	61.52	$NaClO_3 + NaI \cdot 2H_2O$	0	64.80	»
4.28	61.74	»			

№ 3863 SODIUM CHLORATE–SODIUM IODATE–WATER [2397]

$NaClO_3—NaIO_3—H_2O$

Comp. of saturated soln., wt %			t	d	Comp. of solid phase
Sodium Chlorate	Sodium Iodate	Water			
50.14	0.00	49.86	25	—	$NaClO_3$
49.52	1.16	49.32	25	1.444	»
49.38	1.29	49.33	25	1.444	$NaClO_3 + NaIO_3 \cdot H_2O$
49.19	1.30	49.51	25	1.440	$NaIO_3 \cdot H_2O$
48.13	1.33	50.53	25	1.425	»
46.27	1.39	52.34	25	1.404	»
44.56	1.46	53.98	25	1.396	»
42.99	1.52	55.49	25	—	»
38.66	1.69	59.65	25	1.332	»
32.57	1.97	65.46	25	1.273	»
24.67	2.43	72.90	25	1.204	»
16.50	3.14	80.36	25	1.146	»
8.36	4.51	87.13	25	1.098	»
0.00	8.57	91.43	25	1.075	»
55.74	0.00	44.26	50	—	$NaClO_3$
54.98	1.26	43.76	50	—	»
54.71	1.69	43.60	50	—	$NaClO_3 + NaIO_3$
53.83	1.75	44.42	50	—	$NaIO_3$
47.86	2.14	50.00	50	—	»
45.0	2.2	52.8	50	—	$NaIO_3 + NaIO_3 \cdot H_2O$
54.59	1.87	43.54	50	—	$NaClO_3 + NaIO_3 \cdot H_2O$
48.95	2.12	48.93	50	—	$NaIO_3 \cdot H_2O$
43.71	2.41	53.88	50	—	»
33.33	3.23	63.44	50	—	»
16.56	5.69	77.75	50	—	»
10.02	7.67	82.31	50	—	»
0.00	13.49	86.51	50	—	»

№ 3864 SODIUM CHLORATE–POTASSIUM CHLORATE–WATER [328]

$NaClO_3$—$KClO_3$—H_2O

Comp. of saturated soln., wt %			t	d	Comp. of solid phase
Potassium Chlorate	Sodium Chlorate	Water			
7.46	0.46	92.08	24.2	1.045	$KClO_3$
6.93	6.49	86.58	24.2	1.086	»
5.20	20.44	74.36	24.2	1.189	»
4.08	27.89	68.03	24.2	1.241	»
3.28	31.10	65.62	24.2	1.260	»
3.15	36.31	60.54	24.2	1.326	»
3.34	36.91	59.75	24.2	1.329	»
2.82	43.99	53.19	24.2	1.392	»
1.53	49.16	49.31	24.2	1.441	$KClO_3+NaClO_3$
0.00	49.65	50.35	24.2	1.429	$NaClO_3$
12.43	0.00	87.57	40	1.073	$KClO_3$
8.02	10.94	81.04	40	—	»
4.50	31.14	64.36	40	1.268	»
4.22	35.42	60.36	40	1.308	»
3.40	48.00	48.60	40	1.453	»
3.23	51.97	44.80	40	1.483	$KClO_3+NaClO_3$
0.00	53.79	46.21	40	1.462	$NaClO_3$

№ 3865 [1502]

SODIUM CHLORATE – POTASSIUM CHLORATE – WATER

$NaClO_3$—$KClO_3$—H_2O

$t = 20$

Comp. of saturated soln., wt %			Comp. of solid phase
Potassium Chlorate	Sodium Chlorate	Water	
6.75	0.0	93.25	$KClO_3$
4.6	5.0	90.4	»
3.8	10.0	86.2	»
3.4	15.0	81.6	»
3.1	20.0	76.9	»
2.7	30.0	67.3	»
2.5	40.0	57.5	»
2.5	42.0	55.5	$KClO_3+NaClO_3$
2.0	42.5	55.5	$NaClO_3$
1.0	45.0	54.0	»
0.33	47.5	52.17	»
0.0	49.5	50.5	»

№ 3866 SODIUM CHLORATE–POTASSIUM CHLORATE–WATER [1859]

$NaClO_3$—$KClO_3$—H_2O

$t = 25$

Concn. of Sodium Chlorate in aqueous soln., g/l	Solubility of Potassium Chlorate in aqueous soln. of Sodium Chlorate, g/l	Concn. of Sodium Chlorate in aqueous soln., g/l	Solubility of Potassium Chlorate in aqueous soln. of Sodium Chlorate, g/l	Concn. of Sodium Chlorate in aqueous soln., g/l	Solubility of Potassium Chlorate in aqueous soln. of Sodium Chlorate, g/l
0.0	83.030	42.4	63.805	212	40.425
10.6	76.974	53.0	60.027	318	32.340
21.2	71.959	79.5	53.208	424	22.228
31.8	67.655	106	49.170	530	11.385

№ 3867 [2232]

SODIUM CHLORATE–POTASSIUM CHLORATE–WATER

$NaClO_3$—$KClO_3$—H_2O

Concn. in aqueous soln., wt %		t	Comp. of solid phase
Potassium Chlorate	Sodium Chlorate		
0.44	44.21	0	$NaClO_3 + KClO_3$
3.41	51.57	40	»

№ 3868 SODIUM CHLORATE–POTASSIUM CHLORATE–WATER [2238]

$NaClO_3$—$KClO_3$—H_2O

Comp. of saturated soln., g/100 g Water		d	t	Comp. of solid phase
Sodium Chlorate	Potassium Chlorate			
65.9	0.665	1.356	−17.85	$NaClO_3 + KClO_3 +$ ice
71.7	1.00	1.3735	−9.8	$NaClO_3 + KClO_3$
35.3	0.822	1.217	−9.8	$KClO_3 +$ ice
14.62	1.16	1.095	−4.0	»
88.1	2.38	1.4173	10.0	$NaClO_3 + KClO_3$
106.1	5.14	1.4625	30.0	»
127.3	9.97	1.510	50.0	»
153.9	18.0	1.566	70.0	»
212.0	40.6	1.661	100.0	»

№ 3869 SODIUM CHLORATE–SILVER CHLORATE–WATER [2404]

$NaClO_3—AgClO_3—H_2O$

$t = 25$

Concn. in aqueous soln., wt %		t	Concn. in aqueous soln., wt %		t
Silver Chlorate	Sodium Chlorate		Silver Chlorate	Sodium Chlorate	
14.46	0.00	1.126	2.14	46.57	1.415
10.02	8.11	1.143	2.15	46.54	1.413
7.48	17.49	1.194	2.15	46.55	1.417
5.56	27.53	1.270	1.66	47.52	1.419
4.23	34.39	1.317	0.56	49.23	1.428
2.85	41.78	1.375	0.00	50.04	1.424

№ 3870 [1502]

SODIUM CHLORATE–BARIUM CHLORATE–WATER

$NaClO_3—Ba(ClO_3)_2—H_2O$

$t = 20$

Comp. of saturated soln., wt %			Comp. of solid phase
Sodium Chlorate	Barium Chlorate	Water	
0.00	25.16	74.84	$Ba(ClO_3)_2 \cdot H_2O$
4.52	17.92	78.56	»
8.50	10.90	81.60	»
15.52	8.53	75.95	»
25.32	6.49	68.19	»
29.52	5.01	65.47	»
36.50	3.49	60.01	»
43.20	2.89	54.91	$Ba(ClO_3)_2 \cdot H_2O + NaClO_3$
45.00	1.11	53.89	$NaClO_3$
49.70	0.00	50.30	»

№ 3871 [1711]

SODIUM CHLORATE–THALLIUM (I) CHLORIDE–WATER

$NaClO_3—TlCl—H_2O$

$t = 25$

Concn. in aqueous soln., g/l			
Thallium Chloride	Sodium Chlorate	Thallium Chloride	Sodium Chlorate
5.555	53.25	7.909	319.5
6.433	106.5	9.215	426
7.326	213		

№ 3872 [2379]

SODIUM PERCHLORATE–SODIUM HEXAHYDROXOSTANNATE–WATE

$NaClO_4—Na[Sn(OH)_6]—H_2O$

$t = 18$

Concn. of Sodium Perchlorate in aqueous soln., g/l	Solubility of Sodium Hexahydroxo-stannate in aqueous soln. oi Sodium Perchlor: g/l
240.2	301

№ 3873 SODIUM PERCHLORATE–SODIUM IODIDE–SILVER IODIDE–WATER [1972]

$NaClO_4—NaI—AgI—H_2O$

$t = 24—26$

Concn. in aqueous soln., g/l					
Sodium Iodide	Sodium Perchlorate	Silver Iodide	Sodium Iodide	Sodium Perchlorate	Silver Iodide
17.6870	230.4321	0.0178	95.4799	166.8857	0.8428
18.2866	229.9423	0.0150	113.9164	151.8256	1.4415
20.0853	228.4730	0.0200	139.6975	130.7659	2.4651
25.7811	223.8203	0.0326	156.7849	116.8078	3.3337
27.5798	222.3510	0.0373	172.3735	104.0740	4.8128
29.0787	221.1266	0.0376	195.0069	85.5856	6.9022
35.0743	216.2290	0.0589	196.0561	84.7285	7.6066
41.5195	210.9641	0.0902	207.2979	75.5455	8.6630
42.5688	210.1070	0.0979	220.7880	64.5259	10.9872
54.1103	200.6792	0.1791	247.7682	42.4867	15.8000
54.8597	200.0670	0.1953	252.7145	38.4462	15.0253
65.3520	191.4962	0.2958	258.4104	33.7934	17.8660
68.9494	188.5576	0.3428	266.2046	27.4266	19.6972
73.7459	184.6395	0.4132	273.3994	21.5494	20.2841
87.0861	173.7424	0.6339	298.1312	1.3468	28.4072

№ 3874 SODIUM PERCHLORATE–ALUMINUM PERCHLORATE–WATER [1434]

$NaClO_4—Al(ClO_4)_3—H_2O$

$t = 30$

Comp. of saturated soln., wt %			Comp. of solid phase
Sodium Perchlorate	Aluminum Perchlorate	Water	
0.00	52.40	47.60	$Al(ClO_4)_3 \cdot H_2O$
14.03	39.21	46.76	»
19.92	34.65	45.43	$Al(ClO_4)_3 \cdot H_2O + NaClO_4 \cdot H_2O$
22.70	32.48	44.82	$NaClO_4 \cdot H_2O$
40.98	18.51	40.51	»
45.92	14.69	39.39	»
51.45	10.55	38.00	»
68.25	0.00	31.75	»

№ 3875 [1261]

SODIUM PERCHLORATE–POTASSIUM PERCHLORATE–WATER

$NaClO_4—KClO_4—H_2O$

$t = 25$

Concn. in aqueous soln., wt %	
Potassium Perchlorate	Sodium Perchlorate
1.512	1.222
0.994	3.61
0.681	7.087

S.O.I.C. — C*

№ 3876 [359]

SODIUM PERCHLORATE–POTASSIUM PERCHLORATE–WATER

$NaClO_4—KClO_4—H_2O$

$t = 25$

Comp. of saturated soln., wt %			Comp. of solid phase
Potassium Perchlorate	Sodium Perchlorate	Water	
2.02	0.00	97.98	$KClO_4$
2.18	2.38	95.44	»
2.21	4.11	93.68	»
2.03	6.62	91.35	»
1.36	10.00	88.64	»
1.34	11.95	86.71	»
1.60	16.74	81.66	»
1.60	18.39	80.01	»
1.62	26.08	72.30	»
1.45	32.38	66.17	»
1.30	40.82	57.88	»
1.26	49.94	48.80	»
1.33	58.37	40.30	»
1.34	63.26	35.40	»
0.47	66.32	33.21	$KClO_4+NaClO_4 \cdot H_2O$
0.00	67.82	32.18	$NaClO_4 \cdot H_2O$

№ 3877 **SODIUM PERCHLORATE–POTASSIUM PERCHLORATE–WATER** [362]

$NaClO_4—KClO_4—H_2O$

Comp. of saturated soln., wt %			t	Comp. of solid phase
Potassium Perchlorate	Sodium Perchlorate	Water		
0.75	0.00	99.25	0	$KClO_4$
0.63	5.15	93.22	0	»
0.55	6.94	92.51	0	»
0.45	9.23	90.32	0	»
0.49	11.71	87.80	0	»
0.48	11.85	87.67	0	»
0.51	19.54	79.95	0	»
0.50	21.56	77.94	0	»
0.49	31.82	67.69	0	»
0.45	36.58	62.97	0	»
0.46	41.24	58.30	0	»
0.51	50.00	49.49	0	»
0.52	55.37	44.11	0	$KClO_4+NaClO_4$
0.51	55.32	44.17	0	»
0.51	55.35	44.11	0	»
0.54	55.32	44.14	0	»

Table № 3877 (continuation)

Comp. of saturated soln., wt %			t	Comp. of solid phase
Potassium Perchlorate	Sodium Perchlorate	Water		
0.42	58.23	41.35	0	$nKClO_4 \cdot mNaClO_4$
0.28	60.15	39.57	0	»
0.16	61.46	38.38	0	»
0.00	62.89	37.11	0	$NaClO_4 \cdot H_2O$
2.30	0.00	97.70	25	$KClO_4$
2.18	2.38	95.44	25	»
2.21	4.11	93.68	25	»
2.03	6.52	91.35	25	»
1.36	10.00	88.64	25	»
1.34	11.95	86.71	25	»
1.60	16.74	81.66	25	»
1.60	18.39	80.01	25	»
1.62	26.08	72.30	25	»
1.45	32.38	66.17	25	»
1.30	40.82	57.88	25	»
1.26	49.94	48.80	25	»
1.33	58.37	40.30	25	»
1.34	63.26	35.40	25	»
0.47	66.32	33.21	25	$NaClO_4 \cdot H_2O$
0.00	67.82	32.18	25	»

№ 3878 [306]

SODIUM PERCHLORATE–THORIUM OXOCARBONATE HYDRATE–WATER

$NaClO_4—ThOCO_3 \cdot 8H_2O—H_2O$

$t = 20.0$

Content of Sodium Perchlorate in soln., mol/l	Solubility of Thorium Oxo-carbonate, in conversion to Thorium, in aqueous soln. of Sodium Perchlorate, mg/l
3.000	24.9 ±1.2

$Na_2CrO_4—Na_2MoO_4—H_2O$

Comp. of saturated soln., wt %		t	Comp. of solid phase
Sodium Chlorate	Sodium Molybdate		
23.50	0	0	$Na_2CrO_4 \cdot 10H_2O + Na_2MoO_4 \cdot 10H_2O$
18.73	7.31	0	»
11.24	19.24	0	»
8.83	23.30	0	»
8.35	23.86	0	»
5.04	26.81	0	»
0	30.35	0	»
30.50	0	8	»
25.49	7.16	8	»
22.44	11.56	8	»
19.36	16.45	8	»
13.07	25.35	8	»
20.59	19.64	8	»
15.22	24.27	8	»
9.44	29.25	8	»
3.21	33.97	8	»
0	36.38	8	»
37.78	0	15	»
33.57	5.80	15	»
25.97	16.50	15	»
21.23	20.66	15	$Na_2MoO_4 \cdot 2H_2O$
10.62	29.54	15	»
0	39.15	15	»
43.09	0	19	$Na_2CrO_4 \cdot 10H_2O + Na_2MoO_4 \cdot 10H_2O$
40.85	3.20	19	»
38.55	6.45	19	»
43.82	0	19	$Na_2CrO_4 \cdot 6H_2O + Na_2MoO_4 \cdot 6H_2O$
41.98	2.14	19	»
37.76	6.80	19	»
44.80	0.35	19	$Na_2CrO_4 \cdot 4H_2O + Na_2MoO_4 \cdot 4H_2O$
43.68	1.37	19	»
42.42	2.82	19	»
38.00	6.96	19	$Na_2MoO_4 \cdot 2H_2O$
24.71	17.31	19	»
0	39.28	19	»
44.98	0	22	$Na_2CrO_4 \cdot 6H_2O + Na_2MoO_4 \cdot 6H_2O$
42.98	2.07	22	»
40.42	5.10	22	»
40.64	5.08	22	$Na_2CrO_4 \cdot 4H_2O + Na_2MoO_4 \cdot 4H_2O$
40.15	5.70	22	»
40.18	5.95	22	»
45.85	0	25	$Na_2CrO_4 \cdot 6H_2O + Na_2MoO_4 \cdot 4H_2O$
45.38	0.60	25	»
44.55	1.40	25	»
44.53	1.33	25	$Na_2CrO_4 \cdot 4H_2O + Na_2MoO_4 \cdot 4H_2O$
42.85	3.09	25	»
40.30	5.89	25	»
41.12	4.89	25	$Na_2MoO_4 \cdot 2H_2O$
20.20	21.93	25	»
0	39.56	25	»

№ 3880 [2379]

SODIUM CHROMATE–SODIUM HEXAHYDROXOSTANNATE (IV)–WATER

$Na_2CrO_4—Na_2[Sn(OH)_6]—H_2O$

$t = 18$

Concn. of Sodium Chromate in aqueous soln., g/l	Solubility of Sodium Hexahydroxostannate in aqueous soln. of Sodium Chromate, g/l
159.57	375

№ 3881 SODIUM CHROMATE–POTASSIUM CHROMATE–WATER [530]

$Na_2CrO_4—K_2CrO_4—H_2O$

$t = 25$

Concn. in aqueous soln., wt %					
Potassium Chromate	Sodium Chromate	Potassium Chromate	Sodium Chromate	Potassium Chromate	Sodium Chromate
0.0	45.76	15.27	25.15	30.82	8.24
5.49	41.49	16.86	22.97	32.04	8.08
11.75	35.00	21.40	16.49	35.77	4.25
13.20	34.35	25.62	12.33	39.52	0.0
14.92	27.21	30.00	9.46		

№ 3882 SODIUM CHROMATE–CALCIUM CHROMATE–WATER [696]

$Na_2CrO_4—CaCrO_4—H_2O$

Comp. of saturated soln., wt %			t	Comp. of solid phase
Sodium Chromate	Calcium Chromate	Water		
45.76	0.0	54.24	25	$Na_2CrO_4 \cdot 4H_2O$
0.0	2.68	97.32	25	$CaCrO_4$
46.76	0.08	53.16	25	$Na_2CrO_4 \cdot 4H_2O + CaCrO_4$
28.77	0.90	70.33	25	$CaCrO_4$
18.78	1.49	79.73	25	»
51.2	0.0	48.8	50	$Na_2CrO_4 \cdot 4H_2O$
0.0	1.49	98.51	50	$CaCrO_4$
50.86	0.13	49.01	50	$Na_2CrO_4 \cdot 4H_2O + CaCrO_4$
41.25	0.31	58.44	50	$CaCrO_4$
28.89	0.46	70.65	50	»
13.87	0.95	85.18	50	»
55.25	0.00	44.75	75	Na_2CrO_4
0.00	0.81	99.19	75	$CaCrO_4$
55.34	0.06	44.60	75	$Na_2CrO_4 + CaCrO_4$
41.93	0.15	57.92	75	$CaCrO_4$
20.53	0.47	79.00	75	»
3.58	0.68	95.74	75	»

Table № 3882 (continuation)

Comp. of saturated soln., wt %			t	Comp. of solid phase
Sodium Chromate	Calcium Chromate	Water		
55.7	0.0	44.3	95	Na_2CrO_4
0.00	0.64	99.36	95	$CaCrO_4$
54.60	0.12	45.28	95	$Na_2CrO_4+CaCrO_4$
35.89	0.24	63.87	95	$CaCrO_4$
10.32	0.36	89.32	95	»

№ 3883 SODIUM CHROMATE–RUBIDIUM CHROMATE–WATER [953]

$Na_2CrO_4—Rb_2CrO_4—H_2O$

Comp. of saturated soln., wt %			t	Comp. of solid phase
Sodium Chromate	Rubidium Chromate	Water		
45.62	0.00	54.38	25.0	$Na_2CrO_4 \cdot 4H_2O$
44.76	1.78	53.46	25.0	»
44.34	2.96	52.70	25.0	»
43.08	4.88	52.04	25.0	»
42.30	7.30	50.40	25.0	$Na_2CrO_4 \cdot 3Rb_2CrO_4+Na_2CrO_4 \cdot 4H_2O$
38.70	8.00	53.30	25.0	$Na_2CrO_4 \cdot 3Rb_2CrO_4$
35.60	9.40	55.00	25.0	»
31.81	13.00	55.19	25.0	»
29.98	12.51	57.51	25.0	»
28.90	14.20	56.90	25.0	»
26.07	17.56	56.37	25.0	»
19.23	23.18	57.59	25.0	»
15.35	27.58	57.07	25.0	»
10.30	32.92	56.78	25.0	»
7.51	36.46	56.03	25.0	»
6.50	38.11	55.39	25.0	»
5.72	39.02	55.26	25.0	$Na_2CrO_4 \cdot 3Rb_2CrO_4+Rb_2CrO_4$
2.78	41.14	56.08	25.0	Rb_2CrO_4
1.30	42.20	56.50	25.0	»
0.00	43.27	56.73	25.0	»
50.97	0.00	49.03	50.0	$Na_2CrO_4 \cdot 4H_2O$
49.52	0.60	49.88	50.0	»
49.35	2.52	48.13	50.0	»
48.38	4.50	47.12	50.0	»
47.27	6.27	46.46	50.0	$Na_2CrO_4 \cdot 3Rb_2CrO_4+Na_2CrO_4 \cdot 4H_2O$
42.31	7.76	49.93	50.0	»
30.44	16.36	53.20	50.0	»
23.86	19.25	56.89	50.0	»
21.42	22.85	55.73	50.0	»
11.92	32.71	55.37	50.0	»
10.33	34.65	55.02	50.0	»
5.24	40.63	54.13	50.0	»
4.60	43.15	52.25	50.0	$Na_2CrO_4 \cdot 3Rb_2CrO_4+Rb_2CrO_4$
3.97	43.27	52.76	50.0	Rb_2CrO_4
3.00	44.79	52.21	50.0	»
0.00	46.66	53.34	50.0	»

№ 3884 SODIUM CHROMATE–CESIUM CHROMATE–WATER [952]

$Na_2CrO_4—Cs_2CrO_4—H_2O$

Comp. of saturated soln., wt %			t	Comp. of solid phase
Sodium Chromate	Cesium Chromate	Water		
45.91	0.00	54.09	25	$Na_2CrO_4 \cdot 4H_2O$
43.80	2.10	54.10	25	»
42.94	3.09	53.97	25	»
42.62	6.45	46.93	25	»
41.56	8.09	50.35	25	»
41.20	9.00	49.80	25	$Na_2CrO_4 \cdot 4H_2O + Na_2CrO_4 \cdot CsCrO_4 \cdot H_2O$
38.73	11.00	50.27	25	$Na_2CrO_4 \cdot CsCrO_4 \cdot 4H_2O$
37.39	11.87	50.74	25	»
30.91	18.20	50.89	25	»
28.05	19.82	52.13	25	»
26.72	23.71	49.57	25	»
23.20	29.30	47.5	25	$Na_2CsO_4 \cdot Cs_2CrO_4 \cdot 4H_2O + Cs_2CrO_4$
22.90	28.97	48.13	25	Cs_2CrO_4
16.81	33.10	50.09	25	»
12.20	36.30	51.50	25	»
7.90	40.06	52.04	25	»
0.00	45.11	54.89	25	»
50.97	0.00	49.03	50	$Na_2CrO_4 \cdot H_2O$
49.78	1.94	48.28	50	»
49.25	2.52	48.23	50	»
48.61	4.34	47.15	50	»
46.74	7.54	45.72	50	»
45.50	9.80	44.70	50	$Na_2CrO_4 \cdot 4H_2O + Na_2CrO_4 \cdot Cs_2CrO_4 \cdot 4H_2O$
43.52	12.98	43.50	50	$Na_2CrO_4 \cdot CsCrO_4 \cdot 4H_2O$
39.10	16.00	44.90	50	»
33.96	21.18	44.86	50	»
28.94	27.07	44.99	50	»
26.15	31.12	42.73	50	$Na_2CrO_4 \cdot Cs_2CrO_4 \cdot 4H_2O + Cs_2CO_4$
23.61	32.90	43.49	50	Cs_2CrO_4
18.67	38.83	42.50	50	»
11.65	43.30	45.05	50	»
6.39	46.49	47.12	50	»
0.00	53.10	46.90	50	»

№ 3885 SODIUM DICHROMATE–POTASSIUM DICHROMATE–WATER [2421]

$Na_2Cr_2O_7—K_2Cr_2O_7—H_2O$

Comp. of saturated soln., wt %			t	Comp. of solid phase
Potassium Dichromate	Sodium Dichromate	Water		
13.1	0.0	86.9	25	$K_2Cr_2O_7$
12.2	4.38	86.42	25	»
8.68	16.1	75.22	25	»
8.97	20.0	71.03	25	»
7.70	26.9	65.40	25	»
6.52	33.6	59.88	25	»

Comp. of saturated soln., wt %			t	Comp. of solid phase
Potassium Dichromate	Sodium Dichromate	Water		
5.46	43.2	51.34	25	$K_2Cr_2O_7$
5.16	55.5	39.34	25	»
5.25	61.4	33.35	25	$K_2Cr_2O_7+Na_2Cr_2O_7$
0.0	65.5	34.5	25	$Na_2Cr_2O_7$
27.4	0.0	72.6	50	$K_2Cr_2O_7$
24.2	6.5	69.3	50	»
19.3	16.3	64.4	50	»
17.8	21.0	61.2	50	»
13.6	29.3	57.1	50	»
11.3	43.6	45.1	50	»
9.47	48.5	42.03	50	»
9.36	52.5	38.14	50	»
7.92	63.3	29.78	50	$K_2Cr_2O_7+Na_2Cr_2O_7$
0.0	70.3	29.7	50	$Na_2Cr_2O_7$
47.2	0.0	52.8	100	$K_2Cr_2O_7$
42.1	6.3	51.6	100	»
36.5	14.1	49.4	100	»
29.0	23.0	48.0	100	»
24.2	29.6	46.2	100	»
16.4	44.4	39.2	100	»
13.4	56.1	30.5	100	»
12.4	60.6	27.0	100	»
9.96	70.8	20.24	100	$K_2Cr_2O_7+Na_2Cr_2O_7$
0.0	80.1	19.9	100	$Na_2Cr_2O_7$

№ 3886 SODIUM DICHROMATE–CESIUM DICHROMATE–WATER [954]

$Na_2Cr_2O_7—Cs_2Cr_2O_7—H_2O$

Comp. of saturated soln., wt %			t	Comp. of solid phase
Sodium Dichromate	Cesium Dichromate	Water		
0.00	3.75	96.25	25	$Cs_2Cr_2O_7$
17.80	2.00	80.20	25	»
20.63	2.60	76.77	25	»
26.17	2.23	71.60	25	»
34.88	4.80	60.32	25	»
37.87	5.87	56.26	25	»
41.33	5.57	53.10	25	$Na_2Cr_2O_7 \cdot Cs_2Cr_2O_7+Cs_2Cr_2O_7$
43.98	3.08	52.94	25	$Na_2Cr_2O_7 \cdot Cs_2Cr_2O_7$
47.60	1.70	51.70	25	»
51.23	1.57	47.20	25	»
56.31	1.30	42.39	25	»
63.50	1.80	35.70	25	»
64.28	2.09	33.63	25	$Na_2Cr_2O_7 \cdot Cs_2Cr_2O_7+Na_2Cr_2O_7 \cdot H_2O$
64.79	1.62	33.59	25	$Na_2Cr_2O_7 \cdot 2H_2O$
64.51	0.97	34.52	25	»
64.90	0.00	35.10	25	»
0.00	10.99	89.01	50	$Cs_2Cr_2O_7$
9.80	11.45	78.75	50	»

Comp. of saturated soln., wt %			t	Comp. of solid phase
Sodium Dichromate	Cesium Dichromate	Water		
13.03	11,52	75.45	50	$Cs_2Cr_2O_7$
24.05	11.50	74.45	50	»
27.40	11.32	61.58	50	»
31.01	10.80	58.18	50	»
36.93	9.88	53.19	50	$Na_2Cr_2O_7 \cdot Cs_2Cr_2O_7 + Cs_2Cr_2O_7$
41.20	9.86	48.94	50	$Na_2Cr_2O_7 \cdot Cs_2Cr_2O_7$
42.40	9.20	48.40	50	»
48.10	5.40	46.50	50	»
50.79	4.40	44.81	50	»
54.20	2.90	42.90	50	»
64.50	2.30	33.20	50	»
66.03	1.91	32.06	50	»
66.50	1.85	31.65	50	»
68.61	2.17	29.22	50	$Na_2Cr_2O_7 \cdot Cs_2Cr_2O_7 + Na_2Cr_2O_7 \cdot 2H_2O$
69.00	0.51	30.49	50	$Na_2Cr_2O_7 \cdot 2H_2O$
69.50	0.00	30.50	50	»

№ 3887 [1719]

SODIUM DIHYDROGEN ORTHOARSENATE–ARSENIC TRIOXIDE–WATER

$NaH_2AsO_4—As_2O_3—H_2O$

Comp. of saturated soln., wt %			t
Sodium Dihydrogen Orthoarsenate	Arsenic Trioxide	Water	
16.175	2.154	81.671	25
56.256	1.548	42.196	25
15.961	4.361	79.678	60
55.700	3.027	61.273	60
15.602	8.630	75.768	100

№ 3888 [2439]

SODIUM SELENITE–MERCURY (II) SELENITE–WATER

$Na_2SeO_3—HgSeO_3—H_2O$

$t = 25$

Concn. of Sodium Selenite in aqueous soln., g/l	Solubility of Mercury Selenite in aqueous soln. of Sodium Selenite g/l
0.2162	1.8
0.4324	3.2
0.8648	5.3
1.7296	7.0
3.4592	13.9
6.9184	27.3

№ 3889 [2191]

SODIUM SELENATE–MAGNESIUM SELENATE–WATER

$Na_2SeO_4—MgSeO_4—H_2O$

$t = 25$

Comp. of saturated soln., wt %			Comp. of solid phase
Magnesium Selenate	Sodium Selenate	Water	
27.5	0.0	72.5	$MgSeO_4 \cdot 6H_2O$
20.20	12.32	67.48	»
19.90	14.36	65.74	»

Comp. of saturated soln., wt %			Comp. of solid phase
Magnesium Selenate	Sodium Selenate	Water	
18.30	17.20	64.50	$Mg_2SeO_4 \cdot 6H_2O$
16.98	20.11	62.91	»
14.36	24.32	61.32	»
15.18	28.76	56.06	»
10.66	31.22	58.12	$MgSeO_4 \cdot 6H_2O + Na_2SeO_4 \cdot 10H_2O$
8.22	33.24	58.54	$Na_2SeO_4 \cdot 10H_2O$
0.0	36.4	63.6	»

№ 3890 [2191]

SODIUM SELENATE–POTASSIUM SELENATE–WATER

Na_2SeO_4—K_2SeO_4—H_2O

$t = 25$

Comp. of saturated soln., wt %			Comp. of solid phase
Potassium Selenate	Sodium Selenate	Water	
53.20	0.0	46.70	K_2SeO_4
45.20	7.21	47.59	»
37.72	12.76	50.52	»
30.02	19.90	50.08	»
26.80	25.36	57.84	»
23.24	28.52	48.24	»
21.86	31.29	46.85	$K_2SeO_4 + Na_2SeO_4$
19.87	31.78	48.35	Na_2SeO_4
17.42	32.34	50.24	»
0.0	36.40	63.60	»

№ 3891 [2191]

SODIUM SELENATE–CALCIUM SELENATE–WATER

Na_2SeO_4—$CaSeO_4$—H_2O

$t = 25$

Comp. of saturated soln., wt %			Comp. of solid phase
Sodium Selenate	Calcium Selenate	Water	
0.00	6.88	93.12	$CaSeO_4 \cdot 2H_2O$
10.55	5.77	83.68	»
16.17	4.94	78.89	»
17.51	4.79	77.70	»
20.59	4.25	75.16	»
27.98	3.33	68.69	$CaSeO_4 \cdot 2H_2O + CaNa_2(SeO_4)_2 \cdot 2H_2O$
30.01	2.86	67.13	$CaNa_2(SeO_4)_2 \cdot 2H_2O$
33.62	2.32	64.06	»
35.61	2.17	62.22	$CaNa_2(SeO_4)_2 \cdot 2H_2O + Na_2SeO_4$
36.4	0.0	63.6	Na_2SeO_4

№ 3892 SODIUM SELENATE–LEAD (II) SELENATE–WATER [797]

$Na_2SeO_4—PbSeO_4—H_2O$

$t = 25$

Concn. of Sodium Selenate in aqueous soln., mol/l	Solubility of Lead Selenate in aqueous soln. of Sodium Selenate, mol/l	Concn. of Sodium Selenate in aqueous soln., mol/l	Solubility of Lead Selenate in aqueous soln. of Sodium Selenate, mol/l
$0.94 \cdot 10^{-4}$	$4.24 \cdot 10^{-4}$	$1.00 \cdot 10^{-2}$	$0.128 \cdot 10^{-4}$
$4.7 \cdot 10^{-4}$	$2.98 \cdot 10^{-4}$	$0.94 \cdot 10^{-1}$	$0.124 \cdot 10^{-4}$
$0.94 \cdot 10^{-3}$	$1.74 \cdot 10^{-4}$		

№ 3893 SODIUM BROMIDE–SODIUM BROMATE–WATER [2390]

$NaBr—NaBrO_3—H_2O$

Comp. of saturated soln., wt %			t	d	Comp. of solid phase
Sodium Bromide	Sodium Bromate	Water			
45.80	0.00	54.20	10	1.492	$NaBr \cdot 2H_2O$
44.51	2.58	52.91	10	1.517	$NaBr \cdot 2H_2O + NaBrO_3$
43.09	2.83	44.08	10	1.498	$NaBrO_3$
39.40	3.55	57.05	10	1.452	»
11.10	14.46	74.44	10	1.240	»
5.33	18.73	75.94	10	1.220	»
0.00	23.24	76.76	10	1.211	»
48.41	0.00	51.59	25	1.530	$NaBr \cdot 2H_2O$
47.37	1.90	50.73	25	1.546	»
46.82	2.94	50.24	25	1.555	$NaBr \cdot 2H_2O + NaBrO_3$
45.62	3.15	51.23	25	1.542	$NaBrO_3$
39.24	4.61	56.15	25	1.462	»
38.66	4.78	56.56	25	1.457	»
29.83	7.86	62.31	25	1.377	»
21.27	12.04	66.69	25	1.320	»
13.82	16.72	69.46	25	1.282	»
6.46	22.38	71.16	25	1.270	»
0.00	28.29	71.71	25	1.257	»
52.55	0.00	47.45	45	—	$NaBr \cdot 2H_2O$
50.68	3.51	45.81	45	—	$NaBr \cdot 2H_2O + NaBrO_3$
44.39	3.72	51.89	45	—	$NaBrO_3$
28.69	11.17	60.14	45	—	»
7.91	26.65	65.44	45	—	»

№ 3894 SODIUM BROMIDE–SODIUM BROMATE–WATER [386]

$NaBr—NaBrO_3—H_2O$

$t = 35$

Concn. in aqueous soln., wt %		Comp. of solid phase	Concn. in aqueous soln., wt %		Comp. of solid phase
Sodium Bromide	Sodium Bromate		Sodium Bromide	Sodium Bromate	
50.57	0.3	$NaBr \cdot 2H_2O$	18.18	15.30	$NaBrO_3$
47.92	3.31	$NaBr \cdot 2H_2O + NaBrO_3$	8.73	22.59	»
39.69	4.94	$NaBrO_3$	0.0	31.95	»
27.94	9.69	»			

№ 3895 [2379]

SODIUM BROMIDE–SODIUM HEXAHYDROXOSTANNATE (IV)–WATER

$NaBr—Na_2[Sn(OH)_6]—H_2O$

$t = 18$

Concn. of Sodium Bromide in aqueous soln., g/l	Solubility of Sodium Hexahydroxostannate in aqueous soln. of Sodium Bromide, g/l
202.74	370

№ 3896 SODIUM BROMIDE–SODIUM IODIDE–WATER [2494]

$NaBr— NaI —H_2O$

$t = 18$

Comp. of saturated soln., wt %			Comp. of solid phase
Sodium Iodide	Sodium Bromide	Water	
51.10	0.0	48.90	$NaI \cdot 2H_2O$
63.2	2.1	34.70	»
62.2	3.0	34.8	»
60.2	5.0	34.8	»
59.2	5.8	35.0	»
58.8	6.55	34.65	»
57.6	8.15	34.25	$NaI \cdot 2H_2O + NaI \cdot NaBr \cdot 2H_2O$
56.65	8.4	34.95	$NaI \cdot NaBr \cdot 2H_2O$
55.75	9.5	34.75	»
54.90	10.3	34.80	$NaI \cdot NaBr \cdot 2H_2O + NaBr \cdot 2H_2O$
54.0	10.9	35.10	$NaBr \cdot 2H_2O$
51.70	12.1	36.20	»
49.2	13.6	37.2	»
48.2	14.45	37.35	»

Table № 3896 (continuation)

Comp. of saturated soln., wt %			Comp. of solid phase
Sodium Iodide	Sodium Bromide	Water	
45.5	16.3	38.2	$NaBr \cdot 2H_2O$
43.4	17.4	39.2	»
40.2	19.7	40.1	»
38.1	21.3	40.6	»
33.7	24.2	42.1	»
31.5	25.2	43.3	»
30.5	26.4	43.1	»
24.75	30.6	44.65	»
19.3	34.8	45.9	»
15.4	37.5	47.1	»
12.7	39.4	47.9	»
9.6	41.8	48.6	»
7.7	43.1	49.2	»
5.1	45.1	49.8	»
0.0	48.6	51.4	»

№ 3897 SODIUM BROMIDE–SODIUM IODIDE–WATER [253]

$NaBr—NaI—H_2O$

Comp. of saturated soln., wt %			t	Comp. of solid phase
Sodium Bromide	Sodium Iodide	Water		
0.0	95.0	5.0	−0.75	Section I
9.1	85.9	5.0	−4.4	»
16.7	78.3	5.0	−8.2	»
23.0	72.0	5.0	−12.2	»
28.6	66.4	5.0	−17.0	»
33.3	61.7	5.0	−23.0	»
34.6	60.4	5.0	−24.6	»
35.4	59.6	5.0	−25.9	»
36.7	58.3	5.0	−27.4	»
37.5	57.5	5.0	−25.6	»
38.0	57.0	5.0	−24.3	»
39.3	55.7	5.0	−18.6	»
41.2	53.8	5.0	−8.0	»
42.8	52.2	5.0	2.0	»
44.4	50.6	5.0	12.0	»
45.9	49.1	5.0	20.4	»
47.3	47.7	5.0	28.8	»
0.0	90.0	10.0	−2.3	Section II
9.1	80.9	10.0	−6.4	»
16.7	73.3	10.0	−10.4	»
23.0	67.0	10.0	−15.2	»
28.6	61.4	10.0	−21.0	»
33.3	56.7	10.0	−27.2	»
34.6	55.4	10.0	−26.6	»
35.4	54.6	10.0	−24.3	»

Table № 3897 (continuation)

Comp. of saturated soln., wt %			t	Comp. of solid phase
Sodium Bromide	Sodium Iodide	Water		
36.3	53.7	10.0	—22.6	Section II
37.5	52.5	10.0	—16.8	»
39.3	50.7	10.0	— 5.0	»
41.2	48.8	10.0	5.8	»
42.8	47.2	10.0	15.2	»
44.4	45.6	10.0	24.2	»
45.9	44.1	10.0	32.4	»
0.0	80.0	20.0	— 5.9	Section III
9.1	70.9	20.0	—10.6	»
16.7	63.3	20.0	—15.2	»
33.0	47.0	20.0	—20.8	»
25.9	54.1	20.0	—24.0	»
28.0	52.0	20.0	—27.4	»
29.6	50.4	20.0	—26.6	»
30.6	49.4	20.0	—24.8	»
31.5	48.5	20.0	—23.0	»
32.4	47.6	20.0	—18.6	»
34.2	45.8	20.0	— 8.8	»
35.1	44.9	20.0	— 3.2	»
35.9	44.1	20.0	1.2	»
36.7	43.3	20.0	5.6	»
38.3	41.7	20.0	14.4	»
40.1	59.9	20.0	24.5	»
41.8	38.2	20.0	33.6	»
0.0	70.0	30.0	—11.1	Section IV
9.1	60.9	30.0	—17.2	»
16.7	53.3	30.0	—23.8	»
20.0	50.0	30.0	—27.4	»
21.8	48.2	30.0	—30.0	»
22.4	47.6	30.0	—29.0	»
23.0	47.0	30.0	—27.0	»
23.6	46.4	30.0	—26.0	»
24.2	45.8	30.0	—24.8	»
26.4	43.6	30.0	—15.2	»
28.6	41.4	30.0	— 3.0	»
30.0	40.0	30.0	4.6	»
32.0	38.0	30.0	15.2	»
33.8	36.2	30.0	24.0	»
35.4	34.6	30.0	32.2	»
0.0	66.0	34.0	—13.8	Section V
9.1	56.9	34.0	—21.0	»
13.8	52.2	34.0	—25.6	»
16.7	49.3	34.0	—28.6	»
18.0	48.0	34.0	—30.2	»
19.4	46.6	34.0	—29.6	»
20.0	46.0	34.0	—28.0	»
21.2	44.8	34.0	—26.0	»
24.2	41.8	34.0	—13.6	»
27.0	39.0	34.0	1.8	»
29.0	37.0	34.0	12.0	»
31.6	34.4	34.0	26.6	»

Table № 3897 (continuation)

Comp. of saturated soln., wt %			t	Comp. of solid phase
Sodium Bromide	Sodium Iodide	Water		
0.0	64.0	36.0	—15.6	Section VI
5.7	58.3	36.0	—20.0	»
10.7	53.3	36.0	—24.6	»
15.3	48.7	36.0	—30.4	»
16.7	47.3	36.0	—32.4	»
18.0	46.0	36.0	—32.0	»
18.7	45.2	36.0	—28.6	»
20.0	44.0	36.0	—26.0	»
22.5	41.5	36.0	—15.3	»
25.4	38.6	36.0	0.2	»
27.0	37.0	36.0	7.6	»
29.0	35.0	36.0	17.0	»
31.5	32.5	36.0	29 4	»
0.0	61.0	39.0	—18.0	Section VII
4.76	56.24	39.0	—22.0	»
9.1	51.9	39.0	—26.4	»
13.0	48.0	39.0	—31.2	»
14.5	46.5	39.0	—28.6	»
15.2	45.8	39.0	—26.6	»
16.7	44.3	39.0	—24.4	»
17.4	43.6	39.0	—24.0	»
20.0	41.0	39.0	—15.6	»
23.0	38.0	39.0	— 0.6	»
24.8	36.2	39.0	8.0	»
26.5	34.5	39.0	17.0	»
29.6	31.5	39.0	31.6	»
0.0	55.0	45.0	—23.8	Section VIII
1.96	53.04	45.0	—25.7	»
3.8	51.2	45.0	—28.0	»
5.7	49.3	45.0	—26.8	»
7.4	47.6	45.0	—23.8	»
9.1	45.9	45.0	—21.8	»
10.7	44.3	45.0	—20.4	»
13.8	41.2	45.0	—18.4	»
16.7	38.3	45.0	—11.0	»
20.0	35.0	45.0	5.0	»
23.0	32.0	45.0	19.0	»
25.3	29.7	45.0	30.6	»
0.0	50.0	50.0	—25.5	Section IX
1.96	48.04	50.0	—23.8	»
2.9	47.1	50.0	—23.0	»
6.5	43.5	50.0	—19.8	»
9.1	40.9	50.0	—18.2	»
11.5	38.5	50.0	—16.8	»
14.5	35.5	50.0	— 4.4	»
16.7	33.3	50.0	7.0	»
18.7	31.3	50.0	17.0	»
21.3	28.7	50.0	29.5	»
0.0	44.0	56.0	—16.6	Section X
3.84	40.16	56.0	—15.6	»
7.40	36.6	56.0	—14.6	»

Table № 3897 (continuation)

Comp. of saturated soln., wt %			t	Comp. of solid phase
Sodium Bromide	Sodium Iodide	Water		
9.90	34.1	56.0	— 5.2	Section X
12.3	31.7	56.0	8.0	»
13.8	30.2	56.0	16.0	»
15.2	28.8	56.0	22.6	»
16.7	27.3	56.0	31.2	»
0.0	40.0	60.0	—13.6	Section XI
1.96	38.04	60.0	—11.6	»
3.84	36.16	60.00	— 7.2	»
5.7	34.3	60.0	— 2.6	»
7.4	32.6	60.0	1.6	»
9.1	30.9	60.0	5.6	»
10.7	29.3	60.0	12.4	»
12.3	27.7	60.0	20.4	»
13.8	26.2	60.0	26.2	»
15.3	24.7	60.0	33.4	»
0.0	38.0	62.0	2.2	Section XII
1.96	36.04	62.0	5.8	»
3.84	34.16	62.0	9.6	»
5.70	32.3	62.0	13.0	»
7.40	30.6	62.0	16.6	»
9.1	28.9	62.0	20.0	»
10.7	27.3	62.0	30.0	»
11.2	26.8	62.0	40.0	»
95.0	0.0	5.0	— 1.5	Section XIII
85.9	9.1	5.0	— 3.8	»
78.3	16.7	5.0	— 7.0	»
72.0	23.0	5.0	— 9.9	»
61.7	33.3	5.0	—16.6	»
59.1	35.9	5.0	—18.8	»
53.5	41.5	5.0	—24.6	»
49.4	45.6	5.0	—27.6	»
48.0	47.0	5.0	—25.4	»
45.3	49.7	5.0	—21.6	»
42.5	52.1	5.0	—18.8	»
40.7	54.3	5.0	—16.5	»
37.8	57.2	5.0	—14.3	»
35.0	60.0	5.0	— 2.2	»
33.5	61.5	5.0	7.8	»
32.2	62.8	5.0	16.0	»
31.0	64.0	5.0	24.0	»
29.8	65.2	5.0	32.0	»
90.0	0.0	10.0	— 3.6	Section XIV
80.9	9.1	10.0	— 6.4	»
73.3	16.7	10.0	— 9.4	»
61.4	28.6	10.0	—16.4	»
52.5	37.5	10.0	—24.0	»
49.5	40.5	10.0	—27.8	»
45.6	44.4	10.0	—24.2	»
42.7	47.3	10.0	—20.3	»
40.0	50.0	10.0	—17.4	»
36.5	53.5	10.0	—14.8	»

Table № 3897 (continuation)

Comp. of saturated soln., wt %			t	Comp. of solid phase
Sodium Bromide	Sodium Iodide	Water		
33.5	56.5	10.0	— 8.0	Section XIV
31.0	59.0	10.0	7.0	»
30.0	60.0	10.0	14.0	»
29.0	61.0	10.0	20.2	»
27.5	62.5	10.0	29.0	»
0.0	56.0	44.0	—16.6	Section XV
2.8	56.0	41.2	—18.2	»
4.6	56.0	39.4	—19.2	»
6.7	56.0	37.3	—20.4	»
9.7	56.0	34.3	—22.0	»
11.2	56.0	32.8	—23.0	»
12.4	56.0	31.6	—23.6	»
13.7	56.0	30.3	—24.5	»
16.4	56.0	27.6	—26.6	»
18.2	56.0	25.8	—27.0	»
19.1	56.0	24.9	—26.0	»
19.7	56.0	24.3	—25.0	»
58.0	0.0	42.0	—18.6	Section XVI
37.5	20.5	42.0	— 6.6	»
34.2	23.8	42.0	— 5.2	»
31.6	26.4	42.0	— 4.2	»
28.3	29.7	42.0	— 3.6	»
25.5	32.5	42.0	— 3.4	»
22.7	35.3	42.0	— 3.0	»
20.9	37.1	42.0	— 3.2	»
19.3	38.7	42.0	— 3.4	»
16.6	41.4	42.0	— 4.2	»
14.4	43.6	42.0	— 5.0	»
11.5	46.5	42.0	— 7.2	»
9.1	48.9	42.0	—10.0	»
8.6	49.4	42.0	—10.6	»
6.8	51.2	42.0	—13.2	»
5.3	52.7	42.0	—10.2	»
3.9	54.1	42.0	6.4	»
2.7	54.3	42.0	— 5.8	»
52.0	0.0	48.0	21.8	Section XVII
39.2	12.8	48.0	22.4	»
34.5	17.5	48.0	22.0	»
31.7	20.3	48.0	21.6	»
26.1	25.9	48.0	20.8	»
22.7	29.3	48.0	19.8	»
19.6	32.4	48.0	18.8	»
15.6	36.4	48.0	17.2	»
12.3	39.7	48.0	15.0	»
9.8	42.2	48.0	13.2	»
7.3	44.7	48.0	10.6	»
4.8	47.2	48.0	7.6	»
2.6	49.4	48.0	4.2	»
1.3	50.7	48.0	1.8	»
0.8	51.2	48.0	1.4	»

Table № 3897 (continuation)

Section No.	Temperature of points of intersection	Comp. of solid phase
I	—28.2	Ice + $NaBr \cdot 5H_2O$ + $NaBr \cdot 2H_2O$
I	—23.6	$NaBr \cdot 5H_2O$ + $NaBr \cdot 2H_2O$
II	—28.4	Ice + $NaBr \cdot 5H_2O$ + $NaBr \cdot 2H_2O$
II	—23.8	$NaBr \cdot 5H_2O$ + $NaBr \cdot 2H_2O$
III	—29.0	Ice + $NaBr \cdot 5H_2O$ + $NaBr \cdot 2H_2O$
III	—24.0	$NaBr \cdot 5H_2O$ + $NaBr \cdot 2H_2O$
IV	—30.6	Ice + $NaBr \cdot 5H_2O$ + $NaBr \cdot 2H_2O$
IV	—24.4	$NaBr \cdot 5H_2O$ + $NaBr \cdot 2H_2O$
V	—32.2	Ice + $NaBr \cdot 5H_2O$ + $NaBr \cdot 2H_2O$
V	—24.6	$NaBr \cdot 5H_2O$ + $NaBr \cdot 2H_2O$
VI	—33.2	Ice + $NaBr \cdot 5H_2O$ + $NaBr \cdot 2H_2O$
VI	—24.8	$NaBr \cdot 5H_2O$ + $NaBr \cdot 2H_2O$
VII	—33.0	Ice + $NaI \cdot 5H_2O$ + $NaBr \cdot 2H_2O$
VII	—23.0	$NaI \cdot 5H_2O$ + $NaBr \cdot 2H_2O$
VIII	—29.0	Ice + $NaI \cdot 5H_2O$ + $NaBr \cdot 2H_2O$
VIII	—18.0	$NaI \cdot 5H_2O$ + $NaBr \cdot 2H_2O$
IX	—16.2	»
X	—14.4	»
XI	—13.6	$NaI \cdot 5H_2O$ + $NaI \cdot 2H_2O$
XI	6.2	$NaI \cdot 2H_2O$ + $NaBr \cdot 2H_2O$
XII	22.2	»
XIII	—28.8	Ice + $NaI \cdot 5H_2O$ + $NaI \cdot 2H_2O$
XIII	—13.6	$NaI \cdot 5H_2O$ + $NaI \cdot 2H_2O$
XIV	—29.2	Ice + $NaI \cdot 5H_2O$ + $NaI \cdot 2H_2O$
XIV	—13.8	$NaI \cdot 5H_2O$ + $NaI \cdot 2H_2O$
XV	—27.8	$NaI \cdot 2H_2O$ + $NaBr \cdot 5H_2O$
XVI	—13.9	$NaI \cdot 2H_2O$ + $NaBr \cdot 2H_2O$
XVII	— 2.4	»

№ 3898 SODIUM BROMIDE–SODIUM IODIDE–WATER [258]

$NaBr—NaI—H_2O$

Comp. of saturated soln., g-mol			t	Comp. of solid phase
Sodium Bromide	Sodium Iodide	Water		
23.8	76.2	853	—30	Ice + $NaI \cdot 5H_2O$
43.4	56.6	760	—30	$NaI \cdot 5H_2O$ + $NaBr \cdot 5H_2O$
63.0	37.0	822	—30	Ice + $NaBr \cdot 5H_2O$
48.6	51.4	705	—25	$NaI \cdot 5H_2O$ + $NaBr \cdot 5H_2O$ + $NaBr \cdot 2H_2O$
41.2	58.8	646	—20	$NaBr \cdot 2H_2O$ + $NaI \cdot 5H_2O$
24.0	76.0	549	—15	»
18.6	81.4	508	—10	$NaBr \cdot 2H_2O$ + $NaI \cdot 2H_2O$
18.9	81.1	492	— 5	»
19.2	80.8	473	0	»
19.8	80.2	464	5	»
19.8	80.2	453	10	»
20.3	79.6	432	15	»
20.7	79.3	421	20	»
21.4	78.6	404	25	»
21.7	78.3	383	30	»

№ 3899 [679]

SODIUM BROMIDE–SODIUM IODIDE–WATER

$NaBr—NaI—H_2O$

$t = 25$

Comp. of saturated soln., wt %			Comp. of solid phase
Sodium Iodide	Sodium Bromide	Water	
69.29	0.00	30.71	$NaI \cdot 2H_2O$
67.09	2.96	29.95	»
65.49	4.79	29.72	»
64.65	5.79	29.56	$NaI \cdot 2H_2O + NaBr$
64.51	6.03	29.46	»
64.33	6.08	29.59	»
61.31	8.01	30.68	NaBr
55.55	11.50	32.95	»
47.72	16.68	35.60	»
39.49	22.79	37.72	»
31.03	29.37	39.60	»
21.98	36.56	41.46	»
13.56	43.06	43.38	»
8.05	47.47	44.48	»
4.12	50.62	45.26	»
0.00	53.87	46.13	»

№ 3900 [253]

SODIUM BROMIDE–SODIUM IODIDE–POTASSIUM BROMIDE–WATER

$NaBr—NaI—KBr—H_2O$

Comp. of saturated soln., wt %				t	Comp. of solid phase
Sodium Bromide	Sodium Iodide	Potassium Bromide	Water		
4.1	5.9	0.0	90.0	— 2.6	Section I
4.1	5.9	9.1	80.9	— 5.6	»
4.1	5.9	16.7	73.3	— 8.9	»
4.1	5.9	23.0	67.0	—11.7	»
4.1	5.9	26.5	63.5	—13.3	»
4.1	5.9	28.5	61.5	— 6.0	»
4.1	5.9	29.6	60.4	1.4	»
4.1	5.9	30.5	59.5	5.0	»
4.1	5.9	32.0	58.0	12.0	»
4.1	5.9	38.3	51.7	18.0	»
4.1	5.9	35.5	54.5	29.6	»
8.1	11.9	0.0	80.0	— 6.8	Section II
8.1	11.9	9.1	70.9	—10.2	»
8.1	11.9	16.7	63.3	—13.8	»
8.1	11.9	21.9	58.1	—16.1	»
8.1	11.9	22.5	57.5	—14.0	»
8.1	11.9	23.0	57.0	— 7.8	»
8.1	11.9	24.2	55.8	— 2.2	»
8.1	11.9	24.8	54.2	1.2	»
8.1	11.9	27.0	53.0	7.8	»

Table № 3900 (continuation)

Comp. of saturated soln., wt %				t	Comp. of solid phase
Sodium Bromide	Sodium Iodide	Potassium Bromide	Water		
8.1	11.9	29.0	51.0	18.0	Section II
8.1	11.9	32.0	48.0	32.8	»
12.2	17.8	0.0	70.0	—13.0	Section III
12.2	17.8	9.1	60.0	—17.2	»
12.2	17.8	15.2	54.8	—20.3	»
12.2	17.8	16.7	53.3	—17.0	»
12.2	17.8	18.0	52.0	— 6.7	»
12.2	17.8	19.3	50.7	— 1.4	»
12.2	17.8	22.5	47.5	14.0	»
12.2	17.8	24.2	45.8	23.6	»
12.2	17.8	24.8	45.2	27.8	»
12.2	17.8	25.9	44.1	33.0	»
16.2	23.8	0.0	60.0	—22.7	Section IV
16.2	23.8	2.9	57.1	—24.1	»
16.2	23.8	5.7	54.3	—25.5	»
16.2	23.8	9.1	50.9	—27.1	»
16.2	23.8	10.7	39.3	—25.6	»
16.2	23.8	11.5	38.5	—17.0	»
16.2	23.8	14.5	35.5	1.0	»
16.2	23.8	15.9	34.1	10.6	»
16.2	23.8	17.4	32.6	20.0	»
16.2	23.8	18.7	31.3	28.8	»
19.0	28.0	0.0	53.0	—31.6	Section V
19.0	28.0	1.0	52.0	—32.7	»
19.0	28.0	1.96	51.04	—29.6	»
19.0	28.0	3.85	49.15	—27.0	»
19.0	28.0	5.70	47.3	—26.0	»
19.0	28.0	8.2	44.8	—16.4	»
19.0	28.0	9.1	43.9	— 9.6	»
19.0	28.0	10.7	42.3	4.0	»
19.0	28.0	12.3	40.7	17.0	»
19.0	28.0	13.0	40.0	24.0	»
19.0	28.0	13.8	39.2	30.0	»
20.25	29.75	0.0	50.0	—23.7	Section VI
20.25	29.75	1.96	48.04	—18.8	»
20.25	29.75	2.91	47.09	—17.3	»
20.25	29.75	4.76	45.24	—14.8	»
20.25	29.75	5.70	44.30	—14.0	»
20.25	29.75	8.2	41.8	0	»
20.25	29.75	9.9	40.1	14.6	»
20.25	29.75	12.3	47.7	32.6	»
21.5	31.5	0.0	47.0	— 0.6	Section VII
21.5	31.5	2.9	44.1	4.0	»
21.5	31.5	3.85	43.15	5.2	»
21.5	31.5	5.70	41.3	8.0	»
21.5	31.5	6.50	40.5	9.4	»
21.5	31.5	7.4	39.6	15.0	»
21.5	31.5	9.1	37.9	26.0	»
21.5	31.5	9.9	37.1	33.0	»
22.3	32.7	0.0	45.0	12.6	Section VIII
22.3	32.7	1.96	43.04	15.4	»

Table № 3900 (continuation)

Comp. of saturated soln., wt %				t	Comp. of solid phase
Sodium Bromide	Sodium Iodide	Potassium Bromide	Water		
22.3	32.7	3.85	41.15	18.4	Section VIII
22.3	32.7	4.76	40.24	19.6	»
22.3	32.7	5.70	39.30	21.1	»
22.3	32.7	7.0	38.0	24.8	»
22.3	32.7	7.4	37.0	27.8	»
22.3	32.7	8.2	36.8	33.0	»
0.0	0.0	10.0	90.0	— 2.8	Section X
6.8	9.9	10.0	73.3	— 9.0	»
11.7	16.9	10.0	61.4	—16.8	»
15.4	22.1	10.0	52.5	—25.8	»
17.0	24.5	10.0	48.5	—30.8	»
17.7	25.5	10.0	46.8	—30.6	»
17.9	25.9	10.0	46.2	—28.0	»
18.4	26.6	10.0	45.0	—25.0	»
18.8	27.1	10.0	44.1	—22.0	»
19.3	27.9	10.0	42.8	—14.0	»
19.8	28.4	10.0	41.2	— 6.0	»
20.1	29.1	10.0	40.8	1.2	»
20.9	30.1	10.0	39.0	12.2	»
21.5	30.9	10.0	37.6	20.0	»
22.1	31.8	10.0	36.1	27.0	»
22.5	32.4	10.0	35.1	32.0	»
0.0	0.0	5.0	95.0	— 1.3	Section IX
6.8	9.9	5.0	78.3	— 6.7	»
11.7	16.9	5.0	66.4	—14.0	»
15.4	22.1	5.0	57.5	—22.5	»
17.5	25.3	5.0	52.2	—29.8	»
18.3	26.4	5.0	50.3	—32.7	»
18.6	26.7	5.0	49.8	—30.4	»
18 9	27.3	5.0	48.8	—26.4	»
19.4	27.8	5.0	47.8	—24.6	»
19.8	28.4	5.0	46.8	—19.0	»
20.3	29.2	5.0	45.5	— 9.6	»
20.5	29.5	5.0	45.0	— 6.6	»
22.0	29.0	5.0	44.0	1.0	»
21.5	30.9	5.0	42.6	9.7	»
21.8	31.4	5 0	41.8	13.8	»
22.3	32.2	5.0	40.5	20.6	»
22.8	32.7	5.0	39.5	25.0	»
23.2	33.3	5.0	38.5	29.4	»
4.1	5.9	0.0	90.0	— 2.6	Section XI
8.2	11.8	0.0	80.0	— 6.8	»
12.3	18.7	0.0	70.0	—13.0	»
16.4	24.6	0.0	60.0	—22.8	»
18.8	27.1	0.0	54.1	—30.0	»
19.8	28.4	0.0	51.8	—28.2	»
19.9	28.6	0.0	51.5	—26.8	»
20.2	29.0	0.0	50.8	—25.0	»
22.0	29.0	0.0	49.0	—17.4	»
21.4	30.7	0.0	47.9	—10.0	»
21.8	31.4	0.0	46.8	1.0	»

Table № 3900 (continuation)

Comp. of saturated soln., wt %				t	Comp. of solid phase
Sodium Bromide	Sodium Iodide	Potassium Bromide	Water		
22.5	32.4	0.0	45.1	10.4	Section XI
23.2	33.3	0.0	43.5	19.6	»
23.5	33.9	0.0	42.6	23.2	»
23.9	34.4	0.0	41.7	28.4	»
24.3	35.0	0.0	40.7	33.0	»

Table № 3900 (continuation)

Section No.	Temperature of points of intersection	Comp. of solid phase
I	—14.0	Ice+αKBr
I	9.2	αKBr+βKBr
II	—16.3	Ice +αKBr
II	6.4	αKBr+βKBr
III	—20.7	Ice+αKBr
III	4.0	αKBr+βKBr
IV	—27.8	Ice+αKBr
IV	+ 2.0	KBr+βKBr
V	—32.7	Ice+$NaBr \cdot 5H_2O$
V	—25.6	$NaBr \cdot 5H_2O$+αKBr
VI	—13.6	$NaBr \cdot 2H_2O$+αKBr
VII	8.8	»
VIII	22.0	»
IX	—33.0	Ice+$NaBr \cdot 5H_2O$
IX	—23.8	$NaBr \cdot 5H_2O$+$NaBr \cdot 2H_2O$
X	—33.8	Ice +$NaBr \cdot 5H_2O$
X	—24.0	$NaBr \cdot 5H_2O$+$NaBr \cdot 2H_2O$
XI	—32.6	Ice+$NaBr \cdot 5H_2O$
XI	—23.6	$NaBr \cdot 5H_2O$+$NaBr \cdot 2H_2O$

№ 3901 SODIUM BROMIDE–SODIUM IODIDE–POTASSIUM BROMIDE–WATER [257]

$NaBr—NaI—KBr—H_2O$

Comp. of saturated soln., g-mol				t	Comp. of solid phase
Sodium Bromide	Sodium Iodide	Potassium Bromide	Water		
42.0	42.0	16:0	701.0	—30	$NaBr \cdot 5H_2O$+αKBr
39.6	39.6	20.8	782.0	—30	Ice+αKBr
43.2	43.2	13.6	634.0	—20	$NaBr \cdot 2H_2O$+αKBr
28.7	28.7	42.6	994.0	—20	Ice+αKBr
43.7	43.7	12.6	593.0	—10	$NaBr \cdot 2H_2O$+αKBr
49.7	43.7	42.6	565.0	0	»
43.9	43.9	12.2	532.0	10	$NaBr \cdot 2H_2O$+βKBr
8.0	8.0	84.0	1046.0	10	αKBr+βKBr
44.0	44.0	12.0	506.0	20	$NaBr \cdot 2H_2O$+βKBr
0.6	0.6	98.8	1014.0	20	αKBr+βKBr

Comp. of saturated soln., g-mol				t	Comp. of solid phase
Sodium Bromide	Sodium Iodide	Potassium Bromide	Water		
38.4	38.4	32.2	686.0	—25	αKI+αKBr
33.7	33.7	32.6	821.0	—25	Ice+αKBr
38.4	38.4	23.2	666.0	—20	αKI+αKBr
27.0	27.0	46.0	926.0	—20	Ice+αKBr
38.5	38.5	23.0	633.0	—10	αKI+αKBr
38.6	38.6	22.8	603.0	0	»
38.8	38.8	22.4	580.0	10	βKI+βKBr
8.0	8.0	84.0	1016.1	10	αKBr+βKBr
38.9	38.9	22.2	542.0	20	βKI+βKBr
0.7	0.7	98.6	1008.0	20	αKBr+βKBr
54.6	11.6	33.8	738.0	—30	NaBr · $5H_2O$+NaI · $5H_2O$
23.8	4.2	72.0	820.0	—30	Ice+NaBr+$5H_2O$
53.7	9.4	36.9	592.0	—20	NaBr · $2H_2O$+NaI · $5H_2O$
65.9	11.6	22.5	502.0	—10	NaBr·$2H_2O$+αKBr
76.0	13.4	10.6	514.0	—10	αKI+αKBr
65.9	11.5	22.6	468.0	0	NaBr · $2H_2O$+αKBr
76.1	13.5	10.4	496.0	0	αKI+αKBr
66.2	11.6	22.2	447.0	10	NaBr · $2H_2O$+βKBr
75.3	13.2	11.5	467.0	10	βKI+βKBr
66.4	11.5	22.1	424.0	20	NaBr · $2H_2O$+βKBr
76.0	13.4	10.6	439.0	20	βKI+βKBr
66.5	11.7	21.8	390.0	30	NaBr · $2H_2O$+βKBr
78.9	13.7	7.4	417.0	30	βKI+βKBr
55.6	18.6	25.8	688.0	—30	NaI · $5H_2O$+αKI
42.5	14.2	43.3	770.0	—30	Ice+αKI
58.8	19.6	21.6	581.0	—20	NaI · $5H_2O$+αKI
66.3	22.1	11.6	505.0	—10	NaBr · $2H_2O$+αKBr
54.9	18.3	26.8	582.0	—10	αKI+αKBr
66.3	22.1	11.6	468.0	0	NaBr · $2H_2O$+αKBr
54.8	18.2	27.0	543.0	0	αKI+αKBr
66.5	22.2	11.3	447.0	10	NaBr · $2H_2O$+βKBr
54.7	18.2	27.1	515.0	10	βKI+βKBr
4.3	1.4	94.3	668.0	10	αKI+βKI
66.4	22.2	11.4	425.0	20	NaBr · $2H_2O$+βKBr
54.5	18.1	27.4	480.0	20	βKI+βKBr
66.5	22.2	11.3	394.0	30	NaBr · $2H_2O$+βKBr
54.5	18.2	27.3	451.0	30	βKI+βKBr

№ 3902 SODIUM BROMIDE–SODIUM IODIDE–POTASSIUM BROMIDE–POTASSIUM IODIDE–WATER [254]

NaBr—NaI—KBr—KI—H_2O

Comp. of saturated soln., g-mol					t	Comp. of solid phase
Sodium Bromide	Potassium Bromide	Sodium Iodide	Potassium Iodide	Water		
0.00	0.00	97.3	2.7	847.0	- -30	Ice+NaI · $5H_2O$
0.0	0.0	78.9	21.1	729.0	—30	αKI+NaI · $5H_2O$
0.0	0.0	60.3	39.7	797.0	—30	Ice+αKI

Comp. of saturated soln., g-mol					t	Comp. of solid phase
Sodium Bromide	Potassium Bromide	Sodium iodide	Potassium Iodide	Water		
23.8	0.0	76.2	0.0	853.0	—30	Ice + NaI · $5H_2O$
43.4	0.0	56.6	0.0	760.0	—30	NaBr · $5H_2O$ + NaI · $5H_2O$
63.0	0.0	37.0	0.0	822.0	—30	Ice + NaBr · $5H_2O$
88.5	11.5	0.0	0.0	865.0	—30	Ice + αKBr
91.0	9.0	0.0	0.0	845.0	—30	αKBr + NaBr · $5H_2O$
92.0	8.0	0.0	0.0	870.0	—30	Ice + NaBr · $5H_2O$
0.0	4.2	95.8	0.0	845.0	—30	Ice + NaI · $5H_2O$
0.0	26.1	73.9	0.0	663.0	—30	αKBr + NaI · $5H_2O$
0.0	35.4	64.6	0.0	754.0	—30	Ice + αKBr
90.1	0.0	0.0	9.9	845.0	—30	Ice + NaBr · $5H_2O$
88.0	0.0	0.0	12.0	807.0	—30	αKBr + NaBr · $5H_2O$
86.3	0.0	0.0	13.7	836.0	—30	Ice + αKBr
42.0	16.0	42.0	0.0	701.0	—30	αKBr + NaBr · $5H_2O$
39.6	20.8	39.6	0.0	782.0	—30	Ice + αKBr
72.0	0.0	23.8	4.2	820.0	—30	Ice + NaBr · $5H_2O$
33.8	0.0	54.6	11.6	738.0	—30	NaBr · $5H_2O$ + NaI · $5H_2O$
14.2	0.0	42.5	43.3	770.0	—30	Ice + αKI
18.6	0.0	55.6	25.8	688.0	—30	αKI + NaI · $5H_2O$
0.0	0.0	80.3	19.7	668.0	—25	»
0.0	0.0	34.6	65.4	829.0	—25	Ice + αKI
48.6	0.0	51.4	0.0	705.0	—25	NaBr · $5H_2O$ + NaI · $5H_2O$
91.0	9.0	0.0	0.0	797.0	—25	NaBr · $5H_2O$ + αKBr
81.5	18.5	0.0	0.0	945.0	—25	Ice + αKBr
0.0	25.0	75.0	0.0	625.0	—25	αKBr + NaI · $5H_2O$
0.0	44.7	55.3	0.0	834.0	—25	Ice + αKBr
88.0	0.0	0.0	12.0	766.0	—25	NaBr · $5H_2O$ + αKBr
70.4	0.0	0.0	29.6	875.0	—25	Ice + αKBr
30.0	0.0	0.0	70.0	816.0	—25	»
23.1	0.0	0.0	76.9	794.0	—25	αKI + αKBr
19.2	0.0	0.0	80.8	812.0	—25	Ice + αKI
42.5	15.0	42.5	0.0	680.0	—25	NaBr · $5H_2O$ + αKBr
34.6	30.8	34.6	0.0	864.0	—25	Ice + αKI
0.0	23.2	38.4	38.4	686.0	—25	αKI + αKBr
0.0	32.6	33.7	33.7	821.0	—25	Ice + αKBr
38.7	0.0	52.2	9.1	665.0	—25	NaBr · $5H_2O$ + NaI · $5H_2O$
6.9	0.0	20.8	72.3	823.0	—25	Ice + αKI
18.9	0.0	56.5	24.6	633.0	—25	αKI + NaI · $5H_2O$
0.0	0.0	87.6	12.4	532.0	—20	»
0.0	18.8	0.0	81.2	756.0	—20	αKI + αKBr
0.0	62.5	0.0	37.5	1169.0	—20	Ice + αKBr
24.0	0.0	76.0	0.0	549.0	—20	NaBr · $2H_2O$ + NaI · $5H_2O$
91.6	8.4	0.0	0.0	758.0	—20	NaBr · $2H_2O$ + αKBr
32.0	68.0	0.0	0.0	1295.0	—20	Ice + αKBr
0.0	10.2	89.8	0.0	505.0	—20	αKBr + NaI · $5H_2O$
0.0	78.0	22.0	0.0	1196.0	—20	Ice + αKBr
87.2	0.0	0.0	12.8	703.0	—20	NaBr · $2H_2O$ + αKBr
22.5	0.0	0.0	77.5	738.0	—20	αKBr + αKI
43.6	12.8	43.6	0.0	620.0	—20	NaBr · $2H_2O$ + αKBr
14.3	71.4	14.3	0.0	1213.0	—20	Ice + αKBr
0.0	22.8	38.6	38.6	646.0	—20	αKI + αKBr
0.0	72.4	13.8	13.8	1187.0	—20	Ice + αKBr

Table № 3902 (continuation)

Comp. of saturated soln., g-mol					t	Comp. of solid phase
Sodium Bromide	Potassium Bromide	Sodium Iodide	Potassium Iodide	Water		
22.4	0.0	66.0	11.6	524.9	—20	$NaBr \cdot 2H_2O + \alpha KBr$
10.8	0.0	75.6	13.6	521.0	—20	$\alpha KI + \alpha KBr$
22.0	0.0	66.2	11.8	528.0	—20	$NaBr \cdot 2H_2O + \alpha KBr$
18.2	0.0	54.8	27.0	598.0	—20	$\alpha KI + \alpha KBr$
0.0	0.0	90.3	9.7	501.0	—10	$NaI \cdot 2H_2O + \alpha KI$
0.0	19.7	0.0	80.3	732.0	—10	$\alpha KI + \alpha KBr$
18.6	0.0	81.4	0.0	508.0	—10	$NaBr \cdot 2H_2O + NaI \cdot 2H_2O$
91.1	8.9	0.0	0.0	729.0	—10	$NaBr \cdot 2H_2O + \alpha KBr$
0.0	10.1	89.9	0.0	492.0	—10	$NaI \cdot 2H_2O + \alpha KBr$
86.9	0.0	0.0	13.1	676.0	—10	$NaBr \cdot 2H_2O + \alpha KBr$
22.9	0.0	0.0	77.1	720.0	—10	$\alpha KI + \alpha KBr$
43.7	12.6	43.7	0.0	593.0	—10	$NaBr \cdot 2H_2O + \alpha KBr$
0.0	23.0	38.5	38.5	633.0	—10	$\alpha KI + \alpha KBr$
22.5	0.0	65.9	11.6	502.0	—10	$NaBr \cdot 2H_2O + \alpha KBr$
10.6	0.0	76.0	13.4	514.0	—10	$\alpha KI + \alpha KBr$
22.1	0.0	66.3	11.6	505.0	—10	$NaBr \cdot 2H_2O + \alpha KBr$
18.3	0.0	54.9	26.8	582.0	—10	$\alpha KI + \alpha KBr$
0.0	0.0	90.2	9.8	427.0	— 5	$NaI \cdot 2H_2O + \alpha KI$
0.0	20.3	0.0	79.7	698.0	— 5	$\alpha KI + \alpha KBr$
18.9	0.0	81.1	0.0	492.0	— 5	$NaBr \cdot 2H_2O + NaI \cdot 2H_2O$
91.0	9.0	0.0	0.0	697.0	— 5	$NaBr \cdot 2H_2O + \alpha KBr$
0.0	10.0	90.0	0.0	480.0	— 5	$NaI \cdot 2H_2O + \alpha KBr$
86.9	0.0	0.0	13.1	656.0	— 5	$NaBr \cdot 2H_2O + \alpha KBr$
23.2	0.0	0.0	76.8	695.0	— 5	$\alpha KI + \alpha KBr$
43.7	12.6	43.7	0.0	583.0	— 5	$NaBr \cdot 2H_2O + \alpha KBr$
0.0	23.0	38.5	38.5	617.0	— 5	$\alpha KI + \alpha KBr$
23.3	0.0	66.2	11.5	491.0	— 5	$NaBr \cdot 2H_2O + \alpha KBr$
10.3	0.0	76.3	13.4	508.0	— 5	$\alpha KI + \alpha KBr$
22.2	0.0	66.5	11.3	495.0	—10	$NaBr \cdot 2H_2O + \alpha KBr$
18.4	0.0	54.9	26.7	568.0	—10	$\alpha KI + \alpha KBr$
0.0	0.0	89.8	10.2	475.0	0	$NaI \cdot 2H_2O + \alpha KI$
0.0	21.0	0.0	79.0	683.0	0	$\alpha KI + \alpha KBr$
19.2	0.0	80.8	0.0	473.0	0	$NaBr \cdot 2H_2O + NaI \cdot 2H_2O$
90.9	9.1	0.0	0.0	680.0	0	$NaBr \cdot 2H_2O + \alpha KBr$
0.0	10.0	90.0	0.0	466.0	0	$NaI \cdot 2H_2O + \alpha KBr$
86.7	0.0	0.0	13.3	635.0	0	$NaBr \cdot 2H_2O + \alpha KBr$
22.9	0.0	0.0	77.1	673.0	0	$\alpha KI + \alpha KBr$
43.7	12.6	43.7	0.0	565.0	0	$NaBr \cdot 2H_2O + \alpha KBr$
0.0	22.8	38.6	38.6	603.0	0	$\alpha KI + \alpha KBr$
22.6	0.0	65.9	11.5	468.0	0	$KBr + NaBr \cdot 2H_2O$
10.4	0.0	76.1	13.5	496.0	0	$\alpha KI + \alpha KBr$
22.1	0.0	66.3	11.6	468.0	0	$NaBr \cdot 2H_2O + \alpha KBr$
12.3	0.0	55.0	26.7	542.0	0	$\alpha KI + \alpha KBr$
0.0	23.0	40.0	37.0	759.0	—30	Ice $+ \alpha KI + \alpha KBr$
0.0	22.5	72.1	5.4	662.0	—30	$NaI \cdot 5H_2O + \alpha KI + \alpha KBr$
20.3	20.5	59.2	0.0	672.0	—30	$NaBr \cdot 5H_2O + NaI \cdot 5H_2O + \alpha KBr$
0.0	23.0	15.4	61.6	812.0	—25	Ice $+ \alpha KI + \alpha KBr$
0.0	22.0	73.0	5.0	620.0	—25	$NaI \cdot 5H_2O + \alpha KI + \alpha KBr$
21.7	19.5	58.8	0.0	650.0	—25	$NaBr \cdot 5H_2O + NaI \cdot 5H_2O + \alpha KBr$
0.0	22.0	74.1	3.9	558.0	—20	$NaI \cdot 5H_2O + \alpha KI + \alpha KBr$
22.4	17.0	60.6	0.0	588.0	—20	$NaBr \cdot 2H_2O + NaI \cdot 5H_2O + \alpha KBr$

Comp. of saturated soln., g-mol					t	Comp. of solid phase
Sodium Bromide	Potassium Bromide	Sodium Iodide	Potassium Iodide	Water		
0.0	8.5	89.2	2.3	508.0	−15	$NaI \cdot 5H_2O + \alpha KI + \alpha KBr$
2.3	9.5	88.2	0.0	507.0	−15	$NaI \cdot 2H_2O + NaI \cdot 5H_2O + \alpha KBr$
2.3	9.5	88.2	0.0	493.0	−10	$NaBr \cdot 2H_2O + NaI \cdot 2H_2O + \alpha KBr$
0.0	8.0	89.7	2.3	492.0	−10	$NaI \cdot 2H_2O + \alpha KI + \alpha KBr$
2.3	9.5	88.2	0.0	481.0	− 5	$NaBr \cdot 2H_2O + NaI \cdot 2H_2O + \alpha KBr$
0.0	8.0	89.7	2.3	481.0	− 5	$NaI \cdot 2H_2O + \alpha KI + \alpha KBr$
2.3	8.5	89.2	0.0	468.0	0	$NaBr \cdot 2H_2O + NaI \cdot 2H_2O + \alpha KBr$
0.0	8.0	89.2	2.8	470.0	0	$NaI \cdot 2H_2O + \alpha KI + \alpha KBr$

№ 3903 SODIUM BROMIDE–SODIUM IODIDE–POTASSIUM BROMIDE–POTASSIUM IODIDE–WATER [255]

$NaBr—NaI—KBr—KI—H_2O$

Comp. of saturated soln., g-mol					t	Comp. of solid phase
Sodium Bromide	Potassium Bromide	Sodium Iodide	Potassium Iodide	Water		
1.8	9.5	88.7	0.0	458.0	5	$NaBr \cdot 2H_2O + NaI \cdot 2H_2O + \beta\text{-}KBr$
0.0	9.0	89.6	1.4	458.0	5	$NaI \cdot 2H_2O + \beta\text{-}KI + \beta\text{-}KBr$
1.4	9.0	89.6	0.0	445.0	10	$NaBr \cdot 2H_2O + NaI \cdot 2H_2O + \beta\text{-}KBr$
0.0	8.5	90.1	1.4	446.0	10	$NaI \cdot 2H_2O + \beta\text{-}KI + \beta\text{-}KBr$
2.3	9.2	88.5	0.0	431.0	15	$NaBr \cdot 2H_2O + NaI \cdot 2H_2O + \beta\text{-}KBr$
0.0	8.5	90.1	1.4	432.0	15	$NaI \cdot 2H_2O + \beta\text{-}KI + \beta\text{-}KBr$
1.8	9.5	88.7	0.0	424.0	20	$NaBr \cdot 2H_2O + NaI \cdot 2H_2O + \beta\text{-}KBr$
0.0	8.0	90.2	1.8	427.0	20	$NaI \cdot 2H_2O + \beta\text{-}KBr + \beta\text{-}KI$
2.7	9.0	88.3	0.0	406.0	25	$NaBr \cdot 2H_2O + NaI \cdot 2H_2O + \beta\text{-}KBr$
0.0	7.5	89.7	2.8	410.0	25	$NaI \cdot 2H_2O + \beta\text{-}KI + \beta\text{-}KBr$
3.1	9.0	87.9	0.0	389.0	30	$NaBr \cdot 2H_2O + NaI \cdot 2H_2O + \beta\text{-}KBr$
0.0	5.0	89.8	5.2	397.0	30	$NaI \cdot 2H_2O + \beta\text{-}KI + \beta\text{-}KBr$

№ 3904 SODIUM BROMIDE–SODIUM IODIDE–POTASSIUM IODIDE–WATER [252]

$NaBr—NaI—KI—H_2O$

Comp. of saturated soln., wt %				$t_{m.p.}$	Comp. of solid phase
Sodium Iodide	Potassium Iodide	Sodium Bromide	Water		
21.0	4.0	0.0	75.0	− 7.8	Section I
21.0	4.0	9.1	65.9	−13.6	»
21.0	4.0	16.7	58.3	−19.8	»
21.0	4.0	23.1	51.9	−26.4	»
21.0	4.0	27.0	48.0	−29.2	»
21.0	4.0	28.6	46.4	−25.6	»
21.0	4.0	29.5	45.5	−24.3	»
21.0	4.0	30.5	44.5	−21.0	»
21.0	4.0	31.5	43.5	−12.0	»

Comp. of saturated soln., wt %				$t_{m.p.}$	Comp. of solid phase
Sodium Iodide	Potassium Iodide	Sodium Bromide	Water		
21.0	4.0	32.0	43.0	− 7.8	Section I
21.0	4.0	32.4	42.6	− 4.0	»
21.0	4.0	32.9	42.1	0.0	»
21.0	4.0	33.3	41.7	4.6	»
21.0	4.0	33.8	41.2	9.6	»
21.0	4.0	34.2	40.8	13.0	»
21.0	4.0	34.6	40.4	17.0	»
21.0	4.0	35.0	40.0	20.0	»
21.0	4.0	36.3	38.7	30.7	»
35.2	6.8	0.0	58.0	−20.5	Section II
35.2	6.8	3.85	54.15	−24.0	»
35.2	6.8	7.40	50.6	−27.4	»
35.2	6.8	10.70	47.3	−31.0	»
35.2	6.8	13.00	45.0	−30.5	»
35.2	6.8	13.80	44.2	−28.5	»
35.2	6.8	16.00	42.0	−25.8	»
35.2	6.8	18.7	39.3	−24.2	»
35.2	6.8	21.3	37.7	−12.6	»
35.2	6.8	23.0	35.0	1.0	»
35.2	6.8	23.7	34.3	6.0	»
35.2	6.8	24.8	33.2	16.0	»
35.2	6.8	25.4	32.6	20.2	»
35.2	6.8	26.5	31.5	30.0	»
40.2	7.8	0.0	52.0	−27.8	Section III
40.2	7.8	1.96	50.04	−29.4	»
40.2	7.8	3.85	48.15	−31.8	»
40.2	7.8	5.70	46.3	−29.0	»
40.2	7.8	7.40	44.6	−26.4	»
40.2	7.8	11.50	40.5	−22.0	»
40.2	7.8	15.20	36.8	−20.2	»
40.2	7.8	16.70	35.3	−13.0	»
40.2	7.8	17.40	34.6	− 7.2	»
40.2	7.8	18.0	34.0	− 2.0	»
40.2	7.8	19.4	32.6	10.0	»
40.2	7.8	20.6	31.4	22.0	»
40.2	7.8	21.3	30.7	28.0	»
43.6	8.4	0.0	48.0	−28.2	Section IV
43.6	8.4	2.90	45.1	−26.4	»
43.6	8.4	4.76	43.24	−23.4	»
43.6	8.4	6.50	41.5	−21.8	»
43.6	8.4	8.30	39.7	−20.4	»
43.6	8.4	10.70	37.3	−18.8	»
43.6	8.4	12.30	35.7	−18.0	»
43.6	8.4	13.80	34.2	−17.0	»
43.6	8.4	14.5	33.5	−11.0	»
43.6	8.4	15.2	32.8	− 6.0	»
43.6	8.4	16.0	32.0	0.6	»
43.6	8.4	16.7	31.3	6.0	»
43.6	8.4	17.4	30.6	12.0	»
43.6	8.4	18.0	30.0	17.0	»
43.6	8.4	18.7	29.3	28.4	»

Table № 3904 (continuation)

Comp. of saturated soln., wt %				$t_{m.p.}$	Comp. of solid phase
Sodium Iodide	Potassium Iodide	Sodium Bromide	Water		
48.7	9.3	0.0	42.0	—17.8	Section V
48.7	9.3	1.96	40.04	—17.3	»
48.7	9.3	2.90	39.10	—17.0	»
48.7	9.3	3.84	38.16	—16.8	»
48.7	9.3	5.70	36.30	—15.4	»
48.7	9.3	7.40	34.6	—10.8	»
48.7	9.3	8.20	33.8	— 8.7	»
48.7	9.3	9.10	32.9	— 6.0	»
48.7	9.3	10.70	31.3	— 0.6	»
48.7	9.3	12.3	29.7	8.4	»
48.7	9.3	13.0	29.0	12.0	»
48.7	9.3	13.8	28.2	16.4	»
48.7	9.3	14.5	27.5	20.4	»
48.7	9.3	15.2	26.8	24.2	»
48.7	9.3	16.0	26.0	27.6	»
48.7	9.3	16.7	25.3	31.2	»
48.7	9.3	17.4	24.6	38.8	»
48.7	9.3	18.0	24.0	49.0	»
50.4	9.6	0.0	40.0	—10.8	Section VI
50.4	9.6	1.96	38.04	— 7.2	»
50.4	9.6	2.90	37.1	— 5.2	»
50.4	9.6	5.70	34.3	0.4	»
50.4	9.6	7.40	32.6	3.8	»
50.4	9.6	9.10	30.9	7.4	»
50.4	9.6	10.7	29.3	11.6	»
50.4	9.6	12.3	27.7	18.6	»
50.4	9.6	13.0	27.0	21.6	»
50.4	9.6	13.8	26.2	25.4	»
50.4	9.6	14.5	25.5	29.0	»
50.4	9.6	15.2	24.8	32.0	»
50.4	9.6	15.6	24.4	36.0	»
50.4	9.6	16.0	24.0	42.0	»
52.0	10.0	0.0	38.0	3.2	Section VII
52.0	10.0	2.90	35.1	8.4	»
52.0	10.0	4.76	33.24	10.8	»
52.0	10.0	6.54	31.46	14.0	»
52.0	10.0	8.30	29.7	17.2	»
52.0	10.0	9.90	28.10	20.5	»
52.0	10.0	11.50	26.5	24.0	»
52.0	10.0	12.30	25.7	27.4	»
52.0	10.0	13.50	24.5	33.0	»
52.0	10.0	14.20	23.8	36.0	»
52.0	10.0	14.50	23.5	38.4	»
53.7	10.3	0.0	36.0	18.2	Section VIII
53.7	10.3	0.99	35.01	20.2	»
53.7	10.3	1.96	34.04	21.8	»
53.7	10.3	3.84	32.16	24.6	»
53.7	10.3	5.70	30.3	28.2	»
53.7	10.3	6.54	29.46	29.6	»
53.7	10.3	7.40	28.6	31.0	»
53.7	10.3	9.10	26.9	33.0	»

Table № 3904 (continuation)

Comp. of saturated soln., wt %				$t_{m.p.}$	Comp. of solid phase
Sodium Iodide	Potassium Iodide	Sodium Bromide	Water		
53.7	10.3	10.70	25.3	34.6	Section VIII
53.7	10.3	11.40	24.6	37.0	»
53.7	10.3	12.50	23.5	45.0	»
0.0	0.0	10.0	90.0	— 3.4	Section IX
14.0	2.7	10.0	73.3	— 9.6	»
23.5	4.5	10.0	62.0	—16.0	»
28.0	5.3	10.0	56.7	—20.0	»
31.5	6.0	10.0	52.5	—24.6	»
34.8	6.7	10.0	48.5	—29.0	»
38.5	7.4	10.0	44.1	—28.6	»
39.8	7.6	10.0	42.6	—25.4	»
40.7	7.8	10.0	41.5	—23.6	»
42.0	8.0	10.0	40.0	—22.0	»
44.0	8.4	10.0	37.6	—19.2	»
45.7	8.8	10.0	35.5	—17.0	»
47.5	9.0	10.0	33.5	— 9.6	»
48.1	9.3	10.0	32.6	— 4.0	»
48.9	9.4	10.0	31.7	2.6	»
49.3	9.4	10.0	31.3	5.4	»
50.0	9.5	10.0	30.5	11.0	»
50.4	9.6	10.0	30.0	14.5	»
51.0	9.8	10.0	29.2	20.0	»
51.6	9.9	10.0	28.5	24.8	»
52.0	10.0	10.0	28.0	28.0	»
52.4	10.0	10.0	27.6	30.4	»
0.0	0.0	15.0	85.0	— 5.8	Section X
14.0	2.7	15.0	68.3	—12.7	»
23.1	5.4	15.0	56.5	—20.4	»
31.5	6.0	15.0	47.5	—29.0	»
34.6	6.6	15.0	43.8	—32.0	»
37.3	7.1	15.0	40.6	—25.0	»
39.8	7.4	15.0	37.6	—21.3	»
40.7	7.8	15.0	36.5	—20.0	»
42.0	8.0	15.0	35.0	—18.5	»
44.0	8.4	15.0	32.6	—16.8	»
45.7	8.8	15.0	30.5	— 7.8	»
46.1	8.9	15.0	30.0	— 4.6	»
46.6	8.9	15.0	29.5	— 2.0	»
47.5	9.0	15.0	28.5	4.6	»
48.1	9.3	15.0	27.6	10.0	»
48.9	9.4	15.0	26.7	16.4	»
46.6	8.9	15.0	26.0	20.6	»
50.4	9.6	15.0	25.0	27.0	»
0.0	0.0	22.0	78.0	—10.5	Section XI
7.6	1.5	22.0	68.9	—14.2	»
14.0	2.7	22.0	61.3	—18.4	»
19.4	3.6	22.0	65.0	—22.4	»
25.7	4.8	22.0	47.5	—28.6	»
26.9	5.1	22.0	46.0	—30.2	»
26.8	6.1	22.0	45.1	—31.0	»
29.1	5.5	22.0	43.4	—30.6	»

Table № 3904 (continuation)

Comp. of saturated soln., wt %				$t_{m.p.}$	Comp. of solid phase
Sodium Iodide	Potassium Iodide	Sodium Bromide	Water		
29.8	5.7	22.0	42.5	−29.0	Section XI
31.5	5.8	22.0	41.7	−28.0	»
30.8	5.9	22.0	41.3	−27.4	»
31.9	6.0	22.0	40.1	−26.0	»
32.4	6.4	22.0	39.4	−25.0	»
33.1	6.3	22.0	38.6	−24.0	»
32.3	8.5	22.0	37.2	−22.6	»
36.0	6.8	22.0	35.2	−20.5	»
38.3	7.3	22.0	32.4	−18.2	»
40.7	7.8	22.0	29.5	−12.6	»
42.0	8.0	22.0	28.0	− 6.0	»
43.1	8.6	22.0	26.1	2.0	»
45.2	8.7	22.0	24.1	10.4	»
46.6	8.9	22.0	22.5	19.0	»
48.0	9.2	22.0	20.8	27.0	»
48.7	9.3	22.0	20.0	31.0	»
0.0	0.0	25.0	75.0	−12.2	Section XII
8.3	1.6	25.0	65.1	−16.8	»
16.8	3.2	25.0	55.0	−23.0	»
22.3	4.2	25.0	48.5	−28.4	»
24.5	5.4	25.0	45.4	−31.0	»
26.0	5.0	25.0	44.0	−30.2	»
27.6	5.2	25.0	42.2	−27.4	»
28.8	5.0	25.0	41.2	−26.5	»
29.8	5.7	25.0	72.5	−25.4	»
32.2	6.1	25.0	36.7	−22.8	»
33.8	6.3	25.0	34.9	−21.2	»
34.8	6.7	25.0	33.5	−19.6	»
36.3	6.9	25.0	31.8	−18.4	»
38.3	7.0	25.0	29.7	−14.0	»
39.2	7.4	25.0	28.4	− 8.6	»
40.5	7.7	25.0	27.8	− 2.2	»
42.0	8.0	25.0	25.0	4.8	»
44.2	8.4	25.0	22.4	15.2	»
45.6	8.8	25.0	20.6	22.4	»
47.1	9.0	25.0	18.9	29.0	»
47.5	9.0	25.0	18.5	31.0	»
48.2	9.0	25.0	17.8	34.0	»
0.0	0.0	0.0	100.0	0.0	Section XIII
7.6	1.5	0.0	90.9	− 1.8	»
20.8	4.0	0.0	75.2	− 7.6	»
28.0	5.3	0.0	66.7	−13.2	»
35.2	6.7	0.0	58.1	−20.5	»
40.7	7.8	0.0	51.5	−28.6	»
43.1	8.6	0.0	48.1	−28.2	»
45.1	8.6	0.0	46.3	−24.6	»
47.5	9.0	0.0	43.5	−20.0	»
48.9	9.4	0.0	51.7	−17.2	»
50.4	9.6	0.0	40.0	−10.0	»
52.0	10.0	0.0	38.0	3.2	»

Table № 3904 (continuation)

Comp. of saturated soln., wt %				$t_{m.p.}$	Comp. of solid phase
Sodium Iodide	Potassium Iodide	Sodium Bromide	Water		
53.0	10.0	0.0	37.0	11.0	Section XIII
53.8	10.2	0.0	36.0	18.6	»
55.4	10.5	0.0	34.1	30.0	»

Table № 3904 (continuation)

Section No.	Temperature of points of intersection	Comp. of solid phase
I	−30.3	Ice + $NaBr \cdot 5H_2O$
I	−23.6	$NaBr \cdot 5H_2O + NaBr \cdot 2H_2O$
II	−33.0	Ice + $NaBr \cdot 5H_2O$
II	−23.8	$NaBr \cdot 5H_2O + NaBr \cdot 2H_2O$
III	−32.2	Ice + $NaI \cdot 5H_2O$
III	−20.0	$NaI \cdot 5H_2O + NaBr \cdot 2H_2O$
IV	−17.6	»
V	−16.2	$NaI \cdot 5H_2O + \beta\text{-}KI$
V	− 3.0	$\beta\text{-}KI + NaBr \cdot 2H_2O$
V	32.0	$NaBr \cdot 2H_2O + NaBr$
VI	− 2.0	$\alpha\text{-}KI + \beta\text{-}KI$
VI	10.6	$\beta\text{-}KI + NaBr \cdot 2H_2O$
VI	33.0	$NaBr \cdot 2H_2O + NaBr$
VII	11.4	$\beta\text{-}KI + \beta\text{-}KBr$
VII	22.0	$\beta\text{-}KBr + NaBr \cdot 2H_2O$
VII	34.8	$NaBr \cdot 2H_2O + NaBr$
VIII	25.2	$\beta\text{-}KI + \beta\text{-}KBr$
VIII	34.7	$\beta\text{-}KBr + NaBr$
VIII	39.0	$NaBr \cdot 2H_2O + NaBr$
IX	−32.6	Ice + $NaI \cdot 5H_2O$
IX	−16.0	$NaI \cdot 5H_2O + \alpha\text{-}KI$
IX	− 1.6	$\alpha\text{-}KI + \beta\text{-}KI$
X	27.0	$\beta\text{-}KI + \beta\text{-}KBr$
X	−33.0	Ice + $NaI \cdot 5H_2O$
X	−16.4	$NaI \cdot 5H_2O + \alpha\text{-}KI$
XI	−32.2	Ice + $NaBr \cdot 5H_2O$
XI	−27.0	$NaBr \cdot 5H_2O + NaI \cdot 5H_2O$
XI	−17.0	$NaI \cdot 5H_2O + NaBr \cdot 2H_2O$
XI	8.4	$NaBr \cdot 2H_2O + \beta\text{-}KBr$
XII	−31.8	Ice + $NaBr \cdot 5H_2O$
XII	−25.0	$NaBr \cdot 5H_2O + NaI \cdot 5H_2O$
XII	−17.8	$NaI \cdot 5H_2O + NaBr \cdot 2H_2O$
XII	29.8	$NaBr \cdot 2H_2O + \beta\text{-}KBr$
XIII	−31.8	Ice + $NaI \cdot 5H_2O$
XIII	−15.8	$NaI \cdot 5H_2O + \alpha\text{-}KI$

Table № 3904 (continuation)

Comp. of saturated soln., wt %				$t_{m.p.}$	Comp. of solid phase
Sodium Iodide	Sodium Bromide	Potassium Iodide	Water		
7.6	2.4	0.0	90.0	− 2.6	Section I
7.6	2.4	9.1	80.9	− 5.0	»
7.6	2.4	16.7	73.3	− 7.4	»
7.6	2.4	23.0	67.0	− 9.8	»
7.6	2.4	33.3	56.7	−14.5	»
7.6	2.4	39.4	50.6	−18.4	»
7.6	2.4	44.4	45.6	−22.0	»
7.6	2.4	47.4	42.6	−19.0	»
7.6	2.4	50.0	40.0	− 6.0	»
7.6	2.4	52.0	38.0	4.5	»
7.6	2.4	53.5	36.5	13.0	»
7.6	2.4	54.5	35.5	19.0	»
7.6	2.4	55.5	34.5	25.0	»
7.6	2.4	56.5	33.5	32.0	»
22.8	7.2	0.0	70.0	−12.4	Section II
22.8	7.2	9.1	60.9	−15.4	»
22.8	7.2	16.7	53.3	−19.0	»
22.8	7.2	26.0	44.0	−24.3	»
22.8	7.2	28.5	41.5	−26.0	»
22.8	7.2	31.5	38.5	−26.0	»
22.8	7.2	34.2	35.8	−16.0	»
22.8	7.2	36.7	33.3	−16.0	»
22.8	7.2	38.3	31.7	0.0	»
22.8	7.2	41.2	28.8	14.2	»
22.8	7.2	42.8	27.2	23.4	»
22.8	7.2	44.4	25.6	32.6	»
38.0	12.0	0.0	50.0	−25.6	Section III
38.0	12.0	4.7	45.3	−23.0	»
38.0	12.0	7.4	42.6	−22.2	»
38.0	12.0	9.9	40.1	−21.5	»
38.0	12.0	12.3	37.7	−21.2	»
38.0	12.0	14.5	35.5	−18.0	»
38.0	12.0	16.0	34.0	−12.0	»
38.0	12.0	17.3	32.7	− 6.8	»
38.0	12.0	20.0	30.0	5.4	»
38.0	12.0	23.0	27.0	19.0	»
38.0	12.0	24.2	25.8	25.0	»
38.0	12.0	25.9	24.1	33.0	»
39.5	12.5	0.0	48.0	−22.8	Section IV
39.5	12.5	7.4	40.6	−20.2	»
39.5	12.5	12.3	35.7	−19.4	»
39.5	12.5	14.5	33.5	−12.0	»
39.5	12.5	18.0	30.0	3.8	»
39.5	12.5	20.0	28.0	14.0	»
39.5	12.5	21.8	26.2	23.4	»
39.5	12.5	23.0	25.0	30.0	»
45.6	14.0	0.0	40.0	− 0.8	Section V
45.6	14.4	4.7	35.3	4.8	»
45.6	14.4	7.4	32.6	7.8	»
45.6	14.4	9.9	30.1	15.6	»
45.6	14.4	11.5	28.5	24.0	»

Comp. of saturated soln., wt %				$t_{m.p.}$	Comp. of solid phase
Sodium Iodide	Sodium Bromide	Potassium Iodide	Water		
45.6	14.4	13.0	27.0	31.6	Section V
0.0	0.0	5.0	95.0	− 1.0	Section VI
6.9	2.2	5.0	85.9	− 3.8	»
12.7	4.0	5.0	78.3	− 6.6	»
19.8	6.2	5.0	69.0	−11.8	»
25.3	8.0	5.0	61.7	−17.0	»
31.4	9.8	5.0	53.8	−25.2	»
33.7	10.7	5.0	50.6	−30.0	»
35.6	11.2	5.0	48.2	−27.0	»
37.5	11.7	5.0	45.8	−23.2	»
38.8	12.2	5.0	44.0	−21.0	»
41.4	13.0	5.0	40.6	−17.4	»
43.0	13.5	5.0	38.5	−11.0	»
44.1	13.9	5.0	37.0	− 1.2	»
45.0	14.0	5.0	36.0	5.6	»
45.8	14.5	5.0	34.7	14.0	»
46.8	14.7	5.0	33.5	22.0	»
47.5	15.0	5.0	32.5	28.6	»
0.0	0.0	10.0	90.0	− 2.0	Section VII
6.9	2.2	10.0	80.9	− 5.0	»
12.7	4.0	10.0	73.3	− 8.4	»
19.8	6.2	10.0	64.0	−13.6	»
25.3	8.0	10.0	56.7	−19.4	»
29.7	9.3	10.0	51.0	−25.6	»
32.1	10.1	10.0	47.8	−30.4	»
33.7	10.7	10.0	45.6	−28.6	»
35.6	11.2	10.0	43.2	−34.4	»
37.7	11.8	10.0	40.5	−20.8	»
39.5	12.5	10.0	38.0	−18.2	»
41.4	13.0	10.0	35.6	−14.0	»
42.7	13.4	10.0	33.9	− 4.0	»
44.1	13.9	10.0	32.0	7.4	»
45.4	14.3	10.0	30.3	17.7	»
46.8	14.7	10.0	28.5	28.4	»
0.0	0.0	20.0	80.0	− 4.7	Section VIII
6.9	2.2	20.0	70.9	− 8.4	»
12.7	4.0	20.0	63.3	−12.4	»
19.8	6.2	20.0	54.0	−18.5	»
24.3	7.7	20.0	48.0	−24.2	»
27.0	8.5	20.0	44.5	−28.4	»
29.0	9.2	20.0	41.8	−32.2	»
30.5	9.6	20.0	39.9	−30.0	»
32.2	10.0	20.0	37.8	−25.8	»
33.7	10.7	20.0	35.6	−22.6	»
36.0	11.2	20.0	32.8	−19.3	»
37.5	11.7	20.0	30.8	−16.0	»
39.3	22.4	20.0	28.3	− 4.4	»
41.5	13.0	20.0	25.5	7.4	»
43.0	13.5	20.0	23.5	17.0	»
43.6	13.8	20.0	22.6	22.0	»
44.3	14.0	20.0	21.7	28.0	»

Table № 3904 (continuation)

Comp. of saturated soln., wt %				$t_{m.p.}$	Comp. of solid phase
Sodium Iodide	Sodium Bromide	Potassium Iodide	Water		
0.0	0.0	0.0	100.0	0.0	Section IX
6.9	2.2	0.0	90.9	− 2.6	»
12.7	4.0	0.0	83.3	− 5.0	»
19.8	6.2	0.0	74.0	− 9.8	»
25.3	8.0	0.0	66.7	−14.4	»
31.4	9.8	0.0	58.8	−22.2	»
35.6	11.2	0.0	53.2	−30.0	»
37.0	11.7	0.0	51.3	−28.0	»
39.0	12.2	0.0	48.8	−24.0	»
40.6	12.9	0.0	46.5	−20.8	»
42.2	13.3	0.0	44.5	−18.6	»
43.6	13.8	0.0	42.6	−16.6	»
44.3	14.0	0.0	41.7	−14.4	»
45.5	14.2	0.0	40.3	− 3.0	»
45.9	14.5	0.0	39.6	3.2	»
47.2	15.0	0.0	37.8	17.0	»
48.0	15.0	0.0	37.0	24.0	»
48.2	15.4	0.0	36.4	28.6	»

Table № 3904 (continuation)

Section No.	Temperature of points of intersection	Comp. of solid phase
I	−23.6	Ice + α-KI
I	8.6	α-KI + β-KI
II	−28.0	Ice + α-KI
II	4.2	α-KI + β-KI
III	−21.0	$NaI \cdot 5H_2O$ + α-KI
III	− 3.0	α-KI + α-KBr
IV	−19.4	$NaI \cdot 5H_2O$ + α-KI
IV	9.0	β-KI + β-KBr
V	9.0	$NaBr \cdot 2H_2O$ + β-KBr
VI	−31.5	Ice + $NaI \cdot 5H_2O$
VI	−16.3	$NaI \cdot 5H_2O + NaBr \cdot 2H_2O$
VII	−32.0	Ice + $NaI \cdot 5H_2O$
VII	−16.6	$NaI \cdot 5H_2O + NaBr \cdot 2H_2O$
VIII	−33.6	Ice + $NaI \cdot 5H_2O$
VIII	−18.2	$NaI \cdot 5H_2O$ + α-KI
VIII	14.0	$NaBr \cdot 2H_2O$ + β-KBr
IX	−31.2	Ice + $NaI \cdot 5H_2O$
IX	−16.0	$NaI \cdot 5H_2O + NaBr \cdot 2H_2O$

$NaBr—NaIO_3—H_2O$

Comp. of saturated soln., wt %: Sodium Bromide	Sodium Iodate	Water	t	d	Comp. of solid phase
45.08	0.00	54.92	5	1.489	$NaBr \cdot 2H_2O$
45.04	0.08	54.88	5	1.492	$NaBr + 3NaBr \cdot 2NaIO_3 \cdot 15H_2O$
43.99	0.08	55.93	5	1.473	$3NaBr \cdot 2NaIO_3 \cdot 15H_2O$
39.98	0.12	59.90	5	1.415	»
29.50	0.58	69.92	5	1.287	»
24.56	1.09	74.35	5	1.238	»
23.51	1.28	75.21	5	1.230	$3NaBr \cdot 2NaIO_3 \cdot 15H_2O + NaIO_3 \cdot 5H_2O$
22.95	1.25	75.80	5	1.225	$NaIO_3 \cdot 5H_2O$
14.13	1.09	84.78	5	1.132	»
4.97	1.14	93.89	5	1.052	»
0.00	3.30	96.70	5	1.027	»
48.41	0.00	51.59	25	1.530	$NaBr \cdot 2H_2O$
48.21	0.42	51.37	25	1.537	$NaBr \cdot 2H_2O + 3NaBr \cdot 2NaIO_3 \cdot 15H_2O$
46.73	0.45	52.82	25	1.509	$3NaBr \cdot 2NaIO_3 \cdot 15H_2O$
39.55	0.86	59.59	25	1.417	»
35.23	1.51	63.26	25	1.367	»
32.68	2.16	65.16	25	1.343	$3NaBr \cdot 2NaIO_3 \cdot 15H_2O + NaIO_3 \cdot H_2O$
26.39	2.35	71.26	25	1.266	$NaIO_3 \cdot H_2O$
16.40	3.00	80.60	25	1.172	»
7.78	4.46	87.76	25	1.104	»
0.00	8.57	91.43	25	1.075	»
51.55	0.00	48.45	40	—	$NaBr \cdot 2H_2O$
50.84	1.37	47.79	40	—	$NaBr \cdot 2H_2O + 3NaBr \cdot 2NaIO_3 \cdot 10H_2O$
49.38	1.53	49.09	40	—	$3NaBr \cdot 2NaIO_3 \cdot 10H_2O$
47.95	1.77	50.28	40	—	$3NaBr \cdot 2NaIO_3 \cdot 10H_2O + 3NaBr \cdot 2NaIO_3 \cdot 15H_2O$
46.82	1.89	51.29	40	—	$3NaBr \cdot 2NaIO_3 \cdot 15H_2O$
44.86	2.13	53.01	40	—	»
43.77	2.48	53.75	40	—	»
41.96	2.69	55.35	40	—	$3NaBr \cdot 2NaIO_3 \cdot 15H_2O + NaIO_3$
40.05	2.63	57.32	40	—	$NaIO_3$
39.50	2.65	57.85	40	—	$NaIO_3 + NaIO_3 \cdot H_2O$
39.06	2.68	58.26	40	—	$NaIO_3 \cdot H_2O$
32.08	2.77	65.15	40	—	»
20.65	3.65	75.70	40	—	»
0.00	11.70	88.30	40	—	»
53.63	0.00	46.37	50	—	$NaBr \cdot 2H_2O$
53.0	1.57	45.43	50	—	$NaBr \cdot 2H_2O + NaBr$
52.57	2.37	45.06	50	—	$NaBr \cdot 2H_2O + 3NaBr \cdot 2NaIO_3 \cdot 10H_2O$
51.40	2.49	46.11	50	—	$3NaBr \cdot 2NaIO_3 \cdot 10H_2O$
50.93	2.63	46.44	50	—	$3NaBr \cdot 2NaIO_3 \cdot 10H_2O + NaIO_3$
44.74	2.50	52.76	50	—	$NaIO_3$
32.41	3.41	64.18	50	—	»
21.46	5.15	73.39	50	—	$NaIO_3 + NaIO_3 \cdot H_2O$
13.28	6.28	80.44	50	—	$NaIO_3 \cdot H_2O$
6.63	9.03	84.34	50	—	»
0.00	13.49	86.51	50	—	»

№ 3906 [761]

SODIUM BROMIDE–MAGNESIUM BROMIDE–WATER

$NaBr—MgBr_2—H_2O$

$t = 25$

Content in aqueous soln., wt %		Comp. of solid phase
Sodium Bromide	Magnesium Bromide	
31.43	16.20	$NaBr \cdot 2H_2O$
3.10	48.24	$NaBr \cdot 2H_2O + MgBr_2 \cdot 6H_2O$

№ 3907 [593]

SODIUM BROMIDE–MAGNESIUM BROMIDE–POTASSIUM BROMIDE–WATER

$NaBr—MgBr_2—KBr—H_2O$

$t = 25$

Content in aqueous soln., wt %			Comp. of solid phase
Sodium Bromide	Potassium Bromide	Magnesium Bromide	
30.34	6.12	13.55	$NaBr \cdot 2H_2O + KBr$
6.58	3.53	37.86	$NaBr \cdot 2H_2O + KBr + KBr \cdot MgBr_2 \cdot 6H_2O$
2.92	0.20	48.40	$NaBr \cdot 2H_2O + KBr \cdot MgBr_2 \cdot 6H_2O + MgBr_2 \cdot 6H_2O$

№ 3908 [1656]

SODIUM BROMIDE–SULFUR DIOXIDE–WATER

$NaBr—SO_2—H_2O$

$t = 25$

Comp. of Sodium Bromide in aqueous soln., g/l	Solubility of Sulfur Dioxide cm^3 (at t=0 and p=760)/cm^3 solvent
51.45	33.76
102.9	34.54
154.35	35.27
205.8	26.26
257.25	36.84
308.7	37.74

№ 3909 [1130]

SODIUM BROMIDE–SULFUR DIOXIDE–WATER

$NaBr—SO_2—H_2O$

$t = 25$

Comp. of Sodium Bromide in aqueous soln., g/l	Solubility of Sulfur Dioxide cm^3 (at t=0 and p=760)/cm^3 solvent
51.45	30.93

№ 3910 SODIUM BROMIDE–POTASSIUM BROMIDE–WATER [593]

$NaBr—KBr—H_2O$

$t = 25$

Content in aqueous soln., wt %		Comp. of solid phase	Content in aqueous soln., wt %		Comp. of solid phase
Sodium Bromide	Potassium Bromide		Sodium Bromide	Potassium Bromide	
48.67	0	$NaBr \cdot 2H_2O$	28.68	16.59	KBr
43.93	7.67	$NaBr \cdot 2H_2O + KBr$	0	19.66	»

№ 3911 SODIUM BROMIDE–POTASSIUM BROMIDE–WATER [183]

$NaBr—KBr—H_2O$

Concn. in aqueous soln., wt %		*t*	Comp. of solid phase
Potassium Bromide	Sodium Bromide		
0.0	32.8	−20	Ice
1.8	31.0	−20	»
6.5	28.2	−20	»
13.3	22.4	−20	KBr+Ice
10.7	26.9	−20	KBr
6.1	35.5	−20	»
4.4	38.0	−20	»
3.5	38.0	−20	$KBr + NaBr \cdot 2H_2O$
2.1	40.2	−20	$NaBr \cdot 2H_2O$
0.0	41.7	−20	»
0.0	21.6	−10	Ice
3.0	12.4	−10	»
16.0	8.3	−10	»
26.6	0.0	−10	»
31.4	0.0	−10	KBr
25.6	7.4	−10	»
18.6	16.1	−10	»
12.2	26.4	−10	»
7.6	35.0	−10	»
4.6	39.6	−10	»
4.5	39.8	−10	$KBr + NaBr \cdot 2H_2O$
4.0	40.3	−10	$NaBr \cdot 2H_2O$
2.2	41.5	−10	»
0.0	43.2	−10	»
0.0	44.3	0	$NaBr \cdot 2H_2O$
2.3	43.0	0	»
4.7	41.5	0	»
4.8	41.2	0	$NaBr \cdot 2H_2O + KBr$
6.1	39.4	0	KBr
9.1	34.3	0	»
14.0	25.9	0	»
21.0	15.8	0	»
28.2	7.1	0	»
34.0	0.0	0	»

Table № 3911 (continuation)

Concn. in aqueous soln., wt %		t	Comp. of solid phase
Potassium Bromide	Sodium Bromide		
0.0	45.9	10	$NaBr \cdot 2H_2O$
2.4	44.4	10	»
5.0	43.0	10	»
5.4	42.6	10	$NaBr \cdot 2H_2O + KBr$
5.7	42.3	10	KBr
7.9	38.6	10	»
10.6	33.8	10	»
15.6	25.4	10	»
23.2	15.3	10	»
30.7	6.8	10	»
36.7	0.0	10	»
0.0	47.6	20	$NaBr \cdot 2H_2O$
2.5	46.4	20	»
3.6	45.3	20	»
6.3	43.8	20	$NaBr \cdot 2H_2O + KBr$
7.5	41.6	20	KBr
9.4	38.0	20	»
11.7	33.4	20	»
17.5	24.9	20	»
25.6	14.8	20	»
33.2	6.5	20	»
39.4	0.0	20	»
0.0	48.6	25	$NaBr \cdot 2H_2O$
2.6	47.1	25	»
5.2	45.6	25	»
6.7	44.4	25	$NaBr \cdot 2H_2O + KBr$
8.2	41.2	25	KBr
10.1	37.7	25	»
12.4	33.2	25	»
18.4	24.6	25	»
26.8	14.6	25	»
34.4	6.4	25	»
10.6	0.0	25	»
0.0	49.4	30	$NaBr \cdot 2H_2O$
2.6	47.9	30	»
5.3	46.3	30	$NaBr \cdot 2H_2O + KBr$
7.1	45.0	30	KBr
7.0	43.6	30	»
9.0	40.8	30	»
10.7	37.5	30	»
13.0	33.0	30	»
19.3	24.3	30	»
27.9	14.4	30	»
35.8	6.2	30	»
39.0	0.0	30	»
0.0	53.7	50	NaBr
6.78	49.38	50	»
8.18	48.15	50	»
13.48	37.64	50	KBr
14.42	36.94	50	»
17.42	34.28	50	»

Concn. in aqueous soln., wt %		t	Comp. of solid phase
Potassium Bromide	Sodium Bromide		
18.60	33.10	50	KBr
17.20	32.76	50	»
31.18	14.57	50	»
32.42	13.39	50	»
44.5	0.0	50	»

№ 3912 SODIUM BROMIDE–POTASSIUM BROMIDE–WATER [184]

$NaBr—KBr—H_2O$

Content in aqueous soln., wt %		t	Comp. of solid phase
Sodium Bromide	Potassium Bromide		
36.5	4.0	—32	$NaBr \cdot 5H_2O$+α-KBr+Ice
36.2	4.6	—30.8	α-KBr+Ice
37.4	2.0	—29.6	$NaBr \cdot 5H_2O$+Ice
40.3	0	—28.2	»
38.0	3.6	—26.0	$NaBr \cdot 5H_2O$+$NaBr \cdot 2H_2O$+α-KBr
39.2	2.1	—24.0	$NaBr \cdot 5H_2O$+$NaBr \cdot 2H_2O$
41.2	0	—23.8	»
27.0	10.3	—22.6	α-KBr+Ice
32.8	0	—20	Ice
31.0	1.8	—20	»
28.2	6.5	—20	»
22.4	13.3	—20	Ice+α-KBr
26.9	10.7	—20	α-KBr
35.5	6.1	—20	»
38.0	4.4	—20	»
38.0	3.5	—20	α-KBr+$NaBr \cdot 2H_2O$
40.2	2.1	—20	$NaBr \cdot 2H_2O$
41.7	0	—20	»
16.5	17.1	—17.0	α-KBr+Ice
7.5	24.5	—14.4	»
0	30.6	—12.2	»
21.6	0	—10	Ice
12.4	3.0	—10	»
8.3	16.0	—10	»
0	26.6	—10	»
0	31.4	—10	α-KBr
7.4	25.6	—10	»
16.1	18.6	—10	»
26.4	12.2	—10	—
35.0	7.6	—10	—
39.6	4.6	—10	—
39.8	4.5	—10	α-KBr+$NaBr \cdot 2H_2O$
30.3	4.0	—10	$NaBr \cdot 2H_2O$
41.5	2.2	—10	»
43.2	0	—10	»
44.3	0	0	»
43.0	2.3	0	»

Table № 3912 (continuation)

Content in aqueous soln., wt%		t	Comp. of solid phase
Sodium Bromide	Potassium Bromide		
41.5	4.7	0	$NaBr \cdot 2H_2O$
41.2	4.8	0	$NaBr \cdot 2H_2O + \alpha\text{-}KBr$
39.4	6.1	0	$\alpha\text{-}KBr$
36.3	9.1	0	»
25.9	14.0	0	»
15.8	21.0	0	»
7.1	28.2	0	»
0	34.0	0	»
41.4	5.4	7	$NaBr \cdot 2H_2O + \alpha\text{-}KBr + \beta\text{-}KBr$
45.9	0	10	$NaBr \cdot 2H_2O$
44.4	2.4	10	»
43.0	5.0	10	»
42.6	5.4	10	$NaBr \cdot 2H_2O + \beta\text{-}KBr$
42.3	5.7	10	$\beta\text{-}KBr$
38.6	7.9	10	»
33.8	10.6	10	$\alpha\text{-}KBr + \beta\text{-}KBr$
25.4	15.6	10	$\alpha\text{-}KBr$
15.3	23.2	10	»
6.8	30.7	10	»
0	36.7	10	»
25.5	16.5	12	$\alpha\text{-}KBr + \beta\text{-}KBr$
14.8	25.2	16	»
6.7	23.3	19	»
47.6	0	20	$NaBr \cdot 2H_2O$
46.4	2.5	20	»
45.3	3.6	20	»
43.8	6.3	20	$NaBr \cdot 2H_2O + \beta\text{-}KBr$
41.6	7.5	20	$\beta\text{-}KBr$
38.0	9.4	20	»
33.4	11.7	20	»
24.9	17.5	20	»
14.8	25.6	20	»
6.5	33.2	20	»
0	39.4	20	$\alpha\text{-}KBr$
0	40.0	22	$\alpha\text{-}KBr + \beta\text{-}KBr$
48.6	0	25	$NaBr \cdot 2H_2O$
47.1	2.6	25	»
45.6	5.2	25	»
44.4	6.7	25	$NaBr \cdot 2H_2O + \beta\text{-}KBr$
41.2	8.2	25	$\beta\text{-}KBr$
37.7	10.1	25	»
33.2	12.4	25	»
24.6	18.4	25	»
14.6	26.8	25	»
6.4	34.4	25	»
0	40.6	25	»
49.4	0	30	$NaBr \cdot 2H_2O$
47.9	2.6	30	»
46.3	5.3	30	»
45.0	7.1	30	$NaBr \cdot 2H_2O + \beta\text{-}KBr$
43.6	7.0	30	$\beta\text{-}KBr$

Content of aqueous soln., wt %		t	Comp. of solid phase
Sodium Bromide	Potassium Bromide		
40.8	9.0	30	β-KBr
37.5	10.7	30	»
33.0	13.0	30	»
24.3	19.3	30	»
14.4	27.9	30	»
6.2	35.8	30	»
0	39.0	30	»
47.0	7.8	46	$NaBr \cdot 2H_2O$+NaBr+ β-KBr
53.7	0	50	NaBr
49.38	6.78	50	»
48.15	8.18	50	NaBr+ β-KBr
37.64	13.48	50	β-KBr
36.94	14.42	50	»
34.28	17.42	50	»
33.10	18.60	50	»
32.76	17.20	50	»
14.57	31.18	50	»
13.39	32.42	50	»
0	44.5	50	»

№ 3913 SODIUM BROMIDE–POTASSIUM BROMIDE–WATER [4]

$NaBr—KBr—H_2O$

Comp. of saturated soln., wt %			t	d	Comp. of solid phase, wt %		
Potassium Bromide	Sodium Bromide	Water			Potassium Bromide	Sodium Bromide	Water
44.78	0.00	55.22	50	1.4106	—	—	—
41.29	4.66	54.05	50	1.4239	86.60	1.52	11.88
36.04	10.03	53.93	50	1.4288	90.29	1.64	8.07
29.78	18.06	52.16	50	1.4353	96.49	1.10	2.41
19.23	31.43	49.36	50	1.5021	87.79	4.74	7.47
12.77	43.22	44.01	50	1.5694	90.45	4.61	4.94
8.91	48.22	42.87	50	1.6280	68.36	19.48	12.16
7.58	49.23	43.19	50	1.6180	2.50	90.00	7.50
4.96	50.90	44.14	50	1.5950	1.81	81.39	16.80
0.00	53.79	46.21	50	1.5710	—	—	—
47.77	0.00	52.23	75	1.5561	—	—	—
39.96	9.76	52.28	75	1.4638	83.77	1.28	14.95
37.72	12.59	46.69	75	1.4844	91.24	2.67	6.09
29.19	22.69	48.12	75	1.5060	94.53	1.48	3.99
18.49	38.59	42.92	75	1.5686	92.05	4.15	3.80
16.28	42.49	41.23	75	1.6550	81.62	7.46	10.92
12.39	47.33	40.28	75	1.6578	87.38	5.64	6.98
11.81	48.19	40.00	75	1.6886	26.60	58.38	15.02
11.75	48.38	39.87	75	1.6736	5.24	82.86	11.90
9.80	49.35	40.85	75	1.6687	3.08	88.35	8.57
9.13	50.54	40.33	75	1.6551	4.54	70.37	25.09
0.00	54.35	45.65	75	1.6168	—	—	—

№ 3914 SODIUM BROMIDE–POTASSIUM IODIDE–WATER [258]

$NaBr—KI—H_2O$

Comp. of saturated soln., g-mol			t	Comp. of solid phase
Sodium Bromide	Potassium Iodide	Water		
90.1	9.9	845	—30	Ice + $NaBr \cdot 5H_2O$
88.0	12.0	807	—30	α-KBr · $NaBr \cdot 5H_2O$
86.3	13.7	836	—30	Ice + α-KBr
88.0	12.0	766	—25	α-KBr + $NaBr \cdot 5H_2O$
70.4	29.6	875	—25	Ice + α-KBr
30.0	70.0	816	—25	»
23.1	76.9	794	—25	α-KI + α-KBr
19.2	80.8	812	—25	Ice + α-KI
88.0	12.0	726	—20	α-KBr + $NaBr \cdot 2H_2O$
23.1	76.9	766	—20	α-KI + α-KBr
87.2	12.8	703	—15	α-KBr + $NaBr \cdot 2H_2O$
22.5	77.5	738	—15	α-KI + α-KBr
86.9	13.1	676	—10	α-KBr + $NaBr \cdot 2H_2O$
22.9	77.1	720	—10	α-KI + α-KBr
86.9	13.1	656	— 5	α-KBr + $NaBr \cdot 2H_2O$
23.2	76.8	695	— 5	α-KI + α-KBr
86.7	13.3	635	0	α-KBr + $NaBr \cdot 2H_2O$
22.9	77.1	673	0	α-KI + α-KBr
86.7	13.3	615	5	α-KBr + $NaBr \cdot 2H_2O$
22.8	77.2	648	5	β-KI + α-KBr
19.0	81.0	654	5	α-KI + β-KI
86.5	13.5	587	10	β-KBr + $NaBr \cdot 2H_2O$
23.0	77.0	629	10	β-KI + β-KBr
3.4	96.6	667	10	α-KI + β-KI
86.3	13.7	570	15	β-KBr + $NaBr \cdot 2H_2O$
22.8	77.2	603	15	β-KI + β-KBr
86.3	13.7	550	20	β-KBr + $NaBr \cdot 2H_2O$
22.4	77.6	587	20	β-KI + β-KBr
86.1	13.9	531	25	β-KBr + $NaBr \cdot 2H_2O$
22.8	77.2	564	25	β-KI + β-KBr
86.3	13.7	517	30	β-KBr + $NaBr \cdot 2H_2O$
22.7	77.3	544	30	β-KI + β-KBr

№ 3915 [1940]

SODIUM BROMIDE–CALCIUM HYDROXIDE–WATER

$NaBr—Ca(OH)_2—H_2O$

$t = 25$

Content in aqueous soln., mol/100 g Water		Comp. of solid phase	Content in aqueous soln., mol/100 g Water		Comp. of solid phase
Sodium Bromide	Calcium Hydroxide		Sodium Bromide	Calcium Hydroxide	
2.90	0.02478	$Ca(OH)_2$	0.721	0.02805	$Ca(OH)_2$
1.77	0.02780	»	0.348	0.02641	»
1.30	0.02841	»	0.189	0.02485	»
0.979	0.02847	»	0	0.01976	»

№ 3916 [2519]

SODIUM BROMIDE–ARSENIC TRIOXIDE–WATER

$NaBr—As_2O_3—H_2O$

$t = 30$

Comp. of saturated soln., wt %			Comp. of solid phase
Arsenic Trioxide	Sodium Bromide	Water	
2.19	5.57	92.24	As_2O_3
2.09	10.89	87.02	»
1.88	20.79	77.33	»
1.63	30.39	67.98	»
1.50	35.75	62.75	»
1.20	39.24	59.56	$(As_2O_3)_3 \cdot NaBr$
0.953	43.64	55.407	»
0.852	45.99	53.158	»
0.719	50.25	49.031	$(As_2O_3)_3 \cdot NaBr + NaBr \cdot H_2O$
0.0	49.5	50.5	$NaBr \cdot H_2O$

№ 3917 [2011]

SODIUM BROMIDE–SILVER BROMIDE–WATER

$NaBr—AgBr—H_2O$

$t = 20$

Concn. in aqueous soln., g/l					
Sodium Bromide	Silver Bromide	Sodium Bromide	Silver Bromide	Sodium Bromide	Silver Bromide
343.17	1.878	142.00	0.1127	43.73	0.00751
295.84	1.127	120.70	0.0751	29.84	0.00376
264.97	0.751	94.87	0.0376	20.06	0.00187
213.00	0.376	68.63	0.0187	15.44	0.001127
165.67	0.188	53.51	0.01127	11.83	0.000751

№ 3918 [2077]

SODIUM BROMIDE–SILVER BROMIDE–WATER

$NaBr—AgBr—H_2O$

$t = 18$

Concn. of Sodium Bromide in aqueous soln., g/l	Solubility of Silver Bromide in aqueous soln. of Sodium Bromide, g/l	Concn. of Sodium Bromide in aqueous soln., g/l	Solubility of Silver Bromide in aqueous soln. of Sodium Bromide, g/l	Concn. of Sodium Bromide in aqueous soln., g/l	Solubility of Silver Bromide in aqueous soln. of Sodium Bromide, g/l
0.000575	$9.76 \cdot 10^{-6}$	0.0513	$2.8 \cdot 10^{-6}$	5.13	$1.39 \cdot 10^{-4}$
0.001032	$5.62 \cdot 10^{-6}$	0.1028	$3.79 \cdot 10^{-6}$	10.28	$3.99 \cdot 10^{-4}$
0.00205	$3.94 \cdot 10^{-6}$	0.205	$5.34 \cdot 10^{-6}$	20.5	$1.295 \cdot 10^{-3}$
0.00513	$2.707 \cdot 10^{-6}$	0.513	$1.06 \cdot 10^{-5}$	51.3	$8.66 \cdot 10^{-3}$
0.01028	$2.3 \cdot 10^{-6}$	1.028	$2.11 \cdot 10^{-5}$	102.8	$4.26 \cdot 10^{-2}$
0.0205	$2.24 \cdot 10^{-6}$	2.05	$4.17 \cdot 10^{-5}$	205	$3.31 \cdot 10^{-1}$

№ 3919 SODIUM BROMIDE–CESIUM BROMIDE–WATER [778]

$NaBr—CsBr—H_2O$

Comp. of saturated soln., wt %			t	Comp. of solid phase
Sodium Bromide	Cesium Bromide	Water		
48.82	0.00	51.18	25	$NaBr \cdot 2H_2O$
44.18	3.48	52.34	25	»
43.01	7.98	49.01	25	»
42.90	11.42	45.68	25	$NaBr \cdot 2H_2O + NaBr \cdot 3CsBr$
41.07	12.77	46.16	25	$NaBr \cdot 3CsBr$
39.16	13.82	47.02	25	»
35.61	16.33	48.06	25	»
34.97	16.60	48.43	25	»
30.39	22.23	47.38	25	$NaBr \cdot 3CsBr + CsBr$
29.63	25.30	45.07	25	CsBr
24.43	30.02	45.55	25	»
23.52	31.08	45.40	25	»
19.81	34.62	45.57	25	»
10.89	43.60	45.51	25	»
7.54	46.97	45.49	25	»
2.10	52.57	45.33	25	»
0.00	55.20	44.80	25	»
53.56	0.00	46.44	50	NaBr
52.90	1.31	45.79	50	»
50.86	4.66	44.48	50	»
49.08	7.43	43.49	50	»
48.88	8.34	42.78	50	$NaBr \cdot 3CsBr + NaBr$
46.61	9.86	43.53	50	$NaBr + 3CsBr$
44.0	11.58	44.42	50	»
40.06	14.00	45.94	50	»
33.99	23.49	42.52	50	»
32.13	26.50	41.37	50	$NaBr \cdot 3CsBr + CsBr$
25.52	31.64	42.84	50	CsBr
24.20	36.11	39.69	50	»
18.82	41.74	39.44	50	»
14.51	45.64	39.85	50	»
6.94	54.35	38.69	50	»
0.0	62.34	37.66	50	»

№ 3920 SODIUM BROMIDE–MERCURY (II) BROMIDE–WATER [1815]

$NaBr—HgBr_2—H_2O$

$t = 25$

Concn. in aqueous soln., g/l			
Mercury Bromide	Sodium Bromide	Mercury Bromide	Sodium Bromide
28.1	12.14	460	252
102.5	61.3	831	540
194	117.5		

№ 3921 [1415]

SODIUM BROMIDE–MERCURY (II) BROMIDE–WATER

$NaBr—HgBr_2—H_2O$

$t = 15$

Concn. in aqueous soln., wt %		Comp. of solid phase
Mercury Bromide	Sodium Bromide	
23	10	$HgBr_2$
53	20	»
60	22	$HgBr_2 + 2HgBr_2 \cdot NaBr \cdot 2H_2O$
60	22.5	$2HgBr_2 \cdot NaBr \cdot 2H_2O + HgBr_2 \cdot NaBr \cdot 2H_2O$
54	28	$HgBr_2 \cdot NaBr \cdot 2H_2O + NaBr$
52	30	$NaBr + NaBr \cdot 2H_2O$
30	35	$NaBr \cdot 2H_2O$

Note: Data taken from a graph

№ 3922 [291]

SODIUM BROMIDE–LEAD (II) SULFATE–WATER

$NaBr—PbSO_4—H_2O$

$t = 25$

Concn. of Sodium Bromide in aqueous soln., mol/l	Solubility of Lead Sulfate in aqueous soln. of Sodium Bromide, mol/l
$1 \cdot 10^{-3}$	$1.69 \cdot 10^{-4}$
$5 \cdot 10^{-3}$	$2.02 \cdot 10^{-4}$
$1 \cdot 10^{-2}$	$2.51 \cdot 10^{-4}$
$5 \cdot 10^{-2}$	$5.50 \cdot 10^{-4}$
$1 \cdot 10^{-1}$	$9.85 \cdot 10^{-4}$
$5 \cdot 10^{-1}$	$1.87 \cdot 10^{-3}$
1.00	$4.82 \cdot 10^{-3}$
2.00	$1.27 \cdot 10^{-2}$

№ 3923 [1817]

SODIUM BROMIDE–LEAD (II) CHLORIDE–WATER

$NaBr—PbCl_2—H_2O$

$t = 25$

Concn. of Sodium Bromide in aqueous soln., g/l	Solubility of Lead Chloride in aqueous soln. of Sodium Bromide, g/l
49.40	1.94
99.83	3.37
199.6	16.50
290.2	39.81
389.0	88.62

№ 3924 [1817]

SODIUM BROMIDE–LEAD (II) BROMIDE–WATER

$NaBr—PbBr_2—H_2O$

$t = 25$

Concn. of Sodium Bromide in aqueous soln., g/l	Solubility of Lead Bromide in aqueous soln. of Sodium Bromide, g/l	Concn. of Sodium Bromide in aqueous soln., g/l	Solubility of Lead Bromide in aqueous soln. of Sodium Bromide, g/l
0.0	9.63	301.79	71.86
75.19	3.17	378.01	144.4
151.41	8.25	453.2	269.3
226.6	25.85		

№ 3925 **SODIUM BROMIDE–LEAD (II) BROMIDE–WATER** [2424]

$NaBr—PbBr_2—H_2O$

$t = 25$

Concn. in aqueous soln., wt %		d_4	Comp. of solid phase
Sodium Bromide	Lead Bromide		
0.0	0.975	1.0049	$PbBr_2$
0.221	0.733	—	»
0.650	0.502	—	»
1.45	0.310	—	»
3.12	0.249	—	»
4.95	0.332	—	»
6.85	0.364	—	»
9.10	0.543	—	»
11.23	0.715	—	»
22.94	4.80	1.264	»
27.77	17.18	1.525	»
29.38	27.15	1.778	»
29.34	30.78	1.891	»
30.76	36.23	2.107	»
31.09	38.34	2.179	»
31.13	38.70	2.206	$PbBr_2+4PbBr_2 \cdot 3NaBr \cdot 12H_2O$
31.25	36.43	—	$4PbBr_2 \cdot 3NaBr \cdot 12H_2O$
31.90	35.43	2.136	»
33.35	36.76	2.196	»
34.28	38.04	2.299	$4PbBr_2 \cdot 3NaBr \cdot 12H_2O+NaBr$
34.68	37.37	2.275	$NaBr \cdot 2H_2O$
35.98	35.40	2.234	»
38.33	29.82	—	»
42.13	19.46	1.861	»
48.81	0.0	1.543	»

№ 3926 [2403]

SODIUM BROMATE–SODIUM MOLYBDATE–WATER

$NaBrO_3—Na_2MoO_4—H_2O$

$t = 25$

Concn. in aqueous soln., wt %		d	Comp. of solid phase
Sodium Molybdate	Sodium Bromate		
39.38	0.00	1.432	$Na_2MoO_4 \cdot 2H_2O$
38.30	1.80	1.442	»
37.09	3.86	1.453	»
35.57	6.33	1.466	$Na_2MoO_4 \cdot 2H_2O+NaBrO_3$
35.58	6.29	1.468	»
35.60	6.28	1.470	»
35.58	6.30	1.468	»
32.64	7.49	1.440	$NaBrO_3$
27.53	9.86	1.398	»
22.44	12.56	1.363	»
16.18	16.35	1.326	»
11.47	19.40	1.304	»
4.85	24.42	1.278	»
0.00	28.29	1.264	»

№ 3927 **SODIUM BROMATE–SODIUM IODIDE–WATER** [2390]

$NaBrO_3—NaI—H_2O$

$t = 25$

Comp. of saturated soln., wt %			d	Comp. of solid phase
Sodium Bromate	Sodium Iodide	Water		
28.29	0.00	71.71	1.257	$NaBrO_3$
16.57	17.32	66.11	1.332	»
8.92	32.21	58.87	1.438	»
5.78	40.76	53.46	1.521	»
3.62	48.11	48.27	1.619	»
2.23	54.89	42.88	1.727	»
1.44	60.65	37.91	1.836	»
1.30	62.13	36.57	1.874	»
1.17	64.00	34.83	1.914	$NaBrO_3+NaI$
0.00	64.71	35.29	1.904	NaI

№ 3928 **SODIUM BROMATE–POTASSIUM BROMATE–SILVER BROMATE–WATER** [480]

$NaBrO_3—KBrO_3—AgBrO_3—H_2O$

$t = 25$

Concn. in aqueous soln., mol/l		Solubility of Silver Bromate in aqueous soln. of Sodium Bromate and Potassium Bromate, mol/l	Concn. in aqueous soln., mol/l		Solubility of Silver Bromate in aqueous soln. of Sodium Bromate and Potassium Bromate, mol/l
Potassium Bromate	Sodium Bromate		Potassium Bromate	Sodium Bromate	
0.01	0.000	$3.99 \cdot 10^{-3}$	0.01	0.030	$5.49 \cdot 10^{-3}$
0.01	0.001	$4.82 \cdot 10^{-3}$	0.01	0.100	$6.37 \cdot 10^{-3}$
0.01	0.003	$4.92 \cdot 10^{-3}$	0.01	0.300	$7.55 \cdot 10^{-3}$
0.01	0.010	$5.08 \cdot 10^{-3}$	0.01	1.000	$1.12 \cdot 10^{-2}$

№ 3929 **SODIUM BROMATE–SILVER BROMATE–WATER** [480]

$NaBrO_3—AgBrO_3—H_2O$

$t = 25$

Concn. of Sodium Bromate in aqueous soln., mol/l	Solubility of Silver Bromate in aqueous soln. of Sodium Bromate, mol/l	Concn. of Sodium Bromate in aqueous soln., mol/l	Solubility of Silver Bromate in aqueous soln. of Sodium Bromate, mol/l
0.001	$8.22 \cdot 10^{-3}$	0.100	$1.008 \cdot 10^{-2}$
0.003	$8.31 \cdot 10^{-3}$	0.300	$1.138 \cdot 10^{-2}$
0.010	$8.50 \cdot 10^{-3}$	1.000	$1.48 \cdot 10^{-2}$
0.030	$9.27 \cdot 10^{-3}$		

№ 3930 [2403]

SODIUM MOLYBDATE–SODIUM IODATE–WATER

$Na_2MoO_4—NaIO_3—H_2O$

$t = 25$

Concn. in aqueous soln., wt %		d	Comp. of solid phase
Sodium Molybdate	Sodium Iodate		
39.38	0.00	1.432	$Na_2MoO_4 \cdot 2H_2O$
39.16	0.58	1.437	»
38.63	1.69	1.450	»
38.46	2.20	1.453	$Na_2MoO_4 \cdot 2H_2O + NaIO_3 \cdot H_2O$
38.51	2.18	1.452	»
38.43	2.21	1.451	»
38.43	2.21	1.454	»
38.47	2.18	1.455	»
38.46	2.20	1.453	»
37.23	2.24	1.436	$NaIO_3 \cdot H_2O$
31.49	2.54	1.368	»
24.24	3.08	1.277	»
17.89	3.42	1.204	»
11.41	4.16	1.143	»
5.57	5.67	1.009	»
0.00	8.49	1.074	»

№ 3931 [2405]

SODIUM MOLYBDATE–MAGNESIUM MOLYBDATE–WATER

$Na_2MoO_4—MgMoO_4—H_2O$

$t = 25$

Concn. in aqueous soln., wt %		d	Comp. of solid phase
Magnesium Molybdate	Sodium Molybdate		
13.71	4.84	1.177	$MgMoO_4 \cdot 5H_2O$
12.09	9.14	1.210	»
11.22	11.38	1.220	»
11.00	12.10	1.230	»
10.69	12.99	1.238	»
10.39	13.92	1.248	»
10.17	14.68	1.252	»
10.03	14.86	1.250	»
9.84	15.29	1.255	»
9.44	16.45	1.268	»
9.26	17.18	1.270	»
9.28	17.31	1.272	$MgMoO_4 \cdot 5H_2O + MgMoO_4 \cdot Na_2MoO_4 \cdot 2H_2O$
9.26	17.29	1.272	»
9.27	17.27	1.271	»
8.57	17.91	1.272	$MgMoO_4 \cdot Na_2MoO_4 \cdot 2H_2O$
7.34	19.15	1.267	»
6.94	19.85	1.273	»
4.23	23.56	1.285	»
1.62	29.30	1.321	»

Concn. in aqueous soln., wt %		d	Comp. of solid phase
Magnesium Molybdate	Sodium Molybdate		
1.57	29.44	1.322	$MgMoO_4 \cdot Na_2MoO_4 \cdot H_2O$
0.48	34.74	1.373	»
0.39	35.79	1.387	»
0.23	37.41	1.401	»
0.10	38.26	1.419	»
0.13	39.25	1.424	$MgMoO_4 \cdot Na_2MoO_4 \cdot 2H_2O + Na_2MoO_4 \cdot 2H_2O$
0.12	39.32	1.429	»
0.11	39.31	1.429	»

№ 3932 SODIUM MOLYBDATE–MOLYBDENUM OXIDE–WATER [2401]

$Na_2MoO_4—MoO_3—H_2O$

$t = 25$

Concn. in aqueous soln., wt %		Comp. of solid phase	Concn. in aqueous soln., wt %		Comp. of solid phase
Sodium Molybdate	Molybdenum Oxide		Sodium Molybdate	Molybdenum Oxide	
39.42	0.0	$Na_2MoO_4 \cdot 2H_2O$	36.66	0.01	$Na_2Mo_2O_7 \cdot 6H_2O$
39.50	0.0	$Na_2MoO_4 \cdot 2H_2O + Na_2Mo_2O_7 \cdot 6H_2O$	36.38	0.0	»
			33.44	0.026	»
39.46	0.0	»	32.54	0.018	»
39.48	0.0	$Na_2Mo_2O_7 \cdot 6H_2O$	30.85	0.047	»
38.78	0.0	»	28.81	0.083	»
37.66	0.0	»	24.82	0.320	»

№ 3933 SODIUM MOLYBDATE–LANTHANUM MOLYBDATE–WATER [1357]

$Na_2MoO_4—La_2(MoO_4)_3—H_2O$

$t = 25$

Comp. of saturated soln., wt %			Comp. of solid phase
Sodium Molybdate	Lanthanum Molybdate	Water	
0.77	0.01	99.22	$La_2(MoO_4)_3 \cdot H_2O + La_2(MoO_4)_3 \cdot Na_2MoO_4 \cdot 2H_2O$
2.44	0.02	97.54	$La_2(MoO_4)_3 \cdot Na_2MoO_4 \cdot 2H_2O$
2.92	0.03	97.05	»
7.56	0.08	92.36	»
11.50	0.13	88.37	»
11.54	0.13	88.33	$La_2(MoO_4)_3 \cdot Na_2MoO_4 \cdot 2H_2O + La_2(MoO_4)_3 \cdot 2Na_2MoO_4 \cdot 3H_2O$
11.52	0.13	88.35	$La_2(MoO_4)_3 \cdot 2Na_2MoO_4 \cdot 3H_2O$
15.50	0.18	84.32	»
16.19	0.19	83.62	»
20.02	0.25	79.73	»
22.35	0.28	77.37	»
25.26	0.34	74.40	»
25.27	0.34	74.39	$La_2(MoO_4)_3 \cdot 2Na_2MoO_4 \cdot 3H_2O + Na_2MoO_4 \cdot 2H_2O$
25.30	0.34	74.36	»

№ 3934 SODIUM MOLYBDATE–SILVER MOLYBDATE–WATER [2405]

$Na_2MoO_4—Ag_2MoO_4—H_2O$

$t = 25$

Concn. of Sodium Molybdate in aqueous soln., wt %	Comp. of solid phase	Concn. of Sodium Molybdate in aqueous soln., wt %	Comp. of solid phase
5.21	Ag_2MoO_4	37.71	Ag_2MoO_4
13.49	»	39.37	$Ag_2MoO_4 + Na_2MoO_4 \cdot 2H_2O$
21.52	»	39.36	»
30.22	»	39.38	$Na_2MoO_4 \cdot 2H_2O$

Note: The concn. of Silver Molybdate was less than 4.10^{-6} wt %

№ 3935 [2379]

SODIUM HEXAHYDROXOSTANNATE–SODIUM IODIDE–WATER

$Na_2[Sn(OH)_6]— NaI —H_2O$

$t = 18$

Concn. of Sodium Iodide in aqueous soln., g/l	Solubility of Sodium Hexahydroxostannate in aqueous soln. of Sodium Iodide, g/l
295.3	330

№ 3936 [2379]

SODIUM HEXAHYDROXOSTANNATE – SODIUM TUNGSTATE – WATER

$Na_2[Sn(OH)_6]—Na_2WO_4—H_2O$

$t = 18$

Concn. of Sodium Tungstate in aqueous soln., g/l	Solubility of Sodium Hexahydroxostannate in aqueous soln. of Sodium Tungstate, g/l
289.3	350

№ 3937 SODIUM IODIDE–SODIUM IODATE–WATER [2390]

$NaI — NaIO_3—H_2O$

Comp. of saturated soln., wt %			t	d	Comp. of solid phase
Sodium Iodide	Sodium Iodate	Water			
30.43	1.58	67.99	20	1.318	$3NaI \cdot 2NaIO_3 \cdot 20H_2O$
38.26	0.513	61.227	20	1.409	»
54.96	0.062	44.978	20	1.686	»
61.36	0.042	38.598	20	1.811	»
0.0	8.569	91.431	25	1.075	$NaIO_3 \cdot H_2O$
31.57	2.42	66.01	25	—	$NaIO_3 \cdot H_2O + 3NaI \cdot 2NaIO_3 \cdot 20H_2O$
33.06	1.96	65.98	25	1.355	$3NaI \cdot 2NaIO_3 \cdot 20H_2O$
34.29	1.62	64.09	25	—	»
36.56	1.18	62.26	25	1.381	»
40.42	0.704	58.876	24	1.443	»
43.91	0.455	55.635	25	1.494	$3NaI \cdot 2NaIO_3 \cdot 20H_2O + 3NaI \cdot 2NaIO_3 \cdot 15H_2O$
46.06	0.314	53.626	25	—	$3NaI \cdot 2NaIO_3 \cdot 15H_2O$
54.43	0.102	45.478	25	1.605	»
63.71	0.069	36.221	25	1.884	»
64.70	0.075	35.225	25	—	$3NaI \cdot 2NaIO_3 \cdot 15H_2O + NaI \cdot 2H_2O$
64.71	0.0	35.29	25	1.904	$NaI \cdot 2H_2O$

Table № 3937 (continuation)

Comp. of saturated soln., wt %			t	d	Comp. of solid phase
Sodium Iodide	Sodium Iodate	Water			
0.0	11.70	88.30	40	—	$NaIO_3 \cdot H_2O$
40.84	3.02	59.14	40	—	$NaIO_3 + 3NaI \cdot 2NaIO_3 \cdot 15H_2O$
48.20	0.95	50.85	40	—	$3NaI \cdot 2NaIO_3 \cdot 15H_2O$
56.16	0.36	43.48	40	—	»
62.52	0.30	37.18	40	—	»
65.39	0.32	34.29	40	—	$3NaI \cdot 2NaIO_3 \cdot 15H_2O + 3NaI \cdot 2NaIO_3 \cdot 10H_2O$
66.40	0.31	33.29	40	—	$3NaI \cdot 2NaIO_3 \cdot 10H_2O$
66.89	0.30	32.81	40	—	»
67.10	0.30	32.60	40	—	$3NaI \cdot 2NaIO_3 \cdot 10H_2O + NaI \cdot 2H_2O$
67.35	0.0	32.65	40	—	$NaI \cdot 2H_2O$
60.97	0.82	38.21	55	—	$3NaI \cdot 2NaIO_3 \cdot 10H_2O$
63.59	0.67	35.74	55	—	»
66.78	0.60	32.62	55	—	»
68.03	0.61	31.36	55	—	»
69.11	0.59	30.30	55	—	»

№ 3938 SODIUM IODIDE–SODIUM IODATE–WATER [1009]

$NaI — NaIO_3 — H_2O$

Comp. of saturated soln., wt %			t	Comp. of solid phase
Sodium Iodide	Sodium Iodate	Water		
0.00	2.50	97.50	0	$NaIO_3 \cdot 5H_2O$
5.00	2.20	92.80	0	»
10.00	2.00	88.00	0	»
18.60	1.50	79.90	0	$NaIO_3 + 3NaIO_3 \cdot 4NaI \cdot 23H_2O$
20.00	1.30	78.70	0	$3NaIO_3 \cdot 4NaI \cdot 23H_2O$
28.00	0.30	71.70	0	»
30.00	0.20	69.80	0	»
38.30	0.10	61.60	0	»
40.00	0.09	59.91	0	»
50.00	0.07	49.93	0	»
60.00	0.05	39.95	0	»
61.30	0.05	38.65	0	$3NaIO_3 \cdot 4NaI \cdot 23H_2O + NaI \cdot 2H_2O$
61.50	0.00	38.50	0	$NaI \cdot 2H_2O$
0.00	8.30	91.70	21	$NaIO_3 \cdot H_2O$
5.00	6.40	88.60	21	»
10.00	5.05	84.95	21	»
18.60	3.80	77.60	21	»
20.00	3.70	76.30	21	»
28.00	2.80	69.20	21	$NaIO_3 \cdot H_2O + 3NaIO_3 \cdot 4NaI \cdot 23H_2O$
30.00	2.64	67.36	21	$3NaIO_3 \cdot 4NaI \cdot 23H_2O$
38.30	0.90	60.80	21	»
40.00	0.67	59.33	21	»
50.00	0.18	49.82	21	»
60.00	0.10	39.90	21	»
61.30	0.095	38.605	21	»
61.50	0.095	38.405	21	»

Table № 3938 (continuation)

Comp. of saturated soln., wt %			t	Comp. of solid phase
Sodium Iodide	Sodium Iodate	Water		
64.00	0.090	35.910	21	$3NaIO_3 \cdot 4NaI \cdot 23H_2O + NaI \cdot 2H_2O$
64.30	0.00	35.70	21	$NaI \cdot 2H_2O$
0.00	14.00	86.00	50	$NaIO_3 \cdot H_2O$
5.00	10.10	84.90	50	»
10.00	7.00	83.00	50	»
18.60	4.80	76.60	50	»
20.00	4.50	75.50	50	»
28.00	3.6	68.40	50	»
30.0	3.5	66.50	50	»
38.3	2.8	58.90	50	$NaIO_3 \cdot H_2O + 3NaIO_3 \cdot 4NaI \cdot 23H_2O$
40.0	2.7	57.30	50	$3NaIO_3 \cdot 4NaI \cdot 23H_2O$
50.0	1.2	48.80	50	»
60.0	0.47	39.53	50	»
61.3	0.45	38.55	50	»
61.5	0.45	38.05	50	»
64.0	0.42	35.58	50	»
64.3	0.42	35.28	50	»
66.9	0.40	32.70	50	$3NaIO_3 \cdot 4NaI \cdot 23H_2O + NaI \cdot 2H_2O$
69.50	0.00	30.50	50	$NaI \cdot 2H_2O$
0.00	21.00	79.00	80	$NaIO_3$
5.00	17.50	77.50	80	»
10.00	13.70	76.30	80	»
18.60	9.80	71.60	80	»
20.00	7.60	72.40	80	»
28.00	5.00	67.00	80	»
30.00	4.90	65.10	80	»
38.30	4.00	57.70	80	»
40.00	3.80	56.20	80	»
50.00	2.60	47.40	80	»
60.00	2.55	37.45	80	»
61.30	2.50	36.20	80	»
61.50	2.50	36.00	80	»
64.00	2.45	33.55	80	»
64.30	2.45	33.25	80	»
69.50	2.35	28.15	80	»
72.40	2.30	25.30	80	$NaIO_3 + NaI$
74.70	0.00	25.30	80	NaI

№ 3939 SODIUM IODIDE–POTASSIUM BROMIDE–WATER [258]

$NaI-KBr-H_2O$

Comp. of saturated soln., wt %			t	Comp. of solid phase
Sodium Iodide	Potassium Bromide	Water		
95.8	4.2	845	—30	$Ice + NaI \cdot 5H_2O$
73.9	26.1	663	—30	α-$KBr + NaI \cdot 5H_2O$
64.6	35.4	754	—30	$Ice + \alpha$-KBr
75.0	25.0	625	—25	α-$KBr + NaI \cdot 5H_2O$

Table № 3939 (continuation)

Comp. of saturated soln., wt %			t	Comp. of solid phase
Sodium Iodide	Potassium Bromide	Water		
55.3	44.7	884	−25	Ice + α-KBr
75.4	24.6	558	−20	α-KBr + $NaI \cdot 5H_2O$
42.7	57.3	958	−20	Ice + α-KBr
89.8	10.2	505	−15	α-KBr + $NaI \cdot 5H_2O$
22.0	78.0	1196	−15	Ice + α-KBr
89.9	10.1	492	−10	α-KBr + $NaI \cdot 2H_2O$
90.0	10.0	466	0	»
90.3	9.7	457	5	β-KBr + $NaI \cdot 2H_2O$
32.9	67.1	920	5	α-KBr + β-KBr
90.3	9.7	445	10	β-KBr + $NaI \cdot 2H_2O$
13.3	86.7	1024	10	α-KBr + β-KBr
89.7	10.3	427	15	β-KBr + $NaI \cdot 2H_2O$
3.1	96.9	1053	15	α-KBr + β-KBr
90.1	9.9	428	20	β-KBr + $NaI \cdot 2H_2O$
0.6	99.4	1010	20	α-KBr + β-KBr
90.5	9.5	412	25	β-KBr + $NaI \cdot 2H_2O$
90.5	9.5	398	30	»

№ 3940 **SODIUM IODIDE–POTASSIUM IODIDE–POTASSIUM BROMIDE–WATER** [253]

$NaI-KI-KBr-H_2O$

Comp. of saturated soln., wt %				t	Comp. of solid phase
Sodium Iodide	Potassium Iodide	Potassium Bromide	Water		
4.7	5.3	0.0	90.0	− 2.0	Section I
4.7	5.3	9.1	80.9	− 4.6	»
4.7	5.3	16.7	73.3	− 7.5	»
4.7	5.3	23.0	67.0	−10.8	»
4.7	5.3	26.5	63.5	−12.7	»
4.7	5.3	28.6	61.4	− 9.0	»
4.7	5.3	29.6	60.4	− 3.6	»
4.7	5.3	31.0	59.0	2.6	»
4.7	5.3	32.0	58.0	6.4	»
4.7	5.3	33.3	56.7	11.6	»
4.7	5.3	34.6	55.4	16.6	»
4.7	5.3	37.5	52.5	30.0	»
9.4	10.6	0.0	80.0	− 5.4	Section II
9.4	10.6	6.1	73.9	− 8.2	»
9.4	10.6	16.7	63.3	−11.6	»
9.4	10.6	23.0	57.0	−15.0	»
9.4	10.6	24.8	55.2	−11.2	»
9.4	10.6	25.9	54.1	− 4.8	»
9.4	10.6	26.5	53.5	− 1.2	»
9.4	10.6	27.5	52.5	3.0	»
9.4	10.6	28.6	51.4	7.8	»
9.4	10.6	29.6	50.4	12.0	»
9.4	10.6	31.0	49.0	18.0	»
9.4	10.6	33.3	46.7	28.0	»
14.1	15.9	0.0	70.0	− 9.8	Section III

Table № 3940 (continuation)

Comp. of saturated soln., wt %				t	Comp. of solid phase
Sodium Iodide	Potassium Iodide	Potassium Bromide	Water		
14.1	15.9	9.1	60.9	—13.2	Section III
14.1	15.9	16.7	53.3	—16.8	»
14.1	15.9	19.3	50.7	—18.2	»
14.1	15.9	20.6	49.4	—13.0	»
14.1	15.9	21.8	48.2	— 6.6	»
14.1	15.9	23.0	47.0	0.6	»
14.1	15.9	24.8	45.2	8.4	»
14.1	15.9	26.5	43.5	16.0	»
14.1	15.9	28.0	42.0	23.0	»
14.1	15.9	29.0	41.0	27.6	»
14.1	15.9	30.1	39.9	32.4	»
18.8	21.2	0.0	60.0	—16.0	Section IV
18.8	21.2	3.85	56.15	—17.6	»
18.8	21.2	9.1	50.9	—20.0	
18.8	21.2	11.5	48.5	—21.2	»
18.8	21.2	13.8	46.2	—22.6	»
18.8	21.2	16.0	44.0	—15.0	»
18.8	21.2	16.7	43.3	—20.0	»
18.8	21.2	18.0	42.0	2.6	»
18.8	21.2	21.9	38.1	17.0	»
18.8	21.2	23.0	37.0	22.4	»
18.8	21.2	24.2	35.8	28.6	»
18.8	21.2	25.3	34.7	33.0	»
21.1	23.9	0.0	55.0	—20.0	Section V
21.1	23.9	3.85	51.15	—21.6	»
21.1	23.9	5.7	49.3	—22.6	»
21.1	23.9	9.1	45.9	—24.4	»
21.1	23.9	11.5	43.5	—26.0	»
21.1	23.9	13.0	42.0	—19.0	»
21.1	23.9	13.8	41.2	—14.0	»
21.1	23.9	15.2	39.8	5.0	»
21.1	23.9	18.0	37.0	10.0	»
21.1	23.9	19.2	35.8	15.4	»
21.1	23.9	21.9	33.1	28.8	»
21.1	23.9	23.0	32.0	33.2	»
24.4	27.6	0.0	48.0	—25.6	Section VI
24.4	27.6	3.85	44.15	—18.4	»
24.4	27.6	6.5	41.5	—14.2	»
24.4	27.6	11.5	36.5	— 0.6	»
24.4	27.6	13.8	34.2	13.0	»
24.4	27.6	16.7	31.2	28.6	»
24.4	27.6	17.4	30.6	33.0	»
25.4	28.6	0.0	46.0	—14.8	Section VII
25.4	28.6	3.85	42.15	— 7.6	»
25.4	28.6	6.5	39.5	— 1.6	»
25.4	28.6	7.4	38.6	0.2	»
25.4	28.6	9.1	36.9	3.8	»
25.4	28.6	11.5	34.5	12.0	»
25.4	28.6	13.0	33.0	20.8	»
25.4	28.6	14.5	31.5	29.0	»
25.8	29.2	0.0	45.0	— 9.2	Section VIII

Table № 3940 (continuation)

Comp. of saturated soln., wt %				t	Comp. of solid phase
Sodium Iodide	Potassium Iodide	Potassium Bromide	Water		
25.8	29.2	1.96	43.04	— 5.4	Section VIII
25.8	29.2	3.85	41.15	— 1.4	»
25.8	29.2	7.4	37.6	6.0	»
25.8	29.2	9.9	35.1	11.2	»
25.8	29.2	11.5	33.5	19.6	»
25.8	29.2	12.3	32.7	24.4	»
25.8	29.2	13.0	32.0	29.0	»
0.0	0.0	10.0	90.0	— 2.8	Section IX
4.3	4.8	10.0	80.9	— 4.8	»
7.9	8.8	10.0	73.3	— 7.4	»
10.2	11.6	10.0	68.2	— 9.4	»
13.4	15.2	10.0	61.4	—12.8	»
17.6	19.9	10.0	52.6	—18.4	»
21.1	23.9	10.0	45.0	—25.6	»
22.0	24.8	10.0	43.2	—27.6	»
22.4	25.4	10.0	42.2	—24.4	»
22.8	25.7	10.0	41.5	—20.4	»
24.0	27.0	10.0	39.0	— 7.2	»
24.8	28.0	10.0	37.2	2.6	»
25.0	28.3	10.0	36.7	6.0	»
25.6	28.9	10.0	35.6	17.0	»
26.3	29.6	10.0	34.1	26.2	»
26.6	29.9	10.0	33.5	32.0	»
0.0	0.0	14.0	86.0	— 4.3	»
4.3	4.8	14.0	76.9	— 6.4	»
7.9	8.8	14.0	69.3	— 9.0	»
10.8	22.2	14.0	53.0	—11.6	»
13.4	15.2	14.0	57.4	—14.6	»
17.6	19.9	14.0	48.5	—20.3	»
20.3	22.9	14.0	42.8	—25.8	»
21.6	24.3	14.0	40.1	—24.6	»
22.0	24.8	14.0	39.2	—18.0	»
22.4	25.4	14.0	38.2	—13.0	»
23.3	26.2	14.0	36.5	— 5.6	»
24.0	27.0	14.0	35.0	4.4	»
24.4	27.5	14.0	34.1	10.0	Section X
24.8	28.0	14.0	33.2	16.4	»
25.3	28.6	14.0	32.1	25.0	»
25.8	29.2	14.0	31.0	33.4	»
4.7	5.3	0.0	90.0	— 2.0	Section XI
9.4	10.6	0.0	80.0	— 5.4	»
14.1	15.9	0.0	70.0	— 9.8	»
18.8	21.2	0.0	60.0	—16.2	»
21.1	23.9	0.0	55.0	—20.0	»
23.5	26.5	0.0	50.0	—25.8	»
24.4	27.0	0.0	48.0	—14.8	»
25.4	28.6	0.0	46.0	—25.6	»
25.8	29.2	0.0	45.0	— 9.2	»
27.0	30.4	0.0	42.6	3.0	»
27.7	31.3	0.0	41.0	18.0	»
28.5	32.1	0.0	39.4	31.6	»

Table № 3940 (continuation)

Section No.	Temperature of points of intersection	Comp. of solid phase
I	−13.6	Ice+αKBr
I	10.8	αKBr+βKBr
II	−15.6	Ice+αKBr
II	5.8	αKBr+βKBr
III	−18.5	Ice+αKBr
III	3.0	αKBr+βKBr
IV	−23.2	Ice+αKBr
V	−26.4	»
VI	− 6.6	αKI+αKBr
VII	− 1.0	αKI+βKBr
VII	7.2	βKI+βKBr
VIII	2.4	αKI+βKI
VIII	13.2	βKI+βKBr
IX	−28.2	Ice+αKI
IX	3.8	αKI+βKI
X	−29.0	Ice+αKI
X	− 1.0	αKI+βKI
XI	−27.6	Ice+αKI
XI	4.8	αKI+βKI

№ 3941 SODIUM IODIDE–POTASSIUM IODIDE–WATER [1841]

$NaI-KI-H_2O$

Comp. of saturated soln., wt %			t	Comp. of solid phase
Potassium Iodide	Sodium Iodide	Water		
57.33	0.00	42.67	8	KI
37.54	19.97	42.49	8	»
19.20	40.55	40.25	8	»
8.36	55.58	36.06	8	»
7.44	57.13	35.43	8	$KI+NaI\cdot 2H_2O$
5.82	58.23	35.95	8	$NaI\cdot 2H_2O$
0.0	62.49	37.51	8	»
59.78	0.0	40.22	25	KI
50.32	9.29	40.39	25	»
42.08	17.60	40.32	25	»
29.79	30.55	39.66	25	»
21.43	40.11	38.46	25	»
15.90	46.91	37.19	25	»
10.60	54.49	34.91	25	»
7.59	59.35	33.16	25	$KI+NaI\cdot 2H_2O$
7.06	59.71	33.23	25	$NaI\cdot 2H_2O$
6.80	59.97	33.23	25	»
5.44	60.74	33.82	25	»
4.78	61.35	33.87	25	»
2.17	63.15	34.68	25	»
0.00	64.72	35.28	25	»
61.73	0.00	38.27	40	KI
48.04	13.47	38.49	40	»
35.55	26.20	38.25	40	»

Comp. of saturated soln., wt %			t	Comp. of solid phase
Potassium Iodide	Sodium Iodide	Water		
21.33	42.04	36.63	40	KI
9.68	57.85	32.47	40	»
7.23	62.34	30.43	40	KI+NaI · $2H_2O$
4.81	63.92	31.27	40	NaI · $2H_2O$
0.00	67.35	32.65	40	»

№ 3942 SODIUM IODIDE–POTASSIUM IODIDE–WATER [258]

NaI — KI — H_2O

Comp. of saturated soln., mol			t	Comp. of solid phase
Sodium Iodide	Potassium Iodide	Water		
97.3	2.7	847	—30	Ice+NaI · $5H_2O$
78.9	21.1	729	—30	α-KI+NaI · $5H_2O$
60.3	39.7	797	—30	Ice+α-KI
80.3	19.7	668	—25	α-KI+NaI · $5H_2O$
34.6	65.4	829	—25	Ice+α-KI
82.0	18.0	616	—20	α-KI+NaI · $5H_2O$
87.6	12.4	532	—15	»
90.3	9.7	501	—10	α-KI+NaI · $2H_2O$
90.2	9.8	487	— 5	»
89 8	10.2	475	0	»
89.8	10.2	463	5	β-KI+NaI · $2H_2O$
36.0	64.0	649	5	α-KI+β-KI
89.5	10.5	452	10	β-KI+NaI · $2H_2O$
12.8	87.2	671	10	α-KI+β-KI
89.7	10.3	438	15	β-KI+NaI · $2H_2O$
89.7	10.3	427	20	»
89.5	10.5	410	25	»
89.5	10.5	393	30	»

№ 3943 SODIUM IODIDE–POTASSIUM IODIDE–WATER [387]

NaI — KI — H_2O

Comp. of saturated soln., wt %			t	Comp. of solid phase
Potassium Iodide	Sodium Iodide	Water		
61.4	0.0	38.6	0	NaI · $2H_2O$
54.5	7.4	38.1	0	NaI · $2H_2O$+KI
42.0	15.1	42.9	0	KI
30.0	25.8	44.2	0	»
24.1	31.1	44.8	0	»
13.8	42.5	43.7	0	»
7.1	49.1	43.8	0	»
0.0	56.2	43.8	0	»

Table № 3943 (continuation)

Comp. of saturated soln., wt %			t	Comp. of solid phase
Potassium Iodide	Sodium Iodide	Water		
64.8	0.0	35.2	25.0	$NaI \cdot 2H_2O$
58.7	7.5	33.8	25.0	$NaI \cdot 2H_2O + KI$
42.5	22.0	35.5	25.0	KI
34.0	27.8	38.2	25.0	»
30.0	32.0	38.0	25.0	»
24.0	38.1	37.9	25.0	»
15.0	45.8	39.2	25.0	»
6.8	54.1	39.1	25.0	»
0.0	59.7	40.3	25.0	»
69.5	0.0	30.5	50.0	$NaI \cdot 2H_2O$
67.5	4.3	28.2	50.0	»
63.9	7.6	28.5	50.0	$NaI \cdot 2H_2O + KI$
56.9	13.8	29.3	50.0	KI
46.9	22.0	31.1	50.0	»
40.0	27.1	32.9	50.0	»
24.5	41.1	34.4	50.0	»
16.1	47.2	36.7	50.0	»
7.6	55.3	37.1	50.0	»
0.0	62.7	37.3	50.0	»

№ 3944 SODIUM IODIDE–CALCIUM HYDROXIDE–WATER [1940]

$NaI—Ca(OH)_2—H_2O$

$t = 25$

Concn. in aqueous soln., mol/1000 g Water		Comp. of solid phase	Concn. in aqueous soln., mol/1000g Water		Comp. of solid phase
Sodium Iodide	Calcium Hydroxide		Sodium Iodide	Calcium Hydroxide	
4.21	0.01871	$Ca(OH)_2$	0.904	0.02702	$Ca(OH)_2$
2.86	0.02228	»	0.605	0.02678	»
2.11	0.02466	»	0.356	0.02590	»
1.83	0.02541	»	0.159	0.02410	»
1.60	0.02605	»	0.085	0.02261	»
1.33	0.02662	»	0.000	0.01976	»

№ 3945 [2021]

SODIUM IODIDE–SILVER IODIDE–WATER

$NaI— AgI —H_2O$

$t = 25$

Comp. of saturated soln., wt %			Comp. of solid phase
Sodium Iodide	Silver Iodide	Water	
32.85	11.75	55.40	AgI
34.43	14.55	51.02	»
40.19	29.83	29.98	»

Table № 3945 (continuation)

Comp. of saturated soln., wt %			Comp. of solid phase
Sodium Iodide	Silver Iodide	Water	
41.13	32.66	26.21	AgI
41.83	34.90	23.27	AgI + AgI · NaI · 3½ H_2O
45.53	31.27	23.20	AgI · NaI · 3½ H_2O
50.19	27.37	22.44	»
50.57	27.05	22.38	AgI · NaI · 3½ H_2O + NaI
51.22	25.78	23.00	NaI
52.94	22.40	24.66	»
56.11	16.37	27.52	»
59.73	9.95	30.72	»
64.96	0.00	35.04	»

№ 3946 [2537]

SODIUM IODIDE–SILVER IODIDE–WATER

NaI — AgI — H_2O

$t = 20$

Concn. in aqueous soln., g/l	
Sodium Iodide	Silver Iodide
6.45	0.002348
25.48	0.02348
74.95	0.2348
160.39	2.348

№ 3947 SODIUM IODIDE–SILVER IODIDE–WATER [2077]

NaI — AgI — H_2O

$t = 18$

Concn. of Sodium Iodide in aqueous soln., g/l	Solubility of Silver Iodide in aqueous soln. of Sodium Iodide, g/l	Concn. of Sodium Iodide in aqueous soln., g/l	Solubility of Silver Iodide in aqueous soln. of Sodium Iodide, g/l
0.00001492	$1.017 \cdot 10^{-9}$	0.1498	$4.35 \cdot 10^{-6}$
0.0000299	$6.63 \cdot 10^{-8}$	0.299	$9.7 \cdot 10^{-6}$
0.0000747	$4.3 \cdot 10^{-8}$	0.747	$3.03 \cdot 10^{-5}$
0.0001498	$3.69 \cdot 10^{-8}$	1.498	$8.25 \cdot 10^{-5}$
0.000299	$3.99 \cdot 10^{-8}$	2.99	$2.68 \cdot 10^{-4}$
0.000747	$5.07 \cdot 10^{-8}$	7.47	$1.51 \cdot 10^{-3}$
0.001498	$8.79 \cdot 10^{-8}$	14.98	$5.85 \cdot 10^{-3}$
0.00290	$1.1 \cdot 10^{-7}$	2.99	$2.96 \cdot 10^{-2}$
0.00747	$2.25 \cdot 10^{-7}$	74.6	$2.82 \cdot 10^{-1}$
0.01498	$4.27 \cdot 10^{-7}$	148.4	1.97
0.0299	$8.74 \cdot 10^{-7}$	288	17.6
0.0747	$2.09 \cdot 10^{-6}$		

№ 3948 SODIUM IODIDE–ANTIMONY TRIIODIDE–WATER [1659]

$NaI—SbI_3—H_2O$

Comp. of saturated soln., wt %			t	Comp. of solid phase
Antimony Triiodide	Sodium Iodide	Water		
0.00	63.50	36.50	15	NaI
3.47	64.02	32.51	15	$NaI+SbI_3\cdot2NaI\cdot8H_2O$
8.00	52.00	40.00	15	$SbI_3\cdot2NaI\cdot8H_2O$
20.00	40.00	40.00	15	»
23.38	36.29	40.33	15	$SbI_3\cdot2NaI\cdot8H_2O+SbI_3\cdot NaI\cdot6H_2O$
30.43	26.08	43.49	15	$SbI_3\cdot NaI\cdot6H_2O$
0.00	66.32	33.68	35	NaI
7.69	61.53	30.78	35	$NaI+SbI_3\ 2NaI\cdot8H_2O$
14.54	49.09	36.37	35	$SbI_3\cdot2NaI\cdot8H_2O$
25.00	39.28	35.72	35	»
32.20	33.89	33.91	35	$SbI_3\cdot2NaI\cdot8H_2O+SbI_3\cdot NaI\cdot6H_2O$
36.84	28.07	35.09	35	$SbI_3\cdot NaI\cdot6H_2O$
0.00	71.98	28.02	60	NaI
12.10	63.68	24.22	60	»
18.68	59.34	21.98	60	$NaI+SbI_3\cdot2NaI\cdot8H_2O$
18.98	55.69	25.33	60	$SbI_3\cdot2NaI\cdot8H_2O$
28.16	43.66	28.18	60	»
35.13	37.83	27.04	60	»
40.74	34.56	24.70	60	$SbI_3\cdot2NaI\cdot8H_2O+SbI_3\cdot NaI\cdot6H_2O$
43.58	30.76	25.66	60	$SbI_3\cdot NaI\cdot6H_2O$
49.38	25.92	24.70	60	$SbI_3\cdot NaI\cdot6H_2O+SbI_3$
44.77	25.37	29.86	60	SbI_3
40.00	20.00	29.86	60	»
30.55	13.88	55.57	60	»

№ 3949 SODIUM IODIDE–ANTIMONY TRIIODIDE–WATER [1660]

$NaI—SbI_3—H_2O$

Comp. of saturated soln., wt %			t	Comp. of solid phase
Sodium Iodide	Antimony Triiodide	Water		
63.4	0.0	36.6	15	$NaI\cdot2H_2O$
61.40	3.51	35.09	15	$NaI\cdot2H_2O+2NaI\cdot SbI_3\cdot8H_2O$
57.25	4.58	38.17	15	$2NaI\cdot SbI_3\cdot8H_2O$
49.01	11.76	39.23	15	»
40.00	20.00	40.00	15	»
35.29	25.49	39.22	15	$2NaI\cdot SbI_3\cdot8H_2O+NaI\cdot SbI_3\cdot6H_2O$
31.91	25.53	42.56	15	$NaI\cdot SbI_3\cdot6H_2O$
66.33	0.0	33.67	35	$NaI\cdot2H_2O$
63.49	4.28	32.23	35	»
62.12	7.58	30.30	35	$NaI\cdot2H_2O+2NaI\cdot SbI_3\cdot8H_2O$
55.00	11.66	33.34	35	$2NaI\cdot SbI_3\cdot8H_2O$
45.61	19.29	35.10	35	»
37.93	27.58	34.49	35	»
34.42	32.78	32.80	35	$2NaI\cdot SbI_3\cdot8H_2O+NaI\cdot SbI_3\cdot6H_2O$
31.03	34.48	34.49	35	$NaI\cdot SbI_3\cdot6H_2O$
71.92	0.0	28.08	60	$NaI\cdot2H_2O$

Table № 3949 (continuation)

Comp. of saturated soln., wt %			t	Comp. of solid phase
Sodium Iodide	Antimony Triiodide	Water		
63.41	12.20	24.39	60	$NaI \cdot 2H_2O$
59.78	18.47	21.75	60	$NaI \cdot 2H_2O + 2NaI \cdot SbI_3 \cdot 8H_2O$
57.14	19.04	23.82	60	$2NaI \cdot SbI_3 \cdot 8H_2O$
47.22	25.00	27.78	60	»
35.89	38.46	25.65	60	»
33.73	42.16	24.11	60	$2NaI \cdot SbI_3 \cdot 8H_2O + NaI \cdot SbI_3 \cdot 6H_2O$
29.62	45.67	24.71	60	$NaI \cdot SbI_3 \cdot 6H_2O$
26.82	48.79	24.39	60	$NaI \cdot SbI_3 \cdot 6H_2O + SbI_3$
23.33	43.33	33.34	60	SbI_3

№ 3950 SODIUM IODIDE–CESIUM IODIDE–WATER [779]

$NaI—CsI—H_2O$

Comp. of saturated soln., wt %			t	Comp. of solid phase
Sodium Iodide	Cesium Iodide	Water		
64.69	0.00	35.31	25	$NaI \cdot 2H_2O$
62.85	2.87	34.28	25	»
62.20	3.70	34.10	25	»
61.34	4.75	33.91	25	»
61.10	5.00	33.90	25	$NaI \cdot 2H_2O + CsI$
60.15	5.49	34.36	25	CsI
57.51	5.74	36.75	25	»
56.40	6.18	37.42	25	»
42.00	13.18	44.82	25	»
32.81	17.79	49.40	25	»
27.94	20.78	51.28	25	»
27.15	21.15	51.70	25	»
18.16	27.78	54.06	25	»
16.00	29.84	54.16	25	»
7.97	38.00	54.03	25	»
1.57	44.50	53.93	25	»
0.00	46.10	53.90	25	»
69.45	0.00	30.55	50	$NaI \cdot 2H_2O$
66.36	4.19	29.45	50	»
65.53	5.05	29.42	50	»
65.15	7.32	27.53	50	$NaI \cdot 2H_2O + CsI$
56.50	10.35	33.15	50	CsI
51.13	13.98	34.89	50	»
46.24	15.00	38.76	50	»
38.82	20.93	40.25	50	»
35.13	23.62	41.15	50	»
32.23	25.87	41.90	50	»
19.28	37.34	43.38	50	»
12.15	46.61	41.24	50	»
2.20	55.62	42.18	50	»
0.00	58.30	41.70	50	»

№ 3951 [1815]

SODIUM IODIDE–MERCURY (II) IODIDE–WATER

$NaI—HgI_2—H_2O$

$t = 25$

Concn. in aqueous soln., g/l	
Mercury Iodide	Sodium Iodide
187.4	118.6
282.7	206.5
429	331.3

№ 3952 [2424]

SODIUM IODIDE–LEAD (II) IODIDE–WATER

$NaI—PbI_2—H_2O$

Concn. in aqueous soln., wt %		d_4^t	t	Comp. of solid phase
Sodium Iodide	Lead Iodide			
0.041	0.0	—	0	PbI_2
0.0006	2.27	—	0	»
32.94	41.04	—	0	$PbI_2 + PbI_2 \cdot NaI \cdot 4H_2O$
30.85	43.59	—	0	$PbI_2 \cdot NaI \cdot 4H_2O + PbI_2 \cdot 2NaI \cdot 6H_2O$
24.96	47.92	—	0	$PbI_2 \cdot 2NaI \cdot 6H_2O + NaI \cdot 2H_2O$
0.0	61.59	—	0	$NaI \cdot 2H_2O$
0.0	0.077	—	25	PbI_2
0.0546	0.0555	—	25	»
0.1180	0.0288	—	25	»
0.2127	0.0209	—	25	»
0.2881	0.0071	—	25	»
0.4729	0.0042	—	25	»
1.045	0.0022	—	25	»
3.16	0.0026	1.023	25	»
5.12	0.0054	—	25	»
8.29	0.0262	—	25	»
33.44	5.43	1.417	25	»
36.82	11.65	1.630	25	»
39.21	22.00	1.909	25	»
40.44	27.70	2.132	25	»
41.61	30.42	2.272	25	PbI_2
41.91	32.88	2.427	25	»
42.57	34.59	2.542	25	$PbI_2 + PbI_2 \cdot NaI \cdot 4H_2O$
42.82	34.09	2.526	25	$PbI_2 \cdot NaI \cdot 4H_2O$
43.29	32.02	2.428	25	»
43.56	31.82	2.433	25	»
44.38	31.34	2.475	25	»
45.61	31.38	2.499	25	$PbI_2 \cdot NaI \cdot 4H_2O + PbI_2 \cdot 2NaI \cdot 6H_2O$
45.92	31.04	2.496	25	$PbI_2 \cdot 2NaI \cdot 6H_2O$
46.62	30.49	2.490	25	»

Table № 3952 (continuation)

Concn. in aqueous soln., wt %		d_4^t	t	Comp. of solid phase
Sodium Iodide	Lead Iodide			
47.05	29.98	2.481	25	$PbI_2 \cdot 2NaI \cdot 6H_2O$
47.71	26.69	2.492	25	»
47.75	29.41	2.495	25	$PbI_2 \cdot 2NaI \cdot 6H_2O + NaI \cdot 2H_2O$
49.79	26.38	2.426	25	$NaI \cdot 2H_2O$
52.45	21.18	2.303	25	»
58.40	11.22	2.119	25	»
61.23	5.97	2.014	25	»
64.58	0.0	1.919	25	»

№ 3953 SODIUM IODIDE–THALLIUM (III) IODIDE–WATER [430]

$NaI—TlI_3—H_2O$

$t = 25$

Concn. of Sodium Iodide in aqueous soln., mol/l	Solubility of Thallium Iodide in aqueous soln. of Sodium Iodide, (mol/l)·10^5	Concn. of Sodium Iodide in aqueous soln., mol/l	Solubility of Thallium Iodide in aqueous soln. of Sodium Iodide, (mol/l)·10^5
0.012	0.90	1.240	1.23
0.014	0.87	1.761	1.82
0.024	0.48	2.106	2.15
0.056	0.29	2.668	5.13
0.062	0.28	2.730	5.22
0.134	0.29	3.324	8.51
0.268	0.34	3.849	12.45
0.358	0.40	4.455	20.55
0.715	0.66	5.105	35.02
1.072	1.09		

№ 3954 SODIUM IODATE–MAGNESIUM IODATE–WATER [1839]

$NaIO_3—Mg(IO_3)_2—H_2O$

Comp. of saturated soln., wt %			t	d	Comp. of solid phase
Magnesium Iodate	Sodium Iodate	Water			
4.37	0.00	95.63	5	1.038	$Mg(IO_3)_2 \cdot 10H_2O$
3.74	1.45	94.81	5	1.047	»
3.33	2.53	94.14	5	1.055	$Mg(IO_3)_2 \cdot 10H_2O + NaIO_3 \cdot 5H_2O$
2.19	2.68	95.13	5	1.043	$NaIO_3 \cdot 5H_2O$
0.00	3.28	96.72	5	1.028	»
8.54	0.00	91.46	25	1.078	$Mg(IO_3)_2 \cdot 4H_2O$
7.24	2.80	89.96	25	1.091	»
6.05	6.16	87.79	25	1.113	»

Comp. of saturated soln., wt %			t	d	Comp. of solid phase
Magnesium Iodate	Sodium Iodate	Water			
5.86	7.49	86.65	25	1.125	$Mg(IO_3)_2 \cdot 4H_2O + NaIO_3 \cdot H_2O$
5.73	7.66	86.61	25	1.128	»
5.57	8.79	85.64	25	1.139	$Mg(IO_3)_2 \cdot 4H_2O + NaIO_3$
5.42	8.74	85.84	25	1.134	$NaIO_3$
5.40	7.52	87.08	25	1.121	$NaIO_3 \cdot H_2O$
5.33	7.50	87.12	25	1.122	»
4.76	7.77	87.47	25	1.119	»
2.99	7.94	89.07	25	1.102	»
1.18	8.30	90.52	25	1.081	»
0.00	8.57	91.43	25	1.076	»
11.97	0.00	88.03	50	—	$Mg(IO_3)_2 \cdot 4H_2O$
9.95	4.59	85.46	50	—	»
8.41	8.94	82.65	50	—	»
7.67	11.97	80.36	50	—	$Mg(IO_3)_2 \cdot 4H_2O + NaIO_3 \cdot H_2O$
4.93	12.42	82.65	50	—	$NaIO_3 \cdot H_2O$
3.73	13.05	83.22	50	—	»
3.05	13.26	83.69	50	—	»
2.64	13.27	84.09	50	—	»
1.55	13.54	84.91	50	—	»
0.00	13.49	86.52	50	—	»

№ 3955 SODIUM IODATE–POTASSIUM IODATE–WATER [1839]

$NaIO_3—KIO_3—H_2O$

Comp. of saturated soln., wt %			t	d	Comp. of solid phase
Potassium Iodate	Sodium Iodate	Water			
5.16	0.00	94.84	5	1.043	KIO_3
4.71	1.41	93.88	5	1.051	»
4.72	2.17	93.11	5	1.060	$KIO_3 + NaIO_3 \cdot 5H_2O$
3.19	2.48	94.33	5	1.046	$NaIO_3 \cdot 5H_2O$
0.0	3.28	96.72	5	1.028	»
8.45	0.00	91.55	25	1.071	KIO_3
7.09	4.26	88.65	25	1.098	»
6.73	7.13	86.14	25	1.126	$KIO_3 + NaIO_3 \cdot H_2O$
3.79	7.79	88.42	25	1.103	$NaIO_3 \cdot H_2O$
0.00	8.57	91.43	25	1.074	»
13.21	0.00	86.79	50	—	KIO_3
11.92	3.92	84.16	50	—	»
11.14	7.70	81.16	50	—	»
10.61	10.92	78.47	50	—	$KIO_3 + NaIO_3 \cdot H_2O$
7.93	11.41	80.66	50	—	$NaIO_3 \cdot H_2O$
4.24	12.55	83.21	50	—	»
0.00	13.49	86.51	50	—	»

№ 3956 SODIUM IODATE–CALCIUM IODATE–WATER [1840]

$NaIO_3-Ca(IO_3)_2-H_2O$

$t = 25$

Concn. in aqueous soln., wt%		d	Comp. of solid phase
Sodium Iodate	Calcium Iodate		
0.00	0.306	—	$Ca(IO_3)_2 \cdot 6H_2O$
0.522	0.084	1.000	»
5.29	0.000	1.041	»
8.58	0.000	1.074	$Ca(IO_3)_2 \cdot 6H_2O + NaIO_3 \cdot H_2O$
8.58	0.000	—	$NaIO_3 \cdot H_2O$

№ 3957 [2669]

SODIUM IODATE–COBALT (II) IODATE–WATER

$NaIO_3-Co(IO_3)_2-H_2O$

$t = 30$

Concn. in aqueous soln., wt%		d	Comp. of solid phase
Sodium Iodate	Cobalt Iodate		
0.00	0.476	1.0006	$Co(IO_3)_2 \cdot 2H_2O$
0.087	0.417	1.0006	»
0.150	0.380	1.0007	»
0.172	0.368	1.0008	»
0.251	0.327	1.0011	»
0.344	0.282	1.0014	»
0.404	0.258	1.0017	»
0.554	0.204	1.0026	»
0.713	0.164	1.0037	»
0.956	0.123	1.0056	»
1.978	0.041	1.0135	»

№ 3958 SODIUM IODATE–LANTHANUM IODATE–WATER [1779]

$NaIO_3-La(IO_3)_3-H_2O$

$t = 25$

Concn. in aqueous soln., g/l		d_4	Concn. in aqueous soln., g/l		d_4
Lanthanum Iodate	Sodium Iodate		Lanthanum Iodate	Sodium Iodate	
0.6353	0.01808	1.00060	0.3993	0.36155	1.00065
0.5646	0.09029	1.00059	0.1973	0.72329	1.00069
0.5083	0.18077	1.00065	0.1339	0.89745	1.00083

№ 3959 SODIUM TUNGSTATE – CERIUM TUNGSTATE – WATER [1357]

$Na_2WO_4-Ce_2(WO_4)_3-H_2O$

$t = 25$

Concn. in aqueous soln., wt%		Comp. of solid phase
Sodium Tungstate	Cerium Tungstate	
2.65	0.62	$Ce_2(WO_4)_3$ hydrate + $Ce_2(WO_4)_3 \cdot Na_2WO_4 \cdot 11H_2O$
4.00	0.60	$Ce_2(WO_4)_3 \cdot Na_2WO_4 \cdot 11H_2O$
6.13	0.52	»
10.71	0.56	»
13.28	0.48	»
14.24	0.60	»
15.03	0.57	$Ce_2(WO_4)_3 \cdot Na_2WO_4 \cdot 11H_2O + Ce_2(WO_4)_3 \cdot 3Na_2WO_4 \cdot 16H_2O$
17.98	0.42	$Ce_2(WO_4)_3 \cdot 3Na_2WO_4 \cdot 16H_2O$
20.03	0.39	»
23.32	0.40	»
25.35	0.32	$Ce_2(WO_4)_3 \cdot 3Na_2WO_4 \cdot 16H_2O + Ce_2(WO_4)_3 \cdot 5Na_2WO_4 \cdot 23H_2O$
27.64	0.24	$Ce_2(WO_4)_3 \cdot 5Na_2WO_4 \cdot 23H_2O$
30.24	0.15	»
35.07	Traces	»
37.42	—	»
39.90	—	»
42.14	—	»
42.62	—	$Ce_2(WO_4)_3 \cdot 5Na_2WO_4 \cdot 23H_2O + Na_2WO_4 \cdot 2H_2O$

№ 3960 [2579]

SODIUM TUNGSTATE – TUNGSTEN TRIOXIDE – WATER

$Na_2WO_4-WO_3-H_2O$

$t = 25$

Comp. of saturated soln., wt%			d	Comp. of solid phase
Sodium Tungstate	Tungsten Trioxide	Water		
42.60	0.00	57.40	1.565	$Na_2WO_4 \cdot 2H_2O$
42.53	0.00	57.47	1.562	»
42.56	0.00	57.44	1.564	$Na_2WO_4 \cdot 2H_2O + Na_2WO_4 \cdot WO_3 \cdot 5H_2O$
41.52	0.00	58.48	1.543	$Na_2WO_4 \cdot WO_3 \cdot 5H_2O$
38.33	0.00	61.67	1.484	»
34.97	0.00	65.03	1.425	»
30.80	0.00	69.20	1.355	»
25.13	0.00	74.87	1.273	»
22.43	0.00	77.57	1.239	»
20.20	0.00	79.80	1.209	»
18.48	0.00	81.52	1.187	»
15.78	0.00	84.22	1.158	»
14.72	0.00	85.28	1.144	»

Comp. of saturated soln., wt%			d	Comp. of solid phase
Sodium Tungstate	Tungsten Trioxide	Water		
12.11	0.26	87.63	1.118	$5Na_2WO_4 \cdot 7WO_3 \cdot 28H_2O$
6.97	0.92	92.11	1.071	»
5.05	2.26	92.69	1.066	»
17.84	0.00	82.16	1.176	»
17.70	0.00	82.30	1.179	$5Na_2WO_4 \cdot 7WO_3 \cdot 28H_2O + Na_2WO_4 \cdot WO_3 \cdot 5H_2O$

№ 3961 MAGNESIUM OXIDE–MOLYBDENUM OXIDE–WATER [2638]

$MgO—MoO_3—H_2O$

$t = 25$

Comp. of saturated soln., wt%			Comp. of solid phase
Molybdenum Oxide	Magnesium Oxide	Water	
0.64	0.23	99.13	$Mg(OH)_2 + MoO_3 \cdot 8MgO \cdot 48H_2O$
1.88	0.55	97.57	$MoO_3 \cdot 8MgO \cdot 48H_2O + MoO_3 \cdot 4MgO \cdot 12H_2O$
13.15	3.85	83.00	$MoO_3 \cdot 4MgO \cdot 12H_2O + MoO_4Mg \cdot 5H_2O$
12.55	3.50	85.95	$MoO_4Mg \cdot 5H_2O$
40.00	5.45	54.55	$MoO_4Mg \cdot 5H_2O + Mo_3O_{10}Mg \cdot 10H_2O$
1.65	0.145	98.205	$Mo_3O_{10}Mg \cdot 10H_2O$
10.12	0.71	89.17	$Mo_3O_{10}Mg \cdot 10H_2O + MoO_3$
0.128	0.00	99.872	MoO_3

№ 3962 [2422]

MAGNESIUM OXIDE–MAGNESIUM CHLORIDE–WATER

$MgO—MgCl_2—H_2O$

$t = 25$

Concn. in aqueous soln., wt%		d_{25}	Comp. of solid phase
Magnesium Chloride	Magnesium Oxide		
2.36	0.00008	1.019	Undetermined solid soln.
4.47	0.00028	1.038	»
6.79	0.00048	1.056	»
9.02	0.0008	1.075	»
13.14	0.00115	1.111	»
17.53	0.0024	1.141	$2MgO \cdot HCl \cdot 5H_2O$
18.52	0.0025	1.162	»
22.04	0.00245	1.192	»
26.88	0.0025	1.245	»
29.80	0.0024	1.274	»
34.22	0.0030	1.321	»

№ 3963 [1302]

MAGNESIUM OXIDE–MAGNESIUM CHLORIDE–WATER
$MgO—MgCl_2—H_2O$

$t = 25$

Concn. in aqueous soln., wt%		Comp. of solid phase
Magnesium Chloride	Magnesium Oxide	
2.12	Traces	$Mg(OH)_2$
2.65	»	»
6.44	»	»
8.61	»	»
9.45	»	»
10.36	»	»
10.52	»	»
10.88	»	$Mg(OH)_2+3MgO \cdot MgCl_2 \cdot 11H_2O$
10.90	»	»
10.98	»	$3MgO \cdot MgCl_2 \cdot 11H_2O$
14.41	»	»
16.43	»	»
17.36	»	»
21.12	»	»
21.38	»	»
24.21	»	»
26.59	»	»
26.86	»	»
31.14	»	»
33.69	»	»
34.77	»	»
35.70	»	$3MgO \cdot MgCl_2 \cdot 11H_2O+MgCl_2 \cdot 6H_2O$
35.72	»	»
35.72	»	$MgCl_2 \cdot 6H_2O$

№ 3964 [88]

MAGNESIUM OXIDE–PHOSPHORUS PENTOXIDE–WATER
$MgO—P_2O_5—H_2O$

Comp. of saturated soln., wt%			t	Comp. of solid phase
Magnesium Oxide	Phosphorus Pentoxide	Water		
1.9	6.3	91.8	0	$MgHPO_4 \cdot 3H_2O$
4.6	16.5	78.9	0	»
8.0	29.5	62.5	0	$MgHPO_4 \cdot 3H_2O+Mg(H_2PO_4)_2 \cdot 4H_2O$
4.4	49.6	46.0	0	$Mg(H_2PO_4)_2 \cdot 4H_2O$
3.9	52.5	43.6	0	»
4.1	53.2	42.7	0	»
3.1	58.4	38.5	0	$Mg(H_2PO_4)_2 \cdot 4H_2O+Mg(H_2PO_4)_2$
2.2	59.7	38.1	0	$Mg(H_2PO_4)_2$
0.4	62.6	37.0	0	»

Table № 3964 (continuation)

Comp. of saturated soln., wt%			t	Comp. of solid phase
Magnesium Oxide	Phosphorus Pentoxide	Water		
0.4	64.8	34.8	0	$Mg(H_2PO_4)_2$
1.8	6.2	92.0	10	$MgHPO_4 \cdot 3H_2O$
4.7	16.8	78.5	10	»
8.2	31.0	60.8	10	$MgHPO_4 \cdot 3H_2O + Mg(H_2PO_4)_2 \cdot 4H_2O$
4.4	50.5	45.1	10	$Mg(H_2PO_4)_2 \cdot 4H_2O$
3.9	54.5	41.6	10	»
3.4	57.5	39.1	10	»
3.2	58.9	37.9	10	$Mg(H_2PO_4)_2 \cdot 4H_2O + Mg(H_2PO_4)_2 \cdot 2H_2O + Mg(H_2PO_4)_2$
0.4	62.0	37.6	10	$Mg(H_2PO_4)_2$
0.3	63.8	35.9	10	»
0.4	1.3	98.3	50	$MgHPO_4 \cdot 3H_2O$
1.6	6.0	92.4	50	»
2.6	9.3	88.1	50	»
4.6	16.8	78.6	50	»
7.0	27.2	65.8	50	»
8.9	37.3	53.8	50	$MgHPO_4 \cdot 3H_2O + Mg(H_2PO_4)_2 \cdot 4H_2O$
8.3	39.6	52.1	50	$Mg(H_2PO_4)_2 \cdot 4H_2O$
7.2	41.8	51.0	50	$Mg(H_2PO_4)_2 \cdot 4H_2O + Mg(H_2PO_4)_2 \cdot 2H_2O$
5.0	53.4	41.6	50	$Mg(H_2PO_4)_2 \cdot 2H_2O$
3.9	57.5	38.6	50	$Mg(H_2PO_4)_2 \cdot 2H_2O + Mg(H_2PO_4)_2$
2.2	62.7	35.1	50	$Mg(H_2PO_4)_2$
1.5	63.7	34.8	50	»
0.4	1.3	98.3	58	$MgHPO_4 \cdot 3H_2O$
1.6	6.0	92.4	58	»
2.6	9.3	88.1	58	»
4.6	16.8	78.6	58	»
7.0	27.2	65.8	58	»
9.5	38.5	52.0	58	$MgHPO_4 \cdot 3H_2O + Mg(H_2PO_4)_2 \cdot 4H_2O + Mg(H_2PO_4)_2 \cdot 2H_2O$
6.3	50.0	43.7	58	$Mg(H_2PO_4)_2 \cdot 2H_2O$
5.1	55.0	39.9	58	»
4.5	58.0	37.5	58	$Mg(H_2PO_4)_2 \cdot 2H_2O + Mg(H_2PO_4)_2$
3.3	60.0	36.7	58	$Mg(H_2PO_4)_2$
2.1	63.0	34.9	58	»
0.8	3.0	96.2	130	$MgHPO_4 \cdot 3H_2O$
3.1	10.6	86.3	130	»
4.9	18.5	76.6	130	»
6.8	26.1	67.1	130	»
8.0	31.3	60.7	130	»
9.6	39.4	51.0	130	»
12.7	54.0	33.3	130	$MgHPO_4 \cdot 3H_2O + Mg(H_2PO_4)_2 \cdot 2H_2O + Mg(H_2PO_4)_2$
6.2	60.0	33.8	130	$Mg(H_2PO_4)_2$
4.1	63.0	32.9	130	»

$MgO—P_2O_5—H_2O$

Comp. of saturated soln., wt%			t	Comp. of solid phase
Magnesium Oxide	Phosphorus Pentoxide	Water		
2.0	6.5	91.5	25	$MgHPO_4 \cdot 3H_2O$
4.1	14.4	81.5	25	»
7.6	29.5	62.9	25	»
8.3	33.1	58.6	25	$MgHPO_4 \cdot 3H_2O + Mg(H_2PO_4)_2 \cdot 4H_2O$
7.6	33.6	58.8	25	$Mg(H_2PO_4)_2 \cdot 4H_2O$
7.2	35.3	57.5	25	»
6.3	38.0	55.7	25	»
5.7	41.5	52.8	25	»
5.3	43.1	51.6	25	»
4.6	48.6	46.8	25	»
4.6	47.6	47.8	25	»
4.6	48.5	46.9	25	»
4.4	52.8	42.8	25	»
4.6	53.3	42.1	25	$Mg(H_2PO_4)_2 \cdot 4H_2O + Mg(H_2PO_4)_2 \cdot 2H_2O$
3.9	54.7	41.4	25	$Mg(H_2PO_4)_2 \cdot 2H_2O$
3.7	55.6	40.7	25	»
3.8	57.2	39.0	25	»
3.6	57.4	39.0	25	»
3.2	59.6	37.2	25	$Mg(H_2PO_4)_2 \cdot 2H_2O + Mg(H_2PO_4)_2$
3.2	59.6	37.2	25	»
3.2	59.6	37.2	25	»
0.9	63.3	35.8	25	$Mg(H_2PO_4)_2$
0.7	64.0	35.3	25	$Mg(H_2PO_4)_2$
0.5	65.6	33.9	25	»
0.04	68.8	31.16	25	»
0.04	70.0	29.96	25	»
0.5	1.5	98.0	80	$MgHPO_4 \cdot 3H_2O$
0.8	3.0	96.2	80	»
2.2	6.8	91.0	80	»
3.1	10.6	86.3	80	»
4.0	13.8	82.2	80	»
4.9	18.5	76.6	80	»
6.1	23.5	70.4	80	»
6.8	26.1	67.1	80	»
8.0	31.3	60.7	80	»
8.9	35.6	55.5	80	»
9.6	39.4	51.0	80	»
11.0	44.8	44.2	80	$MgHPO_4 \cdot 3H_2O + Mg(H_2PO_4)_2 \cdot 2H_2O$
10.7	44.5	44.8	80	$Mg(H_2PO_4)_2 \cdot 2H_2O$
8.8	47.1	44.1	80	»
8.2	48.0	43.8	80	»
7.8	49.6	42.6	80	»
6.5	53.8	39.7	80	»
6.4	54.9	38.7	80	»
6.0	56.5	37.5	80	$Mg(H_2PO_4)_2 \cdot 2H_2O + Mg(H_2PO_4)_2$
5.0	58.3	36.7	80	$Mg(H_2PO_4)_2$
4.0	60.8	35.2	80	»
2.9	61.8	35.3	80	»
1.9	64.4	33.7	80	»

№ 3966 [89]

MAGNESIUM OXIDE–PHOSPHORUS PENTOXIDE–CALCIUM OXIDE–WATER

$MgO—P_2O_5—CaO—H_2O$

$t = 25$

Comp. of saturated soln., wt%				Comp. of solid phase
Magnesium Oxide	Calcium Oxide	Phosphorus Pentoxide	Water	
0.0	5.7	36.2	58.1	$Ca(H_2PO_4)_2 \cdot H_2O + CaHPO_4$
2.1	4.2	37.8	55.9	»
4.4	2.8	39.2	53.6	»
5.0	1.9	40.1	53.0	»
6.6	1.6	40.8	51.0	»
7.3	1.3	42.3	49.1	»
8.2	1.1	43.6	47.1	»
8.4	1.2	44.6	45.8	»
9.3	0.9	44.5	45.3	»
10.6	0.7	45.0	43.7	$Ca(H_2PO_4)_2 \cdot H_2O + CaHPO_4 + MgHPO_4 \cdot 3H_2O$
0.3	0.0	1.4	98.3	$CaHPO_4 + MgHPO_4 \cdot 3H_2O$
0.7	0.0	3.0	96.3	»
1.8	0.1	7.1	91.0	»
2.9	0.2	10.3	86.6	»
4.1	0.3	16.2	79.4	»
4.4	0.3	16.6	78.7	»
6.8	0.4	26.7	66.1	»
7.5	0.5	30.5	61.5	»
9.2	0.4	36.9	53.5	»
10.0	0.6	41.2	48.2	»
10.6	0.7	45.0	43.7	$Ca(H_2PO_4)_2 \cdot H_2O + CaHPO_4 + MgHPO_4 \cdot 3H_2O$
10.7	0.6	45.0	43.7	$MgHPO_4 \cdot 3H_2O + Ca(H_2PO_4)_2 \cdot H_2O + Mg(H_2PO_4)_2 \cdot 2H_2O$
7.8	0.5	48.5	43.2	$Ca(H_2PO_4)_2 \cdot H_2O + Mg(H_2PO_4)_2 \cdot 2H_2O$
7.6	0.3	49.8	42.3	»
7.4	0.4	50.1	42.1	»
7.1	0.4	50.3	42.2	»
5.8	0.3	55.4	38.5	$Ca(H_2PO_4)_2 \cdot H_2O + Mg(H_2PO_4)_2 \cdot 2H_2O + Mg(H_2PO_4)_2$
5.2	0.4	56.5	37.9	$Ca(H_2PO_4)_2 \cdot H_2O + Mg(H_2PO_4)_2$
4.8	0.4	57.3	37.5	»
4.2	0.3	59.2	36.3	»
3.3	0.3	61.6	34.8	»
3.1	0.3	60.5	36.1	»
0.2	4.4	39.8	55.6	$Ca(H_2PO_4)_2 \cdot H_2O$
0.6	4.3	39.3	55.8	»
1.0	4.0	41.0	54.0	»
1.2	3.8	41.0	54.0	»
1.8	3.2	41.0	54.0	»
2.4	2.6	42.0	53.0	»
3.9	1.6	43.2	51.3	»
4.6	1.4	43.4	50.6	»
6.2	0.7	45.0	48.1	»
9.0	0.6	46.5	43.9	»
8.8	0.2	46.2	44.8	»

Table 3966 (continuation)

Comp. of saturated soln., wt%				Comp. of solid phase
Magnesium Oxide	Calcium Oxide	Phosphorus Pentoxide	Water	
9.1	0.4	46.9	43.6	$Ca(H_2PO_4)_2 \cdot H_2O$
9.2	0.4	47.2	43.2	»
8.8	0.5	46.5	44.2	»
3.6	13.8	21.3	61.3	$CaHPO_4$
2.6	23.9	24.0	49.5	»
2.0	9.5	25.2	63.3	»

№ 3967 [90]

MAGNESIUM OXIDE–PHOSPHORUS PENTOXIDE–CALCIUM OXIDE–WATER

$MgO—P_2O_5—CaO—H_2O$

Comp. of saturated soln., wt%				t	Comp. of solid phase
Magnesium Oxide	Calcium Oxide	Phosphorus Pentoxide	Water		
3.0	0.2	12.0	84.8	10	$CaHPO_4 \cdot 2H_2O + MgHPO_4 \cdot 3H_2O$
5.0	0.5	19.5	75.0	10	»
7.0	0.7	28.0	64.3	10	»
7.5	0.9	31.5	60.1	10	$CaHPO_4 \cdot 2H_2O + MgHPO_4 \cdot 3H_2O + Ca(H_2PO_4)_2 \cdot H_2O$
7.7	0.7	30.0	61.6	10	$MgHPO_4 \cdot 3H_2O + Ca(H_2PO_4)_2 \cdot H_2O + Mg(H_2PO_4)_2 \cdot 4H_2O$
3.2	0.1	59.5	37.2	10	$Ca(H_2PO_4)_2 \cdot H_2O + Mg(H_2PO_4)_2 \cdot 4H_2O + Mg(H_2PO_4)_2 \cdot 2H_2O + Mg(H_2PO_4)_2$
0.0	6.2	20.9	72.9	10	$CaHPO_4 \cdot 2H_2O + Ca(H_2PO_4)_2 \cdot H_2O$
2.0	5.5	20.5	72.0	10	»
5.0	2.8	24.3	67.9	10	»
7.0	1.7	28.0	63.3	10	»
7.5	0.9	31.5	60.1	10	$CaHPO_4 \cdot 2H_2O + MgHPO_4 \cdot 3H_2O + Ca(H_2PO_4)_2 \cdot H_2O$
0.7	0.1	2.8	96.4	25	$CaHPO_4 + MgHPO_4 \cdot 3H_2O$
3.7	0.7	14.3	81.3	25	»
6.5	0.9	24.7	67.9	25	»
7.7	0.9	32.3	59.1	25	$CaHPO_4 + MgHPO_4 \cdot 3H_2O + Ca(H_2PO_4)_2 \cdot H_2O$
7.8	0.7	32.8	58.7	25	$MgHPO_4 \cdot 3H_2O + Ca(H_2PO_4)_2 \cdot H_2O + Mg(H_2PO_4)_2 \cdot 4H_2O$
6.8	0.6	35.1	57.5	25	$Ca(H_2PO_4)_2 \cdot H_2O + Mg(H_2PO_4)_2 \cdot 4H_2O$
6.8	0.6	34.6	58.0	25	»
5.6	0.3	39.6	54.5	25	»
5.0	0.2	45.6	49.2	25	»
4.3	0.1	48.9	46.7	25	»
3.2	0.1	58.0	38.7	25	$Ca(H_2PO_4)_2 \cdot H_2O + Mg(H_2PO_4)_2 \cdot 2H_2O$
3.0	0.1	59.0	37.9	25	$Ca(H_2PO_4)_2 \cdot H_2O + Mg(H_2PO_4)_2 \cdot 2H_2O + Mg(H_2PO_4)_2$
1.0	0.2	63.0	35.8	25	$Ca(H_2PO_4)_2 \cdot H_2O + Mg(H_2PO_4)_2$

Comp. of saturated soln., wt%				t	Comp. of solid phase
Magnesium Oxide	Calcium Oxide	Phosphorus Pentoxide	Water		
0.0	5.8	24.2	70.0	25	$Ca(H_2PO_4)_2 \cdot H_2O+CaHPO_4$
2.0	4.8	24.5	68.7	25	»
4.2	3.2	26.5	66.1	25	»
6.8	1.6	30.3	61.3	25	»
7.7	0.9	32.3	59.1	25	$Ca(H_2PO_4)_2 \cdot H_2O+CaHPO_4+$ $+MgHPO_4 \cdot 3H_2O$
6.5	0.9	24.7	67.9	25	$CaHPO_4+MgHPO_4 \cdot 3H_2O$
7.7	0.9	32.3	59.1	25	$CaHPO_4+MgHPO_4 \cdot 3H_2O+$ $+Ca(H_2PO_4)_2 \cdot H_2O$
7.8	0.7	32.8	58.7	25	$MgHPO_4 \cdot 3H_2O+Ca(H_2PO_4)_2 \cdot H_2O+$ $+Mg(H_2PO_4)_2 \cdot 4H_2O$
6.8	0.6	35.1	57.5	25	$Ca(H_2PO_4)_2 \cdot H_2O+Mg(H_2PO_4)_2 \cdot 4H_2O$
6.8	0.6	34.6	58.0	25	»
5.6	0.3	39.6	54.5	25	»
5.0	0.2	45.6	49.2	25	»
4.3	0.1	48.9	46.7	25	»
4.1	0.04	52.4	43.46	25	$Ca(H_2PO_4)_2 \cdot H_2O+Mg(H_2PO_4)_2 \cdot 4H_2O+$ $+Mg(H_2PO_4)_2 \cdot 2H_2O$
0.1	0.2	1.0	98.7	50	$CaHPO_4+MgHPO_4 \cdot 3H_2O$
1.2	0.3	5.6	92.9	50	»
3.1	0.7	13.0	83.2	50	»
5.5	0.8	22.4	71.3	50	»
7.5	0.7	31.6	60.2	50	»
8.4	0.8	36.5	54.3	50	$CaHPO_4+MgHPO_4 \cdot 3H_2O+$ $+Ca(H_2PO_4)_2 \cdot H_2O$
8.6	0.6	37.3	53.5	50	$MgHPO_4 \cdot 3H_2O+Ca(H_2PO_4)_2 \cdot H_2O+$ $+Mg(H_2PO_4)_2 \cdot 4H_2O$
7.9	0.5	39.7	51.9	50	$Ca(H_2PO_4)_2 \cdot H_2O+Mg(H_2PO_4)_2 \cdot 4H_2O$
7.0	0.30	43.3	49.4	50	$Ca(H_2PO_4)_2 \cdot H_2O+Mg(H_2PO_4)_2 \cdot 4H_2O+$ $+Mg(H_2PO_4)_2 \cdot 2H_2O$
5.0	0.20	53.0	41.8	50	$Ca(H_2PO_4)_2 \cdot H_2O+Mg(H_2PO_4)_2 \cdot 2H_2O$
4.1	0.14	55.5	40.26	50	»
3.6	0.2	56.8	39.4	50	$Ca(H_2PO_4)_2 \cdot H_2O+$ $+Mg(H_2PO_4)_2 \cdot 2H_2O+Mg(H_2PO_4)_2$
2.4	0.3	60.8	36.5	50	$Ca(H_2PO_4)_2 \cdot H_2O+Mg(H_2PO_4)_2$
1.4	0.3	62.4	35.9	50	»
0.0	5.7	29.6	64.7	50	$Ca(H_2PO_4)_2 \cdot H_2O+CaHPO_4$
1.7	4.4	31.1	62.8	50	»
3.4	3.1	32.6	60.9	50	»
5.7	1.9	34.4	58.0	50	»
8.4	0.8	36.5	54.3	50	$Ca(H_2PO_4)_2 \cdot H_2O+CaHPO_4+$ $+MgHPO_4 \cdot 3H_2O$
3.0	0.2	11.5	85.3	58	$CaHPO_4+MgHPO_4 \cdot 3H_2O$
5.0	0.3	20.0	74.7	58	»
7.0	0.5	28.3	64.2	58	»
8.9	0.8	38.7	51.6	58	$CaHPO_4+MgHPO_4 \cdot 3H_2O+$ $+Ca(H_2PO_4)_2 \cdot H_2O$
8.9	0.6	38.5	52.0	58	$Ca(H_2PO_4)_2 \cdot H_2O+MgHPO_4 \cdot 3H_2O+$ $+Mg(H_2PO_4)_2 \cdot 4H_2O+$ $+Mg(H_2PO_4)_2 \cdot 2H_2O$

Table № 3967 (continuation)

Comp. of saturated soln., wt%					Comp. of solid phase
Magnesium Oxide	Calcium Oxide	Phosphorus Pentoxide	Water		
3.9	0.2	56.5	39.4	58	$Ca(H_2PO_4)_2 \cdot H_2O+$ $+Mg(H_2PO_4)_2 \cdot 2H_2O+Mg(H_2PO_4)_2$
0.0	5.7	32.0	62.3	58	$CaHPO_4+Ca(H_2PO_4)_2 \cdot H_2O$
2.0	4.5	32.5	61.0	58	»
5.0	2.4	35.3	57.3	58	»
7.0	1.5	37.0	54.5	58	»
8.9	0.8	38.7	51.6	58	$CaHPO_4+MgHPO_4 \cdot 3H_2O+$ $+Ca(H_2PO_4)_2 \cdot H_2O$
3.0	0.2	12.0	84.8	130	$CaHPO_4+MgHPO_4 \cdot 3H_2O$
5.0	0.2	20.5	74.3	130	»
7.0	0.3	28.5	64.1	130	»
9.6	0.4	39.5	50.5	130	»
12.2	0.7	52.0	35.1	130	$CaHPO_4+MgHPO_4 \cdot 3H_2O+$ $+Ca(H_2PO_4)_2 \cdot H_2O$
12.7	0.0	54.0	33.3	130	$MgHPO_4 \cdot 3H_2O+$ $+Mg(H_2PO_4)_2 \cdot 2H_2O+Mg(H_2PO_4)_2$
12.3	0.6	52.0	35.1	130	$Ca(H_2PO_4)_2 \cdot H_2O+MgHPO_4 \cdot 3H_2O+$ $+Mg(H_2PO_4)_2 \cdot 2H_2O+Mg(H_2PO_4)_2$
0.0	5.7	48.2	46.1	130	$CaHPO_4+Ca(H_2PO_4)_2 \cdot H_2O$
2.0	3.0	50.5	44.5	130	»
5.0	1.5	51.8	41.7	130	»
7.0	1.0	52.3	39.7	130	»
12.2	0.7	52.0	35.1	130	$CaHPO_4+Ca(H_2PO_4)_2 \cdot H_2O+$ $+MgHPO_4 \cdot 3H_2O$

№ 3968 MAGNESIUM OXIDE–ARSENIC PENTOXIDE–WATER [2157]

$MgO—As_2O_5—H_2O$

Content in aqueous soln., wt%		d	t	Comp. of solid phase
Arsenic Pentoxide	Magnesium Oxide			
70.5	0.38	2.18	20	$MgO \cdot 2As_2O_5 \cdot 5H_2O$
67.0	1.37	2.2	20	»
64.66	2.25	2.06	20	»
63.3	6.55	—	20	»
61.46	5.17	1.99	20	»
60.5	7.97	—	20	$MgO \cdot As_2O_5 \cdot 2H_2O \cdot 2H_2O$
60.0	6.05	—	20	»
59.3	5.58	—	20	»
58.8	5.66	—	20	»
56.77	6.24	—	20	»
55.35	6.25	—	20	»
56.09	7.18	—	20	»
55.8	6.2	1.87	20	»

Table № 3968 (continuation)

Content in aqueous soln., wt%		d	t	Comp. of solid phase
Arsenic Pent-oxide	Mag-nesium Oxide			
55.7	6.03	—	20	$MgO \cdot As_2O_5 \cdot 2H_2O \cdot 2H_2O$
55.6	6.26	—	20	»
55.65	7.03	—	20	»
55.2	7.59	—	20	»
53.8	6.0	—	20	»
53.5	7.52	—	20	»
52.3	6.64	—	20	»
52.15	6.28	1.83	20	»
51.72	6.7	—	20	»
49.1	6.47	1.76	20	»
49	6.4	—	20	»
47.6	7.07	—	20	»
47.6	6.0	—	20	»
47	7.42	—	20	»
41.77	6.76	—	20	»
38.3	6.66	1.58	20	»
36.22	6.08	—	20	»
33.44	6.23	1.5	20	»
29.9	5.12	—	20	»
28.4	5.0	—	20	»
29.72	5.38	1.42	20	$2MgO \cdot As_2O_5 \cdot H_2O \cdot 14H_2O$
22.14	3.97	1.27	20	»
17.94	3.14	1.22	20	»
17.6	3.07	1.2	20	»
15.8	2.9	—	20	»
13.52	2.45	1.15	20	»
11.9	2.28	1.14	20	»
10.9	2.01	1.1	20	»
7.05	1.29	1.07	20	»
6.65	1.28	1.06	20	»
1.8	0.36	1.01	20	»
1.67	0.34	1.004	20	»
1.36	0.26	—	20	»
0.94	0.19	1.01	20	»
0.87	0.18	—	20	»
1.6	0.38	1.01	20	Mesophase
1.21	0.238	—	20	»
1.116	0.224	—	20	»
0.905	0.181	—	20	»
0.672	0.142	—	20	»
0.535	0.121	—	20	$3MgO \cdot As_2O_5 \cdot 10H_2O$
0.444	0.105	—	20	»
0.3	0.07	—	20	»
0.266	0.067	1	20	»
0.223	0.06	—	20	»
0.108	0.03	—	20	»
0.103	0.032	1	20	»
70.5	0.71	2.13	60	$MgO \cdot 2As_2O_5 \cdot 5H_2O$
69.71	1.5	2.19	60	»
66.44	3.8	2.14	60	»

Table № 3968 (continuation)

Content in aqueous soln., wt%		d	t	Comp. of solid phase
Arsenic Pent-oxide	Mag-nesium Oxide			
64.89	3.87	2.12	60	$MgO \cdot 2As_2O_5 \cdot 5H_2O$
63.80	4.37	2.11	60	»
62.5	5.2	—	60	»
61.8	5.2	—	60	»
62.2	5.4	—	60	$2MgO \cdot As_2O_5 \cdot H_2O \cdot 2H_2O$
59	5.1	—	60	»
58.34	5.05	1.92	60	»
54.64	5.34	1.89	60	»
52.28	5.68	1.87	60	»
48.46	5.6	1.75	60	»
47.27	5.79	1.72	60	»
44	5.67	—	60	»
41.5	5.77	—	60	»
40	5.27	1.59	60	»
36.73	5.25	1.47	60	»
34.42	5.4	1.48	60	»
33.35	4.64	—	60	»
28.6	4.11	—	60	»
26.31	4.03	1.33	60	»
24.95	4.17	—	60	»
23.1	3.88	1.3	60	»
20.5	3.46	—	60	»
18.85	3.12	1.22	60	»
17.9	2.91	1.2	60	»
16.62	2.4	1.17	60	»
14.95	2.53	1.17	60	»
14.5	2.61	1.15	60	»
10.03	1.86	1.11	60	$2MgO \cdot As_2O_5 \cdot H_2O \cdot 2H_2O + 2MgO \cdot As_2O_5 \cdot H_2O \cdot 5H_2O$
8.35	1.45	—	60	»
7.13	1.24	1.07	60	»
6.72	1.24	1.07	60	»
2.19	0.394	—	60	$2MgO \cdot As_2O_5 \cdot H_2O \cdot 5H_2O$
2.1	0.36	—	60	»
1.84	0.35	—	60	»
1.64	0.299	—	60	»
1.48	0.267	—	60	»
1.2	0.214	—	60	Mesophase
0.885	0.16	—	60	»
0.81	0.155	—	60	»
0.644	0.122	—	60	»
0.552	0.108	—	60	»
0.36	0.068	—	60	»
0.3	0.06	—	60	»
0.258	0.052	—	60	»
0.232	0.054	—	60	»
0.181	0.035	—	60	$3MgO \cdot As_2O_5 \cdot 8H_2O$
0.126	0.038	—	60	»
0.12	0.026	—	60	»
0.0632	0.010	—	60	»

Content in aqueous soln., wt%		d	t	Comp. of solid phase
Arsenic Pentoxide	Magnesium Oxide			
0.046	0.011	—	60	$3MgO \cdot As_2O_5 \cdot 8H_2O$
0.011	0.005	—	60	»
0.008	0.004	—	60	»
0.0057	0.003	—	60	»

№ 3969 MAGNESIUM HYDROXIDE–MAGNESIUM SULFATE–WATER [1469]

$Mg(OH)_2—MgSO_4—H_2O$

$t = 20$

Content in aqueous soln., mol/1000 mol Water		pH	Content in aqueous soln., mol/1000 mol Water		pH
Magnesium Sulfate	Magnesium Hydroxide		Magnesium Sulfate	Magnesium Hydroxide	
50.50	0.0682	8.61	12.66	0.01014	9.26
36.88	—	8.85	0	0.0036	11.1
25.63	—	9.03			

Note. Solid phase $Mg(OH)_2$.

№ 3970 [2234]

MAGNESIUM HYDROXIDE–MAGNESIUM SULFATE–MAGNESIUM CHLORIDE–POTASSIUM SULFATE–POTASSIUM CHLORIDE–WATER

$Mg(OH)_2—MgSO_4—MgCl_2—K_2SO_4—KCl—H_2O$

$t = 50$

Content in saturated soln., mol/1000 mol Water					Comp. of solid phase
K_2^{2+}	Mg^{2+}	Cl_2^{2-}	SO_4^{2-}	$(OH)_2^{2-}$	
20.56	14.84	12.50	22.82	0.0815	$K_2SO_4 + K_2SO_4 \cdot 2MgSO_4 \cdot Mg(OH)_2 \cdot 2H_2O$
33.04	19.44	43.69	8.764	0.0287	»
38.96	23.18	55.04	7.068	0.0364	$KCl + K_2SO_4 + K_2SO_4 \cdot 2MgSO_4 \cdot Mg(OH)_2 \cdot 2H_2O$
14.54	58.20	67.96	4.739	0.0395	$KCl + K_2SO_4 \cdot 2MgSO_4 \cdot Mg(OH)_2 \cdot 2H_2O$
26.08	37.61	57.92	5.738	0.0346	»
7.591	80.82	85.36	3.371	0.0625	$KCl + KCl \cdot MgCl_2 \cdot 6H_2O + K_2SO_4 \cdot 2MgSO_4 \cdot Mg(OH)_2 \cdot 2H_2O$
4.712	84.58	86.36	2.720	0.2051	$KCl \cdot MgCl_2 \cdot 6H_2O + K_2SO_4 \cdot 2MgSO_4 \cdot Mg(OH)_2 \cdot 2H_2O$
2.392	92.78	90.99	4.132	0.0699	$MgSO_4 \cdot H_2O + KCl \cdot MgCl_2 \cdot 6H_2O + K_2SO_4 \cdot 2MgSO_4 \cdot Mg(OH)_2 \cdot 2H_2O$

Content in saturated soln., mol/1000 mol Water					Comp. of solid phase
K_2^{2+}	Mg^{2+}	Cl_2^{2-}	SO_4^{2-}	$(OH)_2^{2-}$	
2.012	85.86	80.70	6.640	0.1600	$MgSO_4 \cdot H_2O + K_2SO_4 \cdot 2MgSO_4 \cdot Mg(OH)_2 \cdot 2H_2O + MgCl_2 \cdot 3Mg(OH)_2 \cdot 8H_2O$
1.744	74.43	59.66	16.03	0.1695	»
2.718	78.39	51.14	29.89	0.0101	$MgSO_4 \cdot 6H_2O + K_2SO_4 \cdot 2MgSO_4 \cdot Mg(OH)_2 \cdot 2H_2O + MgCl_2 \cdot 3Mg(OH)_2 \cdot 8H_2O$
4.096	74.20	38.81	39.39	0.0101	$MgSO_4 \cdot 6H_2O + K_2SO_4 \cdot 2MgSO_4 \cdot Mg(OH)_2 \cdot 2H_2O + MgSO_4 \cdot 5Mg(OH)_2 \cdot 3H_2O$
3.837	71.39	18.49	56.68	0.0630	»
18.70	32.57	6.351	44.87	0.0286	$K_2SO_4 + K_2SO_4 \cdot MgSO_4 \cdot 4H_2O + K_2SO_4 \cdot 2MgSO_4 \cdot Mg(OH)_2 \cdot 2H_2O$
22.58	29.40	27.65	24.35	0.0101	»
32.52	33.39	53.80	11.71	0.0122	$KCl + K_2SO_4 + K_2SO_4 \cdot MgSO_4 \cdot 4H_2O + K_2SO_4 \cdot 2MgSO_4 \cdot Mg(OH)_2 \cdot 2H_2O$
22.04	48.41	57.90	12.54	0.0062	$KCl + K_2SO_4 \cdot MgSO_4 \cdot 4H_2O + K_2SO_4 \cdot 2MgSO_4 \cdot Mg(OH)_2 \cdot 2H_2O$
17.00	60.25	63.01	14.02	0.0	$KCl + K_2SO_4 \cdot MgSO_4 \cdot 4H_2O + KCl \cdot MgSO_4 \cdot 4H_2O$
15.86	61.35	64.26	12.85	0.0048	$KCl + KCl \cdot MgSO_4 \cdot 4H_2O + K_2SO_4 \cdot 2MgSO_4 \cdot Mg(OH)_2 \cdot 2H_2O$
10.51	72.08	76.78	5.554	0.0127	»
7.014	82.45	85.65	3.506	0.0307	$KCl + KCl \cdot MgCl_2 \cdot 6H_2O + KCl \cdot MgSO_4 \cdot 4H_2O + K_2SO_4 \cdot 2MgSO_4 \cdot Mg(OH)_2 \cdot 2H_2O$
3.671	90.05	88.23	5.303	0.0	$MgSO_4 \cdot H_2O + KCl \cdot MgCl_2 \cdot 6H_2O + KCl \cdot MgSO_4 \cdot 4H_2O$
3.241	90.30	76.97	4.988	0.0	$MgSO_4 \cdot H_2O + KCl \cdot MgCl_2 \cdot 6H_2O$
4.997	79.69	60.82	7.718	0.0	$MgSO_4 \cdot H_2O + KCl \cdot MgSO_4 \cdot 4H_2O$
6.417	73.68	48.25	18.98	0.0095	»
6.636	76.92	31.72	35.14	0.0048	$MgSO_4 \cdot H_2O + MgSO_4 \cdot 6H_2O + KCl \cdot MgSO_4 \cdot 4H_2O + K_2SO_4 \cdot 2MgSO_4 \cdot Mg(OH)_2 \cdot 2H_2O$
6.776	71.39	27.68	46.30	0.0111	$MgSO_4 \cdot 6H_2O + K_2SO_4 \cdot MgSO_4 \cdot 4H_2O + K_2SO_4 \cdot 2MgSO_4 \cdot Mg(OH)_2 \cdot 2H_2O$
6.466	73.00	9.874	51.68	0.0074	»
6.226	73.35	14.14	65.70	0.0090	»
6.533	74.42	9.874	71.01	0.0122	»
11.79	63.97	55.41	20.25	0.00	$MgSO_4 \cdot K_2SO \cdot 4H_2O + KCl \cdot MgSO_4 \cdot 3H_2O$
8.933	69.47	51.24	27.20	0.00	»
6.892	75.42	45.23	37.30	0.00	$MgSO_4 \cdot 6H_2O + KCl \cdot MgSO_4 \cdot 3H_2O + K_2SO_4 \cdot MgSO_4 \cdot 4H_2O$
1.166	98.25	97.67	1.803	0.0	$MgSO_4 \cdot H_2O + MgCl_2 \cdot KCl \cdot 6H_2O$
0.458	111.28	110.74	0.996	0.0	$MgSO_4 \cdot H_2O + KCl \cdot MgCl_2 \cdot 6H_2O + MgCl_2 \cdot 6H_2O$

№ 3971 [2160]

MAGNESIUM HYDROXIDE–MAGNESIUM CHLORIDE–WATER

$Mg(OH)_2—MgCl_2—H_2O$

$t = 50$

Content of Magnesium Chloride in aqueous soln., wt%	Comp. of solid phase
37.00	$MgCl_2 \cdot 6H_2O + MgCl_2 \cdot 3MgO \cdot 12H_2O$
14.01	$MgCl_2 \cdot 3MgO \cdot 12H_2O + MgO \cdot H_2O$
14.01—20.95	$MgO \cdot H_2O$ (metastable)
12.45—14.01	$MgCl_2 \cdot 3MgO \cdot 12H_2O$ (metastable)

№ 3972 [1792]

MAGNESIUM HYDROXIDE–MAGNESIUM CHLORIDE–WATER

$Mg(OH)_2—MgCl_2—H_2O$

$t = 100$

Content in aqueous soln., mol/l	
Magnesium Chloride	Magnesium Hydroxide
6.5	0.006
5	0.014
3.5	0.010

№ 3973 [3973]

MAGNESIUM HYDROXIDE–MAGNESIUM CHLORIDE–WATER

$Mg(OH)_2—MgCl_2—H_2O$

$t = 20$

Content in aqueous soln., mol/1000 mol Water		pH	Comp. of solid phase
Magnesium Hydroxide	Magnesium Chloride		
81.34	0.2292	6.85	$MgCl_2(OH)_3 \cdot 4H_2O$
63.12	0.0856	7.34	»
37.80	0.0424	8.14	»
18.46	0.0153	8.71	$Mg(OH)_2$
1.81	0.0049	9.34	»
0	0.0036	11.1	»

№ 3974 [656]

MAGNESIUM HYDROXIDE–POTASSIUM CHLORIDE–WATER

$Mg(OH)_2—KCl—H_2O$

$t = 25$

Content in aqueous soln., wt%		Comp. of solid phase	Content in aqueous soln., wt%		Comp. of solid phase
Potassium Chloride	Magnesium Hydroxide		Potassium Chloride	Magnesium Hydroxide	
25.961	0.040	$KCl + Mg(OH)_2$	18.440	0.050	$Mg(OH)_2$
22.838	0.042	$Mg(OH)_2$	17.210	0.045	»
21.817	0.043	»	12.770	0.050	»
20.208	0.052	»	11.880	0.051	»

Table № 3974 (continuation)

Content in aqueous soln., wt%		Comp. of solid phase	Content in aqueous soln., wt%		Comp. of solid phase
Potassium Chloride	Magnesium Hydroxide		Potassium Chloride	Magnesium Hydroxide	
10.010	0.047	$Mg(OH)_2$	2.593	0.046	$Mg(OH)_2$
9.340	0.044	»	1.536	0.041	»
8.510	0.050	»	0.713	0.035	»
7.190	0.030	»	0	0.039	»
5.588	0.052	»			

№ 3975 [2733]

MAGNESIUM HYDROXIDE–POTASSIUM CHLORIDE–WATER

$Mg(OH)_2—KCl—H_2O$

$t = 25$

Content in aqueous soln., mol/l		Comp. of solid phase
Potassium Chloride	Magnesium Hydroxide	
2	0.00050	$Mg(OH)_2$
0.1	0.00025	»
0.01	0.00038	»

№ 3976 [1729]

MAGNESIUM HYDROGEN CARBONATE–MAGNESIUM SULFATE–POTASSIUM HYDROGEN CARBONATE–POTASSIUM SULFATE–WATER

$Mg(HCO_3)_2—MgSO_4—KHCO_3—K_2SO_4—H_2O$

$t = 25$

Content in aqueous soln., mol/1000 mol Water				Comp. of solid phase	P_{CO_2}, atm
K_2SO_4	$2KHCO_3$	$MgSO_4$	$Mg(HCO_3)_2$		
13.12	1.79	0.0	3.93	$K_2SO_4 + KHCO_3 \cdot MgCO_3 \cdot 4H_2O$	1
13.74	0.84	24.66	3.52	$K_2SO_4 + K_2SO_4 \cdot MgSO_4 \cdot 6H_2O + KHCO_3 \cdot MgCO_3 \cdot 4H_2O$	1
11.45	1.00	29.38	3.55	$K_2SO_4 \cdot MgSO_4 \cdot 6H_2O + KHCO_3 \cdot MgCO_3 \cdot 4H_2O + MgCO_3 \cdot 3H_2O$	1
8.93	1.30	38.07	3.64	$K_2SO_4 \cdot MgSO_4 \cdot 6H_2O + MgCO_3 \cdot 3H_2O$	1
8.23	1.10	40.17	3.66	»	1
6.08	0.0	56.27	3.11	$K_2SO_4 \cdot MgSO_4 \cdot 6H_2O + MgSO_4 \cdot 7H_2O + MgCO_3 \cdot 3H_2O$	1
8.55	2.03	16.08	3.26	$KHCO_3 \cdot MgCO_3 \cdot 4H_2O + MgCO_3 \cdot 3H_2O$	1
1.17	2.83	2.27	2.84	»	1
0.0	2.86	1.00	2.79	»	1

№ 3977 [1025]

MAGNESIUM HYDROGEN CARBONATE–MAGNESIUM SULFATE–CALCIUM HYDROGEN CARBONATE–CALCIUM SULFATE–WATER

$Mg(HCO_3)_2—MgSO_4—Ca(HCO_3)_2—CaSO_4—H_2O$

$t = 25$

Content in aqueous soln., (mol/1000g soln). 10^3				Comp. of solid phase
Mg^{2+}	Ca^{2+}	SO_4^{2+}	$2HCO_3^-$	
17.31	21.08	22.44	15.95	$MgCO_3 + MgCO_3 \cdot CaCO_3 + CaSO_4 \cdot 2H_2O$
7.54	21.58	17.02	12.10	$MgCO_3 \cdot CaCO_3 + CaSO_4 \cdot 2H_2O$
1.07	21.10	13.14	9.03	$MgCO_3 \cdot CaCO_3 + Ca_4SO_4 \cdot 2H_2O + CaCO_3$

№ 3978 [1729]

MAGNESIUM HYDROGEN CARBONATE–MAGNESIUM CHLORIDE–POTASSIUM HYDROGEN CARBONATE–POTASSIUM CHLORIDE–WATER

$Mg(HCO_3)_2—MgCl_2—KHCO_3—KCl—H_2O$

$t = 25$

Content in aqueous soln., wt%				Comp. of solid phase	P_{CO_2}, atm
Potassium Chloride	Potassium Hydrogen Carbonate	Magnesium Chloride	Magnesium Hydrogen Carbonate		
15.17	0.0	8.80	1.14	$KHCO_3 \cdot MgCO_3 \cdot 4H_2O + MgCO_3 \cdot 3H_2O$	1
9.38	0.0	5.31	1.77	»	1
3.43	1.30	1.23	2.14	»	1
1.88	2.06	0.58	2.14	»	1
1.07	2.58	0.27	2.14	»	1
0.0	3.04	0.0	2.40	»	1
1.89	1.34	30.09	0.0	$KCl \cdot MgCl_2 \cdot 6H_2O + MgCl_2 \cdot MgCO_3 \cdot 6H_2O + MgCO_3 \cdot 3H_2O$	1

№ 3979 [1029]

MAGNESIUM HYDROGEN CARBONATE–CALCIUM HYDROGEN CARBONATE–WATER

$Mg(HCO_3)_2—Ca(HCO_3)_2—H_2O$

Concn. in aqueous soln., mol/l		t
Calcium Hydrogen Carbonate	Magnesium Hydrogen Carbonate	
5.34	5.40	0
3.21	3.28	25
3.02	3.06	55

№ 3980 [1023]

MAGNESIUM CARBONATE–MAGNESIUM SULFATE–CALCIUM CARBONATE–CALCIUM SULFATE–WATER

$MgCO_3$—$MgSO_4$—$CaCO_3$—$CaSO_4$—H_2O

$t = 25$

Concn. in aqueous soln., wt%				Comp. of solid phase
Ca	Mg	$(HCO_3)_2$	SO_4	
0.0129	0.0079	0.0792	0.0000	$CaCO_3 \cdot MgCO_3$
0.0100	0.0274	0.1679	0.0000	$CaCO_3 \cdot MgCO_3 + MgCO_3$
0.0325	0.0029	0.1132	0.0000	»
0.0846	0.0026	0.1102	0.0126	$CaCO_3 \cdot MgCO_3 \cdot CaCO_3 + CaSO_4 \cdot 2H_2O$
0.0865	0.0183	0.1476	0.0163	$CaCO_3 \cdot MgCO_3 + CaSO_4 \cdot 2H_2O$
0.0845	0.0421	0.1946	0.0215	$CaCO_3 \cdot MgCO_3 + CaSO_4 \cdot 2H_2O + MgCO_3$
0.0241	0.3720	0.1360	0.1420	$MgCO_3 + CaSO_4 \cdot 2H_2O$

№ 3981 [1027]

MAGNESIUM CARBONATE–MAGNESIUM SULFATE–CALCIUM CARBONATE–CALCIUM SULFATE–WATER

$MgCO_3$—$MgSO_4$—$CaCO_3$—$CaSO_4$—H_2O

Content in saturated soln., mmol/1000g soln.				t	Comp. of solid phase	P_{CO_2}, atm
Ca^{2+}	Mg^{2+}	$(HCO_3)_2^-$	SO_4^{2-}			
15.08	0.00	15.08	0.00	0	$CaCO_3$	1
5.34	5.40	10.74	0.00	0	$CaCO_3 \cdot MgCO_3$	1
13.64	1.32	14.96	0.00	0	$CaCO \cdot MgCO + CaCO_3$	1
0.00	22.52	22.52	0.00	0	$MgCO_3$	1
3.40	12.27	15.67	0.00	0	$CaCO_3 \cdot MgCO_3 + MgCO_3$	1
24.20	0.00	13.40	10.80	0	$CaCO_3 + CaSO_4 \cdot 2H_2O$	1
25.88	1.63	15.29	12.22	0	$CaCO_3 \cdot MgCO_3 + CaCO_3 + CaSO_4 \cdot 2H_2O$	1
20.65	7.17	14.60	13.22	0	$CaCO_3 \cdot MgCO_3 + CaSO_4 \cdot 2H_2O$	1
18.96	10.48	15.26	14.18	0	»	1
20.08	11.47	15.66	15.89	0	»	1
19.70	12.64	16.08	16.26	0	$CaCO_3 \cdot MgCO_3 + CaSO_4 \cdot 2H_2O + MgCO_3$	1
6.09	0.00	6.09	0.00	55	$CaCO_3$	1
3.02	3.06	6.08	0.00	55	$CaCO_3 \cdot MgCO_3$	1
4.23	2.17	6.40	0.00	55	$CaCO_3 \cdot MgCO_3 + CaCO_3$	1
0.00	15.59	15.59	0.00	55	$MgCO_3$	1
2.15	7.39	9.54	0.00	55	$MgCO_3 + CaCO_3 \cdot MgCO_3$	1
20.55	0.00	5.23	15.32	55	$CaCO_3 + CaSO_4 \cdot 2H_2O$	1
19.94	2.28	5.74	16.48	55	$CaCO_3 \cdot MgCO_3 + CaSO_4 \cdot 2H_2O + CaCO_3$	1
19.80	2.45	5.65	16.60	55	»	1
19.34	3.11	5.89	16.56	55	$CaCO_3 \cdot MgCO_3 + CaSO_4 \cdot 2H_2O$	1
17.72	13.00	8.18	22.54	55	$CaCO_3 \cdot MgCO_3 + CaSO_4 \cdot 2H_2O + MgCO_3$	1
14.60	17.62	8.50	23.72	55	$MgCO_3 + CaSO_4 \cdot 2H_2O$	1
13.75	50.95	10.58	54.12	55	»	1
13.49	79.24	12.78	79.95	55	»	1
12.56	149.91	15.55	146.93	55	»	1
12.08	188.15	15.35	184.88	55	»	1

№ 3982 [1032]

MAGNESIUM CARBONATE–MAGNESIUM SULFATE–CALCIUM CARBONATE–CALCIUM SULFATE–WATER

$MgCO_3—MgSO_4—CaCO_3—CaSO_4—H_2O$

$t = 25$

Content in saturated soln., mmol/1000g soln.				Comp. of solid phase
Ca^{2+}	Mg^{2+}	$(HCO_3^-)_2$	SO_4^{2-}	
8.10	1.18	9.28	0.00	$CaCO_3 \cdot MgCO_3 + CaCO_3$
0.39	0.94	1.36	0.00	»
21.10	1.07	9.03	13.14	$CaSO_4 \cdot 2H_2O + CaCO_3 \cdot MgCO_3 + CaCO_3$
15.33	2.58	0.39	17.52	»
2.50	11.26	13.76	0.00	$CaCO_3 \cdot MgCO_3 + MgCO_3$
0.32	1.20	1.52	0.00	»
21.08	17.31	15.95	22.44	$CaCO_3 \cdot MgCO_3 + MgCO_3 + CaSO_4 \cdot 2H_2O$
15.47	3.53	0.47	18.53	»
20.47	0.00	7.52	12.95	$CaSO_4 \cdot 2H_2O + CaCO_3$
15.97	0.00	0.38	15.59	»

№ 3983 [1736]

MAGNESIUM CARBONATE–MAGNESIUM CHLORIDE–WATER

$MgCO_3—MgCl_2—H_2O$

$t = 0$

Content in aqueous soln., g/100g Water	
$MgCl_2 \cdot 6H_2O$	$MgCO_3$
2.55	0.0353
1.05	0.0395
0.51	0.0470
0	0.0944

№ 3984 [1116]

MAGNESIUM CARBONATE–POTASSIUM HYDROGEN CARBONATE–WATER

$MgCO_3 \cdot 3H_2O—KHCO_3—H_2O$

Concn. in aqueous soln., g/l		t	Comp. of solid phase
Magnesium Carbonate	Potassium Hydrogen Carbonate		
1.31	0.00	15	$MgCO_3 \cdot 3H_2O$
1.81	9.93	15	»
2.31	19.45	15	»
2.92	39.96	15	»

Table № 3984 (continuation)

Concn. in aqueous soln., g/l		t	Comp. of solid phase
Magnesium Carbonate	Potassium Hydrogen Carbonate		
2.66	26.84	15	$MgCO_3 \cdot 3H_2O + MgCO_3 \cdot KHCO_3 \cdot 4H_2O$
1.34	52.48	15	$MgCO_3 \cdot KHCO_3 \cdot 4H_2O$
1.02	67.99	15	»
0.39	98.20	15	»
1.20	0.00	25	$MgCO_3 \cdot 3H_2O$
1.59	9.86	25	»
2.06	22.12	25	»
2.50	34.37	25	»
3.00	49.90	25	»
2.71	39.10	25	$MgCO_3 \cdot 3H_2O + MgCO_3 \cdot KHCO_3 \cdot 4H_2O$
1.77	58.99	25	$MgCO_3 \cdot KHCO_3 \cdot 4H_2O$
1.62	64.12	25	»
0.84	112.6	25	»
0.98	0.00	35	$MgCO_3 \cdot 3H_2O$
1.36	10.93	35	»
1.96	28.14	35	»
2.45	48.52	35	»
2.74	58.13	35	»
2.54	50.93	35	$MgCO_3 \cdot 3H_2O + MgCO_3 \cdot KHCO_3 \cdot 4H_2O$
2.12	62.37	35	$MgCO_3 \cdot KHCO_3 \cdot 4H_2O$
1.65	85.44	35	»

№ 3985 [1133]

MAGNESIUM CARBONATE–CALCIUM CARBONATE–WATER

$MgCO_3$—$CaCO_3$—H_2O

$t = 17$

Content in aqueous soln., g/l		Comp. of solid phase
Magnesium Carbonate	Calcium Carbonate	
0.1505	0	$MgCO_3$ (Magnesite)
0.0948	0.0953	$MgCO_3 + MgCO_3 \cdot CaCO_3$ (Dolomite)
0.0093	0.6339	$MgCO_3 \cdot CaCO_3 + CaCO_3$
0	1.00	$CaCO_3$ (Calcite)

№ 3986 MAGNESIUM CARBONATE–CALCIUM CARBONATE–WATER [1026]

$MgCO_3$—$CaCO_3$—H_2O

Concn. in aqueous soln., wt%		t	Comp. of solid phase	P_{CO_2}, atm
Ca	Mg			
0.0604	0.0000	0	$CaCO_3$	1
0.0547	0.0032	0	$CaCO_3 + CaCO_3 \cdot MgCO_3$	1
0.0214	0.0131	0	$CaCO_3 \cdot MgCO_3$	1

Concn. in aqueous soln., wt%		t	Comp. of solid phase	P_{CO_2}, atm
Ca	Mg			
0.0136	0.0298	0	$CaCO_3 \cdot MgCO_3 + MgCO_3$	1
0.0000	0.0547	0	$MgCO_3$	1
0.0360	0.0000	25	$CaCO_3$	1
0.0325	0.0029	25	$CaCO_3 + CaCO_3 \cdot MgCO_3$	1
0.0129	0.0080	25	$CaCO_3 \cdot MgCO_3$	1
0.0100	0.0274	25	$CaCO_3 \cdot MgCO_3 + MgCO_3$	1
0.0000	0.0401	25	$MgCO_3$	1
0.0032	0.0000	25	$CaCO_3$	0.0012
0.0016	0.0023	25	$CaCO_3 + CaCO_3 \cdot MgCO_3$	0.0012
0.0016	0.0025	25	$CaCO_3 \cdot MgCO_3$	0.0012
0.0013	0.0029	25	$CaCO_3 \cdot MgCO_3 + MgCO_3$	0.0012
0.0000	0.0040	25	$MgCO_3$	0.0012
0.0244	0.0000	55	$CaCO_3$	1
0.0170	0.0053	55	$CaCO_3 + CaCO_3 \cdot MgCO_3$	1
0.0121	0.0074	55	$CaCO_3 \cdot MgCO_3$	1
0.0086	0.0180	55	$CaCO_3 \cdot MgCO_3 + MgCO_3$	1
0.0000	0.0379	55	$MgCO_3$	1
0.0138	0.0000	70	$CaCO_3$	1
0.0081	0.0062	70	$CaCO_3 + CaCO_3 \cdot MgCO_3$	1
0.0062	0.0143	70	$CaCO_3 \cdot MgCO_3 + MgCO_3$	1

Table № 3986 (continuation)

Solubility of $CaCO_3 \cdot MgCO_3$, wt%	t	P_{CO_2}, atm
0.1981	0	1
0.0262	25	0.0012
0.1197	25	1
0.1121	55	1
0.0845	75	1

№ 3987 MAGNESIUM CARBONATE–CALCIUM CARBONATE–WATER [1037]

$MgCO_3—CaCO_3—H_2O$

Concn., mmol/1000g soln.,		d	t	Comp. of solid phase	P_{CO_2}, atm
Calcium salts in conversion to $Ca(HCO_3)_2$	Magnesium salts in conversion to $Mg(HCO_3)_2$				
15.40	0.00	1.0021	0	$CaCO_3$	1
10.24	11.90	1.0033	0	»	1
8.01	37.10	1.0063	0	»	1
7.69	42.70	1.0067	0	»	1
7.36	84.40	1.0123	0	»	1
6.50	122.60	1.0159	0	»	1
5.92	182.70	1.0232	0	»	1

Table № 3987 (continuation)

Concn., mmol/1000g soln.		d	t	Comp. of solid phase	p_{CO_2}, atm
Calcium salts in conversion to $Ca(HCO_3)_2$	Magnesium salts in conversion to $Mg(HCO_3)_2$				
4.91	253.10	1.0319	0	$CaCO_3$	1
4.20	298.90	1.0366	0	»	1
3.58	332.40	1.0399	0	»	1
3.14	365.30	1.0434	0	»	1
3.20	376.00	1.0452	0	»	1
2.78	394.20	1.0473	0	$CaCO_3+MgCO_3 \cdot 3H_2O$	1
2.76	396.00	1.0474	0	»	1
11.08	9.30	0.000	0	$CaCO_3$	1
9.05	17.40	1.0035	0	»	1
3.23	77.50	1.0099	0	»	1
1.92	195.80	1.0243	0	»	1
0.55	330.40	1.0388	0	$CaCO_3+MgCO_3 \cdot 5H_2O$	1
10.11	0.00	1.0017	20	$CaCO_3$	1
6.66	16.70	1.0023	20	»	1
2.50	54.00	1.0122	20	»	1
1.40	104.80	1.0185	20	»	1
0.99	171.20	0.000	20	»	1
0.39	241.10	1.0276	20	$CaCO_3+MgCO_3 \cdot 3H_2O$	1
7.05	0.00	1.0010	40	$CaCO_3$	1
4.90	9.90	1.0022	40	»	1
2.95	29.70	1.0040	40	»	1
1.77	53.10	1.0069	40	»	1
0.43	103.40	1.0119	40	»	1
0.23	145.50	1.0150	40	$CaCO_3+MgCO_3 \cdot 3H_2O$	1
6.09	0.00	0.000	55	$CaCO_3$	1
0.91	23.10	1.0028	55	»	1
0.37	48.70	1.0055	55	»	1
0.09	102.20	1.0097	55	$CaCO_3+4MgCO_3 \cdot Mg(OH)_2 \cdot 4H_2O$	1
3.45	0.00	0.000	70	$CaCO_3$	1
1.24	10.10	0.000	70	»	1
0.15	23.60	1.0029	70	»	1
0.07	42.30	1.0049	70	$CaCO_3+4MgCO_3 \cdot Mg(OH)_2 \cdot 4H_2O$	1
1.82	0.00	0.9808	90	$CaCO_3$	1
0.05	15.90	1.0018	90	$CaCO_3+4MgCO_3 \cdot Mg(OH)_2 \cdot 4H_2O$	1

№ 3988 MAGNESIUM CARBONATE–CALCIUM CARBONATE–WATER [1036]

$MgCO_3—CaCO_3—H_2O$

Concn. in aqueous soln., mmol/l			t	d_4^t	Comp. of solid phase
$(HCO_3)^{2-}$	Mg^{2+}	Ca^{2+}			
330.95	330.40	0.55	0	1.0388	$MgCO_3 \cdot 5H_2O+CaCO_3$
241.49	241.10	0.39	20	1.0276	$MgCO_3 \cdot 3H_2O+CaCO_3$
145.73	145.50	0.23	40	1.0150	»
102.29	102.20	0.09	55	1.0097	$4MgCO_3 \cdot Mg(OH)_2 \cdot 4H_2O+CaCO_3$
42.38	42.31	0.07	70	1.0049	»
15.95	15.90	0.05	90	1.0018	»

№ 3989 MAGNESIUM CARBONATE–CALCIUM SULFATE–WATER [1035]

$MgCO_3—CaSO_4—H_2O$

$t = 70$

Content in aqueous soln., mol/1000g soln.				Comp. of solid phase	P_{CO_2}, atm
Ca	Mg	$(HCO_3)_2$	SO_4		
3.45	0.0	3.45	0.0	$CaCO_3$	1
2.03	2.55	4.58	0.0	$CaCO_3 + CaCO_3 \cdot MgCO_3$	1
2.05	2.44	4.49	0.0	»	1
1.55	5.89	7.44	0.0	$CaCO_3 \cdot MgCO_3 + MgCO_3$	1
17.11	0.0	3.26	13.85	$CaCO_3 + CaSO_4$	1
15.87	5.23	3.92	17.18	$CaCO_3 \cdot MgCO_3 + CaCO_3 + CaSO_4$	1
15.63	5.26	3.90	16.99	»	1
15.01	9.68	4.12	20.57	$CaCO_3 \cdot MgCO_3 + CaSO_4 + MgSO_4$	1
14.04	16.07	4.33	25.78	$CaSO_4 + MgCO_3$	1
13.79	30.88	4.52	40.13	»	1

№ 3990 MAGNESIUM CARBONATE (NESQUEHONITE)–SEAWATER [2719]

$MgCO_3 \cdot 3H_2O—\ —H_2O$

$t = 25$

Content of Magnesium Carbonate in aqueous soln., (mol Mg CO_3/l) .10^3	pH	Content of Magnesium Carbonate in aqueous soln., (mol Mg CO_3/l) .10^3	pH
217.7	7.03	79.6	7.74
218.8	7.05	71.5	7.81
142.9	7.36	61.8	7.90
137.8	7.35	32 6	8.25
133.5	7.36	28.7	8.35
111.9	7.46	28.5	8.39
107.6	7.52	21.6	8.46
92.6	7.62	18.5	8.52
		3.08	9.50

№ 3991 [1772]

POTASSIUM MAGNESIUM CARBONATE–POTASSIUM CHLORIDE–WATER

$K_2Mg(CO_3)_2—KCl—H_2O$

$t = 17$

Concn. in aqueous soln., mol/1000g Water		Comp. of solid phase
Magnesium Carbonate	Potassium Chloride	
0.0381	0.274	$K_2Mg(CO_3)_2 \cdot 4H_2O$
0.0352	0.790	»
0.0268	1.490	»
0.0191	2.365	»
0.0139	3.38	»

№ 3992 MAGNESIUM NITRATE–MAGNESIUM SULFATE–WATER [1201]

$Mg(NO_3)_2—MgSO_4—H_2O$

$t = 25$

Comp. of saturated soln., wt%			d	Comp. of solid phase
Magnesium Nitrate	Magnesium Sulfate	Water		
42.89	0.00	57.11	1.397	$Mg(NO_3)_2 \cdot 6H_2O$
41.10	0.99	57.91	1.396	»
40.65	1.77	57.58	1.399	»
40.05	2.28	57.67	1.401	»
39.79	2.85	57.36	1.403	$Mg(NO_3)_2 \cdot 6H_2O + MgSO_4 \cdot 7H_2O$
33.55	3.90	62.55	1.345	$MgSO_4 \cdot 7H_2O$
28.49	5.50	66.01	1.316	»
19.06	10.00	70.94	1.277	»
11.25	15.90	72.85	1.271	»
10.71	16.36	72.93	1.271	»
6.68	20.03	73.29	1.280	»
0.00	26.68	73.32	1.304	»

№ 3993 MAGNESIUM NITRATE–MAGNESIUM SULFATE–WATER [1203]

$Mg(NO_3)_2—MgSO_4—H_2O$

$t = 0$

Comp. of saturated soln., wt%			d	Comp. of solid phase
Magnesium Nitrate	Magnesium Sulfate	Water		
38.58	0.00	61.42	1.368	$Mg(NO_3)_2 \cdot 6H_2O$
37.65	0.99	61.36	1.368	»
37.03	1.38	61.59	1.368	$Mg(NO_3)_2 \cdot 6H_2O + MgSO_4 \cdot 7H_2O$
36.99	1.48	61.53	1.369	»
37.04	1.56	61.40	1.370	»
36.80	1.62	61.58	1.372	»
28.69	2.52	61.79	1.290	$MgSO_4 \cdot 7H_2O$
17.82	6.41	75.77	1.230	»
10.83	10.83	78.34	1.216	»
9.17	11.98	78.85	1.214	»
3.94	16.58	79.48	1.222	»
2.73	17.98	79.29	1.224	»
1.09	19.70	79.21	1.234	»
0.00	20.82	79.18	1.236	»

$Mg(NO_3)_2$—$MgCl_2$—H_2O

Comp. of saturated soln., wt%			t	d	Comp. of solid phase
Magnesium Chloride	Magnesium Nitrate	Water			
35.0	0.0	65.0	15	1.336	$MgCl_2 \cdot 6H_2O$
32.50	4.29	63.21	15	1.352	»
29.90	8.24	61.86	15	1.371	»
27.7	12.2	60.1	15	1.386	»
26.1	14.9	69.0	15	1.397	$MgCl_2 \cdot 6H_2O + Mg(NO_3)_2 \cdot 6H_2O$
20.0	19.5	60.5	15	1.379	$Mg(NO_3)_2 \cdot 6H_2O$
15.0	24.2	60.8	15	1.373	»
9.54	29.90	60.56	15	1.372	»
5.04	34.40	60.56	15	1.373	»
0.0	40.5	59.5	15	1.377	»
35.5	0.0	64.5	25	1.338	$MgCl_2 \cdot 6H_2O$
33.10	4.06	62.84	25	1.353	»
30.70	8.25	61.05	25	1.372	»
28.6	11.5	59.9	25	1.386	»
24.9	17.8	57.3	25	1.412	$MgCl_2 \cdot 6H_2O + Mg(NO_3)_2 \cdot 6H_2O$
18.3	22.7	59.0	25	1.392	$Mg(NO_3)_2 \cdot 6H_2O$
13.7	27.1	59.2	25	1.384	»
8.72	32.70	58.58	25	1.376	»
4.60	36.8	58.6	25	1.382	»
0.0	41.9	58.1	25	1.387	»
37.0	0.0	63.0	50	1.345	$MgCl_2 \cdot 6H_2O$
33.3	6.5	60.2	50	1.372	»
26.4	18.5	55.1	50	1.424	»
23.0	24.4	52.6	50	1.452	»
21.6	26.5	51.9	50	1.462	$MgCl_2 \cdot 6H_2O + Mg(NO_3)_2 \cdot 6H_2O$
18.2	28.6	53.2	50	1.443	$Mg(NO_3)_2 \cdot 6H_2O$
13.0	32.7	54.3	50	1.425	»
9.75	35.90	54.35	50	1.419	»
6.16	39.30	54.54	50	1.414	»
3.32	42.40	54.28	50	1.415	»
0.0	46.0	54.0	50	1.418	»
39.1	0.0	60.9	75	1.360	$MgCl_2 \cdot 6H_2O$
34.6	7.3	50.1	75	1.380	»
33.4	9.4	57.2	75	1.390	»
28.5	17.6	53.9	75	—	»
24.3	28.2	47.5	75	—	»
25.8	30.2	44.0	75	—	$MgCl_2 \cdot 6H_2O + MgCl_2 \cdot 4H_2O$
23.0	37.5	39.5	75	1.575	$MgCl_2 \cdot 4H_2O$
17.1	46.2	36.7	75	—	»
13.1	54.0	32.9	75	—	$MgCl_2 \cdot 4H_2O + Mg(NO_3)_2 \cdot 2H_2O$
11.6	56.1	32.3	75	—	$Mg(NO_3)_2 \cdot 2H_2O$
4.9	63.1	32.0	75	1.679	»
3.9	65.3	30.8	75	—	»
1.8	67.8	30.4	75	—	»
0.0	68.8	31.2	75	—	»
0.0	50.6	49.4	75	1.461	$Mg(NO_3)_2 \cdot 6H_2O$
9.9	43.2	46.9	75	1.528	»
9.0	48.0	43.0	75	—	»

Comp. of saturated soln., wt%			t	d	Comp. of solid phase
Magnesium Chloride	Magnesium Nitrate	Water			
6.5	53.3	40.2	75	1.581	$Mg(NO_3)_2 \cdot 6H_2O$
7.0	57.7	35.3	75	—	»
0.0	63.4	36.6	75	—	»
44.3	0.0	55.7	110	1.40	$MgCl_2 \cdot 6H_2O$
43.95	3.65	52.40	110	—	$MgCl_2 \cdot 6H_2O + MgCl_2 \cdot 4H_2O$
28.3	29.1	42.6	110	—	$MgCl_2 \cdot 4H_2O$
23.1	41.7	35.2	110	—	»
22.4	45.2	32.4	110	—	»
23.3	46.0	30.7	110	—	»
22.2	47.4	30.4	110	—	$MgCl_2 \cdot 4H_2O + MgCl_2 \cdot 2H_2O$
20.1	50.8	29.1	110	1.60	$MgCl_2 \cdot 2H_2O$
19.3	51.8	28.9	110	—	»
17.5	54.0	28.5	110	1.58	»
16.3	56.9	26.8	110	—	»
16.6	57.8	25.6	110	—	»
16.0	57.8	26.2	110	—	»
15.9	58.7	25.4	110	—	»
14.9	58.5	26.6	110	—	»
14.2	59.3	26.5	110	—	»
15.0	61.0	24.0	110	—	$MgCl_2 \cdot 2H_2O + Mg(NO_3)_2 \cdot 2H_2O$
9.95	65.0	25.05	110	—	$Mg(NO_3)_2 \cdot 2H_2O$
0.0	73.5	26.5	110	—	»

№ 3995 MAGNESIUM NITRATE–MAGNESIUM IODATE–WATER [1837

$Mg(NO_3)_2—Mg(IO_3)_2—H_2O$

Comp. of saturated soln., wt%			t	d	Comp. of solid phase
Magnesium Iodate	Magnesium Nitrate	Water			
4.39	0.00	96.61	5	1.034	$Mg(IO_3)_2 \cdot 10H_2O$
3.93	2.49	93.58	5	1.053	»
3.92	5.45	90.63	5	1.076	»
4.03	8.86	87.11	5	1.105	»
4.26	13.10	82.64	5	1.149	»
4.55	17.0	78.45	5	1.184	$Mg(IO_3)_2 \cdot 10H_2O + Mg(IO_3)_2 \cdot 4H_2O$
4.44	18.52	77.04	5	1.197	$Mg(IO_3)_2 \cdot 4H_2O$
3.95	24.08	71.97	5	1.244	»
2.82	34.21	62.97	5	1.333	»
2.39	38.10	59.51	5	1.368	$Mg(IO_3)_2 \cdot 4H_2O + Mg(NO_3)_2 \cdot 6H_2O$
0.46	39.02	60.52	5	1.352	$Mg(NO_3)_2 \cdot 6H_2O$
0.00	39.25	60.75	5	1.347	»
8.55	0.00	91.45	25	1.075	$Mg(IO_3)_2 \cdot 4H_2O$
7.35	3.49	89.16	25	1.089	»
6.66	7.31	86.03	25	1.116	»
5.66	14.60	79.74	25	1.171	»

Comp. of saturated soln., wt%			t	d	Comp. of solid phase
Magnesium Iodate	Magnesium Nitrate	Water			
5.14	18.48	76.38	25	1.202	$Mg(IO_3)_2 \cdot 4H_2O$
4.20	25.81	69.99	25	1.261	»
3.25	33.50	63.25	25	1.325	»
2.46	41.00	56.54	25	1.398	$Mg(IO_3)_2 \cdot 4H_2O + Mg(NO_3)_2 \cdot 6H_2O$
0.90	41.60	57.50	25	1.380	$Mg(NO_3)_2 \cdot 6H_2O$
0.00	42.03	57.97	25	1.378	»
12.05	0.00	87.95	50	1.115	$Mg(IO_3)_2 \cdot 4H_2O$
9.00	7.29	83.71	50	1.141	»
7.31	15.17	77.52	50	1.186	»
5.58	24.35	69.07	50	1.253	»
4.36	32.15	63.49	50	1.321	»
3.57	38.82	57.61	50	1.368	»
3.13	44.41	52.46	50	—	$Mg(IO_3)_2 \cdot 4H_2O + Mg(NO_3)_2 \cdot 6H_2O$
1.52	45.27	53.21	50	—	$Mg(NO_3)_2 \cdot 6H_2O$
0.00	46.09	53.91	50	—	»

№ 3996 [1679]

MAGNESIUM NITRATE–MAGNESIUM NEODYMIUM NITRATE–WATER

$Mg(NO_3)_2 \cdot 6H_2O—3Mg(NO_3)_2 \cdot 2Nd(NO_3)_3—H_2O$

Concn. of Magnesium Nitrate in aqueous soln., g/l	Solubility of Magnesium Neodymium Nitrate in aqueous soln. of Magnesium Nitrate, wt%	t
51.70	59.23	24.8
130.9	50.31	24.4
Saturated solution $Mg(NO_3)_2$	2.92	24.2

№ 3997 [1206]

MAGNESIUM NITRATE–POTASSIUM NITRATE–WATER

$Mg(NO_3)_2—KNO_3—H_2O$

Comp. of saturated soln., wt%			t	d	Comp. of solid phase
Potassium Nitrate	Magnesium Nitrate	Water			
11.7	0.0	88.3	0	1.079	KNO_3
11.5	2.26	86.24	0	1.091	»
8.35	5.45	86.20	0	1.106	»
6.48	11.06	82.46	0	1.146	»
4.29	17.55	78.16	0	1.174	»
4.35	27.29	69.36	0	1.206	»
4.40	33.42	62.18	0	1.350	$KNO_3 + Mg(NO_3)_2 \cdot 6H_2O$
3.26	37.60	59.14	0	1.389	$Mg(NO_3)_2 \cdot 6H_2O$

Table № 3997 (continuation)

Comp. of saturated soln., wt%			t	d	Comp. of solid phase
Potassium Nitrate	Magnesium Nitrate	Water			
1.09	38.45	60.46	0	1.382	$Mg(NO_3)_2 \cdot 6H_2O$
0.0	38.58	61.42	0	1.369	»
27.49	0.0	72.51	25	1.194	KNO_3
21.40	5.98	72.62	25	1.202	»
17.84	10.06	72.10	25	1.221	»
12.41	18.73	68.86	25	1.265	»
10.18	26.09	63.73	25	1.320	»
8.63	37.20	54.17	25	1.438	$Mg(NO_3)_2 \cdot 6H_2O$
7.67	39.30	53.03	25	1.462	»
3.98	40.49	55.53	25	1.431	»
0.0	42.0	58.0	25	1.394	»
46.10	0.0	53.90	50	1.315	KNO_3
39.37	4.44	56.19	50	1.318	»
28.36	14.12	57.52	50	1.316	»
19.36	26.18	54.46	50	1.361	»
15.45	35.86	48.69	50	1.455	»
15.40	37.03	47.57	50	1.459	»
15.02	40.47	44.51	50	1.505	$KNO_3+Mg(NO_3)_2 \cdot 6H_2O$
8.32	42.50	49.18	50	1.462	$Mg(NO_3)_2 \cdot 6H_2O$
0.0	45.44	54.56	50	1.416	»
60.53	0.0	39.47	75	1.470	KNO_3
54.17	3.78	42.05	75	1.445	»
40.68	13.52	45.80	75	1.432	»
30.76	23.30	45.94	75	1.426	»
24.04	34.07	41.89	75	1.510	»
22.30	40.41	37.29	75	1.510	»
22.0	44.58	33.42	75	1.645	$KNO_3+Mg(NO_3)_2 \cdot 6H_2O$
16.96	45.69	37.35	75	1.599	$Mg(NO_3)_2 \cdot 6H_2O$
10.15	48.01	41.84	75	1.537	»
0.0	51.00	49.0	75	1.461	»
70.8	0.0	29.2	99.5	1.529	KNO_3
54.52	12.14	33.34	99.5	1.532	»
42.90	23.19	33.91	99.5	1.555	»
32.05	37.93	30.02	99.5	1.672	»
30.93	38.46	30.61	99.5	1.690	»

№ 3998 [407]

MAGNESIUM NITRATE–POTASSIUM NITRATE–THALLIUM CHROMATE–WATER

$Mg(NO_3)_2—KNO_3—Tl_2CrO_4—H_2O$

$t = 18—20$

Content in soln., g-eq/l		Solubility of Thallium Chromate g/l	Content in soln., g-eq/l		Solubility of Thallium Chromate g/l
Magnesium Nitrate	Potassium Nitrate		Magnesium Nitrate	Potassium Nitrate	
0.100	0.000	$5.9 \cdot 10^2$	0.025	0.075	$5.4 \cdot 10^2$
0.075	0.025	$5.8 \cdot 10^2$	0.000	0.000	$5.3 \cdot 10^2$
0.050	0.050	$5.6 \cdot 10^2$			

$Mg(NO_3)_2$—$Ca(NO_3)_2$—H_2O

Comp. of saturated soln., wt%			t	Comp. of solid phase
Calcium Nitrate	Magnesium Nitrate	Water		
7.30	0.00	92.70	− 1.7	Ice
6.63	5.56	87.81	− 4.8	»
6.07	10.21	83.72	− 7.2	»
5.62	13.97	80.41	− 9.6	»
5.19	17.55	77.26	−12.8	»
4.72	21.43	73.85	−16.7	»
4.39	24.22	71.39	−20.7	»
4.07	26.89	69.04	−25.0	»
3.81	29.02	67.17	−28.3	»
3.47	31.86	64.67	−29.0	$Mg(NO_3)_2 \cdot 9H_2O$
3.39	32.48	64.13	−26.7	»
3.31	33.13	63.56	−24.2	»
3.26	33.56	63.18	−22.7	»
3.20	34.06	62.74	−20.7	»
3.14	34.56	62.30	−19.7	»
3.08	35.12	61.80	−18.2	$Mg(NO_3)_2 \cdot 6H_2O$
2.94	36.23	60.83	−11.5	»
2.84	37.08	60.08	− 5.9	»
2.63	38.83	58.54	5.6	»
2.32	41.40	56.28	20.0	»
3.07	38.15	61.78	−18.2	$Mg(NO_3)_2 \cdot 9H_2O + Mg(NO_3)_2 \cdot 6H_2O$
3.58	30.90	65.52	−32.1	$Mg(NO_3)_2 \cdot 9H_2O$ + Ice
25.30	0.00	74.70	−10.0	Ice
23.45	4.44	72.11	−12.7	»
20.30	11.99	67.71	−18.9	»
17.88	17.79	64.33	−25.5	»
16.03	22.24	61.73	−30.9	»
14.72	25.38	59.90	−29.9	$Mg(NO_3)_2 \cdot 9H_2O$
14.59	25.70	59.71	−28.8	»
14.21	26.61	59.18	−26.0	»
14.10	26.88	59.02	−25.5	»
13.90	27.36	58.74	−24.2	»
13.43	28.49	58.08	−20.7	»
13.26	28.89	57.85	−19.7	$Mg(NO_3)_2 \cdot 6H_2O$
12.98	29.56	57.46	− 7.7	»
10.95	34.42	54.63	8.8	»
10.52	35.47	54.01	14.0	»
10.01	36.68	53.31	20.0	»
13.29	28.80	57.91	−20.0	$Mg(NO_3)_2 \cdot 6H_2O + Mg(NO_3)_2 \cdot 9H_2O$
15.26	24.10	60.64	−34.3	$Mg(NO_3)_2 \cdot 9H_2O$ + Ice
34.90	0.00	65.10	−18.2	Ice
33.31	2.76	63.93	−20.5	»
32.46	4.24	63.30	−21.7	»
31.15	6.52	62.33	−24.2	»
29.79	8.89	61.32	−26.7	»
29.92	11.96	60.02	−30.9	»
25.75	15.92	58.33	−23.3	$Ca(NO_3)_2 \cdot 4H_2O$
24.47	18.14	57.39	−19.7	»
22.24	22.02	55.74	−14.1	»
20.81	24.51	54.68	−11.5	»

Table № 3999 (continuation)

Comp. of saturated soln., wt%			t	Comp. of solid phase
Calcium Nitrate	Magnesium Nitrate	Water		
20.81	24.51	54.68	−18.2	$Mg(NO_3)_2 \cdot 6H_2O$
19.93	26.03	54.04	−10.3	$Ca(NO_3)_2 \cdot 4H_2O$
18.72	28.15	53.13	0.00	$Mg(NO_3)_2 \cdot 6H_2O$
18.38	28.73	52.89	3.5	»
17.53	30.21	58.26	9.2	»
16.88	31.34	51.78	14.7	»
16.27	32.41	51.32	18.5	»
15.63	33.52	50.85	23.2	»
19.83	26.20	53.97	−10.2	$Mg(NO_3)_2 \cdot 6H_2O + Ca(NO_3)_2 \cdot 4H_2O$
27.83	12.30	59.87	−30.9	$Ca(NO_3)_2 \cdot 4H_2O$+Ice
38.00	0.00	62.00	−21.7	Ice
36.12	3.00	60.88	−24.5	»
34.44	5.68	59.88	−27.1	»
30.44	12.08	57.48	−19.3	$Ca(NO_3)_2 \cdot 4H_2O$
29.38	13.76	56.86	−16.3	»
27.11	17.39	55.50	−11.5	»
22.61	24.58	52.81	− 6.8	»
19.88	28.94	51.18	5.5	$Mg(NO_3)_2 \cdot 6H_2O$
18.20	31.63	50.17	15.8	»
17.66	32.49	49.85	20.5	»
21.44	26.45	52.11	− 6.4	$Mg(NO_3)_2 \cdot 6H_2O + Ca(NO_3)_2 \cdot 4H_2O$
32.89	8.15	58.96	−29.4	$Ca(NO_3)_2 \cdot 4H_2O$+Ice
45.20	0.00	54.80	−19.2	$Ca(NO_3)_2 \cdot 4H_2O$
44.60	0.80	54.60	−18.2	»
43.77	1.92	54.31	−16.6	»
42.74	3.31	53.95	−14.4	»
41.34	5.18	53.48	−11.8	»
39.75	7.32	52.93	− 9.0	»
39.03	8.28	52.69	− 7.8	»
37.71	10.06	52.23	− 5.8	»
36.39	11.83	51.78	− 3.6	»
35.38	13.19	51.43	− 2.4	»
34.40	14.51	51.09	− 1.0	»
33.53	15.67	50.80	− 0.2	»
32.15	17.52	50.33	1.4	»
31.06	18.99	49.95	2.4	»
29.67	20.85	49.48	7.8	$Ca(NO_3)_2 \cdot 3H_2O$
29.06	21.67	49.27	9.2	»
28.30	22.69	49.01	11.1	»
27.25	24.10	48.65	13.5	»
26.46	25.16	48.38	15.0	»
25.55	26.39	48.06	14.8	$Mg(NO_3)_2 \cdot 3H_2O$
25.02	27.10	47.88	17.5	»
24.59	27.68	47.73	19.6	»
23.77	28.78	47.45	23.6	»
23.20	29.54	47.26	26.2	»
25.61	26.30	48.09	14.4	$Mg(NO_3)_2 \cdot 6H_2O + Ca(NO_3)_2 \cdot 3H_2O$
31.05	19.00	49.95	2.4	$Ca(NO_3)_2 \cdot 3H_2O + Ca(NO_3)_2 \cdot 4H_2O$
44.70	0.00	52.30	− 9.6	$Ca(NO_3)_2 \cdot 4H_2O$
45.22	3.15	51.63	− 4.8	»
42.85	6.17	50.98	− 2.2	»

Table № 3999 (continuation)

Comp. of saturated soln., wt%			t	Comp. of solid phase
Calcium Nitrate	Magnesium Nitrate	Water		
38.71	11.44	49.85	2.7	$Ca(NO_3)_2 \cdot 4H_2O$
37.55	12.92	49.53	3.8	»
36.07	14.80	49.13	5.1	»
34.49	16.81	48.70	6.3	»
32.60	19.22	48.18	10.7	$Ca(NO_3)_2 \cdot 3H_2O$
31.08	21.15	47.77	13.8	»
30.18	22.29	47.53	15.8	»
29.10	23.67	47.23	17.9	»
28.22	24.79	46.99	19.6	»
27.44	25.78	46.78	21.1	»
26.62	26.83	46.55	22.8	»
25.59	28.13	46.28	26.2	$Mg(NO_3)_2 \cdot 6H_2O$
24.93	28.98	46.09	29.2	»
27.19	26.10	46.71	22.8	$Ca(NO_3)_2 \cdot 4H_2O + Ca(NO_3)_2 \cdot 3H_2O$
34.34	17.00	48.66	6.4	$Ca(NO_3)_2 \cdot 3H_2O + Mg(NO_3)_2 \cdot 4H_2O$
50.80	0.00	49.20	1.0	$Ca(NO_3)_2 \cdot 4H_2O$
46.59	5.03	48.38	5.5	»
45.24	6.64	48.12	7.0	»
42.97	9.35	47.68	8.7	»
40.91	11.81	47.28	10.3	»
39.35	13.68	46.97	11.2	»
36.55	17.02	46.43	14.2	$Ca(NO_3)_2 \cdot 3H_2O$
34.46	19.53	46.01	17.5	»
32.75	21.51	45.74	20.2	»
31.72	22.80	45.48	22.0	»
30.49	24.27	44.24	24.4	»
29.41	25.56	45.03	26.0	»
28.38	26.79	44.83	27.5	»
38.25	15.00	46.75	11.4	$Ca(NO_3)_2 \cdot 4H_2O + Ca(NO_3)_2 \cdot 3H_2O$
0.00	34.40	65.60	−30	$Mg(NO_3)_2 \cdot 9H_2O$
3.51	31.50	64.99	−30	»
14.75	25.30	59.95	−30	»
16.70	24.40	58.90	−30	$Mg(NO_3)_2 \cdot 9H_2O + Ca(NO_3)_2 \cdot 4H_2O$
27.60	12.70	59.70	−30	$Ca(NO_3)_2 \cdot 4H_2O$
30.20	10.40	59.40	−30	$Ca(NO_3)_2 \cdot 4H_2O$ + Ice
28.23	11.60	60.17	−30	Ice
16.34	21.50	62.16	−30	»
3.71	29.85	66.44	−30	»
0.00	33.20	66.80	−30	»
0.00	36.90	63.80	−20	$Mg(NO_3)_2 \cdot 6H_2O$
3.17	34.35	62.48	−20	»
13.29	28.80	57.91	−20	»
15.70	26.90	57.40	−20	$Mg(NO_3)_2 \cdot 6H_2O + Ca(NO_3)_2 \cdot 4H_2O$
24.61	17.90	57.49	−20	$Ca(NO_3)_2 \cdot 4H_2O$
30.60	11.80	57.60	−20	»
45.00	0.00	55.00	−20	»
36.50	0.00	63.50	−20	Ice
33.64	2.2	64.16	−20	»
19.80	13.20	67.00	−20	»
4.42	23.90	71.68	−20	»
0.00	27.60	72.40	−20	»

Comp. of saturated soln., wt%			t	Comp. of solid phase
Calcium Nitrate	Magnesium Nitrate	Water		
0.00	38.80	61.20	—10	$Mg(NO_3)_2 \cdot 6H_2O$
2.92	36.45	60.63	—10	»
12.46	30.80	56.74	—10	»
19.81	26.25	53.94	—10	»
19.90	26.20	53.90	—10	$Mg(NO_3)_2 \cdot 6H_2O + Ca(NO_3)_2 \cdot 4H_2O$
26.04	19.10	54.86	—10	$Ca(NO_3)_2 \cdot 4H_2O$
40.36	6.50	53.14	—10	»
47.60	0.00	52.40	—10	»
25.40	0.00	74.60	—10	Ice
5.58	14.30	80.12	—10	»
0.00	19.50	80.50	—10	»
0.00	40.00	60.00	0	$Mg(NO_3)_2 \cdot 6H_2O$
2.74	37.90	59.36	0	»
11.65	32.75	55.60	0	»
18.66	28.25	53.09	0	»
20.53	27.90	51.57	0	»
22.85	26.40	50.75	0	$Mg(NO_3)_2 \cdot 6H_2O + Ca(NO_3)_2 \cdot 3H_2O$
28.60	20.60	50.80	0	$Ca(NO_3)_2 \cdot 3H_2O + Ca(NO_3)_2 \cdot 4H_2O$
33.46	15.90	50.74	0	$Ca(NO_3)_2 \cdot 4H_2O$
41.18	8.30	50.52	0	»
50.50	0.00	49.50	0	»
0.00	41.30	58.70	10	$Mg(NO_3)_2 \cdot 6H_2O$
2.57	39.35	58.08	10	»
10.83	34.70	54.47	10	»
17.42	30.40	52.18	10	»
19.09	30.20	50.71	10	»
24.75	26.40	48.85	10	$Mg(NO_3)_2 \cdot 6H_2O + Ca(NO_3)_2 \cdot 3H_2O$
28.74	22.10	49.16	10	$Ca(NO_3)_2 \cdot 3H_2O$
32.77	19.00	48.23	10	»
37.00	15.80	47.20	10	$Ca(NO_3)_2 \cdot 3H_2O + Ca(NO_3)_2 \cdot 4H_2O$
41.01	11.70	47.29	10	$Ca(NO_3)_2 \cdot 4H_2O$
53.40	0.00	46.60	10	»
0.00	42.58	57.42	20	$Mg(NO_3)_2 \cdot 6H_2O$
2.38	40.90	56.72	20	»
10.00	36.70	53.30	20	»
16.10	32.70	51.20	20	»
17.65	32.50	49.85	20	»
14.50	27.80	47.70	20	»
26.65	26.10	47.25	20	$Mg(NO_3)_2 \cdot 6H_2O + Ca(NO_3)_2 \cdot 3H_2O$
28.05	25.00	46.95	20	$Ca(NO_3)_2 \cdot 3H_2O$
33.06	25.20	45.74	20	»
46.50	11.30	42.20	20	$Ca(NO_3)_2 \cdot 3H_2O + Ca(NO_3)_2 \cdot 4H_2O$
56.40	0.00	43.60	20	$Ca(NO_3)_2 \cdot 4H_2O$
0.00	49.16	50.84	60	$Mg(NO_3)_2 \cdot 6H_2O$
40.74	31.89	27.37	60	$Mg(NO_3)_2 \cdot 6H_2O + Ca(NO_3)_2$
78.20	0.00	21.80	60	$Ca(NO_3)_2$

№ 4000 MAGNESIUM NITRATE–CALCIUM SULFATE–WATER [1317]

$Mg(NO_3)_2—CaSO_4—H_2O$

$t = 25$

Concn. in aqueous soln., g/l		d_4	Concn. in aqueous soln., g/l		d_4
Magnesium Nitrate	Calcium Sulfate		Magnesium Nitrate	Calcium Sulfate	
0	2.08	0.9981	200	13.34	1.1498
25	5.77	1.0205	300	14.00	1.2190
50	7.88	1.0398	400	14.68	1.2821
100	9.92	1.0786	514	15.04	1.3553

№ 4001 [1503]

MAGNESIUM NITRATE–MANGANESE (II) NITRATE–WATER

$Mg(NO_3)_2—Mn(NO_3)_2—H_2O$

$t = 20$

Comp. of saturated soln., wt%			Comp. of solid phase
Magnesium Nitrate	Manganese Nitrate	Water	
43.68	0.00	56.32	$Mg(NO_3)_2 \cdot 6H_2O$
41.25	2.88	55.87	»
38.00	7.15	54.85	»
37.15	9.90	52.95	»
35.40	12.41	52.19	»
31.60	18.85	49.55	»
20.60	31.30	48.10	»
29.92	26.25	43.83	»
29.0	27.40	43.60	»
27.78	28.36	43.86	»
26.50	29.15	44.35	$Mg(NO_3)_2 \cdot 6H_2O + Mn(NO_3)_2 \cdot 5H_2O$
24.35	32.00	43.65	$Mn(NO_3)_2 \cdot 5H_2O$
16.17	39.92	43.91	»
10.05	44.30	45.65	»
6.15	49.76	44.09	»
0.00	56.81	43.19	»

№ 4002 [2582]

MAGNESIUM NITRATE–SILVER CHROMATE–WATER

$Mg(NO_2)_2—Ag_2CrO_4—H_2O$

$t = 20$

Concn. of Magnesium Nitrate in aqueous soln., g/l	Solubility of Silver Chromate in aqueous soln. of Magnesium Nitrate, wt%
0.74	0.00385
1.48	0.00420
2.97	0.00466
7.42	0.00568

№ 4003 [856]

MAGNESIUM NITRATE–SILVER BROMATE–WATER

$Mg(NO_3)_2—AgBrO_3—H_2O$

$t = 25$

Concn. in aqueous soln., g/l		Concn. in aqueous soln., g/l	
Magnesium Nitrate	Silver Bromate	Magnesium Nitrate	Silver Bromate
0.00000	1.9475	4.4505	2.3130
0.14835	2.0183	14.835	2.6172
0.44505	2.0560	44.505	2.9944
1.4835	2.1880	148.35	4.0790

№ 4004 [481]

VARIOUS METAL NITRATES–SILVER BROMATE–WATER

$—AgBrO_3—H_2O$

$t = 25$

Metal Nitrate			Solubility of Silver Bromate in aqueous soln. of metal Nitrate, (mol/l) · 10²
Name	Formula	Concn. in aqueous soln., mol/l	
Magnesium Nitrate	$Mg(NO_3)$	1.0	1.73
Calcium Nitrate	$Ca(NO_3)_2$	1.0	1.81
Strontium Nitrate	$Sr(NO_3)_2$	1.0	1.94

№ 4005 [406]

MAGNESIUM NITRATE–CESIUM HEXANITROCOBALTATE–WATER

$Mg(NO_3)_2—Cs_3[Co(NO_2)_6]—H_2O$

$t = 20$

Content in aqueous soln.	
Magnesium Nitrate, mol/l	Cesium Hexanitro-cobaltate, (mol/l) · 10⁵
0.0	6.0
0.25	34
0.5	48
1.0	60
2.0	96

№ 4006 [1777]

MAGNESIUM NITRATE–BARIUM BROMATE–WATER

$Mg(NO_3)_2—Ba(BrO_3)_2—H_2O$

$t = 25$

Content in aqueous soln., g/l		d
Magnesium Nitrate	Barium Bromate	
0.00	7.93	1.0038
7.42	8.196	1.0114

№ 4007 [1503]

MAGNESIUM NITRATE–LANTHANUM NITRATE–WATER

$Mg(NO_3)_2—La(NO_3)_3—H_2O$

$t = 20$

Comp. of saturated soln., wt%			Comp. of solid phase
Lanthanum Nitrate	Magnesium Nitrate	Water	
60.13	0.0	39.87	$La(NO_3)_3 \cdot 6H_2O$
58.91	2.20	38.89	»
55.06	6.21	38.73	»
50.11	9.83	40.06	»
42.98	17.40	39.62	»
37.02	19.25	43.73	»
26.0	24.0	50.0	»
22.90	25.48	51.62	$Mg(NO_3)_3 \cdot 6H_2O$
15.35	30.93	53.72	»
6.89	36.30	56.81	»
3.20	39.25	57.55	»
0.0	43.68	56.32	»

№ 4008 [904]

MAGNESIUM NITRATE–LANTHANUM NITRATE–WATER

$Mg(NO_3)_2—La(NO_3)_3—H_2O$

$t = 25$

Comp. of saturated soln., wt%			Comp. of solid phase
Lanthanum Nitrate	Magnesium Nitrate	Water	
60.28	0.00	39.72	$La(NO_3)_3 \cdot 6H_2O$
59.22	1.53	39.25	Solid solutions
58.92	2.47	38.61	»
57.94	3.46	38.60	»
53.99	7.53	38.48	$2La(NO_3)_3 \cdot 3Mg(NO_3)_2 \cdot 24H_2O$
44.67	10.52	44.81	»
37.46	13.73	48.81	»
29.20	19.12	51.68	»
23.66	23.47	52.87	»
15.64	28.58	58.78	»
10 35	32.00	57.65	»
8.35	35.00	56.65	»
4.49	38.69	56.82	»
2.52	39.54	57.94	»
1.65	40.00	58.35	Solid solutions
0.00	43.06	56.94	»
0.00	43.55	56.45	$Mg(NO_3)_2 \cdot 6H_2O$

№ 4009 [1677, 2039]

MAGNESIUM NITRATE–LANTHANUM IODATE–WATER

$Mg(NO_3)_2—La(IO_3)_3—H_2O$

$t = 25$

Concn. of Magnesium Nitrate in aqueous soln., g/l	Solubility of Lanthanum Iodate in aqueous soln. of Magnesium Nitrate, g/l
7.416	1.172

№ 4010 [407]

VARIOUS SALTS–THALLIUM CHROMATE–WATER

$— Tl_2CrO_4—H_2O$

$t = 18—20$

Salt			Solubility of Thallium Chromate in aqueous soln., of salt g/l
Name	Formula	Concn. in aqueous soln., g-eq/l	
Magnesium Nitrate	$Mg(NO_3)_2$	0.000	440
»	»	0.025	460
»	»	0.100	590
»	»	0.250	740
»	»	0.500	1100
Strontium Nitrate	$Sr(NO_3)_2$	0.000	440
»	»	0.050	560
»	»	0.100	630
»	»	0.250	770
»	»	0.500	910
Zinc Nitrate	$Zn(NO_3)_2$	0.000	440
»	»	0.050	530
»	»	0.100	640
»	»	0.500	950
»	»	1.000	1540
»	»	2.000	2380
Zinc Sulfate	$ZnSO_4$	0.000	440
»	»	0.025	530
»	»	0.050	590
»	»	0.250	1010
»	»	0.500	1590

№ 4011 [2039]

MAGNESIUM NITRATE–THALLIUM IODATE–WATER

$Mg(NO_3)_2—TlIO_3—H_2O$

$t = 25$

Concn. of Magnesium Nitrate in aqueous soln., g/l	Solubility of Thallium Iodate in aqueous soln. of Magnesium Nitrate, g/l
2.874	0.930
4.316	0.991

№ 4012 [1011]

MAGNESIUM NITRATE–URANYL NITRATE–WATER

$Mg(NO_3)_2—UO_2(NO_3)_2—H_2O$

Comp. of saturated soln., wt%			t	Comp. of solid phase
Uranyl Nitrate	Magnesium Nitrate	Water		
0.00	41.94	58.06	25	$Mg(NO_3)_2 \cdot 6H_2O$
1.15	41.30	57.55	25	»
7.10	40.50	52.40	25	»
9.2	38.8	52.0	25	»
11.9	37.4	50.7	25	»
13.2	37.2	49.6	25	»
16.0	37.0	47.0	25	$Mg(NO_3)_2 \cdot 6H_2O + UO_2(NO_3)_2 \cdot 6H_2O$
18.4	35.4	46.2	25	$UO_2(NO_3)_2 \cdot 6H_2O$
19.2	33.0	47.8	25	»
20.0	33.6	46.4	25	»
21.0	31.0	48.0	25	»
24.8	28.0	47.2	25	»
25.4	27.6	47.0	25	»
25.8	27.0	47.2	25	»
26.7	25.3	48.0	25	»
30.0	22.5	47.5	25	»
31.8	21.6	46.6	25	»
36.8	17.6	45.6	25	»
37.0	17.4	45.6	25	»
39.6	15.3	45.1	25	»
43.4	11.2	45.4	25	»
47.2	8.0	44.8	25	»
49.8	5.2	45.0	25	»
52.8	3.0	44.2	25	»
55.95	0.0	44.05	25	»
0.0	38.8	61.2	0	$Mg(NO_3)_2 \cdot 6H_2O$
2.6	38.0	59.4	0	»
4.6	37.0	58.4	0	»
5.5	37.2	57.3	0	$Mg(NO_3)_2 \cdot 6H_2O + UO_2(NO_3)_2 \cdot 6H_2O$
7.2	36.2	56.6	0	$UO_2(NO_3)_2 \cdot 6H_2O$
8.2	34.9	56.9	0	»
10.8	31.4	57.8	0	»
12.6	28.7	58.7	0	»

Table № 4012 (continuation)

Comp. of saturated soln., wt%			t	Comp. of solid phase
Uranyl Nitrate	Magnesium Nitrate	Water		
17.0	25.6	57.4	0	$UO_2(NO_3)_2 \cdot 6H_2O$
21.5	22.0	56.5	0	»
24.6	19.7	55.7	0	»
26.0	18.4	55.6	0	»
27.7	17.4	54.9	0	»
29.6	15.7	54.7	0	»
32.0	14.2	53.8	0	»
35.4	11.6	53.0	0	»
35.8	10.7	53.5	0	»
36.8	10.2	53.0	0	»
38.1	9.5	52.4	0	»
40.3	7.6	52.1	0	»
44.5	3.8	51.7	0	»
49.5	0.0	50.5	0	»

№ 4013 [828]

MAGNESIUM FLUORIDE–POTASSIUM FLUORIDE–WATER

MgF_2—KF—H_2O

$t = 20$

Comp. of saturated soln., wt%			Comp. of solid phase
Potassium Fluoride	Magnesium Fluoride	Water	
46.82	0.00	53.18	$KMgF_3$
39.21	0.00	60.79	»
39.19	0.00	60.81	»
30.56	0.00	69.44	»
22.44	0.00	77.56	»
10.12	0.00	89.88	»
5.817	0.00	94.18	»
2.893	0.00	97.11	»
1.164	Traces	98.84	»
$5.687 \cdot 10^{-1}$	»	99.43	»
$4.064 \cdot 10^{-1}$	$6.2 \cdot 10^{-5}$	99.59	»
$4.004 \cdot 10^{-1}$	$6.9 \cdot 10^{-5}$	99.60	»
$2.949 \cdot 10^{-1}$	$7.48 \cdot 10^{-5}$	99.71	MgF_2
$1.711 \cdot 10^{-1}$	$1.37 \cdot 10^{-4}$	99.83	»
$6.582 \cdot 10^{-2}$	$7.73 \cdot 10^{-4}$	99.93	»
$5.88 \cdot 10^{-2}$	$9.17 \cdot 10^{-4}$	99.94	»
$5.90 \cdot 10^{-2}$	$8.24 \cdot 10^{-4}$	99.94	»
$4.061 \cdot 10^{-2}$	$1.40 \cdot 10^{-4}$	99.95	»
$2.94 \cdot 10^{-2}$	$2.00 \cdot 10^{-3}$	99.97	»
$1.761 \cdot 10^{-2}$	$3.69 \cdot 10^{-3}$	99.98	»
$1.774 \cdot 10^{-2}$	$3.66 \cdot 10^{-3}$	99.98	»
$5.94 \cdot 10^{-3}$	$6.29 \cdot 10^{-3}$	99.99	»
$2.90 \cdot 10^{-3}$	$7.05 \cdot 10^{-3}$	99.99	»
0.000	$8.47 \cdot 10^{-3}$	99.99	»

№ 4014 [834]

MAGNESIUM FLUORIDE–SILVER FLUORIDE–WATER

MgF_2—AgF—H_2O

$t = 25$

Comp. of saturated soln., wt%			Comp. of solid phase
Silver Fluoride	Magnesium Fluoride	Water	
46.40	0.00	53.60	AgF+MgF_2
21.87	0.00	78.13	»
7.018	0.00	92.922	»
4.800	0.00	95.200	»
1.245	0.00	98.755	»
0.5142	$1.213 \cdot 10^{-4}$	99.485	»
0.1315	$3.895 \cdot 10^{-4}$	99.868	MgF_2
0.09208	$5.413 \cdot 10^{-4}$	99.907	»
0.03436	$4.365 \cdot 10^{-3}$	99.965	»

№ 4015 [799]

MAGNESIUM SULFITE–MAGNESIUM HYDROGEN SULFITE–MAGNESIUM SULFATE–WATER

$MgSO_3$—$Mg(HSO_3)_2$—$MgSO_4$—H_2O

Content in aqueous soln., wt%		Solubility of Magnesium Sulfite, wt%	d	t
Magnesium Hydrogen Sulfite	Magnesium Sulfate			
0.40	10.0	0.751	1.0968	35
0.40	10.0	0.867	1.0959	40
0.40	10.0	1.076	1.0954	45
0.40	10.0	1.205	1.0948	50
0.40	10.0	1.351	1.0937	55
0.40	10.0	1.520	1.0939	60
0.40	10.0	2.078	1.0968	70
1.14	10.0	0.679	1.1032	35
1.14	10.0	0.837	1.1023	40
1.14	10.0	0.927	1.1020	45
1.14	10.0	1.340	1.1017	50
1.14	10.0	1.729	1.1014	55
1.14	10.0	2.050	1.1017	60
1.14	10.0	2.676	1.1018	70
2.21	10.0	0.643	1.1100	35
2.21	10.0	0.901	1.1092	40
2.21	10.0	0.964	1.1075	45
2.21	10.0	1.073	1.1070	50
2.21	10.0	1.437	1.1070	55
2.21	10.0	1.787	1.1094	60
2.21	10.0	2.661	1.1107	70
2.91	10.0	0.493	1.1178	35
2.91	10.0	0.716	1.1166	40

Table № 4015 (continuation)

Content in aqueous soln., wt%		Solubility of Magnesium Sulfite, wt%	d	t
Magnesium Hydrogen Sulfite	Magnesium Sulfate			
2.91	10.0	1.015	1.1155	45
2.91	10.0	1.072	1.1146	50
2.91	10.0	1.390	1.1153	55
2.91	10.0	1.596	1.1153	60
2.91	10.0	2.599	1.1174	70
3.86	10.0	0.397	1.1211	35
3.86	10.0	0.680	1.1194	40
3.86	10.0	0.932	1.1196	45
3.86	10.0	1.091	1.1186	50
3.86	10.0	1.388	1.1185	55
3.86	10.0	1.725	1.1211	60
3.86	10.0	2.627	1.1226	70
4.42	10.0	0.578	1.1284	35
4.42	10.0	0.739	1.1272	40
4.42	10.0	0.903	1.1265	45
4.42	10.0	1.198	1.1280	50
4.42	10.0	1.579	1.1280	55
4.42	10.0	2.014	1.1282	60
4.42	10.0	2.764	1.1287	70
0.53	0.0	0.595	1.0092	35
0.53	0.0	0.752	1.0087	40
0.53	0.0	0.939	1.0084	45
0.53	0.0	1.060	1.0081	50
0.53	0.0	1.418	1.0077	55
0.53	0.0	1.825	1.0077	60
0.53	0.0	1.968	1.0073	70
1.03	0.0	0.470	1.0166	35
1.03	0.0	0.659	1.0163	40
1.03	0.0	0.838	1.0159	45
1.03	0.0	0.976	1.0150	50
1.03	0.0	1.307	1.0142	55
1.03	0.0	1.535	1.0138	60
1.03	0.0	1.891	1.0097	70
2.09	0.0	0.448	1.0178	35
2.09	0.0	0.561	1.0173	40
2.09	0.0	0.796	1.0165	45
2.09	0.0	0.940	1.0162	50
2.09	0.0	1.246	1.0156	55
2.09	0.0	1.391	1.0144	60
2.09	0.0	1.844	1.0143	70
3.06	0.0	0.437	1.0218	35
3.06	0.0	0.538	1.0212	40
3.06	0.0	0.651	1.0208	45
3.06	0.0	0.800	1.0185	50
3.06	0.0	1.199	1.0180	55
3.06	0.0	1.368	1.0175	60
3.06	0.0	1.549	1.0169	70
3.88	0.0	0.417	1.0298	35
3.88	0.0	0.504	1.0283	40
3.88	0.0	0.612	1.0280	45

Table № 4015 (continuation)

Content in aqueous soln., wt%		Solubility of Magnesium Sulfite, wt%	d	t
Magnesium Hydrogen Sulfite	Magnesium Sulfate			
3.88	0.0	0.736	1.0271	50
3.88	0.0	1.084	1.0277	55
3.88	0.0	1.289	1.0197	60
3.88	0.0	1.538	1.0188	70
4.71	0.0	0.411	1.0358	35
4.71	0.0	0.463	1.0351	40
4.71	0.0	0.581	1.0344	45
4.71	0.0	0.724	1.0329	50
4.71	0.0	0.978	1.0324	55
4.71	0.0	1.154	1.0320	60
4.71	0.0	1.512	1.0313	70

№ 4016 [693]

MAGNESIUM SULFITE–MAGNESIUM HYDROGEN SULFITE–MAGNESIUM SULFATE–SULFUR DIOXIDE–WATER

$MgSO_3$—$Mg(HSO_3)_2$—$MgSO_4$—SO_2—H_2O

Content in aqueous soln., g/l		P_{SO_2}, mmHg	t	Comp. of solid phase
Total Magnesium Sulfite and Magnesium Hydrogen Sulfite in conversion to SO_2	Magnesium Sulfate			
7.04	54.75	0.030	30	$MgSO_3$
8.20	51.20	0.038	40	»
10.3	49.38	0.051	50	»
11.9	51.75	0.060	60	»
11.0	52.00	0.041	30	»
11.84	51.85	0.048	40	»
12.8	54.50	0.057	50	»
12.7	53.65	0.069	60	»
18.2	51.67	0.052	30	»
18.5	52.63	0.065	40	»
21.6	49.85	0.095	50	»
23.6	50.20	0.111	60	»
25.0	51.30	0.073	30	»
25.9	51.13	0.089	40	»
28.8	51.90	0.110	50	»
31.6	51.75	0.141	60	»
9.72	80.25	0.031	30	»
10.5	81.00	0.042	40	»
13.3	80.05	0.052	50	»
13.6	77.80	0.066	60	»
12.4	81.25	0.041	30	»

Table № 4016 (continuation)

Content in aqueous soln., g/l		P_{SO_2}, mmHg	t	Comp. of solid phase
Total Magnesium Sulfite and Magnesium Hydrogen Sulfite in conversion to SO_2	Magnesium Sulfate			
14.1	81.32	0.051	40	$MgSO_3$
16.1	80.45	0.063	50	»
19.3	79.23	0.062	30	»
20.2	81.10	0.079	40	»
22.0	80.25	0.096	50	»
19.9	80.35	0.126	60	»
25.4	80.03	0.122	30	»
28.3	79.00	0.167	40	»
27.6	81.00	0.220	50	»
27.2	81.70	0.295	60	»
8.95	101.26	0.046	40	»
9.56	100.45	0.050	50	»
10.6	100.60	0.059	60	»
10.9	99.90	0.043	30	»
18.1	106.50	0.073	30	»
18.8	105.50	0.102	40	»
21.7	106.80	0.119	50	»
23.6	105.50	0.133	60	»
23.7	107.31	0.130	30	»
24.5	106.89	0.130	40	»
7.87	124.50	0.039	30	»
10.0	119.20	0.048	40	»
10.1	126.10	0.049	30	»
11.5	126.30	0.058	40	»
12.3	122.35	0.068	50	»
17.9	120.62	0.081	30	»
20.0	120.83	0.110	40	»
21.0	120.25	0.158	50	»
24.9	123.15	0.150	30	»
25.66	122.70	0.291	50	»

№ 4017 [2098]

MAGNESIUM SULFATE–MAGNESIUM CHLORIDE–WATER

$MgSO_4 \cdot 6H_2O—MgCl_2 \cdot 6H_2O—H_2O$

$t = 25$

Comp. of saturated soln., mol %		
Magnesium Chloride	Magnesium Sulfate	Water
9.30	1.25	89.44

Table № 4017 (continuation)

$MgSO_4 \cdot 6H_2O—MgCl_2 \cdot 7H_2O—H_2O$

Comp. of saturated soln., mol %		
Magnesium Chloride	Magnesium Sulfate	Water
6.71	1.38	91.91

№ 4018 MAGNESIUM SULFATE–MAGNESIUM CHLORIDE–WATER [2025]

$MgSO_4—MgCl_2—H_2O$

Comp. of saturated soln., mol %			t	Comp. of solid phase
Magnesium Chloride	Magnesium Sulfate	Water		
9.07	0.00	90.93	0	$MgCl_2 \cdot 6H_2O$
8.87	0.31	90.82	0	$MgCl_2 \cdot 6H_2O + MgSO_4 \cdot 7H_2O$
7.73	0.30	91.97	0	$MgSO_4 \cdot 7H_2O$
7.23	0.31	92.46	0	»
6.68	0.32	93.00	0	»
6.00	0.39	93.61	0	»
4.07	0.77	95.16	0	»
2.28	1.61	96.11	0	»
0.89	2.66	96.45	0	»
0.0	3.74	96.26	0	»
9.49	0.0	90.51	25	$MgCl_2 \cdot 6H_2O$
9.02	0.84	90.14	25	$MgCl_2 \cdot 6H_2O + MgSO_4 \cdot 6H_2O$
8.32	0.82	90.86	25	$MgSO_4 \cdot 6H_2O$
7.35	0.84	91.81	25	$MgSO_4 \cdot 6H_2O + MgSO_4 \cdot 7H_2O$
6.33	0.99	92.68	25	$MgSO_4 \cdot 7H_2O$
5.63	1.12	93.25	25	»
3.71	1.79	94.50	25	»
1.94	2.98	95.08	25	»
1.64	3.29	95.07	25	»
0.76	4.27	94.97	25	»

№ 4019 [2159]

MAGNESIUM SULFATE–MAGNESIUM CHLORIDE–WATER

$MgSO_4—MgCl_2—H_2O$

$t = 105$

Content in aqueous soln., wt%		Comp. of solid phase
Magnesium Chloride	Magnesium Sulfate	
42.44	0	$MgCl_2 \cdot 6H_2O$
42.04	0.57	$MgCl_2 \cdot 6H_2O + MgSO_4 \cdot H_2O$
39.12	0.80	$MgSO_4 \cdot H_2O$
0	33.85	»

№ 4020 [2627]

MAGNESIUM SULFATE–MAGNESIUM CHLORIDE–WATER

$MgSO_4$—$MgCl_2$—H_2O

$t = 25$

Comp. of saturated soln., wt%			Comp. of solid phase
Magnesium Chloride	Magnesium Sulfate	Water	
35.86	0.0	64.14	$MgCl_2 \cdot 6H_2O$
34.80	1.45	63.75	»
33.89	2.99	63.12	»
34.03	2.96	63.01	$MgCl_2 \cdot 6H_2O + MgSO_4 \cdot 4H_2O$
33.56	2.81	63.63	$MgSO_4 \cdot 4H_2O$
33.09	2.82	64.09	»
32.97	2.91	64.12	$MgSO_4 \cdot 5H_2O$
32.48	3.04	64.48	»
30.79	3.87	65.34	»
32.36	3.85	63.79	$MgSO_4 \cdot 6H_2O$
32.14	3.77	64.09	»
31.39	3.75	64.86	»
30.91	3.76	65.33	»
30.08	3.80	66.12	»
29.72	4.14	67.14	»
28.21	4.35	67.44	»
27.84	4.29	67.87	»
26.36	4.90	68.74	»
26.06	5.02	68.92	$MgSO_4 \cdot 6H_2O + MgSO_4 \cdot 7H_2O$
23.53	5.44	71.03	$MgSO_4 \cdot 7H_2O$
16.22	10.14	73.64	»
14.77	11.08	74.15	»
14.05	10.68	75.27	»
10.51	14.02	75.47	»
9.42	14.89	75.59	»
5.31	19.63	75.06	»
1.62	24.43	73.95	»
0.0	26.68	73.32	»

№ 4021 [1683]

MAGNESIUM SULFATE–MAGNESIUM CHLORIDE–WATER

$MgSO_4 \cdot 7H_2O$—$MgCl_2$—H_2O

Comp. of saturated soln., wt%			t	Comp. of saturated soln., wt%			t
Magnesium Sulfate	Magnesium Chloride	Water		Magnesium Sulfate	Magnesium Chloride	Water	
21.0	3.8	75.2	20	12.0	3.6	84.4	160
25.5	3.7	70.8	40	5.0	3.7	91.3	180
30.5	3.6	65.9	60	14.0	8.5	77.5	20
34.0	3.5	62.5	80	21.0	8.0	71.0	40
36.0	3.4	60.6	100	26.5	7.7	65.8	60
30.0	3.4	66.6	120	30.5	7.4	62.1	80
20.0	3.5	76.5	140	30.5	7.1	62.4	100

Table № 4021 (continuation)

Comp. of saturated soln., wt%			t	Comp. of saturated soln., wt%			t
Magnesium Sulfate	Magnesium Chloride	Water		Magnesium Sulfate	Magnesium Chloride	Water	
23.0	7.1	69.9	120	13.0	21.3	65.7	60
15.5	7.5	77.0	140	14.5	21.2	64.3	80
8.2	8.2	83.6	160	7.5	23.0	69.5	100
2.5	10.0	87.5	180	3.8	24.0	72.2	120
5.0	24.0	71.0	20	3.0	24.0	73.0	140
10.0	22.2	67.8	40	2.5	23.3	24.2	160

№ 4022 [199]

MAGNESIUM SULFATE–MAGNESIUM CHLORIDE–WATER

$MgSO_4$—$MgCl_2$—H_2O

$t = 35$

Content in aqueous soln., wt%		d	Comp. of solid phase
Magnesium Chloride	Magnesium Sulfate		
36.14	0	1.3363	$MgCl_2 \cdot 6H_2O$
34.77	2.45	1.3546	$MgCl_2 \cdot 6H_2O + MgSO_4 \cdot 6H_2O$
25.16	6.20	1.3252	$MgSO_4 \cdot 6H_2O$
21.21	8.32	1.2817	$MgSO_4 \cdot 6H_2O + MgSO_4 \cdot 7H_2O$
18.30	9.93	1.2715	$MgSO_4 \cdot 7H_2O$
13.31	13.65	1.2664	»
0	29.22	1.3286	»

№ 4023 [295]

MAGNESIUM SULFATE–MAGNESIUM CHLORIDE–WATER

$MgSO_4$—$MgCl_2$—H_2O

$t = 7$

Content in aqueous soln., wt%		d_4	Comp. of solid phase
Magnesium Chloride	Magnesium Sulfate		
34.66	0	1.3296	$MgCl_2 \cdot 6H_2O$
33.77	2.12	—	$MgCl_2 \cdot 6H_2O + MgSO_4 \cdot 7H_2O$
33.02	2.11	—	$MgSO_4 \cdot 7H_2O$
6.04	14.85	—	»
0	22.94	1.2593	»

№ 4024 [332]

MAGNESIUM SULFATE–MAGNESIUM CHLORIDE–WATER

$MgSO_4—MgCl_2—H_2O$

Content in aqueous soln., wt%		$t_{b.p.}$	Comp. of solid phase
Magnesium Chloride	Magnesium Sulfate		
55.56	0	193.0	$MgCl_2 \cdot 2H_2O$
55.42	0.12	193.0	»
55.30	0.48	186.0	$MgCl_2 \cdot 2H_2O + MgSO_4 \cdot H_2O$
40.23	0.70	140.0	$MgSO_4 \cdot H_2O$
38.54	0.75	139.0	»
36.66	0.76	129.5	»
33.59	1.12	125.5	»
30.93	1.54	118.5	»
18.77	11.42	110.5	»
14.98	16.07	109.5	»
13.23	16.90	109.0	»
8.54	22.51	106.0	»
0.26	39.93	104.7	»
0	40.80	104.0	»

№ 4025 [1949]

MAGNESIUM SULFATE–MAGNESIUM CHLORIDE–WATER

$MgSO_4—MgCl_2—H_2O$

$t = 55$

Content in aqueous soln., g/100g Water		Comp. of solid phase
Magnesium Chloride	Magnesium Sulfate	
59.15	1.6	$MgSO_4 \cdot H_2O$ ($5/4H_2O$)
58.8	1.45	»
43.3	3.7	»
32.55	9.7	»
57.35	7.75	$MgSO_4 \cdot 4H_2O$ (metastable)
33.6	15.25	»
34.4	16.85	$MgSO_4 \cdot 6H_2O$
23.7	23.0	»
23.05	22.75	»
9.9	32.05	»
9.25	32.5	»
11.1	35.45	$MgSO_4 \cdot 7H_2O$ (metastable)
10.15	36.8	»
10.1	36.65	»
8.35	30.1	»

№ 4026 [1335]

MAGNESIUM SULFATE–MAGNESIUM CHLORIDE–WATER

$MgSO_4—MgCl_2—H_2O$

$t = 100$

Comp. of saturated soln., wt%			Comp. of solid phase
Magnesium Chloride	Magnesium Sulfate	Water	
42.4	0.0	57.6	$MgCl_2 \cdot 6H_2O$
41.20	1.08	57.72	$MgSO_4$
41.50	1.23	57.27	»
40.25	0.87	58.88	»
40.60	0.60	58.80	»
33.16	0.37	66.47	»
29.0	0.99	70.01	»
37.60	2.12	60.28	»
36.90	2.81	60.29	»
37.35	2.82	59.83	$MgSO_4 + MgSO_4 \cdot H_2O$
31.0	3.30	65.70	$MgSO_4 \cdot H_2O$
20.85	9.05	70.10	»
19.85	8.65	71.50	»
12.50	16.30	71.20	»
8.42	21.20	70.38	»
2.83	29.50	67.67	»
0.00	33.50	66.50	»

№ 4027 [444]

MAGNESIUM SULFATE–MAGNESIUM CHLORIDE–WATER

$MgSO_4—MgCl_2—H_2O$

$t = 25$

Content in aqueous soln., wt%		d_4	Comp. of solid phase
Magnesium Chloride	Magnesium Sulfate		
34.36	2.14	1.3545	$MgCl_2 \cdot 6H_2O + MgSO_4 \cdot 4H_2O$
34.15	2.28	—	$MgSO_4 \cdot 4H_2O$
32.88	2.73	—	»
32.09	3.06	1.3398	$MgSO_4 \cdot 4H_2O + MgSO_4 \cdot 5H_2O$
30.47	3.84	1.3313	$MgSO_4 \cdot 5H_2O$
29.89	3.93	1.3239	$MgSO_4 \cdot 5H_2O + MgSO_4 \cdot 6H_2O$
27.09	4.71	1.3039	$MgSO_4 \cdot 6H_2O$
27.17	4.55	1.3021	»
26.06	4.92	1.2966	$MgSO_4 \cdot 6H_2O + MgSO_4 \cdot 7H_2O$
25.82	5.06	1.2955	$MgSO_4 \cdot 7H_2O$
24.80	5.28	1.2850	»
24.29	5.27	1.2823	»
23.85	5.47	1.2801	»
23.21	5.93	1.2751	»
22.48	6.12	1.2712	»

№ 4028 MAGNESIUM SULFATE–MAGNESIUM CHLORIDE–WATER [444]

$MgSO_4$—$MgCl_2$—H_2O

Content in aqueous soln., wt%		t	d^{25}_{25}	Comp. of solid phase
Magnesium Chloride	Magnesium Sulfate			
34.16	2.50	25	—	$MgCl_2 \cdot 6H_2O + MgSO_4 \cdot 4H_2O$
33.36	2.60	25	1.3468	$MgSO_4 \cdot 3.9H_2O$
33.15	3.33	25	1.3546	$MgSO_4 \cdot 4.28H_2O$
32.64	3.04	25	—	$MgSO_4 \cdot 3.94H_2O$
31.93	3.59	25	—	$MgSO_4 \cdot 4H_2O + MgSO_4 \cdot 5H_2O$
32.11	3.88	25	1.3521	$MgSO_4 \cdot 5H_2O$
31.85	4.08	25	1.3507	»
31.70	3.74	25	—	»
30.91	3.62	25	1.3399	$MgSO_4 \cdot 4.84H_2O$
31.67	4.02	25	—	$MgSO_4 \cdot 5.43H_2O$
31.13	4.10	25	1.3388	$MgSO_4 \cdot 5.99H_2O$
30.65	4.25	25	1.3466	$MgSO_4 \cdot 6H_2O$
30.33	4.50	25	1.3372	»
29.69	3.80	25	—	$MgSO_4 \cdot 5.87H_2O$
29.42	3.63	25	—	$MgSO_4 \cdot 5.98H_2O$
29.37	5.45	25	1.3165	$MgSO_4 \cdot 5.79H_2O$
26.95	4.77	25	—	$MgSO_4 \cdot 6H_2O$
26.08	5.27	25	—	»
26.56	4.79	25	—	$MgSO_4 \cdot 7H_2O$
24.43	5.21	25	1.2822	»
23.92	6.01	25	1.2868	$MgSO_4 \cdot 6.86H_2O$
34.57	2.64	35	—	$MgCl_2 \cdot 6H_2O + MgSO_4 \cdot 4H_2O$
33.54	2.78	35	—	$MgSO_4 \cdot 4H_2O$
32.63	3.78	35	—	$MgSO_4 \cdot 4.02H_2O$
30.27	4.80	35	—	$MgSO_4 \cdot 5H_2O + MgSO_4 \cdot 6.09H_2O$
29.42	5.64	35	—	$MgSO_4 \cdot 6H_2O$
27.12	6.36	35	—	»

№ 4029 [101]

MAGNESIUM SULFATE–MAGNESIUM CHLORIDE–WATER

$MgSO_4$—$MgCl_2$—H_2O

Content in aqueous soln., wt%		t	Comp. of solid phase
Magnesium Chloride	Magnesium Sulfate		
35.65	0	25	$MgCl_2 \cdot 6H_2O$
34.16	2.50	25	$MgCl_2 \cdot 6H_2O + MgSO_4 \cdot 4H_2O$
36.80	0	45	$MgCl_2 \cdot 6H_2O$
36.26	0.43	45	$MgCl_2 \cdot 6H_2O + MgSO_4 \cdot H_2O$
35.52	2.89	45	$MgCl_2 \cdot 6H_2O + MgSO_4 \cdot 4H_2O$
38.95	0	70	$MgCl_2 \cdot 6H_2O$
38.34	0.40	70	$MgCl_2 \cdot 6H_2O + MgSO_4 \cdot H_2O$
40.96	0	90	$MgCl_2 \cdot 6H_2O$
40.63	0.24	90	$MgCl_2 \cdot 6H_2O + MgSO_4 \cdot H_2O$
43.90	0	110	$MgCl_2 \cdot 6H_2O$
43.54	0.18	110	$MgCl_2 \cdot 6H_2O + MgSO_4 \cdot H_2O$

№ 4030 MAGNESIUM SULFATE–MAGNESIUM CHLORIDE–WATER [424]

$MgSO_4$—$MgCl_2$—H_2O

Content in aqueous soln., wt%		d_4^t	t	Comp. of solid phase
Magnesium Chloride	Magnesium Sulfate			
26.06	4.92	1.2966	25	$MgSO_4 \cdot 7H_2O + MgSO_4 \cdot 6H_2O$
29.89	3.93	1.3239	25	$MgSO_4 \cdot 6H_2O + MgSO_4 \cdot 5H_2O$
32.09	3.06	1.3398	25	$MgSO_4 \cdot 5H_2O + MgSO_4 \cdot 4H_2O$
34.36	2.13	1.3545	25	$MgSO_4 \cdot 4H_2O + MgSO_4 \cdot 6H_2O$
0.0	29.32	1.3399	35	$MgSO_4 \cdot 7H_2O$
5.70	22.21	1.3000	35	»
12.55	14.93	—	35	»
19.22	9.79	1.2846	35	»
19.54	9.62	—	35	$MgSO_4 \cdot 7H_2O + MgSO_4 \cdot 6H_2O$
19.56	9.54	1.2872	35	$MgSO_4 \cdot 6H_2O$
20.79	8.97	1.2896	35	»
21.95	8.14	1.2889	35	»
24.72	6.56	—	35	»
25.33	6.45	1.3104	35	»
25.61	6.44	1.3122	35	$MgSO_4 \cdot 6H_2O + MgSO_4 \cdot 5H_2O$
25.61	6.44	1.3122	35	$MgSO_4 \cdot 5.5H_2O$
25.71	6.26	—	35	$MgSO_4 \cdot 5H_2O$
26.14	6.08	—	35	$MgSO_4 \cdot 6H_2O$
26.14	6.08	—	35	»
26.82	5.77	—	35	$MgSO_4 \cdot 6H_2O + MgSO_4 \cdot 5H_2O$
27.01	5.84	—	35	$MgSO_4 \cdot 5.5H_2O$
30.28	5.33	—	35	$MgSO_4 \cdot 5H_2O$
30.21	4.47	—	35	$MgSO_4 \cdot 4.2H_2O$
30.58	3.88	—	35	$MgSO_4 \cdot 4H_2O$
30.58	3.88	—	35	$MgSO_4 \cdot 3.77H_2O$
31.56	3.49	—	35	$MgSO_4 \cdot 3.85H_2O$
35.09	2.82	—	35	$MgSO_4 + 4H_2O + MgCl_2 \cdot 6H_2O$
35.09	2.82	—	35	$MgSO_4 \cdot 3.77H_2O$
34.79	2.58	—	35	$MgSO_4 \cdot 3.9H_2O$
35.20	1.72	1.3570	35	$MgSO_4 \cdot H_2O + MgCl_2 \cdot 6H_2O$
36.14	0.0	1.3363	35	$MgCl_2 \cdot 6H_2O$

№ 4031 [981]

MAGNESIUM SULFATE–MAGNESIUM CHLORIDE–WATER

$MgSO_4$—$MgCl_2$—H_2O

$t = 25$

Content in aqueous soln., mol/1000 mol Water		Comp. of solid phase
Magnesium Chloride	Magnesium Sulfate	
103.9	0	$MgCl_2 \cdot 6H_2O$
102.4	5.2	$MgCl_2 \cdot 6H_2O + MgSO_4 \cdot H_2O$
84.6	8.6	$MgSO_4 \cdot H_2O + MgSO_4 \cdot 6H_2O$
102.7	5.5	$MgCl_2 \cdot 6H_2O + MgSO_4 \cdot 4H_2O$

Table № 4031 (continuation)

Content in aqueous soln., mol/1000 mol Water		Comp. of solid phase
Magnesium Chloride	Magnesium Sulfate	
89.7	7.7	$MgSO_4 \cdot 4H_2O + MgSO_4 \cdot 5H_2O$
83.6	9.0	$MgSO_4 \cdot 5H_2O + MgSO_4 \cdot 6H_2O$
72.2	10.6	$MgSO_4 \cdot 6H_2O + MgSO_4 \cdot 7H_2O$
0	54.4	$MgSO_4 \cdot 7H_2O$

№ 4032 [759]

MAGNESIUM SULFATE–MAGNESIUM CHLORIDE–WATER

$MgSO_4—MgCl_2—H_2O$

$t = 25$

Content in aqueous soln., mol/1000 mol Water		Comp. of solid phase
Magnesium Chloride	Magnesium Sulfate	
105.1	0	$MgCl_2 \cdot 6H_2O$
102.93	5.41	$MgCl_2 \cdot 6H_2O + MgSO_4 \cdot 4H_2O$
101.83	5.35	»
100.00	9.4	$MgCl_2 \cdot 6H_2O + MgSO_4 \cdot 6H_2O$ (metastable)
93.1	7.06	$MgSO_4 \cdot 4H_2O + MgSO_4 \cdot 5H_2O$
89.9	7.18	$MgSO_4 \cdot 5H_2O$
84.86	8.6	$MgSO_4 \cdot 5H_2O + MgSO_4 \cdot 6H_2O$
97.23	9.15	$MgSO_4 \cdot 6H_2O$ (metastable)
86.67	8.57	»
53.81	16.96	»
39.74	25.45	»
16.18	45.86	»
7.15	54.58	»
0	60.94	»
71.34	10.50	$MgSO_4 \cdot 6H_2O + MgSO_4 \cdot 7H_2O$
84.62	9.81	$MgSO_4 \cdot 7H_2O$ (metastable)
47.18	15.72	$MgSO_4 \cdot 7H_2O$
30.32	23.44	»
24.67	27.82	»
17.66	34.29	»
0	54.9	»

№ 4033 [1022]

MAGNESIUM SULFATE–MAGNESIUM CHLORIDE–WATER

$MgSO_4—MgCl_2—H_2O$

$t = 55$

Content in aqueous soln., wt%		Comp. of solid phase
Magnesium Chloride	Magnesium Sulfate	
37.60	0	$MgCl_2 \cdot 6H_2O$
37.33	0.44	$MgCl_2 \cdot 6H_2O + MgSO_4 \cdot H_2O$
14.90	18.09	$MgSO_4 \cdot H_2O + MgSO_4 \cdot 6H_2O$
0	34.40	$MgSO_4 \cdot 6H_2O$

№ 4034

MAGNESIUM SULFATE–MAGNESIUM CHLORIDE–WATER

[669]

$MgSO_4—MgCl_2—H_2O$

Content in aqueous soln., wt%		t	d^t	Comp. of solid phase
Magnesium Chloride	Magnesium Sulfate			
33.51	1.83	3.7	1.325 (20)	$MgSO_4 \cdot 6H_2O + MgSO_4 \cdot 7H_2O + MgCl_2 \cdot 6H_2O$
32.53	2.33	7.5	1.319 (22)	$MgSO_4 \cdot 6H_2O + MgSO_4 \cdot 7H_2O$
31.68	2.52	10.0	1.313 (20)	»
31.13	2.87	12.5	1.310 (22)	»
30.04	3.07	15.0	1.304 (20)	»
29.40	3.51	17.5	1.299 (22)	»
28.15	3.91	20.0	1.293 (22)	»
27.23	4.91	22.5	1.288 (23)	»
25.99	4.75	25.0	1.287 (25.5)	»
24.82	5.54	27.5	1.280 (28)	»
23.42	6.34	30.0	1.278 (31)	»
21.51	7.67	32.5	1.274 (33)	»
19.61	9.45	35.0	1.273 (35.5)	»

Note. The temperature at which the specific gravity of the soln. was determined is shown in parentheses.

№ 4035 [512]

MAGNESIUM SULFATE–MAGNESIUM CHLORIDE–WATER

$MgSO_4—MgCl_2—H_2O$

$t = 75$

Content in aqueous soln., wt%			Comp. of solid phase
Magnesium Sulfate	Magnesium Chloride	Water	
38.00	0.00	62.00	$MgSO_4 \cdot 6H_2O$ (metastable)
37.75	0.60	61.65	»
36.08	2.06	61.86	$MgSO_4 \cdot 6H_2O + MgSO_4 \cdot 4H_2O$ (metastable)
35.89	2.58	61.53	$MgSO_4 \cdot 6H_2O$ (metastable)
34.24	3.29	62.47	»
31.10	4.92	63.98	»
29.92	6.17	63.91	»
27.85	7.86	64.29	»
25.48	10.37	64.15	»
23.76	11.20	65.04	$MgSO_4 \cdot 6H_2O$ (metastable)
17.94	13.92	68.14	$MgSO_4 \cdot H_2O$
10.19	19.39	70.42	»
8.71	21.31	69.98	»
3.10	28.56	68.34	»
0.89	33.15	65.96	»
0.79	36.34	62.87	»
0.40	38.86	60.74	$MgSO_4 \cdot H_2O + MgCl_2 \cdot 6H_2O$
0.30	38.82	60.88	»
0.00	39.10	60.80	$MgCl_2 \cdot 6H_2O$
28.48	4.80	66.72	$MgSO_4 \cdot H_2O$
27.00	6.16	66.84	»
33.15	2.60	64.25	»
23.00	9.46	67.54	»

№ 4036 [449]

MAGNESIUM SULFATE–MAGNESIUM CHLORIDE–POTASSIUM SULFATE–POTASSIUM CHLORIDE–WATER

$MgSO_4—MgCl_2—K_2SO_4—KCl—H_2O$

$t = 25$

Content in aqueous soln., wt%				Comp. of solid phase
Potassium Chloride	Potassium Sulfate	Magnesium Chloride	Magnesium Sulfate	
22.95	0.0	2.61	1.45	$KCl + K_2SO_4$
18.55	0.0	5.78	2.65	»
17.73	2.92	6.15	0.0	»
15.12	0.0	7.98	4.4	$KCl + K_2SO_4 + K_2SO_4 \cdot MgSO_4 \cdot 6H_2O$
11.83	0.0	3.48	7.97	$K_2SO_4 + K_2SO_4 \cdot MgSO_4 \cdot 6H_2O$
0.0	11.71	4.97	5.44	»
0	10.88	0	12.5	»
9.17	0.0	14.17	5.2	$KCl + K_2SO_4 \cdot MgSO_4 \cdot 6H_2O$
4.96	0.0	19.69	7.11	$KCl + K_2SO_4 \cdot MgSO_4 \cdot 6H_2O + MgSO_4 \cdot 7H_2O$

Content in aqueous soln., wt%				Comp. of solid phase
Potassium Chloride	Potassium Sulfate	Magnesium Chloride	Magnesium Sulfate	
4.14	0.0	4.63	19.65	$K_2SO_4 \cdot MgSO_4 \cdot 6H_2O + MgSO_4 \cdot 7H_2O$
3.97	0.0	22.24	6.13	$KCl + MgSO_4 \cdot 7H_2O$
3.45	0.0	23.68	5.79	$KCl + KCl \cdot MgCl_2 \cdot 6H_2O + MgSO_2 \cdot 7H_2O$
3.36	0.0	25.49	2.66	$KCl + KCl \cdot MgCl_2 \cdot 6H_2O$
4.63	0.0	12.66	11.37	$K_2SO_4 \cdot MgSO_4 \cdot 6H_2O + MgSO_4 \cdot 7H_2O$
10.56	5.35	3.6	0.0	K_2SO_4
4.56	8.07	1.56	0.0	»
5.46	6.38	0.0	1.51	»
4.33	10.39	8.86	0.0	»
11.55	0.48	0.0	7.41	»
7.72	4.56	0.0	4.94	»
4.66	6.76	0.0	2.95	»
5.4	5.60	0.0	8.08	»
0.61	10.26	0.0	0.94	»
0.0	10.8	0.87	1.05	»
8.69	0.0	11.45	6.58	$KCl \cdot MgCl_2 \cdot 6H_2O$
8.86	0.0	4.8	9.78	»
0.0	8.47	7.23	7.91	»
0.0	7.38	9.86	7.77	»
0.0	5.12	4.6	17.62	»
0.0	7.45	3.12	14.69	»

№ 4037 [454]

MAGNESIUM SULFATE–MAGNESIUM CHLORIDE–POTASSIUM SULFATE–POTASSIUM CHLORIDE–WATER

$MgSO_4—MgCl_2—K_2SO_4—KCl—H_2O$

$t = 25$

Content in aqueous soln., mol/1000 mol Water				Comp. of solid phase
2KCl	K_2SO_4	$MgCl_2$	$MgSO_4$	
42.0	1.6	0	0	$KCl + K_2SO_4$
39.2	0.0	1.1	5.1	»
31.9	0.0	12.0	10.3	»
29.9	0.0	15.3	11.4	»
26.2	0.0	22.0	11.7	$KCl + K_2SO_4 + K_2SO_4 \cdot MgSO_4 \cdot 6H_2O$
21.0	0.0	30.0	13.1	$KCl + K_2SO_4 \cdot MgSO_4 \cdot 6H_2O$
15.5	0.0	39.9	13.6	»
10.0	0.0	51.2	14.8	»
8.7	0.0	53.4	15.9	»
8.0	0.0	57.6	14.8	$KCl + K_2SO_4 \cdot MgSO_4 \cdot 6H_2O + MgSO_4 \cdot 7H_2O$
7.2	0.0	61.9	13.6	$KCl + MgSO_4 \cdot 7H_2O$
6.1	0.0	66.2	12.2	»
5.3	0.0	72.0	11.2	$KCl + MgSO_4 \cdot 7H_2O + MgSO_4 \cdot 6H_2O$
3.8	11.6	0.0	24.1	$KCl + KCl \cdot MgCl_2 \cdot 6H_2O + MgSO_4 \cdot 6H_2O$
12.2	5.5	0.0	24.0	$K_2SO_4 + K_2SO_4 \cdot MgSO_4 \cdot 6H_2O$
22.8	0.0	7.2	21.5	»
23.8	0.0	12.0	18.2	»

Table № 4037 (continuation)

Content in aqueous soln., mol/1000 mol Water				Comp. of solid phase
2KCl	K_2SO_4	$MgCl_2$	$MgSO_4$	
8.9	0.1	0.9	52.3	$K_2SO_4 \cdot MgSO_4 \cdot 6H_2O + MgSO_4 \cdot 7H_2O$
11.0	0.1	9.2	45.6	»
11.6	0.1	26.9	31.4	»
10.9	0.1	37.8	24.1	»
3.1	0.1	81.1	8.2	$KCl \cdot MgCl_2 \cdot 6H_2O + MgSO_4 \cdot 6H_2O + MgSO_4 \cdot 5H_2O$
2.3	0.1	88.4	7.1	$KCl \cdot MgCl_2 \cdot 6H_2O + MgSO_4 \cdot 5H_2O + MgSO_4 \cdot 4H_2O$
0.3	0.1	102.5	4.6	$KCl \cdot MgCl \cdot 6H_2O + MgCl_2 \cdot 6H_2O + MgSO_4 \cdot 4H_2O$
11.4	0.1	48.1	25.6	$KCl + KCl \cdot MgSO_4 \cdot 3H_2O + K_2SO_4 \cdot MgSO_4 \cdot 6H_2O$
11.2	0.1	46.9	17.2	$KCl \cdot MgSO_4 \cdot 3H_2O + K_2SO_4 \cdot MgSO_4 \cdot 6H_2O + K_2SO_4 \cdot MgSO_4 \cdot 4H_2O$
9.8	0.1	45.0	20.0	$K_2SO_4 \cdot MgSO_4 \cdot 6H_2O + K_2SO_4 \cdot MgSO_4 \cdot 4H_2O + MgSO_4 \cdot 7H_2O$
8.1	0.1	54.2	16.8	$KCl \cdot MgSO_4 \cdot 3H_2O + K_2SO_4 \cdot MgSO_4 \cdot 4H_2O + MgSO_4 \cdot 7H_2O$
4.0	0.1	67.5	11.5	$KCl \cdot MgSO_4 \cdot 3H_2O + MgSO_4 \cdot 7H_2O + MgSO_4 \cdot 6H_2O$
5.4	0.1	72.1	9.6	$KCl + KCl \cdot MgCl_2 \cdot 6H_2O + KCl \cdot MgSO_4 \cdot 3H_2O$
3.5	0.1	78.4	8.9	$KCl \cdot MgCl_2 \cdot 6H_2O + KCl \cdot MgSO_4 \cdot 3H_2O + MgSO_4 \cdot 6H_2O$
2.4	0.1	82.8	8.1	$KCl \cdot MgCl_2 \cdot 6H_2O + MgSO_4 \cdot 6H_2O + MgSO_4 \cdot H_2O$
0.3	0.1	103.2	4.5	$KCl \cdot MgCl_2 \cdot 6H_2O + MgCl_2 \cdot 6H_2O + MgSO_4 \cdot H_2O$

№ 4038 [485]

MAGNESIUM SULFATE–MAGNESIUM CHLORIDE–POTASSIUM SULFATE–POTASSIUM CHLORIDE–WATER

$MgSO_4—MgCl_2—K_2SO_4—KCl—H_2O$

$t = 35$

Content in aqueous soln., wt%				Comp. of solid phase
Potassium Chloride	Potassium Sulfate	Magnesium Chloride	Magnesium Sulfate	
27.55	1.17	0	0	$KCl + K_2SO_4$
22.32	0.0	3.69	2.71	»
18.37	0.0	6.32	3.97	»
15.38	0.0	8.54	5.41	$KCl + K_2SO_4 +$ Sch
12.95	0.0	11.12	5.81	KCl+Sch
12.00	0.0	11.64	5.51	»
11.72	0.0	12.00	5.33	KCl+Sch+Leo
9.13	0.0	14.84	6.21	KCl+Leo
7.75	0.0	16.80	5.84	KCl+Leo+Kai

Content in aqueous soln., wt%				Comp. of solid phase
Potassium Chloride	Potassium Sulfate	Magnesium Chloride	Magnesium Sulfate	
7.34	0.0	19.00	6.42	KCl+Kai
4.98	0.0	21.65	5.30	»
3.89	0.0	24.28	3.97	KCl+Kai+Car
4.06	0.0	27.41	0	KCl+Car
2.93	0.0	26.81	3.00	Kai+Car
1.32	0.0	29.59	2.50	»
0.85	0.0	31.95	2.39	Kai+Car+$MgSO_4 \cdot 4H_2O$
0.18	0.0	34.85	1.50	Car+$MgCl_2 \cdot 6H_2O$+$MgSO_4 \cdot 4H_2O$
0.26	0	36.26	0	Car+$MgCl_2 \cdot 6H_2O$
14.22	—	5.91	6.95	K_2SO_4+Sch
11.94	—	1.50	11.34	»
2.54	9.31	—	14.01	»
0	12.14	0	14.94	»
0.92	0.0	30.91	3.21	Kai+$MgSO_4 \cdot 4H_2O$+$MgSO_4 \cdot 5H_2O$
0.82	0.0	27.57	4.70	Kai+$MgSO_4 \cdot 5H_2O$
1.61	0.0	24.80	6.35	Kai+$MgSO_4 \cdot 5H_2O$+$MgSO_4 \cdot 6H_2O$
2.27	0.0	22.13	8.20	Kai+$MgSO_4 \cdot 6H_2O$
3.64	0.0	19.58	9.87	»
3.70	0.0	17.69	10.80	Kai+Leo+$MgSO_4 \cdot 6H_2O$
3.43	0.0	16.70	11.50	Leo+$MgSO_4 \cdot 6H_2O$+$MgSO_4 \cdot 7H_2O$
3.58	0.0	15.65	12.20	Leo+$MgSO_4 \cdot 7H_2O$
3.94		12.65	14.41	»
4.05	0.0	12.44	15.05	Leo+Sch+$MgSO_4 \cdot 7H_2O$
4.79	0.0	11.47	15.70	Sch+$MgSO_4 \cdot 7H_2O$
4.24	0.0	9.87	17.37	»
4.15	0.0	6.20	21.77	»
0	4.23	0	29.49	»

Note. Kai — Kainite ($KCl \cdot MgCl_2 \cdot 3H_2O$); Car — Carnallite ($KCl \cdot MgCl_2 \cdot 6H_2O$); Leo — Leonite ($K_2SO_4 \cdot MgSO_4 \cdot 4H_2O$); Sch-Schoenite ($K_2SO_4 \cdot MgSO_4 \cdot 6H_2O$).

№ 4039 [816]

MAGNESIUM SULFATE–MAGNESIUM CHLORIDE–POTASSIUM SULFATE–POTASSIUM CHLORIDE–WATER

$MgSO_4$—$MgCl_2$—K_2SO_4—KCl—H_2O

Content in aqueous soln., wt%					t	Comp. of solid phase
Magnesium Sulfate	Potassium Sulfate	Magnesium Chloride	Potassium Chloride	Water		
7.18	0.00	19.81	4.77	68.24	25	$K_2SO_4 \cdot MgSO_4 \cdot 6H_2O$+KCl+$MgSO_4 \cdot 7H_2O$
6.80	0.00	19.15	5.03	69.02	25	$K_2SO_4 \cdot MgSO_4 \cdot 6H_2O$+KCl
6.75	0.00	19.19	5.28	68.78	25	»

Table № 4039 (continuation)

Content in aqueous soln., wt%					t	Comp. of solid phase
Magnesium Sulfate	Potassium Sulfate	Magnesium Chloride	Potassium Chloride	Water		
6.05	0.00	17.62	5.84	70.49	25	$K_2SO_4 \cdot MgSO_4 \cdot 6H_4O+KCl$
5.79	0.00	16.84	6.88	70.49	25	»
5.56	0.00	15.74	8.24	70.46	25	»
5.18	0.00	14.14	9.23	71.45	25	»
4.95	0.00	13.35	9.92	71.78	25	»
4.85	0.00	12.33	10.80	72.02	25	»
4.76	0.00	11.81	11.06	72.37	25	»
4.64	0.00	10.89	11.82	72.65	25	»
4.51	0.00	9.16	14.09	72.24	25	»
4.56	0.00	9.05	14.43	71.96	25	»
4.47	0.00	8.62	14.30	72.61	25	»
4.40	0.00	8.25	14.76	72.59	25	»
4.44	0.00	8.08	14.91	72.57	25	$K_2SO_4 \cdot MgSO_4 \cdot 6H_2O+KCl+K_2SO_4$
6.54	0.00	5.12	12.63	75.71	25	»
6.54	0.00	5.12	12.60	75.74	25	»
6.73	0.00	3.94	12.47	76.86	25	»
7.17	0.00	4.39	12.28	76.16	25	»
8.55	0.00	2.70	11.57	77.18	25	»
9.04	0.00	2.29	11.53	77.14	25	»
10.32	0.00	1.03	11.09	77.56	25	»
11.18	0.00	0.21	10.96	77.65	25	»
11.07	0.00	0.21	10.86	77.86	25	»
10.97	0.59	0.00	9.99	78.45	25	»
11.33	3.69	0.00	7.19	77.79	25	»
11.67	7.19	0.00	4.06	77.08	25	»
11.62	7.26	0.00	4.04	77.08	25	»
12.62	11.15	0.00	0.00	76.23	25	»
26.12	4.22	0.00	0.00	69.66	25	$K_2SO_4 \cdot MgSO_4 \cdot 6H_2O+MgSO_4 \cdot 7H_2O$
24.74	0.51	0.00	3.39	71.36	25	»
23.57	0.00	1.70	4.09	70.64	25	»
21.30	0.00	3.54	4.15	71.01	25	»
19.09	0.00	5.16	4.10	71.65	25	»
17.24	0.00	6.63	4.09	72.04	25	»
15.35	0.00	8.45	4.12	72.08	25	»
14.46	0.00	9.47	4.21	71.86	25	»
12.80	0.00	11.29	4.31	71.60	25	»
11.39	0.00	12.40	4.24	71.97	25	»
10.76	0.00	13.42	4.31	71.51	25	»
9.42	0.00	15.14	4.37	71.06	25	»
8.04	0.00	17.69	4.62	69.65	25	»
7.21	0.00	19.48	4.77	68.54	25	»
7.18	0.00	19.81	4.77	68.24	25	$K_2SO_4 \cdot MgSO_4 \cdot 6H_2O+MgSO_4 \cdot 7H_2O+KCl$
4.13	0.00	7.39	15.87	72.61	25	$KCl+K_2SO_4$
4.96	0.00	7.12	14.35	73.57	25	$K_2SO_4 \cdot MgSO_4 \cdot 6H_2O+K_2SO_4$
6.29	0.00	6.57	12.05	75.09	25	»
8.57	0.00	6.15	9.44	75.84	25	»
10.70	0.00	5.59	7.60	76.11	25	$K_2SO_4 \cdot MgSO_4 \cdot 6H_2O$

Content in aqueous soln., wt%					t	Comp. of solid phase
Magnesium Sulfate	Potassium Sulfate	Magnesium Chloride	Potassium Chloride	Water		
12.41	0.00	5.02	6.64	75.93	25	$K_2SO_4 \cdot MgSO_4 \cdot 6H_2O$
14.33	0.00	4.12	6.25	75.30	25	»
14.35	0.00	4.31	6.01	75.33	25	»
10.64	0.00	0.98	10.24	78.14	25	»
9.04	0.00	2.32	11.39	77.25	25	$K_2SO_4 \cdot MgSO_4 \cdot 6H_2O+K_2SO_4$
7.85	0.00	3.62	11.83	76.70	25	»
7.09	0.00	5.40	11.69	75.82	25	$K_2SO_4 \cdot MgSO_4 \cdot 6H_2O$
6.58	0.00	7.25	11.09	75.08	25	»
6.55	0.00	7.31	11.12	75.02	25	»
5.93	0.00	18.06	5.06	70.95	25	$K_2SO_4 \cdot MgSO_4 \cdot 6H_2O+K_2SO_4$
4.68	0.00	12.43	9.96	72.93	25	»
5.94	0.00	8.25	11.12	74.69	25	$K_2SO_4 \cdot MgSO_4 \cdot 6H_2O$
7.49	0.00	4.95	11.60	75.96	25	»
8.30	0.00	2.90	11.77	77.03	25	$K_2SO_4 \cdot MgSO_4 \cdot 6H_2O+K_2SO_4$
9.31	0.00	1.44	11.50	77.75	25	K_2SO_4
16.40	0.00	7.50	4.21	71.89	25	$K_2SO_4 \cdot MgSO_4 \cdot 6H_2O+MgSO_4 \cdot 7H_2O$
11.64	0.00	7.98	5.80	74.58	25	$K_2SO_4 \cdot MgSO_4 \cdot 6H_2O$
8.01	0.00	7.29	8.99	75.71	25	»
6.81	0.00	6.55	11.29	75.35	25	»
6.17	0.00	6.04	12.89	74.90	25	»
5.86	0.00	5.51	13.96	74.67	25	K_2SO_4
5.85	0.00	5.55	13.88	74.72	25	»
9.01	0.00	12.95	5.31	72.73	25	$K_2SO_4 \cdot MgSO_4 \cdot 6H_2O$
9.04	0.00	12.90	5.44	72.62	25	»
7.59	0.00	12.48	6.56	73.37	25	»
6.45	0.00	11.77	8.65	73.13	25	»
6.43	0.00	11.88	8.53	73.16	25	»
5.45	0.00	10.51	10.81	73.23	25	»
5.47	0.00	10.55	10.69	73.29	25	»
5.43	0.00	10.67	10.62	73.28	25	»
4.60	0.00	8.27	14.33	72.80	25	»
4.39	0.00	7.13	16.05	72.43	25	K_2SO_4
9.55	0.00	15.23	4.21	71.01	25	$MgSO_4 \cdot 7H_2O+K_2SO_4 \cdot MgSO_4 \cdot 6H_2O$
8.78	0.00	15.06	4.53	71.63	25	$K_2SO_4 \cdot MgSO_4 \cdot 6H_2O$
7.39	0.00	14.29	6.40	71.92	25	»
6.26	0.00	12.79	8.75	72.20	25	»
7.58	0.00	17.73	5.08	69.61	25	»
6.05	0.00	17.07	6.16	70.72	25	»
5.53	0.00	15.63	7.77	71.07	25	$K_2SO_4 \cdot MgSO_4 \cdot 6H_2O+KCl$
6.63	0.00	17.43	6.67	69.27	25	$K_2SO_4 \cdot MgSO_4 \cdot 6H_2O$
6.58	0.00	17.47	6.56	69.39	25	»
5.87	0.00	18.77	4.64	70.72	15	$K_2SO_4 \cdot MgSO_4 \cdot 6H_2O+MgSO_4 \cdot 7H_2O+KCl$
5.84	0.00	18.72	4.67	70.77	15	»
5.39	0.00	17.51	5.55	71.55	15	$K_2SO_4 \cdot MgSO_4 \cdot 6H_2O+KCl$
4.72	0.00	15.18	7.52	72.58	15	»

Table № 4039 (continuation)

Comp. of saturated soln. wt %					t	Comp. of solid phase
Magnesium Sulfate	Potassium Sulfate	Magnesium Chloride	Potassium Chloride	Water		
4.25	0.00	12.42	9.66	73.67	15	$K_2SO_4 \cdot MgSO_4 \cdot 6HO_2 + KCl$
4.04	0.00	10.74	11.18	74.02	15	»
3.90	0.00	9.17	12.72	74.21	15	»
3.83	0.00	8.45	13.44	74.28	15	»
3.84	0.00	8.44	13.44	74.28	15	»
3.81	0.00	7.71	14.14	74.34	15	»
3.78	0.00	7.62	14.18	74.42	15	$K_2SO_4 \cdot MgSO_4 \cdot 6H_2O + KCl + K_2SO_4$
3.80	0.00	7.65	14.12	74.43	15	»
23.79	0.00	4.29	0.00	71.92	15	$K_2SO_4 \cdot MgSO_4 \cdot 6H_2O + MgSO_4 \cdot 7H_2O$
23.75	0.00	4.30	0.00	71.95	15	»
23.31	0.00	1.15	2.62	72.92	15	»
23.50	0.00	1.18	2.62	72.30	15	»
21.92	0.91	0.00	3.95	73.22	15	»
21.93	0.85	0.00	3.95	73.27	15	»
18.88	3.23	0.00	4.16	73.73	15	»
16.17	5.41	0.00	4.19	74.23	15	»
13.61	7.40	0.00	4.10	74.89	15	»
11.24	10.49	0.00	4.55	73.72	15	»
9.07	12.99	0.00	4.67	73.27	15	»
7.27	15.71	0.00	4.73	72.29	15	»
6.28	17.35	0.00	4.77	71.60	15	»
5.87	18.74	0.00	4.64	70.72	15	$K_2SO_4 \cdot MgSO_4 \cdot 6H_2O + MgSO_4 \cdot 7H_2O + KCl$
5.84	18.72	0.00	4.67	70.77	15	»
11.92	8.80	0.00	0.00	79.28	15	$K_2SO_4 \cdot MgSO_4 \cdot 6H_2O + K_2SO_4$
10.24	4.56	0.00	4.91	80.29	15	»
10.14	1.74	0.00	7.93	80.19	15	»
9.60	0.00	0.39	9.89	80.12	15	»
9.54	0.00	0.44	9.84	80.18	15	»
8.71	0.00	0.21	10.41	79.67	15	»
8.75	0.00	1.18	10.45	79.62	15	»
7.94	0.00	2.02	10.57	79.47	15	»
7.35	0.00	2.71	10.96	78.98	15	»
7.36	0.00	2.67	11.00	78.97	15	»
6.09	0.00	4.19	11.71	78.01	15	»
5.87	0.00	4.59	11.84	77.70	15	»
5.82	0.00	4.63	11.84	77.71	15	»
5.31	0.00	5.04	11.86	77.79	15	»
4.55	0.00	5.91	12.57	76.97	15	»
3.84	0.00	6.95	13.15	76.06	15	»
3.80	0.00	7.65	14.12	74.43	15	$K_2SO_4 \cdot MgSO_4 \cdot 6H_2O + KCl + K_2SO_4$
3.81	0.00	7.64	14.11	74.44	15	»
3.80	0.00	7.66	14.13	74.41	15	»

№ 4040 [512]

MAGNESIUM SULFATE–MAGNESIUM CHLORIDE–POTASSIUM SULFATE–POTASSIUM CHLORIDE–WATER

$MgSO_4$—$MgCl_2$—K_2SO_4—KCl—H_2O

$t = 75$

Comp. of saturated soln., wt %.					Comp. of solid phase
Potassium Chloride	Magnesium Chloride	Magnesium Sulfate	Potassium Sulfate	Water	
32.58	0.00	0.00	1.45	65.97	$KCl + K_2SO_4$
29.34	2.15	0.00	2.18	66.33	»
25.32	4.50	0.00	3.21	66.97	»
21.92	7.05	0.00	4.17	66.86	»
15.01	10.92	0.00	7.18	66.89	»
13.62	12.06	0.00	8.62	65.70	$KCl + K_2SO_4 + K_2SO_4 \cdot 4H_2O$
13.28	12.10	0.00	8.93	65.69	»
11.56	13.17	0.00	9.29	65.45	$KCl + K_2SO_4 \cdot MgSO_4 \cdot 4H_2O$
8.83	15.93	0.00	9.94	65.30	»
14.65	12.80	5.92	0.00	66.63	$KCl + K_2SO_4 \cdot MgSO_4 \cdot 4H_2O +$ $+ K_2SO_4 \cdot 2MgSO_4$
15.11	13.00	6.08	0.00	65.81	»
12.45	16.78	4.47	0.00	66.30	$KCl + K_2SO_4 \cdot 2MgSO_4$
10.70	18.75	4.13	0.00	66.42	$KCl + K_2SO_4 \cdot 2MgSO_4 + MgCl_2 \cdot 6H_2O$
10.78	19.30	3.77	0.00	66.15	»
5.05	24.80	0.00	3.81	66.34	$KCl + MgCl_2 \cdot 6H_2O$
7.57	23.80	2.38	0.00	66.25	»
4.25	29.71	0.00	1.72	64.29	$KCl + MgCl_2 \cdot 6H_2O + MgSO_4 \cdot H_2O$
5.57	29.25	0.00	0.00	65.18	$KCl + MgSO_4 \cdot H_2O$
0.00	0.00	17.32	16.32	66.28	$K_2SO_4 + K_2SO_4 \cdot MgSO_4 \cdot 4H_2O$
2.66	0.00	16.48	12.34	68.52	»
6.07	0.00	16.20	9.11	68.62	»
9.90	0.00	15.29	4.70	70.11	»
14.10	0.00	14.90	0.91	70.09	»
16.21	0.94	13.30	0.00	69.55	»
16.57	3.50	10.30	0.00	69.63	»
0.00	0.00	31.20	6.99	61.80	$K_2SO_4 \cdot MgSO_4 \cdot 4H_2O +$ $+ K_2SO_4 \cdot 2MgSO_4$
3.11	0.00	30.15	3.72	63.02	»
5.67	0.00	28.50	1.51	64.32	»
7.77	0.87	25.44	0.00	65.92	»
8.28	3.23	21.40	0.00	67.09	»
10.13	6.39	16.15	0.00	67.90	»
12.45	10.34	10.07	0.00	67.14	»
13.83	12.31	6.91	0.00	66.95	»
0.00	0.00	35.56	3.67	60.77	$K_2SO_4 \cdot 2MgSO_4 + MgSO_4 \cdot H_2O$
0.00	1.92	31.05	3.94	63.09	»
3.40	3.41	27.25	0.00	65.94	»
3.00	5.69	24.68	0.00	66.63	»
3.20	7.64	22.04	0.00	67.12	»
3.16	10.56	18.47	0.00	67.81	»
4.01	15.66	12.84	0.00	67.49	»
4.38	19.66	7.86	0.00	68.10	»
4.19	22.03	6.20	0.00	67.58	»

Table № 4040 (continuation)

Comp. of saturated soln., wt %					Comp. of solid phase
Potassium Chloride	Magnesium Chloride	Magnesium Sulfate	Potassium Sulfate	Water	
4.74	23.33	4.67	0.00	68.66	$K_2SO_4 \cdot 2MgSO_4 + MgSO_4 \cdot H_2O + KCl \cdot MgSO_4 \cdot 3H_2O$
4.55	24.28	4.12	0.00	67.05	»
3.54	30.95	1.31	0.00	64.20	$KCl \cdot MgCl_2 \cdot 6H_2O + KCl \cdot MgSO_4 \cdot 3H_2O + MgSO_4 \cdot H_2O$
0.00	38.86	0.40	0.00	60.74	$MgSO_4 \cdot H_2O$ + Bischofite
0.32	38.64	1.01	0.00	0.03	$KCl \cdot MgCl_2 \cdot 6H_2O + MgSO_4 \cdot H_2O$ + Bischofite
0.32	38.85	0.00	0.00	60.83	$KCl \cdot MgCl_2 \cdot 6H_2O$ + Bischofite
6.00	21.49	5.01	0.00	67.50	$K_2SO_4 \cdot 2MgSO_4 + KCl \cdot MgSO_4 \cdot 3H_2O$

№ 4041 [1912]

MAGNESIUM SULFATE–MAGNESIUM CHLORIDE–POTASSIUM SULFATE–POTASSIUM CHLORIDE–WATER

$MgSO_4$—$MgCl_2$—K_2SO_4—KCl—H_2O

$t = 55$

Comp. of saturated soln., wt %					d	Comp. of solid phase
Potassium chloride	Magnesium Chloride	Potassium sulphate	Magnesium Sulfate	Water		
12.41	11.76	7.53	0.00	68.30	—	$K_2SO_4 \cdot MgSO_4 \cdot 4H_2O + K_2SO_4 + KCl$
2.86	20.53	9.19	0.00	67.42	—	$K_2SO_4 \cdot MgSO_4 \cdot 4H_2O + KCl \cdot MgSO_4 \cdot 3H_2O + KCl$
0.00	21.71	10.72	0.18	67.39	1.2996	$K_2SO_4 \cdot MgSO_4 \cdot 4H_2O + K_2SO_4 \cdot 2MgSO_4 + KCl \cdot MgSO_4 \cdot 3H_2O$
0.00	18.28	7.46	6.71	67.55	—	$K_2SO_4 \cdot MgSO_4 \cdot 4H_2O + K_2SO_4 \cdot 2MgSO_4$
0.00	12.90	4.22	17.58	65.30	1.3628	$K_2SO_4 \cdot MgSO_4 \cdot 4H_2O + K_2SO_4 \cdot 2MgSO_4 + MgSO_4 \cdot H_2O$
0.00	10.71	4.37	19.90	65.02	1.3753	$K_2SO_4 \cdot MgSO_4 \cdot 4H_2O + MgSO_4 \cdot H_2O + MgSO_4 \cdot 6H_2O$
0.00	14.90	0.00	18.09	67.01	1.3503	$K_2SO_4 \cdot 6H_2O + MgSO_4 \cdot H_2O$
0.00	15.55	3.99	14.29	66.17	—	$K_2SO_4 \cdot 2MgSO_4 + MgSO_4 \cdot H_2O$
0.00	19.30	2.98	10.82	66.90	—	»
0.00	22.44	3.29	6.66	67.61	1.3140	$K_2SO_4 \cdot 2MgSO_4 + MgSO_4 \cdot H_2O + KCl \cdot MgSO_4 \cdot 3H_2O$
0.00	27.03	3.42	2.56	66.99	—	$KCl \cdot MgSO_4 \cdot 3H_2O + MgSO_4 \cdot H_2O$
0.00	21.75	6.47	3.74	68.04	—	$K_2SO_4 \cdot 2MgSO_4 + KCl \cdot MgSO_4 \cdot 3H_2O$
0.20	36.15	0.00	1.16	62.49	—	$KCl \cdot MgCl_2 \cdot 6H_2O + MgSO_4 \cdot H_2O + MgCl_2 \cdot 6H_2O$

№ 4042 [2689]

MAGNESIUM SULFATE–MAGNESIUM CHLORIDE–POTASSIUM CHLORIDE–WATER

$MgSO_4$—$MgCl_2$—KCl—H_2O

$t = 25$

Content in aqueous soln., mol/1000 mol Water			Comp. of solid phase
2KCl	$MgCl_2$	$MgSO_4$	
25	21	11	KCl+K_2SO_4+Sch
9	55	16	KCl+Sch+$MgSO_4 \cdot 7H_2O$
8	62	15	KCl+$MgSO_4 \cdot 7H_2O$+$MgSO_4 \cdot 6H_2O$
2	98	12	KCl+Car+$MgSO_4 \cdot 6H_2O$
2	99	12	Car+$MgSO_4 \cdot 6H_2O$+$MgSO_4 \cdot 6H_2O$
25	21	11	KCl+K_2SO_4+Sch
4.7	70.0	13.7	KCl+Car+$MgSO_4 \cdot 6H_2O$

Note. Car — Carnallite KCl · $MgCl_2 \cdot 6H_2O$; Sch — Schoenite $K_2SO_4 \cdot MgSO_4 \cdot 6H_2O$.

№ 4043 [2756]

MAGNESIUM SULFATE–MAGNESIUM CHLORIDE–POTASSIUM CHLORIDE–WATER

$MgSO_4$—$MgCl_2$—KCl—H_2O

$t = 25$

Content in aqueous soln., mol/1000 mol Water			Comp. of solid phase
Potassium Chloride	Magnesium Chloride	Magnesium Sulfate	
24.97	20.45	9.84	KCl+K_2SO_4+$K_2SO_4 \cdot 6H_2O$
8	57.5	15.78	KCl+$K_2SO_4 \cdot MgSO_4 \cdot 6H_2O$+$MgSO_4 \cdot 7H_2O$
8.15	60.45	15.7	KCl+$MgSO_4 \cdot 7H_2O$+$MgSO_4 \cdot 6H_2O$
4.8	72.4	13.3	KCl+KCl · $MgSO_4 \cdot 6H_2O$+$MgSO_4 \cdot 6H_2O$
2.01	98.6	12.16	KCl · $MgSO_4 \cdot 6H_2O$+$MgSO_4 \cdot 6H_2O$+$MgCl_2 \cdot 6H_2O$

№ 4044 [1462]

MAGNESIUM SULFATE–MAGNESIUM CHLORIDE–POTASSIUM CHLORIDE–WATER

$MgSO_4$—$MgCl_2$—KCl—H_2O

Content in aqueous soln., mol/1000 mol Water			t	Comp. of solid phase
2KCl	$MgCl_2$	$MgSO_4$		
19.8	18.8	5.0	0	KCl+K_2SO_4+Sch
8.1	42.8	8.5	0	KCl+Sch+$MgSO_4 \cdot 7H_2O$
4.2	67.0	4.4	0	KCl+Car+$MgSO_4 \cdot 7H_2O$

Table № 4044 (continuation)

Content in aqueous soln., mol/1000 mol Water			t	Comp. of solid phase
2KCl	$MgCl_2$	$MgSO_4$		
5.4	69.2	7.5	12.5	$KCl+Sch+MgSO_4 \cdot 7H_2O$
10.6	52.1	14.6	25	$KCl+Kai+Sch$
6.4	67.5	13.9	25	$KCl+Car+Kai$
26.7	23.5	11.3	35	$KCl+K_2SO_4+Sch$
13.8	45.9	12.8	35	$KCl+Kai+Leo$
6.7	22.6	46.6	35	$Sch+Leo+MgSO_4 \cdot 7H_2O$
4.3	56.9	18.9	35	$Kai+Leo+MgSO_4 \cdot 6H_2O$
33.3	21.2	11.4	55	$KCl+K_2SO_4+Leo$
28.7	31.6	13.2	55	$KCl+Kai+Leo$
15.4	41.4	13.0	55	$Kai+Leo+Lb$
6.9	32.5	46.1	55	$Leo+Lb+MgSO_4 \cdot H_2O$
6.2	80.3	4.3	55	$KCl+Car+Kai$
3.6	50.9	33.1	55	$Kai+Lb+MgSO_4 \cdot H_2O$
37.7	21.0	13.3	70	$KCl+K_2SO_4+Lb$
24.4	39.9	15.4	70	$KCl+Kai+Lb$
46.4	17.4	13.8	70	$KCl+K_2SO_4+Lb$
13.1	88.0	5.0	90	$KCl+Lb+MgSO_4 \cdot H_2O$

Note. Kai — Kainite $KCl \cdot MgSO_4 \cdot 3H_2O$; Car — Carnallite· KCl $MgCl_2$ $6H_2O$; Lb — Langbeinite $K_2SO_4 \cdot 2MgSO_4$; Leo —Leonite $K_2SO_4 \cdot MgSO_4 \cdot 4H_2O$; Sch — Schoe $K_2SO_4 \cdot MgSO_4 \cdot 4H_2O$.

№ 4045 [1914]

MAGNESIUM SULFATE–MAGNESIUM CHLORIDE–POTASSIUM CHLORIDE–WATER

$MgSO_4—MgCl_2—KCl—H_2O$

Content in aqueous soln., mol/1000 mol Water			t	Comp. of solid phase
2KCl	$MgCl_2$	$MgSO_4$		
20.5	19.3	7.8	0	$KCl+K_2SO_4+K_2SO_4 \cdot MgSO_4$ $6H_2O$
17.2	22.2	6.9	0	»
23.5	20.9	9.1	15	»
5.8	55.1	11.6	15	$KCl+KCl \cdot MgSO_4 \cdot 3H_2O+$ $+K_2SO_4 \cdot MgSO_4 \cdot 6H_2O$
5.2	57.8	11.6	15	$K_2SO_4 \cdot MgSO_4 \cdot 6H_2O+MgSO_4 \cdot 7H_2O$
25	24.9	9.9	25	$KCl+K_2SO_4 \cdot MgSO_4 \cdot 6H_2O$
10.4	33.8	9.6	25	»
5.3	38.8	13.7	25	—
4.1	43.2	12.9	25	$K_2SO_4 \cdot MgSO_4 \cdot 6H_2O+MgSO_4 \cdot H_2O$
3.6	56.5	3.9	25	—
1.4	56.4	7.6	25	—
27.1	20.7	9.3	35	—
18.2	22.2	9.6	35	—
5.4	47.5	29.0	35	$K_2SO_4 \cdot MgSO_4 \cdot xH_2O+MgSO_4 \cdot xH_2O$
4.3	57.3	18.4	35	—
4.6	54.0	4.6	35	$KCl+KCl \cdot MgCl_2 \cdot 6H_2O$
2.0	58.3	6.9	35	—

Table № 4045 (continuation)

Content in aqueous soln., mol/1000 mol Water			t	Comp. of solid phase
2KCl	$MgCl_2$	$MgSO_4$		
30.3	22.7	11.0	45	$KCl+K_2SO_4+K_2SO_4 \cdot MgSO_4 \cdot 4H_2O$
17.1	47.0	13.4	45	$KCl+KCl \cdot MgSO_4 \cdot 3H_2O+$ $+K_2SO_4 \cdot MgSO_4 \cdot 4H_2O$
15.6	48.1	15.0	45	»
32.1	23.0	12.0	55	$KCl+K_2SO_4+K_2SO_4 \cdot MgSO_4 \cdot 4H_2O$
25.1	40.1	14.8	55	$KCl \cdot MgSO_4 \cdot 3H_2O$
15.1	50.0	18.2	55	—
13.7	56.3	8.8	55	—
4.8	58.5	22.1	55	$KCl \cdot MgSO_4 \cdot 3H_2O+K_2SO_4 \cdot 2MgSO_4+$ $+MgSO_4 \cdot H_2O$
35.7	20.7	11.6	65	$K_2SO_4+K_2SO_4 \cdot 2MgSO_4+$ $+K_2SO_4 \cdot MgSO_4 \cdot 4H_2O$
27.7	35.0	14.9	65	$KCl+KCl \cdot MgSO_4 \cdot 3H_2O+$ $+K_2SO_4 \cdot MgSO_4 \cdot 4H_2O$
2.2	66.9	19.4	65	$K_2SO_4 \cdot MgSO_4 \cdot 4H_2O+MgSO_4 \cdot xH_2O$
1.8	98.5	4.9	65	$KCl \cdot MgCl_2 \cdot 6H_2O+MgSO_4 \cdot xH_2O+$ $+MgCl_2 \cdot 6H_2O$
29.0	35.6	16.0	71	$KCl+K_2SO_4 \cdot 2MgSO_4$
56.8	9.3	8.0	100	$KCl+K_2SO_4$
3.7	138.1	5.3	100	—

№ 4046 [979]

MAGNESIUM SULFATE–MAGNESIUM CHLORIDE–POTASSIUM CHLORIDE–WATER

$MgSO_4$—$MgCl_2$—KCl—H_2O

Content in aqueous soln., mol/1000 mol Water			t	Comp. of solid phase
2KCl	$MgCl_2$	$MgSO_4$		
13.7	50.1	14.2	35	$KCl+K_2SO_4 \cdot MgSO_4 \cdot 4H_2O$
8.6	74.6	4.3	35	$KCl+KCl \cdot MgCl_2 \cdot 6H_2O$
12.0	50.7	20.0	35	$K_2SO_4 \cdot MgSO_4 \cdot 4H_2O+MgSO_4 \cdot 6H_2O$
4.2	88.4	7.6	35	$KCl \cdot MgCl_2 \cdot 6H_2O+MgSO_4 \cdot H_2O$
2.8	91.2	6.9	35	»
5.1	81.1	14.5	35	$MgSO_4 \cdot 6H_2O+MgSO_4 \cdot H_2O$
3.4	81.7	12.4	35	»
25.83	44.02	14.18	55	$KCl+K_2SO_4 \cdot MgSO_4 \cdot 4H_2O$
10.81	82.63	3.21	55	$KCl+KCl \cdot MgCl_2 \cdot 6H_2O$
8.59	94.98	4.46	55	$KCl \cdot MgCl_2 \cdot 6H_2O+MgSO_4 \cdot H_2O$
16.70	49.28	16.30	55	$K_2SO_4 \cdot MgSO_4 \cdot 4H_2O+K_2SO_4 \cdot 2MgSO_4$
5.10	63.40	22.20	55	$K_2SO_4 \cdot 2MgSO_4+MgSO_4 \cdot H_2O$

№ 4047 [1471]

MAGNESIUM SULFATE–MAGNESIUM CHLORIDE–POTASSIUM CHLORIDE–WATER

$MgSO_4—MgCl_2—KCl—H_2O$

Content in aqueous soln., mol/1000 mol Water			t	Comp. of solid phase
2KCl	$MgCl_2$	$MgSO_4$		
32.4	12.5	5.6	25	$KCl+K_2SO_4$
27.6	17.7	7.8	25	»
24.2	22.5	10.2	25	$KCl+K_2SO_4 \cdot MgSO_4 \cdot 6H_2O$
15.7	36.2	10.6	25	»
9.8	52.3	15.4	25	$KCl+K_2SO_4 \cdot MgSO_4 \cdot 6H_2O$ (metastable)
9.3	54.2	15.9	25	»
9.2	53.5	15.6	25	»
41.9	5.2	2.9	35	$KCl+K_2SO_4$
28.8	20.2	9.4	35	»
31.8	15.6	10.2	35	$KCl+K_2SO_4 \cdot MgSO_4 \cdot 6H_2O$ (metastable)
16.9	36.2	12.7	35	»
17.5	36.4	12.75	35	»
9.8	53.8	21.7	35	»
17.35	37.0	11.85	35	$KCl+K_2SO_4 \cdot MgSO_4 \cdot 4H_2O$
10.4	53.1	14.0	35	$KCl+K_2SO_4 \cdot MgSO_4 \cdot 4H_2O$ (metastable)
10.5	53.7	13.9	35	»
9.2	58.2	15.2	35	»
10.5	55.3	10.0	35	$KCl+KCl \cdot MgSO_4 \cdot 3H_2O$
7.2	67.0	5.9	35	»
6.6	72.6	5.0	35	»
6.8	73.0	4.1	35	»
8.4	60.2	19.2	35	$KCl+MgSO_4 \cdot 6H_2O$ (metastable)
37.7	15.9	7.4	55	$KCl+K_2SO_4$
27.5	29.4	12.2	55	$KCl+K_2SO_4 \cdot MgSO_4 \cdot 4H_2O$
19.0	43.0	13.8	55	»
18.0	45.1	14.5	55	$KCl+K_2SO_4 \cdot MgSO_4 \cdot 4H_2O$ (metastable)

№ 4048 [1472]

MAGNESIUM SULFATE–MAGNESIUM CHLORIDE–POTASSIUM CHLORIDE–WATER

$MgSO_4—MgCl_2—KCl—H_2O$

Content in aqueous soln., g/100g Water			t	Comp. of solid phase
Potassium Chloride	Magnesium Chloride	Magnesium Sulfate		
33.85	0.0	1.34	25	$KCl+K_2SO_4$
27.56	5.29	3.14	25	»
21.36	10.57	6.28	25	»
20.86	11.10	6.75	25	$KCl+K_2SO_4+K_2SO_4 \cdot MgSO_4 \cdot 6H_2O$
15.98	15.86	6.62	25	$KCl+K_2SO_4 \cdot MgSO_4 \cdot 6H_2O$
11.59	21.14	7.68	25	»

Content in aqueous soln., g/100g Water			t	Comp. of solid phase
Potassium Chloride	Magnesium Chloride	Magnesium Sulfate		
8.53	26.43	9.56	25	$KCl+K_2SO_4 \cdot MgSO_4 \cdot 6H_2O$
7.12	29.60	10.96	25	$KCl+K_2SO_4 \cdot MgSO_4 \cdot 6H_2O+$ $+MgSO_4 \cdot 7H_2O$ (metastable)
6.13	31.72	12.03	25	$KCl+K_2SO_4 \cdot MgSO_4 \cdot 6H_2O$ (metastable)
8.03	27.49	9.96	25	$KCl+K_2SO_4 \cdot MgSO_4 \cdot 6H_2O+$ $+KCl \cdot MgSO_4 \cdot 3H_2O$
6.13	31.72	7.22	25	$KCl+KCl \cdot MgSO_4 \cdot 3H_2O$
3.97	37.00	4.88	25	»
5.63	32.77	10.02	25	$KCl+MgSO_4 \cdot 6H_2O+MgSO_4 \cdot 7H_2O$ (metastable)
6.13	31.72	10.36	25	$KCl+MgSO_4 \cdot 7H_2O$ (metastable)
3.97	37.00	9.02	25	»
3.89	38.27	4.54	25	—
30.30	5.29	3.01	35	$KCl+K_2SO_4$
23.59	10.57	6.15	35	»
37.50	0.0	1.34	35	»
21.25	12.42	7.48	35	$KCl+K_2SO_4+K_2SO_4 \cdot MgSO_4 \cdot 6H_2O+$ $+K_2SO_4 \cdot MgSO_4 \cdot 4H_2O$
30.30	5.29	6.55	35	$KCl+K_2SO_4 \cdot MgSO_4 \cdot 6H_2O$ (metastable)
23.50	10.57	7.22	35	»
17 71	15.86	8.01	35	»
13.16	21.14	9.22	35	»
9.68	26.43	12.36	35	»
17.71	15.86	7.68	35	$KCl+K_2SO_4 \cdot MgSO_4 \cdot 4H_2O$
13.16	21.14	8.15	35	»
9.68	26.43	9.09	35	$KCl+K_2SO_4 \cdot MgSO_4 \cdot 4H_2O$ (metastable)
7.28	31.72	10.42	35	»
11.17	24.26	8.55	35	$KCl+K_2SO_4 \cdot MgSO_4 \cdot 4H_2O+$ $+KCl \cdot MgSO_4 \cdot 3H_2O$
9.68	26.43	7.48	35	$KCl+KCl \cdot MgSO_4 \cdot 3H_2O$
7.28	31.72	5.21	35	»
5.63	37.00	3.41	35	»
5.38	38.59	3.14	35	»

№ 4049 [815]

MAGNESIUM SULFATE–MAGNESIUM CHLORIDE–POTASSIUM CHLORIDE–WATER

$MgSO_4$—$MgCl_2$—KCl—H_2O

$t = 25$

Content in aqueous soln., wt%			Comp. of solid phase
Potassium Chloride	Magnesium Chloride	Magnesium Sulfate	
15.73	7.64	3.41	$KCl+K_2SO_4$
15.81	7.64	3.77	»
15.75	7.75	3.91	»

Table № 4049 (continuation)

Content in aqueous soln., wt%			Comp. of solid phase
Potassium Chloride	Magnesium Chloride	Magnesium Sulfate	
15.39	7.64	3.83	$KCl + K_2SO_4$
15.41	7.71	3.92	»
15.36	7.67	4.10	»
14.91	7.99	4.21	»
14.91	7.95	4.44	»
14.32	8.38	4.55	$KCl + K_2SO_4 + K_2SO_4 \cdot MgSO_4 \cdot 6H_2O$

№ 4050 [514]

MAGNESIUM SULFATE–MAGNESIUM CHLORIDE–POTASSIUM CHLORIDE–WATER

$MgSO_4—MgCl_2—KCl—H_2O$

Comp. of saturated soln., wt %				t	Comp. of solid phase
Magnesium Sulfate	Magnesium Chloride	Potassium Chloride	Water		
14.08	16.05	3.41	66.46	43	Langbeinite +Kieserite+Leonite+ +Kainite
1.49	27.99	6.48	64.04	85	KCl+Kieserite+Kainite+Langbeinite
0.95	28.23	6.32	64.50	90	KCl+Kieserite + Langbeinite
6.54	7.71	20.85	64.90	81	$KCl + K_2SO_4$+Leonite + Langbeinite
4.13	8.25	21.98	65.64	85	$KCl + K_2SO_4$+ Langbeinite
7.27	18.71	4.45	69.57	23	Leonite+Kainite+Schoenite+Epsomm:
5.17	8.46	15.54	70.83	35.5	$KCl + K_2SO_4$+Leonite + Schoenite
7.04	14.05	13.01	65.90	69	KCl+Kainite+Leonite+Langbeinite
5.37	16.60	12.11	65.92	70	Kainite + Langbeinite +KCl
6.20	18.87	7.24	67.69	33	KCl+Leonite+Kainite+ Schoenite

№ 4051 [690]

MAGNESIUM SULFATE–MAGNESIUM CHLORIDE–CALCIUM SULFATE–CALCIUM CHLORIDE–WATER

$MgSO_4—MgCl_2—CaSO_4—CaCl_2—H_2O$

Comp. of saturated soln., wt%					t	Comp. of solid phase
Magnesium Chloride	Magnesium Sulfate	Calcium Sulfate	Calcium Chloride	Water		
0.00	27.30	0.077	0.00	72.623	25	$MgSO_4 \cdot 7H_2O + CaSO_4 \cdot 2H_2O$
2.335	24.62	0.108	0.00	62.837	25	»
3.37	23.17	0.136	0.00	73.324	25	»
23.30	5.12	0.162	0.00	71.418	25	»
26.06	4.92	0.00	0.00	69.02	25	$MgSO_4 \cdot 7H_2O + MgSO_4 \cdot 6H_2O$

Table № 4051 (continuation)

Comp. of saturated soln., wt%					t	Comp. of solid phase
Magnesium Chloride	Magnesium Sulfate	Calcium Sulfate	Calcium Chloride	Water		
25.92	4.80	0.158	0.00	69.122	25	$MgSO_4 \cdot 7H_2O + MgSO_4 \cdot 6H_2O + CaSO_4 \cdot 2H_2O$
29.89	3.93	0.00	0.00	66.18	25	$MgSO_4 \cdot 6H_2O + MgSO_4 \cdot 5H_2O$
29.87	3.90	0.151	0.00	66.279	25	$MgSO_4 \cdot 6H_2O + MgSO_4 \cdot 5H_2O + CaSO_4 \cdot 2H_2O$
31.96	3.78	0.169	0.00	64.091	25	»
32.09	3.06	0.00	0.00	64.85	25	$MgSO_4 \cdot 4H_2O + MgSO_4 \cdot 5H_2O$
32.00	2.98	0.154	0.00	64.866	25	$MgSO_4 \cdot 4H_2O + MgSO_4 \cdot 5H_2O + CaSO_4 \cdot 2H_2O$
33.13	2.95	0.08	0.00	62.84	25	$MgSO_4 \cdot 4H_2O + CaSO_4 \cdot 2H_2O$
33.24	2.50	0.15	0.00	64.11	25	»
34.36	2.14	0.00	0.00	63.50	25	$MgSO_4 \cdot 4H_2O + MgCl_2 \cdot 6H_2O$
34.32	2.10	0.12	0.00	63.46	25	$MgSO_4 \cdot 4H_2O + MgCl_2 \cdot 6H_2O + CaSO_4 \cdot 2H_2O$
22.53	0.00	0.072	20.54	56.858	25	$MgCl_2 \cdot 6H_2O + CaSO_4 \cdot 2H_2O$
16.81	0.00	0.008	27.01	56.172	25	»
14.95	0.00	0.00	30.50	54.55	25	$MgCl_2 \cdot 6H_2O + MgCl_2 \cdot 2CaCl_2 \cdot 12H_2O$
14.91	0.00	0.012	30.65	54.428	25	$MgCl_2 \cdot 6H_2O + MgCl_2 \cdot 2CaCl_2 \cdot 12H_2O + CaSO_4 \cdot 2H_2O$
9.29	0.00	0.00	38.65	52.06	25	$MgCl_2 \cdot 2CaCl_2 \cdot 12H_2O + CaCl_2 \cdot 6H_2O$
9.27	0.00	0.015	38.24	52.475	25	$MgCl_2 \cdot 2CaCl_2 \cdot 12H_2O + CaSO_4 \cdot 2H_2O$
0.00	0.00	0.006	39.70	60.294	25	»
0.00	34.18	0.01	0.00	65.81	55	$MgSO_4 \cdot 6H_2O + CaSO_4$
3.28	31.05	0.24	0.00	65.43	55	»
7.35	26.21	0.37	0.00	66.07	55	»
9.84	25.01	0.32	0.00	64.83	55	»
15.56	18.10	0.77	0.00	65.57	55	$MgSO_4 \cdot 6H_2O + CaSO_4 + MgSO_4 \cdot H_2O$
13.49	12.56	0.22	0.00	63.73	55	$CaSO_4 + MgSO_4 \cdot H_2O$
27.83	8.89	0.28	0.0	63.00	55	»
37.42	0.76	0.21	0.00	64.81	55	$MgCl_2 \cdot 6H_2O + CaSO_4 + MgSO_4 \cdot H_2O$
35.30	0.00	0.16	8.21	56.33	55	$MgCl_2 \cdot 6H_2O + CaSO_4$
29.10	0.00	0.18	13.95	56.77	55	$MgCl_2 \cdot 6H_2O + CaSO_4 + MgCl_2 \cdot 2CaCl_2 \cdot 12H_2O$
5.00	0.00	0.11	53.32	41.57	55	$MgCl_2 \cdot 2CaCl_2 \cdot 12H_2O + CaSO_4 + CaCl_2 \cdot 2H_2O$
2.14	0.00	0.09	53.82	43.95	55	$CaSO_4 + CaSO_4 + CaCl_2 \cdot 2H_2O$
0.00	0.00	0.02	54.03	45.95	55	»

№ 4052 [2405]

MAGNESIUM SULFATE–MAGNESIUM MOLYBDATE–WATER

$MgSO_4$—$MgMoO_4$—H_2O

$t = 25$

Concn. in aqueous soln., wt%		d_4^{25}	Comp. of solid phase
Magnesium Sulfate	Magnesium Molybdate		
0.0	15.90	1.159	$MgMoO_4 \cdot 5H_2O$
2.91	14.70	1.176	»
6.06	13.28	1.198	»
8.79	12.08	1.218	»
11.54	10.76	1.230	»
14.64	9.20	1.254	»
18.03	7.63	1.271	»
20.72	6.42	1.292	»
23.76	5.06	1.313	$MgMoO_4 \cdot 5H_2O + MgSO_4 \cdot 7H_2O$
24.02	4.51	1.305	$MgSO_4 \cdot 7H_2O$
24.65	3.70	1.296	»
24.78	3.20	1.295	»
25.33	2.26	1.291	»
26.67	0.0	1.278	»

№ 4053 MAGNESIUM SULFATE–MAGNESIUM BROMIDE–WATER [8]

$MgSO_4$—$MgBr_2$—H_2O

$t = 35$

Comp. of saturated soln., wt%			t	Comp. of solid phase
Magnesium Bromide	Magnesium Sulfate	Water		
0.00	30.65	69.35	1.335	$MgSO_4 \cdot 7H_2O$
2.768	26.27	70.96	1.334	»
5.333	23.96	70.71	1.335	»
8.964	20.43	70.61	1.336	»
12.360	17.76	69.88	1.340	»
16.39	14.73	68.88	1.352	»
19.73	12.13	68.14	1.367	»
24.30	9.55	66.15	1.390	»
32.29	6.067	61.14	1.446	»
39.87	5.12	55.01	1.534	$MgSO_4 \cdot 6H_2O$
41.85	3.53	54.62	1.565	»
51.29	1.809	46.70	1.656	»
52.40	1.452	46.15	1.655	$MgBr_2 \cdot 6H_2O$
52.53	1.407	46.06	1.652	»
0.00	53.43	46.57	1.650	»

№ 4054 [1168]

MAGNESIUM SULFATE – ALUMINUM SULFATE – WATER

$MgSO_4$—$Al_2(SO_4)_3$—H_2O

$t = 25$

Comp. of saturated soln., wt%			Comp. of solid phase
Magnesium Sulfate	Aluminum Sulfate	Water	
27.60	0.00	72.40	$MgSO_4 \cdot 7H_2O$
22.69	5.40	71.91	»
19.30	10.18	70.52	»
15.24	16.09	68.67	$MgSO_4 \cdot 7H_2O + MgAl_2(SO_4)_4 \cdot 22H_2O$
14.63	16.38	68.99	$MgAl_2(SO_4)_4 \cdot 22H_2O$
12.71	18.19	69.10	»
11.72	19.12	69.12	»
10.29	20.50	69.21	$MgAl_2(SO_4)_4 \cdot 22H_2O +$ $+ Al_2(SO_4)_3 \cdot 16H_2O$
10.18	20.66	69.16	$Al_2(SO_4)_3 \cdot 16H_2O$
8.39	21.92	69.69	»
3.45	25.48	71.07	»
0.00	28.20	71.80	»

№ 4055 [1713]

MAGNESIUM SULFATE–POTASSIUM SULFATE–WATER

$MgSO_4$—K_2SO_4—H_2O

$t = 83$

Content in aqueous soln., g/100g		Comp. of solid phase
Potassium Sulfate	Magnesium Sulfate	
22.4	0	K_2SO_4
24.67	28.30	$K_2SO_4 + K_2SO_4 \cdot MgSO_4 \cdot 4H_2O$
12.39	44.69	$K_2SO_4 \cdot MgSO_4 \cdot 4H_2O + K_2SO_4 \cdot 2MgSO_4$
2.86	60.62	$K_2SO_4 \cdot 2MgSO_4 + MgSO_4 \cdot H_2O$
0	67.29	$MgSO_4 \cdot H_2O$

№ 4056 [1462]

MAGNESIUM SULFATE–POTASSIUM SULFATE–WATER

$MgSO_4$—K_2SO_4—H_2O

Content in aqueous soln., mol/1000 mol Water		t	Comp. of solid phase
Potassium Sulfate	Magnesium Sulfate		
9.4	16.0	0	$K_2SO_4 + K_2SO_4 \cdot MgSO_4 \cdot 6H_2O$
4.6	41.9	0	$K_2SO_4 \cdot MgSO_4 \cdot 6H_2O + MgSO_4 \cdot 7H_2O$
21.5	38.9	45	$K_2SO_4 + K_2SO_4 \cdot MgSO_4 \cdot 6H_2O$
10.7	56.2	45	$K_2SO_4 \cdot MgSO_4 \cdot 6H_2O + K_2SO_4 \cdot MgSO_4 \cdot 4H_2O$
7.6	74.6	45	$K_2SO_4 \cdot MgSO_4 \cdot 4H_2O + MgSO_4 \quad 7H_2O$
23.7	43.1	55	$K_2SO_4 + K_2SO_4 \cdot MgSO_4 \cdot 4H_2O$
10.2	78.2	70	$K_2SO_4 \cdot MgSO_4 \cdot 4H_2O + K_2SO_4 \cdot 2MgSO_4$
6.4	86.7	70	$K_2SO_4 \cdot 2MgSO_4 + MgSO_4 \cdot H_2O$
27.2	42.9	90	$K_2SO_4 + K_2SO_4 \cdot MgSO_4 \cdot 4H_2O$

№ 4057 [2686]

MAGNESIUM SULFATE–POTASSIUM SULFATE–WATER

$MgSO_4$—K_2SO_4—H_2O

$t = 25$

Comp. of saturated soln., wt%			Comp. of solid phase
Potassium Sulfate	Magnesium Sulfate	Water	
0.0	26.76	73.24	$MgSO_4 \cdot 7H_2O$
1.68	26.67	71.65	»
2.34	26.57	71.09	»
3.76	26.36	69.88	»
4.02	26.39	69.59	$MgSO_4 \quad 7H_2O + MgK_2(SO_4)_2 \cdot 6H_2O$
7.02	18.76	74.22	$MgK_2(SO_4)_2 \cdot 6H_2O$
8.43	16.36	75.21	»
9.63	14.27	76.10	»
10.34	13.26	76.40	»
10.51	12.88	76.61	»
10.70	12.68	76.62	$MgK_2(SO_4)_2 \cdot 6H_2O + K_2SO$
10.77	12.06	77.17	K_2SO_4
10.84	10.69	78.47	»
11.10	7.8	81.1	»
11.03	4.0	84.97	»
10.77	0.0	89.23	»

№ 4058 [1914]

MAGNESIUM SULFATE–POTASSIUM SULFATE–WATER

$MgSO_4—K_2SO_4—H_2O$

Content in aqueous soln., mol/100 mol water		t	Comp. of solid phase
Potassium Sulfate	Magnesium sulfate		
12.3	20.5	0	$K_2SO_4 + K_2SO_4 \cdot MgSO_4 \cdot 6H_2O$
3.3	40.5	0	$MgSO_4 \cdot 7H_2O$
14.8	22.2	15	—
3.2	50.8	15	$MgSO_4 \cdot 7H_2O$
32.3	53.9	100	K_2SO_4

№ 4059 [2729]

MAGNESIUM SULFATE–POTASSIUM SULFATE–WATER

$MgSO_4—K_2SO_4—H_2O$

$t = 30$

Comp. of saturated soln., wt%			Comp. of solid phase
Magnesium Sulfate	Potassium Sulfate	Water	
29.03	0.00	70.97	$MgSO_4 \cdot 7H_2O$
27.69	3.2	69.10	Mixed crystals
27.48	4.36	68.16	»
26.20	5.98	67.82	»
26.28	6.60	67.12	»
26.05	6.93	67.02	»
18.08	8.59	73.33	$MgSO_4 \cdot K_2SO_4 \cdot 6H_2O$
13.19	11.66	75.15	»
12.88	12.47	74.65	»
9.49	12.64	77.87	K_2SO_4
3.45	11.98	84.57	»
0.08	11.20	88.72	»

№ 4060 [2072]

MAGNESIUM SULFATE–POTASSIUM SULFATE–WATER

$MgSO_4—K_2SO_4—H_2O$

Comp. of saturated soln., wt%			t	Comp. of solid phase
Potassium Sulfate	Magnesium Sulfate	Water		
3.42	21.09	75.49	0.0	$MgSO_4 \cdot 12H_2O + MgSO_4 \cdot K_2SO_4 \cdot 6H_2O$
3.83	25.34	70.83	19.8	$MgSO_4 \cdot 7H_2O + MgSO_4 \cdot K_2SO_4 \cdot 6H_2O$
4.26	26.45	69.29	26.1	»
4.13	29.80	66.07	37.8	»
7.86	8.73	83.41	0	$K_2SO_4 + MgSO_4 \cdot K_2SO_4 \cdot 6H_2O$

Comp. of saturated soln., wt%			t	Comp. of solid phase
Potassium Sulfate	Magnesium Sulfate	Water		
10.10	12.08	77.82	20.4	$K_2SO_4+MgSO_4\cdot K_2SO_4\cdot 6H_2O$
11.66	14.38	73.96	34.4	»
12.08	17.00	70.92	44.7	»
12.46	18.09	69.45	49.6	»
12.86	18.38	68.76	55.2	$K_2SO_4+MgSO_4\cdot K_2SO_4\cdot 4H_2O$
14.35	17.20	68.45	63.4	»

№ 4061 [1202]

MAGNESIUM SULFATE–POTASSIUM SULFATE–WATER

$MgSO_4—K_2SO_4—H_2O$

Content in aqueous soln., wt%		d	t	Comp. of solid phase
Potassium Sulfate	Magnesium Sulfate			
6.68	0	1.025	0.2	K_2SO_4
7.03	1.26	1.061	0.2	»
7.51	4.17	1.108	0.2	»
7.77	4.59	1.115	0.2	»
7.86	8.52	1.164	0.2	$K_2SO_4+K_2SO_4\cdot MgSO_4\cdot 6H_2O$
7.53	8.94	1.165	0.2	$K_2SO_4\cdot MgSO_4\cdot 6H_2O$
6.76	11.79	1.192	0.2	»
3.03	20.78	1.265	0.2	$K_2SO_4\cdot MgSO_4\cdot 6H_2O+$ $+MgSO_4\cdot 7H_2O$
7.93	20.74	1.271	0.2	$MgSO_4\cdot 7H_2O$ (metastable)
4.34	20.68	1.274	0.2	»
2.17	20.94	1.259	0.2	$MgSO_4\cdot 7H_2O$
0	20.82	1.236	0.2	»
19.25	0	1.107	99.5	K_2SO_4
19.46	7.34	1.217	99.5	»
18.75	16.25	1.339	99.5	K_2SO_4 (metastable)
18.84	18.00	—	99.5	»
18.9	14.50	1.300	99.5	$K_2SO_4+K_2SO_4\cdot 2MgSO_4$
14.26	18.12	1.304	99.5	$K_2SO_4\cdot 2MgSO_4$
9.04	22.63	1.319	99.5	»
5.54	27.44	1.332	99.5	»
3.31	31.00	1.361	99.5	$K_2SO_4\cdot 2MgSO_4+MgSO_4\cdot H_2O$
0	33.27	1.367	99.5	$MgSO_4\cdot H_2O$

№ 4062 MAGNESIUM SULFATE–POTASSIUM SULFATE–WATER [2605]

$MgSO_4—K_2SO_4—H_2O$

Comp. of saturated soln., wt%			t	Comp. of solid phase
Potassium Sulfate	Magnesium Sulfate	Water		
11.59	0.0	88.41	30	K_2SO_4
11.96	4.03	84.01	30	»
11.98	9.45	78.57	30	»
11.66	12.21	76.13	30	»
11.52	13.31	75.17	30	$K_2SO_4+K_2SO_4 \cdot MgSO_4 \cdot 6H_2O$
10.86	13.92	75.22	30	$K_2SO_4 \cdot MgSO_4 \cdot 6H_2O$
8.21	19.03	72.76	30	»
6.32	22.59	71.09	30	»
4.97	25.88	69.15	30	»
4.47	27.24	68.29	30	$K_2SO_4 \cdot MgSO_4 \cdot 6H_2O+MgSO_4 \cdot 7H_2O$
3.82	27.04	69.14	30	$MgSO_4 \cdot 7H_2O$
3.99	27.08	68.93	30	»
0.0	28.10	71.90	30	»
18.25	6.08	75.67	85	K_2SO_4
17.71	13.60	68.69	85	»
16.94	17.33	65.73	85	$K_2SO_4+K_2SO_4 \cdot MgSO_4 \cdot 4H_2O$
14.86	19.97	65.17	85	$K_2SO_4 \cdot MgSO_4 \cdot 4H_2O$
17.16	25.49	57.35	85	$K_2SO_4 \cdot MgSO_4 \cdot 4H_2O+K_2SO_4 \cdot 2MgSO_4$
9.56	24.66	65.78	85	$K_2SO_4 \cdot 2MgSO_4$
6.58	27.74	65.68	85	»
3.27	32.01	64.72	85	$K_2SO_4 \cdot 2MgSO_4 \cdot MgSO_4 \cdot H_2O$
0.0	32.9	67.1	85	$MgSO_4 \cdot H_2O$
19.53	0.0	80.47	100	K_2SO_4
19.83	8.69	71.48	100	»
19.03	15.64	65.33	100	»
18.68	17.33	63.99	100	»
19.00	13.75	67.25	100	$K_2SO_4 \cdot K_2SO_4 \cdot 2MgSO_4$
14.98	16.49	68.53	100	$K_2SO_4 \cdot 2MgSO_4$
10.40	20.71	68.89	100	»
6.50	25.60	67.90	100	»
3.69	29.90	66.41	100	$K_2SO_4 \cdot 2MgSO_4+MgSO_4 \cdot H_2O$
0.0	31.75	68.25	100	$MgSO_4 \cdot H_2O$

№ 4063 [2606]

MAGNESIUM SULFATE–POTASSIUM SULFATE–WATER

$MgSO_4—K_2SO_4—H_2O$

Content in aqueous soln., wt%		t	Comp. of solid phase
Potassium Sulfate	Magnesium Sulfate		
18.06	0	85	K_2SO_4
18.24	6.08	85	»
17.95	10.84	85	»

Table № 4063 (continuation)

Content in aqueous soln., wt%		t	Comp. of solid phase
Potassium Sulfate	Magnesium Sulfate		
18.12	11.25	85	K_2SO_4
17.71	13.60	85	»
17.54	13.91	85	»
17.06	17.07	85	»
16.94	17.33	85	$K_2SO_4 + K_2SO_4 \cdot MgSO_4 \cdot 4H_2O$
14.86	19.96	85	$K_2SO_4 \cdot MgSO_4 \cdot 4H_2O$
14.50	20.43	85	»
13.73	21.41	85	»
13.90	21.77	85	$K_2SO_4 \cdot MgSO_4 \cdot 4H_2O$ (supersatd.)
13.72	22.10	85	»
12.00	23.65	85	$K_2SO_4 \cdot MgSO_4 \cdot 4H_2O$ (metastable)
12.09	23.85	85	»
10.87	25.83	85	»
9.21	28.61	85	»
13.36	22.08	85	$K_2SO_4 \cdot MgSO_4 \cdot 4H_2O + K_2SO_4 \cdot 2MgSO_4$
10.86	23.90	85	$K_2SO_4 \cdot 2MgSO_4$
10.42	24.06	85	»
9.56	24.66	85	»
6.58	27.74	85	»
4.84	30.63	85	$K_2SO_4 \cdot 2MgSO_4$ (supersatd.)
2.94	34.54	85	$K_2SO_4 \cdot 2MgSO_4$ (metastable)
3.27	32.01	85	$K_2SO_4 \cdot 2MgSO_4 + MgSO_4 \cdot H_2O$
0	32.9	100	$MgSO_4 \cdot H_2O$
19.52	0	100	K_2SO_4
19.93	3.24	100	»
20.10	5.94	100	»
19.83	8.69	100	»
19.49	11.11	100	»
19.18	13.42	100	»
19.18	14.17	100	K_2SO_4 (metastable)
19.03	15.64	100	»
18.63	17.33	100	»
18.99	13.75	100	$K_2SO_4 + K_2SO_4 \cdot 2MgSO_4$
18.87	13.83	100	$K_2SO_4 \cdot 2MgSO_4$
17.26	15.14	100	»
14.98	16.49	100	»
12.80	18.43	100	»
10.40	20.71	100	»
8.57	23.08	100	»
6.50	25.61	100	»
4.58	28.66	100	»
3.22	30.85	100	$K_2SO_4 \cdot 2MgSO_4$ (metastable)

$MgSO_4$—K_2SO_4—H_2O

Comp. of saturated soln., wt%			t	d	Comp. of solid phase
Potassium Sulfate	Magnesium Sulfate	Water			
6.68	0.0	93.32	0	1.025	K_2SO_4
7.03	1.26	91.71	0	1.061	»
7.51	4.17	88.32	0	1.108	»
7.77	4.59	87.64	0	1.115	$K_2SO_4 + K_2SO_4 \cdot MgSO_4 \cdot 6H_2O$
7.85	8.60	83.55	0	1.163	$K_2SO_4 \cdot MgSO_4 \cdot 6H_2O$
7.53	8.94	83.53	0	1.165	»
6.76	11.79	81.45	0	1.192	»
3.03	20.78	76.10	0	1.265	$K_2SO_4 \cdot MgSO_4 \cdot 6H_2O + MgSO_4 \cdot 7H_2O$
7.93	20.74	71.33	0	1.271	$MgSO_4 \cdot 7H_2O$
4.34	20.68	74.98	0	1.274	»
2.17	20.94	76.89	0	1.259	»
0.0	20.82	79.18	0	1.236	»
10.75	0.0	89.25	25	1.088	K_2SO_4
10.86	2.60	86.54	25	1.123	»
10.85	5.05	84.10	25	1.149	»
10.99	12.61	76.40	25	1.230	$K_2SO_4 + K_2SO_4 + MgSO_4 \cdot 6H_2O$
9.89	14.28	75.83	25	1.249	$K_2SO_4 \cdot MgSO_4 \cdot 6H_2O$
9.18	17.02	73.80	25	1.267	»
7.32	20.32	72.36	25	1.294	»
4.90	21.02	69.08	25	1.347	$K_2SO_4 \cdot MgSO_4 \cdot 6H_2O + MgSO_4 \cdot 7H_2O$
2.11	26.61	71.28	25	1.324	$MgSO_4 \cdot 7H_2O$
0.0	26.65	73.35	25	1.305	»
14.14	0.0	85.86	50	1.087	K_2SO_4
14.13	5.05	80.82	50	1.148	»
13.47	13.86	72.67	50	1.255	»
13.10	16.96	69.94	50	1.292	$K_2SO_4 + K_2SO_4 \cdot MgSO_4 \cdot 2H_2O$
13.06	17.36	69.58	50	1.290	$K_2SO_4 \cdot MgSO_4 \cdot 2H_2O$
10.32	21.88	67.80	50	1.322	»
8.14	24.39	67.47	50	1.337	»
5.98	28.59	65.43	50	1.370	»
4.30	32.35	63.35	50	1.394	$K_2SO_4 \cdot MgSO_4 \cdot 2H_2O + MgSO_4 \cdot 6H_2O$
2.60	32.76	64.64	50	1.380	$MgSO_4 \cdot 6H_2O$
0.0	33.50	66.50	50	1.383	»
19.42	0.0	80.58	100	1.119	K_2SO_4
19.46	7.34	73.20	100	1.217	»
18.75	16.25	65.00	100	1.339	»
18.84	18.00	63.16	100	—	»
18.64	14.47	66.89	100	1.305	$K_2SO_4 + K_2SO_4 \cdot 2MgSO_4$
14.26	18.12	67.62	100	1.304	$K_2SO_4 \cdot 2MgSO_4$
9.04	22.63	68.33	100	1.319	»
5.54	27.44	67.02	100	1.332	»
3.31	31.00	65.69	100	1.361	$K_2SO_4 \cdot 2MgSO_4 + MgSO_4 \cdot H_2O$
0.0	33.27	66.73	100	1.367	$MgSO_4 \cdot H_2O$
16.16	0.00	83.84	66	—	K_2SO_4
15.01	7.73	77.26	66	—	»
14.37	14.01	71.62	66	—	»
14.46	17.08	68.46	66	—	$K_2SO_4 + K_2SO_4 \cdot MgSO_4 \cdot 2H_2O$
13.04	19.33	67.63	66	—	$K_2SO_4 \cdot MgSO_4 \cdot 2H_2O$

Table № 4064 (continuation)

Comp. of saturated soln., wt%			t	d	Comp. of solid phase
Potassium Sulfate	Magnesium Sulfate	Water			
7.67	26.97	65.36	66	—	$K_2SO_4 \cdot MgSO_4 \cdot 2H_2O$
5.86	32.53	61.61	66	—	»
4.48	33.89	61.63	66	—	»
4.01	35.66	60.33	66	—	$K_2SO_4 \cdot MgSO_4 \cdot 2H_2O + MgSO_4 \cdot H_2O$
2.08	35.40	62.52	66	—	$MgSO_4 \cdot H_2O$
0.00	35.52	64.48	66	—	»
17.05	0.00	82.95	75	1.095	K_2SO_4
16.13	8.51	75.36	75	1.151	»
15.70	12.70	71.60	75	1.225	»
15.22	17.18	67.60	75	1.302	$K_2SO_4 + K_2SO_4 \cdot MgSO_4 \cdot 2H_2O$
14.09	18.12	67.79	75	1.320	$K_2SO_4 \cdot MgSO_4 \cdot 2H_2O$
11.14	23.32	65.54	75	1.351	»
8.70	26.44	64.86	75	1.402	»
7.71	30.70	61.59	75	1.419	$K_2SO_4 \cdot MgSO_4 \cdot 2H_2O + K_2SO_4 \cdot 2MgSO_4$
5.44	31.66	62.90	75	1.421	$K_2SO_4 \cdot 2MgSO_4$
4.41	32.13	63.46	75	1.424	»
4.02	35.54	60.44	75	1.449	$K_2SO_4 \cdot 2MgSO_4 + MgSO_4 \cdot H_2O$
2.30	35.23	42.47	75	1.437	$MgSO_4 \cdot H_2O$
1.25	37.11	61.64	75	—	»
0.00	37.30	62.70	75	—	»

№ 4065 MAGNESIUM SULFATE–POTASSIUM SULFATE–WATER [1262]

$MgSO_4—K_2SO_4—H_2O$

$t = 35$

Comp. of saturated soln., wt%			d	Comp. of solid phase
Potassium Sulfate	Magnesium Sulfate	Water		
12.40	5.04	82.56	1.148	K_2SO_4
12.28	7.16	80.56	—	»
11.67	13.13	75.20	1.242	»
11.90	14.40	73.70	1.250	$K_2SO_4 + K_2SO_4 \cdot MgSO_4 \cdot 6H_2O$
10.20	16.15	73.65	—	$K_2SO_4 \cdot MgSO_4 \cdot 6H_2O$
9.33	17.82	72.75	1.283	»
7.55	20.37	72.08	1.284	»
5.61	24.60	69.79	—	»
4.35	28.45	67.20	1.360	$K_2SO_4 \cdot MgSO_4 \cdot 6H_2O + MgSO_4 \cdot 7H_2O$
1.78	28.41	69.81	1.336	$MgSO_4 \cdot 7H_2O$
0.0	29.20	70.80	1.332	»

№ 4066 [1335]

MAGNESIUM SULFATE–POTASSIUM SULFATE–WATER

$MgSO_4—K_2SO_4—H_2O$

$t = 100$

Content in aqueous soln., wt%		Comp. of solid phase
Potassium Sulfate	Magnesium Sulfate	
19.00	0	K_2SO_4
17.70	9.25	»
15.40	13.90	»
14.00	21.90	$K_2SO_4 + K_2SO_4 \cdot 2MgSO_4$
11.70	24.00	$K_2SO_4 \cdot 2MgSO_4$
8.00	27.60	»
6.75	28.50	»
3.40	34.00	$K_2SO_4 \cdot 2MgSO_4 + MgSO_4 \cdot H_2O$
0	33.50	$MgSO_4 \cdot H_2O$

№ 4067 MAGNESIUM SULFATE–POTASSIUM SULFATE–WATER [1921]

$MgSO_4—K_2SO_4—H_2O$

Content in aqueous soln., mol/100 mol total salts		t	Comp. of solid phase
Potassium Sulfate	Water		
11.6	2620	−5.2	$K_2SO_4 \cdot MgSO_4 \cdot 6H_2O + MgSO_4 \cdot 12H_2O +$ ice
36.9	4100	−3	$K_2SO_4 + K_2SO_4 \cdot MgSO_4 \cdot 6H_2O +$ ice
9.2	2200	1	$K_2SO_4 \cdot MgSO_4 \cdot 6H_2O + MgSO_4 \cdot 12H_2O + MgSO_4 \cdot 7H_2O$
8.5	1270	41	$K_2SO_4 \cdot MgSO_4 \cdot 6H_2O + K_2SO_4 \cdot MgSO_4 \cdot 4H_2O + MgSO_4 \cdot 7H_2O$
8.6	1190	47.2	$K_2SO_4 \cdot MgSO_4 \cdot 4H_2O + MgSO_4 \cdot 7H_2O + MgSO_4 \cdot 6H_2O$
34.1	1820	47.2	$K_2SO_4 + K_2SO_4 \cdot MgSO_4 \cdot 6H_2O + K_2SO_4 \cdot MgSO_4 \cdot 4H_2O$
7.0	1060	61.0	$K_2SO_4 \cdot MgSO_4 \cdot 4H_2O + MgSO_4 \cdot 6H_2O + MgSO_4 \cdot H_2O$
7.5	1050	66.5	$K_2SO_4 \cdot MgSO_4 \cdot 4H_2O + K_2SO_4 \cdot MgSO_4 + MgSO_4 \cdot H_2O$
33.3	1420	86	$K_2SO_4 \cdot MgSO_4 \cdot 4H_2O + K_2SO_4 \cdot 2MgSO_4$
40.5	1500	89	$K_2SO_4 + K_2SO_4 \cdot MgSO_4 \cdot 4H_2O + K_2SO_4 \cdot 2MgSO_4$
50	1700	100	$K_2SO_4 + K_2SO_4 \cdot 2MgSO_4$
33.3	1690	100	$K_2SO_4 \cdot 2MgSO_4$
8	1330	100	$K_2SO_4 \cdot 2MgSO_4 + MgSO_4 \cdot H_2O$
7.9	1650	120	»
62.0	1810	129	$K_2SO_4 + K_2SO_4 \cdot 2MgSO_4$

№ 4068 [454]

MAGNESIUM SULFATE–POTASSIUM SULFATE–WATER

$MgSO_4$—K_2SO_4—H_2O

$t = 25$

Content in aqueous soln., mol/1000 mol Water		Comp. of solid phase
Potassium Sulfate	Magnesium Sulfate	
12.3	0	K_2SO_4
13.8	16.0	»
14.1	23.1	»
14.6	24.4	$K_2SO_4+K_2SO_4 \cdot MgSO_4 \cdot 6H_2O$
12.5	29.3	$K_2SO_4 \cdot MgSO_4 \cdot 6H_2O$
9.7	39.2	»
6.8	50.4	»
5.6	57.7	$K_2SO_4 \cdot MgSO_4 \cdot 6H_2O+MgSO_4 \cdot 7H_2O$
2.8	55.9	$MgSO_4 \cdot 7H_2O$
0	54.4	»

№ 4069 [1427]

MAGNESIUM SULFATE–POTASSIUM SULFATE–WATER

$MgSO_4$—K_2SO_4—H_2O

Content of Potassium Sulfate in aqueous soln., mol/100 mol total salts	t	Comp. of solid phase
100	−1.5	K_2SO_4+ ice
50	−2.85	$K_2SO_4 \cdot MgSO_4 \cdot 6H_2O$+ice (metastable)
25	−2.97	$K_2SO_4+K_2SO_4 \cdot MgSO_4 \cdot 6H_2O$+ice
13	−5.12	$K_2SO_4 \cdot MgSO_4 \cdot 6H_2O+MgSO_4 \cdot 11H_2O$+ ice
0	−3.9	$MgSO_4 \cdot 11H_2O$+ice

№ 4070 [1427]

MAGNESIUM SULFATE–POTASSIUM SULFATE–WATER

$MgSO_4$—K_2SO_4—H_2O

Content of Potassium Sulfate in aqueous soln., mol/100 mol total salts	t	Comp. of solid phase
11	−6.02	$K_2SO_4 \cdot MgSO_4 \cdot 6H_2O$+ +$MgSO_4 \cdot 9H_2O$+ ice (metastable)
0	−4.87	$MgSO_4 \cdot 9H_2O$+ice (metastable)
10.5	−6.3	$K_2SO_4 \cdot MgSO_4 \cdot 6H_2O$+ +$MgSO_4 \cdot 7H_2O$+ice (metastable)
0	−5.2	$MgSO_4 \cdot 7H_2O$+ ice (metastable)

№ 4071 [1174]

MAGNESIUM SULFATE–POTASSIUM SULFATE–WATER

$MgSO_4$—K_2SO_4—H_2O

$t = 35$

Concn. in aqueous soln., wt%		Comp. of solid phase
Potassium Sulfate	Magnesium Sulfate	
12.13	0.00	K_2SO_4
12.18	3.82	»
12.64	7.83	»
12.10	13.66	$K_2SO_4 + K_2SO_4 \cdot MgSO_4 \cdot 6H_2O$
9.67	17.63	$K_2SO_4 \cdot MgSO_4 \cdot 6H_2O$
7.40	21.75	»
5.53	25.74	»
4.40	28.44	$K_2SO_4 \cdot MgSO_4 \cdot 6H_2O + MgSO_4 \cdot 7H_2O$
2.31	29.63	$MgSO_4 \cdot 7H_2O$
0.00	29.43	»

№ 4072 [485]

MAGNESIUM SULFATE–POTASSIUM SULFATE–WATER

$MgSO_4$—K_2SO_4—H_2O

$t = 35$

Concn. in aqueous soln., wt%		Comp. of solid phase
Potassium Sulfate	Magnesium Sulfate	
12.30	0	K_2SO_4
12.28	4.98	»
12.26	8.38	»
12.14	14.94	$K_2SO_4 + K_2SO_4 \cdot MgSO_4 \cdot 6H_2O$
11.83	15.62	$K_2SO_4 \cdot MgSO_4 \cdot 6H_2O$
7.88	21.04	»
4.98	28.28	»
4.23	29.49	$K_2SO_4 \cdot MgSO_4 \cdot 6H_2O + MgSO_4 \cdot 7H_2O$
3.89	29.20	$MgSO_4 \cdot 7H_2O$
0	29.40	»

№ 4073 [512]

MAGNESIUM SULFATE–POTASSIUM SULFATE–WATER

$MgSO_4$—K_2SO_4—H_2O

$t = 75$

Comp. of saturated soln., wt%			Comp. of solid phase
Potassium Sulfate	Magnesium Sulfate	Water	
0.00	37.70	62.30	$MgSO_4 \cdot H_2O$
1.31	37.59	62.10	»
2.76	36.28	60.96	»
3.67	35.56	60.77	$MgSO_4 \cdot H_2O + K_2SO_4 \cdot 2MgSO_4$
3.56	35.33	61.11	$K_2SO_4 \cdot 2MgSO_4$
3.69	34.45	61.86	»
4.67	33.20	62.13	»
5.30	32.57	62.13	»
6.39	31.70	61.91	»
6.99	31.21	61.80	$K_2SO_4 \cdot 2MgSO_4 + K_2SO_4 \cdot MgSO_4 \cdot 4H_2O$

№ 4074 [512]

MAGNESIUM SULFATE–POTASSIUM SULFATE–WATER

$MgSO_4$—K_2SO_4—H_2O

$t = 75$

Comp. of saturated soln., wt%			Comp. of solid phase
Potassium Sulfate	Magnesium Sulfate	Water	
9.20	27.80	63.00	$K_2SO_4 \cdot MgSO_4 \cdot 4H_2O$
12.59	22.05	65.36	»
16.32	17.32	66.36	$K_2SO_4 \cdot MgSO_4 \cdot 4H_2O + K_2SO_4$
16.73	17.18	66.09	»
16.24	16.60	67.16	K_2SO_4
17.30	0.00	82.70	»

№ 4075 MAGNESIUM SULFATE–POTASSIUM SULFATE–WATER [26]

$MgSO_4$—K_2SO_4—H_2O

$t = 15$

Comp. of saturated soln., wt%			Comp. of solid phase
Potassium Sulfate	Magnesium Sulfate	Water	
9.17	0.00	90.83	K_2SO_4
8.90	1.42	89.68	»
9.18	2.68	88.14	»
9.45	6.13	84.42	»

Table № 4075 (continuation)

Comp. of saturated soln., wt%			Comp. of solid phase
Potassium Sulfate	Magnesium Sulfate	Water	
9.67	8.16	82.17	K_2SO_4
9.37	11.58	79.05	»
9.78	11.81	78.41	$K_2SO_4+K_2SO_4\cdot MgSO_4\cdot 6H_2O$
9.89	11.56	78.55	»
8.43	12.48	79.09	$K_2SO_4\cdot MgSO_4\cdot 6H_2O$
6.78	15.85	77.37	»
5.34	20.30	74.36	»
6.46	23.53	70.01	»
5.92	28.47	70.61	$K_2SO_4\cdot MgSO_4\cdot 6H_2O+MgSO_4\cdot 7H_2O$
5.94	24.23	69.83	»
5.83	23.25	70.42	»
4.50	24.25	71.25	$MgSO_4\cdot 7H_2O$
3.92	24.29	71.79	»
2.12	24.56	73.32	»
1.76	24.56	73.68	»
0.00	24.74	75.26	»

№ 4076 [1713]

MAGNESIUM SULFATE–POTASSIUM SULFATE–CALCIUM SULFATE–WATER

$MgSO_4—K_2SO_4—CaSO_4—H_2O$

$t = 83$

Content in aqueous soln., g/100g Water		Comp. of solid phase
Potassium Sulfate	Magnesium Sulfate	
22.31	22.47	$K_2SO_4+K_2SO_4\cdot MgSO_4\cdot 2CaSO_4\cdot 2H_2O+$ $+K_2SO_4\cdot CaSO_4\cdot H_2O$
9.06	3.11	$K_2SO_4\cdot MgSO_4\cdot 2CaSO_4\cdot 2H_2O+K_2SO_4\cdot$ $\cdot CaSO_4\cdot H_2O+K_2SO_4\cdot 5CaSO_4\cdot H_2O$
5.38	5.59	$K_2SO_4\cdot MgSO_4\cdot 2CaSO_4\cdot 2H_2O+$ $+K_2SO_4\cdot MgSO_4\cdot 4CaSO_4\cdot 2H_2O+$ $+K_2SO_4\cdot 5CaSO_4\cdot H_2O$
3.50	4.24	$K_2SO_4\cdot MgSO_4\cdot 4CaSO_4\cdot 2H_2O+$ $+K_2SO_4\cdot 5CaSO_4\cdot H_2O+CaSO_4$
0.44	62.87	$K_2SO_4\cdot MgSO_4\cdot 4CaSO_4\cdot 2H_2O+$ $+MgSO_4\cdot H_2O+CaSO_4$

№ 4077 [2686]

MAGNESIUM SULFATE – POTASSIUM SULFATE – CALCIUM SULFATE – WATER

$MgSO_4—K_2SO_4—CaSO_4—H_2O$

$t = 25$

Content in aqueous soln., wt%		Comp. of solid phase
Potassium Sulfate	Magnesium Sulfate	
2.74	25.50	$K_2SO_4 \cdot MgSO_4 \cdot 2CaSO_4 \cdot 2H_2O + K_2SO_4 \cdot CaSO_4 \cdot H_2O + CaSO_4 \cdot 2H_2O$
4.23	26.21	$K_2SO_4 \cdot MgSO_4 \cdot 2CaSO_4 \cdot 2H_2O + K_2SO_4 \cdot CaSO_4 \cdot H_2O + MgSO_4 \cdot 7H_2O$
1.71	26.64	$K_2SO_4 \cdot MgSO_4 \cdot 2CaSO_4 \cdot 2H_2O + MgSO_4 \cdot 7H_2O + CaSO_4 \cdot 2H_2O$

№ 4078 [1412]

MAGNESIUM SULFATE–POTASSIUM SULFATE–CALCIUM SULFATE–WATER

$MgSO_4—K_2SO_4—CaSO_4—H_2O$

$t = 100$

Content in aqueous soln., g/100g Water			Comp. of solid phase
Potassium Sulfate	Magnesium Sulfate	Calcium Sulfate	
12.11	0.99	0.098	Pol+Sy
12.84	1.59	0.100	»
13.45	2.17	0.058	»
15.79	2.26	0.066	»
16.05	1.97	0.068	»
16.55	1.96	0.071	»
17.15	1.88	0.073	»
11.60	0	0.075	Sy+C.P.
12.00	0.46	0.095	»
12.15	1.22	0.085	Pol+Sy+C.P.
8.43	1.06	0.134	Pol+C.P.
6.62	1.10	0.080	»
5.85	2.20	0.089	»
4.40	2.22	0.092	»
3.86	3.70	0.113	»
2.55	6.38	0.129	»

Table № 4078 (continuation)

Content in aqueous soln., g/100g Water			Comp. of solid phase
Potassium Sulfate	Magnesium Sulfate	Calcium Sulfate	
6.0	0	0.07	C.P.+$CaSO_4$
5.24	0.95	0.10	»
4.6	1.33	0.92	»
3.77	3.52	0.126	»
3.1	4.8	0.12	C.P. +$CaSO_4$
2.45	4.9	0.12	»
1.23	1.85	0.11	»
7.5	1.7	0.075	Pol+C.P.+$CaSO_4$
10.26	1.5	0.090	Pol+$CaSO_4$
9.32	1.27	0.087	»
8.35	1.43	0.074	»
6.73	2.10	0.074	»
3.92	9.10	0.016	»
3.47	14.06	0.006	»
10.7	0.8	0.093	Pol+Sy+$CaSO_4$
10.0	0	0.092	Sy+$CaSO_4$
5.7	0	0.21	Sy+$CaSO_4 \cdot 2H_2O$
1.1	0	0.137	C.P.+$CaSO_4 \cdot 2H_2O$

Note. Pol— Polyhalite $K_2SO_4 \cdot MgSO_4 \cdot 2CaSO_4 \cdot 2H_2O$; C.P.— Calcium Pentasulfate $K_2SO_4 \cdot 5CaSO_4 \cdot H_2O$; Sy — Syngenite $K_2SO_4 \cdot CaSO_4 \cdot H_2O$.

№ 4079 [148]

MAGNESIUM SULFATE–POTASSIUM SULFATE–CALCIUM SULFATE–WATER

$MgSO_4$—K_2SO_4—$CaSO_4$—H_2O

$t = 55$

Comp. of saturated soln., wt%				Comp. of solid phase
Potassium Sulfate	Magnesium Sulfate	Calcium Sulfate	Water	
5.40	0.00	0.114	94.586	$K_2SO_4 \cdot CaSO_4 \cdot H_2O + K_2SO_4 \cdot 5CaSO_4 \cdot H_2O$
5.33	0.84	0.117	93.813	»
5.23	1.61	0.124	93.036	»
5.25	2.42	0.120	92.200	»
5.30	3.20	0.118	91.382	»
5.28	3.61	0.126	90.984	»
5.30	4.11	0.118	90.472	$K_2SO_4 \cdot CaSO_4 \cdot H_2O + K_2SO_4 \cdot 5CaSO_4 \cdot H_2O + K_2SO_4 \cdot MgSO_4 \cdot 2CaSO_4 \cdot 2H_2O$

Table № 4079 (continuation)

Comp. of saturated soln., wt%				Comp. of solid phase
Potassium Sulfate	Magnesium Sulfate	Calcium Sulfate	Water	
2.15	0.00	0.148	97.702	$K_2SO_4 \cdot 5CaSO_4 \cdot H_2O + CaSO_4$
2.27	0.93	0.153	96.647	»
2.65	2.59	0.155	94.605	»
14.87	0.00	0.020	75.110	$K_2SO_4 \cdot CaSO_4 \cdot H_2O + K_2SO_4$
14.44	2.28	0.020	83.260	»
14.52	2.88	0.025	82.575	»
14.57	4.60	0.019	80.811	»
14.70	5.79	0.020	79.480	»
14.43	7.14	0.020	78.410	»
14.52	11.28	0.020	74.180	»
14.45	16.20	0.018	69.332	»
13.75	18.10	0.018	68.132	$K_2SO_4 \cdot CaSO_4 \cdot H_2O + K_2SO_4 + K_2SO_4 \cdot MgSO_4 \cdot 4H_2O$
5.61	5.67	0.113	88.607	$K_2SO_4 \cdot MgSO_4 \cdot 2CaSO_4 \cdot 2H_2O + K_2SO_4 \cdot CaSO_4 \cdot H_2O$
5.79	5.02	0.102	89.088	»
6.74	6.44	0.095	86.725	»
6.59	8.05	0.097	85.263	»
7.46	11.53	0.055	80.955	»
7.78	13.75	0.030	78.440	»
8.31	14.63	0.030	77.030	»
9.85	15.04	0.040	74.070	»
10.50	17.81	0.016	71.674	»
10.73	18.31	0.000	70.960	$K_2SO_4 \cdot CaSO_4 \cdot H_2O$
7.80	26.50	0.000	65.700	$K_2SO_4 \cdot MgSO_4 \cdot 4H_2O + K_2SO_4 \cdot MgSO_4 \cdot 2CaSO_4 \cdot 2H_2O + CaSO_4$
8.67	24.05	0.000	67.280	»
9.48	23.02	0.000	67.500	»
10.02	22.62	0.000	67.360	»
10.22	21.87	0.000	67.910	»
3.80	6.25	0.160	89.790	$K_2SO_4 \cdot MgSO_4 \cdot 2CaSO_4 \cdot 2H_2O + CaSO_4$
3.34	7.96	0.153	88.547	»
3.51	9.82	0.148	86.382	»
3.40	15.42	0.123	11.057	»
3.50	18.20	0.120	78.180	»
2.00	26.20	0.000	71.800	»
3.80	5.30	0.160	81.740	$CaSO_4 \cdot K_2SO_4 \cdot 5CaSO_4 \cdot H_2O + K_2SO_4 \cdot MgSO_4 \cdot 2CaSO_4 \cdot 2H_2O$
0.00	0.00	0.161	99.839	$CaSO_4$
15.05	0.00	0.00	84.950	K_2SO_4

Table № 4079 (continuation)

Comp. of saturated soln., wt%				Comp. of solid phase
Potassium Sulfate	Magnesium Sulfate	Calcium Sulfate	Water	
0.00	34.94	0.00	65.060	$MgSO_4 \cdot 6H_2O$
13.80	17.97	0.00	68.230	$K_2SO_4 + K_2SO_4 \cdot MgSO_4 \cdot 4H_2O$
4.25	34.06	0.00	61.69	$K_2SO_4 \cdot MgSO_4 \cdot 4H_2O + MgSO_4 \cdot 6H_2O$

№ 4080 [486]

MAGNESIUM SULFATE–POTASSIUM SULFATE–CALCIUM SULFATE–WATER

$MgSO_4—K_2SO_4—CaSO_4—H_2O$

$t = 35$

Comp. of saturated soln., wt%				Comp. of solid phase
Potassium Sulfate	Magnesium Sulfate	Calcium Sulfate	Water	
11.820	0.462	0.0197	87.698	$K_2SO_4 + K_2SO_4 \cdot CaSO_4 \cdot H_2O$
11.606	2.396	0.0183	85.980	»
12.060	6.190	0.0180	81.570	»
12.080	9.805	0.0177	78.097	»
11.904	14.013	0.0170	74.066	»
11.475	15.484	0.01005	73.021	$K_2SO_4 + K_2SO_4 \cdot CaSO_4 \cdot H_2O +$ $+ K_2SO_4 \cdot MgSO_4 \cdot 6H_2O$
11.900	15.454	0.0025	72.643	$K_2SO_4 + K_2SO_4 \cdot MgSO_4 \cdot 6H_2O$
10.505	16.479	0.0100	73.006	$K_2SO_4 \cdot MgSO_4 \cdot 6H_2O + K_2SO_4 \cdot CaSO_4 \cdot H_2O$
8.622	18.573	0.0084	72.797	»
7.762	20.498	0.0075	71.732	»
6.087	22.775	0.0044	71.134	»
5.046	25.265	0.0041	69.685	$K_2SO_4 \cdot MgSO_4 \cdot 6H_2O + K_2SO_4 \cdot CaSO_4 \cdot$ $\cdot H_2O + K_2SO_4 \cdot MgSO_4 \cdot 2CaSO_4 \cdot 2H_2O$
4.652	26.275	0.0037	69.069	$K_2SO_4 \cdot MgSO_4 \cdot 6H_2O +$ $+ K_2SO_4 \cdot MgSO_4 \cdot 2CaSO_4 \cdot 2H_2O$
4.111	27.190	0.0030	68.696	»
3.801	28.497	0.0030	67.699	$K_2SO_4 \cdot MgSO_4 \cdot 6H_2O + MgSO_4 \cdot 7H_2O +$ $+ K_2SO_4 \cdot MgSO_4 \cdot 2CaSO_4 \cdot 2H_2O$
4.191	28.967	0.0015	66.840	$K_2SO_4 \cdot MgSO_4 \cdot 6H_2O + MgSO_4 \cdot 7H_2O$
2.540	28.730	0.0037	68.726	$MgSO_4 \cdot 7H_2O +$ $+ K_2SO_4 \cdot MgSO_4 \cdot 2CaSO_4 \cdot 2H_2O$
1.544	18.972	0.009	69.475	»
1.107	29.017	0.014	69.862	$MgSO_4 \cdot 7H_2O + CaSO_4 \cdot 2H_2O +$ $+ K_2SO_4 \cdot MgSO_4 \cdot 2CaSO_4 \cdot 2H_2O$
0.499	29.185	0.020	70.296	$MgSO_4 \cdot 7H_2O + CaSO_4 \cdot 2H_2O$
1.196	27.492	0.013	71.299	$CaSO_4 \cdot 2H_2O +$ $+ K_2SO_4 \cdot MgSO_4 \cdot 2CaSO_4 \cdot 2H_2O$
1.651	26.007	0.016	72.326	»
2.509	24.765	0.020	72.706	»
3.703	23.042	0.031	73.224	»
4.287	20.557	0.037	75.119	$K_2SO_4 \cdot CaSO_4 \cdot H_2O + CaSO_4 \cdot 2H_2O +$ $+ K_2SO_4 \cdot MgSO_4 \cdot 2CaSO_4 \cdot 2H_2O$
4.904	21.968	0.020	73.108	$K_2SO_4 \cdot CaSO_4 \cdot H_2O +$ $+ K_2SO_4 \cdot MgSO_4 \cdot 2CaSO_4 \cdot 2H_2O$
4.908	23.993	0.007	71.092	»

Table № 4080 (continuation)

Comp. of saturated soln., wt%				Comp. of solid phase
Potassium Sulfate	Magnesium Sulfate	Calcium Sulfate	Water	
4.195	19.196	0.041	76.568	$K_2SO_4 \cdot CaSO_4 \cdot H_2O + CaSO_4 \cdot 2H_2O$
4.189	18.112	0.048	77.651	»
3.997	15.993	0.071	79.939	»
3.895	13.900	0.102	82.103	»
3.797	12.000	0.129	84.074	»
3.667	11.217	0.160	85.956	»
3.600	9.044	0.166	87.190	»
3.600	7.004	0.170	89.226	»
3.549	5.000	0.156	91.295	»
3.540	4.133	0.153	92.174	»
3.636	3.391	0.149	92.824	»
3.605	1.985	0.149	94.261	»
3.580	0.896	0.153	95.371	»
3.592	0.500	0.156	95.752	$K_2SO_4 \cdot CaSO_4 \cdot H_2O +$ $+ K_2SO_4 \cdot 5CaSO_4 \cdot H_2O + CaSO_4 \cdot 2H_2O$

№ 4081 **MAGNESIUM SULFATE–POTASSIUM CHLORIDE–WATER** [1033]

$MgSO_4$—KCl—H_2O

$t = 0$

Comp. of saturated soln., wt%					Comp. of solid phase
Potassium Chloride	Magnesium Chloride	Potassium Sulfate	Magnesium Sulfate	Water	
0.00	3.56	8.70	3.77	83.97	$K_2SO_4 \cdot MgSO_4 \cdot 6H_2O + K_2SO_4 \cdot H_2O$
0.00	6.58	9.81	0.60	83.01	»
8.92	9.76	4.51	0.00	76.81	$K_2SO_4 \cdot MgSO_4 \cdot 6H_2O + K_2SO_4 \cdot H_2O + KCl$
5.19	13.57	4.90	0.00	76.34	$K_2SO_4 \cdot MgSO_4 \cdot 6H_2O + KCl$
0.00	19.72	6.05	0.00	74.23	$K_2SO_4 \cdot MgSO_4 \cdot 6H_2O + KCl + MgSO_4 \cdot 7H_2O$
0.00	7.46	3.99	10.52	78.03	$K_2SO_4 \cdot MgSO_4 \cdot 6H_2O + MgSO_4 \cdot 7H_2O$
17.32	3.36	1.74	0.00	77.58	$K_2SO_4 \cdot H_2O + KCl$
2.45	24.95	0.00	2.05	70.55	$KCl + KCl \cdot MgCl_2 \cdot 6H_2O + MgSO_4 \cdot 7H_2O$
0.15	32.85	0.00	2.35	64.65	$KCl \cdot MgCl \cdot 6H_2O + MgSO_4 \cdot 7H_2O +$ $+ MgCl_2 \cdot 6H_2O$

№ 4082 [824]

MAGNESIUM SULFATE–CALCIUM CARBONATE–WATER

$MgSO_4$—$CaCO_3$—H_2O

Content in aqueous soln.		pH	*t*	Comp. of solid phase
Magnesium Sulfate, wt%	Calcium Carbonate, mg-eq/l			
25	3.48	8.0	16—13	$CaCO_3$
20	3.57	8.2	16—18	»

Table № 4082 (continuation)

Content in aqueous soln.		pH	t	Comp. of solid phase
Magnesium Sulfate, wt%	Calcium Carbonate, mg-eq/l			
15	4.24	8.2	16—18	$CaCO_3$
8	4.76	8.4	16—18	»
3.5	4.41	8.3	16—18	»
1	3.48	8.2	16—18	»
25	9.34	6.8	29—30	»
20	7.53	8.0	29—30	»
15	7.07	8.2	29—30	»
8	5.54	8.4	29—30	»
3.5	3.84	8.5	29—30	»
1	2.93	8.3	29—30	»

№ 4083 MAGNESIUM SULFATE–CALCIUM CARBONATE–WATER [1031]

$MgSO_4—CaCO_3—H_2O$

$t = 25$

Content in aqueous soln., mmol/1000g soln.				Comp. of solid phase
Ca^{2+}	Mg^{2+}	$(HCO_3)_2^{2-}$	SO_4^{2-}	
0.81	0.00	0.81	0.00	$CaCO_3$
0.00	1.65	1.65	0.00	$MgCO_3$
0.40	1.02	1.42	0.00	$CaCO_3 \cdot MgCO_3 + CaCO_3$
0.39	0.94	1.33	0.00	»
0.32	1.20	1.52	0.00	$CaCO_3 \cdot MgCO_3 + MgCO_3$
15.97	0.00	0.38	15.59	$CaCO_3 + CaSO_4$
15.33	2.58	0.39	17.52	$CaCO_3 + CaSO_4 + CaCO_3 + MgCO_3$
15.28	2.85	0.45	17.68	$CaCO_3 \cdot MgCO_3 + CaSO_4 + CaCO_3$
15.47	3.53	0.47	18.53	$CaCO_3 \cdot MgCO_3 + CaSO_4 + MgCO_3$

Note. $P_{CO_2} = 0.0012$ atm.

№ 4084 MAGNESIUM SULFATE–CALCIUM SULFATE–WATER [1325]

$MgSO_4—CaSO_4—H_2O$

$t = 25$

Concn. in aqueous soln., g/l		d_{25}	Concn. in aqueous soln., g/l		d_{25}
Magnesium Sulfate	Calcium Sulfate		Magnesium Sulfate	Calcium Sulfate	
0	2.046	1.0032	21.36	1.478	1.0226
3.20	1.620	1.0055	42.68	1.558	1.0419
6.39	1.507	1.0090	64.14	1.608	1.0626
10.64	1.471	1.0118	85.67	1.617	1.0833

Table № 4084 (continuation)

Concn. in aqueous soln., g/l		d_{25}	Concn. in aqueous soln., g/l		d_{25}
Magnesium Sulfate	Calcium Sulfate		Magnesium Sulfate	Calcium Sulfate	
128.28	1.627	1.1190	232.1	1.254	1.2095
149.67	1.597	1.1377	265.6	1.070	1.2382
165.7	1.549	1.1479	298.0	0.860	1.2624
171.2	1.474	1.1537	330.6	0.647	1.2877
198.8	1.422	1.1813	355.0	0.501	1.3023

№ 4085 [1505]

MAGNESIUM SULFATE–CALCIUM SULFATE–WATER

$MgSO_4—CaSO_4—H_2O$

$t = 25$

Content in aqueous soln., mol/l		Comp. of solid phase
Magnesium Sulfate	Calcium Sulfate	
2.475	0.0093	$CaSO_4 \cdot 2H_2O$
1.243	0.0084	»
0.533	0.0078	»
0.0884	0.0074	»
0	0.0153	»

№ 4086 [1781]

MAGNESIUM SULFATE–CALCIUM SULFATE–WATER

$MgSO_4—CaSO_4—H_2O$

$t = 25$

Concn. in aqueous soln., wt%		d_4	Comp. of solid phase
Magnesium Sulfate	Calcium Sulfate		
0.0	0.20854	0.99911	$CaSO_4 \cdot 2H_2O$
0.06029	0.19565	0.99960	»
0.12167	0.1848	1.00010	»
0.18339	0.1777	1.00067	»
0.0	0.20658	0.99905	Selenite plates
0.13413	0.17886	1.0004	»
0.33161	0.16469	1.0018	»
0.63148	0.15592	1.0044	»
1.2385	0.15912	1.0099	»

№ 408. [148]

MAGNESIUM SULFATE–CALCIUM SULFATE–WATER

$MgSO_4$—$CaSO_4$—H_2O

$t = 55$

Comp. of saturated soln., wt%			Comp. of solid phase
Magnesium Sulfate	Calcium Sulfate	Water	
0.00	0.212	99.788	$CaSO_4 \cdot 2H_2O$
0.52	0.183	99.297	»
0.98	0.166	98.854	»
2.10	0.173	97.727	»
3.89	0.179	95.931	»
6.35	0.186	93.464	»
9.61	0.199	90.191	»
10.14	0.200	89.650	»
12.60	0.180	87.220	»
17.00	0.115	82.885	»
18.78	0.089	81.131	»
20.64	0.070	79.290	»
22.50	0.036	77.463	»
34.70	0.020	65.270	$CaSO_4 \cdot 2H_2O + MgSO_4 \cdot 6H_2O$
34.94	0.00	65.060	$MgSO_4 \cdot 6H_2O$
0.00	0.161	99.839	$CaSO_4$
1.07	0.133	98.797	»
2.38	0.153	97.467	»
5.72	0.165	94.115	»
9.85	0.160	39.990	»
17.10	0.082	82.818	»

№ 4088 [630]

MAGNESIUM SULFATE–CALCIUM SULFATE–WATER

$MgSO_4$—$CaSO_4$—H_2O

$t = 35$

Comp. of saturated soln., wt%			*d*	Comp. of solid phase
Calcium Sulfate	Magnesium Sulfate	Water		
0.21	0.00	99.79	1.002	$CaSO_4 \cdot 2H_2O$
0.19	0.15	99.66	1.002	»
0.16	0.68	99.16	1.010	»
0.16	1.15	99.69	1.015	»
0.17	2.19	97.64	1.024	»
0.17	3.25	96.58	1.036	»
0.18	5.44	94.38	1.057	»
0.18	8.24	91.58	1.085	»
0.18	10.62	89.2	1.112	»
0.17	11.45	88.38	1.124	»
0.16	14.95	84.89	1.160	»
0.14	17.86	82.0	1.194	»
0.09	23.51	76.4	1.275	»
0.03	29.26	70.71	1.347	$CaSO_4 \cdot 2H_2O + MgSO_4 \cdot 7H_2O$
0.02	29.28	79.70	1.342	$MgSO_4 \cdot 7H_2O$
0.00	29.40	70.60	1.324	»

№ 4089 MAGNESIUM SULFATE–CALCIUM SULFATE–WATER [397]

$MgSO_4—CaSO_4—H_2O$

$t = 25$

Comp. of saturated soln., wt%			d	Comp. of solid phase
Magnesium Sulfate	Calcium Sulfate	Water		
0.000	0.209	99.791	1.0015	$CaSO_4 \cdot 2H_2O$
0.180	0.198	99.622	1.0089	»
0.870	0.163	98.967	1.0103	»
1.460	0.155	98.385	1.0165	»
5.400	0.165	94.435	1.0581	»
13.490	0.145	86.365	1.1404	»
19.850	0.117	80.033	1.2188	»
27.050	0.040	72.910	1.3070	$CaSO_4 \cdot 2H_2O + MgSO_4 \cdot 7H_2O$
27.080	0.011	72.909	1.3069	$MgSO_4 \cdot 7H_2O$
27.080	0.007	72.910	1.3069	»
27.100	0.000	72.900	1.3071	»

№ 4090 [398]

MAGNESIUM SULFATE–CALCIUM SULFATE–WATER

$MgSO_4—CaSO_4—H_2O$

$t = 25$

Content in saturated soln., wt%		d	Comp. of solid phase
Magnesium Sulfate	Calcium Sulfate		
0.00	0.209	1.0015	$CaSO_4 \cdot 2H_2O$
0.180	0.198	1.0039	»
0.870	0.163	1.0103	»
1.46	0.155	1.0165	»
5.40	0.165	1.0581	»
13.49	0.145	1.1404	»
19.85	0.117	1.2188	»
27.08	0.007	1.3069	$MgSO_4 \cdot 7H_2O$
27.05	0.040	1.3070	$CaSO_4 \cdot 2H_2O + MgSO_4 \cdot 7H_2O$

№ 4091 MAGNESIUM SULFATE–CALCIUM IODATE–WATER [1388]

$MgSO_4—Ca(IO_3)_2—H_2O$

$t = 25$

Concn. of Magnesium Sulfate in aqueous soln., g/l	Solubility of Calcium Iodate in aqueous soln. of Magnesium Sulfate, wt%	Concn. of Magnesium Sulfate in aqueous soln., g/l	Solubility of Calcium Iodate in aqueous soln. of Magnesium Sulfate, wt%
0.0	0.303	2.408	0.4090
0.2408	0.3219	6.02	0.4952
0.602	0.3397	12.04	0.5768
1.204	0.3663		

№ 4092 MAGNESIUM SULFATE–MANGANESE (II) SULFATE–WATER [1212]

$MgSO_4—MnSO_4—H_2O$

Comp. of saturated soln., wt%			t	Comp. of solid phase
Magnesium Sulfate	Manganese Sulfate	Water		
18.19	3.56	78.25	0	$MgSO_4 \cdot 7H_2O$
17.68	6.17	76.15	0	»
14.61	10.66	74.73	0	»
12.77	14.21	73.02	0	»
11.0	17.83	71.17	0	»
8.37	23.10	68.53	0	$MgSO_4 \cdot 7H_2O + MnSO_4 \cdot 7H_2O$
6.81	25.11	67.98	0	$MnSO_4 \cdot 7H_2O$
3.97	29.25	66.78	0	»
1.88	32.0	66.12	0	»
25.4	1.10	73.5	23	$MgSO_4 \cdot 7H_2O$
25.3	1.83	72.83	23	»
21.16	8.65	70.19	23	»
17.86	14.27	67.87	23	»
14.50	20.40	65.10	23	»
11.88	26.30	61.82	23	»
10.36	28.10	61.54	23	»
9.36	28.50	62.14	23	$MgSO_4 \cdot 7H_2O + MnSO_4 \cdot 5H_2O$ and $MgSO_4 \cdot 6H_2O + MnSO_4 \cdot 5H_2O$
8.43	30.00	61.57	23	$MnSO_4 \cdot 5H_2O$
5.00	33.70	61.30	23	»
3.06	35.20	61.74	23	»
1.04	37.70	61.26	23	»
32.00	2.24	65.76	50	$MgSO_4 \cdot 6H_2O$
31.30	3.15	65.55	50	»
27.60	8.43	63.97	50	»
25.50	12.40	62.10	50	»
24.8	13.04	62.16	50	$MnSO_4 \cdot H_2O + MgSO_4 \cdot 6H_2O$
24.4	13.33	62.27	50	$MnSO_4 \cdot H_2O$
21.2	16.00	62.8	50	»
17.08	19.30	63.62	50	»
10.46	26.00	63.54	50	»
6.00	30.70	63.30	50	»
2.36	33.80	63.84	50	»
34.8	0.30	64.9	90	$MgSO_4 \cdot H_2O + MnSO_4 \cdot H_2O$
33.60	1.64	64.76	90	»
31.30	3.41	65.29	90	»
29.50	5.72	64.78	90	»
27.2	7.55	65.25	90	»
23.5	10.55	65.95	90	»
20.40	13.68	65.92	90	»
13.45	17.86	68.69	90	»
9.43	20.10	71.47	90	»
2.38	25.90	71.71	90	»

$MgSO_4—NiSO_4—H_2O$

Comp. of saturated soln., wt%			t	Comp. of solid phase
Nickel Sulfate	Magnesium Sulfate	Water		
2.18	19.09	78.73	0	$NiSO_4 \cdot MgSO_4 \cdot 7H_2O$
4.14	17.04	78.82	0	»
6.43	15.33	78.24	0	»
10.30	11.66	78.04	0	»
12.31	9.85	77.84	0	»
13.88	8.15	77.97	0	»
16.11	5.83	78.06	0	»
18.57	3.84	77.59	0	»
0.86	28.2	70.94	35	»
2.20	27.2	70.60	35	»
6.0	24.3	69.70	35	»
7.24	23.1	69.66	35	»
9.88	20.6	69.52	35	»
12.89	17.87	69.24	35	»
15.92	15.07	69.01	35	»
18.40	12.70	68.90	35	»
21.6	9.55	68.85	35	»
24.7	6.79	68.51	35	»
25.4	6.36	68.24	35	»
25.9	5.77	68.33	35	»
26.6	5.25	68.15	35	»
27.1	4.54	68.36	35	»
27.6	4.26	68.14	35	α-$NiSO_4 \cdot MgSO_4 \cdot 6H_2O$
3.48	27.0	69.52	40	$NiSO_4 \cdot MgSO_4 \cdot 7H_2O$
6.93	24.1	68.97	40	»
10.69	21.1	68.21	40	»
14.08	18.14	67.78	40	»
16.78	15.86	67.36	40	»
19.43	13.21	67.36	40	»
21.0	11.74	67.26	40	»
22.7	10.09	67.21	40	»
24.0	9.03	66.97	40	»
25.0	8.05	66.95	40	$NiSO_4 \cdot MgSO_4 \cdot 7H_2O+$ $+\alpha$-$NiSO_4 \cdot MgSO_4 \cdot 6H_2O$
25.3	7.55	67.15	40	α-$NiSO_4 \cdot MgSO_4 \cdot 6H_2O$
26.8	5.97	67.23	40	»
28.4	3.31	68.29	40	»
30.1	2.53	67.37	40	»
2.20	29.8	68.00	45	$NiSO_4 \cdot MgSO_4 \cdot 7H_2O$
5.82	26.7	67.48	45	»
9.52	22.8	67.68	45	»
12.90	20.2	66.90	45	»
14.27	19.03	66.70	45	»
14.96	18.34	66.70	45	»
15.59	17.61	66.80	45	»
16.16	17.38	66.46	45	»
16.53	16.91	66.56	45	$NiSO_4 \cdot MgSO_4 \cdot 7H_2O+$ $+\beta$-$NiSO_4 \cdot MgSO_4 \cdot 6H_2O$
19.74	13.89	66.37	45	β-$NiSO_4 \cdot MgSO_4 \cdot 6H_2O$
23.0	10.63	66.37	45	»

Table № 4093 (continuation)

Comp. of saturated soln., wt%			t	Comp. of solid phase
Nickel Sulfate	Magnesium Sulfate	Water		
26.3	7.35	66.35	45	β-$NiSO_4 \cdot MgSO_4 \cdot 6H_2O$
29.0	4.45	66.55	45	»
6.95	28.3	64.75	60	»
9.35	23.4	67.25	60	»
14.09	22.1	63.81	60	»
15.20	18.40	66.40	60	»
18.49	17.47	64.04	60	»
21.7	14.11	64.19	60	»
25.0	11.13	63.87	60	»
28.2	8.01	63.79	60	»
31.6	4.43	63.97	60	»
0.18	36.3	63.52	75	$MgSO_4 \cdot NiSO_4 \cdot H_2O$
0.30	35.3	64.40	75	»
0.87	35.8	63.33	75	»
1.42	35.6	62.98	75	»
2.74	34.5	62.76	75	»
4.03	33.8	62.17	75	β-$NiSO_4 \cdot MgSO_4 \cdot 6H_2O$
7.94	30.1	61.96	75	»
11.09	26.6	62.31	75	»
11.95	26.2	61.85	75	»
15.53	22.7	61.77	75	»
18.92	18.22	62.86	75	»
22.6	15.88	61.52	75	»
26.4	11.97	61.63	75	»
29.8	8.25	61.95	75	»
33.9	4.84	61.26	75	»
0.96	34.5	64.54	80	$MgSO_4 \cdot NiSO_4 \cdot H_2O$
1.81	33.9	64.29	80	»
3.12	32.9	63.98	80	»
9.17	28.7	62.13	80	»
14.04	25.3	60.66	80	β-$NiSO_4 \cdot MgSO_4 \cdot 6H_2O$
16.44	22.8	60.76	80	»
18.63	21.4	59.97	80	»
20.3	19.16	60.54	80	»
24.4	15.15	60.45	80	»
30.2	9.73	60.07	80	»
35.5	4.22	60.28	80	»
39.3	0.70	60.00	80	»
4.08	30.1	65.82	100	$NiSO_4 \cdot MgSO_4 \cdot H_2O$
9.01	26.1	64.89	100	»
12.75	24.5	62.75	100	»
16.56	22.0	61.44	100	»
17.22	20.1	62.68	100	»
22.7	15.70	61.60	100	»
25.8	13.14	61.06	100	»
32.6	7.12	60.28	100	»
37.2	3.11	59.69	100	»

№ 4094 [1500]

MAGNESIUM SULFATE–COPPER (II) SULFATE–WATER

$MgSO_4—CuSO_4—H_2O$

$t = 0$

Comp. of saturated soln., wt%			Comp. of solid phase
Copper Sulfate	Magnesium Sulfate	Water	
0.0	20.9	79.1	$MgSO_4 \cdot 6H_2O$
2.05	20.2	77.75	»
3.65	19.4	76.95	»
6.8	17.6	75.6	$MgSO_4 \cdot 6H_2O + CuSO_4 \cdot 5H_2O$
9.42	12.30	78.28	$CuSO_4 \cdot 5H_2O$
11.12	7.08	81.8	»
13.03	0.0	86.97	»

№ 4095 [1885]

MAGNESIUM SULFATE–COPPER (II) SULFATE–WATER

$MgSO_4—CuSO_4—H_2O$

$t = 30$

Concn. in aqueous soln., wt%		Concn. in aqueous soln., wt%		Concn. in aqueous soln., wt%	
Copper Sulfate	Magnesium Sulfate	Copper Sulfate	Magnesium Sulfate	Copper Sulfate	Magnesium Sulfate
1.57	26.63	8.28	20.98	13.24	12.06
2.19	25.94	8.46	20.06	14.76	8.93
6.04	23.17	10.07	19.31	16.35	5.84
6.07	23.07	10.28	18.82	17.79	3.13
7.92	21.79	11.28	16.54	18.13	1.94

№ 4096 [2596]

MAGNESIUM SULFATE–ZINC HYDROXIDE–WATER

$MgSO_4—Zn(OH)_2—H_2O$

$t = 16—20$

Comp. of solvent, wt%		Solubility of Zinc Hydroxide in aqueous soln. of Magnesium Sulfate, g/l
Magnesium Sulfate	Water	
1	99	0.027

№ 4097 MAGNESIUM SULFATE–RUBIDIUM SULFATE–WATER [964]

$MgSO_4—Rb_2SO_4—H_2O$

$t = 35$

Comp. of saturated soln., wt%			d	Comp. of solid phase
Rubidium Sulfate	Magnesium Sulfate	Water		
0.00	29.63	70.37	1.225	$MgSO_4 \cdot 7H_2O$
1.30	30.42	68.28	1.282	»
2.29	29.67	68.04	—	»
4.78	28.47	66.75	—	»
5.18	29.09	65.73	1.275	$Rb_2SO_4 \cdot MgSO_4 \cdot 6H_2O$
6.00	27.80	66.20	—	»
7.82	21.95	70.23	1.216	»
7.66	19.97	72.37	—	»
9.44	16.11	74.45	—	»
9.88	11.45	78.67	—	»
12.42	8.08	79.50	1.109	»
16.27	4.55	79.18	1.116	»
24.68	1.50	73.82	1.171	»
34.40	0.13	65.47	1.264	»
34.64	0.20	65.16	1.270	$Rb_2SO_4 \cdot MgSO_4 \cdot 6H_2O + Rb_2SO_4$
36.93	0.00	63.07	1.276	Rb_2SO_4

№ 4098 MAGNESIUM SULFATE–SILVER SULFATE–WATER [2081]

$MgSO_4—Ag_2SO_4—H_2O$

Content of Magnesium Sulfate in aqueous soln., mol/l	Solubility of Silver Sulfate in aqueous soln. of Magnesium Sulfate, mol/l	t	Content of Magnesium Sulfate in aqueous soln., mol/l	Solubility of Silver Sulfate in aqueous soln. of Magnesium Sulfate, mol/l	t
0.1	0.0247	25	0.1	0.0506	154
0.1	0.0359	64	0.1	0.0512	160
0.1	0.0422	85	0.1	0.0506	175
0.1	0.0441	104	0.5	0.0440	64

Table № 4098 (continuation)

Content of Magnesium Sulfate in aqueous soln., mol/l	Solubility of Silver Sulfate in aqueous soln. of Magnesium Sulfate, mol/l	t	Content of Magnesium Sulfate in aqueous soln., mol/l	Solubility of Silver Sulfate in aqueous soln. of Magnesium Sulfate, mol/l	t
0.5	0.0520	85	1.0	0.0656	95
0.5	0.0664	129	1.0	0.0718	103
0.5	0.0732	147	1.0	0.0874	124
0.5	0.0801	156	1.0	0.0990	144
1.0	0.0520	64			

№ 4099 [2405]

MAGNESIUM SULFATE–SILVER SULFATE–WATER

$MgSO_4—Ag_2SO_4—H_2O$

$t = 35$

Concn. in aqueous soln., wt%		Comp. of solid phase
Silver Sulfate	Magnesium Sulfate	
0.929	0.00	Ag_2SO_4
0.906	0.320	»
0.952	5.44	»
1.050	11.64	»
1.065	14.50	»
1.052	18.63	»
0.99	24.06	»
0.837	29.35	$Ag_2SO_4+MgSO_4 \cdot 7H_2O$
0.840	29.33	»
0.842	29.31	»
0.840	29.33	»
0.633	29.35	$MgSO_4 \cdot 7H_2O$
0.00	29.77	»

№ 4100 [2344]

MAGNESIUM SULFATE–SILVER CHLORIDE–WATER

$MgSO_4—AgCl—H_2O$

$t = 25$

Concn. in aqueous soln., g/l		Concn. in aqueous soln., g/l		Concn. in aqueous soln., g/l	
Magnesium Sulfate	Silver Chloride	Magnesium Sulfate	Silver Chloride	Magnesium Sulfate	Silver Chloride
0.0	0.00183	0.06101	0.00192	0.3621	0.00208
0.00077	0.00183	0.1212	0.00197	0.6269	0.00213
0.00375	0.00184	0.1815	0.00200	0.9160	0.00218
0.01282	0.00186	0.3010	0.00206	1.205	0.00221
0.02656	0.00189				

№ 4101 MAGNESIUM SULFATE–SILVER IODATE–WATER [1173]

$MgSO_4—AgIO_3—H_2O$

$t = 75$

Content of Magnesium Sulfate in aqueous soln wt%	Solubility of Silver Iodate, mol/l	Content of Magnesium Sulfate in aqueous soln wt%	Solubility of Silver Iodate, mol/l
0.0	0.8417	2.0	0.9111
0.2	0.8522	5.0	0.9629
0.5	0.8698	10.0	1.0201
1.0	0.8855	20.0	1.0928

№ 4102 MAGNESIUM SULFATE–CADMIUM SULFATE–WATER [2707]

$MgSO_4—CdSO_4—H_2O$

Comp. of saturated soln., wt%			t	Comp. of saturated soln., wt%			t
Cadmium Sulfate	Magnesium Sulfate	Water		Cadmium Sulfate	Magnesium Sulfate	Water	
6.5	23.8	69.7	25	2.6	29.6	67.8	40
12.5	20.9	66.6	25	5.0	28.8	66.2	40
17.2	18.3	64.5	25	7.6	27.8	64.6	40
17.4	18.9	63.7	25	9.2	28.0	62.8	40
21.7	17.0	61.3	25	9.6	27.8	62.6	40
25.5	15.2	59.3	25	10.2	27.5	62.3	40
25.7	14.9	59.4	25	9.6	28.3	62.1	40
25.8	15.4	58.8	25	12.1	24.4	63.5	40
26.0	15.0	59.0	25	17.2	20.0	62.8	40
26.1	15.0	58.9	25	14.5	22.4	63.1	40
26.2	15.3	58.5	25	20.0	17.6	62.4	40
26.2	15.1	58.7	25	20.9	17.1	62.0	40
26.4	15.0	58.6	25	24.2	14.3	61.5	40
26.4	15.2	58.4	25	28.2	12.2	59.6	40
26.9	14.9	58.2	25	33.5	10.4	56.1	40
27.0	13.9	59.1	25	38.8	9.2	52.0	40
29.1	12.5	58.4	25	39.6	9.2	51.2	40
30.5	11.2	58.3	25	40.1	8.8	51.1	40
30.5	11.0	58.5	25	40.9	7.7	51.4	40
34.0	8.0	58.0	25	42.7	4.5	52.8	40
38.6	4.3	57.1	25				

№ 4103 [2279]

MAGNESIUM SULFATE–CADIUM SULFATE–WATER

$MgSO_4—CdSO_4—H_2O$

Concn. in aqueous soln., wt%		t	Comp. of solid phase
Cadmium Sulfate	Magnesium Sulfate		
6.5	23.8	25	$MgSO_4 \cdot 7H_2O$
12.5	20.9	25	»
17.2	18.3	25	»

Table № 4103 (continuation)

Concn. in aqueous soln., wt%		t	Comp. of solid phase
Cadmium Sulfate	Magnesium Sulfate		
17.4	18.9	25	$MgSO_4 \cdot 7H_2O$
21.7	17.0	25	»
25.5	15.2	25	$MgSO_4 \cdot 7H_2O + CdSO_4 \cdot 8/3H_2O$
25.7	14.9	25	»
25.8	15.4	25	»
26.0	15.0	25	»
26.1	15.0	25	»
26.2	15.3	25	»
26.2	15.1	25	»
26.4	15.0	25	»
26.4	15.2	25	»
26.9	14.9	25	»
27.0	13.9	25	$CdSO_4 \cdot 8/3H_2O$
29.1	12.5	25	»
30.5	11.2	25	»
30.5	11.0	25	»
34.0	8.0	25	»
38.6	4.3	25	»
2.6	29.6	40	$MgSO_4 \cdot 7H_2O$
5.0	28.8	40	»
7.6	27.8	40	»
9.2	28.0	40	$MgSO_4 \cdot 7H_2O + CdSO_4 \cdot MgSO_4 \cdot 14H_2O$
9.6	27.8	40	»
10.2	27.5	40	»
9.6	28.3	40	»
12.1	24.4	40	$CdSO_4 \cdot MgSO_4 \cdot 14H_2O$
17.2	20.0	40	»
14.5	22.4	40	»
20.0	17.6	40	»
20.9	17.1	40	»
24.2	14.3	40	»
28.2	12.2	40	»
33.5	10.4	40	»
38.8	9.2	40	»
39.6	9.2	40	$CdSO_4 \cdot MgSO_4 \cdot 14H_2O + CdSO_4 \cdot 8/3H_2O$
40.1	8.8	40	»
40.9	7.7	40	$CdSO_4 \cdot 8/3H_2O$
42.7	4.5	40	»

№ 4104 [968]

MAGNESIUM SULFATE–CESIUM SULFATE–WATER

$MgSO_4$—Cs_2SO_4—H_2O

$t = 35$

Comp. of saturated soln., wt%			Comp. of solid phase
Magnesium Sulfate	Cesium Sulfate	Water	
29.58	0.00	70.42	$MgSO_4 \cdot 7H_2O$
29.54	0.99	69.47	»
29.28	3.63	67.09	»
29.06	7.54	63.40	»
28.00	9.00	63.00	»
27.10	13.50	59.40	$MgSO_4 \cdot 7H_2O + Cs_2SO_4 \cdot MgSO_4 \cdot 6H_2O$
25.74	12.62	61.64	$Cs_2SO_4 \cdot MgSO_4 \cdot 6H_2O$
16.10	21.13	62.77	»
7.68	31.86	90.47	»
3.58	41.80	54.62	»
1.88	54.60	43.52	»
0.86	60.03	39.11	»
0.46	61.37	38.17	$MgSO_4 \cdot Cs_2SO_4 \cdot 6H_2O + Cs_2SO_4$
0.53	63.22	36.23	Cs_2SO_4
0.00	66.70	33.30	»

№ 4105 [1677, 2039]

MAGNESIUM SULFATE–LANTHANUM IODATE–WATER

$MgSO_4$—$La(IO_3)_3$—H_2O

$t = 25$

Concn. of Magnesium Sulfate in aqueous soln., g/l	Solubility of Lanthanum Iodate in aqueous soln. of Magnesium Sulfate, g/l
6.016	1.9977
6.016	1.989

№ 4106 [2313]

MAGNESIUM SULFATE–LANTHANUM IODATE–WATER

$MgSO_4$—$La(IO_3)_3$—H_2O

$t = 25$

Concn. of Magnesium Sulfate in aqueous soln., wt%	Solubility of Lanthanum Iodate in aqueous soln. of Magnesium Sulfate, g/l	*d*	Concn. of Magnesium Sulfate in aqueous soln., wt%	Solubility of Lanthanum Iodate in aqueous soln. of Magnesium Sulfate, g/l	*d*
0.120	0.561	0.99766	60.09	3.891	1.06116
0.601	0.802	0.99839	120.3	5.441	1.12180
1.203	0.961	0.99921	180.4	6.431	1.17865
6.01	1.582	1.00496	240.6	7.007	1.23098
12.03	2.005	1.01167			

№ 4107 MAGNESIUM SULFATE–CERIUM IODATE–WATER [1387]

$MgSO_4—Ce(IO_3)_2—H_2O$

$t = 25$

Concn. in aqueous soln., wt%		Concn. in aqueous soln., wt%	
Magnesium Sulfate	Cerium Iodate	Magnesium Sulfate	Cerium Iodate
0.00	0.1729	0.2393	0.3546
0.0240	0.2045	0.5955	0.4645
0.0600	0.2434	1.182	0.5918
0.1199	0.2905		

№ 4108 MAGNESIUM SULFATE–THALLIUM SULFATE–WATER [1207]

$MgSO_4—Tl_2SO_4—H_2O$

Comp. of saturated soln., wt%			t	Comp. of solid phase
Magnesium Sulfate	Thallium Sulfate	Water		
22.37	2.39	75.24	0	$MgSO_4 \cdot 7H_2O + MgSO_4 \cdot Tl_2SO_4 \cdot 6H_2O$
12.66	3.02	84.32	0	$MgSO_4 \cdot Tl_2SO_4 \cdot 6H_2O$
7.42	3.76	88.82	0	$MgSO_4 \cdot TlSO_4 \cdot 6H_2O + Tl_2SO_4$
24.64	3.36	72.00	17.6	$MgSO_4 \cdot 7H_2O + MgSO_4 \cdot Tl_2SO_4 \cdot 6H_2O$
17.52	4.88	77.60	17.6	$MgSO_4 \cdot Tl_2SO_4 \cdot 6H_2O$
13.76	5.75	80.49	17.6	»
12.0	6.40	81.60	17.6	$MgSO_4 \cdot Tl_2SO_4 \cdot 6H_2O + Tl_2SO_4$
31.0	2.59	66.41	40.5	$MgSO_4 \cdot 7H_2O$
30.05	5.06	64.89	40.5	$MgSO_4 \cdot 7H_2O + MgSO_4 \cdot Tl_2SO_4 \cdot 6H_2O$
25.12	6.86	68.02	40.5	$MgSO_4 \cdot Tl_2SO_4 \cdot 6H_2O$
19.66	9.42	70.92	40.5	»
15.62	10.78	73.60	40.5	$MgSO_4 \cdot Tl_2SO_4 \cdot 6H_2O + Tl_2SO_4$
15.13	10.98	73.89	40.5	Tl_2SO_4
3.77	8.36	87.87	40.5	»
1.92	7.74	90.34	40.5	»
35.08	4.70	60.22	64.5	$MgSO_4 \cdot 6H_2O$
34.53	6.47	59.00	64.5	$MgSO_4 \cdot 6H_2O + MgSO_4 \cdot Tl_2SO_4 \cdot 6H_2O$
26.98	12.22	60.80	64.5	$MgSO_4 \cdot Tl_2SO_4 \cdot 6H_2O$
22.65	15.20	62.15	64.5	$MgSO_4 \cdot Tl_2SO_4 \cdot 6H_2O + Tl_2SO_4$
18.40	15.91	65.69	64.5	Tl_2SO_4
4.67	12.52	82.81	64.5	»
30.34	9.41	60.25	10	$MgSO_4 \cdot H_2O$
25.4	24.7	49.9	10	$MgSO_4 \cdot H_2O + Tl_2SO_4$
18.55	23.65	57.80	10	Tl_2SO_4

№ 4109 [2367]

MAGNESIUM SULFATE–THALLIUM CHLORIDE–WATER

$MgSO_4—TlCl—H_2O$

$t = 25$

Concn. in aqueous soln., wt%		Concn. in aqueous soln., wt%	
Thallium Chloride	Magnesium Sulfate	Thallium Chloride	Magnesium Sulfate
0.384	0.0	0.524	0.743
0.457	0.202	0.588	1.520
0.485	0.401	0.615	2.328
0.500	0.522	0.657	4.046

№ 4110 [2702]

MAGNESIUM SULFATE–THALLIUM CHLORIDE WATER

$MgSO_4—TlCl—H_2O$

$t = 20$

Concn in aqueous soln., g/l	
Thallium Chloride	Magnesium Sulfate
3.511	0.602
3.650	1.204
3.856	2.408
4.124	4.816

№ 4111 [2581]

MAGNESIUM SULFATE–THALLIUM BROMIDE–WATER

$MgSO_4—TlBr—H_2O$

$t = 20$

Concn. in aqueous soln., g/l	
Thallium Bromide	Magnesium Sulfate
0.4945	0.241
0.517	0.602
0.530	0.902
0.542	1.20
0.576	2.41
0.620	4.81

№ 4112 [2702]

MAGNESIUM SULFATE–THALLIUM IODIDE–WATER

$MgSO_4—TlI—H_2O$

$t = 20$

Concn. in aqueous soln., g/l	
Thallium Iodide	Magnesium Sulfate
0.0658	0.1204
0.0725	0.903
0.0786	2.408
0.0845	4.816

№ 4113 [2039]

MAGNESIUM SULFATE–THALLIUM IODATE–WATER

$MgSO_4—TlIO_3—H_2O$

$t = 25$

Concn. of Magnesium Sulfate in aqueous soln., g/l	Solubility of Thallium Iodate in aqueous soln. of Magnesium Sulfate g/l
0.601	0.784
1.203	0.824
3.008	0.905
6.016	1.001
6.016	0.995

№ 4114 [1122]

MAGNESIUM DITHIONATE–BARIUM DITHIONATE–WATER

$MgS_2O_6—BaS_2O_6—H_2O$

Comp. of saturated soln., wt%			t	Comp. of solid phase
Magnesium Dithionate	Barium Dithionate	Water		
0.00	15.75	84.25	20	$BaS_2O_6 \cdot 2H_2O$
6.93	10.61	82.46	20	»
31.14	1.29	67.57	20	»
33.95	0.69	65.36	20	$BaS_2O_6 \cdot 2H_2O + MgS_2O_6 \cdot 6H_2O$
33.51	0.0	66.09	20	$MgS_2O_6 \cdot 6H_2O$
7.05	13.37	79.58	30	$BaS_2O_6 \cdot 2H_2O$
18.91	5.85	75.24	30	»

Table № 4114 (continuation)

Comp. of saturated soln., wt%			t	Comp. of solid phase
Magnesium Dithionate	Barium Dithionate	Water		
29.16	1.92	68.92	30	$BaS_2O_6 \cdot 2H_2O$
34.57	0.90	64.53	30	$BaS_2O_6 \cdot 2H_2O + MgS_2O_6 \cdot 6H_2O$
35.24	0.00	64.76	30	$MgS_2O_6 \cdot 6H_2O$

№ 4115 [1253]

MAGNESIUM CHLORIDE–MAGNESIUM BROMIDE–WATER

$MgCl_2$—$MgBr_2$—H_2O

Content in aqueous soln., wt%		Comp. of solid phase	Content in aqueous soln., wt%		Comp. of solid phase
Cl^-	Br^-		Cl^-	Br^-	
23.75	4.56	$Mg(Cl, Br)_2 \cdot 6H_2O$	11.37	25.50	$Mg(Cl, Br)_2 \cdot 6H_2O$
21.31	8.92	»	8.51	30.45	»
18.87	13.14	»	6.27	33.63	»
16.37	17.40	»	3.04	38.77	»
13.62	21.69	»	0	43.66	$MgBr_2 \cdot 6H_2O$

№ 4116 [615]

MAGNESIUM CHLORIDE–MAGNESIUM BROMIDE–WATER

$MgCl_2$—$MgBr_2$—H_2O

$t = 25$

Content in aqueous soln., wt%		d_4	Comp. of solid phase
Magnesium Chloride	Magnesium Bromide		
36.4	0	1.350	$MgCl_2 \cdot 6H_2O$
28.67	10.59	1.3923	$Mg(Cl\ Br)_2 \cdot 6H_2O$
25.34	15.48	—	»
24.60	16.90	1.4344	»
14.07	31.32	1.5271	»
10.84	36.20	1.5598	»
9.78	37.72	1.5711	»
0	50.60	1.642	$MgBr_2 \cdot 6H_2O$

№ 4117 **[615]**

MAGNESIUM CHLORIDE–MAGNESIUM BROMIDE–POTASSIUM CHLORIDE–POTASSIUM BROMIDE–WATER

$MgCl_2$—$MgBr_2$—KCl—KBr—H_2O

Content in aqueous soln wt%						
Potassium Chloride	Potassium Bromide	Magnesium Chloride	Magnesium Bromide	*t*	*d*	Comp. of solid phase
3.90	0	25.35	0	10	1.2711	KCl + Car_{Cl}
2.67	1.34	24.83	—	10	—	—
0	5.05	13.16	19.46	10	1.4253	K (Cl, Br) + $Car_{Cl, Br}$ (tetragonal)
0	2.13	0	42.90	10	1.5582	K (Cl, Br) + Car_{Br} (rhombic)
1.86	1.94	26.82	—	25	1.2777	K (Cl, Br) + $Car_{Cl, Br}$
0	4.96	25.38	2.25	25	1.3021	K (Cl, Br) + $Car_{Cl, Br}$ (rhombic)
0	5.05	23.50	4.90	25	1.3178	»
0	5.02	19.73	11.78	25	1.3629	K (Cl, Br) + $Car_{Cl, Br}$ (tetragonal + rhombic)
0	5.05	16.84	16.16	25	1.3889	»
0	4.75	13.94	21.44	25	1.4170	—
0	4.66	10.02	28.00	25	1.4499	K (Cl, Br) + $Car_{Cl, Br}$ (tetragonal)
0	2.77	0.55	42.38	25	1.5442	K (Cl, Br) + $Car_{Cl, Br}$ (rhombic)
0	0.46	27.05	12.45	25	1.4057	$Car_{Cl, Br}$ + Mg (Cl, Br)$_2$ · 6H_2O
0	0.43	19.52	23.92	25	1.4893	»
0	0.27	15.50	29.70	25	1.5142	» »
4.10	0	27.40	0	35	1.2935	KCl + Car_{Cl}
0	6.54	25.78	1.32	35	1.3255	K (Cl, Br) + $Car_{Cl, Br}$ (rhombic)
0	4.57	19.12	12.51	35	—	—
0	4.11	18.57	13.94	35	1.4160	K (Cl, Br) + $Car_{Cl, Br}$ (rhombic + tetragonal)
0	4.99	12.14	14.42	35	1.4276	K (Cl, Br) + $Car_{Cl, Br}$ (tetragonal)
0	6.21	10.17	20.44	35	1.4345	»
0	6.94	7.95	25.51	35	1.4773	» »
0	6.59	6.04	29.31	35	1.4983	» »
0	5.69	5.01	35.72	35	1.5193	» »
0	5.78	2.00	40.86	35	1.5575	K (Cl, Br) + $Car_{Cl, Br}$ (rhombic)
0	6.24	0	44.05	35	1.5784	KBr + Car_{Br}
4.79	0	26.89	0	55	—	KCl + Car_{Cl} (rhombic)
0	9.07	22.44	6.42	55	—	K (Cl, Br) + $Car_{Cl, Br}$ (rhombic)
0	8.92	17.04	16.25	55	—	»
0	8.82	14.80	19.10	55	—	» »
0	8.43	10.41	26.66	55	—	K (Cl, Br) + $Car_{Cl, Br}$ (tetragonal)
0	9.66	6.48	30.59	55	—	»
0	6.97	4.43	36.66	55	—	K (Cl, Br) + $Car_{Cl, Br}$ (tetragonal + rhombic)
0	5.79	0	43.18	55	—	KBr + Car_{Br} (rhombic)

Note. Car_{Cl} — Carnallite Chloride KCl · $MgCl_2$ · 6H_2O; Car_{Br} — Carnallite Bromide KBr · $MgBr_2$ · 6H_2O; $Car_{Cl, Br}$ — Carnallite Chlorobromide K (Cl, Br) · Mg (Cl, Br)$_2$ · 6H_2O.

№ 4118 [2405]

MAGNESIUM CHLORIDE–MAGNESIUM MOLYBDATE–WATER

$MgCl_2—MgMoO_4—H_2O$

$t=25$

Concn. in aqueous soln., wt%		d	Comp. of solid phase
Magnesium Molybdate	Magnesium Chloride		
15.90	0.00	1.159	$MgMoO_4 \cdot 5H_2O$
12.84	2.03	1.143	»
10.07	4.29	1.128	»
7.14	7.54	1.131	»
4.06	11.58	1.141	»
2.75	15.36	1.156	»
1.55	19.30	1.179	»
0.91	24.20	1.216	»
0.83	24.25	—	»
0.78	25.63	1.231	»
0.67	26.34	1.229	»
0.53	26.96	1.231	$MgMoO_4 \cdot 5H_2O + MgMoO_4 \cdot 2H_2O$
0.58	26.73	1.238	»
0.61	26.80	1.233	»
0.60	26.82	1.238	»
0.58	26.83	1.235	»
0.63	27.52	1.234	$MgMoO_4 \cdot 2H_2O$
0.52	28.03	—	»
0.39	29.29	1.258	»
0.31	30.37	—	»
0.30	31.25	1.278	»
0:08	32.29	1.293	»
0.26	33.34	1.297	»
0.12	34.53	1.312	»
0.10	35.32	—	»
0.14	35.47	1.325	$MgMoO_4 \cdot 2H_2O + MgCl_2 \cdot 6H_2O$
0.09	35.56	—	»
0.17	35.53	1.323	»
0.16	35.54	—	»
0.17	35.53	—	»
0.15	35.53	1.324	»
0.00	35.71	1.323	$MgCl_2 \cdot 6H_2O$

№ 4119 [1253]

MAGNESIUM CHLORIDE–MAGNESIUM IODIDE–WATER

$MgCl_2—MgI_2—H_2O$

$t=25$

Content in aqueous soln., wt%		Comp. of solid phase
Cl^-	I^-	
25.12	2.93	$MgCl_2 \cdot 6H_2O$
23.56	5.91	»
22.81	8.29	»

№ 4120 [2635]

MAGNESIUM CHLORIDE–MAGNESIUM POTASSIUM PHOSPHATE HYDRATE–POTASSIUM PHOSPHATE–WATER

$MgCl_2—MgKPO_4 \cdot 6H_2O—K_3PO_4—H_2O$

$t=25$

Content in saturated soln., mol/l				pH	Comp. of solid phase
Mg	K	P	Cl		
0.075	35.1	13.9	2.5	10.63	$MgKPO_4 \cdot 6H_2O+Mg_3(PO_4)_2 \cdot 22H_2O+$ $+Mg_3(PO_4)_2 \cdot 8H_2O$
0.055	36.9	15.5	2.5	10.53	$MgKPO_4 \cdot 6H_2O+Mg_3(PO_4)_2 \cdot 8H_2O$
0.070	39.4	16.2	2.5	10.51	»
0.13	10.0	3.84	2.0	10.87	$MgKPO_4 \cdot 6H_2O+Mg_3(PO_4)_2 \cdot 22H_2O$
0.058	30.2	12.0	2.0	10.82	$MgKPO_4 \cdot 6H_2O+Mg_3(PO_4)_2 \cdot 8H_2O$
0.053	33.5	13.6	2.0	10.63	$MgKPO_4 \cdot 6H_2O+Mg_3(PO_4)_2 \cdot 22H_2O+$ $+Mg_3(PO_4)_2 \cdot 8H_2O$
0.058	35.3	14.2	2.0	10.64	$MgKPO_4 \cdot 6H_2O+Mg_3(PO_4)_2 \cdot 8H_2O$
0.09	10.5	5.20	0.0	10.70	$MgKPO_4 \cdot 6H_2O+Mg_3(PO_4)_2 \cdot 22H_2O$
0.058	33.7	13.9	0.0	10.67	$MgKPO_4 \cdot 6H_2O+Mg_3(PO_4)_2 \ 8H_2O$
0.053	35.6	15.8	0.0	10.42	»
0.074	37.9	16.5	0.0	10.42	»

№ 4121 **MAGNESIUM CHLORIDE–POTASSIUM NITRATE–WATER** [103]

$MgCl_2—KNO_3—H_2O$

Concn. in aqueous soln., wt%		t	Comp. of solid phase
Potassium Nitrate	Magnesium Chloride		
31.3	0.0	30	KNO_3
20.6	7.8	30	»
14.0	17.0	30	»
11.7	21.4	30	»
10.8	23.2	30	»
10.4	24.0	30	KNO_3+KCl
5.2	27.6	30	KCl
3.8	28.4	30	»
0.4	33.5	30	$KCl+MgCl_2 \cdot 6H_2O$
24.2	0.0	20	KNO_3
15.8	8.2	20	»
10.4	17.6	20	»
8.39	21.9	20	»
8.3	24.0	20	»
8.2	24.5	20	KNO_3+KCl
7.6	24.8	20	KCl

Table № 4121 (continuation)

Concn. in aqueous soln., wt%		t	Comp. of solid phase
Potassium Nitrate	Magnesium Chloride		
3.8	27.6	20	KCl
3.0	28.3	20	»
0.3	33.8	20	$KCl+MgCl_2 \cdot 6H_2O$
17.7	0.0	10	KNO_3
12.0	8.8	10	»
8.0	18.0	10	»
7.8	19.2	10	»
6.8	19.2	10	KNO_3
6.3	21.4	10	»
6.2	22.4	10	KNO_3+KCl
3.6	25.0	10	KCl
1.6	26.6	10	»
0.2	30.0	10	$KCl+MgCl_2 \cdot 6H_2O$
11.5	0.0	0	KNO_3
8.8	9.0	0	»
6.0	18.4	0	»
5.2	22.9	0	»
4.6	24.8	0	»
4.4	25.4	0	KNO_3+KCl
3.6	26.0	0	KCl
1.0	28.7	0	»
1.05	34.0	0	»
0.0	11.6	—10	Ice
1.6	11.2	—10	»
3.3	10.6	—10	»
5.8	9.6	—10	$Ice+KNO_3$
4.8	15.4	—10	KNO_3
4.3	18.8	—10	»
3.8	23.2	—10	»
3.5	25.0	—10	»
3.3	25.4	—10	KNO_3+KCl
0.6	29.0	—10	$KCl+MgCl_2 \cdot 8H_2O$
0.0	32.6	—10	$MgCl_2 \cdot 8H_2O$
0.0	16.8	—20	Ice
2.2	16.2	—20	»
3.4	15.6	—20	$Ice+KNO_3$
2.9	21.4	—20	KNO_3
3.0	19.9	—20	»
2.7	23.5	—20	»
2.4	25.8	—20	KNO_3+KCl
2.35	26.0	—20	KCl
0.6	28.2	—20	$KCl+MgCl_2 \cdot 12H_2O$
0.0	26.8	—20	$MgCl_2 \cdot 12H_2O$
0.0	19.4	—30	Ice
2.0	19.8	—30	$Ice+KNO_3$
2.0	22.4	—30	$KNO_3+MgCl_2 \cdot 12H_2O$
1.4	22.0	—30	$MgCl_2 \cdot 12H_2O$
0.0	22.0	—30	»
2.1	20.2	—33.5	$Ice+KNO_3+MgCl_2 \cdot 12H_2O$
2.0	25.8	—23.2	$KNO_3+KCl+MgCl_2$

№ 4122 MAGNESIUM CHLORIDE–POTASSIUM CHLORIDE–WATER [3251]

$MgCl_2$—KCl—H_2O

Comp. of solvent wt%		Solubility of Potassium Chloride in aqueous soln. of Magnesium Chloride wt%	t	Comp. of solvent wt%		Solubility of Potassium Chloride in aqueous soln. of Magnesium Chloride wt%	t
Magnesium Chloride	Water			Magnesium Chloride	Water		
11	89	14.3	10	21.2	78.8	5.3	10
11	89	15.9	20	21.2	78.8	6.5	20
11	89	17.5	30	21.2	78.8	7.6	30
11	89	19.0	40	21.2	78.8	8.8	40
11	89	20.5	50	21.2	78.8	10.0	50
11	89	21.9	60	21.2	78.8	11.2	60
11	89	24.5	80	21.2	78.8	13.6	80
11	89	25.8	90	21.2	78.8	14.7	90
11	89	27.1	100	21.2	78.8	15.9	100
15	85	9.9	10	30	70	1.9	10
15	85	11.3	20	30	70	2.6	20
15	85	12.7	30	30	70	3.4	30
15	85	14.2	40	30	70	4.2	40
15	85	15.6	50	30	70	5.0	50
15	85	17.0	60	30	70	5.8	60
15	85	19.5	80	30	70	7.3	80
15	85	20.8	90	30	70	8.1	90
15	85	22.1	100	30	70	8.9	100

№ 4123 MAGNESIUM CHLORIDE–POTASSIUM CHLORIDE–WATER [2690]

$MgCl_2$—KCl—H_2O

Comp. of saturated soln., wt%			t	Comp. of solid phase
Magnesium Chloride	Potassium Chloride	Water		
0.00	19.74	80.26	−11.1	Ice + KCl
20.64	0.00	79.36	−33.6	Ice + $MgCl_2 \cdot 12H_2O$
18.31	1.00	80.69	−34.3	Ice + $KCl \cdot MgCl_2 \cdot 12H_2O$
25.48	1.48	73.04	−21	$MgKCl_3 \cdot 6H_2O + MgCl_2 \cdot 12H_2O + KCl$
25.62	2.18	72.20	0	$MgKCl_3 \cdot 6H_2O + KCl$
26.82	3.32	69.96	25	»
28.34	4.16	67.50	50	»
28.43	4.80	66.77	61.5	»
36.47	7.83	55.70	154.5	»
42.90	8.40	48.70	167.5	»
35.50	0.52	63.98	25	$MgKCl_3 + MgCl_2 \cdot 6H_2O$
37.04	0.31	62.65	50	»
39.10	0.75	60.15	80	»
45.71	0.80	53.49	115.7	$MgKCl_3 + MgCl_2 \cdot 6H_2O + MgCl_2 \cdot 4H_2O$
49.01	4.61	46.38	152.5	$MgKCl_3 + MgCl_2 \cdot 4H_2O + KCl$
52.03	6.96	41.01	176	$MgCl_2 \cdot 4H_2O + MgCl_2 \cdot 2H_2O + KCl$
50.15	10.32	39.53	186	$MgCl_2 \cdot 2H_2O + KCl$

№ 4124 MAGNESIUM CHLORIDE–POTASSIUM CHLORIDE–WATER [1961]

$MgCl_2—KCl—H_2O$

Comp. of saturated soln., wt%			t	Comp. of solid phase
Magnesium Chloride	Potassium Chloride	Water		
35.10	0.00	64.90	15	$MgCl_2 \cdot 6H_2O$
27.59	2.76	69.65	15	$KCl+MgCl_2 \cdot KCl \cdot 6H_2O$
23.83	5.49	70.68	15	KCl
16.57	10.73	72.70	15	»
7.66	17.66	74.68	15	»
35.60	0.00	64.40	25	$MgCl \cdot 6H_2O$
35.14	0.50	64.36	25	$MgCl_2 \cdot 6H_2O+MgCl_2 \cdot KCl \cdot 6H_2O$
27.70	3.70	68.60	25	$KCl+MgCl_2 \cdot KCl \cdot 6H_2O$
26.81	4.56	68.63	25	KCl
24.58	5.22	70.20	25	»
23.10	6.06	70.84	25	»
16.40	10.36	73.24	25	»
15.64	11.54	72.92	25	»
7.68	19.50	72.92	25	»
37.55	0.00	62.45	55	$MgCl_2 \cdot 6H_2O$
37.38	0.21	62.41	55	»
36.88	0.62	62.52	55	$MgCl_2 \cdot 6H_2O+MgCl_2 \cdot KCl \cdot 6H_2O$
28.74	4.43	66.83	55	$KCl+MgCl_2 \cdot KCl \cdot 6H_2O$
22.72	9.79	67.49	55	KCl
15.14	14.79	70.07	55	»
7.24	23.40	69.36	55	»
39.80	0.00	60.20	83	$MgCl_2 \cdot 6H_2O$
37.80	1.01	61.19	83	$MgCl_2 \cdot 6H_2O+MgCl_2 \cdot KCl \cdot 6H_2O$
36.99	2.52	60.49	83	$MgCl_2 \cdot KCl \cdot 6H_2O$
29.45	5.80	64.75	83	$MgCl_2 \cdot KCl \cdot 6H_2O+KCl$
20.53	12.75	66.72	83	KCl
14.63	18.31	67.06	83	»
17.77	18.27	63.96	83	»
7.15	26.85	66.0	83	»
43.47	0.0	56.53	105	$MgCl_2 \cdot 6H_2O$
40.75	1.07	58.18	105	$MgCl_2 \cdot 6H_2O+MgCl_2 \cdot KCl \cdot 6H_2O$
36.51	3.62	59.87	105	$MgCl_2 \cdot KCl \cdot 6H_2O$
30.82	7.00	62.18	105	$MgCl_2 \cdot KCl \cdot 6H_2O+KCl$
21.44	14.83	63.73	105	KCl
14.08	21.72	64.20	105	»
13.97	21.87	64.16	105	»
6.52	29.90	63.58	105	»
0.0	36.45	63.55	105	»

№ 4125 MAGNESIUM CHLORIDE–POTASSIUM CHLORIDE–WATER [2061]

$MgCl_2—KCl—H_2O$

$t=25$

Comp. of saturated soln., wt%			d_{25}	Comp. of solid phase
Magnesium Chloride	Potassium Chloride	Water		
35.54	0.00	64.46	1.341	$MgCl_2 \cdot 6H_2O$
35.14	0.53	64.33	—	$MgCl_2 \cdot 6H_2O+MgCl_2 \cdot KCl \cdot 6H_2O$
35.13	0.38	64.49	—	»
35.47	0.52	64.01	—	»

Table № 4125 (continuation)

Comp. of saturated soln., wt%			d_{25}	Comp. of solid phase
Magnesium Chloride	Potassium Chloride	Water		
26.81	3.33	69.86	—	KCl
26.79	3.20	70.01	—	$KCl + MgCl_2 \cdot KCl \cdot 6H_2O$
26.66	3.19	70.15	—	»
19.83	7.90	72.27	1.234	KCl
12.11	13.56	74.33	1.201	»
0.00	26.74	73.26	1.182	»

№ 4126 MAGNESIUM CHLORIDE–POTASSIUM CHLORIDE–WATER [1684]

$MgCl_2—KCl—H_2O$

Content in aqueous soln., g/100g Water		t	d	Comp. of solid phase
Potassium Chloride	Magnesium Chloride			
28.5	0	0	1.156	KCl
3.7	39.1	0	1.262	$KCl + KCl \cdot MgCl_2 \cdot 6H_2O$
0.6	52.8	0	1.330	$KCl \cdot MgCl_2 \cdot 6H_2O + MgCl_2 \cdot 6H_2O$
0	52.9	0	—	$MgCl_2 \cdot 6H_2O$
34.4	0	20	1.177	KCl
4.6	40.2	20	1.274	$KCl + KCl \cdot MgCl_2 \cdot 6H_2O$
0.7	54.3	20	1.332	$KCl \cdot MgCl_2 \cdot 6H_2O + MgCl_2 \cdot 6H_2O$
0	54.6	20	1.332	$MgCl_2 \cdot 6H_2O$
40.2	0	40	1.189	KCl
5.7	41.3	40	1.284	$KCl + KCl \cdot MgCl_2 \cdot 6H_2O$
0.9	56.3	40	1.336	$KCl \cdot MgCl_2 \cdot 6H_2O + MgCl_2 \cdot 6H_2O$
0	57.5	40	1.337	$MgCl_2 \cdot 6H_2O$
45.8	0	60	1.199	KCl
7.1	43.3	60	1.292	$KCl + KCl \cdot MgCl_2 \cdot 6H_2O$
1.2	59.1	60	1.343	$KCl \cdot MgCl_2 \cdot 6H_2O + MgCl_2 \cdot 6H_2O$
0	61.0	60	1.346	$MgCl_2 \cdot 6H_2O$
51.1	0	80	1.207	KCl
8.8	45.6	80	1.298	$KCl + KCl \cdot MgCl_2 \cdot 6H_2O$
1.5	62.7	80	1.354	$KCl \cdot MgCl_2 \cdot 6H_2O + MgCl_2 \cdot 6H_2O$
0	66.1	80	1.362	$MgCl_2 \cdot 6H_2O$

№ 4127 MAGNESIUM CHLORIDE–POTASSIUM CHLORIDE–WATER [37]

$MgCl_2—KCl—H_2O$

Comp. of saturated soln., wt%			t	d	Comp. of solid phase
Potassium Chloride	Magnesium Chloride	Water			
27.39	6.60	66.01	100	—	KCl
14.47	19.73	65.80	100	—	»
6.72	31.09	62.19	100	—	»
0.58	40.93	58.49	100	—	»
6.39	31.49	62.12	100	1.307	$KCl + KCl \cdot MgCl_2 \cdot 6H_2O$

Table № 4127 (continuation)

Comp. of saturated soln., wt%			t	d	Comp. of solid phase
Potassium Chloride	Magnesium Chloride	Water			
0.52	41.74	57.74	100	1.384	$KCl \cdot MgCl_2 \cdot 6H_2O + MgCl_2 \cdot 6H_2O$
31.07	6.32	62.61	125	—	KCl
17.72	18.98	63.30	125	—	»
9.63	30.15	60.22	125	—	»
4.60	39.28	56.12	125	—	»
5.58	38.54	55.88	125	1.335	$KCl + KCl \cdot MgCl_2 \cdot 6H_2O$
1.51	46.40	52.09	125	1.420	$KCl \cdot MgCl_2 \cdot 6H_2O + MgCl_2 \cdot 4H_2O$
33.65	6.03	60.32	150	—	KCl
21.06	18.21	60.73	150	—	»
12.48	29.17	58.35	150	—	»
7.42	38.12	54.46	150	—	»
5.23	44.88	49.89	150	—	»
4.61	44.22	51.17	150	1.381	$KCl + KCl \cdot MgCl_2 \cdot 6H_2O$
1.77	50.40	47.83	150	1.435	$KCl \cdot MgCl_2 \cdot 6H_2O + MgCl_2 \cdot 4H_2O$
2.07	52.90	45.03	175	1.455	$KCl + MgCl_2 \cdot 4H_2O$
38.44	5.59	55.97	200	—	KCl
25.88	17.10	57.02	200	—	»
16.38	27.87	55.75	200	—	»
10.33	36.91	52.76	200	—	»
7.85	43.64	48.51	200	—	»
2.30	55.0	42.70	200	1.472	$KCl + MgCl_2 \cdot 3H_2O$

№ 4128 [660]

MAGNESIUM CHLORIDE–POTASSIUM CHLORIDE–WATER

$MgCl_2$—KCl—H_2O

$t=0$

Content in aqueous soln., wt%		d	Comp. of solid phase
Potassium Chloride	Magnesium Chloride		
22.05	0	1.154	KCl
16.90	4.18	—	»
12.46	8.41	1.164	»
11.95	8.68	—	»
8.19	15.22	1.191	»
6.06	16.50	—	»
4.68	18.77	—	»
3.70	23.76	1.245	»
3.60	21.70	—	»
2.50	26.50	—	$KCl + KCl \cdot MgCl_2 \cdot 6H_2O$
2.40	26.53	1.263	»
0.91	28.67	1.272	$KCl \cdot MgCl_2 \cdot 6H_2O$
0.23	33.37	1.318	»
0.08	34.40	1.324	$KCl \cdot MgCl_2 \cdot 6H_2O + MgCl_2 \cdot 6H_2O$
0	34.49	1.329	$MgCl_2 \cdot 6H_2O$

№ 4129 MAGNESIUM CHLORIDE–POTASSIUM CHLORIDE–WATER [446]

$MgCl_2—KCl—H_2O$

Content in aqueous soln., wt%		t	d	Comp. of solid phase
Potassium Chloride	Magnesium Chloride			
23.84	0	10	1.165	KCl
15.85	6.86	10	1.172	»
11.88	10.98	10	1.182	»
6.73	17.87	10	1.209	»
2.65	26.45	10	1.264	$KCl+KCl \cdot MgCl_2 \cdot 6H_2O$
2.06	27.08	10	1.267	$KCl \cdot MgCl_2 \cdot 6H_2O$
0.09	34.84	10	1.334	$KCl \cdot MgCl_2 \cdot 6H_2O+MgCl_2 \cdot 6H_2O$
0	34.97	20	1.333	$MgCl_2 \cdot 6H_2O$
25.55	0	20	1.174	KCl
16.04	8.30	20	1.185	»
3.16	26.67	20	1.271	$KCl+KCl \cdot MgCl_2 \cdot 6H_2O$
1.83	28.43	20	1.275	$KCl \cdot MgCl_2 \cdot 6H_2O$

№ 4130 [1335]

MAGNESIUM CHLORIDE–POTASSIUM CHLORIDE–WATER

$MgCl_2—KCl—H_2O$

$t=100$

Comp. of saturated soln., wt%			Comp. of solid phase
Potassium Chloride	Magnesium Chloride	Water	
36.0	0.0	64.0	KCl
29.80	4.35	65.85	»
23.59	9.93	66.48	»
17.50	16.35	66.15	»
13.45	20.15	66.40	»
9.60	24.80	65.60	»
5.15	29.89	64.96	$KCl+KCl \cdot MgCl_2 \cdot 6H_2O$
1.88	40.0	58.12	$KCl \cdot MgCl_2 \cdot 6H_2O$
1.64	40.0	58.36	»
0.73	41.0	58.27	$KCl \cdot MgCl_2 \cdot 6H_2O+MgCl_2 \cdot 6H_2O$

№ 4131 [662]

MAGNESIUM CHLORIDE–POTASSIUM CHLORIDE–WATER

$MgCl_2—KCl—H_2O$

Content in aqueous soln., g/100g Water		t	Comp. of solid phase
Potassium Chloride	Magnesium Chloride		
25.55	0	—10	KCl
19.28	5.19	—10	»
11.69	10.42	—10	»
6.07	21.33	—10	»

Table № 4131 (continuation)

Content in aqueous soln., g/100g Water		t	Comp. of solid phase
Potassium Chloride	Magnesium Chloride		
4.35	24.47	−10	KCl
3.50	29.05	−10	»
2.65	35.07	−10	$KCl + KCl \cdot MgCl_2 \cdot 6H_2O$
2.00	37.78	−10	$KCl \cdot MgCl_2 \cdot 6H_2O$
0.90	48.18	−10	$MgCl_2 \cdot 8H_2O$
0	48.20	−10	»
25.89	0	− 5	KCl
19.00	5.18	− 5	»
13.31	10.91	− 5	»
6.95	21.35	− 5	»
4.79	24.48	− 5	»
3.95	29.06	− 5	»
3.10	35.57	− 5	$KCl + KCl \cdot MgCl_2 \cdot 6H_2O$
2.10	37.09	− 5	$KCl \cdot MgCl_2 \cdot 6H_2O$
0.45	43.15	− 5	»
0	51.37	− 5	$MgCl_2 \cdot 8H_2O$
28.26	0	0	KCl
21.44	5.20	0	»
15.05	10.90	0	»
7.82	21.37	0	»
6.11	24.52	0	»
4.83	29.09	0	»
3.55	36.01	0	$KCl + KCl \cdot MgCl_2 \cdot 6H_2O$
0	52.94	0	$MgCl_2 \cdot 6H_2O$

№ 4132 MAGNESIUM CHLORIDE–POTASSIUM CHLORIDE–WATER [447]

$MgCl_2—KCl—H_2O$

Content in aqueous soln., wt%		t	Comp. of solid phase
Potassium Chloride	Magnesium Chloride		
3.74	16.52	−24	KCl+ ice
3.44	18.64	−24	KCl
1.42	24.20	−24	$KCl + MgCl_2 \cdot 12H_2O$
1.13	29.15	−24	$KCl \cdot MgCl_2 \cdot 6H_2O + MgCl_2 \cdot 8H_2O$ (metastable)
0.88	33.60	−24	$KCl \cdot MgCl_2 \cdot 6H_2O + MgCl_2 \cdot 6H_2O$ (metastable)
0	24.5	−24	$MgCl_2 \cdot 12H_2O$
0	29.7	−24	$MgCl_2 \cdot 8H_2O$ (metastable)
0	33.8	−24	$MgCl_2 \cdot 6H_2O$ (metastable)
2.05	17.37	−24	Ice
0	18.21	−24	»
7.10	11.92	−17	KCl+ice
4.97	15.33	−17	KCl
3.09	19.42	−17	»
1.67	25.78	−17	$KCl + KCl \cdot MgCl_2 \cdot 6H_2O$
0.77	27.55	−17	$KCl \cdot MgCl_2 \cdot 6H_2O$
0.17	31.77	−17	$KCl \cdot MgCl_2 \cdot 6H_2O + MgCl_2 \cdot 8H_2O$
0	31.70	−17	$MgCl_2 \cdot 8H_2O$
2.01	74.59	−17	Ice

Table № 4132 (continuation)

Content in aqueous soln., wt%		t	Comp. of solid phase
Potassium Chloride	Magnesium Chloride		
0	15.64	−17	Ice
20.2	0	−10	KCl
11.65	7.52	−10	»
6.05	15.0	−10	»
2.14	26.00	−10	$KCl+KCl\cdot MgCl_2\cdot 6H_2O$
1.15	27.13	−10	$KCl\cdot MgCl_2\cdot 6H_2O$
0.38	29.83	−10	»
0.08	34.60	−10	$KCl\cdot MgCl_2\cdot 6H_2O+MgCl_2\cdot 6H_2O$
0	34.5	−10	$MgCl_2\cdot 6H_2O$
19.2	0	−10	Ice
13.99	4.24	−10	»
5.10	9.26	−10	»
0	11.5	−10	»

№ 4133 [1881]

MAGNESIUM CHLORIDE–POTASSIUM CHLORIDE–WATER

$MgCl_2$—KCl—H_2O

$t=0$

Comp. of saturated soln., wt%			Comp. of solid phase
Potassium Chloride	Magnesium Chloride	Water	
21.87	0.00	78.13	KCl
15.21	5.88	78.91	»
11.70	9.49	78.81	»
6.31	16.33	77.36	»
2.35	25.24	72.41	»
2.12	26.79	71.09	$KCl+KCl\cdot MgCl_2\cdot 6H_2O$
1.68	27.03	71.29	$KCl\cdot MgCl_2\cdot 6H_2O$
0.49	29.70	69.81	»
0.026	34.68	65.294	$KCl\cdot MgCl_2\cdot 6H_2O+MgCl_2\cdot 6H_2O$
0.00	34.63	65.37	$MgCl_2\cdot 6H_2O$

№ 4134 [1470]

MAGNESIUM CHLORIDE–POTASSIUM CHLORIDE–WATER

$MgCl_2$—KCl—H_2O

$t=0$

Content in aqueous soln., wt%		Comp. of solid phase
Potassium Chloride	Magnesium Chloride	
21.87	0	KCl
15.21	5.88	»
11.70	9.49	»
6.31	16.33	»

Content in aqueous soln., mol/1000 mol Water			t	Comp. of solid phase
2KCl	$MgCl_2$	$CaCl_2$		
2	34.5	122	25	$CaCl_2 \cdot 6H_2O + 2MgCl_2 \cdot CaCl_2 \cdot 12H_2O + KCl \cdot MgCl_2 \cdot 6H_2O$
0.5	0.5	239	83	»
9.5	5	142	25	$CaCl_2 \cdot 6H_2O + KCl + KCl \cdot MgCl_2 \cdot 6H_2O$
10.5	27	216	83	»

№ 4139 [1881]

MAGNESIUM CHLORIDE–POTASSIUM CHLORIDE–CALCIUM CHLORIDE–WATER

$MgCl_2—KCl—CaCl_2—H_2O$

$t=0$

Content in aqueous soln., wt%			Comp. of solid phase
Potassium Chloride	Magnesium Chloride	Calcium Chloride	
2.13	26.78	0	$KCl + KCl \cdot MgCl_2 \cdot 6H_2O$
2.15	20.54	7.65	»
2.15	19.17	9.33	»
2.15	15.32	14.55	»
2.13	14.19	16.05	»
2.10	10.37	21.13	»
2.02	6.55	27.66	»
1.87	3.30	33.39	$KCl + KCl \cdot MgCl_2 \cdot 6H_2O + CaCl_2 \cdot 6H_2O$
1.82	2.98	33.84	$KCl + CaCl_2 \cdot 6H_2O$
1.87	2.61	34.51	»
1.91	1.64	35.10	»
1.93	1.33	35.53	»
1.96	0	37.27	»
1.88	3.68	32.99	$KCl \cdot MgCl_2 \cdot 6H_2O + CaCl_2 \cdot 6H_2O$
0.59	8.13	28.75	»
0.32	11.77	25.33	»
0.031	23.77	13.99	»
0.036	23.79	14.04	»
0.028	29.30	6.88	$KCl \cdot MgCl_2 \cdot 6H_2O + MgCl_2 \cdot 6H_2O$
0.024	34.68	0	»
0.019	23.88	14.01	$MgCl_2 \cdot 6H_2O + CaCl_2 \cdot 6H_2O$
0	23.94	14.00	»

№ 4140 [2084]

MAGNESIUM CHLORIDE–POTASSIUM CHLORIDE–CALCIUM CHLORIDE–WATER

$MgCl_2$—KCl—$CaCl_2$—H_2O

$t=35$

Content in aqueous soln., wt%			Comp. of solid phase
Calcium Chloride	Magnesium Chloride	Potassium Chloride	
0.00	27.33	3.81	$KCl \cdot MgCl_2 \cdot 6H_2O + KCl$
9.12	20.61	3.69	»
23.57	10.59	3.47	»
31.66	5.83	3.37	»
35.51	4.07	3.44	»
35.60	4.01	3.43	»
41.00	2.31	3.80	»
45.85	1.30	4.71	»
48.34	1.15	5.69	»
48.89	1.17	6.04	»
49.76	1.02	6.46	$KCl \cdot MgCl_2 \cdot 6H_2O + KCl + \alpha\text{-}CaCl_2 \cdot 4H_2O$
0.00	36.17	0.14	$KCl \cdot MgCl_2 \cdot 6H_2O + MgCl_2 \cdot 6H_2O$
12.97	26.66	0.16	»
26.74	18.07	0.23	$KCl \cdot MgCl_2 \cdot 6H_2O + MgCl_2 \cdot 6H_2O + CaCl_2 \cdot 2MgCl_2 \cdot 12H_2O$
32.01	14.23	0.32	$KCl \cdot MgCl_2 \cdot 6H_2O + CaCl_2 \cdot 2MgCl_2 \cdot 12H_2O$
32.60	13.83	0.33	»
42.75	7.28	0.71	»
43.27	6.98	0.76	»
44.44	6.15	0.97	»
45.30	5.91	0.99	$KCl \cdot MgCl_2 \cdot 6H_2O + \alpha\text{-}CaCl_2 \cdot 4H_2O + CaCl_2 \cdot 2MgCl_2 \cdot 12H_2O$
46.65	4.42	1.33	$KCl \cdot MgCl_2 \cdot 6H_2O + CaCl_2 \cdot 4H_2O$
48.40	2.69	2.28	»
49.25	1.72	3.92	»
49.54	1.36	5.10	»
50.45	0.0	6.48	$KCl + \alpha\text{-}CaCl_2 \cdot 4H_2O$
50.06	0.71	6.46	»
49.83	0.93	6.45	»
45.03	6.20	0.0	$CaCl_2 \cdot 2MgCl_2 \cdot 12H_2O + \alpha\text{-}CaCl_2 \cdot 4H_2O$
45.18	6.13	0.33	»
45.43	5.82	0.70	»
26.68	18.18	0.0	$CaCl_2 \cdot 2MgCl_2 \cdot 12H_2O + MgCl_2 \cdot 6H_2O$

Table № 4143 (continuation)

Comp. of saturated soln., wt%				t	Comp. of solid phase
Potassium Chloride	Calcium Chloride	Magnesium Chloride	Water		
4.00	44.20	1.30	50.50	25	KCl+Carnallite +$CaCl_2 \cdot 6H_2O$
1.00	42.47	2.23	54.30	25	Carnallite +$CaCl_2 \cdot 6H_2O$
1.00	40.97	4.53	53.50	25	»
0.85	38.50	9.30	53.10	25	Carnallite+$CaCl_2 \cdot 6H_2O$+Tachhydrite
0.40	34.88	11.62	53.10	25	Carnallite+Tachhydrite
0.25	30.50	14.80	54.45	25	Carnallite Tachhydrite+$MgCl_2 \cdot 6H_2O$
0.50	26.10	17.40	56.00	25	Carnallite +$MgCl_2 \cdot 6H_2O$
0.30	15.92	23.88	59.90	25	»
0.25	3.62	32.58	63.55	25	»
4.00	2.80	25.20	68.00	55	KCl+Carnallite
4.00	11.88	17.82	66.30	55	»
4.20	18.25	12.16	64.64	55	»
4.10	26.25	8.75	60.90	55	»
4.30	36.30	3.70	58.70	55	»
4.30	34.39	1.81	59.50	55	»
5.90	49.90	2.10	42.10	55	KCl+ Carnallite +Hydrophilite
4.10	54.20	1.50	40.20	55	Hydrophilite+ Carnallite +$CaCl_2 \cdot 2H_2O$
2.50	54.15	2.85	40.50	55	Carnallite+$CaCl_2 \cdot 2H_2O$
1.80	53.10	5.02	40.08	55	$CaCl_2 \cdot 2H_2O$+ Carnallite+Tachhydrite
1.70	51.30	5.80	41.20	55	Tachhydrite+Carnallite
1.50	38.63	12.87	47.00	55	»
1.20	28.20	18.80	51.80	55	»
1.00	17.52	26.28	55.20	55	»
0.80	13.80	29.20	56.20	55	Tachhydrite+ Carnallite+$MgCl_2 \cdot 6H_2O$
0.70	3.90	35.10	60.30	55	Carnallite +$MgCl_2 \cdot 6H_2O$

№ 4144 [1599]

MAGNESIUM CHLORIDE–POTASSIUM CHLORIDE–RUBIDIUM CHLORIDE–WATER

$MgCl_2$— KCl— RbCl— H_2O

$t=20$

Content in aqueous soln., g/l			Content in aqueous soln., g/l		
Potassium Chloride	Rubidium Chloride	Magnesium Chloride	Potassium Chloride	Rubidium Chloride	Magnesium Chloride
125	390	50	35	130	250
85	375	100	32	40	300
60	325	150	31	1.5	330
40	250	200			

№ 4145 [1467]

MAGNESIUM CHLORIDE–POTASSIUM CHLORIDE–RUBIDIUM CHLORIDE–WATER

$MgCl_2—KCl—RbCl—H_2O$

$t=25$

Comp. of saturated soln., wt%				Comp. of solid phase
Magnesium Chloride	Rubidium Chloride	Potassium Chloride	Water	
90.83	Traces	0.0	9.17	$MgCl_2 \cdot 6H_2O + MgCl_2 \cdot RbCl \cdot 6H_2O$
53.96	34.52	0.0	11.52	$MgCl_2 \cdot RbCl \cdot 6H_2O$
34.16	53.15	11.21	1.48	$MgCl_2 \cdot RbCl \cdot 6H_2O + RbCl$
36.07	52.17	0.0	11.76	$MgCl_2 \cdot RbCl \cdot 6H_2O$
82.95	0.0	4.96	12.09	$KCl + KCl \cdot MgCl_2 \cdot 6H_2O$
70.01	9.61	7.55	12.83	(K, Rb)Cl + (K, Rb)Cl + $MgCl \cdot 6H_2O$
59.02	19.25	9.39	12.34	»
45.14	38.41	5.74	10.71	»
47.03	47.90	5.06	0.01	»
80.65	0.60	6.43	12.92	(K, Rb)Cl
81.90	0.86	5.23	12.01	(K, Rb)Cl $\cdot MgCl \cdot 6H_2O$
52.31	28.13	7.75	11.81	»
36.64	44.88	7.17	11.31	(K, Rb)Cl
48.27	37.30	2.94	11.49	(K, Rb)Cl $\cdot MgCl_2 \cdot 6H_2O$
55.27	25.10	8.08	11.55	»
55.58	26.36	6.07	11.99	»
54.10	27.99	5.72	12.19	»
54.34	30.63	3.45	11.58	»

№ 4146 [539]

MAGNESIUM CHLORIDE–POTASSIUM CHLORIDE–RUBIDIUM CHLORIDE–WATER

$MgCl_2—KCl—RbCl—H_2O$

$t=100$

Content in aqueous soln., wt%			Comp. of solid phase
Potassium Chloride	Rubidium Chloride	Magnesium Chloride	
6.40	0	31.50	$KCl + KCl \cdot MgCl \cdot 6H_2O$
8.93	3.49	24.64	(K, Rb)Cl + (K, Rb)Cl $\cdot MgCl_2 \cdot 6H_2O$
9.09	8.08	20.50	»
8.05	16.32	17.25	»
7.07	22.07	17.58	»
8.42	22.79	16.43	»
7.32	24.43	15.87	»
6.59	26.39	16.07	»
0	38.75	13.06	$RbCl + RbCl \cdot MgCl_2 \cdot 6H_2O$
21.06	2.56	11.08	(K, Rb)Cl
17.69	3.06	14.4	»

№ 4153 [2699]

MAGNESIUM CHLORIDE–CALCIUM SULFATE–WATER

$MgCl_2—CaSO_4—H_2O$

$t=25$

Concn. in aqueous soln., g/l		Concn. in aqueous soln., g/l	
Magnesium Chloride	Calcium Sulfate	Magnesium Chloride	Calcium Sulfate
0.474	2.315	1.423	2.682
0.948	2.505	1.897	2.832

№ 4154 [755]

MAGNESIUM CHLORIDE–CALCIUM SULFATE–CALCIUM CHLORIDE–WATER

$MgCl_2—CaSO_4—CaCl_2—H_2O$

$t=25$

Comp. of saturated soln., wt%				Comp. of solid phase
Calcium Chloride	Magnesium Chloride	Calcium Sulfate	Water	
8.81	8.34	0.533	82.317	$CaSO_4 \cdot 2H_2O$
10.15	12.63	0.390	77.230	»
9.56	3.37	0.060	86.710	»
12.30	5.48	0.054	82.170	»
8.39	3.30	0.095	86.215	»
3.41	6.77	0.580	89.240	»
8.10	7.00	0.245	84.647	»
13.38	2.37	0.049	83.301	»
4.18	30.74	0.006	65.076	$CaSO_4 \cdot 2H_2O+MgCl_2 \cdot 6H_2O$
12.28	22.03	0.014	65.680	»
17.10	20.14	0.007	62.753	»
26.78	15.60	0.015	57.600	»
41.71	4.03	0.030	53.230	»
31.61	9.36	0.009	59.020	$CaSO_4 \cdot 2H_2O+CaCl_2 \cdot 6H_2O$
32.10	9.46	0.010	58.440	$2MgCl_2 \cdot CaCl_2 \cdot 12H_2O+CaSO_4 \cdot 2H_2O$
43.60	4.05	0.023	52.330	$CaCl_2 \cdot 6H_2O$
32.40	7.50	0.002	60.100	»
27.63	14.61	0.004	57.765	$2MgCl_2 \cdot CaCl_2 \cdot 12H_2O+CaSO_4 \cdot 2H_2O+$ $+MgCl_2 \cdot 6H_2O$
36.32	4.95	0.014	58.720	$CaCl_2 \cdot 6H_2O+CaSO_4 \cdot 2H_2O$

№ 4155 MAGNESIUM CHLORIDE–CALCIUM CHLORIDE–WATER [2697]

$MgCl_2—CaCl_2—H_2O$

Comp. of saturated soln., wt%			t	Comp. of solid phase
Calcium Chloride	Magnesium Chloride	Water		
41.2	31.6	27.2	16.7	$MgCl_2 \cdot 6H_2O + CaCl_2 \cdot 6H_2O$
57.1	26.0	16.9	21.95	$MgCl_2 \cdot 6H_2O + CaCl_2 \cdot 6H_2O + 2MgCl_2CaCl_2 \cdot 12H_2O$
54.4	28.4	17.1	28.2	$2MgCl_2CaCl_2 \cdot 12H_2O + MgCl_2 \cdot 6H_2O$
0.0	85.63	14.37	116.7	$2MgCl_2CaCl_2 \cdot 12H_2O + MgCl_2 \cdot 6H_2O + MgCl_2 \cdot 4H_2O$
32.3	17.9	49.8	25	$2MgCl_2CaCl_2 \cdot 12H_2O + CaCl_2 \cdot 6H_2O + CaCl_2 \cdot 4H_2O$
80.1	16.1	3.8	28.2	$2MgCl_2CaCl_2 \cdot 12H_2O + CaCl_2 \cdot 4H_2O$
88.7	7.24	4.04	28.2	$CaCl_2 \cdot 6H_2O + CaCl_2 \cdot 4H_2O$

№ 4156 MAGNESIUM CHLORIDE–CALCIUM CHLORIDE–WATER [2061]

$MgCl_2—CaCl_2—H_2O$

$t=25$

Comp. of saturated soln., wt%			d_{25}	Comp. of solid phase
Magnesium Chloride	Calcium Chloride	Water		
0.00	45.06	54.94	1.470	$CaCl_2 \cdot 6H_2O$
4.06	41.87	54.07	1.465	»
7.93	38.95	53.12	1.472	»
9.43	38.70	51.87	1.486	$CaCl_2 \cdot 6H_2O + CaCl_2 \cdot 2MgCl_2 \cdot 12H_2O$
10.78	36.37	52.85	1.473	$CaCl_2 \cdot 2MgCl_2 \cdot 12H_2O$
13.55	32.82	53.63	1.460	»
14.54	31.17	54.29	1.455	$CaCl_2 \cdot 2MgCl_2 \cdot 12H_2O + MgCl_2 \cdot 6H_2O$
16.31	28.12	55.57	1.441	$MgCl_2 \cdot 6H_2O$
18.13	25.09	56.78	1.428	»
23.33	16.05	60.62	1.391	»
27.61	10.33	62.06	1.371	»
35.54	0.0	64.46	1.341	»

№ 4157 MAGNESIUM CHLORIDE–CALCIUM CHLORIDE–WATER [61]

$MgCl_2—CaCl_2—H_2O$

$t=0$

Content in aqueous soln., wt%		Comp. of solid phase
Magnesium Chloride	Calcium Chloride	
32.80	2.32	$MgCl_2 \cdot 6H_2O$
29.20	6.88	»
25.53	11.73	»
23.81	14.05	$MgCl_2 \cdot 6H_2O + CaCl_2 \cdot 6H_2O$
10.77	26.34	$CaCl_2 \cdot 6H_2O$
9.80	27.28	»

Table № 4164 (continuation)

Concn. in aqueous soln., wt%		t	Comp. of solid phase
Calcium Chloride	Magnesium Chloride		
2.0	41.4	110	$CaCl_2 \cdot 2MgCl_2 \cdot 12H_2O + MgCl_2 \cdot 6H_2O$
13.6	30.8	110	$CaCl_2 \cdot 2MgCl_2 \cdot 12H_2O$
18.2	28.0	110	»
26.9	21.2	110	»
40.0	14.0	110	»
47.8	10.0	110	»
49.0	9.5	110	$CaCl_2 \cdot 2MgCl_2 \cdot 12H_2O + 2CaCl_2 \cdot MgCl_2 \cdot 6H_2O$
51.0	8.0	110	$2CaCl_2 \cdot MgCl_2 \cdot 6H_2O$
53.5	6.2	110	»
55.7	4.7	110	$2CaCl_2 \cdot MgCl_2 \cdot 6H_2O + CaCl_2 \cdot 2H_2O$
56.8	4.0	110	$CaCl_2 \cdot 2H_2O$
62.3	0.0	110	»

№ 4165 [687]

MAGNESIUM CHLORIDE–CALCIUM CHLORIDE–WATER

$MgCl_2—CaCl_2—H_2O$

$t=55$

Comp. of saturated soln., wt %			Comp. of solid phase
Calcium Chloride	Magnesium Chloride	Water	
57.20	0.00	42.80	$CaCl_2 \cdot 2H_2O$
55.22	1.98	42.80	»
53.71	4.44	41.85	»
53.10	5.02	41.88	$CaCl_2 \cdot 2H_2O + MgCl_2 \cdot 2CaCl_2 \cdot 12H_2O$
49.87	6.50	43.62	$MgCl_2 \cdot 2CaCl_2 \cdot 12H_2O$
48.23	7.15	44.62	»
15.70	27.46	56.84	»
13.80	29.20	57.00	$MgCl_2 \cdot 2CaCl_2 \cdot 12H_2O + MgCl_2 \cdot 6H_2O$
5.38	34.17	60.45	$MgCl \cdot 6H_2O$
0.00	37.50	62.50	»

№ 4166 [1388]

MAGNESIUM CHLORIDE–CALCIUM IODATE–WATER

$MgCl_2—Ca(IO_3)_2—H_2O$

$t=25$

Concn. of Magnesium Chloride in aqueous soln., g/l	Solubility of Calcium Iodate in aqueous soln. of Magnesium Chloride, wt%	Concn. of Magnesium Chloride in aqueous soln., g/l	Solubility of Calcium Iodate in aqueous soln. of Magnesium Chloride, wt%
0.0	0.303	1.905	0.3736
0.1905	0.3121	4.761	0.4331
0.4761	0.3262	9.523	0.4924
0.9523	0.3444		

№ 4167 [2735]

MAGNESIUM CHLORIDE–IRON (II) HYDROXIDE–WATER

$MgCl_2—Fe(OH)_2—H_2O$

$t=25$

Concn. of Magnesium Chloride in aqueous soln., g/l	Solubility of Iron Hydroxide in aqueous soln. of Magnesium Chloride, g/l	Concn. of Magnesium Chloride in aqueous soln., g/l	Solubility of Iron Hydroxide in aqueous soln. of Magnesium Chloride, g/l	Concn. of Magnesium Chloride in aqueous soln., g/l	Solubility of Iron Hydroxide in aqueous soln. of Magnesium Chloride, g/l
0.0	0.00375	71.4	0.012	143.0	0.022
23.8	0.0075	95.3	0.015	190.5	0.025
47.65	0.010	119.1	0.020	214.2	0.024

№ 4168 [1002]

MAGNESIUM CHLORIDE–IRON (II) CHLORIDE–WATER

$MgCl_2—FeCl_2—H_2O$

$t=15$

Comp. of saturated soln., wt%			Comp. of solid phase
Iron Chloride	Magnesium Chloride	Water	
37.78	0.00	62.22	$FeCl_2 \cdot 4H_2O$
29.81	6.24	63.95	Solid soln. based on $FeCl_2 \cdot 4H_2O$
23.51	11.45	65.04	»
20.50	15.70	63.80	»
12.67	21.08	66.25	»
10.26	23.87	65.87	»
5.85	29.63	64.52	»
4.80	31.49	63.71	»
4.69	31.71	63.60	Solid soln. based on $MgCl_2 \cdot 6H_2O$
4.41	32.38	65.21	»
0.00	34.45	65.55	»

№ 4169 [1160]

MAGNESIUM CHLORIDE–COBALT (II) CHLORIDE–WATER

$MgCl_2—CoCl_2—H_2O$

$t=25$

Comp. of saturated soln., wt%			Comp. of solid phase
Magnesium Chloride	Cobalt Chloride	Water	
0.00	35.87	64.13	$CoCl_2 \cdot 6H_2O$
6.20	28.78	65.02	»
15.40	19.16	65.44	»
22.70	12.56	64.74	»

Table № 4169 (continuation)

Comp. of saturated soln., wt%			Comp. of solid phase
Magnesium Chloride	Cobalt Chloride	Water	
28.78	11.09	60.13	$CoCl_2 \cdot 6H_2O$
29.04	12.08	58.88	»
28.89	12.43	58.68	$CoCl_2 \cdot 6H_2O + CoCl_2 \cdot MgCl_2 \cdot 8H_2O$
29.72	11.61	58.67	$CoCl_2 \cdot MgCl_2 \cdot 8H_2O$
30.54	10.54	58.92	»
31.00	10.00	59.00	»
31.66	8.95	59.39	$CoCl_2 \cdot MgCl_2 \cdot 8H_2O + MgCl_2 \cdot 6H_2O$
31.95	8.64	59.41	$MgCl_2 \cdot 6H_2O$
32.52	7.10	60.38	»
32.80	6.65	60.55	»
33.55	4.97	61.48	»
34.30	3.93	61.77	»
34.83	2.57	62.60	»
36.80	0.00	63.20	»

№ 4170 [1159]

MAGNESIUM CHLORIDE–COBALT (II) CHLORIDE–WATER

$MgCl_2—CoCl_2—H_2O$

$t=25$

Comp. of saturated soln., wt%			Comp. of solid phase
Cobalt Chloride	Magnesium Chloride	Water	
35.87	0.00	64.13	$[Co(H_2O)_6]Cl_2$
28.78	6.20	65.02	»
19.16	15.40	65.44	»
12.56	22.70	64.74	»
11.09	28.78	60.13	»
12.08	29.04	58.88	»
12.43	28.89	58.68	$[Co(H_2O)_6]Cl_2 + [Mg\{(H_2O)_2\}_4]CoCl_4$
11.61	29.72	58.67	$[Mg\{(H_2O)_2\}_4] \cdot CoCl_4$
10.54	30.53	58.92	»
10.00	31.00	59.00	»
8.95	31.66	59.39	$[Mg\{(H_2O)_2\}_4]CoCl_4 + [Mg(H_2O)_6]Cl_2$
8.64	31.95	59.41	$[Mg(H_2O)_6]Cl_2$
7.10	32.52	60.38	»
6.65	32.80	60.55	»
4.97	33.55	61.48	»
4.44	34.00	61.56	»
3.93	34.30	61.77	»
2.57	34.83	62.60	»
0.00	36.20	63.80	»

№ 4171 [2596]

MAGNESIUM CHLORIDE–ZINC HYDROXIDE–WATER

$MgCl_2$—$Zn(OH)_2$—H_2O

$t=16—20$

Comp. of solvent, wt%		Solubility of Zinc Hydroxide in aqueous soln. of Magnesium Chloride, g/l
Magnesium Chloride	Water	
1	99	0.065

№ 4172 MAGNESIUM CHLORIDE–RUBIDIUM CHLORIDE–WATER [1599]

$MgCl_2$—$RbCl$—H_2O

$t=20$

Content in aqueous soln., g/l		Comp. of solid phase	Content in aqueous soln., g/l		Comp. of solid phase
Rubidium Chloride	Magnesium Chloride		Rubidium Chloride	Magnesium Chloride	
625	50	RbCl	168	250	RbCl
550	100	»	68	300	»
450	150	»	19	350	»
310	200	»	6	400	»

Note. In the solid phase formation of $RbCl \cdot MgCl_2 \cdot 6H_2O$ may also take place.

№ 4173 [1467]

MAGNESIUM CHLORIDE–RUBIDIUM CHLORIDE–WATER

$MgCl_2$—$RbCl$—H_2O

$t=25$

Content in aqueous soln., mol/1000 mol Water		Comp. of solid phase
2RbCl	$MgCl_2$	
36.7	32.2	$RbCl + RbCl \cdot MgCl_2 \cdot 6H_2O$
24.8	49.2	$RbCl \cdot MgCl_2 \cdot 6H_2O$
Traces	104	$RbCl \cdot MgCl_2 \cdot 6H_2O + RbCl$

№ 4174 [539]

MAGNESIUM CHLORIDE–RUBIDIUM CHLORIDE–WATER

$MgCl_2—RbCl—H_2O$

$t=100$

Content in aqueous soln., wt%		Comp. of solid phase
Rubidium Chloride	Magnesium Chloride	
58.14	0	RbCl
48.06	7.37	»
38.84	13.09	$RbCl + RbCl \cdot MgCl_2 \cdot 6H_2O$
38.71	14.28	$RbCl \cdot MgCl_2 \cdot 6H_2O$
36.80	14.90	»
30.91	17.11	»
26.58	18.81	»
26.25	19.86	»
25.26	20.28	»
19.82	22.34	»
13.60	25.56	»
8.55	28.19	»
5.33	31.02	»
2.60	33.78	»
0.96	37.86	»
0.57	41.63	»
0.24	42.04	$MgCl_2 \cdot 6H_2O$
0	42.05	»

№ 4175 [2703]

MAGNESIUM CHLORIDE–STRONTIUM SULFATE–WATER

$MgCl_2—SrSO_4—H_2O$

$t=25$

Comp. of solvent, wt%		Solubility of Strontium Sulfate in aqueous soln. of Magnesium Chloride, wt%
Magnesium Chloride	Water	
1.59	98.41	0.198
4.03	95.97	0.204
13.63	86.37	0.240

№ 4176 MAGNESIUM CHLORIDE–STRONTIUM CHLORIDE–WATER [1111]

$MgCl_2—SrCl_2—H_2O$

Comp. of saturated soln., wt%			t	Comp. of solid phase
Strontium Chloride	Magnesium Chloride	Water		
1.2	34.8	64.0	18.9	$SrCl_2 \cdot 6H_2O$
1.3	34.7	64.0	19.4	$SrCl_2 \cdot 6H_2O + SrCl_2 \cdot 2H_2O$
1.3	34.8	63.9	19.9	$SrCl_2 \cdot 2H_2O$
1.7	39.0	59.3	80.0	»
1.7	39.2	59.1	81.0	$SrCl_2 \cdot H_2O$
1.7	41.2	57.1	100.0	»
1.5	42.0	56.5	100.0	$MgCl_2 \cdot 6H_2O$
1.7	39.1	59.2	80.0	$SrCl_2 \cdot 2H_2O + SrCl_2 \cdot H_2O$

№ 4177 MAGNESIUM-CHLORIDE–SILVER CHLORIDE–WATER [1965]

$MgCl_2—AgCl—H_2O$

$t=25$

Concn. in aqueous soln., g/l		Concn. in aqueous soln., g/l		Concn. in aqueous soln., g/l	
Magnesium Chloride	Silver Chloride	Magnesium Chloride	Silver Chloride	Magnesium Chloride	Silver Chloride
61.75	0.083	99.75	0.335	110.7	0.477
67.92	0.110	102.12	0.362	116.6	0.578
79.32	0.167	108.8	0.449	127.3	0.753
91.44	0.254				

№ 4178 MAGNESIUM CHLORIDE–CADMIUM CHLORIDE–WATER [1170]

$MgCl_2—CdCl_2—H_2O$

$t=25$

Comp. of saturated soln., wt%			Comp. of solid phase
Cadmium Chloride	Magnesium Chloride	Water	
0.0	36.2	63.8	$MgCl_2 \cdot 6H_2O$
0.4	35.7	63.9	$MgCl_2 \cdot 6H_2O + CdCl_2 \cdot 2MgCl_2 \cdot 12H_2O$
0.4	35.3	64.3	$CdCl_2 \cdot 2MgCl_2 \cdot 12H_2O$
2.3	31.6	66.1	»
5.4	30.0	64.6	»
14.7	26.2	59.1	»
24.9	22.4	52.7	$CdCl_2 \cdot 2MgCl_2 \cdot 12H_2O + 2CdCl_2 \cdot MgCl_2 \cdot 12H_2O$
26.6	19.2	54.2	$2CdCl_2 \cdot MgCl_2 \cdot 12H_2O$
33.4	14.2	52.4	»
37.7	12.6	49.7	»
40.7	11.5	47.8	»
47.0	8.9	44.1	»
48.1	8.6	43.3	$2CdCl_2 \cdot MgCl_2 \cdot 12H_2O + CdCl_2 \cdot 2.5H_2O$
50.6	3.2	46.2	$CdCl_2 \cdot 2.5H_2O$
54.62	0	45.38	»
48.5	8.8	42.70	$2CdCl_2 \cdot MgCl_2 \cdot 12H_2O + CdCl_2 \cdot H_2O$

№ 4179 MAGNESIUM CHLORIDE–INDIUM CHLORIDE–WATER [1568]

$MgCl_2—InCl_3—H_2O$

$t=20$

Concn. in aqueous soln., g/l		Comp. of solid phase	Concn. in aqueous soln., g/l		Comp. of solid phase
Indium Chloride	Magnesium Chloride		Indium Chloride	Magnesium Chloride	
1318	133	$MgCl_2 \cdot 2InCl_3 \cdot 10H_2O$	646	380	$MgCl_2 \cdot InCl_3 \cdot 7H_2O$
1120	180	»	530	431	$2MgCl_2 \cdot InCl_3 \cdot 2H_2O$
1090	227	$MgCl_2 \cdot InCl_3 \cdot 7H_2O$	516	440	»
734	364	»	501	469	»

№ 4180 MAGNESIUM CHLORIDE–CESIUM CHLORIDE–WATER [1467]

$MgCl_2—CsCl—H_2O$

$t=25$

Concn. in aqueous soln., wt%		Comp. of solid phase
Cesium Chloride	Magnesium Chloride	
65.75	28.80	$CsCl+CsCl \cdot MgCl_2 \cdot 6H_2O$
48.20	44.70	$CsCl \cdot MgCl_2 \cdot 6H_2O$
5.58	84.50	»
1.52	90.20	$CsCl \cdot MgCl_2 \cdot 6H_2O+MgCl_2 \cdot 6H_2O$

№ 4181 MAGNESIUM CHLORIDE–BARIUM CHLORIDE–WATER [1662]

$MgCl_2—BaSO_4—H_2O$

$t=20—25$

Content in aqueous soln., g/l		Content in aqueous soln., g/l		Content in aqueous soln., g/l	
Magnesium Chloride	Barium Sulfate	Magnesium Chloride	Barium Sulfate	Magnesium Chloride	Barium Sulfate
1.0	0.030	10	0.033	50	0.050
2.5	0.030	25	0.050	100	0.050
5.0	0.033				

№ 4182 MAGNESIUM CHLORIDE – BARIUM CHLORIDE – WATER [1114]

$MgCl_2—BaCl_2—H_2O$

Comp. of saturated soln., wt%			Comp. of solid phase	t
Magnesium Chloride	Barium Chloride	Water		
35.7	0.5	63.8	$BaCl_2 \cdot H_2O$	25
35.7	0.5	63.8	$BaCl_2 \cdot 2H_2O$	25
36.4	0.5	63.1	$BaCl_2 \cdot H_2O$	36
36.4	0.5	63.1	$BaCl_2 \cdot 2H_2O$	36

Table № 4182 (continuation)

Comp. of saturated soln., wt%			Comp. of solid phase	t
Magnesium Chloride	Barium Chloride	Water		
36.9	0.5	62.6	$BaCl_2 \cdot H_2O$	42
36.9	0.5	62.6	$BaCl_2 \cdot 2H_2O$	42
37.0	0.5	62.5	$BaCl_2 \cdot H_2O$	45
35.2	0.5	62.3	»	47
35.2	0.2	64.6	»	70
30.5	0.4	69.1	»	70
29.6	0.5	69.9	$BaCl_2 \cdot 2H_2O$	70
26.2	1.1	72.7	»	70
39.80	0.15	60.05	$BaCl_2 \cdot H_2O$	81
39.90	0.15	59.95	$BaCl_2$	83
39.90	0.15	59.95	$BaCl_2 \cdot H_2O$	83
40.0	0.15	59.85	$BaCl_2$	85
37.40	0.00	62.60	$MgCl_2 \cdot 6H_2O$	50
36.90	0.05	63.05	$BaCl_2 \cdot H_2O + MgCl_2 \cdot 6H_2O$	50
36.70	0.05	63.25	$BaCl_2 \cdot H_2O$	50
34.10	0.15	65.75	»	50
32.50	0.20	67.30	$BaCl_2 \cdot 2H_2O$	50
26.80	0.40	72.80	»	50
20.70	2.30	77.0	»	50
12.20	10.60	77.20	»	50
0.00	30.40	69.60	»	50
42.2	0.00	57.8	$MgCl_2 \cdot 6H_2O$	100
40.9	0.2	58.9	$BaCl_2$	100
37.6	0.2	62.2	»	100
31.5	0.9	67.6	»	100
24.0	2.9	73.1	»	100
18.4	8.0	73.6	»	100
13.9	14.0	72.1	»	100
12.7	15.7	71.6	»	100
11.3	17.7	71.0	$BaCl_2 \cdot H_2O$	100
9.9	20.2	69.9	»	100
4.2	29.7	66.1	»	100
2.4	33.2	64.4	»	100
0.0	37.3	62.7	$BaCl_2 \cdot 2H_2O$	100

№ 4183 [1677, 2039]

MAGNESIUM CHLORIDE–LANTHANUM IODATE–WATER

$La(IO_3)_3$—$MgCl_2$—H_2O

$t=25$

Concn. of Magnesium Chloride in aqueous soln., g/l	Solubility of Lanthanum Iodate in aqueous soln. of Magnesium Chloride, g/l
47.66	1.173
4.766	1.154

№ 4184 MAGNESIUM CHLORIDE–LANTHANUM IODATE–WATER [2313]

$MgCl_2$—$La(IO_3)_3$—H_2O

$t=25$

Concn. of Magnesium Chloride in aqueous soln., g/l	Solubility of Lanthanum Iodate in aqueous soln. of Magnesium Chloride, g/l	d	Concn. of Magnesium Chloride in aqueous soln., g/l	Solubility of Lanthanum Iodate in aqueous soln. of Magnesium Chloride, g/l	d
0.0953	0.466	0.99757	47.70	1.846	1.03666
0.477	0.542	0.99798	95.30	2.328	1.07255
0.953	0.603	0.99841	190.60	2.793	1.13805
4.77	0.867	1.00180	286.0	3.079	1.19716
9.53	1.062	1.00589			

№ 4185 [1387]

MAGNESIUM CHLORIDE–CERIUM IODATE–WATER

$MgCl_2$—$Ce(IO_3)_2$—H_2O

$t=25$

Concn. in aqueous soln., wt%	
Magnesium Chloride	Cerium Iodate
0.00	0.1729
0.0189	0.1866
0.0475	0.1988
0.0949	0.2162
0.1897	0.2416
0.4724	0.2960
0.9400	0.3487

№ 4186 [1815]

MAGNESIUM CHLORIDE–MERCURY (II) CHLORIDE–WATER

$MgCl_2$—$HgCl_2$—H_2O

$t=25$

Concn. in aqueous soln., wt%	
Mercury Chloride	Magnesium Chloride
101.6	16.03
195.3	39.5
337	54.35
507	95.15
698	125.8
870	165

№ 4187 [1162]

MAGNESIUM CHLORIDE–MERCURY (II) CHLORIDE–WATER

$MgCl_2$—$HgCl_2$—H_2O

$t=25$

Comp. of saturated sol., wt%			Comp. of solid phase
Mercury Chloride	Magnesium Chloride	Water	
6.90	0.0	93.10	$HgCl_2$
33.47	5.4	61.13	»
47.72	8.28	44.00	»
56.03	9.86	34.11	»
57.46	10.68	31.86	»

Table № 4187 (continuation)

Comp. of saturated sol., wt%			Comp. of solid phase
Mercury Chloride	Magnesium Chloride	Water	
59.23	12.48	28.29	$HgCl_2 + [Mn(H_2O)_6]Hg_3Cl_8$
55.90	14.44	29.66	$[Mg(H_2O)_6]Hg_2Cl_8$
55.78	14.38	29.84	»
54.53	15.24	30.23	»
53.15	16.30	30.55	»
51.88	17.27	30.85	»
50.08	18.77	31.15	$[Mg(H_2O)_6]Hg_3Cl_8 + [Mg(H_2O)_6]HgCl_4$
49.85	19.23	30.92	$[Mg(H_2O)_6]Hg_3Cl_8$
49.47	19.07	31.46	$[Mg(H_2O)_6]HgCl_4$
48.95	19.22	31.83	»
48.78	19.45	31.77	»
47.32	20.01	32.67	»
45.45	20.90	33.65	»
44.93	21.12	33.95	»
43.20	21.95	34.85	$[Mg(H_2O)_6]HgCl_4 + [Mg(H_2O)_6]Cl_2$
41.66	22.43	35.91	$[Mg(H_2O)_6]Cl_2$
31.92	25.29	42.79	»
24.29	27.64	48.07	»
11.76	31.72	56.52	»
0.58	35.48	63.94	»
0.0	35.70	64.30	»

№ 4188 [2264]

MAGNESIUM CHLORIDE–THALLIUM CHLORIDE–WATER

$MgCl_2—TlCl—H_2O$

$t=25$

Concn. of Magnesium Chloride in aqueous soln., g/l	Solubility of Thallium Chloride in aqueous soln. of Magnesium Chloride, g/l
1.191	2.168
2.383	1.482
4.766	0.990
9.532	0.660

№ 4189 MAGNESIUM CHLORIDE–THALLIUM CHLORIDE–WATER [1197]

$MgCl_2$—$TlCl$—H_2O

$t=25$

Concn. in aqueous soln., wt%		Comp. of solid phase	Concn. in aqueous soln., wt%		Comp. of solid phase
Thallium Chloride	Magnesium Chloride		Thallium Chloride	Magnesium Chloride	
0.091	0.57	TlCl	0.058	23.06	TlCl
0.053	2.703	»	0.105	27.89	»
0.042	5.211	»	0.097	28.68	$3TlCl \cdot MgCl_2$
0.029	10.46	»	0.059	30.76	»
0.034	14.66	»	0.013	33.94	»
0.046	16.65	»	0.007	35.12	»

№ 4190 [2039]

MAGNESIUM CHLORIDE–THALLIUM IODATE–WATER

$MgCl_2$—$TlIO_3$—H_2O

$t=25$

Concn. of Magnesium Chloride in aqueous soln., g/l	Solubility of Thallium Iodate in aqueous soln. of Magnesium Chloride, g/l
1.992	0.965
2.991	1.041
2.991	1.045

№ 4191 [2262]

MAGNESIUM CHLORIDE–LEAD (II) CHLORIDE–WATER

$MgCl_2$—$PbCl_2$—H_2O

$t=25$

Concn. in aqueous soln., g/l		Concn. in aqueous soln., g/l	
Magnesium Chloride	Lead Chloride	Magnesium Chloride	Lead Chloride
0	10.79	3	5.7
1	7.7	4	5.2
2	6.5	6	4.4

№ 4192 [1965]

MAGNESIUM CHLORIDE–LEAD (II) CHLORIDE–WATER

$MgCl_2$—$PbCl_2$—H_2O

$t=25$

Concn. in aqueous soln., g/l		Comp. of solid phase
Lead Chloride	Magnesium Chloride	
1.51	5.70	$MgCl_2 \cdot PbCl_2 \cdot 6H_2O$
1.11	12.83	»
0.87	28.50	»
0.86	40.38	»
0.96	75.05	»
1.01	86.92	»

Table № 4192 (continuation)

Concn. in aqueous soln., g/l		Comp. of solid phase
Lead Chloride	Magnesium Chloride	
1.27	124.4	$MgCl_2 \cdot PbCl_2 \cdot 6H_2O$
1.67	168.2	»
1.84	183.8	»
2.17	207.1	»
2.54	230.8	»
3.20	257.4	»
4.03	282.2	»
4.85	302.1	»
5.91	322.1	»
6.74	336.3	»
8.27	360.1	»
9.74	372.05	»
10.23	385.7	»

№ 4193 [2191]

MAGNESIUM SELENATE–POTASSIUM SELENATE–WATER

$MgSeO_4—K_2SeO_4—H_2O$

$t=25$

Comp. of saturated soln., wt%			Comp. of solid phase
Potassium Solenate	Magnesium Solenate	Water	
53.30	0.0	46.70	K_2SeO_4
51.13	0.56	48.31	»
50.58	0.63	48.79	»
47.31	0.89	51.80	$K_2SeO_4+K_2SeO_4 \cdot MgSeO_4 \cdot 6H_2O$
38.50	2.12	59.38	$K_2SeO_4 \cdot MgSeO_4 \cdot 6H_2O$
30.24	5.44	64.32	»
21.83	11.40	66.77	»
21.40	11.48	67.12	$K_2SeO_4 \cdot MgSeO_4 \cdot 6H_2O+MgSeO_4 \cdot 6H_2O$
19.95	12.90	67.15	$MgSeO_4 \cdot 6H_2O$
19.59	13.37	67.04	»
5.82	24.36	69.82	»
0.0	27.5	72.5	»

№ 4194 **[1852]**

MAGNESIUM SELENATE–POTASSIUM SELENATE–WATER

$MgSeO_4—K_2SeO_4—H_2O$

$t=25$

Comp. of saturated soln., wt%			Comp. of solid phase
Magnesium Selenate	Potassium Selenate	Water	
0.00	53.26	46.74	K_2SeO_4
0.191	52.82	46.989	$K_2SeO_4+MgK_2(SeO_4)_2 \cdot 6H_2O$
0.191	52.94	46.869	»
0.35	50.48	49.17	$MgK_2(SeO_4)_2 \cdot 6H_2O$
1.52	41.54	56.94	»
5.68	29.55	64.77	»
13.91	18.30	67.79	»
20.04	12.76	67.20	»
24.08	9.79	66.13	»
31.11	5.77	63.12	»
34.33	3.52	62.15	$MgK_2(SeO_4)_2 \cdot 6H_2O+MgSeO_4 \cdot 6H_2O$
34.42	3.63	61.95	»
35.74	0.00	64.26	$MgSeO_4 \cdot 6H_2O$

№ 4195 **[1253]**

MAGNESIUM BROMIDE–POTASSIUM BROMIDE–WATER

$MgBr_2—KBr—H_2O$

$t=25$

Content in aqueous soln., mol/1000 mol Water		Comp. of solid phase
2KBr	$MgBr_2$	
52.1	0	KBr
4.35	78.6	$KBr+KBr \cdot MgBr_2 \cdot 6H_2O$
0.2	99.0	$KBr \cdot MgBr_2 \cdot 6H_2O+MgBr_2 \cdot 6H_2O$
0	99.1	$MgBr_2 \cdot 6H_2O$

№ 4196 **[615]**

MAGNESIUM BROMIDE–POTASSIUM BROMIDE–WATER

$MgBr_2—KBr—H_2O$

$t=25$

Content in aqueous soln., wt%		d_4	Comp. of solid phase
Potassium Bromide	Magnesium Bromide		
40.81	0	—	KBr
35.29	4.77	—	»
30.23	9.23	—	»
18.14	20.67	—	»

Table № 4196 (continuation)

Content in aqueous soln., wt%		d_4	Comp. of solid phase
Potassium Bromide	Magnesium Bromide		
12.01	28.01	—	KBr
6.13	36.35	—	»
5.17	37.86	—	»
2.68	43.24	1.5631	$KBr + KBr \cdot MgBr_2 \cdot 6H_2O$
0.91	48.23	—	—
0.14	50.43	1.6411	$KBr \cdot MgBr_2 \cdot 6H_2O + MgBr_2 \cdot 6H_2O$
0	50.60	—	$MgBr_2 \cdot 6H_2O$

№ 4197 [116]

MAGNESIUM BROMIDE–RUBIDIUM BROMIDE–WATER

$MgBr_2—RbBr—H_2O$

$t=25$

Concn. in aqueous soln., wt%		Comp. of solid phase
Magnesium Bromide	Rubidium Bromide	
0.0	53.35	RbBr
10.10	41.51	»
20.50	28.20	»
30.40	17.0	»
40.31	9.6	»
50.40	Traces	$RbBr + MgBr_2 \cdot 6H_2O$
50.38	0.0	$MgBr_2 \cdot 6H_2O$

№ 4198 [1568]

MAGNESIUM BROMIDE–INDIUM BROMIDE–WATER

$MgBr_2—InBr_3—H_2O$

$t=20$

Concn. in aqueous soln., g/l		Comp. of solid phase	Concn. in aqueous soln., g/l		Comp. of solid phase
Indium Bromide	Magnesium Bromide		Indium Bromide	Magnesium Bromide	
1653	311	$MgBr_2 \cdot 2InBr_3 \cdot 7H_2O$	778	659	$MgBr_2 \cdot InBr_3 \cdot 7H_2O$
1551	378	»	766	674	»
927	611	$MgBr_2 \cdot InBr_3 \cdot 7H_2O$	639	747	$MgBr_2 \cdot 6H_2O$
859	628	»	449	830	»

№ 4199 MAGNESIUM IODIDE–INDIUM IODIDE–WATER [1568]

$MgI_2—InI_3—H_2O$

$t=20$

Concn. in aqueous soln., g/l		Comp. of solid phase	Concn. in aqueous soln., g/l		Comp. of solid phase
Indium Iodide	Mag-nesium Iodide		Indium Iodide	Mag-nesium Iodide	
816	686	$MgI_2 \cdot 2InI_3 \cdot 11H_2O$	97	1064	$MgI_2 \cdot 6H_2O$
708	750	$3MgI_2 \cdot 4InI_3 \cdot 26H_2O$	64	1075	»
225	890	$MgI_2 \cdot InI_3 \cdot 6H_2O$	51	1138	»
200	961	»	34	1167	»
114	1071	»			

№ 4200 MAGNESIUM IODIDE–THALLIUM IODIDE–WATER [431]

$MgI_2—TlI—H_2O$

$t=25$

Content in aqueous soln.	
Magnesium Iodide, g-eq/l	Thallium Iodide, (mol/l) · 10³
0.41	0.45
0.81	0.63
1.22	0.99
1.62	1.48
2.01	2.13

№ 4201 MAGNESIUM IODATE–CALCIUM IODATE–WATER [2314]

$Mg(IO_3)_2—Ca(IO_3)_2—H_2O$

$t=17.9$

Concn. in aqueous soln., g/l Water	
Calcium Iodate	Magnesium Iodate
2.217	0.0
1.984	0.3775
1.766	0.7524
1.571	1.127
1.238	1.871
0.6905	3.742

№ 4202 ALUMINUM OXIDE–ALUMINUM SULFATE–WATER [235]

$Al_2O_3—Al_2(SO_4)_3—H_2O$

$t=18$

Concn. in aqueous soln., wt%		d	pH	Comp. of solid phase
Aluminum Sulfate	Aluminum Oxide			
2.04	10.50	1.266	3.26	$2Al_2O_3 \cdot SO_3 \cdot xH_2O + Al_2(SO_4)_3 \cdot 18H_2O$
1.41	7.31	1.148	3.45	$2Al_2O_3 \cdot SO_3 \cdot xH_2O$
0.88	4.60	1.089	3.65	»
0.64	3.29	1.052	3.71	»
0.263	1.38	1.023	3.84	»
0.071	0.37	1.011	3.98	»
0.027	0.14	1.005	4.11	»
0.014	0.07	1.002	4.20	»
0.00	0.00	1.000	4.43	$Al_2O_3 \cdot xH_2O$

№ 4203 ALUMINUM OXIDE–PHOSPHORUS PENTOXIDE–WATER [1753]

$Al_2O_3—P_2O_5—H_2O$

Comp. of saturated soln., wt%			t	Comp. of solid phase
Aluminum Oxide	Phosphorus Pentoxide	Water		
6.8	31.2	63.0	60	$Al_2O_3 \cdot P_2O_5 \cdot 7H_2O + Al_2O_3 \cdot P_2O_5 \cdot 4H_2O$
7.9	40.7	51.4	60	$Al_2O_3 \cdot P_2O_5 \cdot 4H_2O + 2Al_2O_3 \cdot 3P_2O_5 \cdot 10H_2O$
8.8	52.7	38.5	60	$2Al_2O_3 \cdot 3P_2O_5 \cdot 10H_2O + Al_2O_3 \cdot 2P_2O_5 \cdot 4H_2O$
8.7	54.5	36.8	60	$Al_2O_3 \cdot 2P_2O_5 \cdot 4H_2O + Al_2O_3 \cdot 3P_2O_5 \cdot 6H_2O$
7.0	54.5	38.5	60	$Al_2O_3 \cdot 3P_2O_5 \cdot 6H_2O$
0.10	2.90	97.0	80	$Al_2O_3 \cdot P_2O_5 \cdot 4H_2O$
0.16	4.69	95.15	80	»
0.20	5.51	94.29	80	»
0.24	7.12	92.64	80	»
0.44	10.35	89.21	80	»
0.55	11.16	88.29	80	»
1.10	15.55	83.35	80	»
1.79	19.08	79.13	80	»
2.49	22.18	75.33	80	»
2.85	24.26	72.89	80	»
5.13	30.08	64.79	80	»

№ 4204 [1292]

ALUMINUM OXIDE–PHOSPHORUS PENTOXIDE–WATER

$Al_2O_3—P_2O_5—H_2O$

$t=25$

Comp. of saturated soln., wt%			Comp. of solid phase
Aluminum Oxide	Phosphorus Pentoxide	Water	
0.04	1.06	98.90	$AlPO_4 \cdot xH_2O$
0.46	5.09	94.45	»
1.06	8.46	90.48	»
1.07	8.68	90.25	»
2.01	11.90	86.09	»
2.19	12.37	85.44	»
2.24	14.02	83.74	»
3.03	16.81	80.16	»
4.50	22.57	72.93	»
4.23	23.21	72.56	»
4.66	23.74	71.60	»
5.46	26.32	68.22	»
6.17	28.18	65.65	$AlPO_4 \cdot H_3PO_4 \cdot 3H_2O$
6.15	29.94	63.91	—
6.57	35.75	57.68	$AlPO_4 \cdot H_3PO_4 \cdot 3H_2O$
6.41	37.17	56.42	»
6.49	38.47	55.04	»
6.45	38.85	54.70	»
6.29	42.79	50.92	»

Table № 4204 (continuation)

Comp. of saturated soln., wt%			Comp. of solid phase
Aluminum Oxide	Phosphorus Pentoxide	Water	
6.28	43.09	50.63	$AlPO_4 \cdot H_3PO_4 \cdot 3H_2O$
6.14	45.84	48.02	»
5.93	46.78	47.29	»
6.03	49.44	44.53	»
6.73	53.15	40.12	»
7.25	54.57	38.18	»

№ 4205 ALUMINUM OXIDE–PHOSPHORUS PENTOXIDE–WATER [1292]

$Al_2O_3—P_2O_5—H_2O$

Comp. of saturated soln., wt%			t	Comp. of solid phase
Aluminum Oxide	Phosphorus Pentoxide	Water		
0.01	0.05	99.94	50	$AlPO_4 \cdot xH_2O$
0.10	3.60	96.30	50	»
0.36	6.35	93.29	50	»
0.65	8.62	90.73	50	»
1.14	11.39	87.50	50	»
1.73	14.77	83.50	50	»
7.47	37.69	54.84	50	$AlPO_4 \cdot H_3PO_4 \cdot 3H_2O$
7.74	38.82	53.44	50	»
7.47	43.71	48.82	50	»
7.55	46.27	46.18	50	»
7.52	47.18	45.30	50	»
7.90	50.86	41.24	50	»
8.42	52.98	38.60	50	»
8.63	54.22	37.15	60	$AlPO_4 \cdot H_3PO_4 \cdot 3H_2O + AlPO_4 \cdot 2H_3PO_4$
6.26	55.60	38.14	50	$AlPO_4 \cdot 2H_3PO_4$
2.09	60.05	37.86	50	»
0.05	3.77	96.18	75	$AlPO_4 \cdot xH_2O$
0.17	6.35	93.48	75	»
0.54	10.36	89.10	75	»
1.24	15.21	83.75	75	»
2.65	22.76	74.59	75	»
3.16	24.86	71.98	75	»
4.14	29.02	66.84	75	»
4.58	30.72	64.70	75	»
5.65	34.67	59.68	75	»
5.95	35.01	59.04	75	»
6.35	36.15	57.50	75	»
8.13	41.77	50.10	75	»
9.39	44.37	46.24	75	$AlPO_4 \cdot xH_2O + AlPO_4 \cdot H_3PO_4 \cdot 3H_2O$
9.38	46.16	44.46	75	$AlPO_4 \cdot H_3PO_4 \cdot 3H_2O$
9.37	49.10	41.53	75	»
8.26	50.52	41.22	75	$AlPO_4 \cdot 2H_3PO_4$
7.36	53.03	39.61	75	»
6.44	55.04	38.52	75	»
6.02	56.29	37.69	75	»
4.09	58.29	37.62	75	»

№ 4206 [1910]

ALUMINUM OXIDE–PHOSPHORUS PENTOXIDE–WATER

Al_2O_3—P_2O_5—H_2O

t=25

Concn. in aqueous soln., wt%		Comp. of solid phase
Phosphorus Pentoxide	Aluminum Oxide	
1.063	0.079	$Al_2O_3 \cdot P_2O_5 \cdot 7H_2O$
3.016	0.291	»
4.832	0.558	»
8.111	1.124	»
11.23	1.984	»
14.86	2.845	»
15.14	3.354	»
16.42	3.803	»
17.89	4.054	$Al_2O_3 \cdot P_2O_5 \cdot 4H_2O$
20.21	4.270	»
23.18	4.516	»
27.12	5.150	»
28.23	5.382	»
30.96	5.08	$2Al_2O_3 \cdot 3P_2O_5 \cdot 10H_2O$
34.88	4.84	»
38.45	4.85	»
40.03	5.06	»
44.16	5.60	»
47.67	6.17	»
49.00	6.78	»
49.96	6.89	»
51.86	7.15	$Al_2O_3 \cdot 3P_2O_5 \cdot 6H_2O$
53.77	5.75	»
53.34	5.57	»
57.21	4.64	»
59.13	4.30	»

№ 4207 [159]

ALUMINUM OXIDE–PHOSPHORUS PENTOXIDE–WATER

Al_2O_3—P_2O_5—H_2O

t=80

Comp. of saturated soln., wt%			Comp. of solid phase
Phosphorus Pentoxide	Aluminum Oxide	Water	
2.90	0.10	97.00	$AlPO_4 \cdot 2H_2O$
4.69	0.16	95.15	»
5.51	0.20	94.29	»
7.12	0.24	92.64	»
10.35	0.44	89.21	»
11.16	0.55	88.29	»

Table № 4207 (continuation)

Comp. of saturated soln., wt%			Comp. of solid phase
Phosphorus Pentoxide	Aluminum Oxide	Water	
15.55	1.10	83.35	$AlPO_4 \cdot 2H_2O$
19.08	1.79	79.13	»
22.18	2.49	75.33	»
24.26	2.85	72.89	»
30.08	5.13	64.79	»
1.57	0.07	98.36	$2Al_2O_3 \cdot 3P_2O_5 \cdot nH_2O$
3.80	0.98	95.22	»
4.05	1.09	94.86	»
4.10	1.13	94.77	»
5.12	1.40	93.48	»
5.47	1.44	93.09	»
6.12	1.46	92.42	»
7.19	1.61	91.20	»
7.24	1.63	91.13	»
7.38	1.77	90.85	»
9.00	1.99	89.01	»
12.80	2.90	84.30	»
13.90	3.62	82.48	»
14.00	3.94	82.06	»
14.27	4.43	80.30	»

№ 4208 [787]

ALUMINUM OXIDE–PHOSPHORUS PENTOXIDE–WATER

$Al_2O_3—P_2O_5—H_2O$

$t=90$

Comp. of saturated soln., wt%			Comp. of solid phase
Phosphorus Pentoxide	Aluminum Oxide	Water	
2.62	0.063	97.317	$AlPO_4 \cdot 2H_2O$
7.82	0.64	91.54	»
9.77	0.39	89.84	»
14.79	1.11	84.10	»
17.73	1.206	81.064	»
19.18	1.99	78.83	»
28.20	4.11	67.69	»
29.43	3.87	66.70	»
31.41	4.85	63.74	»
33.80	5.57	60.63	»
34.59	7.26	58.15	»
37.50	7.65	54.85	»
39.20	7.18	53.62	$2Al_2(HPO_4)_3 \cdot 5H_2O$
41.38	9.17	49.45	»
45.86	10.06	44.08	»
44.20	9.17	46.63	»
52.08	12.14	35.78	»

$Al_2O_3—P_2O_5—H_2O$

$t=60$

Comp. of saturated soln., wt%			d	Comp. of solid phase
Aluminum Oxide	Phosphorus Pentoxide	Water		
0.6	73.0	26.4	—	$Al_2O_3 \cdot 3P_2O_5 \cdot 6H_2O$
1.2	67.5	31.3	1.78	»
1.7	63.0	35.3	—	»
1.9	61.9	36.2	1.725	»
2.5	59.9	37.6	1.73	»
3.1	58.6	38.3	1.73	»
4.2	57.7	38.1	—	»
4.7	56.8	38.5	1.76	»
5.7	56.5	37.8	1.77	»
6.3	56.6	37.1	1.77	»
7.3	56.1	36.6	1.78	$Al_2O_3 \cdot 3P_2O_5 \cdot 6H_2O + Al_2O_3 \cdot 2P_2O_5 \cdot 5H_2O$
7.6	55.9	36.5	1.79	$Al_2O_3 \cdot 3P_2O_5 \cdot 5H_2O$
7.9	54.9	37.2	—	»
8.1	54.5	33.4	—	»
8.3	54.3	37.4	1.81	»
8.5	54.3	37.2	1.80	$Al_2O_3 \cdot 2P_2O_5 \cdot 5H_2O + Al_2O_3 \cdot 2P_2O_5 \cdot 5H_2O$
8.4	54.0	37.6	—	$Al_2O_3 \cdot 2P_2O_5 \cdot 9H_2O$
8.2	52.6	39.2	—	»
8.5	52.4	39.1	1.775	»
8.5	50.5	41.0	1.73	»
8.3	49.9	41.8	—	»
8.0	46.4	45.6	1.66	»
7.6	43.2	49.2	1.59	»
7.8	42.6	49.6	—	»
7.5	42.2	50.3	—	»
7.7	40.9	51.4	1.57	»
7.5	36.2	56.3	1.51	»
9.2	35.4	55,4	1.55	»
8.4	41.4	50.2	1.57	$Al_2O_3 \cdot 2P_2O_5 \cdot 9H_2O + Al_2O_3 \cdot P_2O_5 \cdot 4H_2O$
7.7	40.0	52.3	1.56	»
7.5	37.0	55.5	1.50	»
8.1	38.6	53.3	1.54	$Al_2O_3 \cdot P_2O_5 \cdot 4H_2O$
6.4	33.8	59.8	1.41	»
6.5	33.2	60.3	1.42	»
6.1	33.0	60.9	1.41	»
5.8	32.2	62.0	1.39	»
6.8	31.2	62.0	1.42	»
5.3	30.1	64.6	1.36	»
4.3	27.4	68.3	1.30	»
5.15	25.9	68.95	1.30	»
3.4	23.4	73.2	1.23	»
3.35	20.4	66.25	—	»
2.2	17.6	80.2	1.16	»
1.45	15.0	83.55	1.14	»
2.55	14.5	82.95	1.16	»
1.4	13.3	85.3	1.12	»
1.2	13.0	85.8	1.12	»

Table № 4209 (continuation)

Comp. of saturated soln., wt%			d	Comp. of solid phase
Aluminum Oxide	Phosphorus Pentoxide	Water		
0.9	11.4	87.7	—	$Al_2O_3 \cdot P_2O_5 \cdot 4H_2O$
0.95	11.0	82.05	1.09	»
0.5	9.1	90.4	1.07	»
0.4	7.1	92.5	1.04	»
0.1	1.9	98.0	1.005	»
0.05	1.5	98.45	1.002	»
Traces	0.05	99.95	1.00	»
0	<0.01	>99.99	1.00	$Al_2O_3 \cdot P_2O_5 \cdot 4H_2O + Al_2O_3$

№ 4210 ALUMINUM OXIDE–PHOSPHORUS PENTOXIDE–WATER [806]

$Al_2O_3—P_2O_5—H_2O$

Phosphorus Pentoxide	Aluminum Oxide	Water	t	Phosphorus Pentoxide	Aluminum Oxide	Water	t
4.27	16.88	78.85	40	16.02	13.67	70.31	60
4.96	15.83	79.21	40	17.31	13.55	69.14	60
5.93	15.94	78.13	40	18.01	14.80	67.19	60
6.81	17.36	75.83	40	19.25	15.41	64.34	60
8.10	16.03	75.87	40	20.43	17.74	61.83	60
8.31	15.98	75.71	40	22.99	11.80	65.21	60
10.88	19.44	69.68	40	25.00	9.76	65.24	60
11.34	19.05	69.61	40	26.90	9.90	63.20	60
12.48	18.63	68.89	40	33.16	12.25	54.59	60
12.71	17.27	70.02	40	33.68	11.11	55.21	60
13.47	18.09	68.44	40	34.04	11.54	54.42	60
18.64	18.48	62.88	40	38.81	12.37	48.82	60
23.90	20.84	55.26	40	41.82	15.29	42.89	60
25.16	19.19	55.65	40	45.20	11.53	43.27	60
26.00	18.00	55.00	40	59.19	6.97	33.84	60
27.02	17.25	55.73	40	2.87	15.25	81.88	80
53.32	11.74	34.94	40	11.10	20.74	68.16	80
59.72	8.77	31.51	40	12.03	22.76	65.21	80
61.10	7.34	31.56	40	12.84	19.54	67.62	80
64.85	9.28	25.87	40	13.21	17.63	65.16	80
66.71	10.17	23.12	40	13.42	19.41	67.17	80
73.86	4.80	21.34	40	13.55	21.25	65.20	80
3.41	18.00	78.59	60	15.58	8.68	75.74	80
4.84	17.41	75.75	60	22.40	11.81	65.79	80
5.96	17.70	76.34	60	22.91	11.50	65.59	80
6.34	17.69	75.97	60	31.81	14.33	53.86	80
7.90	14.59	77.51	60	32.50	14.81	52.69	80
8.72	13.60	77.68	60	39.50	10.61	49.89	80
10.00	13.23	76.77	60	47.66	10.91	41.43	80
11.29	13.24	75.47	60	53.40	17.43	29.17	80
13.96	12.50	73.54	60				

(Column groups: Comp. of saturated soln., wt% — Phosphorus Pentoxide, Aluminum Oxide, Water; t — repeated for both halves.)

№ 4211 [1228]

ALUMINUM OXIDE–CALCIUM OXIDE–WATER

Al_2O_3—CaO—H_2O

t=25

Concn. in aqueous soln., wt%		Comp. of solid phase
Aluminum Oxide	Calcium Oxide	
0.0014	0.0123	$Al_2O_3 \cdot 3H_2O$
0.0017	0.0119	»
0.0023	0.0252	»
0.0022	0.0265	$Al_2O_3 \cdot 3H_2O + Al_2O_3 \cdot 3CaO \cdot 6H_2O$
0.0021	0.0395	$Al_2O_3 \cdot 3H_2O$
0.0019	0.0485	»
0.0	0.108	$Ca(OH)_2$

№ 4212 [2322]

ALUMINUM OXIDE–CALCIUM OXIDE–WATER

Al_2O_3—CaO—H_2O

Concn. in saturated soln., g/l		t	Comp. of solid phase
Aluminum Oxide	Calcium Oxide		
0.133	0.223	50	$3CaO \cdot Al_2O_3 \cdot 6H_2O$
0.089	0.317	50	»
0.055	0.254	50	»
0.095	0.233	50	»
0.060	0.386	50	»
0.036	0.232	50	»
0.019	0.325	50	»
0.013	0.319	50	»
0.007	0.408	50	»
0.006	0.404	50	»
0.007	0.416	50	»
0.007	0.552	50	»
0.008	0.559	50	»
0.010	0.678	50	»
0.007	0.671	50	»
0.003	0.747	50	$3CaO \cdot Al_2O_3 \cdot 6H_2O$ with impurities
0.005	0.737	50	»
0.006	0.805	50	»
0.004	0.717	50	»
0.002	0.887	50	»
0.003	0.968	50	$3CaO \cdot Al_2O_3 \cdot 6H_2O + Ca(OH)_2$
0.003	0.956	50	»
0.005	0.930	50	»
0.007	0.980	50	»
0.002	0.924	50	»
0.006	0.947	50	»
0.001	0.077	50	$Al_2O_3 \cdot 3H_2O$
0.002	0.031	50	»

Table № 4212 (continuation)

Concn. in saturated soln., g/l		t	Comp. of solid phase
Aluminum Oxide	Calcium Oxide		
0.001	0.053	50	$Al_2O_3 \cdot 3H_2O$
0.006	0.197	50	»
0.002	0.171	50	»
0.003	0.088	50	»
0.014	0.276	50	»
0.035	0.250	50	»
0.039	0.106	50	»
0.029	0.379	50	»
0.059	0.368	50	»
0.067	0.286	50	»
0.030	0.447	50	»
0.027	0.467	50	»
0.069	0.385	50	»
0.007	0.587	50	»
0.034	0.580	50	»
0.043	0.584	50	»
0.040	0.460	50	»
0.047	0.197	50	»
0.032	0.761	50	»
0.040	0.712	50	»
0.036	0.660	50	»
0.020	0.600	50	»
0.077	0.352	50	»
0.109	0.678	50	»
0.124	0.642	50	»
0.133	0.606	50	»
0.080	0.500	50	»
0.082	0.675	50	»
0.113	0.641	50	»
0.118	0.590	50	»
0.088	0.480	50	»
0.083	0.820	50	»
0.100	0.787	50	»
0.080	0.700	50	»
0.105	0.520	50	»
0.094	0.846	50	»
0.134	0.696	50	»
0.100	0.507	50	»
0.085	0.070	120	»
0.195	0.176	120	»
0.337	0.323	120	»
0.320	0.294	120	»
0.336	0.293	120	»
0.315	0.308	120	$Al_2O_3 \cdot 3H_2O + 3CaO \cdot Al_2O_3 \cdot 6H_2O$
0.323	0.316	120	$Al_2O_3 \cdot 3H_2O$
0.320	0.320	120	»
0.347	0.266	120	»
0.296	0.293	120	»
0.332	0.338	120	$Al_2O_3 \cdot 3H_2O + 3CaO \cdot Al_2O_3 \cdot 6H_2O$
0.330	0.300	120	»
0.300	0.170	120	$3CaO \cdot Al_2O_3 \cdot 6H_2O$

Table № 4212 (continuation)

Concn. in saturated soln., g/l		t	Comp. of solid phase
Aluminum Oxide	Calcium Oxide		
0.327	0.132	120	$3CaO \cdot Al_2O_3 \cdot 6H_2O$
0.335	0.132	120	»
0.359	0.115	120	»
0.432	0.070	120	$3CaO \cdot Al_2O_3 \cdot 6H_2O + Ca(OH)_2$
0.408	0.050	120	»
0.428	0.054	120	»
0.450	0.040	120	»
0.415	0.052	120	»
0.408	0.038	120	»
0.433	0.038	120	»
0.402	0.043	120	»
0.072	0.034	150	$Al_2O_3 \cdot H_2O$
0.160	0.084	150	»
0.240	0.150	150	»
0.250	0.153	150	$Al_2O_3 + 3CaO \cdot Al_2O_3 \cdot 6H_2O$
0.277	0.166	150	»
0.281	0.176	150	»
0.259	0.142	150	»
0.261	0.147	150	»
0.256	0.208	150	$3CaO \cdot Al_2O_3 \cdot 6H_2O$
0.235	0.204	150	»
0.270	0.130	150	»
0.300	0.090	150	»
0.290	0.070	150	»
0.340	0.046	150	$3CaO \cdot Al_2O_3 \cdot 6H_2O + Ca(OH)_2$
0.342	0.050	150	»
0.345	0.054	150	»
0.324	0.060	150	»
0.342	0.061	150	»
0.331	0.042	150	»
0.267	0.154	150	»
0.001	0.022	200	$Al_2O_3 \cdot H_2O$
0.001	0.003	200	»
0.004	0.023	200	»
0.002	0.018	200	»
0.002	0.020	200	»
0.064	0.093	200	$Al_2O_3 \cdot H_2O + 3CaO \cdot Al_2O_3 \cdot 6H_2O$
0.051	0.100	200	»
0.090	0.112	200	»
0.091	0.200	200	$3CaO \cdot Al_2O_3 \cdot 6H_2O + Ca(OH)_2$
0.080	0.258	200	»
0.091	0.200	200	»
0.148	0.175	200	$3CaO \cdot Al_2O_3 \cdot 6H_2O$
0.000	0.047	200	$Ca(OH)_2$
0.000	0.037	250	»
0.045	0.078	250	$Al_2O_3 \cdot H_2O + 4CaO \cdot 3Al_2O_3 \cdot 3H_2O +$ $+ 3CaO \cdot Al_2O_3 \cdot 6H_2O$ (traces)
0.022	0.039	250	»
0.036	0.072	250	»
0.018	0.065	250	$Ca(OH)_2 + CaO \cdot 3Al_2O_3 \cdot 3H_2O$
0.025	0.057	250	»

№ 4213 [1307]

ALUMINUM OXIDE–CALCIUM OXIDE–WATER

$Al_2O_3—CaO—H_2O$

$t=5$

Content in saturated soln., g/l		Comp. of solid phase
Calcium Oxide	Aluminum Oxide	
1.28	<0.002	$Ca(OH)_2+3CaO \cdot Al_2O_3 \cdot 6H_2O$
0.4	0.02	$3CaO \cdot Al_2O_3 \cdot 6H_2O+$ $+3CaO \cdot Al_2O_3 \cdot CaCO_3 \cdot 11H_2O \cdot Ca(OH)_2$
1.28	<0.002	$Ca(OH)_2+4CaO \cdot Al_2O_3 \cdot 19H_2O$
0.59	0.04	$4CaO \cdot Al_2O_3 \cdot 19H_2O+$ $+3CaO \cdot Al_2O_3 \cdot CaCO_3 \cdot 11H_2O \cdot Ca(OH)_2$
0.53	0.06	$4CaO \cdot Al_2O_3 \cdot 19H_2O+2CaO \cdot Al_2O_3 \cdot 10H_2O$
0.51	0.07	$4CaO \cdot Al_2O_3 \cdot 19H_2O+4CaO \cdot Al_2O_3 \cdot 8H_2O$
0.95	0.17	$2CaO \cdot Al_2O_3 \cdot 8H_2O+Al_2O_3$
0.14	0.09	$CaO \cdot Al_2O_3 \cdot 6H_2O+Al_2O_3$
0.22	0.12	$3CaO \cdot Al_2O_3 \cdot 6H_2O+Al_2O_3$
0.25	0.07	$3CaO \cdot Al_2O_3 \cdot 6H_2O+CaO \cdot Al_2O_3 \cdot 10H_2O$

№ 4214 [1356]

ALUMINUM OXIDE–CALCIUM OXIDE–WATER

$Al_2O_3—CaO—H_2O$

$t=1$

Content in saturated soln., g/l		Comp. of solid phase
Aluminum Oxide	Calcium Oxide	
0.070	0.169	$CaO \cdot Al_2O_3 \cdot 10H_2O+CaCO_3+Al_2O_3 \cdot 3H_2O$
0.119	0.222	$CaO \cdot Al_2O_3 \cdot 10H_2O+$ $+3CaO \cdot Al_2O_3 \cdot CaCO_3 \cdot 11H_2O$
0.118	0.290	»
0.130	0.368	$CaO \cdot Al_2O_3 \cdot 10H_2O+2CaO \cdot Al_2O_3 \cdot 8H_2O$
0.161	0.339	$2CaO \cdot Al_2O_3 \cdot 8H_2O+$ $+3CaO \cdot Al_2O_3 \cdot CaCO_3 \cdot 11H_2O$
0.182	0.324	»
0.081	0.419	$2CaO \cdot Al_2O_3 \cdot 8H_2O+CaO \cdot Al_2O_3 \cdot 10H_2O$
0.049	0.537	$2CaO \cdot Al_2O_3 \cdot 8H_2O$
0.024	0.630	$2CaO \cdot Al_2O_3 \cdot 8H_2O+4CaO \cdot Al_2O_3 \cdot 13H_2O$
0.040	0.550	»
0.14	0.693	»
0.008	0.889	»
0.001	1.098	»
0.020	0.314	$4CaO \cdot Al_2O_3 \cdot 13H_2O$
0.028	0.396	»
0.026	0.442	»
0.024	0.489	»
0.026	0.565	»

Table № 4214 (continuation)

Content in saturated soln., g/l		Comp. of solid phase
Aluminum Oxide	Calcium Oxide	
0.022	0.610	$4CaO \cdot Al_2O_3 \cdot 13H_2O$
0.009	0.758	»
0.008	0.868	»
0.001	0.971	»
0.002	1.103	»
0.117	0.235	$3CaO \cdot Al_2O_3 \cdot 6H_2O + 3CaO \cdot Al_2O_3 \cdot CaCO_3 \cdot 11H_2O$
0.134	0.268	»
0.166	0.360	$2CaO \cdot Al_2O_3 \cdot 8H_2O + 3CaO \cdot Al_2O_3 \cdot 6H_2O + 3CaO \cdot Al_2O_3 \cdot CaCO_3 \cdot 11H_2O$
0.169	0.349	»
0.100	0.338	$3CaO \cdot Al_2O_3 \cdot 6H_2O + 3CaO \cdot Al_2O_3 \cdot CaCO_3 \cdot 11H_2O$
0.028	0.319	»
0.040	0.515	$3CaO \cdot Al_2O_3 \cdot 6H_2O + 4CaO \cdot Al_2O_3 \times 13H_2O + 3CaO \cdot Al_2O_3 \cdot CaCO_3 \cdot 11H_2O$
0.010	0.155	$3CaO \cdot Al_2O_3 \cdot 6H_2O + 3CaO \cdot Al_2O_3 \cdot CaCO_3 \cdot 11H_2O$
0.018	0.822	$3CaO \cdot Al_2O_3 \cdot 6H_2O + 3CaO \cdot Al_2O_3 \times CaCO_3 \cdot 11H_2O + 4CaO \cdot Al_2O_3 \cdot 13H_2O$
0.003	1.055	$3CaO \cdot Al_2O_3 \cdot 6H_2O + 3CaO \cdot Al_2O_3 \cdot CaCO_3 \cdot 11H_2O$
0.001	1.371	$Ca(OH)_2 + 3CaO \cdot Al_2O_3 \cdot 6H_2O + 4CaO \cdot Al_2O_3 \cdot 13H_2O$

№ 4215 ALUMINUM OXIDE–ARSENIC PENTOXIDE–WATER [2143]

$Al_2O_3—As_2O_5—H_2O$

Comp. of saturated soln., wt%			d	t	Comp. of solid phase
Aluminum Oxide	Arsenic Pentoxide	Water			
0.4	74.3	25.3	—	60	$3As_2O_5 \cdot 5H_2O + Al_2O_3 \cdot 3As_2O_5 \cdot 10H_2O$
0.4	72.9	26.7	—	60	$Al_2O_3 \cdot 3As_2O_5 \cdot 10H_2O$
0.4	71.3	28.3		60	»
0.15	71.0	28.5	2.15	60	»
0.6	70.3	29.1	2.09	60	»
0.6	67.3	32.1	—	60	»
0.6	66.3	33.1	2.06	60	»
0.6	65.2	34.2	2.01	60	»
1.5	63.6	34.9	1.95	60	»
1.05	63.2	35.75	1.96	60	»
2.3	60.4	37.3	1.96	60	»
2.5	58.9	38.6	1.93	60	»
1.5	64.5	34.0	1.98	60	$Al_2O_3 \cdot 3As_2O_5 \cdot 10H_2O + Al_2O_3 \cdot 3As_2O_5 \cdot 16H_2O$
1.05	63.2	35.75	1.97	60	»
1.4	61.5	37.1	1.93	60	»

Table № 4215 (continuation)

Comp. of saturated soln., wt%			d	t	Comp. of solid phase
Aluminum Oxide	Arsenic Pentoxide	Water			
1.2	63.05	35.75	1.91	60	$Al_2O_3 \cdot 3As_2O_5 \cdot 16H_2O$
2.0	58.5	39.5	1.83	60	»
1.9	56.1	42.0	1.84	60	»
2.0	54.8	43.2	1.78	60	»
2.2	54.1	43.7	—	60	»
2.6	54.2	43.2	1.70	60	»
2.7	49.7	47.6	—	60	»
4.3	46.5	49.2	—	60	»
5.5	45.7	48.8	1.65	60	»
3.9	52.7	43.4	—	60	$Al_2O_3 \cdot 3As_2O_5 \cdot 16H_2O+$ $+Al_2O_3 \cdot As_2O_5 \cdot 4H_2O$
3.5	50.8	45.7	—	60	»
3.9	48.5	47.6	1.65	60	»
6.1	55.5	38.4	1.83	60	$Al_2O_3 \cdot As_2O_5 \cdot 4H_2O$
4.0	52.1	43.9	1.72	60	»
4.3	45.0	50.7	1.65	60	»
2.6	42.1	55.3	—	60	»
2.3	41.0	56.7	1.49	60	»
3.9	38.8	57.3	1.47	60	»
2.7	38.8	58.5	1.47	60	»
1.1	32.4	66.5	1.35	60	»
0.7	27.1	72.2	1.26	60	»
0.8	24.1	75.1	1.23	60	»
0.5	17.6	81.9	—	60	»
0.25	17.2	82.55	—	60	»
0.25	14.9	84.85	1.10	60	»
0.3	8.5	91.2	—	60	»
0.25	6.7	93.05	—	60	»
0.05	4.5	95.45	—	60	»
Traces	0.2	99.8	—	60	$Al_2O_3 \cdot As_2O_5 \cdot 4H_2O+Al_2O_3$
0.0	Traces	100	—	60	»
Traces	70.3	29.7	2.16	20	$Al_2O_3 \cdot 3As_2O_5 \cdot 16H_2O$
0.0	66.7	33.3	2.03	20	»
0.4	65.8	33.8	2.03	20	»
Traces	65.7	34.3	2.07	20	»
0.4	64.6	35.0	1.99	20	»
0.3	63.7	36.0	—	20	»
0.3	61.25	38.45	1.77	20	»
0.3	59.5	40.2	—	20	»
0.35	58.6	41.05	—	20	»
1.0	57.65	41.35	1.80	20	»
0.6	57.3	42.1	1.76	20	»
0.4	54.6	45.0	—	20	»
0.25	53.1	46.65	1.67	20	
0.65	49.1	50.25	1.54	20	»
0.5	47.9	51.6	—	20	»
0.8	41.85	57.35	1.47	20	»
1.6	32.6	65.8	1.32	20	
3.6	30.4	66.0	1.36	20	
1.8	29.0	69.2	—	20	

Comp. of saturated soln., wt%			d	t	Comp. of solid phase
Aluminum Oxide	Arsenic Pentoxide	Water			
3.3	15.1	81.6	—	20	
2.1	27.85	70.05	—	20	$Al_2O_3 \cdot 3As_2O_5 \cdot 16H_2O + Al_2O_3 \cdot As_2O_5 \cdot 4H_2O$
2.5	27.7	69.8	—	20	»
2.8	33.6	63.6	—	20	$Al_2O_3 \cdot As_2O_5 \cdot 4H_2O$
2.1	26.9	71.0	1.30	20	»
1.2	17.7	81.1	1.14	20	»
1.9	15.8	82.3	1.13	20	»
2.0	13.1	84.9	—	20	»
0.8	10.8	88.4	—	20	»
0.6	8.65	90.75	—	20	»
Traces	2.55	97.45	—	20	»
0.0	1.2	98.8	1.0	20	»
0.0	0.2	99.8	1.1	20	$Al_2O_3 + Al_2O_3 \cdot As_2O_5 \cdot 4H_2O$

№ 4216 [2120]

ALUMINUM OXIDE–BARIUM OXIDE–WATER

Al_2O_3—BaO—H_2O

$t=20$

Concn. in aqueous soln., wt%		Comp. of solid phase
Barium Oxide	Aluminum Oxide	
0.0100	0.0020	$Al_2O_3 \cdot xH_2O$
0.5187	0.0373	»
0.9016	0.1179	»
1.205	0.2105	$Al_2O_3 \cdot xH_2O + Al_2O_3 \cdot BaO \cdot 6H_2O$
1.330	0.279	$Al_2O_3 \cdot BaO \cdot 6H_2O$
1.654	0.345	»
1.732	0.378	»
1.873	0.456	»
2.134	0.630	$Al_2O_3 \cdot B_2O \cdot 6H_2O + Al_2O_3 \cdot 2BaO \cdot 5H_2O$
2.211	0.641	$Al_2O_3 \cdot 2BaO \cdot 5H_2O$
3.511	0.671	$Al_2O_3 \cdot 2BaO \cdot 5H_2O + Ba(OH)_2 \cdot 8H_2O$
3.450	0.430	$Ba(OH)_2 \cdot 8H_2O$
3.390	0.002	»
3.320	0.001	»

№ 4217 [1355]

ALUMINUM OXIDE–BARIUM OXIDE–WATER

Al_2O_3—BaO—H_2O

$t=30$

Concn. in aqueous soln., g/l		Comp. of solid phase
Barium Oxide	Aluminum Oxide	
4.8	0.20	$Al_2O_3 \cdot 3H_2O$
9.7	0.39	»
14.2	0.70	»
19.0	0.88	»
23.9	1.06	»
29.1	1.40	»
34.2	1.69	»
38.8	2.00	»
44.7	2.56	»
49.6	2.69	»
52.0	2.80	$2BaO \cdot Al_2O_3 \cdot 5H_2O + Al_2O_3 \cdot 3H_2O$
53.7	2.75	$2BaO \cdot Al_2O_3 \cdot 5H_2O$
55.4	2.70	$2BaO \cdot Al_2O_3 \cdot 5H_2O + Ba(OH)_2 \cdot 8H_2O$
54.1	1.42	$Ba(OH)_2 \cdot 8H_2O$
52.9	0.0	»

№ 4218 [1930]

CALCIUM SULFOALUMINATE–CALCIUM OXIDE–WATER

$Al_2O_3 \cdot 3CaO \cdot 3CaSO_4 \cdot 32H_2O$—$CaO$—$H_2O$

$t=25$

Concn. in aqueous soln., wt%		Concn. in aqueous soln., wt%	
Calcium Sulfoaluminate	Calcium Oxide	Calcium Sulfoaluminate	Calcium Oxide
0.0375	0.00365	0.0115	0.0168
0.0255	0.0056	0.0080	0.0224
0.0165	0.0112	0.0030	0.067

№ 4219 [2765]

ALUMINUM HYDROXIDE–POTASSIUM HYDROXIDE–WATER

$Al(OH)_3$—KOH—H_2O

$t=30$

Comp. of saturated soln., wt%			Comp. of solid phase
Potassium Hydroxide	Aluminum Hydroxide	Water	
30.21	15.58	54.21	$K_2O \cdot Al_2O_3 \cdot 3H_2O$
30.87	12.51	56.62	»
31.65	10.90	57.45	»

Table № 4219 (continuation)

Comp. of saturated soln., wt%			Comp. of solid phase
Potassium Hydroxide	Aluminum Hydroxide	Water	
32.03	9.79	58.18	$K_2O \cdot Al_2O_3 \cdot 3H_2O$
32.40	8.62	58.98	»
35.67	4.68	59.65	»
38.38	3.93	57.69	»
39.40	3.01	57.59	»
41.25	2.20	56.55	»
43.86	1.73	54.41	»
44.74	1.26	54.00	»
45.41	1.80	52.79	»
46.06	1.25	52.69	»
47.83	0.92	51.25	»

№ 4220 [2012]

ALUMINUM HYDROXIDE–ALUMINUM SULFATE–WATER
$Al(OH)_3—Al_2(SO_4)_3—H_2O$

Comp. of saturated soln., wt%			t	Comp. of solid phase
Aluminum Sulfate	Aluminum Hydroxide	Water		
2.31	0.15	97.54	20	$Al_2O_3 \cdot SO_3 \cdot 9H_2O$
4.74	0.28	94.98	20	»
6.50	0.60	92.90	20	»
8.24	1.18	90.58	20	Transition point
8.99	1.10	89.91	20	$Al_2O_3 \cdot 2SO_3 \cdot 12H_2O$
12.93	0.89	86.18	20	»
16.47	1.15	82.38	20	»
19.62	1.88	78.50	20	»
22.44	2.77	74.79	20	»
23.27	3.09	73.64	20	Transition point
24.31	2.02	73.67	20	$Al_2O_3 \cdot 3SO_3 \cdot 16H_2O$
25.60	0.68	73.72	20	»
4.90	1.25	93.85	40	$Al_2O_3 \cdot SO_3 \cdot 9H_2O$
8.00	1.64	90.36	40	$Al_2O_3 \cdot 2SO_3 \cdot 12H_2O$
8.95	1.48	89.57	40	»
12.88	1.20	85.92	40	»
16.37	1.76	81.87	40	»
19.41	2.95	77.64	40	»
21.22	4.32	74.46	40	Transition point
22.23	3.24	74.43	40	$Al_2O_3 \cdot 3SO_3 \cdot 16H_2O$
25.62	1.17	73.21	40	»
32.75	0.40	66.85	40	»
3.11	0.72	96.17	60	$Al_2O_3 \cdot SO_3 \cdot 9H_2O$
7.93	2.27	89.80	60	$Al_2O_3 \cdot 2SO_3 \cdot 12H_2O$
11.06	1.61	87.33	60	»
18.92	2.47	78.61	60	»
23.12	3.58	73.30	60	»
28.56	4.05	67.39	60	$Al_2O_3 \cdot 3SO_3 \cdot 16H_2O$
32.95	0.94	66.11	60	»

№ 4221 [1674]

ALUMINUM HYDROXIDE–POTASSIUM HYDROXIDE–WATER

$Al(OH)_3—KOH—H_2O$

Comp. of saturated soln., wt%			t	Comp. of solid phase
K_2O	Al_2O_3	H_2O		
7.77	0.95	91.28	30	$Al(OH)_3$
21.00	2.62	76.38	30	»
28.80	12.71	58.49	30	»
29.10	16.70	54.20	30	»
29.63	16.60	53.77	30	Potassium Aluminate
37.21	4.08	58.71	30	»
7.92	1.90	90.18	60	$Al(OH)_3$
21.60	10.24	68.16	60	»
25.64	15.92	58.44	60	»
27.81	19.42	52.77	60	»
28.27	21.55	50.18	60	»
38.33	7.08	54.59	60	Potassium Aluminate

№ 4222 [2681]

ALUMINUM THIOCYANATE–SILVER CHLORIDE–WATER

$Al(CNS)_3—AgCl—H_2O$

Comp. of solvent, wt%		Solubility of Silver Chloride, wt%	t
Aluminum Thiocyanate	Water		
9.09	90.91	1.98	25

№ 4223 [1544]

ALUMINUM NITRATE–ALUMINUM FLUORIDE–WATER

$Al(NO_3)_3—AlF_3—H_2O$

$t=25$

Comp. of saturated soln., wt%			Comp. of solid phase
Aluminum Fluoride	Aluminum Nitrate	Water	
0.0	39.43	60.57	$Al(NO_3)_3 \cdot 9H_2O$
0.17	38.83	61.00	»
2.65	38.53	58.82	»
3.95	38.33	57.72	»
5.17	37.92	56.91	»
5.35	37.92	56.73	$Al(NO_3)_3 \cdot 9H_2O+AlF_3 \cdot 3H_2O$
5.41	37.88	56.71	»
5.29	31.05	63.66	$AlF_3 \cdot 3H_2O$
5.20	24.47	70.33	»
5.09	19.46	75.45	»
4.51	15.20	70.29	»
3.99	10.79	85.33	»
2.78	5.78	91.44	»
0.41	0.0	99.59	»

№ 4224 [2122]

ALUMINUM NITRATE–POTASSIUM NITRATE–WATER

$Al(NO_3)_3—KNO_3—H_2O$

$t=25$

Comp. of saturated soln., wt%			Comp. of solid phase
Potassium Nitrate	Aluminum Nitrate	Water	
0.00	39.25	60.75	$Al(NO_3)_3 \cdot 9H_2O$
4.83	37.93	57.24	»
9.41	36.15	54.44	»
13.94	35.23	50.83	$Al(NO_3)_3 \cdot 9H_2O+KNO_3$
18.97	29.81	51.22	KNO_3
20.73	25.19	54.08	»
22.12	20.05	57.83	»
25.91	11.43	62.66	»
26.89	8.75	64.36	»
27.31	0.00	72.69	»

№ 4225 [300]

ALUMINUM NITRATE–POTASSIUM NITRATE–WATER

$Al(NO_3)_3—KNO_3—H_2O$

Comp. of saturated soln., wt%			t	d_4^t	Comp. of solid phase
Potassium Nitrate	Aluminum Nitrate	Water			
6.47	36.0	57.53	0	1.406	$Al(NO_3)_3 \cdot 9H_2O+KNO_3$
9.80	34.8	55.40	20	1.412	»
14.21	36.5	49.29	40	—	»
19.53	38.5	41.97	60	—	»

№ 4226 [407]

ALUMINUM NITRATE–POTASSIUM NITRATE–THALLIUM IODATE–WATER

$Al(NO_3)_3—KNO_3—TlIO_3—H_2O$

$t=18—20$

Content in soln., g-eq/l		Solubility of Thallium Iodate, g/l
Potassium Nitrate	Aluminum Nitrate	
0.100	0.000	0.81
0.075	0.025	0.84
0.050	0.050	0.88
0.025	0.075	0.92
0.000	0.100	0.94

№ 4227 [860]

ALUMINUM NITRATE–CALCIUM FLUORIDE–WATER

$Al(NO_3)_3—CaF_2—H_2O$

$t=25$

Concn. of Aluminum Nitrate in aqueous soln., mol/l	Solubility of Calcium Fluoride in aqueous soln. of Aluminum Nitrate, mol/l
0.3	0.0524
0.1	0.0317
0.01	0.00681
0.003	0.00244
0.001	0.00096

№ 4228 ALUMINUM NITRATE–IRON (III) NITRATE–WATER [2127]

$Al(NO_3)_3—Fe(NO_3)_3—H_2O$

Comp. of saturated soln., wt%			t	Comp. of solid phase
Iron Nitrate	Aluminum Nitrate	Water		
0.00	37.81	62.19	0	$Al(NO_3)_3 \cdot 9H_2O$
7.31	30.25	62.44	0	»
17.57	17.22	65.21	0	»
23.66	12.33	64.01	0	»
25.09	11.02	63.89	0	$Al(NO_3)_3 \cdot 9H_2O + Fe(NO_3)_3 \cdot 9H_2O$
34.25	2.10	63.65	0	$Fe(NO_3)_3 \cdot 9H_2O$
40.15	0.00	59.85	0	»
0.00	39.25	60.75	25	$Al(NO_3)_3 \cdot 9H_2O$
8.77	24.28	66.95	25	»
10.03	22.85	67.12	25	»
17.39	19.95	62.66	25	»
23.87	16.09	60.04	25	»
29.28	15.21	55.51	25	$Al(NO_3)_3 \cdot 9H_2O + Fe(NO_3)_3 \cdot 9H_2O$
30.09	12.15	57.76	25	$Fe(NO_3)_3 \cdot 9H_2O$
36.01	7.12	56.87	25	»
42.13	2.97	54.90	25	»
44.09	1.15	54.76	25	»
46.57	0.00	53.43	25	»
0.00	46.25	53.75	40	$Al(NO_3)_3 \cdot 9H_2O$
5.22	39.83	54.95	40	»
15.13	31.22	53.65	40	»
23.06	23.30	53.64	40	»
32.69	17.12	50.19	40	$Al(NO_3)_3 \cdot 9H_2O + Fe(NO_3)_3 \cdot 9H_2O$
37.66	12.16	50.18	40	$Fe(NO_3)_3 \cdot 9H_2O$
39.79	8.39	51.82	40	»
51.18	0.00	48.82	40	»

№ 4229 ALUMINUM NITRATE–IRON (III) NITRATE–WATER [304]

$Al(NO_3)_3—Fe(NO_3)_3—H_2O$

Comp. of saturated soln., wt%			d	t	Comp. of solid phase
Aluminum Nitrate	Iron Nitrate	Water			
37.81	0.00	62.19	—	0	$Al(NO_3)_3 \cdot 9H_2O$
32.27	4.71	63.02	1.362	0	$Al(NO_3)_3 \cdot 9H_2O + Fe(NO_3)_3 \cdot 9H_2O$
26.65	9.55	63.80	—	0	»
23.46	12.87	63.67	1.375	0	»
21.10	15.57	63.33	1.378	0	»
17.34	20.42	62.24	—	0	»
15.24	23.02	61.74	—	0	»

Table № 4229 (continuation)

Comp. of saturated soln., wt%			d	t	Comp. of solid phase
Aluminum Nitrate	Iron Nitrate	Water			
11.60	26.66	61.74	1.397	0	$Al(NO_3)_3 \cdot 9H_2O + Fe(NO_3)_3 \cdot 9H_2O$
7.94	30.64	61.42	1.405	0	»
4.40	34.08	61.52	1.412	0	»
0.00	40.15	59.85	—	0	$Fe(NO_3)_3 \cdot 9H_2O$
38.56	0.00	61.00	—	20	$Al(NO_3)_3 \cdot 9H_2O$
33.75	6.26	59.99	—	20	$Al(NO_3)_3 \cdot 9H_2O + Fe(NO_3)_3 \cdot 9H_2O$
24.36	15.97	59.67	—	20	»
18.43	23.00	58.57	—	20	»
12.92	28.57	58.51	—	20	»
11.72	30.34	57.94	—	20	»
10.56	31.85	57.59	—	20	»
6.35	37.36	56.29	—	20	»
2.40	40.80	56.80	—	20	»
1.05	42.60	56.35	—	20	»
0.00	45.20	54.80	—	20	$Fe(NO_3)_3 \cdot 9H_2O$
46.25	0.00	53.75	—	40	$Al(NO_3)_3 \cdot 9H_2O$
37.40	8.05	54.55	—	40	$Al(NO_3)_3 \cdot 9H_2O + Fe(NO_3)_3 \cdot 9H_2O$
32.74	12.84	54.42	—	40	»
22.13	23.51	54.56	—	40	»
17.55	29.28	53.17	—	40	»
15.04	31.20	53.76	—	40	»
10.63	37.01	52.36	—	40	»
10.10	38.60	51.30	—	40	»
9.13	39.85	51.02	—	40	»
8.31	41.00	50.69	—	40	»
5.32	45.19	49.49	—	40	»
1.42	40.03	49.55	—	40	»
0.00	51.18	48.82	—	40	$Fe(NO_3)_3 \cdot 9H_2O$

№ 4230 ALUMINUM NITRATE–SILVER SULFATE–WATER [2079]

$Al(NO_3)_3—Ag_2SO_4—H_2O$

$t=25$

Comp. of saturated soln., wt%			Comp. of saturated soln., wt%		
Aluminum Nitrate	Silver Sulfate	Water	Aluminum Nitrate	Silver Sulfate	Water
0.11	0.91	98.92	2.06	1.68	96.26
0.22	0.98	98.80	5.95	2.34	91.71
0.63	1.20	98.17	8.99	2.81	88.20
1.05	1.37	97.58	15.27	3.01	81.72

№ 4231 [407]

VARIOUS METAL NITRATES–THALLIUM IODATE–WATER

—$TlIO_3$—H_2O

t=18—20

Nitrate			Solubility of Thallium Iodate in aqueous soln. of Nitrate, g/l
Name	Formula	Concn. in aqueous soln., g-eq/l	
Potassium Nitrate	KNO_3	0.000	0.59
»	»	0.050	0.69
»	»	0.100	0.81
»	»	0.500	1.22
»	»	1.000	1.47
Zinc Nitrate	$Zn(NO_3)_2$	0.000	0.59
»	»	0.050	0.80
»	»	0.125	0.94
»	»	0.250	1.09
»	»	0.500	1.30
Aluminum Nitrate	$Al(NO_3)_2$	0.000	0.59
»	»	0.050	0.84
»	»	0.100	0.94
»	»	0.500	1.44
»	»	1.000	1.93

№ 4232 [55]

ALUMINUM NITRATE HYDRATE–LEAD (II) IODIDE–WATER

$Al(NO_3)_3 \cdot 9H_2O$—PbI_2—H_2O

t=20

Concn. of Aluminum Nitrate Hydrate in aqueous soln., mol/l	Solubility of Lead Iodide in aqueous soln. of Aluminum Nitrate Hydrate, mol/l
0.20	$4.30 \cdot 10^{-3}$
0.60	$6.07 \cdot 10^{-3}$
1.00	$6.65 \cdot 10^{-3}$

№ 4233 [460]

ALUMINUM NITRATE–THORIUM NITRATE–WATER

$Al(NO_3)_3$—$Th(NO_3)_4$—H_2O

t=20

Comp. of saturated soln., wt%			d_4	Comp. of solid phase
Aluminum Nitrate	Thorium Nitrate	Water		
40.63	0.00	59.37	1.400	$Al(NO_3)_3 \cdot 9H_2O$
33.29	11.28	55.43	0.00	»
22.88	29.13	47.99	1.550	»
15.62	43.42	40.96	1.731	$Al(NO_3)_3 \cdot 9H_2O + Th(NO_3)_4 \cdot 6H_2O$
0.00	64.06	35.94	1.915	$Th(NO_3)_4 \cdot 6H_2O$

№ 4234 [1544]

ALUMINUM FLUORIDE–ALUMINUM SULFATE–WATER

$AlF_3—Al_2(SO_4)_3—H_2O$

$t=25$

Comp. of saturated soln., wt%			Comp. of solid phase
Aluminum Fluoride	Aluminum Sulfate	Water	
0.0	28.22	71.78	$Al_2(SO_4)_3 \cdot 17H_2O$
0.68	28.46	70.86	$Al_2(SO_4)_3 \cdot 17H_2O+Al_2(SO_4)_3 \cdot AlF_3 \cdot 15H_2O$
0.80	28.59	70.61	»
0.80	28.63	70.57	$Al_2(SO_4)_3 \cdot AlF_3 \cdot 15H_2O$
1.18	26.61	72.21	»
1.47	24.85	73.68	»
1.93	22.55	75.52	»
3.05	20.07	76.88	»
3.75	18.95	77.30	»
4.60	17.82	77.58	»
5.33	17.45	77.22	»
5.97	17.12	76.91	$Al_2(SO_4)_3 \cdot AlF_3 \cdot 15H_2O+AlF_3 \cdot 3H_2O$
5.98	17.08	76.94	»
5.51	14.97	79.52	$AlF_3 \cdot 3H_2O$
4.46	10.00	85.54	»
3.43	6.18	90.39	»
0.41	0.0	99.59	»

№ 4235 [15 44]

ALUMINUM FLUORIDE–ALUMINUM CHLORIDE–WATER

$AlCl_3—AlF_3—H_2O$

$t=25$

Concn. in aqueous soln., wt%		Comp. of solid phase
Aluminum Chloride	Aluminum Fluoride	
31.10	0.00	$AlCl_3 \cdot 6H_2O$
30.48	1.50	»
30.37	2.64	$AlCl_3 \cdot 6H_2O+AlF_3 \cdot 5AlCl_3 \cdot 38H_2O$
30.21	3.46	»
30.22	3.46	$AlF_3 \cdot 5AlCl_3 \cdot 38H_2O$
29.81	4.37	»
29.47	4.93	»
28.89	5.62	»
28.78	6.03	»
28.72	6.46	»
28.58	6.90	»

Table № 4235 (continuation)

Concn. in aqueous soln., wt%		Comp. of solid phase
Aluminum Chloride	Aluminum Fluoride	
28.47	6.97	$AlF_3 \cdot 5AlCl_3 \cdot 38H_2O + AlF_3 \cdot 3H_2O$
28.45	7.01	»
24.07	6.30	$AlF_3 \cdot 3H_2O$
21.04	5.96	»
14.33	5.52	»
10.68	4.93	»
5.50	3.56	»
0.00	0.41	»

№ 4236 ALUMINUM FLUORIDE – POTASSIUM FLUORIDE – WATER [857]

$AlF_3—KF—H_2O$

$t=25$

Concn. in aqueous soln., g/l		Comp. of solid phase	Concn. in aqueous soln., g/l		Comp. of solid phase
Potassium Fluoride	Aluminum Fluoride		Potassium Fluoride	Aluminum Fluoride	
0.852	1.588	$AlF_3 \cdot 2KF \cdot H_2O$	8.0	Traces	$AlF_3 \cdot 2KF \cdot H_2O$
0.7592	0.6128	»	11.145	»	$AlF_3 \cdot 2.3KF$
1.075	0.1153	»	13.782	»	$AlF_3 \cdot 2.4KF$
1.328	0.026	»	17.228	»	$AlF_3 \cdot 2.7KF$
2.385	0.042	»	24.244	»	$AlF_3 \cdot 3KF$
2.488	0.006	»	37.336	»	»
3.732	Traces	»	50.6	»	»
4.985	»	»			

№ 4237 ALUMINUM SULFATE–POTASSIUM SULFATE–WATER [2132]

$Al_2(SO_4)_3 \cdot 18H_2O—K_2SO_4—H_2O$

Comp. of saturated soln., wt%			t	Comp. of solid phase
Aluminum Sulfate 18-hydrate	Potassium Sulfate	Water		
19.23	1.85	78.92	0	$K_2Al_2(SO_4)_4 \cdot 24H_2O + Al_2(SO_4)_3$
44.43	1.67	53.90	20	»
46.81	1.81	51.38	35	»
53.98	2.59	43.43	50	»
58.80	4.39	36.81	65	»
61.26	6.01	32.73	77	»

Table № 4237 (continuation)

Comp. of saturated soln., wt%			t	Comp. of solid phase
Aluminum Sulfate 18-hydrate	Potassium Sulfate	Water		
0.47	7.01	92.52	0	$K_2Al_2(SO_4)_4 \cdot 24H_2O + K_2SO_4$
0.79	6.93	92.28	0.5	»
1.46	7.78	90.76	5	»
1.66	8.65	89.69	10	»
1.82	9.67	88.51	15	»
3.33	12.45	84.22	30	»
5.97	13.18	80.85	40	»
9.53	14.78	75.69	50	»
16.77	16.04	66.19	60	»
28.55	17.47	53.98	70	»
40.76	20.20	39.04	80	»

№ 4238 **[1282]**

ALUMINUM SULFATE–POTASSIUM SULFATE–WATER

$Al_2(SO_4)_3—K_2SO_4—H_2O$

$t=25$

Comp. of saturated soln., wt%			Comp. of solid phase
Potassium Sulfate	Aluminum Sulfate	Water	
0.00	27.70	72.30	$Al_2(SO_4)_3 \cdot 18H_2O$
0.98	23.02	76.00	»
1.93	30.27	67.80	$Al_2(SO_4)_3 \cdot 18H_2O + K_2Al_2(SO_4)_4 \cdot 24H_2O$
1.94	27.86	70.20	$K_2Al_2(SO_4)_4 \cdot 24H_2O$
1.46	15.98	82.56	»
1.45	11.28	87.27	»
1.50	10.22	88.28	»
1.72	8.43	89.85	»
1.78	7.07	91.15	»
1.87	6.13	92.00	»
2.05	4.74	93.21	»
2.27	4.47	93.26	»
2.99	3.58	93.43	»
3.61	2.99	93.40	»
5.73	2.03	92.24	»
7.25	1.85	90.90	»
10.60	1.45	87.95	»
11.47	1.35	87.18	$K_2Al_2(SO_4)_3 \cdot 24H_2O + K_2SO_4$
10.69	0.76	87.55	K_2SO_4
10.75	0.00	89.25	»

№ 4245 [2274

ALUMINUM SULFATE–IRON (II) SULFATE–WATER

$Al_2(SO_4)_3$—$FeSO_4$—H_2O

$t=25$

Comp. of saturated soln., wt%			Comp. of solid phase
Iron Sulfate	Aluminum Sulfate	Water	
2.11	26.75	71.14	$Al_2(SO_4)_3 \cdot 18H_2O$
3.85	25.64	70.51	»
4.13	25.41	70.46	$Al_2(SO_4)_3 \cdot 18H_2O + Al_2(SO_4)_3 \cdot FeSO_4 \cdot 24H_2O$
4.19	25.21	70.60	$Al_2(SO_4)_3 \cdot FeSO_4 \cdot 24H_2O$
5.57	23.90	71.53	»
6.60	23.20	70.20	»
7.12	22.64	70.24	»
8.06	21.62	70.32	»
9.36	20.80	70.84	»
10.17	20.16	69.67	$Al_2(SO_4)_3 \cdot FeSO_4 \cdot 24H_2O + FeSO_4 \cdot 7H_2O$
10.60	19.37	70.03	$FeSO_4 \cdot 7H_2O$
15.05	12.29	72.66	»
15.57	11.51	72.92	»
18.68	6.98	74.34	»
19.14	6.06	74.80	»
22.98	0.00	77.02	»

№ 4246 [2592]

ALUMINUM SULFATE–IRON (II) SULFATE–WATER

$Al_2(SO_4)_3$—$FeSO_4$—H_2O

$t=25$

Concn. in aqueous soln., wt%		Concn. in aqueous soln., wt%		Concn. in aqueous soln., wt%	
Aluminum Sulfate	Iron Sulfate	Aluminum Sulfate	Iron Sulfate	Aluminum Sulfate	Iron Sulfate
27.85	0.00	26.96	1.268	25.69	3.198
27.33	0.5447	26.33	2.415	25.28	3.908

№ 4247 [2751]

ALUMINUM SULFATE–IRON (III) SULFATE–WATER

$Al_2(SO_4)_3$—$Fe_2(SO_4)_3$—H_2O

Comp. of saturated soln., wt%			Comp. of solid phase
Iron Sulfate	Aluminum Sulfate	Water	
0.00	27.82	72.18	$Al_2(SO_4)_3 \cdot 18H_2O$
6.064	26.01	67.926	»
9.819	24.21	65.971	»
13.02	21.64	65.34	»
23.28	15.22	61.50	»

Table № 4247 (continuation)

Comp. of saturated soln., wt%			Comp. of solid phase
Iron Sulfate	Aluminum Sulfate	Water	
31.90	10.46	57.64	$Al_2(SO_4)_3 \cdot 18H_2O + Fe_2(SO_4)_3 \cdot 9H_2O$
32.42	10.03	57.55	$Fe_2(SO_4)_3 \cdot 9H_2O$
34.02	8.819	57.161	»
35.82	6.626	57.554	»
38.83	5.200	55.970	»
42.44	2.342	55.218	»
44.97	0.00	55.03	»

№ 4248 ALUMINUM SULFATE–IRON (III) SULFATE–WATER [1687]

$Al_2(SO_4)_3$—$Fe_2(SO_4)_3$—H_2O

Comp. of saturated soln., wt%			t	Comp. of solid phase
Aluminum Sulfate	Iron Sulfate	Water		
26.69	0.00	73.31	20	$Al_2(SO_4)_3 \cdot 18H_2O$
26.10	1.868	72.03	20	»
23.76	5.463	70.80	20	»
20.10	8.575	71.32	20	»
17.05	11.38	71.57	20	»
14.08	14.88	71.04	20	$Al_2(SO_4)_3 \cdot 16H_2O$
13.32	15.56	71.12	20	»
9.442	16.89	73.67	20	»
7.330	18.13	74.54	20	»
4.701	22.71	72.59	20	$Al_2(SO_4)_3 \cdot 16H_2O + Fe_2(SO_4)_3 \cdot 9H_2O$
5.703	20.63	73.67	20	»
3.658	25.88	70.46	20	$Fe_2(SO_4)_3 \cdot 9H_2O$
3.118	27.23	69.65	20	»
1.119	35.38	63.50	20	»
28.97	0.000	71.03	30	$Al_2(SO_4)_3 \cdot 18H_2O$
26.68	3.590	69.73	30	»
21.80	7.803	70.40	30	»
17.95	12.50	69.55	30	$Al_2(SO_4)_3 \cdot 18H_2O + Al_2(SO_4)_3 \cdot 16H_2O$
11.31	16.97	71.72	30	$Al_2(SO_4)_3 \cdot 16H_2O$
7.501	20.50	72.00	30	»
5.613	23.69	70.70	30	»
4.024	26.23	69.75	30	$Al_2(SO_4)_3 \cdot 16H_2O + Fe_2(SO_4)_3 \cdot 9H_2O$
4.037	25.18	70.78	30	$Fe_2(SO_4)_3 \cdot 9H_2O$
5.661	35.21	59.13	30	»
3.735	26.25	70.01	30	»
3.796	27.95	68.25	30	»

$Al_2(SO_4)_3$—$NiSO_4$—H_2O

Comp. of saturated soln., wt%			t	Comp. of solid phase
Nickel Sulfate	Aluminum Sulfate	Water		
0.00	27.80	72.20	25	$Al_2(SO_4)_3 \cdot 18H_2O$
5.58	25.20	69.22	25	»
11.71	21.49	66.80	25	$Al_2(SO_4)_3 \cdot 18H_2O + NiSO_4 \cdot 7H_2O$
11.62	21.51	66.87	25	»
12.51	21.64	65.85	25	$NiSO_4 \cdot 7H_2O$
13.56	21.53	64.91	25	»
16.48	14.96	68.56	25	»
25.30	4.84	69.86	25	»
28.80	0.00	71.20	25	»
0.00	31.30	68.70	40	$Al_2(SO_4)_3 \cdot 18H_2O$
9.27	25.50	65.23	40	»
11.86	23.85	64.29	40	»
15.18	21.24	63.58	40	»
15.57	20.74	63.69	40	»
16.28	20.42	63.30	40	$Al_2(SO_4)_3 \cdot 18H_2O + NiSO_4 \cdot 6H_2O\alpha$
16.44	20.83	62.73	40	$NiSO_4 \cdot 6H_2O\alpha$
20.21	15.86	63.93	40	»
26.67	7.29	66.04	40	»
32.48	0.00	67.52	40	»
0.00	36.98	63.02	60	$Al_2(SO_4)_3 \cdot 18H_2O$
7.08	30.50	62.42	60	»
12.84	25.98	61.18	60	»
16.80	22.97	60.23	60	»
18.60	21.30	60.10	60	»
19.47	20.48	60.25	60	$Al_2(SO_4)_3 \cdot 18H_2O + NiSO_4 \cdot 6H_2O\beta$
20.20	20.95	58.85	60	$NiSO_4 \cdot 6H_2O\beta$
28.30	10.10	61.60	60	»
34.68	0.00	63.52	60	»
0.00	37.10	62.90	65	$Al_2(SO_4)_3 \cdot 18H_2O$
11.87	27.10	61.03	65	»
15.90	23.89	60.21	65	»
17.96	22.73	59.31	65	»
19.76	20.04	60.20	65	$Al_2(SO_4)_3 \cdot 18H_2O + NiSO_4 \cdot 6H_2O\beta$
20.58	19.47	59.95	65	$NiSO_4 \cdot 6H_2O\beta$
21.08	18.71	60.21	65	»
29.50	9.68	60.82	65	»
37.60	0.00	62.40	65	»

№ 4253 **ALUMINUM SULFATE–COPPER (II) SULFATE–WATER** [1374]

$Al_2(SO_4)_3$—$CuSO_4$—H_2O

Comp. of saturated soln., wt%			t	Comp. of solid phase
Copper Sulfate	Aluminum Sulfate	Water		
0.0	27.80	72.20	25	$Al_2(SO_4)_3 \cdot 18H_2O$
2.50	26.60	70.90	25	»
3.83	25.64	70.53	25	»
5.18	24.86	69.96	25	$Al_2(SO_4)_3 \cdot 18H_2O + CuSO_4 \cdot 5H_2O$
5.32	24.65	70.03	25	$CuSO_4 \cdot 5H_2O$
6.11	23.18	70.71	25	»
8.18	18.83	72.99	25	»
11.33	13.05	75.62	25	»
14.69	7.50	77.81	25	»
16.59	3.79	79.62	25	»
18.49	0.00	81.51	25	»
0.00	28.86	71.14	30	$Al_2(SO_4) \cdot 18H_2O$
3.39	26.83	69.78	30	»
5.74	25.22	69.04	30	»
6.30	24.64	69.06	30	$CuSO_4 \cdot 5H_2O$
7.67	22.00	70.33	30	»
9.79	17.65	72.56	30	»
12.13	13.80	74.07	30	»
16.70	5.97	77.33	30	»
18.65	2.75	78.60	30	»
20.33	0.00	79.67	30	»

№ 4254 **ALUMINUM SULFATE–COPPER (II) SULFATE–WATER** [2274]

$Al_2(SO_4)_3$—$CuSO_4$—H_2O

$t = 25$

Comp. of saturated soln., wt%			Comp. of saturated soln., wt%		
Aluminum Sulfate	Copper Sulfate	Water	Aluminum Sulfate	Copper Sulfate	Water
27.80	0.0	72.20	23.18	6.11	70.71
26.60	2.50	70.90	18.83	8.18	72.99
25.66	3.75	70.59	18.10	8.75	73.15
25.64	3.83	70.53	13.05	11.33	75.62
24.86	5.18	69.96	7.50	14.69	77.81
24.65	5.32	70.03	3.79	16.59	79.62
24.32	5.36	70.32	0.0	18.49	81.51

Table № 4277 (continuation)

Concn. in aqueous soln., wt%		t	d	Comp. of solid phase
Calcium Oxide	Phosphorus Pentoxide			
2.307	39.97	25	1.4216	$Ca(H_2PO_4)_2 \cdot H_2O$
2.008	41.48	25	1.4360	»
2.007	41.48	25	—	»
1.708	43.05	25	1.4506	»
1.708	43.02	25	—	»
1.365	45.03	25	1.4700	»
1.064	46.97	25	—	»
0.695	50.26	25	1.5307	»
0.475	53.04	25	1.5676	»
0.320	56.44	25	1.6143	»
0.249	59.21	25	1.6549	»
0.201	61.66	25	1.6932	$Ca(H_2PO_4)_2+Ca(H_2PO_4)_2 \cdot H_2O$
0.190	62.56	25	1.7090	$Ca(H_2PO_4)_2$
0.090	65.55	25	1.7522	»
0.091	65.54	25	1.7518	»
0.051	70.24	25	—	»
0.054	70.42	25	1.8406	»
4.760	21.48	40	1.2480	$CaHPO_4$
5.785	26.47	40	—	$Ca(H_2PO_4)_2 \cdot H_2O+CaHPO_4$
3.922	35.08	40	—	$Ca(H_2PO_4)_2 \cdot H_2O$
2.825	39.76	40	1.4206	»
1.376	47.55	40	1.4998	»
1.371	47.69	40	1.4964	»
0.923	50.89	40	1.5400	»
0.920	50.89	40	1.5390	»
0.737	52.25	40	1.5604	»
0.621	53.90	40	1.5823	»
0.441	57.40	40	1.6338	»
0.437	57.64	40	1.6365	»
0.414	58.32	40	1.6376	»
0.205	62.30	40	1.6960	$Ca(H_2PO_4)_2 \cdot H_2O+Ca(H_2PO_4)_2$
0.214	61.94	40	1.6917	»
0.195	62.64	40	1.7016	»
0.123	65.62	40	1.7512	$Ca(H_2PO_4)_2$
0.120	65.76	40	1.7537	»
0.115	65.82	40	1.7548	»
0.086	69.30	40	1.8183	»
0.089	69.34	40	1.8216	»
0.084	69.35	40	1.8206	»
0.748	3.23	75	1.007	$CaHPO_4$
1.972	10.15	75	—	»
1.963	10.14	75	1.080	»
3.165	16.96	75	1.159	»
4.337	23.95	75	1.246	»
5.387	32.47	75	1.357	»
4.980	37.21	75	1.399	$Ca(H_2PO_4)_2 \cdot H_2O$
4.289	40.18	75	—	»
4.276	40.18	75	1.428	»
3.964	41.57	75	1.440	»
3.676	42.71	75	1.449	»
3.356	44.03	75	1.461	»

Table № 4277 (continuation)

Concn. in aqueous soln., wt%		t	d	Comp. of solid phase
Calcium Oxide	Phosphorus Pentoxide			
2.986	45.91	75	1.482	$Ca(H_2PO_4)_2 \cdot H_2O$
2.622	47.72	75	1.496	»
2.152	50.31	75	1.526	»
1.740	53.33	75	—	»
1.731	53.16	75	1.557	»
1.408	55.97	75	1.595	»
0.999	58.25	75	1.624	$Ca(H_2PO_4)_2$
0.693	60.68	75	1.654	»
0.578	61.85	75	1.668	»
0.379	64.96	75	1.718	»
0.380	65.07	75	1.718	»
0.267	70.12	75	1.806	»
0.266	70.11	75	1.806	»
0.871	5.04	100	1.0143	$CaHPO_4$
2.407	14.90	100	1.107	»
3.958	24.70	100	1.224	»
4.996	32.89	100	1.332	»
5.117	33.88	100	1.342	»
5.538	40.27	100	—	$CaHPO_4 + Ca(H_2PO_4)_2 \cdot H_2O$
5.529	40.15	100	1.432	»
5.432	40.80	100	1.438	$Ca(H_2PO_4)_2 \cdot H_2O$
5.183	41.67	100	1.446	»
4.301	45.37	100	1.476	»
3.411	49.76	100	1.519	»
2.822	52.91	100	1.554	»
2.719	53.57	100	1.564	»
2.479	55.30	100	1.583	»
2.462	55.48	100	1.584	»
2.411	55.98	100	—	»
1.871	57.15	100	1.599	$Ca(H_2PO_4)_2$
1.703	58.00	100	1.610	»
1.569	58.83	100	1.619	»
1.489	59.22	100	1.623	»
1.442	59.55	100	—	»
1.127	61.48	100	1.651	»
0.717	65.35	100	1.708	»
0.719	65.36	100	1.708	»
0.584	70.03	100	1.791	»
0.584	70.04	100	1.791	»

№ 4278 PHOSPHORUS PENTOXIDE–CALCIUM OXIDE–WATER [1473]

P_2O_5—CaO—H_2O

$t=25$

Concn. in aqueous soln., g/l		Comp. of solid phase	Concn. in aqueous soln., g/l		Comp. of solid phase
Ca	PO_4		Ca	PO_4	
0.4809	2.2508	$CaHPO_4 \cdot 2H_2O$	0.3607	1.6810	$CaHPO_4 \cdot 2H_2O$
0.4609	2.1178	»	0.3527	1.6430	»
0.3888	1.8899	»	0.2966	1.4056	»

Table № 4278 (continuation)

Concn. in aqueous soln., g/l		Comp. of solid phase	Concn. in aqueous soln., g/l		Comp. of solid phase
Ca	PO_4		Ca	PO_4	
0.2926	1.3866	$CaHPO_4 \cdot 2H_2O$	0.3567	1.6335	$CaHPO_4$
0.2685	1.2441	»	0.2725	1.2631	»
0.2004	1.0162	»	0.2944	0.9212	»
0.1804	0.8452	»	0.1242	0.5774	Apatite
0.1082	0.4654	»	0.1134	0.5214	»
0.0922	0.3419	»	0.0946	0.4559	»
0.0681	0.3134	»	0.0649	0.3077	»
0.0721	0.2849	»	0.0400	0.2555	»
0.0317	0.0722	»	0.0413	0.2327	»
0.4609	2.2318	$CaHPO_4$	0.0345	0.2042	»
0.4329	2.0703	»	0.0248	0.1766	»
0.3928	1.8614	»	0.0092	0.1159	»
0.3487	1.6525	»			

№ 4279 PHOSPHORUS PENTOXIDE–CALCIUM OXIDE–WATER [419]

P_2O_5—CaO—H_2O

Comp. of saturated soln., wt%			t	Comp. of solid phase
Calcium Oxide	Phosphorus Pentoxide	Water		
1.45	5.09	93.46	40	$CaHPO_4$
2.78	11.00	86.22	40	»
3.61	15.42	80.97	40	»
4.55	20.18	75.27	40	»
5.78	26.47	67.75	40	$CaHPO_4 + Ca(H_2PO_4)_2 \cdot H_2O$
5.51	27.85	66.64	40	$Ca(H_2PO_4)_2 \cdot H_2O$
4.66	31.78	63.60	40	»
4.35	33.00	62.65	40	»
4.30	35.54	60.16	40	»
2.82	40.00	57.18	40	»
1.90	44.34	53.76	40	»
1.80	45.00	53.20	40	»
1.40	47.45	51.15	40	»
1.38	48.26	50.36	40	»
1.15	49.69	49.16	40	»
1.12	50.00	48.88	40	»
0.55	55.00	44.45	40	»
0.27	61.00	38.73	40	»
0.16	63.70	36.14	40	$Ca(H_2PO_4)_2$
0.75	3.23	96.02	75	$CaHPO_4$
1.97	10.15	87.82	75	»
3.17	16.96	78.87	75	»
4.34	23.95	71.67	75	$CaHPO_4$
5.39	32.47	62.14	75	»
4.98	37.21	57.81	75	$Ca(H_2PO_4)_2 \cdot H_2O$
4.28	40.18	55.54	75	»
3.68	42.71	53.51	75	»
2.99	45.91	51.50	75	»

Table № 4279 (continuation)

Comp. of saturated soln., wt%			t	Comp. of solid phase
Calcium Oxide	Phosphorus Pentoxide	Water		
2.63	47.72	50.35	75	$Ca(H_2PO_4)_2 \cdot H_2O$
2.15	50.31	47.54	75	»
1.73	53.16	45.11	75	»
1.41	55.97	42.62	75	»
1.00	58.25	40.75	75	$Ca(H_2PO_4)_2$
0.69	60.68	38.63	75	»
0.58	61.85	37.57	75	»
0.42	64.00	35.58	75	»
2.41	14.90	82.69	100	$CaHPO_4$
3.96	24.70	71.34	100	»
5.12	33.88	61.00	100	»
5.53	40.15	54.32	100	$CaHPO_4 + Ca(H_2PO_4)_2 \cdot H_2O$
5.43	40.80	53.77	100	$Ca(H_2PO_4)_2 \cdot H_2O$
5.18	41.67	53.15	100	»
4.30	45.37	51.33	100	»
4.00	47.00	49.00	100	»
3.65	48.50	47.85	100	»
3.41	49.76	46.83	100	»
2.82	52.91	44.27	100	»
2.72	53.57	43.51	100	»
2.48	55.30	42.12	100	»
2.00	56.70	41.30	100	$Ca(H_2PO_4)_2$
1.70	58.00	40.30	100	»
1.49	59.22	39.29	100	»
1.13	61.48	37.39	100	»
0.90	63.50	35.60	100	»

№ 4280 [944]

PHOSPHORUS PENTOXIDE–CALCIUM OXIDE–WATER

P_2O_5—CaO—H_2O

$t=40$

Comp. of saturated soln., wt%			Comp. of solid phase
Calcium Oxide	Phosphorus Pentoxide	Water	
1.91	5.80	92.29	$CaHPO_4 \cdot 2H_2O$
2.75	8.70	88.55	»
3.48	11.23	85.28	»
3.91	12.74	83.35	»
4.52	14.83	80.65	»
2.82	11.57	85.61	$CaHPO_4$
3.01	12.48	84.51	»
3.17	13.37	83.46	»
3.58	14.66	81.76	»
3.57	14.85	81.58	»
3.54	14.93	81.53	»

№ 4281 PHOSPHORUS PENTOXIDE–CALCIUM OXIDE–WATER [1171]

P_2O_5—CaO—H_2O

Comp. of saturated soln., wt%			t	Comp. of solid phase
Calcium Oxide	Phosphorus Pentoxide	Water		
0.00	57.50	42.50	0	$H_8P_2O_9$
0.112	57.20	42.688	0	$H_8P_2O_9 + CaH_4P_2O_8 \cdot H_2O$
0.716	46.94	52.254	0	$CaH_4P_2O_8 \cdot H_2O$
2.859	35.40	61.741	0	»
4.482	27.96	67.558	0	»
6.008	20.55	73.442	0	$CaH_4P_2O_8 \cdot H_2O + CaHPO_4 \cdot 2H_2O$
4.822	15.47	79.708	0	$CaHPO_4 \cdot 2H_2O$
2.23	6.365	91.405	0	»
0.204	63.02	36.776	25	$H_8P_2O_9 + CaH_4P_2O_8 \cdot H_2O$
5.885	24.40	69.715	25	$CaH_4P_2O_8 \cdot H_2O + CaHPO_4$
6.58	22.62	70.80	25	$CaH_4P_2O_8 \cdot H_2O + CaHPO_4 \cdot 2H_2O$
5.770	29.56	63.86	50	$CaH_4P_2O_8 \cdot H_2O + CaHPO_4$
3.235	9.513	87.252	— 2.6	$CaHPO_4 \cdot 2H_2O$
5.251	16.25	78.499	— 5.5	»
5.819	18.33	75.851	— 7.0	»
6.10	19.62	74.28	— 7.5	$CaHPO_4 \cdot 2H_2O + CaH_4P_2O_8 \cdot H_2O$
6.07	19.50	74.43	— 7.5	»
5.883	19.86	74.257	— 8.0	$CaH_4P_2O_8 \cdot H_2O$
5.47	21.10	73.43	— 9.5	»
3.638	28.95	67.312	— 20.0	»
1.657	35.40	62.943	— 40.0	»
Traces	46.2	53.80	—100.0	$CaH_4P_2O_8 \cdot H_2O + H_8P_2O_9$

№ 4282 PHOSPHORUS PENTOXIDE–CALCIUM SULFATE–WATER [876]

P_2O_5—$CaSO_4$—H_2O

Concn. in aqueous soln., wt%		t	d_4	Comp. of solid phase
Phosphorus Pentoxide	Calcium Sulfate			
0.00	0.27	25	1.006	$CaSO_4$
8.24	0.73	25	1.071	»
21.05	0.75	25	1.175	»
24.82	0.66	25	1.201	»
31.33	0.51	25	1.273	»
35.04	0.39	25	1.310	»
42.40	0.27	25	0.000	»
44.77	0.23	25	1.449	»
46.57	0.16	25	1.470	»
48.80	0.12	25	1.478	»
55.56	0.02	25	1.914	»
63.51	Traces	25	0.000	»
36.80	0.80	25	0.000	$CaSO_4 \cdot ½H_2O$
41.13	0.57	25	1.377	»
47.00	0.32	25	1.445	»

Table № 4282 (continuation)

Concn. in aqueous soln., wt%		t	d_4	Comp. of solid phase
Phosphorus Pentoxide	Calcium Sulfate			
51.20	0.20	25	1.519	$CaSO_4 \cdot {}^1/_2H_2O$
56.06	0.09	25	1.520	»
61.15	0.09	25	1.638	»
0.00	0.21	25	1.004	$CaSO_4 \cdot 2H_2O$
0.00	0.19	25	1.010	»
4.70	0.51	25	1.035	»
4.77	0.46	25	1.035	»
8.24	0.58	25	1.070	»
10.07	0.66	25	1.079	»
10.07	0.58	25	1.079	»
19.99	0.68	25	1.150	»
18.37	0.70	25	1.148	»
18.80	0.70	25	1.164	»
19.43	0.58	25	1.171	»
25.09	0.66	25	1.215	»
25.39	0.61	25	1.220	»
29.65	0.56	25	1.273	»
31.54	0.53	25	1.292	»
34.14	0.49	25	1.314	»
39.40	0.39	25	1.385	»
44.48	0.32	25	1.444	»
46.70	0.29	25	1.490	»
47.60	0.28	25	1.500	»
51.22	0.22	25	1.520	»
55.00	0.17	25	1.530	»
0.00	0.17	80	1.000	$CaSO_4$
1.97	0.44	80	1.004	»
2.92	0.51	80	1.009	»
3.91	0.58	80	1.024	»
4.93	0.58	80	1.029	»
5.88	0.68	80	1.038	»
6.78	0.88	80	1.050	»
8.08	1.00	80	1.065	»
16.00	1.09	80	1.127	»
30.65	0.85	80	1.274	»
31.49	0.92	80	1.283	»
32.24	0.80	80	1.292	»
40.80	0.51	80	1.395	»
41.30	0.58	80	1.400	»
41.60	0.54	80	1.404	»
54.70	0.17	80	1.585	»
56.43	0.10	80	1.615	»
61.63	Traces	80	1.693	»
24.80	1.615	80	1.200	$CaSO_4 \cdot {}^1/_2H_2O$
28.50	1.52	80	1.239	»
29.00	1.53	80	1.250	»
31.48	1.42	80	1.269	»
33.86	1.29	80	1.299	»
38.30	1.06	80	1.360	»
40.01	0.90	80	1.365	»

Table № 4282 (continuation)

Concn. in aqueous soln., wt%		t	d_4	Comp. of solid phase
Phosphorus Pentoxide	Calcium Sulfate			
40.00	0.91	80	1.365	$CaSO_4 \cdot 1/2H_2O$
42.00	0.85	80	1.399	»
0.00	0.19	80	1.000	$CaSO_4 \cdot 2H_2O$
1.00	0.44	80	1.000	»
1.97	0.58	80	1.004	»
2.15	0.56	80	1.005	»
2.92	0.66	80	1.008	»
3.91	0.75	80	1.024	»
4.98	0.83	80	1.028	»
5.88	0.88	80	1.038	»
6.78	0.95	80	1.050	»
7.78	1.02	80	1.060	»
8.11	1.17	80	1.065	»
10.40	1.22	80	1.077	»
12.53	1.34	80	1.090	»
14.02	1.31	80	1.098	»
18.18	1.40	80	1.140	»
20.00	1.44	80	1.160	»
20.50	1.41	80	1.160	»
22.60	1.41	80	1.180	»
25.00	1.28	80	1.210	»
25.20	1.40	80	1.210	»
30.34	1.31	80	1.260	»
30.48	1.35	80	1.260	»
34.46	1.26	80	1.320	»
35.50	1.26	80	1.220	»
40.01	1.11	80	1.380	»
44.46	0.95	80	1.430	»

№ 4283 PHOSPHORUS PENTOXIDE–CALCIUM SULFATE–WATER [877]

$P_2O_5—CaSO_4—H_2O$

Concn. in aqueous soln., wt%		t	d_4^t	Comp. of solid phase
Phosphorus Pentoxide	Calcium Sulfate			
30	0.67	40	—	$CaSO_4$
35	0.55	40	—	»
40	0.36	40	—	»
45	0.27	40	—	»
50	0.18	40	—	»
30	1.07	40	1.2753	$CaSO_4 \cdot ½H_2O$
35	0.85	40	1.3312	»
40	0.63	40	1.3836	»
45	0.47	40	1.4424	»
50	0.30	40	1.5130	»

Table № 4283 (continuation)

Concn. in aqueous soln., wt%		t	d_4^t	Comp. of solid phase
Phosphorus Pentoxide	Calcium Sulfate			
30	0.80	40	1.2753	$CaSO_4 \cdot 2H_2O$
35	0.68	40	1.3312	»
40	0.53	40	1.3836	»
45	0.46	40	1.4424	»
50	0.39	40	1.5130	»
25	0.85	60	—	$CaSO_4$
30	0.75	60	—	»
35	0.62	60	—	»
40	0.48	60	—	»
45	0.32	60	—	»
50	0.26	60	—	»
25	1.44	60	1.2342	$CaSO_4 \cdot \frac{1}{2}H_2O$
50	1.23	60	1.2841	»
33	1.02	60	1.3274	»
40	0.70	60	1.3786	»
45	0.58	60	1.4412	»
48.5	0.39	60	1.5007	»
50	0.35	60	1.5104	»
26	1.05	60	1.2337	$CaSO_4 \cdot 2H_2O$
30	0.98	60	1.2841	»
35	0.88	60	1.3274	»
40	0.79	60	1.3786	»
45	0.64	60	1.4412	»
50	0.54	60	1.5104	»
30	0.88	90	—	$CaSO_4$
35	0.75	90	—	»
40	0.60	90	—	»
50	0.28	90	—	»
30	1.40	90	1.2888	$CaSO_4 \cdot \frac{1}{2}H_2O$
35	1.20	90	1.3664	»
40	1.00	90	1.4154	»
50	0.48	90	1.5145	»

№ 4284

PHOSPHORUS PENTOXIDE–VANADIUM PENTOXIDE–WATER [2453]

$P_2O_5—V_2O_5—H_2O$

$t=25$

Concn. in aqueous soln., wt%		Comp. of solid phase	Concn. in aqueous soln., wt %		Comp. of solid phase
Vanadium Pentoxide	Phosphorus Pentoxide		Vanadium Pentoxide	Phosphorus Pentoxide	
0.6	5.7	$V_2O_5 \cdot P_2O_5 \cdot 6H_2O$	0.05	38.4	$2V_2O_5 \cdot 3P_2O_5 \cdot 13H_2O$
0.5	12.6	»	0.07	39.8	»
0.2	18.2	»	0.07	45.8	»
0.4	19.0	»	0.07	52.0	»
0.35	28.6	»	0.17	64.6	»
0.25	28.7	»	0.12	64.6	»
0.3	30.9	»			

Table № 4287 (continuation)

Comp. of saturated soln., wt%			t	d_4^t	Comp. of solid phase
Phosphorus Pentoxide	Manganese Oxide	Water			
1.260	0.619	98.121	37	1.011	$MnHPO_4 \cdot 3H_2O$
3.206	1.475	95.319	37	—	»
5.96	2.580	89.48	37	1.058	»
10.57	4.56	84.84	37	—	»
14.26	6.05	79.69	37	—	»
19.67	8.13	72.20	37	1.26	»
22.66	9.50	67.84	37	—	»
28.20	11.70	60.10	37	—	$Mn(H_2PO_4)_2 \cdot 2\frac{1}{2}H_2O$
29.30	10.98	59.72	37	—	»
33.60	8.20	58.20	37	—	»
38.30	5.84	55.86	37	—	»
39.00	5.39	55.61	37	—	»
40.00	4.86	55.14	37	—	»
41.90	4.43	53.67	37	—	»
43.10	3.92	52.98	37	—	»
48.00	2.88	49.12	37	—	$Mn(H_2PO_4)_2 \cdot 2\frac{1}{2}H_2O$
50.90	2.35	46.75	37	—	»
52.40	2.04	45.56	37	—	»
56.90	1.69	41.41	37	—	»
59.60	1.94	38.46	37	—	»
60.60	1.19	38.21	37	—	$Mn(H_2PO_4)_2 \cdot H_3PO_4$
61.70	0.71	37.59	37	—	»
62.10	0.78	37.12	37	—	»
63.0	0.31	36.69	37	—	»
63.50	0.52	35.98	37	—	»
64.0	0.29	35.71	37	—	»
64.40	0.28	35.32	37	—	»
65.0	0.25	34.75	37	—	»
1.25	0.57	98.19	50	1.005	$Mn_3(PO_4)_2 \cdot 2MnHPO_4 \cdot 5H_2O$
2.95	1.21	95.84	50	1.027	»
4.40	1.76	93.84	50	1.046	»
4.73	1.85	93.42	50	—	»
6.89	2.94	90.17	50	—	»
7.81	3.05	89.14	50	1.082	»
9.28	3.56	87.16	50	1.10	»
10.50	4.04	85.46	50	1.12	»
20.20	7.99	71.81	50	1.27	»
21.04	8.28	70.68	50	—	»
24.20	9.41	66.39	50	—	»
25.30	9.65	65.05	50	—	»
28.70	11.44	59.86	50	—	»
33.40	9.98	56.62	50	—	$Mn(H_2PO_4)_2 \cdot 2\frac{1}{2}H_2O$
33.50	9.36	57.14	50	—	»
34.10	8.87	57.03	50	—	»
41.40	5.70	52.90	50	—	»
42.70	5.25	52.05	50	—	»
51.10	3.24	45.66	50	—	»
53.50	2.73	43.77	50	—	»
55.40	2.54	42.06	50	—	»
56.80	2.24	40.96	50	—	»

Table № 4287 (continuation)

Comp. of saturated soln., wt%			t	d_4^t	Comp. of solid phase
Phosphorus Pentoxide	Manganese Oxide	Water			
60.20	2.26	37.52	50	—	$Mn(H_2PO_4)_2 \cdot H_3PO_4$
61.0	2.08	36.92	50	—	»
62.80	0.99	36.21	50	—	»
63.40	0.87	35.73	50	—	»
4.29	1.46	94.25	75	—	$Mn_3(PO_4)_2 \cdot 2MnHPO_4 \cdot 5H_2O$
4.41	1.49	91.10	75	—	»
8.25	2.58	89.23	75	—	»
14.29	5.16	80.55	75	—	»
16.91	5.95	77.15	75	—	»
21.10	7.58	71.32	75	—	»
27.80	10.17	62.03	75	—	»
30.40	11.04	58.56	75	—	»
34.50	13.02	52.48	75	—	»
35.50	12.45	52.05	75	—	»
35.70	12.89	51.41	75	—	»
46.10	6.63	47.27	75	—	$Mn(H_2PO_4)_2 \cdot 2\frac{1}{2}H_2O$
47.70	5.73	46.57	75	—	»
52.60	5.35	42.05	75	—	»
57.60	3.86	38.54	75	—	»
58.60	3.33	38.07	75	—	$Mn(H_2PO_4)_2$
59.80	2.98	37.22	75	—	»
60.20	2.53	37.27	75	—	»
60.90	2.21	37.70	75	—	»
61.60	2.20	36.20	75	—	»
62.20	1.65	36.15	75	—	»
63.60	1.58	34.82	75	—	$Mn(H_2PO_4)_2$
63.90	1.70	34.40	75	—	»
2.58	0.78	96.64	85	—	$Mn_3(PO_4)_2 \cdot 2MnHPO_4 \cdot 5H_2O$
3.46	1.02	95.52	85	—	»
5.92	1.82	92.26	85	—	»
11.56	3.48	84.96	85	—	»
22.22	7.18	70.60	85	—	»
27.32	9.68	63.00	85	—	»

№ 4288 [216]

PHOSPHORUS PENTOXIDE–MANGANESE (II) OXIDE–WATER

P_2O_5—MnO—H_2O

$t=25$

Comp. of saturated soln., wt%			Comp. of solid phase
Manganese Oxide	Phosphorus Pentoxide	Water	
0.07	0.03	99.83	$MnHPO_4 \cdot 3H_2O$
0.60	12.30	87.10	»
1.60	28.00	70.40	»
2.00	32.40	65.60	»
2.20	41.50	56.30	»

№ 4292 [1910]

PHOSPHORUS PENTOXIDE–IRON (III) OXIDE–WATER

$P_2O_5—Fe_2O_3—H_2O$

$t=25$

Comp. of saturated soln., wt%		d	Comp. of solid phase
Phosphorus Pentoxide	Iron Oxide		
0.350	0.003	1.00	$Fe_2O_3 \cdot P_2O_5 \cdot 5H_2O$
1.312	0.015	1.00	»
2.32	0.033	1.00	»
3.34	0.053	1.03	»
4.39	0.075	1.01	»
4.61	0.081	1.04	»
6.40	0.139	1.05	»
7.90	0.212	1.05	»
9.23	0.268	1.07	»
11.50	0.479	1.07	»
15.02	1.009	1.07	»
17.23	1.92	1.07	»
19.96	3.12	1.10	»
22.13	4.20	1.15	»
24.50	6.35	1.19	»
26.61	7.92	1.32	$Fe_2O_3 \cdot 2P_2O_5 \cdot 8H_2O$
28.10	7.00	—	»
29.90	5.72	—	»
33.90	4.50	1.42	»
37.00	4.11	—	»
43.00	4.09	—	»
45.20	4.46	1.61	»
45.80	4.51	—	»
48.10	4.78	1.70	$Fe_2O_3 \cdot 3P_2O_5 \cdot 10H_2O$
50.30	4.05	1.76	»
51.50	3.61	—	»
53.80	3.14	—	»
57.30	2.62	—	»
59.10	1.98	—	$Fe_2O_3 \cdot 3P_2O_5 \cdot 6H_2O$
60.0	1.44	1.86	»
61.50	1.114	—	»
64.0	0.919	—	»
66.30	0.907	—	»

№ 4293 PHOSPHORUS PENTOXIDE–IRON (III) OXIDE–WATER [805]

P_2O_5—Fe_2O_3—H_2O

Comp. of saturated soln., wt%			t	Comp. of saturated soln., wt%		
Phosphorus Pentoxide	Iron Oxide	Water		Phosphorus Pentoxide	Iron Oxide	Water
58.45	1.52	40.03	40	63.21	10.70	26.09
32.75	4.23	63.02	40	62.08	7.55	30.27
23.89	2.49	73.62	40	62.54	8.46	29.00
19.12	3.00	77.88	40	60.00	10.45	29.55
17.06	3.76	79.18	40	67.65	28.14	4.21
15.77	4.16	80.07	40	65.31	34.69	0.00
15.17	4.97	79.86	40	65.52	34.48	0.00
13.59	6.34	80.07	40	70.95	29.05	0.00
12.00	8.00	80.00	40	60.29	30.69	9.02
11.70	10.80	78.50	40	60.74	33.26	6.00
10.63	13.41	75.96	40	54.54	31.37	14.09
9.14	13.43	77.43	40	55.24	30.64	14.12
4.50	15.90	79.60	40	10.00	62.72	27.28
3.83	17.00	79.17	40	5.87	37.02	57.11
3.72	17.20	79.08	40	8.32	53.48	38.20
1.78	19.00	79.22	40	12.46	56.64	30.90
48.68	7.94	43.38	60	71.19	8.37	20.44
44.24	6.93	48.83	60	51.61	20.93	27.46
41.17	5.09	53.74	60	—	—	—
38.32	5.78	55.90	60	43.30	13.32	43.38
35.90	6.37	57.73	60	—	—	—
30.18	5.00	64.82	60	—	—	—
29.21	4.83	65.96	60	—	—	—
26.22	5.14	68.04	60	58.84	28.72	12.44
24.55	9.39	66.06	60	57.39	25.41	17.20
21.39	10.46	68.15	60	52.77	22.52	27.41
17.31	20.29	62.40	60	46.47	18.50	35.03
13.65	18.09	68.26	60	25.19	37.76	37.05
11.63	19.29	69.08	60	10.39	37.82	51.79
5.19	20.25	74.56	60	—	—	—
4.39	23.65	71.96	60	8.53	41.56	49.91
2.16	29.58	68.26	60	6.05	41.97	51.98
17.83	2.12	24.05	80	—	—	—
73.62	1.60	24.78	80	78.78	4.01	17.21
70.41	3.77	25.82	80	71.97	6.54	21.49
68.78	4.01	27.21	80	72.56	2.03	25.41
67.70	7.69	24.61	80	62.76	25.66	11.58
64.36	7.06	28.58	80	65.48	14.00	20.52
57.06	11.18	31.76	80	60.17	20.12	19.71
54.30	14.35	31.35	80	59.26	20.00	20.74
51.45	17.15	31.40	80	58.72	25.98	15.30
34.49	13.00	52.51	80	41.48	21.45	37.07
20.96	14.80	64.24	80	47.41	40.45	12.14
19.21	16.44	64.35	80	40.24	33.01	26.75
8.59	19.56	71.85	80	7.99	40.04	51.97
6.55	21.57	71.88	80	8.78	40.27	50.35
6.45	23.06	70.49	80	10.10	46.01	43.89
4.14	23.43	72.43	80	—	—	—
3.38	24.73	71.89	80	8.62	61.70	29.68
2.82	27.68	69.50	80	1.75	49.58	48.67

№ 4294 [1448]

PHOSPHORUS PENTOXIDE–TETRAMMINECOBALT (III) HYDROGEN PHOSPHATE OXIDE WATER

P_2O_5—$[Co(NH_3)_4HPO_4]_2O$—H_2O

t=25

Concn. in aqueous soln., wt%		Concn. in aqueous soln., wt%		Concn. in aqueous soln., wt%	
Tetrammine-cobalt Hydrogen phosphate Oxide	Phosphorus Pentoxide	Tetrammine-cobalt Hydrogen phosphate Oxide	Phosphorus Pentoxide	Tetrammine-cobalt Hydrogen phosphate Oxide	Phosphorus Pentoxide
56.68	4.11	56.68	11.86	57.86	17.45
56.13	5.28	55.63	12.66	57.35	19.62
56.72	6.52	55.74	14.72	58.49	20.83
55.79	8.43	56.18	16.88	58.57	22.57
55.98	8.92				

№ 4295 [1448]

PHOSPHORUS PENTOXIDE–AQUOPENTAMMINECOBALT (III) TRIOXIDE–WATER

P_2O_5—$[Co(NH_3)_5 \cdot H_2O]_2O_3$—$H_2O$

Concn. in aqueous soln., wt%		t	Concn. in aqueous soln., wt%		t
Aquopentammine-cobalt Trioxide	phosphorus Pentoxide		Aquopentammine-cobalt Trioxide	phosphorus Pentoxide	
22.30	22.87	25	26.30	25.48	0
23.07	23.12	25	28.30	27.91	0
23.96	24.34	25	31.40	31.67	0
27.92	27.60	25	32.70	32.68	0
27.84	28.48	25	34.20	34.27	0
29.98	30.30	25	32.28	35.12	0
34.42	35.60	25	30.41	35.23	0
35.87	37.23	25	26.53	38.48	0
35.67	38.67	25	22.98	41.20	0
29.42	42.09	25	21.31	43.13	0
22.99	47.93	25	17.56	47.32	0
18.21	54.64	25	15.36	49.60	0
3.795	3.59	0	14.21	50.91	0
11.54	11.31	0	11.47	56.00	0
15.21	16.26	0	9.98	57.82	0
16.12	16.70	0	8.87	59.85	0
24.00	23.20	0	8.92	60.97	0
25.10	24.33	0			

№ 4296 [1148]

PHOSPHORUS PENTOXIDE–HEXAMMINECOBALT (III) TRIOXIDE–WATER

P_2O_5—$[Co(NH_3)_6]_2O_3$—H_2O

$t=25$

Concn. in aqueous soln., wt%		Concn. in aqueous soln., wt%		Concn. in aqueous soln., wt%	
Hexammine-cobalt Trioxide	Phosphorus Pentoxide	Hexammine-cobalt Trioxide	Phosphorus Pentoxide	Hexammine-cobalt Trioxide	Phosphorus Pentoxide
0.663	0.533	26.88	27.13	11.85	45.41
2.269	3.036	28.46	29.90	9.98	47.90
6.85	8.03	30.75	30.62	8.34	51.33
7.95	8.78	26.63	32.54	7.68	53.16
11.53	13.01	24.11	34.88	6.23	57.71
16.20	18.21	21.24	36.35	5.33	58.77
18.50	19.97	18.12	38.60	2.60	65.04
23.48	24.90	13.79	42.36		

№ 4297 [1754]

PHOSPHORUS PENTOXIDE–COPPER (II) OXIDE–WATER

P_2O_5—CuO—H_2O

$t=60$

Comp. of saturated soln., wt%			Comp. of solid phase
Copper Oxide	Phosphorus Pentoxide	Water	
>3.6	>59.2	<37.2	$CuO \cdot P_2O_5 \cdot 2H_2O$
3.6	59.2	37.2	$CuO \cdot P_2O_5 \cdot 2H_2O + 2CuO \cdot P_2O_5 \cdot 3H_2O$
0.56	2.15	97.29	$2CuO \cdot P_2O_5 \cdot 3H_2O + 4CuO \cdot P_2O_5 \cdot H_2O$
<0.56	>2.15	>97.29	$4CuO \cdot P_2O_5 \cdot H_2O$

№ 4298 [414]

PHOSPHORUS PENTOXIDE–COPPER (II) OXIDE–WATER

P_2O_5—CuO—H_2O

$t=25$

Comp. of saturated soln., wt%			Comp. of solid phase
Copper Oxide	Phosphorus Pentoxide	Water	
0.72	1.65	97.63	$Cu_3(PO_4)_2 \cdot 3H_2O$
3.88	9.1	87.02	»
6.48	14.9	78.62	»
7.04	16.07	76.89	$Cu_3(PO_4)_2 \cdot 3H_2O + CuHPO_4 \cdot H_2O$
7.1	17.03	75.87	$CuHPO \cdot H_2O$
6.8	19.89	73.31	»
6.0	26.6	67.4	»

Table № 4298 (continuation)

Comp. of saturated soln., wt%			Comp. of solid phase
Copper Oxide	Phosphorus Pentoxide	Water	
5.9	35.4	58.7	$CuHPO \cdot H_2O$
4.72	43.37	51.91	»
5.2	47.88	46.92	»
5.0	53.26	41.74	»
6.1	53.9	40.0	$CuHPO_4 \cdot H_2O + Cu(H_2PO_4)_2$
2.05	57.86	40.09	$Cu(H_2PO_4)_2$
0.6	63.5	35.9	»

№ 4299 PHOSPHORUS PENTOXIDE–ZINC OXIDE–WATER [1539]

P_2O_5—ZnO—H_2O

Comp. of saturated soln., wt%			t	Comp. of solid phase
Zinc Oxide	Phosphorus Pentoxide	Water		
2.38	5.08	92.54	25	$Zn_3(PO_4)_2 \cdot 4H_2O$
4.65	9.76	85.59	25	»
6.13	12.42	81.45	25	»
6.56	13.52	79.92	25	»
6.74	14.00	78.26	25	»
6.97	14.37	78.66	25	»
7.37	14.83	77.80	25	»
7.71	15.98	76.31	25	$ZnHPO_4 \cdot 3H_2O$
8.26	17.15	74.59	25	»
8.73	18.33	72.94	25	»
10.74	22.75	66.51	25	»
12.47	26.48	61.05	25	»
13.48	28.70	57.82	25	»
14.16	30.09	55.75	25	»
15.40	32.55	52.05	25	»
15.82	33.79	50.39	25	»
17.30	37.15	45.55	25	»
18.04	39.61	42.35	25	»
16.14	42.05	41.81	25	$Zn(H_2PO_4)_2 \cdot 2H_2O$
13.20	44.53	42.27	25	»
9.58	48.70	41.32	25	»
7.64	52.25	40.11	25	»
7.23	55.97	36.80	25	»
2.08	4.87	93.05	37	$Zn_3(PO_4)_2 \cdot 4H_2O$
4.12	9.46	86.42	37	»
6.27	13.60	90.13	37	»
8.78	18.13	73.09	37	»
9.66	19.48	70.86	37	»
10.16	20.32	69.52	37	»

Comp. of saturated soln., wt%			t	Comp. of solid phase
Zinc Oxide	Phosphorus Pentoxide	Water		
10.88	21.96	67.16	37	$ZnHPO_4 \cdot 3H_2O$
13.26	26.75	59.99	37	»
14.77	29.65	55.58	37	»
17.06	33.39	49.55	37	»
17.92	34.58	47.50	37	»
16.00	36.13	47.87	37	$ZnHPO_4 \cdot 4H_2O$
15.78	37.80	46.42	37	»
16.12	39.93	43.95	37	»
15.81	42.42	41.77	37	»
16.82	42.65	40.53	37	»
17.83	44.89	37.28	37	»
18.05	46.11	35.84	37	$Zn(H_2PO_4)_2 \cdot 2H_2O$
14.74	46.41	38.85	37	»
12.55	48.99	38.46	37	»
11.26	51.35	37.19	37	»
11.12	51.92	36.96	37	»
10.82	54.32	34.86	37	»

№ 4300 PHOSPHORUS PENTOXIDE–ZINC OXIDE–WATER [2459]

$$P_2O_5—ZnO—H_2O$$

Comp. of saturated soln., wt%			t	d_4^t	Comp. of solid phase
Phosphorus Pentoxide	Zinc Oxide	Water			
0.435	0.231	99.334	25	1.003	$Zn_3(PO_4)_2 \cdot 4H_2O$
0.489	0.267	99.244	25	1.004	»
0.714	0.373	98.913	25	1.007	»
0.935	0.484	98.581	25	1.010	»
1.16	0.586	98.254	25	1.010	»
3.13	1.52	95.35	25	1.040	»
4.07	1.95	93.98	25	1.040	»
4.09	1.95	93.96	25	1.040	»
8.72	4.26	87.02	25	1.11	»
11.60	5.59	82.81	25	—	»
14.92	7.28	77.8	25	1.23	»
17.87	8.63	73.5	25	1.28	$ZnHPO_4 \cdot 3H_2O$
19.38	9.30	71.32	25	1.30	»
21.03	10.13	68.84	25	1.33	»
26.40	12.64	60.96	25	—	»
35.50	16.60	47.90	25	—	$ZnHPO_4 \cdot H_2O$
37.30	15.95	46.75	25	1.63	»
39.80	15.91	44.29	25	—	»
40.60	15.75	43.65	25	—	»
44.90	13.06	42.04	25	1.67	$Zn(H_2PO_4)_2 \cdot 2H_2O$

Table № 4300 (continuation)

Comp. of saturated soln., wt%			t	d_4^t	Comp. of solid phase
Phosphorus Pentoxide	Zinc Oxide	Water			
49.60	9.03	41.37	25	—	$Zn(H_2PO_4)_2 \cdot 2H_2O$
53.70	7.66	38.64	25	—	»
55.30	7.23	37.47	25	1.67	»
57.00	7.51	35.49	25	1.72	»
57.30	7.75	34.95	25	—	»
59.10	5.25	35.65	25	—	$Zn(H_2PO_4)_2 \cdot 2H_3PO_4$
60.30	3.72	35.98	25	—	»
62.10	2.45	35.45	25	1.69	»
63.50	1.55	34.95	25	1.69	»
0.467	0.236	99.297	37	1.00	$Zn_3(PO_4)_2 \cdot 4H_2O$
0.907	0.442	91.551	37	1.01	»
0.984	0.508	98.508	37	1.01	»
4.10	1.85	94.05	37	1.05	»
7.69	3.50	88.81	37	1.10	»
7.77	3.49	88.79	37	1.09	»
13.66	6.49	86.34	37	1.19	»
14.49	6.84	78.67	37	1.20	»
17.04	8.26	74.70	37	1.25	»
18.02	8.76	73.22	37	1.27	»
18.02	8.70	73.28	37	1.26	»
19.56	9.70	70.74	37	1.30	$ZnHPO_4 \cdot 3H_2O$
21.07	10.68	68.25	37	1.34	»
24.13	12.27	63.60	37	1.39	»
52.70	11.94	35.36	37	—	$Zn(H_2PO_3)_2 \cdot 2H_2O$
55.10	10.66	34.34	37	—	»
0.84	0.40	98.76	50	1.001	$Zn_3(PO_4)_2 \cdot 4H_2O$
1.61	0.68	97.71	50	1.01	»
4.93	2.01	93.06	50	1.05	»
6.64	2.78	90.58	50	1.07	»
7.67	3.22	89.11	50	1.08	»
10.58	4.62	84.80	50	1.12	»
13.86	6.23	79.91	50	1.14	»
17.56	8.27	74.16	50	1.24	»
17.91	8.51	73.98	50	1.26	»
18.31	8.68	73.01	50	1.27	»
22.04	10.98	66.98	50	1.33	»
22.61	11.40	65.99	50	1.36	»
23.56	11.90	64.54	50	1.37	»
27.40	14.35	58.25	50	1.46	»
27.50	14.41	58.09	50	1.46	»
28.30	14.74	56.96	50	1.45	»
29.60	15.60	54.80	50	1.52	»
30.0	15.97	54.03	50	1.52	»
31.30	16.34	52.36	50	1.54	$ZnHPO_4 \cdot 3H_2O$
34.40	17.48	48.18	50	1.64	»
35.30	17.49	47.21	50	1.62	»
38.70	17.30	44.0	50	1.65	$ZnHPO_4 \cdot H_2O$
41.30	17.73	40.97	50	—	»
41.40	18.01	40.59	50	—	»
43.50	17.34	39.16	50	—	»

Table № 4300 (continuation)

Comp. of saturated soln., wt%			t	d_4^t	Comp. of solid phase
Phosphorus Pentoxide	Zinc Oxide	Water			
45.90	15.98	38.12	50	—	$ZnHPO_4 \cdot H_2O$
49.30	17.20	33.50	50	—	$Zn(H_2PO_4)_2 \cdot 2H_2O$
50.90	15.36	33.74	50	—	»
53.40	12.62	33.98	50	—	»
56.70	11.93	31.37	50	—	»
3.98	1.54	94.48	60	1.02	$Zn_3(PO_4)_2 \cdot 4H_2O$
7.93	3.20	88.87	60	1.08	»
10.43	4.42	85.15	60	1.12	»
16.21	7.52	76.27	60	1.22	»
23.10	11.64	65.26	60	1.36	»
26.60	13.79	59.61	60	1.44	»
28.60	15.37	56.03	60	1.49	»
32.70	16.81	50.49	60	1.58	$ZnHPO_4 \cdot 3H_2O$
36.20	18.42	45.38	60	1.65	»
1.56	0.58	97.86	75	—	$Zn_3(PO_4)_2 \cdot 4H_2O$
7.27	2.68	90.05	75	—	»
7.32	2.66	90.02	75	—	»
13.57	5.71	80.72	75	—	»
18.77	8.72	72.51	75	—	»
24.42	12.33	63.25	75	—	»
24.73	12.59	62.68	75	—	»
25.0	12.80	62.20	75	—	»
25.70	13.27	61.03	75		»
26.30	13.78	59.92	75		»
26.70	13.96	59.34	75		»
27.20	14.25	58.55	75	—	»
29.80	16.32	53.88	75	—	»
31.30	17.49	51.21	75	—	$ZnHPO_4 \cdot H_2O$
35.70	18.47	45.83	75	—	»
45.30	25.20	29.50	75		»
0.77	0.26	98.97	86.3	0.99	$Zn_3(PO_4)_2 \cdot 4H_2O$
1.52	0.50	97.98	86.3	0.99	»
7.36	2.61	90.03	86.3	1.05	»
13.63	5.67	80.70	86.3	1.16	»
15.43	6.66	77.91	86.3	1.16	»
18.06	8.29	73.65	86.3	1.18	»
20.90	10.05	69.05	86.3	1.26	»
22.83	11.38	65.79	86.3	1.31	»
23.51	11.79	64.70	86.3	1.32	»
23.84	11.99	64.17	86.3	—	»
24.02	12.17	63.81	86.3	—	»
26.40	13.60	60.00	86.3	—	»
29.10	15.65	55.25	86.3	—	»
32.90	18.20	48.90	86.3	—	$ZnHPO_4 \cdot H_2O$
38.50	20.09	41.41	86.3	—	»
7.37	2.52	90.11	100	—	$Zn_3(PO_4)_2 \cdot 4H_2O$
13.94	5.81	80.25	100	—	»
17.82	8.08	74.10	100	—	»
21.29	10.33	68.38	100	—	»
21.93	10.88	67.19	100	—	»

№ 4301 PHOSPHORUS PENTOXIDE–STRONTIUM OXIDE–WATER [2630]

$P_2O_5—SrO—H_2O$

$t=25$

Comp. of saturated soln., wt%			d	Comp. of solid phase
Strontium Oxide	Phosphorus Pentoxide	Water		
0.05	61.97	37.98	1.687	$SrH_4(PO_4)_2$
0.13	56.49	43.38	1.594	»
0.44	51.07	48.49	1.527	»
1.78	43.80	54.42	1.455	»
3.24	38.73	58.03	1.418	»
4.61	35.90	59.49	1.410	»
6.52	31.65	61.83	1.394	»
8.08	27.63	64.29	1.379	»
8.42	25.92	65.66	1.360	»
8.88	24.71	66.41	1.357	»
9.25	23.58	67.17	1.351	»
7.84	20.79	71.37	1.294	Sr_2HPO_4
7.52	19.81	72.67	1.278	»
6.53	16.66	76.81	1.230	»
6.32	13.41	80.27	1.194	»
5.11	12.70	82.19	1.169	»
4.13	7.80	88.07	1.112	»
3.18	6.02	90.80	1.083	»
2.67	5.68	91.35	1.074	»

№ 4302 PHOSPHORUS PENTOXIDE–SILVER OXIDE–WATER [1611]

$P_2O_5—Ag_2O—H_2O$

$t=25$

Content in aqueous soln., wt%		Comp. of solid phase	Content in aqueous soln., wt%		Comp. of solid phase
Silver Oxide	Phosphorus Pentoxide		Silver Oxide	Phosphorus Pentoxide	
0.0648	0.0424	Ag_3PO_4	3.84	14.03	Ag_3PO_4
0.0670	0.0428	»	3.99	14.59	»
0.123	0.0834	»	4.18	15.42	»
0.122	0.0832	»	4.52	16.85	»
0.248	0.196	»	5.32	19.56	»
0.249	0.196	»	6.08	22.20	»
0.349	0.315	»	6.09	22.37	»
0.346	0.314	»	6.81	24.60	»
0.638	0.852	»	7.40	26.10	»
0.634	0.857	»	7.97	27.54	$Ag_3PO_4+Ag_2HPO_4$
1.00	1.98	»	8.49	29.44	Ag_2HPO_4
1.47	3.71	»	9.61	33.25	»
1.94	5.76	»	10.51	35.91	»
2.41	7.91	»	12.42	40.14	»
2.84	9.80	»	12.53	40.26	»
3.28	11.75	»	13.34	41.38	»
3.81	13.93	»	15.10	43.24	»

Table № 4302 (continuation)

Content in aqueous soln., wt%		Comp. of solid phase	Content in aqueous soln., wt%		Comp. of solid phase
Silver Oxide	Phosphorus Pentoxide		Silver Oxide	Phosphorus Pentoxide	
15.38	43.84	Ag_2HPO_4	18.43	47.90	AgH_2PO_4
16.17	44.13	»	17.96	49.80	»
16.92	44.70	»	17.71	50.86	»
17.92	45.21	»	17.68	51.43	»
19.41	45.99	»	17.63	51.97	»
19.68	45.89	$Ag_2HPO_4+AgH_2PO_4$	17.72	52.88	»
19.64	45.90	»	17.87	53.37	»
19.61	45.91	»	18.84	54.89	»
20.45	46.17	Ag_2HPO_4	19.99	55.39	»
24.68	46.86	»	24.66	55.60	»
24.97	47.24	»	15.24	61.63	H_3PO_4
19.43	46.30	AgH_2PO_4			

№ 4303 PHOSPHORUS PENTOXIDE–BARIUM OXIDE–WATER [2630]

P_2O_5—BaO—H_2O

$t=25$

Comp. of saturated soln., wt%			d_{25}	Comp. of solid phase
Barium Oxide	Phosphorus Pentoxide	Water		
2.90	2.96	94.14	1.040	$BaHPO_4$
3.66	5.95	90.39	1.085	»
4.33	7.61	88.06	1.107	»
4.80	8.25	86.95	1.118	»
6.31	11.19	82.50	1.165	»
6.09	11.36	82.55	1.166	»
6.84	12.35	80.81	1.188	»
8.50	16.51	74.99	1.246	»
10.86	23.46	65.68	1.354	$BaHPO_4+BaH_4(PO_4)_2$
9.14	28.36	62.50	1.327	»
7.41	32.68	59.91	1.420	»
6.93	33.74	59.33	1.425	»
2.72	44.34	52.94	1.483	»
1.50	49.22	49.28	1.524	»
1.02	52.61	46.37	1.561	»
0.85	53.79	45.36	1.580	»

№ 4304 PHOSPHORUS PENTOXIDE–BARIUM OXIDE–WATER [1752]

BaO—P_2O_5—H_2O

$t=25$

Comp. of saturated soln., wt%			d	Comp. of solid phase
Barium Oxide	Phosphorus Pentoxide	Water		
2.00	2.96	95.04	1.040	$BaHPO_4$
3.66	5.95	90.39	1.085	»
4.33	7.61	88.06	1.107	»

Table № 4304 (continuation)

Comp. of saturated soln., wt%			d	Comp. of solid phase
Barium Oxide	Phosphorus Pentoxide	Water		
4.80	8.25	86.95	1.118	$BaHPO_4$
6.31	11.19	82.50	1.165	»
6.09	11.36	82.55	1.166	»
6.84	12.35	80.81	1.188	»
8.50	16.51	74.99	1.246	»
10.86	23.46	65.68	1.354	$BaHPO_4+Ba(H_2PO_4)_2$
9.14	28.36	62.50	1.397	$Ba(H_2PO_4)_2$
7.41	32.68	59.91	1.420	»
6.93	33.74	59.33	1.425	»
2.72	44.34	52.94	1.483	»
1.50	49.22	49.28	1.524	»
1.02	52.61	46.37	1.561	»
0.85	53.79	45.36	1.580	»

№ 4305 PHOSPHORUS PENTOXIDE–BARIUM OXIDE–WATER [1860]

P_2O_5—BaO—H_2O

$t=38$

Concn. in aqueous soln., g/l		pH	Concn. in aqueous soln., g/l		pH
Barium Oxide	Phosphorus Pentoxide		Barium Oxide	Phosphorus Pentoxide	
0.6502	0.5954	4.76	0.1345	0.1143	5.94
0.4922	0.4499	4.84	0.1076	0.0816	6.33
0.4401	0.3974	4.79	0.0787	0.0560	6.69
0.3772	0.3286	5.28	0.0446	0.0275	6.95
0.2239	0.2399	5.23	0.0396	0.0141	7.25
0.2745	0.2342	5.41			

№ 4306 PHOSPHORUS PENTOXIDE–LANTHANUM PHOSPHATE–WATER [179]

P_2O_5— $LaPO_4$—H_2O

Comp. of saturated soln., wt%			t	Comp. of solid phase
Phosphorus Pentoxide	Lanthanum Phosphate	Water		
10.290	0.060	89.65	20—30	$LaPO_4$
21.430	0.164	78.406	20—30	»
32.300	0.241	67.459	20—30	»
42.700	0.281	57.019	20—30	»
10.390	0.015	89.595	80	»
21.020	0.031	78.949	80	»
31.720	0.053	68.227	80	»
42.050	0.079	57.871	80	»

№ 4307 [1586]

PHOSPHORUS PENTOXIDE–LEAD (II) OXIDE–WATER

$P_2O_5—PbO—H_2O$

$t=25$

Concn. in aqueous soln., g/l		d	Comp. of solid phase
Lead Oxide	Phosphorus Pentoxide		
1.38	52.73	1.036	$PbHPO_4$
2.45	106.75	1.074	»
3.66	160.60	1.113	»
5.87	216.84	1.151	»
8.21	272.63	1.190	»
9.54	328.09	1.227	»
10.01	391.84	1.264	»
11.75	453.35	1.302	»
14.95	549.44	1.375	»
21.42	662.31	1.448	»
19.71	769.06	1.516	$PbH_4(PO_4)_2$
15.47	821.90	1.549	»
11.90	864.06	1.576	»
7.74	956.96	1.635	»

№ 4308 [1466]

PHOSPHORUS PENTOXIDE–THORIUM OXIDE–WATER

$P_2O_5—ThO_2—H_2O$

$t=25$

Comp. of saturated soln., wt%			Comp. of solid phase
Thorium Oxide	Phosphorus Pentoxide	Water	
Traces	4.5	95.5	$Th_3(PO_4)_4$
0.1	9.61	90.29	»
0.2	16.0	83.8	»
0.5	20.0	79.5	$Th_3(PO_4)_4+Th(HPO_4)_2 \cdot H_2O$
3.3	27.1	69.6	$Th(HPO_4)_2 \cdot H_2O$
5.0	30	65.0	$ThHPO_4 \cdot 2H_2PO_4 \cdot 2H_2O$
9.34	46.1	44.56	»
13.3	49.0	37.7	»
19.1	50.8	30.1	»
19.8	52.0	28.2	»

SO_2—SO_3—H_2O

Comp. of solvent, wt%		Solubility of Sulfur Dioxide wt%	t	Comp. of solvent, wt%		Solubility of Sulfur Dioxide wt%	t
Sulfur Trioxide	Water			Sulfur Trioxide	Water		
80.21	19.79	3.85	20	95.02	4.98	27.58	20
80.37	19.63	3.81	20	95.10	4.90	28.34	20
80.64	19.36	3.66	20	95.23	4.77	28.16	20
80.81	19.19	3.65	20	56.75	43.25	1.798	40
81.63	18.37	3.88	20	56.90	43.10	1.747	40
82.03	17.97	3.95	20	57.0	43.0	1.712	40
82.07	17.93	3.85	20	57.05	42.95	1.702	40
82.12	17.88	3.91	20	64.99	35.01	1.602	40
82.13	17.87	3.96	20	65.04	34.96	1.406	40
82.17	17.83	3.98	20	65.13	34.87	1.431	40
82.53	17.47	4.66	20	65.87	34.13	1.406	40
82.75	17.25	4.18	20	73.64	26.36	1.628	40
82.80	17.20	4.58	20	73.70	26.30	1.634	40
82.85	17.15	4.33	20	78.15	21.85	1.920	40
82.90	17.10	4.60	20	78.20	21.80	1.792	40
82.90	17.10	4.48	20	78.20	21.80	1.928	40
82.91	17.09	4.28	20	79.38	20.62	1.981	40
83.65	16.35	4.96	20	79.39	20.61	2.035	40
84.57	15.43	5.62	20	79.52	20.48	2.050	40
85.77	14.23	6.18	20	80.80	19.20	2.020	40
85.94	14.06	6.27	20	80.95	19.05	2.05	40
86.47	13.53	6.77	20	82.70	17.30	2.468	40
86.49	13.51	5.99	20	85.0	15.0	3.176	40
86.65	13.35	5.99	20	85.0	15.0	3.132	40
86.71	13.29	7.06	20	57.75	42.25	0.983	60
86.93	13.07	7.52	20	57.90	42.10	0.986	60
87.11	12.89	6.47	20	57.91	42.09	0.988	60
87.90	12.10	9.06	20	65.88	34.12	0.855	60
88.30	11.70	9.17	20	66.03	33.97	0.834	60
88.60	11.40	9.58	20	66.14	33.86	0.848	60
88.90	11.10	9.41	20	66.23	33.77	0.858	60
90.65	9.35	11.98	20	73.75	26.25	0.949	60
90.70	9.30	12.05	20	73.82	26.18	0.953	60
91.10	8.90	12.17	20	73.87	26.13	0.990	60
91.30	8.70	12.14	20	78.22	21.78	1.196	60
93.16	6.84	18.73	20	78.31	21.69	1.172	60
93.30	6.70	18.93	20	79.50	20.50	1.346	60
93.30	6.70	20.42	20	79.57	20.43	1.337	60
93.34	6.66	20.28	20	79.60	20.40	1.359	60
94.20	5.80	23.31	20	79.90	20.10	1.371	60
94.60	5.40	22.13	20	84.30	15.70	1.625	60
95.22	4.78	25.78	20	84.40	15.60	1.610	60

№ 4310 SULFUR DIOXIDE–POTASSIUM THIOCYANATE–WATER [1656]

SO_2—KCNS—H_2O

Concn. of Potassium Thiocyanate in aqueous soln., g/l	Solubility of Sulfur Dioxide, cm^3/cm^3 Solvent	t	Concn. of Potassium Thiocyanate in aqueous soln., g/l	Solubility of Sulfur Dioxide, cm^3/cm^3 Solvent	t
48.55	37.57	25	48.55	25.63	35
97.1	42.38	25	97.1	28.79	35
145.65	47.02	25	145.65	32.03	35
194.2	51.81	25	194.2	35.05	35
242.75	55.87	25	242.75	38.13	35
291.30	61.26	25	291.30	42.94	35

№ 4311 [1130]

SULFUR DIOXIDE–POTASSIUM THIOCYANATE–WATER

SO_2—KCNS—H_2O

$t=25$

Concn. of Potassium Thiocyanate in aqueous soln., g/l	Solubility of Sulfur Dioxide, cm^3(at t = 0 and P = 760)/cm^3 solvent
48.55	34.42

№ 4312 SULFUR DIOXIDE–POTASSIUM NITRATE–WATER [1656]

SO_2—KNO_3—H_2O

Concn. of Potassium Nitrate in aqueous soln., g/l	Solubility of Sulfur Dioxide, cm^3/cm^3 solvent	t	Concn. of Potassium Nitrate in aqueous soln., g/l	Solubility of Sulfur Dioxide, cm^3/cm^3 solvent	t
50.55	33.80	25	50.55	23.27	35
101.1	34.79	25	101.1	24.03	35
151.65	35.77	25	151.65	24.79	35
202.20	36.66	25	202.20	25.72	35
252.75	37.57	25	252.75	26.54	35
303.30	38.52	25	303.30	27.33	35

№ 4313 [1130]

SULFUR DIOXIDE–POTASSIUM NITRATE–WATER

SO_2—KNO_3—H_2O

$t=25$

Concn. of Potassium Nitrate in aqueous soln., g/l	Solubility of Sulfur Dioxide, cm^3(at t = 0 and P = 760)/cm^3 solvent
50.55	30.97

№ 4314 [1656]

SULFUR DIOXIDE–POTASSIUM SULFATE–WATER

$SO_2—K_2SO_4—H_2O$

t=25

Concn. of Potassium Sulfate in aqueous soln., g/l	Solubility of Sulfur Dioxide, cm^3/cm^3 solvent
43.55	33.20
87.1	33.61

№ 4315 [1130]

SULFUR DIOXIDE–POTASSIUM SULFATE–WATER

$SO_2—K_2SO_4—H_2O$

t=25

Concn. of Potassium Sulfate in aqueous soln., g/l	Solubility of Sulfur Dioxide, cm^3 (t = 0 and P = 760)/ cm^3 solvent
87.1	30.42

№ 4316 SULFUR DIOXIDE–POTASSIUM CHLORIDE–WATER [1656]

$SO_2—KCl—H_2O$

Concn. of Potassium Chloride in aqueous soln., g/l	Solubility of Sulfur Dioxide, cm^3/cm^3 solvent	t	Concn. of Potassium Chloride in aqueous soln., g/l	Solubility of Sulfur Dioxide, cm^3/cm^3 solvent	t
37.3	34.42	25	37.3	23.74	35
74.6	36.05	25	74.6	25.15	35
111.9	37.76	25	111.9	26.54	35
149.2	39.32	25	149.2	27.94	35
186.5	40.96	25	186.5	28.93	35
223.8	42.27	25	223.8	30.02	35

№ 4317 SULFUR DIOXIDE–POTASSIUM CHLORIDE–WATER [1130]

$SO_2—KCl—H_2O$

P=760

Comp. of saturated soln., wt%			t	Comp. of saturated soln., wt%			t
Sulfur Dioxide	Potassium Chloride	Water		Sulfur Dioxide	Potassium Chloride	Water	
13.90	4.49	81.61	10	5.67	21.17	73.16	40
14.42	7.93	77.65	10	4.28	4.61	91.11	48.15
14.89	11.64	73.47	10	4.30	8.81	86.89	48.15
15.38	14.55	70.07	10	4.33	13.06	82.61	48.15
16.35	19.25	64.40	10	4.33	16.67	79.00	48.15
12.18	8.26	79.56	15	4.39	19.17	76.44	48.15
12.82	14.47	72.71	15	4.45	23.55	72.00	48.15
13.60	20.29	66.11	15	3.17	48.73	48.10	60
9.95	4.40	85.65	20	3.08	9.34	87.58	60
10.19	8.39	81.42	20	3.13	16.89	79.98	60
10.49	12.08	77.43	20	3.14	22.91	73.95	60
10.95	17.79	71.26	20	2.51	7.84	89.65	70
20.66	11.32	68.02	20	2.44	17.38	80.18	70
7.24	4.58	88.18	29.9	2.44	24.33	73.23	70
7.40	8.50	84.10	29.9	2.05	7.81	90.14	80
7.68	15.90	76.42	29.9	19.51	17.25	63.24	80
7.98	21.15	70.87	29.9	1.91	22.45	75.64	80
5.36	4.65	89.99	40	1.67	8.31	90.02	90
5.42	8.75	85.83	40	16.21	16.04	67.75	90
5.47	12.64	81.89	40	1.56	24.28	74.16	90
5.56	15.83	78.61	40				

№ 4318 [1875]

SULFUR DIOXIDE–POTASSIUM CHLORIDE–WATER

SO_2—KCl—H_2O

t=25

Concn. of Potassium Chloride in aqueous soln., g/l	Solubility of Sulfur Dioxide, cm^3(at t = 0 and P = 760)/ cm^3 solvent
37.3	31.53

№ 4319 [1130]

SULFUR DIOXIDE–POTASSIUM BROMIDE–WATER

SO_2—KBr—H_2O

t=25

Concn. of Potassium Bromide in aqueous soln., g/l	Solubility of Sulfur Dioxide, cm^3(at t = 0 and P = 760)/ cm^3 solvent
58.5	32.97

№ 4320

SULFUR DIOXIDE–POTASSIUM IODIDE–WATER [1656]

SO_2— KI —H_2O

Concn. of Potassium Iodide in aqueous soln., g/l	Solubility of Sulfur Dioxide, cm^3/cm^3 solvent	t	Concn. of Potassium Iodide in aqueous soln., g/l	Solubility of Sulfur Dioxide, cm^3/cm^3 solvent	t
83	38.66	25	83	26.30	35
166	44.76	25	166	30.25	35
249	50.58	25	249	34.64	35
332	56.75	25	332	38.04	35
415	62.63	25	415	41.87	35
498	68.36	25	498	45.43	35

№ 4321 [1130]

SULFUR DIOXIDE–POTASSIUM IODIDE–WATER

SO_2— KI —H_2O

t=25

Concn. of Potassium Iodide in aqueous soln., g/l	Solubility of Sulfur Dioxide, cm^3(at t = 0 and P = 760)/ cm^3 solvent
83.0	35.42

№ 4322

SULFUR DIOXIDE–CALCIUM OXIDE–WATER [1413]

SO_2—CaO—H_2O

Concn. in aqueous soln., g/l Water			t	P, atm	Concn. in aqueous soln., g/l Water			t	P, atm
Total SO_2	Free SO_2	$CaSO_3$			Total SO_2	Free SO_2	$CaSO_3$		
41.0	22.0	35.6	15	50	98.0	59.0	73.1	15	200
68.0	39.0	54.4	15	100	109	67.0	78.7	15	250
86.0	51.0	65.6	15	150	118	74.0	82.5	15	300

Table № 4322 (continuation)

Concn. in aqueous soln., g/l Water			t	P, atm	Concn. in aqueous soln., g/l Water			t	P, atm
Total SO_2	Free SO_2	$CaSO_3$			Total SO_2	Free SO_2	$CaSO_3$		
128	81.0	88.1	15	350	83.0	52.0	58.0	25	250
138	89.0	91.9	15	400	91.0	57.0	63.8	25	300
147	96.0	95.6	15	450	99.0	63.0	67.5	25	350
155	102	99.4	15	500	106	68.0	71.2	25	400
163	109	101.3	15	550	113	73.0	75.0	25	450
171	115	105.0	15	600	120	78.0	78.7	25	500
178	121	106.9	15	650	127	83.0	82.5	25	550
184	126	108.7	15	700	133	87.0	86.1	25	600
191	132	110.6	15	750	140	92.0	90.0	25	650
27.0	14.0	24.4	25	50	146	97.0	91.9	25	700
47.0	26.0	39.4	25	100	152	102	93.8	25	750
62.0	36.0	48.7	25	150	120.2	79.9	75.5	35	761
73.0	44.0	54.4	25	200					

№ 4323 [2646]

SULFUR DIOXIDE–IRON (II) OXIDE–WATER

$SO_2—FeO—H_2O$

Concn. in aqueous soln., mol %		t	Comp. of solid phase
Iron Oxide	Sulfur Dioxide		
0.5	0.3	15	$FeSO_3 \cdot 5H_2O$
1.0	0.6	15	»
2.0	1.2	15	»
2.5	1.4	15	$FeSO_4 \cdot 5H_2O + Fe(HSO_3)_2$
2.4	2.0	15	$Fe(HSO_3)_2$
2.2	3.5	15	»
2.3	5.0	15	»
2.6	7.0	15	»
3.0	8.1	15	»
0.5	0.75	25	$FeSO_3 \cdot 5H_2O$
1.0	1.3	25	»
1.5	1.6	25	»
1.9	2.0	25	$FeSO_3 \cdot 5H_2O + Fe(HSO_3)_2$
1.5	2.9	25	$Fe(HSO_3)_2$
1.25	4.0	25	»
1.5	5.9	25	»
2.0	7.6	25	»
2.3	8.3	25	»

№ 4324 SULFUR DIOXIDE–ZINC OXIDE–WATER [2646]

$SO_2—ZnO—H_2O$

Comp. of saturated soln., mol %			t	Comp. of solid phase
Zinc Oxide	Sulfur Dioxide	Water		
1.0	2.1	96.9	15	$ZnSO_3 \cdot 2\frac{1}{2}H_2O$
2.0	4.0	94.0	15	»
2.55	5.0	92.45	15	$ZnSO_3 \cdot 2\frac{1}{2}H_2O + Zn(HSO_3)_2$
2.5	6.0	91.5	15	$Zn(HSO_3)_2$
3.0	7.2	89.8	15	»
4.0	9.1	86.9	15	»
5.0	10.75	84.25	15	»
5.6	11.8	82.6	15	»
1.0	2.4	96.6	25	$ZnSO_3 \cdot 2\frac{1}{2}H_2O$
2.0	4.6	93.4	25	»
2.8	6.0	91.2	25	$Zn(SO_3) \cdot 2\frac{1}{2}H_2O + Zn(HSO_3)_2$
2.0	6.4	91.6	25	$Zn(HSO_3)_2$
1.5	7.0	91.5	25	»
1.3	9.0	89.7	25	»
1.5	9.0	89.5	25	»
2.0	10.0	88.0	25	»
3.3	11.8	84.9	25	»

№ 4325 [1900]

SULFUR DIOXIDE–ZINC HYDROXIDE–WATER

$SO_2—Zn(OH)_2—H_2O$

$t=20$

Comp. of saturated soln., wt%			Comp. of solid phase, mol ratio $SO_2 : ZnO$	
Zinc Hydroxide, in conversion to Zinc Oxide	Sulfur Dioxide	Water	Sulfur Dioxide	Zinc Oxide
7.78	14.10	78.12	1	0.978
6.50	11.05	82.45	1	0.974
5.51	9.14	85.35	1	1.00
4.12	6.62	89.26	1	1.00
3.24	5.18	91.58	1	0.990
0.50	0.63	98.87	1	1.05
0.45	0.58	98.97	1	0.995
0.39	0.49	99.12	1	0.994
0.10	0.10	99.80	1	1.58
0.090	0.086	99.824	2	5.03
0.039	0.024	99.937	2	5.28

№ 4326 [1656]

SULFUR DIOXIDE–CADMIUM SULFATE–WATER

$SO_2—CdSO_4—H_2O$

Concn. of Cadmium Sulfate in aqueous soln., g/l	Solubility of Sulfur Dioxide, cm^3/cm^3 solvent	t
52.1	31.11	25
104.2	29.71	25
156.3	28.24	25
208.4	26.58	25
260.5	25.14	25
312.6	23.76	25
52.1	21.45	35
104.2	20.43	35
156.3	19.42	35
208.4	18.31	35
260.5	17.41	35
312.6	16.25	35

№ 4327 [1656]

SULFUR DIOXIDE–CADMIUM CHLORIDE–WATER

$SO_2—CdCl_2—H_2O$

Concn. of Cadmium Chloride in aqueous soln., g/l	Solubility of Sulfur Dioxide, cm^3/cm^3 solvent	t
45.85	31.66	25
91.7	30.55	25
137.55	29.46	25
183.4	28.16	25
229.25	27.09	25
275.10	26.06	25
45.85	21.73	35
91.7	21.23	35
137.55	20.55	35
183.4	20.02	35
229.25	19.23	35
275.10	18.68	35

№ 4328 [1656]

SULFUR DIOXIDE–CADMIUM BROMIDE–WATER

$SO_2—CdBr_2—H_2O$

Concn. of Cadmium Bromide in aqueous soln., g/l	Solubility of Sulfur Dioxide, cm^3/cm^3 solvent	t	Concn. of Cadmium Bromide in aqueous soln., g/l	Solubility of Sulfur Dioxide, cm^3/cm^3 solvent	t
68.05	31.91	25	68.05	21.88	35
136.1	31.01	25	136.1	21.46	35
204.15	30.17	25	204.15	20.81	35
272.2	29.27	25	272.2	20.60	35
340.25	28.15	25	340.25	19.70	35
408.3	27.46	25	408.3	19.17	35

№ 4329 [2646]

SULFUR DIOXIDE–LEAD (II) OXIDE–WATER

$SO_2—PbO—H_2O$

Comp. of saturated soln., mol %			t	Comp. of solid phase
Lead Oxide	Sulfur Dioxide	Water		
0.005	1.0	98.995	15	$PbSO_3 \cdot H_2O$
0.02	3.0	96.98	15	»
0.04	5.0	94.96	15	»
0.07	7.0	92.93	15	»
0.09	8.0	91.91	15	»
0.115	9.0	90.885	15	»
0.12	9.2	90.68	15	$PbSO_4 \cdot H_2O + Pb(HSO_3)_2$

Table № 4329 (continuation)

Comp. of saturated soln., mol %			t	Comp. of solid phase
Lead Oxide	Sulfur Dioxide	Water		
0.09	10.0	89.91	15	$Pb(HSO_3)_2$
0.12	11.0	88.88	15	»
0.005	2.0	97.995	25	$PbSO_3 \cdot H_2O$
0.01	3.0	96.99	25	»
0.02	5.0	94.98	25	»
0.04	7.0	92.96	25	»
0.05	8.0	91.95	25	»
0.06	9.0	90.94	25	»
0.07	9.6	90.3	25	$PbSO_3 \cdot H_2O + Pb(HSO_3)_2$
0.06	10.0	89.94	25	$Pb(HSO_3)_2$
0.065	11.0	88.935	25	»

№ 4330 SULFUR TRIOXIDE–POTASSIUM SULFATE–WATER [1458]

$SO_3—K_2SO_4—H_2O$

$t = 25$

Concn. in aqueous soln., mol/1000g soln.		Comp. of solid phase	Concn. in aqueous soln., mol/1000g soln.		Comp. of solid phase
Potassium Sulfate	Sulfur Trioxide		Potassium Sulfate	Sulfur Trioxide	
0.171	6.42	$KHSO_4$	0.346	8.62	$KH_3(SO_4)_2$
0.190	6.60	»	0.384	8.57	»
0.266	6.91	$KHSO_4 + KH_3(SO_4)_2 \cdot H_2O$	0.412	8.71	»
0.182	7.26	$KH_3(SO_4)_2 \cdot H_2O$	0.583	8.82	»
0.157	7.62	»	0.880	8.65	$KH_3(SO_4)_2 + KHS_2O_7$
0.167	7.88	»	0.882	8.70	KHS_2O_7
0.201	8.00	»	0.561	8.96	»
0.250	8.10	»	0.365	9.80	»
0.352	8.15	»	0.430	9.78	»
0.364	8.16	$KH_3(SO_4)_2 \cdot H_2O + KH_3(SO_4)_2$	0.665	9.80	»
0.341	8.29	$KH_3(SO_4)_2$	0.904	9.66	»
0.322	8.33	»	0.937	9.66	»
0.325	8.45	»			

$SO_3—TiO_2—H_2O$

Comp. of saturated soln., wt%			t	Comp. of solid phase
Titanium Dioxide	Sulfur Trioxide	Water		
0.001	4.73	95.269	100	$TiO_2 \cdot xH_2O$
0.006	9.75	90.244	100	»
0.027	16.21	83.763	100	»
0.079	25.80	74.121	100	»
0.111	27.35	72.539	100	»
0.168	29.35	70.482	100	»
0.260	32.01	67.73	100	»
0.314	34.72	64.966	100	$TiOSO_4 \cdot H_2O$
0.256	36.06	63.684	100	»
0.182	38.04	61.778	100	»
0.113	41.10	58.787	100	»
0.073	43.75	56.177	100	»
0.042	46.92	53.038	100	»
0.040	48.88	51.08	100	»
0.033	50.82	49.147	100	»
0.027	53.33	46.643	100	»
0.024	56.58	43.396	100	»
0.024	58.85	41.126	100	»
0.039	60.78	39.181	100	»
0.051	61.40	38.549	100	»
0.132	62.31	37.558	100	»
0.162	63.52	36.318	100	$TiOSO_4 \cdot H_2SO_4 \cdot 2H_2O$
0.198	65.67	34.132	100	»
0.231	67.18	32.589	100	»
0.320	68.83	30.85	100	$TiOSO_4 \cdot H_2SO_4 \cdot 2H_2O$ (metastabl
0.429	63.10	36.471	100	$TiOSO_4 \cdot H_2SO_4 \cdot 2H_2O$
0.275	65.72	34.005	100	»
0.212	68.38	31.408	100	»
0.211	72.15	27.639	100	»
0.268	74.49	25.242	100	»
0.373	76.53	23.097	100	»
0.601	78.90	20.499	100	»
0.429	78.16	21.411	100	$Ti(SO_4)_2$
0.288	79.10	20.612	100	»
0.238	81.38	18.382	100	»
0.285	82.83	16.885	100	»
0.375	83.35	16.275	100	»
0.374	84.20	15.426	100	»
0.713	34.35	65.937	100	$TiOSO_4 \cdot 2H_2O$ (metastable)
0.354	35.91	63.736	100	»
0.238	37.73	62.032	100	»
0.180	40.03	59.79	100	»
0.122	42.83	57.048	100	»
0.101	47.00	52.899	100	»
0.141	49.48	50.379	100	»
0.292	53.94	45.768	100	»
0.402	55.56	44.038	100	»
0.166	39.92	59.914	125	$TiOSO_4 \cdot H_2O$

Table № 4331 (continuation)

Comp. of saturated soln., wt%			t	Comp. of solid phase
Titanium Dioxide	Sulfur Trioxide	Water		
0.124	42.03	57.846	125	$TiOSO_4 \cdot H_2O$
0.086	44.60	55.314	125	»
0.061	47.13	52.809	125	»
0.048	50.37	49.582	125	»
0.045	53.40	46.555	125	»
0.038	56.07	43.892	125	»
0.041	58.02	41.939	125	»
0.056	60.94	39.004	125	»
0.071	63.02	36.909	125	»
0.117	65.65	34.233	125	»
0.182	67.62	32.198	125	»
0.226	68.36	31.414	125	»
0.470	64.01	35.520	125	$TiOSO_4 \cdot H_2SO_4 \cdot H_2O$
0.314	65.61	34.076	125	»
0.265	67.50	32.235	125	»
0.243	69.22	30.537	125	»
0.290	72.75	26.960	125	»
0.387	75.20	24.413	125	»
0.578	77.76	21.662	125	»
0.414	78.03	21.556	125	$Ti(SO_4)_2$
0.237	79.16	20.603	125	»
0.217	80.35	19.433	125	»
0.203	81.47	18.327	125	»
0.203	81.49	18.307	125	»
0.060	51.80	48.14	150	$TiOSO_4 \cdot H_2O$
0.052	53.72	46.228	150	»
0.054	56.45	43.496	150	»
0.063	59.79	40.147	150	»
0.075	61.90	38.025	150	»
0.090	63.51	36.40	150	$TiOSO_4 \cdot H_2O + TiOSO_4$
0.081	64.73	35.189	150	$TiOSO_4$
0.076	65.91	34.014	150	»
0.082	67.05	34.868	150	»
0.120	68.02	31.860	150	»
0.179	68.73	31.091	150	»
0.330	67.65	32.020	150	$TiOSO_4 \cdot H_2SO_4 \cdot H_2O$
0.313	68.50	31.187	150	»
0.301	70.44	29.259	150	»
0.310	71.70	27.990	150	»
0.327	74.98	24.693	150	»
0.361	76.65	22.989	150	»
0.347	77.78	21.873	150	$Ti(SO_4)_2$
0.238	78.68	21.082	150	»
0.206	80.22	19.574	150	»
0.185	81.43	18.485	150	»
0.098	58.47	41.432	175	$TiOSO_4\ H_2O$
0.094	59.38	40.526	175	»
0.090	60.53	39.380	175	»
0.101	61.45	38.449	175	»
0.113	61.98	37.907	175	$TiOSO_4\ H_2O + TiOSO_4$

Table № 4331 (continuation)

Comp. of saturated soln., wt%			t	Comp. of solid phase
Titanium Dioxide	Sulfur Trioxide	Water		
0.096	62.40	37.504	175	$TiOSO_4$
0.060	65.97	33.97	175	»
0.054	67.60	32.346	175	»
0.060	69.96	29.980	175	»
0.088	72.03	27.882	175	»
0.152	73.94	25.908	175	»
0.262	75.65	24.088	175	»
0.256	76.72	23.024	175	$Ti(SO_4)_2$
0.213	78.05	21.737	175	»
0.193	79.72	20.087	175	»
0.178	80.90	18.922	175	»
0.166	81.65	18.184	175	»
0.050	65.08	34.870	200	$TiOSO_4$
0.051	67.32	32.629	200	»
0.050	69.40	30.550	200	»
0.070	71.76	28.170	200	»
0.092	73.38	26.528	200	»
0.155	75.91	23.935	200	»
0.201	77.52	22.279	200	»
0.233	78.60	21.167	200	»
0.212	79.45	20.338	200	$Ti(SO_4)_2$
0.204	79.92	19.876	200	»
0.182	81.25	18.568	200	»
0.038	69.12	30.842	225	$TiOSO_4$
0.042	70.45	29.508	225	»
0.064	72.56	27.376	225	»
0.074	73.91	26.016	225	»
0.091	75.07	24.839	225	»
0.120	76.60	23.280	225	»
0.178	78.01	21.812	225	»
0.204	78.69	21.106	225	»
0.241	79.55	20.209	225	$Ti(SO_4)_2$
0.229	80.21	19.561	225	»
0.204	81.15	18.646	225	»
0.042	72.10	27.858	250	$TiOSO_4$
0.045	74.17	25.785	250	»
0.048	76.08	23.872	250	»
0.053	77.81	22.137	250	»
0.064	79.88	20.056	250	»
0.099	80.38	19.521	250	»
0.250	81.02	18.730	250	$Ti(SO_4)_2$
0.247	81.32	18.433	250	»
0.038	75.05	24.912	275	$TiOSO_4$
0.044	76.94	23.016	275	»
0.050	80.05	19.900	275	»
0.069	81.55	18.381	275	»
0.038	77.98	21.982	300	»
0.040	80.14	19.820	300	»
0.041	80.88	19.079	300	»
0.056	81.63	18.314	300	»

№ 4332 SULFUR TRIOXIDE–TITANIUM DIOXIDE–WATER [68]

SO_3—TiO_2—H_2O

Comp. of saturated soln., wt%			t	Comp. of solid phase
Titanium Dioxide	Sulfur Trioxide	Water		
0.36	5.75	93.89	25	$TiO_2 \cdot xH_2O$
0.41	13.12	86.47	25	»
0.81	22.07	77.12	25	»
2.61	29.35	68.04	25	»
18.56	26.62	54.82	25	$TiOSO_4 \cdot 2H_2O$ (metastable)
10.11	27.03	62.86	25	»
5.15	29.52	65.33	25	»
1.61	34.98	63.41	25	$TiOSO_4 \cdot 2H_2O$
0.108	40.27	59.622	25	»
0.20	52.44	47.36	25	$TiOSO_4 \cdot 2H_2O$ (metastable)
15.25	27.61	57.14	25	$TiOSO_4 \cdot H_2O$ (metastable)
3.15	34.92	61.93	25	»
1.53	37.44	60.03	25	»
0.25	41.15	58.60	25	»
0.06	43.90	56.04	25	»
0.02	48.46	51.52	25	$TiOSO_4 \cdot H_2O$
0.01	54.91	45.08	25	»
0.04	58.90	41.06	25	»
0.09	62.35	37.56	25	$TiOSO_4 \cdot H_2SO_4 \cdot 2H_2O$
0.06	65.10	34.84	25	»
0.07	69.05	30.88	25	$TiOSO_4 \cdot H_2SO_4 \cdot H_2O$
0.04	74.32	25.64	25	»
0.13	80.46	19.41	25	»
0.39	83.97	15.64	25	$Ti(SO_4)_2$ (metastable)
0.43	87.71	11.86	25	$Ti(SO_4)_2$
0.58	88.80	10.62	25	»
0.26	11.87	87.87	50	$TiO_2 \cdot xH_2O$
0.56	21.35	78.09	50	»
1.07	30.75	68.18	50	»
25.37	27.06	47.57	50	$TiOSO_4 \cdot 2H_2O$ (metastable)
24.18	27.13	48.69	50	»
12.66	27.48	59.86	50	»
4.10	29.36	66.54	50	»
2.05	30.48	67.47	50	$TiOSO_4 \cdot 2H_2O$
0.48	34.48	65.04	50	»
0.18	37.36	62.46	50	»
0.08	41.36	58.56	50	»
0.06	45.51	54.43	50	»
0.17	54.96	44.87	50	$TiOSO_4 \cdot 2H_2O$ (metastable)
1.32	59.30	49.38	50	»
10.90	29.48	59.62	50	$TiOSO_4 \cdot H_2O$ (metastable)
5.94	31.26	62.80	50	»
1.16	37.22	61.62	50	»
0.05	44.94	55.01	50	»
0.04	49.03	50.93	50	$TiOSO_4 \cdot H_2O$
0.02	54.94	45.04	50	»
0.02	57.20	42.78	50	»
0.05	59.95	40.0	50	»
1.04	72.65	26.31	50	$TiOSO_4 \cdot H_2O$ (metastable)
0.15	59.31	40.54	50	$TiOSO_4 \cdot H_2SO_4 \cdot 2H_2O$ (metastable)

Comp. of saturated soln., wt%			t	Comp. of solid phase
Titanium Dioxide	Sulfur Trioxide	Water		
0.10	63.30	36.60	50	$TiOSO_4 \cdot H_2SO_4 \cdot 2H_2O$
0.10	65.90	34.0	50	»
0.18	69.97	29.85	50	$TiOSO_4 \cdot H_2SO_4 \cdot 2H_2O$ (metastable
0.08	69.09	30.83	50	$TiOSO_4 \cdot H_2SO_4 \cdot H_2O$
0.05	71.68	28.27	50	»
0.06	77.20	22.71	50	»
0.14	81.42	18.44	50	»
0.28	83.53	16.19	50	$Ti(SO_4)_2$
0.33	85.71	13.96	50	»
0.12	11.71	88.17	75	$TiO_2 \cdot xH_2O$
0.30	21.27	78.43	75	»
0.62	26.48	72.90	75	»
0.64	33.52	65.84	75	$TiOSO_4 \cdot 2H_2O$
0.18	38.04	61.78	75	»
0.13	53.01	46.86	75	$TiOSO_4 \cdot 2H_2O$ (metastable)
0.44	56.21	43.35	75	»
2.13	58.80	39.07	75	»
0.04	45.66	54.30	75	$TiOSO_4 \cdot H_2O$
0.03	49.38	50.59	75	»
0.04	51.57	48.39	75	»
0.05	58.62	41.33	75	»
0.18	61.20	38.62	75	$TiOSO_4 \cdot H_2SO_4 \cdot 2H_2O$ (metastable
0.14	64.91	34.95	75	$TiOSO_4 \cdot H_2SO_4 \cdot 2H_2O$
0.31	70.53	29.16	75	$TiOSO_4 \cdot H_2SO_4 \cdot 2H_2O$ (metastable
0.14	68.64	31.22	75	$TiOSO_4 \cdot H_2SO_4 \cdot H_2O$
0.13	72.05	27.72	75	»
0.14	75.14	24.72	75	»
0.20	78.26	21.54	75	»
0.23	82.91	16.86	75	$Ti(SO_4)_2$
0.27	84.42	15.31	75	»

№ 4333 [2427]

SULFUR TRIOXIDE–VANADIUM DIOXIDE–WATER

SO_3—VO_2—H_2O

$t=30$

Concn. in aqueous soln., wt%		d	Comp. of solid phase
Vanadium Dioxide	Sulfur Trioxide		
0.0068	0.00	0.996	$VO_2 \cdot H_2O$
3.45	3.33	1.048	»
7.21	6.83	1.096	»
11.51	10.93	1.231	»
13.41	12.78	1.278	»
16.19	15.40	1.361	»
20.21	19.02	1.468	»

Table № 4333 (continuation)

Concn. in aqueous soln., wt%		d	Comp. of solid phase
Vanadium Dioxide	Sulfur Trioxide		
21.14	20.09	1.508	$VO_2 \cdot SO_3 \cdot 5H_2O$
17.24	22.35	1.473	»
17.28	22.50	1.461	»
12.40	26.77	1.427	»
11.24	27.40	1.422	»
8.11	32.35	1.416	»
6.70	35.44	1.436	»
6.52	36.11	1.441	»
17.35	28.09	1.535	$VO_2 \cdot SO_3 \cdot 3H_2O$
17.36	28.09	—	»
14.77	29.71	1.506	»
13.52	30.72	1.496	»
11.73	31.64	—	»
6.04	37.50	1.448	$VO_2 \cdot SO_3 \cdot 3H_2O$
2.27	46.96	1.506	»
2.11	48.17	1.517	»
2.05	50.96	1.662	»
2.19	52.45	1.578	$VO_2 \cdot SO_3 \cdot 3H_2O$
2.80	51.91	1.583	$2VO_2 \cdot 3SO_2 \cdot 4H_2O$
2.51	52.27	1.589	»
0.96	55.24	1.593	»
0.37	58.06	1.622	»
0.275	58.89	1.582	»
0.245	58.58	1.717	»
0.072	65.52	1.754	»
0.080	69.09	1.783	»
0.080	71.16	1.787	»
0.088	72.97	1.795	»
0.097	73.67	1.801	$VO_2 \cdot SO_3$
0.086	74.46	1.844	»
0.081	74.56	1.811	»
0.062	79.17	1.827	»
0.041	80.18	1.827	»
0.035	80.58	1.835	»
0.036	81.02	1.825	»

№ 4334 [2042]

SULFUR TRIOXIDE–VANADIUM PENTOXIDE–WATER

$SO_3—V_2O_5—H_2O$

$t=30$

Concn. in aqueous soln., mol %			Comp. of solid phase
Vanadium Pentoxide	Sulfur Trioxide	Water	
0.011	0.000	0.996	—
0.094	0.684	1.030	—
0.172	1.428	1.066	V_2O_5

Table № 4334 (continuation)

Concn. in aqueous soln., mol %			Comp. of solid phase
Vanadium Pentoxide	Sulfur Trioxide	Water	
0.212	1.742	1.070	V_2O_5
0.270	2.150	1.091	»
0.459	4.232	1.178	»
0.587	5.341	1.219	»
0.850	8.04	1.319	»
1.051	9.81	1.370	»
1.388	11.26	1.440	$V_2O_5+V_2O_5 \cdot 2SO_3 \cdot 8H_2O$
1.060	11.11	1.418	$V_2O_5 \cdot 2SO_3 \cdot 8H_2O$
0.870	11.32	—	»
0.646	12.24	—	»
0.600	13.53	—	»
0.578	13.75	1.443	»
0.620	14.58	1.466	»
0.898	16.46	1.519	»
1.690	20.73	1.661	»
2.030	22.33	1.734	$V_2O_5 \cdot 2SO_3 \cdot 8H_2O+V_2O_5 \cdot 2SO_3 \cdot 3H_2O$
1.920	22.26	1.717	$V_2O_5 \cdot 2SO_3 \cdot 3H_2O$
1.690	22.34	1.694	»
1.580	22.47	1.695	»
1.020	23.20	—	»
0.710	24.34	—	»
0.306	26.72	—	»
0.311	28.15	—	$V_2O_5 \cdot 2SO_3 \cdot 3H_2O$
1.280	28.00	—	$V_2O_5 \cdot 4SO_3 \cdot 4H_2O$ (metastable)
0.601	28.23	1.739	»
0.560	28.28	1.731	»
0.360	28.61	1.708	»
0.056	30.22	1.727	»
0.019	32.78	1.745	»
0.018	36.91	1.790	»
0.015	39.43	1.799	»
0.014	48.93	1.817	»
0.300	28.68	1.714	$V_2O_5 \cdot 4SO_3 \cdot 4H_2O+V_2O_5 \cdot 2SO_3 \cdot 3H_2O$

№ 4335 SULFUR TRIOXIDE–CHROMIUM TRIOXIDE–WATER [1721]

$SO_3—CrO_3—H_2O$

Comp. of saturated soln., wt%			t	d_4	Comp. of solid phase
Chromium Trioxide	Sulfur Trioxide	Water			
25.79	0.00	74.27	25	—	CrO_3
20.69	1.55	77.76	25	1.669	»
17.41	3.34	79.25	25	1.626	»
13.20	5.73	81.07	25	1.579	»
9.70	7.51	82.79	25	1.520	»
6.13	9.99	83.88	25	1.496	»

Table № 4335 (continuation)

Comp. of saturated soln., wt%			t	d_4	Comp. of solid phase
Chromium Trioxide	Sulfur Trioxide	Water			
3.38	12.61	84.01	25	1.460	CrO_3
1.17	15.89	82.94	25	1.479	»
0.14	22.15	77.71	25	1.580	»
0.11	23.22	76.67	25	1.610	»
0.23	26.43	73.34	25	1.680	»
0.28	27.56	72.16	25	1.690	»
0.28	28.17	71.55	25	1.690	»
0.55	29.35	70.10	25	1.728	»
0.97	31.42	67.61	25	1.756	»
1.23	31.93	66.84	25	1.771	»
1.62	32.64	65.74	25	1.786	»
1.58	33.17	65.25	25	1.788	$CrO_3 \cdot SO_3$
1.02	33.14	65.84	25	1.803	»
0.97	34.29	64.74	25	1.806	»
0.80	35.41	63.79	25	1.799	»
0.54	36.14	63.32	25	1.801	»
0.71	38.11	61.18	25	1.818	»
1.12	38.81	60.07	25	1.837	»
0.43	37.79	61.78	25	1.814	»
0.50	38.69	60.81	25	1.812	»
0.36	38.97	60.67	25	1.823	»
0.44	45.80	53.76	25	—	$CrO_3 \cdot SO_3 \cdot H_2O$
0.20	47.01	52.79	25	1.834	»
0.14	51.78	48.08	25	1.841	»
0.40	57.50	42.10	25	1.915	»
0.88	58.94	40.18	25	—	»
13.79	5.25	80.96	45	—	CrO_3
1.00	16.73	82.27	45	—	»
0.17	22.01	77.82	45	—	»
0.16	31.05	68.79	45	1.751	»
0.15	33.55	66.30	45	—	$CrO_3 \cdot SO_3$
1.06	35.40	63.54	45	—	»
0.63	37.07	62.30	45	—	»
0.60	38.10	61.30	45	1.797	»
0.38	42.26	57.36	45	1.814	»
0.30	45.67	54.03	45	—	$CrO \cdot SO_3 \cdot H_2O$

№ 4336 SULFUR TRIOXIDE–CHROMIUM TRIOXIDE–WATER [2366]

$SO_3—CrO_3—H_2O$

Comp. of saturated soln., wt%			t	Comp. of solid phase
Chromium Trioxide	Sulfur Trioxide	Water		
22.67	0.00	77.33	0	CrO_3
10.19	7.36	82.45	0	»
3.32	12.09	84.59	0	»
0.75	16.51	82.74	0	»
0.10	23.22	76.68	0	»
0.22	27.30	72.48	0	»

Table № 4336 (continuation)

Comp. of saturated soln., wt%			t	Comp. of solid phase
Chromium Trioxide	Sulfur Trioxide	Water		
0.65	29.61	69.74	0	CrO_3
2.46	32.43	65.11	0	»
3.28	33.18	63.54	0	»
10.93	35.22	53.85	0	»
1.23	30.97	67.80	0	$CrO_3 \cdot SO_3$
0.82	31.40	67.78	0	»
0.58	33.18	66.24	0	»
0.33	35.52	64.15	0	»
0.17	40.48	59.35	0	»
0.09	44.40	55.51	0	»
23.15	0.00	76.85	20	CrO_3
20.57	1.87	77.56	20	»
18.93	2.68	78.39	20	»
15.72	4.41	79.87	20	»
12.73	6.00	81.27	20	»
10.55	7.38	82.07	20	»
8.08	8.95	82.97	20	»
5.40	10.53	84.07	20	»
3.59	12.19	84.22	20	»
1.90	14.44	83.66	20	»
0.87	16.77	82.36	20	»
0.43	18.56	81.01	20	»
0.22	19.88	79.90	20	»
0.10	23.64	76.26	20	»
0.14	25.73	74.13	20	»
0.41	28.92	70.67	20	»
1.83	32.57	65.60	20	»
4.48	34.91	60.61	20	»
8.48	37.89	53.63	20	»
1.44	31.73	66.83	20	$CrO_3 \cdot SO_3$
1.33	32.05	66.62	20	»
0.69	34.07	65.24	20	»
0.37	37.36	62.27	20	»
0.29	38.97	60.74	20	»
0.11	46.44	53.45	20	»
23.57	0.0	76.43	40	CrO_3
10.95	7.31	81.74	40	»
3.81	12.15	84.04	40	»
1.03	16.51	82.46	40	»
0.12	22.84	77.04	40	»
0.23	26.94	72.83	40	»
1.48	31.83	66.69	40	»
3.58	34.01	62.41	40	»
4.21	34.01	62.41	40	»
6.14	36.52	57.34	40	»
1.66	32.15	66.19	40	$CrO_3 \cdot SO_3$
1.33	33.14	65.53	40	»
1.02	34.25	64.73	40	»
0.62	36.81	62.57	40	»
0.32	42.28	57.40	40	»
0.29	42.83	56.88	40	»

Table № 4336 (continuation)

Comp. of saturated soln., wt%			t	Comp. of solid phase
Chromium Trioxide	Sulfur Trioxide	Water		
2.22	33.66	64.12	75	$CrO_3 \cdot SO_3$
1.46	37.02	61.52	75	»
0.92	40.97	58.11	75	»
0.30	49.89	49.81	75	»
26.38	0.0	73.62	100	CrO_3
13.38	7.13	79.49	100	»
6.98	11.20	81.82	100	»
0.64	20.08	79.28	100	»
0.31	24.01	75.68	100	»
1.67	31.86	66.47	100	»
2.47	32.59	64.94	100	»
3.45	33.79	62.76	100	»
4.97	35.13	59.90	100	»
2.74	36.44	60.82	100	$CrO_3 \cdot SO_3$
2.30	36.83	60.87	100	»
1.57	39.30	59.13	100	»
1.30	41.9	56.80	100	»

№ 4337 SULFUR TRIOXIDE–IRON (II) OXIDE–WATER [226]

SO_3—FeO—H_2O

$t=125$

Comp. of saturated soln., wt%				Comp. of solid phase
Iron Oxide	Iron Oxide (Fe_2O_3)	Sulfur Trioxide	Water	
0.36	0.02	43.57	56.05	$FeSO_4 \cdot H_2O$
0.15	0.05	48.63	51.17	»
0.071	0.02	53.37	46.54	»
0.081	0.01	58.78	41.13	»
0.071	0.02	64.06	35.85	»
0.28	0.00	68.16	31.56	$FeSO_4 \cdot H_2O$+ hexagonal crystals
0.26	0.12	68.23	31.29	hexagonal crystals
0.30	0.02	69.02	30.66	»
0.24	0.03	69.36	30.37	»
0.37	0.00	70.12	29.51	»
0.32	0.00	71.13	28.55	hexagonal crystals+monoclinic crystals
0.35	0.02	71.97	27.64	Monoclinic crystals
0.31	0.00	72.02	27.67	»
0.19	0.14	73.57	26.10	»
0.21	0.00	73.96	25.83	»
0.15	0.00	77.20	22.65	»
0.11	0.03	77.81	22.05	»

Note. The hexagonal and monoclinic crystals represent solid solutions based on $FeSO_4$ H_2SO_4 and $FeS_2O_7 \cdot 2H_2SO_4$, which crystallize in hexagonal form for lower concentrations of SO_3 and in monoclinic form for higher concentrations.

№ 4338 SULFUR TRIOXIDE–IRON (III) OXIDE–WATER [2751]

$SO_3—Fe_2O_3—H_2O$

$t=25$

Comp. of saturated soln., wt%			Comp. of solid phase
Iron Oxide	Sulfur Trioxide	Water	
0.00	71.23	28.77	—
0.24	56.84	43.92	—
3.53	34	62.47	$Fe_2(SO_4)_3 \cdot H_2SO_4 \cdot 9H_2O + Fe_2(SO_4)_3 \cdot H_2SO_4 \cdot 3H_2O$
6.65	32.15	61.20	$Fe_2(SO_4)_3 \cdot H_2SO_4 \cdot 8H_2O$
9.39	31.54	59.07	$Fe_2(SO_4)_3 \cdot H_2SO_4 \cdot 8H_2O + Fe_2(SO_4)_3 \cdot H_2SO_4 \cdot 3H_2O$
12.03	31.51	56.46	$Fe_2(SO_4)_3 \cdot H_2SO_4 \cdot 8H_2O$
13.27	31.84	54.89	»
13.68	31.78	54.54	Unstable
14.49	31.45	54.05	»
15.71	31.88	52.41	»
20.21	31.30	48.49	»
9.39	31.54	59.07	$Fe_2(SO_4)_3 \cdot H_2SO_4 \cdot 8H_2O + Fe_2(SO_4)_3 \cdot 9H_2O$
11.06	29.43	59.51	$Fe_2(SO_4)_3 \cdot 9H_2O$
13.88	28.33	57.79	»
15.23	27.92	56.85	»
16.07	27.98	55.95	»

№ 4339 SULFUR TRIOXIDE–IRON (III) OXIDE–WATER [1086]

$SO_3—F_2O_3—H_2O$

Comp. of saturated soln., wt%			t	Comp. of solid phase
Iron Oxide	Sulfur Trioxide	Water		
0.21	40.64	59.15	18	$Fe_2O_3 \cdot 4SO_3 \cdot 9H_2O$
0.91	36.45	62.64	18	»
6.48	42.43	61.09	18	»
8.00	31.85	60.15	18	»
9.63	31.88	58.49	18	$Fe_2O_3 \cdot 4SO_3 \cdot 9H_2O + Fe_2O_3 \cdot 3SO_3 \cdot 7H_2O$
11.69	30.80	57.51	18	$Fe_2O_3 \cdot 3SO_3 \cdot 7H_2O$
13.88	29.71	56.42	18	»
17.48	29.73	52.79	18	»
18.68	29.64	51.68	18	$Fe_2O_3 \cdot 3SO_3 \cdot 7H_2O +$ + basic solid soln.
17.96	25.42	56.62	18	basic solid soln.
14.00	17.71	68.29	18	»
11.60	13.85	74.55	18	»
6.81	7.60	85.59	18	»
0.27	39.77	59.96	25	$Fe_2O_3 \cdot 4SO_3 \cdot 9H_2O$
0.71	37.22	62.07	25	»
2.38	34.99	62.63	25	»
3.88	33.20	62.92	25	»
8.04	32.06	59.90	25	$Fe_2O_3 \cdot 3SO_3 \cdot 7H_2O$
10.55	30.77	58.68	25	»

Table № 4339 (continuation)

Comp. of saturated soln., wt%			t	Comp. of solid phase
Iron Oxide	Sulfur Trioxide	Water		
13.80	30.02	56.18	25	$Fe_2O_3 \cdot 3SO_3 \cdot 7H_2O$
17.52	29.85	52.63	25	$5(Fe_2O_3 \cdot 3SO_3) \cdot 2Fe_2O_3$
18.56	29.98	51.46	25	»
19.98	29.19	50.83	25	basic solid soln.
19.78	27.90	52.32	25	»
19.55	25.99	54.46	25	»
15.53	17.62	66.85	25	»
13.51	14.58	71.91	25	»
7.91	8.19	83.90	25	»

№ 4340 SULFUR TRIOXIDE–IRON (III) OXIDE–WATER [2345]

$SO_3—Fe_2O_3—H_2O$

$t=50$

Comp. of saturated soln., wt%			Comp. of solid phase
Iron Oxide	Sulfur Trioxide	Water	
0.14	0.39	99.47	$Fe_2O_3 \cdot H_2O$
0.39	0.79	98.82	»
0.90	1.53	97.57	»
1.44	2.30	96.26	$Fe_2O_3 \cdot H_2O+3Fe_2O_3 \cdot 4SO_3 \cdot 9H_2O$
2.55	4.08	93.37	$3Fe_2O_3 \cdot 4SO_3 \cdot 9H_2O$
5.71	9.09	85.20	»
7.19	11.19	81.62	»
15.43	20.08	64.49	»
16.09	20.81	63.10	»
17.96	22.96	59.08	$3Fe_2O_3 \cdot 4SO_3 \cdot 9H_2O+Fe_2O_3 \cdot 2SO_3 \cdot 5H_2O$
20.13	27.18	52.69	$Fe_2O_3 \cdot 2SO_3 \cdot 5H_2O$
20.70	28.40	50.90	$2Fe_2O_3 \cdot 5SO_3 \cdot 17H_2O$
16.78	30.72	52.50	$2Fe_2O_3 \cdot 5SO_3 \cdot 17H_2O+Fe_2O_3 \cdot 3SO_3 \cdot 7H_2O$
10.26	31.91	57.83	$Fe_2O_3 \cdot 3SO_3 \cdot 7H_2O$
8.56	32.52	58.92	»
5.55	33.96	60.49	$Fe_2O_3 \cdot 4SO_3 \cdot 9H_2O$
0.34	41.18	58.48	»
0.10	48.44	51.46	»
0.09	55.34	44.57	»
0.07	59.20	40.73	$Fe_2O_3 \cdot 4SO_3 \cdot 3H_2O$
0.08	62.34	37.58	»
0.07	75.37	24.56	»

№ 4341 [2672]

SULFUR TRIOXIDE–IRON (III) OXIDE–WATER

$SO_3—Fe_2O_3—H_2O$

$t=50$

Concn. in aqueous soln., wt%		Comp. of solid phase
Iron Oxide	Sulfur Trioxide	
0.01	0.10	$Fe_2O_3 \cdot H_2O$
1.95	2.94	$Fe_2O_3 \cdot H_2O + 3Fe_2O_3 \cdot 4SO_3 \cdot 9H_2O$
2.54	3.57	»

№ 4342 [1153]

SULFUR TRIOXIDE–IRON (III) OXIDE–WATER

$SO_3—Fe_2O_3—H_2O$

$t=25$

Comp. of saturated soln., wt%			*d*	Comp. of solid phase
Iron Oxide	Sulfur Trioxide	Water		
2.45	3.55	94.0	1.052	Solid soln.
4.86	6.83	88.31	1.104	»
9.86	12.55	77.59	1.229	»
10.19	12.98	76.83	1.242	»
12.33	18.17	69.50	1.328	»
15.52	19.21	65.27	1.397	»
17.19	23.96	58.85	1.500	»
16.67	24.67	58.66	1.471	»
19.08	25.22	55.70	1.550	»
17.95	28.95	53.10	1.590	»
20.5	26.4	53.10	—	Solid soln. $+ Fe_2O_3 \cdot 2.5SO_3 \cdot 7H_2O$
15.10	29.14	55.76	—	$Fe_2O_3 \cdot 2.5SO_3 \cdot 7H_2O$
15.94	28.88	55.18	—	»
16.70	28.00	55.30	—	»
17.31	27.45	55.24	—	»
18.21	26.85	54.94	—	»
19.06	26.77	54.17	—	»
16.4	28.2	55.4	—	$Fe_2O_3 \cdot 2.5SO_3 \cdot 7H_2O + Fe_2O_3 \cdot 3SO_3 \cdot 9H_2O$
14.49	28.28	57.23	—	$Fe_2O_3 \cdot 3SO_3 \cdot 9H_2O$
12.18	28.88	58.94	—	»
11.06	28.92	60.02	—	»
9.31	29.64	61.05	—	»
7.2	32.1	60.3	—	$Fe_2O_3 \cdot 3SO_3 \cdot 9H_2O + Fe_2O_3 \cdot 4SO_3 \cdot 9H_2O$
6.96	32.06	60.98	—	$Fe_2O_3 \cdot 4SO_3 \cdot 9H_2O$
4.70	32.68	62.62	—	»
2.03	34.33	63.64	—	»
0.26	38.72	61.02	—	»
0.03	56.79	43.18	—	»
8.4	31.8	59.8	—	$Fe_2O_3 \cdot 4SO_3 \cdot 9H_2O + Fe_2O_3 \cdot 3SO_3 \cdot 8H_2O$

Table № 4342 (continuation)

Comp. of saturated soln., wt%			t	Comp. of solid phase
Iron Oxide	Sulfur Trioxide	Water		
10.34	30.96	58.70	—	$Fe_2O_3 \cdot 3SO_3 \cdot 8H_2O$
8.94	31.54	59.52	—	»
14.7	30.3	52.0	—	$Fe_2O_3 \cdot 3SO_3 \cdot 8H_2O + Fe_2O_3 \cdot 2.5SO_3 \cdot 7H_2O$
8.11	31.88	60.01	—	$Fe_2O_3 \cdot 4SO_3 \cdot 9H_2O$
7.96	31.86	60.18	—	Solid soln.

№ 4343 SULFUR TRIOXIDE–IRON (III) OXIDE–WATER [1989]

$SO_3—Fe_2O_3—H_2O$

Comp. of saturated soln., wt%			t	Comp. of solid phase
Iron Oxide	Sulfuric Acid	Water		
2.76	4.12	93.12	10	Solid soln.
7.32	12.59	80.09	10	»
16.30	25.88	57.82	10	»
2.23	16.19	81.58	10	$Fe_2O_3 \cdot 2.44SO_3 + 8.65H_2O$
6.70	31.27	62.03	10	»
0.0	37.23	62.77	10	Solid soln.
0.4	45.00	54.6	10	»
0.27	48.40	51.33	10	»
0.41	50.10	49.49	10	»
0.04	74.50	25.46	10	»
12.46	16.60	70.94	15	»
14.61	21.23	64.16	15	»
15.70	23.05	61.25	15	»
16.63	24.00	59.37	15	»
17.61	26.45	55.94	15	»
16.10	25.40	58.5	15	$Fe_2O_3 \cdot 3SO_3 \cdot 18H_2O$
15.86	24.98	59.16	15	»
14.47	26.15	59.38	15	»
14.20	28.25	57.55	15	»
17.11	31.01	51.88	15	»
18.75	31.94	49.91	15	»
17.75	31.85	50.40	15	»
10.30	30.90	58.80	15	$Fe_2O_3 \cdot 4.46SO_3 \cdot 19.2H_2O$
7.25	32.31	60.44	15	»
2.48	35.23	62.29	15	»
0.80	36.40	62.80	15	»
0.01	37.20	62.79	15	»
3.59	30.60	65.81	15	$Fe_2O_3 \cdot 5.84SO_3 \cdot 30H_2O$
0.77	34.80	64.43	15	»
0.01	37.20	62.79	15	»
7.02	29.39	63.59	15	»
0.98	30.32	68.70	15	»

Table № 4343 (continuation)

Comp. of saturated soln., wt%			t	Comp. of solid phase
Iron Oxide	Sulfur Trioxide	Water		
13.18	32.84	53.98	15	$Fe_2O_3 \cdot 3.3SO_3 \cdot 10.3H_2O$
10.04	39.21	56.75	15	»
2.48	35.23	62.29	15	»
0.80	36.40	62.80	15	»
0.01	9.33	90.66	15	»
14.09	33.80	52.11	15	»
15.65	42.27	42.08	15	»

№ 4344 SULFUR TRIOXIDE–NICKEL (II) OXIDE–WATER [2136]

$SO_3—NiO—H_2O$

$t=300$

Content in aqueous soln., mol/l		Comp. of solid phase	Content in aqueous soln., mol/l		Comp. of solid phase
Sulfur Trioxide	Nickel Oxide		Sulfur Trioxide	Nickel Oxide	
0.403	0.129	$NiSO_4 \cdot H_2O$	0.0989	0.0381	$NiSO_4 \cdot H_2O$
0.375	0.130	»	0.0963	0.0360	»
0.196	0.0662	»	0.0501	0.0205	»
0.193	0.0682	»	0.0500	0.0205	»
0.191	0.0702	»	0.0291	0.0130	»
0.101	0.0392	»	0.0134	0.00642	»

№ 4345 SULFUR TRIOXIDE–NICKEL (II) OXIDE–COPPER (II) OXIDE–URANIUM TRIOXIDE–WATER [2136]

$SO_3—CuO—NiO—UO_3—H_2O$

$t=300$

Content in aqueous soln., mol/l				Content in aqueous soln., mol/l			
Sulfur Trioxide	Uranium Trioxide	Copper Oxide	Nickel Oxide	Sulfur Trioxide	Uranium Trioxide	Copper Oxide	Nickel Oxide
0.999	0.207	0.537	0.0857	0.0942	0.0437	0.0167	0.0216
0.556	0.189	0.218	0.0730	0.0561	0.0227	0.0086	0.0142
0.356	0.129	0.118	0.0524	0.0457	0.0173	0.0071	0.0122
0.207	0.107	0.0388	0.0394	0.0450	0.0170	0.0065	0.0199

Note. Compounds with the following composition form in the solid phase: $UO_3 \cdot UO_2SO_4 \cdot 5H_2O$, $NiSO_4 \cdot H_2O$, $2CuO \cdot CuSO_4 \cdot 2H_2O$

№ 4346 [1929]

SULFUR TRIOXIDE–NICKEL (II) OXIDE–COPPER (II) OXIDE–URANIUM TRIOXIDE–WATER

SO_3—NiO—CuO—UO_3—H_2O

Content in liquid phases in equilibrium, mol/1000g soln.								t
Light phase				Heavy phase				
SO_3	UO_3/SO_3	CuO/SO_3	NiO/SO_3	SO_3	UO_3/SO_3	CuO/SO_3	NiO/SO_3	
0.849	0.649	0.124	0.117	4.10	0.736	0.124	0.118	300
0.926	0.660	0.118	0.113	4.15	0.753	0.121	0.123	300
0.470	0.591	0.047	0.046	6.06	0.850	0.058	0.057	325
1.38	0.657	0.047	0.047	4.40	0.785	0.052	0.051	325
0.434	0.470	0.106	0.108	6.35	0.717	0.120	0.117	325
1.137	0.542	0.101	0.098	4.88	0.678	0.107	0.104	325
0.428	0.425	0.139	0.134	6.43	0.658	0.153	0.147	325
0.196	0.355	0.0827	0.0847	8.31	0.731	0.113	0.118	350
0.427	0.349	0.0895	0.0899	7.67	0.686	0.117	0.112	350
0.799	0.422	0.0910	0.0900	6.68	0.658	0.110	0.106	350
1.141	0.426	0.0892	0.0894	5.72	0.634	0.107	0.099	350
1.661	0.469	0.0890	0.087	4.83	0.613	0.100	0.096	350
0.517	0.306	0.111	0.115	7.32	0.626	0.144	0.138	350
0.824	0.335	0.113	0.114	6.45	0.595	0.138	0.132	350
1.565	0.402	0.120	0.116	5.04	0.557	0.131	0.125	350

№ 4347 [2136]

SULFUR TRIOXIDE–NICKEL (II) OXIDE–URANIUM TRIOXIDE–WATER

SO_3—NiO—UO_3—H_2O

t=300

Content in aqueous soln., mol/l			Content in aqueous soln., mol/l		
Sulfur Trioxide	Uranium Trioxide	Nickel Oxide	Sulfur Trioxide	Uranium Trioxide	Nickel Oxide
0.272	0.1737	0.0721	0.306	0.1632	0.0343
0.1952	0.1330	0.0507	0.217	0.1076	0.0281
0.1879	0.1268	0.0505	0.1922	0.0920	0.0268
0.1564	0.1022	0.0423	0.1281	0.0533	0.0209
0.1281	0.0830	0.0342	0.1075	0.0396	0.0191
0.1031	0.0616	0.0318	0.0819	0.0299	0.0174
0.0734	0.0402	0.0215	0.0696	0.0238	0.0151
0.0517	0.0261	0.0154	0.0513	0.0152	0.0118

Note. Compounds with the following composition form in the solid phase: $UO_3 \cdot UO_2SO_4 \cdot 5H_2O$, $NiSO_4 \cdot H_2O$, $2CuO \cdot CuSO_4 \cdot 2H_2O$

№ 4348 [1929]

SULFUR TRIOXIDE–NICKEL (II) OXIDE–URANIUM TRIOXIDE–WATER

$SO_3—NiO—UO_3—H_2O$

Content in liquid phases in equilibrium, mol/1000g soln.						t
Light phase			Heavy phase			
SO_3	UO_3/SO_3	NiO/SO_3	SO_3	UO_3/SO_3	NiO/SO_3	
0.812	0.751	0.135	4.01	0.844	0.137	300
0.274	0.525	0.042	7.44	0.903	0.061	325
0.420	0.550	0.111	8.65	0.832	0.137	325
1.414	0.645	0.112	4.37	0.762	0.131	325
0.427	0.546	0.160	6.38	0.796	0.178	325
1.265	0.598	0.153	4.61	0.734	0.159	325
2.278	0.626	0.146	2.49	0.625	0.146	325
0.613	0.491	0.034	6.91	0.832	0.051	350
0.760	0.500	0.037	6.40	0.821	0.051	350
1.031	0.527	0.039	5.94	0.802	0.053	350
1.255	0.504	0.040	6.16	0.712	0.048	350
2.010	0.628	0.043	4.47	0.727	0.049	350
0.511	0.420	0.088	7.03	0.773	0.103	350
0.847	0.345	0.188	7.34	0.801	0.144	350

№ 4349 [2346]

SULFUR TRIOXIDE–COPPER (II) OXIDE–WATER

$SO_3—CuO—H_2O$

Comp. of saturated soln., wt%			t	Comp. of solid phase
Copper Oxide	Sulfur Trioxide	Water		
0.00	0.002	99.998	50	$CuO+4CuO \cdot SO_3 \cdot 3H_2O$
0.05	0.05	99.9	50	$4CuO \cdot SO_3 \cdot 3H_2O$
0.13	0.13	99.74	50	»
1.65	1.68	96.67	50	»
3.54	3.58	92.88	50	$4CuO \cdot SO_3 \cdot 3H_2O+3CuO \cdot SO_3 \cdot 2H_2O$
7.53	7.57	84.90	50	$3CuO \cdot SO_3 \cdot 2H_2O$
8.98	9.06	81.96	50	»
10.98	11.08	77.94	50	»
12.44	12.52	75.04	50	$3CuO \cdot SO_3 \cdot 2H_2O+CuO \cdot SO_3 \cdot 5H_2O$
9.34	16.12	74.54	50	$CuO \cdot SO_3 \cdot 5H_2O$
4.74	29.28	65.98	50	»
3.83	34.46	61.71	50	»
3.6	37.5	58.9	50	$CuO \cdot SO_3 \cdot 5H_2O+CuO \cdot SO_3 \cdot 3H_2O$
3.28	38.5	58.22	50	$CuO \cdot SO_3 \cdot 3H_2O$
2.1	40.0	57.9	50	$CuO \cdot SO_3 \cdot 3H_2O+CuO \cdot SO_3 \cdot H_2O$
2.35	43.04	54.61	50	$CuO \cdot SO_3 \cdot H_2O$
1.54	45.40	53.06	50	»

Table № 4349 (continuation)

Comp. of saturated soln., wt%			t	Comp. of solid phase
Copper Oxide	Sulfur Trioxide	Water		
1.09	47.43	51.48	50	$CuO \cdot SO_3 \cdot H_2O$
0.23	56.81	42.96	50	»
0.16	71.42	28.42	50	»
0.17	72.40	27.43	50	$CuO \cdot SO_3 \cdot H_2O + CuO \cdot SO_3$
0.001	0.002	99.997	100	$CuO + 4CuO \cdot SO_3 \cdot 3H_2O$
0.008	0.009	99.983	100	$3CuO \cdot SO_3 \cdot 2H_2O + 4CuO \cdot SO_3 \cdot 3H_2O$
0.02	0.02	99.96	100	»
0.38	0.40	99.22	100	$3CuO \cdot SO_3 \cdot 2H_2O$
1.16	1.24	97.60	100	»
3.69	3.87	92.44	100	»
4.56	4.72	90.72	100	»
5.61	5.81	88.58	100	»
8.62	8.69	82.69	100	»
14.57	14.82	70.51	100	»
21.07	21.44	57.48	100	$3CuO \cdot SO_3 \cdot 2H_2O + 3CuO \cdot 2SO_3 \cdot 5H_2O$
21.14	21.52	57.34	100	$3CuO \cdot 2SO_3 \cdot 5H_2O$
21.37	21.74	56.89	100	$3CuO \cdot 2SO_3 \cdot 5H_2O + CuO \cdot SO_3 \cdot 3H_2O$
17.92	22.73	59.35	100	$CuO \cdot SO_3 \cdot 3H_2O$
15.89	25.44	58.67	100	»
12.60	31.40	56.0	100	$CuO \cdot SO_3 \cdot 3H_2O + CuO \cdot SO_3 \cdot H_2O$
10.49	33.67	55.84	100	$CuO \cdot SO_3 \cdot H_2O$
2.92	45.04	52.04	100	»
1.64	49.20	49.16	100	»
0.84	53.60	45.56	100	»
0.2	70.6	29.2	100	$CuO \cdot SO_3 \cdot H_2O + CuO \cdot SO_3$
0.02	0.05	99.93	200	$3CuO \cdot SO_3 \cdot 2H_2O$
0.17	0.30	99.53	200	»
2.37	2.96	94.67	200	»
4.57	5.58	89.85	200	»
8.58	10.08	81.34	200	»
11.93	13.03	75.04	200	»
17.16	18.94	63.90	200	»
18.46	19.83	61.71	200	$3CuO \cdot SO_3 \cdot 2H_2O + CuO \cdot SO_3 \cdot H_2O$
18.72	21.65	59.63	200	$CuO \cdot SO_3 \cdot H_2O$
14.36	30.99	54.65	200	»
13.36	33.96	52.68	200	»
8.21	43.62	48.17	200	»
6.35	45.81	47.84	200	$CuO \cdot SO_3 \cdot H_2O$
2.14	65.48	32.38	200	»
2.0	67.0	31.0	200	$CuO \cdot SO_3 \cdot H_2O + CuO \cdot SO_3$
1.0	70.0	29.0	200	$CuO \cdot SO_3$

№ 4350 [1233]

SULFUR TRIOXIDE–COPPER (II) OXIDE–WATER

SO_3—CuO—H_2O

$t = 22$

Comp. of saturated soln., wt%			Comp. of solid phase
Copper Oxide	Sulfur Trioxide	Water	
2.22	2.14	95.64	$4CuO \cdot SO_3 \cdot 4H_2O$
5.06	5.01	89.93	»
6.20	6.12	87.68	»
6.70	7.15	86.15	»
13.10	12.67	74.23	»
16.53	17.10	66.37	»
20.90	21.20	57.90	»
23.00	23.30	53.70	»
24.05	24.70	51.25	$4CuO \cdot SO_3 \cdot 4H_2O + CuO \cdot SO_3 \cdot 5H_2O$
20.50	29.70	49.80	$CuO \cdot SO_3 \cdot 5H_2O$
20.90	30.30	48.80	»
19.3	32.1	48.6	»
17.15	34.85	48.0	»
15.1	40.4	44.5	»
12.25	44.0	43.75	»
9.83	48.80	41.37	»
6.97	55.60	37.43	»
4.4	60.7	34.9	»
2.5	68.5	29.0	»
2.23	73.50	24.27	»

№ 4351 [2136]

SULFUR TRIOXIDE–COPPER (II) OXIDE–WATER

SO_3—CuO—H_2O

$t = 300$

Content in aqueous soln., mol/l		Comp. of solid phase	Content in aqueous soln., mol/l		Comp. of solid phase
Sulfur Trioxide	Copper Oxide		Sulfur Trioxide	Copper Oxide	
1.89	1.54	$2CuO \cdot CuSO_4 \cdot 2H_2O$	0.0759	0.0436	$2CuO \cdot CuSO_4 \cdot 2H_2O$
1.16	0.849	»	0.0254	0.0133	»
0.844	0.590	»	0.00730	0.00325	»
0.282	0.177	»	0.00256	0.00970	»

№ 4352 [1233]

SULFUR TRIOXIDE–COPPER (II) OXIDE–WATER

SO_3—CuO—H_2O

$t=22$

Comp. of saturated soln., wt%			Comp. of solid phase
Copper Oxide	Sulfur Trioxide	Water	
1.93	76.60	21.47	$CuO \cdot SO_3 \cdot 3H_2O$
1.4	77.7	20.9	»

№ 4353 [2136]

SULFUR TRIOXIDE–COPPER (II) OXIDE–URANIUM TRIOXIDE–WATER

SO_3—CuO—UO_3—H_2O

$t=300$

Content in aqueous soln., mol/l			Comp. of solid phase
Sulfur Trioxide	Uranium Trioxide	Copper Oxide	
0.323	0.1530	0.1379	$2CuO \cdot CuSO_4 \cdot 2H_2O +$ $+ CuO \cdot 2UO_2SO_4 \cdot 7H_2O$
0.221	0.1271	0.0759	»
0.1950	0.1180	0.0604	»
0.1849	0.1179	0.0561	»
0.1803	0.1156	0.0531	»
0.1800	0.1171	0.0571	—
0.1769	0.1143	0.0538	$2CuO \cdot CuSO_4 \cdot 2H_2O + CuO \cdot 3UO_3$
0.1643	0.0997	0.0538	»
0.1596	0.1024	0.0490	»
0.1257	0.0769	0.0389	»
0.1068	0.0616	0.0323	»
0.1058	0.0630	0.0314	»
0.0851	0.0478	0.0258	»
0.0668	0.0364	0.0206	»
0.0619	0.0321	0.0192	»
0.455	0.295	0.0823	$CuO \cdot 3UO_3$
0.1668	0.1253	0.0321	»
0.1659	0.1232	0.0327	»
0.1577	0.1334	0.01774	»
0.1528	0.1372	0.01045	»
0.1507	0.1392	0.00851	»
0.1411	0.1264	0.01031	»
0.1350	0.1288	0.00377	»
0.1204	0.1246	0.00075	»
0.1111	0.1125	0.00069	»
0.1107	0.1125	0.00072	»
0.1104	0.1118	0.00068	»
0.1104	0.1125	0.00068	»
0.1102	0.1107	0.00068	»
0.1100	0.1130	0.00072	—
0.0940	0.0958	0.00054	$CuO \cdot 3UO_2 + UO_3 \cdot \frac{1}{2}H_2O$
0.0771	0.0769	0.00041	»
0.0500	0.0476	0.00024	»
0.0157	0.0131	0.00010	»

№ 4354 **[1929]**

SULFUR TRIOXIDE–COPPER (II) OXIDE–URANIUM TRIOXIDE–WATER

SO_3—CuO—UO_3—H_2O

Content in liquid phases in equilibrium, mol/1000g soln.						t
Light phase			Heavy phase			
SO_3	UO_3/SO_3	CuO/SO_3	SO_3	UO_3/SO_3	CuO/SO_3	
1.14	0.815	0.0816	1.65	0.835	0.0854	300
1.03	0.772	0.139	3.81	0.846	0.144	300
0.697	0.403	0.333	5.30	0.599	0.379	325
0.344	0.663	0.0369	6.28	0.925	0.0467	325
0.317	0.621	0.0707	6.40	0.889	0.0881	325
0.420	0.645	0.0488	6.28	0.909	0.0622	325
0.415	0.564	0.117	5.95	0.825	0.0141	325
0.418	0.533	0.154	6.42	0.784	0.186	325
0.916	0.532	0.210	5.02	0.695	0.229	325
1.43	0.442	0.346	4.47	0.544	0.365	325
0.179	0.451	0.0335	7.99	0.909	0.0553	350
0.166	0.395	0.113	8.19	0.800	0.165	350
0.195	0.267	0.308	8.49	0.572	0.419	350
0.763	0.456	0.096	6.46	0.742	0.126	350
0.203	0.367	0.192	8.47	0.715	0.250	350
0.396	0.356	0.170	7.62	0.683	0.230	350
0.714	0.385	0.164	6.60	0.655	0.223	350
0.200	0.400	0.125	8.42	0.766	0.197	350
1.14	0.338	0.297	5.66	0.517	0.342	350

№ 4355 **[2140]**

SULFUR TRIOXIDE–COPPER (II) OXIDE–URANIUM TRIOXIDE–WATER

SO_3—CuO—UO_3—H_2O

Content of Sulfur Trioxide in aqueous soln., mol/1000g Water	Solubility, ratio of mol concns. (m)		t	Comp. of solid phase
	Uranium Trioxide m_{UO_3}/m_{SO_3}	Copper Oxide m_{CuO}/m_{SO_3}		
0.0305	0.364	0.308	325	$3CuO \cdot SO_3 \cdot 2H_2O + CuO \cdot 3UO_3$
0.0275	0.375	0.313	325	»
0.0225	0.338	0.329	325	»
0.0308	0.347	0.296	350	»
0.0302	0.371	0.311	350	»
0.0297	0.381	0.311	350	»

№ 4356 [1050]

SULFUR TRIOXIDE–GALLIUM OXIDE–WATER

$SO_3—Ga_2O_3—H_2O$

$t=20$

Comp. of saturated soln., wt%			Comp. of solid phase
Gallium Oxide	Sulfur Trioxide	Water	
0.013	74.65	25.34	$Ga_2(SO_4)_3 \cdot 1.7H_2SO_4 \cdot 6H_2O$
0.06	67.29	32.65	»
0.23	60.26	39.51	»
1.17	50.94	47.89	$Ga_2(SO_4)_3 \cdot H_2SO_4 \cdot 14H_2O$
1.62	45.84	52.54	»
1.96	44.57	53.47	»
2.72	38.11	59.17	»
3.46	35.44	61.10	»
4.52	33.87	61.61	$Ga_2(SO_4)_3 \cdot 20H_2O$
6.79	28.73	64.48	»
10.59	23.86	65.55	»
12.12	22.78	65.10	»
12.88	22.61	64.51	»
15.32	19.68	65.00	»
19.10	16.10	64.80	»
17.05	15.02	67.93	$Ga_2(OH)_{2x}(SO_4)_{3-x} \cdot nH_2O$
17.06	14.63	68.31	$Ga_2(OH)_{1.14} \cdot (SO_4)_{0.93} \cdot 22H_2O$
12.58	12.45	74.97	$Ga(OH)_{1.8} \cdot (SO_4)_{0.6} \cdot 16H_2O$
10.71	10.24	79.05	$Ca(OH)_{2x} \cdot (SO_4)_{3-x} \cdot nH_2O$
8.04	8.73	83.23	$2Ga_2O_3 \cdot SO_3 \cdot 12H_2O$
6.68	6.42	86.90	»
2.72	2.58	94.70	»

№ 4357 [223]

SULFUR TRIOXIDE–NIOBIUM PENTOXIDE–WATER

$SO_3—Nb_2O_5—H_2O$

$t=20$

Comp. of saturated soln., wt%			Comp. of solid phase
Niobium Pentoxide	Sulfur Trioxide	Water	
0.02	3.99	95.99	$Nb_2O_5(xSO_3,\ yH_2O)$
0.02	11.90	88.08	»
0.03	18.96	81.01	»
0.04	22.44	77.52	»
0.05	26.31	73.64	»
0.06	32.99	66.95	»
0.14	36.95	62.91	»
0.22	40.87	58.91	»
0.25	47.70	52.05	$Nb_2O_4SO_4$
0.25	55.11	44.64	»

Table № 4357 (continuation)

Comp. of saturated soln., wt%			Comp. of solid phase
Niobium Pentoxide	Sulfur Trioxide	Water	
0.24	56.31	43.45	$Nb_2O_4(SO_4)_3$
0.22	59.36	40.42	»
0.47	63.49	36.04	»
0.70	68.21	31.09	»
1.01	73.03	25.96	»
1.50	75.30	23.20	$Nb_2O_3(SO_4)_2$
1.78	75.55	22.67	»
2.07	75.74	22.19	»
3.96	76.55	19.49	»
4.60	76.68	18.72	»
4.19	77.12	18.69	»

№ 4358 SULFUR TRIOXIDE–NIOBIUM PENTOXIDE–WATER [225]

SO_3—Nb_2O_5—H_2O

Comp. of saturated soln., wt%			t	Comp. of solid phase
Niobium Pentoxide	Sulfur Trioxide	Water		
0.01	12.74	87.25	100	$2Nb_2O_5 \cdot xSO_3 \cdot yH_2O$
0.20	27.28	72.52	100	»
0.15	31.63	68.22	100	»
0.28	41.70	58.02	100	»
0.86	42.86	56.28	100	»
0.47	44.53	55.00	100	»
0.81	47.99	51.20	100	»
1.42	57.20	41.38	100	»
1.95	59.71	38.34	100	»
0.62	56.55	42.83	150	$Nb_2O_4SO_4$
0.66	57.18	42.16	150	»
0.72	59.33	39.95	150	»
1.91	61.94	36.15	150	»
1.18	63.56	35.26	150	»
2.05	65.92	32.03	150	$Nb_2O_3(SO_4)_3$
2.32	67.54	30.14	150	»
2.05	72.00	26.95	150	»
2.28	73.40	24.32	150	»
2.51	74.40	23.09	150	»
3.14	74.84	22.02	150	»
3.03	75.12	21.85	150	»
3.28	77.22	19.50	150	»
7.46	79.13	13.41	150	$Nb_2O_2(SO_4)_3$
5.92	82.03	12.05	150	»
3.54	82.76	13.70	150	»
3.19	84.73	12.08	150	$Nb_2O(SO_4)_4$
2.40	85.63	11.97	150	»
2.10	85.94	11.96	150	»
2.24	87.37	10.39	150	»

Table № 4358 (continuation)

Comp. of saturated soln., wt%			t	Comp. of solid phase
Niobium Pentoxide	Sulfur Trioxide	Water		
0.43	64.26	35.31	200	$Nb_2O_4SO_4$
0.59	66.22	33.10	200	»
0.83	68.57	30.60	200	»
1.30	69.64	29.06	200	»
1.35	70.48	28.17	200	»
1.46	71.33	27.21	200	»
1.69	72.47	25.84	200	»
2.53	73.18	24.29	200	$Nb_2O_3(SO_4)_2$
2.33	73.95	23.72	200	»
2.16	75.48	22.36	200	»
2.83	77.08	20.09	200	»
2.92	77.63	19.45	200	»
3.15	77.90	18.95	200	»
4.51	78.00	17.15	200	$Nb_2O_3(SO_4)_2+Nb_2O_2(SO_4)_3$
3.27	79.07	17.66	200	$Nb_2O_2(SO_4)_3$
3.24	79.70	17.06	200	»
3.20	79.93	16.87	200	»
3.02	81.00	15.98	200	»
2.64	82.68	14.68	200	»
2.43	84.60	12.97	200	$Nb_2O(SO_4)_4$
2.45	84.96	12.59	200	»
2.40	85.46	12.14	200	»
3.00	85.40	11.60	200	»
2.40	86.40	11.20	200	»

№ 4359 SULFUR TRIOXIDE–TIN (II) OXIDE–WATER [1495]

$SO_3—SnO—H_2O$

Comp. of saturated soln., wt%			t	d_4^t	Comp. of solid phase
Sulfur Trioxide	Tin Oxide	Water			
0.20	0.30	99.50	25	1.004	$SnSO_4 \cdot 2SnO \cdot 2H_2O$
1.36	2.21	96.43	25	1.030	»
1.87	3.07	95.06	25	1.040	»
2.20	3.70	94.10	25	1.053	»
2.76	4.56	92.68	25	1.065	»
3.39	5.72	90.89	25	1.080	»
3.72	6.30	89.98	25	1.090	$SnSO_4 \cdot 2SnO \cdot 2H_2O+SnSO_4 \cdot SnO$
4.85	8.02	87.13	25	1.120	$SnSO_4 \cdot SnO$
5.77	9.68	84.55	25	1.150	»
6.79	11.68	81.53	25	1.185	»
6.90	11.90	81.20	25	1.188	»
8.87	15.15	75.98	25	1.250	»
9.50	16.30	74.20	25	1.270	»
9.53	16.26	74.21	25	1.270	»
10.27	17.50	72.23	25	1.300	$SnSO_4 \cdot SnO+SnSO_4$

Table № 4359 (continuation)

Comp. of saturated soln., wt%			t	d_4^t	Comp. of solid phase
Sulfur Trioxide	Tin Oxide	Water			
10.52	14.64	74.84	25	1.265	$SnSO_4$
10.44	12.78	76.78	25	1.250	»
10.64	11.83	77.53	25	1.240	»
12.85	9.43	77.72	25	1.220	»
15.92	6.30	77.69	25	1.210	»
20.35	3.80	75.85	25	1.223	»
23.00	2.98	74.02	25	1.239	»
31.30	1.50	67.20	25	1.307	»
40.90	0.60	58.50	25	1.404	»
49.63	0.36	50.01	25	1.508	»
0.68	1.04	98.28	50	1.018	$SnSO_4 \cdot 2SnO \cdot 2H_2O$
1.53	2.53	95.94	50	1.037	$SnSO_4 \cdot 2SnO \cdot 2H_2O + SnSO_4 \cdot SnO$
2.17	3.80	94.03	50	1.058	$SnSO_4 \cdot SnO$
5.64	9.33	85.03	50	1.145	»
8.30	14.00	77.70	50	1.230	»
8.70	14.80	76.50	50	1.244	$SnSO_4 \cdot SnO + SnSO_4$
8.78	14.82	76.40	50	1.246	»
14.12	8.74	74.14	50	1.220	$SnSO_4$
19.66	7.42	72.92	50	1.263	»
28.43	5.06	66.51	50	1.326	»

№ 4360 SULFUR TRIOXIDE–TANTALUM PENTOXIDE–WATER [227]

SO_3—Ta_2O_5—H_2O

Comp. of saturated soln., wt%			t	Comp. of solid phase
Tantalum Pentoxide	Sulfur Trioxide	Water		
0.07	10.67	89.26	100	$Ta_2O_4SO_4 \cdot 3H_2O$
0.47	15.00	84.53	100	»
0.51	21.57	87.92	100	»
0.63	25.81	83.56	100	»
0.77	32.60	56.63	100	»
0.42	36.80	62.78	100	»
0.19	45.90	53.91	100	»
0.13	48.80	51.07	100	»
0.05	50.32	49.63	100	»
0.05	55.47	44.48	100	$Ta_2O_5SO_3 \cdot nH_2O$
0.04	62.30	37.66	100	»
0.11	65.01	34.88	100	»
0.60	68.60	30.80	100	»
1.01	71.84	27.15	100	»
0.13	69.95	29.92	200	—
0.13	71.13	28.26	200	—
0.16	72.45	27.39	200	—
0.18	74.42	25.40	200	—

Table № 4360 (continuation)

Comp. of saturated soln., wt%			t	Comp. of solid phase
Tantalum Pentoxide	Sulfur Trioxide	Water		
2.11	75.60	22.29	200	—
6.38	77.20	16.42	200	—
4.95	77.60	17.45	200	—
7.90	80.82	11.28	200	$Ta_2(SO_4)_5$
8.00	82.10	8.55	200	»
7.15	84.30	8.55	200	»
7.45	85.74	16.81	200	»

№ 4361 SULFUR TRIOXIDE–MERCURY (II) OXIDE–WATER [1856]

$SO_3—HgO—H_2O$

Comp. of saturated soln., wt%			t	Comp. of solid phase
Mercury Oxide	Sulfur Trioxide	Water		
0.33	1.24	98.43	25	$3HgO \cdot SO_3$
0.92	2.49	96.59	25	»
1.65	3.93	94.42	25	»
1.85	4.24	93.91	25	$3HgO \cdot SO_3 + 3HgO \cdot 2SO_3 \cdot 2H_2O$
2.12	4.52	93.36	25	»
1.94	4.65	93.41	25	$3HgO \cdot 2SO_3 \cdot 2H_2O$
1.98	5.11	92.91	25	»
2.09	5.58	92.33	25	»
2.08	5.81	92.11	25	»
2.90	5.97	91.13	25	$3HgO \cdot SO_3$
2.05	6.15	91.80	25	$3HgO \cdot 2SO_3 \cdot 2H_2O$
2.13	6.54	91.33	25	»
2.02	6.77	91.21	25	$HgO \cdot SO_3 \cdot H_2O$
1.80	6.90	91.30	25	»
1.01	7.67	91.32	25	»
0.89	7.84	91.27	25	$HgO \cdot SO_3 \cdot H_2O + HgO \cdot SO_3$
0.69	8.36	90.95	25	»
0.53	8.95	90.52	25	»
0.22	10.6	89.18	25	$HgO \cdot SO_3$
Traces	24.2	75.8	25	»
»	60.7	39.3	25	»
0.17	0.96	98.87	50	$3HgO \cdot SO_3$
0.93	3.05	96.02	50	»
1.90	4.92	93.18	50	»
2.09	5.10	92.81	50	»
2.06	5.16	92.78	50	$3HgO \cdot SO_3$
2.12	5.34	92.54	50	»
2.20	5.57	92.23	50	$3HgO \cdot SO_3 + 3HgO \cdot 2SO_3 \cdot 2H_2O$
2.11	5.75	92.14	50	$3HgO \cdot 2SO_3 \cdot 2H_2O$
2.16	5.80	92.04	50	»
2.56	6.27	91.17	50	$3HgO \cdot SO_3 + HgO \cdot SO_3$
2.19	6.34	91.47	50	$3HgO \cdot 2SO_3 \cdot 2H_2O + HgO \cdot SO_3$

Table № 4364 (continuation)

Comp. of liquid phases in equilibrium				
Light liquid phase		Heavy liquid phase		
Content of Sulfur Trioxide mol/1000g soln.	Ratio of concn. of UO_3 to concn. of SO_3, g-mol/1000g soln.	Content of Sulfur Trioxide mol/1000g soln.	Ratio of concn. of UO_3 to concn. of SO_3, g-mol/1000g soln.	t
0.473	0.952	4.33	1.053	300
0.507	0.949	4.24	1.065	300
0.806	0.905	3.68	1.026	300
0.681	0.918	3.87	1.045	300
0.687	0.890	4.11	0.998	300
0.104	0.830	7.20	1.136	325
0.105	0.791	7.18	1.086	325
0.125	0.783	7.24	1.062	325
0.160	0.751	7.20	1.063	325
0.140	0.714	7.32	1.034	325
0.153	0.724	7.11	1.031	325
0.188	0.709	7.03	1.020	325
0.186	0.687	6.80	1.010	325
0.235	0.678	7.05	1.006	325
0.261	0.705	7.00	0.985	325
0.390	0.670	6.57	0.968	325
0.512	0.688	6.16	0.954	325
0.705	0.701	5.27	0.930	325
0.920	0.721	4.85	0.906	325
1.337	0.765	4.29	0.874	325
1.658	0.779	3.99	0.843	325
0.0601	0.806	9.32	1.112	350
0.0603	0.732	9.17	1.088	350
0.0764	0.729	9.14	1.055	350
0.0595	0.622	9.41	1.056	350
0.0730	0.535	9.56	1.021	350
0.074	0.538	10.02	1.016	350
0.083	0.508	9.77	1.007	350
0.071	0.511	9.50	1.002	350
0.107	0.487	9.04	0.995	350
0.112	0.496	9.08	0.987	350
0.172	0.474	8.77	0.962	350
0.196	0.481	8.44	0.956	350
0.277	0.483	7.83	0.942	350
0.371	0.532	7.72	0.928	350
0.471	0.552	7.51	0.909	350
0.528	0.509	7.16	0.897	350
0.520	0.512	6.87	0.892	350
0.728	0.557	6.57	0.872	350
0.793	0.543	6.33	0.862	350
1.473	0.614	5.12	0.814	350
2.36	0.669	3.98	0.768	350

№ 4365 SULFUR TRIOXIDE–URANIUM TRIOXIDE–WATER [2137]

$SO_3—UO_3—H_2O$

Ratio of concns. (mol/l) UO_3/SO_3 in aqueous soln	Temperature at which second liquid (L) or solid (S) phase occurs, for the given SO_3 content (mol/1000g soln.) in the solution:			
	0.2000	0.1000	0.0500	0.0200
1.000	298.0 (L)	308.9 (L)	321.0 (L+S)	$<$ 350 (L)
0.950	301.5 (L)	312.2 (L)	—	—
0.900	306.8 (L)	316.3 (L)	327.0 (L+S)	$<$ 350 (L)
0.800	316.0 (L)	324.8 (L)	333.6 (L)	—
0.750	—	—	—	351.8 (L)
0.700	325.0 (L)	331.5 (L)	340.6 (L)	—
0.600	334.2 (L)	340.0 (L)	348.2 (L)	357.8 (L)
0.500	344.6 (L)	348.8 (L)	356.0 (L)	362.8 (L)
0.400	357.5 (L)	358.3 (L)	363.0 (L)	368.4 (L)
0.300	370.0 (L)	368.5 (L)	370.4 (L)	375.0 (L) *
0.250	—	—	—	374.0 *
0.2000	380 *	377 (L) *	378 *	—
0.1000	—	—	377 *	—
0.0000	380 *	379 *	—	—

* Critical solution temperature

№ 4366 POTASSIUM OXIDE–POTASSIUM SULFITE–WATER [1861]

$K_2O—K_2SO_3—H_2O$

Comp. of saturated soln., wt%			t	Comp. of solid phase
Potassium Sulfate	Potassium Oxide	Water		
47.52	0.0	52.48	0	K_2SO_3
0.43	40.71	58.86	0	$K_2SO_3+KOH \cdot 2H_2O$
0.0	41.01	58.99	0	$KOH \cdot 2H_2O$
48.06	0.0	51.94	11.5	K_2SO_3
0.48	42.41	57.11	11.5	$K_2SO_3+KOH \cdot 2H_2O$
0.0	42.87	57.13	11.5	$KOH \cdot 2H_2O$
49.01	0.0	50.99	25	K_2SO_3
0.47	45.03	54.50	25	$K_2SO_3+KOH \cdot 2H_2O$
0.0	45.51	54.49	25	$KOH \cdot 2H_2O$
49.99	0.0	50.01	33	K_2SO_3
0.37	47.81	51.82	33	$K_2SO_3+KOH \cdot H_2O$
0.0	48.16	51.84	33	$KOH \cdot H_2O+KOH \cdot 1\frac{1}{2}H_2O$
0.0	48.29	51.71	33	$KOH \cdot H_2O+KOH \cdot 2H_2O$
50.37	0.0	49.63	40	K_2SO_3
0.40	48.32	51.28	40	$K_2SO_3+KOH \cdot H_2O$
0.0	48.71	51.29	40	$KOH \cdot H_2O$
52.27	0.0	47.73	65.3	K_2SO_3
0.45	50.14	49.41	65.3	$K_2SO_3+KOH \cdot H_2O$
0.0	50.65	49.35	65.3	$KOH \cdot H_2O$
53.15	0.0	46.85	80	K_2SO_3
0.39	51.50	48.11	80	$K_2SO_3+KOH \cdot H_2O$
0.0	51.81	48.19	80	$KOH \cdot H_2O$
55.53	0.0	44.47	100	K_2SO_3
0.40	54.20	45.31	100	$K_2SO_3+KOH \cdot H_2O$
0.0	54.67	45.33	100	$KOH \cdot H_2O$

№ 4367 POTASSIUM OXIDE–CHROMIUM TRIOXIDE–WATER [2496]

$K_2O—CrO_3—H_2O$

$t=30$

Content in aqueous soln., wt%		Comp. of solid phase	Content in aqueous soln., wt%		Comp. of solid phase
Chromium Trioxide	Potassium Oxide		Chromium Trioxide	Potassium Oxide	
0.00	27.16	K_2CrO_4	10.43	4.91	$K_2Cr_2O_7$
0.00	46.53	»	17.93	3.412	»
0.1775	34.602	»	21.96	3.12	»
1.351	26.602	»	37.14	2.51	»
2.936	22.97	»	43.51	3.01	»
4.69	21.465	»	44.46	3.245	$K_2Cr_2O_7+K_2Cr_3O_{10}$
5.598	20.584	»	46.368	2.823	$K_2Cr_3O_{10}$
9.123	19.771	»	47.651	2.674	»
12.247	19.412	»	49.357	2.353	$K_2Cr_3O_{10}+K_2Cr_4O_{13}$
15.157	19.465	»	50.113	2.146	»
15.407	19.225	»	53.215	1.360	$K_2Cr_4O_3$
20.29	19.07	»	55.829	0.868	»
20.67	19.17	$K_2CrO_4+K_2Cr_2O_7$	62.55	0.796	»
19.096	17.30	$K_2Cr_2O_7$	62.997	0.621	$K_2Cr_4O_3+CrO_3$
16.14	14.02	»	63.289	0.493	»
11.35	7.88	»	62.28	0.0	CrO_3

№ 4368 POTASSIUM OXIDE–CHROMIUM TRIOXIDE–WATER [2007]

$K_2O—CrO_3—H_2O$

Comp. of saturated soln., wt%			t	Comp. of solid phase
Potassium Oxide	Chromium Trioxide	Water		
20.00	5.70	74.30	−25	K_2CrO_4
17.52	13.89	68.59	−13	»
17.12	18.18	64.70	−11.37	»
17.18	18.11	64.71	−11.50	$K_2CrO_4+K_2Cr_2O_4$
8.27	8.01	83.72	− 5	$K_2Cr_2O_7$
1.38	2.93	95.69	− 0.63	»
0.00	6.81	93.19	− 1.78	»
0.00	16.05	83.95	− 5.5	»
0.48	17.25	82.27	− 6.43	»
0.45	23.63	75.92	−10.25	»
0.00	27.26	72.74	−13.22	»
0.00	28.85	71.15	−14.50	»
0.00	35.92	64.08	−22.10	»
0.47	86.14	63.39	−22.11	»
0.88	39.86	59.26	−26.77	»
1.18	42.31	56.51	−30.20	$K_2Cr_2O_7+K_2Cr_3O_{10}$
0.95	43.45	55.60	−34.01	$K_2Cr_3O_{10}$

Table № 4368 (continuation)

Comp. of saturated soln., wt%			t	Comp. of solid phase
Potassium Oxide	Chromium Trioxide	Water		
0.79	45.65	53.56	—39	$K_2Cr_3O_{10}+K_2Cr_4O_{13}$
0.00	49.11	50.89	—49	$K_2Cr_4O_{13}$
0.61	53.57	45.82	—61.5	»
31.18	0.00	68.82	0	$KOH \cdot 2H_2O$
26.06	0.54	73.40	0	K_2CrO_4
19.31	4.27	76.42	0	»
17.06	11.77	71.17	0	»
17.62	18.71	67.67	0	»
17.73	19.04	63.23	0	$K_2CrO_4+K_2Cr_2O_7$
10.90	11.93	77.17	0	$K_2Cr_2O_7$
1.87	3.13	95.0	0	»
0.78	22.38	76.84	0	»
1.47	42.95	55.58	0	$K_2Cr_2O_7+K_2Cr_3O_{10}$
1.25	44.52	54.23	0	$K_2Cr_3O_{10}$
1.17	46.84	51.99	0	»
1.37	47.40	51.23	0	$K_2Cr_3O_{10}+K_2Cr_4O_{13}$
1.24	48.23	50.53	0	$K_2Cr_4O_{13}$
1.16	56.93	41.91	0	»
0.64	61.79	37.57	0	$K_2Cr_4O_{13}+CrO_3$
0.0	61.54	38.46	0	CrO_3
46.8	0.0	33.20	30	$KOH \cdot 2H_2O$
26.89	0.94	72.17	30	K_2CrO_4
22.25	3.06	74.69	30	»
18.65	13.72	67.73	30	»
19.12	20.30	60.58	30	»
19.35	21.0	59.65	30	$K_2CrO_4+K_2Cr_2O_7$
15.04	16.85	68.11	30	$K_2Cr_2O_7$
11.20	13.11	75.69	30	»
2.42	28.21	69.37	30	»
2.50	44.50	53.0	30	$K_2Cr_2O_7+K_2Cr_3O_{10}$
2.35	49.95	47.7	30	$K_2Cr_3O_{10}+K_2Cr_4O_{13}$
1.35	53.39	45.26	30	$K_2Cr_4O_{13}$
0.69	62.81	36.50	30	$K_2Cr_4O_{13}+CrO_3$
0.0	62.52	37.48	30	CrO_3
50.0	0.0	50.0	60	$KOH \cdot 2H_2O$
32.98	0.53	66.49	60	K_2CrO_4
21.05	9.15	69.80	60	»
20.25	14.43	65.32	60	»
20.70	21.97	57.33	60	»
20.61	23.61	55.78	60	$K_2CrO_4+K_2Cr_2O_7$
14.53	20.82	64.63	60	$K_2Cr_2O_7$
10.01	21.21	68.78	60	»
6.86	39.64	53.50	60	»
7.06	49.84	43.10	60	$K_2Cr_2O_7+K_2Cr_3O_{10}$
4.06	54.73	41.21	60	$K_2Cr_3O_{10}$
2.0	60.69	37.31	60	»
1.27	65.77	32.96	60	$K_2Cr_4O_{13}+CrO_3$
0.0	65.12	34.88	60	CrO_3

№ 4369 [2522]

POTASSIUM OXIDE–ARSENIC TRIOXIDE–WATER

$K_2O—As_2O_3—H_2O$

$t=25$

Comp. of saturated soln., wt%			Comp. of solid phase
Potassium Oxide	Arsenic Trioxide	Water	
0.00	2.02	97.98	As_2O_3
5.61	22.63	71.76	»
8.79	36.29	54.92	»
9.69	41.93	48.38	»
11.62	54.02	34.36	»
13.74	58.11	28.15	$K_2As_4O_7=K_2O(As_2O_3)_2$
15.24	53.49	31.27	»
17.28	49.58	33.14	»
18.61	49.21	32.18	»
20.19	49.02	30.79	»
22.01	49.47	28.52	(?)
23.00	49.65	37.35	(?)
24.16	48.96	26.88	$K_2As_4O_9 \cdot 12H_2O= =(K_2O)_3 \cdot (As_2O_3)_2 \cdot 12H_2O$
23.61	44.84	31.55	»
23.49	39.49	37.02	»
24.07	33.60	42.33	»
28.05	24.16	47.79	»
32.93	20.98	46.09	»
35.96	22.47	41.57	»
38.32	29.61	32.07	»
42.48	23.82	33.70	$KOH \cdot H_2O$
43.52	17.57	38.91	»
44.13	11.53	44.34	$KOH \cdot 2H_2O$
45.5	0.0	54.5	»

№ 4370 [1533]

POTASSIUM OXIDE–ARSENIC PENTOXIDE–WATER

$K_2O—As_2O_5—H_2O$

Comp. of saturated soln., wt%			t	Comp. of solid phase
Potassium Oxide	Arsenic Pentoxide	Water		
47.7	0.0	52.3	40	$K_2O \cdot 3H_2O$
46.8	5.6	47.6	40	$5K_2O \cdot As_2O_5 \cdot 18H_2O$
42.6	13.2	44.2	40	$3K_2O \cdot As_2O_5 \cdot 6H_2O$
40.0	19.7	40.3	40	»
32.7	26.7	40.8	40	$3K_2O \cdot As_2O_5 \cdot 14H_2O$
35.1	33.6	31.3	40	$3K_2O \cdot As_2O_5 \cdot 6H_2O$
35.8	34.5	29.7	40	$2K_2O \cdot As_2O_5 \cdot 7H_2O$
30.2	36.2	33.6	40	»
30.1	39.3	30.6	40	$K_2O \cdot As_2O_5 \cdot 2H_2O$
7.4	18.4	74.2	40	»
9.25	65.70	25.05	40	$K_2O \cdot 2As_2O_5 \cdot 5H_2O$

Table № 4370 (continuation)

Comp. of saturated soln., wt%			t	Comp. of solid phase
Potassium Oxide	Arsenic Pentoxide	Water		
8.05	69.80	22.15	40	$K_2O \cdot 3As_2O_5 \cdot 4H_2O$
0.20	73.80	26.00	40	$3As_2O_5 \cdot 5H_2O$
0.00	73.80	26.20	40	»
49.20	0.00	50.80	60	$K_2O \cdot 3H_2O$
47.50	5.53	46.97	60	»
37.7	30.7	31.6	60	$3K_2O \cdot As_2O_5 \cdot 6H_2O$
37.2	37.4	25.4	60	$2K_2O \cdot As_2O_5 \cdot 3H_2O$
34.8	39.3	25.9	60	»
32.7	39.8	27.5	60	$2K_2O \cdot As_2O_5 \cdot 7H_2O$
33.2	42.1	24.7	60	$2K_2O \cdot As_2O_5 \cdot 3H_2O$
32.8	43.3	23.9	60	»
9.15	23.20	67.65	60	$K_2O \cdot As_2O_5 \cdot 2H_2O$
10.40	70.0	19.6	60	$K_2O \cdot 2As_2O_5 \cdot 5H_2O$
9.4	70.6	20.0	60	$K_2O \cdot 3As_2O_5 \cdot 4H_2O$
0.2	74.2	25.6	60	$3As_2O_5 \cdot 5H_2O$
0.0	74.3	25.7	60	»

№ 4371 [579]

POTASSIUM HYDROXIDE–POTASSIUM TETRAHYDROBORATE–WATER

$KOH—KBH_4—H_2O$

Comp. of saturated soln., wt%			t	Comp. of solid phase
Potassium Hydroxide	Potassium Tetrahydro-borate	Water		
0.00	12.66	87.34	0	$KBH_4 \cdot 4H_2O$
0.30	11.86	87.84	0	$KBH_4 \cdot 3H_2O$
1.24	11.22	87.54	0	»
2.04	10.60	87.36	0	$KBH_4 \cdot 3H_2O + KBH_4 \cdot H_2O$
6.90	7.94	85.13	0	$KBH_4 \cdot H_2O$
11.89	6.35	81.76	0	»
17.84	4.01	78.16	0	»
22.39	2.76	74.85	0	»
27.53	1.99	70.48	0	»
31.59	1.37	67.04	0	»
37.08	0.96	61.96	0	»
39.20	0.97	59.83	0	$KBH_4 \cdot H_2O + KBH_4$
44.64	0.60	54.76	0	KBH_4
49.00	0.59	50.41	0	»
50.06	0.60	49.34	0	$KBH_4 + KOH$
50.13	0.64	49.23	0	»
50.14	0.71	49.15	0	»
50.18	0.55	49.27	0	»
51.65	0.48	47.87	0	KBH_4
0.88	15.75	83.37	20	$KBH_4 \cdot 4H_2O$
0.96	15.63	83.91	20	$KBH_4 \cdot 4H_2O + KBH_4 \cdot 3H_2O$
1.06	15.65	83.29	20	$KBH_4 \cdot 3H_2O$

Table № 4371 (continuation)

Comp. of saturated soln., wt%			t	Comp. of solid phase
Potassium Hydroxide	Potassium Tetrahydroborate	Water		
2.83	14.40	82.77	20	$KBH_4 \cdot H_2O$
3.48	13.96	82.56	20	»
5.49	12.68	81.83	20	»
8.45	11.06	80.49	20	»
11.93	9.12	78.95	20	$KBH_4 \cdot H_2O + KBH_4$
12.48	8.86	78.66	20	KBH_4
12.61	8.71	78.68	20	»
17.11	6.63	76.26	20	»
18.93	5.98	75.09	20	»
21.65	4.98	73.37	20	»
27.63	3.30	69.07	20	»
29.87	2.86	67.27	20	»
37.48	1.62	60.90	20	»
40.27	1.43	58.30	20	»
46.01	1.03	52.96	20	»
50.48	0.90	48.62	20	»
54.90	0.80	44.30	20	»
55.06	0.75	44.19	20	$KBH_4 + KOH$
55.10	0.75	44.15	20	»
55.15	0.88	43.97	20	»
55.29	0.86	43.85	20	$KOH \cdot 2H_2O$

№ 4372 [2711]

POTASSIUM HYDROXIDE–POTASSIUM CARBONATE–WATER

$$KOH—K_2CO_3—H_2O$$

$t=30$

Comp. of saturated soln., wt%			Comp. of solid phase
Potassium Carbonate	Potassium Hydroxide	Water	
53.27	0.00	46.73	$K_2CO_3 \cdot 1\frac{1}{2}H_2O$
2.50	53.77	43.73	»
2.05	55.[illegible]4	42.81	$K_2CO_3 \cdot 1\frac{1}{2}H_2O + KOH \cdot 2H_2O$
0.0	55.75	44.25	$KOH \cdot 2H_2O$

№ 4373 [1868]

POTASSIUM HYDROXIDE–POTASSIUM CARBONATE–WATER

$KOH—K_2CO_3—H_2O$

Comp. of saturated soln., wt%			t	Comp. of solid phase
Potassium Carbonate	Potassium Hydroxide	Water		
49.9	2.5	47.6	20	$K_2CO_3 \cdot 1.5H_2O$
47.8	4.4	47.8	20	—
36.9	13.6	49.5	20	—
33.4	16.3	50.3	20	—
33.1	16.8	50.1	20	—
32.8	16.9	50.3	20	$K_2CO_3 \cdot 1.5H_2O$
23.5	24.8	51.7	20	—
19.9	27.8	52.3	20	—
15.3	32.2	52.5	20	—
9.1	38.7	52.2	20	$K_2CO_3 \cdot 1.5H_2O$
6.0	42.2	51.8	20	»
4.6	44.7	50.7	20	—
1.3	50.7	48.0	20	—
48.5	4.8	46.7	40	—
35.1	16.2	48.7	40	—
26.5	23.8	49.7	40	—
22.2	27.2	50.6	40	$K_2CO_3 \cdot 1.5H_2O$
16.5	32.9	50.6	40	—
13.1	36.1	50.8	40	—
9.9	39.7	50.4	40	$K_2CO_3 \cdot 1.5H_2O$
6.8	44.2	49.0	40	—
1.2	56.9	41.9	40	$K_2CO_3 \cdot 1.5H_2O + KOH \cdot 2H_2O$
48.1	6.8	45.1	60	—
38.1	15.6	46.3	60	—
29.7	22.6	47.7	60	—
21.1	30.3	49.6	60	—
14.5	36.8	48.7	60	$K_2CO_3 \cdot 1.5H_2O$
9.1	43.5	47.4	60	—
4.8	51.9	43.3	60	—
2.8	58.4	38.8	60	$K_2CO_3 \cdot 1.5H_2O$
55.0	3.2	41.8	80	—
45.5	11.7	42.8	80	—
41.8	15.0	43.2	80	$K_2CO_3 \cdot 1.5H_2O$
33.8	22.0	44.2	80	—
29.7	25.8	44.5	80	—
25.3	29.7	45.0	80	$K_2CO_3 \cdot 1.5H_2O$
21.1	33.6	45.3	80	—
15.2	39.4	45.4	80	—
9.3	46.8	43.9	80	—
5.0	56.8	38.2	80	$K_2CO_3 \cdot 1.5H_2O$
60.0	0.9	39.1	100	—
58.0	3.0	39.0	100	$K_2CO_3 \cdot 1.5H_2O$
56.0	4.8	39.2	100	—
55.4	5.3	39.3	100	—
51.4	9.0	39.6	100	$K_2CO_3 \cdot 1.5H_2O$
47.0	12.0	41.1	100	—
39.9	22.0	41.3	100	—
32.6	26.1	41.3	100	—

Table № 4373 (continuation)

Comp. of saturated soln., wt%			t	Comp. of solid phase
Potassium Carbonate	Potassium Hydroxide	Water		
26.8	31.5	41.7	100	—
22.1	35.6	42.3	100	$K_2CO_3 \cdot 1.5H_2O$
20.7	37.4	41.9	100	—
19.7	38.4	41.9	100	—
13.5	44.7	41.8	100	—
11.6	47.3	41.1	100	—
8.8	53.6	37.6	100	$K_2CO_3 \cdot 1.5H_2O$
7.7	55.9	36.4	100	—

№ 4374 POTASSIUM HYDROXIDE–POTASSIUM CARBONATE–WATER [1872]

$KOH—K_2CO_3—H_2O$

Comp. of saturated soln., wt%			t	Comp. of solid phase
Potassium Carbonate	Potassium Hydroxide	Water		
49.9	2.5	47.6	20	$K_2CO_3 \cdot 1.5H_2O$
47.8	4.4	47.8	20	—
36.9	13.6	49.5	20	—
33.4	16.3	50.3	20	—
33.1	16.8	50.1	20	$K_2CO_3 \cdot 1.5H_2O$
32.8	16.9	50.3	20	—
23.5	24.8	51.7	20	—
19.9	27.8	52.3	20	—
15.3	32.2	52.5	20	$K_2CO_3 \cdot 1.5H_2O$
9.1	38.7	52.2	20	»
6.0	42.2	51.8	20	—
4.6	44.7	50.7	20	—
1.3	50.7	48.0	20	—
48.5	4.8	46.7	40	—
35.1	16.2	48.7	40	—
26.5	23.8	49.7	40	$K_2CO_3 \cdot 1.5H_2O$
22.2	27.2	50.6	40	—
16.5	32.9	50.6	40	—
13.1	36.1	50.8	40	$K_2CO_3 \cdot 1.5H_2O$
9.9	39.7	50.4	40	—
6.8	44.2	49.0	40	$K_2CO_3 \cdot 1.5H_2O + KOH \cdot 2H_2O$
1.2	56.9	41.9	40	—
48.1	6.8	45.1	60	—
38.1	15.6	46.3	60	—
29.7	22.6	47.7	60	—
21.1	30.3	49.6	60	—
14.5	36.8	48.7	60	$K_2CO_3 \cdot 1.5H_2O$
9.1	43.5	47.4	60	—
4.8	51.9	43.3	60	—
2.8	58.4	38.8	60	$K_2CO_3 \cdot 1.5H_2O$
55.0	3.2	41.8	80	—
45.5	11.7	42.8	80	—
41.8	15.0	43.2	80	$K_2CO_3 \cdot 1.5H_2O$
33.8	22.0	44.2	80	—

Table № 4374 (continuation)

Comp. of saturated soln., wt%			t	Comp. of solid phase
Potassium Carbonate	Potassium Hydroxide	Water		
29.7	25.8	44.5	80	—
25.3	29.7	45.0	80	$K_2CO_3 \cdot 1.5H_2O$
21.1	33.6	45.3	80	—
15.2	39.4	45.4	80	—
9.3	46.8	43.9	80	—
5.0	56.8	38.2	80	$K_2CO_3 \cdot 1.5H_2O$
60.0	0.9	39.1	100	—
58.0	3.0	39.0	100	$K_2CO_3 \cdot 1.5H_2O$
56.0	4.8	39.2	100	—
55.4	5.3	39.3	100	—
51.4	9.0	39.6	100	$K_2CO_3 \cdot 1.5H_2O$
47.0	12.8	40.2	100	—
39.9	22.0	41.1	100	—
32.6	26.1	41.3	100	—
26.8	31.5	41.7	100	—
22.1	35.6	42.3	100	$K_2CO_3 \cdot 1.5H_2O$
20.7	37.4	41.9	100	—
19.7	38.4	41.9	100	—
13.5	44.7	41.8	100	—
11.6	47.3	41.1	100	—
8.8	53.6	37.6	100	$K_2CO_3 \cdot 1.5H_2O$
7.7	55.9	36.4	100	—

№ 4375 POTASSIUM HYDROXIDE–POTASSIUM CARBONATE–WATER [2044]

$KOH—K_2CO_3—H_2O$

Comp. of saturated soln., wt%			t	Comp. of solid phase
Potassium Hydroxide	Potassium Carbonate	Water		
0.00	52.80	47.20	25	$K_2CO_3 \cdot {}^3/_2H_2O$
10.77	39.45	49.78	25	»
14.41	34.84	50.75	25	»
22.83	24.87	52.30	25	»
28.93	18.05	53.12	25	»
37.02	11.10	51.88	25	»
46.33	4.28	49.39	25	»
50.41	2.91	46.68	25	»
52.46	2.41	45.13	25	$K_2CO_3 \cdot {}^3/_2H_2O + KOH \cdot 2H_2O$
52.50	2.36	45.14	25	»
0.00	51.90	48.10	10	$K_2CO_3 \cdot {}^3/_2H_2O$
9.76	39.15	51.09	10	»
20.08	26.93	52.99	10	»
29.51	16.40	54.09	10	»
35.82	10.05	54.13	10	»
42.55	5.10	52.35	10	»
48.92	2.18	48.90	10	$K_2CO_3 \cdot {}^3/_2H_2O + KOH \cdot 2H_2O$
49.03	2.20	48.77	10	»
0.00	51.25	48.75	0	$K_2CO_3 \cdot {}^3/_2H_2O$
6.94	42.50	50.56	0	»

Table № 4375 (continuation)

Comp. of saturated soln., wt%			t	Comp. of solid phase
Potassium Hydroxide	Potassium Carbonate	Water		
13.20	34.30	52.50	0	$K_2CO_3 \cdot {}^3/_2H_2O$
18.20	28.45	53.35	0	»
23.10	22.47	54.43	0	»
29.60	15.13	55.27	0	»
35.43	9.04	55.53	0	»
44.04	3.24	52.72	0	»
47.51	2.09	50.40	0	$K_2CO_3 \cdot {}^3/_2H_2O + KOH \cdot 2H_2O$
47.45	2.11	50.44	0	»
11.05	1.72	87.23	−10	Ice
9.61	4.67	85.72	−10	»
5.68	11.80	82.60	−10	»
0.00	23.60	76.40	−10	»
0.00	49.40	50.60	−10	»
0.00	49.37	50.61	−10	$K_2CO_3 \cdot 6H_2O$
3.88	45.16	50.96	−10	»
6.90	41.90	51.20	−10	»
0.00	50.50	49.50	−10	$K_2CO_3 \cdot 6H_2O + K_2CO_3 \cdot {}^3/_2H_2O$
0.00	50.80	49.20	−10	$K_2CO_3 \cdot {}^3/_2H_2O$
5.09	44.21	50.70	−10	»
13.45	33.60	52.95	−10	»
16.40	29.72	53.88	−10	»
20.00	25.45	54.55	−10	»
21.75	23.50	44.25	−10	»
36.70	7.32	55.98	−10	»
42.92	3.33	53.75	−10	»
45.58	2.01	52.41	−10	$K_2CO_3 \cdot {}^3/_2H_2O + KOH \cdot 2H_2O$
45.70	1.90	52.40	−10	»
46.40	0.92	52.68	−10	$KOH \cdot 2H_2O$
17.39	1.14	81.47	−20	Ice
13.68	7.63	78.69	−20	»
6.84	19.85	73.31	−20	»
0.00	32.52	67.48	−20	»
0.00	44.62	55.38	−20	$K_2CO_3 \cdot 6H_2O$
7.47	36.24	56.29	−20	»
14.68	29.50	55.82	−20	»
17.52	27.41	55.07	−20	$K_2CO_3 \cdot 6H_2O + K_2CO_3 \cdot {}^3/_2H_2O$
22.80	21.83	55.37	−20	$K_2CO_3 \cdot {}^3/_2H_2O$
30.40	12.86	56.74	−20	»
40.54	4.09	55.37	−20	»
44.60	1.73	53.67	−20	$K_2CO_3 \cdot {}^3/_2H_2O + KOH \cdot 2H_2O$
44.90	1.42	53.68	−20	»
24.84	1.22	73.94	−40	Ice
22.04	5.40	72.56	−40	»
18.71	10.65	70.64	−40	»
15.12	17.16	67.72	−40	»
9.62	25.65	64.73	−40	»
7.99	28.62	63.39	−40	$Ice + K_2CO_3 \cdot 6H_2O$
9.88	26.60	63.52	−40	$K_2CO_3 \cdot 6H_2O$
13.12	22.95	63.93	−40	»
20.68	15.15	64.17	−40	»
28.26	11.30	60.44	−40	»

Table № 4375 (continuation)

Comp. of saturated soln., wt%			t	Comp. of solid phase
Potassium Hydroxide	Potassium Carbonate	Water		
31.75	8.87	59.38	—40	$KOH \cdot 4H_2O$
33.40	5.92	60.68	—40	»
35.50	1.78	62.72	—40	»
29.22	1.30	69.48	—60	Ice
28.50	2.48	69.02	—60	»
26.75	5.43	67.82	—60	»
27.50	5.83	66.67	—60	$KOH \cdot 4H_2O$
29.42	3.35	67.23	—60	»
31.36	1.23	67.42	—60	»

№ 4376 POTASSIUM HYDROXIDE–POTASSIUM CARBONATE–WATER [1354]

$KOH—K_2CO_3—H_2O$

Comp. of saturated soln., wt%			t	Comp. of solid phase
Potassium Carbonate	Potassium Hydroxide	Water		
44.0	0.00	56.00	—22	$K_2CO_3 \cdot 5H_2O$
38.50	5.19	56.31	—22	»
36.92	6.44	56.64	—22	»
33.66	9.88	56.46	—22	»
32.80	10.50	55.70	—22	»
30.00	13.80	56.20	—22	»
27.90	17.90	54.20	—22	»
44.34	4.63	51.03	—22	$K_2CO_3 \cdot 1.5H_2O$
43.20	5.48	51.32	—22	»
40.20	7.90	51.90	—22	»
27.20	17.80	55.00	—22	»
13.15	30.15	56.70	—22	»
5.35	38.25	56.40	—22	»
2.60	43.77	53.63	—22	$K_2CO_3 \cdot 1.5H_2O + KOH \cdot 2H_2O$
2.20	44.60	53.20	—22	»
1.02	44.83	54.15	—22	$KOH \cdot 2H_2O$
0.00	45.60	54.40	—22	»
30.50	0.00	69.50	—22	Ice
28.72	2.30	68.98	—22	»
25.91	3.95	70.14	—22	»
22.20	5.90	71.90	—22	»
18.15	8.35	73.50	—22	»
12.00	11.90	76.10	—22	»
4.20	16.39	79.41	—22	»
0.00	18.25	81.75	—22	»
47.70	0.00	52.30	—12	$K_2CO_3 \cdot 5H_2O$
42.90	4.65	52.45	—12	»
41.50	6.00	52.50	—12	»
40.90	6.80	52.30	—12	»
40.33	7.32	52.35	—12	»

Table № 4376 (continuation)

Comp. of saturated soln., wt%			t	Comp. of solid phase
Potassium Carbonate	Potassium Hydroxide	Water		
40.00	8.10	51.90	—12	$K_2CO_3 \cdot 5H_2O$
39.60	8.70	51.70	—12	»
30.80	11.80	57.40	—12	»
43.20	5.90	50.90	—12	$K_2CO_3 \cdot 1.5H_2O$
36.45	11.30	52.25	—12	»
33.00	14.00	53.00	—12	»
31.34	15.20	53.46	—12	»
31.30	15.40	53.30	—12	»
30.80	15.80	53.40	—12	»
26.71	19.20	54.09	—12	»
19.00	25.30	55.70	—12	»
7.00	36.10	56.90	—12	»
4.70	40.20	55.10	—12	»
2.10	45.50	52.40	—12	$K_2CO_3 \cdot 1.5H_2O + KOH \cdot 2H_2O$
2.07	45.50	52.43	—12	»
1.20	46.20	52.60	—12	$KOH \cdot 2H_2O$
0.00	47.50	52.50	—12	»
20.90	0.00	79.10	—12	Ice
18.00	3.20	78.90	—12	»
17.17	3.91	78.92	—12	»
14.55	5.39	80.06	—12	»
12.10	6.60	21.30	—12	»
8.30	9.00	82.70	—12	»
4.84	10.36	84.80	—12	»
0.00	12.50	87.50	—12	»
51.25	0.00	48.75	0.45	$K_2CO_3 \cdot 1.5H_2O$
47.20	3.10	49.70	0.45	»
42.15	6.95	50.90	0.45	»
40.60	8.39	51.01	0.45	»
37.86	10.35	51.79	0.45	»
31.55	15.70	52.75	0.45	»
25.45	20.42	54.13	0.45	»
23.86	21.83	54.31	0.45	»
17.70	27.30	55.00	0.45	»
11.10	33.60	55.30	0.45	»
5.03	41.08	54.89	0.45	»
2.84	44.84	52.32	0.45	»
2.36	46.10	51.54	0.45	»
2.31	46.75	50.94	0.45	»
1.89	47.70	50.41	0.45	$K_2CO_3 \cdot 1.5H_2O + KOH \cdot 2H_2O$
1.87	47.60	50.53	0.45	»
1.25	48.00	50.75	0.45	$KOH \cdot 2H_2O$
0.00	48.95	51.05	0.45	»
53.31	0.00	46.69	30.7	$K_2CO_3 \cdot 1.5H_2O$
43.00	8.20	48.80	30.7	»
40.76	9.90	49.34	30.7	»
31.80	17.44	50.76	30.7	»
31.93	17.29	50.78	30.7	»
30.40	18.80	50.80	30.7	»
22.40	25.40	52.20	30.7	»
18.10	29.20	52.70	30.7	»

Table № 4376 (continuation)

Comp. of saturated soln., wt%			t	Comp. of solid phase
Potassium Carbonate	Potassium Hydroxide	Water		
11.67	35.90	52.43	30.7	$K_2CO_3 \cdot 1.5H_2O$
9.66	37.77	52.57	30.7	»
9.22	38.00	52.78	30.7	»
8.31	39.80	51.89	30.7	»
6.95	42.00	51.05	30.7	»
5.50	44.00	50.50	30.7	»
5.30	44.40	50.30	30.7	»
2.50	53.85	43.65	30.7	»
2.38	55.21	42.31	30.7	»
2.38	55.03	42.59	30.7	»
2.40	54.70	42.90	30.7	$K_2CO_3 \cdot 1.5H_2O + KOH \cdot 2H_2O$
2.38	54.58	43.04	30.7	»
2.37	54.60	43.03	30.7	»
2.39	54.74	42.87	30.7	»
1.90	54.93	43.17	30.7	$KOH \cdot 2H_2O$
1.30	55.20	43.50	30.7	»
0.84	55.63	43.53	30.7	»
1.00	55.38	43.62	30.7	»
0.00	56.50	43.50	30.7	»
2.38	55.51	42.11	30.7	$K_2CO_3 \cdot 1.5H_2O + KOH \cdot H_2O$
2.37	55.60	42.03	30.7	»
1.09	56.40	42.51	30.7	$KOH \cdot H_2O$
0.89	56.62	42.49	30.7	»
0.68	56.98	42.34	30.7	»
0.80	56.66	42.54	30.7	»
0.00	58.00	42.00	30.7	»

№ 4377 [2041]

POTASSIUM HYDROXIDE–POTASSIUM CARBONATE–POTASSIUM MANGANATE–WATER

$KOH—K_2CO_3—K_2MnO_4—H_2O$

Content in aqueous soln.		Solubility of Potassium Manganate, mol/l	t	Content in aqueous soln.		Solubility of Potassium Manganate, mol/l	t
Potassium Hydroxide g-eq/l	Potassium Carbonate mol/l			Potassium Hydroxide g-eq/l	Potassium Carbonate mol/l		
3.0	0.33	1.16	20	8.0	0.33	0.090	40
4.0	0.33	0.696	20	8.0	0.33	0.114	50
6.0	0.33	0.207	20	8.0	0.66	0.047	20
3.0	0.66	0.95	20	8.0	0.66	0.050	25
4.0	0.66	0.560	20	8.0	0.66	0.056	30
6.0	0.66	0.186	20	8.0	0.66	0.064	40
3.0	1.0	0.76	20	8.0	0.66	0.077	50
4.0	1.0	0.462	20	8.0	1.0	0.027	20
6.0	1.0	0.144	20	8.0	1.0	0.031	25
8.0	0.33	0.066	20	8.0	1.0	0.037	30
8:0	0.33	0.072	25	8.0	1.0	0.050	40
8.0	0.33	0.081	30	8.0	1.0	0.064	50

№ 4378 [2041]

POTASSIUM HYDROXIDE–POTASSIUM CARBONATE–POTASSIUM PERMANGANATE–WATER

$KOH—K_2CO_3—KMnO_4—H_2O$

Content in aqueous soln.		Solubility of Potassium Permanganate mol/l	t
Potassium Hydroxide g-eq/l	Potassium Carbonate mol/l		
1.0	0.0	0.120	20
1.0	0.0	0.212	30
1.0	0.0	0.444	40
1.0	0.0	0.49	51
1.0	0.16	0.105	20
1.0	0.16	0.171	30
1.0	0.16	0.33	40
1.0	0.16	0.45	51
1.0	0.33	0.093	20
1.0	0.33	0.148	30
1.0	0.33	0.239	40
1.0	0.33	0.398	51
1.0	0.66	0.080	20
1.0	0.66	0.123	30
1.0	0.66	0.195	40
1.0	0.66	0.30	51
1.0	0.66	0.51	61
1.0	1.0	0.063	20
1.0	1.0	0.105	30
1.0	1.0	0.160	40
1.0	1.0	0.25	51
1.0	1.0	0.42	61
3.0	0.0	0.062	20
3.0	0.0	0.099	30
3.0	0.0	0.154	40
3.0	0.0	0.27	51
3.0	0.0	0.48	61
3.0	0.165	0.056	20
3.0	0.165	0.090	30
3.0	0.165	0.140	40
3.0	0.165	0.24	51
3.0	0.165	0.41	61
3.0	0.33	0.053	20
3.0	0.33	0.084	30
3.0	0.33	0.131	40
3.0	0.33	0.22	51
3.0	0.33	0.35	61
3.0	0.66	0.048	20
3.0	0.66	0.075	30
3.0	0.66	0.116	40
3.0	0.66	0.20	51
3.0	0.66	0.36	61
6.0	0.0	0.028	20
6.0	0.0	0.044	30
6.0	0.0	0.066	40
6.0	0.0	0.106	51
6.0	0.0	0.155	61
6.0	0.165	0.027	20
6.0	0.165	0.043	30
6.0	0.165	0.062	40
6.0	0.165	0.095	51
6.0	0.165	0.150	61
6.0	0.33	0.025	20
6.0	0.66	0.038	30
6.0	0.66	0.057	40
6.0	0.66	0.089	51
6.0	0.66	0.130	61
6.0	1.0	0.024	20
6.0	1.0	0.035	30
6.0	1.0	0.054	40
6.0	1.0	0.081	51
6.0	1.0	0.125	61

№ 4379 [1158]

POTASSIUM HYDROXIDE–POTASSIUM CYANIDE–WATER

$KOH—KCN—H_2O$

$t=25$

Comp. of saturated soln., wt%		
Potassium Cyanide	Potassium Hydroxide	Water
41.56	0.11	58.33
41.67	0.69	57.64
38.92	3.30	57.78
34.72	7.39	57.89
33.58	8.96	57.46
26.39	16.29	57.32

№ 4380 [1746]

POTASSIUM HYDROXIDE–POTASSIUM HEXACYANOFERRATE (III)–WATER

$KOH—K_3Fe(CN)_6—H_2O$

$t=25$

Concn. of Potassium Hydroxide in aqueous soln., g/l	Solubility of Potassium Hexacyanoferrate in aqueous soln. of Potassium Hydroxide g/l
26.3	342.7
54.0	302.3
109.2	215.1

№ 4381 [1746]

POTASSIUM HYDROXIDE–
POTASSIUM HEXAÇYANOFERRATE (III)–
POTASSIUM HEXACYANOFERRATE (II)–WATER

$KOH—K_3Fe(CN)_6—K_4Fe(CN)_6—H_2O$

$t=25$

Concn. of Potassium Hydroxide in aqueous soln., g/l	Concn. in aqueous soln. of Potassium Hydroxide, g/l		Comp. of solid phase
	Potassium Hexa-cyano-ferrate (III)	Potassium Hexa-cyano-ferrate (II)	
26.3	309	66.64	$K_3Fe(CN)_6 + K_4Fe(CN)_6 \cdot 3H_2O$
54.0	275.3	55.19	»
109.2	200.8	35.95	»

№ 4382 [1746]

POTASSIUM HYDROXIDE–POTASSIUM HEXACYANOFERRATE (II)–WATER

$KOH—K_4Fe(CN)_6—H_2O$

$t=25$

Concn. of Potassium Hydroxide in aqueous soln., g/l	Solubility of Potassium Hexacyano-ferrate in aqueous soln. of Potassium Hydroxide, g/l	Comp. of solid phase	Concn. of Potassium Hydroxide in aqueous soln., g/l	Solubility of Potassium Hexacyano-ferrate in aqueous soln. of Potassium Hydroxide, g/l	Comp. of solid phase
5.605	308.5	$K_4Fe(CN)_6 \cdot 3H_2O$	52.8	184.8	$K_4Fe(CN)_6 \cdot 3H_2O$
14.0	283.5	»	78.17	132.1	»
27.8	247.1	»	105.5	86.12	»
39.47	217.4	»			

№ 4383 POTASSIUM HYDROXIDE–POTASSIUM NITRATE–WATER [1290]

$KOH—KNO_3—H_2O$

$t=20$

Concn. in aqueous soln., g/l					
Potassium Nitrate	Potassium Hydroxide	Potassium Nitrate	Potassium Hydroxide	Potassium Nitrate	Potassium Hydroxide
85.632	264.23	36.800	527.901	24.365	786.522
46.00	443.19	30.128	614.295	23.455	833.085
45.495	445.995	27.398	683.859	23.556	842.622

№ 4383a [610]

POTASSIUM HYDROXIDE–POTASSIUM NITRATE–WATER

$KOH—KNO_3—H_2O$

$t=25$

Comp. of saturated soln., wt%			Comp. of solid phase
Potassium Hydroxide	Potassium Nitrate	Water	
5.11	20.36	74.53	KNO_3
12.25	13.31	74.44	»
17.38	9.59	72.03	»
29.65	4.60	65.75	»
20.82	2.49	57.69	»
49.79	1.14	49.07	»
50.85	1.05	48.10	KNO_3+KOH
54.18	0.00	45.82	KOH

№ 4384 [1510]

POTASSIUM HYDROXIDE–POTASSIUM FLUORIDE–WATER

$KOH—KF—H_2O$

$t=20$

Content in aqueous soln., g/100g Water					
Potassium Fluoride	Potassium Hydroxide	Potassium Fluoride	Potassium Hydroxide	Potassium Fluoride	Potassium Hydroxide
98.29	0	69.98	26.57	67.24	43.44
85.48	7.34	68.27	30.59	67.63	43.72
77.12	13.01	68.13	31.57	64.89	44.53
71.62	22.30	67.64	33.01	60.99	47.05
70.52	24.20	66.98	34.02	44.11	60.45
70.21	25.47	67.11	40.49	32.07	70.10

№ 4385 [849]

POTASSIUM HYDROXIDE–POTASSIUM FLUROIDE–WATER

$KOH—KF—H_2O$

Comp. of saturated soln., wt%			t	Comp. of solid phase.
Potassium Fluoride	Potassium Hydroxide	Water		
3.18	15.36	81.46	−21.10	Ice
2.12	10.24	87.64	−10.55	»
1.06	5.12	93.72	− 4.40	»
0.53	2.56	96.91	− 2.00	»
0.27	1.28	98.45	− 0.95	»
6.54	12.60	80.86	−21.0	»
4.36	8.40	87.24	−10.50	»
2.18	4.20	93.62	− 4.25	»
1.09	2.10	96.81	− 1.95	»
0.55	1.05	98.40	− 1.0	»
11.97	3.84	84.19	−11.30	»
7.98	2.58	89.44	− 7.50	»
5.32	1.72	92.96	− 4.50	»
2.66	0.86	96.48	− 2.10	»
1.33	0.43	98.24	− 1.10	»

№ 4386 [167]

POTASSIUM HYDROXIDE–POTASSIUM FLUORIDE–WATER

KOH—KF—H_2O

$t=25$

Comp. of saturated soln., wt%			Comp. of solid phase
Potassium Fluoride	Potassium Hydroxide	Water	
50.9	0.0	49.1	KF · $2H_2O$
49.0	1.2	49.8	»
45.7	4.4	49.1	»
40.3	8.5	50.9	»
39.1	9.9	51.0	»
34.3	14.3	51.4	»
32.1	17.8	50.1	»
30.2	21.3	48.5	»
30.1	22.0	47.9	KF
23.3	28.7	48.0	»
19.0	33.8	48.2	»
16.5	37.3	47.2	»
14.0	41.1	44.9	»
12.2	43.2	44.6	»
10.2	45.8	44.0	»
8.3	49.5	42.2	KF+KOH · $2H_2O$
8.3	49.5	42.2	KOH · $2H_2O$
6.0	49.6	44.4	»
1.7	52.7	45.6	»
0.6	54.4	45.0	»

№ 4387 [956]

POTASSIUM HYDROXIDE–ALUMINUM POTASSIUM OXIDE–POTASSIUM SULFATE–WATER

KOH—$KAlO_2$—K_2SO_4—H_2O

Comp. of saturated soln., wt%				t	d
Aluminum Potassium Oxide	Potassium Hydroxide	Potassium Sulfate	Water		
3.75	5.42	11.05	79.78	95	1.133
6.90	10.20	4.62	78.28	95	1.154
8.00	11.87	3.35	76.78	95	1.174
10.00	14.80	1.97	73.23	95	1.209
12.13	18.50	1.05	68.32	95	1.257
16.30	24.40	0.27	59.03	95	1.367
18.57	27.75	0.14	53.56	95	1.435
19.71	28.76	0.09	51.44	95	1.460
20.88	31.23	0.07	47.82	95	1.504
22.19	32.82	0.02	44.97	95	1.545
23.26	34.66	0.02	42.06	95	1.576
23.76	35.14	0.02	41.08	95	1.600
2.94	4.34	10.60	82.32	75	1.121
6.90	10.28	3.61	79.21	75	1.156
8.01	11.91	2.41	77.67	75	1.178

Table № 4387 (continuation)

Comp. of saturated soln., wt%				t	d
Aluminum Potassium Oxide	Potassium Hydroxide	Potassium Sulfate	Water		
10.56	15.45	1.15	72.84	75	1.230
11.92	17.50	0.76	69.82	75	1.261
14.47	28.85	0.33	64.35	75	1.314
15.58	23.10	0.22	61.10	75	1.358
17.30	25.00	0.11	57.59	75	1.395
19.41	28.25	0.07	52.27	75	1.459
20.40	28.85	0.04	50.71	75	1.479
22.50	31.90	0.01	45.59	75	1.547
22.60	32.65	0.01	44.75	75	1.557
23.00	33.50	0.02	43.48	75	1.571

№ 4388 [2010]

POTASSIUM HYDROXIDE–POTASSIUM SULFIDE BISMUTH SULFIDE–WATER

$KOH—K_2S—Bi_2S_3—H_2O$

$t=25$

Concn. in aqueous soln., g/l		Solubility of Bismuth Sulfide in aqueous soln. of Potassium Hydroxide and Potassium Sulfide, g/l
Potassium Sulfide	Potassium Hydroxide	
55	0	0.043
110	0	0.337
165	0	0.639
55	56	0.240
110	56	1.230
137.5	70	2.354

№ 4389 [1457]

POTASSIUM HYDROXIDE–POTASSIUM SULFATE–WATER

$KOH—K_2SO_4—H_2O$

$t=25$

Comp. of saturated soln., wt%		
Potassium Sulfate	Potassium Hydroxide	Water
10.74	0.0	89.26
7.53	2.89	89.58
4.87	4.85	90.28
2.38	12.66	84.96
0.60	32.03	67.37
0.16	38.30	61.54

№ 4390 [930]

POTASSIUM HYDROXIDE–POTASSIUM SULFATE–WATER

$KOH—K_2SO_4—H_2O$

$t=25$

Content in aqueous soln., wt%		d	Comp. of solid phase
Potassium Sulfate	Potassium Hydroxide		
10.80	0	1.084	K_2SO_4
9.90	0.752	1.080	»
9.82	0.964	1.085	»
8.73	2.01	1.086	»
6.49	4.15	1.086	»
4.43	6.42	1.089	»
1.80	11.64	1.120	»
0.75	16.75	1.160	»
0.35	22.23	1.213	»

$KOH—K_2SO_4—H_2O$

Comp. of saturated soln., wt%			t	Comp. of solid phase
Potassium Hydroxide	Potassium Sulfate	Water		
48.64	0.00	51.36	0	$KOH \cdot 2H_2O$
48.42	0.00	51.58	0	$KOH \cdot 2H_2O + K_2SO_4$
45.45	0.00	54.55	0	K_2SO_4
44.73	0.00	55.27	0	»
42.47	0.00	57.53	0	»
36.54	0.03	63.43	0	»
25.79	0.09	74.12	0	»
23.66	0.12	76.22	0	»
20.51	0.16	79.33	0	»
16.62	0.40	82.98	0	»
10.33	1.39	88.28	0	»
7.21	1.99	90.80	0	»
6.25	2.44	91.31	0	»
3.21	3.98	92.81	0	»
0.00	6.88	93.12	0	»
53.27	0.00	46.73	25	$KOH \cdot 2H_2O$
52.84	0.00	47.16	25	$KOH \cdot 2H_2O + K_2SO_4$
49.98	0.00	50.02	25	K_2SO_4
46.62	0.00	53.38	25	»
42.15	0.00	57.85	25	»
41.53	0.01	58.46	25	»
34.41	0.05	65.54	25	»
25.49	0.22	74.29	25	»
18.56	0.49	80.95	25	»
14.98	1.20	83.82	25	»
13.59	1.49	84.92	25	»
11.36	2.52	86.12	25	»
8.06	3.49	88.45	25	»
0.00	10.51	89.49	25	»
57.57	0.00	42.43	50	$KOH \cdot H_2O$
58.17	0.00	41.83	50	$KOH \cdot H_2O + K_2SO_4$
56.71	0.00	43.29	50	K_2SO_4
48.69	0.02	51.29	50	»
44.45	0.05	55.50	50	»
40.95	0.06	58.99	50	»
30.77	0.25	68.98	50	»
27.01	0.46	72.53	50	»
19.96	0.81	79.23	50	»
18.01	1.02	80.97	50	»
13.12	2.45	84.43	50	»
8.00	5.61	86.39	50	»
2.51	10.81	86.68	50	»
1.34	12.45	86.21	50	»
0.00	14.23	85.77	50	»
59.89	0.00	40.11	75	$KOH \cdot H_2O$
59.53	0.00	40.47	75	$KOH \cdot H_2O + K_2SO_4$
52.23	0.00	47.77	75	K_2SO_4
41.90	0.00	58.10	75	»

Table № 4391 (continuation)

Comp. of saturated soln., wt%			t	Comp. of solid phase
Potassium Hydroxide	Potassium Sulfate	Water		
38.27	0.10	61.63	75	K_2SO_4
26.17	0.52	73.31	75	»
18.85	1.83	79.32	75	»
7.47	6.76	85.77	75	»
0.00	17.60	82.40	75	»

№ 4392 [1931]

POTASSIUM HYDROXIDE–POTASSIUM SULFATE–CALCIUM HYDROXIDE– CALCIUM SULFATE–WATER

$KOH—K_2SO_4—Ca(OH)_2—CaSO_4—H_2O$

$t=25$

Content in aqueous soln., mol/1000g soln.			Comp. of solid phase
$3K_2O$	3CaO	$CaSO_4$	
0.0295	0.00356	0.0331	$CaSO_4 \cdot 2H_2O$
0.0296	0.00381	0.0304	»
0.0297	0.00412	0.0289	»
0.0296	0.00555	0.0223	$CaSO_4 \cdot 2H_2O+Ca(OH)_2$
0.0297	0.00241	0.01176	$Ca(OH)_2$
0.0297	0.00086	0	»

№ 4393 [213]

POTASSIUM HYDROXIDE–POTASSIUM SULFATE–CALCIUM SULFATE–WATER

$KOH—K_2SO_4—CaSO_4—H_2O$

$t=19$

Concn. in aqueous soln., g/l		Solubility of Calcium Sulfate, g/l	Concn. in aqueous soln., g/l		Solubility of Calcium Sulfate, g/l
Potassium Sulfate	Potassium Hydroxide		Potassium Sulfate	Potassium Hydroxide	
0.0	0.56	2.169	2.788	2.85	2.235
0.349	0.56	2.358	5.576	2.85	2.197
0.697	0.56	2.398	11.153	2.85	2.155
1.394	0.56	2.292	22.305	2.85	2.156
2.788	0.56	2.265	0.0	5.61	1.474
5.576	0.56	2.097	0.349	5.61	1.488
11.153	0.56	2.163	0.697	5.61	1.542
22.305	0.56	2.199	1.394	5.61	1.461
0.0	2.85	2.428	2.788	5.61	1.680
0.349	2.85	2.352	5.576	5.61	1.849
0.697	2.85	2.305	11.153	5.61	1.883
1.394	2.85	2.293	22.305	5.61	1.827

№ 4394 [213]

POTASSIUM HYDROXIDE–POTASSIUM SULFATE–LEAD (II) SULFATE–WATER

$KOH—K_2SO_4—PbSO_4—H_2O$

$t=18—20$

Concn. in aqueous soln., g/l		Solubility of Lead Sulfate in aqueous soln. of Potassium Hydroxide and lead sulfate, g/l	Concn. in aqueous soln., g/l		Solubility of Lead Sulfate in aqueous soln. of Potassium Hydroxide and lead sulfate, g/l
Potassium Sulfate	Potassium Hydroxide		Potassium Sulfate	Potassium Hydroxide	
0.000	0.561	0.07006	0.000	2.805	0.42674
0.348	0.561	0.06278	0.348	2.805	0.37397
0.697	0.561	0.05732	0.697	2.805	0.34367
1.394	0.561	0.06582	1.394	2.805	0.37433
2.788	0.561	0.08674	2.788	2.805	0.46526
5.576	0.561	0.10434	5.576	2.805	0.47648
11.153	0.561	0.14589	11.153	2.805	0.4864

№ 4395 [1510]

POTASSIUM HYDROXIDE–POTASSIUM CHLORIDE–WATER

$KOH—KCl—H_2O$

$t=18$

Content in aqueous soln., g/100g Water		Comp. of solid phase	Content in aqueous soln., g/100g Water		Comp. of solid phase
Potassium Chloride	Potassium Hydroxide		Potassium Chloride	Potassium Hydroxide	
30.26	1.76	KCl	10.77	25.16	KCl
28.89	2.31	»	7.08	34.09	»
24.73	6.87	»	6.80	34.56	»
22.15	9.70	»	3.93	45.91	»
17.46	15.04	»	2.19	52.74	»
12.30	21.65	»	2.05	53.64	»

№ 4396 [2748]

POTASSIUM HYDROXIDE–POTASSIUM CHLORIDE–WATER

$KOH—KCl—H_2O$

$t=20$

Content in aqueous soln., g/l		d	Comp. of solid phase	Content in aqueous soln., g/l		d	Comp. of solid phase
Potassium Chloride	Potassium Hydroxide			Potassium Chloride	Potassium Hydroxide		
293	10	1.185	KCl	211	100	1.210	KCl
285	20	1.185	»	205	110	1.210	»
276	30	1.190	»	199	120	1.215	»
265	40	1.192	»	192	130	1.215	»
255	50	1.195	»	185	140	1.220	»
245	60	1.200	»	178	150	1.225	»
236	70	1.200	»	171	160	1.225	»
226	80	1.205	»	165	170	1.230	»
219	90	1.205	»	159	180	1.235	»

Table № 4396 (continuation)

Content in aqueous soln., g/l		d	Comp. of solid phase	Content in aqueous soln., g/l		d	Comp. of solid phase
Potassium Chloride	Potassium Hydroxide			Potassium Chloride	Potassium Hydroxide		
153	190	1.240	KCl	33	530	1.415	KCl
148	200	1.245	»	31	540	1.420	»
142	210	1.250	»	29	550	1.425	»
137	220	1.255	»	27	560	1.430	»
133	230	1.260	»	25	570	1.435	»
128	240	1.265	»	24	580	1.440	»
124	250	1.270	»	23	590	1.445	»
120	260	1.275	»	22	600	1.450	»
115	270	1.280	»	21	610	1.455	»
112	280	1.285	»	20	620	1.460	»
108	290	1.290	»	18	630	1.465	»
104	300	1.295	»	17	640	1.470	»
100	310	1.300	»	16	650	1.475	»
96	320	1.305	»	15	660	1.480	»
93	330	1.310	»	15	670	1.485	»
89	340	1.315	»	15	680	1.490	»
85	350	1.320	»	15	690	1.495	»
81	360	1.325	»	14	700	1.500	»
78	370	1.330	»	14	710	1.505	»
74	380	1.335	»	13	720	1.510	»
71	390	1.340	»	13	730	1.515	»
68	400	1.345	»	13	740	1.520	»
64	410	1.350	»	14	750	1.525	»
61	420	1.355	»	12	760	1.530	»
58	430	1.360	»	12	770	1.535	»
55	440	1.365	»	12	780	1.540	»
53	450	1.370	»	11	790	1.545	»
50	460	1.375	»	11	800	1.550	»
47	470	1.380	»	10	810	1.560	»
44	480	1.385	»	10	820	1.565	»
42	490	1.390	»	9	830	1.570	»
40	500	1.397	»	9	840	1.575	»
38	510	1.405	»	9	850	1.580	»
35	520	1.410	»				

№ 4397 POTASSIUM HYDROXIDE–POTASSIUM CHLORIDE–WATER [1290]

$KOH—KCl—H_2O$

$t=20$

Concn. in aqueous soln., g/l		Concn. in aqueous soln., g/l		Concn. on aqueous soln., g/l	
Potassium Chloride	Potassium Hydroxide	Potassium Chloride	Potassium Hydroxide	Potassium Chloride	Potassium Hydroxide
119.733	264.231	18.874	622.71	10.145	786.522
53.115	443.19	14.248	683.859	9.026	833.085
30.213	527.901	12.533	724.812	8.803	842.622
20.366	614.295	10.295	776.424		

№ 4398 POTASSIUM HYDROXIDE–POTASSIUM CHLORIDE–WATER [1082]

$KOH—KCl—H_2O$

Comp. of saturated soln., wt%			d	t
Potassium Hydroxide	Potassium Chloride	Water		
5	0.0	95	—	−21.3
10	0.0	90	—	−21.3
15	7.9	77.1	1.204	−21.3
20	5.6	74.4	1.239	−21.3
25	3.6	71.4	1.280	−21.3
30	2.2	67.8	1.326	−21.3
35	1.2	63.8	1.376	−21.3
40	0.7	59.3	1.430	−21.3
45	0.4	54.6	1.486	−21.3
50	0.0	50	—	−21.3
55	0.0	45	—	−21.3
5	15.4	79.6	1.180	−11.1
10	11.7	78.3	1.189	−11.1
15	8.8	76.2	1.206	−11.1
20	6.3	73.7	1.238	−11.1
25	4.2	70.8	1.278	−11.1
30	2.6	67.4	1.323	−11.1
35	1.5	63.5	1.372	−11.1
40	0.9	59.1	1.425	−11.1
45	0.5	54.5	1.478	−11.1
50	0.0	50		−11.1
55	0.0	45		−11.1
5	17.6	77.4	1.185	0
10	13.5	76.5	1.194	0
15	9.9	75.1	1.219	0
20	7.2	72.8	1.249	0
25	4.9	70.1	1.283	0
30	3.0	67.0	1.322	0
35	1.8	63.2	1.370	0
40	1.0	59.0	1.419	0
45	0.6	54.4	1.471	0
50	0.0	50	—	0
55	0.0	45	—	0
5	19.2	75.8	1.184	10
10	15.0	75.0	1.203	10
15	11.3	73.7	1.227	10
20	8.0	72.0	1.256	10
25	5.4	69.6	1.285	10
30	3.6	66.4	1.324	10
35	2.2	62.8	1.366	10
40	1.3	58.7	1.415	10
45	0.7	44.3	1.469	10
50	0.5	49.5	1.525	10
55	0.0	45	—	10
5	20.8	74.2	1.180	20
10	16.4	73.6	1.201	20
15	12.3	72.7	1.226	20
20	8 8	71.2	1.252	20
25	6.2	68.8	1.283	20
30	4.2	65.8	1.318	20

Table № 4398 (continuation)

Comp. of saturated soln., wt%			d	t
Potassium Hydroxide	Potassium Chloride	Water		
35	2.6	62.4	1.362	20
40	1.5	58.5	1.411	20
45	0.9	44.1	1.465	20
50	0.6	49.4	1.521	20
55	0.0	45	—	20
5	22.2	72.8	1.182	30
10	17.9	72.1	1.204	30
15	13.6	71.4	1.230	30
20	10.2	69.8	1.258	30
25	7.4	67.6	1.289	30
30	5.0	65.0	1.323	30
35	3.2	61.8	1.364	30
40	1.8	58.2	1.409	30
45	1.1	53.9	1.460	30
50	0.8	49.2	1.515	30
55	0.5	44.5	1.571	30
5	24.7	70.3	1.186	50
10	20.2	69.8	1.208	50
15	16.0	69.0	1.236	50
20	12.3	67.7	1.263	50
25	9.1	65.9	1.293	50
30	6.3	63.7	1.324	50
35	4.1	60.9	1.358	50
40	2.5	57.5	1.404	50
45	1.6	53.4	1.458	50
50	1.1	48.9	1.511	50
55	0.9	44.1	1.568	50
60	0.0	40	—	50
65	0.0	35	—	50
70	0.0	30	—	50
5	27.3	67.7	1.196	70
10	22.7	67.3	1.220	70
15	18.4	66.6	1.243	70
20	14.4	65.6	1.270	70
25	10.8	64.2	1.299	70
30	7.7	62.3	1.330	70
35	5.3	59.7	1.364	70
40	3.7	56.3	1.404	70
45	2.5	52.5	1.451	70
50	1.7	48.3	1.505	70
55	1.1	43.9	1.566	70
60	0.0	40	—	70
65	0.0	35	—	70
70	0.0	30	—	70
5	29.7	65.3	1.214	90
10	24.8	65.2	1.231	90
15	20.0	65.0	1.250	90
20	15.9	64.1	1.272	90
25	12.1	62.9	1.298	90
30	9.1	60.9	1.330	90
35	6.6	58.4	1.367	90

Table № 4398 (continuation)

Comp. of saturated soln., wt%			d	t
Potassium Hydroxide	Potassium Chloride	Water		
40	4.5	55.5	1.408	90
45	3.1	51.9	1.451	90
50	2.2	47.8	1.500	90
55	1.8	43.2	1.556	90
60	1.4	38.6	1.618	90
65	0.0	35	—	90
70	0.0	30	—	90
5	33.0	62.0	—	120
10	27.9	62.1	—	120
15	22.9	62.1	—	120
20	18.0	62.0	—	120
25	14.5	60.5	—	120
30	11.3	58.7	—	120
35	8.8	56.2	—	120
40	6.7	53.3	—	120
45	5.0	50.0	—	120
50	3.3	46.7	—	120
55	2.5	42.5	—	120
60	2.0	38.0	—	120
65	1.7	33.3	—	120
70	0.0	30	—	120
5	36.0	59.0	—	150
10	30.8	59.2	—	150
15	25.8	59.2	—	150
20	21.2	58.8	—	150
25	17.0	58.0	—	150
30	13.8	56.2	—	150
35	10.5	54.5	—	150
40	8.0	52.0	—	150
45	6.0	49.0	—	150
50	4.4	45.6	—	150
55	3.4	41.6	—	150
60	2.7	37.3	—	150
65	2.5	32.5	—	150
70	2.7	27.3	—	150
5	38.3	56.7	—	180
10	33.1	56.9	—	180
15	28.1	56.9	—	180
20	23.5	56.5	—	180
25	19.1	55.9	—	180
30	15.3	54.7	—	180
35	12.1	52.9	—	180
40	9.3	50.7	—	180
45	7.1	47.9	—	180
50	5.5	44.5	—	180
55	4.7	40.3	—	180
60	4.0	36.0	—	180
65	3.9	31.1	—	180

№ 4399 POTASSIUM HYDROXIDE–POTASSIUM CHLORIDE–WATER [2130]

$KOH—KCl—H_2O$

Content in aqueous soln., wt%		$t_{b.p.}$	d	Comp. of solid phase
Potassium Chloride	Potassium Hydroxide			
36.5	0	108.5	—	KCl
31.9	5	109.2	1.227	»
27.4	10	110.0	1.245	»
22.9	15	111.5	1.266	»
18.7	20	113.5	1.292	»
14.9	25	116.5	1.320	»
11.9	30	120.7	1.351	»
9.6	35	125.7	1.382	»
7.7	40	131.5	1.415	»
6.4	45	139.3	1.448	»
5.4	50	151.5	1.8	»
4.7	55	166.3	—	»

№ 4400 POTASSIUM HYDROXIDE–POTASSIUM CHLORIDE–WATER [438]

$KOH—KCl—H_2O$

Content in aqueous soln., wt%			d_4^t	Comp. of solid phase
Potassium Chloride	Potassium Hydroxide			
25.31	1.25	25	1.1821	KCl
23.45	3.23	25	1.1879	»
18.92	8.21	25	1.2050	»
15.78	12.03	25	1.2205	»
11.12	18.25	25	1.2501	»
7.53	24.49	25	1.2852	»
3.80	31.57	25	1.3346	»
29.09	1.05	50	1.1966	»
25.25	5.03	50	1.2083	»
18.80	12.10	50	1.2318	»
10.68	22.43	50	1.2783	»
7.81	27.09	50	1.3051	»
3.40	36.25	50	1.3696	»
17.55	1.81	−11.40	—	KCl+ ice
17.66	1.80	−11.30	—	Ice
16.40	1.67	−10.35	—	»
13.99	1.42	− 8.45	—	»
8.96	0.91	− 5.05	—	»
5.12	0.52	− 2.75	—	»
16.52	3.47	−12.10	—	KCl+ ice
16.74	3.15	−12.05	—	Ice
16.37	3.08	−11.75	—	»
14.50	2.73	−10.05	—	»
8.82	1.66	− 5.55	—	»
4.03	0.76	− 2.35	—	»
11.03	10.37	−16.75	—	KCl+ ice
10.70	9.66	−15.85	—	Ice

Table № 4400 (continuation)

Content in aqueous soln., wt%		t	d_4^t	Comp. of solid phase
Potassium Chloride	Potassium Hydroxide			
10.16	9.18	−14.55	—	Ice
8.69	7.86	−11.65	—	»
6.46	5.83	− 7.85	—	»
2.78	2.51	− 2.95	—	»
2.99	25.04	−44.80	—	KCl+ ice
2.92	22.90	−36.8	—	Ice
2.50	19.69	−27.0	—	»
1.77	13.98	−14.7	—	»
0.52	4.05	− 2.9	—	»

№ 4401 POTASSIUM HYDROXIDE–POTASSIUM CHLORIDE–WATER [1073]

$KOH—KCl—H_2O$

$t=25$

Comp. of saturated soln., wt%			Comp. of saturated soln., wt%		
Potassium Chloride	Potassium Hydroxide	Water	Potassium Chloride	Potassium Hydroxide	Water
22.34	4.21	73.45	4.34	30.56	65.10
22.22	4.34	73.44	3.53	33.37	63.10
18.53	8.34	73.13	2.90	36.07	61.03
18.31	8.69	73.00	2.36	37.90	59.74
15.63	11.79	72.58	1.86	40.51	57.63
14.97	12.76	72.27	1.55	42.28	56.17
13.09	15.92	70.99	1.26	44.18	54.56
12.07	16.44	71.49	1.11	46.04	52.85
9.64	20.37	69.99	0.94	47.74	51.32
9.57	20.64	69.79	0.76	50.43	48.81
7.43	23.62	68.95	0.70	51.64	47.66
8.08	23.90	68.02	0.63	53.32	46.05
6.21	27.01	66.78	0.63	53.21	46.16
5.51	25.41	69.08			

№ 4402 [2267]

POTASSIUM HYDROXIDE–POTASSIUM CHLORIDE–SILVER CHLORIDE–WATER

$KOH—KCl—AgCl—H_2O$

$t=25$

Concn. in aqueous soln., g/l		
Potassium Chloride	Potassium Hydroxide	Silver Chloride
0.255	10.05	0.4896
0.0446	2.00	0.0828

№ 4403 [1313]

POTASSIUM HYDROXIDE–POTASSIUM CHLORATE–WATER

$KOH—KClO_3—H_2O$

$t=25$

Comp. of solvent, g/l		Solubility of Potassium chlorate in aqueous soln. of Potassium Hydroxide, g/l
Potassium Hydroxide	Water	
7.02	92.98	76.60
14.02	85.98	70.23

№ 4404 [1239]

POTASSIUM–HYDROXIDE–POTASSIUM CHLORATE–WATER

$KOH—KClO_3—H_2O$

$t=14—15$

Comp. of saturated soln., wt %		
Potassium Chlorate	Potassium Hydroxide	Water
4.47	1.43	94.10

№ 4405 **POTASSIUM HYDROXIDE–POTASSIUM CHLORATE–WATER** [1290]

$KOH—KClO_3—H_2O$

$t=20$

Concn. in aqueous soln., g/l		Concn. in aqueous soln., g/l		Concn. in aqueous soln., g/l	
Potassium Chlorate	Potassium Hydroxide	Potassium Chlorate	Potassium Hydroxide	Potassium Chlorate	Potassium Hydroxide
11.33	264.3	5.026	482.5	2.636	786.5
10.81	283.8	4.303	527.9	2.390	833.1
7.47	356.2	3.518	614.3	2.341	842.6
5.46	445.9	3.114	683.8		

№ 4406 **POTASSIUM HYDROXIDE–POTASSIUM PERMAGANATE–WATER** [2455]

$KOH—KMnO_4—H_2O$

Concn. of Potassium Hydroxide in aqueous soln., g/l	Solubility of Potassium Permanganate in aqueous soln., of Potassium Hydroxide g/l	t	Concn. of Potassium Hydroxide in aqueous soln., g/l	Solubility of Potassium Permanganate in aqueous soln., of Potassium Hydroxide g/l	t
56.1	7.901	0	336.6	3.635	0
56.1	17.690	10	336.6	6.637	10
56.1	28.287	20	336.6	11.694	19
56.1	49.937	32	336.6	18.015	30
56.1	69.375	40	336.6	25.443	40
56.1	100.823	50	336.6	34.608	50
56.1	185.211	70	336.6	61.63	70
56.1	239.099	80	336.6	79.015	80
112.2	4.899	0	336.6	102.561	90
112.2	10.746	10	448.8	2.686	0
112.2	18.805	20	448.8	4.425	10
112.2	33.66	32	448.8	5.057	20
112.2	48.357	40	448.8	9.798	32
112.2	73.009	50	448.8	13.274	40
112.2	137.328	70	448.8	17.541	50
112.2	188.056	80	448.8	29.709	70
224.4	4.267	0	448.8	36.505	80
224.4	7.585	10	448.8	46.935	90
224.4	12.484	20	561	1.896	0
224.4	23.546	32	561	2.528	10
224.4	33.345	40	561	4.582	20
224.4	48.041	50	561	6.32	30
224.4	90.393	70	561	8.217	40
			561	12.958	70

№ 4407 POTASSIUM HYDROXIDE–POTASSIUM MANGANATE–WATER [2041]

$KOH—K_2MnO_4—H_2O$

Content of Potassium Hydroxide in aqueous soln., g-eq/l	Solubility of Potassium Manganate in aqueous soln. of Potassium Hydroxide, g/l		Content of Potassium Hydroxide in aqueous soln., g-eq/l	Solubility of Potassium Manganate in aqueous soln. of Potassium Hydroxide, g/l	t
2.0	1.69	20	5.0	0.74	50
2.0	1.87	25	6.0	0.24	20
2.0	2.09	30	6.0	0.272	25
2.0	2.57	40	6.0	0.301	30
2.0	2.95	50	6.0	0.360	40
3.0	1.29	20	6.0	0.427	50
3.0	1.37	25	7.0	0.16	20
3.0	1.55	30	7.0	0.172	30
3.0	1.90	40	7.0	0.251	50
3.0	2.21	50	8.0	0.075	20
4.0	0.83	20	8.0	0.080	25
4.0	0.93	25	8.0	0.106	30
4.0	1.04	30	8.0	0.106	40
4.0	1.26	40	8.0	0.132	50
4.0	1.55	50	10.0	0.0160	20
5.0	0.45	20	10.0	0.0184	25
5.0	0.477	25	10.0	0.0204	30
5.0	0.55	30	10.0	0.300	40
5.0	0.62	40	10.0	0.460	50

№ 4408 POTASSIUM HYDROXIDE–POTASSIUM MANGANATE–POTASSIUM PERMANGANATE–WATER [2041]

$KOH—K_2MnO_4—KMnO_4—H_2O$

$t = 20$

Content in aqueous soln.		Solubility of Potassium Permanganate, mol/l	Content in aqueous soln.		Solubility of Potassium Permanganate, mol/l
Potassium Manganate, mol/l	Potassium Hydroxide, g-eq/l		Potassium Manganate, mol/l	Potassium Hydroxide, g-eq/l	
0.070	3.0	0.059	0.167	3.0	0.056
0.082	3.0	0.058	0.175	3.0	0.055
0.113	3.0	0.058	0.205	3.0	0.054
0.145	3.0	0.056			

№ 4409 [2455]

POTASSIUM HYDROXIDE–POTASSIUM MANGANATE–WATER

$KOH—K_2MnO_4—H_2O$

Concn. of Potassium Hydroxide in aqueous soln., g/l	Solubility of Potassium Manganate in aqueous soln. of Potassium Hydroxide, g/l	t	Concn. of Potassium Hydroxide in aqueous soln., g/l	Solubility of Potassium Manganate in aqueous soln. of Potassium Hydroxide, g/l	t
112.2	143.33	0	224.4	107.6	17
112.2	160.08	10	224.4	115.8	25
112.2	180.2	20	224.4	121.9	30
112.2	197.8	30	224.4	134.6	40
112.2	225.0	45	224.4	140.5	45
224.4	87.55	0	224.4	148.2	50

Table № 4409 (continuation)

Concn. of Potassium Hydroxide in aqueous soln., g/l	Solubility of Potassium Manganate in aqueous soln. of Potassium Hydroxide, g/l	t	Concn. of Potassium Hydroxide in aqueous soln., g/l	Solubility of Potassium Manganate in aqueous soln. of Potassium Hydroxide, g/l	t
224.4	158.5	60	448.8	1.517	30
224.4	169.7	70	448.8	18.805	40
224.4	180.6	80	448.8	22.44	50
336.6	24.5	0	448.8	26.391	60
336.6	35.4	15	448.8	30.973	70
336.6	41.2	23	448.8	35.083	80
336.6	47.88	30	561	2.291	0
336.6	57.2	40	561	2.402	10
336.6	61.3	45	561	2.528	20
336.6	74.12	60	561	3.397	30
336.6	83.44	70	561	4.820	40
336.6	92.76	80	561	7.301	50
448.8	9.956	0	561	9.798	60
448.8	11.062	10	561	11.06	70
448.8	12.326	20	561	13.116	80

№ 4410 [2041]

POTASSIUM HYDROXIDE–POTASSIUM PERMANGANATE–WATER

$KOH—KMnO_4—H_2O$

Content of Potassium Hydroxide in aqueous soln., g-eq/l	Solubility of Potassium Permanganate in aqueous soln. of Potassium Hydroxide, mol/l	t	Content of Potassium Hydroxide in aqueous soln., g-eq/l	Solubility of Potassium Permanganate in aqueous soln. of Potassium Hydroxide, mol/l	t
0.00	0.374	20	4.0	0.045	20
0.00	0.550	30	4.0	0.068	30
0.00	0.775	40	4.0	0.115	40
0.00	1.15	50	4.0	0.168	51
0.7	0.260	20	4.0	0.272	61
0.7	0.320	25	5.0	0.036	20
0.7	0.380	30	5.0	0.055	30
0.7	0.440	35	5.0	0.081	40
0.7	0.500	40	5.0	0.120	50
0.7	0.655	50	5.0	0.205	60
1.0	0.120	20	6.0	0.033	20
1.0	0.167	25	6.0	0.049	30
1.0	0.208	30	6.0	0.070	40
1.0	0.255	35	6.0	0.112	51
1.0	0.344	40	6.0	0.168	61
1.0	0.380	45	7.0	0.023	20
1.0	0.475	50	7.0	0.039	30
2.0	0.094	20	7.0	0.060	40
2.0	0.145	30	7.0	0.088	50
2.0	0.240	40	7.0	0.089	51
2.0	0.375	50	7.0	0.130	61
2.0	0.525	60	8.0	0.0198	20
3.0	0.065	20	8.0	0.033	30
3.0	0.099	30	8.0	0.048	40
3.0	0.154	40	8.0	0.069	51
3.0	0.247	50	8.0	0.110	61
3.0	0.480	61			

№ 4411 POTASSIUM HYDROXIDE–POTASSIUM BROMIDE–WATER [1509]

$KOH—KBr—H_2O$

$t=17$

Comp. of saturated soln., wt %					
Potassium Bromide	Potassium Hydroxide	Water	Potassium Bromide	Potassium Hydroxide	Water
35.01	2.28	62.71	16.26	18.19	65.55
28.02	7.33	64.65	8.68	28.29	63.03
23.32	11.54	65.14	3.94	35.28	60.78
18.59	15.28	66.13	1.81	43.88	54.31

№ 4412 POTASSIUM HYDROXIDE–POTASSIUM BROMIDE–WATER [1290]

$KOH—KBr—H_2O$

$t=20$

Concn. in aqueous soln., g/l					
Potassium Bromide	Potassium Hydroxide	Potassium Bromide	Potassium Hydroxide	Potassium Bromide	Potassium Hydroxide
120.4	443.2	41.4	683.8	29.3	786.5
82.4	527.9	36.4	724.8	25.5	832.1
60.9	614.3	29.4	776.4	24.9	842.6
53.6	622.7				

№ 4413 POTASSIUM HYDROXIDE–POTASSIUM TETRATHIOANTIMONATE–WATER [1522]

$KOH—K_3SbS_4—H_2O$

$t=30$

Comp. of saturated soln., wt %			Comp. of solid phase
Potassium Tetra-thioantimonate	Potassium Hydroxide	Water	
75.0	0.0	25.0	$K_3SbS_4 \cdot 5H_2O$
68.4	3.4	28.2	$K_3SbS \cdot 3H_2O$
56.8	11.0	32.2	»
50.9	16.1	33.0	K_3SbS_4
37.7	25.5	36.8	»
19.8	40.5	39.7	»
11.5	49.9	38.6	$K_3SbS_4+KOH \cdot 2H_2O$
9.4	49.9	40.7	$KOH \cdot 2H_2O$
0.0	56.3	43.7	»

№ 4414 POTASSIUM HYDROXIDE–POTASSIUM IODIDE–WATER [1510]

$KOH—KI—H_2O$

$t=19$

Content in aqueous soln., g/100 g Water					
Potassium Iodide	Potassium Hydroxide	Potassium Iodide	Potassium Hydroxide	Potassium Iodide	Potassium Hydroxide
115.94	2.71	51.49	29.21	12.08	76.38
82.54	12.25	30.94	46.48	11.05	76.61
61.05	23.22	19.46	61.78		

№ 4415 **[1290]**

POTASSIUM HYDROXIDE–POTASSIUM IODIDE–WATER

$KOH—KI—H_2O$

$t=20$

Concn. in aqueous soln., g/l					
Potassium Iodide	Potassium Hydroxide	Potassium Iodide	Potassium Hydroxide	Potassium Iodide	Potassium Hydroxide
285.52	527.901	154.878	683.859	111.552	786.522
204.18	614.295	136.784	724.812	92.628	842.061
195.216	622.71				

№ 4416 **[1975]**

POTASSIUM HYDROXIDE–POTASSIUM IODIDE–WATER

$KOH—KI—H_2O$

$t=20$

Concn. in aqueous soln., g/l		d	Concn. in aqueous soln., g/l		d
Potassium Hydroxide	Potassium Iodide		Potassium Hydroxide	Potassium Iodide	
803.6	92.30	1.593	325.7	439.9	1.553
743.7	108.07	1.576	266.2	522.1	—
700.6	125.9	1.555	230.2	578.5	—
618.8	162.8	1.539	224.0	585.2	1.582
589.7	183.8	1.529	134.7	739.4	1.613
493.6	255.8	1.524	104.2	799.8	—
547.1	287.7	—	76.05	851.9	1.653
394.8	360.4	1.525	43.85	919.5	1.703
370.9	380.5	1.531	0.0	1010.8	—
361.4	393.2	1.533	0.0	1014.09	1.716
327.8	437.2	—			

№ 4417 **[1290]**

POTASSIUM HYDROXIDE–POTASSIUM IODATE–WATER

$KOH—KIO_3—H_2O$

$t=20$

Concn. in aqueous soln., g/l					
Potassium Iodate	Potassium Hydroxide	Potassium Iodate	Potassium Hydroxide	Potassium Iodate	Potassium Hydroxide
8.346	264.23	3.082	527.901	2.889	724.812
7.747	283.866	2.782	614.295	3.296	786.522
5.480	356.235	2.739	622.71	4.152	833.085
3.831	445.995	2.803	683.859		

№ 4418 [1833]

POTASSIUM HYDROXIDE–POTASSIUM PERIODATE–WATER

$KOH—KIO_4—H_2O$

$t=25$

Comp. of saturated soln., wt %			d	Comp. of solid phase
Potassium Periodate	Potassium Hydroxide	Water		
0.51	0.00	99.49	1.000	KIO_4
4.12	1.01	94.87	1.044	»
8.03	1.99	89.98	1.087	»
10.32	2.55	87.13	1.116	$KIO_4 + K_4I_2O_9 \cdot 9H_2O$
13.15	4.55	82.30	1.165	$K_4I_2O_9 \cdot 9H_2O$
16.12	6.03	77.85	1.221	»
20.8	8.0	71.2	1.326	»
24.1	9.7	66.2	1.377	»
28.9	11.9	59.2	1.484	»
35.2	14.5	50.3	1.64	»
39.0	16.4	44.6	—	»
44.6	19.1	36.3	—	»

№ 4419 **POTASSIUM HYDROXIDE–CALCIUM OXIDE–WATER** [1661]

$KOH—CaO—H_2O$

Concn. in aqueous soln., g/l		t	Concn. in aqueous soln., g/l		t
Calcium Oxide	Potassium Hydroxide		Calcium Oxide	Potassium Hydroxide	
0.852	1.156	20	0.151	11.900	20
0.611	2.312	20	0.690	1.122	40
0.381	4.186	20	0.280	4.135	40
0.275	5.836	20	0.202	5.680	40
0.185	8.624	20	0.010	11.090	40

№ 4420 [213]

POTASSIUM HYDROXIDE–CALCIUM SULFATE–WATER

$KOH—CaSO_4—H_2O$

$t=19$

Concn. of Potassium Hydroxide in aqueous soln., g/l	Solubility of Calcium Sulfate in aqueous soln. of Potassium Hydroxide, g/l	Concn. of Potassium Hydroxide in aqueous soln., g/l	Solubility of Calcium Sulfate in aqueous soln. of Potassium Hydroxide, g/l
0.000	2.169	2.805	2.428
0.561	2.169	5.610	1.474

№ 4421 POTASSIUM HYDROXIDE–CALCIUM IODATE–WATER [1184]

$KOH—Ca(IO_3)_2—H_2O$

Concn. in aqueous soln., g/l		t	Concn. in aqueous soln., g/l		t
Potassium Hydroxide	Calcium Iodate		Potassium Hydroxide	Calcium Iodate	
0.000	0.903	0	1.001	3.487	25
0.774	1.059	0	1.099	3.532	25
1.286	1.138	0	1.256	3.605	25
1.645	1.192	0	1.519	3.706	25
2.152	1.241	0	2.012	3.863	25
2.482	1.290	0	3.046	4.179	25
3.043	1.340	0	0.000	5.092	40
3.830	1.446	0	0.885	5.646	40
0.000	3.056	25	1.524	6.028	40
0.321	3.211	25	1.814	6.121	40
0.390	3.243	25	2.063	6.309	40
0.453	3.171	25	2.791	6.636	40
0.600	3.339	25	3.660	6.956	40
0.821	3.417	25			

№ 4422 [1117]

POTASSIUM HYDROXIDE–TITANIUM DIOXIDE–WATER

$KOH—TiO_2—H_2O$

$t=18—20$

Comp. of solvent, wt %		Solubility of Titanium Dioxide g/l
Potassium Hydroxide	Water	
10	90	0.30—0.45
40	60	0.70—0.90

№ 4423 POTASSIUM HYDROXIDE–COPPER (II) OXIDE–WATER [2166]

$KOH—CuO—H_2O$

$t=25$

Concn. of aqueous soln. of Potassium Hydroxide, g/l aqueous soln.	Solubility of Copper Oxide in aqueous soln. of Potassium Hydroxide, (g/l).10^5	Concn. of aqueous soln. of Potassium Hydroxide, g/l aqueous soln.	Solubility of Copper Oxide in aqueous soln. of Potassium Hydroxide, (g/l).10^5	Concn. of aqueous soln. of Potassium Hydroxide, g/l aqueous soln.	Solubility of Copper Oxide in aqueous soln. of Potassium Hydroxide, (g/l).10^5
2.335	5.56	14.75	103.3	159.5	7380
3.28	12.25	17.7	308.0	178.0	9500
4.2	15.25	30.4	407.0	233.0	11680
5.225	32.8	42.1	902.0	236.0	12700
6.45	25.1	56.0	1280	284.0	13900
7.76	30.6	91.5	2650	294.0	15250
9.9	41.8	110.0	5430	339.0	23600
12.1	67.5	130.5	6460	470.0	34500

№ 4424 [1979]

POTASSIUM HYDROXIDE–ZINC HYDROXIDE–WATER

$KOH—Zn(OH)_2—H_2O$

$t=18—20$

Concn. in aqueous soln., g/l			
Zinc Hydroxide	Potassium Hydroxide	Zinc Hydroxide	Potassium Hydroxide
9.938	50.4	32.795	113.12
20.770	79.52	36.572	127.68
22.162	87.36	40.249	133.28
26.435	91.28	53.665	155.68
27.031	101.36		

№ 4425 [1695]

POTASSIUM HYDROXIDE–MERCURY (II) OXIDE (YELLOW)–WATER

$KOH—HgO—H_2O$

$t=25$

Concn. in aqueous soln., wt %	
Mercury Oxide	Potassium Hydroxide
0.00516	0.603
0.00517	1.67
0.00523	4.38

№ 4426 [2069]

POTASSIUM HYDROXIDE–THALLIUM CHROMATE–WATER

$KOH—Tl_2CrO_4—H_2O$

Comp. of solvent, wt %		Solubility of Thallium Chromate wt %
Potassium Hydroxide	Water	
31	69	1.8

№ 4427 [1726]

POTASSIUM HYDROXIDE–LEAD (II) OXIDE–WATER

$KOH—PbO—H_2O$

$t=25$

Concn. of Potassium Hydroxide in aqueous soln., g/l	Solubility of Lead Oxide in aqueous soln. of Potassium Hydroxide, g/l
6.603	1.991
56.016	14.084

№ 4428 **POTASSIUM HYDROXIDE–LEAD (II) CHROMATE–WATER** [2031]

$KOH—PbCrO_4—H_2O$

Concn. in aqueous soln., g/l		*t*	Concn. in aqueous soln., g/l		*t*
Lead Chromate	Potassium Hydroxide		Lead Chromate	Potassium Hydroxide	
11.9	23.08	15	26.1	23.08	80
16.2	23.08	60	38.5	23.08	102

№ 4429 **POTASSIUM HYDROXIDE–BISMUTH HYDROXIDE–WATER** [2222]

$KOH—Bi(OH)_3—H_2O$

Concn. in aqueous soln., g/l		*t*	Concn. in aqueous soln., g/l		t
Potassium Hydroxide	Bismuth Hydroxide		Potassium Hydroxide	Bismuth Hydroxide	
28	0	20	560	0.174	20
50	Следы	20	28	0.188	100
112	0.037	20	50	0.249	100
168	0.074	20	112	0.373	100
224	0.100	20	224	0.622	100
280	0.124	20	280	0.622	100
336	0.137	20	448	1.494	100
448	0.137	20	560	2.054	100

[2637

№ 4430 **POTASSIUM TETRABORATE–POTASSIUM PENTABORATE–WATER**

$K_2B_4O_7—KB_5O_8—H_2O$

$t=35$

Content in aqueous soln., g/100 g Water		Comp. of solid phase
Potassium Tetraborate	Potassium Pentaborate	
21.3	0	$K_2B_4O_7 \cdot 4H_2O$
23.8	8.0	$K_2B_4O_7 \cdot 4H_2O + K_2O \cdot 5B_2O_3 \cdot$
0	5.0	$K_2O \cdot 5B_2O_3 \cdot 8H_2O$

№ 4431 **POTASSIUM TETRABORATE–POTASSIUM PENTABOR**

POTASSIUM CHLORIDE–WATER

$K_2B_4O_7—K_2B_{10}O_{16}—KCl—H_2O$

$t=35$

Comp. of saturated soln., wt %				Comp. of
Potassium Tetraborate	Potassium Pentaborate	Potassium Chloride	Water	
0.00	0.00	28.05	71.95	K_2B
17.55	0.00	0.00	72.45	K_2B_1
0.00	4.76	0.00	95.24	KCl+I
4.11	0.0	26.05	69.84	KCl+K
0.00	2.12	27.00	70.88	$K_2B_4O_7 \cdot 4H$
18.05	6.07	0.00	75.88	KCl+
4.55	2.38	25.00	68.07	+K

№ 4432 **POTASSIUM TETRABORATE–POTASS CARBONATE–WATER**

$K_2B_4O_7—K_2CO_3—H_2O$

$t=30$

Comp. of saturated soln., wt %			Comp. o
Potassium Tetraborate	Potassium Carbonate	Water	
0.00	53.34	46.66	K_2C
0.51	52.85	46.64	
1.14	52.69	46.17	
1.35	52.43	46.22	
2.02	52.02	45.96	$K_2CO_3 \cdot 3.$
2.48	51.89	45.63	K
2.50	51.63	45.87	
2.52	50.46	47.02	
2.84	44.80	52.36	
3.34	39.18	57.48	
4.66	28.62	66.72	
6.02	21.23	72.75	
7.27	14.67	78.06	
9.13	8.23	82.64	
[illegible]	7.18	83.23	
1259	2.99	84.78	
15.3	0.00	84.69	

№ 4433 [2763]

POTASSIUM TETRABORATE–POTASSIUM CARBONATE–POTASSIUM SULFATE–WATER

$K_2B_4O_7$—K_2CO_3—K_2SO_4—H_2O

$t=30$

Comp. of saturated soln., wt %				Comp. of solid phase
Potassium Tetraborate	Potassium Carbonate	Potassium Sulfate	Water	
11.34	7.59	0.00	81.07	$K_2SO_4+K_2B_4O_7 \cdot 4H_2O$
8.10	4.30	8.19	79.41	»
7.70	3.73	9.90	78.67	»
6.80	2.45	14.95	75.80	»
6.12	1.82	18.82	73.24	»
5.20	1.05	25.28	68.47	»
5.08	1.01	25.97	67.94	»
4.23	0.47	32.32	62.98	»
4.05	0.45	33.61	61.89	»
3.50	0.19	38.50	57.81	»
3.10	0.12	42.35	54.43	»
2.70	0.05	46.83	50.42	»
2.57	0.03	50.69	46.71	»
2.46	0.02	51.84	45.68	$K_2SO_4+K_2B_4O_7 \cdot 4H_2O+K_2CO_3 \cdot 3.5H_2O$
2.44	0.02	51.87	45.67	»
1.64	0.02	52.32	46.02	$K_2CO_3 \cdot 3.5H_2O+K_2SO_4$
0.93	0.02	52.76	46.29	»
0.00	0.02	53.16	46.82	$K_2CO_3 \cdot 3.5H_2O+K_2B_4O_7 \cdot 4H_2O$
2.48	0.00	51.89	45.63	»

№ 4434 [2763]

POTASSIUM TETRABORATE–POTASSIUM SULFATE–WATER

$K_2B_4O_7$—K_2SO_4—H_2O

$t=30$

Comp. of saturated soln., wt%			Comp. of solid phase
Potassium Tetraborate	Potassium Sulfate	Water	
11.45	0.00	88.55	K_2SO_4
10.65	2.05	87.30	»
9.86	4.06	86.08	»
9.40	5.53	85.07	»
9.14	6.16	84.70	»
8.51	8.43	83.06	»
7.96	9.98	82.06	»
7.67	11.04	81.29	»
7.58	11.33	81.09	$K_2SO_4+K_2B_4O_7 \cdot 4H_2O$
7.58	11.32	81.10	»
7.59	11.35	81.06	»
7.59	11.35	81.06	$K_2B_4O_7 \cdot 4H_2O$
7.41	11.65	80.94	»
6.20	11.92	81.88	»
5.25	12.40	82.35	»
5.06	12.46	82.48	»
3.91	13.03	83.06	»
3.73	13.12	83.15	»
2.34	13.96	83.70	»
0.00	15.31	84.69	»

№ 4435 [2637]

POTASSIUM TETRABORATE–POTASSIUM CHLORIDE–WATER

$K_2B_4O_7$—KCl—H_2O

$t=35$

Content in aqueous soln., soln., g/100 g Water		Comp. of solid phase
Potassium Chloride	Potassium Tetra-borate	
39.0	0	KCl
37.3	5.9	$KCl+K_2B_4O_7 \cdot 4H_2O$
0	21.3	$K_2B_4O_7 \cdot 4H_2O$

№ 4436 [2637]

POTASSIUM PENTABORATE–POTASSIUM CHLORIDE–WATER

KB_5O_8—KCl—H_2O

$t=35$

Content in aqueous soln., soln., g/100 g Water		Comp. of solid phase
KCl	KB_5O_8	
30.0	0	KCl
38.1	3.0	$KCl+K_2O \cdot 5B_2O_3 \cdot 8H_2O$
0	5.0	$K_2O \cdot 5B_2O_3 \cdot 8H_2O$

№ 4437 [2371]

POTASSIUM TETRAFLUOROBORATE–POTASSIUM PERMANGANATE–WATER

KBF_4—$KMnO_4$—H_2O

$t=25$

Concn. in aqueous soln., wt %		Comp. of solid phase
Potassium Tetra-fluoroborate	Potassium Permanganate	
0.0	6.90	$KMnO_4$
0.28	6.76	»
0.33	6.32	»
0.22	6.03	$KMnO_4+KBF_4 \cdot 6KMnO_4$
0.37	5.42	$KBF_4 \cdot 6KMnO_4$
0.48	4.78	»
0.50	4.08	»
0.41	3.19	»
0.22	2.07	$KBF_4+KBF_4 \cdot 6KMnO_4$
0.46	1.58	KBF_4
0.54	1.12	»
0.57	0.0	»

№ 4438 [2372]

POTASSIUM TETRAFLUOROBORATE–POTASSIUM PERCHLORATE–WATER

KBF_4—$KClO_4$—H_2O

$t=25$

Concn. in aqueous soln., wt %		Comp. of solid phase
Potassium Tetra-fluoroborate	Potassium Per-chlorate	
0.574	0.0	KF_4B
0.562	0.213	»
0.569	0.276	»
0.589	0.328	»
0.588	0.393	$KF_4B+8KF_4B \cdot KClO_4$
0.588	0.432	$8KF_4B \cdot KClO_4$
0.476	0.549	»
0.433	0.684	»

Table № 4438 (continuation)

Concn. in aqueous soln., wt %		Comp. of solid phase
Potassium Tetra-fluoroborate	Potassium Per-chlorate	
0.420	0.814	$8KF_4B \cdot KClO_4$
0.427	0.943	»
0.437	1.012	$8KF_4B \cdot KClO_4 + 2KF_4B \cdot KClO_4$
0.406	1.077	$2KF_4B \cdot KClO_4$
0.385	1.165	»
0.337	1.433	»
0.243	1.616	$KClO_4$
0.0	1.971	»

№ 4439 POTASSIUM TETRAFLUOROBORATE–POTASSIUM PERIODATE–WATER [2373]

$KBF_4—KIO_4—H_2O$

$t=35$

Concn. in aqueous soln., wt %		Comp. of solid phase	Concn. in aqueous soln., wt %		Comp. of solid phase
Potassium Tetra-fluoroborate	Potassium Periodate		Potassium Tetra-fluoroborate	Potassium Periodate	
0.92	0.00	KF_4B	0.60	0.48	KIO_4
0.91	0.20	»	0.40	0.55	»
0.90	0.40	KF_4B+KIO_4	0.20	0.68	»
0.80	0.42	KIO_4	0.00	0.80	»

№ 4440 POTASSIUM TETRAFLUOROBORATE–CESIUM TETRAFLUOROBORATE–WATER [564]

$KBF_4—CsBF_4—H_2O$

$t=25$

Comp. of saturated soln., wt %			Comp. of solid phase
Potassium Tetra-fluoroborate	Cesium Tetra-fluoroborate	Water	
0.55	0.00	99.45	KBF_4
0.50	0.20	99.30	»
0.50	0.46	99.04	»
0.52	0.80	98.68	»
0.52	1.20	98.28	»
0.45	1.42	98.13	Solid solution
0.42	1.45	98.13	»
0.42	1.47	98.11	»
0.40	1.50	98.10	»
0.35	1.54	98.11	»
0.20	1.63	98.17	»
0.00	1.75	98.25	$CsBF_4$

№ 4441 [1085]

POTASSIUM HYDROGEN CARBONATE–POTASSIUM CARBONATE–WATER

$KHCO_3—K_2CO_3—H_2O$

$t=18$

Content in aqueous soln., g/l		Comp. of solid phase
Potassium Hydrogen Carbonate	Potassium Carbonate	
280.6	13.3	$KHCO_3$
282.5	14.85	»
272.4	25.5	»
262.3	47.5	»
237.0	103.2	»
230.7	114.1	»
227.6	125.1	»
219.0	145.7	»
203.2	166.8	»
181.0	220.0	»
158.9	289.4	»
151.3	309.2	»
108.4	453.5	»
68.9	607.7	»
64.1	643.0	»
54.6	704.3	»
43.1	800.2	$KHCO_3+2KHCO_3 \cdot K_2CO_3 \cdot 3H_2O$
40.9	805.1	$2KHCO_3 \cdot K_2CO_3 \cdot 3H_2O$
35.0	822.8	»
33.1	825.8	$2KHCO_3 \cdot K_2CO_3 \cdot 3H_2O+K_2CO_3 \cdot 2H_2O$
12.2	836.2	$K_2CO_3 \cdot 2H_2O$
5.9	840.8	»

№ 4442 [1831]

POTASSIUM HYDROGEN CARBONATE–POTASSIUM CARBONATE–WATER

$KHCO_3—K_2CO_3—H_2O$

Content in aqueous soln., wt %		*t*	*d*	Comp. of solid phase
Potassium Hydrogen Carbonate	Potassium Carbonate			
26.55	0	25	1.180	$KHCO_3$
19.31	10.00	25	1.228	»
15.45	16.98	25	1.272	»
12.19	23.36	25	1.316	»
7.35	34.71	25	1.402	»
4.44	43.89	25	1.484	»
2.21	51.79	25	1.558	$KHCO_3+K_2CO_3 \cdot 1.5H_2O$
0	52.74	25	1.555	$K_2CO_3 \cdot 1.5H_2O$
29.93	0	36	—	$KHCO_3$
14.83	22.92	36	1.325	»
7.86	37.57	36	1.437	»
3.88	51.44	36	1.565	$KHCO_3+K_2CO_3 \cdot 1.5H_2O$
0	53.66	36	1.557	$K_2CO_3 \cdot 1.5H_2O$

№ 4443 POTASSIUM HYDROGEN CARBONATE–POTASSIUM CARBONATE–WATER [2623]

$KHCO_3—K_2CO_3—H_2O$

Comp. of saturated soln., wt %			t	d	% Carbon Dioxide in atmosphere	Comp. of solid phase
Potassium Oxide	Carbon Monoxide	Water				
38.00	17.78	44.22	50	1.570	0	$K_2CO_3 \cdot 2H_2O$
37.90	18.07	44.03	50	1.569	16	»
37.43	17.90	44.67	50	1.570	18	$K_2CO_3 \cdot 2H_2O + KHCO_3$
35.81	17.27	46.92	50	1.512	20	$KHCO_3$
31.79	15.80	52.41	50	1.494	22	»
27.25	14.60	58.15	50	1.401	50	»
23.32	14.20	62.48	50	1.352	60	»
19.54	14.10	66.36	50	1.282	80	»
16.95	14.62	68.43	50	1.241	99.5	»
39.65	18.48	41.87	70	1.590	0	$K_2CO_3 \cdot 2H_2O$
39.34	19.41	41.25	70	1.588	22	$K_2CO_3 \cdot 2H_2O + KHCO_3$
39.02	19.16	41.82	70	1.575	26	»
36.36	18.44	45.20	70	1.550	50	$KHCO_3$
29.47	16.85	53.68	70	1.430	55	»
25.99	16.37	57.64	70	1.315	60	»
22.65	16.18	61.17	70	1.335	80	»
20.45	16.87	62.68	70	1.301	99.5	»

№ 4444 POTASSIUM HYDROGEN CARBONATE–POTASSIUM CARBONATE–WATER [2637]

$KHCO_3—K_2CO_3—H_2O$

$t=35$

Content in aqueous soln., g/100 g Water		Comp. of solid phase
Potassium Hydrogen Carbonate	Potassium Carbonate	
40.7	0	$KHCO_3$
10.9	105.0	$KHCO_3 + 2KHCO_3 \cdot K_2CO_3 \cdot 3H_2O$
9.6	113.1	$2KHCO_3 \cdot K_2CO_3 \cdot 3H_2O + K_2CO_3 \cdot 1.5H_2O$
0	115.2	$K_2CO_3 \cdot 1.5H_2O$

№ 4445 POTASSIUM HYDROGEN CARBONATE–POTASSIUM CARBONATE–WATER [1836]

$KHCO_3—K_2CO_3—H_2O$

Comp. of saturated soln., wt %			t	d	Comp. of solid phase
Potassium Carbonate	Potassium Hydrogen Carbonate	Water			
51.33	0.00	48.47	5	1.543	$K_2CO_3 \cdot 1\frac{1}{2}H_2O$
50.59	1.72	47.69	5	1.555	$KCO_3 \cdot 1\frac{1}{2}H_2O + K_2CO_3 \cdot 2KHCO_3 \cdot 1\frac{1}{2}H_2O$
50.03	1.77	48.20	5	—	$K_2CO_3 \cdot 2KHCO_3 \cdot 1\frac{1}{2}H_2O$
49.91	1.90	48.19	5	1.543	»
49.65	1.96	48.39	5	1.542	$K_2CO_3 \cdot 2KHCO_3 \cdot 1\frac{1}{2}H_2O + KHCO_3$
46.38	2.55	51.07	5	1.509	$KHCO_3$
0.00	20.15	79.85	5	1.137	»

Table № 4445 (continuation)

Comp. of saturated soln., wt %			t	d	Comp. of solid phase
Potassium Carbonate	Potassium Hydrogen Carbonate	Water			
52.77	0.00	47.23	25	1.559	$K_2CO_3 \cdot 1\frac{1}{2}H_2O$
51.23	2.64	46.13	25	1.563	$K_2CO_3 \cdot 1\frac{1}{2}H_2O + K_2CO_3 \cdot 2KHCO_3 \cdot 1\frac{1}{2}H_2O$
50.77	2.85	46.38	25	1.562	$K_2CO_3 \cdot 2KHCO_3 \cdot 1\frac{1}{2}H_2O$
50.28	3.02	46.70	25	1.554	»
49.77	3.22	47.01	25	1.549	»
49.48	3.33	47.19	25	1.545	$K_2CO_3 \cdot 2KHCO_3 \cdot 1\frac{1}{2}H_2O + KHCO_3$
48.14	3.56	48.30	25	1.538	$KHCO_3$
42.82	4.67	52.51	25	1.485	»
34.71	7.35	57.94	25	1.402	»
23.36	12.19	74.45	25	1.316	»
16.98	15.45	67.57	25	1.272	»
10.00	19.31	70.69	25	1.228	»
0.00	26.78	73.22	25	1.187	»
53.56	0.00	46.44	35	1.560	$K_2CO_3 \cdot 1\frac{1}{2}H_2O$
51.68	3.26	45.06	35	1.573	$K_2CO_3 \cdot 1\frac{1}{2}H_2O + K_2CO_3 \cdot 2KHCO_3 \cdot 1\frac{1}{2}H_2O$
50.99	3.48	45.53	35	1.563	$K_2CO_3 \cdot 2KHCO_3 \cdot 1\frac{1}{2}H_2O$
50.58	3.62	45.80	35	1.564	»
50.00	3.92	46.08	35	1.556	»
49.51	4.27	46.22	35	1.556	$K_2CO_3 \cdot 2KHCO_3 \cdot 1\frac{1}{2}H_2O + KHCO_3$
48.93	4.27	46.80	35	1.548	$KHCO_3$
37.73	7.50	54.77	35	1.441	»
23.04	14.2	62.75	35	1.329	»
10.04	22.10	67.86	35	1.252	»
0.97	22.82	76.21	35	1.208	»
0.0	29.60	70.40	35	1.200	»
54.8	0.0	45.2	50	—	$K_2CO_3 \cdot 1\frac{1}{2}H_2O$
52.85	3.83	43.32	50	1.584	$K_2CO_3 \cdot 1\frac{1}{2}H_2O + K_2CO_3 \cdot 2KHCO_3 \cdot 1\frac{1}{2}H_2O$
51.97	4.40	43.63	50	1.581	$K_2CO_3 \cdot 2KHCO_3 \cdot 1\frac{1}{2}H_2O$
50.92	4.97	44.11	50	1.569	»
49.65	5.77	44.58	50	1.565	$K_2CO_3 \cdot 2KHCO_3 \cdot 1\frac{1}{2}H_2O + KHCO_3$
0.0	34.2	65.8	50	—	$KHCO_3$

№ 4446 [822]

POTASSIUM HYDROGEN CARBONATE–POTASSIUM CARBONATE–WATER

$KHCO_3$—K_2CO_3—H_2O

$t=42$

Content in aqueous soln., wt %		d	Comp. of solid phase
Potassium Hydrogen Carbonate	Potassium Carbonate		
31.70	0	1.234	$KHCO_3$
28.47	2.64	1.234	»
26.03	5.22	1.238	»
24.65	7.07	1.248	»
21.56	10.53	1.248	»
19.30	14.54	1.278	»
16.55	19.69	1.306	»

Table № 4446 (continuation)

Content in aqueous soln., wt %		d	Comp. of solid phase
Potassium Hydrogen Carbonate	Potassium Carbonate		
15.36	22.15	1.332	$KHCO_3$
12.71	26.34	1.352	»
10.96	30.72	1.388	»
9.03	34.14	1.402	»
7.20	37.42	1.442	»
5.20	43.90	1.490	»
3.90	46.90	1.526	»
2.46	51.10	1.610	$KHCO_3 + K_2CO_3 \cdot 1.5H_2O$
0	55.5	1.598	$K_2CO_3 \cdot 1.5H_2O$

№ 4447 POTASSIUM HYDROGEN CARBONATE–POTASSIUM CARBONATE–WATER [506]

$KHCO_3—K_2CO_3—H_2O$

$t=50$

Comp. of saturated soln., wt %			d	Comp. of solid phase
Potassium Carbonate	Potassium Hydrogen Carbonate	Water		
0.00	34.50	65.50	1.2420	$KHCO_3$
1.93	34.00	64.07	1.2471	»
2.90	31.60	65.50	1.2612	»
4.60	31.10	64.30	1.2643	»
14.40	25.16	60.44	1.3062	»
31.14	16.34	52.52	1.4004	»
40.00	11.00	49.00	—	»
42.70	9.44	47.86	1.4892	»
43.50	8.50	48.00	1.5314	»
45.50	7.80	46.70	—	»
49.50	7.10	43.40	1.5784	$KHCO_3 + 2KHCO_3 \cdot K_2CO_3 \cdot 1.5H_2O$
50.20	6.30	43.50	1.5790	$2KHCO_3 \cdot K_2CO_3 \cdot 1.5H_2O$
51.04	5.07	43.89	1.5798	»
52.03	5.04	42.93	—	»
52.30	4.58	43.12	1.5804	$K_2CO_3 \cdot 1.5H_2O + 2KHCO_3 \cdot K_2CO_3 \cdot 1.5H_2O$
52.28	4.60	43.12	—	»
52.30	4.62	43.08	1.5800	»
54.50	0.00	45.50	1.5938	$K_2CO_3 \cdot 1.5H_2O$

№ 4448 POTASSIUM HYDROGEN CARBONATE–POTASSIUM CARBONATE–POTASSIUM SULFATE–WATER [506]

$KHCO_3—K_2CO_3—K_2SO_4—H_2O$

$t=50$

Comp. of saturated soln., wt %				Comp. of solid phase
Potassium Carbonate	Potassium Sulfate	Potassium Hydrogen Carbonate	Water	
54.10	0.08	0.00	45.82	$K_2CO_3 \cdot 1.5H_2O + K_2SO_4$
52.50	0.08	3.00	44.42	»
53.00	0.06	5.00	41.94	$K_2CO_3 \cdot 1.5H_2O + 2KHCO_3 \cdot K_2CO_3 \cdot 1.5H_2O + K_2SO_4$

Comp. of saturated soln., wt %				Comp. of solid phase
Potassium Carbonate	Potassium Sulfate	Potassium Hydrogen Carbonate	Water	
52.30	0.00	4.60	42.10	$K_2CO_3 \cdot 1.5H_2O + 2KHCO_3 \cdot K_2CO_3 \cdot 1.5H_2O$
53.08	0.06	4.98	41.88	$K_2CO_3 \cdot 1.5H_2O +$ $+ 2KHCO_3 \cdot K_2CO_3 \cdot 1.5H_2O + K_2SO_4$
49.50	0.00	7.10	43.40	$KHCO_3 + 2KHCO_3 \cdot K_2CO_3 \cdot 1.5H_2O$
51.90	0.15	5.26	42.69	$KHCO_3 + K_2SO_4 + 2KHCO_3 \cdot K_2CO_3 \cdot 1.5H_2O$
0.00	2.30	33.48	64.22	$KHCO_3 + K_2SO_4$
8.00	1.10	28.78	62.12	»
22.90	0.80	17.90	58.40	»
30.20	0.48	12.60	56.72	»
37.46	0.40	10.80	51.34	»
36.80	0.30	10.10	52.80	»
38.50	0.20	9.40	51.90	»
41.40	0.16	8.80	49.64	»
42.50	0.15	8.20	49.15	»
46.89	0.15	6.53	46.43	»
51.88	0.14	5.29	42.69	$K_2SO_4 + KHCO_3 + 2KHCO_3 \cdot K_2CO_3 \cdot 1.5H_2O$
51.90	0.13	5.05	42.92	$K_2SO_4 + 2KHCO_3 \cdot K_2CO_3 \cdot 1.5H_2O$
52.90	0.07	5.01	42.02	$K_2CO_3 \cdot 1.5H_2O + K_2SO_4 +$ $+ 2KHCO_3 \cdot K_2CO_3 \cdot 1.5H_2O$

№ 4449 POTASSIUM HYDROGEN CARBONATE–POTASSIUM CARBONATE–POTASSIUM SULFATE–WATER [750]

$KHCO_3—K_2CO_3—K_2SO_4—H_2O$

$t=35$

Comp. of saturated soln., wt %				d	Comp. of solid phase
Potassium Carbonate	Potassium Hydrogen Carbonate	Potassium Sulfate	Water		
52.83	2.47	0.00	44.70	1.5850	$K_2CO_3 \cdot {}^3/_2H_2O +$ $+ K_2CO_3 \cdot 2KHCO_3 \cdot {}^3/_2H_2O$
50.24	4.36	0.00	45.40	1.5687	$KHCO_3 + K_2CO_3 \cdot 2KHCO_3 \cdot {}^3/_2H_2O$
0.00	28.88	2.51	68.82	1.2302	$KHCO_3 + K_2SO_4$
53.60	0.00	0.03	46.37	1.5729	$K_2CO_3 \cdot {}^3/_2H_2O + K_2SO_4$
52.03	3.47	0.03	44.57	1.5873	$K_2CO_3 \cdot {}^3/_2H_2O +$ $+ K_2CO_3 \cdot 2KHCO_3 \cdot {}^3/_2H_2O + K_2SO_4$
60.09	4.56	0.04	45.31	1.5679	$KHCO_3 + K_2CO_3 \cdot 2KHCO_3 \times$ $\times {}^3/_2H_2O + K_2SO_4$

№ 4450 POTASSIUM HYDROGEN CARBONATE–POTASSIUM NITRATE–WATER [2569]

$KHCO_3—KNO_3—H_2O$

Concn. in aqeuous soln., g/l		t	Concn. in aqueous soln., g/l		t
Potassium Nitrate	Potassium Hydrogen Carbonate		Potassium Nitrate	Potassium Hydrogen Carbonate	
236	0.0	14.5	332	0.0	25
220	39.0	14.5	287	89	25
205	76.0	14.5	268	133	25
194	116.0	14.5	249	191	25
183	115.0	14.5			

№ 4451 [1729]

POTASSIUM HYDROGEN CARBONATE–POTASSIUM SULFATE–WATER

$KHCO_3$—K_2SO_4—H_2O

$t=25$

Content in aqueous soln., wt %		Comp. of solid phase
Potassium Sulfate	Potassium Hydrogen Carbonate	
5.00	13.37	K_2SO_4
4.03	16.96	$K_2SO_4+KHCO_3$

№ 4452 **POTASSIUM HYDROGEN CARBONATE–POTASSIUM SULFATE–WATER** [138]

$KHCO_3$—K_2SO_4—H_2O

$t=25$

Comp. of saturated soln., wt %			Comp. of solid phase
Potassium Sulfate	Potassium Hydrogen Carbonate	Water	
10.76	0.00	89.24	K_2SO_4
9.99	1.69	88.32	»
9.53	2.61	87.86	»
7.37	6.52	86.11	»
6.29	9.52	84.19	»
4.71	14.11	81.18	»
4.02	17.56	78.42	»
3.17	20.17	76.66	»
2.43	22.52	75.05	»
2.23	24.72	73.05	$K_2SO_4+KHCO_3$
0.00	26.72	73.28	$KHCO_3$

№ 4453 **POTASSIUM HYDROGEN CARBONATE–POTASSIUM SULFATE–WATER** [505]

$KHCO_3$—K_2SO_4—H_2O

$t=50$

Comp. of saturated soln., wt %			d	Comp. of solid phase
Potassium Sulfate	Potassium Hydrogen Carbonate	Water		
14.25	0.00	85.75	1.1145	K_2SO_4
13.00	2.41	84.59	1.1189	»
11.61	4.92	83.47	1.1245	»
9.53	11.31	79.16	1.1425	»
7.52	14.51	77.97	1.1516	»
6.42	18.81	74.77	1.1621	»
6.02	17.91	76.07	1.1660	»
4.50	24.50	71.00	1.1820	»
3.20	31.50	65.30	1.2195	»
2.25	33.50	64.25	1.2339	$K_2SO_4+KHCO_3$
2.26	33.46	64.28	—	»

Table № 4453 (continuation)

Comp. of saturated soln., wt %			d	Comp. of solid phase
Potassium Sulfate	Potassium Hydrogen Carbonate	Water		
2.30	33.48	64.22	—	$K_2SO_4+KHCO_3$
0.70	34.00	65.30	1.2390	$KHCO_3$
0.00	34.50	65.50	1.2420	»

№ 4454 POTASSIUM HYDROGEN CARBONATE–POTASSIUM SULFATE–WATER [750]

$KHCO_3—K_2SO_4—H_2O$

$t=35$

Comp. of saturated soln., wt %			d	Comp. of solid phase
Potassium Hydrogen Carbonate	Potassium Sulfate	Water		
29.60	0.00	70.40	—	$KHCO_3$
29.27	1.60	69.13	1.2295	»
28.77	2.51	68.72	1.2302	$KHCO_3+K_2SO_4$
21.91	3.72	74.37	1.1895	K_2SO_4
21.34	3.85	74.81	1.1849	»
19.15	4.32	76.53	1.1724	»
9.98	7.44	82.58	—	»
0.00	12.36	87.64	—	»

№ 4455 POTASSIUM HYDROGEN CARBONATE–POTASSIUM CHLORIDE–WATER [2301]

$KHCO_3—KCl—H_2O$

Content in aqueous soln., g/100 g Water		t	d	Comp. of solid phase
Potassium Hydrogen Carbonate	Potassium Chloride			
27.78	0.0	10	1.154	$KHCO_3$
20.87	9.08	10	1.164	»
17.82	15.08	10	1.175	»
13.75	23.93	10	1.195	»
12.55	27.45	10	1.204	$KHCO_3+KCl$
5.17	29.54	10	1.181	KCl
0.0	31.12	10	1.164	»
33.19	0.0	20	1.177	$KHCO_3$
25.34	10.05	20	1.184	»
22.03	15.33	20	1.180	»
19.35	20.14	20	1.198	»
17.33	24.31	20	1.207	»
15.22	29.69	20	1.218	$KHCO_3+KCl$
10.08	31.11	20	1.204	KCl
5.12	32.00	20	1.180	»
0.0	34.35	20	1.174	»
39.23	0.0	29.85	1.197	$KHCO_3$
18.24	31.30	29.85	1.234	$KHCO_3+KCl$
0.0	37.08	29.85	1.181	KCl
45.51	0.0	40	1.217	$KHCO_3$
22.00	33.03	40	1.249	$KHCO_3+KCl$
0.0	40.15	40	1.189	KCl

№ 4456 [1059]

POTASSIUM HYDROGEN CARBONATE–POTASSIUM CHLORIDE–WATER

$KHCO_3$—KCl—H_2O

Comp. of saturated soln., wt %			t	Comp. of solid phase
Potassium Chloride	Potassium Hydrogen Carbonate	Water		
26.40	0.00	73.60	25	KCl
25.80	1.22	72.98	25	»
24.87	2.93	72.20	25	»
24.02	4.65	71.33	25	»
22.94	6.79	70.27	25	»
21.97	9.28	68.93	25	»
20.84	11.40	67.76	25	KCl+$KHCO_3$
16.74	13.79	69.47	25	KCO_3
12.48	16.60	70.92	25	»
8.40	19.69	71.91	25	»
5.12	22.19	72.69	25	»
2.38	24.66	72.96	25	»
0.00	26.79	73.21	25	»
28.64	0.00	71.36	40	KCl
27.34	2.61	70.05	40	»
23.02	11.24	65.74	40	»
21.83	13.70	64.47	40	»
21.40	14.36	64.24	40	KCl+$KHCO_3$
20.96	14.65	64.39	40	$KHCO_3$
17.04	17.35	65.61	40	»
13.91	19.59	66.50	40	»
10.50	22.30	67.20	40	»
7.25	25.16	67.59	40	»
3.28	28.80	67.92	40	»
0.00	31.93	68.07	40	»

№ 4457 [1117]

POTASSIUM HYDROGEN CARBONATE–TITANIUM DIOXIDE–WATER

$KHCO_3$—TiO_2—H_2O

t=18—20

Comp. of solvent, wt %		Solubility of Titanium Dioxide g/l
Potassium Hydrogen Carbonate	Water	
12	88	0.50
25	75	2.75
Saturated solution		7.00

№ 4458 POTASSIUM CARBONATE–POTASSIUM NITRATE–WATER [2659]

$K_2CO_3—KNO_3—H_2O$

Concn. in aqueous soln., g/l		t	Concn. in aqueous soln., g/l		t
Potassium Nitrate	Potassium Carbonate		Potassium Nitrate	Potassium Carbonate	
225	0.0	14.5	265	81.6	25
188	66.4	14.5	199	186.7	25
141	172.9	14.5	148	290.5	25
87	356.9	14.5	115	373.6	25
65	544.9	14.5	80	495.1	25
326	0.0	25			

№ 4459 POTASSIUM CARBONATE–POTASSIUM NITRATE–WATER [2013]

$K_2CO_3—KNO_3—H_2O$

$t=24.2$

Comp. of saturated soln., wt %			Comp. of solid phase
Potassium Nitrate	Potassium Carbonate	Water	
27.37	0.00	72.63	KNO_3
20.14	9.21	70.65	»
10.70	23.07	66.23	»
9.37	24.56	66.07	»
4.14	39.07	56.79	»
2.02	45.81	52.17	»
1.45	51.89	46.66	$KNO_3+K_2CO_3$

№ 4460 POTASSIUM CARBONATE–POTASSIUM METASILICATE–WATER [51]

$K_2CO_3—K_2SiO_3—H_2O$

Comp. of saturated soln., wt %			t	Comp. of solid phase
Potassium Metasilicate	Potassium Carbonate	Water		
5.4	46.5	48.1	0	$K_2CO_3 \cdot 1.5H_2O$
9.0	44.5	46.5	0	»
12.0	39.2	48.8	0	»
18.65	33.0	48.35	0	»
20.99	31.62	47.39	0	»
22.00	29.65	48.35	0	»
22.50	28.60	48.90	0	»
26.70	25.40	47.90	0	»
26.41	24.77	48.82	0	»
32.00	21.20	46.80	0	»
55.00	3.20	41.80	0	»

Table № 4460 (continuation)

Comp. of saturated soln., wt %			t	Comp. of solid phase
Potassium Metasilicate	Potassium Carbonate	Water		
2.71	51.01	46.28	20	$K_2CO_3 \cdot 1.5H_2O$
13.80	39.63	46.57	20	»
15.87	37.07	47.06	20	»
20.96	33.10	45.94	20	»
23.37	31.35	45.28	20	»
27.10	27.80	45.10	20	»
30.40	24.30	45.30	20	»
38.00	18.00	44.00	20	»
41.70	18.50	39.80	20	»
49.66	7.98	42.36	20	»
50.17	8.26	41.57	20	»
52.45	5.70	41.85	20	»
54.78	4.18	41.04	20	»

№ 4461 [1084]
POTASSIUM CARBONATE–POTASSIUM DIHYDROGEN ORTHOPHOSPHATE–WATER

$K_2CO_3—KH_2PO_4—H_2O$

$t=25$

Comp. of saturated soln., wt %		
Potassium Dihydrogen Orthophosphate	Potassium Carbonate	Water
23.01	1.66	35.33
31.85	5.74	62.41

№ 4462 POTASSIUM CARBONATE–POTASSIUM SULFATE–WATER [2197]

$K_2CO_3—K_2SO_4—H_2O$

Content in aqueous soln., wt %		t	d_4^{25}	Comp. of solid phase
Potassium Sulfate	Potassium Carbonate			
10.76	0	25	1.0884	K_2SO_4
6.76	5.85	25	1.1103	»
3.92	12.6	25	—	»
2.48	17.81	25	1.193	»
1.72	22.1	25	1.222	»
0.886	28.5	25	—	»
0.023	53.13	25	1.574	$K_2SO_4+K_2CO_3 \cdot 2H_2O$
0	53.2	25	1.575	$K_2CO_3 \cdot 2H_2O$
7.1	15.33	80	—	K_2SO_4
9.42	13.6	100	—	»

№ 4463 POTASSIUM CARBONATE–POTASSIUM SULFATE–WATER [1835]

$K_2CO_3—K_2SO_4—H_2O$

$t=25$

Comp. of saturated soln., wt %			t	Comp. of solid phase
Potassium Carbonate	Potassium Sulfate	Water		
52.8	0.0	47.2	1.557	$K_2CO_3 \cdot 1\frac{1}{2}H_2O$
52.8	0.03	47.17	1.557	$K_2CO_3+K_2SO_4$
48.0	0.03	51.97	1.506	K_2SO_4
45.1	0.08	54.9	1.465	»
35.0	0.3	64.7	1.348	»
23.2	1.5	75.3	1.230	»
16.4	2.8	80.8	1.170	»
9.4	4.9	85.7	1.122	»
5.5	6.5	88.0	1.103	»
0.0	10.7	89.3	1.083	»

№ 4464 [2637]

POTASSIUM CARBONATE–POTASSIUM SULFATE–WATER

$K_2CO_3—K_2SO_4—H_2O$

Content in aqueous soln., g/100 g Water		t	Comp. of solid phase
Potassium Sulfate	Potassium Carbonate		
14.1	0	35	K_2SO_4
Traces	112.5	35	$K_2SO_4+K_2CO_3 \cdot 1.5H_2O$
0	115.1	35	$K_2CO_3 \cdot 1.5H_2O$
17.1	0	50	K_2SO_4
Traces	121.2	50	$K_2SO_4+K_2CO_3 \cdot 1.5H_2O$
0	121.2	50	$K_2CO_3 \cdot 1.5H_2O$
20.6	0	75	K_2SO_4
Traces	135.8	75	$K_2SO_4+K_2CO_3 \cdot 1.5H_2O$
0	136.4	75	$K_2CO_3 \cdot 1.5H_2O$

№ 4465 POTASSIUM CARBONATE–POTASSIUM SULFATE–WATER [169]

$K_2CO_3—K_2SO_4—H_2O$

Content in aqueous soln., wt %		t	d	Comp. of solid phase
Potassium Sulfate	Potassium Carbonate			
6.78	0	0	1.064	K_2SO_4
4.71	6.88	0	1.105	»
3.33	11.66	0	1.126	»
0.15	45.6	0	1.512	»
Traces	51.3	0	1.564	$K_2SO_4+K_2CO_3 \cdot 2H_2O$
0	51.7	0	1.566	$K_2CO_3 \cdot 2H_2O$
11.48	0	30	1.089	K_2SO_4
4.99	10.64	30	1.105	»
4.21	13.63	30	1.191	»
0.74	33.1	30	1.279	»

Table № 4465 (continuation)

Content in aqueous soln., wt %		t	d	Comp. of solid phase
Potassium Sulfate	Potassium Carbonate			
Traces	53.0	30	1.577	$K_2SO_4+K_2CO_3\cdot 2H_2O$
0	53.3	30	1.579	$K_2CO_3\cdot 2H_2O$
14.7	0	50	1.112	K_2SO_4
5.58	12.71	50	1.163	»
2.89	22.29	50	1.233	»
0.98	35.3	50	1.365	»
Traces	54.7	50	1.586	$K_2SO_4+K_2CO_3\cdot 2H_2O$
0	54.8	50	1.586	$K_2CO_3\cdot 2H_2O$
16.9	0	70	1.120	K_2SO_4
9.73	7.02	70	1.123	»
4.28	19.58	70	1.201	»
1.26	35.39	70	1.333	»
Traces	57.1	70	1.598	$K_2SO_4+K_2CO_3\cdot 2H_2O$
0	57.1	70	1.598	$K_2CO_3\cdot 2H_2O$

№ 4466 POTASSIUM CARBONATE–POTASSIUM SULFATE–WATER [505]

$K_2CO_3—K_2SO_4—H_2O$

$t=50$

Comp. of saturated soln., wt %			d	Comp. of solid phase
Potassium Carbonate	Potassium Sulfate	Water		
0.00	14.25	85.75	1.1145	K_2SO_4
11.12	6.74	82.14	1.1863	»
16.70	4.50	78.80	1.2192	»
21.09	2.96	75.95	1.2516	»
23.90	2.50	73.60	1.2790	»
39.40	1.08	59.52	1.4116	»
44.94	0.50	54.56	1.4712	»
54.12	0.08	45.80	1.5900	$K_2SO_4+K_2CO_3$
54.09	0.08	45.83	1.5908	»
54.21	0.06	45.73	—	K_2CO_3
54.52	0.03	45.45	1.5925	»
54.82	0.00	45.18	1.5938	»

№ 4467 [342]

POTASSIUM CARBONATE–POTASSIUM SULFATE–WATER

$K_2CO_3—K_2SO_4—H_2O$

$t=150$

Comp. of saturated soln., wt %			Comp. of solid phase
Potassium Sulfate	Potassium Carbonate	Water	
21.8	0.0	78.2	K_2SO_4
9.6	18.6	71.8	»
3.4	32.5	64.1	»
0.2	62.9	36.9	»
0.2	68.6	31.2	$K_2SO_4+K_2CO_3\cdot 1.5H_2O$
0.0	70.7	29.3	$K_2CO_3\cdot 1.5H_2O$

№ 4468 POTASSIUM CARBONATE–POTASSIUM SULFATE–WATER [750]

$K_2CO_3—K_2SO_4—H_2O$

$t=35$

Comp. of saturated soln., wt %			d	Comp. of solid phase
Potassium Carbonate	Potassium Sulfate	Water		
0.00	12.36	87.64	—	K_2SO_4
53.60	0.03	46.37	1.5729	$K_2SO_4+K_2CO_3 \cdot {}^3/_2H_2O$
52.94	Traces	47.06	—	»
53.53	0.00	46.47	—	$K_2CO_3 \cdot {}^3/_2H_2O$

№ 4469 [2763]

POTASSIUM CARBONATE–POTASSIUM SULFATE–WATER

$K_2CO_3—K_2SO_4—H_2O$

$t=30$

Comp. of saturated soln., wt %			Comp. of solid phase
Potassium Sulfate	Potassium Carbonate	Water	
11.45	0.00	88.55	K_2SO_4
5.03	10.50	84.47	»
2.47	18.94	78.59	»
1.83	27.67	70.50	»
0.45	38.63	60.92	»
0.15	47.16	52.69	$K_2SO_4+K_2CO_3 \cdot 3.5H_2O$
0.02	53.15	46.83	»
0.00	53.34	46.66	$K_2CO_3 \cdot 3.5H_2O$

№ 4470 [2637]

POTASSIUM CARBONATE–POTASSIUM SULFATE–POTASSIUM CHLORIDE–WATER

$K_2CO_3—K_2SO_4—KCl—H_2O$

Comp. of saturated soln., wt %				t	Comp. of solid phase
Potassium Chloride	Potassium Sulfate	Potassium Carbonate	Water		
28.05	0.00	0.00	71.95	35	KCl
0.00	12.35	0.00	87.65	35	K_2SO_4
0.00	0.00	53.50	46.50	35	$K_2CO_3 \cdot 1\frac{1}{2}H_2O$
27.40	1.21	0.00	71.39	35	$KCl+K_2SO_4$
1.43	0.00	52.53	46.04	35	$KCl+K_2CO_3 \cdot 1\frac{1}{2}H_2O$
0.00	Traces	52.94	47.06	35	$K_2CO_3 \cdot 1\frac{1}{2}H_2O+K_2SO_4$
1.33	»	52.27	46.40	35	$K_2CO_3 \cdot 1\frac{1}{2}H_2O+K_2SO_4+KCl$
37.12	0.00	0.00	62.88	50	KCl

Table 4470 (continuation)

Comp. of saturated soln., wt %				t	Comp. of solid phase
Potassium Chloride	Potassium Sulfate	Potassium Carbonate	Water		
0.00	14.60	0.00	85.40	50	K_2SO_4
0.00	0.00	54.74	45.26	50	$K_2CO_3 \cdot 1\frac{1}{2}H_2O$
29.30	1.25	0.00	69.45	50	K_2SO_4+KCl
1.60	0.00	53.93	44.47	50	$K_2CO_3 \cdot 1\frac{1}{2}H_2O+KCl$
0.00	Traces	54.81	45.19	50	$K_2CO_3+K_2SO_4$
1.60	»	53.91	44.49	50	$K_2CO_3 \cdot 1\frac{1}{2}H_2O+KCl$
33.20	0.00	0.00	66.80	75	KCl
0.00	17.09	0.00	82.91	75	K_2SO_4
0.00	0.00	57.70	42.30	75	$K_2CO_3 \cdot 1\frac{1}{2}H_2O$
32.24	1.27	0.00	66.49	75	$KCl+K_2SO_4$
2.18	0.00	56.18	41.64	75	$K_2CO_3 \cdot 1\frac{1}{2}H_2O+KCl$
0.00	Traces	57.60	42.40	75	$K_2CO_3+K_2SO_4$
2.19	»	55.92	41.89	75	$K_2CO_3+K_2SO_4+KCl$

№ 4471 [170]

POTASSIUM CARBONATE–POTASSIUM SULFATE–POTASSIUM CHLORIDE–WATER

K_2CO_3—K_2SO_4—KCl—H_2O

Comp. of saturated soln., wt %				t	Comp. of solid phase
Potassium Chloride	Potassium Sulfate	Potassium Carbonate	Water		
21.90	0.00	0.00	78.10	0	KCl
21.60	0.93	0.00	77.47	0	$KCl+K_2SO_4$
12.07	0.62	17.33	69.98	0	»
9.11	0.43	23.8	66.66	0	»
0.75	Traces	50.76	48.49	0	$KCl+K_2SO_4+K_2CO_3 \cdot 2H_2O$
0.00	0.00	51.70	48.30	0	$K_2CO_3 \cdot 2H_2O$
27.20	0.00	0.00	72.80	30	KCl
26.22	1.13	0.00	72.65	30	$KCl+K_2SO_4$
18.67	0.79	13.34	67.20	30	»
11.36	0.36	26.19	62.09	30	»
1.04	Traces	52.10	46.86	30	$KCl+K_2SO_4+K_2CO_3 \cdot 2H_2O$
0.00	0.00	53.30	46.70	30	$K_2CO_3 \cdot 2H_2O$
30.00	0.00	0.00	70.0	50	KCl
28.80	1.27	0.00	69.93	50	$KCl+K_2SO_4$
18.93	0.57	14.92	55.58	50	»
10.60	0.24	29.87	59.29	50	»
1.44	Traces	54.80	43.76	50	$KCl+K_2SO_4+K_2CO_3 \cdot 2H_2O$
0.00	0.00	54.80	45.20	50	$K_2CO_3 \cdot 2H_2O$
32.60	0.00	0.00	67.40	70	KCl
31.30	1.39	0.00	67.31	70	$KCl+K_2SO_4$
21.82	0.84	13.33	64.01	70	»
16.48	0.77	19.06	63.69	70	»
1.89	Traces	55.40	42.71	70	$KCl+K_2SO_4+K_2CO_3 \cdot H_2O$
0.00	0.00	57.10	42.90	70	$K_2CO_3 \cdot 2H_2O$

№ 4472 [2711]

POTASSIUM CARBONATE–POTASSIUM CHLORIDE–WATER

$K_2CO_3—KCl—H_2O$

$t=30$

Comp. of saturated soln., wt %			Comp. of solid phase
Potassium Carbonate	Potassium Chloride	Water	
53.27	0.00	46.73	$K_2CO_3 \cdot 1\frac{1}{2}H_2O$
52.22	1.03	46.27	$K_2CO_3 \cdot 1\frac{1}{2}H_2O+KCl$
51.66	1.07	47.27	KCl
1.64	26.22	72.14	»
0.00	28.01	71.99	»

№ 4473 [1242]

POTASSIUM CARBONATE–POTASSIUM CHLORIDE–WATER

$K_2CO_3—KCl—H_2O$

$t=25$

Content in aqueous soln., g/100 g Water		Comp. of solid phase
Calcium Chloride	Calcium Carbonate	
36.97	0	KCl
2.07	110.68	$KCl+K_2CO_3 \cdot 1.5H_2O$
0	113.57	$K_2CO_3 \cdot 1.5H_2O$

№ 4474 [2637]

POTASSIUM CARBONATE–POTASSIUM CHLORIDE–WATER

$KCl—K_2CO_3—H_2O$

Content in aqueous soln., g/100 g Water		t	Comp. of solid phase
Potassium Chloride	Potassium Carbonate		
39.0	0	35	KCl
3.1	114.1	35	$KCl+K_2CO_3 \cdot 1.5H_2O$
0	115.1	35	$K_2CO_3 \cdot 1.5H_2O$
43.1	0	50	KCl
3.6	121.3	50	$KCl+K_2CO_3 \cdot 1.5H_2O$
0	121.2	50	$K_2CO_3 \cdot 1.5H_2O$
49.7	0	75	KCl
5.2	134.9	75	$KCl+K_2CO_3 \cdot 1.5H_2O$
0	136.4	75	$K_2CO_3 \cdot 1.5H_2O$

№ 4475 [95]

POTASSIUM CARBONATE–POTASSIUM CHLORIDE–WATER

$K_2CO_3—KCl—H_2O$

Concn. in aqueous soln., wt %		t	Comp. of solid phase
Potassium Carbonate	Potassium Chloride		
51.9	0.8	20	K_2CO_3+KCl
52.22	1.03	30	»
53.63	1.40	50	»
56.06	2.00	70	»
57.16	2.21	84	»
59.44	2.48	100	»
65.33	2.68	124.5	»

№ 4476 [169]

POTASSIUM CARBONATE–POTASSIUM CHLORIDE–WATER

$K_2CO_3—KCl—H_2O$

Content in aqueous soln., wt %		t	d	Comp. of solid phase
Potassium Chloride	Potassium Carbonate			
21.9	0	0	1.154	KCl
17.4	9.38	0	1.224	»
13.37	14.38	0	1.252	»
7.90	26.4	0	1.331	»
6.46	32.46	0	1.378	»
2.31	40.4	0	1.463	»
0.96	48.4	0	1.552	»
0.79	49.4	0	1.562	»
0.76	50.99	0	1.568	$KCl+K_2CO_3 \cdot 2H_2O$
0	51.7	0	1.566	$K_2CO_3 \cdot 2H_2O$
27.2	0	30	1.188	KCl
20.46	10.21	30	1.228	»
17.15	15.05	30	1.261	»
14.92	19.02	30	1.287	»
12.45	24.04	30	1.324	»
8.04	33.81	30	1.394	»
3.54	41.2	30	1.475	»
1.14	51.3	30	1.565	»
1.05	52.2	30	1.585	$KCl+K_2CO_3 \cdot 2H_2O$
0	53.3	30	1.579	$K_2CO_3 \cdot 2H_2O$
30.0	0	50	1.198	KCl
23.62	10.01	50	1.248	»
21.84	12.08	50	1.260	»
15.58	22.23	50	1.325	»
11.02	30.4	50	1.376	»
3.12	47.13	50	1.517	»
2.49	52.4	50	1.584	»
1.44	53.6	50	1.592	$KCl+K_2CO_3 \cdot 2H_2O$
0	54.8	50	1.586	$K_2CO_3 \cdot 2H_2O$
32.6	0	70	1.211	KCl
20.35	17.23	70	1.293	»
18.34	19.46	70	1.302	»

Table № 4476 (continuation)

Content in aqueous soln., wt %		t	d	Comp. of solid phase
Potassium Chloride	Potassium Carbonate			
16.16	22.15	70	1.316	KCl
10.66	33.5	70	1.410	»
2.57	53.1	70	1.567	»
1.89	55.5	70	1.605	$KCl+K_2CO_3 \cdot 2H_2O$
0	57.1	70	1.598	$K_2CO_3 \cdot 2H_2O$

№ 4477 POTASSIUM CARBONATE–POTASSIUM CHLORATE–WATER [17]

$K_2CO_3—KClO_3—H_2O$

Comp. of solvent, wt %		Solubility of Potassium Chlorate, g/l	t	Comp. of solvent, wt %		Solubility of Potassium Chlorate, g/l	t
Potassium Carbonate	Water			Potassium Carbonate	Water		
5	95	60	20	30	70	25	20
5	95	65	30	30	70	35	30
5	95	85	40	30	70	45	40
5	95	95	50	30	70	60	50
10	90	50	20	40	60	20	20
10	90	60	30	40	60	25	30
10	90	80	40	40	60	45	40
10	90	90	50	40	60	60	50
20	80	40	20	50	50	10	20
20	80	50	30	50	50	25	30
20	80	65	40	50	50	35	40
20	80	75	50	50	50	40	50

№ 4478 POTASSIUM CARBONATE–POTASSIUM CHLORATE–WATER [328]

$K_2CO_3—KClO_3—H_2O$

Comp. of saturated soln., wt %			t	d	Comp. of solid phase
Potassium Carbonate	Potassium Chlorate	Water			
50.58	0.00	49.42	24.2	1.550	$K_2CO_3 \cdot 2H_2O$
52.60	0.16	47.24	24.2	—	$K_2CO_3 \cdot 2H_2O+KClO_3$
49.24	0.25	50.51	24.2	—	$KClO_3$
49.09	0.18	50.73	24.2	—	»
48.63	0.28	51.09	24.2	1.500	»
36.57	0.69	62.74	24.2	1.362	»
35.83	0.73	63.44	24.2	1.362	—
30.55	1.09	68.36	24.2	1.302	$KClO_3$
23.44	1.65	74.91	24.2	1.220	»
22.24	1.82	75.94	24.2	1.218	»
18.52	2.29	79.19	24.2	—	»

Table № 4478 (continuation)

Comp. of saturated soln., wt %			t	d	Comp. of solid phase
Potassium Carbonate	Potassium Chlorate	Water			
17.79	2.39	79.82	24.2	1.180	$KClO_3$
8.81	3.92	87.27	24.2	1.100	»
0.00	7.50	92.50	24.2	1.045	»
54.50	0.00	45.60	40	1.558	$K_2CO_3 \cdot 2H_2O$
51.79	1.03	47.18	40	1.539	$K_2CO_3 \cdot 2H_2O + KClO_3$
29.20	1.45	69.35	40	1.316	$KClO_3$
28.78	4.23	76.99	40	1.195	»
0.00	12.44	87.56	40	1.073	»

№ 4479 POTASSIUM CARBONATE–POTASSIUM PERMANGANATE–WATER [2455]

$K_2CO_3—KMnO_4—H_2O$

Concn. of Potassium Carbonate in aqueous soln., g/l	Solubility of Potassium Permanganate in aqueous soln. of Potassium Carbonate, g/l	t	Concn. of Potassium Carbonate in aqueous soln., g/l	Solubility of Potassium Permanganate in aqueous soln. of Potassium Carbonate, g/l	t
6.910	23.104	0	138.2	7.048	0
6.910	69.138	25	138.2	55.61	40
6.910	116.626	40	276.4	4.267	0
69.10	9.94	0	276.4	14.696	25
69.10	40.914	25	414.6	2.465	0
69.10	79.125	40			

№ 4480 [1145]

POTASSIUM CARBONATE–CALCIUM CARBONATE–WATER

$K_2CO_3—CaCO_3—H_2O$

Content of Potassium Carbonate in aqueous soln., g/100 g Water	t	Comp. of solid phase
59.25	19	$K_2CO_3 \cdot CaCO_3 \cdot CaCO_3$
60.21	50	»
61.00	75	»
61.78	75	»

№ 4481 [1117]

POTASSIUM CARBONATE–TITANIUM DIOXIDE–WATER

$K_2CO_3—TiO_2—H_2O$

$t=18—20$

Comp. of solvent, wt %: Potassium Carbonate	Comp. of solvent, wt %: Water	Solubility of Titanium Dioxide, g/l
30	70	0.02
Saturated solution		0.30

№ 4482 [2596

POTASSIUM CARBONATE–ZINC HYDROXIDE–WATER

$K_2CO_3—Zn(OH)_2—H_2O$

$t=16—20$

Comp. of solvent, wt %: Potassium Carbonate	Comp. of solvent, wt %: Water	Solubility of Zinc Hydroxide, g/l
1	99	0.067

№ 4483 [984]

POTASSIUM CARBONATE–THORIUM OXIDE–WATER

$K_2CO_3—ThO_2—H_2O$

$t=25$

Concn. of Potassium Carbonate in aqueous soln., g/l	Solubility of Thorium Oxide in aqueous soln. of Potassium Carbonate, g/l
69.1	<0.00002

№ 4484 [984]

POTASSIUM CARBONATE–THORIUM HYPOPHOSPHATE–WATER

$K_2CO_3—ThP_2O_6 \cdot 11H_2O—H_2O$

$t=25$

Concn. of Potassium Carbonate in aqueous soln., g/l	Solubility of Thorium Hypophosphate in aqueous soln. of Potassium Carbonate, g/l
69.1	0.250

№ 4485 [586]

POTASSIUM CARBONATE–PLUTONIUM HYDROXIDE–WATER

$K_2CO_3—Pu(OH)_4—H_2O$

$t=20$

Concn. of Potassium Carbonate in aqueous soln., g/l	Solubility of Plutonium Hydroxide in aqueous soln. of Potassium Carbonate, g/l	Concn. of Potassium Carbonate in aqueous soln., g/l	Solubility of Plutonium Hydroxide in aqueous soln. of Potassium Carbonate, g/l	Concn. of Potassium Carbonate in aqueous soln., g/l	Solubility of Plutonium Hydroxide in aqueous soln. of Potassium Carbonate, g/l
50.0	0.0546	300.0	0.352	451.0	0.444
100.0	0.0795	399.0	0.395	500.0	0.508
200.0	0.1787				

№ 4486 [1417]

POTASSIUM CYANIDE–NICKEL (II) CYANIDE–WATER

$KCN—Ni(CN)_2—H_2O$

$t=25$

Comp. of saturated soln., wt %			Comp. of solid phase
Potassium Cyanide	Nickel Cyanide	Water	
39.85	1.00	59.15	KCN
38.79	1.88	59.33	»
37.64	4.69	57.67	»
39.09	5.95	54.96	$KCN+K_2Ni(CN)_4 \cdot H_2O$
30.94	7.46	61.60	$K_2Ni(CN)_4 \cdot H_2O$
23.97	10.82	65.21	»
23.14	10.34	66.52	»
16.82	14.29	68.89	»
16.82	13.04	69.24	$K_2Ni(CN)_4 \cdot H_2O+Ni(CN)_2$
10.58	9.13	80.29	$Ni(CN)_2$
0.28	0.23	99.49	»

№ 4487 [1158]

POTASSIUM CYANIDE–COPPER (I) CYANIDE–WATER

$KCN—CuCN—H_2O$

$t=25$

Comp. of saturated soln., wt %			Comp. of solid phase
Potassium Cyanide	Copper Cyanide	Water	
41.70	0.0	58.3	KCN
40.53	2.16	57.31	»
40.31	4.20	55.49	»
41.70	5.84	52.46	$K_3Cu(CN)_4$
35.96	10.86	53.18	»
33.86	15.30	50.84	»
41.94	6.13	51.93	$K_3Cu(CN)_4 \cdot H_2O$
36.24	12.92	50.84	»
34.82	18.47	46.71	»
35.28	22.52	42.20	»
34.77	22.59	42.64	$KCu(CN)_2$
22.32	14.71	62.97	»
19.95	13.78	66.27	»
12.10	8.84	79.06	»
9.97	7.52	82.51	$KCu_2(CN)_3 \cdot H_2O$
6.05	3.93	90.02	»
6.03	3.88	90.09	»
4.21	1.87	93.92	»
3.27	1.68	95.05	»
1.34	0.60	98.06	»
0.42	Traces	99.58	$KCu_2(CN)_6 \cdot H_2O+CuCN$

№ 4488 [1417]

POTASSIUM CYANIDE–ZINC CYANIDE–WATER

$KCN—Zn(CN)_2—H_2O$

$t=25$

Comp. of saturated soln., wt %			Comp. of solid phase
Potassium Cyanide	Zinc Cyanide	Water	
41.7	0.0	58.3	KCN
41.15	Traces	58.85	$KCN+K_2Zn(CN)_4$
39.04	»	60.96	$K_2Zn(CN)_4$
37.15	0.12	62.73	»
34.46	0.17	65.37	»
32.65	0.20	67.15	»
23.56	0.67	75.77	»
18.22	0.47	81.31	»
14.17	1.50	84.33	»
7.65	5.75	86.60	»
6.64	5.86	87.50	»
6.77	6.97	86.26	$K_2Zn(CN)_4+Zn(CN)_2$
3.81	3.33	92.86	$Zn(CN)_2$
2.50	2.30	95.20	»

№ 4489 [1417]

POTASSIUM CYANIDE–CADMIUM CYANIDE–WATER

$KCN—Cd(CN)_2—H_2O$

$t=25$

Concn. in aqueous soln., wt %		Comp. of solid phase	Concn. in aqueous soln., wt %		Comp. of solid phase
Potassium Cyanide	Cadmium Cyanide		Potassium Cyanide	Cadmium Cyanide	
41.7	0.0	KCN	11.65	16.15	$K_2Cd(CN)_4$
40.67	0.82	$K_2Cd(CN)_4$	9.47	19.14	$K_2Cd(CN)_4+Cd(CN)_2$
28.62	4.65	»	7.82	15.73	$Cd(CN)_2$
17.79	10.70	»	4.65	10.21	»
15.16	12.53	»	Traces	0.36	»

№ 4490 [1158]

POTASSIUM CYANIDE–SILVER CYANIDE–WATER

$KCN—AgCN—H_2O$

$t=25$

Comp. of saturated soln., wt%			Comp. of solid phase
Potassium Cyanide	Silver Cyanide	Water	
41.70	0.00	58.30	KCN
40.77	6.52	52.71	»

№ 4491 POTASSIUM CYANIDE–SILVER CYANIDE–WATER [1158]

$KCN—AgCN—H_2O$

$t=25$

Concn. in aqueous soln., wt %		Comp. of solid phase	Concn. in aqueous soln., wt %		Comp. of solid phase
Potassium Cyanide	Silver Cyanide		Potassium Cyanide	Silver Cyanide	
2.16	Traces	AgCN	20.14	21.66	$KAg(CN)_2$
1.26	2.31	»	26.67	24.71	»
1.64	3.27	»	28.43	26.37	»
2.36	4.42	»	35.19	25.18	$K_3Ag(CN)_4 \cdot H_2O$
6.75	13.53	»	37.76	18.92	»
8.93	17.73	$KAg_2(CN)_3 \cdot H_2O$	40.44	13.71	»
8.05	15.76	»	40.24	10.93	KCN
8.56	16.11	$KAg(CN)_2$	39.91	8.14	»

№ 4492 POTASSIUM CYANIDE–SILVER IODIDE–WATER [2369]

$KCN—AgI—H_2O$

$t=25$

Concn. in aqueous soln., wt %					
Potassium Cyanide	Silver Iodide	Potassium Cyanide	Silver Iodide	Potassium Cyanide	Silver Iodide
0.120	0.332 *	0.134	0.238	1.21	2.07
0.237	0.414 *	0.262	0.445	1.62	2.85
0.355	0.620 *	0.536	0.905	1.96	3.28
0.471	0.822 *	0.926	1.57	2.42	4.04

*Silver Iodide precipitated. In other cases it is crystalline.

№ 4493 POTASSIUM CYANIDE–GOLD (I) CYANIDE–WATER [1158]

$KCN—AuCN—H_2O$

$t=25$

Comp. of saturated soln., wt %			Comp. of solid phase
Potassium Cyanide	Gold Cyanide	Water	
41.70	0.00	58.30	KCN
42.02	Traces	57.98	»
41.98	0.17	57.85	$KCN+KAu(CN)_2$
38.98	0.35	60.67	$KAu(CN)_2$
24.04	0.76	75.20	»

Table № 4493 (continuation)

Comp. of saturated soln., wt %			Comp. of solid phase
Potassium Cyanide	Gold Cyanide	Water	
23.79	1.27	74.94	$KAu(CN)_2$
13.70	2.55	83.75	»
9.66	2.97	87.37	»
7.02	8.34	84.64	»
5.19	11.02	83.79	»
4.78	15.47	79.75	$KAu(CN)_2 + AuCN$
2.60	7.27	90.13	AuCN
2.32	7.32	90.36	»
2.37	6.12	91.51	»

№ 4494 [1158]

POTASSIUM CYANIDE–THALLIUM CYANIDE–WATER

$KCN—TlCN—H_2O$

$t=25$

Comp. of saturated soln., wt %			Comp. of solid phase
Potassium Cyanide	Thallium Cyanide	Water	
41.7	0.0	58.3	KCN
40.83	0.71	58.46	»
39.76	1.65	58.59	»
37.00	3.17	59.83	$KCN + KTl(CN)_2$
37.50	3.11	59.39	$KTl(CN)_2$
36.47	4.16	59.37	»
36.52	4.12	59.36	»
35.93	5.78	58.29	»
36.00	4.99	59.01	TlCN
35.27	4.98	59.75	»
34.45	4.91	60.64	»
34.36	4.90	60.74	»
35.09	4.96	59.95	»
31.25	4.46	64.29	»
30.21	5.00	64.79	»
28.07	4.47	67.46	»
21.00	4.36	74.64	»
14.75	5.46	79.79	»
9.12	7.17	83.71	»
0.51	16.20	83.29	»
0.34	16.12	83.54	»
0.00	16.51	83.49	»

№ 4495 [2563]

POTASSIUM CYANIDE–MERCURY (II) CYANIDE–WATER

$KCN—Hg(CN)_2—H_2O$

$t=25$

Concn. in aqueous soln., g/l	
Mercury Cyanide	Potassium Cyanide
122.6	3.21
135.2	6.41
158.4	12.83

№ 4496 [1417]

POTASSIUM CYANIDE–MERCURY (II) CYANIDE–WATER

$KCN—Hg(CN)_2—H_2O$

$t=25$

Comp. of saturated soln., wt%			Comp. of solid phase
Mercury Cyanide	Potassium Cyanide	Water	
0.00	41.7	58.3	KCN
0.49	39.7	59.81	$KCN+K_2Hg(CN)_4$
2.34	29.47	68.17	$K_2Hg(CN)_4$
11.09	17.34	71.57	»
15.54	14.54	69.92	»
20.65	12.54	66.81	»
31.18	12.51	56.30	»
39.96	12.81	47.23	»
41.85	11.37	46.78	$Hg(CN)_2$
20.34	3.97	55.69	»
13.53	1.09	85.88	»
11.83	0.56	87.61	»
10.89	0.28	88.83	»
10.00	0.00	90.00	»

№ 4497 **POTASSIUM THIOCYANATE–POTASSIUM NITRITE–WATER** [144]

$KCNS—KNO_2—H_2O$

$t=25$

Comp. of saturated soln., wt %			Comp. of solid phase
Potassium Nitrite	Potassium Thiocyanate	Water	
76.16	0.00	23.84	KNO_2
71.56	5.61	22.83	»
67.14	10.87	21.99	»

Table № 4497 (continuation)

Comp. of saturated soln., wt %			Comp. of solid phase
Potassium Nitrite	Potassium Thiocyanate	Water	
62.47	15.89	21.64	KNO_2
59.46	20.51	20.03	»
54.50	27.38	18.12	»
50.82	33.20	15.98	»
45.21	40.52	14.27	KNO_2+KSCN
40.17	43.87	15.96	KSCN
32.46	48.93	18.61	»
27.85	51.82	20.33	»
22.44	55.34	22.22	»
18.12	58.73	23.15	»
13.22	62.20	24.58	»
8.41	65.05	26.54	»
3.68	68.08	28.24	»
0.00	70.83	29.17	»

№ 4498 [136]

POTASSIUM THIOCYANATE–POTASSIUM NITRATE–WATER

$KCNS—KNO_3—H_2O$

$t=25$

Comp. of saturated soln., wt %			Comp. of solid phase
Potassium Nitrate	Potassium Thiocyanate	Water	
27.16	0.00	72.84	KNO_3
25.60	2.04	72.36	»
23.56	4.96	71.48	»
21.70	8.66	69.64	»
19.78	12.33	67.89	»
18.69	15.13	66.18	»
17.60	19.76	62.64	»
16.87	21.73	61.40	»
15.89	25.64	58.47	»
15.16	27.94	56.90	»
14.85	29.40	55.75	»
13.77	33.14	53.09	»
12.30	38.80	48.90	»
11.68	45.49	42.83	»
10.40	54.20	35.40	»
9.94	57.77	32.29	»
9.45	65.20	25.35	KNO_3+KCNS
6.35	66.64	27.01	KCNS
1.93	69.10	28.97	»
0.00	70.70	29.30	»

№ 4499 [139]

POTASSIUM THIOCYANATE–POTASSIUM NITRATE–WATER

$KCNS—KNO_3—H_2O$

$t=25$

Comp. of saturated soln., wt %			Comp. of solid phase
Potassium Nitrate	Potassium Thiocyanate	Water	
27.16	0.00	72.84	KNO_3
25.60	2.04	72.36	»
21.70	8.66	69.64	»
18.69	15.13	66.18	»
16.87	21.73	61.40	»
14.85	29.40	55.75	»
12.30	38.80	48.90	»
11.68	45.49	42.83	»
9.94	57.77	32.29	»
9.45	65.20	25.35	KNO_3+KCNS
6.35	66.64	27.01	KCNS
0.00	70.70	29.30	»

№ 4500 [142]

POTASSIUM THIOCYANATE–POTASSIUM NITRATE–WATER

$KCNS—KNO_3—H_2O$

$t=25$

Comp. of saturated soln., wt %			Comp. of solid phase
Potassium Nitrate	Potassium Thiocyanate	Water	
27.16	0.00	72.84	KNO_3
25.60	2.04	72.36	»
23.56	4.96	71.48	»
21.70	8.66	69.64	»
19.78	12.33	67.89	»
18.69	15.13	66.18	»
17.12	21.00	61.88	»
15.89	25.64	58.47	»
15.11	30.34	54.55	»
13.77	33.14	53.09	»
12.38	38.80	48.82	»
11.65	45.19	43.16	»
10.40	54.20	35.40	»
9.94	57.77	32.29	»
9.45	65.20	25.35	KNO_3+KCNS
6.35	66.64	27.01	KCNS
1.93	69.10	28.97	»
0.00	70.70	29.30	»

$KCNS—K_2SO_4—H_2O$

Comp. of saturated soln., wt %			t	Comp. of solid phase
Potassium Thiocyanate	Potassium Sulfate	Water		
0.00	10.76	89.24	25	K_2SO_4
1.18	9.71	89.11	25	»
3.03	8.42	88.55	25	»
4.94	7.13	87.93	25	»
6.63	6.20	87.17	25	»
8.68	5.24	86.08	25	»
11.80	4.22	83.98	25	»
15.04	3.05	81.91	25	»
18.20	2.14	79.66	25	»
24.07	1.45	74.48	25	»
32.50	0.60	66.90	25	»
38.57	0.31	61.12	25	»
46.15	0.20	53.65	25	»
52.51	0.10	47.39	25	»
70.60	0.04	29.36	25	K_2SO_4+KCNS
70.70	0.00	29.30	25	KCNS
0.00	13.10	86.90	40	K_2SO_4
0.48	12.64	86.88	40	»
2.10	11.87	86.03	40	»
3.89	9.50	86.61	40	»
6.42	8.05	85.53	40	»
8.12	6.41	85.47	40	»
10.54	5.61	83.79	40	»
11.94	5.17	82.89	40	»
14.42	4.13	81.45	40	»
19.31	3.05	77.64	40	»
24.46	1.69	73.85	40	»
25.75	1.28	72.97	40	»
30.06	1.02	68.92	40	»
32.83	0.85	66.32	40	»
36.75	0.79	62.46	40	»
39.63	0.44	59.93	40	»
46.30	0.22	53.48	40	»
55.90	0.13	43.97	40	»
74.24	0.03	25.73	40	K_2SO_4+KCNS
74.31	0.00	25.69	40	KCNS

№ 4502 [145]

POTASSIUM THIOCYANATE–POTASSIUM CHLORIDE–WATER

KCNS—KCl—H_2O

$t=25$

Comp. of saturated soln., wt %			Comp. of solid phase
Potassium Thiocyanate	Potassium Chloride	Water	
0.00	26.56	73.44	KCl
4.39	24.09	71.52	»
8.01	22.07	69.92	»
9.58	21.25	69.17	»
13.05	19.66	67.29	»
20.14	15.84	64.02	»
24.30	13.80	61.90	»
29.18	11.96	58.86	»
33.95	10.01	56.04	»
37.87	8.49	53.64	»
43.47	6.64	49.89	»
49.09	4.90	46.01	»
52.67	4.15	43.18	»
55.71	3.44	40.85	»
56.09	3.35	40.56	»
59.17	2.85	37.98	»
62.48	2.39	35.13	»
65.47	2.00	32.53	»
69.59	1.62	28.79	KCl+KCNS
71.10	0.00	28.90	KCNS

№ 4503 [145]

POTASSIUM THIOCYANATE–POTASSIUM IODIDE–WATER

KCNS—KI—H_2O

$t=25$

Comp. of saturated soln., wt %			Comp. of solid phase
Potassium Thiocyanate	Potassium Iodide	Water	
0.00	59.42	40.58	KI
5.04	55.29	39.67	»
13.65	48.08	38.27	»
21.83	41.23	36.94	»
29.51	34.73	35.76	»
37.00	29.49	33.51	»
44.00	24.25	31.75	»
49.83	20.95	29.22	»
53.16	18.58	28.26	»
55.70	16.97	27.33	»
58.37	15.55	26.08	KI+KCNS
60.46	12.87	26.67	KCNS
63.45	9.38	27.17	»
65.61	6.72	27.67	»
67.06	4.72	28.22	»
71.10	0.00	29.90	»

$KCNS—KBr—H_2O$

Comp. of saturated soln., wt %			t	Comp. of solid phase
Potassium Bromide	Potassium Thiocyanate	Water		
40.87	0.00	59.13	25	KBr
36.58	5.70	57.72	25	»
32.50	11.20	56.30	25	»
27.80	17.55	54.65	25	»
23.60	24.00	52.40	25	»
19.30	31.20	49.50	25	»
14.00	40.80	45.20	25	»
11.30	45.70	43.00	25	»
10.00	48.30	41.70	25	»
8.70	50.19	41.11	25	»
7.30	56.30	36.40	25	»
5.60	62.70	31.70	25	»
4.00	67.30	28.70	25	KBr+KCNS
0.00	71.10	28.90	25	KCNS
43.17	0.00	56.83	40	KBr
40.26	3.97	55.77	40	»
36.95	8.17	54.88	40	»
32.72	13.89	53.39	40	»
29.12	19.00	51.88	40	»
26.11	23.13	50.76	40	»
23.13	27.40	49.47	40	»
20.75	31.47	47.78	40	»
19.28	33.51	47.21	40	»
16.87	38.52	44.61	40	»
14.14	42.41	43.45	40	»
12.06	45.27	42.67	40	»
10.73	48.88	40.39	40	»
9.47	52.22	38.31	40	»
8.71	55.00	36.29	40	»
7.73	57.06	35.21	40	»
6.93	59.93	33.14	40	»
6.12	62.50	31.38	40	»
5.47	65.29	29.24	40	»
4.79	67.96	27.25	40	»
4.29	70.76	24.95	40	KBr+KCNS
1.98	72.37	25.65	40	KCNS
0.00	74.38	25.62	40	»

№ 4505 [931]

POTASSIUM THIOCYANATE–POTASSIUM IODIDE–COPPER (I) THIOCYANATE–WATER

$KCNS—KI—CuCNS—H_2O$

$t=25$

Concn. in aqueous soln., g/l					
Potassium Iodide	Potassium Thiocyanate	Copper Thiocyanate	Potassium Iodide	Potassium Thiocyanate	Copper Thiocyanate
1.5	1.025	0.0209	1.50	3.477	0.210
1.5	1.473	0.0460	1.98	3.16	0.177
2.0	1.00	0.0252	2.5	2.53	0.166
1.5	1.995	0.0638	2.93	2.0	0.139
2.0	1.50	0.05	3.57	1.54	0.128
2.5	1.075	0.0346	1.5	4.035	0.2715
1.5	2.52	0.0994	2.50	3.03	0.2039
2.0	2.0	0.0955	3.60	1.80	0.1910
2.5	1.514	0.0668	2.50	3.364	0.289
1.5	3.045	0.1071	1.50	5.109	0.571
2.5	2.0	0.0932	2.50	4.04	0.4873
3.6	1.10	0.0965	3.6	3.0	0.3735
0.978	3.912	0.218	2.5	4.809	0.762
1.235	3.705	0.205	1.5	6.377	0.991

№ 4506 [2167]

POTASSIUM THIOCYANATE–CALCIUM THIOCYANATE–WATER

$KCNS—Ca(CNS)_2—H_2O$

$t=25$

Concn. in aqueous soln., wt %		Comp. of solid phase
Calcium Thiocyanate	Potassium Thiocyanate	
0.00	70.76	KCNS
4.01	66.16	»
10.51	59.32	»
16.78	53.38	»
21.41	49.34	»
26.61	45.57	»
29.16	44.09	»
31.39	43.75	»
32.21	43.77	$KCNS+3KCNS \cdot Ca(CNS)_2 \cdot 3H_2O$
31.66	43.56	$3KCNS \cdot Ca(CNS)_2 \cdot 3H_2O$
32.11	42.84	»
32.81	42.07	»
33.69	41.12	»
34.22	40.22	»
35.22	39.06	»
37.18	36.89	»
37.47	36.28	$3KCNS \cdot Ca(CNS)_2 \cdot 3H_2O+$ $+2KCNS \cdot Ca(CNS) \cdot 3H_2O$

Table № 4506 (continuation)

Concn. in aqueous soln., wt %		Comp. of solid phase
Calcium Thiocyanate	Potassium Thiocyanate	
37.95	35.39	$2KCNS \cdot Ca(CNS)_2 \cdot 3H_2O$
40.17	32.56	»
44.66	26.55	»
46.65	24.00	»
51.82	18.62	»
53.18	17.77	$2KCNS \cdot Ca(CNS)_2 \cdot 3H_2O+ + \alpha\text{-}Ca(CNS)_2 \cdot 4H_2O$
53.19	17.54	»
54.39	13.81	»
55.68	10.46	»
56.67	7.90	»
57.71	5.23	»
58.99	2.10	»
60.20	0.00	»

№ 4507 POTASSIUM THIOCYANATE–COPPER (I) THIOCYANATE–WATER [931]

$KCNS—CuCNS—H_2O$

$t=25$

Concn. in aqueous soln., mol/l					
Potassium Thiocyanate	Copper Thiocyanate	Potassium Thiocyanate	Copper Thiocyanate	Potassium Thiocyanate	Copper Thiocyanate
0.730	0.0054	2.080	0.0292	5.60	0.4484
1.481	0.0139	3.928	0.1472	6.8	0.8339

№ 4508 POTASSIUM THIOCYANATE– ZINC NITRATE TRIHYDROXIDE–WATER [629]

$KCNS—3Zn(OH)_2 \cdot Zn(NO_3)_2 \cdot xH_2O—H_2O$

$t=25$

Concn. of Potassium Thiocyanate in aqueous soln., mol/l	Solubility of Zinc Nitrate Trihydroxide, mol/l	Concn. of Potassium Thiocyanate in aqueous soln., mol/l	Solubility of Zinc Nitrate Trihydroxide, mol/l	Concn. of Potassium Thiocyanate in aqueous soln., mol/l	Solubility of Zinc Nitrate Trihydroxide, mol/l
0.1	$0.7 \cdot 10^{-2}$	0.8	$2.18 \cdot 10^{-2}$	1.3	$3.05 \cdot 10^{-2}$
0.2	$0.96 \cdot 10^{-2}$	0.9	$2.30 \cdot 10^{-2}$	1.4	$3.90 \cdot 10^{-2}$
0.3	$1.15 \cdot 10^{-2}$	1.0	$2.50 \cdot 10^{-2}$	1.5	$3.15 \cdot 10^{-2}$
0.4	$1.30 \cdot 10^{-2}$	1.1	$2.63 \cdot 10^{-2}$	1.6	$4.35 \cdot 10^{-2}$
0.5	$1.50 \cdot 10^{-2}$	1.2	$2.90 \cdot 10^{-2}$	1.7	$4.90 \cdot 10^{-2}$
0.6	$1.79 \cdot 10^{-2}$				

№ 4509 [1630]

POTASSIUM THIOCYANATE–SILVER THIOCYANATE–WATER

KCNS—AgCNS—H_2O

$t=25$

Concn. in aqueous soln., wt %		Comp. of solid phase
Potassium Thiocyanate	Silver Thiocyanate	
23.86	16.07	KCNS · AgCNS + AgCNS
24.68	16.41	KCNS · AgCNS
32.51	18.34	»
49.43	20.32	»
50.68	20.43	2KCNS · AgCNS + KCNS
53.21	17.53	2KCNS · AgCNS
58.34	13.55	»
61.25	11.76	»
64.47	10.62	»
66.55	9.32	KCNS · 2KCNS · AgCNS
70.53	0.0	KCNS

№ 4510 [2276]

POTASSIUM THIOCYANATE–SILVER THIOCYANATE–WATER

KCNS—AgCNS—H_2O

$t=25$

Concn. in aqueous soln., wt %		Comp. of solid phase
Silver Thiocyanate	Potassium Thiocyanate	
0.00	70.89	KCNS
5.12	68.77	»
9.19	66.88	KCNS + 2KCNS · AgCNS
9.22	66.31	2KCNS · AgCNS
11.78	61.39	»
13.06	59.29	»
16.12	55.21	»
18.74	52.13	»
20.57	50.53	»
20.58	50.01	KCNS · AgCNS
20.28	47.72	»
20.04	43.02	»
19.24	35.88	»
16.68	25.40	»
16.10	23.15	KCNS · AgCNS + AgCNS
11.39	20.98	AgCNS
6.69	16.97	»
1.29	9.67	»
0.00002	0.00	»

№ 4511 [1795, 2614]

POTASSIUM THIOCYANATE–SILVER THIOCYANATE–WATER

$KCNS—AgCNS—H_2O$

Concn. in aqueous soln., g/l		t	Concn. in aqueous soln., g/l		t
Potassium Thiocyanate	Silver Thiocyanate		Potassium Thiocyanate	Silver Thiocyanate	
0.971	0.001161	20	81.564	6.636	20
7.719	0.01659	20	92.245	9.954	20
10.487	0.03318	20	97.1	16.59	20
14.934	0.06636	20	117.00	16.59	20
17.381	0.09954	20	55.69	2.06	25
19.663	0.1659	20	60.85	2.08	25
26.800	0.3318	20	103.6	14.01	25
34.956	0.6636	20	108.9	16.18	25
40.782	0.9954	20	116.6	19.93	25
47.094	1.659	20	121.5	22.34	25
62.921	3.318	20			

№ 4512 [1372]

POTASSIUM THIOCYANATE–SILVER THIOCYANATE–WATER

$KCNS—AgCNS—H_2O$

t=25

Comp. of saturated soln., wt %					
Potassium Thiocyanate	Silver Thiocyanate	Water	Potassium Thiocyanate	Silver Thiocyanate	Water
0.05297	0.00003	99.947	3.1477	0.0600	96.792
0.0999	0.00006	99.900	5.1323	0.2394	94.628
0.3985	0.0005	99.601	7.0286	0.6162	92.355
0.8527	0.0023	99.145	9.7636	1.6038	88.633
1.2784	0.0059	98.716	14.082	4.2604	81.658
2.1143	0.0200	97.866	17.945	7.7457	74.309

Note. Data obtained for a solution with a constant ionic strength of 2.2

№ 4513 [2369]

POTASSIUM THIOCYANATE–SILVER BROMIDE–WATER

$KCNS—AgBr—H_2O$

t=25

Concn. in aqueous soln., wt %					
Potassium Thiocyanate	Silver Bromide	Potassium Thiocyanate	Silver Bromide	Potassium Thiocyanate	Silver Bromide
2.376	0.0206	5.343	0.160	7.002	0.573
2.554	0.0225	6.846	0.532	8.578	1.229
4.806	0.178				

№ 4514 POTASSIUM THIOCYANATE–SILVER IODIDE–WATER [2369]

$KCNS—AgI—H_2O$

$t=25$

Concn. in aqueous soln., wt %					
Potassium Thiocyanate	Silver Iodide	Potassium Thiocyanate	Silver Iodide	Potassium Thiocyanate	Silver Iodide
1.92	0.00047	4.63	0.00120	6.91	0.00343
2.96	0.00033	5.58	0.00094	8.93	0.00383
3.96	0.00157	6.45	0.00228		

№ 4515 [2275]

POTASSIUM THIOCYANATE–BARIUM THIOCYANATE–WATER

$KCNS—Ba(CNS)_2—H_2O$

$t=25$

Comp. of saturated soln., wt %			Comp. of solid phase
Barium Thiocyanate	Potassium Thiocyanate	Water	
0.00	62.61	37.39	$Ba(CNS)_2 \cdot 3H_2O$
14.45	55.72	29.83	»
19.46	53.14	27.40	»
19.50	52.84	27.66	»
26.04	46.60	27.36	$2Ba(CNS)_2 \cdot 3KCNS \cdot 5H_2O$
28.95	43.66	27.39	»
34.02	39.45	26.53	»
34.47	39.10	26.43	»
37.18	37.04	25.78	»
43.92	31.73	24.35	»
45.00	30.75	24.25	»
47.67	28.96	23.37	»
48.28	28.68	23.04	$2Ba(CNS)_2 \cdot 3KCNS \cdot 5H_2O + KCNS$
52.42	22.78	24.80	KCNS
62.32	10.24	27.44	»
70.89	0.0	29.11	»

№ 4516 [2147]

POTASSIUM THIOCYANATE–MERCURY (II) THIOCYANATE–WATER

$KCNS—Hg(CNS)_2—H_2O$

$t=25$

Comp. of saturated soln., wt %			Comp. of solid phase
Mercury Thiocyanate	Potassium Thiocyanate	Water	
0.063	0.0	99.937	$Hg(CNS)_2$
4.05	2.05	93.9	$Hg(CNS)_2 + KHg(CNS)_3$
49.1	33.1	17.8	$K_2Hg(CNS)_4 + KHg(CNS)_3$
10.4	66.4	23.2	$K_2Hg(CNS)_4 + KCNS$
0.0	70.5	29.5	KCNS

№ 4517 [2263]

POTASSIUM THIOCYANATE–THALLIUM THIOCYANATE–WATER

$KCNS—TlCNS—H_2O$

$t=25$

Concn. in aqueous soln., g/l	
Thallium Thiocyanate	Potassium Thiocyanate
2.176	2.208

№ 4518 [1263]

POTASSIUM THIOCYANATE–THALLIUM THIOCYANATE–WATER

$KCNS—TlCNS—H_2O$

$t=25$

Comp. of saturated soln., wt %			Comp. of solid phase
Potassium Thiocyanate	Thallium Thiocyanate	Water	
30.94	0.65	68.41	(K, Tl)CNS
49.10	2.00	48.90	»
52.73	1.29	45.98	»
56.71	1.55	41.74	»
63.14	1.64	35.22	»
64.75	1.49	33.76	»
65.90	1.23	32.87	»
66.16	1.24	32.60	»
66.81	1.04	32.15	»
68.28	0.82	30.90	»

№ 4519 **POTASSIUM THIOCYANATE–THALLIUM THIOCYANATE–WATER** [432]

$KCNS—TlCNS—H_2O$

$t=25$

Concn. in aqueous soln., mol/l					
Thallium Thiocyanate	Potassium Thiocyanate	Thallium Thiocyanate	Potassium Thiocyanate	Thallium Thiocyanate	Potassium Thiocyanate
(33 ±1)	0.10	(54 ±2)	2.5	(770 ±20)	8.0
(22 ±1)	0.25	(77 ±2)	3.0	(880 ±40)	9.0
(18 ±1)	0.50	(146 ±3)	4.0	(848 ±10)	8.5
(19 ±1)	0.75	(250 ±2)	5.0	(660 ±10)	9.65
(21 ±1)	1.0	(410 ±10)	6.0		
(38 ±2)	2.0	(590 ±8)	7.0		

№ 4520 [219]

POTASSIUM SELENOCYANATE–LEAD (II) SELENOCYANATE–WATER

$KCNSe—Pb(CNSe)_2—H_2O$

$t=20$

Concn. in aqueous soln., mol/l					
Potassium Selenocyanate	Lead Selenocyanate	Potassium Selenocyanate	Lead Selenocyanate	Potassium Selenocyanate	Lead Selenocyanate
0.7326	$1.68 \cdot 10^{-3}$	1.4045	$1.76 \cdot 10^{-2}$	2.1698	$7.36 \cdot 10^{-2}$
0.9914	$4.69 \cdot 10^{-3}$	1.8412	$4.11 \cdot 10^{-2}$	2.4070	$9.86 \cdot 10^{-2}$
1.2117	$9.33 \cdot 10^{-3}$				

№ 4521 [1746]

POTASSIUM HEXACYANOFERRATE (III)–POTASSIUM HEXACYANOFERRATE (II)–WATER

$K_3Fe(CN)_6—K_4Fe(CN)_6—H_2O$

$t=25$

Concn. in aqueous soln., g/l		Comp. of solid phase
Potassium Hexacyano-ferrate (III)	Potassium Hexacyano-ferrate (II)	
338.1	79.02	$K_3Fe(CN)_6+$ $+K_4Fe(CN)_6 \cdot 3H_2O$

№ 4522 [2385, 2386]

POTASSIUM HEXACYANOFERRATE (III)–POTASSIUM HEXACYANOFERRATE (II)–WATER

$K_3Fe(CN)_6—K_4Fe(CN)_6—H_2O$

Comp. of saturated soln., wt %			*t*	Comp. of solid phase
Potassium Hexacyano-ferrate (II)	Potassium Hexacyano-ferrate (III)	Water		
19.65	0.0	80.35	15	$K_4Fe(CN)_6 \cdot 3H_2O$
18.16	1.86	79.98	15	»
15.65	4.88	79.47	15	»
14.08	7.75	78.17	15	»
13.71	7.61	78.68	15	»
12.35	10.08	77.57	15	»
10.22	14.10	75.68	15	»
10.14	14.16	75.70	15	»
9.39	14.78	75.83	15	»
8.51	16.90	74.59	15	»

Table № 4522 (continuation)

Comp. of saturated soln., wt %			t	Comp. of solid phase
Potassium Hexacyano-ferrate (III)	Potassium Hexacyano-ferrate (II)	Water		
7.27	19.97	72.76	15	»
6.95	20.52	72.53	15	»
5.72	22.78	72.50	15	»
4.64	25.76	69.60	15	$K_4Fe(CN)_6 \cdot 3H_2O + K_3Fe(CN)_6$
2.49	27.77	69.74	15	$K_3Fe(CN)_6$
2.43	27.62	69.95	15	$K_3Fe(CN)_6$
0.0	29.69	70.31	15	»
32.55	0.0	67.45	50	$K_4Fe(CN)_6 \cdot 3H_2O$
30.54	2.82	66.64	50	»
25.21	8.46	66.33	50	»
23.85	10.11	66.04	50	»
20.88	13.53	65.59	50	»
17.32	18.02	64.66	50	»
16.70	19.21	64.09	50	»
14.67	21.69	63.64	50	»
11.53	27.14	61.33	50	»
10.92	28.46	60.62	50	»
10.45	28.70	60.85	50	»
7.87	32.52	59.61	50	$K_4Fe(CN)_6 \cdot 3H_2O + K_3Fe(CN)_6$
5.81	38.40	59.79	50	$K_3Fe(CN)_6$
5.66	34.53	59.81	50	»
2.14	37.19	60.67	50	»
0.00	39.52	60.48	50	»

№ 4523 POTASSIUM HEXACYANOFERRATE (III)–POTASSIUM HEXACYANOFERRATE (II)–WATER [2385, 2386]

$K_3Fe(CN)_6—K_4Fe(CN)_6—H_2O$

Comp. of saturated soln., wt %			t	Comp. of solid phase
Potassium Hexacyano-ferrate (II)	Potassium Hexacyano-ferrate (III)	Water		
37.00	0.00	63.00	65.0	$K_4Fe(CN)_6 \cdot 3H_2O$
28.49	9.73	61.78	65.0	»
20.85	18.28	60.87	65.0	»
20.72	18.30	60.98	65.0	»

Table № 4523 (continuation)

Comp. of saturated soln., wt %			t	Comp. of solid phase
Potassium Hexacyano-ferrate (II)	Potassium Hexacyano-ferrate (III)	Water		
13.22	28.29	58.49	65	»
9.37	34.00	56.63	65	»
9.22	34.14	56.64	65	$K_4Fe(CN)_6 \cdot 3H_2O + K_3Fe(CN)_6$
4.22	38.45	57.33	65	$K_3Fe(CN)_6$
3.21	39.28	57.51	65	»
2.49	40.49	57.02	65	»
0.00	42.19	57.81	65	»
40.05	0.00	59.95	80	$K_4Fe(CN)_6 \cdot 3H_2O$
27.43	13.65	58.92	80	»
22.24	20.17	57.69	80	»
20.41	22.40	57.19	80	»
16.69	26.95	56.36	80	»
11.10	35.00	53.90	80	$K_4Fe(CN)_6 \cdot 3H_2O + K_3Fe(CN)_6$
5.74	40.00	54.26	80	$K_3Fe(CN)_6$
2.75	42.40	54.85	80	»
0.00	45.27	54.73	80	»

№ 4524 [1259]

POTASSIUM HEXACYANOFERRATE (II)–POTASSIUM SULFATE–WATER

$K_4Fe(CN)_6—K_2SO_4—H_2O$

Comp. of saturated soln., wt %			t	Comp. of solid phase
Potassium Hexacyano-ferrate	Potassium Sulfate	Water		
11.58	0.00	88.42	−1.6	Ice $+ K_4Fe(CN)_6 \cdot 3H_2O$
9.07	4.62	86.31	−2.05	Ice $+ K_4Fe(CN)_6 \cdot 3H_2O + K_2SO_4$
9.80	4.83	85.37	0	$K_4Fe(CN)_6 \cdot 3H_2O + K_2SO_4$
12.48	0.0	87.52	0	$K_4Fe(CN)_6 \cdot 3H_2O$
0.00	6.82	93.18	0	K_2SO_4
25.94	0.00	74.06	30	$K_4Fe(CN)_6 \cdot 3H_2O$
24.17	0.902	74.928	30	»
22.97	2.231	74.799	30	»
22.23	3.902	73.868	30	$K_4Fe(CN)_6 \cdot 3H_2O + K_2SO_4$
16.45	5.535	78.015	30	K_2SO_4
11.08	7.126	81.794	30	»

Table № 4524 (continuation)

Comp. of saturated soln., wt %			t	Comp. of solid phase
Potassium Hexacyano-ferrate	Potassium Sulfate	Water		
8.54	7.818	83.642	30	»
5.01	9.59	85.40	30	»
3.03	10.205	86.765	30	»
0.00	11.53	88.47	30	»
29.28	0.00	70.72	40	$K_4Fe(CN)_6 \cdot 3H_2O$
28.58	1.288	70.132	40	»
27.50	2.468	70.032	40	»
25.66	3.982	70.358	40	$K_4Fe(CN)_6 \cdot 3H_2O+K_2SO_4$
19.63	5.495	74.875	40	K_2SO_4
15.21	7.031	77.759	40	»
10.36	8.557	81.583	40	»
8.56	8.930	82.510	40	»
5.88	10.111	84.009	40	»
3.30	11.306	85.664	40	»
0.00	12.90	87.10	40	»
38.81	0.0	61.19	65	$K_4Fe(CN)_6 \cdot 3H_2O$
33.645	3.215	63.140	65	$K_4Fe(CN)_6 \cdot 3H_2O+K_2SO_4$
0.0	16.0	84.0	65	K_2SO_4
38.75	0.0	61.25	74	$K_4Fe(CN)_6 \cdot 3H_2O$
36.20	3.295	60.505	74	$K_4Fe(CN)_6 \cdot 3H_2O+K_2SO_4$
0.0	16.92	83.09	74	K_2SO_4
40.72	0.0	59.28	85	$K_4Fe(CN)_6 \cdot 3H_2O$
38.715	2.975	58.310	85	$K_4Fe(CN)_6 \cdot 3H_2O+K_2SO_4$
0.0	18.11	81.89	85	K_2SO_4
39.03	2.80	58.17	86.3	$K_4Fe(CN)_6 \cdot 3H_2O+K_4Fe(CN)_6+K_2SO_4$
41.38	0.0	58.62	90	$K_4Fe(CN)_6$
40.14	1.030	58.83	90	»
39.75	1.410	58.84	90	»
39.30	2.770	57.93	90	»
39.165	2.831	58.004	90	$K_4Fe(CN)_6+K_2SO_4$
32.11	4.30	63.59	90	K_2SO_4
30.36	5.26	64.38	90	»
17.20	9.92	72.88	90	»
14.67	11.09	74.24	90	»
5.46	15.42	79.12	90	»
0.0	18.60	81.40	90	»
44.77	0.0	55.23	104.1	$K_4Fe(CN)_6$
39.356	3.344	57.30	104.3	$K_4Fe(CN)_6+K_2SO_4$
0.0	19.50	80.50	101.4	K_2SO_4

№ 4525 POTASSIUM HEXACYANOFERRATE (III)–POTASSIUM SULFATE–WATER [781]

$K_3[Fe(CN)_6]$—K_2SO_4—H_2O

$t=20$

Comp. of saturated soln., wt %			Comp. of solid phase
Potassium Sulfate	Potassium Hexacyano-ferrate	Water	
10.06	0.00	89.94	K_2SO_4
8.70	2.89	88.41	»
7.27	4.37	88.36	»
6.40	7.45	86.15	»
5.00	9.95	85.05	»
4.52	12.97	82.51	»
3.83	15.43	80.74	»
2.43	18.70	78.87	$K_2SO_4+K_4Fe(CN)_6$
2.44	18.67	78.89	»
2.44	18.68	78.88	»
2.43	18.69	78.88	»
2.42	18.72	78.86	»
1.92	19.24	78.84	$K_4Fe(CN)_6$
0.00	22.40	77.60	»

№ 4526 [2699]

POTASSIUM HEXACYANOFERRATE (III)–CALCIUM SULFATE–WATER

$K_3Fe(CN)_6$—$CaSO_4$—H_2O

$t=25$

Concn. in aqueous soln., g/l	
Potassium Hexacyanoferrate	Calcium Sulfate
0.550	2.219
1.096	2.342
2.196	2.560

№ 4527 [747]

POTASSIUM HEXACYANOFERRATE (II)–BARIUM CHLORIDE–WATER

$K_4Fe(CN)_6$—$BaCl_2$—H_2O

$t=25$

Comp. of saturated soln., wt %			Comp. of solid phase
Barium Chloride	Potassium Hexacyano-ferrate	Water	
27.16	0.00	72.84	—
20.05	0.24	79.71	$BaCl_2 \cdot 2H_2O+BaK_2Fe(CN)_6 \cdot 3H_2O$
19.40	0.30	80.30	»
14.96	0.30	84.74	»
8.72	1.62	89.66	»

Table № 4527 (continuation)

Comp. of saturated soln., wt %			Comp. of solid phase
Barium Chloride	Potassium Hexacyano-ferrate	Water	
4.80	3.00	92.20	$BaK_2Fe(CN)_6 \cdot 3H_2O$
0.44	5.47	94.09	»
0.23	8.60	91.17	$BaK_2Fe(CN)_6 \cdot 3H_2O + K_4Fe(CN)_6 \cdot 3H_2O$
0.14	10.66	89.20	»
0.15	14.80	85.05	»
0.20	16.70	83.10	»
0.25	18.50	81.25	»
0.00	23.96	76.04	»

№ 4528 POTASSIUM NITRITE–POTASSIUM NITRATE–WATER [1299]

$KNO_2—KNO_3—H_2O$

Comp. of saturated soln., wt %			t	Comp. of solid phase
Potassium Nitrite	Potassium Nitrate	Water		
68.70	8.53	22.77	25	KNO_2+KNO_3 (mixed crystals)
66.45	11.20	22.35	25	»
64.15	13.05	22.80	25	»
61.5	15.10	23.40	25	β-KNO_3
55.8	14.95	29.25	25	»
35.85	15.05	49.10	25	α-KNO_3
19.90	18.75	61.35	25	»
64.80	14.80	70.40	98.5	KNO_2+KNO_3 (mixed crystals)
57.0	13.80	39.20	98.5	»
51.10	35.15	13.75	98.5	»
46.70	39.15	14.15	98.5	»
39.70	43.80	16.50	98.5	KNO_3
20.45	55.40	24.15	98.5	»

№ 4529 POTASSIUM NITRITE–POTASSIUM NITRATE–WATER [132]

$KNO_2—KNO_3—H_2O$

Comp. of saturated soln., wt %			t	Comp. of solid phase
Potassium Nitrate	Potassium Nitrite	Water		
24.6	0.0	75.4	20	KNO_3
21.0	7.0	72.0	20	»
18.5	16.5	70.0	20	»
15.5	34.0	50.5	20	»
15.0	52.0	33.0	20	»
16.0	60.0	24.0	20	»
16.5	63.0	21.5	20	»
17.0	64.0	19.0	20	KNO_3+KNO_2

Table № 4529 (continuation)

Comp. of saturated soln., wt %			t	Comp. of solid phase
Potassium Nitrate	Potassium Nitrite	Water		
14.0	66.0	20.0	20	KNO_2
9.5	69.0	21.5	20	»
0.0	75.5	24.5	20	»
46.5	0.0	53.5	50	KNO_3
42.0	6.0	52.0	50	»
37.5	12.5	50.0	50	»
30.5	27.5	42.0	50	»
25.5	44.5	30.0	50	»
23.0	53.5	24.5	50	»
22.0	58.5	19.5	50	»
21.0	63.0	16.0	50	KNO_3+KNO_2
17.0	66.0	17.0	50	KNO_2
12.5	68.5	19.0	50	»
8.5	71.5	20.0	50	»
0.0	77.0	23.0	50	»

№ 4530 [144]

POTASSIUM NITRITE–POTASSIUM CHLORIDE–WATER

KNO_2—KCl—H_2O

$t=25$

Comp. of saturated soln., wt %			Comp. of solid phase
Potassium Nitrite	Potassium Chloride	Water	
0.00	26.44	73.56	KCl
5.04	23.77	71.19	»
9.63	21.63	68.74	»
13.56	19.29	67.15	»
20.60	17.00	62.40	»
25.44	14.66	59.90	»
30.59	12.49	56.92	»
35.98	9.71	54.31	»
44.08	8.03	47.89	»
50.71	6.31	42.98	»
55.57	5.12	39.31	»
59.85	4.28	35.87	»
66.20	3.41	30.39	»
66.60	3.16	30.24	$KCl+K(NO_2, Cl)$
67.70	2.66	29.64	$K(NO_2, Cl)$
69.72	1.86	28.42	»
72.15	1.24	26.61	»
74.23	0.69	25.08	»
75.65	0.00	24.31	KNO_2

№ 4531 POTASSIUM NITRITE–STRONTIUM NITRITE–WATER [1000]

$KNO_2—Sr(NO_2)_2—H_2O$

$t=25$

Comp. of saturated soln., wt %			Comp. of solid phase
Strontium Nitrite	Potassium Nitrite	Water	
41.50	0.00	58.50	$Sr(NO_2)_2 \cdot H_2O$
35.28	12.00	52.72	»
33.70	17.76	48.54	»
30.50	25.67	43.83	»
29.10	31.30	39.60	»
28.30	32.30	39.40	$Sr(NO_2)_2 \cdot H_2O + 2KNO_2 \cdot Sr(NO_2)_2 \cdot 4H_2O$
28.00	33.00	39.00	»
28.50	32.50	39.00	»
27.54	33.00	39.46	$2KNO_3 \cdot Sr(NO_2)_2 \cdot 4H_2O$
26.25	34.47	39.28	»
25.20	36.70	38.10	»
21.05	42.50	36.45	»
14.16	51.90	33.93	»
11.50	57.00	31.50	»
10.00	59.00	31.00	»
9.00	64.00	27.00	$KNO_2 + 2KNO_2 \cdot Sr(NO_2)_2 \cdot 4H_2O$
8.60	63.76	27.64	»
8.40	64.00	27.60	»
4.02	68.00	27.98	KNO_2
0.00	75.07	24.93	»

№ 4532 [2298]

POTASSIUM NITRITE–SILVER NITRITE–WATER

$KNO_2—AgNO_2—H_2O$

Concn. in aqueous soln., wt %		t	Comp. of solid phase
Potassium Nitrite	Silver Nitrite		
14.95	1.96	13.5	$AgNO_2 + K_2Ag_2(NO_2)_4 \cdot H_2O$
17.99	4.13	25	»
68.60	6.54	13.5	$KNO_2 + K_2Ag_2(NO_2)_4 \cdot H_2O$
66.70	9.39	25	»

№ 4533 POTASSIUM NITRITE–SILVER NITRITE–WATER [1435]

$KNO_2—AgNO_2—H_2O$

$t=25$

Concn. in aqueous soln., g/l					
Potassium Nitrite	Silver Nitrite	Potassium Nitrite	Silver Nitrite	Potassium Nitrite	Silver Nitrite
0.00	4.135	0.50	3.820	2.00	3.119
0.22	3.974	1.00	3.560	4.00	2.765

№ 4534 [2239]

POTASSIUM NITRITE–SILVER NITRITE–WATER

$KNO_2—AgNO_2—H_2O$

$t=25$

Comp. of saturated soln., wt %			Comp. of solid phase
Silver Nitrite	Potassium Nitrite	Water	
0.75	5.83	93.42	$AgNO_2$
1.99	13.19	84.82	»
4.93	21.73	75.34	$AgNO_2+KAg(NO_2)_2 \cdot {}^7/_2H_2O$
4.89	24.57	70.54	$KAg(NO_2)_2 \cdot ½H_2O$
4.88	28.09	67.03	»
5.01	32.90	62.09	»
5.09	35.11	59.80	»
5.33	38.15	56.52	»
5.57	42.02	52.41	»
6.30	48.83	44.87	»
6.78	52.55	40.67	»
7.28	55.60	37.12	»
7.88	57.47	34.65	»
8.10	60.08	31.82	»
8.31	60.90	30.89	»
8.59	61.02	30.39	»
10.53	73.14	16.33	$KAg(NO_2)_2 \cdot ½H_2O+KNO_2$
4.14	74.15	21.71	KNO_2
0.00	76.22	23.78	»

№ 4535 [718]

POTASSIUM NITRITE–BARIUM NITRITE–WATER

$KNO_2—Ba(NO_2)_2—H_2O$

Comp. of saturated soln., wt %			t	Comp. of solid phase
Potassium Nitrite	Barium Nitrite	Water		
75.18	0.00	24.82	25	$2KNO_2 \cdot Ba(NO_2)_2$
74.04	1.14	24.82	25	»
74.04	1.15	24.81	25	»
69.61	0.80	29.59	25	»
64.06	1.38	34.56	25	»
54.04	2.87	43:09	25	»
51.21	3.10	45.69	25	»
45.10	5.73	49.17	25	»
42.98	6.88	50.14	25	»
37.02	11.06	51.92	25	»
34.89	13.19	51.92	25	»
32.34	14.89	52.77	25	»
30.64	17.78	51.58	25	»
26.81	22.94	50.25	25	»
22.98	28.67	48.35	25	»
20.00	33.83	46.17	25	»
18.36	37.27	44.37	25	$2KNO_2 \cdot Ba(NO_2)_2+Ba(NO_2)_2 \cdot H_2O$

Table № 4535 (continuation)

Comp. of saturated soln., wt %			t	Comp. of solid phase
Potassium Nitrite	Barium Nitrite	Water		
18.30	37.41	44.29	25	$Ba(NO_2)_2 \cdot H_2O$
13.62	39.00	47.38	25	»
6.65	41.81	51.54	25	»
0.00	44.72	55.28	25	»
77.04	0.00	22.96	50	KNO_2
75.31	1.72	22.95	50	$KNO_2+2KNO_2 \cdot Ba(NO_2)_2$
75.36	1.74	22.90	50	»
72.34	2.29	25.37	50	$2KNO_2 \cdot Ba(NO_2)_2$
66.80	2.87	30.33	50	»
62.25	3.84	33.91	50	»
52.84	6.95	40.21	50	»
45.12	13.01	41.87	50	»
39.14	18.93	41.93	50	»
35.14	23.52	41.34	50	»
29.98	30.89	39.13	50	»
25.10	37.85	37.05	50	»
20.42	45.30	34.28	50	»
18.80	48.11	33.09	50	$2KNO_2 \cdot Ba(NO_2)_2+Ba(NO_2)_2 \cdot H_2O$
18.72	48.17	33.11	50	»
15.32	48.86	35.82	50	$Ba(NO_2)_2 \cdot H_2O$
10.64	51.04	38.32	50	»
6.08	52.76	41.16	50	»
0.00	55.30	44.70	50	»

№ 4536 [213]

POTASSIUM NITRATE–POTASSIUM FLUORIDE–LEAD (II) FLUORIDE–WATER

$KNO_3—KF—PbF_2—H_2O$

t=18—20

Concn. in aqueous soln., g/l		Solubility of Lead Fluoride in aqueous soln. of Potassium Nitrate, g/l	Concn. in aqueous soln., g/l		Solubility of Lead Fluoride in aqueous soln. of Potassium Nitrate, g/l
Potassium Fluoride	Potassium Nitrate		Potassium Fluoride	Potassium Nitrate	
0.00	10.10	0.527	3.68	20.20	0.148
0.46	10.10	0.352	7.36	20.20	0.101
0.92	10.10	0.283	14.72	20.20	0.116
1.84	10.10	0.176	20.99	20.20	0.126
3.68	10.10	0.084	0.00	50.50	0.840
7.36	10.10	0.055	0.46	50.50	0.691
14.72	10.10	0.059	0.92	50.50	0.665
20.99	10.10	0.077	1.84	50.50	0.216
0.00	20.20	0.576	3.68	50.50	0.179
0.46	20.20	0.497	7.36	50.50	0.107
0.92	20.20	0.374	14.72	50.50	0.124
1.84	20.20	0.243	20.99	50.50	0.129

№ 4537 [1084]

POTASSIUM NITRATE–POTASSIUM DIHYDROGEN ORTHOPHOSPHATE–WATER

$KNO_3—KH_2PO_4—H_2O$

$t=25$

Comp. of saturated soln., wt %		
Potassium Dihydrogen Orthophosphate	Potassium Nitrate	Water
16.33	5.56	78.11
14.03	12.03	73.94

№ 4538 **POTASSIUM NITRATE–POTASSIUM DIHYDROGEN ORTHOPHOSPHATE–WATER** [104]

$KNO_3—KH_2PO_4—H_2O$

Concn. in aqueous soln., wt %		t	Comp. of solid phase
Potassium Nitrate	Potassium Dihydrogen Orthophosphate		
11.0	0.0	0	KNO_3
9.9	3.6	0	»
9.3	6.0	0	»
8.5	9.0	0	»
8.4	9.8	0	$KNO_3+KH_2PO_4$
4.2	10.5	0	KH_2PO_4
0.0	11.8	0	»
16.6	0.0	10	KNO_3
15.3	3.4	10	»
13.2	8.6	10	»
12.8	10.2	10	$KNO_3+KH_2PO_4$
8.9	11.2	10	KH_2PO_4
5.0	12.6	10	»
0.0	15.0	10	»
23.0	0.0	20	KNO_3
21.6	3.0	20	»
19.0	8.0	20	»
17.9	10.4	20	$KNO_3+KH_2PO_4$
16.1	10.8	20	KH_2PO_4
8.7	13.2	20	»
5.8	14.6	20	»
0.0	18.2	20	»
31.0	0.0	30	KNO_3
28.8	2.8	30	»
25.6	7.4	30	»
23.2	10.9	30	$KNO_3+KH_2PO_4$
15.8	12.2	30	KH_2PO_4
8.4	15.8	30	»
6.8	17	30	»
0.0	21.4	30	»

№ 4539 [1580]

POTASSIUM NITRATE–POTASSIUM SULFATE–WATER

$KNO_3—K_2SO_4—H_2O$

Comp. of saturated soln., wt %			t	d
Potassium Nitrate	Potassium Sulfate	Water		
18.58	4.37	77.05	15	1.165
22.18	4.11	73.71	20	—
25.49	3.93	70.58	25	1.210

№ 4540 POTASSIUM NITRATE–POTASSIUM SULFATE–WATER [1886]

$KNO_3—K_2SO_4—H_2O$

Comp. of saturated soln., wt %			t	Comp. of solid phase
Potassium Nitrate	Potassium Sulfate	Water		
27.64	0.00	72.36	25	KNO_3
27.33	1.64	71.03	25	»
26.01	1.98	72.01	25	»
25.91	3.00	71.09	25	»
25.37	3.95	70.68	25	$KNO_3+K_2SO_4$
23.99	4.45	71.56	25	»
14.52	5.89	79.59	25	K_2SO_4
11.05	6.89	82.06	25	»
5.84	8.12	86.04	25	»
4.82	8.85	86.33	25	»
34.76	0.00	65.24	35	KNO_3
32.61	1.17	66.22	35	»
30.76	3.42	65.82	35	$KNO_3+K_2SO_4$
16.29	6.47	77.24	35	K_2SO_4
0.00	12.17	87.83	35	»
65.92	0.63	33.45	90	$K_2SO_4+KNO_3$

№ 4541 POTASSIUM NITRATE–POTASSIUM SULFATE–WATER [1801]

$KNO_3—K_2SO_4—H_2O$

Comp. of saturated soln., wt %			t	d	Comp. of solid phase
Potassium Nitrate	Potassium Sulfate	Water			
8.96	4.09	86.95	−3.32	—	Ice $+KNO_3+K_2SO_4$
11.74	0.0	88.26	0	—	KNO_3
10.69	3.07	86.24	0	—	»
10.41	4.23	85.36	0	—	$KNO_3+K_2SO_4$
9.48	4.35	86.17	0	—	K_2SO_4
5.71	5.12	89.17	0	—	»
0.00	6.83	93.17	0	—	»
14.64	4.41	80.95	8.5	—	$KNO_3+K_2SO_4$
17.66	4.49	77.85	17.5	1.176	»
2.76	5.27	91.97	20	1.187	»

Table № 4541 (continuation)

Comp. of saturated soln., wt %			t	d	Comp. of solid phase
Potassium Nitrate	Potassium Sulfate	Water			
27.16	0.00	72.84	25	—	KNO_3
25.80	3.31	70.89	25	1.205	»
25.21	4.14	70.65	25	1.210	$KNO_3+K_2SO_4$
23.40	4.37	72.23	25	1.199	K_2SO_4
18.97	5.14	75.89	25	1.170	»
14.31	6.18	79.51	25	1.144	»
8.63	7.61	83.76	25	1.115	»
0.0	10.80	89.20	25	1.086	»
8.99	3.71	87.30	30	1.234	$KNO_3+K_2SO_4$
36.74	3.01	60.25	40	1.287	»
44.66	1.72	53.62	50	1.339	KNO_3
44.11	2.42	53.47	50	1.341	$KNO_3+K_2SO_4$
39.65	2.62	57.73	50	1.301	K_2SO_4
24.43	5.54	70.03	50	1.203	»
18.56	6.86	74.58	50	1.171	»
13.99	8.24	77.77	50	1.152	»
0.0	14.15	85.85	50	1.104	»
59.29	1.31	39.40	75	1.467	$KNO_3+K_2SO_4$
55.85	1.64	42.51	75	1.433	K_2SO_4
50.23	2.13	47.64	75	1.380	»
45.04	2.70	52.26	75	1.335	»
36.11	4.19	59.70	75	1.268	»
26.26	6.55	67.19	75	1.197	»
13.52	10.49	75.99	75	1.148	»
0.0	17.07	82.93	75	1.116	»
70.17	0.84	28.99	100	1.572	$KNO_3+K_2SO_4$
69.78	0.82	29.40	100	1.566	K_2SO_4
69.78	0.82	29.40	100	1.567	»
63.75	1.12	35.13	100	1.497	»
61.10	1.26	37.64	100	1.467	»
53.49	1.16	44.35	100	1.394	»
48.88	1.72	48.40	100	1.352	»
31.52	5.92	62.56	100	1.229	»
18.16	11.05	71.79	100	1.163	»
0.0	19.42	80.58	100	1.119	»

№ 4542 POTASSIUM NITRATE–POTASSIUM SULFATE–WATER [1223]

$KNO_3—K_2SO_4—H_2O$

$t=310$

Comp. of saturated soln., wt %					
Potassium Nitrate	Potassium Sulfate	Water	Potassium Nitrate	Potassium Sulfate	Water
2.9	25.7	71.4	40.0	10.0	50.0
6.1	24.4	69.5	63.0	7.0	30.0
17.0	17.0	66.0	92.20	4.85	2.95

№ 4543 POTASSIUM NITRATE–POTASSIUM SULFATE–WATER [1575]

$KNO_3—K_2SO_4—H_2O$

$t=25$

Content of Potassium Nitrate in aqueous soln., wt %	Solubility of Potassium Sulfate in aqueous soln. of Potassium Nitrate, wt %	Content of Potassium Nitrate in aqueous soln., wt %	Solubility of Potassium Sulfate in aqueous soln. of Potassium Nitrate, wt %
4.95	8.62	14.62	5.85
9.75	7.05	19.60	4.90

№ 4544 POTASSIUM NITRATE–POTASSIUM SULFATE–WATER [208]

$KNO_3—K_2SO_4—H_2O$

Comp. of saturated soln., wt %			t	Comp. of solid phase
Potassium Sulfate	Potassium Nitrate	Water		
4.0	9.0	87.0	−2.7	$H_2O+KNO_3+mK_2SO_4 \cdot nKNO_3 \cdot xH_2O$
6.7	5.0	88.3	9.4	$K_2SO_4+K_2SO_4 \cdot H_2O+mK_2SO_4 \cdot nKNO_3 \cdot xH_2O$
4.4	9.7	85.9	0.9	$KNO_3+K_2SO_4+mK_2SO_4 \cdot nKNO_3 \cdot xH_2O$
4.8	6.4	88.8	−1.6	$H_2O+K_2SO_4 \cdot H_2O+mK_2SO_4 \cdot nKNO_3 \cdot xH_2O$

№ 4545 POTASSIUM NITRATE–POTASSIUM SULFATE–POTASSIUM CHLORIDE–WATER [2017]

$KNO_3—K_2SO_4—KCl—H_2O$

Concn. in aqueous soln., g/100 g Water			d	t	Comp. of solid phase
Potassium Nitrate	Potassium Sulfate	Water			
8.8	1.15	27.4	—	0	$KCl+KNO_3+K_2SO_4$
7.42	1.45	34.9	1.212	25	$KCl+K_2SO_4$
14.4	1.41	34.5	1.236	25	»
23.1	1.40	34.2	1.266	25	$KCl+K_2SO_4+KNO_3$
53.6	1.57	37.6	1.353	50	»
16.7	2.08	46.6	1.258	75	$KCl+K_2SO_4$
40.8	1.93	44.4	1.322	75	»
72.9	1.80	41.8	1.393	75	»
110.3	1.77	39.1	1.461	75	$KCl+K_2SO_4+KNO_3$
204.0	1.86	39.8	1.573	100	»
23.84	0.70	35.23	—	25	$KCl+KNO_3$
12.48	1.45	35.82	—	25	$KCl+K_2SO_4$
21.66	1.38	31.22	—	25	$KCl+K_2SO_4+KNO_3$
29.77	2.66	12.57	—	25	$KNO_3+K_2SO_4$

№ 4546 [2017]

POTASSIUM NITRATE–POTASSIUM SULFATE–POTASSIUM CHLORIDE–WATER

$KNO_3—K_2SO_4—KCl—H_2O$

$t=25$

Concn. in aqueous soln., g/100 g Water			Comp. of solid phase
Potassium Chloride	Potassium Nitrate	Water	
35.23	23.84	0.70	$KCl+KNO_3$
12.57	29.77	2.66	$KNO_3+K_2SO_4$
35.82	12.48	1.45	$KCl+K_2SO_4$
31.22	21.63	1.38	$KCl+KNO_3+K_2SO_4$

№ 4547 [1575]

POTASSIUM NITRATE–POTASSIUM SULFATE–POTASSIUM CHLORIDE–WATER

$KNO_3—K_2SO_4—KCl—H_2O$

$t=25$

Content in aqueous soln., wt %		Solubility of Potassium Sulfate in aqueous soln. of Potassium Nitrate and Potassium Chloride, wt %	Content in aqueous soln., wt %		Solubility of Potassium Sulfate in aqueous soln. of Potassium Nitrate and Potassium Chloride, wt %
Potassium Chloride	Potassium Nitrate		Potassium Chloride	Potassium Nitrate	
4.46	4.46	5.94	4.32	13.04	4.32
4.35	8.69	5.07	8.71	5.17	3.98

№ 4548 [1575]

POTASSIUM NITRATE–POTASSIUM SULFATE–POTASSIUM CHLORIDE–POTASSIUM BROMATE–WATER

$KNO_3—K_2SO_4—KCl—KBrO_3—H_2O$

$t=25$

Content in aqueous soln., wt %			Solubility of Potassium Sulfate in aqueous soln. of Potassium Chloride, Potassium Nitrate and Potassium Bromate, wt %
Potassium Chloride	Potassium Nitrate	Potassium Bromate	
4.47	4.48	0.90	5.73
4.28	4.25	1.56	5.70
4.20	9.79	0.77	4.82
8.28	4.84	0.76	4.01

№ 4549 [1575]

POTASSIUM NITRATE–POTASSIUM SULFATE–POTASSIUM BROMATE–WATER

$KNO_3—K_2SO_4—KBrO_3—H_2O$

$t=25$

Content in aqueous soln., wt %		Solubility of Potassium Sulfate in aqueous soln. of Potassium Nitrate and Potassium Bromate, wt %	Content in aqueous soln., wt %		Solubility of Potassium Sulfate in aqueous soln. of Potassium Nitrate and Potassium Bromate, wt %
Potassium Nitrate	Potassium Bromate		Potassium Nitrate	Potassium Bromate	
4.83	0.97	8.36	9.70	0.92	6.85
5.00	1.91	8.08	9.67	1.57	6.69
5.01	2.43	7.89	14.46	0.79	5.74

№ 4550 [407]

POTASSIUM NITRATE–POTASSIUM SULFATE–THALLIUM CHROMATE–WATER

$KNO_3—K_2SO_4—Tl_2CrO_4—H_2O$

$t=18—20$

Content in soln., g-eq/l		Solubility of Thallium Chromate in aqueous soln. of Potassium Sulfate and Potassium Nitrate, g/l	Content in soln., g-eq/l		Solubility of Thallium Chromate in aqueous soln. of Potassium Sulfate and Potassium Nitrate, g/l
Potassium Sulfate	Potassium Nitrate		Potassium Sulfate	Potassium Nitrate	
1.00	0.00	$13.9 \cdot 10^2$	0.25	0.75	$12.2 \cdot 10^2$
0.75	0.25	$13.6 \cdot 10^2$	0.00	1.00	$11.3 \cdot 10^2$
0.50	0.50	$12.7 \cdot 10^2$			

№ 4551 [407]

POTASSIUM NITRATE–POTASSIUM SULFATE–THALLIUM IODATE–WATER

$KNO_3—K_2SO_4—TlIO_3—H_2O$

$t=18—20$

Content in soln., g-eq/l		Solubility of Thallium Iodate in aqueous soln. of Potassium Nitrate and Potassium Sulfate, g/l	Content in soln., g-eq/l		Solubility of Thallium Iodate in aqueous soln. of Potassium Nitrate and Potassium Sulfate, g/l
Potassium Nitrate	Potassium Sulfate		Potassium Nitrate	Potassium Sulfate	
0.100	0.000	0.81	0.025	0.075	0.89
0.075	0.025	0.83	0.000	0.100	0.92
0.050	0.050	0.86			

№ 4552 [213]

POTASSIUM NITRATE–POTASSIUM SULFATE–LEAD (II) SULFATE–WATER

$KNO_3—K_2SO_4—PbSO_4—H_2O$

$t=18—20$

Concn. in aqueous soln., g/l		Solubility of Lead Sulfate in aqueous soln. of Potassium Sulfate and Potassium Nitrate, g/l	Concn. in aqueous soln., g/l		Solubility of Lead Sulfate in aqueous soln. of Potassium Sulfate and Potassium Nitrate, g/l
Potassium Sulfate	Potassium Nitrate		Potassium Sulfate	Potassium Nitrate	
0.000	10.11	0.1234	2.788	20.22	0.05429
0.697	10.11	0.04428	5.576	20.22	0.05581
1.394	10.11	0.040946	11.153	20.22	0.07097
2.788	10.11	0.037609	0.000	101.1	0.69425
5.576	10.11	0.06187	0.697	101.1	0.4140
11.153	10.11	0.06491	1.394	101.1	0.24931
0.000	20.22	0.13982	2.788	101.1	0.1650
0.697	20.22	0.08614	5.576	101.1	0.2117
1.394	20.22	0.06278	11.153	101.1	0.2226

№ 4553 [2251, 2448]

POTASSIUM NITRATE–POTASSIUM CHLORIDE–WATER

$KNO_3—KCl—H_2O$

$t=20$

Concn. in aqueous soln., wt %					
Potassium Nitrate	Potassium Chloride	Potassium Nitrate	Potassium Chloride	Potassium Nitrate	Potassium Chloride
0.0	25.65	23.72	0.0	13.92	17.14
4.02	24.48	19.25	6.27	12.70	20.71
11.22	22.27	15.98	11.95		

№ 4554 [1248]

POTASSIUM NITRATE–POTASSIUM CHLORIDE–WATER

$KNO_3—KCl—H_2O$

Content in aqueous soln., g/l		d	t	Content in aqueous soln., g/l		d	t
Potassium Nitrate	Potassium Chloride			Potassium Nitrate	Potassium Chloride		
0.0	293.9	1.173	17.5	167.9	249.6	1.240	20.5
65.8	275.0	1.198	17.5	171.1	221.7	1.226	20.5
88.8	273.4	1.210	17.5	176.7	196.1	1.215	20.5
124.8	265.3	1.225	17.5	189.6	151.5	1.198	20.5
148.3	259.8	1.236	17.5	202.3	122.3	1.189	20.5
152.2	259.6	1.239	17.5	224.4	77.4	1.176	20.5
154.9	259.5	1.238	17.5	243.9	47.2	1.170	20.5
153.3	262.4	1.241	17.5	276.8	0.0	1.162	20.5

№ 4555 POTASSIUM NITRATE–POTASSIUM CHLORIDE–WATER [2597]

KNO_3—KCl—H_2O

Content in aqueous soln., g/1000g Water		t	Content in aqueous soln., g/1000g Water		t
Potassium Nitrate	Potassium Chloride		Potassium Nitrate	Potassium Chloride	
0.0	345.2	20	256.8	82.9	20
56.18	342.15	20	311.1	0.0	20
168.54	334.39	20	501.6	310.8	20
202.0	248.7	20	225.8	341.3	25
221.7	165.8	20	1175	402	80

№ 4556 POTASSIUM NITRATE–POTASSIUM CHLORIDE–WATER. [2658]

KNO_3—KCl—H_2O

Content in aqueous soln., g/l		t	Content in aqueous soln., g/l		t
Potassium Nitrate	Potassium Chloride		Potassium Nitrate	Potassium Chloride	
0.0	288.3	14.5	196.9	226.8	25.2
20.64	284.2	14.5	203.9	207.3	25.2
32.18	282.1	14.5	224.4	155.2	25.2
62.23	276.8	14.5	254.0	100.7	25.2
82.77	273.5	14.5	288.7	49.22	25.2
115.9	270.7	14 5	312.3	19.39	25.2
119.1	268.3	14.5	325.5	0.0	25.2
123.4	267.2	14.5	137.1	236.9	14.5
0.0	311.8	25.2	144.0	196.5	14.5
13.76	306.6	25.2	153.3	164.4	14.5
32.18	303.6	25.2	159.5	132.6	14.5
91.26	293.2	25.2	185.2	65.64	14.5
122.7	287.2	25.2	208.2	31.63	14.5
141.4	284.2	25.2	219.8	13.58	14.5
182.7	276	25.2	225.4	0.0	14.5

№ 4557 POTASSIUM NITRATE–POTASSIUM CHLORIDE–WATER [1098]

KNO_3—KCl—H_2O

Content of Potassium Nitrate in aqueous soln., g/1000g Water	Solubility of Potassium Chloride in aqueous soln. of Potassium Nitrate, wt %	t	Content of Potassium Nitrate in aqueous soln., g/1000g Water	Solubility of Potassium Chloride in aqueous soln. of Potassium Nitrate, wt %	t
0.0	22.10	0	0.0	26.73	25
25.27	21.71	0	25.27	26.26	25
50.55	21.25	0	50.55	25.61	25
101.11	20.70	0	101.11	24.58	25
151.66	0.0	0	151.66	23.57	25

№ 4558 POTASSIUM NITRATE–POTASSIUM CHLORIDE–WATER [2056]

$KNO_3—KCl—H_2O$

Concn. in aqueous soln., g/100 g Water		d	t	Comp. of solid phase
Potassium Nitrate	Potassium Chloride			
0.0	37.58	1.186	30	KCl
8.05	36.72	1.219	30	»
19.36	36.19	1.251	30	»
26.83	35.42	1.281	30	$KCl+KNO_3$
29.19	28.71	1.258	30	KNO_3
32.34	19.35	1.241	30	»
38.10	9.44	1.225	30	»
0.0	40.60	1.194	40	KCl
16.86	39.11	1.252	40	»
35.45	37.08	1.305	40	»
39.71	37.49	1.319	40	$KCl+KNO_3$
41.52	32.22	1.312	40	KNO_3
46.31	22.63	1.297	40	»
52.66	11.58	1.279	40	»
0.0	53.58	1.222	91	KCl
52.75	47.85	1.344	91	»
114.6	43.30	1.486	91	»
162.9	39.90	1.552	91	$KCl+KNO_3$
165.6	33.25	1.544	91	KNO_3
181.1	15.56	1.575	91	»
202.8	0.0	1.552	91	»

№ 4559 [1134]

POTASSIUM NITRATE–POTASSIUM CHLORIDE–WATER

$KNO_3—KCl—H_2O$

$t=30$

Comp. of saturated soln., wt %			Comp. of solid phase
Potassium Nitrate	Potassium Chloride	Water	
0.0	27.20	72.80	KCl
11.63	23.43	64.94	»
13.42	22.78	63.80	»
16.39	22.18	61.43	»
16.65	21.90	61.45	$KCl+KNO_3$
16.80	21.30	61.90	KNO_3
17.25	20.25	62.50	»
22.0	10.48	67.52	»
26.36	5.48	68.16	»
31.40	0.0	68.60	»

№ 4560 POTASSIUM NITRATE–POTASSIUM CHLORIDE–WATER [1800

KNO_3—KCl—H_2O

Concn. in aqueous soln., g/100 g Water		d	t	Comp. of solid phase
Potassium Nitrate	Potassium Chloride			
0.0	27.9	—	0	KCl
8.72	27.7	—	0	KCl+KNO_3
13.30	0.0	—	0	KNO_3
0.0	35.9	1.179	25	KCl
16.30	34.8	1.238	25	»
23.0	34.5	1.261	25	KCl+KNO_3
23.3	32.8	1.256	25	KNO_3
28.0	15.9	1.213	25	»
32.2	8.0	1.197	25	»
38.3	0.0	1.189	25	»
0.0	42.4	1.194	50	KCl
54.1	38.2	1.350	50	KCl+KNO_3
61.3	23.7	1.333	50	KNO_3
71.5	11.9	1.329	50	»
85.7	0.0	1.332	50	»
0.0	49.6	1.203	75	KCl
16.8	47.5	1.252	75	»
40.6	45.3	1.316	75	»
72.7	42.5	1.388	75	»
111.0	39.4	1.457	75	KCl+KNO_3
122.0	25.0	1.454	75	KNO_3
139.0	10.2	1.457	75	»
153.0	0.0	1.462	75	»
0.0	55.5	1.211	100	KCl
65.0	49.3	1.365	100	»
132.0	43.5	1.479	100	»
204.0	42.1	1.568	100	KCl+KNO_3
209.0	31.0	1.565	100	KNO_3
224.0	15.2	1.565	100	»
244.0	0.0	1.569	100	»

№ 4561 POTASSIUM NITRATE–POTASSIUM CHLORIDE–WATER [1858

KNO_3—KCl—H_2O

$t=18.5$

Content in aqueous soln., g/l		d	Content in aqueous soln., g/l		d
Potassium Nitrate	Potassium Chloride		Potassium Nitrate	Potassium Chloride	
0.0	298.50	1.174	163.74	207.94	1.214
17.74	294.84	1.182	184.74	139.16	1.188
35.46	292.32	1.189	218.96	67.76	1.168
53.28	288.86	1.198	235.66	40.38	1.162
88.86	283.16	1.215	245.84	26.80	1.160
133.66	274.62	1.235	254.64	13.36	1.157
148.84	271.24	1.242	265.92	0.0	1.156

№ 4562 POTASSIUM NITRATE–POTASSIUM CHLORIDE–WATER [2254]

$KNO_3—KCl—H_2O$

Comp. of saturated soln., wt %			t	Comp. of solid phase
Potassium Nitrate	Potassium Chloride	Water		
0.0	26.5	73.5	25	KCl
4.30	25.05	70.05	25	»
7.92	23.75	68.33	25	»
11.41	22.95	65.64	25	»
14.72	21.94	63.34	25	$KCl+KNO_3$
17.20	15.64	67.16	25	KNO_3
19.52	10.65	69.53	25	»
22.30	6.74	70.96	25	»
24.02	4.50	71.48	25	»
27.86	0.0	72.14	25	»
6.38	20.44	73.22	0.2	$KCl+KNO_3$
11.00	22.23	66.77	17.0	»
21.98	21.28	56.74	40.0	»
34.12	18.56	47.32	60	$KCl+KNO_3$
46.29	15.47	38.24	80	»
53.88	14.96	31.16	100	»

№ 4563 POTASSIUM NITRATE–POTASSIUM CHLORIDE–WATER [1222]

$KNO_3—KCl—H_2O$

Comp. of saturated soln., wt. %			t	Comp. of solid phase
Potassium Chloride	Potassium Nitrate	Water		
37.6	5.6	56.8	150	KCl
29.8	21.8	58.4	150	»
23.2	37.2	39.6	150	»
18.70	47.4	33.9	150	»
14.15	58.9	26.95	150	»
9.50	71.3	19.2	150	»
8.35	75.3	16.35	150	»
3.69	79.5	16.81	150	KNO_3
1.18	81.9	16.92	150	»
40.9	7.04	52.06	200	KCl
38.0	12.60	49.4	200	»
36.2	16.85	46.95	200	»
33.5	22.5	44.0	200	»
28.0	33.1	38.9	200	»
25.8	36.1	38.1	200	»
21.4	48.3	30.3	200	»
11.95	70.9	17.15	200	»
8.02	81.6	10.38	200	»
7.72	82.0	10.28	200	»
3.19	87.0	9.81	200	KNO_3
44.9	5.9	49.2	250	KCl
41.6	13.9	44.5	250	»

Table № 4572 (continuation)

Comp. of saturated soln., wt %			Comp. of solid phase
Potassium Chlorate	Potassium Nitrate	Water	
14.90	2.90	82.20	$n KClO_3 \cdot m KNO_3$
16.40	2.30	81.30	»
18.90	2.50	78.60	»
21.40	2.10	76.50	»
23.21	2.09	74.70	»
24.90	1.80	73.30	»
27.50	1.70	70.80	»
28.60	0.00	71.40	KNO_3

№ 4573 [1261]

POTASSIUM NITRATE–POTASSIUM PERCHLORATE–WATER

$KNO_3—KClO_4—H_2O$

$t=25$

Concn. in aqueous soln., wt %	
Potassium Perchlorate	Potassium Nitrate
1.58	0.956
1.118	2.86
0.826	6.279

[359]

№ 4574

POTASSIUM NITRATE–POTASSIUM PERCHLORATE–WATER

$KNO_3—KClO_4—H_2O$

$t=25$

Comp. of saturated soln., wt %			Comp. of solid phase
Potassium Perchlorate	Potassium Nitrate	Water	
2.02	0.00	97.98	$KClO_4$
2.25	3.07	94.68	$n KClO_4 \cdot m KNO_3$
1.57	5.73	92.70	»
1.00	10.00	89.00	»
0.87	15.50	83.63	»
0.86	19.50	79.64	»
0.87	22.25	76.88	»
1.07	26.87	72.06	»
0.00	27.64	72.36	KNO_3

№ 4575 [731]

POTASSIUM NITRATE–POTASSIUM CHROMATE–WATER

$KNO_3—K_2CrO_4—H_2O$

Concn. in aqueous soln., wt %		t	Comp. of solid phase
Potassium Chromate	Potassium Nitrate		
35.34	2.31	−12.1	Ice + KNO_3 + K_2CrO_4
18.87	5.73	− 6.2	KNO_3 + ice
18.33	5.57	− 5.8	Ice
15.23	4.62	− 4.6	»
8.71	2.64	− 2.6	»
19.53	5.93	− 4.8	KNO_3
21.48	6.52	− 1.8	»
23.83	7.24	2.0	»
24.97	7.58	4.4	»
9.54	9.20	0	»
19.98	6.94	0	»
32.56	5.50	0	»
34.66	4.72	0	KNO_3 + K_2CrO_4

[2018]

se

$\cdot 4H_2O$

KNO_3

$\cdot KNO_3$

[1102]

t

19.85
23.87
23.87

Table № 4575 (continuation)

Concn. in aqueous soln., wt %		t	Comp. of solid phase
Potassium Chromate	Potassium Nitrate		
35.68	3.01	0	K_2CrO_4
8.25	22.26	25	KNO_3
17.76	17.79	25	»
29.32	12.54	25	»
32.31	10.28	25	$KNO_3+K_2CrO_4$
35.32	6.01	25	K_2CrO_4
6.39	40.45	50	KNO_3
13.92	34.17	50	»
26.46	25.48	50	$KNO_3+K_2CrO_4$
32.77	14.44	50	K_2CrO_4
36.82	7.23	50	»
10.64	56.43	96.5	$KNO_3+K_2CrO_4$
5.49	71.15	116	»

№ 4576 [680]

POTASSIUM NITRATE–POTASSIUM CHROMATE–YTTRIUM NITRATE–YTTRIUM CHROMATE–WATER

$KNO_3—K_2CrO_4—Y(NO_3)_3—Y_2(CrO_4)_3—H_2O$

$t=25$

Comp. of saturated soln., wt %					Comp. of solid phase
Potassium Chromate	Yttrium Chromate	Potassium Nitrate	Yttrium Nitrate	Water	
1.47	0.03	5.73	0.00	92.77	$Y_2(CrO_4)_3 \cdot K_2CrO_4 \cdot 6H_2O$
0.00	0.74	4.55	3.09	91.62	»
0.00	0.48	6.24	1.11	92.17	»
2.62	0.05	4.78	0.00	92.55	»
0.00	0.58	5.00	2.37	92.05	»
5.68	0.00	3.00	0.00	91.32	$Y_2(CrO_4)_3 \cdot 4K_2CrO_4 \cdot nH_2O$
8.65	0.12	1.50	0.00	89.73	»
1.00	0.07	4.60	0.00	94.33	$Y_2(CrO_4)_3 \cdot 3K_2CrO_4 \cdot nH_2O$
7.11	0.12	2.52	0 00	90.25	»

№ 4577 [681]

POTASSIUM NITRATE–POTASSIUM CHROMATE–PRASEODYMIUM NITRATE–PRASEODYMIUM CHROMATE–WATER

$KNO_3—K_2CrO_4—Pr(NO_3)_3—Pr_2(CrO_4)_3—H_2O$

$t=25$

Comp. of saturated soln., wt %					Comp. of solid phase
Potassium Chromate	Praseodymium Chromate	Potassium Nitrate	Praseodymium Nitrate	Water	
0.000	0.000	4.242	4.513	91.25	$Pr_2(CrO_4)_3 \cdot K_2CrO_4 \cdot nH_2O$
0.000	0.000	5.757	1.831	92.41	»
0.000	0.000	0.788	6.867	92.35	»
0.000	0.000	5.999	3.532	90.47	»

Table № 4577 (continuation)

Comp. of saturated soln., wt %					Comp. of solid phase
Potassium Chromate	Praseodymium Chromate	Potassium Nitrate	Praseodymium Nitrate	Water	
5.530	0.063	2.969	0.000	91.44	$Pr_2(CrO_4)_3 \cdot 3.5K_2CrO_4 \cdot nH_2O$
8.555	0.000	1.818	0.000	89.63	»
4.190	0.000	3.454	0.000	92.36	»
7.217	0.063	2.545	0.000	90.18	»
2.794	0.000	3.878	0.000	93.33	$Pr_2(CrO_4)_3 \cdot 3.5K_2CrO_4 \cdot nH_2O +$ $+ Pr(CrO) \cdot K_2CrO_4 \cdot nH_2O$
4.190	0.000	3.454	0.000	92.36	»
0.000	0.063	2.727	7.979	89.23	$Pr_2(CrO_4)_3 \cdot 8H_2O +$ admixture of $Pr_2(CrO_4)_3 \cdot K_2CrO_4 \cdot nH_2O$
0.000	0.063	2.060	9.875	88.01	»

№ 4578 POTASSIUM NITRATE–POTASSIUM BROMIDE–WATER [2658]

$KNO_3—KBr—H_2O$

Concn. in aqueous soln., g/l		t	Concn. in aqueous soln., g/l		t
Potassium Nitrate	Potassium Bromide		Potassium Nitrate	Potassium Bromide	
225.4	0.0	14.5	272	110.8	25.2
205.0	42.4	14.5	252.2	163.1	25.2
185.7	93.4	14.5	224.3	143.8	25.2
175	130	14.5	198.1	341.8	25.2
160.6	187.8	14.5	182.8	422.8	25.2
142.2	302.7	14.5	152.2	521.1	25.2
132.2	421.1	14.5	118.4	528.8	25.2
124.9	469.1	14.5	110.3	533.0	25.2
71.4	487.4	14.5	72.9	540.8	25.2
36.6	494.9	14.5	53.3	549.1	25.2
0.0	515.9	14.5	13.3	561.0	25.2
325.5	0.0	25.2	0.0	566.2	25.2
306.2	45.3	25.2			

№ 4579 [139]

POTASSIUM NITRATE–POTASSIUM BROMIDE–WATER

$KNO_3—KBr—H_2O$

$t=25$

Comp. of saturated soln., wt %			Comp. of solid phase
Potassium Nitrate	Potassium Bromide	Water	
27.16	0.00	72.84	KNO_3
23.36	7.82	68.82	»
18.47	16.47	65.06	»
16.08	22.56	61.36	»
14.12	28.08	57.80	»
12.05	35.69	52.26	KNO_3+KBr
10.31	36.46	53.23	KBr
7.51	37.87	54.62	»
3.32	39.63	25.05	»
0.00	40.87	59.13	»

№ 4580 [142]

POTASSIUM NITRATE–POTASSIUM BROMIDE–WATER

KNO_3—KBr—H_2O

$t=25$

Comp. of saturated soln., wt %			Comp. of solid phase
Potassium Nitrate	Potassium Bromide	Water	
27.16	0.00	72.84	KNO_3
23.70	6.06	70.24	»
20.70	12.50	66.80	»
18.47	16.47	65.06	»
17.30	19.90	62.80	»
16.08	22.56	61.36	»
15.25	23.77	60.98	»
14.12	28.08	57.80	»
13.23	30.34	56.43	»
12.05	35.69	52.26	KNO_3+KBr
10.31	36.46	53.23	KBr
7.51	37.87	54.62	»
3.32	39.68	57.05	»
0.00	40.87	59.13	»

№ 4581 [751]

POTASSIUM NITRATE–POTASSIUM BROMIDE–WATER

KNO_3—KBr—H_2O

$t=25$

Comp. of saturated soln., wt %			*d*	Comp. of solid phase
Potassium Bromide	Potassium Nitrate	Water		
0.00	21.97	72.03	1.1917	KNO_3
6.91	23.55	69.54	1.2232	»
18.81	17.63	63.56	1.2954	»
28.39	14.04	57.57	1.3721	»
35.61	11.97	52.42	1.4430	KNO_3+KBr
36.57	9.93	53.50	1.4324	KBr
37.61	7.40	54.99	1.4195	»
39.39	3.24	57.37	1.3976	»
40.76	0.00	59.24	1.3805	»

№ 4582 POTASSIUM NITRATE–POTASSIUM BROMATE–WATER [2390]

$KNO_3—KBrO_3—H_2O$

$t=25$

Comp. of saturated soln., wt %			d	Comp. of solid phase
Potassium Bromate	Potassium Nitrate	Water		
5.61	5.05	89.34	1.074	$KBrO_3$
4.64	11.10	84.26	1.110	»
4.23	16.98	78.79	1.148	»
4.00	23.17	72.83	1.193	»
3.90	27.01	69.09	1.225	$KBrO_3+KNO_3$
2.64	27.27	70.09	1.211	KNO_3
0.00	27.71	72.29	1.193	»

№ 4583 [2410]

POTASSIUM NITRATE–POTASSIUM MOLYBDATE–WATER

$KNO_3—K_2MoO_4—H_2O$

$t=25$

Concn. in aqueous soln., wt %		d	Comp. of solid phase
Potassium Molybdate	Potassium Nitrate		
64.57	0.0	1.800	K_2MoO_4
63.08	3.14	1.821	»
62.67	3.80	1.827	»
62.49	4.30	1.831	»
61.98	5.17	—	»
62.03	5.30	1.843	$K_2MoO_4+KNO_3$
62.15	5.22	1.843	»
62.01	5.25	—	»
62.06	5.26	1.843	»
61.28	5.19	1.827	KNO_3
57.91	5.08	1.753	»
46.08	6.23	1.565	»
35.43	9.35	1.433	»
25.32	13.25	1.353	»
14.39	18.65	1.270	»
0.0	27.71	1.193	»

№ 4584 [143]

POTASSIUM NITRATE–POTASSIUM IODIDE–WATER

$KNO_3—KI—H_2O$

$t=25$

Comp. of saturated soln., wt %			Comp. of solid phase
Potassium Iodide	Potassium Nitrate	Water	
59.88	0.00	40.12	KI
58.60	2.56	38.84	»
57.10	4.97	37.93	»
56.03	6.40	37.57	»
55.06	7.80	37.14	$KI+KNO_3$
48.08	9.60	42.32	KNO_3
43.93	9.90	46.17	»
39.20	10.76	50.04	»
36.79	12.08	51.13	»
31.53	13.80	54.67	»
28.36	14.80	56.84	»
24.42	16.12	59.46	»
21.53	16.76	61.71	»
17.70	18.30	64.00	»
14.56	19.82	65.62	»
11.62	21.40	66.98	»
8.52	22.80	68.68	»
3.50	25.32	71.18	»
0.00	27.16	72.84	»

№ 4585 [751]

POTASSIUM NITRATE–POTASSIUM IODIDE–WATER

$KNO_3—KI—H_2O$

$t=25$

Comp. of saturated soln., wt %			d	Comp. of solid phase
Potassium Iodide	Potassium Nitrate	Water		
59.90	0.00	40.10	1.7215	KI
58.73	2.07	39.20	1.7294	»
57.57	4.03	38.40	1.7370	»
56.49	5.78	37.73	1.7431	»
55.20	7.95	36.85	1.7519	$KI+KNO_3$
47.44	9.36	43.20	1.6231	KNO_3
36.04	12.14	51.82	1.4758	»
18.82	18.35	62.83	1.3111	»
0.00	27.97	72.03	1.1917	»

№ 4586 [213]

POTASSIUM NITRATE–POTASSIUM IODIDE–LEAD (II) IODIDE – WATER

$KNO_3—KI—PbI_2—H_2O$

t=18—20

Concn. in aqueous soln., wt %		Solubility of Lead Iodide in aqueous soln. of Potassium Iodide and Potassium Nitrate, g/l	Concn. in aqueous soln., wt %		Solubility of Lead Iodide in aqueous soln. of Potassium Iodide and Potassium Nitrate, g/l
Potassium Iodide	Potassium Nitrate		Potassium Iodide	Potassium Nitrate	
0.00	0.00	0.665	0.00	50.50	1.06
1.33	0.00	0.160	1.33	50.50	0.777
2.66	0.00	0.071	2.66	50.50	0.660
5.31	0.00	0.095	5.31	50.50	0.438
10.62	0.00	0.132	10.62	50.50	0.342
21.25	0.00	0.142	21.25	50.50	0.269
42.50	0.00	0.191	42.50	50.50	0.329
56.44	0.00	0.218	56.44	50.50	0.00
0.00	10.10	0.738	0.00	101.0	1.21
1.33	10.10	0.256	1.33	101.0	0.905
2.66	10.10	0.151	2.66	101.0	0.780
5.31	10.10	0.098	5.31	101.0	0.592
10.62	10.10	0.149	10.62	101.0	0.398
21.25	10.10	0.222	21.25	101.0	0.342
42.50	10.10	0.282	42.50	101.0	0.288
56.44	10.10	0.00	56.44	101.0	0.299
0.00	20.20	0.885	0.00	151.5	1.493
1.33	20.20	0.398	1.33	151.5	1.36
2.66	20.20	0.281	2.66	151.5	1.22
5.31	20.20	0.213	5.31	151.5	0.910
10.62	20.20	0.176	10.62	151.5	0.612
21.25	20.20	0.232	21.25	151.5	0.424
42.50	20.20	0.301	42.50	151.5	0.414
56.44	20.20	0.00	56.44	151.5	0.415

№ 4587 [1840]

POTASSIUM NITRATE–POTASSIUM IODATE–WATER

$KNO_3—KIO_3—H_2O$

Comp. of saturated soln., wt %			t	d	Comp. of solid phase
Potassium Iodate	Potassium Nitrate	Water			
5.25	0.00	94.75	5	1.043	KIO_3
3.29	5.36	91.35	5	1.060	»
2.93	10,53	86,54	5	1.090	»
2.89	13.53	83.58	4	1.110	»
2.87	14.14	82.99	5	1.120	KIO_3+KNO_3
1.08	14.26	84.66	5	1.100	KNO_3
0.0	14.43	85.57	5	1.097	»
8.46	0.00	91.54	25	1.072	KIO_3
5.92	5.48	88.60	25	1.084	»
5.25	8.51	86.24	25	1.110	»
4.57	16.77	78.66	25	1.156	»
4.48	18.19	77.33	25	1.160	»

Table № 4587 (continuation)

Comp. of saturated soln., wt %			t	d	Comp. of solid phase
Potassium Iodate	Potassium Nitrate	Water			
4.24	26.84	68.02	25	1.232	KIO_3+KNO_3
2.65	27.25	70.10	25	1.215	KNO_3
2.06	27.30	70.64	25	1.210	»
0.00	27.79	72.21	25	1.192	»
13.35	0.0	86.65	50.4	1.110	KIO_3
7.32	17.42	75.26	50.4	1.167	»
5.79	31.43	62.78	50.4	1.259	»
5.17	39.17	55.66	50.4	1.33	»
4.91	42.42	52.67	50.4	1.35	»
4.75	44.23	51.02	50.4	1.37	KIO_3+KNO_3
2.44	45.44	52.12	50.4	1.34	KNO_3
0.00	46.57	53.43	50.4	1.326	»

№ 4588 [1134]

POTASSIUM NITRATE–CALCIUM NITRATE–WATER

$KNO_3—Ca(NO_3)_2—H_2O$

$t=30$

Comp. of saturated soln., wt %			Comp. of solid phase
Potassium Nitrate	Calcium Nitrate	Water	
0.0	59.7	40.30	$Ca(NO_3)_2 \cdot 4H_2O$
10.3	57.0	32.7	»
16.86	58.67	24.47	»
17.2	58.9	23.9	»
17.7	61.5	20.8	»
14.2	64.4	21.4	»
13.0	66.0	21.0	$Ca(NO_3)_2 \cdot 4H_2O+Ca(NO_3)_2 \cdot H_2O$
22.4	58.3	19.3	$Ca(NO_3)_2 \cdot H_2O$
31.8	50.6	17.6	$Ca(NO_3)_2 \cdot H_2O+KNO_3$
30.4	52.0	17.6	KNO_3
25.1	47.87	27.03	»
22.4	37.22	40.38	»
22.15	31.9	45.95	»
22.8	24.8	52.4	»
24.0	18.15	47.85	»
27.0	8.7	64.3	»

№ 4589 POTASSIUM NITRATE–CALCIUM NITRATE–WATER [1685]

KNO_3—$Ca(NO_3)_2$—H_2O

Comp. of saturated soln., wt %			t	Comp. of solid phase
Potassium Nitrate	Calcium Nitrate	Water		
0.00	48.98	51.02	0	$Ca(NO_3)_2 \cdot 4H_2O$
3.00	48.13	48.87	0	»
6.83	47.94	45.23	0	»
9.48	47.21	43.31	0	»
13.67	47.00	39.33	0	»
15.22	46.22	38.54	0	$Ca(NO_3)_2 \cdot 4H_2O + KNO_3$
14.27	44.71	41.02	0	KNO_3
13.75	42.60	43.65	0	»
12.56	39.68	47.76	0	»
11.16	36.46	52.38	0	»
10.81	32.02	57.17	0	»
10.00	27.06	62.94	0	»
9.44	20.14	70.42	0	»
12.12	0.00	87.88	0	»
0.00	56.04	43.96	20	$Ca(NO_3)_2 \cdot 4H_2O$
3.73	54.59	41.68	20	»
8.70	52.45	38.85	20	»
12.54	50.79	36.67	20	»
17.10	48.50	34.40	20	»
19.77	47.01	33.22	20	»
21.70	46.25	32.05	20	$Ca(NO_3)_2 \cdot 4H_2O + KNO_3$
20.62	42.35	37.03	20	KNO_3
18.84	43.83	37.33	20	»
18.44	36.86	44.70	20	»
17.67	29.89	52.44	20	»
17.21	25.66	57.13	20	»
19.90	10.98	69.12	20	»
24.12	0.00	75.88	20	»

№ 4590 [1774]

POTASSIUM NITRATE–CALCIUM NITRATE–WATER

KNO_3—$Ca(NO_3)_2$—H_2O

$t=25$

Comp. of saturated soln., wt %			Comp. of solid phase
Potassium Nitrate	Calcium Nitrate	Water	
0.00	57.98	42.02	$Ca(NO_3)_2 \cdot 4H_2O$
2.64	56.86	40.50	»
7.37	54.82	37.81	»
10.72	53.26	36.02	»
19.06	49.16	31.78	$Ca(NO_3)_2 \cdot 4H_2O + KNO_3$
19.22	40.34	40.44	KNO_3
19.45	31.95	48.60	»
20.32	19.14	60.54	»
20.70	17.34	61.96	»
21.53	11.48	66.99	»
22.19	7.37	70.44	»
23.83	4.96	71.21	»
24.85	2.12	73.03	»
27.30	0.00	72.70	»

№ 4591 POTASSIUM NITRATE–CALCIUM SULFATE–WATER [1321]

$KNO_3—CaSO_4—H_2O$

$t=25$

Concn. in aqueous soln., g/l		d	Concn. in aqueous soln., g/l		d
Potassium Nitrate	Calcium Sulfate		Potassium Nitrate	Calcium Sulfate	
0	2.08	0.9981	100	6.86	1.0625
12.5	3.28	1.0081	150	7.91	1.0924
25.0	4.08	1.0154	200	8.69	1.1224
50	5.26	1.0321			

№ 4592 POTASSIUM NITRATE–CALCIUM SULFATE–WATER [1781]

$KNO_3—CaSO_4—H_2O$

$t=25$

Concn. in aqueous soln., wt %		d_4	Concn. in aqueous soln., wt %		d_4
Potassium Nitrate	Calcium Sulfate		Potassium Nitrate	Calcium Sulfate	
0.00	0.2085	0.9991	0.5311	0.2727	1.0029
0.2782	0.2454	1.0013	1.0330	0.2862	1.0088

№ 4593 [1134]

POTASSIUM NITRATE–CALCIUM CHLORIDE–WATER

$KNO_3—CaCl_2—H_2O$

$t=30$

Comp. of saturated soln., wt %			Comp. of solid phase
Potassium Nitrate	Calcium Chloride	Water	
3.67	50.4	45.93	$CaCl_2 \cdot 4H_2O$
5.1	50.7	44.2	»
5.2	50.7	44.1	$CaCl_2 \cdot 4H_2O+KCl$
5.2	50.3	44.5	KCl
5.42	42.5	52.08	»
5.82	38.3	55.88	»
6.4	34.8	58.8	»
7.64	31.1	61.26	»
10.3	29.9	59.8	»
11.9	28.6	59.5	»
12.4	27.8	59.8	»
13.5	36.9	49.6	»
14.2	26.2	59.6	»
17.6	24.2	58.2	»
19.8	22.7	57.5	»
21.5	21.8	56.7	»
24.6	20.3	55.1	»
25.3	19.8	54.9	»
26.0	19.6	54.4	$KCl+KNO_3$

Table № 4593 (continuation)

Comp. of saturated soln., wt %			Comp. of solid phase
Potassium Nitrate	Calcium Chloride	Water	
26.0	19.4	54.6	KNO_3
25.9	17.8	56.3	»
26.3	14.4	59.3	»
27.0	11.3	61.7	»
27.6	9.5	62.9	»

№ 4594 **POTASSIUM NITRATE–CALCIUM IODATE–WATER** [1388

$KNO_3—Ca(IO_3)_2—H_2O$

$t=25$

Concn. of Potassium Nitrate in aqueous soln., g/l	Solubility of Calcium Iodate in aqueous soln. of Potassium Nitrate, wt %	Concn. of Potassium Nitrate in aqueous soln., g/l	Solubility of Calcium Iodate in aqueous soln. of Potassium Nitrate, wt %	Concn. of Potassium Nitrate in aqueous soln., g/l	Solubility of Calcium Iodate in aqueous soln. of Potassium Nitrate, wt %
0.0	0.303	2.022	0.3393	20.22	0.4677
0.505	0.3125	5.055	0.3695	50.55	0.5861
1.011	0.3207	10.11	0.4102		

№ 4595 **POTASSIUM NITRATE–IRON (III) NITRATE–WATER** [2122

$KNO_3—Fe(NO_3)_3—H_2O$

Comp. of saturated soln., wt %			t	Comp. of solid phase
Iron Nitrate	Potassium Nitrate	Water		
40.15	0.00	59.85	0	$Fe(NO_3)_3 \cdot 9H_2O$
38.68	0.94	60.38	0	»
38.55	3.92	57.53	0	$Fe(NO_3)_3 \cdot 9H_2O + KNO_3$
23.87	2.36	73.77	0	KNO_3
21.82	2.26	75.92	0	»
0.00	11.70	89.30	0	»
46.57	0.00	53.43	25	$Fe(NO_3)_3 \cdot 9H_2O$
43.01	6.03	50.96	25	»
41.90	8.75	49.35	25	»
38.11	12.92	48.97	25	»
39.95	11.02	49.03	25	$Fe(NO_3)_3 \cdot 9H_2O + KNO_3$
38.13	15.95	45.92	25	KNO_3
39.85	10.25	49.90	25	»
40.07	6.13	53.80	25	»
37.43	3.22	59.35	25	»
21.27	5.19	73.54	25	»
11.29	15.91	72.80	25	»
0.00	27.31	72.69	25	»
51.18	0.00	48.82	40	$Fe(NO_3)_3 \cdot 9H_2O$
50.45	16.88	32.67	40	»
44.01	25.31	30.68	40	KNO_3
45.83	21.03	33.14	40	»
45.65	17.99	36.36	40	»
45.15	17.04	37.81	40	»
45.10	16.81	38.09	40	»
40.17	7.75	52.08	40	»
0.00	39.00	61.00	40	»

№ 4596 [2596]

POTASSIUM NITRATE–ZINC HYDROXIDE–WATER

$KNO_3—Zn(OH)_2—H_2O$

$t=16—20$

Comp. of solvent, wt %		Solubility of Zinc Hydroxide, g/l
Potassium Nitrate	Water	
1	99	0.0175

№ 4597 [2057]

POTASSIUM NITRATE–STRONTIUM NITRATE–WATER

$KNO_3—Sr(NO_3)_2—H_2O$

$t=25$

Comp. of saturated soln., wt %		
Potassium Nitrate	Strontium Nitrate	Water
21.02	40.89	38.09

№ 4598 POTASSIUM NITRATE–STRONTIUM NITRATE–WATER [1605]

$KNO_3—Sr(NO_3)_2—H_2O$

Comp. of saturated soln., wt %			t	Comp. of solid phase
Potassium Nitrate	Strontium Nitrate	Water		
22.90	5.49	71.61	20	KNO_3
21.70	9.17	69.13	20	»
21.01	17.10	61.89	20	»
19.60	31.24	49.16	20	»
19.49	34.91	45.60	20	»
19.69	39.56	40.75	20	$KNO_3+Sr(NO_3)_2 \cdot 4H_2O$
17.56	40.37	42.07	20	$Sr(NO_3)_2 \cdot 4H_2O$
12.65	41.12	46.23	20	»
10.00	40.70	49.30	20	»
30.26	23.70	46.04	40	KNO_3
26.90	38.52	34.58	40	$KNO_3+Sr(NO_3)_2 \cdot 4H_2O$
22.50	40.22	37.28	40	$Sr(NO_3)_2 \cdot 4H_2O$
11.19	44.19	44.62	40	»
0.0	47.7	52.3	40	»

№ 4599 [1780]

POTASSIUM NITRATE–STRONTIUM CHLORIDE–WATER

$KNO_3—SrCl_2—H_2O$

$t=25$

Comp. of solvent, wt %		Solubility of Strontium Chloride, wt %	d_4
Potassium Nitrate	Water		
0.980	99.020	35.86	1.4107
4.586	95.414	35.90	1.4349

№ 4600 POTASSIUM NITRATE–SILVER NITRATE–WATER [1577]

$KNO_3—AgNO_3—H_2O$

Comp. of saturated soln., wt %			*t*	Comp. of saturated soln., wt %			*t*
Potassium Nitrate	Silver Nitrate	Water		Potassium Nitrate	Silver Nitrate	Water	
13.5	43.0	43.5	0	32.0	54.0	14.0	50
19.0	44.7	36.3	10	33.5	54.8	11.7	60
23.0	47.0	30.0	20	36.2	55.1	8.7	80
25.0	48.0	27.0	25	38.3	55.3	6.4	100
26.8	49.4	23.8	30	40.0	55.6	4.4	120
29.6	51.5	18.9	40	41.5	55.8	2.7	140

№ 4601 POTASSIUM NITRATE–SILVER NITRATE–WATER [1810]

$KNO_3—AgNO_3—H_2O$

$t=25$

Concn. in aqueous soln., g/l

Silver Nitrate	Potassium Nitrate	Silver Nitrate	Potassium Nitrate	Silver Nitrate	Potassium Nitrate
45.9	321.8	259.6	364.0	507.9	387.2
110.7	322.6	305.6	456.4	745.9	398.6
176.8	333.7				

№ 4602 [2500]

POTASSIUM NITRATE–SILVER NITRATE–WATER

$KNO_3—AgNO_3—H_2O$

$t=30$

Comp. of saturated soln., wt %			Comp. of solid phase
Potassium Nitrate	Silver Nitrate	Water	
31.30	0.0	68.70	KNO_3
30.45	11.51	58.04	»
29.22	23.59	47.19	»
26.58	39.09	34.33	»
25.02	46.38	28.60	$KNO_3+AgNO_3 \cdot KNO_3$
17.38	57.85	24.77	$AgNO_3 \cdot KNO_3$
13.44	65.08	21.48	»
11.52	69.01	19.47	$AgNO_3 \cdot KNO_3+AgNO_3$
5.53	71.65	22.82	$AgNO_3$
0.0	73.0	27.0	»

$KNO_3—AgNO_3—H_2O$

Comp. of saturated soln., wt %			t	Comp. of solid phase
Silver Nitrate	Potassium Nitrate	Water		
10.00	0.00	90.00	—1.9	Ice
9.61	3.84	86.55	—2.7	»
9.26	7.41	83.33	—3.3	»
9.04	9.64	81.32	—4.0	»
9.02	9.80	81.18	—4.1	Ice+KNO_3
8.93	10.71	80.36	—1.2	KNO_3
8.52	14.77	76.71	6.5	»
8.15	18.48	73.37	13.1	»
7.65	23.46	68.89	20.2	»
7.35	26.47	66.18	25.0	»
20.00	0.00	80.00	—3.8	Ice
19.23	3.84	76.93	—4.3	»
18.52	7.41	74.07	—4.8	»
18.29	8.54	73.17	—5.1	»
18.21	9.00	72.79	—5.2	Ice+KNO_3
17.65	11.76	70.59	1.2	KNO_3
16.85	15.73	67.42	8.5	»
15.96	20.21	63.83	15.8	»
14.71	26.47	58.82	25.0	»
30.00	0.00	70.00	—4.4	Ice
29.60	1.32	69.08	—4.6	»
28.85	3.85	67.30	—5.0	»
28.12	6.25	65.63	—5.3	»
27.67	7.80	64.53	—5.6	Ice+KNO_3
26.79	10.71	62.50	—0.9	KNO_3
25.57	14.77	59.66	5.2	»
24.86	17.13	58.01	10.0	»
23.68	21.05	55.27	16.6	»
22.73	24.27	53.00	21.8	»
22.06	26.47	51.47	25.0	»
40.00	0.00	60.00	—6.3	Ice
39.47	1.32	59.21	—6.5	»
38.96	2.60	58.44	—6.7	»
38.46	3.85	57.69	—6.9	»
37.97	5.06	52.97	—7.5	Ice+KNO_3
37.04	7.41	55.55	—2.2	KNO_3
35.30	11.76	52.94	4.6	»
33.52	16.20	50.28	11.2	»
31.75	20.63	47.62	17.9	»
30.60	23.46	45.94	22.4	»
30.00	25.00	45.00	25.0	»
55.00	0.00	45.00	—0.3	$AgNO_3$
54.28	1.32	44.40	—1.4	»
53.57	2.60	43.83	—3.3	»
52.88	3.84	43.28	—4.8	»
52.48	4.60	42.92	—5.5	$AgNO_3$+KNO_3
51.56	6.25	42.19	—0.5	KNO_3
50.30	8.54	41.16	4.8	»
47.96	12.79	39.25	12.2	»
44.84	18.48	36.68	19.1	»
43.42	21.05	35.53	22.0	»

Comp. of saturated soln., wt %			t	Comp. of solid phase
Silver Nitrate	Potassium Nitrate	Water		
42.09	23.46	34.45	25.0	KNO_3
65.00	0.00	35.00	12.2	$AgNO_3$
63.31	2.60	34.09	9.4	»
61.71	5.06	33.23	6.2	»
60.72	6.60	32.68	4.2	$AgNO_3+KNO_3$
59.45	8.54	32.01	9.8	KNO_3
57.35	11.76	30.89	16.4	»
55.40	14.77	29.83	21.5	»
53.87	17.12	29.01	25.0	»

№ 4604 POTASSIUM NITRATE–SILVER ORTHOPHOSPHATE–WATER [833]

$KNO_3—Ag_3PO_4—H_2O$

$t=25$

Content of Potassium Nitrate in aqueous soln., mol/l	Solubility of Silver Orthophosphate in aqueous soln. of Potassium Nitrate, mol/l	Content of Potassium Nitrate in aqueous soln., mol/l	Solubility of Silver Orthophosphate in aqueous soln. of Potassium Nitrate, mol/l	Content of Potassium Nitrate in aqueous soln., mol/l	Solubility of Silver Orthophosphate in aqueous soln. of Potassium Nitrate, mol/l
0.0	$5.11 \cdot 10^{-5}$	$1.0 \cdot 10^{-2}$	$6.56 \cdot 10^{-5}$	1.0	$18.47 \cdot 10^{-5}$
$1.0 \cdot 10^{-5}$	$5.11 \cdot 10^{-5}$	$1.0 \cdot 10^{-1}$	$10.03 \cdot 10^{-5}$	2.0	$21.90 \cdot 10^{-5}$
$1.0 \cdot 10^{-4}$	$5.15 \cdot 10^{-5}$	$0.5 \cdot 10^{-1}$	$15.42 \cdot 10^{-5}$	3.3045	$24.46 \cdot 10^{-5}$
$1.0 \cdot 10^{-3}$	$5.44 \cdot 10^{-5}$				

№ 4605 POTASSIUM NITRATE–SILVER SULFATE–WATER [1385]

$KNO_3—Ag_2SO_4—H_2O$

$t=25$

Concn. of Potassium Nitrate in aqueous soln., g/l	Solubility of Silver Sulfate in aqueous soln. of Potassium Nitrate, wt %	Concn. of Potassium Nitrate in aqueous soln., g/l	Solubility of Silver Sulfate in aqueous soln. of Potassium Nitrate, wt %
3.03	0.9186	60.66	1.6475
6.07	0.9781	151.7	2.3837
15.17	1.1348		

№ 4606 POTASSIUM NITRATE–SILVER CHLORIDE–WATER [2344]

$KNO_3—AgCl—H_2O$

$t=25$

Concn. in aqueous soln., g/l					
Potassium Nitrate	Silver Chloride	Potassium Nitrate	Silver Chloride	Potassium Nitrate	Silver Chloride
0.00	0.00183	0.5022	0.00198	2.026	0.00213
0.00129	0.00183	1.003	0.00204	2.765	0.00217
0.02635	0.00186	1.393	0.00208	3.410	0.00220
0.0514	0.00187	1.660	0.00210	4.055	0.00222
0.1015	0.00189				

№ 4607 POTASSIUM NITRATE–SILVER CHLORIDE–WATER [1479]

$KNO_3—AgCl—H_2O$

Concn. in aqueous soln., g/l		t	Concn. in aqueous soln., g/l		t°
Potassium Nitrate	Silver Chloride		Potassium Nitrate	Silver Chloride	
10.1	0.00113	0	50.5	0.00736	30
10.1	0.00120	8	75.75	0.00847	30
10.1	0.00203	18	101.0	0.00991	30
10.1	0.00387	25	10.1	0.00615	40
10.1	0.00493	30	10.1	0.00865	50

№ 4608 [2582]

POTASSIUM NITRATE–SILVER CHROMATE–WATER

$KNO_3—Ag_2CrO_4—H_2O$

$t=20$

Concn. of Potassium Nitrate in aqueous soln., g/l	Solubility of Silver Chromate in aqueous soln. of Potassium Nitrate, wt %
0	0.00301
0.51	0.00342
4.04	0.00412
10.11	0.00479
16.18	0.0052

№ 4609 [2709]

POTASSIUM NITRATE–SILVER BROMIDE–WATER

$KNO_3— AgBr —H_2O$

$t=25$

Concn. in aqueous soln., g/l	
Potassium Nitrate	Silver Bromide
0.0000	1.9499
7.9659	2.2376
16.7334	2.4215
66.9670	3.0746

№ 4610 POTASSIUM NITRATE–SILVER BROMATE–WATER [478]

$KNO_3—AgBrO_3—H_2O$

$t=25$

Content in aqueous soln., mol/l					
Potassium Nitrate	Silver Bromate	Potassium Nitrate	Silver Bromate	Potassium Nitrate	Silver Bromate
0.000	$8.25 \cdot 10^{-3}$	0.010	$8.61 \cdot 10^{-3}$	0.300	$1.240 \cdot 10^{-2}$
0.001	$8.26 \cdot 10^{-3}$	0.030	$9.30 \cdot 10^{-3}$		
0.003	$8.48 \cdot 10^{-3}$	0.100	$1.024 \cdot 10^{-2}$	1.000	$1.630 \cdot 10^{-2}$

№ 4611 POTASSIUM NITRATE–SILVER IODATE–WATER [1999]

KNO_3— $AgIO_3$—H_2O

$t=25$

Content of Silver Nitrate in aqueous soln., mol/l	Concn., mol/l		Content of Silver Nitrate in aqueous soln., mol/l	Concn., mol/l	
	Ag^+	IO_3^-		Ag^+	IO_3^-
0.001301	$1.813 \cdot 10^{-4}$	$1.832 \cdot 10^{-4}$	0.07050	$2.302 \cdot 10^{-4}$	$2.300 \cdot 10^{-4}$
0.003252	$1.863 \cdot 10^{-4}$	$1.877 \cdot 10^{-4}$	0.1998	$2.668 \cdot 10^{-4}$	$2.663 \cdot 10^{-4}$
0.006503	$1.908 \cdot 10^{-4}$	$1.919 \cdot 10^{-4}$	0.4995	$3.345 \cdot 10^{-4}$	$3.284 \cdot 10^{-4}$
0.01410	$1.991 \cdot 10^{-4}$	$1.996 \cdot 10^{-4}$	0.9989	$4.087 \cdot 10^{-4}$	$4.086 \cdot 10^{-4}$

№ 4612 POTASSIUM NITRATE–SILVER IODATE–WATER [2076]

KNO_3— $AgIO_3$—H_2O

Concn. of Potassium Nitrate in aqueous soln., g/l	Solubility of Silver Iodate in aqueous soln. of Potassium Nitrate, g/l	t	Concn. of Potassium Nitrate in aqueous soln., g/l	Solubility of Silver Iodate in aqueous soln. of Potassium Nitrate, g/l	t
0.000	0.0274	10	51.06	0.9417	25
0.1341	0.02833	10	88.34	0.11041	25
0.3299	0.02893	10	0.00	0.06086	30
0.6535	0.02981	10	0.1341	0.06256	30
1.419	0.03122	10	0.3299	0.06400	30
7.109	0.03617	10	0.6535	0.06572	30
0.000	0.04137	20	1.419	0.06841	30
0.1341	0.04265	20	7.109	0.08043	30
0.3299	0.04364	20	12.26	0.08600	30
0.6535	0.04474	20	25.56	0.09694	30
1.419	0.04683	20	51.06	0.11411	35
7.109	0.05495	20	0.00	0.07316	35
0.000	0.05048	25	0.1341	0.07531	35
0.1315	0.05167	25	0.3299	0.07675	35
0.3288	0.05283	25	0.6535	0.07879	35
0.6575	0.05410	25	1.419	0.08258	35
1.426	0.05662	25	7.032	0.09561	35
7.128	0.06516	25	20.25	0.11518	35
12.26	0.06940	25	50.42	0.13583	35
25.56	0.07899	25			

№ 4613 POTASSIUM NITRATE–CADMIUM IODATE–WATER [2456]

KNO_3—$Cd(IO_3)_2$—H_2O

$t=25$

Comp. of saturated soln., wt %			d_4	Comp. of saturated soln., wt %			d_4
Potassium Nitrate	Cadmium Iodate	Water		Potassium Nitrate	Cadmium Iodate	Water	
0.00	0.96	99.04	0.9985	9.34	1.40	89.26	1.0052
0.26	0.99	98.75	0.9987	20.35	1.68	77.97	1.0145
0.51	1.02	98.47	0.9989	33.11	1.97	64.92	1.0287
1.01	1.06	97.93	0.9991	43.90	2.26	53.84	1.0454
4.880	1.24	93.88	1.0019	50.26	2.43	47.31	1.0578

№ 4614 [2438]

POTASSIUM NITRATE–ANTIMONY TRIFLUORIDE–WATER

$KNO_3—SbF_3—H_2O$

$t=0$

Concn. of Potassium Nitrate in aqueous soln., g/l	Solubility of Antimony Trifluoride in aqueous soln. of Potassium Nitrate, wt %	Concn. of Potassium Nitrate in aqueous soln., g/l	Solubility of Antimony Trifluoride in aqueous soln. of Potassium Nitrate, wt %
101.1	82.08	25.275	80.70
50.55	81.88	12.637	80.05

№ 4615 [1580]

POTASSIUM NITRATE–BARIUM NITRATE–WATER

$KNO_3—Ba(NO_3)_2—H_2O$

Comp. of saturated soln., wt %			t	d
Potassium Nitrate	Barium Nitrate	Water		
11.09	5.28	83.63	17	1.120
13.64	6.08	80.28	21.5	—
17.94	7.45	74.61	30	1.191
29.47	10.80	59.73	50	—

№ 4616 [1605]

POTASSIUM NITRATE–BARIUM NITRATE–WATER

$KNO_3—Ba(NO_3)_2—H_2O$

Comp. of saturated soln., wt %			t	Comp. of solid phase
Potassium Nitrate	Barium Nitrate	Water		
0.00	6.25	93.75	9.1	$Ba(NO_3)_2$
8.15	4.20	87.65	9.1	$Ba(NO_3)_2 + 2KNO_3 \cdot Ba(NO_3)_2$
12.02	1.98	86.00	9.1	$2KNO_3 \cdot Ba(NO_3)_2$
16.80	0.98	82.22	9.1	$2KNO_3 \cdot Ba(NO_3)_2 + KNO_3$
16.76	0.0	83.24	9.1	KNO_3
0.0	8.46	91.54	21.1	$Ba(NO_3)_2$
2.12	7.47	90.41	21.1	»
5.98	6.35	87.67	21.1	»
8.47	6.06	85.47	21.1	»
13.24	5.98	80.78	21.1	$Ba(NO_3)_2 + 2KNO_3 \cdot Ba(NO_3)_2$
18.24	3.35	78.41	21.1	$2KNO_3 \cdot Ba(NO_3)_2$
21.47	2.30	76.23	21.1	»
24.86	1.76	73.38	21.1	$2KNO_3 \cdot Ba(NO_3)_2 + KNO_3$
24.77	0.00	75.23	21.1	KNO_3
14.89	6.62	78.49	25	$Ba(NO_3)_2 + 2KNO_3 \cdot Ba(NO_3)_2$
16.30	5.49	78.21	25	$2KNO_3 \cdot Ba(NO_3)_2$
21.99	3.04	75.97	25	»
27.76	2.04	70.20	25	$2KNO_3 \cdot Ba(NO_3)_2 + KNO_3$
0.00	11.39	88.61	35	$Ba(NO_3)_2$
12.99	8.18	78.83	35	»
17.48	8.08	74.44	35	»
19.75	8.42	71.83	35	$Ba(NO_3)_2 + 2KNO_3 \cdot Ba(NO_3)_2$
24.00	5.85	70.15	35	$2KNO_3 \cdot Ba(NO_3)_2$

Table № 4616 (continuation)

Comp. of saturated soln., wt %			t	Comp. of solid phase
Potassium Nitrate	Barium Nitrate	Water		
26.05	5.02	68.93	35	$2KNO_3 \cdot Ba(NO_3)_2$
34.87	3.02	62.11	35	$2KNO_3 \cdot Ba(NO_3)_2 + KNO_3$
34.98	1.77	63.25	35	KNO_3
35.01	0.00	64.99	35	»

№ 4617 POTASSIUM NITRATE–BARIUM NITRATE–WATER [1728

$KNO_3—Ba(NO_3)_2—H_2O$

Comp. of saturated soln., wt %			t	Comp. of solid phase
Potassium Nitrate	Barium Nitrate	Water		
0.00	9.28	90.72	25	$Ba(NO_3)_2$
3.80	7.72	88.48	25	»
9.13	6.57	84.30	25	»
14.80	6.60	78.60	25	$Ba(NO_3)_2 + 2KNO_3 \cdot Ba(NO_3)_2$
17.14	4.88	77.98	25	$2KNO_3 \cdot Ba(NO_3)_2$
27.67	2.44	69.89	25	$2KNO_3 \cdot Ba(NO_3)_2 + KNO_3$
27.39	0.0	72.61	25	KNO_3
0.0	14.63	85.37	50	$Ba(NO_3)_2$
3.93	12.73	83.34	50	»
10.53	10.92	78.55	50	»
18.76	10.14	71.10	50	»
24.14	10.43	65.43	50	»
28.86	10.81	60.33	50	$Ba(NO_3)_2 + 2KNO_3 \cdot Ba(NO_3)_2$
30.08	10.16	59.76	50	$2KNO_3 \cdot Ba(NO_3)_2$
35.75	6.99	57.26	50	»
40.31	5.64	54.05	50	»
44.66	5.11	50.23	50	$2KNO_3 \cdot Ba(NO_3)_2 + KNO_3$
44.88	4.22	50.90	50	KNO_3
45.45	2.34	52.21	50	»

№ 4618 POTASSIUM NITRATE–BARIUM BROMATE–WATER [1777]

$KNO_3—Ba(BrO_3)_2—H_2O$

Content in aqueous soln., g/l		d	Content in aqueous soln., g/l		d
Potassium Nitrate	Barium Bromate		Potassium Nitrate	Barium Bromate	
0.00	7.93	1.0038	5.05	9.91	1.0080
2.52	8.62	1.0059	10.1	10.25	1.0120

№ 4619 **POTASSIUM NITRATE–BARIUM IODATE–WATER** [1778]

KNO_3—$Ba(IO_3)_2$—H_2O

$t=25$

Content in aqueous soln., g/l			
Potassium Nitrate	Barium Iodate	Potassium Nitrate	Barium Iodate
0.202	0.396	5.055	0.643
1.011	0.445	20.22	0.777

№ 4620 **POTASSIUM NITRATE–BARIUM IODATE–WATER** [2164]

KNO_3—$Ba(IO_3)_2$—H_2O

$t=25$

Content in aqueous soln., g/l					
Potassium Nitrate	Barium Iodate	Potassium Nitrate	Barium Iodate	Potassium Nitrate	Barium Iodate
0.101	0.4024	0.505	0.4287	2.022	0.4901
0.202	0.4097	0.757	0.4414	5.055	0.5632
0.354	0.4205	1.011	0.4541	10.11	0.6631

№ 4621 **POTASSIUM NITRATE–BARIUM IODATE–WATER** [2235]

KNO_3—$Ba(IO_3)_2$—H_2O

$t=25$

Concn. in aqueous soln., g/l					
Barium Iodate	Potassium Nitrate	Barium Iodate	Potassium Nitrate	Barium Iodate	Potassium Nitrate
0.4086	0.51	0.5739	0.25	1.116	4.962
0.4354	1.011	0.6655	0.51	1.577	9.925
0.4593	2.481	0.8575	1.01		

№ 4622 [1677, 2039]

POTASSIUM NITRATE–LANTHANUM IODATE–WATER

KNO_3— $La(IO_3)_3$—H_2O

$t=25$

Concn. of Potassium Nitrate in aqueous soln., g/l	Solubility of Lanthanum Iodate in aqueous soln. of Potassium Nitrate, g/l
10.11	1.1143

№ 4623 POTASSIUM NITRATE–CERIUM NITRATE–WATER [298]

$KNO_3—Ce(NO_3)_3—H_2O$

Comp. of saturated soln., wt %			t	Comp. of solid phase
Potassium Nitrate	Cerium Nitrate	Water		
17.7	0.0	82.3	10	KNO_3
11.0	17.8	11.2	10	»
12.5	37.6	49.9	10	»
14.0	43.0	43.0	10	»
16.2	46.5	62.7	10	$KNO_3+Ce(NO_3)_3 \cdot 2KNO_3 \cdot 2H_2O$
15.6	47.0	37.4	10	$Ce(NO_3)_3 \cdot 2KNO_3 \cdot 2H_2O$
12.4	49.8	37.8	10	»
9.0	55.0	36.0	10	»
7.2	58.0	34.8	10	$Ce(NO_3)_3 \cdot 2KNO_3 \cdot 2H_2O+Ce(NO_3)_3 \cdot 6H_2O$
4.3	59.0	36.7	10	$Ce(NO_3)_3 \cdot 6H_2O$
0.0	60.0	40.0	10	»
24.1	0.0	75.9	20	KNO_3
18.0	16.4	65.6	20	»
16.0	36.1	47.9	20	»
18.0	41.0	41.0	20	»
17.8	45.5	36.7	20	$KNO_3+Ce(NO_3)_3 \cdot 2KNO_3 \cdot 2H_2O$
16.5	47.5	36.0	20	$Ce(NO_3)_3 \cdot 2KNO_3 \cdot 2H_2O$
12.6	50.4	37.0	20	»
9.5	55.5	35.0	20	»
7.4	60.0	32.6	20	$Ce(NO_3)_3 \cdot 2KNO_3 \cdot 2H_2O+Ce(NO_3)_3 \cdot 6H_2O$
3.9	60.5	35.6	20	$Ce(NO_3)_3 \cdot 6H_2O$
0.0	61.5	38.5	20	»
31.5	0.0	68.5	30	KNO_3
23.0	15.4	61.6	30	»
20.0	34.4	45.6	30	»
20.0	40.0	40.0	30	»
20.8	45.0	34.2	30	$KNO_3+Ce(NO_3)_3 \cdot 2KNO_3 \cdot 2H_2O$
17.5	48.0	34.5	30	$Ce(NO_3)_3 \cdot 2KNO_3 \cdot 2H_2O$
13.1	51.8	35.1	30	»
10.5	56.0	33.5	30	»
8.0	62.0	30.0	30	$Ce(NO_3)_3 \cdot 2KNO_3 \cdot 2H_2O+Ce(NO_3)_3 \cdot 6H_2O$
3.4	62.5	34.1	30	$Ce(NO_3)_3 \cdot 6H_2O$
0.0	63.0	37.0	30	»

№ 4624 POTASSIUM NITRATE–SERIUM IODATE–WATER [1387]

$KNO_3—Ce(IO_3)_2—H_2O$

$t=25$

Concn. in aqueous soln., wt %					
Potassium Nitrate	Cerium Iodate	Potassium Nitrate	Cerium Iodate	Potassium Nitrate	Cerium Iodate
0.0	0.1729	0.5019	0.2082	4.797	0.2961
0.0504	0.1841	0.9985	0.2388	9.148	0.3598
0.2014	0.1924	1.977	0.2674		

№ 4625 POTASSIUM NITRATE–THALLIUM NITRATE–WATER [1627]

$KNO_3—TlNO_3—H_2O$

$t=25$

Concn. in aqueous soln., g/l		d	Concn. in aqueous soln., g/l		d
Potassium Nitrate	Thallium Nitrate		Potassium Nitrate	Thallium Nitrate	
351.0	0.0	1.2632	321.0	63.60	1.2436
329.0	2.37	1.1903	330.5	86.18	1.2617
332.4	6.15	1.1956	428.3	123.8	1.2950
333.7	17.64	1.2050	245.1	101.3	1.2050
333.3	49.74	1.2196	0.0	116.1	1.0964

№ 4626 [2363]

POTASSIUM NITRATE–THALLIUM NITRATE–WATER

$KNO_3—TlNO_3—H_2O$

$t=58$

Comp. of saturated soln., wt %		
Thallium Nitrate	Potassium Nitrate	Water
17.56	42.06	40.38

№ 4627 POTASSIUM NITRATE–THALLIUM CHLORIDE–WATER [2263]

$KNO_3—TlCl—H_2O$

$t=25$

Concn. in aqueous soln., g/l					
Thallium Chloride	Potassium Nitrate	Thallium Chloride	Potassium Nitrate	Thallium Chloride	Potassium Nitrate
4.070	1.517	5.076	15.920	7.375	101.11
4.286	3.033	6.153	50.55	9.430	202.22
4.597	7.775				

№ 4628 POTASSIUM NITRATE–THALLIUM CHLORIDE–WATER [1270]

$KNO_3—TlCl—H_2O$

$t=25$

Concn. of Potassium Nitrate in aqueous soln., g/l	Solubility of Thallium Chloride in aqueous soln. of Potassium Nitrate, g/l	d_4	Concn. of Potassium Nitrate in aqueous soln., g/l	Solubility of Thallium Chloride in aqueous soln. of Potassium Nitrate, g/l	d_4
2.022	4.116	1.0009	30.33	5.548	1.0194
5.052	4.380	1.0028	101.059	7.369	1.0632
10.108	4.704	1.0063			

№ 4629 POTASSIUM NITRATE–THALLIUM CHLORIDE–WATER [1306]

$KNO_3—TlCl—H_2O$

Concn. of Potassium Nitrate in aqueous soln., g/l	Solubility of Thallium Chloride in aqueous soln. of Potassium Nitrate, g/l	d	t	Concn. of Potassium Nitrate in aqueous soln., g/l	Solubility of Thallium Chloride in aqueous soln. of Potassium Nitrate, g/l	d	t
5.055	1.888	1.0039	0	50.55	6.131	1.0327	25
20.22	2.291	1.0155	0	101.1	7.268	1.0619	25
50.55	2.840	1.0349	0	5.055	8.631	0.9989	50
101.1	3.531	1.0661	0	20.22	9.922	1.0088	50
5.055	4.380	1.0038	25	50.55	11.500	1.0273	50
20.22	5.160	1.0138	25	101.1	13.251	1.0571	50

№ 4630 POTASSIUM NITRATE–THALLIUM BROMIDE–WATER [1958]

$KNO_3—TlBr—H_2O$

$t=25$

Concn. in aqueous soln., g/l					
Thallium Bromide	Potassium Nitrate	Thallium Bromide	Potassium Nitrate	Thallium Bromide	Potassium Nitrate
0.572	0.00	0.627	1.5216	0.691	5.0803
0.596	0.5075	0.639	2.0301	0.722	7.6280
0.614	1.0140	0.650	2.5376	0.749	10.1808

№ 4631 [2039]

POTASSIUM NITRATE–THALLIUM IODATE–WATER

$KNO_3—TlIO_3—H_2O$

$t=25$

Concn. of Potassium Nitrate in aqueous soln., g/l	Solubility of Thallium Iodate in aqueous soln. of Potassium Nitrate, g/l	Concn. of Potassium Nitrate in aqueous soln., g/l	Solubility of Thallium Iodate in aqueous soln. of Potassium Nitrate, g/l
1.011	0.753	10.11	0.945
2.022	0.788	30.33	1.186
5.055	0.861	50.55	1.393
		101.10	1.828

№ 4632 [213]

POTASSIUM NITRATE–LEAD (II) FLUORIDE–WATER

$KNO_3—PbF_2—H_2O$

$t=18—20$

Concn. of Potassium Nitrate in aqueous soln., g-eq/l	Solubility of Lead Fluoride in aqueous soln. of Potassium Nitrate, mmol/l.
0.1	2.148
0.2	2.353
0.5	3.431

№ 4633 POTASSIUM NITRATE–LEAD (II) NITRATE–WATER [1727]

$KNO_3—Pb(NO_3)_2—H_2O$

Comp. of saturated soln., wt %			t	Comp. of solid phase
Potassium Nitrate	Lead Nitrate	Water		
0.00	37.17	62.83	25	$Pb(NO_3)_2$
2.75	37.37	59.88	25	»
4.90	37.95	57.15	25	»

Table № 4633 (continuation)

Comp. of saturated soln., wt %			t	Comp. of solid phase
Potassium Nitrate	Lead Nitrate	Water		
8.42	39.27	51.31	25	$Pb(NO_3)_2$
14.73	40.86	44.41	25	»
24.67	41.93	33.40	25	$Pb(NO_3)_2+KNO_3$
24.92	40.99	34.09	25	KNO_3
25.93	28.05	46.02	25	»
26.93	9.24	63.83	25	»
27.03	15.14	57.83	25	»
27.39	0.00	72.61	25	»
0.00	44.79	55.21	50	$Pb(NO_3)_2$
5.90	44.44	49.66	50	»
11.26	44.75	43.99	50	»
16.16	44.61	39.23	50	»
19.38	44.38	46.24	50	»
27.47	43.45	29.08	50	»
30.98	43.05	25.97	50	»
33.14	42.14	24.72	50	$Pb(NO_3)_2+KNO_3$
36.33	32.32	31.35	50	KNO_3
37.69	26.47	35.84	50	»
41.72	12.03	46.25	50	»
45.51	0.00	54.49	50	»
0.00	55.65	44.35	100	$Pb(NO_3)_2$
3.79	53.44	42.77	100	»
15.98	49.93	34.09	100	»
36.90	44.50	18.60	100	KNO_3
51.24	27.04	21.72	100	»
71.00	0.00	29.0	100	»

№ 4634 [1843]

POTASSIUM NITRATE–LEAD (II) NITRATE–WATER

$KNO_3—Pb(NO_3)_2—H_2O$

$t=0$

Comp. of saturated soln., wt %			Comp. of solid phase
Potassium Nitrate	Lead Nitrate	Water	
12.24	0.00	87.76	KNO_3
13.67	22.95	63.38	»
14.33	26.05	58.61	»
14.89	36.27	48.84	»
14.86	37.14	48.00	$KNO_3+Pb(NO_3)_2$
14.20	37.37	48.43	$Pb(NO_3)_2$
8.95	32.94	58.11	»
4.15	29.63	66.22	»
0.00	26.66	73.34	»

№ 4635 POTASSIUM NITRATE–LEAD (II) BROMATE–WATER [2112

$KNO_3—Pb(BrO_3)_2—H_2O$

$t=25$

Concn. in aqueous soln., g/l

Lead Bromate	Potassium Nitrate	d	Lead Bromate	Potassium Nitrate	d
15.91	0.0	1.0112	45.46	94.42	1.0906
18.17	4.94	1.0156	52.50	119.68	1.1127
20.27	10.04	1.0181	59.86	145.8	1.1327
23.63	19.78	1.0286	71.76	184.9	1.1664
26.84	29.79	1.0378	78.02	205.8	1.1833
29.61	39.44	1.0456	82.51	226.9	1.2011
32.42	49.06	1.0547	83.29	246.9	1.2161
35.08	58.48	1.0608	80.19	267.2	1.2220
40.44	77.47	1.0776			

№ 4636 [213]

POTASSIUM NITRATE–LEAD (II) IODIDE–WATER

$KNO_3—PbI_2—H_2O$

$t=18—20$

Content of Potassium Nitrate in aqueous soln. g-eq/l	Solubility of Lead Iodide in aqueous soln. of Potassium Nitrate, mmol/l	Content of Potassium Nitrate in aqueous soln. g-eq/l	Solubility of Lead Iodide in aqueous soln. of Potassium Nitrate, mmol/l
0.1	1.601	1.0	2.632
0.2	1.917	1.5	3.240
0.5	2.296		

№ 4637 [55]

POTASSIUM NITRATE–LEAD (II) IODIDE–WATER

$KNO_3—PbI_2—H_2O$

$t=20$

Content of Potassium Nitrate in soln., mol/l	Solubility of Lead Iodide in aqueous soln. of Potassium Nitrate, mol/l
0.20	$2.81 \cdot 10^{-3}$
0.60	$3.81 \cdot 10^{-3}$
1.00	$4.74 \cdot 10^{-3}$

№ 4638 [1778]

POTASSIUM NITRATE–LEAD (II) IODATE–WATER

$KNO_3—Pb(IO_3)_2—H_2O$

$t=25$

Concn. in aqueous soln., g/l

Lead Iodate	Potassium Nitrate	Lead Iodate	Potassium Nitrate
0.0318	0.202	0.0567	5.055
0.0363	1.011	0.0708	20.22

№ 4639 [2040]

POTASSIUM NITRATE–LEAD (II) IODATE–WATER

$KNO_3—Pb(IO_3)_2—H_2O$

$t=25$

Concn. of Potassium Nitrate in aqueous soln., g/l	Solubility of Lead Iodate in aqueous soln. of Potassium Nitrate, g/l
10.11	0.0389

№ 4640 [1409]

POTASSIUM NITRATE–URANYL NITRATE–WATER

$KNO_3—UO_2(NO_3)_2—H_2O$

$t=25$

Comp. of saturated soln., wt %			Comp. of solid phase
Potassium Nitrate	Uranyl Nitrate	Water	
27.44	0.00	72.56	KNO_3
23.15	3.16	73.69	»
21.23	6.46	72.31	»
19.94	9.71	70.35	»
12.74	25.80	61.46	»
11.11	40.60	48.29	»
10.36	48.65	40.99	»
10.40	53.88	35.72	$UO_2(NO_3)_2 \cdot 6H_2O$
9.51	53.90	37.59	»
5.30	54.21	40.49	»
3.36	54.68	41.96	»
0.00	56.08	43.92	»

№ 4641 [688]

POTASSIUM NITRATE–URANYL NITRATE–WATER

$KNO_3—UO_2(NO_3)_2—H_2O$

Comp. of saturated soln., wt %			t	Comp. of solid phase
Uranyl Nitrate	Potassium Nitrate	Water		
39.00	1.92	59.08	−14.8	Ice + KNO_3
40.00	0.00	60.00	−14.4	Ice
39.16	1.96	58.88	−13.6	KNO_3
23.80	3.40	72.80	− 6.4	Ice + KNO_3
23.80	4.76	71.44	− 6.0	Ice
25.00	0.00	75.00	− 5.4	»
7.38	7.50	85.12	− 3.2	Ice + KNO_3
7.40	7.40	85.20	− 3.0	Ice
1.43	8.80	89.77	− 2.8	Ice + KNO_3
23.80	4.76	71.44	− 2.4	KNO_3
7.68	3.84	88.48	− 2.2	Ice
1.44	7.40	91.16	− 2.0	»
1.50	3.84	94.66	− 1.2	»
8.00	0.00	92.00	− 1.0	»
1.52	1.96	96.52	− 0.6	»
38.46	3.84	57.70	− 0.4	KNO_3
1.56	0.00	98.44	− 0.2	Ice
0.00	12.24	87.76	0	KNO_3
1.00	11.30	87.70	0	»
7.40	9.00	83.40	0	»
24.00	6.00	79.00	0	»
38.70	4.10	57.20	0	»
48.30	2.69	49.00	0	$KNO_3 + UO_2(NO_3)_2 \cdot 6H_2O$
49.80	0.00	50.20	0	$UO_2(NO_3)_2 \cdot 6H_2O$
48.30	3.90	47.80	0.2	$UO_2(NO_3)_2 \cdot 6H_2O + KNO_3$
48.47	3.84	47.69	0.2	$UO_2(NO_3)_2 \cdot 6H_2O$

Table № 4641 (continuation)

Comp. of saturated soln., wt %			t	Comp. of solid phase
Uranyl Nitrate	Potassium Nitrate	Water		
49.01	1.96	49.03	0.5	$UO_2(NO_3)_2 \cdot 6H_2O$
50.00	0.00	50.00	1.2	»
1.39	10.71	87.90	1.5	KNO_3
7.14	10.71	82.15	3.6	»
0.00	14.40	85.60	5.0	»
0.90	14.30	84.80	5.0	»
7.10	11.40	81.50	5.0	»
23.60	7.60	68.80	5.0	»
38.00	5.10	56.90	5.0	»
48.00	4.50	47.50	5.0	»
49.80	4.80	45.40	5.0	$KNO_3+UO_2(NO_3)_2 \cdot 6H_2O$
51.30	0.00	48.70	5.0	$UO_2(NO_3)_2 \cdot 6H_2O$
1.34	13.79	84.87	6.8	KNO_3
37.73	5.66	56.61	6.8	»
47.17	5.66	47.17	9.0	»
6.89	13.79	79.32	9.8	»
1.32	15.25	83.43	10.0	»
0.00	17.70	82.30	10.0	»
0.80	16.70	82.50	10.0	»
7.00	14.00	79.00	10.0	»
23.00	9.60	67.40	10.0	»
37.30	6.40	56.30	10.0	»
47.60	5.80	46.60	10.0	»
51.50	6.00	42.50	10.0	$KNO_3+UO_2(NO_3)_2 \cdot 6H_2O$
52.00	0.00	48.00	10.0	$UO_2(NO_3)_2 \cdot 6H_2O$
22.72	9.90	67.38	10.4	KNO_3
1.30	16.66	82.04	12.8	»
37.03	7.40	55.57	13.2	»
6.66	16.66	76.68	14.6	»
1.27	18.03	80.70	14.8	»
0.00	20.00	80.00	15.0	»
0.75	19.80	79.45	15.0	»
6.80	17.00	76.20	15.0	»
22.40	11.70	65.90	15.0	»
37.00	7.90	55.10	15.0	»
47.30	7.20	45.50	15.0	»
53.00	7.30	39.70	15.0	$KNO_3+UO_2(NO_3)_2 \cdot 6H_2O$
52.80	0.00	47.20	15.0	$UO_2(NO_3)_2 \cdot 6H_2O$
46.20	7.40	46.31	16.4	KNO_3
6.45	19.35	74.20	18.6	»
36.36	9.90	53.74	19.0	»
21.73	13.04	65.23	19.2	»
1.23	20.63	78.14	19.6	»
0.00	23.95	76.05	20.0	»
0.70	23.60	75.70	20.0	»
6.40	20.15	73.45	20.0	»
22.00	13.60	64.40	20.0	»
36.30	9.40	54.30	20.0	»
46.30	8.80	44.90	20.0	»
53.60	8.90	37.50	20.0	$KNO_3+UO_2(NO_3)_2 \cdot 6H_2O$

Table № 4641 (continuation)

Comp. of saturated soln., wt %			t	Comp. of solid phase
Uranyl Nitrate	Potassium Nitrate	Water		
54.40	0.00	45.60	20.0	$UO_2(NO_3)_2 \cdot 6H_2O$
6.25	21.87	71.88	22.4	KNO_3
45.45	9.90	44.65	23.0	»
1.20	23.07	75.73	27.2	»
35.71	10.71	53.58	24.4	»
0.00	27.64	72.36	25.0	»
0.65	26.70	72.65	25.0	»
6.20	23.50	70.30	25.0	»
21.50	15.40	63.10	25.0	»
35.60	11.00	53.40	25.0	»
45.80	10.60	43.60	25.0	»
54.10	10.90	35.00	25.0	$KNO_3 + UO_2(NO_3)_2 \cdot 6H_2O$
55.90	0.00	44.10	25.0	$UO_2(NO_3)_2 \cdot 6H_2O$
6.06	24.25	69.69	25.6	KNO_3
20.83	16.66	62.51	31.5	»
0.00	46.09	53.91	50.0	»
0.88	43.50	55.62	50.0	»
4.82	40.12	55.06	50.0	»
17.20	32.00	50.80	50.0	»
30.53	23.66	45.81	50.0	»
40.00	20.00	40.00	50.0	»
52.60	19.70	27.70	50.0	»
58.73	19.60	21.67	50.0	$KNO_3 + UO_2(NO_3)_2 \cdot 6H_2O$
65.50	2.00	32.50	50.0	$UO_2(NO_3)_2 \cdot 6H_2O$
67.50	0.00	32.50	50.0	»
0.00	57.90	42.10	70.0	KNO_3
0.98	55.80	43.22	70.0	»
4.20	52.20	43.60	70.0	»
14.35	43.80	41.85	70.0	»
26.00	35.90	38.10	70.0	»
35.22	30.72	34.06	70.0	»
48.34	25.81	25.85	70.0	»
56.80	22.90	20.30	70.0	»
61.30	21.80	16.90	70.0	»
63.00	21.00	16.00	70.0	$KNO_3 + UO_2(NO_3)_2 \cdot 3H_2O$
76.20	2.00	21.80	70.0	$UO_2(NO_3)_2 \cdot 3H_2O$
78.00	0.00	22.00	70.0	»
0.00	66.80	33.20	90.0	KNO_3
0.75	63.50	35.75	90.0	»
4.25	61.00	34.75	90.0	»
11.60	54.00	34.40	90.0	»
21.72	46.30	31.98	90.0	»
29.70	41.25	29.05	90.0	»
43.32	33.41	23.27	90.0	»
52.21	28.80	18.98	90.0	»
58.30	25.75	15.96	90.0	»
66.60	21.60	11.80	90.0	$KNO_3 + UO_2(NO_3)_2 \cdot 3H_2O$
80.20	1.55	18.25	90.0	$UO_2(NO_3)_2 \cdot 3H_2O$
81.00	0.00	19.00	90.0	»

№ 4642 POTASSIUM HYDROGEN FLUORIDE–POTASSIUM CHLORIDE–WATER [695]

$KHF_2—KCl—H_2O$

Comp. of saturated soln., wt %			t	Comp. of solid phase
Potassium Hydrogen Fluoride	Potassium Chloride	Water		
0.00	26.36	73.64	25	KCl
3.96	23.94	72.10	25	»
8.00	21.97	70.03	25	»
12.46	19.93	67.61	25	»
17.36	17.64	65.00	25	»
18.14	17.28	64.58	25	KHF_2+KHCl_2
19.27	15.54	65.19	25	KHF_2
21.48	11.42	67.10	25	»
23.88	7.41	68.71	25	»
26.20	4.24	69.56	25	»
29.75	0.00	70.25	25	»
0.00	33.20	66.80	75	KCl
4.41	30.22	65.37	75	»
9.78	27.26	62.96	75	»
19.22	22.62	58.16	75	»
26.77	19.27	63.96	75	»
33.06	16.65	50.29	75	»
36.33	15.37	48.30	75	»
40.61	14.15	45.24	75	KHF_2+KCl
41.36	12.95	45.69	75	KHF_2
44.50	7.51	47.99	75	»
46.21	5.27	48.52	75	»
46.68	3.96	49.36	75	»
47.66	3.48	48.86	75	»
51.20	0.00	48.80	75	»

№ 4643 POTASSIUM FLUORIDE–POTASSIUM TETRAFLUOROBORATE–WATER [563]

$KF—KBF_4—H_2O$

$t=25$

Content of Potassium Fluoride in aqueous soln., mol/l	Solubility of Potassium Tetra-fluoroborate mol/1000 g Water	Content of Potassium Fluoride in aqueous soln., mol/l	Solubility of Potassium Tetra-fluoroborate mol/1000 g Water	Content of Potassium Fluoride in aqueous soln., mol/l	Solubility of Potassium Tetra-fluoroborate mol/1000 g Water
0.00000	$4.37 \cdot 10^{-2}$	0.24090	$7.86 \cdot 10^{-3}$	1.570	$4.06 \cdot 10^{-3}$
0.03270	$3.27 \cdot 10^{-2}$	0.26670	$6.98 \cdot 10^{-3}$	1.820	$3.16 \cdot 10^{-3}$
0.08778	$2.15 \cdot 10^{-2}$	0.9965	$6.42 \cdot 10^{-3}$	2.241	$1.67 \cdot 10^{-3}$
0.17720	$2.22 \cdot 10^{-2}$	1.501	$5.01 \cdot 10^{-3}$		

№ 4644 POTASSIUM FLUORIDE–POTASSIUM CHLORIDE–WATER [1336]

$KF—KCl—H_2O$

$t=25$

Content in aqueous soln., wt %		Comp. of solid phase	Content in aqueous soln., wt %		Comp. of solid phase
Potassium Fluoride	Potassium Chloride		Potassium Fluoride	Potassium Chloride	
49.5	0	$KF \cdot 2H_2O$	49.0	0.935	$KF \cdot 2H_2O+KCl$
49.8	0.93	$KF \cdot 2H_2O+KCl$	48.3	1.45	»
49.3	0.935	»	44.8	1.48	KCl

Table № 4644 (continuation)

Content in aqueous soln., wt %		Comp. of solid phase	Content in aqueous soln., wt %		Comp. of solid phase
Potassium Fluoride	Potassium Chloride		Potassium Fluoride	Potassium Chloride	
33.0	3.82	KCl	3.13	22.9	KCl
32.0	4.50	»	1.89	24.6	»
17.1	11.43	»	0.77	26.1	»
9.35	17.7	»	0	26.6	»

№ 4645 POTASSIUM FLUORIDE–POTASSIUM CHLORIDE–WATER [695]

$KF—KCl—H_2O$

Comp. of saturated soln., wt %			t	Comp. of solid phase
Potassium Fluoride	Potassium Chloride	Water		
0.00	26.36	73.64	25	KCl
3.94	22.83	73.23	25	»
7.99	19.47	72.54	25	»
17.75	12.24	70.01	25	»
30.07	5.42	64.51	25	»
41.58	1.95	56.47	25	»
48.56	1.07	50.37	25	»
51.07	0.92	48.01	25	$KF \cdot 2H_2O + KCl$
50.41	0.00	49.59	25	$KF \cdot 2H_2O$
0.00	33.20	66.80	75	KCl
4.62	28.75	66.63	75	»
10.40	23.33	66.27	75	»
22.09	15.29	62.62	75	»
31.08	9.11	59.81	75	»
39.81	5.20	54.99	75	»
51.28	2.58	46.14	75	»
61.20	1.90	36.90	75	KF+KCl
61.04	1.49	37.47	75	KF
59.60	0.00	40.40	75	»

№ 4646 [56]

POTASSIUM FLUORIDE–TITANIUM FLUORIDE–WATER

$KF—TiF_2—H_2O$

$t=20$

Comp. of saturated soln., wt %			Comp. of solid phase
Potassium Fluoride	Titanium Fluoride	Water	
11.020	0.041	88.939	$K_2TiF_6 \cdot H_2O$
6.750	0.052	93.198	»
2.937	0.087	96.976	$K_2TiF_6 \cdot 2H_2O$
1.551	0.169	97.280	»
0.211	0.948	98.841	$(K_2TiF_6)_2 \cdot H_2O$
0.735	2.825	96.422	$K_2TiF_6 \cdot H_2O$
0.652	4.880	94.468	»
0.830	6.360	92.810	»
1.050	13.950	85.000	»

№ 4647 [835]

POTASSIUM FLUORIDE–CHROMIUM FLUORIDE–WATER

$KF—CrF_3—H_2O$

$t=25$

Comp. of saturated soln., wt %			Comp. of solid phase
Potassium Fluoride	Chromium Fluoride	Water	
0.00	4.20	95.80	$CrF_3 \cdot 3H_2O$
0.063	1.46	98.473	»
0.12	1.89	97.99	»
0.49	2.03	97.48	$CrF_3 \cdot 3H_2O + K_2CrF_5 \cdot H_2O$
0.48	1.99	98.53	»
0.48	1.95	97.57	»
0.50	1.84	97.66	»
0.55	1.17	98.28	—
0.62	0.89	98.49	$CrF_3 \cdot 3H_2O + K_2CrF_5 \cdot H_2O$
0.94	0.67	98.39	—
0.55	0.31	99.14	—
2.54	0.007	97.453	$CrF_3 \cdot 3H_2O + K_2CrF_5 \cdot H_2O$
9.80	0.00	90.20	$K_2CrF_5 \cdot H_2O$
10.53	0.00	89.47	»
15.11	0.00	84.89	»
19.54	0.00	80.46	»
37.99	0.00	62.01	»
39.48	0.00	60.52	»
42.01	0.00	57.99	»
44.47	0.00	55.53	»
48.75	0.00	51.25	$K_2CrF_5 \cdot H_2O + KF$

№ 4648 [854]

POTASSIUM FLUORIDE–IRON (III) FLUORIDE–WATER

$KF—FeF_3—H_2O$

$t=25$

Concn. in aqueous soln., wt %		Comp. of solid phase	Concn. in aqueous soln., wt %		Comp. of solid phase
Iron Fluoride	Potassium Fluoride		Iron Fluoride	Potassium Fluoride	
2.30	0.52	$FeF_3 \cdot 3H_2O +$ $+ 2KF \cdot FeF_3 \cdot H_2O$	0.002	3.71	$11KF \cdot 4FeF_3 \cdot 12H_2O$
			0.001	3.97	»
1.01	0.56	$2KF \cdot FeF_3 \cdot H_2O$	0.001	4.00	»
0.24	0.62	»	—	4.86	»
0.03	1.37	»	—	7.17	»
0.02	1.57	»	—	8.94	»
0.006	2.40	$11KF \cdot 4FeF_3 \quad 12H_2O$	—	13.34	$3KF \cdot FeF_3 \cdot 3H_2O$
0.005	2.89	»	—	19.80	»
0.002	3.19	»	—	22.80	»
0.002	3.31	»	—	24.07	»
0.002	3.55	»	—	31.75	»
0.002	3.66	»	—	33.40	»

№ 4649 POTASSIUM FLUORIDE–COBALT (II) FLUORIDE–WATER [2027]

$KF—CoF_2—H_2O$

Concn. in aqueous soln., wt %		t	Concn. in aqueous soln., wt %		t
Potassium Fluoride	Cobalt Fluoride		Potassium Fluoride	Cobalt Fluoride	
1.2	0.86	20	32.6	0.00	20
1.6	0.18	20	4.3	0.012	50
2.3	0.11	20	11.0	0.005	50
13.8	0.013	20	15.5	0.00	50
20.8	0.005	20			

№ 4650 POTASSIUM FLUORIDE–NICKEL (II) FLUORIDE–WATER [2027]

$KF—NiF_2—H_2O$

Comp. of saturated soln., wt %			t	Comp. of solid phase
Potassium Fluoride	Nickel Fluoride	Water		
1.21	1.98	98.81	20	Mixed crystals containing between 0.8 and 1.4 mol KF to 1 mol NiF_2
3.52	1.20	95.28	20	»
6.32	0.80	92.88	20	»
9.64	0.52	89.84	20	»
12.9	0.40	86.70	20	»
16.8	0.01	83.19	20	»
20.2	0.00	79.8	20	»
25.5	0.00	74.5	20	»
28.1	0.00	71.9	20	»
33.0	0.00	67.0	20	»
37.5	0.00	62.5	20	»
1.18	2.01	96.81	50	»
4.30	1.12	94.58	50	»
9.25	0.26	90.49	50	»
15.4	0.03	84.57	50	»
19.2	0.01	80.79	50	»
22.1	0.00	77.9	50	»
25.0	0.00	75.0	50	»
27.7	0.00	72.3	50	»
31.3	0.00	68.7	50	»
36.8	0.00	63.2	50	»
41.3	0.00	58.7	50	»

№ 4651 [2027]

POTASSIUM FLUORIDE–COPPER (II) FLUORIDE–WATER

$KF—CuF_2—H_2O$

$t=20$

Comp. of saturated soln., wt %			Comp. of solid phase
Copper Fluoride	Potassium Fluoride	Water	
1.5	1.2	97.3	Mixed crystals
0.16	11.20	88.64	»

№ 4652 [2027]

POTASSIUM FLUORIDE–ZINC FLUORIDE–WATER

$KF—ZnF_2—H_2O$

$t=20$

Comp. of saturated soln., wt%			Comp. of solid phase
Zinc Fluoride	Potassium Fluoride	Water	
0.25	1.5	98.25	Mixed crystals
0.025	10.0	89.975	»
0.00	19.0	81.0	»

№ 4653 [830]

POTASSIUM FLUORIDE–STRONTIUM FLUORIDE–WATER

$KF—SrF_2—H_2O$

$t=25$

Concn. of Potassium Fluoride in aqueous soln., mol/l	Solubility of Strontium Fluoride in aqueous soln. of Potassium Fluoride, mol/l	Concn. of Potassium Fluoride in aqueous soln., mol/l	Solubility of Strontium Fluoride in aqueous soln. of Potassium Fluoride, mol/l	Concn. of Potassium Fluoride in aqueous soln., mol/l	Solubility of Strontium Fluoride in aqueous soln. of Potassium Fluoride, mol/l
0.00	$9.62 \cdot 10^{-4}$	$2.992 \cdot 10^{-3}$	$2.92 \cdot 10^{-4}$	$5.985 \cdot 10^{-3}$	$1.09 \cdot 10^{-4}$
$4.987 \cdot 10^{-4}$	$8.37 \cdot 10^{-4}$	$3.421 \cdot 10^{-3}$	$2.36 \cdot 10^{-4}$	$6.485 \cdot 10^{-3}$	$9.1 \cdot 10^{-5}$
$9.974 \cdot 10^{-4}$	$6.56 \cdot 10^{-4}$	$3.990 \cdot 10^{-3}$	$2.05 \cdot 10^{-4}$	$7.982 \cdot 10^{-3}$	$6.8 \cdot 10^{-5}$
$1.496 \cdot 10^{-3}$	$5.38 \cdot 10^{-4}$	$4.488 \cdot 10^{-3}$	$1.64 \cdot 10^{-4}$	$8.480 \cdot 10^{-3}$	$6.4 \cdot 10^{-5}$
$1.995 \cdot 10^{-3}$	$4.46 \cdot 10^{-4}$	$4.987 \cdot 10^{-3}$	$1.44 \cdot 10^{-4}$	$8.948 \cdot 10^{-3}$	$5.6 \cdot 10^{-5}$
$2.494 \cdot 10^{-3}$	$3.67 \cdot 10^{-4}$	$5.487 \cdot 10^{-3}$	$1.19 \cdot 10^{-4}$	$9.877 \cdot 10^{-3}$	$5.1 \cdot 10^{-5}$

№ 4654 [57]

POTASSIUM FLUORIDE–ZIRCONIUM FLUORIDE–WATER

$KF—ZrF_4—H_2O$

$t=20$

Comp. of saturated soln., wt %			Comp. of solid phase
Potassium Fluoride	Zirconium Fluoride	Water	
9.060	0.025	90.915	K_3ZrF_7
7.520	0.065	92.415	»
4.665	0.100	95.235	»
2.300	0.262	97.438	$K_3ZrF_7+K_2ZrF_6$
1.700	0.306	97.994	K_2ZrF_6
0.786	1.565	97.649	»
0.630	2.250	97.120	$K_2ZrF_6+KZrF_5$
0.642	2.600	96.758	$KZrF_5$
0.315	5.940	93.745	»

№ 4655 [1058]

POTASSIUM FLUORIDE–MOLYBDENUM TRIOXIDE–WATER

$KF—MoO_3—H_2O$

$t=25$

Comp. of saturated soln., wt %			Comp. of solid phase
Potassium Fluoride	Molybdenum Trioxide	Water	
0.63	0.97	98.40	Region of undetermined comp.
2.55	2.55	94.90	»
4.95	2.29	92.76	»
6.91	2.75	90.34	»
8.52	3.27	88.21	»
10.39	2.36	87.25	»
11.84	3.51	84.65	»
12.71	4.21	83.08	»
14.59	4.90	80.51	»
15.65	4.69	79.66	»
18.05	5.94	76.01	»
19.00	6.11	74.89	$K_2Mo_3OF_6$
20.32	3.91	75.77	»
21.59	1.87	76.54	»
22.43	1.33	76.24	»
34.71	0.19	65.10	»
36.40	0.28	63.32	»
39.18	0.16	60.66	»
50.20	0.44	49.36	»

№ 4656 [2027]

POTASSIUM FLUORIDE–CADMIUM FLUORIDE–WATER

$KF—CdF_2—H_2O$

$t=20$

Comp. of saturated soln., wt %					
Cadmium Fluoride	Potassium Fluoride	Water	Cadmium Fluoride	Potassium Fluoride	Water
2.4	3.4	94.2	0.05	33.1	66.85
0.6	13.6	95.8	0.024	41.2	58.78
0.18	32.5	67.32			

№ 4657 [653]

POTASSIUM FLUORIDE–IODINE PENTOXIDE–WATER*

$KF—I_2O_5—H_2O$

$t=25$

Content of Potassium Fluoride in aqueous soln., wt %	Solubility of Iodine Pentoxide in aqueous soln. of Potassium Fluoride, wt %	Content of Potassium Fluoride in aqueous soln., wt %	Solubility of Iodine Pentoxide in aqueous soln. of Potassium Fluoride, wt %
0.00	73.85	0.43	61.87
0.23	67.67	0.62	56.11
0.37	67.36	1.19	49.81

*Compounds with the following composition form in the solid phase: $KH(IO_3)_2{\cdot}H_2O$ and $KH(IO_3)_2{\cdot}KF{\cdot}4H_2O$.

Table № 4657 (continuation)

Content of Potassium Fluoride in aqueous soln., wt %	Solubility of Iodine Pentoxide in aqueous soln. of Potassium Fluoride, wt %	Content of Potassium Fluoride in aqueous soln., wt %	Solubility of Iodine Pentoxide in aqueous soln. of Potassium Fluoride, wt %
1.86	42.02	6.70	7.03
2.66	32.39	7.53	4.97
3.45	27.61	12.80	2.53
3.68	24.71	13.20	2.38
4.99	20.82	20.35	2.45
4.71	12.78	22.15	0.85
5.02	7.58	30.33	0.68
5.22	6.50	32.87	0.50
35.68	0.44	32.87	0.58
38.20	0.34	33.77	0.52
40.32	0.38	34.93	0.53
42.26	0.19	35.27	0.54
45.77	0.18	46.80	0.15
5.58	5.26	49.35	0.17
6.43	6.57	50.03	0.17

№ 4658 POTASSIUM FLUORIDE–BARIUM FLUORIDE–WATER [830]

$KF—BaF_2—H_2O$

$t=25$

Concn. of Potassium Fluoride in aqueous soln., mol/l	Solubility of Barium Fluoride in aqueous soln. of Potassium Fluoride, mol/l	Concn. of Potassium Fluoride in aqueous soln., mol/l	Solubility of Barium Fluoride in aqueous soln. of Potassium Fluoride, mol/l	Concn. of Potassium Fluoride in aqueous soln., mol/l	Solubility of Barium Fluoride in aqueous soln. of Potassium Fluoride, mol/l
0.00	$9.23 \cdot 10^{-3}$	$2.081 \cdot 10^{-2}$	$4.36 \cdot 10^{-3}$	$1.507 \cdot 10^{-1}$	$5.4 \cdot 10^{-4}$
$5.002 \cdot 10^{-4}$	$9.10 \cdot 10^{-3}$	$2.480 \cdot 10^{-2}$	$3.43 \cdot 10^{-3}$	$2.008 \cdot 10^{-1}$	$4.0 \cdot 10^{-4}$
$1.233 \cdot 10^{-3}$	$8.71 \cdot 10^{-3}$	$5.011 \cdot 10^{-2}$	$1.60 \cdot 10^{-3}$	$2.571 \cdot 10^{-1}$	$3.0 \cdot 10^{-4}$
$5.028 \cdot 10^{-3}$	$7.75 \cdot 10^{-3}$	$7.256 \cdot 10^{-2}$	$1.13 \cdot 10^{-3}$	$3.325 \cdot 10^{-1}$	$3.0 \cdot 10^{-4}$
$7.031 \cdot 10^{-3}$	$7.13 \cdot 10^{-3}$	$1.001 \cdot 10^{-1}$	$7.5 \cdot 10^{-4}$	$5.020 \cdot 10^{-1}$	$2.2 \cdot 10^{-4}$
$1.087 \cdot 10^{-2}$	$6.20 \cdot 10^{-3}$	$1.096 \cdot 10^{-1}$	$6.7 \cdot 10^{-4}$		

№ 4659 [842]
POTASSIUM FLUORIDE–CERIUM (IV) FLUORIDE–WATER

$KF—CeF_4—H_2O$

$t=25$

Comp. of saturated soln., wt %			Comp. of solid phase
Potassium Fluoride	Cerium Fluoride	Water	
45.33	0.55	54.12	$3KF \cdot CeF_4$
41.92	0.41	57.67	»
33.36	$0.97 \cdot 10^{-3}$	66.64	$2KF\ CeF_4$
28.40	$1.03 \cdot 10^{-3}$	71.60	»
18.27	$1.20 \cdot 10^{-3}$	81.73	»
18.19	$1.22 \cdot 10^{-3}$	81.81	»
17.47	$1.43 \cdot 10^{-3}$	82.53	»

Table № 4659 (continuation)

Comp. of saturated soln., wt %			Comp. of solid phase
Potassium Fluoride	Cerium Fluoride	Water	
16.60	$1.46 \cdot 10^{-3}$	83.40	$5KF \cdot 3CeF_4$
13.60	$1.51 \cdot 20^{-3}$	86.40	»
9.37	$1.69 \cdot 10^{-3}$	90.63	»
5.18	$1.68 \cdot 10^{-3}$	94.82	»
5.11	$1.64 \cdot 10^{-3}$	94.89	»
3.52	$1.61 \cdot 10^{-3}$	96.48	»
2.94	$2.99 \cdot 10^{-3}$	97.06	$KF \cdot CeF_4 + 5KF \cdot 3CeF_4$
1.92	$0.90 \cdot 10^{-3}$	98.08	$KF \cdot CeF_4$
0.97	$0.54 \cdot 10^{-3}$	99.03	»
0.68	$0.59 \cdot 10^{-3}$	99.32	»
0.48	$0.44 \cdot 10^{-3}$	99.52	»
0.29	$0.45 \cdot 10^{-3}$	99.71	»
0.18	$0.40 \cdot 10^{-3}$	99.82	»
0.08	$0.39 \cdot 10^{-3}$	99.92	»
0.03	$0.41 \cdot 10^{-3}$	99.97	$CeF_4 \cdot H_2O + KF \cdot CeF_4$

№ 4660 [748]

POTASSIUM FLUORIDE–CERIUM (IV) FLUORIDE–WATER

$KF—CeF_4—H_2O$

$t=25.$

Comp. of saturated soln., wt %			Comp. of solid phase
Potassium Fluoride	Cerium Fluoride	Water	
21.48	0.00	78.52	CeF_3
18.55	0.00	81.45	»
14.21	0.00	85.79	»
11.34	Traces	88.66	»
5.46	»	94.54	»
2.85	$13 \cdot 10^{-4}$	97.15	»
1.16	$14 \cdot 10^{-4}$	88.84	»
0.82	$15 \cdot 10^{-4}$	99.18	»
0.55	$20 \cdot 10^{-4}$	99.45	»
0.35	$21 \cdot 10^{-4}$	99.65	»
0.17	$26 \cdot 10^{-4}$	99.83	»
0.066	$28 \cdot 10^{-4}$	99.93	»
0.042	$30 \cdot 10^{-4}$	99.96	»
0.030	$35 \cdot 10^{-4}$	99.97	»
0.020	$35 \cdot 10^{-4}$	99.98	»

№ 4661 [843]

POTASSIUM FLUORIDE–CERIUM (IV) FLUORIDE–WATER

$KF—CeF_4—H_2O$

$t=25$

Comp. of saturated soln., wt %			Comp. of solid phase
Potassium Fluoride	Cerium Fluoride	Water	
33.61	0.21	66.18	$5KF \cdot 2CeF_4$
34.45	0.27	65.28	»
37.15	0.40	62.45	»
38.52	0.36	61.12	»
39.43	0.34	60.23	»
40.11	0.39	59.50	$3KF \cdot CeF_4$
41.92	0.41	57.67	»
42.28	0.52	57.20	»
44.36	0.53	55.11	»
45.33	0.55	54.12	»
46.88	0.47	53.65	»

№ 4662 [839]

POTASSIUM FLUORIDE–LEAD (II) FLUORIDE–WATER

$KF—PbF_2—H_2O$

$t=25$

Content in aqueous soln., wt %		d
Potassium Fluoride	Lead Fluoride	
44.01	$81.0 \cdot 10^{-2}$	1.4800
48.78	$83.0 \cdot 10^{-2}$	1.5240

№ 4663 **POTASSIUM FLUORIDE–LEAD (II) FLUORIDE–WATER** [839]

$KF—PbF_2—H_2O$

$t=25$

Comp. of saturated soln., wt %			d	Comp. of saturated soln., wt %		
Potassium Fluoride	Lead Fluoride	Water		Potassium Fluoride	Lead Fluoride	Water
0.00	$5.51 \cdot 10^{-2}$	99.94	1.0008	—	—	—
0.50	$0.22 \cdot 10^{-2}$	99.50	1.0040	0.40	71.00	28.60
0.80	$0.157 \cdot 10^{-2}$	99.20	1.0073	—	—	—
1.10	$0.166 \cdot 10^{-2}$	98.90	1.0083	0.60	71.07	28.33
5.03	$0.19 \cdot 10^{-2}$	94.97	1.0492	0.82	75.87	23.31
9.99	$0.43 \cdot 10^{-2}$	90.01	1.0865	2.10	79.72	18.18
14.87	$0.74 \cdot 10^{-2}$	85.13	1.1334	2.80	78.08	19.12
20.10	$1.70 \cdot 10^{-2}$	79.88	1.2159	4.17	71.10	24.73

Table № 4663 (continuation)

Comp. of saturated soln., wt %			d	Comp. of saturated soln., wt %		
Potassium Fluoride	Lead Fluoride	Water		Potassium Fluoride	Lead Fluoride	Water
24.68	$3.66 \cdot 10^{-2}$	75.28	1.2303	9.58	63.92	26.50
29.90	$6.70 \cdot 10^{-2}$	70.03	1.2851	10.35	64.13	25.52
34.12	$14.00 \cdot 10^{-2}$	65.74	1.3339	13.06	61.30	25.64
39.57	$55.60 \cdot 10^{-2}$	59.97	1.4068	17.27	57.90	24.83
44.01	$81.00 \cdot 10^{-2}$	55.18	1.4800	20.00	55.00	25.00
48.78	$83.00 \cdot 10^{-2}$	50.84	1.5240	22.50	54.90	22.60

№ 4664 POTASSIUM FLUORIDE–THORIUM NITRATE–WATER [867]

$KF—Th(NO_3)_4—H_2O$

$t=25$

Concn. in aqueous soln., mol/l·10³					
Thorium Nitrate	Potassium Fluoride	Thorium Nitrate	Potassium Fluoride	Thorium Nitrate	Potassium Fluoride
2.50	0.00	2.41	3.40	0.48	1.03
2.50	2.50	1.52	2.58	0.15	0.38

№ 4665 [1104]

POTASSIUM DIHYDROGEN ORTHOPHOSPHATE–POTASSIUM CHLORIDE–WATER

$KH_2PO_4—KCl—H_2O$

$t=0$

Comp. of saturated soln., wt %		d	Comp. of solid phase
Potassium Chloride	Potassium Dihydrogen Orthophosphate		
100.0	0.0	1.1320	KCl
95.8	4.20	1.1690	»
92.10	7.90	1.1750	$KCl+KH_2PO_4$
73.0	27.0	1.1427	KH_2PO_4
62.70	37.30	1.1199	»
44.90	55.10	1.1183	»
31.70	68.30	1.1179	»
18.10	81.90	1.1158	»
0.0	100.0	1.1151	»

№ 4666 POTASSIUM DIHYDROGEN ORTHOPHOSPHATE–POTASSIUM CHLORIDE–WATER [710]

$KH_2PO_4—KCl—H_2O$

$t=-10.8$

Comp. of saturated soln., wt %		
Potassium Chloride	Potassium Dihydrogen Orthophosphate	Water
18.84	2.31	78.85

№ 4667 POTASSIUM DIHYDROGEN ORTHOPHOSPHATE–POTASSIUM CHLORIDE–WATER [1924]

KH_2PO_4—KCl—H_2O

Comp. of saturated soln., wt %					t	Comp. of solid phase
K^+, g	H^+, g	Cl^-, g	PO_4^{3-}, g	H_2O, mol/100 g salts		
46.92	3.08	45.37	4.63	1358	0	KCl
44.27	5.73	41.38	8.62	1271	0	$KCl+KH_2PO_4$
38.53	11.47	32.74	7.26	1627	0	KH_2PO_4
31.06	18.94	41.40	5.60	1888	0	»
30.61	19.39	20.27	29.73	1901	0	»
24.06	75.94	10.96	39.04	1919	0	»
23.83	26.17	10.49	39.51	1925	0	»
16.66	33.34	0.00	50.00	1778	0	»
46.51	3.49	44.81	5.19	1057	25	KCl
43.54	6.46	40.36	9.64	972	25	$KCl+KH_2PO_4$
37.51	12.49	31.01	18.99	1158	25	KH_2PO_4
30.00	20.00	20.00	30.00	1242	25	»
23.12	26.88	9.43	40.57	1177	25	»
16.66	33.34	0.00	50.00	1012	25	»
50.00	0.00	50.00	0.00	967.4	50	KCl
42.46	7.54	38.65	11.35	783.1	50	$KCl+KH_2PO_4$
34.72	15.28	27.14	22.86	885.7	50	KH_2PO_4
25.63	24.37	13.46	36.54	834.0	50	»
16.66	33.34	0.00	50.00	624.4	50	»
45.95	4.05	43.92	6.08	753.0	75	KCl
43.56	6.44	40.33	9.67	704.9	75	»
40.22	9.78	35.30	14.70	629.5	75	$KCl+KH_2PO_4$
35.46	4.54	28.22	21.78	659.5	75	KH_2PO_4
27.82	22.18	16.70	33.30	615.0	75	»
23.15	26.85	9.78	40.22	563.3	75	»
16.66	33.34	0.00	50.00	400.7	75	»

№ 4668 POTASSIUM DIHYDROGEN ORTHOPHOSPHATE–POTASSIUM SULFATE–WATER [1084]

KH_2PO_4—K_2SO_4—H_2O

$t=25$

Comp. of saturated soln., wt %					
Potassium Dihydrogen Orthophosphate	Potassium Sulfate	Water	Potassium Dihydrogen Orthophosphate	Potassium Sulfate	Water
19.46	1.39	79.15	17.02	6.79	76.19
18.24	3.13	78.63	16.74	6.27	76.99
17.70	6.27	76.03			

№ 4669 POTASSIUM DIHYDROGEN ORTHOPHOSPHATE–POTASSIUM SULFATE–WATER [92]

$K_2SO_4—KH_2PO_4—H_2O$

Comp. of saturated soln., wt %			t	Comp. of solid phase
Potassium Dihydrogen Orthophosphate	Potassium Sulfate	Water		
0.00	6.9	93.1	0	K_2SO_4
3.8	5.6	90.6	0	»
8.6	4.4	87.0	0	$K_2SO_4+KH_2PO_4$
11.3	2.6	86.1	0	KH_2PO_4
12.2	0.0	87.8	0	»
0.0	8.4	91.6	10	K_2SO_4
3.8	7.1	89.1	10	»
8.4	5.8	85.8	10	»
11.4	5.1	83.5	10	»
1.3	4.8	81.9	10	$K_2SO_4+KH_2PO_4$
13.4	4.3	82.3	10	KH_2PO_4
14.0	2.4	83.6	10	»
14.9	0.0	85.1	10	»
0.0	9.8	90.2	20	K_2SO_4
3.7	8.5	87.8	20	»
8.3	7.2	84.5	20	»
11.2	6.5	82.3	20	»
16.0	5.2	78.8	20	$K_2SO_4+KH_2PO_4$
16.3	4.2	79.5	20	KH_2PO_4
17.0	2.4	80.6	20	»
17.8	0.0	82.2	20	»
0.0	11.4	88.6	30	K_2SO_4
3.6	10.0	86.4	30	»
8.2	8.4	83.4	30	»
11.1	7.6	81.3	30	»
15.0	6.4	78.6	30	»
18.6	6.6	75.8	30	$K_2SO_4+KH_2PO_4$
19.1	4.0	76.9	30	KH_2PO_4
19.8	2.2	78.0	30	»
21.0	0.0	79.0	30	»

№ 4670 [1719]

POTASSIUM DIHYDROGEN ORTHOPHOSPHATE–ARSENIC TRIOXIDE–WATER

$KH_2PO_4—As_2O_3—H_2O$

Comp. of saturated soln., wt %			t
Potassium Dihydrogen Orthophosphate	Arsenic Trioxide	Water	
12.706	2.283	85.011	25
12.600	4.594	82.806	60
12.483	8.990	78.427	100

№ 4671 POTASSIUM HYDROGEN ORTHOPHOSPHATE–SILVER ORTHOPHOSPHATE–WATER [833]

$K_2HPO_4—Ag_3PO_4—H_2O$

$t=25$

Content of Potassium Hydrogen Orthophosphate in aqueous soln., mol/l	Solubility of Silver Orthophosphate in aqueous soln. Potassium Hydrogen Orthophosphate, mol/l	Content of Potassium Hydrogen Orthophosphate in aqueous soln., mol/l	Solubility of Silver Orthophosphate in aqueous soln. Potassium Hydrogen Orthophosphate, mol/l	Content of Potassium Hydrogen Orthophosphate in aqueous soln., mol/l	Solubility of Silver Orthophosphate in aqueous soln. Potassium Hydrogen Orthophosphate, mol/l
0.00	$5.11 \cdot 10^{-5}$	$1.00 \cdot 10^{-2}$	$7.86 \cdot 10^{-5}$	2.5	$32.74 \cdot 10^{-5}$
$1.00 \cdot 10^{-5}$	$5.11 \cdot 10^{-5}$	$1.00 \cdot 10^{-1}$	$14.04 \cdot 10^{-5}$	4.0	$44.22 \cdot 10^{-5}$
$1.00 \cdot 10^{-4}$	$5.26 \cdot 10^{-5}$	1.00	$31.35 \cdot 10^{-5}$	4.5	$45.59 \cdot 10^{-5}$
$1.00 \cdot 10^{-3}$	$5.76 \cdot 10^{-5}$				

№ 4672 [2409]

POTASSIUM SULFITOAMIDE–POTASSIUM SULFATE–WATER

$KSO_3NH_2—K_2SO_4—H_2O$

$t=25$

Concn. in aqueous soln., wt %		d	Comp. of solid phase
Potassium Sulfito Amide	Potassium Sulfate		
43.72	0.00	1.327	KSO_3NH_2
42.03	2.17	1.335	$KSO_3NH_2+K_2SO_4$
0.00	10.75	1.084	K_2SO_4

№ 4673 POTASSIUM SULFATE–POTASSIUM PEROXODISULFATE–WATER [1127]

$K_2SO_4—K_2S_2O_8—H_2O$

Comp. of saturated soln., wt %			t	Comp. of solid phase
Potassium Peroxodisulfate	Potassium Sulfate	Water		
0.000	9.206	90.794	15	K_2SO_4
0.488	9.050	90.462	15	»
0.810	8.970	90.220	15	»
1.000	8.900	90.100	15	$K_2SO_4+K_2S_2O_8$
1.050	8.450	90.500	15	$K_2S_2O_8$
1.320	5.537	93.143	15	»
2.380	2.318	95.302	15	»
3.645	0.000	96.355	15	»
0.000	10.020	89.080	20	K_2SO_4
0.453	9.862	89.685	20	»
0.784	9.820	89.396	20	»
0.995	9.740	83.265	20	»
1.197	9.690	89.113	20	»
1.323	9.640	89.037	20	»

Table № 4673 (continuation)

Comp. of saturated soln., wt %			t	Comp. of solid phase
Potassium Peroxo-disulfate	Potassium Sulfate	Water		
1.373	9.630	88.997	20	$K_2SO_4+K_2S_2O_8$
1.570	7.300	91.130	20	$K_2S_2O_8$
1.584	6.955	91.461	20	»
1.750	6.090	92.160	20	»
2.080	4.700	93.220	20	»
2.240	4.340	93.420	20	»
2.242	4.500	93.258	20	»
2.370	3.960	93.670	20	»
3.690	1.340	94.97	20	»
3.780	1.208	95.012	20	»
4.604	0.000	95.396	25	»
0.000	10.750	89.250	20	K_2SO_4
0.468	10.630	88.902	25	»
0.982	10.460	88.558	25	»
1.281	10.310	88.409	25	»
1.940	10.085	87.975	25	$K_2SO_4+K_2S_2O_8$
2.250	7.400	90.350	25	$K_2S_2O_8$
2.924	4.845	92.231	25	»
2.980	4.730	92.290	25	»
4.180	2.150	93.670	25	»
5.730	0.000	94.270	25	»
0.000	11.481	88.519	30	K_2SO_4
0.3066	11.4060	88.2874	30	»
1.2775	11.1050	87.6175	30	»
2.535	10.704	86.761	30	$K_2SO_4+K_2S_2O_8$
3.080	7.700	89.220	30	$K_2S_2O_8$
3.982	5.100	90.918	30	»
5.436	2.210	92.354	30	»
7.0751	0.000	92.9249	30	»

№ 4674 POTASSIUM SULFATE–POTASSIUM CHLORIDE–WATER [2353]

K_2SO_4—KCl—H_2O

Comp. of saturated soln., wt %			t	Comp. of saturated soln., wt %			t
Potassium Chloride	Potassium Sulfate	Water		Potassium Chloride	Potassium Sulfate	Water	
23.37	0.99	75.64	10	27.56	1.20	71.24	40
21.48	1.76	76.76	15.8	28.85	1.27	69.88	50
24.77	1.06	74.17	20	30.06	1.33	68.61	60
25.24	2.13	72.63	25	32.49	1.47	66.04	80
26.22	1.14	72.64	30	34.70	1.67	63.63	100

№ 4675 [1238]

POTASSIUM SULFATE–POTASSIUM CHLORIDE–WATER

$K_2SO_4—KCl—H_2O$

$t=12.5$

Concn. of Potassium Chloride in aqueous soln., g/l	Solubility of Potassium Sulfate in aqueous soln. of Potassium Chloride, g/l	Concn. of Potassium Chloride in aqueous soln., g/l	Solubility of Potassium Sulfate in aqueous soln of Potassium Chloride, g/l
0	99	80	46
20	83	100	35
40	70		
60	57		

№ 4676 [2517]

POTASSIUM SULFATE–POTASSIUM CHLORIDE–WATER

$K_2SO_4—KCl—H_2O$

$t=30$

Comp. of saturated soln., wt %		
Potassium Chloride	Potassium Sulfate	Water
26.20	1.09	72.71

№ 4677 [1240]

POTASSIUM SULFATE–POTASSIUM CHLORIDE–WATER

$K_2SO_4—KCl—H_2O$

Content in aqueous soln., g/100 g Water		t	d	Comp. of solid phase
Potassium Chloride	Potassium Sulfate			
28.20	0	0	1.153	KCl
27.88	1.21	0	—	$KCl—K_2SO_4$
0	7.23	0	1.063	K_2SO_4
36.96	0	25	1.187	KCl
36.63	1.53	25	1.190	$KCl—K_2SO_4$
0	12.02	25	1.088	K_2SO_4
43.12	0	50	1.198	KCl
42.24	1.84	50	1.212	$KCl—K_2SO_4$
0	17.09	50	1.110	K_2SO_4
49.70	0	75	1.204	KCl
18.75	2.12	75	—	$KCl—K_2SO_4$
0	20.80	75	1.120	K_2SO_4
56.20	0	109	1.217	KCl
54.43	2.83	100	1.225	$KCl—K_2SO_4$
0	23.44	100	1.134	K_2SO_4

№ 4678 [1800]

POTASSIUM SULFATE–POTASSIUM CHLORIDE–WATER

$K_2SO_4—KCl—H_2O$

Comp. of saturated soln., wt %			t	d	Comp. of solid phase
Potassium Chloride	Potassium Sulfate	Water			
21.81	0.00	78.19	0	—	KCl
21.50	0.84	77.66	0	—	$KCl—K_2SO_4$
0.00	6.87	93.13	0	—	K_2SO_4
26.42	0.00	73.58	25	1.179	KCl
26.02	1.06	72.92	25	1.185	$KCl—K_2SO_4$

Comp. of saturated soln., wt %			t	d	Comp. of solid phase
Potassium Chloride	Potassium Sulfate	Water			
20.97	1.65	77.38	25	1.153	K_2SO_4
14.96	2.80	82.24	25	1.120	»
7.98	5.26	86.76	25	1.092	»
0.00	10.80	89.20	25	1.086	»
30.07	0.0	69.93	50	1.194	KCl
29.40	1.24	69.36	50	1.200	$KCl+K_2SO_4$
0.00	14.16	85.84	50	1.104	K_2SO_4
33.16	0.0	66.84	75	1.203	KCl
32.33	1.42	66.25	75	1.211	$KCl+K_2SO_4$
25.97	2.47	71.56	75	1.171	K_2SO_4
18.67	4.47	76.86	75	1.135	»
9.99	8.77	81.24	75	1.110	»
0.00	17.09	82.91	75	1.116	»
35.70	0.0	64.30	100	1.211	KCl
34.91	1.62	63.47	100	1.216	$KCl+K_2SO_4$
0.00	19.36	80.64	100	1.119	K_2SO_4

№ 4679 [2637]

POTASSIUM SULFATE–POTASSIUM CHLORIDE–WATER

$KCl—K_2SO_4—H_2O$

Content in aqueous soln., g/100 g Water		t	Comp. of solid phase
Potassium Chloride	Potassium Sulfate		
39.0	0	35	KCl
38.4	1.7	35	$KCl+K_2SO_4$
0	14.1	35	K_2SO_4
43.1	0	50	KCl
42.2	1.8	50	$KCl+K_2SO_4$
0	17.1	50	K_2SO_4
49.7	0	75	KCl
48.5	1.9	75	$KCl+K_2SO_4$
0	20.6	75	K_2SO_4

№ 4680 [1335]

POTASSIUM SULFATE–POTASSIUM CHLORIDE–WATER

$K_2SO_4—KCl—H_2O$

$t=100$

Comp. of saturated soln., wt %			Comp. of solid phase
Potassium Chloride	Potassium Sulfate	Water	
35.60	0.0	64.40	KCl
34.64	1.56	63.80	»
35.96	1.63	62.41	$KCl+K_2SO_4$
28.58	2.46	68.96	K_2SO_4

Table № 4680 (continuation)

Comp. of saturated soln., wt %			Comp. of solid phase
Potassium Chloride	Potassium Sulfate	Water	
18.63	5.54	75.83	K_2SO_4
8.58	11.00	80.42	»
0.0	18.76	81.24	»

№ 4681 POTASSIUM SULFATE–POTASSIUM CHLORIDE–WATER [454]

K_2SO_4—KCl—H_2O

$t=25$

Content in aqueous soln., mol/1000 mol Water		Comp. of solid phase	Content in aqueous soln., mol/1000 mol Water		Comp. of solid phase
2KCl	K_2SO_4		2KCl	K_2SO_4	
43.8	0	KCl	12.2	5.9	K_2SO_4
42.6	1.6	KCl+K_2SO_4	0	12.3	»
27.1	2.7	K_2SO_4			

№ 4682 POTASSIUM SULFATE–POTASSIUM CHLORIDE–WATER [29]

K_2SO_4—KCl—H_2O

Content in aqueous soln., wt %		t	d	Comp. of solid phase
Potassium Chloride	Potassium Sulfate			
21.9	0	0	1.154	KCl
21.6	0.93	0	1.163	KCl+K_2SO_4
8.91	3.10	0	1.113	K_2SO_4
5.24	4.02	0	1.087	»
0.86	6.06	0	1.068	»
0	6.78	0	1.064	»
26.6	0	25	1.187	KCl
25.7	1.11	25	1.191	KCl+K_2SO_4
24.6	1.21	25	1.179	K_2SO_4
16.52	2.49	25	1.140	»
0	10.74	25	1.087	»
27.2	0	30	1.189	KCl
26.9	0.35	30	1.190	»
26.22	1.13	30	1.193	KCl+K_2SO_4
24.4	1.61	30	1.182	K_2SO_4
19.55	2.39	30	1.166	»
14.32	4.18	30	1.121	»
8.59	6.41	30	1.105	»
4.73	7.92	30	1.095	»
2.83	9.10	30	1.093	»
0	11.48	30	1.089	»
30.0	0	50	1.198	KCl
28.8	1.27	50	1.204	KCl+K_2SO_4

Table № 4682 (continuation)

Content in aqueous soln., wt %		t	d	Comp. of solid phase
Potassium Chloride	Potassium Sulfate			
18.70	3.53	50	1.143	K_2SO_4
12.80	6.00	50	1.123	»
9.76	7.39	50	1.120	»
2.84	11.89	50	1.115	»
0	14.17	50	1.112	»
32.6	0	70	1.212	KCl
31.3	1.39	70	1.223	$KCl+K_2SO_4$
30.0	1.81	70	1.206	K_2SO_4
22.19	4.09	70	1.160	»
13.40	7.80	70	1.130	»
8.89	10.08	70	1.125	»
3.83	13.65	70	1.121	»
0	16.60	70	1.120	»
36.1	0	100	—	KCl
34.77	1.64	100	—	$KCl+K_2SO_4$
34.4	1.81	100	—	K_2SO_4
26.7	3.77	100	—	»
22.1	5.22	100	—	»
15.7	8.22	100	—	»
9.47	12.26	100	—	»
0	19.4	100	—	»

№ 4683 [1517]

POTASSIUM SULFATE–POTASSIUM CHLORIDE–WATER

$K_2SO_4—KCl—H_2O$

$t=20$

Content in aqueous soln., wt %		d_4	Comp. of solid phase
Potassium Chloride	Potassium Sulfate		
0.00	10.00	1.081	K_2SO_4
6.24	5.06	1.085	»
10.93	4.32	1.100	»
16.20	28.02	1.129	»
22.68	1.52	1.171	»
25.05	2.31	1.181	K_2SO_4+KCl
25.08	1.15	1.179	KCl
25.59	0.0	1.173	»

№ 4684 POTASSIUM SULFATE–POTASSIUM CHLORIDE–WATER [485]

$K_2SO_4—KCl—H_2O$

$t=35$

Content in aqueous soln., wt %		Comp. of solid phase	Content in aqueous soln., wt %		Comp. of solid phase
Potassium Chloride	Potassium Sulfate		Potassium Chloride	Potassium Sulfate	
28.09	0	KCl	9.00	3.53	K_2SO_4
27.55	1.17	$KCl+K_2SO_4$	0	12.30	»
16.73	3.16	K_2SO_4			

№ 4685 POTASSIUM SULFATE–POTASSIUM CHLORIDE–WATER [733]

$K_2SO_4—KCl—H_2O$

Comp. of saturated soln., wt %			t	Comp. of solid phase
Potassium Chloride	Potassium Sulfate	Water		
40.0	1.0	50.0	145	KCl
44.5	1.2	54.3	206	»
48.9	1.2	49.9	248	»
36.2	2.0	61.8	122	K_2SO_4
43.4	2.2	54.4	197	»
50.6	2.7	48.7	242	»
51.0	2.8	46.2	280	KCl
32.0	4.5	64.5	147	K_2SO_4
35.9	4.0	60.1	195	»
43.0	5.7	51.3	260	»
29.8	5.2	65.0	179	»
35.3	6.2	58.5	228	»
41.0	7.2	51.8	283	»
45.8	8.0	36.2	374	»
23.2	7.8	69.0	185	»
29.1	9.7	61.7	268	»
39.6	13.1	47.3	366	»
46.0	15.3	38.7	407	»
18.2	9.8	72.0	187	»
22.7	12.2	65.1	286	»
29.4	15.8	54.8	355	»
33.3	18.0	48.7	380	»
12.4	12.4	75.2	162	»
15.6	15.6	68.8	280	»
19.1	19.2	61.7	345	»
11.9	12.0	56.1	393	»
8.9	14.7	76.4	168	»
9.5	15.7	74.8	208	»
9.9	16.5	73.6	240	»
12.5	20.8	66.7	342	»
13.4	22.2	64.4	376	»
0.00	26.0	74.0	215	»
0.00	26.2	73.8	256	»
0 00	25.5	74.5	306	»
0.00	20.0	80.0	317	»

№ 4686 POTASSIUM SULFATE–POTASSIUM CHLORIDE–WATER [269]

$K_2SO_4—KCl—H_2O$

$t=25$

Comp. of saturated soln., wt %			Comp. of solid phase
Potassium Chloride	Potassium Sulfate	Water	
0.00	10.77	89.23	K_2SO_4
4.61	7.08	88.31	»
10.62	4.15	85.23	»
26.13	1.05	72.82	K_2SO_4+KCl
26.40	0.59	73.01	KCl
26.68	0.00	73.32	»

№ 4687 POTASSIUM SULFATE–POTASSIUM CHLORIDE–WATER [998]

$K_2SO_4—KCl—H_2O$

Comp. of saturated soln., wt %			t	Comp. of solid phase
Potassium Chloride	Potassium Sulfate	Water		
7.990	0.219	91.791	− 3.65	Ice
11.990	0.299	87.711	− 5.80	»
15.990	0.399	83.611	− 8.4	»
18.990	0.410	80.600	−10.5	»
19.250	0.420	80.330	−10.6	»
19.350	0.425	80.225	−10.7	$Ice+KCl \cdot nH_2O$
19.500	0.430	80.070	−10.4	$KCl \cdot nH_2O$
19.700	0.431	79.869	− 9.8	»
20.000	0.432	79.568	− 9.3	»
20.200	0.435	79.365	− 9.1	$KCl \cdot nH_2O+KCl$
20.500	0.440	79.060	− 7.6	KCl
21.000	0.450	78.550	− 5.6	»
21.500	0.460	78.040	− 3.0	»
22.500	0.475	77.025	1.8	»
23.100	0.485	46.415	6.0	»
23.990	0.500	75.510	11.8	»
24.480	0.505	75.015	14.8	»
25.500	0.507	73.993	20.0	»
2.990	0.090	96.920	− 1.25	Ice
10.00	0.50	89.50	− 5.0	»
15.00	0.70	84.30	− 7.6	»
18.50	0.80	80.70	−10.4	»
18.90	0.81	80.29	−10.8	$Ice+KCl \cdot nH_2O$
19.20	0.85	79.95	− 9.8	$KCl \cdot nH_2O$
19.15	0.88	79.62	− 9.3	»
19.800	0.885	79.315	− 9.2	$KCl \cdot nH_2O+KCl$
20.00	0.89	79.11	− 8.0	KCl
20.40	0.90	78.70	− 6.7	»
20.99	0.91	78.10	− 2.5	»
21.50	0.94	77.56	0.0	»
22.46	0.98	76.56	5.2	»
23.46	1.00	75.54	10.5	»
24.50	1.036	74.464	16.0	»
25.50	1.05	73.45	21.5	»

Table № 4687 (continuation)

Comp. of saturated soln., wt %			t	Comp. of solid phase
Potassium Chloride	Potassium Sulfate	Water		
7.990	0.549	91.461	— 3.7	Ice
9.98	0.649	89.371	— 4.9	»
15.98	0.99	83.03	— 8.6	»
17.990	1.079	80.931	—10.0	»
18.490	1.099	80.411	—10.2	Ice+$K_2SO_4 \cdot H_2O$
18.69	2.59	78.72	— 7.6	$K_2SO_4 \cdot H_2O$
18.990	1.149	79.861	— 4.5	»
19.290	1.187	79.623	— 1.25	»
19.600	1.193	79.207	2.5	»
19.800	1.196	79.004	4.0	»
19.990	1.209	78.801	4.5	»
20.190	1.229	78.581	5.4	$K_2SO_4 \cdot H_2O+K_2SO_4$
20.30	1.23	78.47	6.1	K_2SO_4
20.400	1.231	78.369	6.6	»
20.990	1.259	77.751	10.0	»
21.990	1.349	76.661	16.05	»
22.980	1.390	75.630	20.6	»
3.990	0.299	95.711	— 1.65	Ice
7.990	0.699	91.311	— 3.75	»
11.990	0.990	87.020	— 6.0	»
15.990	1.379	82.631	— 8.7	»
16.600	1.449	81.951	— 9.4	Ice+$K_2SO_4 \cdot H_2O$
16.990	1.489	81.512	— 2.25	$K_2SO_4 \cdot H_2O$
17.190	1.490	81.320	— 0.75	»
17.490	1.499	81.011	2.25	»
17.990	1.510	80.500	5.5	»
18.180	1.520	80.300	6.6	»
18.20	1.53	80.17	7.4	$K_2SO_4 \cdot H_2O+K_2SO_4$
18.490	1.548	79.962	8.1	»
18.990	1.668	79.342	10.25	»
19.480	1.669	78.851	13.3	»
19.990	1.700	78.310	16.25	»
20.990	1.779	77.231	22.35	»
10.99	1.70	87.31	— 6.0	Ice
12.49	1.97	85.54	— 6.75	»
12.69	2.09	85.22	— 7.0	Ice+$K_2SO_4 \cdot H_2O$
12.740	2.095	85.165	— 2.5	$K_2SO_4 \cdot H_2O$
12.810	2.100	85.090	3.0	»
12.99	2.11	84.90	5.5	»
13.70	2.12	84.18	7.8	$K_2SO_4 \cdot H_2O+K_2SO_4$
14.00	2.25	83.75	10.0	K_2SO_4
14.69	2.33	82.98	14.5	»
15.00	2.41	82.59	18.3	»
16.00	2.60	81.40	23.8	»
2.99	1.09	95.92	— 1.6	Ice
4.99	1.79	93.22	— 2.55	»
6.99	2.49	90.52	— 3.85	»
7.99	2.99	89.02	— 4.5	Ice+$K_2SO_4 \cdot H_2O$
8.15	3.06	88.79	0	$K_2SO_4 \cdot H_2O$
8.49	3.19	88.32	3.9	»
8.78	3.29	87.93	5.75	»
8.99	3.39	87.62	7.25	»

Table № 4687 (continuation)

Comp. of saturated soln., wt %			t	Comp. of solid phase
Potassium Chloride	Potassium Sulfate	Water		
9.18	3.49	87.33	8.0	$K_2SO_4 \cdot H_2O + K_2SO_4$
9.45	3.60	86.95	9.2	K_2SO_4
9.69	3.65	86.66	12.5	»
10.49	3.99	85.52	16.8	»
10.99	4.19	84.82	20.5	»
1.99	3.19	98.82	— 1.25	Ice
2.69	4.49	92.82	— 2.0	»
3.19	5.19	91.62	— 2.3	Ice $+ K_2SO_4 \cdot H_2O$
3.25	5.35	91.40	1.0	$K_2SO_4 \cdot H_2O$
3.29	5.49	91.22	5.7	»
3.49	5.69	90.82	7.75	»
3.59	5.89	90.52	8.3	$K_2SO_4 \cdot H_2O + K_2SO_4$
3.60	5.99	90.41	9.0	K_2SO_4
3.61	6.10	90.29	11.0	»
3.99	6.59	89.42	15.75	»
4.49	7.39	88.12	21.75	»
5.00	8.25	86.75	24.5	»
19.70	0.00	80.30	—10.6	Ice $+ KCl \cdot nH_2O$
19.350	0.425	80.225	—10.7	»
18.90	0.81	80.29	—10.8	»
18.65	1.15	80.20	—11.0	Ice $+ K_2SO_4 \cdot nH_2O + K_2SO_4 \cdot H_2O$
18.490	1.099	80.411	—10.2	Ice $+ K_2SO_4 \cdot H_2O$
16.600	1.449	81.951	— 9.4	»
12.690	2.090	85.520	— 7.0	»
7.99	2.99	89.02	— 4.5	»
3.19	5.19	91.62	— 2.3	Ice $+ K_2SO_4 \cdot H_2O$
0.00	7.09	92.91	— 1.8	»
20.68	0.00	79.32	— 6.6	$KCl + KCl \cdot nH_2O$
20.200	0.435	79.365	— 9.1	»
19.800	0.885	79.315	— 9.2	»
19.50	1.12	79.38	—10.0	$KCl + KCl \cdot nH_2O + K_2SO_4 \cdot H_2O$
0.00	8.48	91.52	9.7	$K_2SO_4 \cdot H_2O + K_2SO_4$
3.59	5.89	90.52	8.3	»
9.18	3.49	87.33	8.0	»
13.70	2.12	84.88	7.8	»
18.30	1.53	80.17	7.4	»
20.190	1.229	78.581	5.4	»
21.80	1.10	77.10	1.6	$K_2SO_4 \cdot H_2O + K_2SO_4 + KCl$
15.80	0.00	84.20	— 8	Ice
15.50	0.40	84.10	— 8	»
14.90	1.00	84.10	— 8	»
14.65	1.30	84.05	— 8	»
12.70	0.00	87.30	— 6	»
12.50	0.30	87.20	— 6	»
11.60	1.10	87.30	— 6	»
11.00	1.80	87.20	— 6	»
8.80	0.00	91.20	— 4	»
8.20	0.80	91.00	— 4	»
7.60	1.30	91.10	— 4	»
6.90	2.70	90.40	— 4	»
4.7	0.0	95.3	— 2	»
4.1	0.7	90.2	— 2	»

Table № 4687 (continuation)

Comp. of saturated soln., wt %			t	Comp. of solid phase
Potassium Chloride	Potassium Sulfate	Water		
3.8	1.4	94.8	— 2	Ice
2.7	4.5	92.8	— 2	»
21.8	0.0	78.2	0	KCl
21.7	0.5	77.8	0	»
21.6	0.8	77.6	0	»
21.5	0.9	77.6	0	$KCl+K_2SO_4 \cdot H_2O$
19.3	1.2	79.5	0	$K_2SO_4 \cdot H_2O$
17.3	1.4	81.3	0	»
12.9	2.1	85.0	0	»
8.4	3.2	88.4	0	»
3.3	5.5	91.2	0	»
0.0	7.2	92.8	0	»
24.6	0.0	75.4	15	KCl
24.3	0.8	74.9	15	»
24.2	1.1	74.7	15	$KCl+K_2SO_4$
21.7	1.3	77.0	15	K_2SO_4
19.6	1.7	78.7	15	»
14.7	2.5	82.8	15	»
10.1	3.8	86.1	15	»
3.9	6.5	89.6	15	»
0.0	9.0	91.0	15	»
25.5	0.0	74.5	20	KCl
25.4	0.5	74.1	20	»
25.2	0.8	74.0	20	»
25.10	1.15	73.75	20	$KCl+K_2SO_4$
22.80	1.45	75.75	20	K_2SO_4
20.5	1.8	77.7	20	»
15.5	2.5	82.0	20	»
11.0	4.2	84.8	20	»
4.3	7.2	88.5	20	»
0.0	9.7	90.3	20	»

№ 4688 POTASSIUM SULFATE–POTASSIUM CHLORIDE–WATER [1575]

$K_2SO_4—KCl—H_2O$

$t=25$

Content of Potassium Chloride in aqueous soln., wt %	Solubility of Potassium Sulfate in aqueous soln. of Potassium Chloride wt %	Content of Potassium Chloride in aqueous soln., wt %	Solubility of Potassium Sulfate in aqueous soln. of Potassium Chloride wt %
4.90	6.94	13.75	3.04
9.35	4.59	18.62	1.99

№ 4689 [1575]

POTASSIUM SULFATE–POTASSIUM CHLORIDE–POTASSIUM BROMATE–WATER

$K_2SO_4—KCl—KBrO_3—H_2O$

$t=25$

Content in aqueous soln., wt %		Solubility of Potassium Sulfate in aqueous soln. of Potassium Chloride and Potassium Bromate, wt %	Content in aqueous soln., wt %		Solubility of Potassium Sulfate in aqueous soln. of Potassium Chloride and Potassium Bromate, wt %
Potassium Chloride	Potassium Bromate		Potassium Chloride	Potassium Bromate	
4.57	0.94	6.86	8.07	1.48	4.92
4.21	2.39	6.80			
8.72	0.88	4.72	12.18	0.82	3.39

№ 4690 [1452, 1465]

POTASSIUM SULFATE–POTASSIUM CHLORIDE–CALCIUM SULFATE–CALCIUM CHLORIDE–WATER

$K_2SO_4–KCl–CaSO_4–CaCl_2–H_2O$

Content in aqueous soln., mol/1000 mol Water				t	Comp. of solid phase
2KCl	K_2SO_4	$CaCl_2$	$CaSO_4$		
36.56	0	9.26	0.230	25	$KCl+K_2SO_4 \cdot CaSO_4 \cdot H_2O+CaSO_4 \cdot 2H_2O$
11.33	0	0.655	0.706	25	$K_2SO_4 \cdot CaSO_4 \cdot H_2O+CaSO_4 \cdot 2H_2O$
3.015	1.60	0	0.327	25	»
0.253	3.05	0	0.232	25	»
0	3.223	0	0.223	25	»
16.75	0	30.6	0.17	55	$KCl+K_2SO_4 \cdot 5CaSO_4 \cdot H_2O+CaSO_4 \cdot 2H_2O$

№ 4691 [660]

POTASSIUM SULFATE–POTASSIUM CHLORIDE–CALCIUM SULFATE–CALCIUM CHLORIDE–WATER

$K_2SO_4—KCl—CaSO_4—CaCl_2—H_2O$

Content in aqueous soln., wt %					t
K^+	Ca^{2+}	Cl^-	SO_4^{2-}	H_2O	
10.65	0.7298	10.99	0.1007	77.52	0
10.07	0.7181	10.29	0.1672	78.74	0
10.12	0.634	10.30	0.1052	78.84	0
13.52	0.4386	12.98	0.1225	73.94	25
13.31	0.4317	12.79	0.1119	73.35	25
14.50	0.577	14.35	0.09665	70.39	50
14.90	0.4879	14.34	0.1082	70.11	50
16.72	0.407	16.57	0.1181	66.82	82
18.30	0.293	17.13	0.18605	63.99	100
18.45	0.2813	17.06	0.0923	64.11	100
21.79	0.1576	20.04	0.0757	57.93	175

№ 4692 [782]

POTASSIUM SULFATE–POTASSIUM CHLORIDE–CALCIUM SULFATE–CALCIUM CHLORIDE–WATER

K_2SO_4—KCl—$CaSO_4$—$CaCl_2$—H_2O

$t=25$

Content in aqueous soln., g/100 g Water				Comp. of solid phase
Potassium Chloride	Potassium Sulfate	Calcium Chloride	Calcium Sulfate	
35.60	1.02	0.0	0.04	$KCl+K_2SO_4+K_2SO_4 \cdot CaSO_4 \cdot H_2O$
33.88	0.0	0.83	0.17	$KCl+K_2SO_4 \cdot CaSO_4 \cdot H_2O$
28.99	0.0	4.21	0.13	T »
19.46	0.0	12.31	0.05	$KCl+CaSO_4 \cdot 2H_2O$
17.20	0.0	14.29	0.04	»
18.30	2.68	0.0	0.025	$K_2SO_4+K_2SO_4 \cdot CaSO_4 \cdot H_2O$
8.64	5.47	0.0	0.020	»
0.075	0.0	0.0	0.226	$CaSO_4 \cdot 2H_2O$
0.150	0.0	0.0	0.240	»
0.604	0.0	0.0	0.323	»
0.907	0.0	0.0	0.380	»

№ 4693 [783

POTASSIUM SULFATE–POTASSIUM CHLORIDE–CALCIUM SULFATE–CALCIUM CHLORIDE–WATER

K_2SO_4—KCl—$CaSO_4$—$CaCl_2$—H_2O

$t=55$

Content in aqueous soln., wt %				Comp. of solid phase
K^+	Ca^{2+}	Cl^-	SO_4^{2-}	
16.40	0	14.36	0.71	$KCl+K_2SO_4$
15.71	0.007	14.04	0.71	$KCl+K_2SO_4+K_2SO_4 \cdot CaSO_4 \cdot H_2O$
11.69	0.02	10.09	0.27	$KCl+K_2SO_4 \cdot CaSO_4 \cdot H_2O$
16.08	0.45	15.81	0.11	$KCl+CaSO_4$
9.05	4.60	16.00	0.09	»
4.24	18.09	36.71	0.005	$KCl+KCl \cdot CaCl_2+CaSO_4$
6.76	0.005	0	8.32	$K_2SO_4+K_2SO_4 \cdot CaSO_4 \cdot H_2O$
8.81	0.04	5.48	3.52	»
0	0.01	8.35	2.84	»
2.69	0.03	0	3.25	$K_2SO_4 \cdot CaSO_4 \cdot H_2O+K_2SO_4 \cdot 5CaSO_4 \cdot H_2O$
5.44	0.06	4.55	0.81	»
5.65	0.09	4.82	0.65	»
3.21	0.03	1.16	2.37	$K_2SO_4 \cdot CaSO_4 \cdot H_2O$
8.25	0.06	5.38	2.94	»
2.34	20.37	37.90	0	$KCl \cdot CaCl_2+CaCl_2 \cdot 2H_2O$
1.98	19.98	37.15	0.005	$KCl \cdot CaCl_2+CaCl_2 \cdot 2H_2O+CaSO_4$
1.93	0.03	0	2.37	$K_2SO_4 \cdot 3CaSO_4 \cdot H_2O+CaSO_4$
2.65	0.04	1.18	1.77	»
3.19	0.05	1.93	1.42	»
2.50	0.04	0.79	2.10	$K_2SO_4 \cdot 5CaSO_4 \cdot H_2O$

POTASSIUM SULFATE–POTASSIUM CHLORIDE–ZINC SULFATE–ZINC CHLORIDE–WATER

$K_2SO_4—KCl—ZnSO_4—ZnCl_2—H_2O$

Comp. of saturated soln., mol %					Comp. of solid phase
Zinc Sulfate	Potassium Sulfate	Zinc Chloride	Potassium Chloride	Water	
				t=25	
6	0.0	0.0	0.0	94	$ZnSO_4 \cdot 7H_2O$
0.0	0.0	31.61	0.0	68.39	$ZnCl_2 \cdot 1.5H_2O$
0.0	0.0	0.0	4.141	95.859	KCl
0.0	1.234	0.0	0.0	98.766	K_2SO_4
6.164	0.29	0.0	0.0	93.549	$K_2SO_4 \cdot ZnSO_4 \cdot 6H_2O + ZnSO_4 \cdot 6H_2O$
1.326	0.0	8.841	0.0	89.833	$ZnSO_4 \cdot 7H_2O + ZnSO_4 \cdot 6H_2O$
0.822	0.0	11.7	0.0	76.478	$ZnSO_4 \cdot 6H_2O + ZnSO_4 \cdot H_2O$
0.0388	0.0	35.4	0.0	64.561	$ZnSO_4 \cdot H_2O + ZnCl_2 \cdot 1.5H_2O$
0.0	0.0	35.9	2.39	61.71	$ZnCl_2 + K_2Cl_2 + KZnCl_3 \cdot H_2O$
0.0	0.0	35.6	0.039	64.36	$ZnCl_2 \cdot 1.5H_2O + ZnCl_2 \cdot H_2O$
0.0	0.0	35.9	1.06	63.04	$ZnCl_2 \cdot H_2O + ZnCl_2 \cdot K_2Cl_2$
0.0	0.0	29.75	1.656	68.59	$KZnCl_3 \cdot H_2O + KZnCl_3 \cdot 2H_2O$
0.0	0.0	15.75	11.23	73.02	$KZnCl_3 \cdot 2H_2O + K_2ZnCl_4$
0.0	0.0	14.09	13.71	72.2	$K_2ZnCl_4 + KCl$
0.0	0.1435	0.0	4.1	95.756	$KCl + K_2SO_4$
0.26	12.6	0.0	0.0	97.4	$K_2SO_4 + K_2SO_4 \cdot ZnSO_4 \cdot 6H_2O$
1.22	0.0	8.31	0.75	89.72	$ZnSO_4 \cdot 7H_2O + ZnSO_4 \cdot 6H_2O + K_2SO_4 \cdot ZnSO_4 \cdot 6H_2O$
1.267	0.0	9.55	0.955	88.228	$ZnSO_4 \cdot 6H_2O + ZnSO_4 \cdot H_2O + K_2SO_4 \cdot ZnSO_4 \cdot 6H_2O$
0.916	0.0	12.12	2.503	84.461	$ZnSO_4 \cdot H_2O + K_2SO_4 \cdot ZnSO_4 \cdot 6H_2O + K_2SO_4 \cdot 2ZnCl_3 \cdot 5H_2O$
0.701	0.0	13.11	3.68	82.509	$ZnSO_4 \cdot H_2O + K_2SO_4 \cdot ZnSO_4 \cdot 6H_2O + KZnCl_3 \cdot 2H_2O$
0.346	0.0	13.58	13.87	72.204	$KCl + K_2SO_4 + K_2ZnCl_4$
0.397	0.0	14.15	13.14	72.313	$K_2SO_4 + K_2SO_4 \cdot ZnSO_4 \cdot 6H_2O + K_2ZnCl_4$
0.396	0.0	15.6	11.77	72.234	$K_2SO_4 \cdot ZnSO_4 \cdot 6H_2O + K_2ZnCl_4 + KZnCl_4 \cdot 2H_2O$
0.658	0.0	13.16	4.025	82.157	$K_2SO_4 \cdot ZnSO_4 \cdot 6H_2O + K_2SO_4 \cdot 2ZnCl_2 \cdot 5H_2O + KZnCl_3 \cdot 2H_2O$
0.821	0.0	13.27	3.102	82.807	$ZnSO_4 \cdot H_2O + K_2SO_4 \cdot 2ZnCl_2 \cdot 5H_2O + KZnCl_3 \cdot 2H_2O$
0.055	0.0	29.73	1.645	68.57	$ZnSO_4 \cdot H_2O + KZnCl_3 \cdot 2H_2O + KZnCl_3 \cdot H_2O$
0.092	0.0	36.35	2.282	61.276	$ZnSO_4 \cdot H_2O + KZnCl_3 \cdot H_2O + KZnCl_3$
0.063	0.0	35.9	1.071	62.966	$ZnSO_4 \cdot H_2O + KZnCl_3 + ZnCl_2 \cdot H_2O$
0.044	0.0	35.6	0.415	63.941	$ZnSO_4 \cdot H_2O + ZnCl_2 \cdot H_2O + ZnCl_2 \cdot 1.5H_2O$
4.38	0.0	1.907	0.0	93.713	$ZnSO_4 \cdot 7H_2O$
2.611	0.0	4.84	0.0	92.549	»
1.48	0.0	7.825	0.0	90.695	»
0.405	0.0	15.25	0.0	84.345	$ZnSO_4 \cdot H_2O$
0.071	0.0	21.37	0.0	78.569	»
0.045	0.0	25.43	0.0	74.525	»

Table № 4694 (continuation)

Comp. of saturated soln., mol %					Comp. of solid phase
Zinc Sulfate	Potassium Sulfate	Zinc Chloride	Potassium Chloride	Water	
0.0	0.0	25.13	1.09	73.78	$KZnCl_3 \cdot 2H_2O$
0.0	0.0	21.89	1.009	77.101	»
0.0	0.0	19.36	1.01	79.63	»
0.0	0.0	16.43	1.438	82.132	»
0.0	0.0	14.19	2.25	83.56	»
0.0	0.0	13.69	3.083	83.227	»
0.0	0.0	13.30	4.76	81.94	»
0.0	0.0	14.302	7.005	78.093	»
0.0	0.0	20.805	3.698	75.497	K_2ZnCl_4
0.0	0.0	18.06	7.85	74.09	»
0.0	0.0	14.79	12.74	82.47	»
0.0	0.0	11.95	11.8	76.25	KCl
0.0	0.0	9.525	9.83	70.645	»
0.0	0.0	5.715	7.145	87.14	»
0.0	0.0	2.61	5.48	91.91	»
0.0	0.241	0.0	2.904	96.855	K_2SO_4
0.0	0.66	0.0	0.985	98.355	»
0.601	0.817	0.0	0.0	98.582	$K_2SO_4 \cdot ZnSO_4 \cdot 6H_2O$
1.592	0.498	0.0	0.0	97.91	»
4.27	0.363	0.0	0.0	95.367	»
4.557	0.0	2.11	0.279	93.051	$ZnSO_4 \cdot 7H_2O + K_2SO_4 \cdot ZnSO_4 \cdot 6H_2O$
2.075	0.0	6.47	0.5	90.954	»
1.362	0.0	9.3	0.866	88.472	»
1.148	0.0	10.685	0.56	87.607	$ZnSO_4 \cdot 6H_2O + ZnSO_4 \cdot H_2O$
1.021	0.0	11.17	1.772	86.037	$ZnSO_4 \cdot H_2O + K_2SO_4 \cdot ZnSO_4 \cdot 6H_2O$
0.723	0.0	12.8	3.313	83.164	»
0.141	0.0	1.12	4.82	93.919	KCl
0.134	0.0	4.22	6.425	89.221	»
0.141	0.0	7.86	8.89	83.109	»
0.215	0.0	11.3	11.6	76.885	»
0.42	0.283	0.0	1.65	97.647	$K_2SO_4 + K_2SO_4 \cdot ZnSO_4 \cdot 6H_2O$
0.409	0.0	1.425	3.14	95.026	»
0.323	0.0	3.14	4.27	92.267	»
0.277	0.0	4.895	5.49	89.338	»
0.298	0.0	7.33	7.44	84.932	»
0 314	0.0	11.9	10.43	77.356	»
0.343	0.0	14.5	8.98	76.167	$K_2SO_4 \cdot ZnSO_4 \cdot 6H_2O + KZnCl_3 \cdot 2H_2O$
0.511	0.0	13.21	5.13	81.149	»
0.758	0.0	12.63	3.339	83.273	$K_2SO_4 \cdot ZnSO_4 \cdot 6H_2O + K_2SO_4 \cdot 2ZnCl_2 \cdot 5H_2O$
0.814	0.0	13.105	3.047	83.034	$ZnSO_4 \cdot H_2O + K_2SO_4 \cdot 2ZnCl_2 \cdot 5H_2O$
0.407	0.0	14.3	2.125	83.168	$ZnSO_4 \cdot H_2O + KZnCl_3 \cdot 2H_2O$
0.065	0.0	17.99	1.029	80.916	»
0.069	0.0	22.06	1.015	76.856	»
0.044	0.0	25.63	1.134	73.192	»
$t=35$					
6.84	0.0	0.0	0.0	93.16	$ZnSO_4 \cdot 7H_2O$
0.0	0.0	37.4	0.0	62.6	$ZnCl_2$

Table № 4694 (continuation)

Comp. of saturated soln., mol %					Comp. of solid phase
Zinc Sulfate	Potassium Sulfate	Zinc Chloride	Potassium Chloride	Water	
0.0	0.0	0.0	4.45	95.55	KCl
0.0	1.409	0.0	0.0	98.591	K_2SO_4
7.03	0.288	0.0	0.0	92.682	$ZnSO_4 \cdot 7H_2O + K_2SO_4 \cdot ZnSO_4 \cdot 6H_2O$
4.775	0.0	2.8	0.0	92.425	$ZnSO_4 \cdot 7H_2O + ZnSO_4 \cdot 6H_2O$
1.723	0.0	9.51	0.0	88.767	$ZnSO_4 \cdot 6H_2O + ZnSO_4 \cdot H_2O$
0.038	0.0	37.2	0.0	62.762	$ZnSO_4 \cdot H_2O + ZnCl_2$
0.0	0.0	38.62	3.38	58.00	$KZnCl_3 + KZnCl_3 \cdot H_2O$
0.0	0.0	34.01	2.472	63.518	$KZnCl_3 \cdot H_2O + KZnCl_3 \cdot 2H_2O$
0.0	0.0	16.9	10.825	72.275	$KZnCl_3 \cdot 2H_2O + K_2ZnCl_4$
0.0	0.0	14.51	14.38	71.11	$K_2ZnCl_4 + KCl$
0.0	0.153	0.0	4.4	95.447	$KCl + K_2SO_4$
0.344	1.957	0.0	0.0	98.199	$K_2SO_4 + K_2SO_4 \cdot ZnSO_4 \cdot 6H_2O$
5.03	0.0	2.76	0.232	91.978	$ZnSO_4 \cdot 7H_2O + ZnSO_4 \cdot 6H_2O + K_2SO_4 \cdot ZnSO_4 \cdot 6H_2O$
2.121	0.0	8.99	0.864	88.025	$ZnSO_4 \cdot 6H_2O + ZnSO_4 \cdot H_2O + K_2SO_4 \cdot ZnSO_4 \cdot 6H_2O$
1.06	0.0	12.57	3.59	82.77	$ZnSO_4 \cdot H_2O + K_2SO_4 \cdot ZnSO_4 \cdot 6H_2O + K_2SO_4 \cdot 2ZnCl_2 \cdot 5H_2O$
0.966	0.0	14.295	6.195	78.544	$ZnSO_4 \cdot H_2O + K_2SO_4 \cdot ZnSO_4 \cdot 6H_2O + KZnCl_3 \cdot 2H_2O$
0.304	0.0	14.41	14.56	70.726	$KCl + K_2SO_4 + K_2ZnCl_4$
0.692	0.0	15.02	13.83	70.458	$K_2SO_4 + K_2SO_4 \cdot ZnSO_4 \cdot 6H_2O + K_2ZnCl_4$
0.66	0.0	16.03	12.405	70.905	$K_2SO_4 \cdot ZnSO_4 \cdot 6H_2O + K_2ZnCl_4 + KZnCl_3 \cdot 2H_2O$
0.648	0.0	14.03	8.04	77.282	$K_2SO_4 \cdot ZnSO_4 \cdot 6H_2O + K_2SO_4 \cdot 3ZnCl_3 \cdot 5H_2O + KZnCl_3 \cdot 2H_2O$
0.752	0.0	14.73	4.56	79.968	$ZnSO_4 \cdot H_2O + K_2SO_4 \cdot 2ZnCl_2 \cdot 5H_2O + KZnCl_3 \cdot 2H_2O$
0.047	0.0	30.405	2.418	67.13	$ZnSO_4 \cdot H_2O + KZnCl_3 \cdot 2H_2O + KZnCl_3 \cdot H_2O$
0.052	0.0	38.41	3.38	57.158	$ZnSO_4 \cdot H_2O + KZnCl_3 \cdot H_2O + KZnCl_3$
0.728	0.0	14.92	11.50	72.852	$K_2SO_4 \cdot ZnSO_4 \cdot 6H_2O + K_2ZnCl_4 + K_2SO_4 \cdot 2ZnCl_2 \cdot 5H_2O$
5.1	0.0	2.28	0.0	92.62	$ZnSO_4 \cdot 7H_2O$
3.08	0.0	5.66	0.0	91.26	$ZnSO_4 \cdot 6H_2O$
0.694	0.0	13.56	0.0	85.746	$ZnSO_4 \cdot H_2O$
0.152	0.0	19.73	0.0	80.228	»
0.067	0.0	24.9	0.0	75.033	»
0.055	0.0	31.33	0.0	68.615	»
0.0	0.0	38.1	1.7	60.2	$KZnCl_3$
0.0	0.0	36.0	3.138	60.862	$KZnCl_3 \cdot H_2O$
0.0	0.0	24.8	1.8	74.02	$KZnCl_3 \cdot 2H_2O$
0.0	0.0	21.2	1.837	76.963	»
0.0	0.0	17.63	2.319	80.051	»
0.0	0.0	14.92	4.83	80.25	»
0.0	0.0	15.32	6.85	77.83	»
0.0	0.0	21.43	3.93	74.64	»

Table № 4694 (continuation)

Comp. of saturated soln., mol %					Comp. of solid phase
Zinc Sulfate	Potassium Sulfate	Zinc Chloride	Potassium Chloride	Water	
0.0	0.0	19.39	7.18	73.43	K_2ZnCl_4
0.0	0.0	15.76	12.57	71.67	»
0.0	0.0	11.57	11.88	76.55	»
0.0	0.0	7.62	8.75	83.63	KCl
0.0	0.0	3.4	6.1	91.5	»
0.0	0.0	3.36	9.98	80.67	»
0.0	0.29	0.0	2.9	96.81	K_2SO_4
0.0	0.797	0.0	0.984	98.219	»
0.64	1.023	0.0	0.0	98.337	$K_2SO_4 \cdot ZnSO_4 \cdot 6H_2O$
1.516	0.694	0.0	0.0	87.79	»
3.695	0.469	0.0	0.0	95.836	»
5.73	0.0	1.816	0.276	92.178	$ZnSO_4 \cdot 7H_2O + K_2SO_4 \cdot ZnSO_4 \cdot 6H_2O$
2.96	0.0	6.47	0.521	90.049	$ZnSO_4 \cdot 6H_2O + K_2SO_4 \cdot ZnSO_4 \cdot 6H_2O$
1.845	0.0	9.25	0.39	88.515	$ZnSO_4 \cdot 6H_2O + ZnSO_4 \cdot H_2O$
1.472	0.0	10.81	1.99	85.727	$ZnSO_4 \cdot H_2O + ZnSO_4 \cdot ZnSO_4 \cdot 6H_2O$
0.98	0.0	13.65	5.21	80.16	»
0.885	0.0	13.58	4.15	81.385	$ZnSO_4 \cdot H_2O + K_2SO_4 \cdot 2ZnCl_2 \cdot 5H_2O$
0.162	0.0	3.443	6.32	90.075	$K_2SO_4 + KCl$
0.166	0.0	7.7	9.07	83.064	»
0.232	0.0	10.82	11.52	77.428	»
0.763	0.0	0.260	2.405	96.572	$K_2SO_4 \cdot K_2SO_4 \cdot ZnSO_4 \cdot 6H_2O$
0.683	0.0	1.282	3.028	95.007	»
0.521	0.0	2.55	3.77	93.159	»
0.421	0.0	4.79	5.27	89.519	»
0.386	0.0	8.97	8.55	82.094	»
0.628	0.0	14.19	13.07	72.112	»
0.55	0.0	15.22	11.06	73.17	$K_2SO_4 \cdot ZnSO_4 \cdot 6H_2O + KZnCl_3 \cdot 2H_2O$
0.54	0.0	14.3	9.13	76.03	»
0.701	0.0	15.03	10.61	73.659	$K_2SO_4 \cdot ZnSO_4 \cdot 6H_2O +$ $+ K_2SO_4 \cdot 2ZnCl_2 \cdot 5H_2O$
0.758	0.0	13.5	5.93	79.812	»
0.715	0.0	14.135	7.55	77.6	»
0.517	0.0	15.63	3.223	80.63	$ZnSO_4 \cdot H_2O + KZnCl_3 \cdot 2H_2O$
0.126	0.0	18.9	1.904	79.07	»
0.053	0.0	23.17	1.61	75.167	»
0.049	0.0	35.3	3.01	61.641	»
0.042	0.0	37.75	1.677	60.538	$ZnSO_4 \cdot H_2O + KZnCl_3$

№ 4695 [1239]

POTASSIUM SULFATE–POTASSIUM CHLORATE–WATER

$K_2SO_4—KClO_3—H_2O$

$t = 14—15$

Comp. of saturated soln., wt %		
Potassium Chlorate	Potassium Sulfate	Water
4.71	2.23	93.06
3.98	4.46	91.56

№ 4696 POTASSIUM SULFATE–POTASSIUM CHLORATE–WATER [1859]

$K_2SO_4—KClO_3—H_2O$

$t=25$

Concn. of Potassium Sulfate in aqueous soln. g/l	Solubility of Potassium Chlorate in aqueous soln. of Potassium Sulfate, g/l	Concn. of Potassium Sulfate in aqueous soln. g/l	Solubility of Potassium Chlorate in aqueous soln. of Potassium Sulfate, g/l	Concn. of Potassium Sulfate in aqueous soln. g/l	Solubility of Potassium Chlorate in aqueous soln. of Potassium Sulfate, g/l
0.00	83.030	17.4	75.387	43.5	67.004
2.18	82.570	26.1	71.995	65.25	59.389
4.35	80.622	34.8	69.834	87	54.310
8.7	79.103				

№ 4697 POTASSIUM SULFATE–POTASSIUM CHLORATE–WATER [2394]

$K_2SO_4—KClO_3—H_2O$

Comp. of saturated soln., wt %			t	d	Comp. of solid phase
Potassium Chlorate	Potassium Sulfate	Water			
5.676	0.0	94.324	15	1.032	$KClO_3$
3.29	7.86	88.85	15	1.085	$KClO_3+K_2SO_4$
0.0	9.258	90.742	15	1.076	K_2SO_4
7.897	0.0	92.103	25	1.048	$KClO_3$
6.72	2.73	90.55	25	1.063	»
5.77	5.57	89.66	25	1.080	»
5.06	8.19	86.75	25	1.099	»
4.96	8.64	86.40	25	1.100	$KClO_3+K_2SO_4$
3.30	9.43	87.27	25	1.099	K_2SO_4
1.80	9.93	88.27	25	1.089	»
0.0	10.76	89.24	25	1.083	»
13.90	0.0	86.10	45	—	$KClO_3$
9.80	9.13	81.07	45	—	$KClO_3+K_2SO_4$
0.0	13.53	86.47	45	—	K_2SO_4

№ 4698 [2269]

POTASSIUM SULFATE–POTASSIUM PERCHLORATE–WATER

$K_2SO_4—KClO_4—H_2O$

$t=25$

Concn. in aqueous soln., g/l	
Potassium Perchlorate	Potassium Sulfate
20.57	0.0
18.26	4.34
16.40	8.66

№ 4699 [1261]

POTASSIUM SULFATE–POTASSIUM PERCHLORATE–WATER

$K_2SO_4—KClO_4—H_2O$

$t=25$

Concn. in aqueous soln., wt %	
Potassium Perchlorate	Potassium Sulfate
1.614	0.847
1.144	2.537
0.841	4.93

№ 4700 POTASSIUM SULFATE–POTASSIUM CHROMATE–WATER [1627]

$K_2SO_4—K_2CrO_4—H_2O$

$t=25$

Concn. in aqueous soln., g/l		d	Concn. in aqueous soln., g/l		d	Concn. in aqueous soln., g/l		d
Potassium Sulfate	Potassium Chromate		Potassium Sulfate	Potassium Chromate		Potassium Sulfate	Potassium Chromate	
107.7	0.0	1.083	17.54	522.3	1.377	48.64	184.4	1.1700
106.0	20.02	1.092	0.0	553.5	1.398	26.68	285.6	1.2255
59.46	134.5	1.141	127.9	0.0	1.0863	51.61	521.2	1.3688
30.47	290.5	1.231	107.6	20.1	1.0934	0.0	527.8	1.3781
19.30	490.5	1.356	80.72	88.0	1.1235			

№ 4701 POTASSIUM SULFATE–POTASSIUM CHROMATE–WATER [1076]

$K_2SO_4—K_2CrO_4—H_2O$

$t=25$

Comp. of saturated soln., wt %

Potassium Chromate	Potassium Sulfate	Water	Potassium Chromate	Potassium Sulfate	Water
38.50	0.47	61.03	16.45	4.52	79.03
37.76	0.73	61.51	12.03	5.85	82.12
36.44	1.16	62.40	6.69	7.69	85.62
33.60	1.54	64.86	3.80	8.94	87.26
28.37	2.32	69.31	1.72	9.63	88.65
20.69	3.65	75.66	0.00	10.80	89.20

№ 4702 POTASSIUM SULFATE–POTASSIUM CHROMATE–WATER [530]

$K_2SO_4—K_2CrO_4—H_2O$

$t = 25$

Concn. in aqueous soln., wt %

Potassium Chromate	Potassium Sulfate	Potassium Chromate	Potassium Sulfate	Potassium Chromate	Potassium Sulfate
39.52	0.0	30.91	1.88	15.53	5.47
36.11	1.11	25.75	2.66	4.95	8.75
33.91	1.51	23.20	3.49	0.00	10.76

№ 4703 [975]

POTASSIUM SULFATE–POTASSIUM CHROMATE–WATER

$K_2SO_4—K_2CrO_4—H_2O$

$t = 25$

Comp. of saturated soln., wt %			Comp. of solid phase, wt %	
Potassium Chromate	Potassium Sulfate	Water	Potassium Chromate	Potassium Sulfate
0.00	10.80	89.20	0.00	100.00
15.98	4.96	79.06	2.49	97.51
23.86	3.56	72.58	5.49	94.51
32.30	2.04	65.66	29.80	70.20
34.12	2.00	63.88	44.45	55.55
37.39	1.07	61.54	72.00	28.00
39.40	0.00	60.60	100.00	0.00

№ 4704 [134]

POTASSIUM SULFATE–POTASSIUM DICHROMATE–WATER

$K_2SO_4–K_2Cr_2O_7–H_2O$

Comp. of saturated soln., wt %			t	Comp. of solid phase
Potassium Dichromate	Potassium Sulfate	Water		
11.10	0.00	88.90	20	$K_2Cr_2O_7$
10.16	1.02	88.82	20	»
9.91	1.40	88.69	20	»
9.53	1.87	88.60	20	»
8.56	3.38	88.06	20	»
7.80	4.42	87.78	20	»
6.92	6.41	86.67	20	»
6.24	8.23	85.53	20	$K_2Cr_2O_7+K_2SO_4$
5.42	8.48	86.10	20	K_2SO_4
4.20	8.89	86.91	20	»
3.08	9.43	87.49	20	»
1.98	9.64	88.38	20	»
0.45	9.90	89.65	20	»
0.00	10.00	90.00	20	»
13.08	0.00	86.92	25	$K_2Cr_2O_7$
12.09	1.56	86.35	25	»
11.14	2.95	85.91	25	»
10.12	4.15	85.73	25	»
9.15	5.80	85.05	25	»
7.70	8.01	84.29	25	»
7.43	8.33	84.24	25	$K_2Cr_2O_7+K_2SO_4$
6.69	8.65	84.66	25	K_2SO_4
6.15	8.80	85.05	25	»
4.03	9.60	86.37	25	»
3.13	9.92	86.95	25	»
2.28	10.10	87.62	25	»
1.19	10.50	88.31	25	»
0.74	10.70	88.56	25	»
0.00	10.80	89.20	25	»
15.40	0.00	84.60	30	$K_2Cr_2O_7$
13.16	2.86	83.98	30	»
12.20	4.66	83.14	30	»

Comp. of saturated soln., wt %			t	Comp. of solid phase
Potassium Dichromate	Potassium Sulfate	Water		
11.39	5.95	82.66	30	$K_2Cr_2O_7$
10.15	7.75	82.10	30	»
9.75	8.66	81.59	30	»
9.41	9.07	81.52	30	$K_2Cr_2O_7+K_2SO_4$
6.80	9.75	83.45	30	K_2SO_4
4.54	10.15	85.31	30	»
3.48	10.40	86.12	30	»
2.94	10.74	86.32	30	»
1.76	11.15	87.09	30	»
1.01	11.28	87.71	30	»
0.00	11.48	0.00	30	»
20.60	0.00	79.40	40	$K_2Cr_2O_7$
19.45	1.17	79.38	40	»
18.94	1.80	79.36	40	»
17.65	3.54	78.81	40	»
16.89	4.40	78.71	40	»
15.98	5.53	78.49	40	»
15.24	6.63	78.13	40	»
14.61	7.76	77.63	40	»
14.40	8.20	77.40	40	$K_2Cr_2O_7+K_2SO_4$
11.66	8.88	79.46	40	K_2SO_4
9.90	9.60	80.50	40	»
8.30	10.05	81.65	40	»
6.88	10.44	82.68	40	»
5.60	10.80	83.60	40	»
4.45	11.43	84.12	40	»
3.71	11.85	84.44	40	»
2.70	12.20	85.10	40	»
2.46	12.36	85.18	40	»
1.84	12.46	85.70	40	»
1.34	12.76	85.90	40	»
0.34	12.94	86.72	40	»
0.00	13.10	86.90	40	»

№ 4705 POTASSIUM SULFATE–POTASSIUM PERMANGANATE–WATER [2668]

$K_2SO_4—KMnO_4—H_2O$

$t = 25$

Comp. of saturated soln., wt %			d	Comp. of solid phase
Potassium Permanganate	Potassium Sulfate	Water		
7.10	0.00	92.90	1.0454	$KMnO_4$
6.59	0.80	92.61	1.0483	»
5.92	1.98	92.10	1.0537	»
4.52	5.47	90.01	1.0730	»
3.87	7.79	88.34	1.0876	»
3.55	9.26	87.19	1.0979	$KMnO_4+K_2SO_4$
0.00	10.75	89.25	1.0864	$KMnO_4$

№ 4706 POTASSIUM SULFATE–POTASSIUM BROMIDE–WATER [1238]

$K_2SO_4—KBr—H_2O$

$t = 14$

Concn. of Potassium Bromide in aqueous soln., g/l	Solubility of Potassium Sulfate in aqueous soln. of Potassium Bromide, g/l	Concn. of Potassium Bromide in aqueous soln., g/l	Solubility of Potassium Sulfate in aqueous soln. of Potassium Bromide, g/l	Concn. of Potassium Bromide in aqueous soln., g/l	Solubility of Potassium Sulfate in aqueous soln. of Potassium Bromide, g/l
0	101.6	60	74	100	60
20	91	80	66	120	55
40	82				

№ 4707 POTASSIUM SULFATE–POTASSIUM BROMIDE–WATER [10]

$K_2SO_4—KBr—H_2O$

Comp. of saturated soln., wt %			t	d	Comp. of solid phase
Potassium Bromide	Potassium Sulfate	Water			
0.00	6.78	93.22	0	0.9863	K_2SO_4
4.02	4.95	91.03	0	0.0212	»
6.52	4.09	89.39	0	1.0459	»
10.95	3.14	85.91	0	1.0940	»
14.12	2.22	83.06	0	1.1428	»
17.39	1.80	80.81	0	1.1710	»
21.12	1.61	77.27	0	1.2102	»
25.20	1.32	73.48	0	1.2640	»
29.20	1.14	69.66	0	1.2828	»
31.41	0.99	67.60	0	1.3303	»
35.54	0.68	73.78	0	1.3200	K_2SO_4+KBr
33.78	0.32	65.90	0	1.3119	KBr
34.11	0.00	65.89	0	1.3180	»
0.00	10.70	89.30	25	1.0883	K_2SO_4
4.75	8.12	87.13	25	1.0970	»
9.72	5.90	84.38	25	1.1426	»
13.50	4.50	82.00	25	1.1810	»
18.00	3.30	78.70	25	1.2201	»
22.05	2.76	75.19	25	1.2610	»
26.00	2.21	71.79	25	1.2871	»
32.00	1.80	65.20	25	1.3381	»
38.20	1.14	60.66	25	1.3862	»
39.20	1.11	59.69	25	1.3920	K_2SO_4+KBr
39.86	0.75	59.39	25	1.3894	KBr
40.67	0.00	59.33	25	1.3862	»
0.00	14.20	85.80	50	1.1255	K_2SO_4
2.00	12.93	85.07	50	1.1338	»
4.40	11.92	83.68	50	1.1527	»
7.81	9.83	82.36	50	1.1821	»
11.40	8.07	81.53	50	1.2120	»
14.80	6.90	88.50	50	1.2380	»
24.02	4.24	71.74	50	1.3063	»
28.73	3.43	67.84	50	1.3384	»
35.70	2.72	61.58	50	1.3840	»

Table 4707 (continuation)

Comp. of saturated soln., wt %			t	d	Comp. of solid phase
Potassium Bromide	Potassium Sulfate	Water			
39.54	2.17	58.29	50	1.4081	K_2CO_4
41.00	2.00	57.00	50	1.4180	K_2SO_4+KBr
43.29	0.93	55.78	50	1.4237	KBr
44.48	0.00	55.52	50	1.4200	»

№ 4708 [1452]

POTASSIUM SULFATE–POTASSIUM BROMIDE–CALCIUM SULFATE–CALCIUM BROMIDE–WATER

$K_2SO_4—KBr—CaSO_4—CaBr_2—H_2O$

$t = 25$

Content of Potassium Bromide in aqueous soln., g/l	Solubility in aqueous soln. of Potassium Bromide, g/l		Content of Potassium bromide in aqueous soln., g/l	Solubility in aqueous soln. of Potassium Bromide, g/l	
	Calcium Sulfate	Calcium Oxide		Calcium Sulfate	Calcium Oxide
400	2.505	5.90	80	6.012	—
320	3.691	3.080	40	4.489	—
240	8.764	0.364	20	3.617	—
160	7.124	0.030	0	2.550	—
120	6.640	—			

№ 4709 [2390]

POTASSIUM SULFATE–POTASSIUM BROMATE–WATER

$K_2SO_4—KBrO_3—H_2O$

$t = 25$

Comp. of saturated soln., wt %			d	Comp. of solid phase
Potassium Bromate	Potassium Sulfate	Water		
6.08	2.67	91.25	1.066	$KBrO_3$
5.02	5.44	89.54	1.083	»
4.27	8.20	87.53	1.100	»
4.00	9.35	86.65	1.108	$KBrO_3+K_2SO_4$
3.40	9.45	87.15	1.103	K_2SO_4
1.69	10.12	88.19	1.094	»
0.00	10.76	89.24	1.083	»

№ 4710 [1575]

POTASSIUM SULFATE–POTASSIUM BROMATE–WATER

$K_2SO_4—KBrO_3—H_2O$

$t = 25$

Content of Potassium Bromate in aqueous soln., wt %	Solubility of Potassium Sulfate in aqueous soln. of Potassium Bromate, wt %	Content of Potassium Bromate in aqueous soln., wt %	Solubility of Potassium Sulfate in aqueous soln. of Potassium Bromate, wt %
1.05	10.36	3.04	9.68
2.03	10.04	3.64	9.47

№ 4711 POTASSIUM SULFATE–POTASSIUM MOLYBDATE–WATER [1076]

$K_2SO_4—K_2MoO_4—H_2O$

$t = 25$

Comp. of saturated soln., wt %					
Potassium Molybdate	Potassium Sulfate	Water	Potassium Molybdate	Potassium Sulfate	Water
64.87	0.00	35.13	49.49	0.75	49.76
64.26	0.25	35.49	31.00	1.44	67.56
63.73	0.27	36.00	14.39	3.25	82.36
55.74	0.43	43.83	4.18	7.55	88.27
51.49	0.61	47.90	0.00	10.80	89.20

№ 4712 [368]

POTASSIUM SULFATE–POTASSIUM MOLYBDATE–WATER

$K_2SO_4—K_2MoO_4—H_2O$

$t = 25$

Comp. of saturated soln., wt %			Comp. of solid phase
Potassium Molybdate	Potassium Sulfate	Water	
0.00	10.63	89.37	K_2SO_4
7.21	7.06	85.73	»
14.09	4.61	81.30	»
24.89	2.04	73.07	»
36.34	0.51	63.15	»
47.53	0.25	52.22	»
58.63	0.12	41.25	$K_2MoO_4+K_2SO_4$
62.86	0.10	37.04	»
63.46	0.10	36.54	»
63.34	0.09	36.27	»
63.40	0.09	36.51	»
63.43	0.09	36.48	»
63.85	0.05	36.10	K_2MoO_4
64.53	0.00	35.47	»

№ 4713 [675]

POTASSIUM SULFATE–POTASSIUM MOLYBDATE–WATER

$K_2SO_4—K_2MoO_4—H_2O$

$t = 25$

Comp. of saturated soln., wt %			Comp. of solid phase
Potassium Molybdate	Potassium Sulfate	Water	
0.00	10.63	89.37	K_2SO_4
7.21	7.06	85.73	»
14.09	4.61	11.30	»
24.89	2.04	73.07	»
36.34	0.51	63.15	»
47.53	0.25	52.22	»
58.63	0.12	41.25	»

Table 4713 (continuation)

Comp. of saturated soln., wt %			Comp. of solid phase
Potassium Molybdate	Potassium Sulfate	Water	
62.86	0.10	37.04	$K_2SO_4+K_2MoO_4$
63.46	0.10	36.44	»
63.33	0.09	36.578	»
63.40	0.09	36.51	»
63.43	0.09	36.48	»
63.85	0.05	36.10	K_2MoO_4
64.53	0.00	35.47	»
0.00	10.60	89.40	K_2SO_4
2.85	9.06	88.09	»
5.62	7.76	86.62	»
8.30	6.47	84.23	»
10.86	5.54	83.60	»
13.07	4.53	86.40	»
24.06	1.80	74.14	»
32.50	0.79	66.71	»
39.20	0.44	60.36	»
44.68	0.26	55.06	»
49.25	0.17	50.58	»
53.14	0.08	46.78	»
64.60	0.00	35.40	K_2MoO_4
55.47	9.03	35.50	»
59.99	4.70	35.31	»
62.24	2.35	35.41	»

№ 4714 POTASSIUM SULFATE–POTASSIUM IODIDE–WATER [1238]

$K_2SO_4—KI—H_2O$

$t = 12.5$

Concn. of Potassium Iodide in aqueous soln., g/l	Solubility of Potassium Sulfate in aqueous soln. of Potassium Iodide, g/l	Concn. of Potassium Iodide in aqueous soln., g/l	Solubility of Potassium Sulfate in aqueous soln. of Potassium Iodide, g/l	Concn. of Potassium Iodide in aqueous soln., g/l	Solubility of Potassium Sulfate in aqueous soln. of Potassium Iodide, g/l
0	99	60	77	100	66
20	92	80	72	120	60
40	84				

№ 4715 POTASSIUM SULFATE–POTASSIUM IODIDE–WATER [2393]

$K_2SO_4—KI—H_2O$

$t = 25$

Comp. of saturated soln., wt %			d	Comp. of solid phase
Potassium Iodide	Potassium Sulfate	Water		
59.76	0.0	40.24	1.718	KI
59.60	0.08	40.32	1.722	$KI+K_2SO_4$
58.70	0.10	41.20	1.701	K_2SO_4

Table 4715 (continuation)

Comp. of saturated soln., wt %			d	Comp. of solid phase
Potassium Iodide	Potassium Sulfate	Water		
39.57	0.69	59.74	1.399	K_2SO_4
28.81	1.70	69.49	1.273	»
18.57	3.57	77.86	1.185	»
9.13	6.57	84.30	1.127	»
0.0	10.76	89.24	1.083	»

№ 4716 [1452]

POTASSIUM SULFATE–POTASSIUM IODIDE–CALCIUM SULFATE–CALCIUM IODIDE–WATER

K_2SO_4— KI —$CaSO_4$— CaI_2—H_2O

$t=25$

Content in aqueous soln., mol/1000 g Water				Comp. of solid phase
2 KI	K_2SO_4	CaI_2	$CaSO_4$	
13.31	0.0	0.835	0.565	$K_2SO_4 \cdot CaSO_4 \cdot H_2O + CaSO_4 \cdot 2H_2O$
5.62	0.731	0.0	0.443	»

№ 4717 POTASSIUM SULFATE–POTASSIUM IODATE–WATER [1839]

K_2SO_4— KIO_3—H_2O

Comp. of saturated soln., wt %			t	d	Comp. of solid phase
Potassium Iodide	Potassium Sulfate	Water			
5.16	0.00	94.84	5	1.043	KIO_3
3.14	4.07	92.79	5	1.060	»
2.57	7.08	90.35	5	1.081	$KIO_3+K_2SO_4$
1.80	7.25	90.95	5	1.077	K_2SO_4
0.0	7.64	92.36	5	1.062	»
8.45	0.00	91.55	25	1.071	KIO_3
5.66	4.75	89.59	25	1.085	»
4.72	7.74	87.54	25	1.103	»
4.30	9.65	86.05	25	1.117	$KIO_3+K_2SO_4$
2.44	10.10	87.46	25	1.103	K_2SO_4
0.0	10.76	89.24	25	1.083	»
13.21	0.00	86.79	50	—	KIO_3
8.68	7.0	83.42	50	—	»
7.39	11.43	81.18	50	—	»
7.06	12.0	80.94	50	—	$KIO_3+K_2SO_4$
3.78	13.0	83.22	50	—	K_2SO_4
0.00	14.18	85.82	50	—	»

Table 4724 (continuation)

Content in aqueous soln., wt %		t	Comp. of solid phase
Potassium Sulfate	Calcium Sulfate		
22.8	0.031	151	$K_2SO_4 + K_2SO_4 \cdot CaSO_4 \cdot H_2O$
19.3	0.073	168	$K_2SO_4 \cdot CaSO_4 \cdot H_2O + K_2SO_4 \cdot 5CaSO_4 \cdot H_2O$
21.2	0.066	169	»
19.8	0.060	169	»
25.1	0.073	200	$K_2SO_4 + K_2SO_4 \cdot CaSO_4 \cdot H_2O$
24.5	0.081	200	$K_2SO_4 \cdot CaSO_4 \cdot H_2O + K_2SO_4 \cdot 5CaSO_4 \cdot H_2O$
12.0	0.046	200	$K_2SO_4 \cdot 5CaSO_4 \cdot H_2O + CaSO_4$
27.3	0	250	K_2SO_4

№ 4725 POTASSIUM SULFATE–CALCIUM SULFATE–WATER [1842]

K_2SO_4—$CaSO_4$—H_2O

Content in aqueous soln., wt %		t	d	Comp. of solid phase
Potassium Sulfate	Calcium Sulfate			
0.0	0.212	40	0.9960	$CaSO_4 \cdot 2H_2O$
1.027	0.149	40	—	»
2.066	0.147	40	1.0108	»
2.596	0.154	40	1.0152	»
2.917	0.155	40	1.0187	$CaSO_4 \cdot 2H_2O + 5CaSO_4 \cdot K_2SO_4 \cdot H_2O$
3.956	0.143	40	1.0252	$5CaSO_4 \cdot K_2SO_4 \cdot H_2O + CaSO_4 \cdot K_2SO_4 \cdot H_2O$
4.061	0.134	40	1.0258	$CaSO_4 \cdot K_2SO_4 \cdot H_2O$
4.666	0.101	40	1.031	»
4.797	0.094	40	1.032	»
5.754	0.070	40	1.039	»
7.763	0.040	40	1.056	»
9.376	0.029	40	1.069	»
11.284	0.020	40	1.086	»
12.895	0.020	40	1.099	$CaSO_4 \cdot K_2SO_4 \cdot H_2O + K_2SO_4$
12.897	0.010	40	1.099	K_2SO_4
12.86	0.0	40	1.099	»
5.660	0.106	60	—	$5CaSO_4 \cdot K_2SO_4 \cdot H_2O + CaSO_4 \cdot K_2SO_4 \cdot H_2O$
6.172	0.099	60	—	$CaSO_4 \cdot K_2SO_4 \cdot H_2O$
10.044	0.041	60	—	»
15.40	0.019	60	—	$CaSO_4 \cdot K_2SO_4 \cdot H_2O + K_2SO_4$
15.40	0.00	60	—	K_2SO_4
0.0	0.160	60	—	$CaSO_4$
1.007	0.101	60	—	»
2.013	0.107	60	—	»
3.000	0.111	60	—	»
3.562	0.114	60	—	»
4.10	0.116	60	—	$CaSO_4 + 5CaSO_4 \cdot K_2SO_4 \cdot H_2O$
4.035	0.117	60	—	$5CaSO_4 \cdot K_2SO_4 \cdot H_2O$
0.0	0.167	100	—	$CaSO_4 \cdot 2H_2O$
0.260	0.135	100	—	»
0.787	0.133	100	—	»
1.090	0.136	100	—	$CaSO_4 \cdot 2H_2O + 5CaSO_4 \cdot K_2SO_4 \cdot H_2O$
10.62	0.066	100	—	$5CaSO_4 \cdot K_2SO_4 \cdot H_2O + CaSO_4 \cdot H_2O$
11.62	0.055	100	—	$CaSO_4 \cdot K_2SO_4 \cdot H_2O$
13.608	0.042	100	—	»

№ 4726 [1846]

POTASSIUM SULFATE–CALCIUM SULFATE–WATER

$K_2SO_4—CaSO_4—H_2O$

Content in aqueous soln., wt %		t	Comp. of solid phase
Potassium Sulfate	Calcium Sulfate		
0.12	0.188	35	$CaSO_4 \cdot 2H_2O$
0.08	0.195	35	»
0.04	0.204	35	»
0	0.212	35	$CaSO_4$
0.08	0.220	35	»
0.04	0.230	35	»
0	0.242	35	»
0.1	0.185	45	»
0.05	0.192	45	»
0	0.206	45	»

№ 4727 [213]

POTASSIUM SULFATE–CALCIUM SULFATE–WATER

$K_2SO_4—CaSO_4—H_2O$

$t=18—20$

Concn. of Potassium Sulfate in aqueous soln., g/l	Solubility of Calcium Sulfate in aqueous soln. of Potassium Sulfate, g/l	Concn. of Potassium Sulfate in aqueous soln., g/l	Solubility of Calcium Sulfate in aqueous soln. of Potassium Sulfate, g/l	Concn. of Potassium Sulfate in aqueous soln., g/l	Solubility of Calcium Sulfate in aqueous soln. of Potassium Sulfate, g/l
0.00	2.17	1.394	1.47	11.153	1.39
0.348	1.99	2.788	1.31	22.306	1.68
0.697	1.81	5.576	1.19		

№ 4728 [784

POTASSIUM SULFATE–CALCIUM SULFATE–WATER

$K_2SO_4—CaSO_4—H_2O$

$t=55$

Comp. of saturated soln., wt %			Comp. of solid phase
Potassium Sulfate	Calcium Sulfate	Water	
0.00	0.183	99.82	$CaSO_4$
1.30	0.129	98.57	»
9.83	0.033	90.14	$CaSO_4 \cdot K_2SO_4 \cdot H_2O$
5.02	0.108	94.87	$5CaSO_4 \cdot K_2SO_4 \cdot H_2O$
15.09	0.007	84.90	$CaSO_4 \cdot K_2SO_4 \cdot H_2O+K_2SO_4$
5.92	0.095	83.98	$5CaSO_4 \cdot K_2SO_4 \cdot H_2O+CaSO_4 \cdot K_2SO_4 \cdot H_2O$
2.94	0.125	96.94	$CaSO_4$
4.20	0.120	95.68	$CaSO_4+5CaSO_4 \cdot K_2SO_4 \cdot H_2O$

№ 4729 [148]

POTASSIUM SULFATE–CALCIUM SULFATE–WATER

$K_2SO_4—CaSO_4—H_2O$

$t=55$

Concn. in aqueous soln., wt %		Comp. of solid phase
Potassium Sulfate	Calcium Sulfate	
0.00	0.212	$CaSO_4 \cdot 2H_2O$
0.51	0.161	»
0.99	0.154	»
1.30	0.155	»
2.42	0.157	»
3.55	0.168	»
4.73	0.180	$CaSO_4 \cdot 2H_2O+K_2SO_4 \cdot CaSO_4 \cdot H_2O$
5.20	0.131	$K_2SO_4 \cdot CaSO_4 \cdot H_2O$
6.45	0.086	»
9.35	0.048	»
13.01	0.027	»
14.87	0.020	$K_2SO_4+K_2SO_4 \cdot CaSO_4 \cdot H_2O$
15.05	0.00	K_2SO_4
2.57	0.139	$K_2SO_4 \cdot 5CaSO_4 \cdot H_2O$
3.10	0.133	»
4.43	0.124	»
5.40	0.114	$K_2SO_4 \cdot 5CaSO_4 \cdot H_2O+K_2SO_4 \cdot CaSO_4 \cdot H_2O$
2.15	0.148	$K_2SO_4 \cdot 5CaSO_4 \cdot H_2O+CaSO_4$
0.00	0.161	$CaSO_4$
0.92	0.144	»
1.52	0.145	»
3.12	0.153	»
4.75	0.166	$CaSO_4+K_2SO_4 \cdot CaSO_4 \cdot H_2O$
1.75	0.154	$K_2SO_4 \cdot 5CaSO_4 \cdot H_2O+CaSO_4 \cdot 2H_2O$

№ 4730 [486]

POTASSIUM SULFATE–CALCIUM SULFATE–WATER

$K_2SO_4—CaSO_4—H_2O$

$t=35$

Comp. of saturated soln., wt %			Comp. of solid phase
Potassium Sulfate	Calcium Sulfate	Water	
0.000	0.211	99.789	$CaSO_4 \cdot 2H_2O$
0.684	0.166	99.150	»
1.752	0.152	98.096	»
2.088	0.150	97.762	»
2.990	0.154	96.856	»
3.349	0.158	96.493	»
3.540	0.161	96.299	$CaSO_4 \cdot 2H_2O+K_2SO_4 \cdot CaSO_4 \cdot H_2O$
3.681	0.152	96.167	$K_2SO_4 \cdot CaSO_4 \cdot H_2O$
3.400	0.159	96.441	$CaSO_4 \cdot 2H_2O+K_2SO_4 \cdot 5CaSO_4 \cdot H_2O$

Table 4730 (continuation)

Comp. of saturated soln., wt %			Comp. of solid phase
Potassium Sulfate	Calcium Sulfate	Water	
3.433	0.158	96.409	$K_2SO_4 \cdot 5CaSO_4 \cdot H_2O$
3.554	0.158	96.288	»
3.600	0.158	96.242	$K_2SO_4 \cdot 5CaSO_4 + K_2SO_4 \cdot CaSO_4 \cdot H_2O$
3.868	0.128	96.004	$K_2SO_4 \cdot CaSO_4 \cdot H_2O$
4.445	0.108	95.447	»
5.178	0.0856	94.7364	»
5.967	0.0642	93.9688	»
7.689	0.0421	92.2689	»
9.154	0.0296	90.8164	»
9.955	0.0267	90.0183	»
11.764	0.0199	88.2161	»
12.129	0.0187	87.8523	»
12.283	0.0177	87.6993	$K_2SO_4 \cdot CaSO_4 \cdot H_2O + K_2SO_4$
12.296	0.0119	87.6921	K_2SO_4
12.303	0.000	87.697	»

№ 4731 [470]

POTASSIUM SULFATE–CALCIUM SULFATE–WATER

$K_2SO_4—CaSO_4—H_2O$

$t=80$

Concn. in aqueous soln., wt %		Comp. of solid phase
Potassium Sulfate	Calcium Sulfate	
0.00	0.170	$CaSO_4$
1.24	0.160	$CaSO_4 + K_2SO_4 \cdot 5CaSO_4 \cdot H_2O$
8.80	0.078	$K_2SO_4 \cdot CaSO_4 \cdot H_2O +$ $+ K_2SO_4 \cdot 5CaSO_4 \cdot H_2O$
17.80	0.024	$K_2SO_4 \cdot 5CaSO_4 \cdot H_2O + K_2SO_4$
17.80	0.000	K_2SO_4

№ 4732 [1388]

POTASSIUM SULFATE–CALCIUM IODATE–WATER

$K_2SO_4—Ca(IO_3)_2—H_2O$

$t=25$

Concn. of Potassium Sulfate in aqueous soln., g/l	Solubility of Calcium Iodate in aqueous soln. of Potassium Sulfate, wt %
0.0	0.303
0.3485	0.3196
0.8713	0.3389
1.7426	0.3666
3.485	0.4008
8.713	0.5077
17.426	0.6290

№ 4733 POTASSIUM SULFATE–CALCIUM IODATE–WATER [1184]

$K_2SO_4—Ca(IO_3)_2—H_2O$

Concn. in aqueous soln., wt %		t	Concn. in aqueous soln., wt %		t
Potassium Sulfate	Calcium Iodate		Potassium Sulfate	Calcium Iodate	
0.058	0.102	0	0.190	0.371	25
0.102	0.110	0	0.220	0.380	25
0.170	0.118	0	0.259	0.391	25
0.220	0.124	0	0.349	0.410	25
0.269	0.130	0	0.053	0.533	40
0.359	0.135	0	0.104	0.558	40
0.408	0.139	0	0.155	0.578	40
0.059	0.328	25	0.210	0.600	40
0.090	0.340	25	0.245	0.612	40
0.140	0.357	25			

№ 4734 [1377]

POTASSIUM SULFATE–MANGANESE (II) SULFATE–WATER

$K_2SO_4—MnSO_4—H_2O$

$t=0$

Comp. of saturated soln., wt %			Comp. of solid phase
Potassium Sulfate	Manganese Sulfate	Water	
0.00	34.78	65.22	$MnSO_4 \cdot 7H_2O$
1.68	34.49	63.83	»
1.85	34.53	63.62	»
2.81	34.38	62.81	»
3.35	34.30	62.35	$MnSO_4 \cdot 7H_2O + MnSO_4 \cdot K_2SO_4 \cdot 4H_2O$
4.00	30.28	65.72	$MnSO_4 \cdot K_2SO_4 \cdot 4H_2O$
4.48	27.62	67.90	»
5.00	25.50	69.50	»
5.54	23.64	70.82	»
7.69	17.39	74.92	»
8.88	13.68	77.44	$MnSO_4 \cdot K_2SO_4 \cdot 4H_2O + K_2SO_4$
8.48	11.80	79.72	K_2SO_4
7.97	8.21	83.82	»
7.75	6.21	86.04	»
6.82	0.00	93.18	»

№ 4735 POTASSIUM SULFATE–MANGANESE (II) SULFATE–WATER (1204)

K_2SO_4—$MnSO_4$—H_2O

Comp. of saturated soln., wt %			t	Comp. of solid phase
Potassium Sulfate	Manganese Sulfate	Water		
10.85	14.85	74.30	17.5	$K_2SO_4+MnSO_4\cdot 5H_2O$
12.25	15.46	72.29	30	$K_2SO_4+MnSO_4\cdot K_2SO_4\cdot 4H_2O$
5.50	33.64	60.86	30	$MnSO_4\cdot K_2SO_4\cdot 4H_2O+K_2SO_4\cdot 2MnSO_4$
13.10	15.65	71.25	35	$K_2SO_4+MnSO_4\cdot K_2SO_4\cdot 2H_2O$
6.60	30.40	63.00	35	$MnSO_4\cdot K_2SO_4\cdot 4H_2O+K_2SO_4\cdot 2MnSO_4$
13.78	14.26	71.96	40	$K_2SO_4+K_2SO_4\cdot MnSO_4\cdot 2H_2O$
7.22	27.18	65.60	40	$K_2SO_4\cdot 2MnSO_4+K_2SO_4\cdot MnSO_4\cdot 2H_2O$
3.34	34.66	62.00	40	$K_2SO_4\cdot 2MnSO_4+MnSO_4\cdot H_2O$
14.17	0.0	85.83	50	K_2SO_4
15.02	2.77	82.21	50	»
15.11	7.36	77.53	50	»
15.34	11.78	72.88	50	$K_2SO_4+K_2SO_4\cdot MnSO_4\cdot 2H_2O$
13.40	14.43	72.17	50	$K_2SO_4\cdot MnSO_4\cdot 2H_2O$
12.34	16.45	71.21	50	»
11.85	17.17	70.98	50	»
11.55	18.30	70.15	50	»
11.41	18.70	69.89	50	$K_2SO_4\cdot MnSO_4\cdot 2H_2O+K_2SO_4\cdot 2MnSO_4$
9.51	21.52	68.97	50	$K_2SO_4\cdot 2MnSO_4$
7.30	24.82	67.88	50	»
5.85	27.90	66.25	50	»
3 48	35.22	61.29	50	$K_2SO_4\cdot 2MnSO_4+MnSO_4\cdot H_2O$
0.0	37.3	62.7	50	$MnSO_4\cdot H_2O$
15.91	10.50	73.59	55	$K_2SO_4+K_2SO_4\cdot MnSO_4\cdot 2H_2O$
12.83	16.71	70.46	55	$K_2SO_4\cdot 2MnSO_4+K_2SO_4\cdot MnSO_4\cdot 2H_2O$
3.10	33.60	63.30	55	$K_2SO_4\cdot 2MnSO_4+MnSO_4\cdot H_2O$
16.75	8.63	74.62	66	$K_2SO_4+K_2SO_4\cdot MnSO_4\cdot 2H_2O$
14.77	11.35	73.88	66	$K_2SO_4\cdot 2MnSO_4+K_2SO_4\cdot MnSO_4\cdot 2H_2O$
17.92	0.0	82.08	80	K_2SO_4
18.60	3.38	78.02	80	»
18.40	5.41	76.19	80	»
18.40	5.84	75.76	80	$K_2SO_4+K_2SO_4\cdot 2MnSO_4$
17.13	6.16	76.77	80	$K_2SO_4\cdot 2MnSO_4$
13.80	6.92	79.28	80	»
10.82	8.81	80.37	80	»
6.66	14.57	78.77	80	»
6.19	16.65	77.16	80	»
4.85	20.42	74.73	80	»
2.61	29.58	67.81	80	$K_2SO_4\cdot 2MnSO_4+MnSO_4\cdot H_2O$
0.0	30.24	69.76	80	$MnSO_4\cdot H_2O$
19.42	0.0	80.58	97	K_2SO_4
19.78	2.09	78.13	97	»
19.89	3.63	76.48	97	$K_2SO_4+K_2SO_4\cdot 2MnSO_4$
16.84	4.05	79.11	97	$K_2SO_4\cdot 2MnSO_4$
12.80	4.31	82.89	97	»
9.68	5.46	84.86	97	»
7.25	7.08	85.67	97	»
6.00	11.65	83.35	97	»
4.18	17.0	78.82	97	»
3.48	25.92	71.60	97	$K_2SO_4\cdot 2MnSO_4+MnSO_4\cdot H_2O$
0.0	28.49	71.51	97	$MnSO_4\cdot H_2O$

№ 4736 [1333]

POTASSIUM SULFATE–IRON (III) SULFATE–WATER

K_2SO_4—$Fe_2(SO_4)_3$—H_2O

$t=25$

Comp. of saturated soln., wt %			Comp. of solid phase
Iron Sulfate	Potassium Sulfate	Water	
27.80	2.83	69.37	$Fe_2(SO_4)_3 \cdot 2K_2SO_4 \cdot 14H_2O$
24.30	2.94	72.76	»
21.80	3.34	74.86	»
20.80	3.38	75.82	»
13.10	4.56	82.34	»
11.20	5.43	83.37	»
9.02	5.32	85.66	»
6.04	6.21	87.75	»
2.35	8.0	89.65	»
0.85	11.70	87.45	$Fe_2(SO_4)_3 \cdot 2K_2SO_4 \cdot 14H_2O+K_2SO_4$
1.25	12.20	86.55	»
1.53	12.50	85.97	»
1.68	13.10	85.22	»
1.88	13.70	84.42	»
0.69	11.90	87.41	K_2SO_4
0.31	11.80	87.89	»
0.10	11.50	88.40	»
0.02	11.30	88.68	»

№ 4737 [1378]

POTASSIUM SULFATE–COBALT (II) SULFATE–WATER

K_2SO_4—$CoSO_4$—H_2O

$t=25$

Comp. of saturated soln., wt %			Comp. of solid phase
Cobalt Sulfate	Potassium Sulfate	Water	
0.00	10.60	89.40	K_2SO_4
2.02	11.13	86.85	»
3.15	11.60	85.25	$K_2SO_4+K_2SO_4 \cdot CoSO_4$
6.10	7.95	85.95	$K_2SO_4 \cdot CoSO_4$
12.11	5.55	82.34	»
13.75	5.10	81.15	»
16.79	4.41	78.80	»
24.70	3.02	72.28	»
27.60	2.57	69.83	$K_2SO_4 \cdot CoSO_4+CoSO_4$
27.52	2.32	70.16	$CoSO_4$
27.40	0.00	72.60	»

№ 4738 POTASSIUM SULFATE–COBALT (II) SULFATE–WATER [1210]

$K_2SO_4—CoSO_4—H_2O$

Comp. of saturated soln., wt %			t	Comp. of solid phase
Potassium Sulfate	Cobalt Sulfate	Water		
6.70	1.13	92.17	0	$K_2SO_4+CoSO_4 \cdot K_2SO_4 \cdot 6H_2O$
2.00	16.54	81.46	0	$CoSO_4 \cdot 7H_2O+CoSO_4 \cdot K_2SO_4 \cdot 6H_2O$
3.82	3.40	92.78	0	$CoSO_4 \cdot K_2SO_4 \cdot 6H_2O$
10.07	2.74	87.19	25	$K_2SO_4+CoSO_4 \cdot K_2SO_4 \cdot 6H_2O$
1.97	21.23	76.80	25	$CoSO_4 \cdot 7H_2O+CoSO_4 \cdot K_2SO_4 \cdot 6H_2O$
6.74	6.00	87.26	25	$CoSO_4 \cdot K_2SO_4 \cdot 6H_2O$
2.11	24.29	73.60	39	$CoSO_4 \cdot 7H_2O+CoSO_4 \cdot 6H_2O+$ $+CoSO_4 \cdot K_2SO_4 \cdot 6H_2O$
11.76	4.10	84.14	40	$K_2SO_4+CoSO_4 \cdot K_2SO_4 \cdot 6H_2O$
2.52	24.27	73.21	46	$CoSO_4 \cdot 6H_2O+CoSO_4 \cdot K_2SO_4 \cdot 6H_2O$
9.16	8.16	82.68	46	$CoSO_4 \cdot K_2SO_4 \cdot 6H_2O$
12.71	6.57	80.72	55	$K_2SO_4+CoSO_4 \cdot K_2SO_4 \cdot 6H_2O$
10.43	9.28	80.29	60	$CoSO_4 \cdot 6H_2O+CoSO_4 \cdot K_2SO_4 \cdot 6H_2O$
3.21	26.24	70.55	68	$CoSO_4 \cdot 6H_2O+CoSO_4 \cdot H_2O+$ $+CoSO_4 \cdot K_2SO_4 \cdot 6H_2O$
3.79	25.98	70.23	69	$CoSO_4 \cdot H_2O+CoSO_4 \cdot K_2SO_4 \cdot 6H_2O$
7.14	22.24	70.62	80	»
12.27	14.87	72.86	88	$K_2SO_4+CoSO_4 \cdot K_2SO_4 \cdot 6H_2O$
9.61	18.90	71.49	88	$CoSO_4 \cdot H_2O+CoSO_4 \cdot K_2SO_4 \cdot 6H_2O$
11.04	16.66	72.30	94	$CoSO_4 \cdot H_2O+K_2SO_4+CoSO_4 \cdot K_2SO_4 \cdot 6H_2O$
12.49	17.31	70.20	100	$CoSO_4 \cdot H_2O+K_2SO_4$

№ 4739 [1219]

POTASSIUM SULFATE–COBALT (II) SULFATE–WATER

$K_2SO_4—CoSO_4—H_2O$

Concn. in aqueous soln., wt %		t	Comp. of solid phase
Cobalt Sulfate	Potassium Sulfate		
1.32	7.90	0	$K_2SO_4+K_2SO_4 \cdot CoSO_4 \cdot 6H_2O$
3.83	3.81	0	$K_2SO_4 \cdot CoSO_4 \cdot 6H_2O$
21.0	1.66	0	$K_2SO_4 \cdot CoSO_4 \cdot 6H_2O+CoSO_4 \cdot 7H_2O$
4.70	13.31	38	$K_2SO_4+K_2SO_4 \cdot CoSO_4 \cdot 6H_2O$
15.75	6.71	38	$K_2SO_4 \cdot CoSO_4 \cdot 6H_2O$
31.0	3.8	38	$K_2SO_4 \cdot CoSO_4 \cdot 6H_2O+CoSO_4 \cdot 7H_2O$
6.80	14.36	50	$K_2SO_4+K_2SO_4 \cdot CoSO_4 \cdot 6H_2O$
10.32	12.01	50	$K_2SO_4 \cdot CoSO_4 \cdot 6H_2O$
33.3	3.5	50	$K_2SO_4 \cdot CoSO_4 \cdot 6H_2O+CoSO_4 \cdot 6H_2O$
13.51	17.9	75	$K_2SO_4+K_2SO_4 \cdot CoSO_4 \cdot 6H_2O$
17.63	14.5	75	$K_2SO_4 \cdot CoSO_4 \cdot 6H_2O$
34.5	7 32	75	$K_2SO_4 \cdot CoSO_4 \cdot 6H_2O+CoSO_4 \cdot H_2O$
29.4	12.80	90	»
22.4	18.55	96	$K_2SO_4+K_2SO_4 \cdot CoSO_4 \cdot 6H_2O$
24.4	16.57	96	$K_2SO_4 \cdot CoSO_4 \cdot 6H_2O$
27.9	14.95	96	$K_2SO_4 \cdot CoSO_4 \cdot 6H_2O+CoSO_4 \cdot H_2O$
22.9	18.50	99.5	$K_2SO_4+K_2SO_4 \cdot CoSO_4 \cdot 4H_2O$
24.9	17 75	99.5	$K_2SO_4 \cdot CoSO_4 \cdot 4H_2O$
25.6	17.50	99.5	$K_2SO_4 \cdot CoSO_4 \cdot 4H_2O+CoSO_4 \cdot H_2O$

№ 4740 POTASSIUM SULFATE–COBALT (II) CHLORIDE–WATER [1219]

K_2SO_4—$CoCl_2$—H_2O

$t=38$

Concn. in aqueous soln., wt %				Comp. of solid phase
Co	K_2	Cl_2	SO_4	
0.0	15.08	13.20	0.72	$KCl+K_2SO_4$
1.95	13.20	13.60	1.04	»
2.97	12.35	13.75	1.49	»
3.43	12.13	13.83	1.80	»
3.40	11.10	12.52	2.29	»
3.80	11.85	13.85	2.01	$KCl+K_2SO_4+CoSO_4 \cdot K_2SO_4 \cdot 6H_2O$
4.04	11.35	13.78	1.90	$KCl+CoSO_4 \cdot K_2SO_4 \cdot 6H_2O$
4.20	24.80	13.05	1.95	»
4.68	11.10	14.25	2.02	»
7.12	8.93	15.15	2.09	»
6.70	8.28	14.00	2.20	»
7.16	9.10	15.30	2.26	»
1.78	6.00	0.01	10.25	$K_2SO_4+CoSO_4 \cdot K_2SO_4 \cdot 6H_2O$
1.72	6.22	2.27	7.35	»
1.82	7.32	5.03	5.16	»
2.73	10.50	10.90	2.59	»
3.56	11.03	12.93	1.87	»
18.15	4.50	26.7	0.00	$KCl+CoCl_2 \cdot 6H_2O$
18.24	5.34	26.5	0.52	»
18.40	3.45	26.3	1.14	»
18.0	4.93	24.4	2.48	$KCl+CoCl_2 \cdot 6H_2O+CoSO_4 \cdot 6H_2O$
11.7	1.85	0.0	21.3	$CoSO_4 \cdot 7H_2O+CoSO_4 \cdot K_2SO_4 \cdot 6H_2O$
12.08	2.06	7.9	11.68	$CoSO_4 \cdot 7H_2O+CoSO_4 \cdot 6H_2O$
18.30	0.0	19.30	2.91	$CoSO_4 \cdot 6H_2O+CoCl_2 \cdot 6H_2O$

№ 4741 [1376]

POTASSIUM SULFATE–NICKEL (II) SULFATE–WATER

K_2SO_4—$NiSO_4$—H_2O

$t=25$

Comp. of saturated soln., wt %			Comp. of solid phase
Nickel Sulfate	Potassium Sulfate	Water	
27.94	0.00	71.06	$NiSO_4 \cdot 7H_2O$
27.87	3.82	68.31	»
27.92	6.30	65.78	»
27.90	8.26	63.84	$NiSO_4 \cdot 7H_2O+NiSO_4 \cdot K_2SO_4 \cdot 6H_2O$
23.09	6.82	70.08	$NiSO_4 \cdot K_2SO_4 \cdot 6H_2O$
19.33	5.62	75.05	»
15.66	4.75	79.59	»
10.15	4.11	85.74	»
3.90	4.15	91.95	»
1.35	4.93	93.72	»
0.33	11.16	88.51	$NiSO_4 \cdot K_2SO_4 \cdot 6H_2O+K_2SO_4$
0.00	10.59	89.41	K_2SO_4

№ 4742 **POTASSIUM SULFATE–NICKEL (II) SULFATE–WATER** [1210]

K_2SO_4—$NiSO_4$—H_2O

Comp. of saturated soln., wt %			t	Comp. of solid phase
Potassium Sulfate	Nickel Sulfate	Water		
6.80	0.14	93.06	0	$K_2SO_4+K_2SO_4\cdot NiSO_4\cdot 6H_2O$
1.09	21.62	77.29	0	$NiSO_4\cdot 7H_2O+K_2SO_4\cdot NiSO_4\cdot 6H_2O$
1.18	29.52	69.30	25	»
1.30	32.40	66.30	30	$NiSO_4\cdot 7H_2O+K_2SO_4\cdot NiSO_4\cdot 6H_2O+$ $+NiSO_4\cdot 6H_2O$
13.15	0.51	86.34	40	$K_2SO_4+K_2SO_4\cdot NiSO_4\cdot 6H_2O$
2.08	33.35	64.57	40	$NiSO_4\cdot 6H_2O+K_2SO_4\cdot NiSO_4\cdot 6H_2O$
14.76	1.20	84.04	55	$K_2SO_4+K_2SO_4\cdot NiSO_4\cdot 6H_2O$
2.30	35.0	62.70	55	$NiSO_4\cdot 6H_2O+K_2SO_4\cdot NiSO_4\cdot 6H_2O$
2.99	41.19	55.82	88	»
20.21	5.19	74.60	100	$K_2SO_4+K_2SO_4\cdot NiSO_4\cdot 6H_2O$
3.60	42.30	54.10	100	$NiSO_4\cdot 6H_2O+K_2SO_4\cdot NiSO_4\cdot 6H_2O$

№ 4743 [2407]

POTASSIUM SULFATE–COPPER (II) SULFATE–WATER

K_2SO_4—$CuSO_4$—H_2O

$t=25$

Concn. in aqueous soln., wt %		d	Comp. of solid phase
Potassium Sulfate	Copper Sulfate		
11.03	1.29	1.097	$K_2SO_4+K_2SO_4\cdot CuSO_4\cdot 6H_2O$
6.35	3.42	1.085	$K_2SO_4\cdot CuSO_4\cdot 6H_2O$
5.70	4.30	1.089	»
5.38	4.94	1.093	»
3.09	19.16	1.246	$K_2SO_4\cdot CuSO_4\cdot 6H_2O+CuSO_4\cdot 5H_2O$

№ 4744 [2407]

POTASSIUM SULFATE–COPPER (II) SULFATE–THALLIUM SULFATE–WATER

K_2SO_4—$CuSO_4$—Tl_2SO_4—H_2O

$t=25$

Comp. of saturated soln., wt %				d	Comp. of solid phase
Thallium Sulfate	Potassium Sulfate	Copper Sulfate	Water		
2.62	0.42	18.77	78.20	1.244	$CuSO_4\cdot 5H_2O+Tl_2SO_4\cdot CuSO_4\cdot 6H_2O+$ $+K_2SO_4\cdot CuSO_4\cdot 6H_2O$
2.13	1.07	18.90	77.90	1.249	»
1.56	1.70	19.02	77.72	1.251	»
0.56	2.53	19.17	77.74	1.249	»
0.25	2.67	19.22	77.85	1.245	»
0.23	2.72	19.23	77.82	1.247	»
0.51	3.11	15.57	80.81	1.204	$Tl_2SO_4\cdot CuSO_4\cdot 6H_2O+$ $+K_2SO_4\cdot CuSO_4\cdot 6H_2O$

Table № 4744 (continuation)

Comp. of saturated soln., wt %				d	Comp. of solid phase
Thallium Sulfate	Potassium Sulfate	Copper Sulfate	Water		
0.17	3.31	15.90	80.62	1.207	$Tl_2SO_4 \cdot CuSO_4 \cdot 6H_2O + K_2SO_4 \cdot CuSO_4 \cdot 6H_2O$
5.58	0.38	1.91	92.14	1.073	$Tl_2SO_4 + K_2SO_4 + Tl_2SO_4 \cdot CuSO_4 \cdot 6H_2O + K_2SO_4 \cdot CuSO_4 \cdot 6H_2O$
5.04	3.51	1.42	90.04	1.093	»
4.93	7.27	1.04	86.76	1.119	»
4.57	9.41	0.927	85.09	1.131	»
4.50	9.56	0.912	85.12	1.134	»
4.31	9.68	0.911	85.10	1.133	»
4.23	9.71	0.926	85.14	1.132	»
4.21	9.69	0.933	85.17	1.132	»
3.29	0.99	1.005	85.71	1.125	»
2.51	10.26	1.078	86.14	1.119	»

№ 4745 [2596]

POTASSIUM SULFATE-ZINC HYDROXIDE-WATER

$K_2SO_4—Zn(OH)_2—H_2O$

$t=16—20$

Comp. of solvent, wt %		Solubility of Zinc Hydroxide, g/l
Potassium Sulfate	Water	
1	99	0.0375

№ 4746 [1376]

POTASSIUM SULFATE-ZINC SULFATE-WATER

$K_2SO_4—ZnSO_4—H_2O$

$t=25$

Comp. of saturated soln., wt %			Comp. of solid phase
Potassium Sulfate	Zinc Sulfate	Water	
10.59	0.0	89.41	K_2SO_4
10.71	1.11	88.18	»
10.94	1.91	87.15	$K_2SO_4 + K_2SO_4 \cdot ZnSO_4 \cdot 6H_2O$
6.52	4.73	88.75	$K_2SO_4 \cdot ZnSO_4 \cdot 6H_2O$
5.99	5.40	88.61	»
3.19	25.11	71.70	»
2.27	33.01	64.72	»
1.95	35.58	62.47	»
1.82	36.00	62.18	»
1.85	36.39	61.76	$K_2SO_4 \cdot ZnSO_4 \cdot 6H_2O + ZnSO_4 \cdot 7H_2O$
1.44	36.42	62.14	$ZnSO_4 \cdot 7H_2O$
0.0	36.61	63.39	»

№ 4747 POTASSIUM SULFATE–ZINC SULFATE–WATER [1204]
$K_2SO_4—ZnSO_4—H_2O$

Comp. of saturated soln., wt %			t	Comp. of solid phase
Potassium Sulfate	Zinc Sulfate	Water		
6.88	0.00	93.12	0	$K_2SO_4+K_2SO_4 \cdot ZnSO_4 \cdot 6H_2O$
2.91	3.68	93.41	0	$K_2SO_4 \cdot ZnSO_4 \cdot 6H_2O$
0.99	29.83	69.18	0	$K_2SO_4 \cdot ZnSO_4 \cdot 6H_2O+ZnSO_4 \cdot 7H_2O$
9.69	0.37	89.94	12	$K_2SO_4+K_2SO_4 \cdot ZnSO_4 \cdot 6H_2O$
1.82	32.60	65.58	12	$K_2SO_4 \cdot ZnSO_4 \cdot 6H_2O+ZnSO_4 \cdot 7H_2O$
13.78	3.18	83.04	41	$K_2SO_4+K_2SO_4 \cdot ZnSO_4 \cdot 6H_2O$
2.50	40.30	57.20	41	$K_2SO_4 \cdot ZnSO_4 \cdot 6H_2O+ZnSO_4 \cdot 6H_2O$
14.28	4.34	81.38	50	$K_2SO_4+K_2SO_4 \cdot ZnSO_4 \cdot 6H_2O$
2.15	43.20	54.65	50	$K_2SO_4 \cdot ZnSO_4 \cdot 6H_2O+ZnSO_4 \cdot 6H_2O$
4.20	39.90	55.90	68	$K_2SO_4 \cdot ZnSO_4 \cdot 6H_2O+ZnSO_4 \cdot H_2O$
17.55	0.00	82.45	80	K_2SO_4
17.87	2.07	80.06	80	»
18.17	8.48	73.35	80	»
18.24	11.80	69.96	80	»
17.77	13.46	68.77	80	»
18.14	13.67	68.19	80	$K_2SO_4+K_2SO_4 \cdot ZnSO_4 \cdot 6H_2O$
16.91	14.32	68.77	80	$K_2SO_4 \cdot ZnSO_4 \cdot 6H_2O$
14.36	16.29	69.35	80	»
13.15	21.04	65.81	80	»
12.17	23.00	64.83	80	»
10.26	27.30	62.44	80	»
8.60	31.00	60.40	80	»
6.51	36.84	56.65	80	»
6.60	38.38	55.02	80	$K_2SO_4 \cdot ZnSO_4 \cdot 6H_2O+ZnSO_4 \cdot H_2O$
6.02	38.62	55.36	80	$K_2SO_4 \cdot H_2O$
3.59	38.92	57.49	80	»
0.00	40.56	59.54	80	»
19.42	0.00	80.58	100	K_2SO_4
20.50	9.05	70.45	100	»
21.43	10.16	68.41	100	»
20.07	23.18	56.75	100	$K_2SO_4+K_2SO_4 \cdot ZnSO_4 \cdot 6H_2O$
19.13	24.96	55.91	100	$K_2SO_4 \cdot ZnSO_4 \cdot 6H_2O$
18.85	24.40	56.75	100	»
17.69	25.82	56.49	100	»
15.08	30.62	54.30	100	»
13.51	36.70	49.79	100	$K_2SO_4 \cdot ZnSO_4 \cdot 6H_2O+ZnSO_4 \cdot H_2O$
7.15	36.95	55.90	100	$ZnSO_4 \cdot H_2O$
2.23	38.85	58.92	100	»
0.00	37.71	62.29	100	»

№ 4748 POTASSIUM SULFATE–STRONTIUM SULFATE–WATER [1140]
$K_2SO_4—SrSO_4—H_2O$

Content in aqueous soln., wt %		t	Comp. of solid phase
Potassium Sulfate	Strontium Sulfate		
1.27	<0.01	17.5	$K_2SO_4 \cdot SrSO_4+SrSO_4$
1.88	<0.01	50	»
2.71	<0.01	75	»
3.90	<0.01	100	»

№ 4749 [833]

POTASSIUM SULFATE–SILVER ORTHOPHOSPHATE–WATER

$K_2SO_4—Ag_3PO_4—H_2O$

$t=25$

Content of Potassium Sulfate in aqueous soln., mol/l	Solubility of Silver Orthophosphate in aqueous soln. of Potassium Sulfate, mol/l	Content of Potassium Sulfate in aqueous soln., mol/l	Solubility of Silver Orthophosphate in aqueous soln. of Potassium Sulfate, mol/l
0.00	$5.11 \cdot 10^{-5}$	$1 \cdot 10^{-3}$	$7.02 \cdot 10^{-5}$
$1 \cdot 10^{-5}$	$5.15 \cdot 10^{-5}$	$1 \cdot 10^{-2}$	$7.92 \cdot 10^{-5}$
$1 \cdot 10^{-4}$	$5.22 \cdot 10^{-5}$	$1 \cdot 10^{-1}$	$15.18 \cdot 10^{-5}$
$5 \cdot 10^{-4}$	$5.54 \cdot 10^{-5}$	$5 \cdot 10^{-1}$	$26.68 \cdot 10^{-5}$
$1 \cdot 10^{-3}$	$5.83 \cdot 10^{-5}$	0.6755	$29.75 \cdot 10^{-5}$

№ 4750 [1530]

POTASSIUM SULFATE–SILVER SULFATE–WATER

$K_2SO_4—Ag_2SO_4—H_2O$

$t=25$

Concn. in aqueous soln., g/l	
Potassium Sulfate	Silver Sulfate
1.74	7.67
3.49	7.36
8.72	7.20
17.44	7.24

№ 4751 [1143]

POTASSIUM SULFATE–SILVER SULFATE–WATER

$K_2SO_4—Ag_2SO_4—H_2O$

Concn. in aqueous soln., wt %		t	Concn. in aqueous soln., wt %		t
Potassium Sulfate	Silver Sulfate		Potassium Sulfate	Silver Sulfate	
3.22	0.863	33	3.12	1.273	75
5.62	0.940	33	5.73	1.406	75
8.37	1.046	33	8.43	1.554	75
10.41	1.117	33	10.55	1.665	75
11.80	1.177	33	13.17	1.806	75
3.20	1.023	51	17.06	2.021	75
5.61	1.127	51	3.23	1.488	100
8.40	1.247	51	5.60	1.675	100
10.55	1.340	51	8.45	1.890	100
13.16	1.450	51	11.30	2.115	100
14.37	1.524	51	15.07	2.410	100
			18.58	2.677	100

№ 4752 [1385]

POTASSIUM SULFATE–SILVER SULFATE–WATER

$K_2SO_4—Ag_2SO_4—H_2O$

$t=25$

Concn. of Potassium Sulfate in aqueous soln., g/l	Solubility of Silver Sulfate in aqueous soln. of Potassium Sulfate, wt %	Concn. of Potassium Sulfate in aqueous soln., g/l	Solubility of Silver Sulfate in aqueous soln. of Potassium Sulfate, wt %
1.74	0.8032	34.85	0.8063
3.48	0.7800	87.13	1.0183
8.71	0.7551	saturated	1.1652
17.43	0.7577		

№ 4753 [2578]

SILVER SULFATE–POTASSIUM SULFATE–WATER

$AgSO_4—K_2SO_4—H_2O$

$t=25$

Concn. in aqueous soln., wt %		d	Comp. of solid phase
Silver Sulfate	Potassium Sulfate		
0.833	0.000	1.004	Ag_2SO_4
0.79	3.86	1.033	»
0.905	7.29	1.067	»
0.991	9.72	1.084	»
1.009	10.25	1.090	»
1.03	10.92	—	$Ag_2SO_4+K_2SO_4$
1.04	10.92	1.095	»
1.04	10.94	1.097	»
1.04	10.95	1.101	»
1.032	10.93	1.093	»
1.04	10.91	1.102	»
1.04	10.93	1.098	»
1.04	10.93	1.098	»
0.885	10.89	1.094	K_2SO_4
0.568	10.82	1.089	»
0.000	10.80	1.084	»

№ 4754 [2344]

POTASSIUM SULFATE–SILVER CHLORIDE–WATER

$K_2SO_4—AgCl—H_2O$

$t=25$

Concn. in aqueous soln., g/l			
...sium ...te	Silver Chloride	Potassium Sulfate	Silver Chloride
	0.00183	0.5406	0.00202
	0.00183	0.8364	0.00211
	0.00185	1.183	0.00215
	0.00188	1.495	0.00218
	0.00191	1.857	0.00222
	0.00195	2.310	0.00225
	0.00198		

№ 4755 [482]

POTASSIUM SULFATE–SILVER BROMATE–WATER

$K_2SO_4—AgBrO_3—H_2O$

$t=25$

Content in aqueous soln., mol/l	
Potassium Sulfate	Silver Bromate
0.000	$8.25 \cdot 10^{-3}$
0.001	$8.45 \cdot 10^{-3}$
0.003	$8.65 \cdot 10^{-3}$
0.010	$9.42 \cdot 10^{-2}$
0.030	$1.061 \cdot 10^{-2}$
0.100	$1.305 \cdot 10^{-2}$
0.300	$1.762 \cdot 10^{-2}$

№ 4756 POTASSIUM SULFATE–CADMIUM SULFATE–WATER [2117]

K_2SO_4—$CdSO_4$—H_2O

t=25

Comp. of saturated soln., wt %			Comp. of solid phase
Cadmium Sulfate	Potassium Sulfate	Water	
43.53	0.0	56.47	$CdSO_4 \cdot 8/3H_2O$
43.71	0.55	55.74	»
43.21	1.85	54.94	»
43.05	3.15	53.80	$CdSO_4 \cdot 8/3H_2O + CdSO_4 \cdot K_2SO_4$
39.22	4.33	56.45	$CdSO_4 \cdot K_2SO_4$
35.83	5.46	58.71	»
33.81	6.14	60.05	»
31.24	7.20	61.56	»
29.00	8.48	62.52	»
27.02	10.05	63.93	»
24.91	11.18	63.91	»
24.19	11.91	63.90	$Cd_2SO_4 \cdot K_2SO_4 + K_2SO_4$
18.95	11.90	69.15	K_2SO_4
15.73	11.62	72.65	»
12.55	11.40	76.05	»
10.62	11.39	77.99	»
6.60	11.50	81.90	»
4.58	11.13	84.29	»
2.66	10.77	86.57	»
0.0	10.77	89.23	»

№ 4757 POTASSIUM SULFATE–CADMIUM SULFATE–WATER [1208]

K_2SO_4—$CdSO_4$—H_2O

Comp. of saturated soln., wt %			t	Comp. of solid phase
Cadmium Sulfate	Potassium Sulfate	Water		
42.78	2.27	54.95	0	$CdSO_4 \cdot 8/3H_2O + 3CdSO_4 \cdot K_2SO_4 \cdot 5H_2O$
41.04	3.17	55.79	0	$3CdSO_4 \cdot K_2SO_4 \cdot 5H_2O$
39.23	4.22	56.55	0	»
39.08	4.52	56.40	0	$3CdSO_4 \cdot K_2SO_4 \cdot 5H_2O + CdSO_4 \cdot K_2SO_4 \cdot 4H$[illegible]
30.15	6.42	63.43	0	$CdSO_4 \cdot K_2SO_4 \cdot 4H_2O$
27.38	7.20	65.42	0	»
20.61	9.04	70.35	0	$CdSO_4 \cdot K_2SO_4 \cdot 4H_2O + K_2S$[illegible]
15.48	8.60	75.92	0	K_2SO_4
41.86	2.48	55.64	13.3	$CdSO_4 \cdot 8/3H_2O + 3CdSO_4 \cdot$ [illegible]
41.52	2.68	55.80	13.3	$3CdSO_4 \cdot K_2SO$[illegible]
36.93	5.61	57.46	13.3	$3CdSO_4 \cdot K_2SO_4 \cdot 5H_2O$ [illegible] $+CdSO_4 \cdot$ [illegible]
35.58	6.05	58.37	13.3	$CdSO_4$ [illegible]
31.80	7.07	61.13	13.3	[illegible]
21.18	10.64	68.18	13.3	$CdSO_4 \cdot K_2$ [illegible]
15.69	10.39	73.92	13.3	[illegible]
6.69	9.89	83.42	13.3	[illegible]
43.40	2.08	54.52	25	$CdSO_4 \cdot 8/3H_2O$
39.38	3.58	57.04	25	$3CdSO_4 \cdot$ [illegible]
35.22	6.38	58.40	25	$3CdSO_4 \cdot K_2SO_4 \cdot 5H$[illegible]
30.62	7.97	61.41	25	$CdSO_4 \cdot$ [illegible]

Table № 4757 (continuation)

Comp. of saturated soln., wt %			t	Comp. of solid phase
Cadmium Sulfate	Potassium Sulfate	Water		
26.11	9.52	64.37	25	$CdSO_4 \cdot K_2SO_4 \cdot 1\frac{1}{2}H_2O$
22.03	11.12	66.85	25	»
20.52	11.90	67.58	25	$CdSO_4 \cdot K_2SO_4 \cdot 1\frac{1}{2}H_2O + K_2SO_4$
16.35	11.90	71.75	25	K_2SO_4
43.58	2.07	54.35	50	$CdSO_4 \cdot H_2O + 3CdSO_4 \cdot K_2SO_4 \cdot 2H_2O$
41.10	2.83	56.07	50	$3CdSO_4 \cdot K_2SO_4 \cdot 2H_2O$
40.43	3.14	56.43	50	$3CdSO_4 \cdot K_2SO_4 \cdot 2H_2O + 3CdSO_4 \cdot K_2SO_4 \cdot 5H_2O$
35.33	5.78	58.89	50	$3CdSO_4 \cdot K_2SO_4 \cdot 5H_2O$
31.70	8.52	59.78	50	$3CdSO_4 \cdot K_2SO_4 \cdot 5H_2O + CdSO_4 \cdot K_2SO_4 \cdot 1\frac{1}{2}H_2O$
29.20	9.22	61.58	50	$CdSO_4 \cdot K_2SO_4 \cdot 1\frac{1}{2}H_2O$
23.70	11.19	65.11	50	»
17.35	13.77	68.88	50	»
15.14	14.88	69.98	50	$CdSO_4 \cdot K_2SO_4 \cdot 1\frac{1}{2}H_2O + K_2SO_4$
3.20	14.33	82.47	50	K_2SO_4
42.65	1.90	55.45	60	$CdSO_4 \cdot H_2O + 3CdSO_4 \cdot K_2SO_4 \cdot 2H_2O$
37.93	4.12	57.95	60	$3CdSO_4 \cdot K_2SO_4 \cdot 2H_2O$
33.80	6.42	59.88	60	»
31.51	8.07	60.42	60	$3CdSO_4 \cdot K_2SO_4 \cdot 2H_2O + 3CdSO_4 \cdot K_2SO_4 \cdot 5H_2O +$ $+ 3CdSO_4 \cdot 2K_2SO_4 \cdot H_2O$
30.38	9.07	60.55	60	$3CdSO_4 \cdot 2K_2SO_4 \cdot H_2O$
29.50	9.39	61.11	60	$3CdSO_4 \cdot 2K_2SO_4 \cdot H_2O + CdSO_4 \cdot K_2SO_4 \cdot 1\frac{1}{2}H_2O$
30.20	9.20	60.60	66	$3CdSO_4 \cdot K_2SO_4 \cdot 2H_2O + 3CdSO_4 \cdot 2K_2SO_4 \cdot H_2O$
40.88	2.24	56.88	75	$CdSO_4 \cdot H_2O + 3CdSO_4 \cdot K_2SO_4 \cdot 2H_2O$
35.13	4.05	60.82	75	$3CdSO_4 \cdot K_2SO_4 \cdot 2H_2O$
25.68	10.45	63.87	75	»
26.15	10.30	63.55	75	$3CdSO_4 \cdot K_2SO_4 \cdot 2H_2O + 3CdSO_4 \cdot 2K_2SO_4 \cdot H_2O$
23.10	11.65	65.25	75	$3CdSO_4 \cdot 2K_2SO_4 \cdot H_2O$
21.79	12.39	65.82	75	$3CdSO_4 \cdot 2K_2SO_4 \cdot H_2O + CdSO_4 \cdot K_2SO_4 \cdot 1\frac{1}{2}H_2O$
17.32	14.37	68.31	75	$CdSO_4 \cdot K_2SO_4 \cdot 1\frac{1}{2}H_2O$
14.17	15.93	69.90	75	»
11.69	17.54	70.77	75	$CdSO_4 \cdot K_2SO_4 \cdot 1\frac{1}{2}H_2O + K_2SO_4$
6.76	17.06	76.18	75	K_2SO_4
39.42	2.38	58.20	90	$CdSO_4 \cdot H_2O + 3CdSO_4 \cdot K_2SO_4 \cdot 2H_2O$
32.04	4.44	63.52	90	$3CdSO_4 \cdot K_2SO_4 \cdot 2H_2O$
25.71	7.68	66.61	90	$3CdSO_4 \cdot K_2SO_4 \cdot 2H_2O$
20.61	12.83	66.56	90	$3CdSO_4 \cdot K_2SO_4 \cdot 2H_2O + 3CdSO_4 \cdot 2K_2SO_4 \cdot H_2O$
17.24	14.78	67.98	90	$3CdSO_4 \cdot 2K_2SO_4 \cdot H_2O$
15.91	15.79	68.30	90	$3CdSO_4 \cdot 2K_2SO_4 \cdot H_2O + CdSO_4 \cdot K_2SO_4 \cdot 1\frac{1}{2}H_2O$
13.37	16.43	70.20	90	$CdSO_4 \cdot K_2SO_4 \cdot 1\frac{1}{2}H_2O$
10.45	19.45	70.10	90	$CdSO_4 \cdot K_2SO_4 \cdot 1\frac{1}{2}H_2O + K_2SO_4$
6.00	19.42	74.58	90	K_2SO_4
37.58	2.54	59.88	99	$CdSO_4 \cdot H_2O + 3CdSO_4 \cdot K_2SO_4 \cdot 2H_2O$
32.31	3.69	64.00	99	$3CdSO_4 \cdot K_2SO_4 \cdot 2H_2O$
28.73	4.95	66.32	99	»
22.42	7.77	69.81	99	»
20.13	10.58	69.29	99	»
17.38	13.74	68.88	99	$3CdSO_4 \cdot K_2SO_4 \cdot 2H_2O + 3CdSO_4 \cdot 2K_2SO_4 \cdot H_2O$
16.36	14.36	69.28	99	$3CdSO_4 \cdot 2K_2SO_4 \cdot H_2O$
13.17	17.53	69.30	99	$3CdSO_4 \cdot 2K_2SO_4 \cdot H_2O + CdSO_4 \cdot K_2SO_4 \cdot 1\frac{1}{2}H_2O$
10.78	19.29	69.93	99	$CdSO_4 \cdot K_2SO_4 \cdot 1\frac{1}{2}H_2O$
9.50	20.23	70.27	99	$CdSO_4 \cdot K_2SO_4 \cdot 1\frac{1}{2}H_2O + K_2SO_4$
5.29	19.93	74.78	99	K_2SO_4

№ 4758 **[246]**

POTASSIUM SULFATE–INDIUM SULFATE–WATER

K_2SO_4—$In_2(SO_4)_3$—H_2O

$t=20$

Comp. of saturated soln., wt %			Comp. of solid phase
Indium Sulfate	Potassium Sulfate	Water	
54.0	0.0	46.0	$In_2(SO_4)_3 \cdot 5H_2O$
50.6	0.2	49.2	$KIn(SO_4)_2 \cdot 2H_2O$
44.0	0.3	55.7	»
41.8	0.4	57.8	»
38.0	0.5	61.5	»
37.2	0.9	61.9	»
29.3	2.0	68.7	»
23.3	2.0	74.7	»
22.1	2.1	55.8	»
18.7	2.6	78.7	»
17.4	2.8	79.8	»
16.8	2.9	80.3	»
14.6	3.2	82.2	»
13.57	4.5	81.93	»
13.1	4.8	82.1	$K_2In(OH)(SO_4)_2 \cdot nH_2O$
7.6	5.6	86.8	»
4.1	5.5	90.4	»
2.2	5.8	92.0	»
1.2	9.0	89.8	»
0.0	10.0	90.0	K_2SO_4

№ 4759 **[2438]**

POTASSIUM SULFATE–ANTIMONY TRIFLUORIDE–WATER

K_2SO_4—SbF_3—H_2O

$t=0$

Concn. of Potassium Sulfate in aqueous soln., g/l	Solubility of Antimony Trifluoride in aqueous soln. of Potassium Sulfate, wt %
87.05	80.76
43.525	80.33
21.86	80.26

№ 4760 POTASSIUM SULFATE–CESIUM SULFATE–WATER [936]

K_2SO_4—Cs_2SO_4—H_2O

$t=25$

Comp. of saturated soln., wt %			d	Comp. of solid phase
Cesium Sulfate	Potassium Sulfate	Water		
64.36	0.00	35.64	2.016	Cs_2SO_4
63.34	1.13	35.52	2.015	»
61.93	2.97	35.10	2.024	»
61.88	3.10	35.02	—	$Cs_2SO_4+K_2SO_4$
62.00	2.96	35.04	—	»
62.08	2.85	35.07	—	»
61.96	2.97	35.07	—	»
61.99	3.00	35.01	—	»
61.90	3.95	35.05	—	»
62.02	2.89	35.09	—	K_2SO_4
55.41	3.45	41.14	1.946	»
49.86	4.13	46.01	1.841	»
44.91	4.82	50.27	1.742	»
39.14	5.56	55.20	1.621	»
32.24	6.38	61.38	1.502	»
27.67	6.71	65.62	1.446	»
21.12	7.60	71.28	1.348	»
15.84	8.73	75.43	1.260	»
10.80	9.11	80.09	1.195	»
5.11	10.02	84.87	1.133	»
0.00	10.76	89.24	1.091	»

№ 4761 [937]

POTASSIUM SULFATE–CESIUM SULFATE–WATER

K_2SO_4—Cs_2SO_4—H_2O

$t=50$

Comp. of saturated soln., wt %			Comp. of solid phase
Potassium Sulfate	Cesium Sulfate	Water	
0.00	66.13	33.87	Cs_2SO_4
0.97	64.93	34.10	»
2.32	63.97	33.71	»
3.98	62.49	33.53	»
3.84	62.60	33.56	$K_2SO_4+Cs_2SO_4$
3.75	62.76	33.49	»
3.66	62.71	33.63	»
3.91	62.48	33.61	»
3.96	62.67	33.37	K_2SO_4
4.56	56.81	38.63	»
5.64	48.36	46.00	»
6.69	42.88	50.43	»
7.45	37.58	54.97	»
8.57	30.89	60.54	»

Table № 4761 (continuation)

Comp. of saturated soln., wt %			Comp. of solid phase
Potassium Sulfate	Cesium Sulfate	Water	
9.39	25.60	65.01	K_2SO_4
10.43	20.65	68.92	»
11.38	15.62	73.00	»
12.51	8.53	78.96	»
13.02	4.70	82.28	»
14.26	0.00	85.74	»

№ 4762 [1143]

POTASSIUM SULFATE–LANTHANUM SULFATE–WATER

$K_2SO_4—La_2(SO_4)_3—H_2O$

$t=16.5$

Comp. of saturated soln., wt %			Comp. of solid phase
Lanthanum Sulfate	Potassium Sulfate	Water	
2.15	0.00	97.85	$La_2(SO_4)_3 \cdot 9H_2O$
0.72	0.24	99.04	$La_2(SO_4)_3 \cdot K_2SO_4 \cdot 2H_2O$
0.27	0.49	99.24	»
0.18	0.84	98.98	»
0.15	1.02	98.93	$La_2(SO_4)_3 \cdot 5K_2SO_4$
0.02	1.14	98.84	»

№ 4763

POTASSIUM SULFATE–LANTHANUM SULFATE–WATER

[2767]

$K_2SO_4—La_2(SO_4)_3—H_2O$

$t=25$

Comp. of saturated soln., wt %			Comp. of solid phase
Lanthanum Sulfate	Potassium Sulfate	Water	
0.87	0.23	98.90	$La_2(SO_4)_3 \cdot K_2SO_4 \cdot 2H_2O$
0.77	0.23	99.00	»
0.41	0.45	99.14	»
0.28	0.79	98.93	$2La_2(SO_4)_3 \cdot 3K_2SO_4 \cdot 8H_2O$
0.26	0.74	99.00	$2La_2(SO_4)_3 \cdot 3K_2SO_4 \cdot 8H_2O + La_2(SO_4)_3 \cdot 3K_2SO_4$
0.24	0.87	98.89	$La_2(SO_4)_3 \cdot 3K_2SO_4$
0.00	1.62	98.38	$La_2(SO_4)_3 \cdot 3K_2SO_4 + La_2(SO_4)_3 \cdot 4K_2SO_4 \cdot H_2O$
0.00	1.51	98.49	$La_2(SO_4)_3 \; 4K_2SO_4 \cdot H_2O$
0.00	4.24	95.76	»
0.00	4.78	95.22	$La_2(SO_4)_3 \cdot 4\frac{1}{2} K_2SO_4 \cdot 2H_2O$
0.00	10.16	89.84	»
0.00	11.11	88.89	$La_2(SO_4)_3 \cdot 5K_2SO_4 \cdot 2H_2O$
0.00	11.02	88.98	$La_2(SO_4)_3 \cdot 5K_2SO_4 \cdot 2H_2O + K_2SO_4$

№ 4764 POTASSIUM SULFATE–LANTHANUM IODATE–WATER [1677, 2039]

K_2SO_4— $La(IO_3)_3$—H_2O

$t=25$

Concn. of Potassium Sulfate in aqueous soln., g/l	Solubility of Lanthanum Iodate in aqueous soln. of Potassium Sulfate, g/l	Concn. of Potassium Sulfate in aqueous soln., g/l	Solubility of Lanthanum Iodate in aqueous soln. of Potassium Sulfate, g/l	Concn. of Potassium Sulfate in aqueous soln., g/l	Solubility of Lanthanum Iodate in aqueous soln. of Potassium Sulfate, g/l
0.035	0.6084	0.348	0.8066	3.484	1.57126
0.087	0.6468	1.219	1.1184	8.71	2.1152
0.174	0.7011	2.177	1.3454	8.71	2.2366

№ 4765 POTASSIUM SULFATE–LANTHANUM IODATE–WATER [2313]

K_2SO_4— $La(IO_3)_3$—H_2O

$t=25$

Concn. of Potassium Sulfate in aqueous soln., g/l	Solubility of Lanthanum Iodate in aqueous soln. of Potassium Sulfate, g/l	d_4	Concn. of Potassium Sulfate in aqueous soln., g/l	Solubility of Lanthanum Iodate in aqueous soln. of Potassium Sulfate, g/l	d_4
0.174	0.547	0.99790	1.74	0.988	0.99952
0.87	0.806	0.99847	8.64	1.704	1.00554

№ 4766 POTASSIUM SULFATE–CERIUM SULFATE–WATER [1141]

K_2SO_4—$Ce_2(SO_4)_3$—H_2O

$t=16$

Concn. in aqueous soln., wt %			
Potassium Sulfate	Cerium Sulfate	Potassium Sulfate	Cerium Sulfate
0.0	9.70	0.719	0.248
0.176	0.945	1.273	0.041
0.505	0.427		

№ 4767 [2773]

POTASSIUM SULFATE–CERIUM SULFATE–WATER

K_2SO_4—$Ce_2(SO_4)_3$—H_2O

$t=25$

Concn. in aqueous soln., wt %		Comp. of solid phase
Potassium Sulfate	Cerium Sulfate	
0.16	1.74	$Ce_2(SO_4)_3 \cdot K_2SO_4 \cdot 2H_2O + 2Ce_2(SO_4)_3 \cdot 3K_2SO_4 \cdot 8H_2O$
0.79	0.30	$2Ce_2(SO_4)_3 \cdot 3K_2SO_4 \cdot 8H_2O$
1.13	0.15	$2Ce_2(SO_4)_3 \cdot 3K_2SO_4 \cdot 8H_2O + Ce_2(SO_4)_3 \cdot 4\frac{1}{2} K_2SO_4$
1.17	0.13	»

Table № 4767 (continuation)

Concn. in aqueous soln., wt %		Comp. of solid phase
Potassium Sulfate	Cerium Sulfate	
1.26	0.12	$2Ce_2(SO_4)_3 \cdot 3K_2SO_4 \cdot 8H_2O + Ce_2(SO_4)_3 \cdot 4^1/_2K_2SO_4$
1.43	0.05	»
1.77	0.00	»
2.87	0.00	»
4.65	0.00	»
5.56	0.00	$Ce_2(SO_4)_3 \cdot 5K_2SO_4$
6.22	0.00	»
7.72	0.00	»
9.22	0.00	»
9.80	0.00	$Ce_2(SO_4)_3 \cdot 5K_2SO_4 + K_2SO_4$

№ 4768 POTASSIUM SULFATE–CERIUM SULFATE–WATER [2773]

$K_2SO_4—Ce_2(SO_4)_3—H_2O$

$t=25$

Concn. in aqueous soln., wt %		Comp. of solid phase	Concn. in aqueous soln., wt %		Comp. of solid phase
Potassium Sulfate	Cerium Sulfate		Potassium Sulfate	Cerium Sulfate	
0.02	7.95	$Ce_2(SO_4)_3 \cdot K_2SO_4 \cdot 2H_2O$	0.07	4.88	$Ce_2(SO_4)_3 \cdot K_2SO_4 \cdot 2H_2O$
0.04	6.66	»	0.12	3.48	»

№ 4769 POTASSIUM SULFATE–CERIUM SULFATE–WATER [2527]

$K_2SO_4—Ce_2(SO_4)_3—H_2O$

Concn. in aqueous soln., wt %		t	Comp. of solid phase
Potassium Sulfate	Cerium Sulfate		
0.0	13.93	0	$Ce_2(SO_4)_3 \cdot 12H_2O$
0.0	14.06	0	$Ce_2(SO_4)_3 \cdot 8H_2O$
0.012	13.94	0	$Ce_2(SO_4)_3 \cdot K_2SO_4 \cdot 2H_2O$
0.073	9.79	0	»
0.107	6.11	0	»
0.107	5.58	0	$Ce_2(SO_4)_3 \cdot K_2SO_4 \cdot 2H_2O + 2Ce_2(SO_4)_3 \cdot 3K_2SO_4 \cdot 8H_2O$
0.142	4.12	0	$2Ce_2(SO_4)_3 \cdot 3K_2SO_4 \cdot 8H_2O$
0.217	2.04	0	»
0.279	1.10	0	»
0.407	0.631	0	»
0.692	0.190	0	$2Ce_2(SO_4)_3 \cdot 3K_2SO_4 \cdot 8H_2O + Ce_2(SO_4)_3 \cdot 4K_2SO_4$
0.724	0.158	0	$Ce_2(SO_4)_3 \cdot 4K_2SO_4$ (or $5K_2SO_4$)
0.812	0.091	0	»
1.353	0.018	0	»
0.0	7.77	25	$Ce_2(SO_4)_3 \cdot 8H_2O$
0.032	7.27	25	$Ce_2(SO_4)_3 \cdot K_2SO_4 \cdot 2H_2O$
0.062	4.63	25	»
0.083	2.60	25	»
0.142	1.09	25	»
0.202	0.642	25	»

Concn. in aqueous soln., wt %		t	Comp. of solid phase
Potassium Sulfate	Cerium Sulfate		
0.307	0.461	25	$Ce_2(SO_4)_3 \cdot K_2SO_4 \cdot 2H_2O$
0 471	0.356	25	»
0.657	0.321	25	$Ce_2(SO_4)_3 \cdot K_2SO_4 \cdot 2H_2O + 2Ce_2(SO_4)_3 \cdot 3K_2SO_4 \cdot 8H_2O$
0.686	0.285	25	$2Ce_2(SO_4)_2 \cdot 3K_2SO_4 \cdot 8H_2O$
0.934	0.218	25	»
0.983	0.189	25	$2Ce_2(SO_4)_3 \cdot 3K_2SO_4 \cdot 8H_2O$
1.056	0.157	25	$2Ce_2(SO_4)_3 \cdot 3K_2SO_4 \cdot 8H_2O + Ce_2(SO_4)_3 \cdot 4K_2SO_4$
1.148	0.117	25	$Ce_2(SO_4)_3 \cdot 4K_2SO_4$ (or $5K_2SO_4$)
1.395	0.043	25	»
2.641	0.002	25	»
0.0	4.622	50	$Ce_2(SO_4)_3 \cdot 8H_2O$
0.0	3.327	50	$Ce_2(SO_4)_3 \cdot 4H_2O$
0.011	4.278	50	$Ce_2(SO_4)_3 \cdot 4H_2O + Ce_2(SO_4)_3 \cdot K_2SO_4 \cdot 2H_2O$
0.021	3.735	50	$Ce_2(SO_4)_3 \cdot K_2SO_4 \cdot 2H_2O$
0.062	1.904	50	»
0.132	0.776	50	»
0.190	0.460	50	»
0.320	0.279	50	»
0.532	0.160	50	»
0.819	0.083	50	»
1.428	0.078	50	$Ce_2(SO_4)_3 \cdot K_2SO_4 \cdot 2H_2O + Ce_2(SO_4)_3 \cdot 4K_2SO_4$
1.651	0.055	50	$Ce_2(SO_4)_3 \cdot 4K_2SO_4$ (or $5K_2SO_4$)
2.029	0.034	50	»
2.289	0.016	50	»
0.0	1.076	75	$Ce_2(SO_4)_3 \cdot 4H_2O$
0.019	1.150	75	»
0.061	1.356	75	$Ce_2(SO_4)_3 \cdot 4H_2O + Ce_2(SO_4)_3 \cdot K_2SO_4 \cdot 2H_2O$
0.068	1.148	75	$Ce_2(SO_4)_3 \cdot K_2SO_4 \cdot 2H_2O$
0.101	0.697	75	»
0.144	0.384	75	»
0.324	0.134	75	»
0.545	0.065	75	»
1.549	0.040	75	»
2.197	0.019	75	»
2.278	0.023	75	$Ce_2(SO_4)_3 \cdot K_2SO_4 \cdot 2H_2O + Ce_2(SO_4)_3 \cdot 4K_2SO_4$

№ 4770 POTASSIUM SULFATE–CERIUM IODATE–WATER [1386]

K_2SO_4—$Ce(IO_3)_2$—H_2O

$t=25$

Concn. in aqueous soln., wt %			
Potassium Sulfate	Cerium Iodate	Potassium Sulfate	Cerium Iodate
0.00	0.1729	0.3460	0.3486
0.0347	0.2008	0.8599	0.4470
0.0868	0.2362	1.703	0.5532
0.1735	0.2806		

№ 4771 [2444]

POTASSIUM SULFATE–PRASEODYMIUM SULFATE–WATER

$K_2SO_4—PrSO_4—H_2O$

$t=25$

Comp. of saturated soln., wt %			Comp. of solid phase
Potassium Sulfate	Praseodymium Sulfate	Water	
4.07	0.0	95.93	$5K_2SO_4 \cdot Pr_2(SO_4)_3 \cdot H_2O$
4.01	0.0	95.99	»
3.20	0.0	96.80	$4\frac{1}{2}K_2SO_4 \cdot Pr_2(SO_4)_3$
1.47	0.16	98.37	$4K_2SO_4 \cdot Pr_2(SO_4)_3 \cdot H_2O$
1.03	0.18	98.79	$3K_2SO_4 \cdot Pr_2(SO_4)_3 \cdot 2H_2O$
0.88	0.32	98.80	»
0.67	0.40	98.93	»
0.34	0.52	99.14	—
0.27	1.87	97.86	$K_2SO_4 \cdot Pr_2(SO_4)_3 \cdot 2H_2O$
0.19	2.63	97.18	—

№ 4772 POTASSIUM SULFATE–NEODYMIUM SULFATE–WATER [2768]

$K_2SO_4—Nd_2(SO_4)_3—H_2O$

$t=25$

Comp. of saturated soln., wt %			Comp. of solid phase
Neodymium Sulfate	Potassium Sulfate	Water	
2.76	0.00	97.24	$Nd_2(SO_4)_3 \cdot K_2SO_4 \cdot 2H_2O$
2.02	0.15	97.83	»
1.69	0.30	98.01	»
1.62	0.47	97.91	$Nd_2(SO_4)_3 \cdot K_2SO_4 \cdot 2H_2O+2Nd_2(SO_4)_3 \cdot 3K_2SO_4 \cdot 8H_2O$
1.13	0.54	98.33	$2Nd_2(SO_4)_3 \cdot 3K_2SO_4 \cdot 8H_2O$
0.26	1.75	97.99	»
0.17	2.61	97.22	»
0.17	2.80	97.03	$2Nd_2(SO_4)_3 \cdot 3K_2SO_4 \cdot 8H_2O+Na_2(SO_4)_3 \cdot 3K_2SO_4 \cdot 2H_2O$
0.18	2.98	96.84	»
0.18	3.01	96.81	$Nd_2(SO_4)_3 \cdot 3K_2SO_4 \cdot 2H_2O$
0.17	3.40	96.43	$Nd_2(SO_4)_3 \cdot 3K_2SO_4 \cdot 2H_2O+Nd_2(SO_4)_3 \cdot 4K_2SO_4 \cdot H_2O$
0.08	4.06	95.86	$Nd_2(SO_4)_3 \cdot 4K_2SO_4 \cdot H_2O$
0.08	4.94	94.98	»
0.07	6.20	93.73	$Nd_2(SO_4)_3 \cdot 4K_2SO_4 \cdot 2H_2O$
0.04	8.52	91.44	»
0.04	10.01	89.95	»
0.04	10.85	89.11	$Nd_2(SO_4)_3 \cdot 5K_2SO_4 \cdot 2H_2O$
0.00	11.18	88.82	$Nd_2(SO_4)_3 \cdot 5K_2SO_4 \cdot 2H_2O+K_2SO_4$

№ 4773 [2760]

POTASSIUM SULFATE–MERCURY (I) SULFATE–WATER

$K_2SO_4—Hg_2SO_4—H_2O$

$t=25$

Concn. of Potassium Sulfate in aqueous soln., g/l	Solubility of Mercury Sulfate in aqueous soln. of Potassium Sulfate, g/l
34.87	0.45

№ 4774 [2363]

POTASSIUM SULFATE–THALLIUM SULFATE–WATER

$K_2SO_4—Tl_2SO_4—H_2O$

Concn. in aqueous soln., wt%		t
Thallium Sulfate	Potassium Sulfate	
0.0412	0.0895	15
0.0899	0.1282	62
0.1279	0.1810	100

№ 4775 **POTASSIUM SULFATE–THALLIUM SULFATE–WATER** [2704]

$K_2SO_4—Tl_2SO_4—H_2O$

Concn. in aqueous soln., wt%		t	d	Concn. in aqueous soln., wt%		t	d
Thallium Sulfate	Potassium Sulfate			Thallium Sulfate	Potassium Sulfate		
3.568	0.00	10	1.031	4.13	9.42	25	1.118
2.94	7.29	10	1.087	3.79	9.46	25	1.113
0.00	8.42	10	1.066	3.61	9.53	25	1.113
5.222	0.00	25	1.046	2.80	9.88	25	—
5.04	1.51	25	1.056	1.79	10.24	25	1.100
4.98	3.04	25	—	1.79	10.31	25	1.101
4.91	4.38	25	1.080	1.79	10.26	25	1.099
4.84	5.67	25	1.091	1.02	10.46	25	1.093
4.50	8.40	25	1.113	1.03	10.44	25	1.093
4.46	8.64	25	1.113	0.00	10.76	25	1.085
4.26	9.37	25	1.118	7.72	0.00	45	1.064
4.25	9.31	25	1.118	6.40	11.30	45	1.146
4.21	9.43	25	—	0.00	13.58	45	1.098

№ 4776 [1270]

POTASSIUM SULFATE–THALLIUM CHLORIDE–WATER

$K_2SO_4—TlCl—H_2O$

$t=25$

Concn. of Potassium Sulfate in aqueous soln., g/l.	Solubility of Thallium Chloride in aqueous soln., of Potassium Sulfate, g/l	d_4
1.738	4.267	1.0012
4.351	4.658	1.0037
8.700	5.126	1.0074
26.100	6.237	1.0221
86.978	8.194	1.0698

№ 4777 [2581]

POTASSIUM SULFATE–THALLIUM BROMATE–WATER

$K_2SO_4—TlBr—H_2O$

$t=20$

Concn. in aqueous soln., g/l	
Thallium Bromate	Potassium Sulfate
0.4887	0.3485
0.5328	1.7426
0.5882	5.2278

№ 4778 POTASSIUM SULFATE–THALLIUM IODATE–WATER [2039]

$K_2SO_4—TlIO_3—H_2O$

$t=25$

Concn. of Potassium Sulfate in aqueous soln., g/l	Solubility of Thallium Iodate in aqueous soln. of Potassium Sulfate g/l.	Concn. of Potassium Sulfate in aqueous soln., g/l.	Solubility of Thallium Iodate in aqueous soln. of Potassium Sulfate, g/l.	Concn. of Potassium Sulfate in aqueous soln., g/l.	Solubility of Thallium Iodate in aqueous soln. of Potassium Sulfate, g/l.
0.174	0.725	2.264	0.854	11.149	1.091
0.348	0.743	5.226	0.953	26.130	1.335
0.871	0.787	5.801	0.965	43.550	1.528
1.742	0.832	8.710	1.042	87.10	1.945

№ 4779 POTASSIUM SULFATE–LEAD SULFATE–WATER [1879]

$K_2SO_4—PbSO_4—H_2O$

Concn. in aqueous soln., g/l		t	Concn. in aqueous soln., g/l		t	Concn. in aqueous soln., g/l		t
Lead Sulfate	Potassium Sulfate		Lead Sulfate	Potassium Sulfate		Lead Sulfate	Potassium Sulfate	
0.0125	0.1	18	0.0002	30.0	18	0.0055	1.5	30
0.0047	0.5	18	0.0002	100.0	18	0.0036	5.0	30
0.0053	1.0	18	0.0128	0.1	30	0.0047	7.5	30
0.0042	2.5	18	0.0064	0.5	30	0.0007	10.0	30
0.0037	5.0	18	0.0060	0.75	30	0.0011	20.0	30
0.0007	10.0	18						

№ 4780 POTASSIUM SULFATE–LEAD SULFATE–WATER [213]

$K_2SO_4—PbSO_4—H_2O$

$t=18—20$

Concn. of Potassium Sulfate in aqueous soln., g/l	Solubility of Lead Sulfate in aqueous soln. of Potassium Sulfate, g/l	Concn. of Potassium Sulfate in aqueous soln., g/l	Solubility of Lead Sulfate in aqueous soln. of Potassium Sulfate g/l
0.00	0.042	2.788	0.017
0.348	0.028	5.576	0.033
0.697	0.024	11.153	0.034
1.394	0.016		

№ 4781 [1143]

POTASSIUM SULFATE–THORIUM SULFATE–WATER

$K_2SO_4—Th(SO_4)_2—H_2O$

Concn. in aqueous soln., wt%		*t*	Comp. of solid phase
Thorium Sulfate	Potassium Sulfate		
1.37	0.0	16	$Th(SO_4)_2 \cdot 9H_2O$
1.632	0.415	16	$Th(SO_4)_2 \cdot K_2SO_4 \cdot 4H_2O$
2.125	0.973	16	»
3.058	1.104	16	»
2.423	1.179	16	»
2.146	1.238	16	»
1.655	1.308	16	»
1.589	1.337	16	$Th(SO_4)_2 \cdot 2K_2SO_4 \cdot 2H_2O$
0.850	1.452	16	»
0.362	1.804	16	»
0.068	2.987	16	»
0.026	3.891	16	$Th(SO_4)_2 \cdot 3.5K_2SO_4$
0.003	4.602	16	»
0.916	0.0	75	$Th(SO_4)_2 \cdot 9H_2O$
1.114	0.848	75	$Th(SO_4)_2 \cdot K_2SO_4 \cdot 4H_2O$
1.146	1.14	75	»
1.095	1.145	75	»
0.887	1.242	75	»
0.486	1.273	75	»
0.291	1.813	75	»
0.194	3.016	75	$Th(SO_4)_2 \cdot 2K_2SO_4 \cdot 2H_2O$
0.244	4.440	75	»
0.161	5.068	75	»
0.116	5.593	75	»
0.029	6.694	75	$Th(SO_4)_2 \cdot 3.5K_2SO_4$
0.021	8.854	75	»

№ 4782 [1409]

POTASSIUM SULFATE–URANYL SULFATE–WATER

$K_2SO_4—UO_2SO_4—H_2O$

$t=25$

Comp. of saturated soln., wt %			Comp. of solid phase
Potassium Sulfate	Uranyl Sulfate	Water	
10.4	0.0	89.6	K_2SO_4
10.67	9.91	88.42	»
11.02	1.59	87.39	»
11.17	2.39	86.44	»
9.01	2.41	88.58	$K_2(UO_2)(SO_4)_2 \cdot 2H_2O$
7.31	2.43	90.26	»
3.50	4.73	91.77	»
2.69	9.91	87.40	»

Table 4782 (continuation)

Comp. of saturated soln., wt %			Comp. of solid phase
Potassium Sulfate	Uranyl Sulfate	Water	
1.91	31.69	66.40	$K_2(UO_2)(SO_4) \cdot 2H_2O$
1.79	35.97	62.24	»
1.77	39.73	58.50	»
1.71	51.70	46.59	$K_2(UO_2)_2(SO_4)_3 \cdot 5H_2O$
0.54	59.20	40.26	»
0.45	61.50	38.05	»
0.23	61.25	38.52	$UO_2SO_4 \cdot 3H_2O$
0.0	61.18	38.82	»

№ 4783 POTASSIUM DITHIONATE–BARIUM DITHIONATE–WATER [1122]

$K_2S_2O_6—BaS_2O_6—H_2O$

Comp. of saturated soln., wt %			t	Comp. of solid phase
Potassium Dithionate	Barium Dithionate	Water		
1.21	8.02	90.77	0	$BaS_2O_6 \cdot 2H_2O$
2.75	8.88	88.37	0	$BaS_2O_6 \cdot 2H_2O+K_2S_2O_6$
2.60	4.69	92.71	0	$K_2S_2O_6$
2.52	0.00	97.48	0	»
2.69	16.43	80.88	20	$BaS_2O_6 \cdot 2H_2O$
6.03	17.15	76.82	20	$BaS_2O_6 \cdot 2H_2O+K_2S_2O_6$
6.02	11.82	82.16	20	$K_2S_2O_6$
6.23	0.00	93.77	20	»
4.16	20.64	75.20	30	$BaS_2O_6 \cdot 2H_2O$
8.03	20.69	71.28	30	$BaS_2O_6 \cdot 2H_2O+K_2S_2O_6$
8.11	20.50	71.39	30	$K_2S_2O_6$
8.33	11.74	79.93	30	»
8.49	6.02	85.49	30	»
8.54	0.00	91.46	30	»

№ 4784 POTASSIUM TRITHIONATE–POTASSIUM TETRATHIONATE–WATER [2026]

$K_2S_3O_6—K_2S_4O_6—H_2O$

Comp. of saturated soln., wt %			t	Comp. of solid phase
Potassium Tetrathionate	Potassium Trithionate	Water		
12.00	6.68	81.32	0	$K_2S_4O_6+K_2S_3O_6$
21.85	1.80	76.35	20	$K_2S_4O_6$
21.60	2.30	76.10	20	»
18.82	6.86	74.32	20	»
18.90	11.50	69.60	20	»
18.65	13.26	69.09	20	$K_2S_4O_6+K_2S_3O_6$
14.17	14.51	71.32	20	$K_2S_3O_6$
11.94	16.70	71.36	20	»
5.75	16.23	78.02	20	»
2.37	17.67	79.96	20	»
12.00	6.68	81.32	20	»

№ 4785 [2026]

POTASSIUM TRITHIONATE–POTASSIUM PENTATHIONATE–WATER

$K_2S_3O_6—K_2S_5O_6—H_2O$

$t=0$

Comp. of saturated soln., wt %			Comp. of solid phase
Potassium Trithionate	Potassium Pentathionate	Water	
7.60	2.87	89.53	$K_2S_3O_6+K_2S_5O_6 \cdot 1\frac{1}{2}H_2O+S$
6.35	7.20	86.45	$K_2S_5O_6 \cdot 1\frac{1}{2}H_2O+S$
8.13	8.17	83.70	$K_2S_5O_6 \cdot 1\frac{1}{2}H_2O$
10.50	8.72	80.78	$K_2S_5O_6 \cdot 1\frac{1}{2}H_2O+K_2S_3O_6+S$
10.84	9.98	79.18	$K_2S_3O_6+S$
14.04	3.74	82.22	»
14.12	4.37	81.51	»
14.99	2.46	82.55	»

№ 4786 POTASSIUM TETRATHIONATE–POTASSIUM PENTATHIONATE–WATER [2026]

$K_2S_4O_6—K_2S_5O_6—H_2O$

Comp. of saturated soln., wt %			t	Comp. of solid phase
Potassium Tetrathionate	Potassium Pentathionate	Water		
4.71	14.49	80.80	0	$K_2S_2O_4+K_2S_5O_6 \cdot 1\frac{1}{2}H_2O$
2.53	21.39	76.08	20	$K_2S_5O_6 \cdot 1\frac{1}{2}H_2O$
5.86	20.59	73.55	20	»
7.53	20.19	72.28	20	»
10.34	20.86	68.80	20	»
10.88	20.10	69.02	20	$K_2S_5O_6 \cdot 1\frac{1}{2}H_2O+K_2S_4O_6$
10.53	19.81	69.66	20	$K_2S_4O_6$
11.71	17.62	70.67	20	»
14.03	14.30	71.67	20	»
17.39	8.70	73.91	20	»
18.93	6.44	74.63	20	»
22.46	0.82	76.72	20	»

№ 4787 POTASSIUM CHLORIDE–POTASSIUM CHLORATE–WATER [2748]

$KCl—KClO_3—H_2O$

$t=20$

Concn. in aqueous soln., g/l		d	Concn. in aqueous soln., g/l		d	Concn. in aqueous soln., g/l		d
Potassium Chlorate	Potassium Chloride		Potassium Chlorate	Potassium Chloride		Potassium Chlorate	Potassium Chloride	
71.1	0.0	1.050	30.0	80	1.075	21.0	160	1.119
58.0	10	1.050	27.0	100	1.086	20.0	180	1.130
49.0	20	1.050	24.5	120	1.098	20.0	200	1.140
39.5	40	1.054	22.5	140	1.108	20.0	250	1.168
34.0	60	1.064						

№ 4788 [1239]

POTASSIUM CHLORIDE–POTASSIUM CHLORATE–WATER

$KCl—KClO_3—H_2O$

t=14—15

Comp. of saturated soln., wt %		
Potassium Chlorate	Potassium Chloride	Water
4.45	1.91	93.64
3.58	3.82	92.60

№ 4789 POTASSIUM CHLORIDE–POTASSIUM CHLORATE–WATER [1859]

$KCl—KClO_3—H_2O$

t=25

Concn. of Potassium Chloride in aqueous soln., g/l	Solubility of Potassium Chlorate in aqueous soln. of Potassium Chloride, g/l	Concn. of Potassium Chloride in aqueous soln., g/l	Solubility of Potassium Chlorate in aqueous soln. of Potassium Chloride, g/l	Concn. of Potassium Chloride in aqueous soln., g/l	Solubility of Potassium Chlorate in aqueous soln. of Potassium Chloride, g/l
0.00	83.030	29.6	64.191	74	48.142
7.4	76.653	37.0	60.375	148	34.673
14.8	71.484	55.5	54.010	222	27.600
22.2	67.793				

№ 4790 POTASSIUM CHLORIDE–POTASSIUM CHLORATE–WATER [1626]

$KCl—KClO_3—H_2O$

Comp. of saturated soln., wt %			d	t	Comp. of solid phase
Potassium Chloride	Potassium Chlorate	Water			
21.90	0.0	78.10	1.153	0	KCl
21.16	0.71	78.13	1.157	0	$KCl+KClO_3$
16.21	0.82	82.97	1.121	0	$KClO_3$
8.47	1.09	90.44	1.068	0	»
0.0	3.05	96.95	1.022	0	»
25.70	0.0	74.30	1.176	20	KCl
25.17	0.89	73.94	1.183	20	»
24.60	1.55	73.85	1.184	20	$KCl+KClO_3$
25.0	1.56	73.44	1.177	20	»
20.26	1.75	77.99	1.153	20	$KClO_3$
13.47	2.44	84.09	1.106	20	»
7.44	3.74	88.82	1.070	20	»
2.84	5.18	91.98	1.051	20	»
0.0	6.78	93.22	1.044	20	»
27.30	0.0	72.70	1.182	30	KCl
26.64	0.87	72.49	1.190	30	»
25.81	2.29	71.90	1.198	30	$KCl+KClO_3$
17.86	3.12	79.02	1.140	30	$KClO_3$

Table № 4790 (continuation)

Comp. of saturated soln., wt %			d	t	Comp. of solid phase
Potassium Chloride	Potassium Chlorate	Water			
13.06	3.82	83.14	1.108	30	$KClO_3$
11.03	4.38	84.59	1.097	30	»
9.45	4.82	85.73	1.088	30	»
4.67	6.62	88.71	—	30	»
0.0	9.24	90.76	1.058	30	»
28.75	0.0	71.25	1.188	40	KCl
27.74	1.54	70.72	1.196	40	»
26.49	3.15	70.36	1.206	40	$KCl+KClO_3$
20.66	3.88	75.46	1.165	40	$KClO_3$
18.43	4.30	77.27	1.148	40	»
13.16	5.44	81.40	1.116	40	»
10.76	6.39	82.85	1.106	40	»
8.71	7.25	84.04	1.098	40	»
7.66	7.64	84.70	1.092	40	»
4.60	9.20	86.20	1.084	40	»
0.0	11.65	88.35	1.074	40	»
30.18	0.0	69.82	1.194	50	KCl
28.24	3.27	68.49	1.207	50	»
27.45	4.46	68.09	1.214	50	$KCl+KClO_3$
28.30	4.32	67.38	1.211	50	»
18.53	5.76	75.71	1.155	50	$KClO_3$
17.55	6.01	76.44	1.147	50	»
8.51	9.66	81.83	1.105	50	»
0.0	14.76	85.24	1.088	50	»

№ 4791 [1220]

POTASSIUM CHLORIDE–POTASSIUM CHLORATE–WATER

$KCl—KClO_3—H_2O$

Concn. in aqueous soln., wt %		t	Comp. of solid phase
Potassium Chloride	Potassium Chlorate		
34.7	10.40	150	KCl
34.4	10.88	150	»
31.7	17.80	150	»
28.5	23.7	150	»
26.0	29.5	150	»
18.22	37.9	150	$KClO_3$
16.90	38.9	150	»
4.97	50.6	150	»
35.5	13.14	175	KCl
32.3	19.95	175	»
26.6	32.0	175	»
20.5	44.9	175	$KCl+KClO_3$
20.4	46.0	175	»
20.0	46.3	175	»

Table № 4791 (continuation)

Concn. in aqueous soln., wt %		t	Comp. of solid phase
Potassium Chloride	Potassium Chlorate		
9.28	55.9	175	$KClO_3$
38.5	12.3	200	KCl
35.6	16.8	200	»
25.2	36.8	200	»
21.6	45.9	200	»
14.55	58.0	200	»
5.76	66.6	200	$KCl+KClO_3$
3.29	69.4	200	$KClO_3$

№ 4792 POTASSIUM CHLORIDE–POTASSIUM CHLORATE–WATER [2238]

$KCl—KClO_3—H_2O$

Comp. of saturated soln., g/100 g Water		d	t	Comp. of solid phase
Potassium Chloride	Potassium Chlorate			
24.2	0.53	1.139	−10.85	$KCl+KClO_3+$ ice
24.7	0.56	1.143	− 9.8	$KCl+KClO_3$
22.05	0.565	1.1295	− 9.8	$KClO_3+$ ice
8.93	0.968	1.062	− 4.0	»
31.23	1.44	1.1709	10.0	$KCl+KClO_3$
37.00	3.21	1.1937	30.0	»
42.30	6.35	1.215	50.0	»
46.5	11.6	1.240	70.0	»
51.8	25.9	1.280	100.0	»

№ 4793 [2269]

POTASSIUM CHLORIDE–POTASSIUM PERCHLORATE–WATER

$KCl—KClO_4—H_2O$

$t=25$

Concn. in aqueous soln., g/l	
Potassium Perchlorate	Potassium Chloride
20.57	0.0
17.80	3.72
15.60	7.42

№ 4794 [1261]

POTASSIUM CHLORIDE–POTASSIUM PERCHLORATE–WATER

$KCl—KClO_4—H_2O$

$t=25$

Concn. in aqueous soln., g/l	
Potassium Perchlorate	Potassium Chloride
1.539	0.714
1.015	2.162
0.704	4.314

№ 4795 [1220]

POTASSIUM CHLORIDE–POTASSIUM PERCHLORATE–WATER

$KCl—KClO_4—H_2O$

Concn. in aqueous soln., wt %		t	Comp. of solid phase
Potassium Chloride	Potassium Perchlorate		
39.6	3.46	150	KCl
39.1	5.17	150	»
35.0	11.2	150	$KCl+KClO_4$
35.3	11.6	150	»
33.0	12.3	150	$KClO_4$
25.3	17.08	150	»
23.1	17.7	150	»
16.0	23.7	150	»
11.62	26.5	150	»
38.6	5.96	175	KCl
39.1	7.20	175	»
33.3	18.0	175	»
33.0	19.0	175	$KCl+KClO_4$
10.3	37.0	175	$KClO_4$
41.3	5.7	200	KCl
38.8	12.2	200	»
36.2	16.45	200	»
33.9	21.5	200	»
32.2	23.3	200	»
23.9	31.6	200	$KClO_4$
22.0	33.6	200	»
11.6	42.8	200	»
11.1	44.3	200	»
38.7	15.4	225	KCl
29.8	31.8	225	»
28.4	34.7	225	»
26.6	38.0	225	$KCl+KClO_4$
26.4	39.2	225	»
26.0	39.0	225	»
15.08	47.5	225	$KClO_4$
11.34	51.9	225	»
42.6	11.6	250	KCl
41.8	14.6	250	»
35.3	26.5	250	»
2.5	29.2	250	»
24.1	45.6	250	»
24.0	46.3	250	$KCl+KClO_4$
14.52	55.8	250	$KClO_4$

№ 4796 [1223]

POTASSIUM CHLORIDE–POTASSIUM PERCHLORATE–WATER

$KCl—KClO_4—H_2O$

Concn. in aqueous soln., wt %		t	Comp. of solid phase
Potassium Chloride	Potassium Perchlorate		
32.2	1.52	75	$KCl+\alpha\text{-}KClO_4$
43.1	3.36	100	»
34.1	7.80	150	»
35.8	16.7	200	»
38.5	26.8	250	»

№ 4797 **POTASSIUM CHLORIDE–POTASSIUM PERCHLORATE–WATER** [364]

$KCl—KClO_4—H_2O$

$t=20$

Comp. of saturated soln., wt %			Comp. of solid phase, wt %		
Potassium Perchlorate	Potassium Chloride	Water	Potassium Perchlorate	Potassium Chloride	Water
1.72	0.00	98.28	—	—	—
1.69	1.76	96.55	72.63	0.56	26.81
1.37	6.56	92.08	67.58	2.34	70.08
1.19	10.26	88.55	54.33	5.26	40.41
1.02	13.72	85.26	53.12	6.83	40.05
0.96	18.27	80.77	50.49	9.25	40.26
0.88	21.24	77.88	57.97	9.64	32.42
0.69	22.78	76.53	64.82	8.31	26.87
0.51	24.39	75.10	52.86	10.01	37.13
0.39	24.45	75.16	60.18	10.32	29.50
0.36	24.52	75.12	54.26	17.27	28.47
0.27	24.85	74.88	9.60	72.50	17.90
0.24	25.01	74.75	4.75	82.14	15.11
0.20	25.36	74.44	3.85	82.57	13.58
0.00	25.98	74.02	—	—	—

№ 4798 [927]

POTASSIUM CHLORIDE–POTASSIUM PERCHLORATE–WATER

$KCl—KClO_4—H_2O$

$t=35$

Comp. of saturated soln., wt %			Comp. of solid phase
Potassium Perchlorate	Potassium Chloride	Water	
3.25	0.00	96.75	$KClO_4$
2.64	2.63	94.73	»
2.60	6.14	91.80	»
1.58	9.71	88.71	»
1.37	12.94	85.69	»

Table № 4798 (continuation)

Comp. of saturated soln., wt %			Comp. of solid phase
Potassium Perchlorate	Potassium Chloride	Water	
1.05	16.14	82.81	$KClO_4$
0.98	20.28	78.74	»
0.93	24.01	74.46	$KClO_4+KCl$
0.94	24.06	75.00	»
0.89	24.28	74.83	»
0.54	25.85	73.61	»
0.36	26.20	73.44	$nKCl \cdot mKClO_4$
0.26	27.02	72.72	»
0.16	27.34	72.50	»
0.00	28.04	71.96	KCl

№ 4799 POTASSIUM CHLORIDE–POTASSIUM PERCHLORATE–WATER [365]

$KCl—KClO_4—H_2O$

Comp. of saturated soln., wt %			t	Comp. of solid phase
Potassium Perchlorate	Potassium Chloride	Water		
0.75	0.00	99.25	0	$KClO_4$
0.68	2.75	95.57	0	»
0.62	3.89	95.49	0	»
0.64	5.12	94.24	0	»
0.57	7.01	92.42	0	»
0.51	10.31	89.17	0	»
0.52	13.52	85.96	0	»
0.54	16.54	82.92	0	»
0.50	16.86	82.64	0	$KClO_4+KCl$
0.49	17.04	82.47	0	»
0.49	17.25	82.26	0	»
0.47	17.21	82.32	0	$nKClO_4 \cdot mKCl$
0.39	19.79	79.82	0	»
0.32	20.21	79.67	0	»
0.24	21.10	78.66	0	»
0.00	22.20	77.80	0	KCl
1.17	0.00	98.83	10	$KClO_4$
0.98	2.10	96.92	10	»
0.87	6.69	92.44	10	»
0.87	11.21	87.92	10	»
0.85	15.61	83.54	10	»
0.66	20.16	79.18	10	»
0.47	23.35	76.18	10	$KClO_4+KCl$
0.48	23.44	76.08	10	$KClO_4$
0.46	23.49	76.05	10	»
0.46	23.52	76.02	10	»
0.48	23.40	76.12	10	»
0.32	23.90	75.78	10	$nKClO_4 \cdot mKCl$
0.28	23.95	75.77	10	»
0.00	23.97	76.03	10	KCl

№ 4800 POTASSIUM CHLORIDE – POTASSIUM HEXACHLOROPLATINATE – WATER [1093]

$KCl—K_2PtCl_6—H_2O$

$t=20$

Concn. of Potassium Chloride in aqueous soln., g/l	Solubility of Potassium Hexachloro-platinate in aqueous soln. of Potassium Chloride, wt %	Concn. of Potassium Chloride in aqueous soln., g/l	Solubility of Potassium Hexachloro-platinate in aqueous soln. of Potassium Chloride, wt %	Concn. of Potassium Chloride in aqueous soln., g/l	Solubility of Potassium Hexachloro-platinate in aqueous soln. of Potassium Chloride, wt %
37.3	0.0109	149.2	0.0045	298.4	0.0042
74.6	0.0046	223.8	0.0043	Saturated	0.0034

№ 4801 [1094]

POTASSIUM CHLORIDE–POTASSIUM HEXACHLOROPLATINATE–WATER

$KCl—K_2PtCl_6—H_2O$

$t=100$

Concn. of Potassium Chloride in aqueous soln., g/l	Solubility of Potassium Hexachloro-platinate in aqueous soln. of Potassium Chloride, g/l
149.2	50.3

№ 4802 [1093]

POTASSIUM CHLORIDE–POTASSIUM HEXACHLOROPLATINATE–WATER

$KCl—K_2PtCl_6—H_2O$

$t=20$

Concn. of Potassium Chloride in aqueous soln., g/l	Solubility of Potassium Hexachloro-platinate in aqueous soln. of Potassium Chloride, wt %
14.92	0.0236
18.65	0.0207

№ 4803 POTASSIUM CHLORIDE–POTASSIUM METAVANADATE–WATER [2671]

$KCl—KVO_3—H_2O$

$t=25$

Comp. of saturated soln., wt %			Comp. of solid phase, wt %		
Potassium Chloride	Potassium Metavanadate	Water	Potassium Chloride	Potassium Metavanadate	Water
26.2	0.0	73.8	79.8	0.0	20.2
26.2	0.06	73.74	75.2	0.80	24.0
26.2	0.06	73.74	69.9	1.6	28.5
26.2	0.06	73.74	67.3	3.1	29.6
26.2	0.06	73.74	63.2	3.2	33.6
0.0	9.7	90.3	0.0	58.0	42.0
0.3	6.2	93.5	0.2	35.5	64.3
3.0	3.4	93.6	0.7	81.0	18.3
4.7	1.9	93.4	1.3	72.2	26.5
5.9	1.4	92.7	3.0	46.9	50.1
1.1	4.7	94.2	33.4	12.0	54.6
14.4	0.9	84.7	5.8	59.9	34.3
14.5	0.8	84.7	6.5	57.5	36.0
15.4	0.6	84.0	9.0	41.9	49.1
19.0	0.4	80.6	7.7	59.9	32.4
21.9	0.4	77.7	9.2	57.8	33.0
24.2	0.3	75.5	14.0	41.9	34.1
26.4	0.3	73.3	33.4	47.1	19.5
26.4	0.3	73.3	49.7	25.9	24.4

№ 4804 POTASSIUM CHLORIDE–POTASSIUM CHROMATE–WATER [1818]

$KCl—K_2CrO_4—H_2O$

$t=25$

Concn. in aqueous soln., g/l			
Potassium Chromate	Potassium Chloride	Potassium Chromate	Potassium Chloride
250.9	29.82	190.3	128.2
248.6	34.29	169.6	162.7
209.7	97.72	153.3	201.4

№ 4805 POTASSIUM CHLORIDE–POTASSIUM DICHROMATE–WATER [2421]

$KCl—K_2Cr_2O_7—H_2O$

Comp. of saturated soln., wt %			t	Comp. of solid phase
Potassium Dichromate	Potassium Chloride	Water		
10.0	2.1	87.9	25	$K_2Cr_2O_7$
8.3	4.3	87.4	25	»
6.53	6.32	87.15	25	»
5.5	8.2	86.3	25	»
4.55	10.0	85.45	25	»
3.9	11.6	84.5	25	»
3.36	13.0	83.64	25	»
2.6	15.7	81.7	25	»
2.2	18.6	79.2	25	»
1.7	21.2	77.1	25	»
1.57	23.0	75.43	25	»
1.2	26.0	72.8	25	$K_2Cr_2O_7+KCl$
1.05	26.1	72.85	25	KCl
0.0	26.3	73.7	25	»
25.8	2.4	71.8	50	$K_2Cr_2O_7$
21.3	4.6	74.1	50	»
18.9	6.6	74.5	50	»
16.7	8.6	74.7	50	»
14.9	10.6	74.5	50	»
13.2	11.9	74.9	50	»
12.4	13.3	74.3	50	»
11.0	14.8	74.2	50	»
8.8	18.1	73.1	50	»
6.8	21.4	71.8	50	»
5.7	24.0	70.3	50	»
4.1	28.9	67.0	50	$K_2Cr_2O_7+KCl$
3.7	29.0	67.3	50	KCl
1.9	29.8	68.3	50	»
0.0	30.6	69.4	50	»
47.2	0.0	52.8	100	$K_2Cr_2O_7$
42.1	3.5	54.4	100	»
37.1	7.2	55.7	100	»
32.2	11.4	56.4	100	»
27.6	15.6	56.8	100	»
20.6	22.1	55.3	100	»
18.2	24.9	56.9	100	»
16.2	27.6	56.2	100	$K_2Cr_2O_7+KCl$
11.6	29.6	58.8	100	KCl
6.4	31.9	61.7	100	»
0.0	35.9	64.1	100	»

№ 4806 POTASSIUM CHLORIDE–POTASSIUM DICHROMATE–WATER [1818]

$KCl—K_2Cr_2O_7—H_2O$

$t=25$

Concn. in aqueous soln., mol/l			
Potassium Chloride	Potassium Dichromate	Potassium Chloride	Potassium Dichromate
4.15	0.04	4.10	0.16
4.12	0.08	4.05	0.28

№ 4807 POTASSIUM CHLORIDE–POTASSIUM DICHROMATE–WATER [299]

$KCl—K_2Cr_2O_7—H_2O$

Comp. of saturated soln., wt %			t	Comp. of solid phase
Potassium Dichromate	Potassium Chloride	Water		
13.0	0.0	87.0	25	$K_2Cr_2O_7$
7.2	1.6	91.2	25	»
4.0	9.6	86.4	25	»
2.5	14.6	82.9	25	»
1.5	19.7	78.8	25	»
0.7	24.8	74.5	25	»
0.7	26.0	73.3	25	$K_2Cr_2O_7+KCl$
0.0	26.2	73.8	25	KCl
26.2	0.0	73.8	50	$K_2Cr_2O_7$
19.5	4.0	76.5	50	»
14.0	8.6	77.4	50	»
10.5	13.4	76.1	50	»
7.4	18.5	74.1	50	»
4.5	23.9	71.6	50	»
4.0	28.8	69.2	50	»
3.6	28.0	68.4	50	$K_2Cr_2O_7+KCl$
2.1	29.0	68.9	50	KCl
0.7	29.7	69.6	50	»
0.0	30.0	70.0	50	»
38.0	0.0	62.0	75	$K_2Cr_2O_7$
30.5	3.5	66.0	75	»
24.5	7.5	68.0	75	»
20.0	12.0	68.0	75	»
15.0	17.0	68.0	75	»
11.0	22.2	66.8	75	»
9.0	27.3	63.7	75	»
8.6	29.9	61.5	75	$K_2Cr_2O_7+KCl$
7.1	29.0	63.9	75	KCl
3.5	30.5	66.0	75	»
2.1	31.5	66.4	75	»
0.7	32.7	66.4	75	»
0.0	33.5	66.5	75	»

№ 4808 POTASSIUM CHLORIDE–POTASSIUM PERMANGANATE–WATER [2455]

$KCl—KMnO_4—H_2O$

Concn. of Potassium Chloride in aqueous soln., g/l	Solubility of Potassium Permanganate in aqueous soln. of Potassium Chloride, g/l	t	Concn. of Potassium Chloride in aqueous soln., g/l	Solubility of Potassium Permanganate in aqueous soln. of Potassium Chloride, g/l	t
7.46	22.04	0	74.6	8.407	0
7.46	68.18	25	74.6	34.76	25
7.46	116.62	40	74.6	70.16	40
37.3	12.01	0	149.2	5.989	0
37.3	48.35	25	149.2	22.63	25
37.3	92.28	40	149.2	45.51	40

№ 4809 [1818]

POTASSIUM CHLORIDE–POTASSIUM PERMANGANATE–WATER

$KCl—KMnO_4—H_2O$

$t=25$

Concn. in aqueous soln., g-mol/l			
Potassium Chloride	Potassium Permanganate	Potassium Chloride	Potassium Permanganate
4.19	0.0	4.04	0.11
4.02	0.05	4.00	0.40

№ 4810 POTASSIUM CHLORIDE–POTASSIUM BROMIDE–WATER [1578]

$KCl—KBr—H_2O$

Comp. of saturated soln., wt %			t	Comp. of saturated soln., wt %			t
Potassium Bromide	Potassium Chloride	Water		Potassium Bromide	Potassium Chloride	Water	
17.5	10.5	72.0	−20	28.0	11.5	60.5	40
21.5	10.8	67.7	0	30.6	11.8	57.6	60
23.2	11.0	65.8	10	33.4	12.1	54.5	80
24.8	11.2	64.0	20	35.7	12.6	51.7	100
25.5	11.3	63.2	25	38.0	12.9	49.1	120
26.3	11.4	62.3	30	40.6	13.4	46.0	150

№ 4811 POTASSIUM CHLORIDE–POTASSIUM BROMIDE–WATER [1627]

$KCl—KBr—H_2O$

$t=25$

Concn. in aqueous soln., g/l		d	Comp. of solid phase, mol %	
Potassium Bromide	Potassium Chloride		Potassium Chloride	Potassium Bromide
558.1	0.0	1.3756	0.0	100
531.5	23.44	1.3700	0.0	100
503.6	46.57	1.3648	8.23	91.77
454.6	82.62	1.3544	15.68	74.32
379.6	136.6	1.3320	33.66	66.34
324.8	166.9	1.3119	63.51	36.49
218.0	213.9	1.2689	82.29	17.71
140.7	250.9	1.2455	88.04	11.96
47.5	291.7	1.1977	96.98	3.02
0.0	311.3	1.1756	100.0	0.0

№ 4812 POTASSIUM CHLORIDE–POTASSIUM BROMIDE–WATER [2658]

$KCl—KBr—H_2O$

$t=25.2$

Concn. in aqueous soln., g/l					
Potassium Bromide	Potassium Chloride	Potassium Bromide	Potassium Chloride	Potassium Bromide	Potassium Chloride
567.0	0.0	397.7	126.8	156.1	238.0
502.5	50.0	320.4	173.8	101.3	267.1
494.2	60.4	297.7	183.5	62.5	281.6
440.7	100.7	268.0	192.4	58.4	287.2
421.6	110.4	211.9	217.1	0.0	311.8
407.2	120				

№ 4813 [1253]

POTASSIUM CHLORIDE–POTASSIUM BROMIDE–WATER

$KCl—KBr—H_2O$

Content in aqueous soln., wt %		t	Comp. of solid phase
Cl^-	Br^-		
14.69	1.92	25	K(Cl, Br)
9.48	7.94	25	»
7.43	13.34	25	»
6.79	15.09	25	»
5.07	18.70	25	»
3.86	21.00	25	»
1.73	24.39	25	»
1.21	25.16	25	»
1.22	25.37	25	»
0.70	26.08	25	»
0.68	25.95	25	»
0	27.38	25	KBr
12.92	3.39	45	K(Cl, Br)
11.32	6.77	45	»

№ 4814 POTASSIUM CHLORIDE–POTASSIUM BROMIDE–WATER [1075]

$KCl—KBr—H_2O$

$t=25$

Comp. of saturated soln., wt %					
Potassium Bromide	Potassium Chloride	Water	Potassium Bromide	Potassium Chloride	Water
40.64	0.0	59.36	24.19	13.27	62.54
37.12	3.24	59.64	17.45	16.97	65.58
35.03	5.06	59.91	8.98	21.54	69.48
31.80	7.56	60.64	0.0	26.53	73.47
28.00	10.50	61.50			

№ 4815 [1421]

POTASSIUM CHLORIDE–POTASSIUM BROMIDE–WATER

$KCl—KBr—H_2O$

Content in aqueous soln., g/100 g salts		$t_{b.p.}$	Comp. of solid phase
Potassium Chloride	Potassium Bromide		
100	0	108.5	KCl
90	10	108.6	K(Cl, Br)
80	20	108.9	»
70	30	109.25	»
60	40	109.75	»
50	50	110.15	»
40	60	110.6	»
30	70	110.95	»
25	75	111.7	»
20	80	111.25	»
10	90	111.15	»
5	95	111.05	»
0	100	110.9	KBr

№ 4816 POTASSIUM CHLORIDE–POTASSIUM BROMIDE–WATER [615]

$KCl—KBr—H_2O$

$t=25$

Content in aqueous soln., wt %		Comp. of solid phase	Content in aqueous soln., wt %		Comp. of solid phase
Potassium Chloride	Potassium Bromide		Potassium Chloride	Potassium Bromide	
26.72	0	KCl	9.78	29.25	K(Cl, Br)
25.88	1.88	K(Cl, Br)	8.02	31.45	»
20.88	10.00	»	2.08	37.51	»
17.90	15.50	»	0	40.81	KBr
14.28	22.22	»			

№ 4817 POTASSIUM CHLORIDE–POTASSIUM BROMIDE–WATER [977]

$KCl—KBr—H_2O$

Concn. in aqueous soln., wt %		d_4^t	t	Concn. in aqueous soln., wt %		d_4^t	t
Potassium Bromide	Potassium Chloride			Potassium Bromide	Potassium Chloride		
0.539	21.295	1.155	0	8.15	23.08	1.224	35
0.640	21.850	1.154	0	10.68	21.65	1.239	35
3.360	20.360	1.169	0	11.12	21.75	1.242	35
8.184	18.048	1.196	0	13.02	20.67	1.253	35
19.159	12.480	1.264	0	14.63	20.14	1.265	35
22.3	11.048	1.280	0	24.24	14.23	1.320	35
22.2	8.109	1.296	0	25.22	13.76	1.324	35
30.504	4.573	1.312	0	30.18	10.64	1.349	35
0.0	27.95	1.185	35	30.40	10.34	1.350	35
0.18	28.04	1.186	35	32.11	9.19	1.359	35
0.61	27.40	1.186	35	36.50	5.09	1.373	35
1.60	26.79	1.192	35	39.64	2.77	1.387	35
2.91	26.46	1.198	35	42.59	0.00	1.397	35
7.30	23.69	1.221	35				

№ 4818 POTASSIUM CHLORIDE–POTASSIUM BROMIDE–WATER [105]

$KCl—KBr—H_2O$

Comp. of saturated soln., wt %			t	Comp. of solid phase
Potassium Chloride	Potassium Bromide	Water		
20.2	0.0	79.8	—10	KCl+KBr (solid soln.)
15.8	8.4	75.8	—10	»
11.0	17.8	71.2	—10	»
4.9	26.5	68.6	—10	»
0.0	31.4	68.6	—10	»
1.0	0.0	81.0	—10	Ice
12.4	8.8	78.8	—10	»
5.0	19.0	76.0	—10	»
0.0	26.6	73.4	—10	»
12.0	0.0	88.0	— 5	»
4.3	9.6	86.1	— 5	»
0.0	16.0	84.0	— 5	»
20.8	0.0	79.2	— 5	KCl+KBr (solid soln.)
16.6	8.3	75.1	— 5	»
11.7	17.6	70.7	— 5	»
5.8	26.3	67.9	— 5	»
1.3	31.7	67.0	— 5	»
0.0	32.8	67.2	— 5	»
21.8	0.0	78.2	0	»
17.6	8.3	74.1	0	»
12.5	17.4	70.0	0	»
7.1	26.0	66.9	0	»
2.7	31.2	66.1	0	»
0.0	34.0	66.0	0	»
22.6	0.0	74.4	5	»
18.4	8.2	73.4	5	»

Table № 4818 (continuation)

Comp. of saturated soln., wt %			t	Comp. of solid phase
Potassium Chloride	Potassium Bromide	Water		
13.6	17.2	69.2	5	KCl + KBr (solid soln.)
8.2	25.7	66.1	5	»
4.2	30.7	65.1	5	»
0.0	35.2	64.8	5	»
23.4	0.0	76.6	10	»
19.1	8.1	72.8	10	»
14.4	17.1	68.5	10	»
9.6	25.2	65.2	10	»
5.3	30.4	64.4	10	»
0.0	36.6	63.4	10	»
24.2	0.0	75.8	15	»
20.0	8.1	72.0	15	»
15.4	16.9	67.7	15	»
10.7	25.0	64.3	15	»
6.5	30.0	63.5	15	»
0.0	36.8	63.2	15	»
25.0	0.0	75.0	20	»
20.8	8.0	71.2	20	»
16.2	16.8	67.0	20	»
11.7	24.7	63.6	20	»
7.8	29.6	62.6	20	»
0.0	39.2	60.8	20	»
25.8	0.0	74.2	25	»
21.7	7.9	70.4	25	»
16.8	16.7	66.5	25	»
12.4	24.5	63.1	25	»
9.0	29.1	61.9	25	»
0.0	40.6	59.4	25	»
26.6	0.0	73.4	30	»
22.6	7.8	69.6	30	»
17.4	16.5	66.1	30	»
13.0	24.3	62.7	30	»
10.0	28.8	61.2	30	»
0.0	41.9	58.1	30	»

№ 4819 [1607]

POTASSIUM CHLORIDE–POTASSIUM BROMIDE–WATER

$KCl—KBr—H_2O$

$t=25$

Comp. of saturated soln., mol %			Comp. of solid phase
Potassium Chloride	Potassium Bromide	Water	
89.77	0.0	10.23	KCl
69.34	21.18	9.48	KCl+KBr
51.69	39.47	8.84	»
46.13	45.40	9.47	»

Table № 4819 (continuation)

Comp. of saturated soln., mol %			Comp. of solid phase
Potassium Chloride	Potassium Bromide	Water	
36.66	54.98	8.36	KCl + KBr
31.02	60.76	8.21	»
23.28	68.30	8.32	»
16.10	75.38	8.52	»
7.49	83.87	8.64	»
0.00	91.22	8.78	KBr

№ 4820 POTASSIUM CHLORIDE–POTASSIUM BROMIDE–WATER [289]

$KCl—KBr—H_2O$

$t=25$

Concn. in aqueous soln., wt %					
Potassium Bromide	Potassium Chloride	Potassium Bromide	Potassium Chloride	Potassium Bromide	Potassium Chloride
0.0	26.27	23.42	13.31	27.91	10.52
2.85	24.75	24.27	12.77	28.20	10.72
5.66	23.33	25.38	12.19	28.38	10.22
8.51	21.96	26.58	11.83	28.76	9.89
14.48	18.22	26.90	11.32	28.93	9.85
17.12	17.20	27.13	11.08	29.09	9.65
19.31	15.74	27.42	11.03	40.62	0.0
21.85	14.91				

№ 4821 POTASSIUM CHLORIDE–POTASSIUM BROMIDE–WATER [292]

$KCl—KBr—H_2O$

$t=25$

Comp. of saturated soln., wt %			d	Comp. of solid phase
Potassium Chloride	Potassium Bromide	Water		
26.36	0.00	73.64	1.1376	KCl
25.38	1.78	72.84	1.1864	»
22.77	5.24	71.99	1.2070	»
22.66	6.57	70.77	1.2109	»
21.38	9.11	69.51	1.2238	KCl+KBr
21.11	9.33	69.56	1.2244	»
19.51	12.88	67.61	1.2434	»
17.14	17.24	65.62	1.2688	»
16.69	17.60	65.71	1.2718	»
14.99	20.82	64.19	1.2892	»
13.94	23.00	63.06	1.3020	»
12.53	25.18	62.29	1.3148	»
11.33	26.50	62.17	1.3222	»
11.45	26.90	61.65	1.3250	»
10.55	28.32	61.13	1.3309	»
9.12	30.17	60.71	1.3396	»
8.21	31.05	60.74	1.3457	»

Table № 4821 (continuation)

Comp. of saturated soln., wt %			d	Comp. of solid phase
Potassium Chloride	Potassium Bromide	Water		
7.93	31.73	60.34	1.3470	KCl+KBr
7.19	32.40	60.41	1.3506	»
7.16	33.23	59.61	1.3514	»
6.02	33.82	60.16	1.3581	»
5.59	34.93	59.48	1.3602	»
0.00	40.62	59.38	1.4034	KBr

№ 4822 POTASSIUM CHLORIDE–POTASSIUM BROMIDE–POTASSIUM IODIDE–WATER [1577]

$KCl—KBr—KI—H_2O$

Content in aqueous soln., wt %			t	Comp. of solid phase
Potassium Chloride	Potassium Bromide	Potassium Iodide		
0.10	9.7	42.9	—17	KCl+KBr+KI
0.10	9.8	46.2	11	»
0.10	11.0	49.1	30	»
0.10	12.4	52.7	82	»

№ 4823 [289]

POTASSIUM CHLORIDE–POTASSIUM BROMIDE–POTASSIUM IODIDE–WATER

$KCl—KBr—KI—H_2O$

$t=25$

Content in aqueous soln., wt %			Comp. of solid phase
Potassium Chloride	Potassium Bromide	Potassium Iodide	
9.65	29.09	0	K(Cl, Br)
7.77	25.70	5.85	»
7.92	23.71	11.72	»
8.29	22.88	12.04	»
7.45	20.40	17.29	»
6.47	18.35	22.41	»
5.81	16.18	27.67	»
5.05	13.93	32.93	»
4.46	12.28	37.45	»
4.72	10.36	42.34	»
2.88	8.22	48.63	»
3.05	8.06	48.13	K(Cl, Br)+KI
3.10	8.01	48.39	»
3.02	8.06	48.40	»

№ 4824 POTASSIUM CHLORIDE–POTASSIUM BROMATE–WATER [2390]

$KCl—KBrO_3—H_2O$

$t=25$

Comp. of saturated soln., wt %			d	Comp. of solid phase
Potassium Bromate	Potassium Chloride	Water		
7.533	0.00	92.467	1.054	$KBrO_3$
4.63	4.33	89.04	1.058	»
3.24	9.03	87.73	1.082	»
2.44	14.45	83.31	1.112	»
1.97	19.71	78.32	1.147	»
1.65	24.87	73.48	1.183	»
1.61	25.89	72.50	1.192	$KBrO_3+KBr$
1.48	25.93	72.59	1.187	KBr
0.00	26.36	73.64	1.179	»

№ 4825 POTASSIUM CHLORIDE–POTASSIUM MOLYBDATE–WATER [366]

$KCl—K_2MoO_4—H_2O$

$t=25$

Comp. of saturated soln., wt %			d	Comp. of solid phase
Potassium Chloride	Potassium Molybdate	Water		
0.00	26.45	73.55	1.182	KCl
4.25	24.20	71.45	1.202	»
13.25	19.54	67.21	1.260	»
27.74	12.52	59.74	1.366	»
35.40	9.20	55.40	1.439	»
44.57	5.87	49.56	1.536	»
52.57	3.28	44.15	1.640	»
58.37	2.00	39.63	1.728	»
63.06	1.40	35.54	1.812	»
63.78	1.30	34.92	1.821	$KCl+K_2MoO_4$
63.70	1.29	35.01	1.824	»
63.63	1.29	35.08	1.825	»
61.71	1.29	36.00	1.823	»
63.86	1.26	34.88	1.826	»
63.65	1.29	35.06	1.826	»
63.82	1.28	34.90	1.826	»
64.20	0.58	35.22	1.826	K_2MoO_4
64.75	0.00	35.25	1.829	»

№ 4826 POTASSIUM CHLORIDE–POTASSIUM IODIDE–WATER [1577]

$KCl—KI—H_2O$

Concn. in aqueous soln., wt %		t	Concn. in aqueous soln., wt %		t
Potassium Chloride	Potassium Iodide		Potassium Chloride	Potassium Iodide	
3.7	50.5	0	6.2	61.0	100
4.2	53.0	20	7.3	63.7	140
4.7	55.3	40	8.3	65.5	180
5.2	57.5	60	9.4	66.3	220
5.7	59.4	80	10.0	66.5	246

№ 4827 POTASSIUM CHLORIDE–POTASSIUM IODIDE–WATER [1075]

$KCl—KI—H_2O$

t = 25

Concn. in aqueous soln., g/100 g water					
Potassium Chloride	Potassium Iodide	Potassium Chloride	Potassium Iodide	Potassium Chloride	Potassium Iodide
0.0	149.26	11.74	133.90	29.56	23.83
4.06	144.03	15.10	105.91	31.38	14.83
7.63	137.79	19.64	68.22	33.68	7.0
11.36	132.60	23.75	43.89	36.12	0.0

№ 4828 POTASSIUM CHLORIDE–POTASSIUM IODATE–WATER [1839]

$KCl—KIO_3—H_2O$

Comp. of saturated soln., wt %			d	t	Comp. of solid phase
Potassium Iodate	Potassium Chloride	Water			
0.0	22.84	77.16	1.155	5	KCl
1.44	22.64	75.92	1.170	5	$KCl—KIO_3$
1.50	19.56	78.94	1.147	5	KIO_3
1.79	10.13	88.02	1.084	5	»
2.91	3.03	94.06	1.044	5	»
5.16	0.0	94.84	1.043	5	»
0.0	26.36	73.64	1.179	25	KCl
2.10	25.82	72.08	1.197	25	$KCl—KIO_3$
2.40	19.64	77.96	1.153	25	KIO_3
3.01	12.37	84.62	1.109	25	»
4.29	5.93	89.78	1.073	25	»
5.83	2.78	91.39	1.066	25	»
8.45	0.0	91.55	1.071	25	»
0.0	30.03	69.97	—	50	KCl
1.77	29.40	68.83	—	50	»
3.07	29.08	67.85	—	50	$KCl—KIO_3$
3.74	23.17	73.09	—	50	KIO_3
4.71	15.64	79.65	—	50	»
7.42	6.83	85.75	—	50	»
13.21	0.0	86.79	—	50	»

№ 4829 [1785]

POTASSIUM CHLORIDE–POTASSIUM IODIDE–WATER

$KCl—KI—H_2O$

Content in aqueous soln., g/l		t	Comp. of solid phase
Potassium Chloride	Potassium Iodide		
297.4	0.0	20	KCl
66.9	907	20	KCl+KI
50.5	923	20	KI
34.7	953	20	»
18.2	974	20	»
0.0	1003	20	»
318.8	0.0	30	KCl
78.4	929	30	KCl+KI
55.3	968	30	KI
42.1	980	30	»
20.3	1004	30	»
7.8	1032	30	»
339.8	0.0	40	KCl
78.6	959	40	KCl+KI
55.5	979	40	KI
43.9	1012	40	»
20.4	1045	40	»
0.0	1072	40	»

№ 4830 [592]

POTASSIUM CHLORIDE–POTASSIUM IODIDE–WATER

$KCl—KI—H_2O$

$t=25$

Content in aqueous soln., wt %		d_4	Comp. of solid phase
Potassium Chloride	Potassium Iodide		
26.30	0	1.1781	KCl
23.51	5.00	1.2016	»
23.98	6.21	—	»
23.19	5.74	1.2063	»
20.65	12.43	—	»
20.60	13.00	—	»
17.49	19.59	—	»
17.19	19.84	1.3143	»
14.65	26.77	1.3686	»
14.07	29.33	—	»
11.38	34.55	1.4417	»
9.29	40.20	1.5007	»
6.99	47.61	—	»
5.03	54.63	1.6920	KCl+KI
2.11	57.89	1.7086	KI
0	59.78	1.7204	»

POTASSIUM CHLORIDE–POTASSIUM IODIDE–WATER

KCl—KI—H_2O

Concn. in aqueous soln., wt %		t	Comp. of solid phase
Potassium Chloride	Potassium Iodide		
22.20	0.00	0	KCl
21.33	2.45	0	»
20.43	4.65	0	»
19.12	7.54	0	»
18.40	10.08	0	»
17.12	12.51	0	»
16.48	14.89	0	»
15.18	18.17	0	»
13.71	22.21	0	»
12.06	27.02	0	»
10.59	31.51	0	»
9.34	35.83	0	»
8.12	38.97	0	»
7.27	39.75	0	»
6.74	41.75	0	»
5.89	45.01	0	»
5.48	47.59	0	»
4.67	49.56	0	»
4.07	51.89	0	KI+KCl
4.07	51.88	0	»
0.00	57.70	0	KI
26.30	0.0	25	KCl
24.55	3.07	25	»
23.86	4.73	25	»
22.57	7.19	25	»
22.10	8.16	25	»
21.77	8.64	25	»
21.20	11.80	25	»
19.67	13.56	25	»
18.90	15.27	25	»
16.90	19.65	25	»
15.25	25.04	25	»
12.80	29.97	25	»
12.22	31.60	25	»
11.61	33.81	25	»
10.52	36.13	25	»
9.77	38.74	25	»
8.55	41.38	25	»
8.01	43.02	25	»
7.38	45.14	25	»
6.63	47.19	25	»
6.10	49.02	25	»
5.01	53.12	25	»
4.53	54.54	25	KI+KCl
4.57	54.52	25	»
4.60	54.67	25	»
4.52	54.60	25	»
4.53	54.57	25	»
0.0	59.76	25	KI

Table № 4831 (continuation)

Concn. in aqueous soln., wt %		t	Comp. of solid phase
Potassium Chloride	Potassium Iodide		
30.10	0.0	50	KCl
27.49	5.01	50	»
26.34	7.52	50	»
24.85	10.02	50	»
22.89	15.39	50	»
19.98	19.07	50	»
15.79	27.87	50	»
13.15	35.03	50	»
11.13	40.01	50	»
7.75	48.82	50	»
5.73	54.97	50	»
5.04	57.34	50	KI +KCl
4.85	57.40	50	»
4.92	57.46	50	»
4.92	57.40	50	»
0.0	62.70	50	KI
33.20	0.0	75	KCl
31.06	4.00	75	»
27.67	9.29	75	»
25.14	15.04	75	»
23.05	18.36	75	»
22.09	20.02	75	»
19.56	24.98	75	»
17.15	30.13	75	»
14.01	35.85	75	»
12.54	40.93	75	»
9.40	45.02	75	»
8.70	49.84	75	»
6.65	55.86	75	»
5.43	58.00	75	KI +KCl
5.42	59.01	75	»
5.46	59.06	75	»
5.44	59.02	75	»
0.0	65.30	75	KI

№ 4832 POTASSIUM CHLORIDE–POTASSIUM IODIDE–WATER [9]

$KCl-KI-H_2O$

$t=25$

Comp. of saturated soln., wt %			d	Comp. of solid phase
Potassium Chloride	Potassium Iodide	Water		
26.48	0.00	73.52	1.1802	KCl
23.75	5.05	71.20	1.035	»
21.09	10.08	68.83	1.2781	»
18.95	15.11	65.94	1.3253	»
17.06	20.01	62.93	1.3725	»

Table № 4832 (continuation)

Comp. of saturated soln., wt %			d	Comp. of solid phase
Potassium Chloride	Potassium Iodide	Water		
14.84	25.08	60.08	1.4223	KCl
12.96	30.00	57.04	1.4688	»
11.10	35.04	53.86	1.5158	»
9.22	40.10	50.68	1.5635	»
7.23	45.09	47.68	1.6152	»
5.90	49.90	44.20	1.6625	»
4.68	54.51	40.81	1.7012	KCl+KI
4.65	54.56	40.79	1.7000	KI
3.31	55.95	40.74	1.7090	»
3.04	56.20	40.76	1.7090	»
1.93	57.51	40.56	1.7102	»
1.72	57.98	40.30	1.7108	»
1.69	58.00	40.31	1.7108	»
0.00	59.78	40.22	1.7208	»

№ 4833 [2594]

POTASSIUM CHLORIDE–POTASSIUM PERRHENATE–WATER

$KCl—KReO_4—H_2O$

$t=30$

Concn. in aqueous soln., wt %		Comp. of solid phase
Potassium Chloride	Potassium Perrhenate	
1.45	0.00	$KReO_4$
0.414	1.597	»
0.170	1.504	»
0.100	12.14	»
0.052	27.24	$KReO_4$+KCl
0.00	27.23	KCl

№ 4834 [2594]

POTASSIUM CHLORIDE–POTASSIUM PERRHENATE–WATER

$KCl—KReO_4—H_2O$

$t=0$

Concn. in aqueous soln., wt %		Comp. of solid phase
Potassium Chloride	Potassium Perrhenate	
0.358	0.00	$KReO_4$
0.025	8.68	»
0.019	17.09	»
0.017	21.91	$KReO_4$+KCl
0.013	28.93	KCl
0.005	21.90	»
0.00	21.88	»

$KCl—KReO_4—H_2O$

Comp. of saturated soln., wt %			t	Comp. of solid phase
Potassium Perrhenate	Potassium Chloride	Water		
0.35	0	99.65	0	$KReO_4$
0.22	0.10	99.68	0	»
0.10	0.38	99.52	0	»
0.06	0.76	99.14	0	»
0.03	2.64	97.33	0	»
0.02	4.53	95.45	0	»
0.013	7.71	92.16	0	»
0.011	9.93	90.599	0	»
0.011	10.75	89.239	0	»
0.011	12.29	87.699	0	»
0.010	12.55	87.44	0	»
0.008	17.50	82.492	0	»
0.007	21.53	78.463	0	$KReO_4+KCl$
0.00	22.16	77.84	0	KCl
1.19	0.00	98.81	25	$KReO_4$
0.74	0.32	98.94	25	»
0.57	0.47	98.96	25	»
0.52	0.61	98.87	25	»
0.43	0.93	98.64	25	»
0.32	1.26	98.42	25	»
0.30	1.41	98.29	25	»
0.22	2.28	97.50	25	»
0.18	3.01	96.81	25	»
0.094	7.55	92.356	25	»
0.092	7.74	92.168	25	»
0.084	8.70	91.216	25	»
0.074	10.30	89.626	25	»
0.069	11.16	88.771	25	»
0.064	12.03	87.906	25	»
0.063	16.77	83.167	25	»
0.057	18.44	81.503	25	»
0.041	22.48	77.479	25	»
0.038	22.69	77.272	25	»
0.034	25.78	74.186	25	$KReO_4+KCl$
0.00	26.41	73.59	25	KCl
3.12	0	96.88	50	$KReO_4$
2.24	0.44	97.32	50	»
1.52	1.54	96.94	50	»
1.50	1.58	96.92	50	»
0.69	5.28	94.03	50	»
0.56	6.79	92.65	50	»
0.45	7.45	92.10	50	»
0.43	8.84	90.73	50	»
0.35	10.09	89.56	50	»

Comp. of saturated soln., wt %			t	Comp. of solid phase
Potassium Perrhenate	Potassium Chloride	Water		
0.34	11.38	88.28	50	»
0.30	11.68	88.02	50	»
0.29	12.48	87.23	50	»
0.27	13.26	86.47	50	$KReO_4$
0.27	14.09	85.64	50	»
0.25	15.43	84.32	50	»
0.24	15.98	83.78	50	»
0.16	21.02	78.82	50	»
0.15	24.88	74.97	50	»
0.11	28.55	71.34	50	$KReO_4$+KCl
0.00	29.77	70.23	50	KCl
6.92	0.00	93.08	75	$KReO_4$
5.28	0.94	97.78	75	»
3.79	3.03	93.18	75	»
3.15	3.92	92.93	75	»
3.18	4.06	92.76	75	»
2.08	7.24	90.68	75	»
1.33	10.94	87.73	75	»
1.06	15.25	83.69	75	»
0.87	17.02	82.11	75	»
0.58	21.32	78.10	75	»
0.39	31.67	67.94	75	»
0.35	32.53	67.12	75	$KReO_4$+KCl
0.00	33.17	66.83	75	KCl

№ 4836 POTASSIUM CHLORIDE–CALCIUM OXIDE–WATER [1309]

KCl—CaO—H_2O

Concn. in aqueous soln., g/l		t	Concn. in aqueous soln., g/l		t
Potassium Chloride	Calcium Oxide		Potassium Chloride	Calcium Oxide	
0	1.360	0	120	1.606	15
30	1.701	0	240	1.199	15
60	1.725	0	0	0.635	99
120	1.718	0	30	0.788	99
240	1.248	0	60	0.876	99
0	1.310	15	120	0.894	99
30	1.658	15	240	0.617	99
60	1.674	15			

№ 4837 [1751]

POTASSIUM CHLORIDE–CALCIUM OXIDE–ARSENIC PENTOXIDE–WATER

$KCl—CaO—As_2O_5—H_2O$

$t=17$

Concn. of Potassium Chloride in aqueous soln., g/l	Solubility in aqueous soln. of Potassium Chloride, wt %		pH
	Calcium Oxide	Arsenic Pentoxide	
7.46	0.14	0.30	7.4
37.3	0.18	0.39	7.2
74.6	0.19	0.44	7.0
149.2	0.20	0.45	6.8
223.8	0.19	0.41	6.4

№ 4838 POTASSIUM CHLORIDE–CALCIUM HYDROXIDE–WATER [2106]

$KCl—Ca(OH)_2—H_2O$

Content in aqueous soln.		t	Comp. of solid phase	Content in aqueous soln.		t	Comp. of solid phase
Potassium Chloride, wt %	Calcium Oxide, g/l			Potassium Chloride wt. %	Calcium Oxide g/l		
30	1.21	15	$Ca(OH)_2$	10	1.42	50	$Ca(OH)_2$
25	1.42	15	»	5	1.28	50	»
20	1.49	15	»	0	1.09	50	»
15	1.56	15	»	30	0.59	100	»
10	1.84	15	»	25	1.00	100	»
5	1.70	15	»	20	1.00	100	»
0	1.284	15	»	15	1.00	100	»
30	0.85	50	»	10	1.00	100	»
25	1.06	50	»	5	1.13	100	»
20	1.13	50	»	0	0.664	100	»
15	1.21	50	»				

№ 4839 POTASSIUM CHLORIDE–CALCIUM HYDROXIDE–WATER [1967]

$KCl—Ca(OH)_2—H_2O$

$t=50$

Content in aqueous soln., mol/100 g soln.		Comp. of solid phase	Content in aqueous soln., mol/100 g soln.		Comp. of solid phase
Potassium Chloride	Calcium Hydroxide		Potassium Chloride	Calcium Hydroxide	
0.5750	0	$Ca(OH)_2$	0.1894	0.0042	$Ca(OH)_2$
0.3552	0.0017		0.0947	0.0045	»
0.3315	0.0033	»	0.0236	0.0042	»
0.2841	0.0036	»	0	0.0029	»
0.2368	0.0038	»			

№ 4840 POTASSIUM CHLORIDE–CALCIUM HYDROXIDE–WATER [1940]

$KCl—Ca(OH)_2—H_2O$

$t = 25$

Content in aqueous soln., mol/1000 g Water		Comp. of solid phase	Content in aqueous soln., mol/1000g Water		Comp. of solid phase
Potassium Chloride	Calcium Hydroxide		Potassium Chloride	Calcium Hydroxide	
2.87	0.02351	$Ca(OH)_2$	0.527	0.02708	$Ca(OH)_2$
1.96	0.02610	»	0.264	0.02550	»
1.62	0.02696	»	0.076	0.02773	»
1.28	0.02744	»	0.038	0.02157	»
1.08	0.02768	»	0	0.01976	»
0.989	0.02766	»			

Note. $P_{CO_2} < 10^{-}$[illegible] atm.

№ 4841 POTASSIUM CHLORIDE–CALCIUM CARBONATE–WATER [1329]

$KCl—CaCO_3—H_2O$

$t = 25$

Concn. in aqueous soln., wt %		Note	Concn. in aqueous soln., wt %		Note
Potassium Chloride	Calcium Carbonate		Potassium Chloride	Calcium Carbonate	
0.00	0.0013	In contact with air	0.00	0.062	In contact with Carbon Dioxide at $P = 760$ mm Hg
3.90	0.0078		3.90	0.145	
7.23	0.0078		7.23	0.150	
13.82	0.0072		13.82	0.165	
18.21	0.0070		18.21	0.154	
26.00	0.0060		26.00	0.126	

№ 4842 POTASSIUM CHLORIDE–CALCIUM CARBONATE–WATER [1532]

$KCl—CaCO_3—H_2O$

$t \sim 20$

Content in aqueous soln.		Comp. of solid phase	Content in aqueous soln.		Comp. of solid phase
Potassium Chloride, mol/l	0.33 N H_2SO_4, cm^3/l soln.*		Potassium Chloride, mol/l	0.33 N H_2SO_4, cm^3/l soln.*	
3	1.35	$CaCO_3$	0.2	0.75	$CaCO_3$
2	1.30	»	0.125	0.65	»
1	1.20	»	0.5	0.60	»
0.5	1.10	»	0	0.60	»
0.333	0.90	»			

Note. The asterisk indicates the quantity of acid used in titration of soln.

№ 4843 [1134]

POTASSIUM CHLORIDE–CALCIUM NITRATE–WATER

$KCl—Ca(NO_3)_2—H_2O$

$t=30$

Comp. of saturated soln. wt %			Comp. of solid phase
Potassium Chloride	Calcium Nitrate	Water	
4.80	62.40	32.80	$Ca(NO_3)_2 \cdot 4H_2O$
6.37	65.60	28.03	»
9.00	73.50	17.50	$Ca(NO_3)_2 \cdot 4H_2O + Ca(NO_3)_2 \cdot H_2O$
13.85	59.90	26.25	$Ca(NO_3)_2 \cdot H_2O$
16.00	52.35	31.65	»
17.56	47.83	34.61	»
18.20	45.80	36.00	$Ca(NO_3)_2 \cdot H_2O + KNO_3$
16.70	44.33	38.97	KNO_3
16.60	44.20	39.20	»
15.83	39.00	45.17	»
16.00	34.56	49.44	»
17.14	27.18	55.68	»
18.30	23.87	57.83	»
20.54	20.17	59.29	»
23.40	17.31	59.29	»
24.60	16.20	59.20	$KNO_3 + KCl$
24.74	14.65	60.61	KCl
25.80	8.22	65.98	»

№ 4844 [1511]

POTASSIUM CHLORIDE–CALCIUM SULFATE–WATER

$KCl—CaSO_4—H_2O$

$t=21$

Concn. of Potassium Chloride in aqueous soln., g/l	Solubility of Calcium Sulfate in aqueous soln. of Potassium Chloride, g/l	Concn. of Potassium Chloride in aqueous soln., g/l	Solubility of Calcium Sulfate in aqueous soln. of Potassium Chloride, g/l
0	2.05	60	6.6
10	3.6	80	7.2
20	4.5	100	7.5
40	5.8	125	Double salt

№ 4845 [2481]

POTASSIUM CHLORIDE–CALCIUM SULFATE–WATER

$KCl—CaSO_4—H_2O$

Content of Potassium Chloride in aqueous soln., mol/l	Solubility of Calcium Sulfate in aqueous soln. of Potassium Chloride, wt %	t	Comp. of solid phase
1	0.589	25	$CaSO_4 \cdot 2H_2O$
0.5	0.492	25	»
0.1	0.312	25	»
0.05	0.275	25	»
1	0.598	40	»
0.5	0.496	40	»
0.1	0.317	40	»
0.05	0.282	40	»
0.1	0.31	25	$CaSO_4 \cdot 0.5H_2O$
0	0.26	25	»
0.1	0.32	40	»
0	0.23	40	»

№ 4846 [4846]

POTASSIUM CHLORIDE–CALCIUM SULFATE–WATER

$KCl—CaSO_4—H_2O$

$t=15—17$

Content in aqueous soln., g/l		Comp. of solid phase	Content in aqueous soln., g/l		Comp. of solid phase
Potassium Chloride	Calcium Sulfate		Potassium Chloride	Calcium Sulfate	
173	11.1	$KCl+CaSO_4 \cdot 0.5H_2O$	18.5	4.6	$CaSO_4 \cdot 2H_2O$
81.8	12.5	—	9.2	3.2	»
80.6	12.45	—	0	2.6	»

№ 4847 [2699]

POTASSIUM CHLORIDE–CALCIUM SULFATE–WATER

$KCl—CaSO_4—H_2O$

$t=25$

Concn. in aqueous soln., g/l			
Potassium Chloride	Calcium Sulfate	Potassium Chloride	Calcium Sulfate
0.746	4.493	4.474	6.140
1.491	4.765	5.219	6.385
2.237	5.037	8.947	7.502
2.982	5.351		

№ 4848 [1134]

POTASSIUM CHLORIDE–CALCIUM CHLORIDE–WATER

$KCl—CaCl_2—H_2O$

$t=30$

Comp. of saturated soln., wt %			Comp. of solid phase
Potassium Chloride	Calcium Chloride	Water	
0.3	50.0	49.7	$CaCl_2 \cdot 4H_2O$
2.1	49.6	48.3	»
3.35	49.5	47.15	»
4.4	49.2	46.4	$CaCl_2 \cdot 4H_2O+KCl$
4.0	46.8	49.2	KCl
4.85	30.3	64.85	»
7.9	24.0	68.1	»
12.85	16.0	71.15	»
18.5	9.25	72.25	»
24.8	2.33	72.87	»

№ 4849 POTASSIUM CHLORIDE–CALCIUM CHLORIDE–WATER [2061]

$KCl—CaCl_2—H_2O$

$t=25$

Comp. of saturated soln., wt %			d_{25}	Comp. of solid phase
Potassium Chloride	Calcium Chloride	Water		
0.00	45.06	54.94	1.470	$CaCl_2 \cdot 6H_2O$
3.28	44.73	51.99	—	»
3.06	44.72	52.22	1.485	»
3.05	44.66	52.29	1.485	$CaCl_2 \cdot 6H_2O+KCl$
3.15	37.82	59.03	1.402	KCl
3.72	32.34	63.94	1.349	»
7.52	23.15	69.33	1.273	»
11.64	16.55	71.81	1.236	»
17.63	8.53	73.84	1.204	»
26.74	0.00	73.26	1.182	»

4850 POTASSIUM CHLORIDE–CALCIUM CHLORIDE–WATER [2162]

$KCl—CaCl_2—H_2O$

$t=20$

Content in aqueous soln., wt %		Comp. of solid phase	Content in aqueous soln., wt %		Comp. of solid phase
Potassium Chloride	Calcium Chloride		Potassium Chloride	Calcium Chloride	
25.4	0	KCl	2.54	30.70	KCl
19.41	6.21	»	1.19	35.78	»
10.62	14.14	»	0.63	43.8	$KCl+CaCl_2 \cdot 6H_2O$
6.8	19.80	»	0	42.7	$CaCl_2 \cdot 6H_2O$

№ 4851 POTASSIUM CHLORIDE–CALCIUM CHLORIDE–WATER [301]

$KCl—CaCl_2—H_2O$

Content in aqueous soln., wt %		t	d	Comp. of solid phase
Potassium Chloride	Calcium Chloride			
19.75	0	—10	1.136	KCl
5.70	22.20	—10	1.240	»
1.98	35.29	—10	1.372	$KCl+CaCl_2\cdot 6H_2O$
0	35.5	—10	1.353	$CaCl_2\cdot 6H_2O$
18.67	0	—10	1.128	Ice
8.56	8.13	—10	1.133	»
0	14.5	—10	1.134	»
22.19	0	—10	1.154	KCl
12.19	10.93	—10	1.193	»
5.07	23.53	—10	1.265	»
2.52	34.65	—10	1.366	»
2.42	37.00	—10	1.388	$KCl+CaCl_2\cdot 6H_2O$
0	37.57	—10	1.376	$CaCl_2\cdot 6H_2O$
2.73	41.05	15	—	$KCl+CaCl_2\cdot 6H_2O$
0.63	43.79	20	—	»
3.05	44.66	25	—	»

№ 4852 POTASSIUM CHLORIDE–CALCIUM CHLORIDE–WATER [1881]

$KCl—CaCl_2—H_2O$

Comp. of saturated soln., wt %			t	Comp. of solid phase
Potassium Chloride	Calcium Chloride	Water		
0.00	37.55	62.45	0	$CaCl_2\cdot 6H_2O$
0.69	37.41	61.90	0	»
1.96	37.36	60.68	0	$CaCl_2\cdot 6H_2O+KCl$
2.62	33.42	63.96	0	KCl
3.88	25.89	70.23	0	»
6.48	20.03	73.49	0	»
8.38	16.43	75.19	0	»
12.78	10.52	76.70	0	»
16.93	5.54	77.53	0	»
21.87	0.00	78.13	0	»
0.00	42.70	57.30	20	$CaCl_2\cdot 6H_2O$
0.63	43.80	55.57	20	$CaCl_2\cdot 6H_2O+KCl$
1.19	35.78	63.03	20	KCl
2.54	30.70	66.76	20	»
6.80	19.80	73.40	20	»
10.62	14.14	75.24	20	»
19.41	6.21	74.38	20	»
25.40	0.00	74.60	20	»

№ 4853 [510]

POTASSIUM CHLORIDE–CALCIUM CHLORIDE–WATER

$KCl—CaCl_2—H_2O$

Content in aqueous soln., wt %		t	Comp. of solid phase
Potassium Chloride	Calcium Chloride		
1.80	34.50	—10	$KCl+CaCl_2 \cdot 6H_2O$
2.10	36.63	0	»
0	37.25	0	$CaCl_2 \cdot 6H_2O$
3.42	44.53	25	$KCl+CaCl_2 \cdot 6H_2O$
0	44.70	25	$CaCl_2 \cdot 6H_2O$
30.04	0	50	KCl
19.0	11.89	50	»
7.0	28.51	50	»
4.17	42.08	50	$KCl+CaCl_2 \cdot 2H_2O$
4.18	55.11	50	$CaCl \cdot 2H_2O$
0	56.86	50	KCl
36.1	0	100	»
13.92	25.32	100	»
8.98	38.39	100	$KCl+CaCl_2 \cdot 2H_2O$
3.21	59.89	100	$CaCl_2 \cdot 2H_2O$
0	61.05	100	KCl
36.61	0	100	»
4.15	59.11	100	$KCl+CaCl_2 \cdot 2H_2O$
2.65	62.66	100	$CaCl_2 \cdot 2H_2O$
0	63.70	100	»

№ 4854 [2082]

POTASSIUM CHLORIDE–CALCIUM CHLORIDE–WATER

$KCl—CaCl_2—H_2O$

$t=35$

Comp. of saturated soln., wt %			Comp. of solid phase
Calcium Chloride	Potassium Chloride	Water	
0.00	28.04	71.96	KCl
3.12	24.66	72.22	»
7.28	20.77	71.95	»
13 38	15.59	71.03	»
19.75	10.77	69.48	»
26.45	6.89	66.66	»
31.56	4.89	63.55	»
38.37	3.74	57.99	»
43.56	4.25	52.19	»
46.48	4.90	48.62	»
47.65	5.10	47.25	»
50.47	6.48	43.05	$KCl+CaCl_2 \cdot 4H_2O$
50.43	6.48	43.09	»
50.70	3.26	46 04	$CaCl_2 \cdot 4H_2O$
51.02	1.28	47.70	»
51.06	0.82	48.12	»
51.33	0.00	48.67	»

№ 4855 [2083]

POTASSIUM CHLORIDE–CALCIUM CHLORIDE–WATER

$KCl—CaCl_2—H_2O$

Concn. in aqueous soln., wt %		t	Comp. of solid phase
Calcium Chloride	Potassium Chloride		
0.00	33.16	75	KCl
11.73	21.62	75	»
18.27	16.00	75	»
28.47	9.62	75	»
37.65	6.77	75	»
42.84	6.97	75	»
47.64	8.43	75	»
50.19	10.32	75	$KCl+2KCl \cdot CaCl_2 \cdot 2H_2O$
50.18	10.31	75	»
50.21	10.36	75	»
50.22	10.32	75	»
50.34	10.36	75	$2KCl \cdot CaCl_2 \cdot 2H_2O$
50.92	9.36	75	»
53.85	6.21	75	»
54.03	6.08	75	»
54.52	5.45	75	»
54.79	5.55	75	»
56.33	4.51	75	»
56.28	4.33	75	»
56.57	5.20	75	»
56.79	4.00	75	»
57.00	3.92	75	»
57.62	3.60	75	$2KCl \cdot CaCl_2 \cdot 2H_2O+CaCl_2 \cdot 2H_2O$
57.64	3.58	75	»
57.69	3.58	75	»
57.68	3.60	75	»
57 77	2.56	75	$CaCl_2 \cdot 2H_2O$
58.58	0.00	75	»
49.47	12.42	95	$KCl+2KCl \cdot CaCl_2 \cdot 2H_2O$
49.40	12.44	95	»
52.47	7.25	95	$2KCl \cdot CaCl_2 \cdot 2H_2O$
58.26	3.93	95	$2KCl \cdot CaCl_2 \cdot 2H_2O+CaCl_2 \cdot 2H_2O$

№ 4856 [1106]

POTASSIUM CHLORIDE–CALCIUM CHLORIDE–WATER

$KCl—CaCl_2—H_2O$

Content in aqueous soln., wt %		t	Comp. of solid phase
Potassium Chloride	Calcium Chloride		
4.0	47.3	27.0	$KCl+CaCl_2 \cdot 6H_2O+\alpha\text{-}CaCl_2 \cdot 4H_2O$
7.0	51.8	37.8	$KCl+KCl \cdot CaCl_2+\alpha\text{-}CaCl_2 \cdot 4H_2O$
4.3	55.0	43.4	$KCl \cdot CaCl_2+\alpha\text{-}CaCl_2 \cdot 4H_2O+CaCl_2 \cdot 2H_2O$

№ 4857 POTASSIUM CHLORIDE–CALCIUM CHLORIDE–WATER [1107]

$KCl—CaCl_2—H_2O$

Concn. in aqueous soln., wt %		t	Comp. of solid phase
Calcium Chloride	Potassium Chloride		
0.0	29.9	50	KCl
5.9	24.9	50	»
12.7	18.1	50	»
23.8	9.8	50	»
30.9	6.1	50	»
39.0	4.8	50	»
45.7	5.7	50	»
48.9	6.6	50	»
50.85	8.60	50	$KCl+CaCl_2 \cdot KCl$
50.75	8.65	50	»
51.6	7.1	50	$CaCl_2 \cdot KCl$
52.1	6.6	50	»
52.9	6.0	50	»
53.8	5.6	50	»
54.8	4.6	50	»
55.51	4.10	50	$CaCl_2 \cdot KCl+CaCl_2 \cdot 2H_2O$
55.58	4.09	50	$CaCl_2 \cdot 2H_2O$
56.54	0.0	50	»
56.58	0.0	50	»
0.0	36.2	95	KCl
6.4	29.2	95	»
11.3	24.2	95	»
18.0	18.6	95	»
28.0	12.7	95	»
38.4	10.5	95	»
41.3	10.3	95	»
43.2	10.2	95	»
45.3	10.5	95	»
48.0	11.9	95	»
48.6	12.2	95	»
49.6	12.4	95	$KCl+CaCl_2 \cdot KCl$
49.9	12.1	95	$CaCl_2 \cdot KCl$
52.2	9.3	95	»
53.0	8.0	95	»
54.0	7.2	95	»
55.6	5.9	95	»
56.4	5.3	95	»
58.3	3.9	95	»
59.6	2.9	95	$CaCl_2 \cdot KCl+CaCl_2 \cdot 2H_2O$
59.8	2.5	95	$CaCl_2 \cdot 2H_2O$
60.20	0.0	95	»
45.40	3.90	26.5	$KCl+CaCl_2 \cdot 6H_2O$
47.20	4.50	27.0	$CaCl_2 \cdot 6H_2O+\alpha\text{-}CaCl_2 \cdot 4H_2O+KCl$
47.75	4.40	27.0	$CaCl_2 \cdot 6H_2O+\alpha\text{-}CaCl_2 \cdot 4H_2O$
48.10	4.30	27.6	$\alpha\text{-}CaCl_2 \cdot 4H_2O+KCl$
48.40	4.30	28.0	KCl
48.55	4.65	28.0	$\alpha\text{-}CaCl_2 \cdot 4H_2O+KCl$
47.75	4.65	28.0	$\alpha\text{-}CaCl_2 \cdot 4H_2O$
51.40	7.30	37.75	KCl
51.50	7.30	37.75	$KCl+\alpha\text{-}CaCl_2 \cdot 4H_2O$

Table № 4857 (continuation)

Concn. in aqueous soln., wt %		t	Comp. of solid phase
Calcium Chloride	Potassium Chloride		
51.90	7.35	37.75	α-$CaCl_2 \cdot 4H_2O$
51.50	7.35	38.0	KCl
51.60	7.20	38.0	$CaCl_2 \cdot KCl$
51.70	7.15	38.0	α-$CaCl_2 \cdot 4H_2O$
52.00	7.00	38.0	»
54.70	4.55	43.25	$CaCl_2 \cdot KCl$
54.90	4.50	43.25	$CaCl_2 \cdot KCl + \alpha$-$CaCl_2 \cdot 4H_2O$
55.15	4.25	43.25	α-$CaCl_2 \cdot 4H_2O$
54.45	4.40	43.50	$CaCl_2 \cdot KCl$
55.10	4.20	43.50	$CaCl_2 \cdot KCl + CaCl_2 \cdot 2H_2O$
55.20	4.00	43.50	$CaCl_2 \cdot 2H_2O$

№ 4858 [784]

POTASSIUM CHLORIDE–CALCIUM CHLORIDE–WATER

$KCl—CaCl_2—H_2O$

$t=55$

Comp. of saturated soln., wt %			Comp. of solid phase
Calcium Chloride	Potassium Chloride	Water	
55.26	4.63	40.11	$CaCl_2 \cdot 2H_2O$
56.40	4.46	39.14	$CaCl_2 \cdot 2H_2O + KCl \cdot CaCl_2$
49.35	7.79	42.86	$KCl \cdot CaCl_2 + KCl$
40.18	5.11	54.71	KCl
31.89	6.07	62.04	»
31.55	6.35	62.10	»
12.75	18.27	68.98	»
4.82	25.66	69.52	»

№ 4859 [110]

POTASSIUM CHLORIDE–CALCIUM CHLORIDE–WATER

$KCl—CaCl_2—H_2O$

Comp. of saturated soln., wt %			t	Comp. of solid phase
Water	Potassium Chloride	Calcium Chloride		
71.8	0.0	28.2	—40	Ice
71.5	1.5	27.0	—40	Ice+KCl
68.8	1.6	29.6	—40	KCl
68.5	0.4	31.1	—40	$KCl + CaCl_2 \cdot 6H_2O$

Comp. of saturated soln., wt %			t	Comp. of solid phase
Water	Potassium Chloride	Calcium Chloride		
69.0	0.0	31.0	—40	$CaCl_2 \cdot 6H_2O$
75.0	0.0	25.0	—30	Ice
74.0	3.2	22.8	—30	Ice+KCl
72.6	3.1	24.3	—30	KCl
68.6	2.0	29.4	—30	»
66.8	0.7	32.5	—30	$KCl+CaCl_2 \cdot 6H_2O$
67.6	0.0	32.4	—30	$CaCl_2 \cdot 6H_2O$
78.70	0.00	21.30	—20	Ice
77.30	3.40	19.30	—20	»
77.70	6.70	15.60	—20	Ice+KCl
75.20	6.00	18.80	—20	KCl
71.70	4.30	24.00	—20	»
68.40	2.40	29.20	—20	»
65.00	1.00	34.00	—20	$KCl+CaCl_2 \cdot 6H_2O$
66.00	0.00	34.00	—20	$CaCl_2 \cdot 6H_2O$
85.50	0.00	14.50	—10	Ice
82.40	8.60	9.00	—10	»
80.80	19.20	0.00	—10	»
79.70	20.30	0.00	—10	KCl
79.60	19.20	1.20	—10	»
78.70	12.60	8.70	—10	»
74.70	6.70	18.60	—10	»
71.70	4.30	24.00	—10	»
68.00	3.00	29.00	—10	»
63.80	2.00	34.20	—10	»
63.20	1.20	35.60	—10	$KCl+CaCl_2 \cdot 6H_2O$
64.80	0.00	35.20	—10	$CaCl_2 \cdot 6H_2O$
77.90	22.10	0.00	0	KCl
77.80	21.00	1.20	0	»
76.70	14.80	8.50	0	»
73.60	8.00	18.40	0	»
70.90	5.30	23.80	0	»
67.50	3.60	28.90	0	»
63.40	2.60	34.00	0	»
61.00	1.60	37.40	0	$KCl+CaCl_2 \cdot 6H_2O$
63.00	0.00	37.00	0	$CaCl_2 \cdot 6H_2O$
76.40	23.60	0.00	10	KCl
76.00	22.80	1.20	10	»
75.10	16.60	8.30	10	»
72.50	9.30	18.20	10	»
70.40	6.00	23.60	10	»
67.00	4.40	28.60	10	»
62.00	3.20	23.80	10	»
58.80	2.20	39.00	10	$KCl+CaCl_2 \cdot 6H_2O$
60.60	0.00	39.40	10	$CaCl_2 \cdot 6H_2O$
73.80	26.20	0.00	25	KCl
73.60	25.20	1.20	25	»
72.80	19.20	8.00	25	»
71.20	11.00	17.80	25	»
69.50	7.20	23.30	25	»
66.40	5.20	28.40	25	»

Comp. of saturated soln., wt %			t	Comp. of solid phase
Water	Potassium Chloride	Calcium Chloride		
62.40	4.00	33.60	25	KCl
54.80	3.60	41.60	25	»
50.00	4.00	46.00	25	$KCl+CaCl_2 \cdot 6H_2O$
55.60	0.00	44.40	25	$CaCl_2 \cdot 6H_2O$
70.00	30.00	0.00	50	KCl
70.10	28.80	1.10	50	»
69.90	22.30	7.80	50	»
68.40	14.60	17.00	50	»
66.80	10.80	11.40	50	»
64.80	7.60	27.60	50	»
61.20	6.00	32.80	50	»
54.20	5.00	40.80	50	»
68.20	14.40	17.40	50	»
67.40	12.00	20.60	50	»
65.40	10.80	23.80	50	»
40.40	8.60	51.00	50	»
40.40	4.20	55.40	50	α-$KCl \cdot CaCl_2+CaCl_2 \cdot 2H_2O$
64.40	7.40	28.20	50	KCl
41.00	6.00	53.00	50	α-$KCl \cdot CaCl_2$
43.40	0.00	57.60	50	$CaCl_2 \cdot 2H_2O$
63.80	36.20	0.00	100	KCl
62.20	17.00	20.80	100	»
62.00	17.20	20.80	100	»
60.40	14.60	25.00	100	»
57.20	12.40	30.40	100	»
51.00	10.60	38.40	100	»
49.00	10.60	40.40	100	»
37.60	12.60	49.80	100	$KCl+\alpha$-$KCl \cdot CaCl_2$
39.00	6.20	54.80	100	α-$KCl \cdot CaCl_2$
37.10	2.60	60.30	100	α-$KCl \cdot CaCl_2+CaCl_2 \cdot 2H_2O$
38.60	0.00	61.40	100	$CaCl_2 \cdot 2H_2O$
59.40	40.60	0.00	150	KCl
56.80	24.20	19.00	150	»
53.80	20.80	25.40	150	»
49.60	18.60	31.80	150	»
47.00	17.60	35.40	150	»
39.60	17.60	42.80	150	»
35.40	18.40	46.20	150	$KCl+\alpha$-$KCl \cdot CaCl_2$
36.20	13.20	50.60	150	α-$KCl \cdot CaCl_2$
36.80	6.40	56.80	150	»
31.00	2.00	67.00	150	α-$KCl \cdot CaCl_2+CaCl_2 \cdot 2H_2O$
32.40	0.00	67.60	150	$CaCl_2 \cdot 2H_2O$
55.10	44.90	0.00	200	KCl
48.50	35.50	16.00	200	»
41.00	28.00	31.00	200	»
40.20	27.00	32.80	200	»
30.20	29.80	40.00	200	$KCl+\beta$-$KCl \cdot CaCl_2$
32.00	25.00	43.00	200	β-$KCl \cdot CaCl_2$
33.40	19.40	47.20	200	»
34.20	13.60	52.20	200	»
34.00	6.80	59.20	200	»

Comp. of saturated soln., wt %			t	Comp. of solid phase
Water	Potassium Chloride	Calcium Chloride		
20.80	2.20	77.00	200	β-KCl · $CaCl_2$+$CaCl_2 \cdot H_2O$
24.40	0.00	75.60	200	$CaCl_2 \cdot H_2O$
50.50	49.50	0.00	250	KCl
43.80	41.60	14.60	250	»
36.10	36.60	27.30	250	»
28.60	32.60	38.80	250	β-KCl · $CaCl_2$+KCl
28.60	32.20	39.20	250	β-KCl · $CaCl_2$
24.40	26.00	44.60	250	»
31.20	20.00	48.80	250	»
32.40	14.00	53.60	250	»
32.50	6.70	60.80	250	»
19.20	2.40	78.40	250	β-KCl · $CaCl_2$+$CaCl_2 \cdot H_2O$
21.40	0.00	78.60	250	$CaCl_2 \cdot H_2O$
47.50	52.50	0.00	300	KCl
40.20	46.40	13.40	300	»
33.40	41.40	25.20	300	»
26.40	36.00	37.60	300	β-KCl · $CaCl_2$+KCl
26.80	33.00	40.20	300	β-KCl · $CaCl_2$
27.20	26.80	46.00	300	»
28.80	20.80	50.40	300	»
31.00	14.20	54.80	300	»
30.00	7.20	62.80	300	»
18.00	3.70	78.30	300	$CaCl_2$+β-KCl · $CaCl_2$
20.00	0.00	80.00	300	$CaCl_2$
				Triple invariant points
69.86	0.24	29.90	−50.5	Ice+KCl+$CaCl_2 \cdot 6H_2O$
48.30	4.50	47.20	27.0	$CaCl_2 \cdot 6H_2O$+$CaCl_2 \cdot 4H_2O$+KCl
41.25	7.30	51.45	37.8	$CaCl_2 \cdot 4H_2O$+α-KCl · $CaCl_2$+KCl
40.65	4.30	55.05	43.4	$CaCl_2 \cdot 4H_2O$+α-KCl · $CaCl_2$+ +$CaCl_2 \cdot 2H_2O$
30.40	29.60	40.00	196.0	KCl+α-KCl · $CaCl_2$+β-KCl · $CaCl_2$
24.00	2.00	74.00	170.0	$CaCl_2 \cdot 2H_2O$+$CaCl_2 \cdot H_2O$+ +α-KCl · $CaCl_2$
22.30	2.10	75.60	196.0	$CaCl_2 \cdot H_2O$+α+β-KCl · $CaCl_2$
19.20	2.40	78.40	250.0	$CaCl_2 \cdot H_2O$+$CaCl_2$+β-KCl · $CaCl_2$

№ 4860 POTASSIUM CHLORIDE–CALCIUM CHLORIDE–WATER [185]

$KCl—CaCl_2—H_2O$

Comp. of saturated soln., wt %			t	Comp. of solid phase
Calcium Chloride	Potassium Chloride	Water		
31.40	0.00	68.60	−40	$CaCl_2 \cdot 6H_2O$
31.20	0.80	68.00	−40	»
31.00	0.70	68.30	−40	»
30.90	0.75	68.35	−40	$CaCl_2 \cdot 6H_2O$+KCl

Comp. of saturated soln., wt %			t	Comp. of solid phase
Calcium Chloride	Potassium Chloride	Water		
29.30	0.90	69.80	—40	KCl
28.30	1.90	69.80	—40	»
27.10	2.10	70.80	—40	KCl+ice
27.40	1.50	71.10	—40	Ice
27.80	0.60	71.60	—40	»
28.20	0.00	71.80	—40	»
32.60	0.00	67.40	—30	$CaCl_2 \cdot 6H_2O$
32.40	0.80	66.80	—30	»
32.20	1.30	66.50	—30	»
32.10	1.70	66.20	—30	$CaCl_2 \cdot 6H_2O$+KCl
30.90	1.90	67.20	—30	KCl
28.20	2.30	68.80	—30	»
23.30	3.60	73.10	—30	KCl+ice
23.80	2.40	73.80	—30	Ice
24.40	1.40	74.20	—30	»
24.80	0.60	74.60	—30	»
25.00	0.00	75.00	—30	»
34.00	0.00	66.00	—20	$CaCl_2 \cdot 6H_2O$
33.70	0.80	65.50	—20	»
33.60	1.30	65.10	—20	»
33.40	1.80	64.80	—20	$CaCl_2 \cdot 6H_2O$+KCl
32.50	2.00	65.50	—20	KCl
28.00	2.70	69.00	—20	»
18.80	5.80	75.40	—20	»
17.20	6.60	76.20	—20	KCl+ice
19.80	2.00	78.20	—20	Ice
20.50	0.50	79.00	—20	»
20.80	0.00	79.20	—20	»
35.50	0.00	64.50	—10	$CaCl_2 \cdot 6H_2O$
35.40	0.90	63.70	—10	»
35.20	1.30	63.50	—10	»
35.00	2.00	63.00	—10	$CaCl_2 \cdot 6H_2O$+KCl
34.40	2.20	63.40	—10	KCl
28.00	3.20	68.80	—10	»
18.60	7.10	74.30	—10	»
9.20	12.80	78.00	—10	»
0.00	20.10	79.90	—10	»
37.40	0.00	62.60	0	$CaCl_2 \cdot 6H_2O$
37.30	0.90	61.80	0	»
37.10	1.40	61.50	0	»
36.80	2.30	60.90	0	$CaCl_2 \cdot 6H_2O$+KCl
34.20	2.70	63.10	0	KCl
27.70	4.00	68.30	0	»
18.40	8.40	73.20	0	»
9.00	14.60	76.40	0	»
0.00	21.80	78.20	0	»
39.90	0.00	60.10	10	$CaCl_2 \cdot 6H_2O$
39.60	1.00	59.40	10	»
39.40	1.50	59.10	10	»
39.10	2.20	58.70	10	»
37.00	2.90	60.10	10	KCl

Table № 4860 (continuation)

Comp. of saturated soln., wt %			t	Comp. of solid phase
Calcium Chloride	Potassium Chloride	Water		
39.00	2.80	58.20	10	$CaCl_2 \cdot 6H_2O + KCl$
34.00	3.30	62.70	10	KCl
27.50	4.70	67.80	10	»
18.00	9.60	72.40	10	»
8.80	16.40	74.80	10	»
0.00	23.60	76.40	10	»
43.0	0.00	57.00	20	$CaCl_2 \cdot 6H_2O$
42.50	1.60	55.90	20	»
42.10	2.40	55.50	20	»
41.80	3.20	55.00	20	$CaCl_2 \cdot 6H_2O + KCl$
39.60	3.40	57.00	20	KCl
33.80	3.80	62.40	20	»
27.40	5.40	67.20	20	»
17.80	10.90	71.30	20	»
8.60	18.10	73.30	20	»
0.00	25.40	74.60	20	»
50.20	0.00	49.80	30	$CaCl_2 \cdot 4H_2O$
49.60	1.80	48.60	30	»
49.50	2.80	47.60	30	»
49.00	3.80	47.20	30	»
48.65	4.75	46.60	30	$CaCl_2 \cdot 4H_2O + KCl$
46.20	4.20	52.60	30	KCl
39.40	3.80	56.80	30	»
33.60	4.30	62.10	30	»
27.10	6.10	66.80	30	»
17.60	12.20	70.20	30	»
8.40	19.80	71.80	30	»
0.00	26.90	73.10	30	»
44.92	4.60	50.48	35	»
27.30	6.40	66.30	35	»
17.44	12.80	69.76	35	»
8.35	20.70	70.95	35	»
50.20	5.20	44.60	35	$CaCl_2 \cdot 4H_2O + KCl$
53.60	0.00	46.40	40	$CaCl_2 \cdot 4H_2O$
52.50	2.00	45.50	40	»
51.70	3.90	44.40	40	»
51.20	5.20	43.60	40	$CaCl_2 \cdot 4H_2O + KCl$
44.80	4.80	50.40	40	KCl
39.17	4.40	56.43	40	»
33.50	4.80	61.70	40	»
27.00	6.60	66.40	40	»
17.21	13.60	69.19	40	»
8.20	21.80	70.00	40	»
0.00	28.80	71.20	40	»
38.50	0.95	60.55	6.5	$CaCl_2 \cdot 6H_2O$
36.60	0.90	62.50	—4.0	»
35.40	0.87	63.73	—11.0	»
33.00	0.81	66.19	—26.8	»
31.80	0.79	67.41	—35.0	»
31.00	0.77	68.23	—40.4	»
29.90	0.74	69.36	—50.1	$CaCl_2 \cdot 6H_2O +$ ice

Table № 4860 (continuation)

Comp. of saturated soln., wt %			t	Comp. of solid phase
Calcium Chloride	Potassium Chloride	Water		
29.60	0.73	69.67	—45.0	Ice
27.20	0.67	72.13	—37.8	»
25.20	0.62	74.18	—31.0	»
21.80	0.54	77.66	—22.7	»
19.50	0.48	80.02	—18.0	»
14.60	0.36	85.04	—10.2	»
52.20	1.95	45.85	39.0	$CaCl_2 \cdot 4H_2O$
50.80	1.89	47.31	34.8	»
50.00	1.86	48.14	31.7	»
49.50	1.84	48.66	29.1	$CaCl_2 \cdot 4H_2O + CaCl_2 \cdot 6H_2O$
47.60	1.77	50.63	27.8	$CaCl_2 \cdot 6H_2O$
44.80	1.67	53.53	24.5	»
42.40	1.58	56.02	19.2	»
41.00	1.52	57.48	14.4	»
39.00	1.45	59.45	7.0	»
37.00	1.38	61.62	— 1.8	»
35.00	1.30	63.70	—10.3	»
34.00	1.26	64.74	—18.0	»
31.80	1.18	66.02	—34.0	»
31.00	1.15	67.85	—39.8	»
29.60	1.10	69.30	—50.6	$CaCl_2 \cdot 6H_2O$ + ice
29.00	1.08	69.92	—46.0	Ice
27.80	1.03	71.17	—41.0	»
26.50	0.99	72.51	—36.0	»
25.70	0.96	73.34	—33.0	»
23.60	0.88	75.52	—27.8	»
21.00	0.78	78.22	—21.0	»
18.40	0.68	80.92	—16.0	»
13.00	0.48	86.52	— 8.5	»
44.20	2.51	53.29	25.0	$CaCl_2 \cdot 6H_2O$
41.20	2.34	56.46	17.5	»
38.00	2.16	59.84	5.6	»
37.40	2.12	60.48	3.3	»
35.60	2.02	62.38	— 6.4	»
34.40	1.95	63.65	—12.0	»
34.30	1.95	63.75	—15.3	$CaCl_2 \cdot 6H_2O + KCl$
33.60	1.91	64.49	—17.0	KCl
32.20	1.83	65.97	—21.0	»
30.20	1.71	68.09	—35.0	»
29.60	1.68	68.72	—39.4	»
28.50	1.62	69.78	—45.8	Ice+KCl
28.00	1.60	70 40	—42.4	Ice
27.40	1.56	71.04	—40.5	»
24.00	1.36	74.64	—28.6	»
18.20	1.03	80.77	—16.4	»
10.40	0.59	89.01	— 6.4	»
51.00	3.87	45.13	38.0	$CaCl_2 \cdot 4H_2O$
49.80	3.78	46.42	34.8	»
49.10	3.73	47.17	31.0	»
48.90	3.71	47.39	29.1	$CaCl_2 \cdot 4H_2O + CaCl_2 \cdot 6H_2O$
45.80	3.48	50.72	27.0	$CaCl_2 \cdot 6H_2O$

Comp. of saturated soln., wt %			t	Comp. of solid phase
Calcium Chloride	Potassium Chloride	Water		
44.0	3.34	52.66	25.0	$CaCl_2 \cdot 6H_2O$
42.50	3.23	54.27	22.7	»
41.80	3.17	55.03	20.0	»
41.00	3.01	55.89	16.5	$CaCl_2 \cdot 6H_2O+KCl$
39.80	3.02	57.18	14.5	KCl
36.80	2.79	60.41	9.5	»
34.80	2.64	62.56	2.0	»
33.80	2.57	63.63	— 3.5	»
32.60	2.47	64.93	— 8.0	»
31.60	2.40	66.00	—12.0	»
31.40	2.38	66.22	—15.5	»
30.70	2.33	66.97	—19.6	»
29.60	2.25	68.15	—24.6	»
29.10	2.21	68.69	—31.0	»
28.60	2.17	69.33	—32.4	»
27.40	2.08	70.51	—41.0	Ice+KCl
26.20	1.99	71.81	—37.0	Ice
24.40	1.85	73.75	—31.4	»
20.20	1.53	78.27	—21.0	»
15.20	1.15	83.65	—12.0	»
37.00	3.70	59.30	21.6	KCl
35.20	3.52	61.28	13.8	»
32.70	3.27	64.08	3.8	»
30.40	3.04	65.56	— 7.4	»
28.20	2.82	69.98	20.6	»
27.40	2.74	69.86	28.0	»
26.00	2.60	71.40	—36.8	KCl+ice
21.00	2.10	76.90	—23.5	Ice
20.50	2.05	77.45	—21.5	»
19.00	1.90	79.10	—19.0	»
17.00	1.70	81.30	—15.0	»
11.60	1.16	87.24	— 8.4	»
10.70	0.00	89.30	— 5.8	»
10.50	2.00	87.50	— 6.6	»
10.38	3.00	86.61	— 7.0	»
10.21	4.00	85.69	— 7.8	»
10.10	6.00	83.90	— 9.2	»
9.79	8.00	82.21	—10.9	»
9.61	9.00	81.39	—11.8	»
9.47	10.00	80.53	—12.7	»
9.42	11.00	79.58	—13.6	»
9.41	11.00	79.54	—14.0	»
9.40	11.90	78.70	—14.2	Ice +KCl
9.30	12.00	78.70	—13.6	KCl
9.20	12.40	78.40	—12.0	»
9.19	13.00	77.81	— 9.0	»
9.10	13.50	77.40	— 6.0	»
9.00	14.00	77.00	— 3.5	»
8.98	14.50	76.52	— 1.0	»
8.92	15.00	76.08	2.0	»
8.80	16.00	75.20	7.6	»

Table № 4860 (continuation)

Comp. of saturated soln., wt %			t	Comp. of solid phase
Calcium Chloride	Potassium Chloride	Water		
8.70	17.00	74.30	13.8	KCl
8.62	18.00	73.38	20.4	»
8.50	19.00	72.50	25.2	»
20.00	0.00	80.00	—18.0	Ice
19.60	2.10	78.30	—19.2	»
19.37	3.10	77.53	—20.0	»
19.10	4.10	76.80	—20.6	»
19.07	4.60	76.63	—21.0	»
19.05	5.00	75.95	—21.2	»
19.00	5.50	75.50	—21.4	Ice+KCl
18.80	6.00	75.20	—18.5	KCl
18.60	6.60	74.80	—13.6	»
18.72	7.20	74.08	—10.0	»
18.40	8.00	73.60	— 2.6	»
18.18	9.10	72.72	6.0	»
18.00	10.00	72.00	13.0	»
17.75	11.00	71.25	20.4	»
17.70	12.00	70.30	28.8	»
17.40	13.00	69.60	36.6	»
28.90	0.00	71.10	—45.0	Ice
28.80	0.50	70.70	—45.0	»
28.60	1.00	70.40	—45.2	»
28.52	1.50	69.98	—45.6	»
28.50	1.60	69 80	—45.8	Ice+KCl
28.38	2.00	69.62	—35.4	KCl
28.20	2.50	69.30	—24.8	»
28.02	3.00	68.98	—13.2	»
27.70	3.50	68.60	— 6.0	»
27.75	4.00	68.25	1.2	»
27.60	4.50	67.90	8.0	»
27.47	5.00	67.53	15.0	»
27.30	5.50	67.20	22.4	»
27.19	6.00	66.81	28.8	»
35.10	0.00	64.90	—13.2	$CaCl_2 \cdot 6H_2O$
34.50	0.50	64.60	—13.5	»
34.70	1.10	64.20	—14.2	»
34.62	1.50	63.88	—14.8	»
34.60	1.90	63.50	—15.2	$CaCl_2 \cdot 6H_2O+KCl$
34.59	2.00	63.41	—14.4	KCl
34.20	2.50	63.30	— 4.8	»
34.05	3.00	62.95	5.6	»
33.82	3.50	62.68	15.8	»
33.65	4.00	62.35	24.0	»
33.50	4.50	62.00	33.6	»
41.00	0.00	59.00	14.0	$CaCl_2 \cdot 6H_2O$
40.20	0.50	58.70	13.8	»
40.60	1.00	58.40	13.6	»
40.40	1.50	58.10	13.4	»
39.90	2.50	57.60	13.0	»
39.70	2.90	57.40	12.8	$CaCl_2 \cdot 6H_2O+KCl$
39.60	3.20	57.20	17.8	KCl

Table № 4860 (continuation)

Comp. of saturated soln., wt %			t	Comp. of solid phase
Calcium Chloride	Potassium Chloride	Water		
39.40	3.40	57.20	21.0	KCl
39.37	3.80	58.83	28.0	»
39.35	4.00	56.65	32.6	»
39.20	4.40	56.40	41.4	»
47.10	0.00	52.90	26.5	$CaCl_2 \cdot 6H_2O$
46.90	0.50	52.60	26.4	»
46.62	1.00	52.38	26.2	»
46.20	2.00	51.80	26.0	»
45.90	2.50	51.60	25.8	»
45.70	3.00	51.30	25.7	»
45.30	3.90	50.80	25.4	$CaCl_2 \cdot 6H_2O+KCl$
45.00	4.30	50.70	30.0	KCl
44.70	4.60	50.70	36.0	»
44.80	4.80	50.40	39.0	»

№ 4861 **POTASSIUM CHLORIDE–CALCIUM CHLORIDE–STRONTIUM CHLORIDE–WATER** [1110]

$KCl—CaCl_2—SrCl_2—H_2O$

Comp. of saturated soln., wt %				t	Comp. of solid phase
Strontium Chloride	Calcium Chloride	Potassium Chloride	Water		
0.8	39.6	2.6	58.0	18.0	$(Ca, Sr)Cl_2 \cdot 6H_2O$
1.2	38.4	2.6	57.8	18.0	»
6.7	25.4	4.2	63.7	18.0	»
23.4	7.2	6.3	63.1	18.0	»
0.3	45.8	4.0	49.9	26.4	»
0.8	44.4	3.6	51.2	26.4	»
1.6	42.7	3.5	52.2	26.4	»
1.6	41.2	3.5	53 7	26.4	»
2.0	40.5	3.5	54.0	26.4	»
0.3	46.2	3.8	49.7	26.9	»
0.9	45.0	3.7	50.4	26.9	»
1.2	43.9	3.2	51.7	26.9	»
1.6	42.9	3.4	52.1	26.9	$SrCl_2 \cdot 2H_2O+KCl$
1.5	42.4	3.3	52.8	26.9	»
2.1	40.6	3.3	54.0	26.9	»
2.0	40.2	3.4	54.4	26.9	$CaCl_2 \cdot SrCl_2 \cdot 6H_2O+KCl$
0.9	48.0	4.4	46.7	27.4	$CaCl_2 \cdot 4H_2O+KCl$
0.9	47.1	4.2	47.8	27.4	$CaCl_2 \cdot SrCl_2 \cdot 4H_2O+KCl$
0.8	46.2	4.2	48.8	27.4	»
1.3	45.3	4.2	49.2	27.4	$SrCl_2 \cdot 2H_2O+CaCl_2 \cdot SrCl_2 \cdot 6H_2O+KCl$
1.2	43.9	3.6	51.3	27.4	$SrCl_2 \cdot 2H_2O+KCl$
1.7	41.9	3.6	52.8	27.4	»
2.0	41.1	3.6	53.3	27.4	»
2.2	40.5	3.6	53.7	27.4	$CaCl_2 \cdot SrCl \cdot 6H_2O+KCl$
4.0	36.0	3.6	56.4	27.4	»
1.1	47.4	4.7	46.8	28.0	$SrCl_2 \cdot 2H_2O+\alpha\text{-}CaCl_2 \cdot 4H_2O+KCl$

Comp. of saturated soln., wt %				t	Comp. of solid phase
Strontium Chloride	Calcium Chloride	Potassium Chloride	Water		
1.1	47.1	4.1	47.7	28.0	$SrCl_2 \cdot 2H_2O + KCl$
1.1	44.6	2.7	51.6	28.0	»
1.7	42.2	3.0	53.1	28.0	»
1.9	41.1	3.0	54.0	28.0	»
2.9	39.1	3.0	55.0	28.0	$SrCl_2 \cdot 2H_2O + CaCl_2 \cdot SrCl_2 \cdot 6H_2O + KCl$
3.3	35.7	3.3	57.7	28.0	$CaCl_2 \cdot SrCl_2 \cdot 6H_2O + KCl$
13.0	20.0	5.2	61.8	28.0	»
1.4	50.4	7.0	41.2	37.0	$SrCl_2 \cdot 2H_2O + \alpha\text{-}CaCl_2 \cdot 4H_2O$
1.8	50.0	7.3	40.9	37.0	$SrCl_2 \cdot 2H_2O + KCl$
1.4	50.7	7.0	40.9	37.5	$SrCl_2 \cdot 2H_2O + KCl \cdot CaCl_2$
1.3	50.4	7.4	40.0	37.5	$SrCl_2 \cdot 2H_2O + KCl$
1.7	53.5	4.3	40.9	42.6	$SrCl_2 \cdot 2H_2O + \alpha\text{-}CaCl_2 \cdot 4H_2O$
1.4	52.5	6.5	39.6	42.6	$SrCl_2 \cdot 2H_2O + KCl \cdot CaCl_2$
1.2	53.2	4.2	41.4	42.8	$SrCl_2 \cdot 2H_2O + CaCl_2 \cdot 2H_2O$
1.2	52.5	5.2	42.1	42.8	$SrCl_2 \cdot 2H_2O + KCl \cdot CaCl_2$
2.2	55.8	3.6	38.4	74.0	$CaCl_2 \cdot 2H_2O + KCl \cdot CaCl_2$
1.9	56.0	3.6	38.5	74.0	$SrCl_2 \cdot 2H_2O + KCl \cdot CaCl_2$
2.0	56.0	3.8	38.2	75.0	$SrCl_2 \cdot H_2O + KCl \cdot CaCl_2$
3.1	54.5	4.6	37.8	75.0	$SrCl_2 \cdot 2H_2O + KCl \cdot CaCl_2$
3.0	58.3	3.1	35.6	100	$CaCl_2 \cdot 2H_2O + KCl \cdot CaCl_2$
3.2	57.5	3.1	36.2	100	$SrCl_2 \cdot H_2O + KCl \cdot CaCl_2$
1.8	41.8	3.5	52.9	26.6	$SrCl_2 \cdot 2H_2O + CaCl_2 \cdot SrCl_2 \cdot 6H_2O + KCl$
1.0	48.0	4.7	46.3	27.8	$SrCl_2 \cdot 2H_2O + \alpha\text{-}CaCl_2 \cdot 4H_2O +$ $+ CaO_2 \cdot SrCl_2 \cdot 6H_2O + KCl$
1.5	50.4	7.2	40.9	37.2	$SrCl_2 \cdot 2H_2O + \alpha\text{-}CaCl_2 \cdot 4H_2O + KCl +$ $+ KCl \cdot CaCl_2$
1.4	53.0	5.2	40.4	42.7	$SrCl_2 \cdot 2H_2O + \alpha\text{-}CaCl_2 \cdot 4H_2O + CaCl_2 \cdot 2H_2O +$ $+ KCl \cdot CaCl_2$
2.0	56.0	3.5	38.5	74.5	$SrCl_2 \cdot 2H_2O + CaCl_2 \cdot 2H_2O + SrCl_2 \cdot H_2O +$ $+ KCl \cdot CaCl_2$

№ 4862 [1745]

POTASSIUM CHLORIDE–CALCIUM IODATE–WATER

$KCl—Ca(IO_3)_2—H_2O$

$t = 25$

Concn. of Potassium Chloride in aqueous soln., g/l	Solubility of Calcium Iodate in aqueous soln. of Potassium Chloride, g/l	Concn. of Potassium Chloride in aqueous soln., g/l	Solubility of Calcium Iodate in aqueous soln. of Potassium Chloride, g/l
0.0	3.110	37.28	5.911
3.728	3.724	55.92	6.827
7.456	4.133	74.56	7.323
11.184	4.441	111.84	8.871
22.368	5.170	149.12	9.978

№ 4863 [1957]

POTASSIUM CHLORIDE–CALCIUM IODATE–WATER

$KCl—Ca(IO_3)_2—H_2O$

$t = 25$

Concn. in aqueous soln., wt %	
Calcium Iodate	Potassium Chloride
0.304	0.0
0.342	0.186
0.374	0.371
0.404	0.737

№ 4864 POTASSIUM CHLORIDE–MANGANESE (II) CHLORIDE–WATER [2618]

$KCl—MnCl_2—H_2O$

Comp. of saturated soln., wt %			t	Comp. of solid phase
Manganese Chloride	Potassium Chloride	Water		
40.23	0.00	59.77	6	$MnCl_2 \cdot 4H_2O$
35.94	9.41	54 65	6	$MnCl_2 \cdot 4H_2O+MnCl_2 \cdot KCl \cdot 2H_2O+KCl$
0.00	23.06	76.94	6	KCl
44.46	0.00	55.54	28.4	$MnCl_2 \cdot 4H_2O$
43.28	8.66	48.06	28.4	$MnCl_2 \cdot 4H_2O+MnCl_2 \cdot KCl \cdot 2H_2O$
38.65	13.79	47.56	28.4	$MnCl_2 \cdot 4H_2O+MnCl_2 \cdot KCl \cdot 2H_2O+KCl$
0.00	26.91	73.09	28.4	KCl
50.14	6.01	43.85	52.8	$MnCl_2 \cdot 4H_2O+MnCl_2 \cdot 2H_2O+$ $+MnCl \cdot KCl \cdot H_2O$
51.72	0.00	48.28	58.3	$MnCl_2 \cdot 4H_2O+MnCl_2 \cdot 2H_2O$
51.86	0.00	48.14	62.6	$MnCl_2 \cdot 2H_2O$
49.95	6.67	43.38	62.6	$MnCl_2 \cdot 2H_2O+MnCl_2 \cdot KCl \cdot 2H_2O$
44.05	12.49	43.46	62.6	$MnCl_2 \cdot KCl \cdot 2H_2O+MnCl_2 \cdot 2KCl \cdot 2H_2O$
36.85	18.77	44.38	62.6	$MnCl_2 \cdot 2KCl \cdot 2H_2O+MnCl_2 \cdot 4KCl$
0.00	31.57	68.43	62.6	KCl

№ 4865 POTASSIUM CHLORIDE–MANGANESE (II) CHLORIDE–WATER [647]

$KCl—MnCl_2—H_2O$

Comp. of saturated soln., wt %			t	Comp. of solid phase
Potassium Chloride	Manganese Chloride	Water		
62.30	35.60	2.10	—50.2	$MnCl_2 \cdot 6H_2O$
0.00	37.30	62.70	—49.1	»
3.20	32.18	64.62	—39.5	»
6.30	20.95	72.75	—22.6	»
12.80	8.90	78.30	—18.9	»
6.35	36.15	57.50	—16.0	»
19.50	0.00	80.50	—10.5	—
10.80	37.60	51.60	5.0	$MnCl_2 \cdot 4H_2O$
13.00	37.90	49.10	19.3	»

№ 4866 [1854]

POTASSIUM CHLORIDE–IRON (III) CHLORIDE–WATER

$KCl—FeCl_3—H_2O$

$t=21$

Comp. of saturated soln., wt %			Comp. of solid phase
Iron Chloride	Potassium Chloride	Water	
0.00	34.97	65.03	KCl
13.44	24.45	62.11	Mixed crystals
23.18	16.54	60.28	»
28.05	11.69	60.26	»
35.72	11.68	52.60	»
36.62	11.19	52.19	$FeCl_3 \cdot 2KCl \cdot H_2O$
37.35	13.67	48.98	»
51.69	7.54	40.77	»
83.89	0.00	16.11	$FeCl_3$

$KCl—FeCl_3—H_2O$

Comp. of saturated soln., wt %			t	Comp. of solid phase
Iron Chloride	Potassium Chloride	Water		
42.66	0.00	57.34	0	$FeCl_3 \cdot 6H_2O$
42.53	1.07	56.40	0	$FeCl_3 \cdot 6H_2O + FeCl_3 \cdot 2KCl \cdot H_2O$
38.55	2.85	58.60	0	$FeCl_3 \cdot 2KCl \cdot H_2O$
35.67	6.03	58.30	0	$FeCl_3 \cdot 2KCl \cdot H_2O$ · CK
38.12	6.18	55.70	0	M.C.
33.54	6.14	60.32	0	»
28.85	6.22	64.93	0	»
25.78	6.73	67.49	0	»
23.25	7.54	69.21	0	»
16.38	10.19	73.43	0	»
0.00	21.60	78.40	0	KCl
49.76	0.00	50.24	25	$FeCl_3 \cdot 6H_2O$
49.33	1.22	49.45	25	$FeCl_3 \cdot 6H_2O + FeCl_3 \cdot 2KCl \cdot H_2O$
47.27	2.01	50.72	25	$FeCl_3 \cdot 2KCl \cdot H_2O$
43.02	5.65	51.33	25	»
38.02	10.29	51.69	25	»
35.88	11.97	52.15	25	$FeCl_3 \cdot 2KCl \cdot H_2O + KCl$
27.27	12.26	60.47	25	KCl
22.14	14.99	62.87	25	»
14.90	17.40	67.70	25	»
0.00	26.02	73.98	25	»
73.79	0.00	26.21	35	$Fe_2Cl_6 \cdot 5H_2O$
73.03	1.65	25.32	35	$Fe_2Cl_6 \cdot 5H_2O + FeCl_3 \cdot 2KCl \cdot H_2O$
65.81	2.21	31.98	35	$FeCl_3 \cdot 2KCl \cdot H_2O$
57.53	2.35	40.12	35	»
49.57	3.52	46.91	35	»
45.94	5.05	49.01	35	»
40.25	8.60	51.15	35	»
36.02	13.81	50.17	35	$FeCl_3 \cdot 2KCl \cdot H_2O + KCl$
34.44	13.11	52.45	35	KCl
29.21	14.02	56.77	35	»
20.12	16.82	63.06	35	»
10.91	20.21	68.88	35	»
0.0	27.80	72.20	35	»
78.86	0.0	21.14	60	$FeCl_3 \cdot 2H_2O$
77.13	1.90	20.97	60	$FeCl_3 \cdot 2H_2O + FeCl_3 \cdot 2KCl \cdot H_2O$
74.08	2.25	23.67	60	$FeCl_3 \cdot 2KCl \cdot H_2O$
61.70	3.10	35.20	60	»
53.98	4.96	41.06	60	»
36.78	14.41	48.81	60	$FeCl_3 \cdot 2KCl \cdot H_2O + KCl$
28.73	16.30	54.97	60	KCl
0.0	31.30	68.70	60	»

Note. M.C.–Mixed crystals.

№ 4868 [504]

POTASSIUM CHLORIDE–IRON (III) CHLORIDE–WATER

$KCl—FeCl_3—H_2O$

$t=35$

Comp. of saturated soln., wt %			Comp. of solid phase
Potassium Chloride	Iron Chloride	Water	
15.0	17.0	68.0	KCl
11.5	32.0	56.5	$KCl+FeCl_3 \cdot 2KCl \cdot H_2O$
5.0	44.0	51.0	$FeCl_3 \cdot 2KCl \cdot H_2O$
3.5	51.5	45.0	$FeCl_3 \cdot 2KCl \cdot H_2O+FeCl_3 \cdot 6H_2O$

№ 4869 [2161]

POTASSIUM CHLORIDE–COBALT (II) CHLORIDE–WATER

$KCl—CoCl_2—H_2O$

$t=20$

Comp. of saturated soln., wt %			Comp. of solid phase
Cobalt Chloride	Potassium Chloride	Water	
33.52	4.73	61.75	$CoCl_2 \cdot 6H_2O$
33.15	6.81	60.04	$CoCl_2 \cdot 6H_2O+KCl$
31.58	6.79	61.63	KCl
29.17	7.19	63.64	»
28.37	7.70	63.93	»
27.90	8.69	63.41	»
26.45	8.45	65.10	»
20.25	11.56	68.19	»
14.05	15.05	70.90	»
7.40	19.37	73.23	»

№ 4870 [1643]

POTASSIUM CHLORIDE–COBALT (II) CHLORIDE–WATER

$KCl—CoCl_2—H_2O$

$t=25$

Comp. of saturated soln., wt %			Comp. of solid phase
Potassium Chloride	Cobalt Chloride	Water	
0.00	35.67	64.33	$CoCl_2 \cdot 6H_2O$
2.43	35.20	62.37	»
4.43	34.77	60.80	»
6.22	34.53	59.25	»
8.24	34.10	57.66	$CoCl_2 \cdot 6H_2O+KCl$
8.09	33.25	58.66	KCl
10.19	24.71	65.10	»
14.19	16.09	69.72	»
20.29	7.50	72.21	»
26.46	0.00	73.54	»

$KCl—CoCl_2—H_2O$

Comp. of saturated soln., wt %		t	Comp. of solid phase
Cobalt Chloride	Potassium Chloride		
3.95	18.58	0	KCl
12.60	12.33	0	»
16.05	10.22	0	»
28.4	5.44	0	$KCl+CoCl_2 \cdot 6H_2O$
39.3	4.00	38	$CoCl_2 \cdot 6H_2O$
40.2	6.92	38	»
40.0	8.60	38	$KCl+CoCl_2 \cdot 6H_2O$
33.8	9.55	38	KCl
25.5	12.46	38	»
15.3	17.52	38	»
7.67	22.9	38	»
3.71	26.9	50	»
11.95	20.8	50	»
28.9	12.52	50	»
34.1	11.92	50	»
38.6	11.12	50	»
41.1	12.25	50	KCl+ solid soln.
40.5	12.82	50	»
45.9	6.39	50	Solid soln.
47.0	5.16	50	Solid soln.$+CoCl_2 \cdot 4H_2O$
47.9	3.34	50	$CoCl_2 \cdot 4H_2O$
46.9	2.45	50	»
15.2	18.65	51	KCl
30.4	12.25	51	»
36.2	12.39	51	»
40.7	12.56	51	KCl+solid solution
42.6	10.61	51	Solid soln.
45.1	7.70	51	»
45.2	7.55	51	»
46.5	6.01	51	»
47.0	5.20	51	Solid soln. $+CoCl_2 \cdot 4H_2O$
46.8	4.95	51	$CoCl_2 \cdot 4H_2O$
47.3	0.50	51	»
2.87	30.8	75	KCl
7.81	27.5	75	»
22.9	20.2	75	»
29.6	17.65	75	»
35.7	20.6	75	»
37.2	19.5	75	»
38.0	23.1	75	»
40.9	22.9	75	KCl+solid soln.
40.4	28.1	75	»
41.4	23.1	75	Solid soln.
43.3	19.30	75	»
43.5	16.00	75	»
44.6	15.27	75	»
45.8	11.85	75	»
48.0	8.18	75	»
49.0	5.92	75	Solid soln. $+CoCl_2 \cdot 2H_2O$
50.1	1.19	75	$CoCl_2 \cdot 2H_2O$

Concn. in aqueous soln., wt %		t	Comp. of solid phase
Cobalt Chloride	Potassium Chloride		
7.58	30.8	99.5	KCl
15.80	28.8	99.5	»
21.4	27.6	99.5	»
29.2	25.1	99.5	»
28.5	27.8	99.5	»
32.9	27.9	99.5	»
36.7	28.6	99.5	»
40.5	36.1	99.5	KCl+ solid soln.
41.4	31.5	99.5	KCl (blue) +solid soln.
43.0	32.4	99.5	Solid soln.
44.6	26.4	99.5	»
47.0	17.95	99.5	»
49.0	11.74	99.5	»
50.6	7.00	99.5	Solid soln. $+CoCl_2 \cdot 2H_2O$
49.1	4.90	99.5	$CoCl_2 \cdot 2H_2O$
51.1	3.24	99.5	»
51.4	2.85	99.5	»

№ 4872 [1812]
POTASSIUM CHLORIDE–COPPER (II) SULFATE–WATER
$KCl—CuSO_4—H_2O$

$t=25$

Concn. in aqueous soln., g/l	
Copper Sulfate	Potassium Chloride
238.9	41.9
249.2	87.5
290.3	175.0

№ 4873 POTASSIUM CHLORIDE–COPPER (I) CHLORIDE–WATER [1289]
KCl—CuCl—HO

$t=22$

Concn. in aqueous soln., wt %		Comp. of solid phase	Concn. in aqueous soln., wt %		Comp. of solid phase
Potassium Chloride	Copper Chloride		Potassium Chloride	Copper Chloride	
3.87	0.115	CuCl	23.57	13.99	CuCl · 2KCl
6.56	0.405	»	23.50	11.39	»
8.24	0.861	»	23.49	7.35	»
11.33	2.19	»	24.04	4.53	»
15.30	4.80	»	25.03	3.14	»
17.47	7.19	»	26.28	2.20	»
20.31	10.21	»	27.06	1.60	»
21.64	13.32	»	26.68	1.21	KCl
23.84	17.23	»	26.32	0.58	»
25.24	21.47	»	25.68	0.0	»
23.87	15.48	CuCl · 2KCl			

№ 4874 POTASSIUM CHLORIDE–COPPER (I) CHLORIDE–WATER [2683]

$KCl—CuCl—H_2O$

$t=25$

Concn. in aqueous soln., mol %		Comp. of solid phase	Concn. in aqueous soln., mol %		Comp. of solid phase
Copper Chloride	Potassium Chloride		Copper Chloride	Potassium Chloride	
0.27	0.09	CuCl	1.32	1.62	$CuCl \cdot KCl \cdot H_2O+$ $+CuCl \cdot 2KCl$
0.68	0.18	»			
0.96	0.25	»	1.25	1.55	$CuCl \cdot 2KCl$
1.91	0.54	»	1.25	1.50	»
2.10	0.62	»	1.10	1.35	»
2.05	0.90	$CuCl+CuCl \cdot KCl \cdot$ $\cdot H_2O$	1.00	1.22	»
			0.90	1.15	»
1.97	0.95	$CuCl \cdot KCl \cdot H_2O$	0.80	1.05	»
1.90	1.00	»	0.52	0.85	»
1.85	1.05	»	0.40	0.80	»
1.80	1.10	»	0.30	0.75	»
1.75	1.15	»	0.25	0.75	$CuCl \cdot 2KCl+KCl$
1.66	1.20	»	0.25	0.85	KCl
1.65	1.24	»	0.23	0.90	»
1.55	1.25	»	0.20	1.79	»
1.44	1.45	»	0.00	8.00	»

№ 4875 POTASSIUM CHLORIDE–COPPER (II) CHLORIDE–WATER [2195]

$KCl—CuCl_2—H_2O$

Comp. of saturated soln., mol %			t	Comp. of solid phase
Copper Chloride	Potassium Chloride	Water		
4.82	8.60	86.58	39.4	$CuCl_2 \cdot 2KCl \cdot 2H_2O+KCl$
5.42	9.68	84.90	49.9	»
6.35	11.19	82.46	60.4	»
8.55	14.50	76.95	79.1	»
10.70	17.50	71.80	90.5	»
11.39	18.30	70.31	93.7	$CuCl_2 \cdot KCl+KCl$
11.95	19.62	68.43	98.8	»
8.81	1.74	89.45	0	$CuCl_2 \cdot 2KCl \cdot 2H_2O+CuCl_2 \cdot 2H_2O$
10.90	4.60	84.50	39.6	»
11.35	5.72	82.93	50.1	»
12.10	6.23	81.67	52.9	»
12.70	6.83	80.47	60.2	$CuCl_2 \cdot KCl+CuCl_2 \cdot 2H_2O$
13.42	6.68	79.90	72.6	»
11.78	9.18	79.04	64.2	$CuCl_2 \cdot 2KCl \cdot 2H_2O+CuCl_2 \cdot KCl$
11.40	11.53	77.07	72.5	$CuCl_2 \cdot KCl$

POTASSIUM CHLORIDE–COPPER (II) CHLORIDE–WATER

$KCl—CuCl_2—H_2O$

Comp. of saturated soln., wt %			t	d	Comp. of solid phase
Potassium Chloride	Copper Chloride	Water			
0.20	39.9	59.9	—65.0	—	Ice+$CuCl_2 \cdot KCl \cdot 2H_2O$+$CuCl_2 \cdot 2H_2O$
0.70	36.5	62.8	—63.2	—	Ice+$CuCl_2 \cdot KCl \cdot 2H_2O$+$CuCl_2 \cdot 2KCl \cdot 2H_2O$
0.93	39.74	59.33	—30.0	—	$CuCl_2 \cdot KCl \cdot 2H_2O$+$CuCl_2 \cdot 2H_2O$
1.67	36.51	61.82	—30.0	—	$CuCl_2 \cdot KCl \cdot 2H_2O$+$CuCl_2 \cdot 2KCl \cdot 2H_2O$
8.0	21.40	70.60	—21.0	—	Ice+KCl+$CuCl_2 \cdot KCl \cdot 2H_2O$
9.04	21.10	69.86	—18	—	Ice+KCl
3.35	40.97	55.68	0	—	$CuCl_2 \cdot KCl \cdot 2H_2O$+$CuCl_2 \cdot 2H_2O$
3.68	40.17	56.15	0	1.537	$CuCl_2 \cdot KCl \cdot 2H_2O$
4.88	38.23	56.89	0	1.509	$CuCl_2 \cdot KCl \cdot 2H_2O$+$CuCl_2 \cdot 2KCl \cdot 2H_2O$
5.19	35.98	58.83	0	1.478	$CuCl_2 \cdot 2KCl \cdot 2H_2O$
8.32	28.48	63.20	0	—	»
13.22	21.07	65.71	0	1.340	»
13.92	20.15	65.93	0	1.338	$CuCl_2 \cdot 2KCl \cdot 2H_2O$+KCl
15.26	13.11	71.63	0	1.251	KCl
18.26	6.14	75.60	0	—	»
21.85	0.00	78.15	0	—	»
5.90	42.45	51.65	20	1.578	$CuCl_2 \cdot KCl \cdot 2H_2O$+$CuCl_2 \cdot 2H_2O$
7.70	39.88	52.42	20	—	$CuCl_2 \cdot KCl \cdot 2H_2O$+$CuCl_2 \cdot 2KCl \cdot 2H_2O$
17.98	21.06	60.96	20	1.370	$CuCl_2 \cdot 2KCl \cdot 2H_2O$+KCl
0.00	43.60	56.40	25	—	$CuCl_2 \cdot 2H_2O$
2.94	43.45	53.61	25	1.558	»
5.49	43.25	51.26	25	1.581	»
6.27	43.16	50.57	25	1.593	»
6.46	43.15	50.39	25	1.594	$CuCl_2 \cdot 2H_2O$+$CuCl_2 \cdot KCl \cdot 2H_2O$
6.75	42.76	50.49	25	1.589	$CuCl_2 \cdot KCl \cdot 2H_2O$
7.50	41.59	50.91	25	1.583	»
7.94	41.27	50.79	25	1.573	»
8.37	40.43	51.20	25	1.567	$CuCl_2 \cdot KCl \cdot 2H_2O$+$CuCl_2 \cdot 2KCl \cdot 2H_2O$
9.23	38.40	52.37	25	1.541	$CuCl_2 \cdot 2KCl \cdot 2H_2O$
11.37	33.10	55.53	25	1.475	»
14.84	27.90	57.26	25	—	»
17.77	23.24	58.99	25	—	»
19.03	21.30	59.67	25	1.350	$CuCl_2 \cdot 2KCl \cdot 2H_2O$+KCl
21.80	11.21	66.99	25	1.127	KCl
24.01	5.47	70.52	25	1.190	»
26.40	0.00	73.60	25	—	»
8.15	44.10	47.75	35	—	$CuCl_2 \cdot 2KCl \cdot 2H_2O$+$CuCl_2 \cdot 2H_2O$
10.33	41.57	48.10	35	—	$CuCl_2 \cdot 2KCl \cdot 2H_2O$+$CuCl_2 \cdot KCl \cdot 2H_2O$
21.06	22.56	56.38	35	—	$CuCl_2 \cdot 2KCl \cdot 2H_2O$+KCl
11.55	45.67	42.78	50	—	$CuCl_2 \cdot 2KCl \cdot 2H_2O$+$CuCl_2 \cdot 2H_2O$
14.00	42.92	43.08	50	—	$CuCl_2 \cdot 2KCl \cdot 2H_2O$+$CuCl_2 \cdot KCl \cdot 2H_2O$
25.36	26.97	47.67	50	—	$CuCl_2 \cdot 2KCl \cdot 2H_2O$+KCl
0.00	47.24	52.76	54.0	—	$CuCl_2 \cdot 2H_2O$
12.40	46.30	41.30	54.4	—	$CuCl_2$+$CuCl_2 \cdot KCl \cdot 2H_2O$
13.55	45.73	40.72	56.0	—	Mixed crystals +$CuCl_2 \cdot KCl \cdot 2H_2O$

Comp. of saturated soln., wt %			t	d	Comp. of solid phase
Potassium Chloride	Copper Chloride	Water			
12.36	46.61	41.03	58	—	$CuCl_2 \cdot 2H_2O$
12.82	46.85	40.33	58	—	$CuCl$ + mixed crystals
14.36	45.64	40.00	58	—	Mixed crystals
14.94	45.00	40.06	58	—	Mixed crystals + $CuCl_2 \cdot KCl \cdot 2H_2O$
15.47	44.40	40.13	58	—	$CuCl_2 \cdot KCl \cdot 2H_2O$
15.60	44.40	40.00	58.4	—	$CuCl_2 \cdot KCl \cdot 2H_2O$ + mixed crystals + $CuCl_2 \cdot KCl$
12.95	46.81	40.24	59.0	—	Mixed crystals
14.42	45.46	40.12	59.0	—	»
14.85	45.20	39.95	59.0	—	Mixed crystals + $CuCl_2 \cdot KCl$
15.56	44.53	39.91	59.0	—	$CuCl_2 \cdot KCl$
16.0	44.20	39.80	59.2	—	$CuCl_2 \cdot KCl + CuCl_2 \cdot KCl \cdot 2H_2O + CuCl_2 \cdot 2KCl \cdot 2H_2O$
13.0	47.10	39.90	60.0	—	$CuCl_2 \cdot KCl$ + Mixed crystals + $CuCl_2 \cdot 2H_2O$
16.18	44.14	39.68	60.0	—	$CuCl_2 \cdot 2KCl \cdot 2H_2O + CuCl_2 \cdot KCl$
12.93	47.13	39.94	61.0	—	$CuCl_2 \cdot KCl + CuCl_2 \cdot 2H_2O$
0.00	50.40	49.60	75.0	—	$CuCl_2 \cdot 2H_2O$
5.74	49.88	44.38	75	1.711	»
12.45	49.30	38.25	75	1.787	$CuCl_2 \cdot 2H_2O + CuCl_2 \cdot KCl$
15.35	46.70	37.95	75	1.781	$CuCl_2 \cdot KCl$
18.44	44.11	37.45	75	—	»
20.27	42.73	37.00	75	1.761	»
21.95	41.42	36.63	75	1.756	$CuCl_2 \cdot KCl + CuCl_2 \cdot 2KCl \cdot 2H_2O$
25.16	35.75	39.09	75	1.716	$CuCl_2 \cdot 2KCl \cdot 2H_2O$
28.33	31.56	40.11	75	1.629	$CuCl_2 \cdot 2KCl \cdot 2H_2O + KCl$
28.13	23.74	48.13	75	1.499	KCl
29.79	12.63	57.58	75	1.360	»
33.20	0.0	66.80	75	—	»
29.74	38.03	32.23	88	—	$CuCl_2 \cdot KCl + CuCl_2 \cdot 2KCl \cdot 2H_2O$
30.32	36.22	33.46	88	—	$CuCl_2 \cdot 2KCl \cdot 2H_2O + KCl$
31.20	37.50	31.30	93.2	—	$CuCl_2 \cdot 2KCl \cdot 2H_2O + KCl + CuCl_2 \cdot KCl$
0.00	54.60	45.40	100	—	$CuCl_2 \cdot 2H_2O$
5.53	53.69	40.78	100	1.690	»
11.53	53.26	35.21	100	1.853	$CuCl_2 \cdot 2H_2O + CuCl_2 \cdot KCl$
13.14	51.30	35.56	100	1.815	$CuCl_2 \cdot KCl$
25.42	41.80	32.78	100	1.771	»
31.69	38.21	30.10	100	1.783	»
33.29	37.51	29.20	100	1.700	$CuCl_2 \cdot KCl + KCl$
32.20	31.11	36.69	100	1.651	KCl
35.70	0.00	64.30	100	—	»
36.80	0.00	63.20	108.6	—	»
0.00	56.66	43.34	112	—	$CuCl_2 \cdot 2H_2O$
0.00	57.90	42.10	115.8	—	»
35.21	38.17	26.62	116	—	$CuCl_2 \cdot KCl + KCl$
10.10	56.02	33.88	116	—	$CuCl_2 \cdot KCl + CuCl_2 \cdot 2H_2O$
35.50	38.40	26.10	119.9	—	$CuCl_2 \cdot KCl + KCl$
9.90	58.15	31.95	121.5	—	$CuCl_2 \cdot KCl + CuCl_2 \cdot 2H_2O$

№ 4877 [1959]

POTASSIUM CHLORIDE–COPPER (II) IODATE–WATER

$KCl—Cu(IO_3)_2—H_2O$

$t=25$

Concn. in aqueous soln., wt %			
Copper Iodate	Potassium Chloride	Copper Iodate	Potassium Chloride
0.134	0.000	0.164	0.261
0.140	0.037	0.172	0.373
0.145	0.075	0.184	0.558
0.154	0.149	0.194	0.743

№ 4878 [2596]

POTASSIUM CHLORIDE–ZINC HYDROXIDE–WATER

$KCl—Zn(OH)_2—H_2O$

$t=16—20$

Comp. of solvent, wt %		Solubility of Zinc Hydroxide, g/l
Potassium Chloride	Water	
1	99	0.043

№ 4879 [1475]

POTASSIUM CHLORIDE–ZINC CHLORIDE–WATER

$KCl—ZnCl_2—H_2O$

$t=34$

Comp. of saturated soln., wt %			Comp. of solid phase
Potassium Chloride	Zinc Chloride	Water	
4.25	0.00	95.75	KCl
6.05	3.36	90.59	»
7.18	5.41	87.41	»
8.95	7.97	83.08	»
13.98	14.34	71.68	»
11.25	16.73	72.02	»

№ 4880 [1467]

POTASSIUM CHLORIDE–RUBIDIUM CHLORIDE–WATER

$KCl—RbCl—H_2O$

$t=25$

Comp. of saturated soln., wt %					
Potassium Chloride	Rubidium Chloride	Water	Potassium Chloride	Rubidium Chloride	Water
15.27	0.00	84.73	6.81	20.61	72.98
12.74	5.54	81.72	5.12	24.22	70.66
10.95	10.19	78.86	2.99	27.65	69.36
8.76	15.73	75.51	0.00	68.03	31.97

№ 4881 [539]

POTASSIUM CHLORIDE–RUBIDIUM CHLORIDE–WATER

$KCl—RbCl—H_2O$

$t=100$

Content in aqueous soln., wt %		Comp. of solid phase	Content in aqueous soln., wt %		Comp. of solid phase
Potassium Chloride	Rubidium Chloride		Potassium Chloride	Rubidium Chloride	
35.9	0	KCl	15.23	37.99	(K, Rb) Cl
31.15	10.25	(K, Rb)Cl	3.40	39.73	»
27.81	15.77	»	8.95	46.29	»
25.07	20.37	»	6.32	50.81	»
21.86	25.70	»	3.64	54.66	»
18.49	29.79	»	0	58.14	RbCl

№ 4882 [2662]

POTASSIUM CHLORIDE–STRONTIUM CARBONATE–WATER

$KCl—SrCO_3—H_2O$

Comp. of solvent, wt %		Solubility of Strontium Carbonate, g/1000 g Water	t	Comp. of solvent, wt %		Solubility of Strontium Carbonate, g/1000 g Water	t
Potassium Chloride	Water			Potassium Chloride	Water		
0.0	100.0	0.008	25	0.0	100.0	0.010	40
0.149	99.851	0.009	25	0.149	99.851	0.014	40
0.371	99.629	0.012	25	0.371	99.629	0.018	40
0.740	99.260	0.016	25	0.740	99.260	0.021	40
1.47	98.53	0.020	25	1.47	98.53	0.025	40
2.19	97.81	0.023	25	2.19	97.81	0.028	40
2.89	97.11	0.025	25	2.89	97.11	0.029	40
3.59	96.41	0.026	25	3.59	96.41	0.030	40
6.94	93.06	0.028	25	6.94	93.06	0.033	40
18.28	81.72	0.032	25	18.28	81.72	0.037	40

№ 4883 [2703]

POTASSIUM CHLORIDE–STRONTIUM SULFATE–WATER

$KCl—SrSO_4—H_2O$

$t=25$

Comp. of solvent, wt %		Solubility of Strontium Sulfate, wt %
Potassium Chloride	Water	
8.22	91.78	0.193
12.54	87.46	0.193
18.08	81.92	0.251

№ 4884 [2227]

POTASSIUM CHLORIDE–STRONTIUM CHLORIDE–WATER

$KCl—SrCl_2—H_2O$

$t=25$

Content in aqueous soln., mol/1000 g Water		d_4	Comp. of solid phase
Potassium Chloride	Strontium Chloride		
0.797	3.382	1.4266	$SrCl_2 \cdot 6H_2O$
0.429	3.441	1.4152	»
0.2165	3.475	1.4085	»
0.0360	3.508	1.4052	»
0	3.517	1.4015	»

№ 4885 [1780]

POTASSIUM CHLORIDE–STRONTIUM CHLORIDE–WATER

$KCl—SrCl_2—H_2O$

$t=25$

Comp. of solvent, wt %		Solubility of Strontium Chloride, wt %	d_4
Potassium Chloride	Water		
0.533	99.467	35.62	1.4032
3.129	96.871	34.80	1.4085
6.013	93.987	33.89	1.4152
10.627	89.373	32.40	1.4266

№ 4886 [1108]

POTASSIUM CHLORIDE–STRONTIUM CHLORIDE–WATER

$KCl—SrCl_2—H_2O$

Comp. of saturated soln., wt %			t	Comp. of solid phase
Strontium Chloride	Potassium Chloride	Water		
0.0	25.4	74.6	18	KCl
14.2	15.7	70.1	18	»
24.8	9.8	65.4	18	»
30.5	7.5	62.0	18	»
31.0	7.2	61.8	18	$KCl+SrCl_2 \cdot 6H_2O$
32.0	4.6	63.4	18	$SrCl_2 \cdot 6H_2O$
34.3	0.0	65.7	18	»
0.0	31.3	68.7	60	KCl
18.5	18.1	63.4	60	»
29.3	12.3	58.4	60	»
39.9	8.0	52.1	60	»
43.5	7.2	49.3	60	$KCl+SrCl_2 \cdot 2H_2O$
43.8	6.7	49.5	60	$SrCl_2 \cdot 2H_2O$
45.1	3.7	51.2	60	»
45.2	3.3	51.5	60	»
45.6	2.7	51.7	60	$SrCl_2 \cdot 2H_2O+SrCl_2 \cdot 2H_2O$

Table № 4886 (continuation)

Comp. of saturated soln., wt %			t	Comp. of solid phase
Strontium Chloride	Potassium Chloride	Water		
45.7	2.6	51.7	60	$SrCl_2 \cdot 6H_2O$
46.2	0.0	53.8	60	»
0.0	36.2	63.8	100	KCl
16.9	24.7	58.4	100	»
28.9	17.4	53.7	100	»
35.9	13.8	50.3	100	»
44.2	11.1	44.7	100	»
46.5	10.9	42.6	100	»
46.7	10.8	42.5	100	$KCl+SrCl_2 \cdot 2H_2O$
46.9	10.1	43.0	100	$SrCl_2 \cdot 2H_2O$
47.9	8.0	44.1	100	»
50.8	0.0	49.2	100	»
31.0	7.2	61.8	18.0	$KCl+SrCl_2 \cdot 6H_2O$
36.4	7.1	56.5	40.0	»
40.1	7.0	52.9	50.0	»
42.1	7.1	50.8	54.7	»
43.1	7.2	49.7	57.5	»
43.3	7.2	49.5	58.7	$KCl+SrCl_2 \cdot 2H_2O$
43.5	7.2	49.3	60.0	»
43.9	8.2	47.9	70.0	»
44.5	8.9	46.6	80.0	»
45.4	9.8	44.8	90.0	»
46.7	10.8	42.5	100.0	»
48.7	11.9	39.4	110.0	»
49.4	12.4	38.2	114.0	»

№ 4887 [93]

POTASSIUM CHLORIDE–OXOZIRCONIUM CHLORIDE–WATER

$KCl—ZrOCl_2—H_2O$

$t=25$

Comp. of saturated soln., wt %			Comp. of solid phase
Potassium Chloride	Oxozirconium Chloride	Water	
26.80	0.00	73.20	KCl
19.50	7.60	72.90	»
16.11	12.83	71.06	»
12.16	18.64	69.20	»
11.89	19.19	68.92	»
8.85	24.37	66.78	»
7.12	27.27	65.61	»
4.45	31.85	63.70	»
4.20	33.47	62.33	»
4.16	34.40	61.44	$KCl+ZrOCl_2 \cdot 8H_2O$
3.92	34.71	61.37	$ZrOCl_2 \cdot 8H_2O$
3.07	34.90	62.03	»
0.00	37.60	62.40	»

№ 4888 POTASSIUM CHLORIDE–SILVER CHLORIDE–WATER [2486]

$KCl—AgCl—H_2O$

$t=15$

Concn. in aqueous soln., wt %			
Potassium Chloride	Silver Chloride	Potassium Chloride	Silver Chloride
10.0	0.000	22.47	0.045
14.29	0.004	24.0	0.072
16.66	0.008	25.0	0.084
20.00	0.020		

№ 4889 POTASSIUM CHLORIDE–SILVER CHLORIDE–WATER [1650]

$KCl—AgCl—H_2O$

Concn. in aqueous soln., g/l					
Potassium Chloride	Silver Chloride	t	Potassium Chloride	Silver Chloride	t
82.77	0.0202	25	212.3	0.2638	25
106.2	0.0336	25	229.5	0.3482	25
127.6	0.0559	25	255.1	0.5151	25
150.6	0.0881	25	286.3	0.8187	25
178.5	0.1501	25	247.7	0.2480	1
195.8	0.1988	25	220.1	0.3984	35

№ 4890 [1730]

POTASSIUM CHLORIDE–SILVER CHLORIDE–WATER

$KCl—AgCl—H_2O$

$t=25$

Concn. in aqueous soln., g/l	
Potassium Chloride	Silver Chloride
0.00236	0.001836
0.00471	0.002178
0.01491	0.003052
0.02984	0.003209

№ 4891 [2330]

POTASSIUM CHLORIDE–SILVER CHLORIDE–WATER

$KCl—AgCl—H_2O$

$t=25$

Concn. in aqueous soln., g/l	
Potassium Chloride	Silver Chloride
0.0746	0.000045
0.7456	0.000057
7.456	0.00039
18.640	0.00110
37.280	0.00347

№ 4892 **POTASSIUM CHLORIDE–SILVER CHLORIDE–WATER** [2331]

$KCl—AgCl—H_2O$

$t=25$

Concn. of Potassium Chloride in aqueous soln., g/l	Solubility of Silver Chloride in aqueous soln. of Potassium Chloride, g/l	Concn. of Potassium Chloride in aqueous soln., g/l	Solubility of Silver Chloride in aqueous soln. of Potassium Chloride, g/l
0.0021	0.00128	0.7456	0.000066
0.0037	0.000684	7.456	0.000403
0.0074	0.000396	37.28	0.00367
0.0745	0.000067	74.56	0.01437
0.1864	0.000052		

№ 4893 [1136]

POTASSIUM CHLORIDE–SILVER CHLORIDE–WATER

$KCl—AgCl—H_2O$

$t=25$

Concn. in aqueous soln., g/l	
Potassium Chloride	Silver Chloride
0.1119	0.0000222
0.746	0.0000574

№ 4894 **POTASSIUM CHLORIDE–CADMIUM CHLORIDE–WATER** [2615]

$KCl—CdCl_2—H_2O$

Comp. of saturated soln., wt %			t	Comp. of solid phase
Potassium Chloride	Cadmium Chloride	Water		
0.00	52.67	47.33	19.3	$CdCl_2 \cdot 2\frac{1}{2}H_2O$
4.03	36.84	59.13	19.3	$CdCl_2 \cdot 2\frac{1}{2}H_2O + CdCl_2 \cdot KCl \cdot H_2O$
8.03	19.54	72.43	19.3	$CdCl_2 \cdot KCl \cdot H_2O$
21.21	8.23	79.56	19.3	$CdCl_2 \cdot KCl \cdot H_2O + CdCl_2 \cdot 4KCl$
25.52	1.06	73.42	19.3	$CdCl_2 \cdot 4KCl + KCl$
25.34	0.00	74.66	19.3	KCl
0.00	56.45	43.55	29.7	$CdCl_2 \cdot 2\frac{1}{2}H_2O$
0.35	49.22	50.43	29.7	»
4.04	38.92	57.04	29.7	$CdCl_2 \cdot 2\frac{1}{2}H_2O + CdCl_2 \cdot KCl \cdot H_2O$
6.30	30.01	63.69	29.7	$CdCl_2 \cdot KCl \cdot H_2O$
7.76	22.42	69.82	29.7	»
11.47	17.28	71.25	29.7	»
18.29	11.26	70.45	29.7	»
22.53	10.38	67.09	29.7	$CdCl_2 \cdot KCl \cdot H_2O + CdCl_2 \cdot 4KCl$
26.88	1.73	71.39	29.7	$CdCl_2 \cdot 4KCl + KCl$
27.12	0.00	72.88	29.7	KCl
0.00	57.24	42.76	40.1	$CdCl_2 \cdot H_2O$

Table № 4894 (continuation)

Comp. of saturated soln., wt %			t	Comp. of solid phase
Potassium Chloride	Cadmium Chloride	Water		
1.46	47.29	51.25	40.1	$CdCl_2 \cdot H_2O + CdCl_2 \cdot KCl \cdot H_2O$
7.04	31.76	61.20	40.1	$CdCl_2 \cdot KCl \cdot H_2O$
9.93	24.76	65.31	40.1	»
14.85	16.71	68.44	40.1	»
22.98	12.28	64.74	40.1	»
23.88	12.64	63.48	40.1	$CdCl_2 \cdot KCl \cdot H_2O + CdCl_2 \cdot 4KCl$
28.20	2.08	69.72	40.1	$CdCl_2 \cdot 4KCl + KCl$
28.75	0.00	71.25	40.1	KCl
0.0	58.31	41.69	54.5	$CdCl_2 \cdot H_2O$
1.13	49.95	48.92	54.5	$CdCl_2 \cdot H_2O + CdCl_2 \cdot KCl \cdot H_2O$
11.32	27.10	61.58	54.5	$CdCl_2 \cdot KCl \cdot H_2O$
25.76	15.38	58.86	54.5	$CdCl_2 \cdot KCl \cdot H_2O + CdCl_2 \cdot 4KCl$
30.40	2.80	66.80	54.5	$CdCl_2 \cdot 4KCl + KCl$
30.07	0.0	69.93	54.5	KCl

№ 4895 POTASSIUM CHLORIDE–CADMIUM CHLORIDE–WATER [1802]

$KCl—CdCl_2—H_2O$

Comp. of saturated soln., wt %			t	d_4^t	Comp. of solid phase
Potassium Chloride	Cadmium Chloride	Water			
0.35	41.6	58.05	− 7.8	—	$3CdCl_2 \cdot KCl \cdot 4H_2O$
19.15	0.9	79.95	− 6.2	—	$CdCl_2 \cdot 4HCl$
15.15	4.4	80.45	− 5.7	—	$CdCl_2 \cdot KCl \cdot H_2O$
3.05	28.5	68.45	− 3.7	—	»
0.2	45.5	54.3	0	1.603	$3CdCl_2 \cdot KCl \cdot 4H_2O$
1.57	35.25	63.18	0	1.433	»
3.45	28.65	67.9	0	1.352	$CdCl_2 \cdot KCl \cdot H_2O \cdot$
3.75	27.25	69.0	0	1.337	»
4.5	21.8	73.7	0	1.268	»
10.05	7.85	82.1	0	1.167	»
16.5	5.15	78.35	0	1.169	»
17.6	4.75	77.65	0	1.172	$CdCl_2 \cdot 4KCl$
19.3	1.9	78.8	0	1.154	»
0.29	52.1	47.61	25	1.722	$3CdCl_2 \cdot KCl \cdot 4H_2O$
1.33	44.05	54.62	25	1.571	»
3.65	38.5	57.85	25	1.496	»
4.96	35.3	59.74	25	1.457	$CdCl_2 \cdot KCl \cdot H_2O$
20.6	9.67	69.73	25	1.240	$CdCl_2 \cdot KCl$
5.6	42.8	51.6	50	1.579	»
26.1	13.0	60.9	50	1.312	$CdCl_2 \cdot 4KCl$
29.55	3.3	67.15	50	1.151	»
3.55	58.8	37.65	100	—	$3CdCl_2 \cdot KCl \cdot 4H_2O$
5.0	58.6	36.4	100	1.881	»
5.5	58.3	36.2	100	1.878	»
7.8	51.45	40.75	100	—	$CdCl_2 \cdot KCl$
12.75	39.4	47.85	100	—	»

Comp. of saturated soln., wt %			t	d_4^t	Comp. of solid phase
Potassium Chloride	Cadmium Chloride	Water			
21.7	27.5	50.8	100	—	$CdCl_2 \cdot KCl$
29.8	25.25	44.95	100	—	»
30.7	23.7	45.6	100	—	$CdCl_2 \cdot 4KCl$
31.6	19.55	48.85	100	1.415	»
34.45	10.15	55.4	100	—	KCl
35.25	6.4	58.35	100	1.275	»
0.26	43.2	56.54	—10.3	—	$CdCl_2 \cdot 4H_2O + 3CdCl_2 \cdot KCl \cdot 4H_2O$ + ice
0.20	44.9	54.90	— 7.2	—	$CdCl_2 \cdot 4H_2O + 3CdCl_2 \cdot KCl \cdot 4H_2O$
0.24	47.0	52.76	— 3.3	—	»
0.18	45.8	54.02	— 5.7	—	$CdCl_2 \cdot 4H_2O + CdCl_2 \cdot 2\frac{1}{2}H_2O + 3CdCl_2 \cdot KCl \cdot 4H_2O$
0.19	47.4	52.41	0	1.639	$CdCl_2 \cdot 2\frac{1}{2}H_2O + 3CdCl_2 \cdot KCl \cdot 4H_2O$
0.19	54.7	45.11	25	1.783	»
0.16	56.9	42.94	32	1.830	»
0.15	57.4	42.45	33.7	1.843	$CdCl_2 \cdot 4H_2O + CdCl_2 \cdot 2\frac{1}{2}H_2O + 3CdCl_2 \cdot KCl \cdot 4H_2O$
0.18	57.5	42.32	35	1.841	$CdCl_2 \cdot 4H_2O + 3CdCl_2 \cdot KCl\ 4H_2O$
0.43	57.6	41.97	50	1.834	»
1.45	58.3	40.25	75	1.844	»
2.75	59.2	38.05	90	1.869	»
4.1	59.9	36.0	100	1.901	»
2.75	30.05	67.2	— 5.8	—	$3CdCl_2 \cdot KCl\ 4H_2O + CdCl_2 \cdot KCl \cdot H_2O$ + ice
3.1	31.5	65.4	0	1.391	»
3.7	34.15	62.15	11	1.434	»
4.45	37.45	58.1	25	1.488	»
4.8	38.3	56.9	28.8	1.503	$3CdCl_2 \cdot KCl \cdot 4H_2O + CdCl_2 \cdot KCl \cdot H_2O + CdCl_2 \cdot KCl$
4.9	39.05	56.05	32	1.516	$3CdCl_2 \cdot KCl \cdot 4H_2O + CdCl_2 \cdot KCl$
5.45	43.6	51.95	50	1.593	»
5.85	50.3	43.85	75	1.715	»
5.9	57.8	36.3	100	1.874	»
15.7	4.0	80.3	— 8.1	—	Ice + $CdCl_2 \cdot KCl \cdot H_2O + CdCl_2 \cdot 4KCl$
17.5	5.05	77.45.	0	1.176	$CdCl_2 \cdot KCl \cdot H_2O + CdCl_2 \cdot 4KCl$
19.6	6.8	73.6	11	1.206	»
20.0	7.0	73.0	12.4	1.212	$CdCl_2 \cdot KCl \cdot H_2O + CdCl_2 \cdot 4KCl + CdCl_2 \cdot KCl$
20.25	7.3	72.45	14	1.215	$CdCl_2 \cdot KCl + CdCl_2 \cdot 4KCl$
22.1	9.35	68.55	25	1.250	»
23.1	10.45	66.45	31	1.268	»
25.8	14.35	59.85	50	1.328	»
27.6	17.5	54.9	60	1.377	»
28.6	19.8	51.6	75	1.410	»
29.45	22.0	48.55	85	1.443	»
30.2	25.2	43.6	100	1.491	»
19.65	0.45	79.9	—10.7	—	Ice + $CdCl_2 \cdot 4KCl$ + KCl
21.95	0.6	77.45	0	1.161	$CdCl_2 \cdot 4KCl$ + KCl
26.5	1.3	72.2	25	1.193	»
30.15	2.7	67.15	50	1.223	»
32.95	5.1	61.95	75	1.259	»
35.35	8.6	56.05	100	1.303	»

№ 4896 POTASSIUM CHLORIDE–CADMIUM CHLORIDE–WATER* [261]

$KCl—CdCl_2—H_2O$

Comp. of saturated soln., wt %: Cadmium Chloride	Potassium Chloride	Water	t	Comp. of saturated soln., wt %: Cadmium Chloride	Potassium Chloride	Water	t
54.64	0.00	45.36	25	56.46	0.00	43.54	45
50.89	1.80	47.31	25	55.07	0.99	43.94	45
41.19	2.82	55.99	25	48.24	1.30	50.46	45
35.74	3.63	60.63	25	43.01	2.24	54.75	45
28.18	3.29	68.53	25	37.34	3.83	58.83	45
21.85	6.92	71.23	25	35.70	4.61	59.69	45
20.87	9.16	69.97	25	27.56	8.48	64.96	45
11.79	10.88	77.33	25	25.10	9.41	65.49	45
11.07	15.51	73.42	25	24.64	10.10	65.26	45
10.49	16.03	73.48	25	18.53	13.78	67.69	45
8.54	16.81	74.65	25	17.45	14.76	67.79	45
6.52	17.91	75.57	25	12.16	20.19	67.65	45
9.09	21.76	69.15	25	11.82	24.18	64.00	45
9.02	23.90	67.08	25	9.71	25.15	55.14	45
6.14	25.41	68.45	25	0.00	29.23	70.77	45
0.00	26.23	73.77	25	23.32	2.37	74.31	45
11.56	5.00	83.44	25	15.36	6.24	78.40	45
16.33	7.36	76.31	25	19.67	8.01	72.32	45
20.87	9.16	69.97	25	10.55	8.57	80.88	45
4.79	7.54	87.67	25	12.08	10.38	77.54	45
9.07	13.98	76.95	25	17.53	13.69	68.78	45
10.49	16.03	73.48	25				

*At 25° the solid phase takes the form of a basic salt of comp. $CdCl_2 \cdot KCl \cdot H_2O$, and at 45° the comp. is $CaCl_2 \cdot KCl$.

№ 4897 [2414]

-'OTASSIUM CHLORIDE–CADMIUM CHLORIDE–VARIOUS CHLORIDES–WATER

$KCl—CdCl_2—\ldots—H_2O$

$t=16$

Chloride: Name	Formula	Concn. in aqueous soln., g/l: Cadmium Chloride	Potassium Chloride	Added Chloride	d_4
Hydrogen Chloride	HCl	6.05	9.84	321.9	1.1403
Lithium Chloride	LiCl	30.43	49.43	190.1	1.1380
Calcium Chloride	$CaCl_2$	49.50	80.52	209.5	1.2333
—	—	92.94	238.2	—	

№ 4898 POTASSIUM CHLORIDE–CADMIUM IODATE–WATER [2279]

$KCl—Cd(IO_3)_2—H_2O$

$t=25$

Concn. in aqueous soln., mol/1000 g Water					
Potassium Chloride	Cadmium Iodate	Potassium Chloride	Cadmium Iodate	Potassium Chloride	Cadmium Iodate
0.02	0.0022	0.08	0.0034	1.5	0.025
0.04	0.0027	0.5	0.0095	2.0	0.035
0.06	0.0031	1.0	0.017	2.5	0.046

Note. Data taken from a graph.

№ 4899 [1686]

POTASSIUM CHLORIDE–TIN (II) CHLORIDE–WATER

$KCl—SnCl_2—H_2O$

$t=25$

Comp. of saturated soln., wt %			Comp. of solid phase
Potassium Chloride	Tin Chloride	Water	
25.78	0.00	74.22	KCl
23.82	2.12	74.06	»
24.61	3.16	72.23	»
23.06	4.32	72.62	$SnCl_2 \cdot 2KCl \cdot 2H_2O$
22.45	4.27	73 28	»
19.84	7.44	72.72	»
17.36	12.25	70.39	»
15.03	18.30	65.67	»
4.27	24.44	71.29	»
11.20	31.20	57.60	$SnCl_2 \cdot KCl \cdot H_2O$
10.12	33.16	56.72	»
9.49	40.71	49.80	»
7.90	47.79	44.31	»
7.28	58.46	34.26	»
6.56	65.58	26.86	»
5.51	72.88	21.61	»
4.80	70.81	24.39	$SnCl_2 \cdot 2H_2O$
2.18	68.66	29.16	»
0.84	68.41	30.75	»
0.00	70.07	29.93	»

№ 4900 [2438]

POTASSIUM CHLORIDE–ANTIMONY TRIFLUORIDE–WATER

$KCl—SbF_3—H_2O$

$t=0$

Concn. of Potassium Chloride in aqueous soln., g/l	Solubility of Antimony Trifluoride in aqueous soln. of Potassium Chloride, wt %
74.6	82.20
37.3	81.76
18.45	81.19
9.325	80.28

№ 4901 [1812]

POTASSIUM CHLORIDE–BARIUM HYDROXIDE–WATER

$KCl—Ba(OH)_2—H_2O$

$t=25$

Content in aqueous soln., g/l	
Potassium Chloride	Barium Hydroxide
259.5	59.3
130.5	56.6
86.0	55.3
0.0	47.6

№ 4902 POTASSIUM CHLORIDE–BARIUM CARBONATE–WATER [1352]

$KCl—BaCO_3—H_2O$

Comp. of solvent, wt %		Solubility of Barium Carbonate, g/l	t	Comp. of solvent, wt %		Solubility of Barium Carbonate, g/l	t
Potassium Chloride	Water			Potassium Chloride	Water		
0.15	99.85	0.0847	b.p.	10.0	90.0	0.2175	10
1.0	99.0	0.1781	b.p.	10.0	90.0	0.2408	20
3.0	97.0	0.2667	b.p.	10.0	90.0	0.2972	40
10.0	90.0	0.4274	b.p.	10.0	90.0	0.3491	60
30.0	70.0	0.5550	b.p.	10.0	90.0	0.4049	70

№ 4903 POTASSIUM CHLORIDE–BARIUM CARBONATE–WATER [2622]

$KCl—BaCO_3—H_2O$

Comp. of solvent, wt %		Solubility of Barium Carbonate, g/100 g Water		Comp. of solvent, wt %		Solubility of Barium Carbonate, g/100 g Water	
Potassium Chloride	Water	$t = 25$	$t = 40$	Potassium Chloride	Water	$t = 25$	$t = 40$
0.00	100.0	0.0180	0.0241	2.19	97.81	0.0528	0.0662
0.149	99.851	0.0213	0.0329	2.89	97.11	0.0593	0.0708
0.371	99.629	0.0262	0.0409	3.59	96.41	0.0648	0.0742
0.74	99.26	0.0335	0.0493	6.94	93.06	0.0821	0.0826
1.47	98.53	0.0446	0.0597	18.28	81.72	0.0924	0.0976

№ 4904 POTASSIUM CHLORIDE–BARIUM SULFATE–WATER [1950]

$KCl—BaSO_4—H_2O$

$t \sim 20$

Content of Potassium Chloride in aqueous soln., mol/l	Solubility of Barium Sulfate in soln. of Potassium Chloride, g/l	Comp. of solid phase	Content of Potassium Chloride in aqueous soln., mol/l	Solubility of Barium Sulfate in soln. of Potassium Chloride, g/l	Comp. of solid phase
1.00	0.0190	$BaSO_4$	0.13	0.0100	$BaSO_4$
0.75	0.0160	»	0.06	0.0075	»
0.50	0.0140	»	0.03	0.0045	»
0.25	0.0115	»	0	0.0043	»

№ 4905 [1631]

POTASSIUM CHLORIDE–BARIUM CHLORIDE–WATER

$KCl—BaCl_2—H_2O$

Comp. of saturated soln., wt %			t
Barium Chloride	Potassium Chloride	Water	
13.83	18.97	67.20	25

№ 4906 [2516]

POTASSIUM CHLORIDE–BARIUM CHLORIDE–WATER

$KCl—BaCl_2—H_2O$

Content in aqueous soln., wt %		t	Comp. of solid phase
Potassium Chloride	Barium Chloride		
28.63	0	40	KCl
23.98	9.15	40	$KCl+BaCl_2 \cdot 2H_2O$
0	28.98	40	$BaCl_2 \cdot 2H_2O$
31.2	0	60	KCl
23.09	14.83	60	$KCl+BaCl_2 \cdot 2H_2O$
0	31.7	60	$BaCl_2 \cdot 2H_2O$

№ 4907 [1229]

POTASSIUM CHLORIDE–BARIUM CHROMATE–WATER

$KCl—BaCrO_4—H_2O$

$t=25$

Concn. in aqueous soln., g/l	
Potassium Chloride	Barium Chromate
4.66	0.00717
18.64	0.01158

№ 4908 [1751]

POTASSIUM CHLORIDE–BARIUM HYDROGEN ORTHOARSENATE–WATER

$KCl—BaHAsO_4—H_2O$

$t=17$

Concn. in aqueous soln., g/l	
Potassium Chloride	Barium Hydrogen Orthoarsenate
149	3.3

№ 4909 [1751]

POTASSIUM CHLORIDE–BARIUM ORTHOARSENATE–WATER

$KCl—Ba_3(AsO_4)_2—H_2O$

$t=17$

Concn. in aqueous soln., g/l	
Potassium Chloride	Barium Orthoarsenate
149	0.45

№ 4910 [1957]

POTASSIUM CHLORIDE–BARIUM BROMATE–WATER

$KCl—Ba(BrO_3)_2—H_2O$

$t=25$

Concn. in aqueous soln., wt %	
Barium Bromate	Potassium Chloride
0.783	0.0
0.832	0.149
0.870	0.299
0.908	0.448
0.941	0.596
0.967	0.745

№ 4911 [2164]

POTASSIUM CHLORIDE–BARIUM IODATE–WATER

$KCl—Ba(IO_3)_2—H_2O$

$t=25$

Content in aqueous soln., g/l			
Potassium Chloride	Barium Iodate	Potassium Chloride	Barium Iodate
0.0	0.3946	0.5592	0.4380
0.0746	0.4029	0.7456	0.4472
0.1491	0.4092	1.491	0.4799
0.2610	0.4185	3.728	0.5442
0.3728	0.4258	7.456	0.6183

№ 4912 [2235]

POTASSIUM CHLORATE–BARIUM IODATE–WATER

$KCl–Ba(IO_3)_2–H_2O$

$t=25$

Concn. in aqueous soln., g/l	
Barium Iodate	Potassium Chloride
0.3984	0.0
0.4072	0.07
0.4374	0.37
0.4529	0.75
0.5496	3.736
0.6212	7.47
0.7630	18.29
0.9232	36.59
1.159	73.19

№ 4913 [1957]

POTASSIUM CHLORIDE–BARIUM IODATE–WATER

$KCl—Ba(IO_3)_2—H_2O$

$t=25$

Concn. in aqueous soln., wt %	
Barium Iodate	Potassium Chloride
0.39	0.0
0.402	0.094
0.410	0.187
0.423	0.347
0.443	0.738
0.451	0.922
0.483	1.82
0.523	3.58
0.553	5.26
0.573	6.89

№ 4914 [1677, 2039]

POTASSIUM CHLORIDE–LANTHANUM IODATE–WATER

$KCl—La(IO_3)_3—H_2O$

$t=25$

Concn. of Potassium Chloride in aqueous soln., g/l	Solubility of Lanthanum Iodate in aqueous soln. of Potassium Chloride, g/l	Concn. of Potassium Chloride in aqueous soln., g/l	Solubility of Lanthanum Iodate in aqueous soln. of Potassium Chloride, g/l
0.0746	0.6048	7.46	1.0937
0.373	0.6470	14.92	1.246
0.746	0.685	37.3	1.7017
3.73	0.8847	74.801	2.027
7.46	1.0428	149.2	2.510

№ 4915 POTASSIUM CHLORIDE–LANTHANUM IODATE–WATER [2313]

$KCl—La(IO_3)_3—H_2O$

$t=25$

Concn. of Potassium Chloride in aqueous soln., g/l	Solubility of Lanthanum Iodate in aqueous soln. of Potassium Chloride, g/l	d_4	Concn. of Potassium Chloride in aqueous soln., g/l	Solubility of Lanthanum Iodate in aqueous soln. of Potassium Chloride, g/l	d_4
0.0	0.440	0.99749	7.45	1.274	1.00243
0.0745	0.454	0.99750	37.25	1.627	1.02090
0.3725	0.489	0.99776	74.50	2.102	1.04277
0.745	0.673	0.99801	149.0	2.587	1.08268
3.725	0.793	1.00000			

№ 4916 POTASSIUM CHLORIDE–CERIUM CHLORIDE–WATER [959]

$KCl—CeCl_3—H_2O$

Comp. of saturated soln., wt %			t	Comp. of solid phase
Cerium Chloride	Potassium Chloride	Water		
0.0	27.2	72.8	30	KCl
7.9	21.0	71.1	30	»
27.9	7.0	65.1	30	»
41.3	5.0	53.7	30	»
47.0	5.0	48.0	30	$KCl+CeCl_3 \cdot 7H_2O$
48.5	2.1	49.4	30	$CeCl_3 \cdot 7H_2O$
49.2	1.0	49.8	30	»
49.2	0.0	50.3	30	»
0.00	30.0	70.0	50	KCl
7.6	23.5	68.9	50	»
27.2	9.2	63.6	50	»
40.9	6.0	53.1	50	»
48.5	6.0	45.5	50	$KCl+CeCl_3 \cdot 7H_2O$
50.7	1.9	47.4	50	$CeCl_3 \cdot 7H_2O$
51.3	0.9	47.8	50	»
51.7	0.0	48.3	50	»

№ 4917 [1818]

POTASSIUM CHLORIDE–MERCURY (II) OXIDE–WATER

$KCl—HgO—H_2O$

$t=25$

Concn. in aqueous soln., wt %	
Mercury Oxide	Potassium Chloride
0.08664	26.1
0.11913	52.2
0.14079	63.3
0.24909	111.2
0.41154	175.4
0.58482	233.3

№ 4918 [54]

POTASSIUM CHLORIDE–MERCURY AMIDOCHLORIDE–WATER

$KCl—HgNH_2Cl—H_2O$

Content of Potassium Chloride in aqueous soln., mol/l	Solubility of Mercury Amidochloride in aqueous soln. of Potassium Chloride, mol/l
0.200	$4.0 \cdot 10^{-4}$
0.600	$5.2 \cdot 10^{-4}$
1.000	$7.2 \cdot 10^{-4}$

Note. Data obtained at room temperature.

№ 4919 [879]

POTASSIUM CHLORIDE–MERCURY (II) CHLORIDE–WATER

$KCl—HgCl_2—H_2O$

$t=20$

Comp. of saturated soln., wt %			Comp. of solid phase
Mercury Chloride	Potassium Chloride	Water	
6.98	0.0	93.02	$HgCl_2$
10.32	0.99	88.69	»
13.32	2.02	84.66	»
17.57	3.21	79.22	»
19.34	3.73	76.93	$HgCl_2+2HgCl_2 \cdot KCl$
19.20	4.28	76.52	$2HgCl_2 \cdot KCl$
19.42	5.06	75.52	»
19.70	5.52	74.78	$2HgCl_2 \cdot KCl+HgCl_2 \cdot KCl$
18.70	5.64	75.66	$HgCl_2 \cdot KCl$
15.45	6.94	77.61	»
16.57	10.87	72.56	»
18.45	12.23	69.32	»
19.43	13.64	66.93	»
21.63	16.33	62.04	»
23.10	17.90	59.00	»
24.20	19.25	56.55	$HgCl_2 \cdot KCl+HgCl_2 \cdot 2KCl$
22.65	19.70	57.65	$HgCl_2 \cdot 2KCl$
20.70	20.25	59.05	»
18.43	21.40	60.17	»
15.0	23.28	61.72	»
12.0	25.70	62.30	$HgCl_2 \cdot 2KCl+KCl$
10.13	25.50	64.37	KCl
6.95	25.60	67.45	»
1.53	25.65	72.82	»

№ 4920 [1633]

POTASSIUM CHLORIDE–MERCURY (II) CHLORIDE–WATER

$KCl—HgCl_2—H_2O$

$t=25$

Comp. of saturated soln., wt %			Comp. of solid phase
Potassium Chloride	Mercury Chloride	Water	
26.46	0.0	73.54	KCl
26.24	15.04	58.72	$KCl+2KCl \cdot HgCl_2 \cdot H_2O$
26.43	15.02	58.55	»
26.33	15.02	58.65	»
26.33	14.92	58.75	»
23.74	18.91	57.35	$2KCl \cdot HgCl_2 \cdot H_2O$
22.36	21.39	56.25	»
21.39	23.88	54.73	»
20.32	27.62	52.06	$2KCl \cdot HgCl_2 \cdot H_2O+KCl \cdot HgCl_2 \cdot H_2O$
20.26	27.38	52.36	»
17.85	25.34	56.81	$KCl \cdot HgCl_2 \cdot H_2O$
9.26	18.95	71.79	»
7.80	19.56	72.64	»

Comp. of saturated soln., wt %			Comp. of solid phase
Potassium Chloride	Mercury Chloride	Water	
6.84	22.81	70.35	$KCl \cdot HgCl_2 \cdot H_2O$
6.66	24.32	69.02	»
6.52	25.13	68.35	$KCl \cdot HgCl_2 \cdot H_2O + KCl \cdot 2HgCl_2 \cdot 2H_2O$
6.64	25.16	68.20	»
6.27	25.11	68.62	$KCl \cdot 2HgCl_2 \cdot 2H_2O$
5.77	24.73	69.50	»
4.68	24.75	70.57	$KCl \cdot 2HgCl_2 \cdot 2H_2O + HgCl_2$
4.66	25.17	70.17	»
4.69	24.82	70.49	»
0.0	6.90	93.10	$HgCl_2$

№ 4921 POTASSIUM CHLORIDE–MERCURY (II) CHLORIDE–WATER [1815]

$KCl—HgCl_2—H_2O$

$t=25$

Concn. in aqueous soln., g/l			
Mercury Chloride	Potassium Chloride	Mercury Chloride	Potassium Chloride
72	0	103.5	24.33
103.5	11.01	137.2	27.53
96.4	19.15	227	75.18

№ 4922 POTASSIUM CHLORIDE–MERCURY (II) CHLORIDE–WATER [2660]

$KCl—HgCl_2—H_2O$

$t=34$

Comp. of saturated soln., wt %			d	Comp. of solid phase
Mercury Chloride	Potassium Chloride	Water		
8.55	0.0	91.45	1.068	$HgCl_2$
19.14	2.6	78.26	1.202	»
27.2	4.2	68.6	—	»
34.1	5.4	60.5	—	»
37.0	6.0	57.0	1.497	»
37.4	6.0	56.6	1.512	$HgCl_2 + 2HgCl_2 \cdot KCl \cdot 2H_2O$
37.8	6.5	55.7	1.525	$2HgCl_2 \cdot KCl \cdot 2H_2O$
37.7	6.7	55.6	1.533	»
37.6	7.3	55.1	1.545	»
37.8	7.45	54.75	1.549	$2HgCl_2 \cdot KCl \cdot 2H_2O + 3HgCl_2 \cdot 2KCl \cdot {}^5/_3H_2O$
37.4	7.8	54.8	1.534	$3HgCl_2 \cdot 2KCl \cdot {}^5/_3H_2O$
37.6	8.0	54.4	1.530	»
36.8	8.2	55.0	1.524	$3HgCl_2 \cdot 2KCl \cdot {}^5/_3H_2O + HgCl_2 \cdot KCl \cdot {}^3/_4H_2O$
35.9	8.2	55.9	1.512	$HgCl_2 \cdot KCl \cdot {}^3/_4H_2O$
34.5	8.5	57.0	1.486	»
27.9	9.6	62.5	1.386	»

Table № 4922 (continuation)

Comp. of saturated soln., wt %			d	Comp. of solid phase
Mercury Chloride	Potassium Chloride	Water		
25.5	11.4	63.1	1.368	$HgCl_2 \cdot KCl \cdot {}^3/_4 H_2O$
26.8	14.9	58.3	1.428	»
30.8	20.0	49.2	1.561	»
32.6	22.0	45.4	1.620	»
33.0	21.9	45.1	1.624	$HgCl_2 \cdot KCl \cdot {}^3/_4 H_2O + HgCl_2 \cdot 2KCl \cdot H_2O$
32.8	21.9	45.3	1.621	$HgCl_2 \cdot 2KCl \cdot H_2O$
29.0	23.0	48.0	—	»
24.5	24.8	50.7	—	$HgCl_2 \cdot 2KCl \cdot H_2O$
20.65	26.0	53.35	1.459	$HgCl_2 \cdot 2KCl \cdot 2H_2O + KCl$
17.8	26.8	55.4	—	KCl
12.2	27.0	60.8	—	»
8.4	27.3	64.3	—	»
4.5	27.5	68.0	—	»
0.0	27.8	72.2	—	»

№ 4923 POTASSIUM CHLORIDE–THALLIUM CHLORIDE–WATER [2264]

$KCl—TlCl—H_2O$

$t=25$

Concn. of Potassium Chloride in aqueous soln., g/l	Solubility of Thallium Chloride in aqueous soln. of Potassium Chloride, g/l	Concn. of Potassium Chloride in aqueous soln., g/l	Solubility of Thallium Chloride in aqueous soln. of Potassium Chloride, g/l
1.865	2.092	14.92	0.635
3.730	1.422	59.68	0.408
7.46	0.957		

№ 4924 POTASSIUM CHLORIDE–THALLIUM CHLORIDE–WATER [1306

$KCl—TlCl—H_2O$

Concn. of Potassium Chloride in aqueous soln., g/l	Solubility of Thallium Chloride in aqueous soln. of Potassium Chloride, g/l	d	t	Concn. of Potassium Chloride in aqueous soln., g/l	Solubility of Thallium Chloride in aqueous soln. of Potassium Chloride, g/l	d	t
3.73	1.406	1.0004	25	3.73	4.402	0.9943	50
7.46	0.9284	1.0003	25	7.46	3.073	0.9957	50
14.92	0.619	1.0033	25	14.92	2.142	0.9996	50
37.3	0.422	1.0063	25	37.3	1.454	1.0121	50

№ 4925 [2039]

POTASSIUM CHLORIDE–THALLIUM IODATE–WATER

$KCl—TlIO_3—H_2O$

$t=25$

Concn. of Potassium Chloride in aqueous soln., g/l	Solubility of Thallium Iodate in aqueous soln. of Potassium Chloride, g/l
0.746	0.760
1.492	0.799
3.730	0.885
7.460	0.995

№ 4926 [2262]

POTASSIUM CHLORIDE–LEAD (II) CHLORIDE–WATER

$KCl—PbCl_2—H_2O$

$t=25$

Concn. in aqueous soln., g/l			
Potassium Chloride	Lead Chloride	Potassium Chloride	Lead Chloride
0	10.79	6	5.3
1	9.3	8	4.5
2	8.2	10	3.9
3	7.2	14	3.1
4	6.5		

№ 4927 **POTASSIUM CHLORIDE–LEAD (II) CHLORIDE–WATER** [1554]

$KCl—PbCl_2—H_2O$

$t=25.2$

Concn. of Potassium Chloride in aqueous soln., g/l	Solubility of Lead Chloride in aqueous soln. of Potassium Chloride, g/l	Concn. of Potassium Chloride in aqueous soln., g/l	Solubility of Lead Chloride in aqueous soln. of Potassium Chloride, g/l	Concn. of Potassium Chloride in aqueous soln., g/l	Solubility of Lead Chloride in aqueous soln. of Potassium Chloride, g/l
0.0	10.791	0.149	8.944	74.533	1.363
0.0746	10.657	4.468	6.291	112.034	1.385
0.186	10.527	7.452	3.309	149.4	1.546
0.365	10.296	37.345	2.058	224.1	2.709
0.738	9.812	52.354	2.052		

№ 4928 **POTASSIUM CHLORIDE–LEAD (II) CHLORIDE–WATER** [1289]

$KCl—PbCl_2—H_2O$

$t=20$

Concn. in aqueous soln., wt %		Comp. of solid phase
Lead Chloride	Potassium Chloride	
0.264	1.455	$PbCl_2$
0.202	2.230	»
0.184	2.797	»
0.172	3.603	»
0.174	3.805	$PbCl_2+2PbCl_2 \cdot KCl$
0.148	4.289	$2PbCl_2 \cdot KCl$
0.132	4.767	»
0.107	6.938	»
0.102	9.131	»
0.109	11.749	»

Table № 4928 (continuation)

Concn. in aqueous soln., wt %		Comp. of solid phase
Lead Chloride	Potassium Chloride	
0.124	14.055	$2PbCl_2 \cdot KCl$
0.142	15.666	»
0.147	16.412	»
0.164	17.083	»
0.181	17.605	$2PbCl_2 \cdot KCl + PbCl_2 \cdot KCl \cdot {}^1/_3H_2O$
0.182	18.277	$PbCl_2 \cdot KCl \cdot {}^1/_3H_2O$
0.194	19.843	»
0.205	20.664	»
0.215	21.709	»
0.230	22.763	»
0.247	23.723	»
0.0317 *	3.295 *	$PbCl_2 \cdot KCl \cdot {}^1/_3H_2O + KCl$

*g per 1000 g Water

№ 4929 **POTASSIUM CHLORIDE–LEAD (II) CHLORIDE–WATER** [1489]

$KCl—PbCl_2—H_2O$

Comp. of saturated soln., wt %			t	Comp. of solid phase
Lead Chloride	Potassium Chloride	Water		
0.91	0.0	99.09	14	$PbCl_2$
0.18	1.37	98.45	14	»
0.14	2.44	97.42	14	»
0.13	3.43	96.44	14	»
0.15	3.79	96.06	14	$PbCl_2 + 2PbCl_2 \cdot KCl$
0.13	4.0	95.87	14	$2PbCl_2 \cdot KCl$
0.08	4.7	95.22	14	»
0.08	12.75	87.17	14	»
0.10	15.65	84.25	14	»
0.11	15.87	84.02	14	»
0.11	16.26	83.63	14	»
0.12	16.83	83.05	14	$2PbCl_2 \cdot KCl + PbCl_2 \cdot KCl \cdot {}^1/_3H_2O$
0.13	17.57	82.30	14	$PbCl_2 \cdot KCl \cdot {}^1/_3H_2O$
0.14	18.64	81.22	14	»
0.15	18.97	80.88	14	»
0.17	22.45	77.38	14	»
0.20	24.61	75.19	14	$PbCl_2 \cdot KCl \cdot {}^1/_3H_2O + KCl$
0.0	24.52	75.48	14	KCl
1.75	0.0	98.25	50	$PbCl_2$
0.65	2.0	97.35	50	»
0.5	4.0	95.5	50	»
0.57	6.0	93.43	50	»
0.62	6.4	92.98	50	$PbCl_2 + 2PbCl_2 \cdot KCl$
0.50	12.0	87.5	50	$2PbCl_2 \cdot KCl$
0.52	14.0	85.48	50	»
0.55	16.0	83.45	50	»
0.62	18.0	81.38	50	»
0.70	20.0	79.3	50	»
0.85	22.0	77.15	50	»
0.97	22.8	76.23	50	$2PbCl_2 \cdot KCl + PbCl_2 \cdot KCl \cdot {}^1/_3H_2O$

Comp. of saturated soln., wt %			t	Comp. of solid phase
Lead Chloride	Potassium Chloride	Water		
1.10	28.0	70.90	50	$PbCl_2 \cdot KCl \cdot {}^1/_3H_2O$
1.27	30.0	68.73	50	»
1.5	31.24	67.26	50	$PbCl_2 \cdot KCl \cdot {}^1/_3H_2O + KCl$
1.0	31.0	68.0	50	KCl
0.5	31.0	68.5	50	»
0.0	31.07	68.93	50	»
3.10	0.0	96.90	100	$PbCl_2$
2.00	2.0	96.0	100	»
1.75	4.0	94.25	100	»
1.78	6.0	92.22	100	»
2.00	8.0	90.0	100	»
2.38	10.0	87.62	100	»
2.57	10.6	86.83	100	$PbCl_2 + 2PbCl_2 \cdot KCl$
2.42	12.0	85.58	100	$2PbCl_2 \cdot KCl$
2.32	14.0	83.68	100	»
2.35	16.0	81.65	100	»
2.50	18.0	79.50	100	»
2.75	20.0	77.25	100	»
3.10	22.0	74.90	100	»
3.50	24.0	72.50	100	»
4.25	26.0	69.75	100	$2PbCl_2 \cdot KCl$
5.00	27.2	68.8	100	$2PbCl_2 \cdot KCl + PbCl_2 \cdot KCl \cdot {}^1/_3H_2O$
4.90	28.0	68.1	100	$PbCl_2 \cdot KCl \cdot {}^1/_3H_2O$
5.10	30.0	64.9	100	»
5.45	32.0	62.55	100	»
5.80	34.0	60.2	100	»
6.05	35.3	58.65	100	$PbCl_2 \cdot KCl \cdot {}^1/_3H_2O + KCl$
3.5	35.0	61.5	100	KCl
1.25	35.2	63.55	100	»
0.0	35.5	64.5	100	»

№ 4930 POTASSIUM CHLORIDE–LEAD (II) CHLORIDE–WATER [1965]

$KCl—PbCl_2—H_2O$

$t=25$

Concn. in aqueous soln., g/l		Comp. of solid phase	Concn. in aqueous soln., g/l		Comp. of solid phase
Lead Chloride	Potassium Chloride		Lead Chloride	Potassium Chloride	
2.57	18.62	$PbCl_2$	1.34	96.85	$KCl \cdot 2PbCl_2$
2.22	27.56	»	1.43	115.48	»
1.98	34.27	»	1.58	134.1	»
1.89	38.74	»	1.79	154.2	»
1.88	40.23	»	1.93	161.7	»
1.77	46.94	$KCl \cdot 2PbCl_2$	2.07	168.4	$KCl \cdot PbCl_2$
1.53	55.13	»	2.06	179.9	»
1.48	58.11	»	2.23	197.4	»
1.36	72.26	»	2.70	222.0	»
1.34	81.95	»	2.73	223.5	»
1.34	89.40	»	3.34	238.4	»

№ 4931 POTASSIUM CHLORIDE–LEAD (II) CHLORIDE–WATER [1300]

$KCl—PbCl_2—H_2O$

$t=25$

Concn. in aqueous soln., wt %		Comp. of solid phase
Lead Chloride	Potassium Chloride	
1.083	0.0	$PbCl_2$
1.008	0.037	»
0.961	0.075	»
0.877	0.152	»
0.747	0.296	»
0.462	0.735	»
0.301	1.469	»
0.205	3.022	»
0.227	4.969	$PbCl_2+KCl\cdot 2PbCl_2$
0.128	7.048	$KCl\cdot 2PbCl_2$
0.127	9.787	»
0.142	12.93	»
0.236	18.74	$KCl\cdot 2PbCl_2+KCl\cdot PbCl_2\cdot {}^1/_3H_2O$
0.303	22.54	$KCl\cdot PbCl_2\cdot {}^1/_3H_2O$
0.386	26.43	$KCl\cdot PbCl_2\cdot {}^1/_3H_2O+KCl$
0.267	26.52	KCl
0.122	26.44	»
0.00	26.53	»

№ 4932 POTASSIUM CHLORIDE–LEAD (II) CHLORIDE–WATER [1074]

$KCl—PbCl_2—H_2O$

$t=25$

Concn. in aqueous soln., wt %		Comp. of solid phase
Lead Chloride	Potassium Chloride	
1.08	0.0	$PbCl_2$
0.46	0.74	»
0.299	1.47	»
0.224	2.906	»
0.214	3.374	»
0.209	3.838	»
0.219	4.479	»
0.226	4.976	»
0.235	5.200	»
0.254	5.557	$PbCl_2+2PbCl_2\cdot KCl$
0.218	6.961	$2PbCl_2\cdot KCl$
0.131	11.48	»
0.172	15.37	»
0.236	18.72	$2PbCl_2\cdot KCl+PbCl_2\cdot KCl\cdot {}^1/_3H_2O$
0.269	20.89	$PbCl_2\cdot KCl\cdot {}^1/_3H_2O$
0.310	22.47	»
0.348	24.72	»
0.384	26.42	$PbCl_2\cdot KCl\cdot {}^1/_3H_2O+KCl$
0.293	26.52	KCl
0.147	26.48	»
0.0735	26.45	»
0.0092	26.41	»

№ 4933 POTASSIUM CHLORIDE–LEAD (II) BROMIDE–WATER [1817]

$KCl—PbBr_2—H_2O$

$t=25$

Concn. of Potassium Chloride in aqueous soln., g/l	Solubility of Lead Bromide in aqueous soln. of Potassium Chloride, g/l	Concn. of Potassium Chloride in aqueous soln., g/l	Solubility of Lead Bromide in aqueous soln. of Potassium Chloride, g/l
0.0	9.63	144.72	2.28
35.81	2.04	217.08	3.88
72.36	2.14	289.45	5.62

№ 4934 [2040]

POTASSIUM CHLORIDE–LEAD (II) IODATE–WATER

$KCl—Pb(IO_3)_2—H_2O$

$t=25$

Concn. of Potassium Chloride in aqueous soln., g/l	Solubility of Lead Iodate in aqueous soln. of Potassium Chloride, g/l
7.46	0.0387

№ 4935 POTASSIUM CHLORATE–POTASSIUM BROMIDE–WATER [1239]

$KClO_3—KBr—H_2O$

Comp. of saturated soln., wt %			t	Comp. of saturated soln., wt %			t
Potassium Chlorate	Potassium Bromide	Water		Potassium Chlorate	Potassium Bromide	Water	
4.49	3.05	92.46	14~15	4.60	2.0	93.40	13
3.60	6.10	90.30	14—15	4.20	3.0	92.80	13
5.18	0.20	94.62	13	4.0	4.0	92.0	13
5.20	0.60	94.20	13	3.46	6.0	90.54	13
5.06	0.80	94.14	13	2.80	8.0	89.20	13
5.04	1.0	93.96	13	2.40	10.0	87.60	13

№ 4936 POTASSIUM CHLORATE–POTASSIUM BROMIDE–WATER [2395]

$KClO_3—KBr—H_2O$

$t=25$

Comp. of saturated soln., wt %			d	Comp. of solid phase
Potassium Bromide	Potassium Chlorate	Water		
40.63	0.00	59.37	1.380	KBr
40.02	1.42	58.56	1.356	$KBr+KClO_3$
39.47	1.42	59.11	1.376	$KClO_3$

Table № 4936 (continuation)

Comp. of saturated soln., wt %			d	Comp. of solid phase
Potassium Bromide	Potassium Chlorate	Water		
31.66	1.87	66.47	1.292	$KClO_3$
24.20	2.41	73.39	1.216	»
16.99	3.21	79.8	1.160	»
9.30	4.59	86.11	1.100	»
0.00	7.905	92.095	1.047	»

№ 4937 POTASSIUM CHLORATE–POTASSIUM BROMATE–WATER [2620]

$KClO_3—KBrO_3—H_2O$

$t=25$

Comp. of saturated soln., wt %			d	Comp. of saturated soln., wt %			d
Potassium Bromate	Potassium Chlorate	Water		Potassium Bromate	Potassium Chlorate	Water	
7.533	0.0	92.467	1.054	4.02	6.75	89.23	1.072
6.46	2.26	91.28	—	2.79	7.08	90.13	1.064
5.75	4.08	90.17	1.067	2.07	7.26	90.67	—
5.63	4.29	90.08	—	1.02	7.60	91.38	1.053
4.945	6.531	88.524	1.078	0.0	7.895	92.105	1.048

№ 4938 [1239]

POTASSIUM CHLORATE–POTASSIUM IODIDE–WATER

$KClO_3—KI—H_2O$

$t=14—15$

Comp. of saturated soln., wt %		
Potassium Chlorate	Potassium Iodide	Water
4.59	4.25	91.16
3.65	8.51	87.74

№ 4939 POTASSIUM CHLORATE–POTASSIUM IODIDE–WATER [2395]

$KClO_3—KI—H_2O$

$t=25$

Comp. of saturated soln., wt %			d	Comp. of solid phase
Potassium Chlorate	Potassium Iodide	Water		
7.905	0.0	92.095	1.047	$KClO_3$
5.04	9.33	85.63	1.103	»
3.35	18.74	77.91	1.178	»

Table № 4939 (continuation)

Comp. of saturated soln., wt %			d	Comp. of solid phase
Potassium Chlorate	Potassium Iodide	Water		
2.30	28.72	68.98	1.275	$KClO_3$
1.60	39.26	59.14	1.400	»
1.10	49.94	48.96	1.555	»
0.82	58.34	40.84	1.702	»
0.83	59.28	39.89	1.724	$KClO_3$+KJ
0.67	59.36	39.97	1.724	KJ
0.00	59.76	40.24	1.718	»

№ 4940 POTASSIUM CHLORATE–POTASSIUM IODATE–WATER [2397]

$KClO_3$—KIO_3—H_2O

Comp. of saturated soln., wt %			t	d	Comp. of solid phase
Potassium Chlorate	Potassium Iodate	Water			
7.90	0.00	92.10	25	1.048	$KClO_3$
7.31	2.92	89.77	25	1.068	»
6.80	5.43	87.77	25	1.090	$KClO_3$+KIO_3
5.31	5.85	88.84	25	1.082	KIO_3
2.31	7.05	90.64	25	1.070	»
0.00	8.45	91.55	25	1.043	»
15.78	0.0	84.22	50	—	$KClO_3$
15.11	2.41	82.48	50	—	»
14.31	5.27	80.42	50	—	»
13.77	7.27	78.96	50	—	$KClO_3$+KIO_3
8.58	8.76	82.66	50	—	KIO_3
3.71	10.87	85.42	50	—	»
0.00	13.21	86.79	50	—	»

№ 4941 [2162]

POTASSIUM CHLORATE–CALCIUM CHLORATE–WATER

$KClO_3$—$Ca(ClO_3)_2$—H_2O

$t=20$

Comp. of saturated soln., wt %			Comp. of solid phase
Potassium Chlorate	Calcium Chlorate	Water	
0.00	66.16	33.84	$Ca(ClO_3)_2 \cdot 2H_2O$
1.10	64.31	34.59	$Ca(ClO_3)_2 \cdot 2H_2O + KClO_3$
2.40	47.52	50.08	$KClO_3$
2.39	36.89	60.72	»
2.45	28.60	68.95	»
2.97	11.15	85.88	»
6.76	0.00	93.24	»

№ 4942 POTASSIUM CHLORATE–RUBIDIUM CHLORATE–WATER [383]

$KClO_3—RbClO_3—H_2O$

$t=25$

Comp. of saturated soln., wt %			Comp. of saturated soln., wt %		
Potassium Chlorate	Rubidium Chlorate	Water	Potassium Chlorate	Rubidium Chlorate	Water
0.00	6.65	93.35	7.81	3.38	83.81
2.15	5.64	92.11	7.86	3.38	88.76
2.41	5.43	92.16	7.55	3.70	88.70
3.53	5.04	91.43	8.15	0.95	90.90
3.91	4.87	91.22	8.22	1.49	90.29
7.16	3.72	89.22	8.59	0.00	91.41
7.25	3.94	88.81	8.60	0.00	91.40

№ 4943 POTASSIUM CHLORATE–CESIUM CHLORATE–WATER [383]

$KClO_3—CsClO_3—H_2O$

$t=25$

Comp. of saturated soln., wt %			Comp. of saturated soln., wt %		
Potassium Chlorate	Cesium Chlorate	Water	Potassium Chlorate	Cesium Chlorate	Water
0.00	7.76	92.24	7.57	4.70	87.73
1.83	6.28	91.95	7.62	4.48	87.90
3.63	5.46	90.91	7.78	3.01	89.21
5.18	5.14	89.68	8.03	0.99	90.98
6.98	4.76	88.31	8.59	0.00	91.61
7.73	4.54	87.73			

№ 4944 [2164]

POTASSIUM CHLORATE–BARIUM IODATE–WATER

$KClO_3—Ba(IO_3)_2—H_2O$

$t=25$

Content in aqueous soln., g/l	
Potassium Chlorate	Barium Iodate
0.613	0.4287
1.226	0.4502
3.498	0.5023
9.192	0.5768

№ 4945 [1711]

POTASSIUM CHLORATE–THALLIUM CHLORIDE–WATER

$KClO_3—TlCl—H_2O$

$t=25$

Concn. in aqueous soln., g/l	
Thallium Chloride	Potassium Chlorate
5.674	61.28

№ 4946 POTASSIUM CHLORATE–THALLIUM CHLORATE–WATER [2436]

$KClO_3$–$TlClO_3$–H_2O

$t=10$

Concn. in aqueous soln., g/l		d	Concn. in aqueous soln., g/l		d
Thallium Chlorate	Potassium Chlorate		Thallium Chlorate	Potassium Chlorate	
25.637	0.00	1.0210	6.706	46.410	1.0357
19.637	6.884	1.0222	6.723	47.109	1.0363
12.001	26.100	1.0278	4.858	47.312	1.0345
9.036	40.064	1.0338	2.769	47.134	1.0330
7.885	46.497	1.0359	0.00	49.925	1.0330
7.935	46.535	1.0360			

№ 4947 [2363]

POTASSIUM CHLORATE–THALLIUM CHLORATE–WATER

$KClO_3$—$TlClO_3$—H_2O

Comp. of saturated soln., wt %			t
Thallium Chlorate	Potassium Chlorate	Water	
2.64	3.11	94.25	0
8.97	1.34	89.69	15
9.83	12.57	77.60	50
27.88	23.45	48.67	100

№ 4948 POTASSIUM PERCHLORATE–POTASSIUM PERMANGANATE–WATER [2233]

$KClO_4$—$KMnO_4$—H_2O

$t=7$

Concn. in aqueous soln., g/l		Concn. in aqueous soln., g/l	
Potassium Permanganate	Potassium Perchlorate	Potassium Permanganate	Potassium Perchlorate
0.00	8.86	20.26	4.68
4.65	7.55	22.86	4.59
10.71	5.93	26.55	4.09
12.50	5.49	28.97	3.49
15.79	5.36	31.30	2.80
19.34	4.77	36.98	3.92
18.84	5.39	41.81	0.00

№ 4949 POTASSIUM PERCHLORATE–POTASSIUM PERMANGANATE–WATER [1742]

$KClO_4—KMnO_4—H_2O$

$t=25$

Concn. in aqueous soln., wt %		Concn. in aqueous soln., wt %	
Potassium Perchlorate	Potassium Permanganate	Potassium Perchlorate	Potassium Permanganate
1.868	0.4712	0.8434	4.573
1.381	2.065	0.7816	4.864
1.230	2.651	0.5872	5.698
1.245	2.628	0.5548	5.732
0.9777	4.009	0.3582	6.333

№ 4950 [1261]

POTASSIUM PERCHLORATE–BARIUM NITRATE–WATER

$KClO_4—Ba(NO_3)_2—H_2O$

$t=25$

Concn. in aqueous soln., wt %	
Potassium Perchlorate	Barium Nitrate
2.148	1.249
2.276	3.709
2.383	7.18

№ 4951 [1261]

POTASSIUM PERCHLORATE–BARIUM CHLORIDE–WATER

$KClO_4—BaCl_2—H_2O$

$t=25$

Concn. in aqueous soln., wt %	
Potassium Perchlorate	Barium Chloride
2.107	1.000
2.154	2.94
2.16	5.752

№ 4952 [2164]

POTASSIUM PERCHLORATE–BARIUM IODATE–WATER

$KClO_4—Ba(IO_3)_2—H_2O$

$t=24.93$

Content of Potassium Perchlorate in aqueous soln., mol/l	Solubility of Barium Iodate in aqueous soln. of Potassium Perchlorate, mmol/l
0.004	0.866
0.008845	0.905
0.035	1.045
0.075	1.164

№ 4953 [2007]

POTASSIUM CHROMATE–POTASSIUM DICHROMATE–WATER

$K_2CrO_4—K_2Cr_2O_7—H_2O$

Comp. of saturated soln., wt %			t
Potassium Oxide	Chromium Trioxide	Water	
12.69	13.38	73.93	—11.5
12.96	13.91	73.13	0
13.78	14.96	71.26	+30
14.29	16.37	69.34	60
15.69	19.70	64.61	106.8

Note. t = −11.5° is the eutetic point; t = 106.8° is the boiling point.

№ 4954 [1818]

POTASSIUM CHROMATE–POTASSIUM BROMIDE–WATER

$K_2CrO_4—KBr—H_2O$

$t=25$

Concn. in aqueous soln., g/l	
Potassium Chromate	Potassium Bromide
244.41	48.75
223.4	97.40
202.38	147.2
177.82	211.82
161.65	260.03
141.61	321.03

№ 4955 POTASSIUM CHROMATE–POTASSIUM MOLYBDATE–WATER [1076]

$K_2CrO_4—K_2MoO_4—H_2O$

$t=25$

Comp. of saturated soln., wt %			Comp. of saturated soln., wt %		
Potassium Chromate	Potassium Molybdate	Water	Potassium Chromate	Potassium Molybdate	Water
39.26	0.0	60.74	2.91	56.26	40.83
30.06	9.32	60.62	2.42	59.64	37.94
21.89	21.83	56.28	1.82	61.18	37.00
18.03	28.68	54.29	0.76	63.90	35.34
6.64	46.38	46.98	0.60	64.28	35.12
4.40	51.90	43.70	0.0	64.87	35.13
4.45	52.09	43.46			

№ 4956 POTASSIUM DICHROMATE–POTASSIUM TRICHROMATE–WATER [2007]

$K_2Cr_2O_7—K_2Cr_3O_{10}—H_2O$

Comp. of saturated soln., wt %			t	Comp. of saturated soln., wt %			t
Potassium Oxide	Chromium Trioxide	Water		Potassium Oxide	Chromium Trioxide	Water	
1.18	42.51	56.31	—30	2.50	44.50	53.00	30
1.47	42.99	55.54	0	7.06	49.84	43.10	60
2.20	43.10	54.70	20	16.80	59.20	24.00	114

Note. t = −30° is the eutetic point; t = 114° is the boiling point.

POTASSIUM DICHROMATE–POTASSIUM BROMIDE–WATER

$K_2Cr_2O_7$—KBr—H_2O

Concn. in aqueous soln., wt %		t	Comp. of solid phase
Potassium Bromide	Potassium Dichromate		
0.0	4.43	0	$K_2Cr_2O_7$
2.90	2.20	0	»
6.90	1.20	0	»
14.90	0.70	0	»
25.76	0.40	0	»
34.30	0.30	0	$K_2Cr_2O_7$+KBr
34.5	0.0	0	KBr
0.0	7.50	10	$K_2Cr_2O_7$
2.85	4.90	10	»
6.78	3.20	10	»
14.80	1.40	10	»
19.82	1.00	10	»
37.90	0.50	10	$K_2Cr_2O_7$+KBr
38.1	0.0	10	KBr
0.0	11.10	20	$K_2Cr_2O_7$
3.96	8.06	20	»
9.92	4.53	20	»
16.96	2.55	20	»
22.10	1.93	20	»
30.81	1.19	20	»
34.75	0.88	20	»
39.15	0.65	20	»
39.40	0.0	20	»
0.0	15.40	30	»
6.04	10.32	30	»
16.53	4.94	30	»
26.47	2.56	30	»
34.33	1.56	30	»
40.94	0.95	30	$K_2Cr_2O_7$+KBr
41.45	0.0	30	KBr
0.0	20.60	40	$K_2Cr_2O_7$
6.84	14.58	40	»
18.70	7.23	40	»
23.50	4.21	40	»
42.42	1.65	40	$K_2Cr_2O_7$+KBr
43.2	0.0	40	KBr

№ 4958 POTASSIUM DICHROMATE–CALCIUM CHLORIDE–WATER [1818]

$K_2Cr_2O_7$—$CaCl_2$—H_2O

$t=25$

Concn. in aqueous soln., g/l		Concn. in aqueous soln., g/l		Concn. in aqueous soln., g/l	
Calcium Chloride	Potassium Dichromate	Calcium Chloride	Potassium Dichromate	Calcium Chloride	Potassium Dichromate
635.5	0.0	633.8	92.28	78.26	131.9
629.4	4.90	13.32	141.2	91.02	128.4
631.6	7.84	25.0	142.7	103.9	125.5
637.1	12.26	51.06	139.2	633.8	92.2
630.5	17.16				

№ 4959 POTASSIUM DICHROMATE–CALCIUM CHLORIDE–WATER [405]

$K_2Cr_2O_7$—$CaCl_2$—H_2O

Comp. of saturated soln., wt %					t	Comp. of solid phase
K_2^{2+}	Ca^{2+}	$Cr_2O_7^{2-}$	Cl_2^{2-}	H_2O		
1.07	13.30	0.0	24.50	61.13	0	$CaCl_2\cdot 6H_2O+KCl$
1.03	13.23	0006	24.34	61.40	0	»
1.10	13.32	0.96	24.25	60.37	0	$CaCl_2\cdot 6H_2O+KCl+K_2Cr_2O_7+(CaCrO_4)$
11.56	0.0	0.28	10.39	77.77	0	$K_2Cr_2O_7+KCl$
9.03	2.01	0.24	11.67	77.05	0	»
3.68	7.33	0.65	16.09	72.75	0	$K_2Cr_2O_7+KCl+(CaCrO_4)$
2.17	10.41	0.56	20.20	66.66	0	»
1.33	12 79	0.71	23.61	61.56	0	»
0.65	8.07	45.30	0.0	45.98	0	$K_2Cr_2O_7+CaCr_2O_7\cdot 6H_2O$
0.71	8.74	38.26	3.55	48.74	0	»
0.63	9.56	30.65	7.42	51.74	0	»
0.46	11.34	16.60	15.03	56.57	0	»
0.61	12.86	8.84	20.40	57.29	0	»
0.65	12.94	8.52	20.69	57.20	0	»
0.80	12.90	8.79	20.67	56.84	0	$K_2Cr_2O_7+CaCr_2O_7\cdot 6H_2O+CaCl_2\cdot 6H_2O+(CaCrO_4)$
0.0	13.30	8.12	20.86	57.72	0	$CaCr_2O_7\cdot 6H_2O+CaCl_2\cdot 6H_2O$
0.23	13.03	7.86	20.69	58.19	0	»
0.90	13.15	1.38	23.63	60.94	0	$CaCl_2\cdot 6H_2O+K_2Cr_2O_7$
0.71	13.22	1.95	23.45	60.67	0	»
0.53	13.14	4.62	22.21	59.50	0	$CaCl_2\cdot 6H_2O+K_2Cr_2O_7+(CaCrO_4)$
0.46	12.94	7.72	20.78	58.10	0	»
1.61	16.28	0.0	30.26	51.85	25	$CaCl_2\cdot 6H_2O+KCl$
2.26	15.23	3.91	27.71	50.89	25	$CaCl_2\cdot 6H_2O+KCl+(CaCrO_4)$
2.48	15.36	4.47	27.95	49.74	25	$CaCl_2\cdot 6H_2O+KCl+K_2Cr_2O_7+(CaCrO_4)$
14.08	0.0	0.89	12.48	72.55	25	$K_2Cr_2O_7+KCl$
9.89	2.98	0.82	13.97	72.34	25	»
5.67	6.64	0.84	16.62	70.23	25	$K_2Cr_2O_7+KCl+(CaCrO_4)$
3.46	16.56	3.81	31.19	44.98	25	»
0.99	8.84	50.37	0.0	39.80	25	$K_2Cr_2O_7+CaCr_2O_7\cdot 5H_2O$
1.00	9.63	42.85	3.88	42.64	25	»

Comp. of saturated soln., wt %					t	Comp. of solid phase
K_2^{2+}	Ca^{2+}	$Cr_2O_7^{2-}$	Cl_2^{2-}	H_2O		
1.03	11.67	29.25	11.98	46.07	25	$K_2Cr_2O_7+CaCr_2O_7 \cdot 5H_2O$
1.08	12.77	22.60	16.16	47.39	25	$K_2Cr_2O_7+CaCr_2O_7 \cdot 5H_2O+(CaCrO_4)$
0.0	15.31	14.35	22.39	47.95	25	$CaCr_2O_7 \cdot 5H_2O+CaCl_2 \cdot 6H_2O+(CaCrO_4)$
1.79	15.03	12.20	23.66	47.32	25	$CaCr_2O_7 \cdot 5H_2O+CaCl_2 \cdot 6H_2O+K_2Cr_2O_7+$ $+(CaCrO_4)$

Note. $(CaCrO_4)$ is a sparingly soluble compound.

№ 4960 POTASSIUM DICHROMATE–CALCIUM DICHROMATE–WATER [404]

$K_2Cr_2O_7—CaCr_2O_7—H_2O$

Comp. of saturated soln., wt %			t	Comp. of solid phase
Calcium Dichromate	Potassium Chromate	Water		
53.47	0.00	46.53	0	$CaCr_2O_7 \cdot 6H_2O$
52.69	0.46	46.85	0	»
52.01	1.50	46.49	0	»
51.57	2.45	45.98	0	$CaCr_2O_7 \cdot 6H_2O+K_2Cr_2O_7$
46.86	2.47	50.67	0	$K_2Cr_2O_7$
36.41	2.60	60.99	0	»
23.96	3.26	72.78	0	»
12.12	3.35	84.53	0	»
0.00	4.58	95.42	0	»
58.96	0.00	41.04	25	$CaCr_2O_7 \cdot 5H_2O$
57.70	1.19	41.11	25	»
57.00	2.18	40.82	25	»
56.49	3.71	39.80	25	$CaCr_2O_7 \cdot 5H_2O+K_2Cr_2O_7$
43.00	4.21	52.79	25	$K_2Cr_2O_7$
35.72	5.22	59.06	25	»
23.77	6.74	69.49	25	»
12.21	9.26	78.53	25	»
0.00	13.13	86.87	25	»
64.35	0.00	35.65	50	$CaCr_2O_7 \cdot 4H_2O$
63.47	1.17	35.36	50	»
63.07	1.80	35.13	50	»
62.68	3.08	34.24	50	»
61.92	3.72	34.36	50	»
61.60	4.25	34.15	50	»
61.60	4.73	33.67	50	»
61.24	5.69	33.07	50	$CaCr_2O_7 \cdot 4H_2O+K_2Cr_2O_7$
59.91	5.91	34.18	50	$K_2Cr_2O_7$
56.61	6.88	36.51	50	»
52.93	7.37	39.70	50	»
44.28	8.57	47.15	50	»
34.57	10.57	54.86	50	»
20.96	15.92	63.12	50	»
10.74	20.06	69.20	50	»
0.00	25.83	74.18	50	»

№ 4961 [1818]

POTASSIUM DICHROMATE–STRONTIUM CHLORIDE–WATER

$K_2Cr_2O_7—SrCl_2—H_2O$

$t=25$

Concn. in aqueous soln., g/l	
Potassium Dichromate	Strontium Chloride
141.69	40.45
142.68	79.32
133.85	160.22
131.89	176.87
128.46	225.25
116.69	275.22

№ 4962 [2007]

POTASSIUM TRICHROMATE–POTASSIUM TETRACHROMATE–WATER

$K_2Cr_3O_{10}—K_2Cr_4O_{13}—H_2O$

Comp. of saturated soln., wt %			t
Potassium Trichromate	Potassium Tetrachromate	Water	
0.79	45.69	53.52	—39
1.37	47.40	51.23	0
2.00	48.46	49.54	20
2.25	49.95	47.80	30
5.01	54.09	40.90	60

Note. t = −39° is the eutectic point.

№ 4963 [2007]

POTASSIUM TETRACHROMATE–CHROMIUM TRIOXIDE–WATER

$K_2Cr_4O_{13}–CrO_3–H_2O$

Comp. of saturated soln., wt %			t
Potassium Oxide	Chromium Trioxide	Water	
0.64	61.79	37.57	0
0.62	62.80	36.58	20
0.69	62.81	36.50	30
1.27	65.77	32.96	60

№ 4964 [1818]

POTASSIUM PERMANGANATE–CALCIUM CHLORIDE–WATER

$KMnO_4—CaCl_2—H_2O$

$t=25$

Concn. in aqueous soln., g/l	
Potassium Permanganate	Calcium Chloride
63.21	94.35
37.92	271.9
28.76	377.4
25.28	424.9

№ 4965 [2233]

POTASSIUM PERMANGANATE–RUBIDIUM PERMANGANATE–WATER

$KMnO_4—RbMnO_4—H_2O$

$t=7$

Concn. in aqueous soln., g/l	
Potassium Permanganate	Rubidium Permanganate
4.28	4.64
11.84	4.54

№ 4966 [1818]

POTASSIUM PERMANGANATE–STRONTIUM CHLORIDE–WATER

$KMnO_4—SrCl_2—H_2O$

$t=25$

Concn. in aqueous soln., g/l	
Strontium Chloride	Potassium Permanganate
494.9	0.0
493.3	0.63

Table № 4965 (continuation)

Concn. in aqueous soln., g/l	
Potassium Permanganate	Rubidium Permanganate
19.03	6.40
29.80	7.97
31.39	8.44
32.56	8.69
35.61	5.32
41.81	0.00

Table № 4966 (continuation)

Concn. in aqueous soln., g/l	
Strontium Chloride	Potassium Permanganate
486.9	12.96
485.4	25.92
467.9	42.98
358.5	48.60
310.9	51.83
196.5	61.31
50.75	71.74

№ 4967 [1719]

POTASSIUM DIHYDROGEN ORTHOARSENATE–ARSENIC TRIOXIDE–WATER

$KH_2AsO_4—As_2O_3—H_2O$

Comp. of saturated soln., wt %			t
Potassium Dihydrogen Orthoarsenate	Arsenic Trioxide	Water	
17.704	2.339	79.957	25
17.199	4.727	78.074	60
16.767	9.360	73.873	100

№ 4968 [2191]

POTASSIUM SELENATE–CALCIUM SELENATE–WATER

$K_2SeO_4—CaSeO_4—H_2O$

$t=25$

Comp. of saturated soln., wt %			Comp. of solid phase
Potassium Selenate	Calcium Selenate	Water	
0.00	6.88	93.12	$CaSeO_4 \cdot H_2O$
13.62	6.67	79.71	»
16.43	6.42	77.15	»
17.90	6.22	75.88	»
19.29	5.56	75.15	»
26.18	5.78	68.04	$CaSeO_4 \cdot 2H_2O$
31.52	5.26	63.22	$CaSeO_4 \cdot 2H_2O+CaK_2(SeO_4)_2 \cdot 2H_2O$
35.42	3.24	61.34	$CaK_2(SeO_4)_2 \cdot 2H_2O$
43.05	1.12	55.83	$CaK_2(SeO_4)_2 \cdot 2H_2O+K_2SeO_4$
53.3	0.0	46.7	K_2SeO_4

№ 4969 POTASSIUM SELENATE–LEAD (II) SELENATE–WATER [797]

$K_2SeO_4—PbSeO_4—H_2O$

$t=25$

Concn. of Potassium Selenate in aqueous soln., mol/l	Solubility of Lead Selenate in aqueous soln. of Potassium Selenate, mol/l	Concn. of Potassium Selenate in aqueous soln., mol/l	Solubility of Lead Selenate in aqueous soln. of Potassium Selenate, mol/l
$0.825 \cdot 10^{-4}$	$3.41 \cdot 10^{-4}$	$0.825 \cdot 10^{-2}$	$0.137 \cdot 10^{-4}$
$4.12 \cdot 10^{-4}$	$3.13 \cdot 10^{-4}$	$1.00 \cdot 10^{-2}$	$0.128 \cdot 10^{-4}$
$0.825 \cdot 10^{-3}$	$1.66 \cdot 10^{-4}$	$0.825 \cdot 10^{-1}$	$0.146 \cdot 10^{-4}$

№ 4970 POTASSIUM BROMIDE–POTASSIUM BROMATE–WATER [211]

$KBr—KBrO_3—H_2O$

Comp. of saturated soln., wt %			t	Comp. of solid phase
Potassium Bromide	Potassium Bromate	Water		
35.30	0.00	64.70	0	KBr
35.08	0.57	64.35	0	»
35.16	0.65	64.19	0	$KBr+KBrO_3$
0.00	2.96	97.04	0	$KBrO_3$
39.4	0.00	60.6	20	KBr
39.26	1.22	59.52	20	$KBr+KBrO_3$
22.20	1.85	75.95	20	$KBrO_3$
11.80	2.73	75.47	20	»
0.00	6.43	93.57	20	»
43.54	0.00	56.46	40	KBr
42.97	0.34	56.69	40	»
42.44	1.28	56.28	40	»
42.34	2.19	55.47	40	$KBr+KBrO_3$
35.97	2.60	61.43	40	$KBrO_3$
20.84	4.32	74.84	40	»
8.50	7.32	74.18	40	»
0.00	11.70	88.30	40	»
46.2	0.00	53.8	60	KBr
45.17	3.70	51.13	60	$KBr+KBrO_3$
49.93	3.75	46.32	60	$KBrO_3$
44.56	3.73	51.71	60	»
31.90	5.71	62.39	60	»
12.69	12.09	75.22	60	»
10.07	12.74	77.19	60	»
9.52	12.83	77.65	60	»
9.37	13.14	77.49	60	»
4.48	14.92	76.60	60	»
4.35	15.58	80.07	60	»
3.94	16.43	79.63	60	»
0.00	18.21	80.79	60	»
49.72	0.00	50.28	80	KBr
46.5	5.62	47.88	80	$KBr+KBrO_3$
45.57	5.60	48.83	80	$KBrO_3$
7.45	20.74	71.81	80	»
0.00	25.53	74.47	80	»

№ 4971 POTASSIUM BROMIDE–POTASSIUM BROMATE–WATER [2390]

$KBr—KBrO_3—H_2O$

$t=25$

Comp. of saturated soln., wt %			d	Comp. of solid phase
Potassium Bromide	Potassium Bromate	Water		
40.62	0.00	59.38	1.381	KBr
40.08	1.20	58.72	1.389	»
40.00	1.43	58.57	1.392	$KBr+KBrO_3$
34.82	1.62	63.56	1.328	$KBrO_3$
26.05	2.06	71.89	1.237	»
17.48	2.73	79.79	1.161	»
7.82	4.29	87.89	1.089	»
0.00	7.533	92.467	1.054	»

№ 4972 POTASSIUM BROMIDE–POTASSIUM IODIDE–WATER [1578]

$KBr—KI—H_2O$

Comp. of saturated soln., wt %			t	Comp. of saturated soln., wt %			t
Potassium Bromide	Potassium Iodide	Water		Potassium Bromide	Potassium Iodide	Water	
9.2	42.5	48.3	—20	11.2	49.6	39.2	40
9.9	45.3	44.8	0	11.9	51.3	36.8	60
10.2	46.6	43.2	10	12.6	52.7	34.7	80
10.5	47.5	42.0	20	13.2	53.8	33.0	100
10.7	48.0	41.3	25	14.0	54.8	31.2	120
10.9	48.6	40.5	30	14.9	55.5	29.6	150

№ 4973 POTASSIUM BROMIDE–POTASSIUM IODIDE–WATER [1075]

$KBr—KI—H_2O$

$t=25$

Comp. of saturated soln., wt %			Comp. of saturated soln., wt %		
Potassium Bromide	Potassium Iodide	Water	Potassium Bromide	Potassium Iodide	Water
28.13	18.99	52.88	11.91	47.86	40.23
20.25	32.41	47.34	8.88	50.99	40.13
14.87	41.55	43.58	8.78	51.08	40.14
12.05	47.88	40.07	7.44	52.42	40.14

№ 4974 POTASSIUM BROMIDE–POTASSIUM IODIDE–WATER [256]

$KBr—KI—H_2O$

Comp. of saturated soln., mol			t	Comp. of solid phase
Potassium Bromide	Potassium Iodide	Water		
18.1	81.9	773	—20	α-KI+α-KBr
27.9	72.1	891	—20	Ice+α-KBr
18.8	81.2	756	—15	α-KI+α-KBr
62.5	37.5	1169	—15	Ice+α-KBr
19.7	80.3	732	—10	α-KI+α-KBr
20.3	79.7	698	— 5	»
21.0	79.0	683	0	»
17.2	82.8	670	5	α-KI+β-KI
21.7	78.3	655	5	β-KI+β-KBr
65.4	34.6	978	5	α-KBr+β-KBr
5.9	94.1	651	10	α-KI+β-KI
22.8	77.2	632	10	β-KI+β-KBr
75.7	24.3	985	10	α-KBr+β-KBr
23.3	76.7	607	15	β-KI+β-KBr
85.3	14.7	1000	15	α-KBr+β-KBr
23.8	76.2	592	20	β-KI+β-KBr
96.4	3.6	1002	20	α-KBr+β-KBr
24.6	75.4	565	25	β-KI+β-KBr
25.3	74.7	546	30	»

№ 4975 POTASSIUM BROMIDE–POTASSIUM IODATE–WATER [2390]

$KBr—KIO_3—H_2O$

Comp. of saturated soln., wt %			t	d	Comp. of solid phase
Potassium Bromide	Potassium Iodate	Water			
36.25	0.00	63.75	5	1.333	KBr
35.71	1.80	62.49	5	1.352	$KBr+KIO_3$
30.50	1.77	67.73	5	1.290	KIO_3
22.59	1.80	75.61	5	1.208	»
14.80	1.95	83.25	5	1.136	»
7.58	2.45	99.97	5	1.080	»
0.00	5.186	94.814	5	1.043	»
40.62	0.00	59.38	25	1.381	KBr
40.28	0.98	58.74	25	1.396	»
39.75	2.36	57.89	25	1.407	$KBr+KIO_3$
34.38	2.47	63.15	25	1.341	KIO_3
25.91	2.73	71.36	25	1.294	»
17.40	3.26	79.34	25	1.168	»
8.35	4.53	77.12	25	1.106	»
0.00	8.452	91.548	25	1.071	»

№ 4976 [2390]

POTASSIUM BROMIDE–POTASSIUM IODATE–WATER

$KBr—KIO_3—H_2O$

$t=50$

Comp. of saturated soln., wt %			Comp. of solid phase
Potassium Bromide	Potassium Iodate	Water	
44.77	0.00	55.23	KBr
43.88	2.22	53.90	»
43.50	3.07	53.43	$KBr+KIO_3$
38.03	3.32	58.65	KIO_3
30.57	3.87	65.56	»
23.19	4.66	72.15	»
15.65	5.80	78.55	»
8.18	8.16	83.66	»
0.00	13.20	86.80	»

№ 4977 [1940]

POTASSIUM BROMIDE–CALCIUM HYDROXIDE–WATER

$KBr—Ca(OH)_2—H_2O$

$t=25$

Content in aqueous soln., mol/1000 g Water		Comp. of solid phase
Potassium Bromide	Calcium Hydroxide	
3.24	0.02058	$Ca(OH)_2$
2.09	0.02441	»
1.55	0.02596	»
1.27	0.02636	»
0.930	0.02670	»
0.759	0.02669	»
0.453	0.02608	»
0.270	0.02509	»
0.162	0.02404	»
0	0.01976	»

№ 4978 [1511]

POTASSIUM BROMIDE–CALCIUM SULFATE–WATER

$KBr—CaSO_4—H_2O$

$t=21$

Concn. of Potassium Bromide in aqueous soln., g/l	Solubility of Calcium Sulfate in aqueous soln. of Potassium Bromide, g/l	Concn. of Potassium Bromide in aqueous soln., g/l	Solubility of Calcium Sulfate in aqueous soln. of Potassium Bromide, g/l
0	2.05	100	6.3
10	3.1	125	6.7
20	3.6	150	7.0
40	4.5	200	7.3
60	5.2	250	Double salt
80	5.9		

№ 4979 POTASSIUM BROMIDE–COPPER (I) IODIDE–WATER [1992]

$KBr—CuI—H_2O$

Concn. in aqueous soln., g/l		t	Concn. in aqueous soln., g/l		t
Copper Iodide	Potassium Bromide		Copper Iodide	Potassium Bromide	
1.467	238	19.5	3.595	357	23
1.558	238	24	7.126	476	22
3.409	357	19.5	6.977	476	22

№ 4980 [2519]

POTASSIUM BROMIDE–ARSENIC TRIOXIDE–WATER

$KBr—As_2O_3—H_2O$

$t=30$

Comp. of saturated soln., wt %			Comp. of solid phase
Arsenic Trioxide	Potassium Bromide	Water	
2.25	0.336	97.414	As_2O_3+D *
0.818	2.51	96.672	D *
0.460	12.78	86.760	»
0.327	22.59	77.083	»
0.290	27.40	71.310	»
0.275	36.98	62.745	»
0.207	39.04	60.553	»
0.166	42.07	57.764	D *+KBr
0.0	41.3	58.7	KBr

*The composition of D changes from $(As_2O_3)_2KBr$ to $(As_2O_3)_7(KBr)_4$.

№ 4981 POTASSIUM BROMIDE–RUBIDIUM BROMIDE–WATER [554]

$KBr—RbBr—H_2O$

$t=25$

Concn. in aqueous soln., mol/1000 g Water		Concn. in aqueous soln., mol/1000 g Water		Concn. in aqueous soln., mol/1000 g Water	
Rubidium Bromide	Potassium Bromide	Rubidium Bromide	Potassium Bromide	Rubidium Bromide	Potassium Bromide
$\sim 10^{-5}$	5.748	3.331	3.689	4.061	3.120
0.842	5.268	3.528	3.576	4.360	2.702
2.233	4.427	3.681	3.354	4.912	2.116
2.936	3.972	3.882	3.260	6.737	10^{-3}
3.054	3.903	3.838	3.348		

№ 4982 [2038]

POTASSIUM BROMIDE–SILVER BROMIDE–WATER

$KBr—AgBr—H_2O$

Comp. of saturated soln., wt %			t	Comp. of solid phase
Potassium Bromide	Silver Bromide	Water		
39.87	1.67	58.46	−13	AgBr+KBr
39.54	1.636	58.824	30	»
40.88	2.24	56.88	40	»
43.09	3.25	53.66	50	»
42.43	3.12	54.45	50	»
44.24	4.20	51.56	60	»
45.48	5.80	48.72	70	»
45.53	5.60	48.87	70	»
46.91	6.72	46.37	80	»
48.15	8.47	43.38	90	»
47.98	9.00	43.02	90	»
48.86	10.55	40.79	100	»

№ 4983 [2710]

POTASSIUM BROMIDE–SILVER BROMIDE–WATER

$KBr—AgBr—H_2O$

$t=25$

Concn. in aqueous soln., g/l		Concn. in aqueous soln., g/l		Concn. in aqueous soln., g/l	
Potassium Bromide	Silver Bromide	Potassium Bromide	Silver Bromide	Potassium Bromide	Silver Bromide
0.0000297.5	0.000751	17.85	0.001878	254.66	0.7512
0.000119	0.0001878	23.205	0.001878	276.08	1.1268
0.0002975	0.0000751	31.535	0.003756	321.3	1.878
0.000476	0.00004695	47.600	0.007512	328.44	2.1973
0.00119	0.00001878	55.335	0.011268	371.04	3.756
0.119	0.00001878	71.400	0.01878	396.51	4.695
1.309	0.0000751	98.77	0.03756	422.21	7.512
3.2725	0.00001878	119.00	0.07512	437.92	7.493
5.712	0.0002817	139.23	0.11268	497.42	13.503
9.8175	0.00011695	164.22	0.1878	528.36	17.972
13.685	0.0007512	805.87	0.3756	578.82	264.61

№ 4984 [519]

POTASSIUM BROMIDE–SILVER BROMIDE–WATER

$KBr—AgBr—H_2O$

$t=20$

Content of Potassium Bromide in aqueous soln., mol/l	Solubility of Silver Bromide in aqueous soln. of Potassium Bromide, mol/l	Content of Potassium Bromide in aqueous soln., mol/l	Solubility of Silver Bromide in aqueous soln. of Potassium Bromide, mol/l
0.099	0.000006	0.385	0.000039
0.198	0.000011	0.784	0.00020

Table № 4984 (continuation)

Content of Potassium Bromide in aqueous soln., mol/l	Solubility of Silver Bromide in aqueous soln. of Potassium Bromide, mol/l	Content of Potassium Bromide in aqueous soln., mol/l	Solubility of Silver Bromide in aqueous soln. of Potassium Bromide, mol/l
1.434	0.00104	0.377	0.000058
2.790	0.0128	0.775	0.00031
4.915	0.132	1.57	0.00195
0.099 *	0.000006	2.78	0.0130
0.197	0.000016		

*All subsequent experiements were carried out with a solution of constant ionic strength of 5 mol/l. for which Potassium Nitrate was added in appropriate quantities to each solution.

№ 4985 POTASSIUM BROMIDE–SILVER BROMIDE–WATER [2348]

$KBr—AgBr—H_2O$

Concn. of Potassium Bromide in aqueous soln., g/l	Solubility of Silver Bromide in aqueous soln. of Potassium Bromide, g/l	t	Concn. of Potassium Bromide in aqueous soln., g/l	Solubility of Silver Bromide in aqueous soln. of Potassium Bromide, g/l	t
0.376	0.004695	20	119.0	0.13338	50
0.376	0.006573	30	119.0	0.16714	60
0.376	0.008451	40	376.0	0.30048	20
0.376	0.011831	50	376.0	0.33804	30
0.376	0.016714	60	376.0	0.41316	40
119.0	0.060096	20	376.0	0.46950	50
119.0	0.08451	30	376.0	0.50706	60
119.0	0.10517	40			

№ 4986 [2486]

POTASSIUM BROMIDE–SILVER IODIDE–WATER

$KBr—AgI—H_2O$

Concn. in aqueous soln., wt %		t
Potassium Bromide	Silver Iodide	
30.77	0.132	15

$KBr—CdBr_2—H_2O$

Comp. of saturated soln., wt %			t	d_4^t	Comp. of solid phase
Potassium Bromide	Cadmium Bromide	Water			
6.1	13.9	80.0	— 2.7	—	Ice
10.9	12.4	76.7	— 3.7	—	»
13.7	12.0	74.3	— 4.5	—	»
10.6	24.4	65.0	— 5.1	—	»
19.5	22.4	58.1	— 7.6	—	»
15.2	34.8	50.0	— 9.2	—	»
24.7	21.8	53.5	—10.1	—	»
25.4	29.0	45.6	—12.0	—	»
32.0	28.1	39.9	—16.8	—	»
7.7	37.9	54.4	0	1.585	$CdBr_2 \cdot 4H_2O$
20.2	34.1	45.7	0	1.708	$CdBr_2 \cdot KBr \cdot H_2O$
35.1	33.5	31.4	0	1.989	»
34.6	21.0	44.4	0	1.669	KBr
34.45	12.25	53.3	0	1.502	»
14.8	35.5	49.7	— 9.4	—	Ice $+ CdBr_2 \cdot 4H_2O + CdBr \cdot KBr \cdot H_2O$
14.1	40.0	45.9	0	1.744	$CdBr_2 \cdot 4H_2O + CdBr_2 \cdot KBr \cdot H_2O$
33.0	29.0	38.0	—18.1	—	ice $+ KBr + CdBr_2 \cdot KBr \cdot H_2O$
35.7	33.7	30.6	0	2.011	$KBr + CdBr_2 \cdot KBr \cdot H_2O$
13.45	43.2	43.35	6.5	1.805	$CdBr_2 \cdot 4H_2O + CdBr_2 \cdot KBr \cdot H_2O$
37.1	36.0	26.9	12.0	2.108	$KBr + CdBr_2 \cdot KBr \cdot H_2O$
12.75	46.4	40.85	13.0	1.872	$CdBr_2 \cdot 4H_2O + CdBr_2 \cdot KBr \cdot H_2O$
12.6	47.5	39.9	15.2	1.896	$CdBr_2 \cdot 4H_2O + 3CdBr_2 \cdot KBr \cdot 4H_2O +$ $+ CdBr_2 \cdot KBr \cdot H_2O$
11.05	48.8	40.15	18.0	1.901	$CdBr_2 \cdot 4H_2O + 3CdBr_2 \cdot KBr \cdot 4H_2O$
12.9	47.8	39.3	18.0	1.911	$3CdBr_2 \cdot KBr \cdot 4H_2O + CdBr_2 \cdot KBr \cdot H_2O$
37.6	37.5	24.9	20.0	2.163	$KBr + CdBr_2 \cdot KBr \cdot H_2O$
37.7	38.3	24.0	22.9	2.185	$KBr + CdBr_2 \cdot KBr \cdot H_2O + CdBr_2 \cdot KBr$
37.9	38.3	23.8	25.0	2.194	»
6.0	52.6	41.4	25	1.892	$CdBr_2 \cdot 4H_2O$
9.7	51.0	39.3	25	1.928	$3CdBr_2 \cdot KBr \cdot 4H_2O$
11.8	49.7	38.5	25	1.936	»
25.0	37.8	37.2	25	1.869	$CdBr_2 \cdot KBr \cdot H_2O$
35.5	37.3	27.2	25	2.093	»
37.9	21.4	40.7	25	1.717	KBr
39.05	10.65	50.3	25	1.527	»
7.5	52.7	39.8	25	1.929	$CdBr_2 \cdot 4H_2O + 3CdBr_2 \cdot KBr \cdot 4H_2O$
13.7	48.7	37.6	25	1.947	$3CdBr_2 \cdot KBr \cdot 4H_2O + CdBr \cdot KBr \cdot H_2O$
4.5	57.6	37.9	32.5	2.081	$CdBr_2 \cdot 4H_2O + 3CdBr_2 \cdot KBr \cdot 4H_2O$
3.7	59.2	37.1	35	2.031	$CdBr_2 \cdot 4H_2O + CdBr_2 + 3CdBr_2 \cdot KBr \cdot 4H_2O$
15.0	50.0	35.0	35	2.007	$3CdBr_2 \cdot KBr \cdot 4H_2O + CdBr_2 \cdot KBr \cdot H_2O$
4.3	58.9	36.8	37	2.033	$CdBr_2 \cdot 4H_2O + 3CdBr_2 \cdot KBr \cdot 4H_2O$
7.05	57.7	35.25	45	2.056	»
15.9	51.5	32.6	45	2.067	$3CdBr_2 \cdot KBr \cdot 4H_2O + CdBr_2 \cdot KBr \cdot H_2O$
39.2	40.5	20.3	50	2.291	$KBr + CdBr_2 \cdot KBr$
16.4	52.2	31.4	50	2.095	$3CdBr_2 \cdot KBr \cdot 4H_2O + CdBr_2 \cdot KBr \cdot H_2O$
16.6	52.5	30.9	52.1	2.111	$3CdBr_2 \cdot KBr \cdot 4H_2O + CdBr_2 \cdot KBr \cdot H_2O +$ $+ 3CdBr_2 \cdot KBr \cdot H_2O$
10.35	56.6	33.05	52.5	2.092	$CdBr_2 \cdot 4H_2O + 3CdBr_2 \cdot KBr \cdot 4H_2O$

Table № 4987 (continuation)

Comp. of saturated soln., wt %			t	d_4^t	Comp. of solid phase
Potassium Bromide	Cadmium Bromide	Water			
10.2	56.8	33.0	54.5	2.088	$CdBr_2$
16.9	52.6	30.5	56	2.119	$2CdBr_2 \cdot KBr \cdot H_2O$
12.5	55.8	31.7	56	2.116	$CdBr_2+3CdBr_2 \cdot KBr \cdot 4H_2O$
14.4	54.6	31.0	56	2.123	$3CdBr \cdot KBr \cdot 4H_2O+2CdBr_2 \cdot KBr_2 \cdot H_2O$
18.0	51.8	30.2	56	2.117	$2CdBr_2 \cdot KBr \cdot H_2O+CdBr_2 \cdot KBr \cdot H_2O$
18.25	51.7	30.05	56.7	2.118	$2CdBr_2 \cdot KBr \cdot H_2O+CdBr_2 \cdot KBr \cdot H_2O+CdBr_2 \cdot KBr$
13.5	55.5	31.0	57.6	2.132	$CdBr_2+3CdBr_2 \cdot KBr \cdot 4H_2O+2CdBr_2 \cdot KBr \cdot H_2O$
14.3	55.2	30.5	60	2.138	$CdBr_2+2CdBr_2 \cdot KBr \cdot H_2O$
19.6	52.5	27.9	60	2.165	$2CdBr_2 \cdot KBr \cdot H_2O+CdBr_2 \cdot KBr$
15.9	54.8	29.3	70	2.148	$CdBr_2$
18.2	53.4	28.4	70	2.163	$2CdBr_2 \cdot KBr \cdot H_2O$
21.0	50.9	28.1	70	2.144	$CdBr_2 \cdot KBr \cdot H_2O$
17.0	54.4	28.6	70	2.166	$CdBr_2+2CdBr_2 \cdot KBr \cdot H_2O$
19.6	52.5	27.9	70	2.165	$2CdBr_2 \cdot KBr \cdot H_2O+CdBr_2 \cdot KBr$
40.1	42.4	17.5	75	2.373	$KBr+CdBr_2 \cdot KBr$
19.65	53.7	26.65	80	2.200	$CdBr_2+2CdBr_2 \cdot KBr \cdot H_2O$
21.1	53.5	25.4	86	2.224	$CdBr_2+2CdBr_2 \cdot KBr \cdot H_2O+CdBr \cdot KBr$
21.6	53.5	23.9	90	2.232	$CdBr_2+CdBr_2 \cdot KBr$
23.1	53.4	23.5	100	2.257	»
7.0	58.3	34.7	100	1.998	$CdBr_2$
15.2	55.3	29.5	100	2.106	»
40.9	44.1	15.0	100	2.446	$KBr+CdBr \cdot KBr \cdot H_2O$
27.6	48.8	23.6	100	2.200	$CdBr_2 \cdot KBr$
30.3	46.5	23.2	100	2.190	»
33.3	44.7	22.0	100	2.207	»
36.1	43.6	20.3	100	2.321	»
40.8	41.4	17.8	100	2.324	KBr
42.2	29.6	28.2	100	1.949	»
47.0	12.8	40.2	100	1.636	»
24.35	53.5	22.15	110	2.287	$CdBr_2+CdBr_2 \cdot KBr \cdot H_2O$

№ 4988 [2438]

POTASSIUM BROMIDE–ANTIMONY TRIFLUORIDE–WATER

$KBr—SbF_3—H_2O$

$t=0$

Concn. of Potassium Bromide in aqueous soln., g/l	Solubility of Antimony Trifluoride in aqueous soln. of Potassium Bromide wt %
119	81.77
59.5	81.81
29.75	82.00
14.875	80.66

№ 4989 [233]

POTASSIUM BROMIDE–ISOMERIC DIAMMINEDIBROMOPLATINATES–WATER

$KBr—Pt(NH_3)_2Br_2—H_2O$

$t=25$

Diamminedibromoplatinate		Content of Potassium Bromide in aqueous soln., mol/l	Solubility of Diamminedibromoplatinate in aqueous soln. of Potassium Bromide, wt %
Name	Formula		
cis-Diamminedibromoplatinate	*cis*- $[Pt(NH_3)_2Br_2]$	0.0	0.100
	»	1.0	0.077
	»	1.5	0.086
	»	2.0	0.089
	»	4.0	0.105
	»	4.5	0.112
trans-Diamminedibromoplatinate	*trans*- $[Pt(NH_3)_2Br_2]$	0.0	0.013
	»	1.0	0.009
	»	4.5	0.009

№ 4990 [1815]

POTASSIUM BROMIDE–MERCURY (II) BROMIDE–WATER

$KBr—HgBr_2—H_2O$

$t=25$

Concn. in aqueous soln., g/l	
Mercury Bromide	Potassium Bromide
6.12	0.0
35.3	23.95
170.2	91.6
490	283.0
696	413

№ 4991 POTASSIUM BROMIDE–MERCURY (II) BROMIDE–WATER [2324]

$KBr—HgBr_2—H_2O$

Comp. of saturated soln., wt %			t	Comp. of solid phase
Mercury Bromide	Potassium Bromide	Water		
0.3	0.0	99.7	0	$HgBr_2$
15.0	7.5	77.5	0	»
24.0	12.0	64.0	0	$HgBr_2 + HgBr_2 \cdot KBr \cdot H_2O$
31.5	20.0	48.5	0	$HgBr_2 \cdot KBr \cdot H_2O$
37.5	25.0	37.5	0	»
43.2	30.0	26.8	0	»
48.2	32.5	19.3	0	$HgBr_2 \cdot KBr \cdot H_2O + KBr$
40.0	32.2	27.8	0	KBr
30.0	31.8	38.2	0	»

Comp. of saturated soln., wt %			t	Comp. of solid phase
Mercury Bromide	Potassium Bromide	Water		
20.0	32.5	47.5	0	KBr
0.0	34.0	66.0	0	»
0.68	0.0	99.32	34	$HgBr_2$
15.0	5.0	80.0	34	»
23.0	10.0	67.0	34	»
38.0	15.0	47.0	34	»
51.0	19.0	30.0	34	$HgBr_2 + HgBr_2 \cdot KBr \cdot H_2O$
48.5	23.0	28.5	34	$HgBr_2 \cdot KBr \cdot H_2O$
48.0	28.0	24.0	34	»
55.0	33.5	11.5	34	$HgBr_2 \cdot KBr \cdot H_2O + KBr$
45.0	34.0	21	34	KBr
30.0	35.0	35	34	»
15.0	40.0	45	34	»
0.0	42.1	57.9	34	»
2.7	0.0	97.3	80	$HgBr_2$
30.0	10.0	60.0	80	»
50.0	15.0	35.0	80	»
75.0	18.5	6.5	80	$HgBr_2 + HgBr_2 \cdot KBr \cdot H_2O$
70.0	21.5	8.5	80	$HgBr_2 \cdot KBr \cdot H_2O$
65.0	25.0	10	80	»
63.0	28.0	9	80	»
62.0	32.0	6	80	$HgBr_2 \cdot KBr \cdot H_2O + KBr$
50.0	35.0	15	80	KBr
26.0	40.0	36	80	»
15.0	44.0	41	80	»
0.0	48.0	52	80	»

№ 4992 [1817]

POTASSIUM BROMIDE–LEAD (II) CHLORIDE–WATER

$KBr—PbCl_2—H_2O$

$t=25$

Concn. of Potassium Bromide in aqueous soln., g/l	Solubility of Lead Chloride in aqueous soln. of Potassium Bromide, g/l
63.075	2.194
127.34	2.469
254.68	7.167
382.02	18.648
509.363	49.338

№ 4993 [1817]

POTASSIUM BROMIDE–LEAD (II) BROMIDE–WATER

$KBr—PbBr_2—H_2O$

$t=25$

Concn. of Potassium Bromide in aqueous soln., g/l	Solubility of Lead Bromide in aqueous soln. of Potassium Bromide, g/l
0.0	9.63
94.01	2.66
188.02	3.21
257.04	9.72
384.37	31.35
509.32	66.69

№ 4994 [1300]

POTASSIUM BROMIDE–LEAD (II) BROMIDE–WATER

$KBr—PbBr_2—H_2O$

$t=25$

Comp. of saturated soln., wt %			Comp. of solid phase
Lead Bromide	Potassium Bromide	Water	
0.969	0.0	99.031	$PbBr_2$
0.946	0.056	98.998	»
0.875	0.131	98.994	»
0.782	0.423	98.795	»
0.347	1.325	98.328	»
0.256	2.292	97.452	»
0.310	4.967	34.723	$PbBr_2+KBr\cdot 2PbBr_2$
0.292	5.609	94.099	$KBr\cdot 2PbBr_2$
0.330	12.94	86.73	»
0.345	13.18	86.475	»
0.427	14.54	85.033	»
0.560	18.27	81.17	»
0.666	19.05	80.284	»
0.753	20.64	78.607	»
0.868	21.61	77.522	»
1.494	25.69	72.816	»
3.160	30.11	66.73	$KBr\cdot 2PbBr_2+KBr\cdot PbBr_2\cdot {}^1/_3H_2O$
3.622	31.97	64.408	$KBr\cdot PbBr_2\cdot {}^1/_3H_2O$
4.248	33.80	61.952	»
5.038	35.68	59.282	»
7.205	38.87	53.925	$KBr\cdot PbBr_2\cdot {}^1/_3H_2O+KBr$
4.929	39.32	55.751	KBr
1.835	40.28	57.885	»
0.0	40.52	59.48	»

№ 4995 [2368]

POTASSIUM BROMIDE–LEAD (II) BROMIDE–WATER

$KBr—PbBr_2—H_2O$

$t=25$

Concn. in aqueous soln., wt %		Concn. in aqueous soln., wt %	
Lead Bromide	Potassium Bromide	Lead Bromide	Potassium Bromide
0.037	0.313	6.83	0.077
0.073	0.310	12.06	0.072
0.183	0.297	12.49	0.073
0.365	0.276	17.87 *	0.069
0.727	0.242	22.77 *	0.068
1.80	0.172	30.54 *	0.094
3.54	0.098		

*Solid phase–$2PbBr_2\cdot KBr$.

№ 4996 POTASSIUM BROMATE–POTASSIUM IODIDE–WATER [2390]

$KBrO_3—KI—H_2O$

$t=25$

Comp. of saturated soln., wt %			d	Comp. of solid phase
Potassium Bromate	Potassium Iodide	Water		
7.533	0.00	92.467	1.054	$KBrO_3$
4.54	8.77	86.69	1.103	»
2.96	18.85	78.19	1.182	»
2.17	28.60	69.23	1.278	»
1.63	38.99	59.38	1.402	»
1.21	50.06	48.73	1.565	»
0.99	58.14	40.87	1.707	»
0.96	59.20	39.84	1.729	$KBrO_3+KI$
0.00	59.76	40.24	1.718	KI

№ 4997 [2262]

POTASSIUM BROMATE–SILVER BROMATE–WATER

$KBrO_3—AgBrO_3—H_2O$

$t=24.5$

Concn. in aqueous soln., g/l	
Silver Bromate	Potassium Bromate
1.911	0.0
1.225	1.42
0.536	5.78

№ 4998 POTASSIUM BROMATE–SILVER BROMATE–WATER [480]

$KBrO_3—AgBrO_3—H_2O$

$t=25$

Concn. of Potassium Bromate in aqueous soln., mol/l	Solubility of Silver Bromate in aqueous soln. of Potassium Bromate, mol/l	Concn. of Potassium Bromate in aqueous soln., mol/l	Solubility of Silver Bromate in aqueous soln. of Potassium Bromate, mol/l
0.000	$8.25 \cdot 10^{-3}$	0.030	$2.49 \cdot 10^{-3}$
0.001	$7.32 \cdot 10^{-3}$	0.100	$1.06 \cdot 10^{-3}$
0.003	$5.29 \cdot 10^{-3}$	0.300	$7.03 \cdot 10^{-4}$
0.010	$3.99 \cdot 10^{-3}$		

№ 4999 POTASSIUM BROMATE–BARIUM BROMATE–WATER [1777]

$KBrO_3$–$Ba(BrO_3)_2$–H_2O

$t=25$

Content in aqueous soln., g/l		d	Content in aqueous soln., g/l		d
Potassium Bromate	Barium Bromate		Potassium Bromate	Barium Bromate	
0.00	7.93	1.0038	8.35	3.415	1.0062
4.18	5.216	1.0046	16.7	1.72	1.0109

№ 5000 POTASSIUM MOLYBDATE–POTASSIUM IODATE–WATER [2410]

$KMoO_4$—KIO_3—H_2O

$t=25$

Concn. in aqueous soln., wt %		d	Comp. of solid phase	Concn. in aqueous soln., wt %		d	Comp. of solid phase
Potassium Molybdate	Potassium Iodate			Potassium Molybdate	Potassium Iodate		
63.94	1.15	1.822	$K_2MoO_4+KIO_3$	43.86	1.63	1.484	KIO_3
63.97	1.15	1.823	»	38.10	1.92	1.406	»
63.93	1.20	1.818	»	30.97	2.33	1.331	»
63.95	1.17	1.821	»	21.75	2.92	1.225	»
60.52	1.13	—	KIO_3	13.93	3.69	1.151	»
56.57	1.13	1.683	»	10.00	4.37	1.115	»
51.08	1.30	1.600	»	0.00	8.45	1.071	»

№ 5001 POTASSIUM IODIDE–POTASSIUM IODATE–WATER [2394]

KI—KIO_3—H_2O

$t=25$

Comp. of saturated soln., wt %			d	Comp. of solid phase
Potassium Iodide	Potassium Iodate	Water		
59.76	0.0	40.24	1.718	KI
59.14	1.10	39.76	1.731	»
58.62	2.25	39.13	—	»
58.51	2.35	39.14	1.751	$KI+KIO_3$
57.02	2.35	40.63	1.722	KIO_3
41.10	2.54	56.36	1.451	»
22.38	3.27	74.35	1.227	»
12.04	4.33	83.63	—	»
2.40	7.15	90.45	1.053	»
0.0	8.449	91.551	1.071	»

№ 5002 POTASSIUM IODIDE–POTASSIUM IODATE–WATER [1009]

$KI—KIO_3—H_2O$

Comp. of saturated soln., wt %			t	Comp. of solid phase
Potassium Iodide	Potassium Iodate	Water		
0.0	4.5	95.5	0	KIO_3
2.0	4.1	93.9	0	»
5.0	3.55	91.45	0	»
10.0	2.50	87.50	0	»
20.0	2.40	77.60	0	»
30.0	2.35	67.65	0	»
40.0	2.30	57.70	0	»
50.0	2.30	47.70	0	»
52.0	2.30	45.70	0	$KI+KIO_3$
55.7	0.00	44.30	0	KI
0.0	7.90	92.10	21	KIO_3
2.0	7.40	90.60	21	»
5.0	6.55	88.45	21	»
10.0	5.30	84.70	21	»
20.0	3.30	76.70	21	»
30.0	2.70	67.30	21	»
40.0	2.60	57.40	21	»
50.0	2.60	47.40	21	»
58.4	2.60	39.00	21	$KI+KIO_3$
55.7	0.00	44.30	21	KI
0.0	18.50	81.50	75	KIO_3
2.0	16.50	81.50	75	»
5.0	13.90	81.10	75	»
10.0	9.90	80.10	75	»
20.0	7.40	72.60	75	»
30.0	5.85	64.15	75	»
40.0	4.50	55.50	75	»
50.0	3.60	46.40	75	»
63.7	3.20	33.10	75	$KI+KIO_3$
65.3	0.0	34.70	75	KI

№ 5003 [213]

POTASSIUM IODIDE–CALCIUM NITRATE–LEAD (II) IODIDE–WATER

$KI—Ca(NO_3)_2—PbI_2—H_2O$

$t=18—20$

Concn. in aqueous soln., g/l		Solubility of Lead Iodide in aqueous soln. of Potassium Iodide and Calcium Nitrate g/l	Concn. in aqueous soln., g/l		Solubility of Lead Iodide in aqueous soln. of Potassium Iodide and Calcium Nitrate g/l
Potassium Iodide	Calcium Nitrate		Potassium Iodide	Calcium Nitrate	
0.00	8.2	1.23	5.31	8.2	0.186
1.33	8.2	0.751	10.62	8.2	0.224
2.66	8.2	0.460	21.25	8.2	0.259

Table № 5003 (continuation)

Concn. in aqueous soln., g/l		Solubility of Lead Iodide in aqueous soln. of Potassium Iodide and Calcium Nitrate g/l	Concn. in aqueous soln., g/l		Solubility of Lead Iodide in aqueous soln. of Potassium Iodide and Calcium Nitrate g/l
Potassium Iodide	Calcium Nitrate		Potassium Iodide	Calcium Nitrate	
42.50	8.2	0.429	5.31	16.4	0.184
56.44	8.2	0.00	10.62	16.4	0.228
0.00	16.4	1.17	21.25	16.4	0.246
1.33	16.4	0.735	42.50	16.4	0.423
2.66	16.4	0.421	56.44	16.4	0.00

№ 5004 POTASSIUM IODIDE–CALCIUM SULFATE–WATER [1511]

$KI—CaSO_4—H_2O$

$t=21$

Concn. of Potassium Iodide in aqueous soln., g/l	Solubility of Calcium Sulfate in aqueous soln. of Potassium Iodide, g/l	Concn. of Potassium Iodide in aqueous soln., g/l	Solubility of Calcium Sulfate in aqueous soln. of Potassium Iodide, g/l
0	2.05	100	5.1
10	2.8	125	5.45
20	3.2	150	5.8
40	3.9	200	5.95
60	4.5	250	6.00
80	4.85	300	Двойная соль

№ 5005 POTASSIUM IODIDE–COPPER (I) THIOCYANATE–WATER [931]

$KI—CuCNS—H_2O$

$t=25$

Concn. in aqueous soln., mol/l		Concn. in aqueous soln., mol/l	
Potassium Iodide	Copper Thiocyanate	Potassium Iodide	Copper Thiocyanate
1.020	0.0094	4.044	0.140
1.065	0.0090	5.00	0.2110
2.036	0.0359	5.10	0.218
3.060	0.0661	6.012	0.399

№ 5006 POTASSIUM IODIDE–COPPER (I) IODIDE–WATER [931]

$KI—CuI—H_2O$

$t=25$

Concn. in aqueous soln., mol/l		Concn. in aqueous soln., mol/l	
Potassium Iodide	Copper Iodide	Potassium Iodide	Copper Iodide
1.014	0.0104	3.959	0.366
2.00	0.061	5.878	1.218
3.050	0.167		

№ 5007 POTASSIUM IODIDE–ARSENIC TRIOXIDE–WATER [2519]

$KI—As_2O_3—H_2O$

$t=30$

Comp. of saturated soln., wt %			Comp. of solid phase
Arsenic Trioxide	Potassium Iodide	Water	
2.26	0.00	97.74	As_2O_3
0.772	1.19	98.038	$(As_2O_3)_2 \cdot KI$
0.296	9.56	90.144	»
0.183	22.89	76.927	»
0.150	34.31	65.540	»
0.119	40.79	59.091	»
0.081	47.07	52.849	»
0.115	53.51	46.375	»
0.134	60.54	39.326	$(As_2O_3)KI + KI$
0.0	61.5	38.5	KI

№ 5008 [1780]

POTASSIUM IODIDE–STRONTIUM CHLORIDE–WATER

$KI—SrCl_2—H_2O$

$t=25$

Comp. of solvent, wt %		Solubility of Strontium Chloride, wt %	d_4
Potassium Iodide	Water		
1.5	98.50	35.45	1.4093
8.23	91.77	33.79	1.4466
9.08	90.92	33.60	1.4513
19.34	80.66	30.90	1.5154

№ 5009 POTASSIUM IODIDE–PALLADIUM IODIDE–WATER [855]

$KI—PdI_2—H_2O$

$t=18$

Concn. in aqueous soln., g/l		Concn. in aqueous soln., g/l		Concn. in aqueous soln., g/l	
Palladium Iodide	Potassium Iodide	Palladium Iodide	Potassium Iodide	Palladium Iodide	Potassium Iodide
0.00505	16.60	0.02307	34.86	0.03966	46.48
0.00901	19.92	0.02740	37.18	0.7715	59.76
0.01658	26.56	0.02884	37.85	0.09012	73.04
0.01875	29.88	0.03172	38.51	0.1694	99.6
0.02019	33.20	0.03389	39.18	0.4686	166.0

№ 5010 POTASSIUM IODIDE–SILVER IODIDE–WATER [2486]

$KI—AgI—H_2O$

$t=15$

Concn. in aqueous soln., wt %		Concn. in aqueous soln., wt %		Concn. in aqueous soln., wt %	
Potassium Iodide	Silver Iodide	Potassium Iodide	Silver Iodide	Potassium Iodide	Silver Iodide
20.0	0.80	33.3	7.33	50.0	25.0
21.74	1.576	40.0	13.0	57.15	40.0
25.0	2.75				

№ 5011 POTASSIUM IODIDE–SILVER IODIDE–WATER [1795]

$KI—AgI—H_2O$

$t=25$

Concn. in aqueous soln., g/l		Concn. in aqueous soln., g/l		Concn. in aqueous soln., g/l	
Potassium Iodide	Silver Iodide	Potassium Iodide	Silver Iodide	Potassium Iodide	Silver Iodide
55.6	0.0853	167.3	3.32	246.7	15.46
97.3	0.512	169.0	3.47	270.6	24.01
121.8	1.032	233.4	12.55	321.5	46.42

№ 5012 POTASSIUM IODIDE–SILVER IODIDE–WATER [2684]

$KI—AgI—H_2O$

Comp. of saturated soln., wt %			t	Comp. of solid phase
Potassium Iodide	Silver Iodide	Water		
56.1	0.0	43.9	0	KI
53.0	9.0	38.0	0	»
51.2	18.0	30.8	0	»
46.6	31.3	22.1	0	»
44.0	37.9	18.1	0	KI+AgI · KI
42.7	37.6	19.7	0	AgI · KI
41.3	38.0	20.7	0	AgI · KI · AgI
36.4	28.1	35.5	0	AgI
34.6	26.6	38.8	0	»
26.1	6.5	67.4	0	»
20.5	1.5	78.0	0	»
9.8	0.2	90.0	0	»
48.7	27.5	23.8	0	AgI · 2KI+KI
50.3	21.0	28.7	0	AgI · 2KI
60.35	0.00	39.65	30	KI
55.5	16.0	28.5	30	»
46.9	35.8	17.3	30	»

Table № 5012 (continuation)

Comp. of saturated soln., wt %			t	Comp. of solid phase
Potassium Iodide	Silver Iodide	Water		
43.9	42.8	13.3	30	KI
43.2	44.1	12.7	30	KI+AgI · KI
40.9	47.7	11.4	30	AgI · KI
38.6	49.7	11.7	30	AgI · KI+AgI
38.8	42.8	18.4	30	AgI
37.6	29.4	33.0	30	»
31.4	10.0	58.6	30	»
10.2	0.1	89.7	30	»
62.6	0.0	37.4	50	KI
59.1	10.7	30.2	50	»
55.5	22.8	21.7	50	»
43.2	45.0	11.8	50	»
37.6	53.4	9.0	50	KI +AgI · KI
37.1	53.5	9.4	50	AgI · KI
36.6	53.5	9.9	50	AgI · KI +AgI
36.5	53.5	10.0	50	AgI
38.1	39.0	23.9	50	»
36.7	28.0	35.3	50	»
33.8	10.0	50.2	50	»
24.8	2.5	72.7	50	»

№ 5013 POTASSIUM IODIDE–SILVER IODIDE–WATER [2537]

$KI—AgI—H_2O$

Concn. in aqueous soln., g/l		t	Concn. in aqueous soln., g/l		t
Silver Iodide	Potassium Iodide		Silver Iodide	Potassium Iodide	
9.392	204.18	20	0.004696	13.45	20
4.696	179.28	20	0.002348	8.96	20
2.348	153.55	20	0.001409	6.64	20
1.409	129.48	20	0.000939	5.40	20
0.9392	116.20	20	0.000470	3.82	20
0.4696	95.45	20	0.000235	2.49	20
0.2348	74.70	20	0.02348	23.2	40
0.1409	62.25	20	0.2348	71.38	40
0.09392	53.95	20	2.348	151.06	40
0.04696	39.01	20	0.02348	20.75	60
0.02348	28.22	20	0.2348	61.42	60
0.01409	22.41	20	2.348	149.4	60
0.009392	18.26	20			

$KI—CdI_2—H_2O$

Comp. of saturated soln., wt %			t	d_4^t	Comp. of solid phase
Cadmium Iodide	Potassium Iodide	Water			
33.0	40.3	26.7	−26.8	—	Ice
39.3	35.7	25.0	−23.2	—	»
29.2	35.8	35.0	−17.6	—	»
41.0	27.9	31.1	−16.7	—	»
36.2	24.6	39.2	−11.4	—	»
24.2	29.5	46.3	−10.9	—	»
29.8	20.3	49.9	− 7.1	—	»
21.0	19.1	59.9	− 5.0	—	»
15.8	19.2	65.0	− 4.8	—	»
18.5	12.6	68.9	− 2.5	—	»
33.5	41.25	25.25	−26.7	—	KI
41.95	36.15	21.90	−24.4	—	$2CdI_2 \cdot 3KI \cdot 4H_2O$
39.2	40.05	20.75	−20.7	—	$CdI_2 \cdot 2KI \cdot 3H_2O$
42.25	35.2	22.55	−20.1	—	$CdI_2 \cdot KI \cdot H_2O$
42.15	28.9	28.95	−14.4	—	$CdI_2 \cdot 4H_2O$
40.5	18.0	41.5	− 8.3	—	CdI_2
44.75	36.45	18.8	0.0	—	$CdI_2 \cdot KI \cdot H_2O$
42.3	15.8	41.9	20	1.810	CdI_2
45.1	31.95	22.95	20	2.353	»
46.95	38.2	14.85	20	2.719	$CdI_2 \cdot KI \cdot H_2O$
41.05	43.0	15.95	20	2.624	KI
26.65	47.25	26.1	20	2.164	»
18.1	50.8	31.1	20	1.986	»
52.05	41.2	6.75	82	3.168	$CdI_2 \cdot 2KI \cdot ½H_2O$
54.75	39.25	6.0	95	—	$CdI_2 \cdot KI \cdot H_2O$
49.2	16.5	34.3	100	1.927	CdI_2
50.4	28.5	21.1	100	2.335	»
55.5	35.8	8.7	100	—	»
56.2	37.4	6.4	100	3.191	$CdI_2 \cdot KI \cdot H_2O$
56.0	38.25	5.75	100	3.235	»
54.95	39.75	5.3	100	—	$CdI_2 \cdot 2KI \cdot ½H_2O$
53.1	41.1	5.8	100	—	»
34.8	47.7	17.5	100	2.417	KI
14.35	58.8	26.85	100	2.009	»
42.4	29.75	27.85	−19.3	—	Ice $+CdI_2 \cdot 4H_2O+CdI_2 \cdot KI \cdot H_2O$
42.6	30.6	26.8	−14.3	—	$CdI_2 \cdot 4H_2O+CdI_2 \cdot KI \cdot H_2O$
43.2	31.25	25.55	−11.7	—	»
43.35	31.4	25.25	−10.9	—	$CdI_2 \cdot 4H_2O+CdI_2+CdI_2 \cdot KI \cdot H_2O$
43.5	31.65	24.85	− 9.85	—	$CdI_2+CdI_2 \cdot KI \cdot H_2O$
44.05	32.45	23.5	− 5.1	—	»
44.7	33.4	21.9	0	2.420	»
47.1	35.0	17.9	20	2.575	»
49.4	35.7	14.9	40	2.706	»
51.4	36.25	12.35	60	2.836	»
53.9	36.5	9.6	80	2.982	»
56.4	36.6	17.0	100	—	»
41.7	35.25	23.05	−25.3	—	Ice $+CdI_2 \cdot KI \cdot H_2O+2CdI_2 \cdot 3KI \cdot 4H_2O$
41.9	35.4	22.7	−24.1	—	$CdI_2 \cdot KI \cdot H_2O+2CdI_2 \cdot 3KI \cdot 4H_2O$

Comp. of saturated soln., wt %			t	d_4^t	Comp. of solid phase
Cadmium Iodide	Potassium Iodide	Water			
43.7	37.0	19.3	−11.8	—	$CdI_2 \cdot KI \cdot H_2O + 2CdI_2 \cdot 3KI \cdot 4H_2O$
45.15	38.2	16.65	0	2.644	»
41.55	36.6	21.85	−27.7	—	Ice $+ 2CdI_2 \cdot 3KI \cdot 4H_2O + CdI_2 \cdot 2KI \cdot 3H_2O$
41.75	36.7	21.55	−27.0	—	$2CdI_2 \cdot 3KI \cdot 4H_2O + CdI_2 \cdot KI \cdot H_2O$
41.8	36.95	21.25	−26.6	—	»
42.0	37.2	20.8	−24.4	—	»
43.0	38.0	19.0	−18.6	—	»
43.3	38.35	18.35	−15.8	—	»
43.3	38.45	18.25	−14.2	—	$2CdI_2 \cdot 3KI \cdot 4H_2O + CdI_2 \cdot 2KI \cdot 2H_2O$
45.05	38.9	16.05	0	2.672	»
46.25	39.15	14.6	10	2.738	»
46.3	39.15	14.55	10	2.740	$CdI_2 \cdot KI \cdot H_2O + 2CdI_2 \cdot 3KI \cdot 4H_2O + CdI_2 \cdot KI \cdot 2H_2O$
47.15	39.45	13.4	20	2.793	$CdI_2 \cdot KI \cdot H_2O + CdI_2 \cdot 2KI \cdot 2H_2O$
48.7	40.05	11.25	40	2.899	»
50.4	40.5	9.1	60	3.022	»
51.9	41.0	7.1	75	3.153	»
52.1	41.05	6.85	76.3	3.165	$CdI_2 \cdot KI \cdot H_2O + CdI_2 \cdot KI \cdot 2H_2O + CdI_2 \cdot 2KI \cdot \frac{1}{2}H_2O$
52.4	40.9	6.7	78	3.172	$CdI_2 \cdot KI \cdot H_2O + CdI_2 \cdot 2KI \cdot \frac{1}{2}H_2O$
53.9	40.15	5.95	90	3.235	»
55 75	39.2	5.05	100	3.307	»
33.15	40.8	24.05	−30.2	—	Ice $+ CdI_2 \cdot 2KI \cdot 3H_2O + KI$
35.7	41.0	23.3	−28.0	—	$CdI_2 \cdot 2KI \cdot 3H_2O + KI$
36.85	41.25	21.9	−23.4	—	»
37.95	41.5	20.55	−18.9	—	»
38.35	41.6	20.05	−17.3	—	$CdI_2 \cdot 2KI \cdot 3H_2O + CdI_2 \cdot 2KI \cdot 2H_2O + KI$
37.4	41.25	21.35	−24.3	—	$CdI_2 \cdot 2KI \cdot 2H_2O + KI$
38.7	41.8	19.5	−13.8	—	»
40.4	42.3	7.3	0	2.583	»
42.5	42.8	14.7	20	2.693	»
44.35	43.2	12.45	40	2.795	»
46.2	43.5	10.3	60	2.907	»
48.6	43.6	7.8	80	—	»
48.9	43.65	7.45	81.4	—	$CdI_2 \cdot 2KI \cdot 2H_2O + CdI_2 \cdot 2KI \cdot \frac{1}{2}H_2O + KI$
48.95	43.65	7.4	83	—	$CdI_2 \cdot 2KI \cdot \frac{1}{2}H_2O + KI$
49.65	43.7	6.65	100	3.120	»

№ 5015 POTASSIUM IODIDE–ANTIMONY (III) IODIDE–WATER [1658, 1660]

$KI—SbI_3—H_2O$

Comp. of saturated soln., wt %			t	Comp. of solid phase
Potassium Iodide	Antimony Iodide	Water		
48.75	26.25	25.00	16	$KI + 2KI \cdot SbI_3 \cdot H_2O$
41.55	32.46	25.99	16	$SbI_3 \cdot 2KI \cdot SbI_3 \cdot H_2O$
48.88	28.88	22.24	32	$KI + 2KI \cdot SbI_3 \cdot H_2O$

Table № 5015 (continuation)

Comp. of saturated soln., wt %			t	Comp. of solid phase
Potassium Iodide	Antimony Iodide	Water		
43.95	34.06	21.99	32	$SbI_3+2KI \cdot SbI_3 \cdot H_2O$
49.0	31.0	20.0	48	$KI+2KI \cdot SbI_3 \cdot H_2O$
42.08	38.61	19.31	48	$SbI_3+2KI \cdot SbI_3 \cdot H_2O$

№ 5016 POTASSIUM IODIDE–MERCURY (II) IODIDE–WATER [2563]

$KI—HgI_2—H_2O$

$t=25$

Concn. in aqueous soln., g/l		Concn. in aqueous soln., g/l	
Potassium Iodide	Mercury Iodide	Potassium Iodide	Mercury Iodide
8.3	11.2	166	227.2
16.6	22.7	249	340.8
33.2	45.4	332	454.5
83	113.6	415	578

№ 5017 POTASSIUM IODIDE–MERCURY (II) IODIDE–WATER [1538]

$KI—HgI_2—H_2O$

Comp. of saturated soln., wt %			t	Comp. of solid phase
Mercury Iodide	Potassium Iodide	Water		
19.3	50.9	29.8	20	KI
32.4	44.4	23.2	20	»
48.0	39.0	13.0	20	»
53.6	37.4	9.0	20	$KI+KHgI_3$
52.6	37.8	9.6	20	$KHgI_3$
52.2	35.1	12.7	20	»
51.2	35.5	13.3	20	$KHgI_3 \cdot H_2O$
50.3	26.7	23.0	20	$KHgI_3 \cdot H_2O+HgI_2$
49.4	26.6	24.0	20	HgI_2
40.2	23.7	36.1	20	»
22.5	14.9	62.6	20	»
0.0	60.6	39.4	30	KI
53.0	40.0	7.0	30	$KI+KHgI_3$
52.7	39.6	7.7	30	$KHgI_3$
52.2	40.0	7.8	30	»
51.2	40.2	8.6	30	»
50.3	39.3	10.4	30	»
49.8	33.7	16.5	30	»
52	33	15	30	»
51.7	31.4	16.9	30	$KHgI_3 \cdot H_2O$
52.2	29.1	18.7	30	»

№ 5018 POTASSIUM IODIDE–MERCURY (II) IODIDE–WATER [2242]

$KI—HgI_2—H_2O$

t=22.5

Comp. of saturated soln., wt %			Comp. of solid phase
Mercury Iodide	Potassium Iodide	Water	
0.01	0.0	99.99	HgI_2
2.02	1.95	96.03	»
10.96	16.16	72.88	»
45.06	24.60	30.34	»
46.93	24.57	28.50	»
49.13	25.10	25.77	$HgI_2+HgI_2 \cdot KI \cdot 1\frac{1}{2}H_2O$
49.06	27.58	23.36	$HgI_2 \cdot KI \cdot 1\frac{1}{2}H_2O$
48.97	28.44	22.59	»
49.18	29.18	21.64	»
49.28	31.18	19.54	»
49.15	32.68	18.17	»
49.67	33.37	16.96	»
50.10	33.92	15.98	»
51.13	33.79	15.08	»
51.74	34.04	14.22	$HgI_2 \cdot KI$
51.58	34.92	13.50	»
51.53	34.84	13.63	»
51.28	36.45	12.27	KI
50.01	36.69	13.30	»
47.04	37.61	15.35	»
37.41	40.41	22.18	»
18.37	49.60	32.03	»
1.60	58.33	40.07	»
0.0	59.20	40.80	»

№ 5019 POTASSIUM IODIDE–MERCURY (II) IODIDE–WATER [2323]

$KI—HgI_2—H_2O$

Comp. of saturated soln., wt %			t	Comp. of solid phase
Mercury Iodide	Potassium Iodide	Water		
19.1	13.3	67.6	0	HgI_2
32.7	21.1	46.2	0	»
39.9	25.0	35.1	0	»
41.8	26.7	31.5	0	$HgI_2 \cdot KI \cdot H_2O$
43.3	29.7	27.0	0	»
45.5	32.1	22.4	0	»
47.8	34.8	17.4	0	»
50.2	37.5	12.3	0	»
47.5	38.1	14.4	0	KI
43.7	38.9	17.4	0	»
39.2	40.0	20.8	0	»
34.0	41.9	24.1	0	»
28.4	43.9	27.7	0	»
24.1	45.6	30.3	0	»

Table № 5019 (continuation)

Comp. of saturated soln., wt %			t	Comp. of solid phase
Mercury Iodide	Potassium Iodide	Water		
19.3	47.4	33.3	0	KI
15.6	49.2	35.2	0	»
19.6	12.0	68.4	10	HgI_2
34.9	22.0	43.1	10	»
41.7	25.5	32.8	10	»
44.8	27.0	28.2	10	$HgI_2 \cdot KI \cdot H_2O_2$
46.6	30.7	22.7	10	»
48.1	33.7	18.2	10	»
49.3	35.7	15.0	10	»
50.8	37.5	11.7	10	»
50.9	37.9	11.2	10	$HgI_2 \cdot KI \cdot H_2O + KI$
48.1	38.3	13.6	10	KI
46.5	38.6	14.9	10	»
44.7	39.0	16.3	10	»
38.2	41.2	20.6	10	»
35.7	42.3	22.0	10	»
10.3	7.4	82.3	20	HgI_2
24.3	16.2	59.5	20	»
34.7	21.3	44.0	20	»
41.3	24.6	34.1	20	»
48.2	27.3	24.5	20	$HgI_2 + HgI_2 \cdot KI \cdot H_3O$
48.0	29.0	23.0	20	$HgI_2 \cdot KI \cdot H_2O$
48.7	31.3	20.0	20	»
49.2	33.4	17.4	20	»
49.9	34.9	15.2	20	»
50.9	36.8	12.3	20	»
50.0	37.9	12.1	20	KI
45.3	39.1	15.6	20	»
31.0	44.4	24.6	20	»
16.7	50.6	22.7	20	»
8.7	53.8	37.5	20	»
3.1	2.2	94.7	34	HgI_2
20.3	12.9	66.8	34	»
47.4	26.2	26.4	34	»
53.4	27.7	18.9	34	$HgI_2 + HgI_2 \cdot KI \cdot H_2O$
52.4	29.3	18.3	34	$HgI_2 \cdot KI \cdot H_2O$
51.9	30.7	17.4	34	»
51.7	34.4	13.9	34	»
52.5	37.6	9.9	34	$HgI_2 \cdot KI \cdot H_2O + KI$
49.8	38.3	11.9	34	KI
46.1	39.3	14.6	34	»
42.3	40.5	17.2	34	»
33.7	44.2	22.1	34	»
15.6	52.9	31.5	34	»
4.1	3.2	92.7	56	HgI_2
27.4	16.6	56.0	56	»
46.1	25.0	28.9	56	»
57.1	27.5	15.4	56	»
57.4	28.8	13.8	56	$HgI_2 \cdot KI \cdot H_2O$
56.4	30.2	13.4	56	»
54.9	32.8	12.3	56	»

Comp. of saturated soln., wt %			t	Comp. of solid phase
Mercury Iodide	Potassium Iodide	Water		
54.2	37.5	8.3	56	$HgI_2+KI\cdot H_2O+KI$
53.0	37.9	9.1	56	»
46.6	39.8	13.6	56	»
41.2	42.0	16.8	56	»
24.7	49.4	25.9	56	»
12.8	56.1	31.1	56	»
3.9	61.4	34.7	56	»
1.2	0.6	98.2	80	HgI_2
19.1	11.3	69.6	80	»
40.8	20.9	38.3	80	»
54.3	24.6	21.1	80	»
65.0	26.2	8.8	80	$HgI_2+HgI_2\cdot KI\cdot H_2O$
63.2	27.6	9.2	80	$HgI_2\cdot KI\cdot H_2O$
61.1	29.4	9.5	80	»
58.2	33.0	8.8	80	»
57.5	36.2	6.3	80	$HgI_2\cdot KI\cdot H_2O+KI$
56.7	36.6	6.7	80	KI
46.5	40.4	13.1	80	»
28.9	48.3	22.8	80	»
15.5	56.2	28.3	80	»
8.0	60.8	31.2	80	»

№ 5020 POTASSIUM IODIDE–THALLIUM IODIDE–WATER [431]

$KI-TlI-H_2O$

Content in aqueous soln.		t	Content in aqueous soln.		t
Potassium Iodide, mol/l	Thallium Iodide, [mol/l]·10[5]		Potassium Iodide, mol/l	Thallium Iodide, [mol/l]·10[5]	
1.48	1.49	20	3.66	16.38	30
1.95	2.15	20	3.88	18.85	30
2.41	3.39	20	4.33	25.74	30
2.83	5.12	20	4.74	34.86	30
3.21	7.35	20	5.11	46.28	30
3.56	9.85	20	5.48	57.08	30
3.84	12.84	20	0.49	1.46	40
4.04	15.91	20	0.97	2.22	40
4.38	17.96	20	1.49	3.05	40
4.76	23.65	20	1.89	4.9	40
5.23	33.21	20	2.31	6.95	40
5.60	44.28	20	2.67	9.90	40
0.49	0.82	30	2.99	13.18	40
0.98	1.76	30	3.24	16.88	40
1.45	2.78	30	3.50	19.90	40
1.92	3.81	30	3.85	25.11	40
2.34	5.76	30	4.27	34.02	40
2.75	7.48	30	4.68	44.65	40
3.11	10.05	30	5.01	59.5	40
3.43	13.02	30	5.37	73.42	40

Table № 5020 (continuation)

Content in aqueous soln.			Content in aqueous soln.		
Potassium Iodide, mol/l	Thallium Iodide, [mol/l]·10^5	t	Potassium Iodide, mol/l	Thallium Iodide, [mol/l]·10^5	t
0.48	2.11	50	3.33	34.32	60
0.96	3.29	50	3.73	45.72	60
1.41	4.99	50	4.10	57.81	60
1.84	7.29	50	4.44	73.84	60
2.23	10.08	50	4.74	91.97	60
2.58	13.50	50	5.03	108.4	60
3.05	21.05	50	0.46	6.20	70
3.36	24.86	50	0.90	8.74	70
3.81	33.68	50	1.31	11.39	70
4.20	44.65	50	1.66	14.8	70
4.58	57.97	50	1.95	18.95	70
4.89	73.80	50	2.40	26.40	70
5.10	88.57	50	2.85	33.10	70
0.48	3.46	60	3.25	47.55	70
0.94	5.07	60	3.76	62.32	70
1.38	7.19	60	3.99	77.29	70
1.77	10.32	60	4.30	92.5	70
2.12	13.74	60	4.57	112.6	70
2.44	17.36	60	4.80	134.9	70
2.89	25.02	60			

№ 5021 POTASSIUM IODIDE–THALLIUM IODIDE–WATER [430]

$KI—TlI_3—H_2O$

$t=25$

Comp. of Potassium Iodide in aqueous soln., mol/l	Solubility of Thallium Iodide, (mol/l)·10^5	Comp. of Potassium Iodide in aqueous soln., mol/l	Solubility of Thallium Iodide, (mol/l)·10^5
0.010	0.90	2.420	4.66
0.020	0.58	2.697	6.06
0.030	0.46	2.862	6.55
0.044	0.38	3.256	9.22
0.065	0.34	3.554	13.58
0.128	0.32	3.860	17.34
0.248	0.33	4.272	22.47
0.500	0.49	4.302	23.55
0.747	0.77	4.760	28.98
1.013	1.28	5.140	39.52
1.425	2.20	5.500	49.87
1.948	3.68		

№ 5022 POTASSIUM IODIDE–LEAD (II) IODIDE–WATER [2508]

$KI—PbI_2—H_2O$

Comp. of saturated soln., wt %			t	Comp. of solid phase
Lead Iodide	Potassium Iodide	Water		
0.00	14.01	85.99	5	Double salt $+PbI_2$
0.71	20.48	78.81	20	»
1.85	24.07	74.08	28	»
3.01	30.05	66.94	39	»
12.71	37.43	49.86	67	»
25.10	40.72	34.18	80	»
22.39	38.99	38.62	80	»
35.88	38.68	25.44	104.5	»
43.34	36.51	20.15	120	»
48.13	35.11	16.76	137	»
58.67	30.85	10.48	175	»
0.0	76.95	23.05	189	»
3.94	55.21	40.85	9	Double salt $+KI$
4.58	55.39	40.03	13	»
6.91	56.00	34.09	23	»
15.34	55.52	29.14	50	»
19.98	54.70	25.32	64	»
24.40	53.58	22.02	83.5	»
26.30	53.02	20.68	92	»
35.76	49.21	15.03	137	»
44.47	44.90	10.63	165	»
91.15	0.0	8.85	218	»
58.72	36.68	4.60	241	»
92.72	0.0	7.28	242	»
93.85	0.0	6.15	250	»
54.60	39.30	6.10	157	$PbI_2 \cdot 2KI \cdot 2\frac{1}{2}H_2O$
50.34	36.30	13.36	172	»
51.56	37.20	11.24	186	»
52.45	37.80	9.75	194	»
53.38	38.50	8.12	201	»

№ 5023 POTASSIUM IODIDE–LEAD (II) IODIDE–WATER [1489]

$KI—PbI_2—H_2O$

Comp. of saturated soln., wt %			t	Comp. of solid phase
Lead Iodide	Potassium Iodide	Water		
0.05	0.00	99.95	13	PbI_2
0.001	2.21	97.789	13	»
0.001	4.18	95.819	13	»
0.002	9.09	90.908	13	»
0.05	16.02	83.93	13	$PbI_2+PbI_2 \cdot KI \cdot 2H_2O$
0.02	17.26	82.72	13	$PbI_2 \cdot KI \cdot 2H_2O$
0.19	21.32	78.49	13	»
0.68	34.45	64.87	13	»
0.77	37.25	61.98	13	»
1.47	45.17	53.36	13	»

Table № 5023 (continuation)

Comp. of saturated soln., wt %			t	Comp. of solid phase
Lead Iodide	Potassium Iodide	Water		
2.91	52.14	44.85	13	$PbI_2 \cdot KI \cdot 2H_2O$
4.54	56.55	38.91	13	$PbI_2 \cdot KI \cdot 2H_2O + KI$
0.00	57.66	42.34	13	KI
0.002	2.84	97.158	50	PbI_2
0.02	9.33	90.65	50	»
0.02	14.48	85.50	50	»
0.12	17.36	82.52	50	»
0.14	18.77	81.09	50	»
0.15	19.50	80.35	50	»
0.27	21.00	79.73	50	»
1.74	29.43	68.83	50	»
1.82	29.51	68.67	50	»
2.45	30.47	67.08	50	»
2.52	31.77	65.71	50	»
3.11	32.12	64.77	50	$PbI_2 + PbI_2 \cdot KI \cdot 2H_2O$
3.10	36.18	60.72	50	$PbI_2 \cdot KI \cdot 2H_2O$
3.99	38.44	58.57	50	»
4.23	42.28	53.49	50	»
5.64	51.93	52.53	50	»
9.59	53.96	36.45	50	»
11.74	56.41	31.85	50	»
14.14	58.49	27.37	50	$PbI_2 \cdot KI \cdot 2H_2O + KI$
9.72	58.49	31.79	50	KI
5.55	60.28	34.37	50	»
1.52	61.98	36.50	50	»
1.48	62.02	36.50	50	»
0.00	62.39	37.61	50	»

№ 5024 POTASSIUM IODIDE–LEAD (II) IODIDE–WATER [1300]

$KI-PbI_2-H_2O$

$t=25$

Concn. in aqueous soln., wt %		Comp. of solid phase
Potassium Iodide	Lead Iodide	
0.083	0.0234	PbI_2
0.166	0.0092	»
0.332	0.0040	»
0.664	0.0020	»
0.830	0.0019	»
1.661	0.0016	»
3.320	0.0020	»
8.307	0.0073	»
13.61	0.0316	»
14.90	0.0578	»
19.45	0.253	»
21.32	0.428	$PbI_2 + PbI_2 \cdot KI \cdot 2H_2O$

Table № 5024 (continuation)

Concn. in aqueous soln., wt %		Comp. of solid phase
Potassium Iodide	Lead Iodide	
26.82	0.553	$PbI_2 \cdot KI \cdot 2H_2O$
36.73	1.221	»
43.80	1.249	»
49.43	3.712	»
51.61	4.654	»
54.60	5.876	»
56.59	7.421	$PbI_2 \cdot KI \cdot 2H_2O + KI$
57.82	5.235	KI
59.13	5.49	»
59.72	0.0	»

№ 5025 POTASSIUM IODIDE–LEAD (II) IODIDE–WATER [2687]

$KI—PbI_2—H_2O$

Comp. of saturated soln., wt %			t	Comp. of solid phase
Potassium Iodide	Lead Iodide	Water		
56.07	0.00	43.93	0	KI
55.65	1.23	43.12	0	»
55.28	2.00	42.72	0	»
55.10	3.10	41.80	0	$KI + KPbI_3 \cdot 2H_2O$
53.10	2.41	44.49	0	$KPbI_2 \cdot 2H_2O$
46.20	1.30	52.50	0	»
40.56	0.724	58.716	0	»
35.20	0.445	64.355	0	»
29.72	0.263	70.017	0	»
25.56	0.146	74.294	0	»
19.60	0.074	80.326	0	»
15.26	0.044	85.696	0	»
11.42	0.037	88.543	0	$KPbI_3 \cdot 2H_2O + PbI_2$
10.35	0.030	89.62	0	PbI_2
7.26	0.012	92.728	0	»
4.55	0.002	95.448	0	»
2.21	0.001	97.789	0	»
0.166	0.003	99.831	0	»
0.0	0.043	99.957	0	»
59.77	0.00	40.23	25	KI
58.53	2.267	39.403	25	»
58.40	3.663	37.937	25	»
57.75	5.22	37.03	25	»
56.60	7.30	36.10	25	$KI + KPbI_3 \cdot 2H_2O$
55.75	5.901	38.349	25	$KPbI_3 \cdot 2H_2O$
51.42	4.491	44.089	25	»
43.48	2.348	54.172	25	»
38.38	1.537	60.083	25	»
30.77	0.801	68.429	25	»
21.29	0.422	78.288	25	$KPbI_3 \cdot 2H_2O + PbI_2$
21.09	0.410	78.50	25	PbI_2

Comp. of saturated soln., wt %			t	Comp. of solid phase
Potassium Iodide	Lead Iodide	Water		
21.03	0.408	78.562	25	PbI_2
19.58	0.256	80.164	25	»
12.52	0.033	87.447	25	»
9.97	0.020	90.01	25	»
0.645	0.001	99.354	25	»
0.159	0.010	99.831	25	»
0.0	0.075	99.925	25	»

№ 5026 POTASSIUM IODIDE–LEAD (II) IODIDE–WATER [213]

$KI—PbI_2—H_2O$

$t=18—20$

Content of Potassium Iodide in aqueous soln., g-eq/l	Solubility of Lead Iodide in aqueous soln. of Potassium Iodide, mol/l	Content of Potassium Iodide in aqueous soln., g-eq/l	Solubility of Lead Iodide in aqueous soln. of Potassium Iodide, mol/l
0.0	1.442	0.064	0.286
0.008	0.347	0.128	0.307
0.016	0.154	0.256	0.413
0.032	0.206	0.340	0.473

№ 5027 [2043]

POTASSIUM IODIDE–LEAD (II) IODIDE–WATER

$KI—PbI_2—H_2O$

$t=25$

Concn. in aqueous soln., g/l	
Lead Iodide	Potassium Iodide
0.7565	0.00
0.1296	1.66
0.0301	4.98
0.0212	9.96
0.0195	1.328
0.0185	16.6
0.0187	20.75
0.0204	24.90
0.0226	29.05
0.0253	33.20
0.0276	37.35
0.0311	41.5
0.0337	45.65
0.0377	49.80
0.0424	53.95
0.0472	58.10
0.0614	66.40

№ 5028 [891]

POTASSIUM IODIDE–LEAD (II) IODIDE–WATER

$KI—PbI_2—H_2O$

$t=18—20$

Content of Potassium Iodide in aqueous soln., mol/l	Solubility of Lead Iodide in aqueous soln. of Potassium Iodide, $(mol/l)\cdot 10^5$
0.010	28.10
0.030	6.53
0.060	4.60
0.080	4.23
0.100	4.01
0.125	4.06
0.150	4.42
0.175	4.90
0.200	5.49
0.225	5.98
0.250	6.75
0.275	7.30
0.300	8.17
0.325	9.20
0.350	10.23

№ 5029 [1005]

POTASSIUM IODATE–SILVER IODATE–WATER

$KIO_3—AgIO_3—H_2O$

$t=20$

Concn. in aqueous soln., g/l	
Potassium Iodate	Silver Iodate
1.070	0.01697
5.350	0.1414
10.70	0.2347

№ 5030 [1778]

POTASSIUM IODATE–BARIUM IODATE–WATER

$KIO_3—Ba(IO_3)_2—H_2O$

$t=25$

Content in aqueous soln., g/l	
Potassium Iodate	Barium Iodate
0.0227	0.368
0.1134	0.303
0.2270	0.229

№ 5031 [1779]

POTASSIUM IODATE–LANTHANUM IODATE–WATER

$KIO_3—La(IO_3)_3—H_2O$

$t=25$

Concn. in aqueous soln., g/l		d_4
Lanthanum Iodate	Potassium Iodate	
0.6290	0.0212	1.00030
0.5634	0.1061	1.00027
0.4970	0.2122	1.00030
0.3738	0.4243	1.00031

№ 5032 [1778]

POTASSIUM IODATE–LEAD (II) IODATE–WATER

$KIO_3—Pb(IO_3)_2—H_2O$

$t=25$

Concn. in aqueous soln., g/l	
Lead Iodate	Potassium Iodate
0.0199	0.0113
0.0122	0.0227

№ 5033 [1764]

POTASSIUM IRON (II) SULFATE–POTASSIUM ZINC SULFATE–WATER

$FeK_2(SO_4)_2 \cdot 6H_2O—ZnK_2(SO_4)_2 \cdot 6H_2O—H_2O$

$t=6.8$

Concn. in aqueous soln., g/l		d	Concn. in aqueous soln., g/l		d
Potassium Iron Sulfate	Potassium Zinc Sulfate		Potassium Iron Sulfate	Potassium Zinc Sulfate	
329.15	0.0	1.2067	142.76	63.87	1.1339
275.03	19.39	1.1859	119.46	67.83	1.1204
246.80	28.82	1.1758	87.69	72.46	1.1085
203.22	47.62	1.1606	68.10	77.78	1.0993
173.59	53.07	1.1474	0.0	113.62	1.0795

№ 5034 [2335]

POTASSIUM COBALT (II) SULFATE–POTASSIUM COPPER (II) SULFATE–WATER

$CoK_2(SO_4)_2 \cdot 6H_2O—CuK_2(SO_4)_2 \cdot 6H_2O—H_2O$

Concn. in aqueous soln., g/l		*t*	*d*	Concn. in aqueous soln., g/l		*t*	*d*
Potassium Cobalt Sulfate	Potassium Copper Sulfate			Potassium Cobalt Sulfate	Potassium Copper Sulfate		
106.81	0.0	0	1.0719	124.47	5.90	7.8	1.0852
87.06	11.40	0	1.0673	110.20	14.79	7.8	1.0825
65.31	25.43	0	1.0612	94.53	25.30	7.8	1.0796
56.68	31.19	0	1.0601	82.36	32.18	7.8	1.0760
44.15	39.48	0	1.0566	73.90	37.08	7.8	1.0730
33.30	46.32	0	1.0538	58.54	46.92	7.8	1.0703
22.09	53.71	0	1.0524	44.02	56.02	7.8	1.0668
13.46	58.33	0	1.0488	29.49	66.14	7.8	1.0639
0.00	67.67	0	1.0462	17.21	74.48	7.8	1.0617
133.80	0.00	7.8	1.0879	0.00	85.99	7.8	1.0587

№ 5035 [1851]

POTASSIUM COPPER (II) SULFATE–POTASSIUM COBALT (II) SULFATE–WATER

$CuK_2(SO_4)_2—CoK_2(SO_4)_2—H_2O$

$t = 25$

Comp. of saturated soln., wt %			Comp. of saturated soln., wt %		
Potassium Cobalt Sulfate	Potassium Copper Sulfate	Water	Potassium Cobalt Sulfate	Potassium Copper Sulfate	Water
11.74	1.312	86.948	7.60	4.61	87.79
11.71	1.323	86.957	5.39	6.28	88.33
9.72	2.961	89.319	5.37	6.30	88.33
9.71	2.965	87.325	2.918	8.20	88.882
7.63	4.60	87.77	2.915	8.21	88.875

№ 5036 [1770]

POTASSIUM NICKEL (II) SULFATE–POTASSIUM COPPER (II) SULFATE–WATER

$K_2SO_4 \cdot NiSO_4 \cdot 6H_2O—K_2SO_4 \cdot CuSO_4 \cdot 6H_2O—H_2O$

Concn. in aqueous soln., g/l		*t*	*d*	Comp. of solid phase, wt %	
Potassium Copper Sulfate	Potassium Nickel Sulfate			Potassium Copper Sulfate	Potassium Nickel Sulfate
70.71	0.0	0	1.0466	100	0.0
65.45	3.04	0	1.0456	88.88	11.12
57.00	9.43	0	1.0454	77.98	22.02
53.04	11.52	0	1.0433	69.02	30.98
45.85	15.83	0	1.0434	54.56	45.44

Table № 5036 (continuation)

Concn. in aqueous soln., g/l		t	d	Comp. of solid phase, wt %	
Potassium Copper Sulfate	Potassium Nickel Sulfate			Potassium Copper Sulfate	Potassium Nickel Sulfate
37.39	21.37	0	1.0395	45.22	54.78
27.89	28.50	0	1.0393	32.35	67.61
14.70	37.64	0	1.0357	14.05	85.95
0.0	48.07	0	1.0224	0.0	100
92.12	0.0	8.5	1.0582	100	0.0
79.73	6.47	8.5	1.0281	80.46	19.54
69.71	13.53	8.5	1.0563	71.41	28.59
60.13	20.53	8.5	1.0563	61.12	38.88
55.38	21.29	8.5	1.0547	47.98	52.02
43.51	29.61	8.5	1.0496	39.83	60.17
35.60	33.79	8.5	1.0472	29.34	70.66
24.46	39.13	8.5	1.0426	21.07	78.93
15.36	47.85	8.5	1.0451	12.32	87.68
0.0	60.18	8.5	1.0397	0.0	100.0

№ 5037 [1851]

POTASSIUM COPPER (II) SULFATE–POTASSIUM NICKEL (II) SULFATE–WATER

$CuK_2(SO_4)_2$—$NiK_2(SO_4)_2$—H_2O

$t=25$

Comp. of saturated soln., wt %	
Potassium Copper Sulfate	Potassium Nickel Sulfate
9.25	0.72
9.26	0.69
7.84	1.505
7.84	1.508
6.56	2.221
6.55	2.218
5.07	3.084
5.06	3.080
2.980	4.33
2.981	4.34

№ 5038 [1851]

POTASSIUM ZINC SULFATE–POTASSIUM NICKEL (II) SULFATE–WATER

$ZnK_2(SO_4)_2$—$NiK_2(SO_4)_2$—H_2O

$t=25$

Comp. of saturated soln., wt %		
Potassium Zinc Sulfate	Potassium Nickel Sulfate	Water
10.64	0.457	88.903
10.64	0.454	88.906
9.35	1.026	89.624
9.36	1.036	89.604
7.91	1.697	90.393
7.91	1.692	90.398
6.01	2.657	91.333
6.01	2.662	91.328
3.99	3.78	92.23

№ 5039 [2338]

POTASSIUM COPPER (II) SULFATE–POTASSIUM ZINC SULFATE–WATER

$CuSO_4 \cdot K_2SO_4 \cdot 6H_2O—ZnSO_4 \cdot K_2SO_4 \cdot 6H_2O—H_2O$

$t=6.8$

Concn. in aqueous soln., g/l		d	Comp. of solid phase, wt %	
Potassium Copper Sulfate	Potassium Zinc Sulfate		Potassium Copper Sulfate	Potassium Zinc Sulfate
85.32	0.0	1.0581	100	0.0
75.83	12.11	1.0595	87.82	12.18
68.79	22.53	1.0620	77.17	22.83
58.14	34.58	1.0634	66.05	33.95
49.28	45.55	1.0644	55.55	44.45
42.93	54.10	1.0658	45.64	54.36
33.40	65.60	1.0672	36.81	63.19
24.74	75.91	1.0682	27.64	72.36
16.99	84.80	1.0690	18.69	71.31
0.0	106.08	1.0707	0.0	100

№ 5040 [1851]

POTASSIUM COPPER (II) SULFATE–POTASSIUM ZINC SULFATE–WATER

$CuK_2(SO_4)_2—ZnK_2(SO_4)_2—H_2O$

$t=25$

Comp. of saturated soln., wt %			Comp. of saturated soln., wt %		
Potassium Zinc Sulfate	Potassium Copper Sulfate	Water	Potassium Zinc Sulfate	Potassium Copper Sulfate	Water
10.36	1.305	88.335	6.65	4.74	88.61
10.38	1.302	88.318	4.74	6.41	88.85
8.54	3.012	88.448	4.73	6.43	88.84
8.52	3.008	88.472	2.617	8.22	89.163
6.63	4.74	88.63	2.616	8.22	89.164

№ 5041 [2407]

POTASSIUM COPPER (II) SULFATE–THALLIUM COPPER (II) SULFATE–WATER

$K_2SO_4CuSO_4{\cdot}6H_2O–CuSO_4{\cdot}Tl_2SO_4{\cdot}6H_2O–H_2O$

$t=25$

Concn. in aqueous soln., wt %			d_4	Concn. in aqueous soln., wt%			d_4
Thallium Sulfate	Potassium Sulfate	Copper Sulfate		Thallium Sulfate	Potassium Sulfate	Copper Sulfate	
5.15	0.47	2.12	1.073	0.98	4.71	4.46	1.095
3.12	2.77	3.55	1.088	0.68	4.93	4.76	1.094
2.48	3.42	3.95	1.092	0.37	5.14	4.87	1.093

$CaO—Ca(NO_3)_2—H_2O$

Concn. in aqueous soln., wt %		t	Comp. of solid phase
Calcium Nitrate	Calcium Oxide		
0.0	0.1150	25	$Ca(OH)_2$
4.836	0.0978	25	»
9.36	0.1074	25	»
13.77	0.1193	25	»
22.46	0.1444	25	»
27.83	0.1650	25	»
32.94	0.1931	25	»
40.66	0.2579	25	»
44.44	0.3060	25	»
45.28	0.2802	25	$Ca_2N_2O_7 \cdot 3H_2O$
47.79	0.2314	25	»
51.07	0.1894	25	»
53.20	0.1659	25	»
55.25	0.1486	25	»
57.72	0.0836	25	$Ca(NO_3)_2 \cdot 4H_2O$
57.98	0.0	25	»
0.0	0.0561	100	$Ca(OH)_2$
2.42	0.0550	100	»
4.91	0.0624	100	»
15.39	0.1110	100	»
16.10	0.1200	100	»
21.86	0.155	100	»
33.03	0.269	100	»
42.26	0.480	100	»
50.94	0.973	100	»
53.75	1.261	100	»
55.40	1.477	100	»
55.43	1.476	100	»
55.05	1.491	100	»
56.89	1.635	100	$Ca(OH)_2 + Ca_2N_2O_7 \cdot 2H_2O$
57.03	1.686	100	»
57.91	1.596	100	$Ca_2N_2O_7 \cdot 2H_2O$
58.67	1.576	100	»
60.44	1.348	100	»
62.82	1.167	100	»
66.44	1.077	100	»
69.12	1.141	100	»
70.60	1.252	100	$Ca_2N_2O_7 \cdot 2H_2O + CaN_2O_7 \cdot \frac{1}{2}H_2O$
70.40	1.203	100	$CaN_2O_7 \cdot \frac{1}{2}H_2O$
71.44	1.103	100	»
73.85	0.937	100	»
75.74	0.849	100	»
76.94	0.815	100	»
77.62	0.804	100	$Ca(NO_3)_2$
77.74	0.412	100	»
78.43	0.0	100	»

№ 5043 CALCIUM OXIDE–CALCIUM SULFATE–WATER [1324]

$CaO—CaSO_4—H_2O$

$t=25$

Concn. in aqueous soln., g/l		Comp. of solid phase
Calcium Sulfate	Calcium Oxide	
0.0	1.166	$Ca(OH)_2$
0.391	1.141	»
0.666	1.150	»
0.955	1.215	»
1.214	1.242	»
1.588	1.222	$Ca(OH)_2+CaSO_4 \cdot 2H_2O$
1.634	0.939	$CaSO_4 \cdot 2H_2O$
1.722	0.610	»
1.853	0.349	»
1.918	0.176	»
2.030	0.062	»
2.126	0.0	»

№ 5044 CALCIUM OXIDE–CALCIUM CHLORIDE–WATER [2766]

$CaO—CaCl_2—H_2O$

Comp. of solvent, wt %		Solubility of Calcium Oxide, g/l solvent	t	Comp. of solvent, wt %		Solubility of Calcium Oxide, g/l solvent	t
Calcium Chloride	Water			Calcium Chloride	Water		
0	100	1.374	20	10	90	1.328	80
0	100	1.162	40	10	90	1.389	100
0	100	1.026	60	15	85	1.993	20
0	100	0.845	80	15	85	1.781	40
0	100	0.664	100	15	85	1.706	60
5	95	1.370	20	15	85	1.736	80
5	95	1.160	40	15	85	1.842	100
5	95	1.020	60	25	75	3.710	100
5	95	0.936	80	30	70	1.630	20
5	95	0.906	100	30	70	3.680	40
10	90	1.6661	20	30	70	3.664	60
10	90	1.419	40	30	70	4.122	80
10	90	1.313	60	30	70	4.922	100

№ 5045 CALCIUM OXIDE–CALCIUM CHLORIDE–WATER [2513]

$CaO—CaCl_2—H_2O$

$t=25$

Comp. of saturated soln., wt %			Comp. of solid phase
Calcium Chloride	Calcium Oxide	Water	
5.02	0.101	94.879	$Ca(OH)_2$
10.00	0.115	89.885	»

Table № 5045 (continuation)

Comp. of saturated soln., wt %			Comp. of solid phase
Calcium Chloride	Calcium Oxide	Water	
15.14	0.140	84.720	$Ca(OH)_2$
18.15	0.148	81.702	$Ca(OH)_2+CaCl_2 \cdot 4CaO \cdot 24H_2O$
18.01	0.152	81.838	$CaCl_2 \cdot 4CaO \cdot 14H_2O$
21.02	0.147	78.833	»
28.37	0.170	71.460	»
32.67	0.225	67.105	»
33.21	0.245	66.545	»
33.72	0.254	66.026	$CaCl_2 \cdot 4CaO \cdot 14H_2O+CaCl_2 \cdot CaO \cdot 2H_2O$
34.36	0.173	65.767	$CaCl_2 \cdot CaO \cdot 2H_2O$
38.61	0.060	61.330	»
41.32	0.048	58.632	»
44.30	0.030	55.670	»
44.61	0.029	55.361	$CaCl_2 \cdot CaO \cdot 2H_2O+CaCl_2 \cdot 6H_2O$
44.77	0.000	55.23	$CaCl_2 \cdot 6H_2O$

№ 5046 CALCIUM OXIDE–CALCIUM CHLORIDE–WATER [2204]

$CaO—CaCl_2—H_2O$

Comp. of saturated soln., wt %			t	Comp. of solid phase
Calcium Oxide	Calcium Chloride	Water		
0.130	0.00	99.870	10	$Ca(OH)_2$
0.102	15.46	84.438	10	$Ca(OH)_2+CaCl_2 \cdot 3CaO \cdot 16H_2O$
0.110	26.30	73.560	10	$CaCl_2 \cdot 3CaO \cdot 16H_2O$
0.115	32.38	67.505	10	»
0.140	34.60	65.260	10	»
0.151	35.44	64.409	10	$CaCl_2 \cdot 3CaO \cdot 16H_2O+CaCl_2 \cdot CaO \cdot 2H_2O$
0.106	38.23	61.664	10	$CaCl_2 \cdot CaO \cdot 2H_2O$
0.126	38.45	61.424	10	»
0.000	39.40	60.60	10	$CaCl_2 \cdot 6H_2O$
0.101	5.02	94.879	25	$Ca(OH)_2$
0.115	10.00	89.885	25	»
0.128	12.94	87.932	25	»
0.140	15.14	84.720	25	»
0.145	17.20	82.655	25	»
0.148	18.15	81.702	25	$Ca(OH)_2+CaCl_2 \cdot 3CaO \cdot 16H_2O$
0.147	21.02	78.833	25	$CaCl_2 \cdot 3CaO \cdot 16H_2O$
0.146	23.80	76.054	25	»
0.147	24.33	75.523	25	»
0.170	28.37	71.460	25	»
0.180	29.54	70.280	25	»
0.225	32.67	67.105	25	»
0.245	33.21	66.545	25	$CaCl_2 \cdot 3CaO \cdot 16H_2O+CaCl_2 \cdot CaO \cdot 2H_2O$
0.173	34.36	65.467	25	$CaCl_2 \cdot CaO \cdot 2H_2O$
0.060	38.61	61.330	25	»
0.048	41.32	58.632	25	»
0.030	44.30	55.670	25	»

Comp. of saturated soln., wt %			t	Comp. of solid phase
Calcium Oxide	Calcium Chloride	Water		
0.022	44.51	55.468	25	$CaCl_2 \cdot CaO \cdot 2H_2O + CaCl_2 \cdot 6H_2O$
0.036	44.64	55.324	25	»
0.00	44.77	55.23	25	$CaCl_2 \cdot 6H_2O$
0.104	0.00	99.896	40	$Ca(OH)_2$
0.455	29.38	70.165	40	»
0.609	31.93	67.461	40	$Ca(OH)_2 + CaCl_2 \cdot CaO \cdot 2H_2O$
0.571	31.90	67.529	40	»
0.580	31.86	68.560	40	»
0.055	49.97	49.975	40	$CaCl_2 \cdot CaO \cdot 2H_2O + CaCl_2 \cdot 4H_2O$
0.000	51.18	48.82	40	$CaCl_2 \cdot 4H_2O$
0.100	0.00	99.900	45	$Ca(OH)_2$
0.621	31.72	68.659	45	$Ca(OH)_2 + CaCl_2 \cdot CaO \cdot 2H_2O$
0.634	31.95	68.416	45	»
0.442	35.87	63.688	45	$CaCl_2 \cdot CaO \cdot 2H_2O$
0.069	38.12	61.811	45	»
0.074	43.31	56.616	45	»
0.047	52.10	47.853	45	»
0.121	59.96	39.919	45	$CaCl_2 \cdot CaO \cdot 2H_2O + CaCl_2 \cdot 2H_2O$
0.00	57.19	42.81	45	$CaCl_2 \cdot 2H_2O$
0.096	0.00	99.904	50	$Ca(OH)_2$
0.112	4.08	95.808	50	»
0.142	15.02	84.838	50	»
0.538	29.41	70.052	50	»
0.436	29.57	70.994	50	»
0.362	29.66	69.978	50	$Ca(OH)_2 + CaCl_2 \cdot CaO \cdot 5H_2O$
0.517	29.60	69.883	50	$CaCl_2 \cdot CaO \cdot 5H_2O$
0.470	31.70	67.830	50	»
0.106	35.20	64.694	50	»
0.113	36.95	62.937	50	$CaCl_2 \cdot CaO \cdot 5H_2O + CaCl_2 \cdot CaO \cdot 2H_2O$
0.146	36.82	63.034	50	»
0.124	36.93	62.946	50	$CaCl_2 \cdot CaO \cdot 2H_2O$
0.069	39.43	60.501	50	»
0.051	45.57	54.379	50	»
0.046	46.42	53.534	50	»
0.034	47.19	52.776	50	»
0.046	48.75	51.204	50	»
0.058	54.05	45.892	50	$CaCl_2 \cdot CaO \cdot 2H_2O + CaCl_2 \cdot 2H_2O$
0.000	56.95	43.05	50	$CaCl_2 \cdot 2H_2O$

№ 5047 CALCIUM OXIDE–CALCIUM CHLORIDE–WATER [2278]

$CaO—CaCl_2—H_2O$

$t=0$

Concn. in aqueous soln., wt%		d_t	Comp. of solid phase
Calcium Oxide	Calcium Chloride		
0.214	2.7	1.025	$Ca(OH)_2$
0.226	5.5	1.048	»
0.216	6.8	1.056	$Ca(OH)_2+CaCl_2 \cdot 3CaO \cdot 16H_2O$
0.212	7.6	1.067	$CaCl_2 \cdot 3CaO \cdot 16H_2O$
0.186	10.2	1.089	»
0.172	12.5	1.111	»
0.164	15.2	1.133	»
0.152	17.4	1.155	»
0.148	20.0	1.186	»
0.145	22.8	1.213	»
0.142	25.6	1.240	»
0.138	28.0	1.265	»
0.152	29.7	1.280	»
0.190	34.9	1.338	»
0.00	37.6	1.370	$CaCl_2 \cdot 6H_2O$

№ 5048 CALCIUM OXIDE–CALCIUM CHLORIDE–WATER [1304]

$CaO—CaCl_2—H_2O$

$t=25$

Concn. in aqueous soln., wt %		Comp. of solid phase
Calcium Oxide	Calcium Chloride	
0.141	14.85	$Mg(OH)_2+Ca(OH)_2$
0.150	18.06	$Mg(OH)_2+Ca(OH)_2+3CaO \cdot CaCl_2 \cdot 16H_2O$
0.145	20.08	$Mg(OH)_2+3CaO \cdot CaCl_2 \cdot 16H_2O$
0.145	23.56	»
0.168	29.05	»
0.170	29.22	»
0.235	32.98	$Mg(OH)_2+3CaO \cdot CaCl_2 \cdot 16H_2O+CaO \cdot CaCl_2 \cdot 2H_2O$
0.098	36.44	$Mg(OH)_2+CaO \cdot CaCl_2 \cdot 2H_2O$
0.083	36.82	»
0.091	37.10	»
0.051	40.85	»
0.034	44.92	$Mg(OH)_2+CaO \cdot CaCl_2 \cdot 2H_2O+CaCl_2 \cdot 6H_2O$

№ 5049 CALCIUM OXIDE–CALCIUM HYPOCHLORITE–WATER [2278]

$CaO—Ca(ClO)_2—H_2O$

$t=0$

Concn. in aqueous soln., wt %		d_4	Comp. of solid phase
Calcium Oxide	Calcium Hypochlorite		
0.0	21.8	1.213	$Ca(Cl)_2 \cdot 3H_2O$
0.106	21.2	1.203	$Ca(ClO)_2 \cdot 2CaO \cdot 2H_2O$
0.106	19.8	1.188	»
0.113	19.2	1.185	»
0.077	15.1	1.142	$Ca(ClO)_2 \cdot 3CaO \cdot 3H_2O$
0.093	13.0	1.123	»
0.093	11.1	1.104	»
0.091	9.9	1.093	»
0.110	8.2	1.077	$Ca(OH)_2$
0.105	5.7	1.053	»
0.106	3.0	1.030	»

№ 5050 CALCIUM OXIDE–CALCIUM BROMIDE–WATER [2205]

$CaO—CaBr_2—H_2O$

$t=25$

Comp. of saturated soln., wt %			Comp. of solid phase
Calcium Oxide	Calcium Bromide	Water	
0.000	60.07	39.93	$CaBr_2 \cdot 6H_2O$
0.209	60.09	39.701	$CaBr_2 \cdot 6H_2O + 3CaBr_2 \cdot 4CaO \cdot 16H_2O$
0.192	59.48	40.328	$3CaBr_2 \cdot 4CaO \cdot 16H_2O$
0.199	58.90	40.901	»
0.235	56.94	42.825	»
0.271	55.29	44.439	»
0.358	54.22	45.422	$3CaBr_2 \cdot 4CaO \cdot 16H_2O + CaBr_2 \cdot 3CaO \cdot 16H_2O$
0.247	53.39	46.363	$CaBr_2 \cdot 4CaO \cdot 16H_2O$
0.132	50.04	49.828	»
0.098	44.21	55.692	»
0.077	41.36	58.563	»
0.080	32.92	67.000	»
0.080	32.50	67.420	»
0.089	28.21	71.701	»
0.089	25.15	74.761	»
0.092	22.39	77.518	»
0.103	21.80	78.097	$CaBr_2 \cdot 3CaO \cdot 16H_2O + Ca(OH)_2$
0.097	20.75	79.153	»
0.115	19.18	80.705	$Ca(OH)_2$
0.101	13.75	86.149	»
0.119	9.28	90.601	»
0.118	0.00	99.882	»

$CaO—CaBr_2—H_2O$

Comp. of saturated soln., wt %			t	Comp. of solid phase wt %		
Calcium Oxide	Calcium Bromide	Water		Calcium Oxide	Calcium Bromide	Water
0.11	0.00	99.89	40	—	—	—
0.045	12.90	87.06	40	31.30	7.66	61.04
0.13	22.50	77.37	40	33.2	12.2	54.6
0.13	28.20	71.67	40	40.5	12.8	46.7
0.18	30.40	69.42	40	15.6	28.9	55.5
0.21	32.30	67.49	40	25.5	25.8	48.7
0.16	36.50	63.34	40	14.8	33.6	51.6
0.16	37.90	61.84	40	11.6	34.5	53.9
0.24	41.60	58.16	40	16.2	35.7	48.1
0.41	50.0	49.59	40	10.8	42.6	46.6
0.55	52.10	47.35	40	12.1	42.8	45.1
0.70	52.50	46.80	40	13.1	43.7	43.2
0.45	54.00	45.55	40	11.0	53.6	35.4
0.35	55.30	44.35	40	11.0	54.5	34.5
0.26	58.50	41.24	40	9.45	56.7	33.8
0.25	59.50	40.25	40	8.33	58.2	33.5
0.20	61.7	38.1	40	5.45	62.70	31.85
0.00	68.01	32.0	40	—	—	—
0.097	0.00	99.90	50	—	—	—
0.045	9.24	90.74	50	36.2	5.5	58.3
0.15	14.0	85.85	50	30.50	7.89	61.61
0.092	15.10	84.81	50	30.5	9.58	59.92
0.12	18.80	81.08	50	29.3	13.8	56.9
0.11	22.80	77.09	50	31.4	13.6	55.0
0.14	34.6	65.26	50	32.9	19.8	47.3
0.10	35.4	64.50	50	46.5	13.8	39.7
0.21	37.20	62.59	50	40.9	17.3	41.8
0.30	40.3	59.4	50	12.6	35.6	51.8
0.14	41.86	58.0	50	13.1	36.6	50.3
0.68	47.70	51.62	50	12.2	40.6	47.2
0.74	49.10	50.16	50	9.91	42.9	47.19
0.89	51.40	47.61	50	14.7	42.1	43.2
1.07	51.50	47.43	50	9.11	49.11	41.78
0.68	52.60	46.72	50	11.5	53.3	35.2
0.46	55.30	44.24	50	10.3	54.6	35.1
0.38	59.10	40.52	50	6.79	61.50	31.71
0.30	59.70	40.00	50	11.2	57.0	31.8
0.00	72.0	28.00	50	—	—	—
0.088	0.00	99.91	60	—	—	—
0.33	24.50	75.17	60	37.9	13.4	48.7
0.28	34.40	65.32	60	32.9	19.5	47.6
0.40	42.3	57.3	60	45.9	16.7	37.4
1.26	51.60	47.14	60	5.98	49.1	44.92
1.20	52.4	46.4	60	8.15	52.30	39.55
0.38	56.50	43.12	60	9.75	55.40	34.85
0.37	59.80	39.83	60	8.41	64.10	27.49
0.27	61.10	38.63	60	5.99	60.00	34.01
0.00	73.5	26.5	60	—	—	—

№ 5052 CALCIUM OXIDE–CALCIUM IODIDE–WATER [2204]

$CaO—CaI_2—H_2O$

$t=25$

Comp. of saturated soln., wt %			Comp. of solid phase
Calcium Iodide	Calcium Oxide	Water	
0.00	0.118	99.882	$Ca(OH)_2$
9.14	0.076	90.784	»
17.42	0.089	82.491	»
25.80	0.087	74.113	»
25.44	0.097	74.463	$Ca(OH)_2+CaI_2 \cdot 3CaO \cdot 16H_2O$
31.33	0.089	68.581	$CaI_2 \cdot 3CaO \cdot 16H_2O$
33.00	0.075	66.925	»
40.30	0.064	59.636	»
46.37	0.054	53.576	»
47.09	0.055	52.855	»
49.70	0.047	50.253	»
59.34	0.139	40.521	»
66.72	0.587	32.693	$CaI_2 \cdot 3CaO \cdot 16H_2O+CaI_2 \cdot 6H_2O$
66.65	0.553	33.897	»
66.80	0.0	33.20	$CaI_2 \cdot 6H_2O$

№ 5053 CALCIUM OXIDE–ARSENIC TRIOXIDE–WATER [2152]

$CaO—As_2O_3—H_2O$

$t=60$

Comp. of saturated soln., wt %			Comp. of solid phase
Calcium Oxide	Arsenic Trioxide	Water	
0.03	4.90	95.07	$CaO \cdot As_2O_3 \cdot nH_2O$
0.019	0.09	99.89	»
0.08	Traces	99.92	$2CaO \cdot As_2O_3 \cdot nH_2O$

№ 5054 CALCIUM OXIDE–ARSENIC TRIOXIDE–WATER [2152]

$CaO—As_2O_3—H_2O$

$t=20$

Comp. of saturated soln., wt %			Comp. of solid phase
Calcium Oxide	Arsenic Trioxide	Water	
0.03	2.00	97.97	$CaO \cdot As_2O_3 \cdot nH_2O$
0.019	0.07	99.91	»
0.11	Traces	99.89	$2CaO \cdot As_2O_3 \cdot nH_2O$

№ 5055 CALCIUM OXIDE–ARSENIC PENTOXIDE–WATER [2585]

$CaO—As_2O_5—H_2O$

$t=35$

Concn. in aqueous soln., g/l		d_4	Comp. of solid phase
Calcium Oxide	Arsenic Pentoxide		
0.956	1.372	0.996	—
7.20	27.79	1.025	$CaHAsO_4 \cdot H_2O$
20.74	87.00	1.090	»
37.09	150.1	1.157	»
57.0	232.3	1.242	»
58.2	252.1	1.261	»
69.4	285.6	1.300	»
92.5	405.4	1.413	$CaH_4(AsO_4)_2$
79.0	518.4	1.480	»
64.7	690.5	1.590	»
50.1	853.6	1.697	»
33.6	995.0	1.795	»
16.2	1230	1.975	»
2.93	1539	2.167	»
2.53	1599	2.217	$CaH_4(AsO_4)_2+3As_2O_5 \cdot 5H_2O$

№ 5056 CALCIUM OXIDE–ARSENIC PENTOXIDE–WATER [1750]

$CaO—As_2O_5—H_2O$

Concn. in aqueous soln., wt %		pH	t	Comp. of solid phase
Calcium Oxide	Arsenic Pentoxide			
6.5	27.5	2.4	17	$CaO \cdot As_2O_5 \cdot 2H_2O+2CaO \cdot As_2O_5 \cdot 3H_2O$
0.09	0.32	6.8	17	$2CaO \cdot As_2O_5 \cdot 3H_2O+5CaO \cdot 2As_2O_5 \cdot 10H_2O$
0.05	0.10	7.8	17	$5CaO \cdot 2As_2O_5 \cdot 10H_2O+3CaO \cdot As_2O_5 \cdot 10H_2O$
0.04	Traces	9.6	17	$3CaO \cdot As_2O_5 \cdot 10H_2O+4CaO \cdot As_2O_5 \cdot 5H_2O$
6.6	28.5	2.4	40	$CaO \cdot As_2O_5 \cdot 2H_2O+2CaO \cdot As_2O_5 \cdot 3H_2O$
0.09	0.23	6.4	40	$2CaO \cdot As_2O_5 \cdot 3H_2O+5CaO \cdot 2As_2O_5 \cdot 5H_2O$
0.03	0.06	7.4	40	$5CaO \cdot 2As_2O_5 \cdot 5H_2O+3CaO \cdot As_2O_5 \cdot 4H_2O$
0.006	0.008	9.0	40	$3CaO \cdot As_2O_5 \cdot 4H_2O+4CaO \cdot As_2O_5 \cdot H_2O$
6.7	29.7	2.4	60	$CaO \cdot As_2O_5 \cdot 2H_2O+2CaO \cdot As_2O_5 \cdot 3H_2O$
0.07	0.2	6.2	60	$2CaO \cdot As_2O_5 \cdot 3H_2O+5CaO \cdot 2As_2O_5 \cdot 5H_2O$
0.03	0.07	6.8	60	$5CaO \cdot 2As_2O_5 \cdot 5H_2O+3CaO \cdot As_2O_5 \cdot 2H_2O$
0.008	0.017	7.8	60	$3CaO \cdot As_2O_5 \cdot 2H_2O+4CaO \cdot As_2O_5 \cdot H_2O$
6.8	32.0	2.4	90	$CaO \cdot As_2O_5 \cdot 2H_2O+2CaO \cdot As_2O_5 \cdot H_2O$
0.06	0.22	5.9	90	$2CaO \cdot As_2O_5 \cdot H_2O+5CaO \cdot 2As_2O_5 \cdot 5H_2O$
0.02	0.07	6.2	90	$5CaO \cdot 2As_2O_5 \cdot 5H_2O+3CaO \cdot As_2O_5 \cdot 2H_2O$
0.009	0.017	8.0	90	$3CaO \cdot As_2O_5 \cdot 2H_2O+4CaO \cdot As_2O_5 \cdot H_2O$

№ 5057 [376]

CALCIUM HYDROXIDE–CALCIUM CARBONATE–WATER
$Ca(OH)_2—CaCO_3—H_2O$

Content in aqueous soln., wt %		t	Comp. of solid phase
Calcium Carbonate	Calcium Hydroxide		
0.0050	0.00	25	$CaCO_3$
Traces	0.13	25	$CaCO_3+Ca(OH)_2$
»	0.15	25	»
»	0.14	25	»
»	0.12	25	$Ca(OH)_2$
0.0026	0.00	75	$CaCO_3$
Traces	0.071	75	$CaCO_3+Ca(OH)_2$
»	0.102	75	»
»	0.103	75	»
»	0.106	75	»
»	0.100	75	»
»	0.104	75	»
»	0.100	75	$Ca(OH)_2$

№ 5058 [1326]

CALCIUM HYDROXIDE–CALCIUM SULFATE–WATER
$Ca(OH)_2—CaSO_4—H_2O$
$t=25$

Content in aqueous soln., g/l		Comp. of solid phase
Calcium Sulfate	Calcium Hydroxide	
1.166	0	$CaSO_4 \cdot 2H_2O$
1.141	0.341	»
1.215	0.955	»
1.242	1.214	»
0.939	1.634	$Ca(OH)_2$
0.319	1.853	»
0.176	1.918	»
0.062	2.032	»
0	2.126	»

№ 5059 [1930]

CALCIUM HYDROXIDE–CALCIUM SULFATE–WATER
$Ca(OH)_2—CaSO_4—H_2O$
$t=25$

Concn. in aqueous soln., wt %		Comp. of solid phase
Calcium Hydroxide	Calcium Sulfate	
0.0	0.209	$CaSO_4 \cdot 2H_2O$
0.137	0.170	$CaSO_4 \cdot 2H_2O+Ca(OH)_2$
0.140	0.169	$Ca(OH)_2$
0.143	0.0	»

№ 5060 [1235]

CALCIUM HYDROXIDE–CALCIUM SULFATE–WATER
$Ca(OH)_2$—$CaSO_4$—H_2O

Content in saturated soln., g-eq/l		t	Content in saturated soln., g-eq/l		t
Calcium Sulfate	Calcium Hydroxide		Calcium Sulfate	Calcium Hydroxide	
0.02340	0.04360	0	0.02583	0.03000	50
0.02440	0.03610	25	0.02290	0.02200	80
0.02694	0.03449	35			

№ 5061 [2761]

CALCIUM HYDROXIDE–CALCIUM CHLORIDE–WATER
$Ca(OH)_2$—$CaCl_2$—H_2O
t=16—20

Content in aqueous soln., g/l		Comp. of solid phase	Content in aqueous soln., g/l		Comp. of solid phase
$CaCl_2$	CaO		$CaCl_2$	CaO	
122.1	1.47	—	12.2	0.98	$Ca(OH)_2$
85.5	1.31	—	9.8	0.99	»
73.2	1.225	—	7.3	1.01	»
61.0	1.18	—	4.9	1.04	»
41.5	1.06	$Ca(OH)_2$	3.0	1.08	»
24.4	1.015	»	0	1.24	»

№ 5062 [614]

CALCIUM HYDROXIDE–CALCIUM CHLORIDE–WATER
$Ca(OH)_2$—$CaCl_2$—H_2O
t=25

Content in aqueous soln., wt %		Comp. of solid phase
Calcium Oxide	Hydrogen Chloride	
10.85	13.85	$CaCl_2 \cdot 3Ca(OH)_2 \cdot 11H_2O$
10.40	21.05	»

№ 5063 [1792]

CALCIUM HYDROXIDE–CALCIUM CHLORIDE–WATER
$Ca(OH)_2$—$CaCl_2$—H_2O
t=100

Content in aqueous soln., mol/l		Comp. of solid phase
Calcium Chloride	Calcium Hydroxide	
5	0.006	Basic salts
3.5	0.006	»

№ 5064 [752]

CALCIUM HYDROXIDE–CALCIUM CHLORIDE–WATER

$Ca(OH)_2—CaCl_2—H_2O$

$t=50$

Content in aqueous soln., wt %		Comp. of solid phase
$CaCl_2$	CaO	
56.70	0	$CaCl_2 \cdot 2H_2O$
53.35	0.052	$CaCl_2 \cdot 2H_2O + CaCl_2 \cdot CaO \cdot 2H_2O$
37.12	0.115	$CaCl_2 \cdot CaO \cdot 2H_2O + CaCl_2 \cdot CaO \cdot 5H_2O$
30.02	0.360	$CaCl_2 \cdot CaO \cdot 5H_2O + Ca(OH)_2$
0	0.098	$Ca(OH)_2$

№ 5065 [545]

CALCIUM HYDROXIDE–CALCIUM CHLORIDE–WATER

$CaCl_2—Ca(OH)_2—H_2O$

Comp. of saturated soln., wt %			t	Comp. of solid phase
Calcium Hydr-oxide	Calcium Chloride	Water		
0.18	0.00	99.82	5	$Ca(OH)_2$
0.15	6.53	93.32	5	»
0.15	7.57	92.28	5	$Ca(OH)_2 + CaCl_2 \cdot 3Ca(OH)_2 \cdot 12H_2O$
0.13	11.06	88.81	5	$CaCl_2 \cdot 3Ca(OH)_2 \cdot 12H_2O$
0.11	13.62	86.27	5	»
0.10	19.25	80.65	5	»
0.11	29.03	70.86	5	»
0.16	34.75	65.09	5	»
0.19	36.58	63.23	5	$CaCl_2 \cdot 3Ca(OH)_2 \cdot 12H_2O + CaCl_2 \cdot Ca(OH)_2 \cdot H_2O$
0.10	37.04	62.86	5	$CaCl_2 \cdot Ca(OH)_2 \cdot H_2O$
0.09	37.90	62.01	5	$CaCl_2 \cdot Ca(OH)_2 \cdot H_2O + CaCl_2 \cdot 6H_2O$
0.00	38.29	61.71	5	$CaCl_2 \cdot 6H_2O$
0.18	0.00	99.82	10	$Ca(OH)_2$
0.14	7.33	92.53	10	»
0.14	8.87	90.99	10	»
0.15	10.26	89.59	10	$Ca(OH)_2 + CaCl_2 \cdot 3Ca(OH)_2 \cdot 12H_2O$
0.14	12.12	87.74	10	$CaCl_2 \cdot 3Ca(OH)_2 \cdot 12H_2O$
0.12	18.27	81.61	10	»
0.12	25.04	74.84	10	»
0.15	30.90	68.95	10	»
0.23	36.09	63.68	10	$CaCl_2 \cdot 3Ca(OH)_2 \cdot 12H_2O + CaCl_2 \cdot Ca(OH)_2 \cdot H_2O$
0.16	36.58	63.26	10	$CaCl_2 \cdot Ca(OH)_2 \cdot H_2O$
0.10	37.90	62.00	10	»
0.08	39.00	60.92	10	$CaCl_2 \cdot Ca(OH)_2 \cdot H_2O + CaCl_2 \cdot 6H_2O$
0.00	39.40	60.60	10	$CaCl_2 \cdot 6H_2O$
0.15	0.00	99.85	30	$Ca(OH)_2$
0.16	12.80	87.04	30	»
0.21	19.61	80.18	30	»
0.23	20.63	79.14	30	$Ca(OH)_2 + CaCl_2 \cdot 3Ca(OH)_2 \cdot 12H_2O$
0.24	23.50	76.26	30	$CaCl_2 \cdot 3Ca(OH)_2 \cdot 12H_2O$
0.26	26.58	73.16	30	»
0.37	32.05	67.58	30	»

Table № 5065 (continuation)

Comp. of saturated soln., wt %			t	Comp. of solid phase
Calcium Hydr-oxide	Calcium Chloride	Water		
0.39	32.76	66.85	30	$CaCl_2 \cdot 3Ca(OH)_2 \cdot 12H_2O$
0.43	32.90	66.67	30	$CaCl_2 \cdot 3Ca(OH)_2 \cdot 12H_2O + CaCl_2 \cdot Ca(OH)_2 \cdot H_2O$
0.22	33.85	65.93	30	»
0.15	35.96	63.89	30	»
0.08	38.74	61.18	30	»
0.07	41.88	58.05	30	»
0.06	46.60	53.34	30	$CaCl_2 \cdot Ca(OH)_2 \cdot H_2O + \alpha\text{-}CaCl_2 \cdot 4H_2O$
0.00	49.25	50.75	30	$\alpha\text{-}CaCl_2 \cdot 4H_2O$
0.14	0.00	99.86	40	$Ca(OH)_2$
0.18	13.02	86.80	40	»
0.23	19.44	80.33	40	»
0.37	26.08	73.55	40	»
0.58	30.00	69.42	40	»
0.67	31.00	68.33	40	»
0.74	32.00	67.26	40	$Ca(OH)_2 + CaCl_2 \cdot Ca(OH)_2 \cdot H_2O$
0.56	32.38	67.16	40	$CaCl_2 \cdot Ca(OH)_2 \cdot H_2O$
0.21	34.47	65.32	40	»
0.10	38.72	61.18	40	»
0.05	44.29	55.66	40	»
0.06	48.25	51.69	40	»
0.06	49.44	50.50	40	$CaCl_2 \cdot Ca(OH)_2 \cdot H_2O + \alpha\text{-}CaCl_2 \cdot 4H_2O$
0.06	49.32	50.62	40	»
0.00	52.95	47.05	40	$\alpha\text{-}CaCl_2 \cdot 4H_2O$
0.12	0.00	99.88	45	$Ca(OH)_2$
0.13	5.45	94.42	45	»
0.27	21.55	78.18	45	»
0.41	28.15	71.44	45	»
0.71	31.18	68.11	45	»
0.76	31.28	67.96	45	$Ca(OH)_2 + CaCl_2 \cdot Ca(OH)_2 \cdot H_2O$
0.32	32.24	67.44	45	$CaCl_2 \cdot Ca(OH)_2 \cdot H_2O$
0.13	36.05	63.82	45	»
0.09	39.59	60.32	45	»
0.06	43.69	56.25	45	»
0.07	50.63	49.30	45	»
0.08	53.48	46.44	45	$CaCl_2 \cdot Ca(OH)_2 \cdot H_2O + CaCl_2 \cdot 2H_2O$
0.00	56.50	43.50	45	$CaCl_2 \cdot 2H_2O$
0.12	0.00	99.88	50	$Ca(OH)_2$
0.20	16.89	82.91	50	»
0.23	19.64	80.13	50	»
0.32	24.31	75.37	50	»
0.62	30.75	68.63	50	»
0.74	31 64	67.62	50	$Ca(OH)_2 + CaCl_2 \cdot Ca(OH)_2 \cdot H_2O$
0.72	31.78	67.50	50	»
0.60	32.40	67.00	50	$CaCl_2 \cdot Ca(OH)_2 \cdot H_2O$
0.44	33.20	66.36	50	»
0.16	36.32	63.52	50	»
0 10	39.13	60.77	50	»
0.09	40.42	59.58	50	»
0.08	49.61	50.31	50	»
0.09	53.81	46.10	50	$CaCl_2 \cdot Ca(OH)_2 \cdot H_2O + CaCl_2 \cdot 2H_2O$

Table № 5065 (continuation)

Comp. of saturated soln., wt %			t	Comp. of solid phase
Calcium Hydroxide	Calcium Chloride	Water		
0.00	56.90	43.10	50	$CaCl_2 \cdot 2H_2O$
0.12	0.00	99.88	55	$Ca(OH)_2$
0.13	6.75	93.12	55	»
0.18	13.35	86.47	55	»
0.25	19.25	80.50	55	»
0.55	28.10	71.35	55	»
0.67	30.09	69.24	55	»
0.76	31.76	67.48	55	$Ca(OH)_2 + CaCl_2 \cdot Ca(OH)_2 \cdot H_2O$
0.53	32.23	67.24	55	$CaCl_2 \cdot Ca(OH)_2 \cdot H_2O$
0.37	33.27	66.36	55	»
0.20	35.13	64.67	55	»
0.12	36.77	63.11	55	»
0.07	43.34	56.59	55	»
0.07	45.48	54.45	55	»
0.07	49.75	50.18	55	»
0.06	52.65	47.29	55	»
0.08	54.92	45.00	55	$CaCl_2 \cdot Ca(OH)_2\ H_2O + CaCl_2 \cdot 2H_2O$
0.00	57.27	42.73	55	$CaCl_2 \cdot 2H_2O$
0.10	0.00	99.90	75	$Ca(OH)_2$
0.12	6.30	93.58	75	»
0.30	19.33	80.37	75	»
0.55	26.03	73.42	75	»
0.72	30.58	68.70	75	$Ca(OH)_2 + CaCl_2 \cdot Ca(OH)_2 \cdot H_2O$
0.30	32.99	66.71	75	$CaCl_2 \cdot Ca(OH)_2 \cdot H_2O$
0.10	46.45	53.45	75	»
0.12	56.98	42.90	75	$CaCl_2 \cdot Ca(OH)_2 \cdot H_2O + CaCl_2 \cdot 2H_2O$
0.00	58.85	41.15	75	$CaCl_2 \cdot 2H_2O$
0.13	0.00	99.87	0.116	Ice $+ Ca(OH)_2$
0.282	6.800	92.918	0.0	$Ca(OH)_2 + CaCl_2 \cdot 3Ca(OH)_2 \cdot 13H_2O$
0.15	7.57	92.28	5.0	»
0.19	36.58	63.23	5.0	$CaCl_2 \cdot 3Ca(OH)_2 \cdot 12H_2O + CaCl_2 \cdot Ca(OH)_2 \cdot H_2O$
0.09	37.90	62.01	5.0	$CaCl_2 \cdot Ca(OH)_2 \cdot H_2O + CaCl_2 \cdot 6H_2O$
0.15	10.26	89.59	10.0	$Ca(OH)_2 + CaCl_2 \cdot 3Ca(OH)_2 \cdot 12H_2O$
0.23	36.09	63.68	10.0	$CaCl_2 \cdot 3Ca(OH)_2 \cdot 12H_2O + CaCl_2 \cdot Ca(OH)_2 \cdot H_2O$
0.20	35.94	63.86	10.0	»
0.08	39.00	60.92	10.0	$CaCl_2 \cdot Ca(OH)_2 \cdot H_2O + CaCl_2 \cdot 6H_2O$
0.195	18.15	81.655	25.0	$Ca(OH)_2 + CaCl_2 \cdot 3Ca(OH)_2 \cdot 13H_2O$
0.323	33.21	66.464	25.0	$CaCl_2 \cdot 3Ca(OH)_2 \cdot 13H_2O + CaCl_2 \cdot Ca(OH)_2 \cdot H_2O$
0.029	44.51	55.461	25.0	$CaCl_2 \cdot Ca(OH)_2 \cdot H_2O + CaCl_2 \cdot 6H_2O$
0.00	53.00	47.00	29.0	$CaCl_2 \cdot 6H_2O + \gamma\text{-}CaCl_2 \cdot 4H_2O$
0.00	52.30	47.70	29.5	$CaCl_2 \cdot 6H_2O + \alpha\text{-}CaCl_2 \cdot 4H_2O$
0.23	20.63	79.14	30.0	$Ca(OH)_2 + CaCl_2 \cdot 3Ca(OH)_2 \cdot 12H_2O$
0.43	32.90	66.67	30.0	$CaCl_2 \cdot 3Ca(OH)_2 \cdot 12H_2O + CaCl_2 \cdot Ca(OH)_2 \cdot H_2O$
0.06	46.60	53.37	30.0	$CaCl_2 \cdot Ca(OH)_2 \cdot H_2O + \alpha\text{-}CaCl_2 \cdot 4H_2O$
0.00	50.1	49.90	30.2	$\alpha\text{-}CaCl_2 \cdot 4H_2O + CaCl_2 \cdot 6H_2O$
0.00	56.0	44.00	38.5	$\gamma\text{-}CaCl_2 \cdot 4H_2O + CaCl_2 \cdot 2H_2O$
0.804	31.930	67.233	40.0	$Ca(OH)_2 + CaCl_2 \cdot Ca(OH)_2 \cdot H_2O$
0.74	32.00	67.26	40.0	»
0.06	49.44	50.50	40.0	$CaCl_2 \cdot Ca(OH)_2 \cdot H_2O + \alpha\text{-}CaCl_2 \cdot 4H_2O$
0.00	53.00	47.00	41.0	$\beta\text{-}CaCl_2 \cdot 4H_2O + CaCl_2 \cdot 2H_2O$

Comp. of saturated soln., wt %			t	Comp. of solid phase
Calcium Hydroxide	Calcium Chloride	Water		
0.837	31.95	67.213	45.0	$Ca(OH)_2+CaCl_2 \cdot Ca(OH)_2 \cdot H_2O$
0.76	31.28	67.96	45.0	»
0.08	53.48	46.44	45.0	$CaCl_2 \cdot Ca(OH)_2 \cdot H_2O+CaCl_2 \cdot 2H_2O$
0.00	56.6	43.40	45.1	$CaCl_2 \cdot 4H_2O+CaCl_2 \cdot 2H_2O$
0.74	31.64	67.62	50.0	$Ca(OH)_2+CaCl_2 \cdot Ca(OH)_2 \cdot H_2O$
0.09	53.81	46.10	50.0	$CaCl_2 \cdot Ca(OH)_2 \cdot H_2O+CaCl_2 \cdot 2H_2O$
0.76	31.76	67.48	55	$CaCl_2 \cdot Ca(OH)_2 \cdot H_2O+Ca(OH)_2$
0.08	54.92	45.00	55	»
0.72	30.58	68.70	75	»
0.12	56.98	42.90	75	$CaCl_2 \cdot Ca(OH)_2 \cdot H_2O+CaCl_2 \cdot 2H_2O$

№ 5066 [191]

CALCIUM HYDROXIDE–CALCIUM CHLORIDE–WATER

$Ca(OH)_2—CaCl_2—H_2O$

$t=-20$

Comp. of saturated soln., wt %			Comp. of solid phase
Calcium Chloride	Calcium Hydroxide	Water	
20.40	0.09	79.51	$H_2O+CaCl_2 \cdot 3Ca(OH)_2 \cdot 12H_2O$
22.20	0.06	77.74	$CaCl_2 \cdot 3Ca(OH)_2 \cdot 12H_2O$
28.30	0.08	71.62	»
32.77	0.09	67.14	—
33.68	0.07	66.25	$CaCl_2 \cdot 3Ca(OH)_2 \cdot 12H_2O$
33.90	0.08	66.02	$CaCl_2 \cdot 3Ca(OH)_2 \cdot 12H_2O+CaCl_2 \cdot 6H_2O$
34.1	0.0	65.9	$CaCl_2 \cdot 6H_2O$

№ 5067 [883]

CALCIUM HYDROXIDE–CALCIUM CHLORIDE–WATER

$Ca(OH)_2—CaCl_2—H_2O$

Comp. of saturated soln., wt %			t	Comp. of solid phase
Calcium Chloride	Calcium Hydroxide	Water		
26.72	0.186	73.094	50	$Ca(OH)_2$
30.07	0.203	68.73	50	$Ca(OH)_2+CaCl_2 \cdot Ca(OH)_2 \cdot H_2O$
33.75	0.206	66.044	50	$CaCl_2 \cdot Ca(OH)_2 \cdot H_2O$
38.44	0.180	61.38	50	»
41.17	0.172	58.66	50	»
46.40	0.122	53.48	50	»
50.21	0.096	49.69	50	»
52.37	0.08	47.55	50	$CaCl_2 \cdot Ca(OH)_2 \cdot H_2O+CaCl_2 \cdot 2H_2O$
56.95	0.00	43.05	50	$CaCl_2 \cdot 2H_2O$
17.40	0.064	82.54	75	$Ca(OH)_2$

Table № 5067 (continuation)

Comp. of saturated soln., wt %			t	Comp. of solid phase
Calcium Chloride	Calcium Hydroxide	Water		
24.38	0.080	75.54	75	$Ca(OH)_2$
31.73	0.125	68.15	75	»
32.20	0.154	67.65	75	$Ca(OH)_2 + CaCl_2 \cdot Ca(OH)_2 \cdot H_2O$
32.37	0.202	67.43	75	$CaCl_2 \cdot Ca(OH)_2 \cdot H_2O + Ca(OH)_2$
40.57	0.172	59.26	75	$CaCl_2 \cdot Ca(OH)_2 \cdot H_2O$
46.90	0.062	53.04	75	»
48.52	0.131	51.35	75	»
51.44	0.160	48.40	75	$2CaCl_2 \cdot Ca(OH)_2 \cdot H_2O$
53.74	0.214	46.05	75	»
54.92	0.08	45.00	75	»
57.35	0.12	42.53	75	$2CaCl_2 \cdot Ca(OH)_2 \cdot H_2O + CaCl_2 \cdot 2H_2O$
58.95	0.00	41.05	75	$CaCl_2$
32.56	0.211	67.23	80	$Ca(OH)_2 + CaCl_2 \cdot Ca(OH)_2 \cdot H_2O$
34.60	0.158	65.24	80	$CaCl_2 \cdot Ca(OH)_2 \cdot H_2O$
41.41	0.102	58.49	80	»
47.03	0.160	52.81	80	»
49.97	0.098	49.93	80	$2CaCl_2 \cdot Ca(OH)_2 \cdot H_2O$
52.37	0.061	47.57	80	»
55.24	0.133	44.63	80	»
57.05	0.072	42.88	80	»

№ 5068 [154]

CALCIUM HYDROXIDE–CALCIUM CHLORIDE–CALCIUM CHLORATE–WATER

$Ca(OH)_2—CaCl_2—Ca(ClO_3)_2—H_2O$

$t = 25$

Comp. of saturated soln., wt %				Comp. of solid phase
Calcium Hydr-oxide	Calcium Chloride	Calcium Chlorate	Water	
0.05	43.00	3.80	53.15	$CaCl_2 \cdot 6H_2O + CaCl_2 \cdot Ca(OH)_2 \cdot H_2O$
0.06	41.38	11.18	47.38	$CaCl_2 \cdot 6H_2O + \alpha\text{-}CaCl_2 \cdot 4H_2O + CaCl_2 \cdot Ca(OH)_2 \cdot H_2O$
0.07	38.03	17.95	43.95	$CaCl_2 \cdot 4H_2O + CaCl_2 \cdot Ca(OH)_2 \cdot H_2O$
0.07	34.54	24.78	40.61	»
0.07	30.55	33.34	36.04	$\alpha\text{-}CaCl_2 \cdot 4H_2O + CaCl_2 \cdot Ca(OH)_2 \cdot H_2O + Ca(ClO_3) \cdot 2H_2O$
0.08	22.88	40.56	36.48	$\alpha\text{-}CaCl_2 \cdot Ca(OH)_2 \cdot H_2O + Ca(ClO_3)_2 \cdot 2H_2O$
0.09	18.35	45.16	36.40	»
0.14	10.70	53.63	35.53	$CaCl_2 \cdot Ca(OH)_2 \cdot H_2O + Ca(ClO_3)_2 \cdot Ca(OH)_2 + Ca(ClO_3)_2 \cdot 2H_2O$
0.16	8.99	55.38	35.47	$Ca(ClO_3)_2 \cdot Ca(OH)_2 \cdot 12H_2O + CaCl_2 \cdot Ca(OH)_2 \cdot H_2O$
0.12	3.99	61.43	34.46	»
0.54	24.30	15.25	59.91	$Ca(OH)_2 + CaCl_2 \cdot 3Ca(OH)_2 \cdot 12H_2O$
0.74	18.78	24.58	55.90	»

Comp. of saturated soln., wt %				Comp. of solid phase
Calcium Hydr-oxide	Calcium Chloride	Calcium Chlorate	Water	
0.22	13.57	11.27	74.94	$Ca(OH)_2 + CaCl_2 \cdot 3Ca(OH)_2 \cdot 12H_2O$
0.26	10.84	21.42	67.48	»
0.62	2.73	46.57	50.08	$Ca(OH)_2 + Ca(ClO_3)_2 \cdot Ca(OH)_2$
0.66	5.66	42.80	50.88	»
0.74	7.96	39.52	51.78	$Ca(OH)_2 + CaCl_2 \cdot 3Ca(OH)_2 \cdot 12H_2O + Ca(ClO_3)_2 \cdot Ca(OH)_2$
0.73	11.11	35.69	52.47	$CaCl_2 \cdot 3Ca(OH)_2 \cdot 12H_2O + Ca(ClO_3)_2 \cdot Ca(OH)_2$
0.73	13.11	33.18	52.98	»
0.73	14.11	31.93	53.23	$CaCl_2 \cdot 3Ca(OH)_2 \cdot 12H_2O + CaCl_2 \cdot Ca(OH)_2 \cdot xH_2O + Ca(ClO_3)_2 \cdot Ca(OH)_2$
0.20	13.74	39.48	46.58	$CaCl_2 \cdot Ca(OH)_2 \cdot H_2O + Ca(ClO_3)_2 \cdot Ca(OH)_2$
0.16	12.10	47.00	40.74	»

№ 5069 [189]

CALCIUM HYDROXIDE–CALCIUM CHLORATE–WATER

$Ca(OH)_2—Ca(ClO_3)_2—H_2O$

$t=75$

Comp. of saturated soln., wt %			Comp. of solid phase
Calcium Chlorate	Calcium Hydroxide	Water	
0.00	0.10	99.90	$Ca(OH)_2$
7.72	0.10	92.18	»
25.95	0.17	73.88	»
31.35	0.21	68.44	»
44.76	0.35	54.89	»
50.90	0.86	48.24	»
52.97	1.18	45.85	$Ca(OH)_2 + Ca(ClO_3)_2 \cdot Ca(OH)_2$
62.56	0.36	34.08	$Ca(ClO_3)_2 \cdot Ca(OH)_2$
69.44	0.29	30.27	»
71.48	0.00	28.52	»
76.71	0.47	22.82	»
77.83	0.00	22.17	$Ca(ClO_3)_2 \cdot 2H_2O$

CALCIUM HYDROXIDE–CALCIUM CHLORATE–WATER

$Ca(OH)_2$—$Ca(ClO_3)_2$—H_2O

Comp. of saturated soln., wt %			t	Comp. of solid phase
Calcium Hydr-oxide	Calcium Chlorate	Water		
0.17	0.00	99.83	0	$Ca(OH)_2$
0.15	20.62	79.23	0	»
0.16	23.07	76.77	0	$Ca(OH)_2 + Ca(ClO_3)_2 \cdot 3Ca(OH)_2 \cdot 12H_2O$
0.16	23.08	76.76	0	»
0.14	24.61	75.25	0	$Ca(ClO_3)_2 \cdot 3Ca(OH)_2 \cdot 12H_2O$
0.14	26.33	73.53	0	»
0.13	27.92	71.95	0	»
0.12	33.06	66.82	0	»
0.15	46.10	53.75	0	»
0.17	47.70	52.13	0	»
0.22	48.60	51.18	0	»
0.22	50.06	49.72	0	»
0.25	51.49	48.26	0	$Ca(ClO_3)_2 \cdot 3Ca(OH)_2 \cdot 12H_2O + Ca(ClO_3)_2 \cdot Ca(OH)_2$
0.15	54.03	45.82	0	$Ca(ClO_3)_2 \cdot Ca(OH)_2$
0.11	56.59	43.30	0	»
0.10	60.43	39.47	0	»
0.10	63.31	36.59	0	$Ca(ClO_3)_2 \cdot Ca(OH)_2 + Ca(ClO_3)_2 \cdot 2H_2O$
0.10	63.40	36.50	0	»
0.00	63.70	36.30	0	$Ca(ClO_3)_2$ $2H_2O$
0.16	0.00	99.84	25	$Ca(OH)_2$
0.13	4.38	95.49	25	»
0.12	9.07	90.81	25	»
0.12	12.29	87.59	25	»
0.13	18.27	81.60	25	»
0.14	23.56	76.30	25	»
0.15	28.35	71.50	25	»
0.18	32.75	67.07	25	»
0.26	42.41	57.33	25	»
0.42	48.13	51.45	25	»
0.57	49.72	49.71	25	$Ca(OH)_2 + Ca(ClO_3) \cdot Ca(OH)_2$
0.33	52.40	47.27	25	$Ca(ClO_3)_2 \cdot Ca(OH)_2$
0.20	56.17	43.63	25	»
0.13	59.79	40.08	25	»
0.12	63.28	36.60	25	»
0.13	66.34	33.53	25	$Ca(ClO_3)_2 \cdot Ca(OH)_2 + Ca(ClO_3) \cdot 2H_2O$
0.00	67.18	32.82	25	$Ca(ClO_3)_2 \cdot 2H_2O$

№ 5071 [1481]

CALCIUM HYDROXIDE–CALCIUM IODATE–WATER

$Ca(OH)_2—Ca(IO_3)_2—H_2O$

$t=25$

Concn. of Calcium Hydroxide in aqueous soln., g/l	Solubility of Calcium Iodate in aqueous soln. of Calcium Hydroxide, g/l	Concn. of Calcium Hydroxide in aqueous soln., g/l	Solubility of Calcium Iodate in aqueous soln. of Calcium Hydroxide, g/l
0.000	3.057	1.057	2.503
0.269	2.835	1.121	2.484
0.318	2.803	1.378	2.429
0.763	2.593	1.551	2.398
0.953	2.527		

№ 5072 [1940]

CALCIUM HYDROXIDE–STRONTIUM CHLORIDE–WATER

$Ca(OH)_2—SrCl_2—H_2O$

$t=25$

Content in aqueous soln., mol/1000 g Water		Comp. of solid phase
Calcium Hydroxide	Strontium Chloride	
0.01976	0	$Ca(OH)_2$
0.02624	0.0653	»
0.03043	0.120	»
0.03702	0.283	»
0.04135	0.445	»
0.04819	0.823	»
0.05336	1.33	»
0.05622	2.19	»
0.05623	2.23	»
0.05537	3.18	»

Note. $P_{CO_2} < 10^{-13}$ atm.

№ 5073 [1940]

CALCIUM HYDROXIDE–CESIUM CHLORIDE–WATER

$Ca(OH)_2—CsCl—H_2O$

$t=25$

Content in aqueous soln., mol/1000 g Water		Comp. of solid phase
Cesium Chloride	Calcium Hydroxide	
3.55	0.02056	$Ca(OH)_2$
1.56	0.02647	»
1.25	0.02703	»
0.862	0.02724	»
0.537	0.02684	»
0.199	0.02451	»
0.073	0.02236	»
0	0.01976	»

$Ca(OH)_2—MeX—H_2O$

$t=25$

Salt		Concn. in aqueous soln., g/l Water	
Name	Formula	Salt	Calcium Hydroxide
—	—	0	1.46
Sodium Chloride	NaCl	0.701	1.53
»	»	3.74	1.68
»	»	7.07	1.78
»	»	21.39	2.01
»	»	44.60	2.16
»	»	70.72	2.20
»	»	102.3	2.17
»	»	125.7	2.12
»	»	130.9	2.10
»	»	164.2	1.98
»	»	219.8	1.79
Potassium Chloride	KCl	2.83	1.60
»	»	5.67	1.68
»	»	19.68	1.89
»	»	39.29	2.01
»	»	73.74	2.05
»	»	80.52	2.05
»	»	95.44	2.03
»	»	120.8	2.00
»	»	146.1	1.93
»	»	214.0	1.74
Cesium Chloride	CsCl	12.29	1.66
»	»	33.50	1.82
»	»	90.41	1.99
»	»	145.1	2.02
»	»	210.5	2.00
»	»	262.7	1.96
»	»	597.7	1.52
Lithium Chloride	LiCl	0.229	1.68
»	»	4.88	1.83
»	»	10.56	2.07
»	»	19.72	2.35
»	»	35.19	2.65
»	»	58.09	2.96
»	»	78.44	3.14
»	»	95.40	3.26
»	»	128.5	3.44
»	»	161.5	3.53
»	»	227.3	3.59
»	»	439.7	3.31
Barium Chloride	$BaCl_2$	3.96	1.69
»	»	7.71	1.83
»	»	13.75	1.99
»	»	23.12	2.18
»	»	33.11	2.34
»	»	59.77	2.65
»	»	88.51	2.90

Table № 5074 (continuation)

Salt		Concn. in aqueous soln., g/l Water	
Name	Formula	Salt	Calcium Hydroxide
Barium Chloride	$BaCl_2$	130.0	3.15
»	»	268.7	3.64
»	»	299.9	3.70
»	»	331.1	3.72
Strontium Chloride	$SrCl_2$	8.40	1.94
»	»	19.02	2.25
»	»	44.87	2.74
»	»	70.55	3.06
»	»	130.5	3.57
»	»	210.9	3.95
»	»	347.2	4.17
»	»	353.5	4.17
»	»	504.2	4.10
Sodium Chlorate	$NaClO_3$	20.12	1.83
»	»	41.09	1.96
»	»	76.22	2.04
»	»	86.65	2.05
»	»	121.3	2.06
»	»	169.3	2.02
»	»	209.7	1.94
»	»	315.1	1.74
»	»	485.4	1.41
Sodium Perchlorate	$NaClO_4$	14.45	1.72
»	»	48.86	1.88
»	»	58.29	1.90
»	»	95.02	1.90
»	»	142.0	1.84
»	»	193.5	1.74
»	»	293.7	1.52
»		437.1	1.20
Sodium Bromide	NaBr	19.45	1.84
»	»	35.81	1.96
»	»	74.20	2.08
»	»	100.7	2.11
»	»	133.8	2.10
»	»	182.2	2.06
»	»	298.4	1.84
Potassium Bromide	KBr	19.28	1.78
»	»	32.13	1.86
»	»	53.91	1.93
»	»	90.32	1.98
»	»	110.7	1.98
»	»	151.1	1.95
»	»	184.4	1.92
»	»	248.7	1.81
»	»	385.6	1.52
Sodium Iodide	NaI	12.74	1.68
»	»	23.84	1.79
»	»	53.37	1.92
»	»	90.70	1.98
»	»	135.5	2.00

Table № 5074 (continuation)

Salt		Concn. in aqueous soln., g/l Water	
Name	Formula	Salt	Calcium Hydroxide
Sodium Iodide	NaI	199.4	1.97
»	»	239.9	1.93
»	»	274.4	1.88
»	»	316.3	1.83
»	»	428.8	1.65
»	»	631.2	1.39
Sodium Nitrate	$NaNO_3$	14.45	1.88
»	»	18.45	1.92
»	»	29.83	2.05
»	»	43.35	2.16
»	»	71.23	2.30
»	»	114.7	2.39
»	»	149.6	2.43
»	»	192.9	2.43
»	»	238.8	2.39
»	»	321.3	2.28
»	»	401.2	2.17
Sodium Nitrite	$NaNO_2$	17.39	1.97
»	»	31.40	2.14
»	»	58.93	2.32
»	»	89.70	2.41
»	»	140.1	2.46
»	»	200.1	2.46
»	»	284.3	2.37
»	»	401.6	2.20

№ 5075 CALCIUM HYDROGEN CARBONATE–CALCIUM SULFATE–WATER [1663]

$Ca(HCO_3)_2—CaSO_4—H_2O$

$t=25$

Concn. in aqueous soln., g/l Water		p_{CO_2}, atm.	Concn. in aqueous soln., g/l Water		p_{CO_2}, atm.
Calcium Hydrogen Carbonate	Calcium Sulfate		Calcium Hydrogen Carbonate	Calcium Sulfate	
0.504	1.965	0.121	1.200	1.850	0.897
0.577	1.939	0.167	1.221	1.824	0.916
0.704	1.905	0.270	1.218	1.796	0.957
0.861	1.835	0.431	1.224	1.796	0.063
1.008	1.804	0.598	1.224	1.767	0.960
1.041	1.807	0.652	1.229	1.793	0.953
1.088	1.826	0.728	1.273	1.159	0.966
1.127	1.833	0.770	1.346	0.562	0.962

№ 5076 CALCIUM HYDROGEN CARBONATE–CALCIUM SULFATE–WATER [752]

$Ca(HCO_3)_2—CaSO_4—H_2O$

$t=40$

Content in aqueous soln., wt %		Comp. of solid phase
Calcium Sulfate	Calcium Hydrogen Carbonate	
0.236	0	$CaSO_4 \cdot 2H_2O$
0.250	0.0059	»
0.271	0.0229	$CaSO_4 \cdot 2H_2O$ + solid soln.
0.244	0.0369	Solid solution
0.237	0.0524	»
0.222	0.0793	»
0.228	0.0938	»
0.216	0.1208	Solid soln. + $CaCO_3$
0.214	0.1190	$CaCO_3$
0.116	0.0980	»
0	0.0971	»

№ 5077 [1398]

CALCIUM CARBONATE–CALCIUM SULFATE–WATER

$CaCO_3—CaSO_4—H_2O$

Concn., g/l				
All solid substances	Ca	SO_4	CO_3	
0.210	0.058	0.131	0.006	159
0.208	0.060	0.128	0.012	159
0.139	0.041	0.076	0.013	178
0.095	0.028	0.046	0.013	195
0.075	0.022	0.0359	0.011	206
0.048	0.0	0.0216	0.009	225
0.040	0.0	0.0146	0.008	241
0.032	0.0	0.0117	0.008	252

№ 5078 [824]

CALCIUM CARBONATE–CALCIUM SULFATE–WATER

$CaCO_3—CaSO_4—H_2O$

Content in aqueous soln.		t	pH	Comp. of solid phase
Calcium Sulfate, wt %	Calcium Carbonate mg-eq/l			
0.200	0.96	18—19	7.5	$CaCO_3$
0.175	0.76	18—19	7.4	»
0.150	0.76	18—19	7.5	»
0.125	0.89	18—19	7.4	»
0.100	1.08	18—19	7.5	»

Table № 5078 (continuation)

Content in aqueous soln.		t	pH	Comp. of solid phase
Calcium Sulfate, wt %	Calcium Carbonate, mg-eq/l			
0.075	0.96	18—19	7.4	$CaCO_3$
0.050	1.15	18—19	7.5	»
0.200	0.69	30	7.8	»
0.175	0.67	30	7.8	»
0.150	0.69	30	7.8	»
0.125	0.75	30	7.8	»
0.100	0.72	30	7.8	»
0.075	0.77	30	7.9	»
0.050	0.88	30	7.9	»

№ 5079 [1736]

CALCIUM CARBONATE–CALCIUM CHLORIDE–WATER

$CaCO_3—CaCl_2—H_2O$

$t=0$

Content in aqueous soln., g/1000 g Water		Content in aqueous soln., g/1000 g Water	
$CaCl_2 \cdot 6H_2O$	CaO	$CaCl_2 \cdot 6H_2O$	CaO
2.75	0.0084	0.55	0.0090
1.1	0.0084	0	0.0174

№ 5080 [1398]

CALCIUM CARBONATE–CALCIUM SULFATE–WATER

$CaCO_3—CaSO_4—H_2O$

Content in aqueous soln., g/1000 g Water			t	Comp. of solid phase
Ca^{2+}	SO_4^{2-}	CO_3^{2-}		
0.059	0.129	0.009	159	$CaSO_4+CaCO_3$
0.041	0.076	0.013	178	»
0.028	0.048	0.011	195	»
0.022	0.036	0.011	206	»
0.00	0.022	0.009	225	»
0.00	0.015	0.008	241	»
0.00	0.012	0.008	252	»

№ 5081 CALCIUM CARBONATE–SEAWATER [2717]

$CaCO_3$— — —H_2O

Solubility of $CaCO_3$ (mg-eq/l) in Seawater											
Pressure of CO_2 (atm.·10^{-4})											
0.5	1	2	3	4	5	6	7	8	9	10	t
0.9	1.2	1.5	1.7	2.0	2.2	2.5	2.3	2.6	2.8	2.9	0
0.9	1.1	1.4	1.7	1.9	2.1	2.2	2.4	2.5	2.6	2.8	2
0.9	1.1	1.4	1.6	1.8	2.0	2.1	2.3	2.4	2.5	2.6	4
0.8	1.0	1.3	1.6	1.7	1.9	2.0	2.2	2.3	2.4	2.5	6
0.8	1.0	1.3	1.5	1.7	1.8	1.9	2.1	2.2	2.3	2.4	8
0.7	0.9	1.2	1.4	1.6	1.7	1.8	1.9	2.1	2.2	2.3	10
0.7	0.9	1.1	1.3	1.5	1.6	1.7	1.8	2.0	2.1	2.2	12
0.7	0.8	1.1	1.3	1.4	1.5	1.6	1.7	1.8	2.0	2.1	14
0.6	0.8	1.0	1.2	1.3	1.5	1.6	1.6	1.7	1.9	2.1	16
0.6	0.8	1.0	1.1	1.3	1.4	1.5	1.6	1.6	1.8	1.9	18
0.6	0.7	0.9	1.1	1.2	1.3	1.4	1.5	1.6	1.7	1.7	20
0.6	0.7	0.9	1.0	1.1	1.2	1.3	1.4	1.5	1.6	1.6	22
0.5	0.6	0.8	1.0	1.1	1.2	1 3	1.3	1.4	1.5	1.6	24
0.5	0.6	0.8	0.9	1.0	1.1	1.2	1.3	1.3	1.4	1.5	26
0.5	0.6	0.7	0.9	1.0	1.0	1.1	1.2	1.3	1.3	1.4	28
0.4	0.5	0.7	0.8	0.9	1.0	1.1	1.1	1.2	1.2	1.3	30

№ 5082 CALCIUM CARBONATE–SEAWATER [2718]

$CaCO_3$— — — H_2O

Solubility of Calcium Carbonate in Seawater, mg/l	P_{CO_2}, atm · 10^{-4}	t	Solubility of Calcium Carbonate in Seawater, mg/l	P_{CO_2}, atm · 10^{-4}	t
60	1	0	35	1	20
75	2	0	45	2	20
80	3	0	50	3	20
100	5	0	60	5	20
135	10	0	83	10	20
45	1	10	25	1	30
60	2	10	35	2	30
65	3	10	35	3	30
78	5	10	45	5	30
105	10	10	60	10	30

Note. Seawater contained 3.5% salts.

№ 5083 CALCIUM CARBONATE–SEAWATER [334]

$CaCO_3$— — — H_2O

Content in aqueous soln., g/l			d	b.p.	Comp. of solid phase
Cl^-	HCO_3^-	$CaCO_3$			
18.4	0.0467	0.0383	1.024	100.6	$CaCO_3$
18.5	0.0347	0.0285	1.025	100.6	»
23.7	0.0233	0.0191	1.029	100.8	»
26.5	0.0196	0.0161	1.034	100.8	»
29.7	0.0191	0.0156	1.037	101.0	»
43.6	0.0189	0.0155	1.055	101.4	»
55.2	0.0190	0.0156	1.067	101.5	»

№ 5084 CALCIUM CARBONATE–SEAWATER [212]

$CaCO_3$— — —H_2O

Solubility of $CaCO_3$ (wt %) in seawater					t
Content of Cl^- (%)					
1.85	3.70	4.50	5.55	6.50	
—	—	—	—	0.0056	—7.4
—	—	0.0082	0.0071	0.0062	—4.8
—	0.0094	0.0086	0.0073	0.0067	—3.5
0.0114	0.0099	0.0093	0.0084	0.0077	—1.8
0.0110	—	—	—	—	0

№ 5085 [2167]

CALCIUM THIOCYANATE–SILVER THIOCYANATE–WATER

$Ca(CNS)_2$—AgCNS—H_2O

Comp. of saturated soln., wt %			t	Comp. of solid phase
Silver Thio-cyanate	Calcium Thio-cyanate	Water		
0.0	60.20	39.80	25	α-$Ca(CNS)_2 \cdot 4H_2O$
4.03	58.80	37.17	25	»
7.49	57.05	35.46	25	»
7.52	57.03	35.45	25	α-$Ca(CNS)_2 \cdot 4H_2O + Ca(CNS)_2 \cdot 2AgCNS \cdot 2H_2O$
8.50	55.21	36.29	25	$Ca(CNS)_2 \cdot 2AgCNS \cdot 2H_2O$
8.26	54.74	37.00	25	»
10.27	50.76	38.97	25	»
13.09	44.99	41.92	25	»
14.96	41.95	43.09	25	»
19.22	37.02	43.76	25	»
25.71	29.12	45.17	25	»
26.74	27.86	45.40	25	»
29.79	25.47	44.74	25	»
31.01	25.15	43.84	25	$Ca(CNS)_2 \cdot 2AgCNS \cdot 2H_2O$+AgCNS
30.70	24.52	44.78	25	AgCNS
22.88	21.72	55.46	25	»
15.42	19.23	65.35	25	»
11.74	17.33	70.93	25	»
8.09	15.01	76.90	25	»
3.42	14.11	82.47	25	»
0.0	59.00	41.00	20	β-$Ca(CNS)_2 \cdot 4H_2O$
4.24	57.27	38.49	20	»
5.19	56.85	37.96	20	»
7.50	56.41	36.09	20	»
8.73	55.44	35.83	20	»

№ 5086 CALCIUM NITRITE–CALCIUM NITRATE–WATER [724]

$Ca(NO_2)_2—Ca(NO_3)_2—H_2O$

$t=25$

Comp. of saturated soln., wt %			Comp. of solid phase
Calcium Nitrite	Calcium Nitrate	Water	
49.80	0.00	50.20	$Ca(NO_2)_2 \cdot 4H_2O$
47.82	2.92	49.26	»
44.80	9.03	46.17	»
43.02	12.97	44.01	$Ca(NO_2)_2 \cdot 4H_2O + Ca(NO_2)_2 \cdot H_2O$
42.90	12.88	44.22	»
39.32	15.81	44.87	$Ca(NO_2)_2 \cdot H_2O$
36.54	19.80	43.66	»
32.21	26.70	41.09	»
30.07	30.03	39.90	»
26.03	35.64	38.33	»
21.53	42.18	36.29	»
18.57	47.70	33.73	»
17.20	50.00	32.80	$Ca(NO_2)_2 \cdot H_2O + Ca(NO_3)_2 \cdot 4H_2O$
17.27	50.31	32.42	»
14.21	50.30	35.49	$Ca(NO_3)_2 \cdot 4H_2O$
9.07	52.20	38.73	»
4.90	54.64	40.46	»
0.00	57.50	42.50	»

№ 5087 CALCIUM NITRITE–SILVER NITRITE–WATER [2298]

$Ca(NO_2)_2—AgNO_2—H_2O$

Comp. of saturated soln., wt %			t
Calcium Nitrite	Silver Nitrite	Water	
45.38	5.50	49.12	14

№ 5088 CALCIUM NITRITE–THALLIUM NITRITE–WATER [723]

$Ca(NO_2)_2—TlNO_2—H_2O$

$t=25$

Comp. of saturated soln., wt %			Comp. of solid phase
Thallium Nitrite	Calcium Nitrite	Water	
0.00	49.01	50.99	$Ca(NO_2)_2 \cdot H_2O$
1.53	46.51	51.96	»
2.30	42.05	55.65	$Ca(NO_2)_2 \cdot H_2O + Ca(NO_2)_2 \cdot TlNO_2 \cdot 3H_2O$
2.49	41.94	55.57	»
6.38	33.71	59.91	$Ca(NO_2)_2 \cdot TlNO_2 \cdot 3H_2O$

Comp. of saturated soln., wt %			Comp. of solid phase
Thallium Nitrite	Calcium Nitrite	Water	
10.65	28.25	61.10	$Ca(NO_2)_2 \cdot TlNO_2 \cdot 3H_2O$
17.71	23.86	58.43	»
21.26	22.42	56.32	»
25.77	19.87	54.36	$Ca(NO_2)_2 \cdot TlNO_2 \cdot 3H_2O + TlNO_2$
25.96	19.59	54.45	»
22.40	17.13	60.47	$TlNO_2$
24.52	12.67	62.81	»
26.20	5.99	67.81	»
32.25	0.00	67.75	»

№ 5089 CALCIUM NITRATE–CALCIUM ORTHOPHOSPHATE–WATER [77]

$Ca(NO_3)_2—Ca_3(PO_4)_2—H_2O$

Concn. in aqueous soln., wt %			d_4^t	t	Comp. of solid phase
Calcium Oxide	Nitrogen Pentoxide	Phosphorus Pentoxide			
5.81	0.0	24.20	—	25	$Ca(H_2PO_4)_2 \cdot H_2O + CaHPO_4$
6.10	1.46	23.00	1.280	25	»
6.53	3.17	21.20	1.298	25	»
7.22	4.78	19.53	1.298	25	»
8.06	7.45	17.07	1.291	25	»
8.72	10.18	14.60	1.320	25	»
9.58	12.77	12.51	1.328	25	»
10.52	16.06	10.45	1.341	25	»
11.86	20.60	8.19	1.366	25	»
12.08	19.00	7.69	1.370	25	»
13.37	22.80	5.65	1.392	25	»
13.95	24.27	4.96	1.408	25	»
15.37	27.46	3.55	1.446	25	»
17.57	31.71	2.02	1.486	25	»
18.70	34.87	1.25	1.539	25	»
19.73	37.26	1.18	1.551	25	$Ca(H_2PO_4)_2 \cdot H_2O + CaHPO_4 + Ca(NO_3)_2 \cdot 4H_2O$
19.68	37.60	0.97	1.551	25	$CaHPO_4 + Ca(NO_3)_2 \cdot 4H_2O$
19.66	37.58	0.36	1.558	25	»
19.79	38.20	0.0	1.563	25	$Ca_3(PO_4)_2 + CaHPO_4 + Ca(NO_3)_2 \cdot 4H_2O$
19.25	36.21	2.05	1.552	25	$Ca(H_2PO_4)_2 \cdot H_2O + Ca(NO_3)_2 \cdot 4H_2O$
18.75	35.47	3.79	1.576	25	»
18.53	34.60	4.45	1.542	25	»
18.23	34.69	5.54	1.568	25	»
17.42	32.60	8.10	1.563	25	»
16.85	32.30	9.92	1.579	25	»
16.14	31.90	10.84	1.571	25	»
16.29	30.92	12.01	—	25	»
15.80	30.94	12.62	1.577	25	»
14.70	29.76	15.69	—	25	»
14.27	29.86	17.79	—	25	»

Concn. in aqueous soln., wt %			d_4^+	t	Comp. of solid phase
Calcium Oxide	Nitrogen Pentoxide	Phosphorus Pentoxide			
14.24	30.36	18.21	1.572	25	$Ca(H_2PO_4)_2 \cdot H_2O + Ca(NO_3)_2 \cdot 4H_2O + Ca(NO_3)_2 \cdot 3H_2O$
12.82	30.76	21.49	—	25	$Ca(H_2PO_4)_2 \cdot H_2O + Ca(NO_3)_2 \cdot 3H_2O$
11.79	30.43	23.94	—	25	»
10.82	30.70	26.74	—	25	»
9.76	29.83	29.90	—	25	»
7.78	26.89	34.63	1.637	25	»
7.27	25.32	35.62	1.656	25	»
6.81	25.62	37.05	1.670	25	»
5.70	23.08	39.25	1.679	25	»
4.99	21.52	41.70	1.684	25	»
4.23	19.34	44.56	1.592	25	»
11.03	51.89	0.0	1.536	65	$Ca(NO_3)_2 \cdot 4H_2O + Ca(NO_3)_2 \cdot 3H_2O$
12 29	49.94	2.84	1.559	25	»
12.10	49.03	3.98	—	65	»
12.78	47.70	4.54	1.566	25	»
12.91	47.86	6.04	—	25	»
13.57	46.05	7.68	1.588	25	»
13.49	44.44	7.74	1.586	25	»
14.34	41.61	9.31	1.605	25	»
14.76	40.52	11.30	1.596	25	»
14.60	37.10	14.03	1.605	25	»
14.11	34.53	14.96	1.605	25	»
9.54	56.28	0.0	1.525	25	$Ca(NO_3)_2 \cdot 3H_2O + Ca(NO_3)_2 \cdot 2H_2O$
9.65	53.03	3.77	1.555	25	»
8.43	51.93	7.13	1.555	25	»
8.07	50.11	10.12	1.571	25	»
7.41	47.35	13.56	1.590	25	»
6.83	44.54	16.52	1.599	25	»
6.20	41.10	21.23	1.612	25	»
5.47	38.44	23.28	1.610	25	»
5.06	38.69	23.68	1.618	25	»
4.89	37.34	25.37	1.630	25	»
3.50	34.15	29.95	1.647	25	»
2.53	31.27	33.50	1.659	25	»
1.80	29.08	37.67	1.653	25	»
1.28	26.85	40.95	1.665	25	»
1.19	25.88	43.15	1.670	25	»
2.79	62.84	0.0	1.459	25	$Ca(NO_3)_2 \cdot 2H_2O + Ca(NO_3)_2$
2.03	61.12	4.21	1.468	25	»
1.56	59.56	6.93	1.487	25	»
1.24	56.75	9.97	1.481	25	»
0.99	54.0	14.03	1.499	25	»
0.58	52.40	17.92	1.509	25	»
0.35	48.89	23.34	1.541	25	»
0.23	44.77	28.51	1.564	25	»
0.0	42.65	30.65	1.560	25	»
5.88	0.0	28.85	1.314	50	$Ca(H_2PO_4)_2 \cdot H_2O + CaHPO_4$
6.82	4.24	24.71	—	50	»
8.15	8.32	20.66	1.326	50	»

Concn. in aqueous soln., wt %			d_4^t	t	Comp. of solid phase
Calcium Oxide	Nitrogen Pentoxide	Phosphorus Pentoxide			
10.81	16.65	15.48	1.374	50	$Ca(H_2PO_4)_2 \cdot H_2O + CaHPO_4$
13.00	22.53	12.23	1.427	50	»
14.77	26.71	11.19	1.482	50	»
15.94	29.56	10.11	—	50	»
17.60	33.58	9.71	—	50	»
19.93	38.62	7.80	—	50	»
21.54	41.65	6.53	—	50	»
23.41	45.81	4.80	—	50	$Ca(H_2PO_4)_2 \cdot H_2O + CaHPO_4 + Ca(NO_3)_2 \cdot 2H_2O$
20.79	43.28	11.12	—	50	$Ca(H_2PO_4)_2 \cdot H_2O + Ca(NO_3)_2 \cdot 2H_2O + Ca(NO_3)_2$
20.24	42.56	11.55	—	50	$Ca(H_2PO_4)_2 \cdot H_2O + Ca(NO_3)_2$
15.43	37.59	18.26	—	50	»
14.79	36.60	19.76	1.622	50	»
10.85	31.10	27.03	1.641	50	»
9.33	29.86	30.87	1.655	50	»
7.30	26.63	36.27	1.653	50	»
5.63	19.40	45.57	1.684	50	»
5.40	18.39	48.11	—	50	»
25.95	51.23	0.0	—	50	$Ca(NO_3)_2 \cdot 2H_2O + Ca(NO_3)_2$
24.36	48.75	3.16	—	50	»
23.11	48.21	4.85	—	50	»
23.64	48.00	5.42	—	50	»
22.69	46.43	7.34	—	50	»
21.77	45.01	9.38	—	50	»
20.79	43.28	11.12	—	50	$Ca(H_2PO_4)_2 \cdot H_2O + Ca(NO_3)_2 \cdot 2H_2O + Ca(NO_3)_2$

№ 5090 CALCIUM NITRATE–CALCIUM SULFATE–WATER [2552]

$Ca(NO_3)_2—CaSO_4—H_2O$

$t=25$

Concn. in aqueous soln., g/l		d_4	Concn. in aqueous soln., g/l		d_4
Calcium Nitrate	Calcium Sulfate		Calcium Nitrite	Calcium Sulfate	
0.00	2.08	0.998	300	0.76	1.204
25	1.24	1.014	400	0.57	1.265
50	1.20	1.032	500	0.40	1.328
100	1.13	1.067	544	0.35	1.352
200	0.93	1.137			

№ 5091 CALCIUM NITRATE–CALCIUM THIOSULFATE–WATER [2015]

$Ca(NO_3)_2—CaS_2O_3—H_2O$

Comp. of saturated soln., wt %			t	Comp. of solid phase
Calcium Thiosulfate	Calcium Nitrate	Water		
5.46	46.02	48.52	9	$Ca(NO_3)_2 \cdot 4H_2O$
6.81	45.68	47.51	9	$Ca(NO_3)_2 \cdot 4H_2O + CaS_2O_3 \cdot 6H_2O$
10.46	?7.92	61.62	9	$CaS_2O_3 \cdot 6H_2O$
22.81	10.49	66.70	9	»
4.27	54.03	41.70	25	$Ca(NO_3)_2 \cdot 4H_2O$
9.10	50.25	40.65	25	»
13.00	45.92	41.08	25	$Ca(NO_3)_2 \cdot 4H_2O + CaS_2O_3 \cdot 6H_2O$
13.83	42.93	43.24	25	$CaS_2O_3 \cdot 6H_2O$
17.09	32.01	50.90	25	»
23.78	19.51	56.71	25	»
29.85	8.15	62.00	25	»

№ 5092 CALCIUM NITRATE–CALCIUM CHLORIDE–WATER [1134]

$Ca(NO_3)_2—CaCl_2—H_2O$

$t=30$

Comp. of saturated soln., wt %			Comp. of solid phase
Calcium Chloride	Calcium Nitrate	Water	
45.00	10.81	44.19	$CaCl_2 \cdot 4H_2O$
44.70	11.45	43.85	»
42.35	17.03	40.62	»
7.57	65.26	27.17	$Ca(NO_3)_2 \cdot 4H_2O$
10.70	54.40	34.90	»
11.05	53.40	35.55	»
7.27	52.90	39.83	»
7.60	55.00	40.40	»

№ 5093 CALCIUM NITRATE–CALCIUM CHLORIDE–WATER [1543]

$Ca(NO_3)_2—CaCl_2—H_2O$

$t=25$

Comp. of saturated soln., wt %			d_{25}	Comp. of solid phase
Calcium Nitrate	Calcium Chloride	Water		
0.00	44.92	55.08	1.453	$CaCl_2 \cdot 6H_2O$
2.95	44.75	52.30	1.474	»
5.43	44.14	50.43	5.524	»
6.74	44.24	49.02	1.523	»

Comp. of saturated soln., wt %			d_{25}	Comp. of solid phase
Calcium Nitrate	Calcium Chloride	Water		
7.60	44.98	47.42	1.541	$CaCl_2 \cdot 6H_2O + CaCl_2 \cdot 4H_2O$
9.03	43.78	47.19	1.556	$CaCl_2 \cdot 4H_2O$
16.86	40.35	42.79	1.619	»
18.07	39.75	42.18	1.630	»
20.55	37.73	41.72	1.630	$CaCl_2 \cdot 4H_2O + CaCl_2 \cdot Ca(NO_3)_2 \cdot H_2O$
20.94	37.36	41.70	1.630	$CaCl_2 \cdot Ca(NO_3)_2 \cdot 4H_2O$
21.80	36.80	41.40	1.624	»
22.29	37.04	40.67	1.623	»
31.63	29.01	39.36	1.651	»
34.20	27.57	48.23	1.660	»
41.92	21.75	36.33	1.692	»
43.02	21.10	35.88	1.700	»
49.28	16.92	33.80	1.733	»
52.21	15.20	32.59	1.749	$CaCl_2 \cdot Ca(NO_3)_2 \cdot 4H_2O + Ca(NO_3)_2 \cdot 4H_2O$
50.86	15.75	33.39	1.790	$Ca(NO_3)_2 \cdot 4H_2O$
51.30	14.45	34.25	1.713	»
48.01	15.06	36.93	1.696	»
46.94	14.57	38.49	1.655	»
46.45	12.91	40.64	1.625	»
46.17	12.02	41.81	1.616	»
48.72	9.19	42.09	1.595	»
48.93	8.24	42.83	1.590	»
48.88	7.78	44.34	1.588	»
55.03	2.26	42.71	1.581	»
58.35	0.00	41.65	1.579	»

№ 5094 [920]

CALCIUM NITRATE–CALCIUM CHLORIDE–BARIUM NITRATE–WATER

$Ca(NO_3)_2—CaCl_2—Ba(NO_3)_2—H_2O$

Comp. of saturated soln., g-mol				t	Comp. of solid phase
Calcium Chloride	Calcium Nitrate	Barium Nitrate	Water		
100.00	0.00	0.00	587.5	80	$CaCl_2 \cdot H_2O$
98.50	1.36	0.14	944.4	80	$BaCl_2 \cdot H_2O$
96.20	3.44	0.36	1058.8	80	»
92.37	6.90	0.73	1077.0	80	»
88.42	10.48	1.10	1118.2	80	»
84.28	14.22	1.50	1084.7	80	»
80.10	17.90	2.00	1130.7	80	$Ba(NO_3)_2$
57.38	38.60	4.02	1768.4	80	»
36.58	57.40	6.02	2450.2	80	»
100.00	0.00	0.00	550.50	100	$CaCl_2$
98.50	1.36	0.14	664.60	100	$BaCl_2$
96.20	3.44	0.36	743.70	100	»
92.37	6.90	0.73	897.00	100	»

Comp. of saturated soln., g-mol				t	Comp. of solid phase
Calcium Chloride	Calcium Nitrate	Barium Nitrate	Water		
88.42	10.48	1.10	998.80	100	$BaCl_2$
84.28	14.22	1.50	1003.90	100	»
80.10	17.90	2.00	1027.40	100	$Ba(NO_3)_2$
57.38	38.60	4.02	1398.50	100	»
36.58	57.40	6.02	1762.40	100	»
100.00	0.00	0.00	512.83	120	$CaCl_2$
98.50	1.36	0.14	469.80	120	$BaCl_2$
96.20	3.44	0.36	538.70	120	»
92.37	6.90	0.73	749.70	120	»
88.42	10.48	1.10	897.60	120	»
88.42	14.22	1.50	932.00	120	$Ba(NO_3)_2$
80.10	17.90	2 00	1172.40	120	»
57.38	38.60	4.02	1408.30	120	»
82.50	15.84	1.66	1060.00	80	$BaCl_2 \cdot H_2O + Ba(NO_3)_2$
82.00	16.29	1.71	1000.00	100	$BaCl_2 + Ba(NO_3)_2$
81.50	16.74	1.76	900.00	120	»

№ 5095 CALCIUM NITRATE–CALCIUM CHROMATE–WATER [696]

$Ca(NO_3)_2—CaCrO_4—H_2O$

Comp. of saturated soln., wt %			t	Comp. of solid phase
Calcium Nitrate	Calcium Chromate	Water		
58.35	0.00	41.65	25	$Ca(NO_3)_2 \cdot 4H_2O$
0.00	2.68	97.32	25	$CaCrO_4$
54.90	0.04	45.06	25	$Ca(NO_3)_2 \cdot 4H_2O + CaCrO_4$
38.87	0.13	61.00	25	$CaCrO_4$
28.99	0.39	70.62	25	»
14.64	1.14	84.22	25	»
73.00	0.00	27.00	50	$Ca(NO_3)_2 \cdot 3H_2O$
0.00	1.49	98.51	50	$CaCrO_4$
67.59	0.01	32.40	50	$Ca(NO_3)_2 \cdot 3H_2O + CaCrO_4$
51.77	0.01	48.22	50	$CaCrO_4$
47.99	0.05	51.96	50	»
26.80	0.28	72.92	50	»
14.48	0.58	84.94	50	»
78.20	0.00	21.80	75	$Ca(NO_3)_2 \cdot 2H_2O$
0.00	0.81	99.19	75	$CaCrO_4$
74.43	0.01	25.56	75	$Ca(NO_3)_2 \cdot 2H_2O + CaCrO_4$
48.67	0.01	51.32	75	$CaCrO_4$
26.43	0.15	73.42	75	»
13.46	0.36	86.18	75	»
78.35	0.00	21.65	95	$Ca(NO_3)_2 \cdot 2H_2O$
0.00	0.64	99.36	95	$CaCrO_4$
76.92	0.01	23.07	95	$Ca(NO_3)_2 \cdot 2H_2O + CaCrO_3$
51.74	0.02	48.24	95	$CaCrO_4$
29.65	0.97	69.38	95	»

№ 5096 CALCIUM NITRATE–STRONTIUM NITRATE–WATER [1984]

$Ca(NO_3)_2—Sr(NO_3)_2—H_2O$

Concn. in aqueous soln., wt %		t	Comp. of solid phase
Calcium Nitrate	Strontium Nitrate		
0.00	44.50	25	$Sr(NO_3)_2 \cdot 4H_2O$
12.23	32.30	25	$Sr(NO_3)_2 \cdot 4H_2O+Sr(NO_3)_2$
17.60	27.70	25	$Sr(NO_3)_2$
25.90	19.90	25	»
39.15	9.25	25	»
44.00	6.54	25	»
46.80	4.85	25	»
47.05	4.71	25	»
48.10	4.58	25	»
51.00	3.15	25	»
56.10	1.55	25	$Sr(NO_3)_2+Ca(NO_3)_2 \cdot 4H_2O$
57.20	0.00	25	$Ca(NO_3)_2 \cdot 4H_2O$
0.00	48.25	60	$Sr(NO_3)_2$
33.04	15.42	60	»
51.00	4.20	60	»
70.20	0.14	60	»
76.00	0.084	60	»
76.96	0.07	60	$Ca(NO_3)_2+Sr(NO_3)_2$
77.15	0.00	60	$Ca(NO_3)_2$

№ 5097 CALCIUM NITRATE–STRONTIUM NITRATE–WATER [823]

$Ca(NO_3)_2—Sr(NO_3)_2—H_2O$

$t=0$

Comp. of saturated soln., wt %			Comp. of solid phase
Strontium Nitrate	Calcium Nitrate	Water	
28.52	0.00	71.48	$Sr(NO_3)_2 \cdot 4H_2O$
22.51	7.88	69.61	»
20.02	10.50	69.48	»
18.75	11.92	69.33	»
18.03	12.12	69.85	»
13.92	18.37	67.71	»
9.45	26.57	63.98	»
8.06	29.06	62.88	»
7.31	31.78	60.91	»
6.06	34.87	59.07	»
5.91	35.58	58.51	»
5.59	37.46	56.95	»
4.95	38.51	56.54	»
4.49	40.97	54.54	»
3.88	44.17	51.95	»
3.61	44.90	51.49	»
3.52	45.50	50.98	»
3.50	47.67	48.83	$Sr(NO_3)_2 \cdot 4H_2O+Ca(NO_3)_2 \cdot 4H_2O$
1.57	48.54	49.89	$Ca(NO_3)_2 \cdot 4H_2O$
1.20	49.27	49.53	»
0.00	50.14	49.86	»

№ 5098 [2364]

CALCIUM NITRATE–STRONTIUM SULFATE–WATER

$Ca(NO_3)_2—SrSO_4—H_2O$

$t=20$

Concn. in aqueous soln., g/l	
Calcium Nitrate	Strontium Sulfate
5	0.483
10	0.619
20	1.081
30	1.275
40	1.488
50	1.688
60	1.955

№ 5099 [856]

CALCIUM NITRATE–SILVER BROMATE–WATER

$Ca(NO_3)_2—AgBrO_3—H_2O$

$t=25$

Concn. in aqueous soln., g/l	
Calcium Nitrate	Silver Bromate
0.00000	1.9475
0.16409	1.8650
0.49227	1.9027
1.6409	2.1574
4.9227	2.3295
16.409	2.6172
49.227	3.1359
164.09	4.2676

№ 5100 [920]

CALCIUM NITRATE–BARIUM NITRATE–WATER

$Ca(NO_3)_2—Ba(NO_3)_2—H_2O$

Comp. of saturated soln., g-mol			t	Comp. of solid phase
Barium Nitrate	Calcium Nitrate	Water		
100.00	0.00	6603.0	80	$Ba(NO_3)_2$
48.57	51.43	4752.0	80	»
29.53	70.47	4022.0	80	»
21.19	78.81	3515.9	80	»
13.58	86.42	2657.4	80	»
6.54	93.46	2049.8	80	»
3.68	96.32	894.4	80	»
0.00	100.00	142.3	80	$Ca(NO_3)_2$
100.00	0.00	4136.0	100	$Ba(NO_3)_2$
41.49	58.51	4067.0	100	»
59.51	40.49	3772.5	100	»
48.57	51.43	3341.0	100	»
29.53	70.47	2807.0	100	»
21.19	78.81	2507.9	100	»
13.58	86.42	1980.7	100	»
6.54	93.46	1384.8	100	»
3.68	96.32	620.1	100	»
0.00	100.00	140.2	100	$Ca(NO_3)_2$
100.00	0.00	3229.3	120	$Ba(NO_3)_2$
59.51	40.49	2863.3	120	»
48.57	51.43	2646.4	120	»
29.53	70.47	2230.8	120	»
21.19	78.81	1981.5	120	»
13.58	86.42	1578.9	120	»
6.54	93.46	1030.1	120	»
3.68	96.32	500.0	120	»
0.00	100.00	137.9	120	$Ca(NO_3)_2$
$<$ 0.10	$>$ 99.90	$<$ 142.3	80	$Ba(NO_3)_2+Ca(NO_3)_2$
$<$ 0.10	$>$ 99.90	$<$ 140.9	100	»
$<$ 0.10	$>$ 99.90	$<$ 137.9	120	»

$Ca(NO_3)_2—BaCl_2—H_2O$

Comp. of saturated soln., wt %			t	Comp. of solid phase
Barium Chloride	Calcium Nitrate	Water		
27.50	0.00	72.50	20	$BaCl \cdot 2H_2O$
27.40	1.09	71.51	20	»
27.20	1.56	71.24	20	»
27.15	1.82	71.03	20	»
27.06	2.55	70.39	20	»
27.00	3.00	70.00	20	»
27.00	3.65	69.35	20	»
86.20	3.80	70.00	20	$Ba(NO_3)_2$
24.00	3.80	72.20	20	»
18.81	3.50	77.69	20	»
10.00	4.50	85.50	20	»
6.50	5.60	87.90	20	»
3.50	7.24	89.26	20	»
2.26	9.50	88.24	20	»
2.26	9.60	88.14	20	»
1.20	14.82	83.98	20	»
0.25	34.91	64.84	20	»
0.00	56.40	43.60	20	$Ca(NO_3)_2 \cdot 4H_2O$
30.50	0.00	69.50	40	$BaCl_2 \cdot 2H_2O$
30.10	1.05	68.85	40	»
30.05	1.40	68.55	40	»
30.00	1.75	68.25	40	»
29.70	2.45	67.85	40	»
29.20	3.54	67.26	40	»
29.00	4.26	66.74	40	»
25.00	5.00	70.00	40	$Ba(NO_3)_2$
23.70	5.20	71.10	40	»
18.48	5.20	76.32	40	»
9.00	6.80	84.20	40	»
5.70	9.43	84.87	40	»
4.40	11.95	83.65	40	»
2.80	14.58	82.62	40	»
2.05	18.00	79.95	40	»
0.56	34.80	64.64	40	»
0.00	65.50	34.50	40	$Ca(NO_3)_2 \cdot 4H_2O$
33.50	0.00	66.50	60	$BaCl_2 \cdot 2H_2O$
33.00	1.01	65.99	60	»
32.80	1.35	65.85	60	»
32.70	1.68	65.62	60	»
32.40	2.37	65.23	60	»
32.10	3.40	64.50	60	»
32.00	4.08	63.92	60	»
31.30	5.15	63.55	60	»
23.39	6.50	70.11	60	»
18.03	7.50	74.47	60	$Ba(NO_3)_2$
11.50	8.85	79.65	60	»
7.00	11.62	82.38	60	»
5.50	14.17	80.33	60	»
1.80	28.00	70.20	60	»
0.99	77.50	22.49	60	$Ca(NO_3)_2$

Table № 5101 (continuation)

Comp. of saturated soln., wt %			t	Comp. of solid phase
Barium Chloride	Calcium Nitrate	Water		
23.50	0.00	76.50	− 7.6	Ice $+BaCl_2 \cdot 2H_2O$
23.15	1.15	75.70	− 7.7	»
23.15	1.53	75.32	− 8.0	»
23.00	1.92	75.08	− 8.3	Ice $+BaCl_2 \cdot 2H_2O+Ba(NO_3)_2$
19.20	1.97	78.83	− 7.6	Ice $+Ba(NO_3)_2$
12.50	2.27	85.23	− 5.0	»
4.00	3.36	92.64	− 3.6	»
2.39	4.10	93.51	− 2.2	»
1.50	4.93	93.57	− 2.0	»
1.25	5.93	92.82	− 5.0	»
1.00	7.43	91.57	− 3.0	»
0.80	9.92	89.28	− 4.5	»
0.60	14.91	84.49	− 6.0	»
0.08	34.97	64.95	−15.0	»
0.00	42.00	58.00	−28.5	Ice $+Ca(NO_3)_2 \cdot 4H_2O$

№ 5102 CALCIUM NITRATE–LEAD (II) BROMATE–WATER [2165]

$Ca(NO_3)_2—Pb(BrO_3)_2—H_2O$

$t=25$

Concn. in aqueous soln., g/l		d	Concn. in aqueous soln., g/l		d
Lead Bromate	Calcium Nitrate		Lead Bromate	Calcium Nitrate	
15.91	0.0	1.0112	43.86	94.53	1.0997
16.51	1.028	1.0099	51.16	127.59	1.1283
16.85	1.830	1.0117	55.28	156.19	1.1457
18.15	3.536	1.0135	66.44	226.40	1.2090
19.90	8.466	1.0195	76.26	293.10	1.2607
24.94	19.926	1.0313	83.94	356.19	1.3074
28.72	32.39	1.0425	90.00	415.92	1.3489
34.71	54.97	1.0643	98.06	523.98	1.4215
40.60	80.69	1.0869			

№ 5103 CALCIUM NITRATE–LEAD (II) IODIDE–WATER [213]

$Ca(NO_3)_2—PbI_2—H_2O$

$t=18—20$

Concn. in aqueous soln., g-eq/l		Solubility of Lead Iodide, mmol/l	Concn. in aqueous soln., g-eq/l		Solubility of Lead Iodide, mmol/l
Potassium Iodide	Calcium Nitrate		Potassium Iodide	Calcium Nitrate	
0.0	0.1	2.662	0.0	0.2	2.534
0.008	0.1	1.632	0.008	0.2	1.594
0.016	0.1	0.995	0.016	0.2	0.912
0.032	0.1	0.404	0.032	0.2	0.400
0.064	0.1	0.485	0.064	0.2	0.494
1.128	0.1	0.562	0.128	0.2	0.532
0.256	0.1	0.930	0.256	0.2	0.918

№ 5104 [55]

CALCIUM NITRATE–LEAD (II) IODIDE–WATER

$Ca(NO_3)_2—PbI_2—H_2O$

$t=20$

Concn. of Calcium Nitrate hydrate in aqueous soln., mol/l	Solubility of Lead Iodide in aqueous soln. of Calcium Nitrate hydrate, mol/l
0.20	$2.93 \cdot 10^{-3}$
0.60	$3.86 \cdot 10^{-3}$
1.00	$4.81 \cdot 10^{-3}$

№ 5105 [1010]

CALCIUM NITRATE–URANYL NITRATE–WATER

$Ca(NO_3)_2—UO_2(NO_3)_2—H_2O$

Comp. of saturated soln., wt %			t	Comp. of solid phase
Uranyl Nitrate	Calcium Nitrate	Water		
49.50	0.00	50.50	0.1	$UO_2(NO_3)_2 \cdot 6H_2O$
43.22	4.16	52.62	0.1	»
37.21	10.40	52.39	0.1	»
21.25	15.52	53.23	0.1	»
26.02	20.49	53.49	0.1	»
23.30	23.15	53.55	0.1	»
22.05	24.95	53.00	0.1	»
20.82	26.16	53.02	0.1	»
12.21	35.82	51.97	0.1	»
10.15	37.49	52.36	0.1	»
8.75	39.43	51.82	0.1	»
8.44	39.95	51.61	0.1	»
6.76	43.32	49.92	0.1	$UO_2(NO_3)_2 \cdot 6H_2O+Ca(NO_3)_2 \cdot 4H_2O$
5.10	42.69	52.21	0.1	$Ca(NO_3)_2 \cdot 4H_2O$
4.55	43.20	52.25	0.1	»
2.55	45.00	52.45	0.1	»
0.00	50.63	49.37	0.1	»
55.92	0.00	44.08	25	$UO_2(NO_3)_2 \cdot 6H_2O$
48.83	4.69	46.48	25	»
45.75	7.00	47.25	25	»
34.75	17.87	47.38	25	»
31.10	22.08	46.82	25	»
29.65	22.40	47.95	25	»
26.02	26.23	47.75	25	»
22.98	29.05	47.97	25	»
17.60	34.31	48.09	25	»
16.40	37.02	47.58	25	»
15.18	39.21	45.61	25	»
13.25	41.05	45.70	25	»
9.66	46.40	43.94	25	»

Table № 5105 (continuation)

Comp. of saturated soln., wt %			t	Comp. of solid phase
Uranyl Nitrate	Calcium Nitrate	Water		
8.72	47.51	43.77	25	$UO_2(NO_3)_2 \cdot 6H_2O$
7.97	48.10	43.93	25	»
7.92	50.48	41.60	25	$UO_2(NO_3)_2 \cdot 6H_2O + Ca(NO_3)_2 \cdot 4H_2O$
7.41	49.59	43.00	25	$Ca(NO_3)_2 \cdot 4H_2O$
6.88	49.20	43.92	25	»
4.18	50.00	45.82	25	»
2.01	52.81	45.18	25	»
0.00	57.97	42.03	25	»

№ 5106 [28]

CALCIUM FLUORIDE–CALCIUM CHLORIDE–WATER

CaF_2—$CaCl_2$—H_2O

Content of Calcium Chloride in aqueous soln., wt %	Solubility of Calcium Fluoride in aqueous soln. of Calcium Chloride, g/100 ml soln.	t	Content of Calcium Chloride in aqueous soln., wt %	Solubility of Calcium Fluoride in aqueous soln. of Calcium Chloride, g/100 ml soln.	t
24.94	0.0041	200	42.20	0.0431	200
25.00	0.0365	250	42.18	0.1415	250
24.90	0.0638	300	42.19	0.2268	300
42.19	0.0209	150			

№ 5107 [868]

CALCIUM FLUORIDE–IRON (III) NITRATE–WATER

CaF_2—$Fe(NO_3)_3$—H_2O

$t=25$

Concn. of Iron Nitrate in aqueous soln., mol/l	Solubility of Calcium Fluoride in aqueous soln. of Iron Nitrate, mol/l
0.003	0.0028
0.01	0.0068
0.03	0.0181
0.1	0.0468
0.3	0.1200
1.0	0.2480

№ 5108 [868]

CALCIUM FLUORIDE–IRON (III) CHLORIDE–WATER

CaF_2—$FeCl_3$—H_2O

$t=25$

Concn. of Iron Chloride in aqueous soln., mol/l	Solubility of Calcium Fluoride in aqueous soln. of Iron Chloride, mol/l
0.003	0.0027
0.01	0.0067
0.03	0.0146
0.1	0.0365
0.3	0.0724
1.0	0.1545

№ 5109 [866]

CALCIUM FLUORIDE–THORIUM NITRATE–WATER

CaF_2—$Th(NO_3)_4$—H_2O

$t=25$

Concn. in aqueous soln., mol/l	
Thorium Nitrate	Calcium Fluoride
0.010	0.006
0.025	0.014
0.050	0.028
0.100	0.051
0.266	0.127
0.333	0.153
0.500	0.225

№ 5110 [870]

CALCIUM FLUORIDE–URANYL NITRATE–WATER

CaF_2—$UO_2(NO_3)_2$—H_2O

$t=25$

Content of Uranyl Nitrate in aqueous soln., mol/1000 g Water	Solubility of Calcium Fluoride in conversion to Fluorine, g-ion/l
0.004	0.0046
0.02	0.02
0.04	0.034
0.08	0.05
0.10	0.05
0.20	0.10
0.40	0.20

№ 5111 CALCIUM HYDROGEN ORTHOPHOSPHATE*–SALTS–WATER [2101]

$CaHPO_4 \cdot 2H_2O$—MeX—H_2O

$t=40$

Salt		Concn. in aqueous soln., g/l	
Name	Formula	Salt	Divalent Calcium Orthophosphate
—	—	0	0.571
Sodium Chloride	NaCl	2.34	0.434
»	»	5.84	0.535
»	»	11.69	0.668
»	»	23.38	0.888
»	»	46.76	1.103
»	»	58.45	1.224
Potassium Chloride	KCl	7.46	0.690
»	»	29.82	0.943
»	»	74.56	1.184
Sodium Sulfate	Na_2SO_4	13.50	0.847
»	»	27.00	1.027
»	»	54.12	1.323
Magnesium Sulfate	$MgSO_4$	2.17	0.873
»	»	5.42	1.342
»	»	10.71	2.323

*Brucite

№ 5112 CALCIUM ORTHOPHOSPHATE–CALCIUM SULFATE–WATER [1341]

$Ca_3(PO_4)_2—CaSO_4—H_2O$

$t=75.3$

Concn. in aqueous soln., wt %			Comp. of solid phase
Calcium Oxide	Phosphorus Pentoxide	Sulfur Trioxide	
0.109	0.294	0.151	$Ca_3(PO_4)_2 \cdot H_2O+CaHPO_4+CaSO_4 \cdot 2H_2O$
0.122	0.310	0.154	»
0.118	0.283	0.157	»
0.281	0.976	0.141	$CaHPO_4+CaSO_4 \cdot 2H_2O+CaSO_4$ (traces)
0.741	3.07	0.157	$CaHPO_4+CaSO_4 \cdot 2H_2O$
2.01	9.49	0.125	»
2.68	13.35	0.091	»
3.49	18.09	0.075	»
3.94	20.93	0.081	»
4.92	27.51	0.052	»
5.27	30.53	0.048	$CaHPO_4+CaSO_4 \cdot 2H_2O+CaSO_4 \cdot 1/2H_2O$
5.37	31.68	0.0	$CaHPO_4+CaSO_4 \cdot H_2O$
5.45	32.72	0.050	$CaHPO_4+CaSO_4 \cdot 1/2H_2O$
5.52	33.54	0.035	»
5.64	35.51	0.033	$CaHPO_4+CaSO_4 \cdot 1/2H_2O+Ca(H_2PO_4)_2 \cdot H_2O$
5.68	35.60	0.028	»
5.38	36.54	0.035	$Ca(H_2PO_4)_2 \cdot H_2O+CaSO_4 \cdot 1/2H_2O$
5.10	37.60	0.026	»
4.35	40.62	0.018	»
3.30	44.63	0.017	»
2.96	46.00	Traces	»
2.40	49.02	0.0	»
1.65	54.24	0.0	»
1.275	57.46	0.0	$Ca(H_2PO_4)_2 \cdot H_2O+CaSO_4 \cdot 1/2H_2O+$ $+Ca(H_2PO_4)_2$ (traces)
1.269	57.38	0.0	»
1.150	56.47	0.0	»

№ 5113 CALCIUM ORTHOPHOSPHATE–SALTS–WATER [1542]

$Ca_3(PO_4)_2—MeX—H_2O$

$t=14$

Salt			Solubility of Calcium Orthophosphate in aqueous soln. of salt, g/l solvent
Name	Formula	Concn. in aqueous soln., wt %	
—	—	0	0.228
Ammonium Chloride	NH_4Cl	31.38	1.371
»	»	Saturated	1.293
Ammonium Sulfate	$(NH_4)_2SO_4$	36.10	2.413
»	»	Saturated	5.885
Magnesium Chloride, 6-water	$MgCl_2 \cdot 6H_2O$	46.50	1.287
»	»	Saturated	1.892
Magnesium Sulfate, 7-water	$MgSO_4 \cdot 7H_2O$	51.29	1.973

Table № 5113 (continuation)

Salt			Solubility of Calcium Orthophosphate in aqueous soln. of salt, g/l solvent
Name	Formula	Concn. in aqueous soln., wt %	
Magnesium Sulfate, 7-water	$MgSO_4 \cdot 7H_2O$	Saturated	3.600
Potassium Magnesium Chloride, 6-water	$MgCl_2 \cdot KCl \cdot 6H_2O$	44.20	1.577
»	»	Saturated	1.154
Potassium Magnesium Sulfate, Magnesium Chloride, 6-water	$MgSO_4 \cdot K_2SO_4 \cdot MgCl_2 \cdot 6H_2O$	41.50	1.777
»	»	Saturated	2.491
Potassium Sulfate	K_2SO_4	42.69	4.904
»	»	Saturated	4.765
Sodium Chloride	NaCl	33.33	1.321
»	»	Saturated	0.641
Sodium Nitrate	$NaNO_3$	42.10	1.583
»	»	Saturated	0.864
Sodium Sulfate, 10-water	$Na_2SO_4 \cdot 10H_2O$	57.93	2.491
»	»	Saturated	3.227

№ 5114 [1316]

CALCIUM SULFATE–CALCIUM CHLORIDE–WATER

$CaSO_4—CaCl_2—H_2O$

$t=25$

Concn. in aqueous soln., g/l		Concn. in aqueous soln., g/l	
Calcium Chloride	Calcium Sulfate	Calcium Chloride	Calcium Sulfate
0.00	2.06	51.33	1.02
7.49	1.24	97.02	0.84
11.96	1.18	192.71	0.47
25.77	1.10	280.30	0.20
32.05	1.08	367.85	0.03

№ 5115 [657]

CALCIUM SULFATE–CALCIUM CHLORIDE–WATER

$CaSO_4—CaCl_2—H_2O$

$t=16—22$

Content in aqueous soln., wt %		Comp. of solid phase
Calcium Chloride	Calcium Sulfate	
40	0.0128	$CaSO_4 \cdot 2H_2O$
35	0.0151	»
30	0.0186	»
25	0.0395	»
20	0.0555	»

Table № 5115 (continuation)

Content in aqueous soln., wt %		Comp. of solid phase
Calcium Chloride	Calcium Sulfate	
15	0.761	$CaSO_4 \cdot 2H_2O$
10	0.0851	»
5	0.1036	»
4	0.1042	»
3	0.1057	»
2	0.1125	»
1	1.1130	»
0.9	0.1189	»
0.8	0.1208	»
0.7	0.1324	»
0.5	0.1404	»
0.1	0.1856	»

№ 5116 CALCIUM SULFATE–CALCIUM CHLORIDE–WATER [2699]

$CaSO_4—CaCl_2—H_2O$

$t=25$

Concn. in aqueous soln., g/l		Concn. in aqueous soln., g/l	
Calcium Chloride	Calcium Sulfate	Calcium Chloride	Calcium Sulfate
0.55	1.865	2.775	1.470
1.11	1.715	3.33	1.416
1.665	1.607	4.44	1.321
2.22	1.525	5.55	1.226

№ 5117 [423]

CALCIUM SULFATE–CALCIUM CHLORIDE–WATER

$CaSO_4—CaCl_2—H_2O$

$t=20$

Comp. of saturated soln., g			Comp. of solid phase
Calcium Chloride	Calcium Sulfate	Water	
25.0	0.017	75.0	$CaSO_4$
20.0	0.029	80.0	»
15.0	0.047	85.0	»
10.0	0.069	90.0	»
5.0	0.099	95.0	»
0	0.208	100	»

№ 5118 CALCIUM SULFATE–CALCIUM CHLORIDE–WATER [784]

$CaSO_4—CaCl_2—H_2O$

Comp. of saturated soln., wt %			t	Comp. of solid phase
Calcium Chloride	Calcium Sulfate	Water		
20.30	0.031	79.669	25	—
26.51	0.081	73.472	25	—
27.53	0.021	72.449	25	—
33.87	0.011	66.119	25	—
38.91	0.008	61.082	25	—
8.23	0.102	91.668	55	$CaSO_4 \cdot 2H_2O$
9.08	0.097	90.823	55	»
13.55	0.075	86.375	55	»
23.00	0.03	76.97	55	»
27.08	0.021	72.899	55	»
31.91	0.016	68.074	55	»
42.50	0.061	57.439	55	»
47.77	0.092	52.138	55	α-$CaSO_4 \cdot ½H_2O$
50.45	0.041	49.509	55	»
51.30	0.010	48.690	55	»
56.94	0.031	43.029	55	α-$CaSO_4 \cdot ½H_2O + CaCl_2 \cdot 2H_2O$

№ 5119 CALCIUM SULFATE–CALCIUM CHLORIDE–WATER [784]

$CaSO_4—CaCl_2—H_2O$

$t = 55$

Comp. of saturated soln., wt %			Comp. of solid phase
Calcium Sulfate	Calcium Chloride	Water	
0.102	8.23	91.67	$CaSO_4 \cdot 2H_2O$
0.097	9.08	80.82	»
0.075	13.55	86.37	»
0.030	23.00	76.97	»
0.020	27.08	72.90	»
0.016	31.91	68.07	»
0.061	42.50	57.44	»
0.090	47.47	52.44	α-$CaSO_4 \cdot ½H_2O$
0.040	50.45	49.51	»
0.010	51.30	48.69	»
0.030	56.94	43.03	$CaCl_2 \cdot 2H_2O + \alpha$-$CaSO_4 \cdot ½H_2O$
0.075	8.42	91.50	$CaSO_4$
0.013	27.00	72.99	»

№ 5120 CALCIUM SULFATE–CALCIUM CHLORIDE–WATER [236]

$CaSO_4—CaCl_2—H_2O$

$t=110$

Comp. of saturated soln., wt %			Comp. of solid phase
Calcium Chloride	Calcium Sulfate	Water	
0.0	0.080	99.92	$CaSO_4$
5.35	0.054	94.60	»
10.80	0.043	89.16	»
16.70	0.032	83.27	»
25.30	0.017	74.68	»
41.40	0.003	58.60	»
50.45	0.001	49.55	»
61.50	0.001	38.50	$CaSO_4+CaCl_2 \cdot 2H_2O$
62.30	0.000	37.70	$CaCl \cdot 2H_2O$

№ 5121 [756]

CALCIUM SULFATE–CALCIUM CHLORIDE–WATER

$CaSO_4—CaCl_2—H_2O$

$t=25$

Comp. of saturated soln., wt %			Comp. of solid phase
Calcium Chloride	Calcium Sulfate	Water	
38.91	0.0080	61.08	$CaSO_4 \cdot 2H_2O$
33.87	0.0110	66.12	»
27.53	0.0210	72.45	»
26.51	0.0180	73.47	»
29.06	0.0025	70.94	»
23.22	0.0168	76.76	»
16.86	0.0407	33.10	»
9.03	0.0783	89.89	»
4.95	0.0977	94.95	»
3.13	0.1055	96.76	»
2.53	0.1075	97.36	»
1.19	0.1172	98.69	»
0.75	0.1239	99.13	»
0.00	0.2060	99.79	»

$CaSO_4—CrO_3—H_2O$

Concn. in aqueous soln., wt %		t	Comp. of solid phase
Chromium Trioxide	Calcium Sulfate		
0.208	0.0177	25	$CaSO_4 \cdot 2H_2O+CrO_3$
0.628	0.0025	25	»
0.0259	0.0066	25	»
0.0628	0.0106	25	»
0.144	0.0179	25	»
0.206	0.0181	25	»
0.210	0.0180	25	»
0.268	0.0160	25	»
0.290	0.0153	25	»
0.440	0.0079	25	»
0.485	0.0064	25	$CaSO_4 \cdot 2H_2O$
0.536	0.0051	25	»
0.568	0.0036	25	»
0.625	0.0025	25	$CaSO_4 \cdot 2H_2O+CrO_3$
0.192	0.0298	60	»
0.211	0.0283	60	»
0.327	0.0236	60	»
0.580	0.0035	60	»
0.130	0.0266	60	$CaSO_4 \cdot 2H_2O$
0.471	0.0093	60	$CaSO_3$
0.636	0.0018	60	$CaSO_4+CrO_3$
0.636	0.0018	60	»
0.0229	0.0096	60	»
0.0357	0.0137	60	»
0.081	0.0211	60	»
0.122	0.0251	60	»
0.181	0.0289	60	$CaSO_4 \cdot CrO_3$
0.233	0.0279	60	»
0.322	0.0245	60	»
0.358	0.0211	60	»
0.413	0.0163	60	$CaSO_4 \cdot 2H_2O$
0.627	0.0019	60	$CaSO_4$
0.192	0.0458	95	»
0.387	0.0286	95	»
0.407	0.0276	95	»
0.492	0.0097	95	»
0.666	0.0017	95	$CaSO_4+CrO_3$
0.668	0.0018	95	»
0.0145	0.0995	95	»
0.0469	0.0232	95	»
0.0923	0.0328	95	$CaSO_4$
0.142	0.0327	95	»
0.197	0.0305	95	»
0.234	0.0281	95	»
0.448	0.0125	95	»
0.513	0.0086	95	»
0.588	0.0050	95	»
0.471	0.0112	95	»
0.230	0.0178	20—26	$CaSO_4 \cdot 2H_2O$

$CaSO_4$—$MnSO_4$—H_2O

Comp. of saturated soln., wt %			t	Comp. of solid phase
Manganese Sulfate	Calcium Sulfate	Water		
39.42	0.00	60.58	25	$MnSO_4 \cdot 5H_2O$
39.21	0.05	60.74	25	»
39.02	0.07	60.91	25	»
38.10	0.12	61.78	25	»
37.05	0.15	62 80	25	»
36.82	0.16	63.02	25	$MnSO_4 \cdot 5H_2O + CaSO_4 \cdot 2H_2O$
34.10	0.19	65.71	25	$CaSO_4 \cdot 2H_2O$
32.10	0.23	67.67	25	»
24.06	0.27	75.67	25	»
12.42	0.28	87.30	25	»
6.05	0.26	93.69	25	»
4.52	0.25	95.23	25	»
1.33	0.21	98.46	25	»
0.00	0.20	99.80	25	»
33.07	0.00	66.93	75	$MnSO_4 \cdot H_2O$
32.81	0.05	67.15	75	»
32.12	0.09	67.79	75	»
31.24	0.14	68.62	75	$MnSO_4 \cdot H_2O + CaSO_4$
31.02	0.15	68.82	75	$CaSO_4$
28.92	0.16	70.92	75	»
27.10	0.166	72.724	75	»
25.82	0.166	74.014	75	»
16.71	0.15	83.14	75	»
12.10	0.14	87.76	75	»
9.62	0.13	90.25	75	»
5.12	0.12	94.76	75	»
0.00	0.11	99.89	75	»

№ 5124 [2699]

CALCIUM SULFATE–
COBALT HEXAMMINECHLORIDE–
WATER

$CaSO_4$–$Co(NH_3)_6Cl_3$–H_2O

$t=25$

Concn. in aqueous soln., g/l	
Cobalt Hexammine-chloride	Calcium Sulfate
0.890	2.383
1.783	2.737
2.673	3.023

№ 5125 CALCIUM SULFATE–NICKEL (II) SULFATE–WATER [1334]

$CaSO_4—NiSO_4—H_2O$

Concn. in aqueous soln., wt %		t	Concn. in aqueous soln., wt %		t
Nickel Sulfate	Calcium Sulfate		Nickel Sulfate	Calcium Sulfate	
0.00	0.211	45	0.0	0.177	90
6.23	0.220	45	7.38	0.222	90
10.61	0.223	45	11.15	0.268	90
20.42	0.243	45	16.23	0.309	90
23.50	0.236	45	21.63	0.332	90
25.45	0.246	45	23.76	0.348	90
26.62	0.246	45	24.93	0.351	90
29.43	0.255	45	29.72	0.359	90
35.03	0.252	45	31.09	0.350	90
37.99	0.257	45	31.36	0.345	90
43.54	0.242	45	34.16	0.340	90
49.31	0.200	45	37.24	0.337	90
48.36	0.0	45	40.06	0.335	90
0.0	0.199	75	43.12	0.327	90
7.69	0.226	75	44.58	0.330	90
19.60	0.273	75	46.61	0.325	90
24.36	0.287	75	48.95	0.291	90
33.32	0.294	75	49.56	0.299	90
40.16	0.321	75	53.91	0.280	90
40.88	0.315	75	57.88	0.265	90
44.59	0.302	75	61.34	0.246	90
49.70	0.276	75	72.40	0.186	90
58.81	0.240	75	72.54	0.183	90
61.25	0.220	75	68.70	0.0	90
60.30	0.0	75			

№ 5126 CALCIUM SULFATE–COPPER (II) SULFATE–WATER [1180]

$CaSO_4—CuSO_4—H_2O$

$t=25$

Concn. in aqueous soln., g/l		d	Concn. in aqueous soln., g/l		d
Copper Sulfate	Calcium Sulfate		Copper Sulfate	Calcium Sulfate	
1.144	2.068	1.002	39.407	1.718	1.041
3.564	1.986	1.005	49.382	1.744	1.051
6.048	1.944	1.007	58.880	1.782	1.061
7.279	1.858	1.009	97.950	1.931	1.098
14.814	1.760	1.016	146.725	2.048	1.146
19.729	1.736	1.021	196.021	2.076	1.192
29.543	1.688	1.030	224.916	2.088	1.218

№ 5127 CALCIUM SULFATE – COPPER (II) SULFATE – WATER [6713]

$CaSO_4—CuSO_4—H_2O$

$t=25$

Concn. in aqueous soln., wt %		d_4	Concn. in aqueous soln., wt %		d_4
Copper Sulfate	Calcium Sulfate		Copper Sulfate	Calcium Sulfate	
0.1996	0.1843	1.0010	3.267	0.1747	1.0324
0.7922	0.1671	1.0069	13.465	0.1944	1.1486
1.5842	0.1661	1.0149			

№ 5128 CALCIUM SULFATE–RUBIDIUM SULFATE–WATER [1452]

$CaSO_4—Rb_2SO_4—H_2O$

Content in aqueous soln., mol/1000 mol Water		t	Comp. of solid phase
Rubidium Sulfate	Calcium Sulfate		
16.9*	—	0	$CaSO_4 \cdot 2H_2O+RbSO_4 \cdot CaSO_4 \cdot H_2O$
18.79*	—	25	$CaSO_4 \cdot 2H_2O+Rb_2SO_4 \cdot 2CaSO_4$
25.52	0.50	25	$Rb_2SO_4 \cdot 2CaSO_4+Rb_2SO_4 \cdot CaSO_4 \cdot H_2O$
39.5*	—	40	»
13.1*	—	40	$CaSO_4 \cdot 2H_2O+Rb_2SO_4 \cdot 2CaSO_4$

Note. The asterisk indicates the total content of both salts.

№ 5129 CALCIUM SULFATE–SILVER SULFATE–WATER [1580]

$CaSO_4—Ag_2SO_4—H_2O$

Concn. in aqueous soln., g/l		t	d_4^t
Calcium Sulfate	Silver Sulfate		
2.31	7.235	17	1.0083
2.61	8.11	25	1.010

№ 5130 CALCIUM SULFATE–SILVER CHLORIDE–WATER [2344]

$CaSO_4—AgCl—H_2O$

$t=25$

Concn. in aqueous soln., g/l		Concn. in aqueous soln., g/l	
Calcium Sulfate	Silver Chloride	Calcium Sulfate	Silver Chloride
0.0	0.00183	0.1385	0.00196
0.00087	0.00183	0.2072	0.00199
0.00431	0.00184	0.3447	0.00205
0.0142	0.00187	0.4135	0.00210
0.0697	0.00192		

№ 5131 CALCIUM SULFATE–CESIUM SULFATE–WATER [1452]

$CaSO_4—Cs_2SO_4—H_2O$

Total content of mixture of salts in aqueous soln., mol/1000 mol Water	t	Comp. of solid phase
14.0	25	$Cs_2SO_4 \cdot 2CaSO_4 + CaSO_4 \cdot 2H_2O$
12.4	60	»

№ 5132 [232]

CALCIUM SULFATE (GYPSUM)–VARIOUS METAL CHLORIDES–WATER

$CaSO_4 \cdot 2H_2O— MeCl—H_2O$

$t=25$

Metal Chloride			Solubility of Gypsum, mol/1000 g Water
Name	Formula	Content in aqueous soln., mol/1000 g Water	
Lanthanum Chloride	$LaCl_3$	0.00	0.015
»	»	0.50	0.124
»	»	1.00	0.175
»	»	1.28	0.176
»	»	1.50	0.174
»	»	2.10	0.161
»	»	2.48	0.145
Praseodymium Chloride	$PrCl_3$	0.00	0.015
»	»	0.50	0.133
»	»	0.64	0.162
»	»	1.05	0.175
»	»	1.16	0.173
»	»	1.52	0.165
»	»	2.00	0.150
Neodymium Chloride	$NdCl_3$	0.00	0.015
»	»	0.50	0.121
»	»	0.96	0.168
»	»	1.25	0.170
»	»	1.59	0.157
»	»	2.38	0.140
Samarium Chloride	$SmCl_3$	0.00	0.015
»	»	0.96	0.163
»	»	1.26	0.161
»	»	1.97	0.135

№ 5133 CALCIUM SULFATE–CERIUM SULFATE–WATER [275]

$CaSO_4—Ce_2(SO_4)_3—H_2O$

$t = 25$

Comp. of saturated soln., wt %			Comp. of solid phase	Comp. of saturated soln., wt %			Comp. of solid phase
Calcium Sulfate	Cerium Sulfate	Water		Calcium Sulfate	Cerium Sulfate	Water	
0.24	3.22	96.54	$mCaSO_4 \cdot nCe_2(SO_4)_3 \cdot xH_2O$	0.34	8.55	91.11	$mCaSO_4 \cdot nCe_2(SO_4)_3 \cdot xH_2O$
0.41	4.29	95.30	»	0.21	9.21	90.58	»
0.33	6.53	93.14	»	0.26	10.76	88.98	»
0.26	8.20	91.54	»	0.18	14.06	85.76	»
0.40	8.83	90.77	»	0.52	14.05	85.43	»
0.14	8.07	91.79	»	0.28	15.19	84.53	»

№ 5134 CALCIUM SULFATE–SEAWATER [2629]

$CaSO_4—$ —— $—H_2O$

Total content of Cl⁻ in aqueous soln., %	Solubility of Calcium Sulfate (%) and specific gravity of soln. at given temperature															
	25°		50°		75°		100°		125°		150°		175°		200°	
	%	d	%	d	%	d	%	d	%	d	%	d	%	d	%	d
0	—	—	0.164	0.9897	0.105	0.9759	0.065	0.9590	0.040	0.9397	0.023	0.9175	0.012	0.8921	0.01	0.863
1	—	—	0.282	1.0041	0.215	0.9902	0.151	0.9761	0.093	0.957	0.060	0.9343	0.033	0.9117	0.02	0.889
2	—	—	0.363	1.0168	0.262	1.0026	0.180	0.9872	0.116	0.9677	0.077	0.9461	0.042	0.9241	0.03	0.901
3	—	—	0.384	1.0291	0.277	1.0150	0.191	0.9983	0.123	0.9790	0.091	0.9584	0.047	0.9367	0.03	0.914
4	—	—	0.388	1.0415	0.281	1.0272	0.194	1.0098	0.123	0.9908	0.080	0.9706	0.046	0.9492	0.03	0.926
5	—	—	0.380	1.0545	0.276	1.0402	0.188	1.0220	0.115	1.0033	0.075	0.9828	0.043	0.9618	0.03	0.939
6	—	—	0.359	1.0681	0.262	1.0534	0.176	1.0353	0.108	1.0169	0.070	0.9966	0.041	0.9757	0.03	0.952
7	—	—	0.326	1.0821	0.238	1.0673	0.161	1.0498	0.101	1.0313	0.065	1.0110	0.039	0.9907	0.03	0.968
8	—	—	0.285	1.0960	0.210	1.0809	0.145	1.0641	0.094	1.0458	0.061	1.0257	0.036	1.0059	0.02	0.984
9	—	—	0.248	1.1096	0.184	1.0951	0.130	1.0783	0.087	1.0605	0.056	1.0413	0.035	1.021	0.02	0.998
10	0.29	1.136	0.217	1.1235	0.163	1.1089	0.116	1.0926	0.079	1.0732	0.052	1.0565	0.033	1.0358	0.02	1.013
11	0.26	1.150	0.191	1.1368	0.146	1.1221	0.105	1.1065	0.073	1.0895	0.049	1.0711	0.032	1.0509	0.02	1.028
12	0.22	1.163	0.167	1.1502	0.130	1.1352	0.095	1.1200	0.066	1.1036	0.047	1.0855	0.031	1.0653	0.02	1.043
13	0.19	1.177	0.143	1.1636	0.114	1.1485	0.085	1.1337	0.061	1.1175	0.044	1.0994	0.030	1.0797	0.02	1.058
14	0.15	1.190	0.120	1.1769	0.097	1.1617	0.075	1.1473	0.056	1.1313	0.042	1.1134	0.029	1.0940	0.02	1.073
15	0.12	1.204	0.099	1.1902	0.081	1.1751	0.065	1.1616	0.054	1.1451	0.040	1.1277	0.028	1.1086	0.02	1.088

№ 5135 CALCIUM SULFATE–SEAWATER [1776]

$CaSO_4$— —— —H_2O

Content in aqueous soln., wt %		t	d	Comp. of solid phase
Cl^-	$CaSO_4$			
1.629	0.394	40	1.0171	$CaSO_4$
1.629	0.348	50	1.0124	»
1.269	0.253	75	0.9984	»
1.629	0.177	100	0.9831	»
1.629	0.112	125	0.9640	»
1.629	0.072	150	0.9414	»
1.629	0.042	175	0.919	»
1.629	0.026	200	0.894	»
4.964	0.450	35	1.0613	»
4.964	0.426	40	1.0591	»
4.964	0.380	50	1.0539	»
4.964	0.273	75	1.0395	»
4.969	0.191	100	1.0215	»
4.964	0.112	125	1.0027	»
4.964	0.077	150	0.9823	»
9.724	0.287	30	1.1300	»
9.724	0.270	35	1.1275	»
9.724	0.248	40	0.1248	»
9.724	0.226	50	1.1194	»
9.724	0.171	75	1.1050	»
9.724	0.120	100	1.0886	»
9.724	0.080	125	1.0712	»
9.724	0.059	150	1.0521	»
14.806	0.103	50	1.1872	»
14.806	0.096	60	1.1819	»
14.806	0.086	75	1.1724	»
14.806	0.073	100	1.1584	»
14.806	0.054	125	1.1423	»

№ 5136 CALCIUM SULFATE–SEAWATER [335]

$CaSO_4$— —— —H_2O

Content in aqueous soln., wt %				b.p.	d	Comp. of solid phase
Sodium Chloride	Magnesium Chloride	Magnesium Sulfate	Calcium Sulfate			
2.53	0.318	0.218	0.109	100.6	1.024	—
10.79	1.364	0.927	0.432	102.5	1.105	$CaSO_4$
11.95	1.518	1.024	0.333	103.0	1.115	»
13.17	1.671	1.130	0.231	103.8	1.130	»
15.08	1.912	1.257	0.204	104.2	1.145	»
18.54	2.350	1.58	0.159	105.6	1 181	»
20.85	2.64	1.78	0.081	107.4	1.208	»
22.35	2.83	1.91	0.035	108.0	1.230	»

Table № 5136 (continuation)

Content in aqueous soln., wt %				b.p.	d	Comp. of solid phase
Sodium Chloride	Magnesium Chloride	Magnesium Sulfate	Calcium Sulfate			
23.34	3.12	2.11	0.028	108.5	1.235	$CaSO_4+NaCl$
23.85	3.20	2.17	0.025	109.0	1.240	»
22.94	3.79	2.56	0.020	109.0	1.254	»
22.09	4.38	2.96	0.019	109.0	1.260	»
21.90	4.61	3.12	0.016	109.0	1.260	»
21.15	5.20	3.52	0.016	109.2	1.263	»
20.45	5.69	3.85	Traces	109.4	1.265	»
19.01	6.81	4.61	»	109.4	1.267	»
17.63	7.85	4.30	»	109.6	1.269	»

№ 5137 CALCIUM SULFATE (GYPSUM)–SEAWATER [2629]

$CaSO_4 \cdot 2H_2O$— — —H_2O

Total Content of Cl^- in aqueous soln., %	Solubility of Gypsum (%) and specific gravity of soln. at given temperature									
	0°		25°		50°		75°		100°	
	%	d	%	d	%	d	%	d	%	d
1	0.290	1.0167	0.360	1.0136	0.375	1.0050	0.360	0.9917	0.34	0.978
2	0.384	1.0315	0.424	1.0269	0.435	1.0175	0.428	0.0043	0.42	0.989
3	0.411	1.0451	0.447	1.0399	0.458	1.0298	0.452	1.0167	—	—
4	0.420	1.0587	0.456	1.0529	0.466	1.0423	0.461	1.0290	—	—
5	0.418	1.0723	0.455	1.0661	0.466	1.0554	0.460	1.0421	—	—
6	0.402	1.0868	0.443	1.0798	0.457	1.0691	0.451	1.0554	—	—
7	0.370	1.1014	0.415	1.0940	0.436	1.0834	0.429	1.0694	—	—
8	0.331	1.1160	0.373	1.1083	0.398	1.0972	0.395	1.0829	—	—
9	0.294	1.1311	0.332	1.1222	0.358	1.1109	0.355	1.0969	—	—
10	0.260	1.1459	0.295	1.1363	0.319	1.1246	0.317	1.105	—	—
11	0.229	1.1602	0.260	1.1500	0.283	1.1378	—	—	—	—
12	0.198	1.1741	0.226	1.1635	0.247	1.1511	—	—	—	—
13	0.168	1.1883	0.192	1.1767	0.211	1.1644	—	—	—	—
14	0.138	1.2018	0.158	1.1900	0.175	1.1774	—	—	—	—
15	0.110	1.2155	0.127	1.2037	0.140	1.1906	—	—	—	—

№ 5138 CALCIUM SULFATE (GYPSUM)–SEAWATER [1776]

$CaSO_4 \cdot 2H_2O$— — — H_2O

Content in aqueous soln., wt %		t	d	Comp. of solid phase
Cl^-	$CaSO_4$			
1.629	0.361	0	1.0261	$CaSO_4 \cdot 2H_2O$
1.629	0.415	30	1.0207	»
1.629	0.419	35	1.0192	»

Table № 5138 (continuation)

Content in aqueous soln., wt %		t	d	Comp. of solid phase
Cl⁻	$CaSO_4$			
1.629	0.421	40	1.0174	$CaSO_4 \cdot 2H_2O$
1.629	0.417	50	1.0130	»
1.629	0.415	75	1.0000	»
1.629	0.400	100	0.9853	$CaSO_4 \cdot 2H_2O$ (metastable
3 263	0.462	50	1.0334	$CaSO_4 \cdot 2H_2O$
4.964	0.418	0	1.0717	»
4.964	0.456	25	1.0654	»
4.964	0.465	35	1.0615	»
4.964	0.463	40	1.0595	»
4.964	0.465	55	1.0524	»
4.964	0.461	75	1.0415	»
6.545	0.450	50	1.0757	»
9.724	0.280	0	1.1419	»
9.724	0.312	25	1.1324	»
9.724	0.309	30	1.1303	»
9.724	0.314	35	1.1280	»
9.724	0.333	50	1.1206	»
9.724	0.332	75	1.1068	»
14.806	0.120	0	1.2125	»
14.806	0.133	25	1.2010	»
14.806	0.150	30	1.1878	»

№ 5139 [317]

CALCIUM THIOSULFATE–CALCIUM CHLORIDE–WATER

$CaS_2O_3—CaCl_2—H_2O$

$t=25$

Concn. in aqueous soln., wt %		Comp. of solid phase
Calcium Chloride	Calcium Thiosulfate	
0.0	34.68	$CaS_2O_3 \cdot 6H_2O$
6.08	28.55	»
10.24	24.10	»
12.75	20.80	»
17.16	16.24	»
19.75	14.20	»
20.76	13.88	»
26.88	10.15	»
29.68	9.15	»
31.10	8.33	»
32.65	7.99	»
33.34	7.49	$CaS_2O_3 \cdot 6H_2O + CaS_2O_3 \cdot H_2O$
38.46	3.45	$CaS_2O_3 \cdot H_2O$
42.30	1.43	»
43.16	1.32	»
45.01	1.03	$CaS_2O_3 \cdot H_2O + CaCl_2 \cdot 6H_2O$
45.65	0.0	$CaCl_2 \cdot 6H_2O$

№ 5140 [2278]

CALCIUM CHLORIDE–CALCIUM HYPOCHLORITE–WATER

$CaCl_2—Ca(ClO)_2—H_2O$

$t=0$

Comp. of saturated soln., wt %			d_4	Comp. of solid phase
Calcium Hypochlorite	Calcium Chloride	Water		
0.0	37.6	62.4	1.370	$CaCl_2 \cdot 6H_2O$
1.6	36.9	61.5	1.382	»
2.6	36.4	61.0	1.388	$CaCl_2 \cdot 6H_2O + Ca(ClO)_2 \cdot 3H_2O$
2.9	34.0	63.1	1.366	$Ca(ClO)_2 \cdot 3H_2O$
4.2	30.7	65.1	1.333	»
4.5	28.5	67.0	1.322	»
7.4	22.2	70.4	1.285	»
8.8	19.4	71.8	1.270	»
11.0	15.4	73.6	1.255	»
14.0	10.8	75.2	1.238	»
16.1	8.4	75.5	1.234	»
18.6	5.3	76.1	1.228	»
19.6	3.1	77.3	1.220	»
21.8	0.0	78.2	1.213	»

№ 5141 [2162]

CALCIUM CHLORIDE–CALCIUM CHLORATE–WATER

$CaCl_2—Ca(ClO_3)_4—H_2O$

$t=20$

Comp. of saturated soln., wt %			Comp. of solid phase
Calcium Chlorate	Calcium Chloride	Water	
0.00	42.7	57.3	$CaCl_2 \cdot 6H_2O$
6.74	39.49	53.77	»
10.56	37.35	52.09	»
15.13	35.83	52.09	»
22.88	33.84	43.28	$CaCl_2 \cdot 6H_2O + CaCl_2 \cdot 4H_2O$
29.23	31.46	39.31	$CaCl_2 \cdot 4H_2O$
33.47	28.29	38.24	$CaCl_2 \cdot 4H_2O + Ca(ClO_3)_2 \cdot 2H_2O$
42.85	19.67	37.48	$Ca(ClO_3)_2 \cdot 2H_2O$
49.61	12.83	37.56	»
53.57	10.34	36.09	»
58.74	7.16	34.10	»
66.16	0.00	33.84	»

№ 5142 CALCIUM CHLORIDE–CALCIUM CHLORATE–WATER [1543]

$CaCl_2—Ca(ClO_3)_2—H_2O$

$t=25$

Comp. of saturated soln., wt %			d_{25}	Comp. of solid phase
Calcium Chlorate	Calcium Chloride	Water		
0.0	44.92	55.08	1.453	$CaCl_2 \cdot 6H_2O$
2.36	43.26	54.38	1.490	»
3.85	44.06	52.09	1.480	»
8.85	42.60	48.55	1.544	$CaCl_2 \cdot 6H_2O+CaCl_2 \cdot 4H_2O$
10.70	41.22	48.08	1.570	$CaCl_2 \cdot 4H_2O$
12.12	41.70	46.18	1.565	»
12.29	40.15	47.56	1.578	»
13.41	40.53	46.06	1.580	»
16.55	38.82	44.63	1.603	»
16.91	37.92	45.17	1.614	»
17.69	38.17	44.14	1.607	»
19.65	37.75	42.60	1.618	»
21.52	36.91	41.57	1.625	»
22.14	35.59	42.27	1.641	»
25.20	34.69	40.11	1.659	»
30.11	31.51	38.38	1.718	»
32.01	31.49	36.50	1.725	»
33.36	30.65	35.99	1.744	$CaCl_2 \cdot 4H_2O+Ca(ClO_3)_2 \cdot 2H_2O$
33.01	30.43	36.56	1.735	$Ca(ClO_3)_2 \cdot 2H_2O$
34.71	29.21	36.08	1.733	»
36.02	28.15	35.83	1.730	»
44.59	19.19	36.22	1.731	»
53.45	10.31	36.24	1.751	»
58.65	5.86	35.49	1.717	»
66.05	0.0	33.95	1.781	»

№ 5143 CALCIUM CHLORIDE–CALCIUM CHLORATE–WATER [541]

$CaCl_2—Ca(ClO_3)_2—H_2O$

Comp. of saturated soln., wt %			t	Comp. of solid phase
Calcium Chlorate	Calcium Chloride	Water		
0.00	57.27	42.73	55	$CaCl_2 \cdot 2H_2O$
10.06	50.44	39.50	55	»
37.92	33.16	28.92	55	»
41.74	29.69	28.57	55	$CaCl_2 \cdot 2H_2O+Ca(ClO_3)_2$
45.98	26.29	27.73	55	—
49.04	23.87	27.09	55	$Ca(ClO_3)_2$
51.78	21.04	27.18	55	$Ca(ClO_3)_2+Ca(ClO_3)_2 \cdot 2H_2O$
56.01	15.59	28.40	55	$Ca(ClO_3)_2 \cdot 2H_2O$
56.23	14.28	29.49	55	»
71.75	0.00	28.25	55	»

Comp. of saturated soln., wt %			t	Comp. of solid phase
Calcium Chlorate	Calcium Chloride	Water		
0.00	58.85	41.15	75	$CaCl_2 \cdot 2H_2O$
7.79	53.49	38.72	75	»
16.89	47.37	35.74	75	»
19.37	45.63	35.00	75	»
23.23	42.51	34.26	75	»
27.29	40.51	22.20	75	»
41.94	34.03	24.03	75	$CaCl_2 \cdot 2H_2O + Ca(ClO_3)_2$
44.22	30.98	24.80	75	$Ca(ClO_3)_2$
50.53	24.00	25.47	75	»
70.09	5.72	24.19	75	»
74.55	2.37	23.08	75	»
77.83	0.00	22.17	75	»

№ 5144 CALCIUM CHLORIDE–CALCIUM DICHROMATE–WATER [405]

$CaCl_2—CaCr_2O_7—H_2O$

Comp. of saturated soln., wt %			t	Comp. of solid phase
Calcium Dichromate	Calcium Chloride	Water		
53.47	0.00	46.53	0	$CaCr_2O_7 \cdot 6H_2O$
47.27	3.94	48.79	0	»
37.69	10.10	52.21	0	»
31.10	14.68	54.22	0	»
24.87	19.05	56.08	0	»
19.33	23.27	57.40	0	»
9.63	32.65	57.72	0	$CaCr_2O_7 \cdot 6H_2O + CaCl_2 \cdot 6H_2O$
9.41	32.45	58.14	0	$CaCl_2 \cdot 6H_2O$
8.45	32.71	58.84	0	»
5.66	34.92	59.42	0	»
2.62	36.22	61.16	0	»
0.00	37.16	62.84	0	»
59.00	0.00	41.00	25	$CaCr_2O_7 \cdot 5H_2O$
47.86	7.33	44.81	25	»
32.47	18.77	48.76	25	»
22.80	27.31	49.89	25	»
21.64	28.36	50.00	25	»
16.88	34.81	48.31	25	$CaCr_2O_7 \cdot 5H_2O + (CaCrO_4)$
17.01	35.04	47.95	25	$CaCr_2O_7 \cdot 5H_2O + CaCl_2 \cdot 6H_2O + (CaCrO_4)$
14.94	36.47	48.59	25	$CaCl_2 \cdot 6H_2O + (CaCrO_4)$
12.92	37.52	49.56	25	»
12.33	38.43	49.24	25	»
10.05	38.93	51.02	25	»
5.87	41.82	52.31	25	»
2.95	42.73	54.32	25	»
1.77	43.93	54.30	25	»
0.00	44.44	55.56	25	$CaCl_2 \cdot 6H_2O$

Table № 5144 (continuation)

Comp. of saturated soln., wt %			t	Comp. of solid phase
Calcium Dichromate	Calcium Chloride	Water		
64.35	0.00	35.65	50	$CaCr_2O_7 \cdot 4H_2O$
61.02	2.63	36.35	50	»
60.14	2.89	36.27	50	»
56.25	6.16	37.59	50	»
53.62	7.97	38.41	50	»
52.57	8.84	38.59	50	$CaCr_2O_7 \cdot 4H_2O$
52.88	9.84	37.28	50	$CaCr_2O_7 \cdot 4H_2O + (CaCrO_4)$
52.53	9.95	37.52	50	$CaCr_2O_7 \cdot 4H_2O$
52.29	11.44	36.27	50	»
47.39	18.39	34.22	50	»
45.75	13.77	40.48	50	»
45.52	22.77	31.71	50	$CaCr_2O_7 \cdot 4H_2O + (CaCrO_4)$
43.25	16.37	40.38	50	»
39.80	30.65	29.55	50	»
37.88	20.25	41.87	50	»
19.79	47.29	32.92	50	$CaCl_2 \cdot 2H_2O + (CaCrO_4)$
14.32	49.55	36.13	50	»
12.83	50.56	36.61	50	»
11.08	51.24	37.68	50	»
10.48	51.51	38.01	50	»
7.30	52.95	39.75	50	»
2.95	55.89	41.16	50	»
0.31	56.80	42.89	50	»
0.16	56.44	43.40	50	»
0.00	56.88	43.12	50	$CaCl_2 \cdot 2H_2O$

Note. ($CaCrO_2$) is a sparingly soluble compound

№ 5145 [1786]

CALCIUM CHLORIDE–MANGANESE TETRAHYDROGEN ORTHOBORATE–WATER

$CaCl_2—MnH_4(BO_3)_2—H_2O$

Concn. in aqueous soln., g/l		t	Concn. in aqueous soln., g/l		t
Manganese Tetrahydrogen Orthoborate	Calcium Chloride		Manganese Tetrahydrogen Orthoborate	Calcium Chloride	
2.81	20	18	2.25	20	60
2.44	20	40	1.35	20	80

№ 5146 [2735]

CALCIUM CHLORIDE–IRON (II) HYDROXIDE–WATER
$CaCl_2—Fe(OH)_2—H_2O$

$t=25$

Concn. of Calcium Chloride in aqueous soln., g/l	Solubility of Iron Hydroxide in aqueous soln. of Calcium Chloride, g/l	Concn. of Calcium Chloride in aqueous soln., g/l	Solubility of Iron Hydroxide in aqueous soln. of Calcium Chloride, g/l
0.0	0.00375	55.5	0.0016
27.8	0.0042	83.3	0.0014
55.5	0.0045	111.1	0.00125
83.3	0.0046	139.8	0.00105
111.1	0.0047	166.4	0.00095
138.9	0.0048	222.0	0.00075
0.0	0.00375	249.8	0.00070
27.8	0.0020		

№ 5147 [1165]

CALCIUM CHLORIDE–COBALT (II) CHLORIDE–WATER
$CaCl_2—CoCl_2—H_2O$

Comp. of saturated soln., wt %			t	Comp. of solid phase
Cobalt Chloride	Calcium Chloride	Water		
0.00	37.00	63.00	0	$CaCl_2 \cdot 6H_2O$
1.64	36.25	62.11	0	»
2.36	35.82	61.82	0	»
2.85	35.48	61.67	0	$CaCl_2 \cdot 6H_2O + CoCl_2 \cdot 6H_2O$
2.82	36.77	60.41	0	$CoCl_2 \cdot 6H_2O$
7.22	26.42	64.36	0	»
29.50	0.00	70.50	0	»
0.00	45.05	54.95	25	$CaCl_2 \cdot 6H_2O$
2.77	43.37	53.86	25	»
8.34	41.13	50.53	25	»
10.36	40.63	49.01	25	$CaCl_2 \cdot 6H_2O + CoCl_2 \cdot 2H_2O$
11.43	38.54	50.03	25	$CoCl_2 \cdot 2H_2O$
13.64	34.90	51.46	25	»
13.51	34.83	51.66	25	$CoCl_2 \cdot 2H_2O + CoCl_2 \cdot 2H_2O$
12.52	33.46	54.02	25	$CoCl_2 \cdot 6H_2O$
11.39	29.92	58.69	25	»
12.44	26.10	61.46	25	»
18.50	18.66	62.84	25	»
23.66	12.61	63.73	25	»
26.88	9.08	64.04	25	»
35.87	0.00	64.13	25	»
0.00	56.70	43.30	50	$CaCl_2 \cdot 2H_2O$
6.75	51.43	41.82	50	»
11.55	47.74	40.71	50	»
12.79	46.63	40.58	50	$CaCl_2 \cdot 2H_2O + CoCl_2 + 2H_2O$
13.72	42.85	43.43	50	$CoCl_2 \cdot 2H_2O$
23.24	25.98	50.78	50	»
27.67	20.87	51.46	50	»

Comp. of saturated soln., wt %			t	Comp. of solid phase
Cobalt Chloride	Calcium Chloride	Water		
35.41	11.88	52.71	50	$CoCl_2 \cdot 2H_2O$
40.22	7.87	51.91	50	»
41.92	6.31	51.77	50	$CoCl_2 \cdot 2H_2O + CoCl_2 \cdot 6H_2O$
43.50	3.34	53.16	50	$CoCl_2 \cdot 6H_2O$
46.00	0.00	54.00	50	»

№ 5148 [2596]

CALCIUM CHLORIDE–ZINC HYDROXIDE–WATER

$CaCl_2—Zn(OH)_2—H_2O$

$t=16—20$

Comp. of solvent, wt %		Solubility of Zinc Hydroxide, g/l
Calcium Chloride	Water	
1	99	0.0575

№ 5149 [947]

CALCIUM CHLORIDE–ZINC CHLORIDE–WATER

$CaCl_2—ZnCl_2—H_2O$

Concn. in aqueous soln., wt %		t	Comp. of solid phase
Calcium Chloride	Zinc Chloride		
52.0	0.0	25	$CaCl_2$
43.33	23.6	25	»
41.35	28.80	25	»
24.19	53.11	25	»
22.29	60.56	25	$CaCl_2+ZnCl_2$
0.0	81.0	25	$ZnCl_2$
57.8	0.0	60	$CaCl_2$
50.42	6.4	60	»
46.99	13.05	60	»
34.41	34.42	60	»
27.55	41.7	60	»
28.30	50.19	60	$CaCl_2+ZnCl_2$
27.09	47.6	60	»
15.92	63.87	60	$ZnCl_2$
6.58	76.09	60	»
3.67	79.89	60	»
2.01	80.49	60	»
0.9	81.52	60	»
0.0	83.0	60	»

$CaCl_2—ZnCl_2—H_2O$

Comp. of saturated soln., wt %			t	Comp. of solid phase
Calcium Chloride	Zinc Chloride	Water		
35.5	0.0	64.5	0	$CaCl_2 . 6H_2O$
37.4	3.9	68.7	0	»
33.3	10.9	55.8	0	»
31.0	19.0	50.0	0	$CaCl_2 . 4H_2O$
30.6	21.9	47.5	0	»
30.0	27.0	43.0	0	»
30.0	29.0	41.0	0	»
30.0	30.0	40.0	0	$CaCl_2 . 4H_2O+CaCl_2 . ZnCl_2 . 5H_2O$
30.6	29.8	39.6	0	»
29.5	30.1	40.4	0	»
29.5	29.1	41.4	0	»
28.0	33.0	39.0	0	$CaCl_2 . ZnCl_2 . 5H_2O$
20.0	43.0	37.0	0	»
20.0	47.0	33.0	0	»
18.0	52.0	30.0	0	»
17.0	54.0	29.0	0	»
0.0	65.5	34.5	0	»
47.3	0.0	52.7	25	$CaCl_2 . 6H_2O$
43.3	6.0	50.7	25	»
38.9	14.3	46.8	25	»
41.7	19.4	38.9	25	$CaCl_2 . 4H_2O$
41.3	19.1	39.6	25	»
36.7	19.6	43.7	25	»
33.7	22.4	43.9	25	»
35.3	27.5	37.2	25	$CaCl_2 . 4H_2O+CaCl_2 . ZnCl_2 . 5H_2O$
35.3	26.4	38.3	25	»
35.3	27.1	37.6	25	»
35.0	27.5	37.5	25	»
34.6	28.8	36.6	25	$CaCl_2 . ZnCl_2 . 5H_2O$
30.9	31.5	37.6	25	
28.8	34.6	36.6	25	»
24.3	45.0	30.7	25	»
23.9	47.6	28.5	25	»
22.4	48.9	28.7	25	»
20.5	53.3	26.2	25	»
19.9	57.8	22.3	25	»
0.0	80.0	20.0	25	$CaCl_2 . ZnCl_2 . 5H_2O+ZnCl_2 . (H_2O)_x$

№ 5151 [2519]

CALCIUM CHLORIDE–ARSENIC TRIOXIDE–WATER

$CaCl_2—As_2O_3—H_2O$

$t=20$

Comp. of saturated soln., wt %			Comp. of solid phase
Arsenic Trioxide	Calcium Chloride	Water	
1.78	0.00	98.22	As_2O_3
1.39	12.66	85.95	»
1.01	23.09	75.90	»
0.865	27.68	71.455	»
0.757	31.85	67.393	»
0.697	36.01	63.293	»
0.675	41.92	57.405	$As_2O_3+CaCl_2 . 6H_2O$
0.0	42.7	57.3	$CaCl_2 . 6H_2O$

№ 5152 [2703]

CALCIUM CHLORIDE–STRONTIUM SULFATE–WATER

$CaCl_2—SrSO_4—H_2O$

$t=25$

Comp. of solvent, wt %		Solubility of Strontium Sulfate, wt %
Calcium Chloride	Water	
8.67	91.33	0.175
16.51	83.49	0.184
33.70	66.30	0.170

№ 5153 [1693]

CALCIUM CHLORIDE–STRONTIUM SULFATE–WATER

$CaCl_2—SrSO_4—H_2O$

Comp. of solvent, wt %		Solubility of Strontium Sulfate, g/l	t
Calcium Chloride	Water		
10	90	0.313	20
10	90	0.358	98

№ 5154 [1109]

CALCIUM CHLORIDE–STRONTIUM CHLORIDE–WATER

$CaCl_2—SrCl_2—H_2O$

Comp. of saturated soln., wt %			t	Comp. of solid phase
Strontium Chloride	Calcium Chloride	Water		
34.15	0.0	65.85	18.0	$SrCl_2 . 6H_2O$
26.9	6.6	66.5	18.0	$SrCl_2 . 6H_2O . CaCl_2 . 6H_2O$
18.0	14.6	65.4	18.0	»
4.0	31.2	64.8	18.0	»
3.0	33.3	63.7	18.0	»
2.1	35.5	62.4	18.0	»
1.9	37.7	60.4	18.0	»
1.4	38.7	59.9	18.0	»
1.1	39.7	59.2	18.0	»
0.7	40.4	58.9	18.0	»
0.5	40.6	58.9	18.0	»

Comp. of saturated soln., wt %			t	Comp. of solid phase
Strontium Chloride	Calcium Chloride	Water		
0.4	41.2	58.4	18.0	$SrCl_2 \cdot 6H_2O \cdot CaCl_2 \cdot 6H_2O$
0.3	41.4	58.3	18.0	»
0.0	42.05	57.95	18.0	$CaCl_2 \cdot 6H_2O$
36.6	0.0	63.4	28.0	$SrCl_2 \cdot 6H_2O$
17.6	17.2	65.2	28.0	$SrCl_2 \cdot 6H_2O \cdot CaCl_2 \cdot 6H_2O$
6.0	30.7	63.3	28.0	»
3.4	35.4	61.2	28.0	»
2.7	40.1	57.2	28.0	»
2.2	41.7	56.1	28.0	»
2.1	42.3	55.6	28.0	»
1.7	43.6	54.7	28.0	»
1.5	44.3	54.2	28.0	»
1.0	45.2	53.8	28.0	»
1.1	45.9	53.0	28.0	»
1.0	46.1	52.9	28.0	»
0.6	46.6	52.8	28.0	»
0.0	47.2	52.8	28.0	$CaCl_2 \cdot 6H_2O$
2.6	40.3	57.1	28.5	$SrCl_2 \cdot 6H_2O \cdot CaCl_2 \cdot 6H_2O$
2.6	40.9	56.5	28.5	»
2.4	42.2	55.4	28.5	$SrCl_2 \cdot 2H_2O$
2.3	42.7	55.0	28.5	»
1.7	44.1	54.2	28.5	»
1.6	44.2	54.2	28.5	$SrCl_2 \cdot 6H_2O \cdot CaCl_2 \cdot 6H_2O$
1.3	44.6	54.1	28.5	»
1.2	45.2	53.6	28.5	»
1.0	46.0	53.0	28.5	»
0.8	46.9	52.3	28.5	»
0.0	47.9	52.1	28.5	$CaCl_2 \cdot 6H_2O$
36.8	0.0	63.2	29.0	$SrCl_2 \cdot 6H_2O$
20.0	16.0	64.0	29.0	$SrCl_2 \cdot 6H_2O \cdot CaCl_2 \cdot 6H_2O$
6.4	30.6	63.0	29.0	»
2.8	39.1	58.1	29.0	»
2.9	40.6	56.5	29.0	»
2.9	41.6	55.5	29.0	$SrCl_2 \cdot 2H_2O$
2.5	42.0	55.5	29.0	»
2.5	42.5	55.0	29.0	»
1.7	44.0	54.3	29.0	»
1.7	44.3	54.0	29.0	»
1.6	44.6	53.8	29.0	$SrCl_2 \cdot 6H_2O \cdot CaCl_2 \cdot 6H_2O$
1.4	45.2	53.4	29.0	»
1.4	45.8	52.8	29.0	»
1.1	46.8	52.1	29.0	»
0.9	47.7	51.4	29.0	»
0.2	48.7	51.1	29.0	»
0.0	49.2	50.8	29.0	$CaCl_2 \cdot 6H_2O$
3.7	38.5	57.7	29.3	$SrCl_2 \cdot 6H_2O \cdot CaCl_2 \cdot 6H_2O$
3.5	39.9	56.6	29.3	»
3.4	40.2	56.4	29.3	$SrCl_2 \cdot 2H_2O + SrCl_2 \cdot 6H_2O \cdot CaCl_2 \cdot 6H_2O$
2.7	41.6	55.7	29.3	$SrCl_2 \cdot 2H_2O$
2.5	42.4	55.1	29.3	»

Table № 5154 (continuation)

Comp. of saturated soln., wt %			t	Comp. of solid phase
Strontium Chloride	Calcium Chloride	Water		
2.3	42.8	54.9	29.3	$SrCl_2 \cdot 2H_2O$
1.9	44.8	53.3	29.3	»
1.7	46.5	51.8	29.3	$SrCl_2 \cdot 6H_2O \cdot CaCl_2 \cdot 6H_2O$
1.4	47.4	51.2	29.3	»
0.7	48.7	50.6	29.3	»
0.3	49.1	50.6	29.3	»
0.0	49.6	50.4	29.3	$CaCl_2 \cdot 6H_2O$
37.0	0.0	63.0	29.7	$SrCl_2 \cdot 6H_2O$
28.7	7.5	63.8	29.7	$SrCl_2 \cdot 6H_2O \cdot CaCl_2 \cdot 6H_2O$
13.5	21.9	64.6	29.7	»
3.1	39.4	57.5	29.7	»
3.0	40.4	56.6	29.7	$SrCl_2 \cdot 2H_2O + SrCl_2 \cdot 6H_2O \cdot CaCl_2 \cdot 6H_2O$
2.0	42.2	55.8	29.7	$SrCl_2 \cdot 2H_2O$
1.4	44.6	54.0	29.7	»
1.3	46.6	52.1	29.7	»
1.4	46.9	51.7	29.7	»
1.2	47.7	51.1	29.7	»
1.2	48.2	50.6	29.7	»
1.0	49.2	49.8	29.7	$SrCl_2 \cdot 2H_2O + \alpha\text{-}CaCl_2 \cdot 4H_2O$
1.0	49.0	50.0	29.7	$\alpha\text{-}CaCl_2 \cdot 4H_2O$
0.8	49.2	50.0	29.7	»
0.0	50.2	49.8	29.7	»
9.6	31.6	58.8	44.3	$SrCl_2 \cdot 6H_2O \cdot CaCl_2 \cdot 6H_2O$
7.2	34.4	58.4	44.3	$SrCl_2 \cdot 2H_2O$
1.0	55.2	43.8	44.3	$\alpha\text{-}CaCl_2 \cdot 4H_2O$
1.0	55.0	44.0	44.7	$SrCl_2 \cdot 2H_2O$
0.9	55.6	43.5	44.7	$\alpha\text{-}CaCl_2 \cdot 4H_2O$
0.9	56.0	43.1	45.3	$SrCl_2 \cdot 2H_2O$
0.9	56.3	42.8	45.7	$CaCl_2 \cdot 2H_2O$
2.0	42.2	45 8	28.2	$SrCl_2 \cdot 2H_2O + SrCl_2 \cdot 6H_2O \cdot CaCl_2 \cdot 6H_2O$
1.0	48.5	50.5	29.4	$SrCl_2 \cdot 2H_2O + \alpha\text{-}CaCl_2 \cdot 4H_2O + SrCl_2 \cdot 6H_2O \cdot CaCl_2 \cdot 6H_2O$
1.0	55.7	43.3	44.9	$SrCl_2 \cdot 2H_2O + \alpha\text{-}CaCl_2 \cdot 4H_2O + CaCl_2 \cdot 2H_2O$

№ 5155 [1056]

CALCIUM CHLORIDE–STRONTIUM CHLORIDE–BARIUM CHLORIDE–WATER
$CaCl_2$—$SrCl_2$—$BaCl_2$—H_2O

Comp. of saturated soln., wt %				t	Comp. of solid phase
Calcium Chloride	Barium Chloride	Strontium Chloride	Water		
57.10	0.00	1.00	41.90	60	$CaCl_2 \cdot 2H_2O + SrCl_2 \cdot 2H_2O$
0.00	0.70	45.84	53.46	60	$BaCl_2 \cdot 2H_2O + SrCl_2 \cdot 2H_2O$
56.35	0.178	0.00	43.47	60	$CaCl_2 \cdot 2H_2O + BaCl_2 \cdot 2H_2O$
53.29	0.156	1.297	45.26	60	$CaCl_2 \cdot 2H_2O + BaCl_2 \cdot 2H_2O + SrCl_2 \cdot 2H_2O$
19.89	0.31	22.08	57.72	60	$BaCl_2 \cdot 2H_2O + SrCl_2 \cdot 2H_2O$
0.00	31.30	0.00	68.70	60	$BaCl_2 \cdot 2H_2O$
59.10	0.00	2.70	38.20	100	$CaCl_2 \cdot 2H_2O + SrCl_2 \cdot H_2O$
0.00	2.29	47.52	50.19	100	$BaCl_2 \cdot 2H_2O + SrCl_2 \cdot H_2O$
58.64	0.39	0.00	40.97	100	$CaCl_2 \cdot 2H_2O + BaCl_2 \cdot 2H_2O$
55.33	0.336	2.74	41.57	100	$CaCl_2 \cdot 2H_2O + SrCl_2 \cdot H_2O + BaCl_2 \cdot 2H_2O$
19.70	1.907	19.85	58.54	100	$SrCl_2 \cdot H_2O + BaCl_2 \cdot 2H_2O$
0.00	37.00	0.00	63.00	100	$BaCl_2 \cdot 2H_2O$
0.00	0.00	58.50	41.50	100	$SrCl_2 \cdot H_2O$

№ 5156 [354]

CALCIUM CHLORIDE–ZIRCONIUM CHLORIDE OXIDE–WATER
$CaCl_2$—$ZrOCl$—H_2O

Comp. of saturated soln., wt %			t	Comp. of solid phase
Calcium Chloride	Zirconium Chloride Oxide	Water		
0.00	36.96	63.04	25	$ZrOCl_2 \cdot 8H_2O$
0.24	36.50	63.26	25	»
3.19	31.95	64.86	25	»
7.78	26.34	65.88	25	»
8.70	24.05	67.25	25	»
12.61	19.31	68.05	25	»
15.80	14.10	70.10	25	»
20.75	8.60	70.65	25	»
23.94	6.68	69.62	25	»
26.91	3.04	70.05	25	»
31.16	1.11	67.73	25	
36.47	0.33	63.20	25	»
44.20	0.07	55.73	25	$ZrOCl_2 \cdot 8H_2O + CaCl_2 \cdot 6H_2O$
45.23	0.00	54.77	25	$CaCl_2 \cdot 6H_2O$
0.00	39.98	60.02	50	$ZrOCl_2 \cdot 8H_2O$
6.56	31.50	61.95	50	»
11.04	25.58	63.48	50	»
23.03	10.55	66.42	50	»
27.24	6.47	66.29	50	»
30.56	4.35	65.09	50	»
33.80	2.40	63.80	50	»
47.00	0.71	52.29	50	»
54.85	0.64	44.51	50	$ZrOCl_2 \cdot 8H_2O + CaCl_2 \cdot 6H_2O$
56.95	0.00	43.05	50	$CaCl_2 \cdot 6H_2O$

№ 5157 [1650

CALCIUM CHLORIDE–SILVER CHLORIDE–WATER

$CaCl_2—AgCl—H_2O$

Concn. in aqueous soln., g/l		t	Concn. in aqueous soln., g/l		t
Calcium Chloride	Silver Chloride		Calcium Chloride	Silver Chloride	
194.9	0.1378	1	207.4	0.3120	25
97.0	0.0413	25	223.8	0.4007	25
122.2	0.0716	25	251.8	0.5970	25
152.1	0.1287	25	277.8	0.8237	25
181.2	0.2092	25	178.8	0.2582	35
184.3	0.2165	25			

№ 5158 [1170]

CALCIUM CHLORIDE–CADMIUM CHLORIDE–WATER

$CaCl_2—CdCl_2—H_2O$

$t=25$

Comp. of saturated soln., wt %			Comp. of solid phase
Cadmium Chloride	Calcium Chloride	Water	
0.00	45.05	54.95	$CaCl_2 \cdot 6H_2O$
1.60	44.60	53.80	$CaCl_2 \cdot 6H_2O + CdCl_2 \cdot 2CaCl_2 \cdot 12H_2O$
4.50	39.80	55.70	$CdCl_2 \cdot 2CaCl_2 \cdot 12H_2O$
8.70	37.50	53.8	»
12.00	35.00	53.00	»
16.50	32.50	51.00	»
17.80	32.20	50.00	»
21.80	30.50	47.70	»
24.30	28.80	46.90	$CdCl_2 \cdot 2CaCl_2 \cdot 12H_2O + CdCl_2 \cdot H_2O + CaCl_2 \cdot 4H_2O$
25.00	28.30	46.70	$CdCl_2 \cdot H_2O + CaCl_2 \cdot 4H_2O$
32.80	21.70	45.50	»
38.30	19.30	42.40	»
43.10	16.70	40.20	»
46.80	14.90	38.30	»
47.80	11.30	40.90	$CdCl_2 \cdot 2.5H_2O + CdCl_2 \cdot H_2O$
48.60	8.30	43.10	$CdCl_2 \cdot 2.5H_2O$
50.00	6.10	43.90	»
54.62	0.00	45.38	»

№ 5159 CALCIUM CHLORIDE–INDIUM CHLORIDE–WATER [1568]

$CaCl_2—InCl_3—H_2O$

$t=20$

Concn. in aqueous soln., g/l		Comp. of solid phase
Indium Chloride	Calcium Chloride	
1439	84	$InCl_3 \cdot 4H_2O$
1353	108	$CaCl_2 \cdot 2InCl_3 \cdot 12H_2O$
1276	136	$3CaCl_2 \cdot 4InCl_3 \cdot 30H_2O$
934	306	$CaCl_2 \cdot JnCl_3 \cdot 8H_2O$
643	458	»
543	545	»
424	601	»
362	673	»

№ 5160 CALCIUM CHLORIDE–CESIUM CHLORIDE–WATER [700]

$CaCl_2—CsCl—H_2O$

Comp. of saturated soln., wt %			Comp. of solid phase
Cesium Chloride	Calcium Chloride	Water	
			$t=25$
0.00	44.70	55.30	$CaCl_2 \cdot 6H_2O$
0.00	44.45	55.55	$CaCl_2 \cdot 6H_2O+CsCl \cdot CaCl_2$
0.00	44.07	55.93	$CsCl \cdot CaCl_2$
0.43	42.62	56.95	»
1.25	41.30	57.45	»
1.99	40.10	57.91	»
3.11	39.41	57.48	»
5.10	38.68	56.22	$CsCl \cdot CaCl_2+2CsCl \cdot CaCl_2 \cdot 2H_2O$
6.49	37.58	55.93	$2CsCl \cdot CaCl_2 \cdot 2H_2O$
9.68	35.42	54.90	»
11.57	33.78	54.65	»
15.18	31.09	53.73	»
18.88	28.05	53.07	»
23.53	25.60	50.87	»
27.68	22.49	49.83	»
36.53	18.44	45.03	»
42.15	15.50	42.55	»
45.49	13.68	40.83	»
48.10	12.16	39.74	»
50.90	11.59	37.51	$2CsCl \cdot CaCl_2 \cdot 2H_2O+5CsCl \cdot CaCl_2$
51.28	11.37	37.35	$5CsCl \cdot CaCl_2$
52.47	9.67	37.86	»
54.98	8.27	36.75	»
55.64	7.86	36.50	»
57.59	6.76	36.65	$5CsCl \cdot CaCl_2+CsCl$
59.68	5.31	35.01	$CsCl$
63.71	1.49	34.80	»
65.77	0.00	34.23	»

Table № 5160 (continuation)

Comp. of saturated soln., wt %			Comp. of solid phase
Cesium Chloride	Calcium Chloride	Water	
			$t=50$
0.00	56.79	43.21	$CaCl_2 \cdot 2H_2O$
0.00	56.79	43.21	$CaCl_2 \cdot 2H_2O + CsCl \cdot CaCl_2$
0.00	56.76	43.24	$CsCl \cdot CaCl_2$
0.00	52.50	47.50	»
0.00	48.71	51.29	»
0.14	47.69	52.17	»
0.57	45.39	54.04	»
1.04	44.31	54.65	»
3.07	41.83	55.10	»
3.66	40.85	55.49	»
5.65	38.96	55.39	»
6.69	38.32	54.99	»
9.26	36.51	54.23	$CsCl \cdot CaCl_2 + 2CsCl \cdot CaCl_2 \cdot 2H_2O$
10.16	96.22	53.62	$2CsCl \cdot CaCl_2 \cdot 2H_2O$
12.75	33.25	54.00	»
17.47	30.02	52.51	»
18.89	29.13	51.98	»
24.40	25.76	49.84	»
28.68	22.52	48.80	»
33.63	20.65	45.72	»
36.50	19.10	44.40	»
38.36	17.86	43.78	»
40.69	16.81	42.50	»
41.78	16.03	42.19	»
46.00	14.00	40.00	»
50.33	12.07	37.60	»
52.03	11.60	36.37	$2CsCl \cdot CaCl_2 \cdot 2H_2O + 5CaCl \cdot CaCl_2$
53.46	10.49	36.05	$5CsCl \cdot CaCl_2$
54.08	9.84	36.08	»
56.12	8.75	35.13	»
60.16	6.23	33.61	»
61.25	5.85	32.90	$5CsCl \cdot CaCl_2 + CsCl$
61.86	5.53	32.61	$CsCl$
65.50	2.50	32.00	»
68.61	0.00	31.39	»
			$t=75$
0.00	58.56	41.44	$CaCl_2 \cdot 2H_2O$
0.00	58.56	41.44	$CaCl_2 \cdot 2H_2O + CsCl \cdot CaCl_2$
0.00	57.25	42.75	$CsCl \cdot CaCl_2$
0.00	52.31	47.69	»
0.15	50.90	48.95	»
2.51	44.59	52.90	»
5.22	40.66	54.12	»
6.51	39.74	53.75	»
11.40	34.92	53.68	»
17.42	32.22	50.36	$CsCl \cdot CaCl_2 + 2CsCl \cdot CaCl_2 \cdot 2H_2O$
17.99	31.88	50.13	$2CsCl \cdot CaCl_2 \cdot 2H_2O$
21.46	29.61	48.93	»
22.32	28.22	49.46	»
27.06	25.93	47.01	»

Comp. of saturated soln., wt %			Comp. of solid phase
Cesium Chloride	Calcium Chloride	Water	
35.99	20.47	43.54	$2CsCl \cdot CaCl_2 \cdot 2H_2O$
46.97	14.37	38.66	»
52.02	11.98	37.00	»
54.08	10.96	34.96	»
55.03	10.39	34.58	$2CsCl \cdot CaCl_2 \cdot 2H_2O + 5CsCl \cdot CaCl_2$
55.99	10.00	34.01	$5CsCl \cdot CaCl_2$
58.07	9.11	32.82	»
59.46	8.72	31.79	»
59.80	8.46	31.74	$5CsCl \cdot CaCl_2 + CsCl$
62.51	6.50	30.99	CsCl
67.95	2.49	29.56	»
70.62	0.00	29.38	»

№ 5161 CALCIUM CHLORIDE–BARIUM NITRATE–WATER [920]

$CaCl_2—Ba(NO_3)_2—H_2O$

Comp. of saturated soln., g-mol			t	Comp. of solid phase
Calcium Chloride	Barium Nitrate	Water		
100.00	0.00	587.5	80	$CaCl_2 \cdot H_2O$
99.02	0.98	1387.0	80	$BaCl_2 \cdot H_2O$
97.45	2.55	1399.9	80	»
94.81	5.19	1554.4	80	»
88.99	11.01	1709.4	80	»
85.88	14.12	1607.5	80	»
82.52	17.48	1651.4	80	$Ba(NO_3)_2$
74.25	24.75	2059.1	80	»
66.96	33.04	2448.8	80	»
57.44	42.56	2760.4	80	»
36.25	63.75	3705.0	80	»
18.37	81.63	4498.8	80	»
0.00	100.00	6603.0	80	»
85.0	15.0	1540.0	80	$Ba(NO_3)_2 + BaCl_2 \cdot H_2O$
100.00	0.00	550.5	100	$CaCl_2$
99.02	0.98	1141.49	100	$BaCl_2$
97.45	2.55	1355.80	100	»
94.81	5.19	1495.14	100	»
88.99	11.01	1570.90	100	»
85.88	14.12	1471.40	100	»
82.52	17.48	1371.49	100	$Ba(NO_3)_2$
74.25	24.75	1631.20	100	»
66.96	33.04	1898.50	100	»
57.44	42.56	2246.60	100	»
36.25	63.75	3126.90	100	»
18.37	81.63	3667.20	100	»
0.00	100.00	4130.00	100	»
82.5	17.5	1340.0	100	$Ba(NO_3)_2 + BaCl_2$
100.00	0.00	519.82	120	$CaCl_2$

Comp. of saturated soln., g-mol			t	Comp. of solid phase
Calcium Chloride	Barium Nitrate	Water		
99.02	0.98	999.41	120	$BaCl_2$
97.45	2.55	1261.90	120	»
94.81	5.19	1430.72	120	»
87.99	11.01	1451.38	120	»
85.88	14.12	1391.20	120	»
82.52	17.48	1298.70	120	»
74.25	24.75	1317.41	120	$Ba(NO_3)_2$
66.96	33.04	1532.08	120	»
57.44	42.56	1843.97	120	»
46.44	53.56	2224.89	120	»
36.25	63.75	2674.98	120	»
18.37	81.63	3043.81	120	»
0.00	100.00	3229.30	120	»
81.5	18.5	1170.0	120	$Ba(NO_3)_2+BaCl_2$

№ 5162 CALCIUM CHLORIDE–BARIUM NITRATE–WATER [921]

$CaCl_2—Ba(NO_3)_2—H_2O$

Comp. of saturated soln., wt %			t	Comp. of solid phase
Calcium Chloride	Barium Nitrate	Water		
0.00	8.50	91.50	20	$Ba(NO_3)_2$
3.98	7.50	88.52	20	»
8.00	7.00	85.00	20	»
12.13	6.00	81.87	20	»
16.25	5.00	78.75	20	»
17.79	3.80	78.41	20	»
20.24	2.00	77.76	20	»
23.07	0.60	76.33	20	»
25.62	0.30	74.08	20	»
27.43	0.25	72.32	20	»
42.70	0.00	57.30	20	$CaCl_2 \cdot 6H_2O$
0.00	12.50	87.50	40	$Ba(NO_3)_2$
3.83	11.00	85.17	40	»
7.74	10.00	82.26	40	»
11.80	8.50	79.70	40	»
15.82	7.50	76.68	40	»
17.11	7.50	75.39	40	»
19.93	3.50	76.57	40	»
22.85	1.60	75.55	40	»
25.52	0.70	73.78	40	»
27.39	0.40	72.21	40	»
29.61	0.09	70.30	40	»
53.50	0.00	46.50	40	$CaCl_2 \cdot 4H_2O$
0.00	16.68	83.32	60	$Ba(NO_3)_2$
3.66	14.80	81.54	60	»
7.48	13.00	79.52	60	»
11.44	11.30	77.26	60	»

Comp. of saturated soln., wt %			t	Comp. of solid phase
Calcium Chloride	Barium Nitrate	Water		
15.39	10.00	74.61	60	$Ba(NO_3)_2$
16.74	9.50	73.76	60	»
19.54	5.40	75.06	60	»
22.50	3.00	74.50	60	»
25.32	1.50	73.18	60	»
27.31	0.70	71.99	60	»
28.93	0.25	70.82	60	»
57.80	0.00	42.20	60	$CaCl_2 \cdot 2H_2O$
0.00	4.75	95.25	—0.70	Ice $+ Ba(NO_3)_2$
4.09	4.74	91.17	—2.5	»
8.22	4.40	86.38	—5.7	»
12.40	3.80	83.80	—8.2	»
16.59	3.00	80.41	—14.5	»
18.22	1.50	80.28	—16.5	»
20.54	0.50	78.96	—17.5	»
30.22	0.00	69.78	—49.8	Ice $+ \alpha$-$CaCl_2 \cdot 4H_2O$
15.50	4.30	81.20	—8.5	Ice $+ CaCl_2 \cdot 6H_2O + Ba(NO_3)_2$
16.10	5.10	78.80	20.0	$CaCl_2 \cdot 6H_2O + Ba(NO_3)_2$
17.11	7.50	75.39	40.0	$CaCl_2 \cdot 4H_2O + Ba(NO_3)_2$
18.50	8.70	72.80	60.0	$CaCl_2 \cdot 2H_2O + Ba(NO_3)_2$

№ 5163 CALCIUM CHLORIDE–BARIUM CHLORIDE–WATER [60]

$CaCl_2—BaCl_2—H_2O$

Content in aqueous soln., g/l		t	d	Concn. in aqueous soln., g/l		t	d
Calcium Chloride	Barium Chloride			Calcium Chloride	Barium Chloride		
310.5	14.2	0	1.2349	85.3	183.6	0	1.2145
237.1	37.4	0	1.2100	274.0	44.6	25	1.2248
159.2	95.1	0	1.1926	223.2	76.1	25	1.2136
136.3	115.3	0	1.1968	159.5	135.9	25	1.2250

№ 5164 CALCIUM CHLORIDE–BARIUM CHLORIDE–WATER [174]

$CaCl_2—BaCl_2—H_2O$

$t=25°$

Content in aqueous soln., wt %		Comp. of solid phase
Calcium Chloride	Barium Chloride	
45.07	0	$CaCl_2 \cdot 6H_2O$
31.00	0.30	$BaCl_2 \cdot 2H_2O$
26.53	0.64	»
25.25	0.92	»
19.37	3.65	»
13.75	8.92	»
9.84	13.35	»
5.60	18.81	»
0	26.95	»

CALCIUM CHLORIDE–BARIUM CHLORIDE–WATER

$CaCl_2$—$BaCl_2$—H_2O

Comp. of saturated soln., g-mol			t	Comp. of solid phase
Calcium Chloride	Barium Chloride	Water		
0.00	100.0	2149.0	80	$BaCl_2 \cdot H_2O$
7.84	92.16	2238.9	80	»
15.19	84.81	2258.2	80	»
22.15	77.85	2298.9	80	»
40.88	59.12	2322.4	80	»
61.78	38.22	2289.5	80	»
79.56	20.44	2213.9	80	»
82.89	17.11	2118.1	80	»
86.62	13.38	2033.3	80	»
93.60	6.40	1838.1	80	»
96.86	3.14	1686.8	80	»
98.73	1.27	1440.7	80	»
100.00	0.00	587.5	80	$CaCl_2 \cdot H_2O$
0.00	100.00	1970.7	100	$BaCl_2$
7.84	92.16	2021.4	100	»
15.19	84.81	2041.5	100	»
20.77	79.23	2077.5	100	»
40.88	59.12	2058.9	100	»
61.78	38.22	2079.4	100	»
71.61	28.39	2037.5	100	»
79.56	20.44	2014.8	100	»
86.62	13.38	1870.9	100	»
93.60	6.40	1706.2	100	»
96.86	3.14	1529.1	100	»
98.73	1.27	1303.2	100	»
100.00	0.00	550.5	100	$CaCl_2$
0.00	100.00	1845.2	120	$BaCl_2$
7.84	92.16	1886.4	120	»
15.19	84.81	1915.0	120	»
22.15	77.85	1941.0	120	»
20.77	79.23	1946.9	120	»
48.92	51.08	1992.7	120	»
61.78	38.22	1970.8	120	»
71.61	28.39	1936.6	120	»
79.56	20.44	1924.5	120	»
82.89	17.11	1875.0	120	»
86.62	13.38	1757.7	120	»
93.60	6.40	1582.7	120	»
96.86	3.14	1395.1	120	»
98.73	1.27	1138.1	120	»
100.00	0.00	512.83	120	$CaCl_2$
$>$99.9	$<$0.1	587.50	80	$CaCl_2 \cdot H_2O + BaCl_2 \cdot H_2O$
$>$99.9	$<$0.1	550.53	100	$CaCl_2 + BaCl_2$
$>$99.9	$<$0.1	512.83	120	»

№ 5166 CALCIUM CHLORIDE–BARIUM CHLORIDE–WATER [218]

$CaCl_2—BaCl_2—H_2O$

Comp. of saturated soln., wt %			t	Comp. of saturated soln., wt %			t
Barium Chloride	Calcium Chloride	Water		Barium Chloride	Calcium Chloride	Water	
21.03	4.23	74.74	24	22.72	4.57	72.71	50
16.71	7.17	76.12	24	21.00	5.74	73.26	50
12.83	10.04	77.13	24	20.22	6.21	73.57	50
9.58	12.74	77.68	24	14.84	10.19	74.97	50
6.61	15.83	77.56	24	12.98	11.90	75.12	50
2.67	20.63	76.70	24	9.74	14.67	75.59	50
1.51	23.68	74.81	24	7.34	17.25	75.41	50
0.80	25.72	74.28	24	3.72	20.30	75.98	50
0.35	28.27	71.38	24	3.80	21.32	74.88	50
0.23	30.28	69.49	24	2.73	23.09	74.18	50
0.09	31.69	68.22	24	1.65	26.12	72.23	50
0.05	32.99	66.96	24	0.96	26.91	72.13	50
27.60	0.00	72.40	30	0.57	29.66	69.77	50
21.59	4.30	74.11	30	0.31	30.94	68.75	50
19.84	5.67	74.49	30	0.22	31.21	68.57	50
17.80	6.91	75.29	30	31.70	0.00	68.30	60
17.37	7.10	75.53	30	24.62	3.43	71.95	60
13.02	10.22	76.76	30	20.85	6.67	72.48	60
10.00	12.71	77.29	30	17.51	9.21	73.28	60
3.26	19.82	76.92	30	12.29	12.34	75.34	60
1.31	24.21	74.48	30	10.95	14.17	74.88	60
0.10	32.00	67.90	30	6.65	18.66	74.69	60
29.0	0.00	71.00	40	5.60	20.00	74.40	60
22.01	4.44	73.55	40	1.18	25.78	73.04	60
20.85	5.20	73.95	40	0.45	28.98	70.57	60
19.04	6.52	74.44	40	0.30	30.62	69.08	60
18.54	6.93	74.53	40	0.02	32.46	67.52	60
13.43	10.32	76.25	40	33.10	0.00	66.90	70
11.21	12.51	76.28	40	25.37	3.77	70.86	70
7.72	15.57	76.71	40	21.94	6.31	71.75	70
5.95	17.36	76.69	40	17.39	9.73	72.88	70
3.79	20.18	76.03	40	12.90	12.63	74.47	70
2.51	22.25	75.24	40	9.81	15.20	74.99	70
0.92	26.13	72.95	40	3.91	20.03	76.06	70
0.67	26.95	72.38	40	2.70	23.42	73.88	70
0.25	29.89	69.86	40	1.47	25.96	72.57	70
0.17	32.35	67.48	40	0.92	27.11	71.97	70
0.13	32.59	67.18	40	0.60	29.75	69.65	70
30.40	0.00	69.60	50	0.20	31.51	68.29	70
25.81	2.99	71.20	50	0.09	32.37	67.54	70

№ 5167 [2164]

CALCIUM CHLORIDE–BARIUM IODATE–WATER

$CaCl_2—Ba(IO_3)_2—H_2O$

Content in aqueous soln., g/l	
Calcium Chloride	Barium Iodate
0.2287	0.4307
0.5507	0.4653
1.102	0.5096

№ 5168 [1001]

CALCIUM CHLORIDE–LANTHANUM CHLORIDE–WATER

$CaCl_2$— $LaCl_3$—H_2O

$t=25$

Comp. of saturated soln., wt %			Comp. of solid phase
Lanthanum Chloride	Calcium Chloride	Water	
49.00	0.00	51.00	$LaCl_3 \cdot 7H_2O$
48.84	0.21	50.95	»
47.88	0.76	51.36	»
45.92	2.72	51.36	»
42.62	4.82	48.56	»
32.43	13.74	53.83	»
25.15	20.18	54.67	»
16.16	28.77	55.07	»
14.04	30.67	55.29	»
11.21	33.98	54.81	»
10.69	35.02	56.29	»
9.23	36.63	54.14	»
7.58	40.10	52.32	$LaCl_3 \cdot 7H_2O + CaCl_2 \cdot 6H_2O$
7.56	40.12	52.32	»
7.56	40.15	52.29	»
6.98	40.31	52.71	$CaCl_2 \cdot 6H_2O$
2.53	43.32	54.15	»
0.00	45.24	54.76	»

№ 5169 [1001]

CALCIUM CHLORIDE–NEODYMIUM CHLORIDE–WATER

$CaCl_2$—$NdCl_3$—H_2O

$t=25$

Comp. of saturated soln., wt %			Comp. of solid phase
Neodymium Chloride	Calcium Chloride	Water	
49.94	0.00	50.06	$NdCl_3 \cdot 6H_2O$
49.40	0.45	51.15	»
47.80	1.42	50.78	»
44.72	3.12	52.16	»
37.98	8.26	53.76	»
30.45	14.28	55.27	»
21.27	22.56	56.17	»
14.81	28.96	56.23	»
8.14	35.44	56.42	»
6.39	39.14	54.47	$NdCl_3 \cdot 6H_2O + CaCl_2 \cdot 2H_2O$
6.40	39.16	54.44	»
4.35	41.23	54.42	$CaCl_2 \cdot 6H_2O$
1.73	43.16	55.11	»
0.00	45.24	54.76	»

№ 5170 [2412]

CALCIUM CHLORIDE–MERCURY (I) CHLORIDE–WATER

$CaCl_2—HgCl—H_2O$

t=25

Concn. in aqueous soln., g/l		*d*	Comp. of solid phase
Mercury Chloride	Calcium Chloride		
0.022	39.96	—	HgCl + 0.1 g Hg
0.033	55.5	—	»
0.081	111	1.064	»
0.118	138.75	1.105	»
0.231	195.36	1.151	»
0.322	257.52	1.205	»
0.430	324.67	1.243	»
0.518	432.9	1.315	»
0.510	499.5	1.358	»

№ 5171 [1815]

CALCIUM CHLORIDE–MERCURY (II) CHLORIDE–WATER

$CaCl_2—HgCl_2—H_2O$

t=25

Concn. in aqueous soln., g/l			
Mercury Chloride	Calcium Chloride	Mercury Chloride	Calcium Chloride
98.8	21.08	491.6	107
208	44.64	719	158.4
301	72.7	898	191.3

№ 5172 [1162]

CALCIUM CHLORIDE–MERCURY (II) CHLORIDE–WATER

$CaCl_2—HgCl_2—H_2O$

t=25

Concn. in aqueous soln., wt %		Comp. of solid phase
Mercury Chloride	Calcium Chloride	
6.90	0.00	$HgCl_2$
38.50	7.25	»
52.66	10.53	»
56.71	11.35	$HgCl_2 + [Ca(H_2O)_6] \cdot [(HgCl_2)_6 \cdot Cl_2]$
56.87	11.43	»
56.85	11.61	$[Ca(H_2O)_6] \cdot [(HgCl_2)_6 \cdot Cl_2]$
56.31	12.60	»
56.20	13.09	»
55.70	14.80	»
55.68	14.70	»
55.50	16.04	»
55.63	16.90	»

Table № 5172 (continuation)

Concn. in aqueous soln., wt %		Comp. of solid phase
Mercury Chloride	Calcium Chloride	
55.58	17.81	$[Ca(H_2O)_6] \cdot [(HgCl_2)_6 \cdot Cl_2]$
55.42	18.77	»
54.96	19.98	»
54.87	20.40	»
53.86	21.67	»
52.67	23.19	»
52.21	23.62	$[Ca(H_2O)_6] \cdot [(HgCl_2)_6] + \alpha\text{-}CaCl_2 \cdot 4H_2O$
52.15	23.53	»
51.97	23.96	»
51.31	24.66	$[Ca(H_2O)_6] \cdot [(HgCl_2)_6Cl_2]$
50.90	25.13	»
50.84	25.19	$[Ca(H_2O)_6] \cdot [(HgCl_2)_6Cl_2] + [Ca(H_2O)_4] \cdot [HgCl_4]$
50.81	25.25	»
50.81	25.30	»
50.78	25.22	»
50.76	25.24	»
50.72	25.33	$[Ca(H_2O)_6] \cdot [(HgCl_2)_6Cl_2] + [Ca(H_2O)_4] \cdot [HgCl_4]$
51.47	24.92	$[Ca(H_2O)_4] \cdot [HgCl_4] + [Ca(H_2O)_6] \cdot [(HgCl_2)_6Cl_2]$
50.99	25.21	»
50.90	25.14	$[Ca(H_2O)_4] \cdot [HgCl_4]$
50.84	25.19	$[Ca(H_2O)_4] \cdot [HgCl_4] + [Ca(H_2O)_6] \cdot [(HgCl_2)_6 \cdot Cl_2]$
50.81	25.25	»
50.81	25.30	»
50.77	25.25	$[Ca(H_2O)_4] \cdot [HgCl_4]$
50.76	25.38	$[Ca(H_2O)_4] \cdot [HgCl_4] + CaCl_2 \cdot 2H_2O$
50.67	25.35	»
50.72	25.33	$[Ca(H_2O)_4] \cdot [HgCl_4] + [Ca(H_2O)_6] \cdot [(HgCl_2)_6Cl_2]$
50.72	25.48	$[Ca(H_2O)_4] \cdot HgCl_4]$
50.78	25.22	$CaCl_2 \cdot 2H_2O + [Ca(H_2O)_6] \cdot [(HgCl_2)_6Cl_2]$
50.76	25.24	»
50.76	25.38	$CaCl_2 \cdot 2H_2O + [Ca(H_2O)_4] \cdot [HgCl_4]$
50.67	25.35	»
50.30	25.50	$CaCl_2 \cdot 2H_2O$
50.21	25.48	»
49.99	25.74	»
49.21	26.04	»
49.16	26.17	»
48.50	26.33	»
47.77	26.81	»
47.01	27.22	»
46.06	27.77	»
45.82	27.90	$CaCl_2 \cdot 2H_2O + \beta\text{-}CaCl_2 \cdot 4H_2O$
45.71	27.81	»
39.24	31.61	$CaCl_2 \cdot 2H_2O + \gamma\text{-}CaCl_2 \cdot 4H_2O$
0.00	55.00	$CaCl_2 \cdot 2H_2O$
39.24	31.61	$CaCl_2 \cdot 2H_2O + \gamma\text{-}CaCl_2 \cdot 4H_2O$
37.14	32.36	$\gamma\text{-}CaCl_2 \cdot 4H_2O$
25.10	37.99	»
24.25	38.50	$\gamma\text{-}CaCl_2 \cdot 4H_2O + CaCl_2 \cdot 6H_2O$
0.00	52.13	$\gamma\text{-}CaCl_2 \cdot 4H_2O$
45.82	27.90	$\beta\text{-}CaCl_2 \cdot 4H_2O + CaCl_2 \cdot 2H_2O$

Concn. in aqueous soln., wt %		Comp. of solid phase
Mercury Chloride	Calcium Chloride	
45.71	27.81	β-$CaCl_2 \cdot 4H_2O$+$CaCl_2 \cdot 2H_2O$
46.03	27.15	β-$CaCl_2 \cdot 4H_2O$
32.73	33.65	»
25.24	37.39	»
24.71	37.73	β-$CaCl_2 \cdot 4H_2O$+$CaCl_2 \cdot 6H_2O$
0.14	51.14	β-$CaCl_2 \cdot 4H_2O$
0.00	51.30	»
51.20	24.46	α-$CaCl_2 \cdot 4H_2O$
51.90	23.93	»
51.97	23.96	α-$CaCl_2 \cdot 4H_2O$+$[Ca(H_2O)_6] \cdot [(HgCl_2)_6Cl_2]$
52.16	23.77	»
52.21	23.62	»
52.15	23.53	»
50.24	24.50	α-$CaCl_2 \cdot 4H_2O$
50.08	24.71	»
49.68	24.65	»
48.65	25.35	»
47.53	25.54	»
47.37	25.74	»
47.19	25.65	»
41.62	27.99	»
39.97	29.01	»
30.70	32.72	»
24.00	36.00	α-$CaCl_2 \cdot 4H_2O$+$CaCl_2 \cdot 6H_2O$
0.00	48.85	α-$CaCl_2 \cdot 4H_2O$
23.00	39.30	$CaCl_2 \cdot 6H_2O$+γ-$CaCl_2 \cdot 4H_2O$
24.71	37.73	$CaCl_2 \cdot 6H_2O$+β-$CaCl_2 \cdot 4H_2O$
24.00	36.00	$CaCl_2 . 6H_2O$+α-$CaCl_2 \cdot 4H_2O$
23.73	35.95	$CaCl_2 \cdot 6H_2O$
22.10	36.32	»
20.08	37.04	»
17.04	38.36	»
0.00	45.05	»

№ 5173 [2648]

CALCIUM CHLORIDE–MERCURY (II) CHLORIDE–WATER

$CaCl_2—HgCl_2—H_2O$

t=25

Concn. in aqueous soln., wt %		d_4	Concn. in aqueous soln., wt %		d_4
Mercury Chloride	Calcium Chloride		Mercury Chloride	Calcium Chloride	
101.82	11.1	1.0821	599.24	222	1.4960
129.79	22.2	1.1074	864.79	333	1.6805
155.58	33.3	1.1318	1130.6	444	1.8429
206.90	55.5	1.1788	1396.7	555	1.9878
336.41	111	1.2907	166.25	666	2.1111

№ 5174 [2264]

CALCIUM CHLORIDE–THALLIUM CHLORIDE–WATER

$CaCl_2—TlCl—H_2O$

$t=25$

Concn. of Calcium Chloride in aqueous soln., g/l	Solubility of Thallium Chloride in aqueous soln. of Calcium Chloride, g/l	Concn. of Calcium Chloride in aqueous soln., g/l	Solubility of Thallium Chloride in aqueous soln. of Calcium Chloride, g/l
1.387	2.156	5.55	1.000
2.775	1.497	11.1	0.681

№ 5175 [1197]

CALCIUM CHLORIDE–THALLIUM CHLORIDE–WATER

$CaCl_2—TlCl—H_2O$

$t=25$

Concn. in aqueous soln., wt %			Comp. of solid phase
Thallium Chloride	Calcium Chloride	Water	
0.038	4.760	95.202	TlCl
0.05	15.60	84.35	»
0.085	22.03	77.885	»
0.109	29.18	70.711	»
0.139	31.13	68.731	»
0.290	40.27	59.440	»
0.461	46.78	52.759	»

№ 5176 [2262]

CALCIUM CHLORIDE–LEAD (II) CHLORIDE–WATER

$CaCl_2—PbCl_2—H_2O$

$t=25$

Concn. in aqueous soln., g/l	
Calcium Chloride	Lead Chloride
0	10.79
1	8.7
2	7.6
3	6.7
4	6.0
6	4.8
8	3.9
10	3.3

№ 5177 [1817]

CALCIUM CHLORIDE–LEAD (II) CHLORIDE–WATER

$CaCl_2—PbCl_2—H_2O$

$t=25$

Concn. of Calcium Chloride in aqueous soln., g/l	Solubility of Lead Chloride in aqueous soln. of Calcium Chloride, g/l	Comp. of solid phase
28.857	1.936	$CaCl_2 \cdot PbCl_2 \cdot 6H_2O$
52.720	1.566	»
105.995	2.144	»
168.149	3.526	»
228.639	5.721	»

№ 5178 [1965]

CALCIUM CHLORIDE–LEAD (II) CHLORIDE–WATER

$CaCl_2—PbCl_2—H_2O$

$t=25$

Concn. in aqueous soln., g/l		Comp. of solid phase
Lead Chloride	Calcium Chloride	
3.34	11.10	$CaCl_2 . PbCl_2 . 6H_2O$
2.98	27.75	»
2.92	38.85	»
2.95	52.17	»
2.93	65.49	»
3.07	84.35	»
3.27	102.1	»
3.45	116.5	»
3.51	121.5	»
3.62	132.1	»
3.90	155.4	»
4.48	190.9	»
5.08	201.4	»
5.08	225.3	»
5.92	255.3	»
6.82	278.6	»
7.23	287.5	»
7.51	292.4	»
8.11	301.3	»
9.04	311.8	»
9.48	315.2	»

№ 5179 [947]

CALCIUM CHLORIDE–LEAD (II) CHLORIDE–WATER

$CaCl_2—PbCl_2—H_2O$

Concn. in aqueous soln., wt %		t	Comp. of solid phase
Lead Chloride	Calcium Chloride		
1.031	0.0	25	$PbCl_2$
0.576	0.35	25	»
0.382	0.653	25	»
0.227	1.25	25	»
0.164	2.44	25	»
0.12	4.83	25	»
0.156	9.16	25	»
0.31	17.18	25	»
2.226	29.9	25	»
6.36	42.3	25	$PbCl_2+CaCl_2$
1.887	0.0	60	$PbCl_2$
1.368	0.35	60	»
1.068	0.67	60	»
0.757	1.28	60	»
0.520	2.56	60	»
0.478	4.17	60	»

Table № 5179 (continuation)

Concn. in aqueous soln., wt %		t	Comp. of solid phase
Lead Chloride	Calcium Chloride		
0.489	10.51	60	$PbCl_2$
0.914	17.43	60	»
3.69	19.46	60	»
7.265	44.48	60	»

№ 5180 [1817]

CALCIUM CHLORIDE–LEAD (II) BROMIDE–WATER

$CaCl_2—PbBr_2—H_2O$

$t=25$

Concn. of Calcium Chloride in aqueous soln., g/l	Solubility of Lead Bromide in aqueous soln. of Calcium Chloride, g/l	Concn. of Calcium Chloride in aqueous soln., g/l	Solubility of Lead Bromide in aqueous soln. of Calcium Chloride, g/l
0.0	9.63	109.34	2.54
27.19	1.70	163.72	4.98
54.39	1.99		

№ 5181 [1817]

CALCIUM CHLORIDE–LEAD (II) BROMIDE–WATER

$CaCl_2—PbBr_2—H_2O$

Concn. of Calcium Chloride in aqueous soln., g/l	Solubility of Lead Bromide, g/l
218.12	8.64

№ 5182 [2519]

CALCIUM BROMIDE–ARSENIC TRIOXIDE–WATER

$CaBr_2—As_2O_3—H_2O$

Comp. of saturated soln., wt %			Comp. of solid phase
Arsenic Trioxide	Calcium Bromide	Water	
1.28	20.13	78.59	As_2O_3

№ 5183 [2519]

CALCIUM BROMIDE–ARSENIC TRIOXIDE–WATER

$CaBr_2—As_2O_3—H_2O$

$t=20$

Comp. of saturated soln., wt %			Comp. of solid phase
Arsenic Trioxide	Calcium Bromide	Water	
1.58	9.65	88.87	As_2O_3
0.912	34.90	64.188	»
0.789	41.00	58.211	»
0.698	47.67	51.632	»
0.513	52.06	47.427	»
0.687	58.22	41.093	$As_2O_3+CaBr_2 \cdot 6H_2O$
0.00	58.20	41.80	$CaBr_2 \cdot 6H_2O$

№ 5184 [1568]

CALCIUM BROMIDE–INDIUM BROMIDE–WATER

$CaBr_2—InBr_3—H_2O$

$t=20$

Concn. in aqueous soln., g/l		Comp. of solid phase
Indium Bromide	Calcium Bromide	
1844	170	$CaBr_2 . 2InBr_2 . 8H_2O$
1801	200	»
803	580	$CaBr_2 . InBr_3 . 8H_2O$
769	765	»
618	780	»
534	822	»
473	834	$3CaBr_2 . 2InBr_3 . 23H_2O$
398	853	»
290	908	$2CaBr_2 . InBr_3 . 14H_2O$
264	938	»

№ 5185 [1815]

CALCIUM BROMIDE–MERCURY (II) WATER

$CaBr_2—HgBr_2—H_2O$

$t=25$

Concn. in aqueous soln., g/l	
Mercury Bromide	Calcium Bromide
42.15	14.4
244.35	128.8
489	378
995	486
1319	751

№ 5186 [1817]

CALCIUM BROMIDE–LEAD (II) CHLORIDE–WATER

$CaBr_2—PbCl_2—H_2O$

$t=25$

Concn. of Calcium Bromide in aqueous soln., g/l	Solubility of Lead Chloride in aqueous soln. of Calcium Bromide, g/l
46.979	1.894
94.957	2.962
190.914	10.532
284.871	26.193
380.828	60.435

№ 5187 [1817]

CALCIUM BROMIDE–LEAD (II) BROMIDE–WATER

$CaBr_2—PbBr_2—H_2O$

$t=25$

Concn. of Calcium Bromide in aqueous soln., g/l	Solubility of Lead Bromide in aqueous soln. of Calcium Bromide, g/l	Concn. of Calcium Bromide in aqueous soln., g/l	Solubility of Lead Bromide in aqueous soln. of Calcium Bromide, g/l
0.0	9.63	208.0	16.07
52.0	2.45	313.0	43.12
104.0	4.42	417.0	190.4

№ 5188 CALCIUM IODIDE–INDIUM IODIDE–WATER [1568]

$CaI_2—InI_3—H_2O$

$t=20$

Concn. in aqueous soln., g/l		Comp. of solid phase
Indium Iodide	Calcium Iodide	
1243	697	$CaI_2 \cdot 2InI_3 \cdot 8H_2O$
225	1071	»
109	1122	$2CaI_2 \cdot 3InI_3 \cdot 13H_2O$
38	1179	$CaI_2 \cdot InI_3 \cdot 6H_2O$
32	1247	$CaI_2 \cdot H_2O$
16	1315	»

№ 5189 CALCIUM IODIDE–MERCURY (II) IODIDE–WATER [1815]

$CaI_2—HgI_2—H_2O$

$t=25$

Concn. in aqueous soln., g/l			
Mercury Iodide	Calcium Iodide	Mercury Iodide	Calcium Iodide
22.7	15.56	199.8	137.5
118.7	74	776	529

№ 5190 CALCIUM IODATE–VARIOUS SUBSTANCES–WATER [1970]

$Ca(IO_3)_2$— —— —H_2O

Substance		Concn. in aqueous soln., g/l		t
Name	Formula	Substance	Calcium Iodate	
—	—	0.0	2.219	18
—	—	0.0	3.057	25
—	—	0.0	3.864	30
Calcium Chloride	$CaCl_2$	0.555	1.864	18
»	»	1.110	1.688	18
»	»	2.775	1.462	18
»	»	5.55	1.287	18
»	»	0.555	2.655	25
»	»	1.110	2.511	25
»	»	2.775	2.172	25
»	»	5.55	1.915	25
»	»	0.555	3.373	30
»	»	1.110	3.205	30
»	»	2.775	2.823	30
»	»	5.55	2.566	30
Calcium Hydroxide	$Ca(OH)_2$	0.80	1.755	18
»	»	0.80	2.495	25
»	»	0.80	3.353	30
Hydrogen Chloride	HCl	3.64	3.665	18

Table № 5190 (continuation)

Substance		Concn. in aqueous soln., g/l		t
Name	Formula	Substance	Calcium Iodate	
Hydrogen Chloride	HCl	36.47	12.360	18
»	»	3.64	4.913	25
»	»	36.47	16.533	25
»	»	3.64	6.005	30
»	»	36.47	20.393	30
Potassium Iodate	KIO_3	3.574	0.792	18
»	»	3.574	1.462	25
»	»	3.574	2.094	30
Magnesium Chloride	$MgCl_2$	4.79	3.279	18
»	»	48.09	5.966	18
»	»	4.79	4.367	25
»	»	48.09	7.798	25
»	»	4.79	5.381	30
»	»	48.09	9.475	30
Sodium Chloride	NaCl	5.833	3.030	18
»	»	14.61	3.603	18
»	»	29.22	4.309	18
»	»	58.45	5.381	18
»	»	5.833	4.094	25
»	»	14.61	4.796	25
»	»	29.22	5.693	25
»	»	58.45	6.980	25
»	»	5.833	5.030	30
»	»	14.61	5.849	30
»	»	29.22	6.902	30
»	»	58.45	8.422	30
Sodium Hydroxide	NaOH	0.40	2.417	18
»	»	1.00	2.768	18
»	»	2.0	3.119	18
»	»	0.40	3.275	25
»	»	1.0	3.665	25
»	»	2.0	4.094	25
»	»	0.40	4.289	30
»	»	1.00	4.562	30
»	»	2.0	5.030	30

№ 5191 **CALCIUM IODATE–SALTS–WATER** [2753]

$Ca(IO_3)_2$—MeX—H_2O

$t=25$

Salt		Concn. in aqueous soln., g/l		d
Name	Formula	Salt	Calcium Iodate	
—	—	0.00	3.057	0.9998
Calcium Chloride	$CaCl_2$	0.694	2.609	1.0001
»	»	2.775	2.121	1.0016
»	»	5.550	1.9106	1.0036
Potassium Hexacyanoferrate (II)	$K_4Fe(CN)_6$	0.460	3.268	1.0007

Table № 5191 (continuation)

Salt		Concn. in aqueous soln., g/l		d
Name	Formula	Salt	Calcium Iodate	
Potassium Hexacyanoferrate (II)	$K_4Fe(CN)_6$	0.921	3.447	1.0013
»	»	1.381	3.677	1.0018
»	»	1.842	3.821	1.0023
Potassium Chloride	KCl	0.932	3.240	1.0008
»	»	1.864	3.404	1.0017
»	»	3.728	3.661	1.0032
»	»	7.456	4.063	1.0057
Magnesium Sulfate	$MgSO_4$	0.752	3.525	1.001
»	»	1.505	3.817	1.002
»	»	2.257	4.063	1.004
»	»	3.010	4.270	1.004
Sodium Chloride	NaCl	0.731	3.232	1.001
»	»	1.461	3.384	1.0015
»	»	2.922	3.622	1.0025
»	»	5.845	3.989	1.0050
Sodium Sulfate	Na_2SO_4	0.888	3.470	1.001
»	»	1.776	3.802	1.002
»	»	2.664	4.075	1.003
»	»	3.551	4.309	1.004

№ 5192 CHROMIUM TRIOXIDE–COPPER (II) OXIDE–WATER [1340]

$CrO_3—CuO—H_2O$

$t=29.9$

Concn. in aqueous soln., wt %		Comp. of solid phase
Copper Oxide	Chromium Trioxide	
0.0	64.8	CrO_3
1.0	64.4	»
5.7	62.1	»
10.9	60.6	»
12.6	61.0	»
13.2	61.8	$CuCr_2O_7 \cdot 2H_2O+CrO_3$
13.1	61.7	»
13.8	60.7	$CuCr_2O_7 \cdot 2H_2O$
14.6	58.6	»
15.0	54.7	»
14.5	53.1	$CuCrO_4$
15.1	47.4	»
15.4	46.9	»
15.0	40.5	»
12.7	31.9	»
10.1	25.4	»
8.8	22.3	»
7.94	19.6	»
6.60	16.6	»
5.80	14.8	»
4.31	11.3	»

Table № 5192 (continuation)

Concn. in aqueous soln., wt %		Comp. of solid phase
Copper Oxide	Chromium Trioxide	
6.80	10.5	$CuCrO_4 \cdot Cu(OH)_2$
5.70	8.53	»
4.74	7.31	$2CuCrO_4 \cdot 3Cu(OH)_2 \cdot H_2O$
4.74	7.32	»
2.40	5.88	»
2.42	5.89	»

№ 5193 CHROMIUM TRIOXIDE–ZINC OXIDE–WATER [1743]

$CrO_3—ZnO—H_2O$

$t=25$

Concn. in aqueous soln., g/l		Comp. of solid phase
Zinc Oxide	Chromium Trioxide	
0.409	0.604	$4ZnO \cdot CrO_3 \cdot 3H_2O$
2.24	4.19	»
5.86	11.5	$4ZnO \cdot CrO_3 \cdot 3H_2O + 3ZnO \cdot CrO_3 \cdot 2H_2O$
10.7	22.2	$3ZnO \cdot CrO_3 \cdot 2H_2O$
26.7	57.5	»
30.4	66.7	$3ZnO \cdot CrO_3 \cdot 2H_2O + 4ZnO \cdot CrO_3 \cdot 3H_2O$
32.2	70.6	$4ZnO \cdot CrO_3 \cdot 3H_2O$
66.1	151	»
83.7	192	$4ZnO \cdot CrO_3 \cdot 3H_2O + 3ZnO \cdot 2CrO_3 \cdot H_2O$
123	285	$3ZnO \cdot 2CrO_3 \cdot H_2O$
193	450	»
196	461	$3ZnO \cdot 2CrO_3 \cdot H_2O + ZnO \cdot CrO_3 \cdot H_2O$
202	475	$ZnO \cdot CrO_3 \cdot H_2O$
389	940	»

№ 5194 CHROMIUM TRIOXIDE–ZINC OXIDE–WATER [2758]

$CrO_3—ZnO—H_2O$

$t=25$

Comp. of saturated soln., wt %			d	Comp. of solid phase, wt %		
Zinc Oxide	Chromium Trioxide	Water		Zinc Oxide	Chromium Trioxide	Water
<0.01	<0.01	99.98	0.998	27.30	6.15	66.55
0.04	0.08	99.88	—	27.60	7.41	66.99
0.21	0.34	99.45	1.000	27.10	7.79	65.11
0.47	0.90	98.63	1.002	23.70	8.18	68.12
0.74	1.37	97.90	—	22.70	8.26	69.04
0.95	1.89	97.16	1.016	20.80	8.32	70.88
1.36	3.40	95.24	1.032	21.00	9.47	69.53
1.62	4.75	93.63	1.052	19.4	10.10	70.50
2.88	5.54	91.58	1.063	24.5	13.7	61.8
2.45	5.20	92.35	1.056	34.3	19.4	46.3
2.60	5.72	91.68	1.066	31.7	20.0	48.3
2.85	6.01	91.14	1.073	29.2	19.6	51.2
3.73	10.00	86.27	1.132	27.6	20.7	51.7

Table № 5194 (continuation)

Comp. of saturated soln., wt %			d	Comp. of solid phase, wt %		
Zinc Oxide	Chromium Trioxide	Water		Zinc Oxide	Chromium Trioxide	Water
6.00	15.15	78.50	1.216	29.3	24.2	46.5
9.45	22.00	68.55	1.323	34.4	28.4	37.2
10.10	24.1	65.8	1.354	26.6	27.8	45.6
10.80	26.0	63.2	1.390	37.0	30.9	32.1
11.50	27.7	60.8	1.429	31.2	30.8	38.0
12.20	29.5	58.3	—	27.9	31.2	40.9
12.6	30.1	57.3	1.490	33.6	33.0	33.4
12.6	29.9	57.5	—	30.0	33.4	36.6
14.1	33.6	52.3	1.550	26.8	34.9	38.3
15.7	35.1	49.2	1.588	39.1	35.7	25.2
14.8	35.8	49.4	—	22.1	35.9	42.0
14.6	35.7	49.7	1.602	31.4	36.5	32.1
15.0	36.6	48.4	1.598	30.8	38.3	30.9
15.4	36.7	47.9	1.608	38.3	42.0	19.7
15.8	37.0	47.2	—	29.2	42.4	28.4
15.4	36.5	48.1	1.620	34.4	44.5	21.1
16.4	39.0	44.6	1.688	33.3	46.2	20.5
17.1	40.7	42.2	1.742	31.8	46.6	21.6
18.9	42.4	38.7	1.738	35.9	47.4	16.7
18.3	43.8	47.9	1.793	31.9	47.0	21.1
18.6	45.4	36.0	1.878	26.8	46.7	26.5
19.4	46.0	34.6	—	28.7	47.0	24.3
19.4	47.0	33.6	1.865	29.6	47.4	23.0
20.5	48.7	30.8	—	27.4	48.5	24.1
21.1	51.0	27.9	1.925	27.5	50.7	21.8
21.1	51.7	27.2	—	26.9	52.7	20.4
21.1	52.0	26.9	—	25.4	52.8	21.8
21.0	52.0	27.0	—	29.5	54.2	16.3
21.9	50.8	27.3	—	23.0	54.4	22.6
20.4	52.9	26.7	2.07	23.2	57.4	19.4
18.8	53.8	27.4	1.965	21.2	56.5	22.3
18.6	55.0	26.6	2.07	21.8	59.1	19.1
17.9	57.6	24.5	—	20.7	59.5	19.8
16.3	60.6	23.1	1.970	19.6	61.2	19.2
16.5	60.7	22.8	2.13	20.5	62.4	17.1
16.4	60.5	23.1	—	10.0	78.7	11.3
16.0	60.8	23.2	2.06	13.3	67.0	29.7
13.4	60.5	26.1	—	3.9	89.0	7.1
12.5	60.6	26.9	—	3.7	88.6	7.7
10.3	61.0	28.7	—	3.9	86.6	9.5
6.9	61.4	31.7	—	2.3	90.0	7.7
3.6	62.3	34.1	—	0.78	91.0	7.22
<0.0	63.4	36.6	—	—	—	—
<0.01	0.004	99.9	—	91.84	4.25	3.91
<0.01	0.006	99.9	—	85.39	7.51	7.10
<0.01	0.005	99.9	—	77.77	13.16	9.07
<0.01	0.002	99.9	—	72.70	16.00	11.30
0.01	0.003	99.9	—	69.38	17.40	13.22
0.288	0.555	99.157	—	67.14	20.91	11.95
0.979	2.047	96.974	—	66.02	22.33	11.65
1.442	2.198	96.360	—	64.70	23.40	11.90
2.191	5.003	92.806	—	61.12	27.95	10.93

Note. The solid phase takes the form of an amorphous mass.

№ 5195 CHROMIUM TRIOXIDE–ARSENIC PENTOXIDE–WATER [2102]

CrO_3–As_2O_5–H_2O

t=25

Comp. of saturated soln., wt %					
Chromium Trioxide	Arsenic Pentoxide	Water	Chromium Trioxide	Arsenic Pentoxide	Water
0.10	0.60	99.30	1.68	11.20	87.12
0.15	0.95	98.90	1.90	11.83	86.27
0.20	1.78	98.02	2.50	13.70	83.80
0.25	2.30	97.45	2.54	15.05	82.41
0.28	2.55	97.17	3.37	16.35	80.28
0.41	3.36	96.23	3.72	17.80	78.48
0.46	3.85	95.69	4.41	20.00	75.59
0.57	5.05	94.38	4.35	21.85	73.80
0.78	6.45	92.77	3.92	22.60	73.48
1.05	7.60	91.35	3.23	24.60	72.17
1.50	9.60	88.90	2.89	27.10	70.01

№ 5196 CHROMIUM TRIOXIDE–RUBIDIUM OXIDE–WATER [2498]

CrO_3—Rb_2O—H_2O

t=30

Comp. of saturated soln., wt %			Comp. of solid phase
Rubidium Oxide	Chromium Trioxide	Water	
60.56	0.00	39.44	RbOH
56.82	0.00	43.18	Rb_2CrO_4
37.88	0.78	61.34	»
34.89	2.89	62.22	»
30.20	4.96	64.84	»
28.17	8.54	63.29	»
27.99	11.98	60.03	»
28.73	15.38	55.89	»
28.55	15.54	65.91	$Rb_2CrO_4+Rb_2Cr_2O_7$
23.87	13.69	62.44	$Rb_2Cr_2O_7$
17.56	9.98	72.46	»
8.47	5.72	85.81	»
7.98	4.58	87.44	»
4.60	4.87	90.53	»
3.57	8.16	88.27	»
3.38	13.91	82.71	»
3.45	15.05	81.50	$Rb_2Cr_2O_7+Rb_2Cr_2O_{10}$
3.59	15.31	81.10	$Rb_2Cr_2O_7$
3.19	15.19	81.62	$Rb_2Cr_2O_{10}$
2.37	18.96	78.67	»
1.66	24.92	73.42	»
1.61	37.34	61.05	»
1.54	48.20	50.26	»
1.67	53.87	44.46	»
1.28	54.29	44.43	$Rb_2Cr_2O_{10}+Rb_2Cr_4O_{13}$
1.07	58.69	40.24	$Rb_2Cr_4O_{13}$

Table № 5196 (continuation)

Comp. of saturated soln., wt %			Comp. of solid phase
Rubidium Oxide	Chromium Trioxide	Water	
0.93	62.38	36.69	$Rb_2Cr_4O_{13}$
0.93	62.74	36.33	»
0.92	63.07	36.01	$Rb_2Cr_4O_{13}+CrO_3$
0.00	62.28	37.72	CrO_3

№ 5197 CHROMIUM TRIOXIDE–SILVER OXIDE–WATER [1340]

$CrO_3—Ag_2O—H_2O$

$t=29.9$

Comp. of saturated soln., wt %			Comp. of solid phase
Silver Oxide	Chromium Oxide	Water	
0.09	0.21	99.70	Ag_2CrO_4+ solid soln.
0.08	0.21	99.71	»
0.11	0.19	99.70	»
0.10	0.18	99.72	»
0.07	0.34	99.59	Solid solution
0.08	2.55	97.37	»
0.1	4.07	95.83	»
0.1	7.11	82.88	»
0.1	8.11	91.79	»
0.1	15.4	84.5	»
0.1	24.0	75.9	»
0.1	35.1	64.8	»
0.1	47.0	52.9	»
0.1	53.6	46.3	»
0.1	58.7	41.2	»
0.1	62.0	37.9	$Ag_2Cr_2O_7+CrO_3$
0.1	61.9	38.0	»

№ 5198 CHROMIUM TRIOXIDE–THORIUM OXIDE–WATER [1284]

$CrO_3—ThO_2—H_2O$

$t\doteq25$

Comp. of saturated soln., wt %			Comp. of solid phase
Thorium Oxide	Chromium Trioxide	Water	
0.044	0.033	99.923	$Th(CrO_4)_2 \cdot 3H_2O$
0.452	1.376	98.172	»
2.09	4.58	93.33	»
3.08	7.15	89.77	»
6.52	13.09	80.39	»
11.27	21.75	66.98	»
15.62	28.00	56.38	»

Comp. of saturated soln., wt %			Comp. of solid phase
Thorium Oxide	Chromium Trioxide	Water	
22.87	37.27	39.86	$Th(CrO_4)_2 \cdot 3H_2O$
23.25	37.45	39.30	»
24.22	39.49	36.29	$Th(CrO_4)_2 \cdot 3H_2O + Th(CrO_4)_2 \cdot CrO_3 \cdot 3H_2O$
23.04	40.23	36.73	$Th(CrO_4)_2 \cdot CrO_3 \cdot 3H_2O$
21.59	41.56	36.85	»
17.08	44.20	38.72	»
7.91	52.60	39.49	»
7.52	56.79	35.69	9
9.09	57.57	33.34	»
17.65	61.38	20.97	$Th(CrO_4)_2 \cdot CrO_3 \cdot 3H_2O + CrO_3$
6.27	61.84	32.89	CrO_3
0.0	62.87	37.13	»

№ 5199 CHROMIUM FLUORIDE–RUBIDIUM FLUORIDE–WATER [840]

$CrF_3—RbF—H_2O$

t=25

Comp. of saturated soln., wt %			Comp. of solid phase
Chromium Fluoride	Rubidium Fluoride	Water	
0.71	0.95	98.34	$CrF_3 \cdot 3H_2O + 2RbF \cdot CrF_3 \cdot H_2O$
1.04	0.43	98.53	»
27.57	0.00	72.43	$2RbF \cdot CrF_3 \cdot H_2O$
32.54	0.00	67.46	»
41.32	0.00	58.68	»
45.94	0.00	54.06	»
47.60	0.00	52.40	»
51.95	0.00	48.05	»
54.24	0.00	45.76	»
60.87	0.00	39.13	»
0.74	8.25	91.01	$CrF_3 \cdot 3H_2O + 2RbF \cdot CrF_3 \cdot H_2O$
0.84	7.89	91.27	»
0.55	7.98	91.47	»

№ 5200 CHROMIUM FLUORIDE–CESIUM FLUORIDE–WATER [840]

$CrF_3—CsF—H_2O$

t=25

Comp. of saturated soln., wt %			Comp. of solid phase
Chromium Fluoride	Cesium Fluoride	Water	
1.30	4.87	98.83	$CrF_3 \cdot 3H_2O + 2CsF \cdot CrF_3 \cdot H_2O$
1.38	1.20	97.42	»
4.31	0.79	94.90	»
5.96	0.73	93.31	»
7.72	0.70	91.58	»
10.06	0.52	89.42	»

Comp. of saturated soln., wt %			Comp. of solid phase
Chromium Fluoride	Cesium Fluoride	Water	
13.74	0.28	85.98	$2CsF \cdot CrF_3 \cdot H_2O$
16.20	0.25	83.55	»
21.77	0.20	78.03	»
26.23	0.19	73.58	»
30.58	0.19	69.23	»
41.07	0.18	58.75	»
49.80	0.25	49.95	»
54.54	0.22	45.24	»
59.34	0.15	40.51	$3CsF \cdot CrF_3$
60.91	0.11	38.98	»
63.40	0.058	36.54	»
67.69	0.018	32.29	»
70.99	0.00	29.01	»
83.70	0.00	16.30	$CsF \cdot H_2O$
7.35	0.73	91.92	$CrF_3 \cdot 3H_2O + 2CsF \cdot CrF_3 \cdot H_2O$
36.10	0.18	63.72	$2CsF \cdot CrF_3 \cdot H_2O$
58.50	0.16	41.34	$2CsF \cdot CrF_3 \cdot H_2O + 3CsF \cdot CrF_3$
71.75	0.00	28.25	$3CsF \cdot CrF_3$
2.38	6.52	91.10	$CrF_3 \cdot 3H_2O + 2CsF \cdot CrF_3 \cdot H_2O$
1.95	6.70	91.35	»

№ 5201 ARSENIC PENTOXIDE–MANGANESE (II) OXIDE–WATER [433]

$As_2O_5—MnO—H_2O$

$t=75$

Comp. of saturated soln., wt %			Comp. of solid phase
Arsenic Pentoxide	Manganese Oxide	Water	
0.22	0.71	99.07	$MnHAsO_4 \cdot 4H_2O$
0.33	0.78	98.89	»
3.37	0.50	96.13	»
3.47	0.43	91.10	»
4.22	0.49	95.29	»
6.36	0.72	92.92	»
6.83	0.72	92.45	»
7.57	0.73	91.70	»
15.65	0.98	83.37	»
17.00	1.13	81.87	»
19.19	1.14	79.67	»
25.57	1.39	73.04	»
30.18	1.73	68.09	»
39.98	2.44	57.58	»

№ 5202 MANGANESE (II) NITRATE–LANTHANUM NITRATE–WATER [1503]

$Mn(NO_3)_2$—$La(NO_3)_3$—H_2O

$t=20$

Comp. of saturated soln., wt %			Comp. of solid phase
Lanthanum Nitrate	Manganese Nitrate	Water	
60.13	0.0	39.87	$La(NO_3)_3 \cdot 6H_2O$
53.90	5.08	41.02	»
49.20	8.91	41.89	»
41.45	13.22	45.33	»
36.82	16.19	46.99	»
29.81	21.00	49.19	»
27.45	21.95	50.60	»
24.70	26.80	48.50	$Mn(NO_3)_2 \cdot 6H_2O$
22.95	27.50	49.55	»
21.25	29.18	49.57	»
14.15	41.10	44.75	»
10.00	45.00	45.00	»
6.22	49.10	44.58	»
0.0	58.81	41.19	»

№ 5203 MANGANESE (II) SULFATE–MANGANESE (II) CHLORIDE–WATER [278]

$MnSO_4$—$MnCl_2$—H_2O

$t=25$

Comp. of saturated soln., wt %			*d*	Comp. of solid phase
Manganese Chloride	Manganese Sulfate	Water		
43.51	0.0	56.49	1.494	$MnCl_2 \cdot 4H_2O$
42.15	2.47	55.38	1.500	»
41.19	3.52	55.29	1.521	»
41.04	3.90	55.06	1.535	»
40.85	4.81	54.34	1.552	$MnCl_2 \cdot 4H_2O + MnSO_4 \cdot H_2O$
37.08	6.45	56.47	1.464	$MnSO_4 \cdot H_2O$
28.78	11.60	59.62	1.450	»
21.46	15.20	63.34	1.430	»
16.87	20.17	62.96	1.431	»
14.85	22.33	62.82	—	»
14.17	22.70	63.13	—	»
11.28	25.48	63.24	1.450	»
7.72	29.68	62.60	1.460	»
4.41	34.85	60.74	—	»
3.55	35.01	61.44	—	»
0.0	39.58	60.42	1.495	»

№ 5204 MANGANESE (II) SULFATE–IRON (II) SULFATE–WATER [2734]

$MnSO_4—FeSO_4—H_2O$

Comp. of saturated soln., wt %			t	Comp. of solid phase
Iron Sulfate	Manganese Sulfate	Water		
12.55	3.07	84.38	0	$FeSO_4 \cdot 7H_2O$ solid soln.
11.44	5.83	82.73	0	»
9.26	10.66	80.08	0	»
6.84	16.52	76.64	0	»
5.69	19.18	75.13	0	»
1.25	29.30	69.45	0	»
0.34	31.28	68.38	0	»
0.14	31.84	68.02	0	»
23.09	0.09	76.82	25	»
21.41	2.55	76.04	25	»
18.94	6.82	74.24	25	»
16.62	10.33	73.05	25	»
14.39	13.57	72.04	25	»
7.78	24.96	67.26	25	»
1.82	36.95	61.23	25	»
1.33	37.83	60.84	25	$FeSO_4 \cdot 7H_2O$ solid soln. + $+MnSO_4 \cdot 5H_2O$ solid soln.
0.85	38.29	60.86	25	$MnSO_4 \cdot 5H_2O$ solid soln.
0.33	39.39	60.28	25	»

№ 5205 MANGANESE (II) SULFATE–COPPER (II) SULFATE–WATER [2608]

$MnSO_4—CuSO_4—H_2O$

$t=18$

Concn. in aqueous soln., mol %		Comp. of solid phase
Copper Sulfate	Manganese Sulfate	
2.23	0.0	Triclinic crystals of $CuMnSO_4 \cdot 5H_2O$
2.17	0.429	»
1.46	3.57	»
1.24	4.44	»
0.995	5.23	»
0.68	5.95	»
0.32	6.55	»
0.995	5.23	Monoclinic crystals of $CuMnSO_4 \cdot 7H_2O$
0.68	5.95	»
0.00	6.87	»

MANGANESE (II) SULFATE–RUBIDIUM SULFATE–WATER [1205]

$MnSO_4$—Rb_2SO_4—H_2O

Comp. of saturated soln., wt %			t	Comp. of solid phase
Manganese Sulfate	Rubidium Sulfate	Water		
33.40	4.80	61.80	0	$MnSO_4 \cdot 7H_2O+MnSO_4 \cdot Rb_2SO_4 \cdot 6H_2O$
9.30	8.95	81.75	0	$MnSO_4 \cdot Rb_2SO_4 \cdot 6H_2O$
6.32	10.83	82.85	0	»
1.69	26.51	71.80	0	$MnSO_4 \cdot Rb_2SO_4 \cdot 6H_2O+Rb_2SO_4$
37.5	3.3	59.2	25	$MnSO_4 \cdot H_2O+MnSO_4 \cdot Rb_2SO_4 \cdot 6H_2O$
26.93	10.67	62.40	25	$MnSO_4 \cdot Rb_2SO_4 \cdot 6H_2O$
3.77	32.73	63.50	25	$MnSO_4 \cdot Rb_2SO_4 \cdot 6H_2O+Rb_2SO_4$
3.03	33.47	63.50	25	»
32.05	5.50	62.45	40	$2MnSO_4 \cdot Rb_2SO_4$
24.02	6.17	60.81	40	»
14.80	13.21	71.99	40	»
7.72	25.34	66.94	40	»
5.11	33.49	61.40	40	»
3.17	39.75	57.08	60	$2MnSO_4 \cdot Rb_2SO_4+Rb_2SO_4$
29.00	3.58	67.42	55	$2MnSO_4 \cdot Rb_2SO_4$
20.56	5.77	73.67	55	»
11.60	10.55	77.85	55	»
5.72	23.95	70.33	55	»
4.08	30.75	65.17	55	»
30.80	2.97	66.23	75	$MnSO_4 \cdot H_2O+2MnSO_4 \cdot Rb_2SO_4$
29.25	2.81	67.94	75	$2MnSO_4 \cdot Rb_2SO_4$
7.59	9.61	82.80	75	»
2.31	26.48	71.21	75	»
2.12	45.56	52.32	75	»
22.65	2.26	75.09	100	»
3.91	9.09	87.00	100	»
1.76	20.74	77.50	100	»
1.34	31.96	66.70	100	»

№ 5207

MANGANESE (II) SULFATE–THALLIUM SULFATE–WATER [1205]

$MnSO_4$—Tl_2SO_4—H_2O

Comp. of saturated soln., wt %			t	Comp. of solid phase
Manganese Sulfate	Thallium Sulfate	Water		
34.40	3.26	62.34	0	$MnSO_4 \cdot 7H_2O+MnSO_4 \cdot Tl_2SO_4 \cdot 6H_2O$
3.88	9.85	86.27	0	$Tl_2SO_4+MnSO_4 \cdot Tl_2SO_4 \cdot 6H_2O$
14.2	6.0	79.8	8	»
38.6	3.2	58.2	25	$MnSO_4 \cdot 5H_2O+2MnSO_4 \cdot Tl_2SO_4$
38.42	2.52	59.06	30	$MnSO_4 \cdot H_2O+2MnSO_4 \cdot Tl_2SO_4$
28.91	7.39	63.70	30	$2MnSO_4 \cdot Tl_2SO_4$
31.21	9.32	59.47	30	$MnSO_4 \cdot Tl_2SO_4 \cdot 6H_2O$
27.14	11.08	61.78	30	$2MnSO_4 \cdot Tl_2SO_4+Tl_2SO_4$
35.41	2.83	61.76	50	$MnSO_4 \cdot H_2O+2MnSO_4 \cdot Tl_2SO_4$

Table № 5207 (continuation)

Comp. of saturated soln., wt %			t	Comp. of solid phase
Manganese Sulfate	Thallium Sulfate	Water		
30.14	4.58	65.28	50	$2MnSO_4 \cdot Tl_2SO_4$
22.45	9.07	68.48	50	»
13.77	13.38	72.85	50	$2MnSO_4 \cdot Tl_2SO_4 + Tl_2SO_4$
8.50	11.37	80.13	50	Tl_2SO_4
4.38	10.08	85.54	50	»
36.78	3.62	59.60	40	$MnSO_4 \cdot H_2O + 2MnSO_4 \cdot Tl_2SO_4$
27.92	5.43	66.65	40	$2MnSO_4 \cdot Tl_2SO_4$
21.85	13.05	65.10	40	$2MnSO_4 \cdot Tl_2SO_4 + Tl_2SO_4$
34.57	2.06	63.37	66	$MnSO_4 \cdot H_2O + 2MnSO_4 \cdot Tl_2SO_4$
20.24	5.84	73.92	66	$2MnSO_4 \cdot Tl_2SO_4$
15.01	7.65	77.34	66	»
9.07	11.81	79.12	66	»
8.85	14.75	76.40	66	$2MnSO_4 \cdot Tl_2SO_4 + Tl_2SO_4$
7.23	12.80	79.97	66	Tl_2SO_4
24.17	2.33	73.50	100	$MnSO_4 \cdot H_2O + 2MnSO_4 \cdot Tl_2SO_4$
7.56	4.75	87.69	100	$2MnSO_4 \cdot Tl_2SO_4$
4.29	7.25	88.46	100	»
2.49	16.06	81.45	100	»
2.14	17.28	80.58	100	$2MnSO_4 \cdot Tl_2SO_4 + Tl_2SO_4$

№ 5208 MANGANESE (II) SULFATE–THORIUM SULFATE–WATER [1303]

$MnSO_4$—$Th(SO_4)_2$—H_2O

t=30

Concn. in aqueous soln., wt %		Comp. of solid phase
Manganese Sulfate	Thorium Sulfate	
0.00	2.14	$Th(SO_4)_2 \cdot 8H_2O$
1.41	3.03	»
2.93	3.94	»
4.54	4.69	»
9.33	6.64	»
13.54	7.75	»
15.49	8.23	»
17.68	8.66	»
13.79	7.91	$Th(SO_4)_2 \cdot 8H_2O + MnSO_4 \cdot Th(SO_4)_2 \cdot 7H_2O$
16.94	6.65	$MnSO_4 \cdot Th(SO_4)_2 \cdot 7H_2O$
19.86	5.82	»
29.75	3.11	»
39.82	2.58	»
38.47	2.17	»
36.78	1.87	»
44.62	1.20	»
45.27	1.38	»
54.59	0.50	»
65.67	0.135	$MnSO_4 \cdot Th(SO_4)_2 \cdot 7H_2O + MnSO_4 \cdot 5H_2O$
66.41	0.00	$MnSO_4 \cdot 5H_2O$

№ 5209 MANGANESE (II) SULFATE–THORIUM SULFATE–WATER [1379]

$MnSO_4—Th(SO_4)_2—H_2O$

$t=30$

Comp. of saturated soln., wt %			Comp. of solid phase
Manganese Sulfate	Thorium Sulfate	Water	
39.91	0.00	60.09	$MnSO_4 \cdot 5H_2O$
39.61	0.08	60.31	$MnSO_4 \cdot 5H_2O + MnSO_4 \cdot Th(SO_4)_2 \cdot 7H_2O$
35.19	0.32	64.49	$MnSO_4 \cdot Th(SO_4)_2 \cdot 7H_2O$
30.86	0.94	68.20	»
26.52	1.35	72.13	»
22.39	2.34	75.27	»
15.80	4.63	79.57	»
13.70	5.38	80.92	»
11.33	6.50	82.17	$MnSO_4 \cdot Th(SO_4)_2 \cdot 7H_2O + Th(SO_4)_2 \cdot 8H_2O$
13.99	6.78	79.23	$Th(SO_4)_2 \cdot 8H_2O$
12.52	6.65	80.83	»
11.16	6.39	82.45	»
8.04	5.73	86.23	»
4.16	3.22	91.62	»
2.74	3.69	93.57	»
1.35	2.90	95.75	»
0.00	2.09	97.91	»

№ 5210 MANGANESE (II) CHLORIDE–CADMIUM CHLORIDE–WATER [1170]

$MnCl_2—CdCl_2—H_2O$

$t=25$

Comp. of saturated soln., wt %			Comp. of solid phase
Cadmium Chloride	Manganese Chloride	Water	
0.0	43.55	56.45	$MnCl_2 \cdot 4H_2O$
7.7	40.40	51.9	»
16.5	36.60	46.9	»
27.5	31.10	41.4	»
31.6	27.50	40.9	$MnCl_2 \cdot 4H_2O + 4CdCl_2 \cdot MnCl_2 \cdot 10H_2O$
36.50	22.20	41.3	$4CdCl_2 \cdot MnCl_2 \cdot 10H_2O$
37.10	21.90	41.0	»
40.90	18.00	41.1	$4CdCl_2 \cdot MnCl_2 \cdot 10H_2O + CdCl_2 \cdot 2.5H_2O$
45.50	11.30	43.2	$CdCl_2 \cdot 2.5H_2O$
51.60	2.90	45.5	»
54.62	0.00	45.38	»
43.00	15.30	41.7	$4CdCl_2 \cdot MnCl_2 \cdot 10H_2O + CdCl_2 \cdot H_2O$

№ 5211 [2264]

MANGANESE (II) CHLORIDE–THALLIUM CHLORIDE–WATER

$MnCl_2—TlCl—H_2O$

$t=25$

Concn. of Manganese Chloride in aqueous soln., g/l	Solubility of Thallium Chloride in aqueous soln. of Manganese Chloride, g/l
1.574	2.154
3.148	1.480
6.297	0.988
12.594	0.686

№ 5212 **MANGANESE (II) CHLORIDE–LEAD (II) CHLORIDE–WATER** [2262]

$MnCl_2—PbCl_2—H_2O$

$t=25$

Concn. in aqueous soln., g/l			
Manganese Chloride	Lead Chloride	Manganese Chloride	Lead Chloride
0	10.79	6	5.0
1	9.5	8	4.1
2	8.3	10	3.5
3	7.3	14	2.8
4	6.3		

№ 5213 **IRON (III) OXIDE–IRON (II) SULFATE–WATER** [235]

$Fe_2O_3—FeSO_4—H_2O$

$t=18$

Concn. in aqueous soln.		d_4	pH	Comp. of solid phase
Iron Sulfate, wt %	Fe, g/l			
19.0	88.0	1.259	6.46	$FeSO_4 \cdot 7H_2O + Fe_2O_3 \cdot 2FeO \cdot SO_3 \cdot xH_2O$
14.5	62.6	1.178	6.62	$Fe_2O_3 \cdot 2FeO \cdot SO_3 \cdot xH_2O$
9.1	36.8	1.103	6.79	»
8.5	34.5	1.100	6.80	»
4.3	16.5	1.046	6.94	»
1.55	5.8	1.015	7.14	»
0.30	1.1	1.004	8.50	»

№ 5214 **IRON (III) OXIDE–ARSENIC PENTOXIDE–WATER** [1787]

$Fe_2O_3—As_2O_5—H_2O$

$t=25$

Concn. in aqueous soln., wt %		Comp. of solid phase
Iron Oxide	Arsenic Pentoxide	
Traces	2.6	$Fe_2O_3 \cdot As_2O_5 \cdot xH_2O$ (where $x < 6$)
»	3.5	»
»	5.4	»
»	5.8	»
»	6.4	»
0.001	7.96	»
0.004	10.11	»
0.010	12.49	»
0.036	16.19	»
0.037	18.82	$Fe_2O_3 \cdot As_2O_5 \cdot xH_2O + Fe_2O_3 \cdot 2As_2O_5 \cdot 8H_2O$
0.059	21.77	»
0.076	23.13	»
0.1	8.0—15.5	$Fe_2O_3 \cdot 2As_2O_5 \cdot 8H_2O$

№ 5215 **IRON (III) OXIDE–ARSENIC PENTOXIDE–WATER** [2624]

$Fe_2O_3—As_2O_5—H_2O$

$t=30$

Comp. of saturated soln., wt %			Comp. of solid phase
Iron Oxide	Arsenic Pentoxide	Water	
Traces	< 2.48	> 97.52	$Fe_2O_3 \cdot As_2O_5 \cdot 9H_2O$
»	2.48	97.52	$Fe_2O_3 \cdot As_2O_5 \cdot 9H_2O + 3As_2O_5 \cdot 22H_2O$
0.12	12.00	87.88	$2Fe_2O_3 \cdot 3As_2O_5 \cdot 22H_2O + Fe_2O_3 \cdot 3As_2O_5 \cdot 17H_2O$
Traces	> 12.00	< 88.00	$Fe_2O_3 \cdot 3As_2O_5 \cdot 17H_2O$

№ 5216 **IRON (II) SULFATE–IRON (III) SULFATE–WATER** [1333]

$FeSO_4—Fe_2(SO_4)_3—H_2O$

Comp. of saturated soln., wt %			t	Comp. of solid phase
Iron (II) Sulfate	Iron (III) Sulfate	Water		
0.38	44.75	54.87	25	BFS
0.82	44.15	55.03	25	»
1.36	43.79	54.85	25	»
2.01	43.32	54.67	25	»
2.61	43.25	54.14	25	»
3.07	42.60	54.33	25	$BFS + FeSO_4 \cdot 7H_2O$
5.30	38.42	56.28	25	»

Table № 5216 (continuation)

Comp. of saturated soln., wt %			t	Comp. of solid phase
Iron (II) Sulfate	Iron (III) Sulfate	Water		
8.16	36.60	55.24	25	$Fe_2(SO_4)_3+FeSO_4 \cdot 7H_2O$
10.47	28.36	61.17	25	»
15.23	16.64	68.13	25	»
20.24	5.73	74.03	25	»
22.50	1.36	76.14	25	»
0.19	48.73	51.08	50	BFS
0.67	46.54	52.79	50	»
3.67	44.50	51.83	50	»
6.18	41.03	52.79	50	$BFS+FeSO_4 \cdot 7H_2O$
8.02	38.87	53.11	50	»
10.55	36.61	52.84	50	»
14.74	32.94	52.32	50	$FeSO_4 \cdot 7H_2O$
19.40	25.35	55.25	50	»
21.76	19.85	58.39	50	»
27.88	9.67	62.45	50	»
30.35	5.40	64.25	50	»
31.99	2.79	65.22	50	»

Note. BFS – Basic Iron (III) Sulfate

№ 5217 IRON (II) SULFATE–COBALT (II) SULFATE–WATER [228]

$FeSO_4—CoSO_4—H_2O$

$t=20$

Comp. of saturated soln., wt %					
$CoSO_4 \cdot 7H_2O$	$FeSO_4 \cdot 7H_2O$	Water	$CoSO_4 \cdot 7H_2O$	$FeSO_4 \cdot 7H_2O$	Water
48.16	0.00	51.48	25.71	15.45	58.84
38.81	3.77	57.42	13.89	25.71	60.40
38.80	3.81	57.39	14.29	25.89	59.82
34.92	8.54	56.54	8.09	29.67	62.24
27.68	13.84	58.48	8.00	29.47	62.53
27.97	13.59	58.44	0.00	37.05	62.95

Note. The solid phase is a solid heptahydrate solution.

№ 5218 IRON (II) SULFATE–NICKEL (II) SULFATE–WATER [228]

$FeSO_4—NiSO_4—H_2O$

$t=20$

Concn. in aqueous soln., wt %			
Fe	Ni	Fe	Ni
0.00	49.24	18.61	29.72
1.29	47.85	21.18	27.23
2.34	46.48	27.70	16.41
5.13	43.82	31.78	9.24
4.87	44.95	34.94	4.50
9.08	42.69	36.40	2.46
16.27	36.40	37.05	0.00
16.75	34.69		

Note. The solid phase is a solid solution of $FeSO_4 \cdot 7H_2O + NiSO_4 \cdot 7H_2O$.

№ 5219 IRON (II) SULFATE–COPPER (II) SULFATE–WATER [1064]

$FeSO_4$—$CuSO_4$—H_2O

Comp. of saturated soln., wt %			t	d	Comp. of solid phase
Copper Sulfate	Iron Sulfate	Water			
28.64	5.03	66.33	25	1.256	$CuSO_4 \cdot 5H_2O$
28.87	7.08	64.05	25	1.275	»
28.26	12.46	59.28	25	1.320	»
27.44	15.09	57.47	25	1.320	»
27.62	17.65	54.73	25	1.343	»
23.23	18.68	58.09	25	1.314	$FeSO_4 \cdot 7H_2O$
22.09	20.28	57.63	25	1.308	»
19.78	23.05	57.17	25	1.308	»
19.84	22.29	57.87	25	1.305	»
14.96	27.56	57.48	25	1.300	»
14.64	28.40	56.96	25	—	»
12.18	29.84	57.98	25	1.291	»
10.76	31.39	57.85	25	1.289	»
5.52	36.29	58.19	25	1.275	»
3.88	37.45	58.67	25	1.266	»
41.92	4.50	53.58	56	1.357	$CuSO_4 \cdot 5H_2O$
41.61	6.28	52.11	56	1.383	»
41.10	10.12	48.78	56	1.408	»
41.24	17.90	40.86	56	1.469	»
39.31	26.55	34.14	56	1.537	»
31.15	34.07	34.78	56	1.512	$FeSO_4 \cdot 7H_2O$
27.62	37.92	34.46	56	1.505	»
21.25	43.49	35.26	56	1.482	»
19.52	45.00	35.48	56	1.472	»
19.81	45.04	35.15	56	1.477	»
15.01	50.31	34.68	56	1.470	»
11.95	52.95	35.10	56	1.460	»
10.73	54.65	34.62	56	1.450	»
5.35	60.21	34.44	56	1.457	»
3.80	60.48	35.72	56	1.437	»

№ 5220 IRON (II) SULFATE–COPPER (II) SULFATE–WATER [1331]

$FeSO_4$—$CuSO_4$—H_2O

$t=30$

Comp. of saturated soln., wt %			Comp. of solid phase
Copper Sulfate	Iron Sulfate	Water	
18.60	2.47	78.93	$CuSO_4 \cdot 5H_2O$
17.53	2.63	79.84	»
16.79	5.34	77.87	»
16.80	8.59	74.61	»
15.34	8.76	75.90	»
14.92	12.47	72.61	$CuSO_4 \cdot 5H_2O$ + solid soln.
13.69	14.15	72.16	Solid solution
11.55	16.40	72.05	»

№ 5221 IRON (II) SULFATE–ZINC SULFATE–WATER [231]

$FeSO_4 \cdot 7H_2O—ZnSO_4 \cdot 7H_2O—H_2O$

$t=0$

Comp. of saturated soln., wt %			Comp. of solid phase
Iron Sulfate Hydrate	Zinc Sulfate Hydrate	Water	
24.35	1.17	74.48	$FeSO_4 \cdot 7H_2O + ZnSO_4 \cdot 7H_2O$
24.30	1.16	74.54	»
23.85	1.71	74.44	»
24.10	1.73	74.17	»
23.82	2.84	73.34	»
23.74	2.76	73.50	»
23.51	3.04	73.45	»
23.61	2.99	73.40	»
22.91	5.32	71.77	»
22.92	5.28	71.80	»
22.70	5.91	71.39	»
22.74	6.05	71.21	»
20.93	11.20	67.87	»
21.19	11.40	67.41	»
20.54	12.60	66.86	»
20.56	12.30	67.14	»
14.43	27.20	58.37	»
13.94	28.40	57.66	»
7.82	39.70	52.48	»
8.01	38.90	53.09	»
7.08	37.30	55.62	»
6.99	36.90	56.11	»

№ 5222 [2608]

IRON (II) SULFATE–CADMIUM SULFATE–WATER

$FeSO_4—CdSO_4—H_2O$

$t=25$

Comp. of saturated soln., wt %		
Cadmium Sulfate	Iron Sulfate	Water
43.19	0.00	56.81
34.24	6.32	59.44
0.00	21.06	78.94

№ 5223 IRON (II) CHLORIDE–IRON (III) CHLORIDE–WATER [1664]

$FeCl_2$—$FeCl_3$—H_2O

$t=25$

Concn. in aqueous soln., wt %		Comp. of solid phase
Iron (III) Chloride	Iron (II) Chloride	
39.2	0.0	$FeCl_2 \cdot 4H_2O$
35.9	4.1	»
31.4	9.9	»
22.8	19.46	»
14.45	30.4	»
8.3	39.8	»
7.05	41.95	»
5.44	46.65	$FeCl_2 \cdot 4H_2O + FeCl_2 \cdot 6H_2O$
3.96	47.01	$FeCl_2 \cdot 6H_2O$
3.49	47.41	»
0.0	49.7	»

№ 5224 IRON (II) CHLORIDE–COBALT (II) CHLORIDE–WATER [1003]

$FeCl_2$—$CoCl_2$—H_2O

$t=15$

Comp. of saturated soln., wt %			Comp. of solid phase
Iron Chloride	Cobalt Chloride	Water	
37.78	0.00	62.22	$FeCl_2 \cdot 4H_2O$
35.75	2.07	62.18	Solid soln. based on $FeCl_2 \cdot 4H_2O$
34.92	2.67	52.41	»
34.36	3.25	61.49	$FeCl_2 \cdot 4H_2O$
33.89	4.11	62.00	»
33.21	4.70	62.09	»
32.47	5.53	62.00	»
32.25	5.55	62.20	»
31.38	5.75	62.87	Solid soln. based on $CoCl_2 \cdot 6H_2O$
30.88	5.65	63.47	»
30.80	5.80	63.40	»
30.12	5.86	64.02	»
29.45	6.52	64.03	»
29.16	6.71	64.13	»
27.53	8.42	64.05	»
26.89	9.60	63.51	»
25.75	10.05	64.20	»
25.31	10.13	64.56	»
25.53	11.74	64.73	»
21.75	13.29	64.96	»
21.61	13.46	64.93	»
20.07	14.93	65.00	»
18.42	17.16	64.42	»
16.04	18.14	65.82	»
15.14	18.96	65.90	»

Table № 5224 (continuation)

Comp. of saturated soln., wt %			Comp. of solid phase
Cobalt Chloride	Iron Chloride	Water	
14.51	19.71	65.78	Solid soln. based on $CoCl_2 \cdot 6H_2O$
12.48	21.61	65.91	»
10.67	24.03	65.30	»
7.15	26.88	65.97	»
3.62	29.71	66.67	»
0.00	33.44	66.56	$CoCl_2 \cdot 6H_2O$

№ 5225 IRON (II) CHLORIDE–NICKEL (II) CHLORIDE–WATER [2292]

$FeCl_2—NiCl_2—H_2O$

$t=25$

Comp. of saturated soln., wt %			Comp. of solid phase
Iron Chloride	Nickel Chloride	Water	
39.61	0.00	60.39	$FeCl_2 \cdot 4H_2O$
35.26	4.44	60.30	Solid solution
29.82	10.52	59.66	»
22.72	17.61	59.67	»
21.56	19.42	59.02	»
16.25	24.03	59.72	»
9.26	30.68	60.06	»
0.00	38.88	61.12	$NiCl_2 \cdot 6H_2O$

№ 5226 IRON (II) CHLORIDE–COPPER (I) CHLORIDE–WATER [2014]

$FeCl_2—CuCl—H_2O$

$t=21.5$

Comp. of saturated soln., wt %			Comp. of solid phase
Iron Chloride	Copper Chloride	Water	
0.00	1.51	98.49	CuCl
5.60	1.24	93.16	»
10.30	1.59	88.11	»
13.65	2.62	83.73	»
19.71	5.33	74.96	»
21.35	5.87	72.78	»
23.21	6.71	70.08	»
28.00	7.95	64.05	»
31.60	9.98	58.42	»
35.30	11.48	53.22	»
37.30	11.80	50.90	$CuCl+FeCl_2 \cdot 4H_2O$
37.10	11.19	51.71	$FeCl_2 \cdot 4H_2O$
38.25	6.58	55.17	»
39.40	0.00	60.60	»

$FeCl_2—ZnCl_2—H_2O$

$t=15$

Comp. of saturated soln., wt %			Comp. of solid phase
Iron Chloride	Zinc Chloride	Water	
37.78	0.00	62.28	$FeCl_2 \cdot 4H_2O$
34.87	5.80	59.33	Solid soln. based on $FeCl_2 \cdot 4H_2O$
34.51	6.74	58.75	»
31.84	11.15	57.01	»
31.92	12.44	55.64	»
30.21	15.00	54.79	»
28.82	19.91	51.27	»
27.77	26.83	45.40	»
27.60	32.73	39.67	»
27.65	34.00	38.35	»
27.52	34.50	37.98	»
27.33	34.98	37.69	Solid soln. based on $FeCl_2 \cdot ZnCl_2 \cdot 6H_2O$
27.27	34.60	38.13	»
23.79	38.88	37.33	»
23.12	38.32	38.56	»
21.94	39.85	38.21	»
19.98	44.06	35.96	»
17.38	46.96	35.66	»
16.01	46.82	37.17	»
14.36	51.05	34.59	»
11.24	55.15	33.61	»
11.01	55.16	33.83	»
10.01	57.58	32.41	»
8.98	60.95	30.07	»
8.19	62.79	29.02	»
7.73	64.92	27.35	»
6.35	67.66	25.99	»
5.52	68.21	26.27	»
4.35	69.34	26.31	»
4.24	69.56	26.20	»
4.31	70.58	25.11	Solid soln. based on $ZnCl_2 \cdot 1.5H_2O$
3.97	71.48	24.55	»
3.85	71.70	25.45	»
3.17	72.02	24.81	»
2.69	72.47	24.84	»
2.20	74.20	23.60	»
1.29	76.14	22.57	»
0.00	77.14	22.86	$ZnCl_2 \cdot 1.5H_2O$

№ 5228 IRON (III) CHLORIDE–COBALT (II) CHLORIDE–WATER [2293]

$FeCl_3—CoCl_2—H_2O$

$t=25$

Comp. of saturated soln., wt %			Comp. of solid phase
Iron Chloride	Cobalt Chloride	Water	
0.0	36.08	63.92	$CoCl_2 \cdot 6H_2O$
7.77	29.36	62.87	»
19.08	20.95	59.97	»
35.52	10.13	54.35	»
41.07	7.99	50.94	»
43.31	7.42	49.27	$CoCl_2 \cdot 6H_2O+FeCl_3 \cdot 6H_2O$
43.95	6.58	49.47	$FeCl_3 \cdot 6H_2O$
44.50	5.99	49.51	»
47.17	2.47	50.36	»
49.42	0.0	50.58	»

№ 5229 IRON (III) CHLORIDE–NICKEL (II) CHLORIDE–WATER [2293]

$FeCl_3—NiCl_2—H_2O$

$t=25$

Comp. of saturated soln., wt %			Comp. of solid phase
Iron Chloride	Nickel Chloride	Water	
49.42	0.00	50.58	$FeCl_3 \cdot 6H_2O$
46.08	4.37	49.55	»
43.39	6.80	49.81	»
41.01	10.01	48.98	»
46.24	7.33	46.43	$NiCl_2 \cdot 4H_2O$
42.22	9.89	47.89	»
40.63	10.71	48.66	$NiCl_2 \cdot 4H_2O+FeCl_3 \cdot 6H_2O$
37.47	11.81	50.72	$NiCl_2 \cdot 4H_2O$
31.35	16.43	52.22	»
29.77	18.11	52.12	»
26.41	19.80	53.79	$NiCl_2 \cdot 4H_2O+NiCl_2 \cdot 6H_2O$
22.68	22.21	55.11	$NiCl_2 \cdot 6H_2O$
10.24	30.89	58.87	»
0.00	38.88	61.12	»

№ 5230 IRON (III) CHLORIDE–CESIUM CHLORIDE–WATER [1854]

$FeCl_3—CsCl—H_2O$

$t=21$

Comp. of saturated soln., wt %			Comp. of solid phase
Iron Chloride	Cesium Chloride	Water	
0.0	65.0	35.0	CsCl
0.45	55.18	44.37	$FeCl_3 \cdot 3CsCl \cdot H_2O$
2.1	52.38	45.52	»
5.24	51.44	43.32	»
7.80	47.70	44.50	$FeCl_3 \cdot 2CsCl \cdot H_2O$
8.93	41.15	49.92	»
15.34	25.25	59.41	»
21.65	14.96	63.39	»
27.96	8.42	63.62	»
48.71	0.94	50.35	»
83.89	0.00	16.11	$FeCl_3$

№ 5231 IRON (III) CHLORIDE–BARIUM SULFATE–WATER [1662]

$FeCl_3—BaSO_4—H_2O$

$t=20—25$

Content in aqueous soln., g/l			
Iron Chloride	Barium Sulfate	Iron Chloride	Barium Sulfate
1.0	0.058	25	0.150
2.5	0.072	50	0.160
5.0	0.115	100	0.170
10	0.123		

№ 5232 IRON (II) BROMIDE–IRON (III) BROMIDE–WATER [336]

$FeBr_2—FeBr_3—H_2O$

Comp. of saturated soln., wt %			t	Comp. of solid phase
Iron (II) Bromide	Iron (III) Bromide	Water		
38.83	27.36	33.81	60	$FeBr_2 \cdot 4H_2O$
38.90	27.85	33.25	60	»
29.61	41.77	28.62	60	»
29.90	42.50	27.60	60	»
24.52	52.25	23.23	60	»
22.82	57.39	19.79	60	»
23.61	57.20	19.19	60	»
23.56	60.88	5.56	60	»
23.30	62.80	13.90	60	$FeBr_2 \cdot 4H_2O + Fe_3Br_8 \cdot 6H_2O$

Table № 5232 (continuation)

Comp. of saturated soln., wt %			t	Comp. of solid phase
Iron (II) Bromide	Iron (III) Bromide	Water		
16.54	67.45	16.01	60	$Fe_3Br_8 \cdot 6H_2O$
18.00	66.86	15.14	60	»
10.03	74.50	15.47	60	»
5.20	80.00	14.80	60	»
2.42	83.49	14.09	60	»
0.00	82.00	18.00	25	$FeBr_3 \cdot 2H_2O$
1.12	81.80	17.09	25	»
2.70	81.65	15.65	25	$FeBr_3 \cdot 2H_2O + Fe_3Br_8 \cdot 6H_2O$
2.04	80.40	17.56	25	$Fe_3Br_8 \cdot 6H_2O$
2.60	81.76	15.64	25	»
3.92	79.68	16.40	25	»
4.62	78.77	16.61	25	»
6.27	76.50	17.23	25	»
6.05	76.60	17.35	25	»
9.72	72.50	17.78	25	»
9.02	73.08	17.90	25	»
12.48	68.87	18.85	25	»
13.40	68.60	18.00	25	»
15.80	65.15	19.05	25	»
18.08	63.90	18.02	25	$Fe_3Br_8 \cdot 6H_2O + FeBr_2 \cdot 4H_2O$
17 40	64.85	17.75	25	»
17.38	64.05	18.57	25	»
19.06	57.94	23.00	25	$FeBr_2 \cdot 4H_2O$
19.18	55.89	24.93	25	»
20.96	52.75	26.29	25	»
21.45	50.15	28.40	25	»
29.10	35.95	34.95	25	»
30.48	36.29	33.23	25	»
33.70	30.88	35.42	25	»
34.20	30.30	35.50	25	»
14.16	17.15	68.69	25	»
53.70	6.30	40.00	25	»
53.33	6.48	40.19	25	»
54.59	5.36	40.05	25	$FeBr_2 \cdot 4H_2O + FeBr_2 \cdot 6H_2O$
54.29	3.62	42.09	25	$FeBr_2 \cdot 6H_2O$
55.43	0.00	46.57	25	»
1.12	63.58	35.30	25	$FeBr_3 \cdot 6H_2O$
1.20	63.79	35.01	25	»
5.64	60.93	33.43	25	»
5.33	61.67	33.00	25	»
10.97	60.56	28.47	25	
11.04	60.72	28.24	25	»
9.93	67.23	22.84	25	»
6.53	70.57	22.90	25	»
6.56	71.20	22.24	25	»
3.75	74.22	22.03	25	»
0.96	76.50	22.54	25	»
53.40	0.00	46.60	15	$FeBr_2 \cdot 6H_2O$
49.60	5.65	44.75	15	»
48.18	6.41	45.41	15	»
44.16	12.50	43.34	15	»

Table № 5232 (continuation)

Comp. of saturated soln., wt %			t	Comp. of solid phase
Iron (II) Bromide	Iron (III) Bromide	Water		
43.70	13.18	43.12	15	$FeBr_2 \cdot 6H_2O$
36.60	21.45	42.95	15	»
37.40	21.75	40.85	15	»
37.18	22.17	40.65	15	»
33.32	26.25	40.43	15	»
32.93	28.20	38.87	15	$FeBr_2 \cdot 6H_2O + FeBr_3 \cdot 6H_2O$
32.23	27.55	40.22	15	»
32.53	27.80	39.67	15	»
30.40	29.60	40.00	15	$FeBr_3 \cdot 6H_2O$
29.60	28.85	41.55	15	»
30.35	29.85	39.80	15	»
27.50	31.70	40.80	15	»
26.55	34.40	39.05	15	»
23.55	34.71	41.74	15	»
13.73	44.40	41.87	15	»
14.05	44.80	41.15	15	»
13.20	44.80	42.00	15	»
12.12	44.87	43.01	15	»
10.03	47.12	42.85	15	»
9.62	47.98	42.40	15	»
7.26	49.85	42.89	15	»
8.36	48.40	43.24	15	»
5.94	49.15	44.91	15	»
1.93	55.40	42.67	15	»
1.77	55.00	43.23	15	»
0.00	56.20	43.80	15	»
50.00	6.25	43.75	20	$FeBr_2 \cdot 6H_2O$
41.20	18.13	40.67	20	»
36.00	24.25	39.75	20	Mixed crystals
51.00	0.00	49.00	0	$FeBr_2 \cdot 6H_2O$
38.40	13.60	48.00	0	»
47.80	4.30	47.90	0	»
38.00	15.00	47.00	0	»
28.30	26.10	45.60	0	»
47.10	4.70	48.20	0	»
40.00	12.80	47.20	0	»
25.5	28.80	45.70	0	$FeBr_3 \cdot 6H_2O + FeBr_2 \cdot 6H_2O$
26.30	28.30	45.40	0	»
20.60	33.80	45.60	0	$FeBr_3 \cdot 6H_2O$
21.00	32.80	46.20	0	»
5.40	48.00	46.60	0	»
5.80	47.00	47.20	0	»
14.40	39.80	45.80	0	»
12.90	41.10	46.00	0	»
1.10	52.50	46.40	0	»
0.00	53.00	47.0	0	»
38.85	11.80	49.35	—10	$FeBr_2 \cdot 6H_2O$
36.00	14.65	49.35	—10	»
32.75	18.40	48.85	—10	»
29.35	21.80	49.85	—10	»
22.90	28.95	48.15	—10	»

Comp. of saturated soln., wt %			t	Comp. of solid phase
Iron (II) Bromide	Iron (III) Bromide	Water		
20.10	32.60	47.30	—10	$FeBr_2 \cdot 6H_2O + FeBr_3 \cdot 6H_2O$
18.80	34.20	47.00	—10	$FeBr_3 \cdot 6H_2O$
15.95	36.40	47.65	—10	»
9.00	43.20	47.80	—10	»
8.00	44.20	47.80	—10	»
Traces	51.05	48.95	—10	»
49.12	0.00	50.88	—10	$FeBr_2 \cdot 6H_2O$
37.80	12.55	49.65	—20	»
35.30	14.35	50.35	—20	»
30.30	19.70	50.00	—20	»
28.80	21.32	49.88	—20	»
22.75	28.75	48.50	—20	»
18.60	33.40	48.00	—20	$FeBr_3 \cdot 6H_2O$
15.65	35.75	48.60	—20	»
9.75	41.20	49.05	—20	»
7.95	43.20	48.85	—20	»
Traces	49.20	50.80	—20	»
48.14	0.00	51.86	—20	$FeBr_2 \cdot 6H_2O$
35.20	14.20	50.60	—25	»
29.45	20.10	50.45	—25	»
28.80	21.40	49.80	—25	»
22.60	28.70	48.70	—25	»
19.00	32.80	48.20	—25	$FeBr_2 \cdot 6H_2O + FeBr_3 \cdot 6H_2O$
16.00	35.40	48.60	—25	$FeBr_3 \cdot 6H_2O$
9.90	40.25	49.85	—25	»
7.80	42.40	49.80	—25	»
Traces	48.30	51.70	—25	»
48.00	0.00	52.00	—25	$FeBr_2 \cdot 6H_2O$
37.30	12.70	51.00	—30	»
35.20	14.32	50.48	—30	»
28.75	20.50	50.75	—30	»
28.70	21.25	50.05	—30	»
22.50	28.41	49.09	—30	»
19.40	32.30	48.30	—30	$FeBr_3 \cdot 6H_2O$
15.35	35.00	59.65	—30	»
10.25	39.20	50.55	—30	»
7.70	41.60	50.70	—30	»
Traces	47.40	52.60	—30	»
31.75	12.90	56.35	—40	$FeBr_2 \cdot 9H_2O$
26.65	19.75	53.60	—40	»
22.65	24.40	52.95	—40	»
22.60	23.80	53.60	—40	»
20.95	26.42	52.63	—40	»
15.08	34.40	50.52	—40	$FeBr_2 \cdot 9H_2O + FeBr_3 \cdot 6H_2O$
10.80	37.25	51.95	—40	$FeBr_3 \cdot 6H_2O$
7.30	40.30	52.40	—40	»
Traces	45.95	54.05	—40	»
43.00	0.00	57.00	—40	$FeBr_2 \cdot 9H_2O$
31.00	12.65	56.35	—45	$FeBr_3 \cdot 9H_2O$
24.60	20.20	55.20	—45	»
15.00	32.50	52.50	—45	»

Comp. of saturated soln., wt %			t	Comp. of solid phase
Iron (II) Bromide	Iron (III) Bromide	Water		
9.85	37.10	53.05	—45	$FeBr_2 \cdot 6H_2O$
7.30	39.70	54.00	—45	»
Traces	45.50	54.50	—45	»
16.00	6.50	77.50	—10	Ice
14.00	10.35	75.65	—10	»
9.90	12.50	78.60	—10	»
6.80	15.40	77.80	—10	»
3.80	19.40	76.80	—10	»
Traces	22.60	77.40	—10	»
22.10	0.00	77.90	—10	»
22.20	9.03	68.73	—20	»
18.55	13.75	67.70	—20	»
13.60	17.17	69.23	—20	»
9.35	21.30	69.35	—20	»
4.85	25.80	69.35	—20	»
Traces	31.40	68.60	—20	»
30.75	0.00	69.25	—20	»
24.10	9.90	66.00	—25	»
20.20	15.00	64.80	—25	»
14.95	18.75	64.30	—25	»
10.30	23.20	66.50	—25	»
5.25	28.70	66.05	—25	»
Traces	34.50	65.50	—25	»
35.20	0.00	64.80	—25	»
25.70	10.42	63.88	—30	»
21.60	16.00	62.40	—30	»
16.10	20.35	63.55	—30	»
11.00	25.05	63.95	—30	»
5.60	31.10	63.30	—30	»
Traces	36.90	63.10	—30	»
37.13	0.00	62.87	—30	»
28.60	11.64	59.76	—40	»
23.42	17.35	59.23	—40	»
17.80	22.48	59.72	—40	»
12.32	28.10	59.58	—40	»
6.40	35.15	58.45	—40	»
Traces	41.30	58.70	—40	»
41.00	0.00	59.00	—40	»
29.50	12.00	58.50	—45	»
24.00	17.90	58.10	—45	»
18.50	23.25	58.25	—45	»
13.20	29.35	57.45	—45	»
6.80	36.80	56.40	—45	»
Traces	43.10	56.90	—45	»

№ 5233 COBALT (II) OXIDE–COBALT (II) SULFATE–WATER [235]

$CoO—CoSO_4—H_2O$

$t=18$

Concn. in aqueous soln., wt %		d_4	pH	Comp. of solid phase
Cobalt Sulfate	Cobalt g/l			
24.8	120.1	1.278	6.60	$CoO \cdot CoSO_4 \cdot xH_2O + CoSO_4 \cdot 7H_2O$
11.7	50.2	1.130	6.99	$CoO \cdot CoSO_4 \cdot H_2O$
6.7	27.4	1.073	7.10	»
2.48	9.7	1.030	7.26	»
1.47	5.7	1.019	7.34	»
0.91	3.5	1.015	7.38	»
0.52	2.0	1.010	7.49	»
0.0	0.0	1.000	8.50	$Co(OH)_2$

№ 5234 COBALT (II) NITRATE–NICKEL (II) NITRATE–WATER [229]

$Co(NO_3)_2—Ni(NO_3)_2—H_2O$

$t=20$

Concn. in aqueous soln., wt %			
$CO(NO_3)_2 \cdot 6H_2O$	$Ni(NO_3)_2 \cdot 6H_2O$	$CO(NO_3)_2 \cdot 6H_2O$	$Ni(NO_3)_2 \cdot 6H_2O$
77.07	0.0	30.81	47.49
75.17	0.88	30.88	47.54
75.55	0.90	26.15	51.43
75.42	0.85	25.98	51.29
72.82	2.66	24.44	58.83
73.60	2.67	24.34	53.84
73.23	3.84	24.70	54.32
72.64	4.10	25.16	53.45
68.94	8.03	14.84	63.49
69.25	8.02	14.53	63.11
69.72	8.03	14.51	62.47
69.40	8.02	8.82	69.02
62.02	13.70	8.78	69.01
61.95	13.67	4.71	73.84
49.71	26.74	4.76	73.80
49.80	25.97	2.08	73.30
38.11	38.85	2.09	72.74
38.84	39.59	0.0	76.21

№ 5235 [229]

COBALT (II) NITRATE HYDRATE–NICKEL (II) NITRATE HYDRATE–WATER

$Co(NO_3)_2 \cdot 6H_2O—Ni(NO_3)_2 \cdot 6H_2O—H_2O$

$t=20$

Comp. of saturated soln., wt %			Comp. of solid phase
Cobalt Nitrate Hydrate	Nickel Nitrate Hydrate	Water	
77.07	0.00	22.93	Solid solutions with the structure of Cobalt Nitrate
75.17	0.88	23.95	»
75.55	0.90	23.55	»

Table № 5235 (continuation)

Comp. of saturated soln., wt %			Comp. of solid phase
Cobalt Nitrate Hydrate	Nickel Nitrate Hydrate	Water	
75.42	0.85	23.73	Solid solutions with the structure of Cobalt Nitrate
72.82	2.66	24.52	»
73.60	2.67	23.73	»
73.23	3.84	22.93	»
72.64	4.10	23.26	»
68.94	8.03	23.03	»
69.25	8.02	22.73	»
69.75	8.03	22.22	»
69.40	8.02	22.58	»
62.02	13.70	24.28	»
61.95	13.67	24.38	»
49.71	26.74	23.55	»
49.80	25.97	24.23	»
38.11	38.85	23.04	»
38.84	39.59	21.57	»
30.81	47.49	21.70	»
30.88	47.54	21.58	»
26.15	51.43	22.42	»
25.98	51.29	22.73	»
24.44	53.83	21.73	»
24.34	43.84	21.82	»
24.70	54.32	20.98	Solid solutions with the structure of Nickel Nitrate
25.16	53.45	21.39	»
14.84	63.49	21.67	»
14.53	63.11	22.36	»
14.51	62.47	23.02	»
8.82	69.02	22.16	»
8.78	69.01	22.21	»
4.71	73.84	21.45	»
4.76	73.80	21.44	»
2.08	73.30	24.62	»
2.09	72.74	25.17	»

№ 5236 COBALT (II) NITRATE–COPPER (II) NITRATE–WATER [2739]

$Co(NO_3)_2$—$Cu(NO_3)_2$—H_2O

Comp. of saturated soln., wt %			t	d_{20}^{20}	Comp. of solid phase
Copper Nitrate	Cobalt Nitrate	Water			
0.0	47.5	52.5	14	—	$Co(NO_3)_2 \cdot 6H_2O$
5.2	44.3	50.5	14	—	»
25.8	28.7	45.5	14	—	»
34.1	22.8	43.1	14	—	»
36.5	19.0	44.5	14	—	$Co(NO_3)_2 \cdot 6Cu(NO_3)_2 \cdot 42H_2O$
34.8	16.7	48.5	14	—	»
38.9	16.0	45.1	14	—	»
41.5	14.0	44.5	14	—	»
45.6	11.1	43.3	14	—	$Cu(NO_3)_2 \cdot 6H_2O$

Table № 5236 (continuation)

Comp. of saturated soln., wt %			t	d_{20}^{20}	Comp. of solid phase
Copper Nitrate	Cobalt Nitrate	Water			
47.0	5.9	47.1	14	—	$Cu(NO_3)_2 \cdot 6H_2O$
52.5	0.0	47.5	14	—	»
0.0	49.3	50.7	20	1.598	$Co(NO_3)_2 \cdot 6H_2O$
9.9	41.2	48.9	20	1.610	»
13.1	38.9	48.0	20	1.613	»
20.3	34.2	45.5	20	1.640	»
33.7	24.5	41.8	20	1.705	»
43.1	17.9	39.0	20	1.773	»
45.0	16.7	38.3	20	—	»
49.4	10.8	39.8	20	—	$Cu(NO_3)_2 \cdot 6H_2O$
49.8	8.9	41.3	20	1.707	»
55.6	0.0	44.4	20	—	»
0.0	52.7	47.3	30	—	$Co(NO_3)_2 \cdot 6H_2O$
9.2	44.7	46.1	30	—	»
27.1	29.2	43.7	30	—	»
37.9	22.2	39.9	30	—	»
45.7	16.5	37.8	30	—	$Cu(NO_3)_2 \cdot 6H_2O$
48.4	13.0	38.6	30	—	»
54.9	5.3	39.8	30	—	»
61.4	0.0	38.6	30	—	»

№ 5237 COBALT (II) NITRATE–COPPER (II) NITRATE–WATER [2747]

$Co(NO_3)_2$—$Cu(NO_3)_2$—H_2O

Concn. in aqueous soln., wt %		t	d_4^t	Comp. of solid phase
Copper Nitrate	Cobalt Nitrate			
52.50	0.00	14	—	$Cu(NO_3)_2 \cdot 6H_2O$
47.00	5.90	14	—	»
45.60	11.10	14	—	»
41.50	14.00	14	—	$Cu(NO_3)_2 \cdot Co(NO_3)_2$
38.90	16.00	14	—	»
38.40	16.70	14	—	»
36.50	19.00	14	—	»
34.10	22.80	14	—	$Co(NO_3)_2 \cdot 6H_2O$
25.80	28.70	14	—	»
5.20	44.30	14	—	»
0.00	47.50	14	—	»
55.60	0.0	20	—	$Cu(NO_3)_2 \cdot 6H_2O$
49.80	8.90	20	1.707	»
49.40	10.80	20	—	»
45.0	16.70	20	—	$Co(NO_3)_2 \cdot 6H_2O$
43.10	17.90	20	1.773	»
33.70	24.50	20	1.705	»
20.30	34.20	20	1.640	»
13.10	38.90	20	1.613	»
9.90	41.20	20	1.610	»
0.00	49.30	20	1.598	»

Table № 5237 (continuation)

Concn. in aqueous soln., wt %		t	d_4^t	Comp. of solid phase
Copper Nitrate	Cobalt Nitrate			
61.40	0.00	30	—	$Cu(NO_3)_2 \cdot 2\frac{1}{2}H_2O$
54.90	5.30	30	—	»
48.40	13.0	30	—	»
45.70	16.50	30	—	»
37.90	22.20	30	—	$Co(NO_3)_2 \cdot 6H_2O$
27.10	29.20	30	—	»
9.20	44.70	30	—	»
0.00	52.70	30	—	»

№ 5238 COBALT (II) SULFATE–COBALT (II) CHLORIDE–WATER [2048]

$CoSO_4—CoCl_2—H_2O$

$t=25$

Comp. of saturated soln., wt %			Comp. of solid phase
Cobalt Chloride	Cobalt Sulfate	Water	
36.0	0.0	64.0	$CoCl_2 \cdot 6H_2O$
33.8	3.6	62.6	»
32.8	4.6	62.6	$CoCl_2 \cdot 6H_2O+CoSO_4 \cdot 7H_2O$
26.8	6.1	67.1	$CoSO_4 \cdot 7H_2O$
23.5	6.7	69.8	»
20.3	9.6	70.1	»
15.5	11.3	73.2	»
9.45	16.4	74.15	»
5.6	20.1	74.3	»
3.0	22.5	74.5	»
1.8	23.8	74.4	»
0.8	25.4	73.8	»
0.0	26.8	73.2	»

№ 5239 COBALT (II) SULFATE–COBALT (II) CHLORIDE–WATER [1219]

$CoSO_4—CoCl_2—H_2O$

Concn. in aqueous soln., wt %		t	Comp. of solid phase
Cobalt Chloride	Cobalt Sulfate		
29.4	4.03	0	$CoSO_4 \cdot 7H_2O+CoCl_2 \cdot 6H_2O$
28.9	2.53	0	»
19.50	4.96	0	$CoSO_4 \cdot 7H_2O$
12.40	8.43	0	»
11.53	8.74	0	»
3.11	16.00	0	»
37.4	3.70	38	$CoCl_2 \cdot 6H_2O$
36.4	4.70	38	$CoCl_2 \cdot 6H_2O+CoSO_4 \cdot 6H_2O$
36.3	4.85	38	»

Table № 5239 (continuation)

Concn. in aqueous soln., wt %		t	Comp. of solid phase
Cobalt Chloride	Cobalt Sulfate		
32.8	7.33	38	$CoSO_4 \cdot 6H_2O$
30.4	7.25	38	»
28.2	7.52	38	»
25.4	9.25	38	»
21.8	10.56	38	»
20.2	12.26	38	»
21.9	11.28	38	»
17.71	13.64	38	»
13.91	17.11	38	$CoSO_4 \cdot 6H_2O + CoSO_4 \cdot 7H_2O$
10.24	18.99	38	$CoSO_4 \cdot 7H_2O$
8.72	21.2	38	»
3.97	26.9	38	»
6.18	26.9	50	$CoSO_4 \cdot 6H_2O$
6.79	26.2	50	»
12.95	19.8	50	»
22.3	13.6	50	»
32.3	8.69	50	»
32.9	9.36	50	$CoSO_4 \cdot 6H_2O + CoSO_4 \cdot H_2O$
39.7	5.95	50	»
43.5	6.09	50	$CoSO_4 \cdot 6H_2O + CoCl_2 \cdot 4H_2O$
43.3	6.35	50	»
21.8	13.38	50	$CoSO_4 \cdot H_2O$
34.9	5.18	50	»
35.7	6.65	50	$CoSO_4 \cdot 6H_2O + CoSO_4 \cdot H_2O$
43.4	3.89	50	$CoSO_4 \cdot H_2O + CoCl_2 \cdot 4H_2O$
45.0	2.80	50	$CoSO_4 \cdot 6H_2O + CoSO_4 \cdot H_2O$
43.9	3.89	51	$CoSO_2 \cdot H_2O$
7.20	20.2	63.5	»
1.35	34.5	75	»
9.03	25.9	75	»
12.79	20.3	75	»
18.45	15.7	75	»
20.0	13.0	75	»
20.8	6.43	75	»
41.5	3.40	75	»
43.1	2.53	75	»
50.2	2.18	75	$CoCl_2 \cdot 2H_2O + CoSO_4 \cdot H_2O$
47.6	2.56	75	»
48.2	1.75	75	$CoCl_2 \cdot 2H_2O$
52.6	0.48	99.5	$CoCl_2 \cdot 2H_2O + CoSO_4 \cdot H_2O$
50.8	1.82	99.5	»
49.2	4.30	99.5	»
39.0	2.37	99.5	$CoSO_4 \cdot H_2O$
26.0	5.85	99.5	»
30.1	4.39	99.5	»
23.2	7.46	99.5	»
12.21	15.58	99.5	»
6.47	20.4	99.5	»
5.57	21.4	99.5	»
2.82	26.8	99.5	»

№ 5240 COBALT (II) SULFATE–COBALT (II) IODATE–WATER [2669]

$CoSO_4—Co(IO_3)_2—H_2O$

$t=30$

Concn. in aqueous soln., wt %		d_4	Comp. of solid phase
Cobalt Sulfate	Cobalt Iodate		
0.00	0.476	1.0006	$Co(IO_3)_2 \cdot 2H_2O$
0.040	0.462	1.0008	»
0.081	0.454	1.0011	»
0.095	0.454	1.0012	»
0.191	0.437	1.0019	»
0.200	0.433	1.0020	»
0.364	0.417	1.0034	»
0.400	0.413	1.0037	»
0.800	0.388	1.0076	»
1.200	0.384	1.0117	»
1.600	0.380	1.0155	»

№ 5241 COBALT (II) SULFATE–NICKEL (II) SULFATE–WATER [811]

$CoSO_4—NiSO_4—H_2O$

$t=20$

Comp. of saturated soln., wt %			Comp. of solid phase, wt %		
Nickel Sulfate	Cobalt Sulfate	Water	Nickel Sulfate	Cobalt Sulfate	Water
27.7	0.0	72.3	55.1	0.00	44.9
22.4	6.5	71.1	45.1	10.0	44.9
20.3	8.9	70.8	40.8	14.3	44.9
17.5	12.1	70.4	36.5	18.6	44.9
15.7	13.9	70.4	34.4	20.7	44.9
15.1	14.7	70.2	33.4	21.7	44.9
15.2	14.7	70.1	33.3	21.8	44.9
14.7	15.0	70.3	16.6	38.5	44.9
13.3	16.4	70.3	15.2	39.9	44.9
10.4	18.4	71.2	11.6	43.5	44.9
7.0	20.8	72.2	7.5	47.6	44.9
0.0	25.6	74.4	0.0	55.1	44.9

№ 5242 COBALT (II) SULFATE–COPPER (II) SULFATE–WATER [1438]

$CoSO_4—CuSO_4—H_2O$

Comp. of saturated soln., wt %			t	Comp. of solid phase
Copper Sulfate	Cobalt Sulfate	Water		
0.00	19.82	80.18	0	$CoSO_4 \cdot 7H_2O$
0.59	19.09	80.32	0	Solid solution
1.57	17.64	80.79	0	»
3.14	15.02	81.84	0	»

Table № 5242 (continuation)

Comp. of saturated soln., wt %			t	Comp. of solid phase
Copper Sulfate	Cobalt Sulfate	Water		
6.66	10.67	82.67	0	Solid solution
9.53	7.89	82.58	0	»
11.95	5.94	82.11	0	Solid soln. $+CuSO_4 \cdot 5H_2O$
12.33	4.05	83.62	0	$CuSO_4 \cdot 5H_2O$
12.55	2.80	84.65	0	»
12.78	1.00	86.22	0	»
12.87	0.0	87.13	0	»
0.00	27.60	72.40	25	$CoSO_4 \cdot 7H_2O$
1.76	25.24	73.00	25	Solid solution
3.71	23.25	73.04	25	»
7.45	18.55	74.00	25	»
11.61	15.03	73.36	25	»
12.91	13.88	73.21	25	Solid soln. $+CuSO_4 \cdot 5H_2O$
14.68	9.70	75.62	25	$CuSO_4 \cdot 5H_2O$
16.66	5.04	78.30	25	»
18.18	1.14	80.68	25	»
18.45	0.00	81.55	25	»

№ 5243 COBALT (II) SULFATE–CADMIUM SULFATE–WATER [1164]

$CoSO_4—CdSO_4—H_2O$

Comp. of saturated soln., wt %			t	Comp. of solid phase
Cadmium Sulfate	Cobalt Sulfate	Water		
0.00	27.05	72.95	25	$CoSO_4 \cdot 7H_2O$
9.97	22.38	67.65	25	Solid soln. of $CoSO_4 \cdot 7H_2O + CdSO_4 \cdot 7H_2O$
17.78	18.89	63.33	25	»
23.90	15.62	60.48	25	»
28.88	13.42	57.70	25	»
31.02	12.35	56.63	25	»
32.64	11.55	55.81	25	»
34.52	9.44	56.04	25	$CdSO_4 \cdot {}^8/_3H_2O$ (solid soln.)
35.85	8.00	56.15	25	»
38.84	5.05	56.11	25	»
40.46	3.43	56.11	25	»
41.84	1.78	56.38	25	»
43.83	0.00	56.17	25	$CdSO_4 \cdot {}^8/_3H_2O$
0.00	39.94	60.06	80	$CoSO_4 \cdot 4H_2O$
5.94	36.60	57.46	80	Solid soln. of $CdSO_4 \cdot 4H_2O + CoSO_4 \cdot 4H_2O$
9.91	34.30	55.79	80	»
14.08	32.01	53.91	80	Solid soln. of $CdSO_4 \cdot 4H_2O + CoSO_4 \cdot 4H_2O$ and solid soln. of $CdSO_4 \cdot {}^8/_3H_2O + CoSO_4 \cdot {}^8/_3H_2O$
16.63	29.66	53.71	80	Solid soln. of $CdSO_4 \cdot {}^8/_3H_2O + CoSO_4 \cdot {}^8/_3H_2O$
17.30	28.77	53.93	80	»
17.83	28.32	53.85	80	»
19.31	26.90	53.79	80	»

Comp. of saturated soln., wt %			t	Comp. of solid phase
Cadmium Sulfate	Cobalt Sulfate	Water		
20.83	25.59	53.58	80	Solid soln. of $CdSO_4 \cdot H_2O + CoSO_4 \cdot H_2O$
21.24	24.63	54.13	80	»
21.85	23.32	54.83	80	»
22.83	22.06	55.11	80	»
26.08	18.12	55.80	80	»
27.94	15.56	56.50	80	»
29.37	14.13	56.50	80	»
34.98	7.20	57.82	80	Solid soln. of β-$CdSO_4 + \alpha$-$CdSO_4$
38.11	3.00	58.89	80	»
40.31	0.00	59.69	80	β-$CdSO_4 \cdot H_2O$
45.80	0.00	54.20	80	α-$CdSO_4 \cdot H_2O$

№ 5244 COBALT (II) SULFATE–RUBIDIUM SULFATE–WATER [1210]

$CoSO_4—Rb_2SO_4—H_2O$

Comp. of saturated soln., wt %			t	Comp. of solid phase
Cobalt Sulfate	Rubidium Sulfate	Water		
1.80	3.89	94.31	0	$CoSO_4 \cdot Rb_2SO_4 \cdot 6H_2O$
16.40	0.99	82.61	0	$CoSO_4 \cdot Rb_2SO_4 \cdot 6H_2O + CoSO_4 \cdot 7H_2O$
0.19	25.08	74.73	25	$CoSO_4 \cdot Rb_2SO_4 \cdot 6H_2O + Rb_2SO_4$
3.57	6.15	90.28	25	$CoSO_4 \cdot Rb_2SO_4 \cdot 6H_2O$
21.24	1.90	76.86	25	$CoSO_4 \cdot Rb_2SO_4 \cdot 6H_2O + CoSO_4 \cdot 7H_2O$
26.25	2.14	71.61	40	$CoSO_4 \cdot Rb_2SO_4 \cdot 6H_2O + CoSO_4 \cdot 7H_2O +$ $+ CoSO_4 \cdot 6H_2O$
5.44	6.34	88.22	44.5	$CoSO_4 \cdot Rb_2SO_4 \cdot 6H_2O$
6.87	11.87	81.26	60	»
0.94	28.76	70.30	69	$CoSO_4 \cdot Rb_2SO_4 \cdot 6H_2O + Rb_2SO_4$
7.69	13.29	79.02	69	$CoSO_4 \cdot Rb_2SO_4 \cdot 6H_2O$
26.03	3.92	70.05	69	$CoSO_4 \cdot Rb_2SO_4 \cdot 6H_2O + CoSO_4 \cdot 6H_2O +$ $+ CoSO_4 \cdot H_2O$
9.57	16.55	73.88	88	$CoSO_4 \cdot Rb_2SO_4 \cdot 6H_2O$
20.70	9.68	69.62	88	$CoSO_4 \cdot Rb_2SO_4 \cdot 6H_2O + CoSO_4 \cdot H_2O$
2.78	31.15	66.07	100	$CoSO_4 \cdot Rb_2SO_4 \cdot 6H_2O + Rb_2SO_4$
10.6	18.2	71.2	100	$CoSO_4 \cdot Rb_2SO_4 \cdot 6H_2O$
18.85	14.20	66.95	100	$CoSO_4 \cdot Rb_2SO_4 \cdot 6H_2O + CoSO_4 \cdot H_2O$

№ 5245 COBALT (II) SULFATE–THALLIUM SULFATE–WATER [1211]

$CoSO_4—Tl_2SO_4—H_2O$

Comp. of saturated soln., wt %			t	Comp. of solid phase
Cobalt Sulfate	Thallium Sulfate	Water		
1.38	3.05	95.57	0	$CoSO_4 \cdot Tl_2SO_4 \cdot 6H_2O + Tl_2SO_4$
16.60	1.96	81.44	0	$CoSO_4 \cdot Tl_2SO_4 \cdot 6H_2O + CoSO_4 \cdot 7H_2O$
0.80	5.10	94.10	25	Tl_2SO_4
2.90	5.49	91.61	25	$Tl_2SO_4 + CoSO_4 \cdot Tl_2SO_4 \cdot 6H_2O$
3.33	5.10	91.57	25	$CoSO_4 \cdot Tl_2SO_4 \cdot 6H_2O$
5.11	4.31	90.58	25	»
18.56	2.48	78.96	25	»
20.53	2.85	76.62	25	$CoSO_4 \cdot Tl_2SO_4 \cdot 6H_2O + CoSO_4 \cdot 7H_2O$
20.72	1.69	77.59	25	$CoSO_4 \cdot 7H_2O$
3.82	6.61	89.57	36	$CoSO_4 \cdot Tl_2SO_4 \cdot 6H_2O + Tl_2SO_4$
23.50	2.98	73.52	36	$CoSO_4 \cdot Tl_2SO_4 \cdot 6H_2O + CoSO_4 \cdot 7H_2O$
6.21	9.53	84.26	50	$CoSO_4 \cdot Tl_2SO_4 \cdot 6H_2O + Tl_2SO_4$
23.65	3.66	72.69	50	$CoSO_4 \cdot Tl_2SO_4 \cdot 6H_2O + CoSO_4 \cdot 6H_2O$
22.15	9.97	67.88	78.5	$CoSO_4 \cdot Tl_2SO_4 \cdot 6H_2O + CoSO_4 \cdot H_2O$
13.66	15.53	70.81	84.5	$CoSO_4 \cdot Tl_2SO_4 \cdot 6H_2O + Tl_2SO_4$
20.8	12.5	66.7	84.5	$CoSO_4 \cdot Tl_2SO_4 \cdot 6H_2O + CoSO_4 \cdot H_2O$
2.31	14.54	83.15	100	Tl_2SO_4
6.13	15.80	78.07	100	»
12.47	18.03	69.50	100	»
14.29	18.15	67.56	100	»
15.25	18.80	65.95	100	$Tl_2SO_4 + CoSO_4 \cdot H_2O$
19.18	9.48	71.34	100	$CoSO_4 \cdot H_2O$
23.23	3.41	73.36	39.5	$CoSO_4 \cdot 7H_2O + CoSO_4 \cdot 6H_2O + CoSO_4 \cdot Tl_2SO_4 \cdot 6H_2O$
26.50	5.17	68.33	69.5	$CoSO_4 \cdot H_2O + CoSO_4 \cdot 6H_2O + CoSO_4 \cdot Tl_2SO_4 \cdot 6H_2O$
16.25	17.25	66.50	94	$CoSO_4 \cdot H_2O + Tl_2SO_4 + CoSO_4 \cdot Tl_2SO_4 \cdot 6H_2O$

№ 5246 COBALT (II) CHLORIDE–NICKEL (II) CHLORIDE–WATER [2293]

$CoCl_2—NiCl_2—H_2O$

$t=25$

Comp. of saturated soln., wt %			Comp. of solid phase
Nickel Chloride	Cobalt Chloride	Water	
0.00	36.08	63.92	$CoCl_2 \cdot 6H_2O$
4.31	32.06	63.63	$(Co, Ni)Cl_2 \cdot 6H_2O$
10.45	26.17	63.38	»
13.39	23.42	63.19	»
22.51	14.87	62.62	»
26.13	11.83	62.04	»
30.63	7.66	61.71	»
31.91	6.28	61.81	»
34.36	4.38	61.26	»
38.88	0.00	61.12	$NiCl_2 \cdot 6H_2O$

№ 5247 COBALT (II) CHLORIDE–COPPER (II) CHLORIDE–WATER [1166]

$CoCl_2—CuCl_2—H_2O$

$t=25$

Comp. of saturated soln., wt %			Comp. of solid phase
Copper Chloride	Cobalt Chloride	Water	
0.0	35.87	64.13	$CoCl_2 \cdot 6H_2O$
5.34	33.19	61.47	»
9.99	30.50	59.51	»
15.08	27.48	57.44	»
18.01	26.06	56.93	$CoCl_2 \cdot 6H_2O + CuCl_2 \cdot 2H_2O$
25.62	17.48	56.90	$CuCl_2 \cdot 2H_2O$
30.33	12.66	57.01	»
34.00	8.99	57.01	»
39.29	4.34	56.37	»
44.00	0.0	56.00	»

№ 5248 COBALT (II) CHLORIDE–ZINC CHLORIDE–WATER [1160]

$CoCl_2—ZnCl_2—H_2O$

$t=25$

Comp. of saturated soln., wt %			Comp. of solid phase
Zinc Chloride	Cobalt Chloride	Water	
0.00	35.87	64.13	$CoCl_2 \cdot 6H_2O$
9.26	32.03	58.71	»
21.75	29.35	48.90	»
26.74	30.86	43.40	»
27.91	31.20	40.89	$CoCl_2 \cdot 6H_2O + CoCl_2 \cdot 2H_2O$
28.43	31.08	40.49	$CoCl_2 \cdot 2H_2O$
32.16	29.77	38.07	»
33.62	29.40	36.98	»
33.31	29.48	37.21	$CoCl_2 \cdot 2H_2O$ + solid soln. A
33.31	28.22	48.47	solid soln. A
37.86	26.09	36.05	»
43.09	22.20	34.71	»
47.00	19.56	33.44	»
50.00	17.45	32.55	»
59.29	12.34	28.37	»
60.10	13.19	26.71	»
60.09	13.51	26.40	Solid soln. A + solid soln. B
59.92	13.77	26.31	Solid soln. B
61.15	12.91	25.94	»
63.08	12.28	24.64	»
63.38	12.40	24.22	»
63.40	12.61	23.99	Solid soln. B + solid soln. C
59.54	15.78	24.68	Solid soln. C
63.40	12.63	23.97	»
64.00	12.16	23.84	»
69.88	7.35	22.77	»

Table № 5248 (continuation)

Comp. of saturated soln., wt %			Comp. of solid phase
Zinc Chloride	Cobalt Chloride	Water	
72.07	5.69	22.24	Solid soln. C
74.50	3.97	21.53	»
76.09	2.76	21.15	»
76.98	2.34	20.68	»
80.19	0.43	19.38	»
81.78	0.0	18.22	$ZnCl_2 \cdot {}^5/_3H_2O$
57.95	17.23	24.82	Solid soln. C + solid soln. D
60.09	15.36	24.55	Solid soln. A + solid soln. D
64.03	12.32	23.65	Solid soln. D
66.98	10.18	22.84	»
67.72	9.59	22.69	»
67.46	9.48	23.06	»
73.07	5.96	20.97	»
73.82	5.37	20.81	»
77.38	4.19	18.43	»
77.67	4.70	17.63	Solid soln. E
80.02	0.61	19.37	Solid soln. C + solid soln. E
80.88	0.0	19.12	$ZnCl_2 \cdot {}^4/_3H_2O$
78.58	2.60	18.82	$ZnCl_2$ (hexagonal plates)
81.5	0.0	18.5	»
79.02	2.45	18.53	$ZnCl_2$ (octahedra)
79.20	2.06	18.74	$ZnCl_2$ (double pyramids)

Note. Solid solution A – $[Zn(H_2O)_6] \cdot [ZnCl_4] + [Co(H_2O)_6] \cdot [ZnCl_4]_2$; solid solution B – $[Zn_2(H_2O)_{10}] \cdot [ZnCl_4] + [Co(H_2O)_{10}] \cdot [ZnCl_4]_2$; solid solution C – $[Zn_2(H_2O)_{10}] \cdot [Zn_2Cl_6]_2 + [Co_2(H_2O)_{10}] \cdot [Zn_2Cl_6]$; solid solution D – $[Zn_2(H_2O)_6] \cdot [ZnCl_4]_2 + [Co_2(H_2O)_6] \cdot [ZnCl_4]_2$; solid solution E – $[Zn(H_2O)_4] \cdot [Zn_2Cl_6] + [Co(H_2O)_4] \cdot [ZnCl_6]$.

№ 5249 COBALT (II) CHLORIDE–ZINC CHLORIDE–WATER [1159]

$CoCl_2—ZnCl_2—H_2O$

$t=25$

Comp. of saturated soln., wt %			Comp. of solid phase
Cobalt Chloride	Zinc Chloride	Water	
35.87	0.00	64.13	$CoCl_2 \cdot 6H_2O$
32.03	9.26	58.71	»
29.35	21.75	48.90	»
30.86	26.74	42.40	»
31.20	27.91	40.89	$CoCl_2 \cdot 6H_2O + CoCl_2 \cdot 2H_2O$
31.08	28.43	40.49	$CoCl_2 \cdot 2H_2O$
29.77	32.16	38.07	»
29.64	33.21	37.15	»
29.40	33.62	36.98	»
29.48	33.31	37.21	$CoCl_2 \cdot 2H_2O + ZnCl_2 \cdot 3H_2O + [Co(H_2O)_6]ZnCl_4$
28.22	34.99	36.79	$ZnCl_2 \cdot 3H_2O + [Co(H_2O)_6]ZnCl_4$

Comp. of saturated soln., wt %			Comp. of solid phase
Cobalt Chloride	Zinc Chloride	Water	
26.09	37.86	36.05	$ZnCl_2 \cdot 3H_2O+[Co(H_2O)_6]ZnCl_4$
22.20	43.19	34.61	»
19.56	47.00	33.44	»
17.45	50.00	32.55	»
13.10	57.15	29.75	»
12.82	59.29	27.89	»
12.34	59.29	28.37	»
12.78	60.29	26.93	»
13.19	60.10	26.71	»
13.51	60.09	26.40	$ZnCl_2 \cdot 3H_2O+[Co(H_2O)_6]ZnCl_4+ZnCl_2 \cdot 2.5H_2O+$ $+[Co(H_2O)_{10}]Zn_2Cl_8$
13.77	59.92	26.31	$ZnCl_2 \cdot 2.5H_2O+[Co(H_2O)_{10}]Zn_2Cl_8$
12.91	61.15	25.94	»
12.72	61.82	25.46	»
12.50	62.59	24.91	»
12.28	63.08	24.64	»
12.40	63.38	24.22	»
12.61	63.40	23.99	$ZnCl_2 \cdot 2.5H_2O+[Co(H_2O)_{10}]Zn_2Cl_8+$ $+ZnCl_2 \cdot 1^2/_3H_2O+[Co_2(H_2O)_{10}](Zn_2Cl_6)_2$
15.78	59.54	24.68	$ZnCl_2 \cdot 1^2/_3H_2O+[Co_2(H_2O)_{10}](Zn_2Cl_6)_2$
12.63	63.40	23.97	»
12.16	64.00	23.84	»
7.35	69.98	22.67	»
5.69	72.35	21.96	»
5.66	72.32	22.02	»
5.64	72.45	21.91	»
3.97	74.50	21.53	»
2.76	76.36	20.88	»
2.34	0.00	97.66	»
0.43	80.19	19.38	»
0.00	81.78	18.22	$[Co_2(H_2O)_{10}] \cdot [Zn_2Cl_6]_2+ZnCl_2 \cdot 1^2/_3H_2O$
17.23	57.95	24.82	$[Co_2(H_2O)_{10}] \cdot [Zn_2Cl_6]_2+ZnCl_2 \cdot 1^2/_3H_2O$
15.36	60.09	24.55	$ZnCl_2 \cdot 3H_2O+[Co(H_2O)_6](ZnCl_4)_2+$ $+ZnCl_2 \cdot 1\frac{1}{2}H_2O+[Co_2(H_2O)_6](ZnCl_4)_2$
12.32	64.03	23.65	$ZnCl_2 \cdot 1\frac{1}{2}H_2O+[Co(H_2O)_6](ZnCl_2)_2$
10.18	66.98	22.84	»
9.59	67.72	22.69	»
9.48	68.0	22.52	»
5.96	73.09	20.95	»
5.37	73.82	20.81	»
4.19	77.38	18.43	»
4.70	74.67	20.63	»
0.61	80.02	19.37	$ZnCl_2 \cdot 1\frac{1}{2}H_2O+[Co_2(H_2O)_6](ZnCl_2)_2+$ $+ZnCl_2 \cdot 1^1/_3H_2O+[Co(H_2O)_4] \cdot [Zn_2Cl_6]$
0.00	81.28	18.72	$[Co(H_2O)_4] \cdot [Zn_2Cl_6]+ZnCl_2 \cdot 1^1/_3H_2O$
0.00	80.90	19.10	»
2.60	78.58	18.82	$ZnCl_2$ (hexagonal)
0.00	81.84	18.16	»
2.45	79.02	18.53	$ZnCl_2$ (octahedral)
0.00	81.44	18.56	»
2.06	79.20	18.74	$ZnCl_2$ (rhombohedral)
0.00	81.52	18.48	»

№ 5250 COBALT (II) CHLORIDE–RUBIDIUM CHLORIDE–WATER [1643]

$CoCl_2$—$RbCl$—H_2O

$t=25$

Comp. of saturated soln., wt %			Comp. of solid phase
Rubidium Chloride	Cobalt Chloride	Water	
0.0	35.67	64.33	$CoCl_2 \cdot 6H_2O$
3.38	35.20	61.42	»
5.98	35.11	58.91	$CoCl_2 \cdot 6H_2O + CoCl_2 \cdot RbCl \cdot 2H_2O$
6.20	34.92	58.88	»
6.83	34.58	58.59	$CoCl_2 \cdot RbCl \cdot 2H_2O$
7.79	34.29	57.92	$CoCl_2 \cdot RbCl \cdot 2H_2O + CoCl_2 \cdot 2RbCl \cdot 2H_2O$
7.56	34.43	58.01	»
9.34	31.74	58.92	$CoCl_2 \cdot 2RbCl \cdot 2H_2O$
12.31	28.89	58.80	»
30.52	16.09	53.39	»
38.83	11.37	49.80	$CoCl_2 \cdot 2RbCl \cdot 2H_2O + RbCl$
38.03	11.62	50.35	»
43.58	5.41	51.01	$RbCl$
48.57	0.00	51.43	»

№ 5251 COBALT (II) CHLORIDE–STRONTIUM CHLORIDE–WATER [1165]

$CoCl_2$—$SrCl_2$—H_2O

Comp. of saturated soln., wt %			t	Comp. of solid phase
Strontium Chloride	Cobalt Chloride	Water		
0.00	35.87	64.13	25	$CoCl_2 \cdot 6H_2O$
4.28	32.57	63.15	25	»
9.00	29.21	61.79	25	»
12.19	26.90	60.91	25	$CoCl_2 \cdot 6H_2O + SrCl_2 \cdot 6H_2O$
12.31	26.57	61.12	25	»
13.69	24.55	61.76	25	$SrCl_2 \cdot 6H_2O$
14.50	24.17	61.33	25	»
24.93	10.83	64.24	25	»
35.07	0.0	64.93	25	»
0.0	49.50	50.50	50	$CoCl_2 \cdot 2H_2O$
10.71	42.16	47.13	50	»
12.25	41.02	46.73	50	»
22.16	31.98	45.86	50	$CoCl_2 \cdot 2H_2O + SrCl_2 \cdot 2H_2O$
22.30	33.77	43.93	50	»
29.24	24.45	46.31	50	$SrCl_2 \cdot 2H_2O$
34.12	18.05	47.83	50	»
47.50	0.00	52.50	50	»

№ 5252 COBALT (II) CHLORIDE–CADMIUM CHLORIDE–WATER [1166]

$CoCl_2—CdCl_2—H_2O$

$t=25$

Comp. of saturated soln., wt %			Comp. of solid phase
Cobalt Chloride	Cadmium Chloride	Water	
0.00	54.62	45.38	$CdCl_2 \cdot 2\frac{1}{2}H_2O$
7.38	48.73	43.99	»
11.71	45.40	42.89	»
12.23	45.20	42.57	$CdCl_2 \cdot 2\frac{1}{2}H_2O + CdCl_2 \cdot H_2O$
13.74	44.25	42.01	$CdCl_2 \cdot H_2O + 4CdCl_2 \cdot CoCl_2 \cdot 10H_2O$
14.01	43.95	42.04	$4CdCl_2 \cdot CoCl_2 \cdot 10H_2O$
15.02	42.57	42.41	»
15.16	42.43	42.41	»
15.80	41.57	42.63	$4CdCl_2 \cdot CoCl_2 \cdot 10H_2O + 2CdCl_2 \cdot CoCl_2 \cdot 12H_2O$
19.26	35.95	44.79	$2CdCl_2 \cdot CoCl_2 \cdot 12H_2O$
22.23	32.10	45.67	»
24.33	30.02	45.65	»
25.00	29.61	45.39	»
27.75	28.22	44.03	$3CdCl_2 \cdot CoCl_2 \cdot 12H_2O + CoCl_2 \cdot 6H_2O$
28.66	23.71	47.63	$CoCl_2 \cdot 6H_2O$
31.00	17.99	51.01	»
32.45	11.20	56.35	»
35.06	5.39	60.55	»
35.87	0.00	64.13	»

№ 5253 COBALT (II) CHLORIDE–CESIUM CHLORIDE–WATER [1643]

$CoCl_2—CsCl—H_2O$

$t=25$

Comp. of saturated soln., wt %			Comp. of solid phase
Cobalt Chloride	Cesium Chloride	Water	
35.67	0.0	64.33	$CoCl_2 \cdot 6H_2O$
35.73	1.65	62.62	$CoCl_2 \cdot 6H_2O + CoCl_2 \cdot CsCl \cdot 2H_2O$
35.60	2.07	62.33	$CoCl_2 \cdot CsCl \cdot 2H_2O$
32.58	3.62	63.80	»
30.82	6.12	63.06	»
31.09	5.03	63.88	$CoCl_2 \cdot CsCl \cdot 2H_2O + CoCl_2 \cdot 2CsCl$
22.57	16.22	61.21	$CoCl_2 \cdot 2CsCl$
12.76	32.60	54.64	»
10.42	37.63	51.95	»
8.98	40.56	50.46	$CoCl_2 \cdot 2CsCl + CoCl_2 \cdot 3CsCl$
7.75	42.76	49.49	$CoCl_2 \cdot 3CsCl$
0.45	64.50	35.05	»
0.25	64.48	35.27	$CoCl_2 \cdot 3CsCl + CsCl$
0.0	65.61	34.39	$CsCl$

№ 5254 COBALT (II) CHLORIDE–BARIUM CHLORIDE–WATER [2161]

$CoCl_2—BaCl_2—H_2O$

$t=20$

Comp. of saturated soln., wt %			Comp. of solid phase
Cobalt Chloride	Barium Chloride	Water	
6.34	16.50	77.16	$BaCl_2 \cdot 2H_2O$
13.14	10.03	76.83	»
19.52	5.48	75.00	»
24.92	2.01	73.07	»
29.57	0.88	69.55	»
33.80	0.38	65.82	»
34.15	0.37	65.48	»
34.43	0.45	65.12	$BaCl_2 \cdot 2H_2O+CoCl_2 \cdot 6H_2O$

№ 5255 COBALT (II) CHLORIDE–MERCURY (II) CHLORIDE–WATER [1160]

$CoCl_2—HgCl_2—H_2O$

$t=25$

Comp. of saturated soln., wt %			Comp. of solid phase
Mercury Chloride	Cobalt Chloride	Water	
0.00	35.87	64.13	$CoCl_2 \cdot 6H_2O$
15.14	30.69	54.17	»
30.68	26.04	43.28	»
41.25	23.38	35.37	»
51.00	21.54	17.46	$CoCl_2 \cdot 6H_2O+CoCl_2 \cdot HgCl_2 \cdot 4H_2O$
51.76	20.96	17.28	$CoCl_2 \cdot HgCl_2 \cdot 4H_2O$
54.34	19.50	26.16	»
56.95	18.34	24.71	»
57.81	18.02	24.17	$CoCl_2 \cdot HgCl_2 \cdot 4H_2O+HgCl_2$
56.54	15.14	28.32	$HgCl_2$
54.92	13.95	31.13	»
37.29	8.59	54.02	»
30.25	6.70	63.05	»
21.56	4.33	74.11	»
6.90	0.00	93.10	»

№ 5256 COBALT (II) CHLORIDE–MERCURY (II) CHLORIDE–WATER [1159]

$CoCl_2—HgCl_2—H_2O$

$t=25$

Comp. of saturated soln., wt %			Comp. of solid phase
Cobalt Chloride	Mercury Chloride	Water	
35.87	0.00	64.13	$[Co(H_2O)_6]Cl_2$
30.69	15.14	54.17	»
26.04	30.68	43.28	»
23.38	41.25	35.37	»

Comp. of saturated soln., wt %			Comp. of solid phase
Cobalt Chloride	Mercury Chloride	Water	
21.54	51.00	27.46	$[Co(H_2O)_6]Cl_2 + [Co(H_2O)_4]HgCl_4$
20.96	51.76	27.28	$[Co(H_2O)_4]HgCl_4$
19.50	54.34	26.16	»
18.34	56.95	24.71	»
18.02	57.81	24.17	$[Co(H_2O)_4HgCl_4] + HgCl_2$
15.14	56.54	28.32	$HgCl_2$
13.95	54.92	31.13	»
12.95	51.43	35.62	»
8.59	37.39	54.02	»
6.70	30.26	63.04	»
4.33	21.56	74.11	»
0.00	6.90	93.10	»

№ 5257 COBALT (II) CHLORIDE–THORIUM CHLORIDE–WATER [1165]

$CoCl_2—ThCl_4—H_2O$

Comp. of saturated soln., wt %			t	Comp. of solid phase
Cobalt Chloride	Thorium Chloride	Water		
0.55	55.20	44.25	0	$CoCl_2 \cdot 6H_2O + ThCl_4 \cdot 8H_2O$
0.0	55.75	44.25	0	$ThCl_4 \cdot 8H_2O$
0.0	55.61	44.39	0	»
35.87	0.0	64.13	25	$CoCl_2 \cdot 6H_2O$
22.47	18.50	59.03	25	»
17.34	26.26	56.40	25	»
7.85	42.46	49.69	25	»
4.21	54.25	41.54	25	$CoCl_2 \cdot 6H_2O + ThCl_4 \cdot 8H_2O$
3.34	55.17	41.49	25	$ThCl_4 \cdot 8H_2O$
1.97	56.25	41.78	25	»
0.0	58.00	42.00	25	»

№ 5258 NICKEL (II) NITRATE–LANTHANUM NITRATE–WATER [904]

$Ni(NO_3)_2—La(NO_3)_3—H_2O$

$t=25$

Comp. of saturated soln., wt %			Comp. of solid phase
Lanthanum Nitrate	Nickel Nitrate	Water	
60.28	0.00	39.72	$La(NO_3)_3 \cdot 6H_2O$
55.26	3.86	40.88	$2La(NO_3)_3 \cdot 3Ni(NO_3)_2 \cdot 24H_2O$
52.34	5.90	41.76	»
42.53	12.12	45.35	»

Table № 5258 (continuation)

Comp. of saturated soln., wt %			Comp. of solid phase
Lanthanum Nitrate	Nickel Nitrate	Water	
35.60	16.50	47.90	$2La(NO_3)_3 \cdot 3Ni(NO_3)_2 \cdot 24H_2O$
29.20	20.60	50.20	»
24.23	23.45	52.32	»
21.00	27.38	51.62	»
16.13	32.00	51.87	»
10.37	36.98	52.65	»
6.12	41.50	52.38	»
2.35	43.54	54.11	»
0.00	48.08	51.92	Solid solutions
0.00	50.28	49.72	»
0.00	51.70	49.30	$Ni(NO_3)_2 \cdot 6H_2O$

№ 5259 NICKEL (II) SULFATE–COPPER (II) SULFATE–WATER [1627]

$NiSO_4—CuSO_4—H_2O$

Concn. in aqueous soln., mol %		*t*	Concn. in aqueous soln., mol %		*t*
Nickel Sulfate	Copper Sulfate		Nickel Sulfate	Copper Sulfate	
98.43	1.57	35	97.35	2.65	67
92.31	7.69	35	91.69	8.31	67
88.34	11.66	35	86.45	13.55	67
83.08	16.92	35	83.61	16.39	67
74.37	25.63	35	75.54	24.46	67
72.10	27.90	35			

№ 5260 NICKEL (II) SULFATE–COPPER (II) SULFATE–WATER [1540]

$NiSO_4—CuSO_4—H_2O$

Comp. of saturated soln., wt %			*t*	Comp. of solid phase
Copper Sulfate	Nickel Sulfate	Water		
8.5	20.3	71.2	15	Solid soln. enriched with Copper + solid soln. enriched with Nickel:
8.9	23.2	67.9	30	
10.3	26.4	63.3	50	
13.6	27.5	58.9	70	$(Cu, Ni)SO_4 \cdot 5H_2O +$ $+(Ni, Cu)SO_4 \cdot 7H_2O$

№ 5261 NICKEL (II) SULFATE–RUBIDIUM SULFATE–WATER [1210]

$NiSO_4—Rb_2SO_4—H_2O$

Comp. of saturated soln., wt %			t	Comp. of solid phase
Nickel Sulfate	Rubidium Sulfate	Water		
22.59	1.84	75.57	0	$NiSO_4 \cdot 7H_2O+NiSO_4 \cdot Rb_2SO_4 \cdot 6H_2O$
7.89	3.25	88.86	0	$NiSO_4 \cdot Rb_2SO_4 \cdot 6H_2O$
0.08	33.70	66.22	25	$Rb_2SO_4+NiSO_4 \cdot Rb_2SO_4 \cdot 6H_2O$
3.96	6.81	89.23	25	$NiSO_4 \cdot Rb_2SO_4 \cdot 6H_2O$
30.90	2.17	66.93	25	$NiSO_4 \cdot 7H_2O+NiSO_4 \cdot Rb_2SO_4 \cdot 6H_2O$
34.60	2.32	63.08	29	$NiSO_4 \cdot 7H_2O+NiSO_4 \cdot Rb_2SO_4 \cdot 6H_2O+$ $+NiSO_4 \cdot 6H_2O$
6.38	11.04	82.58	44.5	$NiSO_4 \cdot Rb_2SO_4 \cdot 6H_2O$
35.75	3.94	60.31	54	$NiSO_4 \cdot Rb_2SO_4 \cdot 6H_2O+NiSO_4 \cdot 6H_2O$
8.45	14.6	76.95	60	$NiSO_4 \cdot Rb_2SO_4 \cdot 6H_2O$
9.73	16.8	73.47	69	»
12.96	22.35	64.69	88	»
0.91	46.99	52.10	100	$NiSO_4 \cdot Rb_2SO_4 \cdot 6H_2O+Rb_2SO_4$
14.88	25.57	59.55	100	$NiSO_4 \cdot Rb_2SO_4 \cdot 6H_2O$
42.80	7.98	49.22	100	$NiSO_4 \cdot Rb_2SO_4 \cdot 6H_2O+NiSO_4 \cdot 6H_2O$

№ 5262 NICKEL (II) SULFATE–THALLIUM SULFATE–WATER [1210]

$NiSO_4—Tl_2SO_4—H_2O$

Comp. of saturated soln., wt %			t	Comp. of solid phase
Nickel Sulfate	Thallium Sulfate	Water		
0.16	2.69	97.15	0	$Tl_2SO_4+NiSO_4 \cdot Tl_2SO_4 \cdot 6H_2O$
0.51	1.68	97.81	0	$NiSO_4 \cdot Tl_2SO_4 \cdot 6H_2O$
21.30	0.63	78.07	0	$NiSO_4 \cdot Tl_2SO_4 \cdot 6H_2O+NiSO_4 \cdot 7H_2O$
0.33	5.15	94.52	25	$NiSO_4 \cdot Tl_2SO_4 \cdot 6H_2O+Tl_2SO_4$
1.01	3.20	95.79	25	$NiSO_4 \cdot Tl_2SO_4 \cdot 6H_2O$
27.92	0.89	71.19	25	$NiSO_4 \cdot Tl_2SO_4 \cdot 6H_2O+NiSO_4 \cdot 7H_2O$
33.85	2.28	63.87	30	$NiSO_4 \cdot Tl_2SO_4 \cdot 6H_2O+NiSO_4 \cdot 7H_2O+$ $+NiSO_4 \cdot 6H_2O$
0.85	8.29	90.86	46	$NiSO_4 \cdot Tl_2SO_4 \cdot 6H_2O+Tl_2SO_4$
1.77	5.77	92.46	46	$NiSO_4 \cdot Tl_2SO_4 \cdot 6H_2O$
32.62	1.55	65.83	46	$NiSO_4 \cdot Tl_2SO_4 \cdot 6H_2O+NiSO_4 \cdot 6H_2O$
2.25	7.30	90.45	55	$NiSO_4 \cdot Tl_2SO_4 \cdot 6H_2O$
33.85	2.28	63.87	55	$NiSO_4 \cdot Tl_2SO_4 \cdot 6H_2O+NiSO_4 \cdot 6H_2O$
3.15	14.65	82.20	80	$NiSO_4 \cdot Tl_2SO_4 \cdot 6H_2O+Tl_2SO_4$
4.15	13.42	82.45	80	$NiSO_4 \cdot Tl_2SO_4 \cdot 6H_2O$
37.89	4.32	57.69	90	$NiSO_4 \cdot Tl_2SO_4 \cdot 6H_2O+NiSO_4 \cdot 6H_2O$
7.34	20.08	72.58	100	$NiSO_4 \cdot Tl_2SO_4 \cdot 6H_2O+Tl_2SO_4$
41.45	7.03	51.52	100	$NiSO_4 \cdot Tl_2SO_4 \cdot 6H_2O+NiSO_4 \cdot 6H_2O$

№ 5263 NICKEL (II) CHLORIDE–CADMIUM CHLORIDE–WATER [1166]

$NiCl_2—CdCl_2—H_2O$

$t=25$

Concn. in aqueous soln., wt %		Comp. of solid phase
Cadmium Chloride	Nickel Chloride	
0.0	39.58	$NiCl_2 \cdot 6H_2O$
4.58	37.39	»
9.44	35.73	»
9.96	35.56	$NiCl_2 \cdot 6H_2O + CdCl_2 \cdot 2NiCl_2 \cdot 12H_2O$
2.47	43.31	$CdCl_2 \cdot 2NiCl_2 \cdot 12H_2O$
11.01	34.91	»
23.35	28.55	»
29.40	25.93	»
30.30	25.72	»
31.37	25.01	$CdCl_2 \cdot 2NiCl_2 \cdot 12H_2O + 2CdCl_2 \cdot NiCl_2 \cdot 12H_2O$
32.05	24.23	$2CdCl_2 \cdot NiCl_2 \cdot 12H_2O$
33.80	22.30	$2CdCl_2 \cdot NiCl_2 \cdot 12H_2O + 4CdCl_2 \cdot NiCl_2 \cdot 10H_2O$
38.16	18.09	$4CdCl_2 \cdot NiCl_2 \cdot 10H_2O$
39.64	16.83	»
43.45	13.99	»
44.51	13.26	»
45.20	12.70	$4CdCl_2 \cdot NiCl_2 \cdot 10H_2O + CdCl_2 \cdot H_2O$
45.34	11.84	$CdCl_2 \cdot H_2O + CdCl_2 \cdot 2.5H_2O$
46.98	7.61	$CdCl_2 \cdot 2.5H_2O$
54.62	0.0	»

№ 5264 COPPER (II) OXIDE–COPPER (II) SULFATE–WATER [235]

$CuO—CuSO_4—H_2O$

$t=18$

Concn. in aqueous soln.		d_4	pH	Comp. of solid phase
Copper Sulfate, wt %	Cu, g/l			
17.4	82.8	1.196	3.80	$3CuO \cdot CuSO_4 \cdot xH_2O + CuSO_4 \cdot 5H_2O$
8.8	39.1	1.119	4.07	$3CuO \cdot CuSO_4 \cdot H_2O$
5.87	24.8	1.060	4.17	»
4.29	17.4	1.038	4.22	»
1.55	6.2	1.014	4.31	»
0.68	2.7	1.007	4.41	»
0.40	1.6	1.004	4.45	»
0.19	0.72	1.002	4.96	»
0.14	0.55	1.001	5.04	»
0.10	0.38	1.001	5.24	»
0.025	0.10	1.000	5.80	»

№ 5265 [235]

COPPER (II) OXIDE–COPPER (II) SULFATE–ZINC OXIDE–ZINC SULFATE–WATER

$CuO—CuSO_4—ZnO—ZnSO_4—H_2O$

$t=18$

Concn. in aqueous soln., g/l		pH	Comp. of solid phase
Zinc	Copper		
185.3	0.38	5.30	$3ZnO \cdot ZnSO_4 \cdot xH_2O + 2CuO \cdot CuSO_4 \cdot xH_2O$
144.5	0.28	5.44	»
108.5	0.17	5.55	»
70.5	0.12	5.70	»
53.0	0.083	5.80	»
30.3	0.045	6.05	»
16.6	Traces	6.18	»
9.4	0.0	6.31	»

№ 5266 [2145]

COPPER (II) OXIDE–ARSENIC PENTOXIDE–WATER

$CuO—As_2O_5—H_2O$

$t=18$

Comp. of saturated soln., wt %			d_4	Comp. of solid phase
Arsenic Pentoxide	Copper Oxide	Water		
0.180	0.0045	99.8155	—	$4CuO \cdot As_2O_5 \cdot H_2O$
0.340	0.083	99.577	—	»
0.490	0.120	99.390	—	»
0.660	0.141	99.199	—	»
0.840	0.180	98.980	—	»
0.890	0.175	98.935	—	»
0.990	0.210	98.800	—	»
1.048	0.206	98.745	—	»
1.055	0.220	98.725	—	$4CuO \cdot As_2O_5 \cdot H_2O + 3CuO \cdot As_2O_5 \cdot 4H_2O$
1.070	0.227	98.703	—	$3CuO \cdot As_2O_5 \cdot 4H_2O$
1.120	0.230	98.650	—	»
0.110	0.032	99.858	—	$5CuO \cdot 2As_2O_5 \cdot 10H_2O$
0.240	0.051	99.709	—	»
0.360	0.084	99.556	—	»
0.440	0.102	99.458	—	»
0.750	0.172	99.078	—	$5CuO \cdot 2As_2O_5 \cdot 10H_2O + 2CuO \cdot As_2O_5 \cdot 3H_2O$
0.700	0.153	99.147	—	»
0.540	0.120	99.34	—	»

Table № 5266 (continuation)

Comp. of saturated soln., wt %			d_4	Comp. of solid phase
Arsenic Pentoxide	Copper Oxide	Water		
0.420	0.095	99.485	—	$2CuO \cdot As_2O_5 \cdot 3H_2O$
0.700	0.140	99.160	—	»
0.950	0.210	98.840	—	»
1.120	0.240	98.640	—	»
3.52	0.56	95.92	1.03	»
6.35	0.79	92.86	1.06	»
10.42	1.12	88.46	1.10	»
44.29	1.40	54.31	1.14	»
28.40	1.51	70.09	1.34	»
34.45	1.76	63.79	1.46	»
40.25	2.00	57.75	1.53	»
50.00	2.78	47.22	1.74	»
59.45	2.95	37.60	1.95	»
60.25	3.18	36.57	1.95	»
61.10	2.80	36.10	1.95	$2CuO \cdot As_2O_5 \cdot 3H_2O + CuO \cdot 2As_2O_5 \cdot 7H_2O$
61.87	2.00	36.13	1.95	$CuO \cdot 2As_2O_5 \cdot 7H_2O$
62.25	1.80	35.95	1.97	»
69.25	1.50	36.25	1.96	»
64.50	1.30	34.20	1.97	»
64.80	1.04	34.16	1.96	»
65.10	0.81	34.09	2.00	$CuO \cdot 2As_2O_5 \cdot 7H_2O + CuO \cdot 2As_2O_5 \cdot 5H_2O$
66.25	0.50	33.75	2.03	$CuO \cdot 2As_2O_5 \cdot 5H_2O$
67.50	0.35	32.15	2.06	»
68.00	0.40	31.60	2.04	»
68.50	0.30	31.20	2.05	»
69.00	0.25	30.75	2.09	»
71.45	0.00	28.55	2.15	$CuO \cdot 2As_2O_5 \cdot 5H_2O + As_2O_5$

№ 5267 COPPER (II) HYDROXIDE–COPPER (II) SELENATE–WATER [1138]

$$Cu(OH)_2—CuSeO_4—H_2O$$

$t=30$

Comp. of saturated soln., mol %			Comp. of solid phase
Copper Hydroxide	Copper Selenate	Water	
1.83	1.92	96.25	$4CuO \cdot SeO_3 \cdot H_2O$
1.03	1.09	97.88	»
0.55	0.56	98.89	»
0.40	0.41	99.19	»
0.32	0.33	99.35	»
0.31	0.31	99.38	»
0.19	0.20	99.61	»
0.14	0.14	99.72	»

№ 5268 COPPER (II) NITRATE–COPPER (II) SULFATE–WATER [2149]

$Cu(NO_3)_2$—$CuSO_4$—H_2O

Comp. of saturated soln., wt %			t	Comp. of solid phase
Copper Nitrate	Copper Sulfate	Water		
54.94	000	45.06	20	$Cu(NO_3)_2 \cdot 6H_2O$
55.28	1.20	43.52	20	»
55.08	1.08	43.84	20	$Cu(NO_3)_2 \cdot 6H_2O + CuSO_4 \cdot 5H_2O$
40.24	1.69	58.07	20	$CuSO_4 \cdot 5H_2O$
34.24	2.42	63.34	20	»
11.51	7.77	80.72	20	»
0.0	17.52	82.48	20	»
60.92	0.00	39.08	35	$Cu(NO_3)_2 \cdot 3H_2O$
60.53	1.27	38.20	35	$Cu(NO_3)_2 \cdot 3H_2O + CuSO_4 \cdot 5H_2O$
59.11	1.02	40.87	35	$CuSO_4 \cdot 5H_2O$
54.67	0.70	44.63	35	»
51.90	0.86	47.24	25	»
43.10	1.62	55.28	35	»
11.37	11.04	77.59	35	»
0.00	21.00	79.0	35	»

№ 5269 COPPER (II) NITRATE–LEAD (II) NITRATE–WATER [1594]

$Cu(NO_3)_2$—$Pb(NO_3)_2$—H_2O

$t=20$

Comp. of saturated soln., wt %				Comp. of solid phase
Lead Nitrate	Copper Nitrate	Water		
35.52	0.0	64.48	1.419	$Pb(NO_3)_2 + Cu(NO_3)_2 \cdot 6H_2O$
26.75	5.24	68.01	1.354	»
19.46	10.52	70.02	1.322	»
13.25	17.14	69.61	1.321	»
9.94	22.47	67.59	1.343	»
8.66	25.13	66.21	1.360	»
4.86	35.79	59.35	1.451	»
2.85	44.09	53.06	1.546	»
1.73	49.20	49.07	1.622	»
1.01	55.51	43.48	1.700	»

№ 5270 [601]

COPPER (II) THIOCYANATE–ZINC SULFATE–WATER

$Cu_2(SCN)_2$—$ZnSO_4$—H_2O

$t=20$

Concn. of Zinc Sulfate in aqueous soln., wt %	Solubility of Copper Thiocyanate in aqueous soln. of Zinc Sulfate, g/l
50	0.112

№ 5271 COPPER (II) SULFATE–COPPER (II) CHLORIDE–WATER [2508]

$CuSO_4—CuCl_2—H_2O$

$t=30$

Comp. of saturated soln., wt %			Comp. of solid phase
Copper Sulfate	Copper Chloride	Water	
20.32	0.00	79.68	$CuSO_4 \cdot 5H_2O$
13.62	6.58	79.80	»
8.93	15.68	75.39	»
4.77	25.67	69.56	»
3.21	39.48	57.31	»
2.90	42.62	54.48	$CuSO_4 \cdot 5H_2O + CuCl_2 \cdot 2H_2O$
1.14	43.25	55.61	$CuCl_2 \cdot 2H_2O$
0.00	43.95	56.05	»

№ 5272 COPPER (II) SULFATE–COPPER (II) ARSENATE–WATER [274]

$CuSO_4—Cu_3(AsO_4)_2—H_2O$

Comp. of saturated soln., wt %			t	Comp. of solid phase
Copper Arsenate	Copper Sulfate	Water		
0.01	0.00	99.99	25	$Cu_3(AsO_4) \cdot 4H_2O$
0.39	2.21	97.40	25	$Cu_3(AsO_4)_2 \cdot 4H_2O + 3Cu_3(AsO_4)_2 \cdot 0.33CuSO_4 \cdot 13H_2O$
0.15	16.55	83.30	25	$CuSO_4$
0.00	18.47	81.53	25	$CuSO_4 \cdot 5H_2O$
0.01	0.00	99.99	50	$Cu(AsO_4)_2 \cdot 4H_2O$
0.17	2.64	97.10	50	$Cu_3(AsO_4)_2 \cdot 4H_2O + 3Cu_3(AsO_4)_2 \cdot 0.33CuSO_4 \cdot 13H_2O$
0.64	20.81	78.55	50	$CuSO_4 \cdot 5H_2O$
0.00	25.20	74.80	50	»

№ 5273 COPPER (II) SULFATE–COPPER (II) ARSENATE–COPPER [282]

$CuSO_4—Cu_3(AsO_4)_2—H_2O$

Comp. of saturated soln., wt %			t	Comp. of solid phase
Copper Arsenate	Copper Sulfate	Water		
0.001	0.000	99.999	25	$Cu_3(AsO_4)_2 \cdot 4H_2O$
0.19	2.23	97.57	25	»
0.40	4.23	95.37	25	$Cu_3(AsO_4)_2 \cdot 4H_2O + 3Cu_3(AsO_4)_2 \cdot 0.33CuSO_4 \cdot 13H_2O$
0.38	7.61	92.01	25	$3Cu_3(AsO_4)_2 \cdot 0.33CuSO_4 \cdot 13H_2O$
0.36	9.30	90.34	25	»
0.34	9.55	90.11	25	»
0.34	12.29	87.37	25	»
0.35	16.55	83.10	25	$3Cu_3(AsO_4)_2 \cdot 0.33CuSO_4 \cdot 13H_2O + CuSO_4 \cdot 5H_2O$
0.29	17.80	81.91	25	»
0.27	18.71	81.02	25	»

Comp. of saturated soln., wt %			t	Comp. of solid phase
Copper Arsenate	Copper Sulfate	Water		
0.20	18.30	81.50	25	$CuSO_4 \cdot 5H_2O$
0.27	19.04	80.69	25	»
0.00	18.49	81.51	25	»
0.001	0.000	99.999	50	$Cu_3(AsO_4)_2 \cdot 4H_2O$
0.29	1.29	98.43	50	»
0.57	2.64	96.79	50	$3Cu_3(AsO_4)_2 \cdot 0.33CuSO_4 \cdot 13H_2O + Cu_3(AsO_4)_2 \cdot 4H_2O$
0.51	5.25	94.24	50	$3Cu_3(AsO_4)_2 \cdot 0.33CuSO_4 \cdot 13H_2O$
0.52	7.71	91.77	50	»
0.57	12.13	87.30	50	»
0.64	20.81	78.55	50	$3Cu_3(AsO_4)_2 \cdot 0.33CuSO_4 \cdot 13H_2O + CuSO_4 \cdot 5H_2O$
0.60	21.27	78.13	50	»
0.58	22.64	76.78	50	»
0.39	22.58	77.03	50	$CuSO_4 \cdot 5H_2O$
0.37	22.40	77.23	50	»
0.15	24.42	75.43	50	»
0.00	25.20	74.80	50	»

№ 5274 COPPER (II) SULFATE–ZINC SULFATE–WATER [2607]

$CuSO_4—ZnSO_4—H_2O$

$t=18$

Comp. of solid soln., mol %			
Copper Sulfate	Zinc Sulfate	Copper Sulfate	Zinc Sulfate
2.23	0	0.48	5.27
1.76	2.00	0.28	5.26
1.34	3.43	0.0	6.05
1.12	4.73	1.12	4.72
1.77	3.20	0.48	5.26
1.16	4.22	0.25	5.46
0.96	4.47	0.00	5.61
0.78	4.76		

№ 5275 [1812]

COPPER (II) SULFATE–RUBIDIUM CHLORIDE–WATER

$CuSO_4—RbCl—H_2O$

$t=25$

Concn. in aqueous soln., g/l	
Copper Sulfate	Rubidium Chloride
223.4	0.0
250.2	132.0

№ 5276 COPPER (II) SULFATE–SILVER SULFATE–WATER [1068]

$CuSO_4—Ag_2SO_4—H_2O$

$t=25$

Concn. of Copper Sulfate in aqueous soln., g/l	Solubility of Silver Sulfate in aqueous soln. of Copper Sulfate, wt %	Concn. of Copper Sulfate in aqueous soln., g/l	Solubility of Silver Sulfate in aqueous soln. of Copper Sulfate, wt %
16.93	0.807	108.4	1.033
34.47	0.845	131.2	1.086
52.06	0.890	150.4	1.127
71.26	0.941	175.1	1.175
89.92	0.988	197.1	1.216

№ 5277 COPPER (II) SULFATE–CERIUM SULFATE–WATER [275]

$CuSO_4—Ce_2(SO_4)_3—H_2O$

$t=25$

Comp. of saturated soln., wt %			Comp. of solid phase
Cerium Sulfate	Copper Sulfate	Water	
3.22	0.24	96.54	—
4.29	0.41	95.30	$mCuSO_4 \cdot nCe_2(SO_4)_3 \cdot xH_2O$
6.53	0.33	93.14	»
8.20	0.26	91.54	»
8.83	0.40	90.77	»
8.07	0.14	91.79	»
8.55	0.34	91.11	»
9.21	0.21	90.58	»
10.76	0.26	88.98	»
14.06	0.18	85.76	»
14.05	0.52	85.43	»
15.19	0.28	84.53	»

№ 5278 COPPER (II) SULFATE–THALLIUM SULFATE–WATER [2407]

$CuSO_4—Tl_2SO_4—H_2O$

$t=25$

Concn. in aqueous soln., wt %		*d*	Comp. of solid phase
Thallium Sulfate	Copper Sulfate		
5.22	0.00	1.046	Tl_2SO_4
5.65	1.86	1.070	$Tl_2SO_4+Tl_2SO_4 \cdot CuSO_4 \cdot 6H_2O$
2.81	18.70	1.237	$Tl_2SO_4 \cdot CuSO_4 \cdot 6H_2O+CuSO_4 \cdot 5H_2O$
0.00	18.50	1.206	$CuSO_4 \cdot 5H_2O$

№ 5279 COPPER (II) SULFATE–COPPER (II) CHLORIDE–WATER [417]

$CuSO_4—CuCl_2—H_2O$

$t=25$

Comp. of saturated soln., wt %			Comp. of solid phase
Copper Sulfate	Copper Chloride	Water	
18.70	0.00	81.30	$CuSO_4 \cdot 5H_2O$
14.80	4.40	80.80	»
12.83	6.75	80.42	»
12.25	9.05	78.70	»
10.80	11.57	77.63	»
9.80	12.83	77.37	»
7.52	17.99	74.49	»
7.08	20.15	72.77	»
7.05	20.18	72.77	$CuSO_4 \cdot 4H_2O$
7.35	20.74	71.91	»
6.67	22.52	70.81	»
4.28	27.77	67.95	»
2.99	32.47	64.54	»
3.16	34.13	62.71	»
2.98	35.39	61.63	»
2.72	36.04	61.24	$CuSO_4 \cdot 3H_2O$
3.46	38.91	57.63	»
2.18	42.33	55.49	»
2.84	41.68	55.48	$CuSO_4 \cdot 3H_2O + CuCl_2 \cdot 2H_2O$
1.92	41.76	56.32	»
1.29	42.26	56.45	»
2.05	41.78	56.17	»
2.39	42.07	55.54	$CuSO_4 \cdot 3H_2O + CuCl_2 \cdot 2H_2O$
2.31	42.56	55.13	»
2.13	41.95	55.92	»
1.46	41.95	56.59	»
1.72	42.02	56.26	»
1.54	41.53	56.93	»
2.27	41.77	55.96	»
2.43	41.87	55.70	»
2.27	41.77	55.96	$CuCl_2 \cdot 2H_2O$
1.26	42.84	55.90	»
0.65	43.02	56.33	»
0.00	43.32	56.68	»

№ 5280 COPPER (II) SULFATE–COPPER (II) CHLORIDE–WATER [279]

$CuSO_4—CuCl_2—H_2O$

$t=25$

Comp. of saturated soln., wt %			Comp. of solid phase
Copper Chloride	Copper Sulfate	Water	
0.00	18.70	81.30	$CuSO_4 \cdot 5H_2O$
4.40	14.80	80.80	»
6.75	12.83	80.42	»
9.05	12.25	78.70	»

Table № 5280 (continuation)

Comp. of saturated soln., wt %			Comp. of solid phase
Copper Chloride	Copper Sulfate	Water	
11.57	10.80	77.63	$CuSO_4 \cdot 5H_2O$
12.83	9.80	77.37	»
17.99	7.52	74.49	»
20.15	7.08	72.77	»
20.18	7.05	72.77	$CuSO_4 \cdot 4H_2O$
20.74	7.35	71.91	»
22.52	6.67	70.81	»
27.77	4.28	67.95	»
32.47	2.99	64.54	»
34.13	3.16	62.71	»
35.39	2.98	61.63	»
36.04	2.72	61.24	$CuSO_4 \cdot 3H_2O$
38.91	3.46	57.63	»
42.33	2.18	55.49	»
41.68	2.84	55.48	$CuSO_4 \cdot 3H_2O + CuCl_2 \cdot 2H_2O$
41.76	1.92	56.32	»
42.26	1.29	56.45	»
41.78	2.05	56.17	»
42.07	2.39	55.54	»
42.56	2.31	55.13	»
41.95	2.13	55.92	»
41.95	1.46	56.59	»
42.02	1.72	56.26	»
41.53	1.54	56.93	»
41.77	2.27	55.96	»
41.87	2.43	55.70	»
41.77	2.27	55.96	$CuCl_2 \cdot 2H_2O$
42.84	1.26	55.90	»
43.02	0.65	56.33	»
43.32	0.00	56.68	»

№ 5281 COPPER (I) CHLORIDE–COPPER (II) CHLORIDE–WATER [1597]

$CuCl—CuCl_2—H_2O$

$t=19$

Concn. in aqueous soln., g/l		Comp. of solid phase
Copper (II) Chloride	Copper (I) Chloride	
134.5	15.8	CuCl
260.0	40.7	»
435.7	63.8	»
536.5	78.7	»
661.9	87.4	$CuCl + CuCl_2 \cdot 2H_2O$

№ 5282 [1780]

COPPER (II) CHLORIDE–STRONTIUM CHLORIDE–WATER

$CuCl_2—SrCl_2—H_2O$

Comp. of solvent, wt %		Solubility of Strontium Chloride, wt %	d_4
Copper Chloride	Water		
4.58	95.42	34.005	1.4200
13.27	86.73	30.40	1.4595

№ 5283 [1170]

COPPER (II) CHLORIDE–CADMIUM CHLORIDE–WATER

$CuCl_2—CdCl_2—H_2O$

$t=25$

Comp. of saturated soln., wt %			Comp. of solid phase
Cadmium Chloride	Copper Chloride	Water	
0.0	43.60	56.40	$CuCl_2 \cdot 2H_2O$
8.20	39.40	52.40	»
13.70	36.80	49.50	»
21.40	32.40	46.20	»
22.40	32.00	45.60	$CuCl_2 \cdot 2H_2O + CdCl_2 \cdot CuCl_2 \cdot 4H_2O$
23.70	31.40	44.90	$CdCl_2 \cdot CuCl_2 \cdot 4H_2O$
24.40	30.70	44.90	»
28.10	27.60	44.30	»
30.50	26.20	43.30	»
30.70	26.40	42.90	»
31.80	25.10	43.10	»
36.40	22.60	41.0	»
38.80	21.10	40.10	$CdCl_2 \cdot CuCl_2 \cdot 4H_2O + CdCl_2 \cdot H_2O$
39.70	18.80	41.50	$CdCl_2 \cdot 2.5H_2O + CdCl_2 \cdot H_2O$
47.40	7.10	45.50	$CdCl_2 \cdot 2.5H_2O$
54.62	0.00	45.38	»

№ 5284 [2500]

COPPER (II) CHLORIDE–BARIUM CHLORIDE–WATER

$CuCl_2—BaCl_2—H_2O$

$t=30$

Comp. of saturated soln., wt %			Comp. of solid phase
Barium Chloride	Copper Chloride	Water	
0.00	43.95	56.05	$CuCl_2 \cdot 2H_2O$
1.25	42.45	56.30	»
2.72	42.36	54.92	$CuCl_2 \cdot 2H_2O + BaCl_2 \cdot 2H_2O$
2.84	41.18	55.98	$BaCl_2 \cdot H_2O$
3.98	37.42	58.60	»
5.49	30.76	63.75	»
10.13	21.76	68.11	»
17.08	11.49	71.43	»
22.76	5.13	78.11	»
27.60	0.00	72.40	»

№ 5285 COPPER (II) CHLORIDE–CERIUM CHLORIDE–WATER [959]

$CuCl_2—CeCl_3—H_2O$

Comp. of saturated soln., wt %			t	Comp. of solid phase
Cerium Chloride	Copper Chloride	Water		
0.0	43.5	56.5	20	$CuCl_2 \cdot 2H_2O$
6.0	36.4	57.6	20	»
8.6	33.9	57.5	20	»
16.4	27.0	56.6	20	»
22.2	22.7	55.1	20	»
26.1	20.1	53.8	20	»
37.4	13.4	49.2	20	»
42.5	11.5	46.0	20	$CeCl_2 \cdot 2H_2O + CeCl_3 \cdot 7H_2O$
43.7	8.5	47.8	20	$CeCl_3 \cdot 7H_2O$
46.0	4.6	49.4	20	»
48.5	0.0	51.5	50	$CuCl_2 \cdot 2H_2O$
0.0	46.3	53.7	50	»
8.0	36.7	55.3	50	»
14.6	31.2	64.2	50	»
15.5	30.5	64.0	50	»
25.0	22.5	52.5	50	»
36.0	15.8	48.2	50	»
41.5	13.4	45.1	50	»
45.0	12.5	42.5	50	$CuCl_2 \cdot 2H_2O + CeCl_3 \cdot 7H_2O$
45.6	10.9	43.5	50	$CeCl_3 \cdot 7H_2O$
47.0	7.2	45.8	50	»
49.3	3.8	46.9	50	»
51.7	0.0	48.3	50	»

№ 5286 COPPER (II) CHLORIDE–MERCURY (II) CHLORIDE–WATER [2514]

$CuCl_2—HgCl_2—H_2O$

$t=35$

Comp. of saturated soln., wt %			Comp. of solid phase
Mercury Chloride	Copper Chloride	Water	
0.00	44.47	55.53	$CuCl_2 \cdot 2H_2O$
21.03	33.50	45.47	»
37.30	26.07	36.63	»
44.47	23.31	32.22	»
50.47	21.50	28.03	$CuCl_2 \cdot 2H_2O + HgCl_2$
52.44	19.40	28.16	$HgCl_2$
52.54	18.46	29.00	»
52.81	18.06	29.13	»
51.03	14.73	34.24	»
49.50	5.94	44.56	»
23.87	2.64	73.49	»
8.51	0.00	91.49	»

№ 5287 COPPER (II) CHLORIDE–MERCURY (II) CHLORIDE–WATER [1162]

$CuCl_2—HgCl_2—H_2O$

Comp. of saturated soln., wt %			t	Comp. of solid phase
Mercury Chloride	Copper Chloride	Water		
0.00	43.32	56.68	25	$CuCl_2 \cdot 2H_2O$
48.75	22.53	28.72	25	»
51.68	21.46	26.86	25	$CuCl_2 \cdot 2H_2O + HgCl_2$
51.50	21.35	27.15	25	$HgCl_2$
51.57	20.95	27.48	25	»
51.83	20.64	27.53	25	»
52.89	18.86	28.25	25	»
51.47	14.86	33.67	25	»
6.90	0.0	93.10	25	»
0.00	44.47	55.53	35	$CuCl_2 \cdot 2H_2O$
49.71	22.50	27.79	35	»
52.13	21.69	26.18	35	$CuCl_2 \cdot 2H_2O + HgCl_2$
52.35	21.09	26.56	35	$HgCl_2$
52.60	20.27	27.13	35	»
53.28	18.70	28.02	35	»
52.36	14.51	33.13	35	»
8.51	0.00	91.49	35	»

№ 5288 COPPER (II) CHLORIDE–THALLIUM CHLORIDE–WATER [2264]

$CuCl_2—TlCl—H_2O$

Concn. of Copper Chloride in aqueous soln., g/l	Solubility of Thallium Chloride in aqueous soln. of Copper Chloride, g/l	Concn. of Copper Chloride in aqueous soln., g/l	Solubility of Thallium Chloride in aqueous soln. of Copper Chloride, g/l
1.682	2.171	6.727	1.012
3.363	1.473	13.454	0.698

№ 5289 COPPER (II) CHLORIDE–THORIUM CHLORIDE–WATER [125]

$CuCl_2—ThCl_4—H_2O$

$t=25$

Comp. of saturated soln., wt %			Comp. of solid phase
Copper Chloride	Thorium Chloride	Water	
44.62	0.00	55.38	$CuCl_2 \cdot 2H_2O$
35.20	8.12	56.68	»
29.50	14.20	56.30	»
25.42	19.14	55.44	»
20.44	25.60	53.96	»
15.50	31.82	52.68	»
9.56	40.63	49.81	»
7.60	44.20	48.20	$CuCl_2 \cdot 2H_2O + ThCl_4 \cdot 8H_2O$
4.45	47.50	48.05	$ThCl_4 \cdot 8H_2O$
2.20	51.48	46.32	»
0.00	55.42	44.58	»

№ 5290 ZINC OXIDE–ZINC SULFATE–WATER [1416]

$ZnO—ZnSO_4—H_2O$

$t=25$

Concn. in aqueous soln., wt %		Comp. of solid phase
Zinc Oxide	Zinc Sulfate	
1.8	1.9	$3Zn(OH)_2 \cdot ZnSO_4 \cdot 4H_2O$
4.4	4.4	»
8.1	8.0	»
10.7	10.7	»
13.6	13.6	»
14.5	14.5	»
17.6	17.6	»
18.3	18.3	$3Zn(OH)_2 \cdot 3ZnSO_4 \cdot 4H_2O + ZnSO_4 \cdot 7H_2O$

№ 5291 ZINC OXIDE–ZINC CHLORIDE–WATER [1529]

$ZnO—ZnCl_2—H_2O$

$t=18—20$

Comp. of saturated soln., wt %			Comp. of solid phase
Zinc Oxide	Zinc Chloride	Water	
0.01	7.59	92.30	$ZnCl_2 \cdot 4ZnO \cdot 6H_2O$
0.11	18.83	81.06	»
0.34	31.37	68.29	»
0.40	33.85	65.75	»
0.46	36.09	63.45	»
0.54	38.38	61.08	»
0.91	48.53	50.56	»
1.41	54.71	43.88	»
1.07	58.51	40.42	»
0.52	66.64	32.84	$ZnCl_2 \cdot ZnO \cdot 1\frac{1}{2}H_2O$

№ 5292 ZINC OXIDE–BARIUM OXIDE–WATER [2216]

$ZnO—BaO—H_2O$

Comp. of saturated soln., wt %			*t*	Comp. of solid phase
Zinc Oxide	Barium Oxide	Water		
0.0	26.1	73.9	70	$Ba(OH)_2 \cdot 8H_2O$
0.3	26.1	73.6	70	»
0.5	27.1	72.4	70	»
0.9	27.6	71.5	70	»
1.2	28.6	70.2	70	»
1.6	28.6	69.8	70	»
1.9	29.5	68.6	70	»
4.6	32.9	62.5	70	»
5.2	34.2	60.6	70	»
7.4	37.1	55.5	70	»
7.5	35.8	56.7	70	»
8.2	37.3	56.5	70	»

Table № 5292 (continuation)

Comp. of saturated soln., wt %			t	Comp. of solid phase
Zinc Oxide	Barium Oxide	Water		
10.9	41.1	48.0	70	$Ba(OH)_2 \cdot 8H_2O$
11.0	41.8	47.2	70	»
13.5	51.8	34.7	70	$2BaO \cdot ZnO \cdot 7H_2O$
13.6	51.6	34.8	70	»
14.1	52.4	39.5	70	»
13.9	52.3	33.8	70	»
13.6	52.1	34.3	70	»
15.2	46.7	38.1	70	$BaO \cdot ZnO \cdot 2H_2O$
14.2	40.1	45.7	70	$BaO \cdot ZnO \cdot 3H_2O$
14.4	40.7	44.9	70	»
10.4	31.3	58.3	70	»
9.3	29.8	60.9	70	»
6.4	24.2	69.4	70	»
4.7	20.6	74.7	70	»
3.3	17.0	79.7	70	»
1.8	13.2	85.0	70	»
0.2	3.3	96.5	70	ZnO
1.1	8.4	90.5	70	»
0	53.8	46.2	90	$Ba(OH)_2 \cdot H_2O$
2.7	52.6	44.7	90	»
3.0	48.0	49.0	90	$2BaO \cdot ZnO \cdot 3H_2O$
3.2	46.1	50.7	90	»
3.7	53.1	43.2	90	»
4.0	46.4	49.6	90	»
4.4	46.2	49.4	90	»
7.8	46.4	45.8	90	$2BaO \cdot ZnO \cdot 7H_2O$
11.5	45.9	42.6	90	»
14.6	51.6	33.8	90	»
16.1	49.8	34.1	90	$BaO \cdot ZnO \cdot 2H_2O$
16.4	48.7	34.9	90	»
17.0	46.7	36.3	90	»
17.3	44.9	37.8	90	»
17.7	44.6	37.7	90	»
17.9	43.1	39.0	90	»
18.2	44.2	37.6	90	»
17.5	41.0	41.5	90	$BaO \cdot ZnO \cdot 3H_2O$
15.3	38.6	46.1	90	»
15.0	39.0	46.0	90	»
13.3	36.0	50.7	90	»
10.1	31.0	58.9	90	»
6.4	25.5	68.0	90	»
5.5	22.5	72.0	90	»
5.3	20.4	74.3	90	»
4.4	20.2	75.4	90	»
3.1	16.9	80.0	90	»
3.0	15.2	81.8	90	»
2.2	14.4	83.4	90	»
2.1	11.5	86.4	90	ZnO
1.3	9.8	88.9	90	»
0.9	6.1	93.0	90	»
0.7	7.0	92.3	90	»
0.3	3.6	96.1	90	»
0.1	1.9	98.0	90	»

№ 5293 [2596]

ZINC HYDROXIDE–BARIUM NITRATE–WATER

$Zn(OH)_2—Ba(NO_3)_2—H_2O$

$t = 16—20$

Comp. of solvent, wt %		Solubility of Zinc Hydroxide, g/l
Barium Nitrate	Water	
1	99	0.025

№ 5294 [2596]

ZINC HYDROXIDE–BARIUM CHLORIDE–WATER

$Zn(OH)_2—BaCl_2—H_2O$

$t = 16—20$

Comp. of solvent, wt %		Solubility of Zinc Hydroxide, g/l
Barium Chloride	Water	
1	99	0.038

№ 5295 [407]

ZINC NITRATE–ZINC SULFATE–THALLIUM CHROMATE–WATER

$Zn(NO_3)_2—ZnSO_4—Tl_2CrO_4—H_2O$

$t = 18—20$

Content in aqueous soln., g-eq/l		Solubility of Thallium Chromate, g/l
Zinc Sulfate	Zinc Nitrate	
0.500	0.000	$15.9 \cdot 10^2$
0.375	0.125	$13.8 \cdot 10^2$
0.250	0.250	$12.8 \cdot 10^2$
0.125	0.375	$11.5 \cdot 10^2$
0.000	0.500	$9.5 \cdot 10^2$

№ 5296 [479, 481]

ZINC NITRATE–SILVER BROMATE–WATER

$Zn(NO_3)_2—AgBrO_3—H_2O$

$t = 25$

Concn. of Zinc Nitrate in aqueous soln. mol/l	Solubility of Silver Bromate in aqueous soln. of Zinc Nitrate, mol/l
0.000	$8.25 \cdot 10^{-3}$
0.001	$8.43 \cdot 10^{-3}$
0.003	$8.75 \cdot 10^{-3}$
0.010	$9.12 \cdot 10^{-3}$

№ 5297 [479, 481]

ZINC NITRATE–SILVER BROMATE–WATER

$Zn(NO_3)_2—AgBrO_3—H_2O$

$t = 25$

Concn. of Zinc Nitrate in aqueous soln. mol/l	Solubility of Silver Bromate in aqueous soln. of Zinc Nitrate, mol/l
0.030	$1.02 \cdot 10^{-2}$
0.100	$1.16 \cdot 10^{-2}$
0.300	$1.47 \cdot 10^{-2}$
1.000	$1.93 \cdot 10^{-2}$

№ 5298 ZINC NITRATE–CADMIUM NITRATE–WATER [2359]

$Zn(NO_3)_2—Cd(NO_3)_2—H_2O$

An eutectic mixture of $Cd(NO_3)_2 \cdot 4H_2O$ and $Zn(NO_3)_2 \cdot 6H_2O$, containing 27.5 mol % $Zn(NO_3)_2 \cdot 6H_2O$, forms at t = 28.

№ 5299 ZINC NITRATE–LANTHANUM NITRATE–WATER [904]

$Zn(NO_3)_2—La(NO_3)_3—H_2O$

$t=25$

Comp. of saturated soln., wt %			Comp. of solid phase
Lanthanum Nitrate	Zinc nitrate	Water	
60.27	0.00	39.73	$La(NO_3)_3 \cdot 6H_2O$
56.40	3.12	40.48	Solid solutions
54.13	5.87	41.00	»
53.12	6.35	40.53	$2La(NO_3)_3 \cdot 3Zn(NO_3)_2 \cdot 24H_2O$
44.30	10.92	44.78	»
41.53	13.90	44.57	»
36.42	16.62	46.96	»
32.97	20.38	46.65	»
27.50	24.05	48.45	»
20.51	30.22	49.27	»
11.42	36.81	51.87	»
6.01	43.72	50.27	»
3.25	46.48	50.27	»
1.64	48.18	50.18	»
1.05	51.13	47.82	Solid solutions
0.55	53.10	46.35	»
0.00	55.98	44.02	$Zn(NO_3)_2 \cdot 6H_2O$

№ 5300 ZINC NITRATE–URANYL NITRATE–WATER [1013]

$Zn(NO_3)_2—UO_2(NO_3)_2—H_2O$

Comp. of saturated soln., wt %			t	Comp. of solid phase
Uranyl Nitrate	Zinc Nitrate	Water		
0.0	48.3	51.7	0	$Zn(NO_3)_2 \cdot 6H_2O$
1.6	48.0	50.4	0	»
2.9	48.0	49.1	0	$Zn(NO_3)_2 \cdot 6H_2O + UO_2(NO_3)_2 \cdot 6H_2O$
3.3	46.3	50.4	0	$UO_2(NO_3)_2 \cdot 6H_2O$
3.8	45.1	51.1	0	»
3.6	44.5	51.9	0	»
6.5	40.3	53.2	0	»
12.9	32.8	54 3	0	»
16.5	28.7	54.8	0	»
17.8	26.2	56.0	0	»
21.9	23.1	55.0	0	»
24.0	21.8	54.2	0	»
28.1	17.0	54.9	0	»
32.3	15.0	52.7	0	»
37.3	9.3	53.4	0	»
38.3	8.2	53.5	0	»
41.6	6.8	51.6	0	»
49.5	0.0	50.5	0	»
0.0	55.9	44.1	25	$Zn(NO_3)_2 \cdot 6H_2O$
2.2	54.2	43.6	25	»
3.9	53.3	42.8	25	»
6.3	51.9	41.8	25	»
8.1	51.1	40.8	25	»

Table № 5300 (continuation)

Comp. of saturated soln., wt %			t	Comp. of solid phase
Uranyl Nitrate	Zinc Nitrate	Water		
8.2	50.3	41.5	25	$Zn(NO_3)_2 \cdot 6H_2O$
10.0	50.0	40.0	25	»
12.0	48.3	39.7	25	$Zn(NO_3)_2 \cdot 6H_2O + UO_2(NO_3)_2 \cdot 6H_2O$
12.2	47.0	40.8	25	$UO_2(NO_3)_2 \cdot 6H_2O$
12.1	45.5	42.4	25	»
11.1	44.8	44.1	25	»
14.6	37.6	47.8	25	»
14.8	37.1	48.1	25	»
19.3	32.0	48.7	25	»
24.0	28.2	47.8	25	»
25.8	25.8	48.4	25	»
34.4	18.3	47.3	25	$UO_3(NO_3)_2 \cdot 6H_2O$
40.6	12.8	46.6	25	»
45.8	8.0	46.2	25	»
48.9	6.8	44.2	25	»
55.9	0.0	44.1	25	»

№ 5301 ZINC SULFATE– SILVER SULFATE–WATER [1068]

$ZnSO_4$—Ag_2SO_4—H_2O

$t=25$

Concn. of Zinc Sulfate in aqueous soln., g/l	Solubility of Silver Sulfate in aqueous soln. of Zinc Sulfate, wt %	Concn. of Zinc Sulfate in aqueous soln., g/l	Solubility of Silver Sulfate in aqueous soln. of Zinc Sulfate, wt %
19.78	0.800	133.8	1.060
40.81	0.847	155.0	1.100
62.24	0.900	186.0	1.156
84.86	0.953	213.4	1.198
109.5	1.008	243.3	1.238

№ 5302 ZINC SULFATE–SILVER BROMATE–WATER [483]

$ZnSO_4$—$AgBrO_3$—H_2O

$t=25$

Concn. of Zinc Sulfate in aqueous soln., g/l	Solubility of Silver Bromate in aqueous soln. of Zinc Sulfate, g/l	Concn. of Zinc Sulfate in aqueous soln., g/l	Solubility of Silver Bromate in aqueous soln. of Zinc Sulfate, g/l
0.161	1.940	16.14	2.67
0.484	2.01	48.42	3.31
1.61	2.10	161.4	4.13
4.84	2.33		

№ 5303 ZINC SULFATE–THALLIUM SULFATE–WATER [1207]

$ZnSO_4—Tl_2SO_4—H_2O$

Comp. of saturated soln., wt %			t	Comp. of solid phase
Thallium Sulfate	Zinc Sulfate	Water		
0.81	26.81	72.38	0	$ZnSO_4 \cdot 7H_2O + Tl_2SO_4 \cdot ZnSO_4 \cdot 6H_2O$
2.74	0.97	96.29	0	$Tl_2SO_4 + Tl_2SO_4 \cdot ZnSO_4 \cdot 6H_2O$
1.37	34.69	63.94	17.6	$ZnSO_4 \cdot 7H_2O + Tl_2SO_4 \cdot ZnSO_4 \cdot 6H_2O$
1.73	26.36	71.91	17.6	$Tl_2SO_4 \cdot ZnSO_4 \cdot 6H_2O$
4.68	2.04	93.28	17.6	$Tl_2SO_4 \cdot ZnSO_4 \cdot 6H_2O + Tl_2SO_4$
2.32	39.58	58.10	40.5	$ZnSO_4 \cdot 6H_2O + Tl_2SO_4 \cdot ZnSO_4 \cdot 6H_2O$
2.64	37.16	60.20	40.5	$Tl_2SO_4 \cdot ZnSO_4 \cdot 6H_2O$
3.2	33.6	63.2	40.5	»
3.91	28.20	67.89	40.5	»
4.39	22.66	72.95	40.5	»
6.12	9.97	83.91	40.5	»
6.82	7.30	82.82	40.5	»
8.34	4.48	87.18	40.5	$Tl_2SO_4 \cdot ZnSO_4 \cdot 6H_2O + Tl_2SO_4$
2.93	41.07	56.00	64.5	$ZnSO_4 \cdot H_2O$
5.66	40.04	54.30	64.5	$ZnSO_4 \cdot H_2O + Tl_2SO_4 \cdot ZnSO_4 \cdot 6H_2O$
6.35	38.30	55.35	64.5	$Tl_2SO_4 \cdot ZnSO_4 \cdot 6H_2O$
8.57	34.45	56.98	64.5	»
10.85	21.39	67.76	64.5	»
14.71	10.43	74.86	64.5	$Tl_2SO_4 \cdot ZnSO_4 \cdot 6H_2O + Tl_2SO_4$
13.95	8.77	77.28	64.5	Tl_2SO_4
3.55	37.10	59.35	100	$ZnSO_4 \cdot H_2O$
11.97	34.58	53.45	100	$ZnSO_4 \cdot H_2O + Tl_2SO_4 \cdot ZnSO_4 \cdot 6H_2O$
24.5	31.3	44.2	100	$Tl_2SO_4 + Tl_2SO_4 \cdot ZnSO_4 \cdot 6H_2O$
23.7	12.11	64.19	100	Tl_2SO_4

№ 5304 ZINC SULFATE–THALLIUM CHLORIDE–WATER [1306]

$ZnSO_4—TlCl—H_2O$

Concn. of Zinc Sulfate in aqueous soln., g/l	Solubility of Thallium Chloride in aqueous soln. of Zinc Sulfate, g/l	d	t	Concn. of Zinc Sulfate in aqueous soln., g/l	Solubility of Thallium Chloride in aqueous soln. of Zinc Sulfate, g/l	d	t
8.069	2.099	1.0105	0	48.414	4.589	1.0515	25
16.138	2.348	1.0188	0	96.731	4.646	1.0993	25
48.414	2.912	1.0674	0	8.069	9.660	1.0046	50
96.731	3.413	1.1015	0	16.138	10.692	1.0249	50
8.069	4.320	1.0097	25	48.414	12.832	1.0459	50
16.138	4.404	1.0184	25	96.731	14.784	1.0952	50

№ 5305 ZINC SULFATE–THALLIUM CHLORIDE–WATER [2039]

$ZnSO_4—TlCl—H_2O$

$t=25$

Concn. in aqueous soln., g/l			
Thallium Chloride	Zinc Chloride	Thallium Chloride	Zinc Chloride
4.939	8.069	6.645	48.414
5.467	16.138	7.683	96.731

№ 5306 [2039]

ZINC SULFATE–THALLIUM IODATE–WATER

$ZnSO_4$—$TlIO_3$—H_2O

$t=25$

Concn. of Zinc Sulfate in aqueous soln., g/l	Solubility of Thallium Iodate in aqueous soln. of Zinc Sulfate, g/l
8.069	1.055

№ 5307 [1650]

ZINC CHLORIDE–SILVER CHLORIDE–WATER

$ZnCl_2$—$AgCl$—H_2O

Concn. in aqueous soln., g/l		t
Zinc Chloride	Silver Chloride	
325.5	0.052	25

№ 5308 [1568]

ZINC CHLORIDE–INDIUM CHLORIDE–WATER

$ZnCl_2$—$InCl_3$—H_2O

$t=20$

Concn. in aqueous soln., g/l		Comp. of solid phase
Indium Chloride	Zinc Chloride	
549	1155	$ZnCl_2 \cdot 2InCl_3 \cdot 8H_2O$
482	1250	$2ZnCl_2 \cdot InCl_3 \cdot 5H_2O$

№ 5309 [1197]

ZINC CHLORIDE–THALLIUM CHLORIDE–WATER

$ZnCl_2$—$TlCl$—H_2O

$t=25$

Comp. of saturated soln., wt %			Comp. of solid phase
Thallium Chloride	Zinc Chloride	Water	
0.374	0.0	99.626	$TlCl$
0.04	1.097	98.863	»
0.007	2.457	97.536	»
0.053	10.02	89.927	»
0.135	27.26	72.605	»
0.440	43.110	56.450	$3TlCl \cdot ZnCl_2$
0.49	47.80	51.71	»
0.86	55.27	43.87	»
1.03	57.40	41.57	»
1.50	58.42	40.08	»
1.76	60.58	37.66	»
2.04	60.81	37.15	$2TlCl \cdot ZnCl_2$
3.55	67.13	29.32	»
5.08	67.48	27.44	»
7.69	69.40	22.91	»
9.08	69.46	21.46	»
10.23	69.02	20.75	»
6.[illegible]9	73.16	20.65	»
4.63	74.92	20.40	»
0.0	79.60	20.40	$ZnCl_2$

№ 5310 [2262]

ZINC CHLORIDE–LEAD (II) CHLORIDE–WATER

$ZnCl_2—PbCl_2—H_2O$

$t=25$

Concn. in aqueous soln., g/l	
Zinc Chloride	Lead Chloride
0	10.79
14	3.0

№ 5311 ZINC CHLORIDE–LEAD (II) CHLORIDE–WATER [947]

$ZnCl_2—PbCl_2—H_2O$

Concn. in aqueous soln., wt %		t	Comp. of solid phase
Lead Chloride	Zinc Chloride		
1.031	0.0	25	$PbCl_2$
0.513	0.49	25	»
0.3	0.8	25	»
0.252	0.727	25	»
0.22	0.846	25	»
0.161	12.54	25	»
0.09	16.72	25	»
0.08	26.00	25	»
0.10	37.21	25	»
0.14	47.91	25	»
0.17	55.33	25	»
0.184	66.68	25	»
0.696	76.68	25	»
0.78	80.05	25	$PbCl_2+ZnCl_2$
1.887	0.0	60	$PbCl_2$
0.69	2.05	60	»
0.54	3.96	60	»
0.48	7.68	60	»
0.45	14.52	60	»
0.31	26.52	60	»
0.147	43.71	60	»
0.195	47.41	60	»
0.215	66.16	60	»
0.345	69.29	60	»
3.86	80.96	60	$PbCl_2+ZnCl_2$

№ 5312 ZINC BROMIDE–INDIUM BROMIDE–WATER [1568]

$ZnBr_2—InBr_3—H_2O$

$t=20$

Concn. in aqueous soln., g/l		Comp. of solid phase
Indium Bromide	Zinc Bromide	
1029	1278	$3ZnBr_2 \cdot 4InBr_3 \cdot 14H_2O$
877	1402	»
649	1646	»
562	1750	$3ZnBr_2 \cdot InBr_3 \cdot 4H_2O$

№ 5313 ZINC IODIDE–INDIUM IODIDE–WATER [1568]

$ZnI_2—InI_3—H_2O$

$t=20$

Concn. in aqueous soln., g/l		Comp. of solid phase
Indium Iodide	Zinc Iodide	
993	1133	$2ZnI_2 \cdot 3InI_3 \cdot 13H_2O$
630	1689	»
397	1806	$ZnI_2 \cdot 3InI_3 \cdot 5H_2O$
335	1913	»
239	2080	»
285	218.7	$ZnI_2 \cdot H_2O$
264	2218	»

№ 5314 ARSENIC TRIOXIDE–STRONTIUM OXIDE–WATER [2152]

$As_2O_3—SrO—H_2O$

Comp. of saturated soln., wt %			t	Comp. of solid phase
Strontium Oxide	Arsenic Trioxide	Water		
0.18	2.0	97.82	20	$SrO \cdot As_2O_3 \cdot nH_2O$
0.14	0.22	99.64		»
0.65	Traces	99.35	20	$2SrO \cdot As_2O_3 \cdot nH_2O$
0.32	5.6	94.08	60	$SrO \cdot As_2O_3 \cdot nH_2O$
0.15	0.21	99.64	60	»
3.01	Traces	96.99	60	$2SrO \cdot As_2O_2 \cdot nH_2O$

№ 5315 ARSENIC TRIOXIDE–STRONTIUM OXIDE–WATER [2153]

$As_2O_3—SrO—H_2O$

Comp. of saturated soln., wt %			t	Comp. of solid phase
Strontium Oxide	Arsenic Trioxide	Water		
0.18	2.01	97.81	20	$As_2O_3+SrO \cdot As_2O_3$
0.18	1.91	97.91	20	$SrO \cdot As_2O_3$
0.16	1.58	97.26	20	»
0.15	1.47	98.38	20	»
0.13	1.01	98.86	20	»
0.13	0.89	98.98	20	»
0.12	0.46	99.42	20	»
0.12	0.38	99.50	20	»
0.14	0.21	98.65	20	$SrO \cdot As_2O_3+2SrO \cdot As_2O_3$
0.15	0.21	98.64	20	»
0.14	0.22	98.64	20	»
0.14	0.15	99.71	20	$2SrO \cdot As_2O_3$
0.22	0.12	99.66	20	»
0.55	0.05	99.40	20	»
0.65	Traces	99.35	20	$2SrO \cdot As_2O_3+Sr(OH)_2 \cdot 8H_2O$
0.21	5.55	94.24	60	As_2O_3
0.32	5.60	94.08	60	$SrO \cdot As_2O_3+As_2O_3$

Comp. of saturated soln., wt%			t	Comp. of solid phase
Strontium Oxide	Arsenic Trioxide	Water		
0.31	5.01	94.68	60	$SrO \cdot As_2O_3$
0.24	3.78	95.98	60	»
0.20	2.30	97.50	60	»
0.16	1.50	98.34	60	»
0.16	0.57	99.27	60	»
0.13	0.49	99.38	60	»
0.15	0.21	99.64	60	»
0.16	0.21	99.63	60	$SrO \cdot As_2O_3 + 2SrO \cdot As_2O_3$
0.15	0.20	99.65	60	»
0.15	0.21	99.64	60	$2SrO \cdot As_2O_3$
0.22	0.13	99.65	60	»
0.74	Traces	99.26	60	»
1.75	»	98.25	60	»
2.11		97.89	60	»
2.60	»	97.40	60	»
3.10	»	96.90	60	$2SrO \cdot As_2O_3 + Sr(OH)_2 \cdot 8H_2O$

№ 5316 ARSENIC TRIOXIDE–STRONTIUM CHLORIDE–WATER [2519]

$As_2O_3—SrCl_2—H_2O$

$t=30$

Comp. of saturated soln., wt%			Comp. of solid phase
Arsenic Trioxide	Strontium Chloride	Water	
2.14	6.27	91.59	As_2O_3
1.92	13.67	84.41	»
1.67	21.29	77.04	»
1.46	27.46	71.08	»
1.28	34.03	64.69	»
1.23	36.16	62.61	$As_2O_3 + SrCl_2 \cdot 6H_2O$
0.0	37.5	62.5	$SrCl_2 \cdot 6H_2O$

№ 5317 [2519]

ARSENIC TRIOXIDE–STRONTIUM BROMIDE–WATER

$As_2O_3—SrBr_2—H_2O$

$t=30$

Comp. of saturated soln., wt%			Comp. of solid phase
Arsenic Trioxide	Strontium Bromide	Water	
1.69	11.69	86.62	As_2O_3
1.74	22.09	76.17	»
1.48	31.98	66.54	»
1.25	41.91	56.84	»
1.07	46.87	52.06	»
0.99	48.91	50.10	$As_2O_3 + SrBr_2 \cdot 6H_2O$
0.00	49.11	50.89	$SrBr_2 \cdot 6H_2O$

As_2O_3—BaO—H_2O

Comp. of saturated soln., wt%			t	Comp. of solid phase
Barium Oxide	Arsenic Trioxide	Water		
0.17	2.94	96.89	20	$2BaO \cdot 3As_2O_3$
0.18	2.58	97.24	20	»
0.16	1.97	97.87	20	$As_2O_3 + 2BaO \cdot 3As_2O_3$
0.15	1.94	97.91	20	»
0.16	1.88	97.96	20	$2BaO \cdot 3As_2O_3$
0.18	1.32	98.50	20	»
0.22	1.15	98.63	20	»
0.17	0.99	98.84	20	»
0.29	0.57	99.14	20	»
0.52	0.73	98.75	20	$2BaO \cdot 3As_2O_3 + BaO \cdot As_2O_3$
0.54	0.74	98.72	20	»
0.54	0.75	99.71	20	»
0.53	0.70	98.77	20	$BaO \cdot As_2O_3$
0.49	0.61	98.99	20	»
0.51	0.58	98.91	20	»
0.52	0.54	98.94	20	»
0.59	0.55	98.86	20	$BaO \cdot As_2O_3 + 2BaO \cdot As_2O_3$
0.60	0.56	98.84	20	»
0.61	0.56	98.13	20	»
0.58	0.52	98.90	20	$2BaO \cdot As_2O_3$
0.91	0.11	98.98	20	»
1.01	0.094	98.90	20	»
1.15	0.097	98.75	20	»
1.51	0.08	98.41	20	»
1.75	0.05	98.20	20	»
2.23	0.049	97.72	20	»
2.54	0.045	97.42	20	»
2.73	0.05	97.22	20	»
3.04	0.05	96.91	20	»
3.40	0.05	96.55	20	»
4.59	0.05	95.36	20	»
0.17	5.30	94.53	60	As_2O_3
0.38	5.60	94.02	60	»
0.42	5.70	93.88	60	$As_2O_3 + 2BaO \cdot 3As_2O_3$
0.45	5.40	94.15	60	$2BaO \cdot 3As_2O_3$
0.44	5.30	94.26	60	»
0.32	4.21	95.47	60	»
0.25	3.85	96.90	60	»
0.23	3.57	96.20	60	»
0.24	3.20	96.56	60	»
0.24	3.01	96.75	60	»
0.23	2.85	96.93	60	»
0.19	1.97	97.84	60	»
0.19	1.82	97.99	60	»
0.20	1.64	98.16	60	»
0.44	0.83	98.73	60	»
0.51	0.75	98.74	60	$2BaO \cdot 3As_2O_3 + BaO \cdot As_2O_3$
0.52	0.75	98.73	60	»
0.50	0.75	98.75	60	»

Table № 5318 (continuation)

Comp. of saturated soln., wt%			t	Comp. of solid phase
Barium Oxide	Arsenic Trioxide	Water		
0.50	0.74	98.68	60	$BaO \cdot As_2O_3$
0.53	0.73	98.74	60	»
0.54	0.71	98.75	60	»
0.58	0.69	98.73	60	$BaO \cdot As_2O_3 + 3BaO \cdot 2As_2O_3$
0.58	0.68	98.74	60	»
0.58	0.61	98.81	60	»
0.55	0.53	98.92	60	»
0.55	0.43	99.02	60	»
0.56	0.41	99.03	60	»
0.63	0.38	98.99	60	»
0.77	0.33	98.90	60	$3BaO \cdot As_2O_3$
0.83	0.32	98.85	60	»
0.88	0.31	98.81	60	»
1.18	0.27	98.55	60	»
1.27	0.27	98.46	60	»
1.52	0.28	98.20	60	$3BaO \cdot 2As_2O_3 + 2BaO \cdot As_2O_3$
1.59	0.28	98.13	60	»
1.52	0.26	98.22	60	»
2.60	0.12	97.28	60	»
2.90	0.14	96.96	60	»
3.02	0.15	96.83	60	»
3.60	0.12	96.28	60	»
5.16	0.11	94.73	60	$2BaO \cdot As_2O_3$
8.00	0.13	91.87	60	»
10.90	0.15	88.95	60	»
11.90	0.13	87.97	60	»
12.90	0.13	86.97	60	»
13.60	0.15	86.25	60	»
17.00	0.19	82.81	60	»
18.30	0.18	81.52	60	$2BaO \cdot As_2O_3 + Ba(OH)_2 \cdot 8H_2O$

№ 5319 ARSENIC TRIOXIDE–BARIUM OXIDE–WATER [2153]

As_2O_3—BaO—H_2O

Content in aqueous soln., wt%		t	Comp. of solid phase
Barium Oxide	Arsenic Trioxide		
0.17	2.94	20	$2BaO \cdot 3As_2O_3$ in the presence of saturated soln.
0.18	2.58	20	»
0.16	1.97	20	$As_2O_3 + 2BaO \cdot 3As_2O_3$
0.15	1.94	20	»
0.16	1.88	20	»
0.18	1.32	20	»
0.22	1.15	20	$2BaO \cdot 3As_2O_3$

Table № 5319 (continuation)

Content in aqueous soln., wt%		t	Comp. of solid phase
Barium Oxide	Arsenic Trioxide		
0.17	0.99	20	$2BaO \cdot 3As_2O_3$
0.29	0.57	20	»
0.52	0.73	20	$2BaO \cdot 3As_2O_3 + BaO \cdot As_2O_3$
0.54	0.74	20	»
0.54	0.75	20	»
0.53	0.70	20	»
0.49	0.61	20	$BaO \cdot As_2O_3$
0.51	0.58	20	»
0.52	0.54	20	»
0.59	0.55	20	»
0.60	0.56	20	$BaO \cdot As_2O_3 + 2BaO \cdot As_2O_3$
0.61	0.56	20	»
0.58	0.52	20	»
0.91	0.11	20	»
1.01	0.094	20	»
1.15	0.097	20	»
1.51	0.08	20	»
1.75	0.05	20	»
2.23	0.049	20	$2BaO \cdot As_2O_3$
2.54	0.045	20	»
2.73	0.05	20	»
3.04	0.05	20	»
3.40	0.05	20	»
4.59	0.05	20	»
0.17	5.3	60	As_2O_3
0.38	5.6	60	»
0.42	5.7	60	$As_2O_3 + 2BaO \cdot 3As_2O_3$
0.45	5.4	60	»
0.44	5.3	60	»
0.32	4.21	60	»
0.25	3.85	60	»
0.23	3.57	60	»
0.24	3.2	60	»
0.24	3.01	60	$2BaO \cdot As_2O_3$
0.23	2.85	60	»
0.19	1.97	60	»
0.19	1.82	60	»
0.20	1.64	60	»
0.44	0.83	60	»
0.51	0.75	60	»
0.52	0.75	60	$2BaO \cdot 3As_2O_3 + BaO \cdot As_2O_3$
0.50	0.75	60	»
0.58	0.82	60	Metastable equilibrium
0.53	0.82	60	»
0.50	0.79	60	»
0.50	0.74	60	$BaO \cdot As_2O_3$
0.53	0.73	60	»
0.54	0.71	60	»
0.62	0.67	60	Metastable equilibrium
0.58	0.67	60	»
0.58	0.69	60	$BaO \cdot As_2O_3 + 3BaO \cdot 2As_2O_3$

Content in aqueous soln., wt%		t	Comp. of solid phase
Barium Oxide	Arsenic Trioxide		
0.58	0.61	60	$BaO \cdot As_2O_3 + 3BaO \cdot 2As_2O_3$
0.55	0.53	60	»
0.55	0.43	60	»
0.56	0.41	60	»
0.63	0.38	60	»
0.77	0.33	60	$3BaO \cdot 2As_2O_3$
0.83	0.32	60	»
0.88	0.31	60	»
1.18	0.27	60	»
1.27	0.27	60	»
1.52	0.28	60	$3BaO \cdot 2As_2O_3 + 2BaO \cdot As_2O_3$
1.59	0.28	60	»
1.52	0.26	60	»
2.60	0.12	60	»
2.90	0.14	60	»
3.02	0.15	60	»
3.60	0.12	60	»
5.16	0.11	60	$2BaO \cdot As_2O_3$
8.0	0.13	60	»
10.9	0.15	60	»
11.9	0.13	60	»
12.9	0.13	60	»
13.6	0.15	60	»
17.0	0.19	60	»
18.3	0.18	60	$2BaO \cdot As_2O_3 + Ba(OH)_2 \cdot 8H_2O$
18.3	0.18	60	»

№ 5320 ARSENIC TRIOXIDE–BARIUM OXIDE–WATER [2152]

As_2O_3—BaO—H_2O

Comp. of saturated soln., wt%			t	Comp. of solid phase
Barium Oxide	Arsenic Trioxide	Water		
0.15	1.97	97.88	20	$2BaO \cdot 3As_2O_3 \cdot nH_2O$
0.54	0.75	98.71	20	$BaO \cdot As_2O_3 \cdot nH_2O$
0.59	0.55	98.86	20	»
3.74	0.05	96.21	20	$2BaO \cdot As_2O_3 \cdot nH_2O$
0.42	5.70	93.88	60	$2BaO \cdot 3As_2O_3 \cdot nH_2O$
0.51	0.75	98.74	60	$BaO \cdot As_2O_3 \cdot nH_2O$
0.58	0.68	98.74	60	»
1.52	0.28	98.20	60	$3BaO \cdot 2As_2O_3 \cdot nH_2O$
17.32	0.12	82.56	60	$2BaO \cdot As_2O_3 \cdot nH_2O$

№ 5321 ARSENIC TRIOXIDE–BARIUM CHLORIDE–WATER [2519]

$As_2O_3—BaCl_2—H_2O$

$t=30$

Comp. of saturated soln., wt%			Comp. of solid phase
Arsenic Trioxide	Barium Chloride	Water	
2.24	3.84	93.92	As_2O_3
2.20	8.72	89.08	»
2.19	8.86	88.95	»
2.15	10.34	87.51	»
1.69	9.55	88.76	$(As_2O_3)_2 \cdot BaCl_2$
1.12	13.62	85.26	»
0.905	16.93	82.165	»
0.737	20.06	79.203	»
0.608	23.87	75.522	»
0.506	26.54	72.954	$(As_2O_3)_2 \cdot BaCl_2 + BaCl_2 \cdot 2H_2O$
0.0	27.6	72.4	$BaCl \cdot 2H_2O$

№ 5322 ARSENIC TRIOXIDE–BARIUM BROMIDE–WATER [2519]

$As_2O_3—BaBr_2—H_2O$

$t=30$

Comp. of saturated soln., wt%			Comp. of solid phase
Arsenic Trioxide	Barium Bromide	Water	
2.09	9.41	88.50	As_2O_3
2.03	16.88	81.09	»
1.97	24.03	74.00	»
1.87	24.41	73.72	»
1.58	23.49	74.93	$(As_2O_3)_2 \cdot BaBr_2$
0.757	29.09	70.153	»
0.678	33.08	66.242	»
0.464	38.19	61.346	»
0.322	43.02	56.658	»
0.277	50.03	49.693	$(As_2O_3)_2 \cdot BaBr_2 + BaBr_2 \cdot 2H_2O$
0.0	50.62	49.38	$BaBr_2 \cdot 2H_2O$

№ 5323 ARSENIC PENTOXIDE–STRONTIUM OXIDE–WATER [2631]

$As_2O_5—SrO—H_2O$

$t=25$

Comp. of saturated soln., wt%			*d*	Comp. of solid phase
Strontium Oxide	Arsenic Pentoxide	Water		
0.14	0.18	99.68	1.001	$SrHAsO_4$
0.48	1.01	98.51	1.010	»
1.56	3.33	95.11	1.042	»
2.43	5.41	92.16	1.080	»
3.84	8.33	87.83	1.126	»

Comp. of saturated soln., wt%			d	Comp. of solid phase
Strontium Oxide	Arsenic Pentoxide	Water		
5.72	12.29	81.99	1.199	$SrHAsO_4$
7.30	17.31	75.39	1.273	$SrHAsO_4 + Sr(H_2AsO_4)_2$
6.81	23.38	69.81	1.339	$Sr(H_2AsO_4)_2$
6.24	26.08	67.68	1.362	»
5.98	33.59	60.43	1.489	»
5.60	37.59	56.81	1.542	»
5.01	39.71	55.28	1.555	»
4.44	42.94	52.62	1.605	»
2.23	55.20	42.57	1.805	»
1.32	57.97	40.71	1.851	»
0.43	64.75	34.82	1.999	»
0.09	71.71	28.20	2.236	»

№ 5324 ARSENIC PENTOXIDE–BARIUM OXIDE–WATER [1749, 1798]

$As_2O_5—BaO—H_2O$

$t=17$

Concn. in aqueous soln., wt%		d_4	Comp. of solid phase
Barium Oxide	Arsenic Pentoxide		
2.92	0.0	1.038	$Ba(OH)_2 \cdot 8H_2O$
2.895	—	1.037	$Ba_3(AsO_4)_2 \cdot H_2O$
2.91	—	1.039	»
2.61	—	1.030	»
1.98	—	1.022	»
1.92	—	1.024	»
1.60	—	1.019	»
0.66	—	1.004	»
0.375	<0.001	1.002	$BaO + As_2O_5$
0.013	0.006	1.0	»
0.012	0.006	1.0	»
0.030	0.025	1.0	»
0.035	0.027	1.0	»
0.031	0.025	1.0	»
0.029	0.024	1.0	»
0.030	0.024	1.0	»
0.031	0.025	1.0	»
0.031	0.025	1.0	»
0.030	0.023	1.0	»
0.035	0.033	—	$Ba_2H_2(AsO_4)_2 \cdot 2H_2O$
0.050	0.053	—	»
0.111	0.156	—	»
0.218	0.296	—	»
0.75	1.24	1.021	»
3.72	2.32	1.051	»
4.08	6.78	1.094	»
7.20	12.27	1.208	»
7.82	13.66	1.230	»
8.32	15.13	1.255	»
10.71	19.08	1.353	»

Table № 5324 (continuation)

Concn. in aqueous soln., wt%		d_4	Comp. of solid phase
Barium Oxide	Arsenic Pentoxide		
12.53	23.63	1.441	$Ba_2H_2(AsO_4)_2 \cdot 2H_2O$
13.12	24.44	1.450	»
13.53	25.14	1.481	»
12.59	27.60	1.502	$BaH_4(AsO_4)_2 \cdot H_2O$
11.68	30.28	1.554	»
11.27	33.59	1.576	»
10.56	38.41	1.636	»
9.93	42.01	1.691	»
9.48	43.80	1.725	»
9.32	45.10	1.750	»
9.23	46.04	1.762	$BaH_4(AsO_4)_2$
8.89	46.63	1.772	»
8.87	47.60	1.802	»
7.45	52.05	1.850	»
6.03	56.07	1.904	»
5.42	57.88	1.941	»
4.80	59.70	1.970	»
4.15	61.52	1.984	»
2.65	66.56	2.077	»
8.82	46.02	1.765	$BaO \cdot 2As_2O_5 \cdot 3H_2O$
4.30	46.15	1.605	»
2.15	46.90	1.572	»
0.35	50.48	1.602	»
0.15	55.71	1.734	»
0.13	59.38	1.819	»
Traces	61.36	1.868	»
—	67.84	2.061	»
—	71.72	2.172	»
—	71.96	2.181	$As_2O_3 \cdot 4H_2O$

$t = 30$

Comp. of saturated soln., wt%			d_{20}^{20}	Comp. of solid phase
Barium Oxide	Arsenic Pentoxide	Water		
0.22	0.002	99.778	1.003	Solid solution
0.11	0.003	99.887	1.002	»
0.03	0.34	99.63	1.001	$BaHSO_4 \cdot H_2O + Ba_2(AsO_4)_2 \cdot xH_2O$
2.63	3.98	93.39	1.067	$BaHAsO_4 \cdot H_2O$
3.60	5.86	90.54	1.094	»
4.36	7.26	88.48	1.117	»
4.04	6.72	89.24	1.108	»
5.49	9.28	85.23	1.151	»
6.25	10.56	83.19	1.176	»
6.62	11.27	82.11	1.188	»
6.98	11.12	81.90	1.281	»
9.46	14.92	75.68	1.271	»
11.67	18.52	69.81	1.357	»
13.59	22.95	63.56	1.474	»
15.37	27.26	57.37	1.592	»

Comp. of saturated soln., wt%			d_{20}^{20}	Comp. of solid phase
Barium Oxide	Arsenic Pentoxide	Water		
16.65	28.01	55.34	—	$BaHAsO_4 \cdot H_2O + BaH_4(AsO_4)_2 \cdot 2H_2O$
15.65	31.17	53.18	1.694	$BaH_4(AsO_4)_2 \cdot 2H_2O$
13.35	35.86	50.79	1.719	»
7.53	47.27	45.20	1.809	»
6.56	48.48	44.96	1.817	»
Traces	68.40	31.60	2.150	»
»	72.04	27.96	2.280	$BaH(AsO_4)_2 \cdot 2H_2O + 3As_2O_5 \cdot 5H_2O$
»	59.44	40.56	1.869	$3As_2O_5 \cdot 5H_2O$

№ 5325 SELENIUM DIOXIDE–BARIUM SELENITE–WATER [2245]

SeO_2—$BaSeO_3$—H_2O

Comp. of saturated soln., wt%			t	Comp. of solid phase
Selenium Dioxide	Barium Selenite	Water		
0.00	0.005	99.995	0	$BaSeO_3$
0.17	0.39	99.44	0	»
0.27	0.64	99.09	0	$BaSeO_3 + BaSeO_5$
1.11	0.65	98.24	0	$BaSeO_5$
2.04	0.58	97.38	0	»
4.57	0.58	94.85	0	»
8.80	0.69	90.51	0	»
17.9	0.89	81.21	0	»
26.6	1.10	72.30	0	»
30.6	1.18	68.22	0	»
35.6	1.44	62.96	0	»
44.3	1.99	53.71	0	»
49.7	2.44	47.86	0	»
57.8	3.42	39.78	0	»
58.6	3.53	37.87	0	$BaSeO_5 + H_2SeO_3$
58.7	2.51	38.79	0	H_2SeO_3
58.8	1.24	39.96	0	»
58.9	0.00	41.1	0	»
0.00	0.005	99.995	25	$BaSeO_3$
0.25	0.58	99.17	25	»
0.42	0.97	98.61	25	$BaSeO_3 + BaSeO_5$
0.41	0.95	98.64	25	$BaSeO_5$
1.27	0.96	97.77	25	»
5.94	0.97	93.09	25	»
13.5	1.09	85.41	25	»
23.3	1.38	75.32	25	»
32.2	1.74	65.06	25	»
44.7	2.42	52.88	25	»
59.9	3.84	36.26	25	»
63.6	4.36	32.04	25	»
67.2	5.06	27.74	25	$BaSeO_5 + H_2SeO_3$
67.3	4.61	28.09	25	H_2SeO_3

Table № 5325 (continuation)

Comp. of saturated soln., wt%			t	Comp. of solid phase
Selenium Dioxide	Barium Selenite	Water		
67.7	3.13	29.17	25	H_2SeO_3
68.3	1.54	30.16	25	»
68.8	0.00	31.2	25	»
0.00	0.005	99.995	50	$BaSeO_3$
0.26	0.67	98.97	50	»
0.52	1.18	98.30	50	$BaSeO_3+BaSeO_5$
1.34	1.16	97.50	50	$BaSeO_5$
4.87	1.28	94.95	50	»
13.7	1.45	84.85	50	»
24.7	1.80	73.50	50	»
37.1	2.38	60.52	50	»
44.7	2.81	52.49	50	»
57.8	3.90	38.30	50	»
68.3	5.32	26.38	50	»
71.8	6.42	21.78	50	»
73.4	6.63	19.97	50	$BaSeO_5+H_2SeO_3$
73.8	5.80	20.40	50	H_2SeO_3
74.8	3.66	21.54	50	»
75.4	2.75	21.85	50	»
76.2	1.51	22.29	50	»
76.6	0.00	23.4	50	»

№ 5326 SELENIUM CHLORIDE OXIDE–BARIUM CHLORIDE–WATER [2752]

$SeOCl_2—BaCl_2—H_2O$

$t=25$

Comp. of solvent, mol %		Solubility of Barium Chloride, wt%	Comp. of solvent, mol %		Solubility of Barium Chloride, wt%
Selenium Chloride Oxide	Water		Selenium Chloride Oxide	Water	
100.00	0.00	3.95	80.00	20.00	2.32
94.12	5.88	3.86	66.67	33.33	1.40
88.89	11.11	3.37	50.00	50.00	0.45

№ 5327 RUBIDIUM CARBONATE–DIOXOAMERICIUM CARBONATE–WATER [2253]

$Rb_2CO_3—(AmO_2)_2CO_3—H_2O$

$t=25$

Comp. of saturated soln., wt%			Comp. of solid phase
Dioxoamericium Carbonate	Rubidium Carbonate	Water	
0.376	[illegible].73	92.894	$RbAmO_2CO_3$
0.0043	5.64	94.3557	»

№ 5328 RUBIDIUM NITRITE–STRONTIUM NITRITE–WATER [725]

$RbNO_2—Sr(NO_2)_2—H_2O$

$t=25$

Comp. of saturated soln., wt%			Comp. of solid phase
Strontium Nitrite	Rubidium Nitrite	Water	
41.50	0.00	58.50	$Sr(NO_2)_2 . H_2O$
40.40	4.40	55.20	»
35.58	13.50	50.92	»
33.38	22.10	44.52	»
30.46	31.70	37.84	»
27.30	43.90	28.80	$Sr(NO_2)_2 . H_2O+2RbNO_2 \cdot Sr(NO_2)_2 . H_2O$
26.92	47.00	26.08	»
26.98	46.94	26.08	»
26.90	47.00	26.10	»
25.50	47.50	27.00	$2RbNO_2 . Sr(NO_2)_2 . H_2O$
16.60	59.00	24.40	»
11.00	68.30	20.20	»
6.99	77.10	15.91	»
5.57	81.26	13.17	»
3.10	87.11	9.79	$RbNO_2+2RbNO_2 . Sr(NO_2)_2 . H_2O$
3.20	87.30	9.50	»
3.14	87.20	9.66	»
0.00	88.70	11.30	$RbNO_2$

№ 5329 RUBIDIUM NITRITE–BARIUM NITRITE–WATER [719]

$RbNO_2—Ba(NO_2)_2—H_2O$

$t=25$

Comp. of saturated soln., wt%			Comp. of solid phase
Rubidium Nitrite	Barium Nitrite	Water	
87.77	0.00	12.23	$RbNO_2$
84.48	2.42	13.10	$RbNO_2+2RbNO_2 . Ba(NO_2)_2$
84.37	2.45	13.18	»
83.04	2.98	13.98	$2RbNO_2 . Ba(NO_2)_2$
79.17	3.54	17.29	»
65.73	8.30	25.97	»
61.67	11.62	26.71	»
53.45	16.66	29.89	»
45.31	23.78	30.91	$2RbNO_2 . Ba(NO_2)_2+RbNO_2 . 2Ba(NO_2)_2$
45.28	23.75	30.97	»
42.57	25.14	32.29	$RbNO_2 . 2Ba(NO_2)_2$
38.83	26.98	34.19	»
36.65	28.55	34.80	»
30.61	32.44	36.95	»
27.39	34.89	37.72	»
23.31	37.63	39.06	$RbNO_2 . 2Ba(NO_2)_2+Ba(NO_2)_2 . H_2O$
23.34	37.61	39.05	»
20.50	38.57	40.93	$Ba(NO_2)_2 . H_2O$
18.19	39.32	42.49	»
13.98	40.96	45.06	»
6.06	42.96	50.98	»
0.00	45.30	54.70	»

№ 5330 RUBIDIUM NITRITE–THALLIUM NITRITE–WATER [722]

$RbNO_2—TlNO_2—H_2O$

$t=25$

Comp. of saturated soln., wt%			Comp. of solid phase
Thallium Nitrite	Rubidium Nitrite	Water	
0.0	90.64	9.36	$RbNO_2$
0.91	89.46	9.63	$RbNO_2+TlNO_2$ (solid solns.)
3.18	88.27	8.55	»
10.89	82.17	6.94	»
14.38	79.74	5.88	»
16.27	77.60	6.13	»
17.46	77.01	5.53	»
18.90	76.32	4.78	Eutonic from solid solns.
18.58	76.36	5.06	»
29.68	68.69	1.63	$TlNO_2+RbNO_2$ (solid solns).
27.67	63.95	8.38	»
29.53	60.65	9.82	»
32.07	58.38	9.55	»
34.05	56.05	9.90	Eutonic from solid solns. $+TlNO_2$
33.99	56.27	9.74	»
34.40	51.90	13.70	$TlNO_2$
32.43	48.89	18.68	»
28.76	44.18	27.06	»
26.16	36.55	37.29	»
24.75	34.11	41.14	»
23.80	25.33	50.87	»
24.52	19.32	56.16	»
25.15	15.29	59.56	»
27.81	7.56	64.63	»
32.25	0.00	67.75	»

№ 5331 RUBIDIUM NITRATE–CERIUM NITRATE–WATER [298]

$RbNO_3—Ce(NO_3)_3—H_2O$

Comp. of saturated soln., wt%			t	Comp. of solid phase
Rubidium Nitrate	Cerium Nitrate	Water		
25.5	0.0	74.5	10	$RbNO_3$
20.2	16.0	63.8	10	»
26.0	23.0	51.0	10	»
28.8	35.5	35.7	10	»
29.5	36.5	34.0	10	$RbNO_3+Ce(NO_3)_3 \cdot 2RbNO_3 \cdot 4H_2O$
27.5	37.5	35.0	10	$Ce(NO_3)_3 \cdot 2RbNO_3 \cdot 4H_2O$
18.5	43.8	37.7	10	»
9.5	55.5	35.0	10	»
7.0	58.2	34.8	10	$Ce(NO_3)_3 \cdot 2RbNO_3 \cdot 4H_2O+Ce(NO_3)_3 \cdot 6H_2O$
3.5	59.0	37.5	10	$Ce(NO_3)_3 \cdot 6H_2O$
0.0	60.0	40.0	10	»
35.0	0.0	65.0	20	$RbNO_3$
30.0	14.0	56.0	20	»
31.7	29.3	39.0	20	»

Table № 5331 (continuation)

Comp. of saturated soln., wt%			t	Comp. of solid phase
Rubidium Nitrate	Cerium Nitrate	Water		
33.2	33.6	33.2	20	$RbNO_3$
34.4	35.6	30.0	20	$RbNO_3 + Ce(NO_3)_3 \cdot 2RbNO_3 \cdot 4H_2O$
29.5	38.5	32.0	20	$Ce(NO_3)_3 \cdot 2PbNO_3 \cdot 4H_2O$
21.3	45.1	33.6	20	»
13.0	54.0	33.0	20	»
7.5	59.9	32.6	20	$Ce(NO_3)_3 \cdot 2RbNO_3 \cdot 4H_2O + Ce(NO_3)_3 \cdot 6H_2O$
6.2	60.3	33.5	20	$Ce(NO_3)_3 \cdot 6H_2O$
3.2	61.2	35.6	20	»
0.0	61.5	38.5	20	»
42.3	0.0	57.7	30	$RbNO_3$
40.4	11.8	47.8	30	»
38.4	26.5	35.1	30	»
39.0	30.5	30.5	30	»
39.5	34.5	25.0	30	$RbNO_3 + Ce(NO_3)_3 \cdot 2RbNO_3 \cdot 4H_2O$
31.3	39.5	29.2	30	$Ce(NO_3)_3 \cdot 2RbNO_3 \cdot 4H_2O$
24.3	45.2	30.5	30	»
16.7	53.0	30.3	30	»
8.5	62.7	28.8	30	$Ce(NO_3)_3 \cdot 2RbNO_3 \cdot 4H_2O + Ce(NO_3)_3 \cdot 6H_2O$
5.4	62.8	31.8	30	$Ce(NO_3)_3 \cdot 6H_2O$
3.0	62.8	34.2	30	»
0.0	63.0	37.0	30	»

№ 5332 [2319]

RUBIDIUM HYDROGEN SULFATE–RUBIDIUM HYDROGEN TELLURATE–WATER

$RbHSO_4$—$RbHTeO_4$—H_2O

$t=25$

Concn. in aqueous soln., g/l			
Rubidium Hydrogen Sulfate	Rubidium Hydrogen Tellurate	Rubidium Hydrogen Sulfate	Rubidium Hydrogen Tellurate
26.67	38.40	59.07	20.18
32.12	31.58	498.3	0.028
42.92	26.76		

№ 5333

RUBIDIUM SULFATE–SILVER SULFATE–WATER

[1516]

Rb_2SO_4—Ag_2SO_4—H_2O

$t=25$

Concn. in aqueous soln., wt%		d	Comp. of solid phase
Silver Sulfate	Rubidium Sulfate		
0.833	0.0	1.004	Ag_2SO_4
0.79	5.85	1.052	»
0.92	12.99	1.126	»
1.10	20.45	1.200	»

Table № 5333 (continuation)

Concn. in aqueous soln., wt%		d	Comp. of solid phase
Silver Sulfate	Rubidium Sulfate		
1.18	24.41	1.252	Ag_2SO_4
1.23	26.13	1.272	»
1.30	28.95	1.316	»
1.33	30.85	1.336	»
1.37	32.93	1.359	»
1.38	33.66	1.375	$Ag_2SO_4+Rb_2SO_4$
1.37	33.61	1.373	»
1.38	33.63	1.378	»
1.38	33.63	1.361	»
1.38	33.63	1.372	»
0.62	33.72	1.355	Rb_2SO_4
0.0	33.77	1.348	»

№ 5334 RUBIDIUM SULFATE–CADMIUM SULFATE–WATER [1208]

Rb_2SO_4—$CdSO_4$—H_2O

Comp. of saturated soln., wt%			t	Comp. of solid phase
Cadmium Sulfate	Rubidium Sulfate	Water		
41.92	3.98	54.10	0	$CdSO_4 \cdot {}^8/_3H_2O+3CdSO_4 \cdot Rb_2SO_4 \cdot 5H_2O$
39.39	4.80	55.81	0	$3CdSO_4 \cdot Rb_2SO_4 \cdot 5H_2O$
36.88	6.25	56.87	0	$3CdSO_4 \cdot Rb_2SO_4 \cdot 5H_2O+CdSO_4 \cdot Rb_2SO_4 \cdot 6H_2O$
27.51	8.09	64.40	0	$CdSO_4 \cdot Rb_2SO_4 \cdot 6H_2O$
18.34	10.00	71.66	0	»
10.86	12.78	76.36	0	»
6.22	16.97	76.81	0	»
4.91	19.20	76.89	0	$CdSO_4 \cdot Rb_2SO_4 \cdot 6H_2O+Rb_2SO_4$
42.27	3.51	54.22	25	$CdSO_4 \cdot {}^8/_3H_2O+3CdSO_4 \cdot Rb_2SO_4 \cdot 5H_2O$
42.14	3.60	54.26	25	$3CdSO_4 \cdot Rb_2SO_4 \cdot 5H_2O$
38.75	5.10	56.15	25	»
32.70	9.27	58.03	25	»
28.41	14.29	57.30	25	»
26.81	16.90	56.29	25	$3CdSO_4 \cdot Rb_2SO_4 \cdot 5H_2O+CdSO_4 \cdot Rb_2SO_4 \cdot 6H_2O$
14.89	24.56	60.55	25	$CdSO_4 \cdot Rb_2SO_4 \cdot 6H_2O$
9.18	30.66	60.16	25	»
6.70	32.75	60.55	25	$CdSO_4 \cdot Rb_2SO_4 \cdot 6H_2O+Rb_2SO_4$
4.39	34.39	61.22	25	Rb_2SO_4
43.48	2.89	53.63	50	$CdSO_4 \cdot H_2O+3CdSO_4 \cdot Rb_2SO_4 \cdot 5H_2O$
39.05	4.90	56.05	50	$3CdSO_4 \cdot Rb_2SO_4 \cdot 5H_2O$
36.32	5.86	57.82	50	»
29.90	11.83	58.27	50	»
23.31	18.09	58.60	50	»
22.68	20.17	57.15	50	»
19.65	25.23	55.12	50	$3CdSO_4 \cdot Rb_2SO_4 \cdot 5H_2O+CdSO_4 \cdot Rb_2SO_4 \cdot 2H_2O$
16.51	27.61	55.88	50	$CdSO_4 \cdot Rb_2SO_4 \cdot 2H_2O$
14.80	29.10	56.10	50	»
13.12	31.10	55.78	50	»
8.82	35.93	56.25	50	»
7.92	37.54	54.54	50	$CdSO_4 \cdot Rb_2SO_4 \cdot 2H_2O+Rb_2SO_4$

Table № 5334 (continuation)

Comp. of saturated soln., wt%			t	Comp. of solid phase
Cadmium Sulfate	Rubidium Sulfate	Water		
22.32	19.10	58.58	62	$3CdSO_4 \cdot Rb_2SO_4 \cdot 5H_2O$
19.90	24.10	56.00	62	$3CdSO_4 \cdot Rb_2SO_4 \cdot 5H_2O + 2CdSO_4 \cdot Rb_2SO_4$
17.47	24.85	57.68	62	$2CdSO_4 \cdot Rb_2SO_4$
15.30	30.03	54.67	62	$2CdSO_4 \cdot Rb_2SO_4 + CdSO_4 \cdot Rb_2SO_4 \cdot 2H_2O$
10.19	35.79	54.02	62	$CdSO_4 \cdot Rb_2SO_4 \cdot 2H_2O$
8.94	38.26	52.80	62	$CdSO_4 \cdot Rb_2SO_4 \cdot 2H_2O + Rb_2SO_4$
40.40	4.38	55.22	75	$CdSO_4 \cdot H_2O + 3CdSO_4 \cdot Rb_2SO_4 \cdot 5H_2O$
39.75	4.70	55.55	75	$3CdSO_4 \cdot Rb_2SO_4 \cdot 5H_2O$
31.06	9.49	59.45	75	»
27.81	11.34	60.85	75	$3CdSO_4 \cdot Rb_2SO_4 \cdot 5H_2O + 2CdSO_4 \cdot Rb_2SO_4$
22.17	15.35	62.48	75	$2CdSO_4 \cdot Rb_2SO_4$
13.93	24.40	61.67	75	»
10.00	30.61	59.39	75	»
8.35	38.30	53.35	75	»
7.61	40.89	52.50	75	$2CdSO_4 \cdot Rb_2SO_4 + Rb_2SO_4$
37.80	6.25	55.95	97.5	$CdSO_4 \cdot H_2O + 2CdSO_4 \cdot Rb_2SO_4$
38.15	5.53	56.32	97.5	$2CdSO_4 \cdot Rb_2SO_4$
30.28	8.96	60.76	97.5	»
14.74	14.89	70.37	97.5	»
9.38	26.50	64.12	97.5	»
5.43	38.82	55.75	97.5	»
2.27	46.25	51.48	97.5	»

№ 5335 RUBIDIUM SULFATE–INDIUM SULFATE–WATER [244]

$Rb_2SO_4—In_2(SO_4)_3—H_2O$

$t=20$

Comp. of saturated soln., wt%			Comp. of solid phase
Indium Sulfate	Rubidium Sulfate	Water	
54.0	0.0	46.0	$In_2(SO_4)_3 \cdot 5H_2O$
52.7	0.1	47.2	$In_2(SO_4)_3 \cdot 5H_2O + RbIn(SO_4)_2 \cdot 3H_2O$
49.9	0.1	50.0	»
44.4	0.39	55.21	$RbIn(SO_4)_2 3H_2O$
39.9	0.6	59.50	»
34.12	1.9	63.98	»
33.70	2.2	64.10	»
29.50	3.2	67.30	»
28.20	3.8	68.0	»
17.70	9.2	73.10	»
13.9	14.3	71.8	»
13.5	14.5	72.0	$RbIn(SO_4)_2 \cdot 3H_2O + Rb_2In(OH)(SO_4)_2 \cdot 2H_2O$
11.6	16.0	72.4	$Rb_2In(OH)(SO_4)_2 \cdot 2H_2O$
11.3	16.5	72.2	»
10.5	17.5	72.0	»
10.0	18.5	71.5	»
9.8	20.2	70.0	»
9.02	21.0	69.98	»
8.9	21.0	70.1	»

Table № 5335 (continuation)

Comp. of saturated soln., wt%			Comp. of solid phase
Indium Sulfate	Rubidium Sulfate	Water	
8.5	22.0	69.5	$Rb_2In(OH)(SO_4)_2 \cdot 2H_2O$
4.0	23.2	72.8	$Rb_2In(OH)_2(SO_4)_2 \cdot nH_2O$
2.0	27.5	70.5	»
0.85	30.0	69.15	»
0.46	30.5	69.04	»
0.095	32.0	69.905	»
0.0	32.5	67.5	Rb_2SO_4

№ 5336 RUBIDIUM SULFATE–PRASEODYMIUM SULFATE–WATER [2388]

$Rb_2SO_4—Pr_2(SO_4)_3—H_2O$

$t=25$

Comp. of saturated soln., wt%			Comp. of solid phase
Praseodymium Sulfate	Rubidium Sulfate	Water	
4.82	0.30	94.88	$Pr_2(SO_4)_3 \cdot 8H_2O$
3.67	0.33	96.00	»
1.98	0.38	97.64	—
1.28	0.41	98.31	—
1.08	0.38	98.54	—
0.06	6.72	93.22	$Pr_2(SO_4)_3 \cdot Rb_2SO_4 \cdot 8H_2O$
0.0	15.22	84.78	»
0.0	17.59	82.41	»
0.0	19.43	80.57	»
0.0	20.84	79.16	—

№ 5337 RUBIDIUM CHLORIDE–RUBIDIUM BROMIDE–WATER [557]

$RbCl—RbBr—H_2O$

Content in aqueous soln., mol/l		t	Content in aqueous soln., mol/l		t
Rubidium Chloride	Rubidium Bromide		Rubidium Chloride	Rubidium Bromide	
7.78	0.00	25	8.90	0.00	25
7.48	0.39	25	8.43	0.46	45
6-94	1.20	25	8.14	0.88	45
5.94	2.43	25	7.86	1.38	45
5.64	2.77	25	7.11	2.23	45
4.99	3.23	25	6.45	3.07	45
4.47	3.58	25	5.97	3.42	45
3.70	4.16	25	5.15	4.19	45
3.34	4.46	25	5.04	4.24	45
2.75	4.82	25	3.59	5.30	45
1.97	5.38	25	3.10	5.61	45
1.68	5.51	25	2.39	6.16	45
0.93	6.04	25	1.73	6.68	45
0.44	6.37	25	1.02	7.18	45
0.15	6.64	25	0.48	7.51	45
0.00	6.74	25	0.00	7.98	45

№ 5338 [1812]

RUBIDIUM CHLORIDE–BARIUM HYDROXIDE–WATER

$RbCl—Ba(OH)_2—H_2O$

$t=25$

Content in aqueous soln., g/l	
Rubidium Chloride	Barium Hydroxide
0.0	47.6
151.1	55.5

№ 5339 RUBIDIUM CHLORIDE–MERCURY (II) CHLORIDE–WATER [1632]

$RbCl —HgCl_2—H_2O$

$t=25$

Comp. of saturated soln., wt%			Comp. of solid phase
Mercury Chloride	Rubidium Chloride	Water	
0.00	48.57	51.43	RbCl
9.18	46.76	44.06	$RbCl+2RbCl \cdot HgCl_2 \cdot H_2O$
9.49	47.54	42.97	»
9.39	47.55	43.06	»
9.47	47.3	43.23	»
10.35	47.65	42.0	$2RbCl \cdot HgCl_2 \cdot H_2O$
19.58	35.16	45.26	»
19.94	34.77	45.29	$2RbCl \cdot HgCl_2 \cdot H_2O+3RbCl \cdot 2HgCl_2 \cdot 2H_2O$
20.10	34.76	45.14	»
20.17	30.27	49.56	$3RbCl \cdot 2HgCl_2 \cdot 2H_2O$
20.55	29.20	50.25	»
20.63	27.38	51.99	»
20.87	26.84	52.29	$3RbCl_2 \cdot 2HgCl_2 \cdot 2H_2O+RbCl \cdot HgCl_2 \cdot H_2O$
20.97	27.09	51.94	»
20.58	26.15	53.27	$RbCl \cdot HgCl_2 \cdot H_2O$
18.71	23.81	57.48	»
14.25	18.10	67.65	»
10.42	10.87	78.71	»
10.56	10.68	78.76	$RbCl \cdot HgCl_2 \cdot H_2O+3RbCl \cdot 4HgCl_2 \cdot H_2O$
10.05	10.50	79.45	»
9.86	10.06	80.08	$3RbCl \cdot 4HgCl_2 \cdot H_2O$
8.71	8.48	82.81	»
8.80	8.46	82.74	»
8.70	5.68	85.62	»
8.33	5.10	86.57	»
8.25	3.43	88.32	$3RbCl \cdot 4HgCl_2 \cdot H_2O+RbCl \cdot 5HgCl_2$
8.00	3.38	88.62	»
7.71	2.98	89.31	»
7.64	1.89	90.47	»
7.55	1.50	90.95	»
7.21	1.10	91.69	»
7.16	0.79	92.05	$RbCl \cdot 5HgCl_2+HgCl_2$
7.42	0.84	91.74	»
6.90	0.00	93.10	$HgCl_2$

№ 5340 RUBIDIUM CHLORATE–CESIUM CHLORATE–WATER [383]

$RbClO_3—CsClO_3—H_2O$

$t=25$

Comp. of saturated soln., wt%					
Rubidium Chlorate	Cesium Chlorate	Water	Rubidium Chlorate	Cesium Chlorate	Water
0.00	7.79	92.21	5.13	4.08	90.79
0.79	7.31	91.90	5.37	3.44	91.19
1.58	6.89	91.53	5.86	2.15	91.99
2.49	6.37	91.14	6.10	1.23	92.67
3.74	5.73	90.53	6.65	0.00	93.35
4.60	5.02	90.38			

№ 5341 RUBIDIUM HYDROGEN TELLURATE–RUBIDIUM HYDROGEN SELENATE–WATER [2319]

$RbHTeO_4—RbHSeO_4—H_2O$

$t=25$

Concn. in aqueous soln., g/l			
Rubidium Hydrogen Selenate	Rubidium Hydrogen Tellurate	Rubidium Hydrogen Selenate	Rubidium Hydrogen Tellurate
76.46	39.51	462.8	5.00
95.82	35.30	859.3	3.40
171.70	22.98		

№ 5342 RUBIDIUM BROMIDE–RUBIDIUM IODIDE–WATER [429]

$RbBr–RbI–H_2O$

$t=25$

Comp. of saturated soln., wt%			Comp. of solid phase
Rubidium Bromide	Rubidium Iodide	Water	
53.70	0.00	46.30	RbBr
53.20	0.92	45.88	»
52.21	1.12	46.67	»
52.47	1.70	45.83	»
52.76	3.01	44.23	»
53.53	3.98	42.49	$RbBr+Rb[IBr_2]$
51.82	5.01	43.17	$Rb[IBr_2]$
51.38	4.02	44.60	»
50.62	5.07	44.31	»
50.02	5.25	44.73	»
46.08	6.19	47.73	»
44.54	7.84	47.62	»
42.32	7.63	50.05	»
40.77	8.59	50.64	»
36.17	10.84	53.03	»
26.86	19.49	53.65	»
23.20	27.02	49.78	»

Table № 5342 (continuation)

Comp. of saturated soln., wt%			Comp. of solid phase
Rubidium Bromide	Rubidium Iodide	Water	
22.14	44.73	33.13	$Rb[IBr_2]$
23.19	48.03	28.76	»
26.38	54.67	18.95	»
21.06	55.04	23.90	»
22.27	56.42	21.31	»

№ 5343 RUBIDIUM IODIDE–CESIUM IODIDE–WATER [556]

$RbI—CsI—H_2O$

$t=25$

Content in aqueous soln., mol/l			
Rubidium Iodide	Cesium Iodide	Rubidium Iodide	Cesium Iodide
7.63	$\approx 10^{-3}$	3.91	2.18
7.50	0.37	2.75	2.42
7.42	0.48	1.88	2.67
7.39	0.64	1.33	2.77
7.25	0.90	0.63	3.05
7.02	1.38	0.55	3.05
7.01	1.42	0.18	3.14
6.88	1.48	$\approx 10^{-3}$	3.32
6.04	1.66		

№ 5344 RUBIDIUM IODIDE–MERCURY (II) IODIDE–WATER [2325]

$RbI—HgI_2—H_2O$

$t=34$

Concn. in aqueous soln., wt%		Comp. of solid phase
Mercury Iodide	Rubidium Iodide	
4.0	5.0	$HgI_2 \cdot RbI \cdot H_2O$
10.7	10.0	»
24.3	20.0	»
36.3	30.0	»
46.0	40.0	»
47.0	41.0	$HgI_2 \cdot RbI \cdot H_2O + 3HgI \cdot 5RbI$
44.7	43.0	$3HgI \cdot 5RbI$
43.7	44.7	»
44.0	45.5	$3HgI \cdot 5RbI + RbI$
40.0	47.0	RbI
30.0	51.0	»
20.0	54.7	»
10.0	59.0	»
0.0	64.0	»

№ 5345 RUBIDIUM IODIDE–THALLIUM TRIIODIDE–WATER [430]

$RbI—TlI_3—H_2O$

$t=25$

Concn. of Rubidium Iodide in aqueous soln., mol/l	Solubility of Thallium Triiodide in aqueous soln. of Rubidium Iodide, (mol/l) · 10⁵	Concn. of Rubidium Iodide in aqueous soln., mol/l	Solubility of Thallium Triiodide in aqueous soln. of Rubidium Iodide, (mol/l) · 10⁵
0.01	0.90	1.68	2.75
0.03	0.40	2.13	4.00
0.06	0.29	2.50	6.02
0.13	0.32	3.13	13.54
0.50	0.55	3.60	22.10
0.87	0.87		

№ 5346 STRONTIUM OXIDE–STRONTIUM CHLORIDE–WATER [2204]

$SrO—SrCl_2—H_2O$

Comp. of saturated soln., wt%			t	Comp. of solid phase
Strontium Chloride	Strontium Oxide	Water		
30.68	0.00	69.32	0	$SrCl_2 \cdot 6H_2O$
30.55	0.31	69.14	0	$SrCl_2 \cdot 6H_2O + SrO \cdot 9H_2O$
29.03	0.30	70.67	0	$SrO \cdot 9H_2O$
23.74	0.29	75.97	0	»
0.0	0.35	99.65	0	»
35.60	0.0	64.40	25	$SrCl_2 \cdot 6H_2O$
35.65	0.85	63.50	25	$SrCl_2 \cdot 6H_2O + SrCl_2 \cdot SrO \cdot 9H_2O$
33.41	1.09	65.50	25	$SrCl_2 \cdot SrO \cdot 9H_2O + SrO \cdot 9H_2O$
22.09	0.75	77.16	25	$SrO \cdot 9H_2O$
12.52	0.86	86.62	25	»
0.00	0.85	99.15	25	»
39.62	0.00	60.38	40	$SrCl_2 \cdot 6H_2O$
39.26	1.36	59.38	40	$SrCl_2 \cdot 6H_2O + SrCl_2 \cdot SrO \cdot 9H_2O$
36.62	1.73	61.65	40	$SrCl_2 \cdot SrO \cdot 9H_2O$
36.08	1.76	63.16	40	»
34.13	2.03	63.84	40	»
32.87	2.10	64.93	40	»
32.07	2.47	65.46	40	$SrCl_2 \cdot SrO \cdot 9H_2O + SrO \cdot 9H_2O$
29.01	1.81	69.08	40	$SrO \cdot 9H_2O$
28.84	1.90	69.26	40	»
18.48	1.42	80.10	40	»
0.00	1.48	98.52	40	»

№ 5347 STRONTIUM OXIDE–STRONTIUM BROMIDE–WATER [2204]

$SrO—SrBr_2—H_2O$

$t=25$

Comp. of saturated soln., wt%			Comp. of solid phase
Strontium Bromide	Strontium Oxide	Water	
49.79	0.00	50.21	$SrBr_2 \cdot 6H_2O$
49.78	0.21	50.01	$SrBr_2 \cdot 6H_2O+SrBr_2 \cdot SrO \cdot 9H_2O$
48.06	0.25	51.69	$SrBr_2 \cdot SrO \cdot 9H_2O$
47.65	0.28	52.07	»
44.24	0.47	55.27	»
42.76	0.61	56.63	»
41.78	0.64	57.58	»
40.66	0.84	58.50	$SrBr_2 \cdot SrO \cdot 9H_2O+SrO \cdot 9H_2O$
38.68	0.79	60.53	$SrO \cdot 9H_2O$
35.83	0.73	64.44	»
27.33	0.66	72.01	»
21.46	0.65	77.89	»
16.27	0.66	83.07	»
0.00	0.85	99.15	»

№ 5348 STRONTIUM OXIDE–STRONTIUM IODIDE–WATER [2204]

$SrO—SrI_2—H_2O$

$t=25$

Comp. of saturated soln., wt%			Comp. of solid phase
Strontium Iodide	Strontium Oxide	Water	
67.70	0.0	32.30	$SrI_2 \cdot 6H_2O$
64.04	0.16	35.80	$SrI_2 \cdot 6H_2O+SrI_2 \cdot 2SrO \cdot 9H_2O$
60.77	0.18	39.05	$SrI_2 \cdot 2SrO \cdot 9H_2O$
55.58	0.46	43.96	»
55.09	0.51	44.40	»
54.16	0.60	45.24	»
53.04	0.72	46.24	»
49.37	0.74	49.89	$SrI_2 \cdot 2SrO \cdot 9H_2O+SrO \cdot 9H_2O$
39.50	0.55	59.95	$SrO \cdot 9H_2O$
29.98	0.53	69.49	»
22.36	0.57	77.07	»
0.0	0.85	99.15	»

№ 5349 STRONTIUM HYDROXIDE–STRONTIUM NITRATE–WATER [2306]

$Sr(OH)_2—Sr(NO_3)_2—H_2O$

$t=25$

Comp. of saturated soln., wt%			d_{25}	Comp. of solid phase
Strontium Hydroxide	Strontium Nitrate	Water		
0.00	44.21	55.79	1.481	$Sr(NO_3)_2$
0.211	44.18	55.609	1.492	»
0.428	44.50	55.072	1.494	»

Table № 5349 (continuation)

Comp. of saturated soln., wt%			d_{25}	Comp. of solid phase
Strontium Hydroxide	Strontium Nitrate	Water		
0.962	44.33	54.708	1.506	$Sr(OH)_2 \cdot 8H_2O$
0.972	42.20	56.828	1.490	»
0.914	38.56	60.526	1.419	»
0.894	35.69	63.426	1.381	»
0.856	31.68	67.464	1.327	»
0.799	27.21	71.991	1.267	»
0.778	22.18	77.042	1.217	»
0.761	19.09	80.149	1.178	»
0.765	15.10	84.135	1.148	»
0.739	11.24	88.021	1.108	»
0.738	8.162	91.100	1.079	»
0.737	5.874	93.389	1.059	»
0.741	4.228	95.031	1.033	»

№ 5350 STRONTIUM HYDROXIDE–STRONTIUM HYDROGEN SULFIDE–WATER [2640]

$Sr(OH)_2—Sr(SH)_2—H_2O$

Comp. of saturated soln., wt%			t	Comp. of saturated soln., wt%			t
Strontium Hydroxide	Strontium Hydrogen Sulfide	Water		Strontium Hydroxide	Strontium Hydrogen Sulfide	Water	
0.41	0.00	99.59	0	0.80	26.20	73.00	40
0.30	1.56	98.14	0	0.42	28.30	71.28	40
0.25	6.68	93.07	0	0.30	29.25	70.45	40
0.22	12.65	87.13	0	0.10	31.60	68.30	40
0.20	14.80	85.00	0	0.00	31.60	68.40	40
0.16	18.20	81.64	0	3.63	0.00	96.37	60
0.15	19.05	70.80	0	3.15	1.60	95.25	60
0.10	23.40	76.50	0	3.04	6.08	90.88	60
0.10	24.80	75.10	0	3.16	11.85	84.99	60
0.20	27.50	72.30	0	3.15	16.72	80.13	60
0.00	27.50	72.50	0	1.75	25.0	73.25	60
0.82	0.0	99.18	20	1.75	26.6	71.65	60
0.72	5.90	93.38	20	0.60	30.8	68.60	60
0.70	11.4	87.90	20	0.10	33.6	66.30	60
0.62	15.8	83.58	20	0.00	33.2	66.80	60
0.60	18.3	81.10	20	7.78	0.00	92.22	80
0.55	20.3	79.15	20	5.60	5.60	88.80	80
1.00	24.0	75.00	20	5.00	9.50	85.50	80
0.25	26.0	73.75	20	4.88	15.50	79.62	80
0.20	26.8	73.2	20	4.10	19.60	76.30	80
0.15	27.5	72.35	20	2.70	27.50	69.80	80
0.00	29.7	70.3	20	2.10	28.00	69.90	80
1.75	0.00	98.25	40	1.40	30.40	68.20	80
1.65	1.40	96.95	40	0.05	35.60	64.35	80
1.60	4.87	93.53	40	0.00	35.60	64.40	80
1.60	10.83	87.57	40	23.20	0.00	76.80	100
1.55	15.93	82.52	40	14.00	7.40	78.60	100
1.50	20.00	78.50	40	0.15	37.80	62.05	100
1.10	24.10	74.80	40	0.00	37.80	62.20	100

№ 5351 [2298]

STRONTIUM NITRITE–SILVER NITRITE–WATER

$Sr(NO_2)_2—AgNO_2—H_2O$

Concn. in aqueous soln., wt%		t
Strontium Nitrite	Silver Nitrite	
41.38	5.76	14

№ 5352 [725]

STRONTIUM NITRITE–CESIUM NITRITE–WATER

$Sr(NO_2)_2—CsNO_2—H_2O$

$t=25$

Comp. of saturated soln., wt%			Comp. of solid phase
Strontium Nitrite	Cesium Nitrite	Water	
41.50	0.00	58.50	$Sr(NO_2)_2 \cdot H_2O$
34.45	8.67	53.88	»
35.90	16.00	48.10	»
32.20	27.00	40.80	»
31.50	32.50	36.00	»
31.30	35.98	32.72	$Sr(NO_2)_2 \cdot H_2O + CsNO_2 \cdot Sr(NO_2)_2 \cdot H_2O$
31.00	36.00	33.00	»
30.98	36.28	32.74	»
31.22	36.60	32.18	»
24.77	39.68	35.55	$CsNO_2 \cdot Sr(NO_2)_2 \cdot H_2O$
16.20	50.80	33.00	»
10.80	60.90	28.30	»
8.38	67.30	24.32	»
5.07	77.00	17.93	»
4.37	81.00	14.63	$CsNO_2 + CsNO_2 \cdot Sr(NO_2)_2 \cdot H_2O$
4.47	81.19	14.34	»
0.00	85.63	14.37	$CsNO_2$

№ 5353 [160]

STRONTIUM NITRITE–THALLIUM NITRITE–WATER

$Sr(NO_2)_2—TlNO_2—H_2O$

$t=25$

Comp. of saturated soln., wt%			Comp. of solid phase
Thallium Nitrite	Strontium Nitrite	Water	
0.00	42.31	57.69	$Sr(NO_2)_2 \cdot H_2O$
4.21	42.25	53.54	»
9.14	39.01	51.85	»
15.26	35.81	48.93	»
17.50	35.75	46.75	»
23.21	33.66	43.13	$Sr(NO_2)_2 \cdot H_2O + Sr(NO_2)_2 \cdot TlNO_2 \cdot 3H_2O$

Table № 5353 (continuation)

Comp. of saturated soln., wt%			Comp. of solid phase
Thallium Nitrite	Strontium Nitrite	Water	
22.99	33.84	43.17	$Sr(NO_2)_2 \cdot H_2O + Sr(NO_2)_2 \cdot TlNO_2 \cdot 3H_2O$
24.98	32.07	42.95	$Sr(NO_2)_2 \cdot TlNO_2 \cdot 3H_2O$
27.88	30.84	41.28	»
29.49	30.42	40.09	»
33.00	28.30	38.70	$Sr(NO_2)_2 \cdot TlNO_2 \cdot 3H_2O + TlNO_2$
33.41	28.25	38.34	»
30.27	27.02	42.71	$TlNO_2$
28.84	23.88	47.28	»
27.62	20.64	51.74	»
26.68	15.30	58.02	»
27.98	7.32	64.70	»
32.25	0.00	67.75	»

№ 5354 STRONTIUM NITRATE–STRONTIUM CHLORIDE–WATER [1780]

$Sr(NO_3)_2—SrCl_2—H_2O$

$t=25$

Comp. of solvent, wt%		Solubility of Strontium Chloride, wt%	d_4
Strontium Nitrate	Water		
1.431	98.569	35.42	1.4113
5.750	94.250	34.47	1.4336
10.415	89.585	33.30	1.4636
25.985	74.015	28.97	1.6664

№ 5355 STRONTIUM NITRATE–STRONTIUM CHLORIDE–WATER [1543]

$Sr(NO_3)_2—SrCl_2—H_2O$

$t=25$

Comp. of saturated soln., wt%			d	Comp. of solid phase
Strontium Chloride	Strontium Nitrate	Water		
0.00	44.28	55.72	1.476	$Sr(NO_3)_2 \cdot 4H_2O$
6.32	48.69	44.99	1.492	»
5.54	38.57	55.89	1.488	»
10.79	35.14	54.07	1.508	»
10.87	34.28	54.84	1.510	»
12.11	33.71	54.18	1.513	$Sr(NO_3)_2$
15.10	30.15	54.75	1.515	»
18.83	27.25	63.92	1.523	»
23.07	22.75	54.18	1.537	»
28.45	19.62	51.93	1.565	»
28.97	18.79	52.24	1.568	$Sr(NO_3)_2 + SrCl_2 \cdot 6H_6O$
29.23	18.00	52.77	1.563	$SrCl_2 \cdot 6H_2O$
30.75	14.10	55.15	1.526	»
32.51	9.97	57.52	1.483	»
35.82	0.00	64.18	1.403	»

№ 5356 STRONTIUM NITRATE–STRONTIUM BROMIDE–WATER [1779]

$Sr(NO_3)_2—SrBr_2—H_2O$

$t=25$

Comp. of saturated soln., wt%			d_4^{24}
Strontium Bromide	Strontium Nitrate	Water	
51.59	0.00	48.41	1.700
51.41	0.37	48.22	—
51.24	0.73	48.03	1.703
50.88	1.47	47.65	1.728
50.07	3.04	46.89	1.737
48.61	5.88	45.51	1.748
43.50	15.96	40.54	1.773

№ 5357 STRONTIUM NITRATE–SILVER BROMATE–WATER [856]

$Sr(NO_3)_2—AgBrO_3—H_2O$

$t=25$

Concn. in aqueous soln., g/l			
Strontium Nitrate	Silver Bromate	Strontium Nitrate	Silver Bromate
0.00000	1.9475	6.3489	2.3578
0.21163	1.9806	21.163	2.8765
0.63489	2.0254	63.489	3.3717
2.1163	2.1126	211.63	4.5741

№ 5358 STRONTIUM NITRATE–LEAD (II) NITRATE–WATER [1627]

$Sr(NO_3)_2—Pb(NO_3)_2—H_2O$

$t=25$

Content in aqueous soln., mol %		n_D^{20}	Content in aqueous soln., mol %		n_D^{20}
Lead Nitrate	Strontium Nitrate		Lead Nitrate	Strontium Nitrate	
62.12	35.88	1.431	20.38	79.62	1.419
53.16	46.84	1.435	10.63	89.37	1.532
38.23	61.77	1.417	10.77	89.23	1.528
52.82	47.18	1.417	14.20	85.80	1.524
8.25	91.75	1.532			

№ 5359 STRONTIUM NITRATE–LEAD (II) BROMATE–WATER [2112]

$Sr(NO_3)_2—Pb(BrO_3)_2—H_2O$

$t=25$

Concn. in aqueous soln., g/l		d	Concn. in aqueous soln., g/l		d
Lead Bromate	Strontium Nitrate		Lead Bromate	Strontium Nitrate	
15.91	0.0	1.0112	41.39	162.75	1.1107
16.58	1.178	1.0117	45.09	122.81	1.1301
17.14	2.182	1.0129	51.39	161.11	1.1627
18.02	4.144	1.0144	57.83	199.88	1.1971
20.63	10.622	1.0226	71.07	290.78	1.2748
23.98	21.04	1.0332	81.72	376.47	1.3438
29.35	41.79	1.0548	90.52	457.03	1.4084
33.80	62.40	1.0741	97.46	533.17	1.4668
37.86	82.59	1.0933	103.16	607.38	1.5261

№ 5360 STRONTIUM NITRATE–URANYL NITRATE–WATER [1010]

$Sr(NO_3)_2—UO_2(NO_3)_2—H_2O$

$t=25$

Comp. of saturated soln., wt%			Comp. of solid phase
Uranyl Nitrate	Strontium Nitrate	Water	
55.87	0.00	44.13	$UO_2(NO_3)_2 \cdot 6H_2O$
52.64	1.09	46.35	»
51.55	2.65	45.80	»
51.39	4.78	43.83	$UO_2(NO_3)_2 \cdot 6H_2O + Sr(NO_3)_2$
49.76	5.55	44.69	$Sr(NO_3)_2$
47.48	7.00	45.52	»
46.05	7.97	45.98	»
45.24	8.02	46.74	»
40.39	11.39	48.22	»
37.33	13.88	48.70	$Sr(NO_3)_2 + Sr(NO_3)_2 \cdot 4H_2O$
34.10	14.42	51.48	$Sr(NO_3)_2 \cdot 4H_2O$
27.75	20.75	51.50	»
20.40	26.06	53.54	»
11.20	34.95	54.75	»
0.00	44.12	55.88	»

№ 5361 STRONTIUM DITHIONATE–LEAD (II) DITHIONATE–WATER [1627]

$SrS_2O_6—PbS_2O_6—H_2O$

$t=25$

Concn. in aqueous soln., mol %		d	Concn. in aqueous soln., mol %		d
Lead Dithionate	Strontium Dithionate		Lead Dithionate	Strontium Dithionate	
0.0	100.0	1.1126	75.75	24.25	1.3252
1.05	98.95	1.1184	78.09	21.91	1.3726
15.31	84.69	1.1503	88.29	11.71	1.4671
46.80	53.20	1.2147	100.0	0.00	1.6817
62.30	37.70	1.2889			

№ 5362 [1650]

STRONTIUM CHLORIDE–SILVER CHLORIDE–WATER

$SrCl_2$—AgCl—H_2O

$t=25$

Concn. in aqueous soln., g/l			
Strontium Chloride	Silver Chloride	Strontium Chloride	Silver Chloride
43.60	0.0047	196.3	0.1068
78.40	0.0132	237.2	0.1790
107.7	0.0247	277.0	0.2886
124.6	0.0337	329.1	0.5139
144.1	0.0498	413.5	1.169
169.6	0.0729	457.8	1.722

№ 5363 **STRONTIUM CHLORIDE–INDIUM CHLORIDE–WATER** [1568]

$SrCl_2$—$InCl_3$—H_2O

$t=20$

Concn. in aqueous soln., g/l		Comp. of solid phase
Indium Chloride	Strontium Chloride	
1349	192	$SrCl_2 \cdot 2InCl_3 \cdot 8H_2O$
1106	258	$SrCl_2 \cdot InCl_3 \cdot 6H_2O$
1045	265	»
1040	280	»
963	289	»
746	367	»
742	371	$SrCl_2 \cdot 6H_2O$

№ 5364 [1950]

STRONTIUM CHLORIDE–BARIUM SULFATE–WATER

$SrCl_2$—$BaSO_4$—H_2O

$t=20$

Content of Strontium Chloride in aqueous soln., mol/l	Solubility of Barium Sulfate in aqueous soln. of Strontium Chloride, g/l	Comp. of solid phase
0.150	0.0225	$BaSO_4$
0.075	0.0150	»
0.038	0.0120	»
0.019	0.0095	»
0.0095	0.0060	»
0.008	0.0050	»
0.0045	0.0040	»
0.0025	0.0045	»
0	0.0043	»

№ 5365 STRONTIUM CHLORIDE–MERCURY (II) CHLORIDE–WATER [1815]

$SrCl_2—HgCl_2—H_2O$

$t=25$

Concn. in aqueous soln., g/l			
Mercury Chloride	Strontium Chloride	Mercury Chloride	Strontium Chloride
85.63	26	364.4	114.9
153	49.35	482.5	166
225.4	82.3	623	219.6

№ 5366 STRONTIUM CHLORIDE–MERCURY (II) CHLORIDE–WATER [1162]

$SrCl_2—HgCl_2—H_2O$

$t=25$

Concn. in aqueous soln., wt%		Comp. of solid phase
Mercury Chloride	Strontium Chloride	
6.90	0.00	$HgCl_2$
34.70	9.30	»
48.21	13.93	»
56.62	16.65	»
57.17	17.34	»
58.30	18.04	$HgCl_2+[Sr(H_2O)_8]\cdot[(HgCl_2)_3Cl_2]$
58.46	18.01	»
58.68	18.47	$HgCl_2+[Sr(H_2O)_6]\cdot[(HgCl_2)_6\cdot Cl_2]$
58.49	19.42	$[Sr(H_2O)_6]\cdot[(HgCl_2)_6\cdot Cl_2]$
58.46	19.53	»
59.49	19.66	»
58.13	18.00	$[Sr(H_2O)_8]\cdot[(HgCl_2)_3Cl_2]$
56.20	19.20	»
55.95	19.52	»
55.45	19.82	»
54.33	20.88	»
52.70	21.90	»
51.20	22.50	»
50.87	22.98	»
48.90	23.69	$[Sr(H_2O)_8]\cdot[(HgCl_2)_3Cl_2]+SrCl_2\cdot 6H_2O$
48.81	24.01	»
48.77	23.90	$[Sr(H_2O)_8]\cdot[(HgCl_2)_3Cl_2]$
0.00	40.00	$[Sr(H_2O)_8]\cdot[(SrCl_2)_3Cl_2]$
48.20	23.60	$SrCl_2\cdot 6H_2O$
46.25	24.02	»
45.32	24.26	»
34.52	26.78	»
31.02	27.50	»
18.82	30.50	»
8.55	33.42	»
0.0	35.84	»

№ 5367 STRONTIUM CHLORIDE–THALLIUM CHLORIDE–WATER [1197]

$SrCl_2—TlCl—H_2O$

$t = 25$

Concn. in aqueous soln., wt%		Comp. of solid phase	Concn. in aqueous soln., wt%		Comp. of solid phase
Thallium Chloride	Strontium Chloride		Thallium Chloride	Strontium Chloride	
0.039	4.760	TlCl	0.053	28.55	TlCl
0.036	9.087	»	0.065	33.31	»
0.037	16.66	»	0.087	34.78	»
0.047	23.06	»			

№ 5368 STRONTIUM CHLORIDE–LEAD (II) CHLORIDE–WATER [1817]

$SrCl_2—PbCl_2—H_2O$

$t=25$

Concn. of Strontium Chloride in aqueous soln., g/l	Solubility of Lead Chloride in aqueous soln. of Strontium Chloride, g/l	Concn. of Strontium Chloride in aqueous soln., g/l	Solubility of Lead Chloride in aqueous soln. of Strontium Chloride, g/l
40.4	1.886	210.9	3.945
80.9	1.780	245.1	4.390
109.4	2.292	327.6	7.506

№ 5369 STRONTIUM CHLORIDE–LEAD (II) CHLORIDE–WATER [1965]

$SrCl_2—PbCl_2—H_2O$

$t=25$

Concn. in aqueous soln., g/l		Comp. of solid phase
Lead Chloride	Strontium Chloride	
3.66	15.86	$SrCl_2 \cdot PbCl_2$
3.48	25.38	»
3.45	47.58	»
3.51	63.45	»
3.64	88.83	»
3.92	115.8	»
4.15	134.8	»
4.54	163.4	»
4.59	168.1	»
5.15	202.2	»
5.52	210.9	»
5.60	222.1	»
5.98	234.8	»
6.99	263.3	»
7.12	266.4	»
7.56	275.2	»
8.36	288.7	»
9.04	299.8	»
9.65	307.7	»

№ 5370 STRONTIUM CHLORIDE–LEAD (II) BROMIDE–WATER [1817]

$SrCl_2—PbBr_2—H_2O$

$t=25$

Concn. of Strontium Chloride in aqueous soln., g/l	Solubility of Lead Bromide in aqueous soln. of Strontium Chloride, g/l	Concn. of Strontium Chloride in aqueous soln., g/l	Solubility of Lead Bromide in aqueous soln. of Strontium Chloride, g/l
0.00	9.63	210.94	5.67
52.34	2.07	245.04	7.34
105.47	2.30	331.47	9.97

№ 5371 STRONTIUM BROMIDE–INDIUM BROMIDE–WATER [1568]

$SrBr_2—InBr_2—H_2O$

$t=20$

Concn. in aqueous soln., g/l		Comp. of solid phase
Indium Bromide	Strontium Bromide	
1795	141	$SrBr_2 \cdot 2InBr_3 \cdot 9H_2O$
1653	282	»
1625	294	»
1591	295	»
1498	378	»
1412	401	»
1400	424	$SrBr_2 \cdot InBr_3 \cdot 10H_2O$
1260	457	»
1248	472	$2SrBr_2 \cdot InBr_3 \cdot 16H_2O$

№ 5372 STRONTIUM BROMIDE–MERCURY (II) BROMIDE–WATER [1815]

$SrBr_2—HgBr_2—H_2O$

$t=25$

Concn. in aqueous soln., g/l	
Mercury Bromide	Strontium Bromide
37.45	15.33
169.4	81.2
325	165.4
637	346.5
806.	464

№ 5373 STRONTIUM BROMIDE–LEAD (II) CHLORIDE–WATER [1817]

$SrBr_2—PbCl_2—H_2O$

$t=25$

Concn. of Strontium Bromide in aqueous soln., g/l	Solubility of Lead Bromide in aqueous soln. of Strontium Bromide, g/l
64.3	1.75
128.6	3.17
257.4	13.56
386.0	38.87
514.7	111.1

№ 5374 STRONTIUM BROMIDE–LEAD (II) BROMIDE–WATER [1817]

$SrBr_2—PbBr_2—H_2O$

$t=25$

Concn. of Strontium Bromide in aqueous soln., g/l	Solubility of Lead Bromide in aqueous soln. of Strontium Bromide, g/l	Concn. of Strontium Bromide in aqueous soln., g/l	Solubility of Lead Bromide in aqueous soln. of Strontium Bromide, g/l
0.0	9.63	257.50	16.03
64.38	2.47	386.26	57.22
128.75	4.67	515.01	208.7

№ 5375 STRONTIUM IODIDE–INDIUM IODIDE–WATER [1568]

$SrI_2—InI_3—H_2O$

$t=20$

Concn. in aqueous soln., g/l		Comp. of solid phase
Indium Iodide	Strontium Iodide	
1209	748	$SrI_2 \cdot InI_3 \cdot 10H_2O$
1023	793	»
950	806	»
932	810	»
591	962	»
130	1157	»
95	1138	$2SrI_2 \cdot InI_3 \cdot 18H_2O$
89	1287	$SrI_2 \cdot H_2O$

№ 5376 STRONTIUM IODIDE–MERCURY (II) IODIDE–WATER [1815]

$SrI_2—HgI_2—H_2O$

$t=25$

Concn. in aqueous soln., g/l			
Mercury Iodide	Strontium Iodide	Mercury Iodide	Strontium Iodide
96.3	86.75	264.6	184.2
145.4	121.4	315.7	207.6

№ 5377 [1906]

YTTRIUM HYDROXIDE–YTTRIUM NITRATE–WATER

$Y(OH)_3—Y(NO_3)_3—H_2O$

$t=25$

Comp. of saturated soln., wt%			d	Comp. of solid phase
Yttrium Nitrate	Yttrium Hydroxide	Water		
3.03	0.01	96.96	1.0260	$Y(OH)_3$
12.17	0.03	87.80	1.1106	»
19.95	0.05	80.00	1.1907	»
24.79	0.12	75.09	1.2517	$Y(OH)_3+3Y_2O_3 \cdot 4N_2O_5 \cdot 2H_2O$

Comp. of saturated soln., wt%			d	Comp. of solid phase
Yttrium Nitrate	Yttrium Hydroxide	Water		
30.69	0.08	69.23	1.3268	$3Y_2O_3 \cdot 4N_2O_5 \cdot 2H_2O$
36.93	0.06	63.01	1.4104	»
42.18	0.04	57.78	1.4867	»
47.08	0.04	52.87	1.5587	»
50.91	0.04	49.05	1.6259	»
55.01	0.03	44.96	1.6931	»
57.80	0.03	42.17	1.7440	$3Y_2O_3 \cdot 4N_2O_5 \cdot 2H_2O+Y(NO_3)_3$
58.60	0.0	41.40	1.7446	$Y(NO_3)_3$

№ 5378 SILVER OXIDE–IODINE HEPTOXIDE–WATER [1762]

$Ag_2O—I_2O_7—H_2O$

Comp. of saturated soln., wt%			t	Comp. of solid phase
Silver Oxide	Iodine Heptoxide	Water		
Traces	Traces	100.00	0	$Ag_2O+5Ag_2O \cdot I_2O_7$
»	0.0235	99.98	0	$5Ag_2O \cdot I_2O_7$
»	0.0252	99.97	0	»
0.0016	0.0306	99.97	0	$5Ag_2O \cdot I_2O_7+2Ag_2O \cdot I_2O_7 \cdot 3H_2O$
0.0025	1.321	98.68	0	$2Ag_2O \cdot I_2O_7 \cdot 3H_2O$
0.0039	13.08	86.93	0	»
0.0064	21.87	78.12	0	»
0.0175	35.48	64.50	0	»
0.0236	39.13	60.85	0	»
0.0460	49.72	50.23	0	$2Ag_2O \cdot I_2O_7 \cdot 3H_2O+$ $+Ag_2O \cdot I_2O_7 \cdot 4H_2O$
0.0398	52.11	47.85	0	$Ag_2O \cdot I_2O_7 \cdot 4H_2O$
0.0409	57.77	42.19	0	»
0.0203	60.95	39.03	0	»
Traces	62.53	37.47	0	$Ag_2O \cdot I_2O_7 \cdot 4H_2O+I_2O_7 \cdot 5H_2O$
»	62.51	37.49	0	$I_2O_7 \cdot 5H_2O$
»	0.00	100.00	35	$Ag_2O+5Ag_2O \cdot I_2O_7$
»	0.00	100.00	35	$5Ag_2O \cdot I_2O_7$
»	0.04	99.96	35	»
»	0.05	99.95	35	»
»	0.06	99.94	35	$5Ag_2O \cdot I_2O_7+2Ag_2O \cdot I_2O_7 \cdot H_2O$
»	0.18	99.82	35	$2Ag_2O \cdot I_2O_7 \cdot H_2O$
»	1.22	98.78	35	»
»	3.30	96.70	35	»
»	15.53	84.47	35	»
»	26.41	73.59	35	»
»	34.50	65.50	35	»
»	50.83	49.17	35	$2Ag_2O \cdot I_2O_7 \cdot H_2O+Ag_2O \cdot I_2O_7$
»	56.56	43.44	35	$Ag_2O \cdot I_2O_7$
»	61.01	38.99	35	»
»	63.73	36.27	35	$Ag_2O \cdot I_2O_7+I_2O_7 \cdot 5H_2O$
»	63.81	36.19	35	$I_2O_7 \cdot 5H_2O$

№ 5379 SILVER CYANIDE–SILVER NITRATE–WATER [1795]

$AgCN—AgNO_3—H_2O$

$t=25$

Concn. in aqueous soln., g/l	
Silver Cyanide	Silver Nitrate
1.216	501

№ 5380 SILVER THIOCYANATE–BARIUM THIOCYANATE–WATER [2247]

$AgCNS—Ba(CNS)_2—H_2O$

$t=25$

Concn. in aqueous soln., wt%		Comp. of solid phase
Barium Thiocyanate	Silver Thiocyanate	
62.61	0.00	$Ba(CNS)_2 \cdot 3H_2O$
61.39	4.29	»
58.91	12.58	»
57.36	17.04	»
55.99	22.58	»
55.91	22.98	$Ba(CNS)_2 \cdot AgCNS \cdot 2H_2O + Ba(CNS)_2 \cdot 3H_2O$
55.20	23.91	$Ba(CNS)_2 \cdot AgCNS \cdot 2H_2O$
54.14	25.01	»
53.23	26.35	»
52.61	27.89	»
52.01	29.17	$Ba(CNS)_2 \cdot AgCNS \cdot 2H_2O + Ba(CNS)_2 \cdot 2AgCNS \cdot 2H_2O$
47.74	29.81	$Ba(CNS)_2 \cdot 2AgCNS \cdot 2H_2O$
44.28	31.67	»
40.61	34.70	»
39.54	35.79	$Ba(CNS)_2 \cdot 2AgCNS \cdot 2H_2O + Ba(CNS)_2 \cdot 3AgCNS \cdot 2H_2O$
39.02	35.79	$Ba(CNS)_2 \cdot 3AgCNS \cdot 2H_2O$
38.41	35.13	»
37.49	34.58	»
37.13	34.28	»
36.78	34.01	»
36.04	33.79	$Ba(CNS)_2 \cdot 3AgCNS \cdot 2H_2O + AgCNS$
34.50	27.99	AgCNS
32.20	23.24	»
27.00	14.67	»
18.18	4.47	»
0.68	0.75	»

№ 5381 SILVER NITRITE–SILVER NITRATE–WATER [2243]

$AgNO_2—AgNO_3—H_2O$

$t=18$

Concn. in aqueous soln., g/l	
Silver Nitrate	Silver Nitrite
0.000	3.184
0.439	3.042
0.878	2.926
1.756	2.601
3.512	2.201
7.024	1.799
14.048	1.480

№ 5382 SILVER NITRITE–SILVER NITRATE–WATER [1435]

$AgNO_2—AgNO_3—H_2O$

$t=25$

Concn. in aqueous soln., g/l	
Silver Nitrate	Silver Nitrite
0.000	4.135
0.438	3.991
1.00	3.735
2.00	3.432
4.00	2.943
8.00	2.498

№ 5383 SILVER NITRITE–BARIUM NITRITE–WATER [2298]

$AgNO_2—Ba(NO_2)_2—H_2O$

$t=13.5$

Comp. of saturated soln., wt%			Comp. of solid phase
Barium Nitrite	Silver Nitrite	Water	
36.74	5.85	57.41	$AgNO_2+Ag_2Ba(NO_2)_4\cdot H_2O$
40.84	5.13	54.03	$Ba(NO_2)_2+Ag_2Ba(NO_2)_4\cdot H_2O$

№ 5384 SILVER NITRITE–BARIUM NITRITE–WATER [2241]

$AgNO_2—Ba(NO_2)_2—H_2O$

$t=21$

Comp. of saturated soln., wt%			Comp. of solid phase
Silver Nitrite	Barium Nitrite	Water	
4.81	34.88	60.31	$AgNO_2$
5.48	37.31	57.21	»
5.92	39.00	55.08	$AgNO_2+Ba(NO_2)_2\cdot AgNO_2\cdot H_2O$
5.93	39.27	54.80	»
5.55	40.85	53.60	$Ba(NO_2)_2\cdot AgNO_2\cdot H_2O$
5.46	42.57	51.97	»
5.27	43.37	51.36	»
5.20	43.67	51.13	»
5.18	44.43	50.39	$Ba(NO_2)_2\cdot AgNO_2\cdot H_2O+Ba(NO_2)_2\cdot H_2O$
5.22	44.79	49.99	»
2.49	43.19	54.32	$Ba(NO_2)_2\cdot H_2O$
0.00	42.31	57.69	»

№ 5385 SILVER NITRATE–SILVER SULFATE–WATER [2079]

$AgNO_3—Ag_2SO_4—H_2O$

$t=25$

Comp. of saturated soln., wt%					
Silver Nitrate	Silver Sulfate	Water	Silver Nitrate	Silver Sulfate	Water
0.167	0.736	99.097	0.453	0.558	98.989
0.295	0.654	99.051	0.755	0.403	98.842

№ 5386 [1795]

SILVER NITRATE–SILVER CHLORIDE–WATER

$AgNO_3—AgCl—H_2O$

Concn. in aqueous soln., g/l		t
Silver Nitrate	Silver Chloride	
510	0.8	25

№ 5387 [2332]

SILVER NITRATE–SILVER CHLORIDE–WATER

$AgNO_3—AgCl—H_2O$

$t=25$

Concn. in aqueous soln., g/l	
Silver Nitrate	Silver Chloride
0.00849	0.0407
0.01699	0.0205
0.05097	0.0090
0.1699	0.0062
0.8495	0.0053
1.699	0.0070
8.495	0.0244
16.99	0.0416
33.98	0.0788
67.96	0.1734
135.7	0.4716
169.6	0.6966

№ 5388 [2078]

SILVER NITRATE–SILVER CHLORIDE–WATER

$AgNO_3—AgCl—H_2O$

$t=20$

Concn. of Silver Nitrate in aqueous soln., g/l	Solubility of Silver Chloride in aqueous soln. of Silver Nitrate, g/l × 10^4
0.1699	0.416
0.3398	0.373
0.8495	0.373
1.6990	0.402
3.3980	0.474
8.495	0.645
16.990	1.020
33.980	2.898
84.950	15.61
169.90	69.8
339.80	436

№ 5389 **SILVER NITRATE–SILVER PERMANGANATE–WATER** [1794]

$AgNO_3—AgMnO_4—H_2O$

$t=20$

Concn. in aqueous soln., g/l			
Silver Nitrate	Silver Permanganate	Silver Nitrate	Silver Permanganate
0.00	9.222	137.4	1.882
10.05	6.078	195.6	1.905
19.88	3.901	287.3	2.064
30.02	3.243	373.2	2.563
30.27	3.130	456.8	2.994
39.89	2.812	537.7	3.402
49.11	2.472	615.4	3.992
59.73	2.268	1029.8	7.802
78.37	2.155	1421.3	12.202
97.65	2.087	1535.3	14.221
118.0	1.996		

№ 5390 **SILVER NITRATE–SILVER BROMIDE–WATER** [1795]

$AgNO_3—AgBr—H_2O$

$t=25$

Concn. in aqueous soln., g/l	
Silver Nitrate	Silver Bromide
509.7	0.4

№ 5391 SILVER NITRATE–SILVER BROMIDE–WATER [2710]

$AgNO_3$—$AgBr$—H_2O

t=20

Concn. in aqueous soln., g/l			
Silver Bromide	Silver Nitrate	Silver Bromide	Silver Nitrate
0.0001878	19.54	0.02817	183.49
0.0002817	24.64	0.04695	219.17
0.0004695	36.53	0.07512	244.66
0.0007512	42.48	0.1127	285.43
0.0011268	53.52	0.1878	341.50
0.001878	71.36	0.2817	382.28
0.002817	81.55	0.4695	468.92
0.004695	96.84	0.3944	509.7
0.007512	117.23	0.7512	535.18
0.011268	127.42	1.1268	586.16
0.01878	158.01	1.878	688.10

№ 5392 [2078]

SILVER NITRATE–SILVER BROMIDE–WATER

$AgNO_3$—$AgBr$—H_2O

t=20

Concn. of Silver Nitrate in aqueous soln., g/l	Solubility of Silver Bromide in aqueous soln. of Silver Nitrate, g/l	Concn. of Silver Nitrate in aqueous soln., g/l	Solubility of Silver Bromide in aqueous soln. of Silver Nitrate, g/l
1.699	$0.864 \cdot 10^{-4}$	33.980	$61.00 \cdot 10^{-4}$
3.388	$1.442 \cdot 10^{-4}$	84.95	$6.06 \cdot 10^{-2}$
8.48	$3.300 \cdot 10^{-4}$	169.90	0.3562
16.990	$10.82 \cdot 10^{-4}$	339.80	1.740

№ 5393 [2262]

SILVER NITRATE–
SILVER BROMATE–WATER

$AgNO_3$—$AgBrO_3$—H_2O

t=24.5

Concn. in aqueous soln., g/l	
Silver Bromate	Silver Nitrate
1.911	0.0
1.203	1.445
0.510	5.882

№ 5394 [2405]

SILVER NITRATE–SILVER MOLYBDATE–WATER

$AgNO_3—Ag_2MoO_4—H_2O$

$t=25$

Concn. of Silver Nitrate in saturated soln., wt%	Comp. of solid phase
12.32	Ag_2MoO_4
23.54	»
34.87	»
50.56	»
57.84	»
68.14	»
71.81	$Ag_2MoO_4+NaNO_3$
71.80	»
71.82	$NaNO_3$

Note. The concn. of Silver Molybdate was less than $4 \cdot 10^{-6}$ wt%.

№ 5395 [2052]

SILVER NITRATE–SILVER IODIDE–WATER

$AgNO_3—AgI—H_2O$

Comp. of solvent	Solubility of Silver Iodide wt%	t
Saturated aqueous soln. of Silver Nitrate	2.25	11
»	10.95	b.p.

№ 5396 [1795]

SILVER NITRATE–SILVER IODIDE–WATER

$AgNO_3—AgI—H_2O$

$t=25$

Concn. in aqueous soln., g/l		Comp. of solid phase
Silver Nitrate	Silver Iodide	
34.0	0.068	AgI
59.5	0.121	»
84.9	0.299	»
118.9	0.850	»
206.4	3.08	Ag_2INO_3
276.9	6.26	»
346.6	10.9	»
431.5	16.1	$Ag_3I(NO_3)_2$
637.1	33.2	»
796.8	53.2	»
1002.4	85.0	»

№ 5397 [2078]

SILVER NITRATE–SILVER IODIDE–WATER

$AgNO_3$—AgI—H_2O

$t=20$

Concn. of Silver Nitrate in aqueous soln., g/l	Solubility of Silver Iodide in aqueous soln of Silver Nitrate, g/l
0.03398	$2.119 \cdot 10^{-7}$
0.08495	$4.46 \cdot 10^{-7}$
0.1699	$1.035 \cdot 10^{-6}$
0.3398	$2.680 \cdot 10^{-6}$
1.699	$5.26 \cdot 10^{-4}$
3.388	$2.03 \cdot 10^{-3}$
8.48	$1.305 \cdot 10^{-3}$
16.990	$6.10 \cdot 10^{-2}$
33.980	0.3192
84.95	3.192
169.90	19.650
339.80	83.60

№ 5398 [2406]

SILVER NITRATE–SILVER IODATE–WATER

$AgNO_3$—$AgIO_3$—H_2O

$t=25$

Comp. of saturated soln., wt%					
Silver Nitrate	Barium Iodate	Water	Silver Nitrate	Silver Iodate	Water
22.65	0.009	77.341	70.43	0.035	29.535
48.47	0.006	51.524	71.84	0.041	28.119

№ 5399 [1121]

SILVER NITRATE–BARIUM NITRATE–WATER

$AgNO_3$—$Ba(NO_3)_2$—H_2O

$t=30$

Comp. of saturated soln., wt%			Comp. of solid phase
Silver Nitrate	Barium Nitrate	Water	
0.00	10.40	89.60	$Ba(NO_3)_2$
8.07	7.36	84.57	»
20.69	5.77	73.54	»
26.17	5.05	68.78	»
41.39	4.69	54.02	»
44.11	4.57	51.32	»
61.87	3.58	34.55	»
71.62	2.57	25.81	$Ba(NO_3)_2+AgNO_3$
73.06	0.68	26.26	$AgNO_3$
73.00	0.00	27.00	»

№ 5400 SILVER NITRATE–LEAD (II) NITRATE–WATER [2240]

$AgNO_3—Pb(NO_3)_2—H_2O$

$t=25$

Comp. of saturated soln., wt%			Comp. of solid phase
Lead Nitrate	Silver Nitrate	Water	
37.17	0.00	62.83	$Pb(NO_3)_2$
34.29	11.21	54.50	»
31.11	23.28	45.61	»
29.17	28.81	41.02	»
27.21	35.28	37.51	»
24.84	41.84	33.32	»
22.79	46.86	30.35	»
22.39	49.25	28.36	»
18.68	56.63	24.69	»
17.43	58.68	23.89	$Pb(NO_3)_2+AgNO_3$
17.41	58.70	23.89	»
17.50	58.53	23.97	»
11.50	63.17	25.33	$AgNO_3$
6.79	66.79	26.42	»
0.00	71.00	29.00	»

№ 5401 SILVER SULFATE–CESIUM SULFATE–WATER [1516]

$Ag_2SO_4—Cs_2SO_4—H_2O$

Concn. in aqueous soln., wt%		d	t	Comp. of solid phase
Silver Sulfate	Cesium Sulfate			
0.833	0.0	1.004	25	Ag_2SO_4
0.76	5.93	1.055	25	»
0.86	13.94	1.136	25	»
0.98	20.52	1.211	25	»
1.07	27.53	1.306	25	»
1.19	35.16	1.411	25	»
1.22	36.19	1.420	25	»
1.28	41.19	1.514	25	»
1.38	46.96	1.623	25	»
1.43	48.11	1.635	25	»
1.43	51.99	1.743	25	»
1.53	57.94	1.871	25	»
1.57	62.74	2.005	25	»
1.57	62.99	2.011	25	»
1.57	63.72	2.048	25	$Ag_2SO_4+Cs_2SO_4$
1.58	63.71	2.049	25	»
1.57	63.80	2.054	25	»
1.57	63.79	2.045	25	»
1.57	63.76	2.049	25	»
0.57	64.21	2.016	25	Cs_2SO_4
1.0	64.53	2.007	25	»
1.80	64.28	2.057	35	$Ag_2SO_4+Cs_2SO_4$
1.80	64.27	2.048	35	»
1.80	64.22	2.068	35	»
1.80	64.31	2.059	35	»
1.80	64.27	2.058	35	»

№ 5402 [2079]

SILVER SULFATE–LANTHANUM SULFATE–WATER

$Ag_2SO_4—La_2(SO_4)_3—H_2O$

$t=25$

Comp. of saturated soln., wt%		
Lanthanum Sulfate	Silver Sulfate	Water
0.11	0.84	99.05
0.28	0.85	98.87
0.54	0.85	98.61

№ 5403 [2619]

SILVER SULFATE–VARIOUS SUBSTANCES–WATER

$Ag_2SO_4—\ ——\ —H_2O$

$t=25$

Substance		Concn. in aqueous soln., g/l	
Name	Formula	Substance	Silver Sulfate
Nitric Acid	HNO_3	0.0	8.41
»	»	1.0	9.33
»	»	2.00	10.18
»	»	4.00	11.83
Sulfuric Acid	H_2SO_4	0.0	8.41
»	»	1.42	8.55
»	»	2.84	8.68
»	»	5.16	8.86
Potassium Sulfate	K_2SO_4	2.36	7.93
»	»	4.73	7.68
Potassium Hydrogen Sulfate	$KHSO_4$	7.16	8.13
»	»	14.32	8.07

№ 5404 [1777]

SILVER SULFATE–SALTS–WATER

$Ag_2SO_4—MeX—H_2O$

$t=25$

Salt		Concn. in aqueous soln., g/l		d_{25}
Name	Formula	Salt	Silver Sulfate	
Potassium Nitrate	KNO_3	0.0	8.344	—
»	»	2.516	8.996	1.0072
»	»	5.027	9.531	1.0092
»	»	10.09	10.435	1.0034
Magnesium Nitrate	$Mg(NO_3)_2$	1.837	9.267	1.0073
»	»	3.675	10.029	1.0094
»	»	7.380	11.334	1.0133
Silver Nitrate	$AgNO_3$	4.989	6.095	1.0065
»	»	9.967	4.487	1.0084
»	»	19.912	2.644	1.0137

Table № 5404 (continuation)

Salt		Concn. in aqueous soln., g/l		d_{25}
Name	Formula	Salt	Silver Sulfate	
Potassium Sulfate	K_2SO_4	2.177	7.899	1.0064
»	»	4.354	6.694	1.0079
»	»	8.700	7.490	1.0112
»	»	17.40	7.531	1.0180
Magnesium Sulfate	$MgSO_4$	1.205	8.140	1.0061
»	»	3.014	7.941	1.0079
»	»	6.020	7.740	1.0105
»	»	12.04	7.733	1.0164

№ 5405 [2344]

SILVER CHLORIDE–LANTHANUM NITRATE–WATER

$AgCl—La(NO_3)_3—H_2O$

$t=25$

Concn. in aqueous soln., g/l	
Lathanum Nitrate	Silver Chloride
0.0	0.00183
0.00139	0.00183
0.0248	0.00187
0.0467	0.00188
0.1878	0.00195
0.3754	0.00201
0.5395	0.00205
0.9123	0.00211
1.243	0.00215
1.657	0.00220
2.154	0.00224

№ 5406 [2344]

SILVER CHLORIDE–LANTHANUM SULFATE–WATER

$AgCl—La_2(SO_4)_3—H_2O$

$t=25$

Concn. in aqueous soln., g/l	
Lanthanum Sulfate	Silver Chloride
0.0	0.00183
0.00121	0.00183
0.00373	0.00187
0.0138	0.00191
0.0585	0.00195
0.1396	0.00198
0.2339	0.00204
0.3786	0.00211
0.5425	0.00216
0.7330	0.00223
1.293	0.00225
1.509	0.00229

№ 5407 [2344]

SILVER CHLORIDE–BARIUM NITRATE–WATER

$AgCl—Ba(NO_3)_2—H_2O$

$t=25$

Concn. in aqueous soln., g/l			
Barium Nitrate	Silver Chloride	Barium Nitrate	Silver Chloride
0.0	0.00183	0.8059	0.00203
0.00167	0.00183	1.151	0.00207
0.00945	0.00185	1.464	0.00210
0.0552	0.00187	2.195	0.00215
0.1847	0.00191	2.925	0.00218
0.3918	0.00196	3.657	0.00221
0.5730	0.00199		

№ 5408 SILVER CHLORIDE–BARIUM CHLORIDE–WATER [1650

$AgCl—BaCl_2—H_2O$

$t=25$

Concn. in aqueous soln., g/l			
Barium Chloride	Silver Chloride	Barium Chloride	Silver Chloride
129.95	0.0266	278.65	0.1822
167.65	0.0485	339.46	0.3383

№ 5409 SILVER CHLORIDE*–MERCURY (I) NITRATE–WATER [2221]

$AgCl—HgNO_3—H_2O$

$t=25$

Concn. in aqueous soln., g/l			
Mercury Nitrate	Silver Chloride	Mercury Nitrate	Silver Chloride
2.626	0.620	13.13	1.310
3.282	0.715	26.26	2.000
6.565	0.990	262.6	6.896

* Identical results were obtained for crystalline and amorphous Silver Chloride. Varying the concn. of the free Nitric Acid between the limits of 0.1N and 2N did not show any effect.

№ 5410 SILVER CHLORIDE–SALTS–WATER [2681]

$AgCl—MeX—H_2O$

Salt			Solubility of Silver Chloride in aqueous soln. of salt, wt%	t
Name	Formula	Concn. in aqueous soln., wt%		
Barium Chloride	$BaCl_2$	27.32	0.057	24.5
»	»	Saturated	0.014	20
Barium Thiocyanate	$Ba(CNS)_2$	9.1	0.20	25
Calcium Thiocyanate	$Ca(CNS)_2$	9.1	0.15	25
Calcium Chloride	$CaCl_2$	41.26	0.571	24.5
»	»	Saturated	0.093	20
Copper (II) Chloride	$CuCl_2$	»	0.053	24.5
Iron (II) Chloride	$FeCl_2$	»	0.169	24.5
Iron (III) Chloride	$FeCl_3$	»	0.006	24.5
Manganese (II) Chloride	$MnCl_2$	»	0.013	24.5
Magnesium Chloride	$MgCl_2$	33.3	0.50	25
»	»	36.35	0.531	24.5
»	»	Saturated	0.171	20
Strontium Chloride	$SrCl_2$	»	0.088	20
Zinc Chloride	$ZnCl_2$	»	0.0134	24.5
Potassium Chloride	KCl	»	0.0475	20
»	»	24.95	0.0776	19.6
Potassium Cyanide	KCN	4.75	2.75	25
Potassium Thiocyanate	KCNS	9.1	0.11	25
Sodium Chloride	NaCl	25.95	0.105	19.6
»	»	Saturated	0.095	20

№ 5411 SILVER CHLORIDE–VARIOUS SUBSTANCES–WATER [1651]

$AgCl - \text{——} - H_2O$

$t = 25$

Substance		Concn in aqueous soln., g/l	
Name	Formula	Substance*	Silver Chloride
Sodium Chloride	$NaCl$	20.46	0.00027
Ammonium Chloride	NH_4Cl	18.71	0.00023
Potassium Chloride	KCl	26.06	0.00025
Rubidium Chloride	$RbCl$	42.31	0.00026
Calcium Chloride	$CaCl_2$	19.42	0.00025
Strontium Chloride	$SrCl_2$	27.74	0.00246
Barium Chloride	$BaCl_2$	36.45	0.00224
Hydrogen Chloride	HCl	12.76	0.00031

* In all cases 0.35 N soln.

№ 5412 SILVER CHLORIDE–VARIOUS SUBSTANCES–WATER [1485]

$AgCl - \text{——} - H_2O$

Comp. of solvent, mol.	Wt of one equivalent of solvent, g	Solubility of Silver Chloride mmol/g-eq solvent	t
$\frac{1}{2}CaCl_2 + 5H_2O$	145.58	4.52	20
»	145.58	5.96	40
»	145.58	7.57	60
»	145.58	9.81	80
$LiCl + 5H_2O$	132.48	2.85	20
»	132.48	4.19	40
»	132.48	5.77	60
»	132.48	7.91	80
$NH_4Cl + 10H_2O$	233.66	1.978	20
»	233.66	2.91	40
»	233.66	4.22	60
»	233.66	6.10	80
$NaCl + 10H_2O$	238.62	1.271	20
»	238.62	1.913	40
»	238.62	2.88	60
»	238.62	4.41	80
$\frac{1}{2}CaCl_2 + 10H_2O$	235.66	1.096	20
»	235.66	1.607	40
»	235.66	2.41	60
»	235.66	3.55	80
$LiCl + 10H_2O$	222.56	0.636	20

Table № 5412 (continuation)

Comp. of solvent, mol.	Wt of one equivalent of solvent, g	Solubility of Silver Chloride mmol/g-eq solvent	t
$LiCl+10H_2O$	222.56	1.087	40
»	222.56	1.775	60
»	222.56	2.83	80
$KCl+10H_2O$	216.63	0.599	20
»	216.63	0.959	40
»	216.63	1.514	60
»	216.63	2.285	80
$KCl+20H_2O$	434.9	0.416	20
»	434.9	0.733	40
»	434.9	1.309	60
»	434.9	2.250	80
$NH_4Cl+20H_2O$	413.8	0.386	20
»	413.8	0.683	40
»	413.8	1.185	60
»	413.8	2.075	80
$NaCl+20H_2O$	418.8	0.269	20
»	418.8	0.504	40
»	418.8	0.938	60
»	418.8	1.668	80
$\frac{1}{2}CaCl_2+20H_2O$	415.8	0.248	20
»	415.8	0.457	40
»	415.8	0.841	60
»	415.8	1.449	80
$LiCl+20H_2O$	402.7	0.171	20
»	402.7	0.360	40
»	402.7	0.712	60
»	402.7	1.304	80
$HCl+20H_2O$	396.8	0.175	20
»	396.8	0.349	40
»	396.8	0.663	60
»	396.8	1.200	80
$KCl+30H_2O$	615.0	0.176	20
»	615.0	0.368	40
»	615.0	0.721	60
»	615.0	1.365	80
$NH_4Cl+30H_2O$	594.0	0.183	20
»	594.0	0.367	40
»	594.0	0.709	60
»	594.0	1.321	80
$\frac{1}{2}CaCl_2+30\ H_2O$	596.0	0.133	20
»	596.0	0.273	40
»	596.0	0.548	60
»	596.0	1.042	80
$LiCl+30H_2O$	582.9	0.100	20
»	582.9	0.228	40
»	582.9	0.481	60
»	582.9	0.955	80
$HCl+30H_2O$	577.0	0.099	20
»	577.0	0.226	40
»	577.0	0.461	60
»	577.0	0.903	80

№ 5413 SILVER CHROMATE–NITRATES–WATER [1359]

Ag_2CrO_4— —— —H_2O

t=100

Nitrate		Comp. of solvent, wt%		Solubility of Silver Chromate g/l
Name	Formula	Nitrate	Water	
—	—	0	100	0.64
Sodium Nitrate	$NaNO_3$	33.3	66.7	0.64
Potassium Nitrate	KNO_3	33.3	66.7	1.92
Ammonium Nitrate	NH_4NO_3	33.3	66.7	3.20
Magnesium Nitrate	$Mg(NO_3)_2$	33.3	66.7	2.56

№ 5414 [1794]

SILVER PERMANGANATE–SILVER PERCHLORATE–WATER

$AgMnO_4$—$AgClO_4$—H_2O

t=20

Concn. in aqueous soln., g/l	
Silver Perchlorate	Silver Permanganate
0.00	9.222
11.71	1.542
128.5	1.111
157.5	1.021
209.3	0.975
232.1	0.816

№ 5415 [2221]

SILVER BROMIDE–MERCURY (I) NITRATE–WATER

$AgBr$—$HgNO_3$—H_2O

t=25

Concn. in aqueous soln., g/l	
Silver Bromide	Mercury Nitrate
0.575	2.63
0.618	3.28
0.863	6.56
1.200	13.13
1.640	26.3
6.878	262.6

Note. The presence of Nitric Acid in quantities from 0.1 N to 2 N did not show any effect. Identical results were obtained also using crystalline and amorphous Silver Bromide.

№ 5416 [1485]

SILVER BROMIDE–BROMIDE COMPOUNDS–WATER

$AgBr$— —— —H_2O

Bromide compound			Solubility of Silver Bromide mg/g-eq solvent	t
Name	Formula	Concn. in aqueous soln., wt%		
Potassium Bromide	KBr	39.80	22.17	20
»	»	39.80	24.06	40
»	»	39.80	27.17	60
»	»	39.80	31.81	80
»	»	24.83	3.04	20
»	»	24.83	3.78	40
»	»	24.83	4.94	60
»	»	24.83	6.75	80

Table № 5416 (continuation)

Bromide compound			Solubility of Silver Bromide mg/g-eq solvent	t
Name	Formula	Concn. in aqueous soln., wt%		
Potassium Bromide	KBr	18.43	1.068	20
»	»	18.43	1.521	40
»	»	18.43	2.24	60
»	»	18.43	3.34	80
Calcium Bromide	$CaBr_2$	35.68	9.37	20
»	»	35.68	11.09	40
»	»	35.68	13.18	60
»	»	35.68	16.01	80
»	»	21.71	1.663	20
»	»	21.71	2.23	40
»	»	21.71	3.19	60
»	»	21.71	4.47	80
»	»	15.61	0.663	20
»	»	15.61	1.021	40
»	»	15.61	1.60	60
»	»	15.61	2.51	80
Hydrogen Bromide	HBr	30.98	6.86	20
»	»	30.98	8.11	40
»	»	30.98	9.72	60
»	»	30.98	12.15	80
»	»	18.33	1.312	20
»	»	18.33	1.87	40
»	»	18.33	2.70	60
»	»	18.33	3.94	80
»	»	13.02	0.605	20
»	»	13.02	0.913	40
»	»	13.02	1.496	60
»	»	13.02	2.31	80

№ 5417 SILVER BROMIDE–SALTS–WATER [2486]

$AgBr—MeX—H_2O$

$t=15$

Salt			Equilibrium Concns., g	
Name	Formula	Concn. in aqueous soln., wt%	Salt	Silver Bromide
Sodium Chloride	NaCl	21	100	0.182
»	»	Saturated solution	100	0.474
Sodium Chloride + Potassium Bromide	KBr	»	100	3.019
	NaCl+KBr	»	95+10	0.075

№ 5418 SILVER BROMIDE–SALTS–WATER [2681]

$AgBr—MeX—H_2O$

Salt			Solubility of Silver Bromide wt%	t
Name	Formula	Concn in aqueous soln., wt%		
Sodium Thiosulfate	$Na_2S_2O_3$	1	0.35	20
»	»	5	1.90	20
»	»	10	3.50	20
»	»	15	4.20	20
»	»	20	5.80	20
Sodium Sulfite	Na_2SO_3	10	0.04	25
»	»	20	0.08	25
Potassium Cyanide	KCN	5	6.55	25
Potassium Thiocyanate	KCNS	10	0.73	25
Ammonium Thiocyanate	NH_4CNS	5	0.21	20
»	»	10	2.04	20
»	»	15	5.30	20
Calcium Thiocyanate	$Ca(CNS)_2$	10	0.53	25
Barium Thiocyanate	$Ba(CNS)_2$	10	0.35	25
Aluminum Thiocyanate	$Al(CNS)_3$	10	4.50	25

№ 5419 SILVER BROMIDE–SALTS–WATER [1406]

$AgBr—MeX—H_2O$

Salt			Solubility of Silver Bromide wt%	t
Name	Formula	Concn. in aqueous soln., wt%		
Sodium Thiosulfate	$Na_2S_2O_3$	1	0.50	20
»	»	5	2.40	20
»	»	10	4.59	20
»	»	15	6.58	20
»	»	20	8.50	20
Potassium Cyanide	KCN	5	6.85	25

№ 5420 SILVER BROMIDE–SALTS–WATER [2175]

$AgBr—MeX—H_2O$

$t=20$

Salt		Solubility of Silver Bromide in 1% aqueous soln. of salt, g/l
Name	Formula	
Sodium Thiosulfate	NaS_2O_3	2.06
Ammonium Thiocyanate	NH_4CNS	0.03
Ammonium Carbonate	NH_4CO_3	0.004
Sodium Sulfate	Na_2SO_4	0.055

№ 5421 [479, 481]

SILVER BROMATE–CADMIUM NITRATE–WATER

$AgBrO_3—Cd(NO_3)_2—H_2O$

$t=25$

Content of Cadmium Nitrate in aqueous soln., mol/l	Solubility of Silver Bromate mol/l
0.000	$8.25 \cdot 10^{-3}$
0.001	$8.13 \cdot 10^{-3}$
0.003	$8.42 \cdot 10^{-3}$
0.010	$8.75 \cdot 10^{-3}$

№ 5422 [479, 481]

SILVER BROMATE–CADMIUM NITRATE–WATER

$AgBrO_3—Cd(NO_3)_2—H_2O$

$t=25$

Concn. of Cadmium Nitrate in aqueous soln., mol/l	Solubility of Silver Bromate in aqueous soln. of Cadmium Nitrate, mol %
0.030	$9.10 \cdot 10^{-3}$
0.100	$1.04 \cdot 10^{-2}$
0.300	$1.32 \cdot 10^{-2}$
1.000	$1.74 \cdot 10^{-2}$

№ 5423 [483]

SILVER BROMATE–CADMIUM SULFATE–WATER

$AgBrO_3—CdSO_4—H_2O$

$t=25$

Concn. of Cadmium Sulfate in aqueous soln., g/l	Solubility of Silver Bromate in aqueous soln. of Cadmium Sulfate, g/l
0.208	1.93
0.624	1.99
2.08	2.09
6.24	2.30
20.8	2.55
62.4	3.14
208	4.06

№ 5424 [856]

SILVER BROMATE–BARIUM NITRATE–WATER

$AgBrO_3—Ba(NO_3)_2—H_2O$

$t=25$

Concn. in aqueous soln., g/l	
Barium Nitrate	Silver Bromate
0.00000	1.9475
0.26135	1.9428
0.78405	1.9782
2.6135	2.1079
7.8405	2.3366
26.135	2.8765
78.405	3.5131

№ 5425 SILVER BROMATE–SALTS–WATER [1447]

$AgBrO_3—MeX—H_2O$

$t=25$

Salt			Solubility of Silver Bromate, wt%	d_4^{25}
Name	Fomula	Concn. in aqueous soln., wt%		
—	—	—	0.187	0.9990
Potassium Perchlorate	$KClO_4$	0.345	0.205	1.0010
»	»	0.687	0.216	1.0033
»	»	1.366	0.228	1.0073
Potassium Sulfate	K_2SO_4	0.192	0.219	1.0009
»	»	0.383	0.236	1.0028
»	»	0.764	0.261	1.0062
Magnesium Nitrate	$Mg(NO_3)_2$	0.185	0.210	1.0003
»	»	0.369	0.222	1.0018
»	»	0.734	0.237	1.0046
Magnesium Sulfate	$MgSO_4$	0.239	0.210	—
»	»	0.608	0.227	—
»	»	1.186	0.242	—
»	»	2.330	0.268	—
Barium Nitrate	$Ba(NO_3)_2$	0.326	0.214	1.0018
»	»	0.649	0.227	1.0046
»	»	1.290	0.244	1.0098
Sodium Sulfate	Na_2SO_4	0.354	0.235	1.0025
»	»	0.705	0.258	1.0059
»	»	6.630	0.437	1.0622
Cadmium Sulfate	$CdSO_4$	1.031	0.245	1.0017
»	»	4.952	0.314	1.0492
Cerium Nitrate	$Ce(NO_3)_2$	0.164	0.209	1.0002
»	»	0.328	0.220	1.0014
Potassium Nitrate	KNO_3	0.402	0.214	—
»	»	1.000	0.233	—
»	»	1.880	0.260	—
»	»	3.826	0.298	—

№ 5426 SILVER IODIDE–MERCURY (II) NITRATE–WATER [2221]

$AgI—Hg(NO_3)_2—H_2O$

$t=25$

Concn. in aqueous soln., g/l			
Mercury Nitrate	Silver Iodide	Mercury Nitrate	Silver Iodide
3.246	0.800	16.23	1.737
4.057	0.841	32.46	2.730
8.115	1.118	324.6	25.160

№ 5427 SILVER IODIDE–SALTS–WATER [2681]

$AgI—MeX—H_2O$

Salt		Comp. of solvent, wt%		Solubility of Silver Iodide, wt%	t
Name	Formula	Salt	Water		
Sodium Thiosulfate	$Na_2S_2O_3$	1.0	99.0	0.03	20
»	»	4.76	95.24	0.15	20
»	»	9.09	90.91	0.30	20
»	»	13.04	86.96	0.40	20
»	»	16.7	83.3	0.60	20
Potassium Cyanide	KCN	4.76	95.24	8.28	25
Sodium Sulfite	Na_2SO_3	9.09	90.91	0.01	25
»	»	16.7	83.3	0.02	25
Ammonium Thiocyanate	NH_4CNS	4.76	95.24	9.02	20
»	»	9.09	90.91	0.08	20
»	»	13.04	86.96	0.13	20
Calcium Thiocyanate	$Ca(CNS)_2$	9.09	90.91	0.03	25
Barium Thiocyanate	$Ba(CNS)_2$	9.09	90.91	0.02	25
Aluminum Thiocyanate	$Al(CNS)_3$	9.09	90.91	0.02	25

№ 5428 SILVER IODATE–IODINE PENTOXIDE–WATER [2406]

$AgIO_3—I_2O_5—H_2O$

Comp. of saturated soln., wt%			d_4^{20}	Comp. of solid phase
Silver Iodate	Iodine Pentoxide	Water		
		$t = 25°$		
3.03	18.96	78.01	1.203	$AgIO_3$
9.99	33.17	56.84	1.455	»
2.99	47.21	49.80	—	»
22	45	33.0	—	$AgIO_3 + AgIO_3 \cdot I_2O_5$
20	46	34.0	—	»
14.96	54.06	30.98	—	$AgIO_3 \cdot I_2O_5$
5.00	55.02	39.98	1.864	»
8.00	58.81	33.19	1.999	»
5.01	62.15	32.84	2.12	»
10.01	61.67	28.32	—	»
5.00	63.57	31.43	2.18	»
5.00	64.48	30.52	2.22	»
5.00	66.39	28.61	—	»
5.00	67.35	27.65	—	»
5.00	68.29	26.71	—	»
5.00	69.24	25.76	2.45	»
5.50	69.25	25.25	—	»

Table № 5428 (continuation)

Comp. of saturated soln., wt%			d_4^{20}	Comp. of solid phase
Silver Iodate	Iodine Pentoxide	Water		
			$t=45°$	
20.01	27.53	52.46	1.410	$AgIO_3$
19.97	32.21	47 82	1.512	»
20.00	36.98	43.02	1.637	»
10.00	42.68	47.32	1.653	»
8.99	44.10	46.91	1.681	»
5.00	46.96	48.04	1.702	»
21	44	35.00	—	$AgIO_3+AgIO_3\cdot I_2O_5$
22	42	36.00	—	»
5.01	49.81	45.18	—	»
5.01	50.75	44.24	1.727	$AgIO_3\cdot I_2O_5$
5.00	51.24	43.76	1.737	»
5.00	51.71	43.29	—	»
15.00	53.11	31.89	—	»
5.00	54.09	40.92	1.820	»
5.00	55.03	39.97	1.849	»
5.01	55.98	39.01	—	»
5.01	56.93	38.06	1.911	»
15.01	58.92	26.07	—	»
15.00	62.59	22.41	—	»
5.01	68.76	26.23	2.42	»
4.98	69.71	25.31	2.47	»
14.95	66.21	18.84	—	»
5.00	71.14	23.86	2.54	»
2.00	74.00	24.00	2.53	$AgIO_3\cdot I_2O_5+I_2O_5\cdot H_2O$

№ 5429 SILVER IODATE–SALTS–WATER [1173]

$AgIO_3—MeX—H_2O$

$t=75$

Salt		Concn. in aqueous soln., g/l Water	
Name	Formula	Salt	Silver Iodate
Without salts	—	—	0.238
Potassium Chlorate	$KClO_3$	0.245	0.245
»	»	0.612	0.252
»	»	1.225	0.259
»	»	2.45	0.269
»	»	6.127	0.288
»	»	12.25	0.308
Potassium Nitrate	KNO_3	0.101	0.242
»	»	0.202	0.250
»	»	0.505	0.251
»	»	1.01	0.259
»	»	2.02	0.271
»	»	5.05	0.293
»	»	10.11	0.318

Table № 5429 (continuation)

Salt		Concn. in aqueous soln., g/l Water	
Name	Formula	Salt	Silver Iodate
Potassium Sulfate	K_2SO_4	0.087	0.242
»	»	0.174	0.248
»	»	0.348	0.255
»	»	0.870	0.272
»	»	1.74	0.290
»	»	3.48	0.314
»	»	8.70	0.366
Barium Nitrate	$Ba(NO_3)_2$	0.130	0.244
»	»	0.261	0.246
»	»	0.522	0.254
»	»	1.31	0.264
»	»	2.61	0.273
Magnesium Sulfate	$MgSO_4$	0.024	0.241
»	»	0.060	0.246
»	»	0.120	0.250
»	»	0.240	0.258
»	»	0.600	0.272
»	»	1.20	0.288
»	»	2.40	0.309

№ 5430 CADMIUM OXIDE–CADMIUM SULFATE–WATER [235]

$CdO—CdSO_4—H_2O$

$t=18$

Concn. in aqueous soln.		d_4	pH	Comp. of solid phase
Cadmium Sulfate, wt%	Cd, g/l			
43.4	368.6	1.580	5.65	$3CdO \cdot CdSO_4 \cdot xH_2O + CdSO_4 \cdot 8/3H_2O$
25.6	178.2	1.292	6.31	$3CdO \cdot CdSO_4 \cdot xH_2O$
16.9	109.3	1.176	6.70	»
10.9	64.8	1.097	7.0	»
6.88	38.7	1.042	7.19	»
2.35	12.8	1.010	7.38	»
2.17	11.8	1.009	7.38	»
0.00	0.0	1.000	8.00	$Cd(OH)_2$

№ 5431 CADMIUM HYDROXIDE–CADMIUM SULFATE–WATER [901]

$Cd(OH)_2—CdSO_4—H_2O$

$t=25$

Comp. of saturated soln., wt%			Comp. of solid phase
Cadmium Hydroxide	Cadmium Sulfate	Water	
$\sim 2 \cdot 10^{-5}$	40.75	59.25	$CdSO_4 \cdot Cd(OH)_2$
$\sim 2 \cdot 10^{-5}$	33.34	66.66	»
$\sim 2 \cdot 10^{-5}$	29.75	70.25	»

Table № 5431 (continuation)

Comp. of saturated soln., wt%			Comp. of solid phase
Cadmium Hydroxide	Cadmium Sulfate	Water	
$\sim 2 \cdot 10^{-5}$	24.53	75.47	$CdSO_4 \cdot Cd(OH)_2$
$\sim 2 \cdot 10^{-5}$	18.66	81.34	»
$\sim 2 \cdot 10^{-5}$	16.46	83.54	»
$\sim 2 \cdot 10^{-5}$	15.22	84.78	»
$\sim 2 \cdot 10^{-5}$	12.76	87.24	»
$\sim 2 \cdot 10^{-5}$	8.92	91.08	»
$\sim 2 \cdot 10^{-5}$	5.18	94.82	»
$\sim 2 \cdot 10^{-5}$	1.72	98.28	»
$\sim 2 \cdot 10^{-5}$	0.92	99.08	»
$\sim 2 \cdot 10^{-5}$	0.24	99.76	$CdSO_4 \cdot 2Cd(OH)_2$
$\sim 2 \cdot 10^{-5}$	0.694	99.306	»
$\sim 2 \cdot 10^{-5}$	0.15	99.85	»

№ 5432 CADMIUM NITRATE–LEAD (II) IODIDE–WATER [891]

$Cd(NO_3)_2—PbI_2—H_2O$

$t=18\div20$

Comp. of aqueous solvent mol/l Cadmium Nitrate	Solubility of Lead Iodide, mol/l · 10^3	Comp. of aqueous solvent mol/l Cadmium Nitrate	Solubility of Lead Iodide, mol/l · 10^3
0.0739	9.20	0.5725	40.80
0.1201	12.80	0.6205	45.90
0.2790	23.90	0.8394	60.20
0.3830	30.20	1.0046	68.80
0.5095	38.30	1.1960	80.80

№ 5433 CADMIUM NITRATE–URANYL NITRATE–WATER [1013]

$Cd(NO_3)_2—UO_2(NO_3)_2—H_2O$

Comp. of saturated soln., wt%			t	Comp. of solid phase
Uranyl Nitrate	Cadmium Nitrate	Water		
0.0	55.6	44.4	0	$Cd(NO_3)_2 \cdot 4H_2O$
2.9	52.6	44.5	0	»
3.9	52.1	44.0	0	»
6.1	49.8	44.1	0	»
6.6	49.5	43.9	0	$Cd(NO_3)_2 \cdot 4H_2O + UO(NO_3)_2 \cdot 6H_2O$
7.6	48.0	44.4	0	$UO_3(NO_3)_2 \cdot 6H_2O$
9.9	49.0	47.1	0	»
12.2	39.0	48.8	0	»
17.8	31.7	50.5	0	»
17.9	32.1	50.0	0	»
19.8	30.1	50.1	0	»
25.6	23.8	50.6	0	»
25.8	23.5	50.7	0	»

Table № 5433 (continuation)

Comp. of saturated soln., wt%			t	Comp. of solid phase
Uranyl Nitrate	Cadmium Nitrate	Water		
27.3	22.3	50.4	0	$UO_3(NO_3)_2 \cdot 6H_2O$
30.7	19.0	50.3	0	»
37.1	11.5	51.4	0	»
38.0	10.6	51.4	0	»
40.6	8.6	50.8	0	»
44.7	3.9	51.4	0	»
49.5	0.0	50.5	0	»
0.0	61.4	38.6	25	$Cd(NO_3)_2 \cdot 4H_2O$
1.2	60.0	38.8	25	»
3.2	58.3	38.5	25	»
5.9	56.0	38.1	25	»
8.4	54.6	37.0	25	»
9.7	53.5	36.8	25	$Cd(NO_3)_2 \cdot 4H_2O + UO_2(NO_3)_2 \cdot 6H_2O$
9.5	52.5	38.0	25	$UO_2(NO_3)_2 \cdot 6H_2O$
9.5	52.2	38.3	25	»
9.9	51.8	38.3	25	»
10.2	51.6	38.2	25	»
10.9	49.1	40.0	25	»
12.9	45.9	41.2	25	»
13.0	46.4	40.6	25	»
15.2	43.4	41.4	25	»
20.2	37.5	42.3	25	»
26.8	29.7	43.5	25	»
33.4	22.3	44.3	25	»
48.1	7.4	44.5	25	»
55.9	0.0	44.1	25	»

№ 5434 [1677, 2039]

CADMIUM SULFATE–LANTHANUM IODATE–WATER

$CdSO_4—La(IO_3)_3—H_2O$

$t=25$

Concn. of Cadmium Sulfate in aqueous soln., g/l	Solubility of Lanthanum Iodate in aqueous of Cadmium Sulfate, g/l
10.420	2.127

№ 5435 CADMIUM SULFATE–THALLIUM SULFATE–WATER [1208]

$CdSO_4—Tl_2SO_4—H_2O$

Comp. of saturated soln., wt%			t	Comp. of solid phase
Cadmium Sulfate	Thallium Sulfate	Water		
41.90	3.40	54.70	0	$CdSO_4 \cdot 8/3H_2O + 3CdSO_4 \cdot Tl_2SO_4 \cdot 5H_2O$
39.48	5.21	55.31	0	$3CdSO_4 \cdot Tl_2SO_4 \cdot 5H_2O$
39.07	6.50	54.43	0	$3CdSO_4 \cdot Tl_2SO_4 \cdot 5H_2O + Tl_2SO_4$

Table № 5435 (continuation)

Comp. of saturated soln., wt%			t	Comp. of solid phase
Cadmium Sulfate	Thallium Sulfate	Water		
32.04	8.13	59.83	0	Tl_2SO_4
42.75	1.80	55.45	25	$CdSO_4 \cdot 8/3H_2O$
42.00	3.60	54.40	25	$CdSO_4 \cdot 8/3H_2O + 3CdSO_4 \cdot Tl_2SO_4 \cdot 5H_2O$
39.43	4.95	55.62	25	$3CdSO_4 \cdot Tl_2SO_4 \cdot 5H_2O$
35.03	8.55	56.42	25	»
31.12	12.95	56.93	25	$3CdSO_4 \cdot Tl_2SO_4 \cdot 5H_2O + Tl_2SO_4$
6.81	7.21	85.98	25	Tl_2SO_4
43.32	4.81	51.87	50	$CdSO_4 \cdot H_2O + 3CdSO_4 \cdot Tl_2SO_4 \cdot 5H_2O$
37.95	6.97	55.08	50	$3CdSO_4 \cdot Tl_2SO_4 \cdot 5H_2O$
38.12	7.38	54.50	50	»
36.00	8.10	55.90	50	»
32.07	12.03	55.90	50	»
28.93	16.77	54.30	50	$3CdSO_4 \cdot Tl_2SO_4 \cdot 5H_2O + Tl_2SO_4$
12.18	12.88	75.94	50	Tl_2SO_4
37.25	7.90	54.85	75	$CdSO_4 \cdot H_2O + 3CdSO_4 \cdot Tl_2SO_4 \cdot 5H_2O$
36.65	8.83	54.52	75	$3CdSO_4 \cdot Tl_2SO_4 \cdot 5H_2O$
32.78	12.70	54.52	75	$3CdSO_4 \; Tl_2SO_4 \cdot 5H_2O + 2CdSO_4 \cdot Tl_2SO_4$
29.28	16.22	54.50	75	$2CdSO_4 \cdot Tl_2SO_4$
26.63	17.42	55.95	75	»
23.06	19.54	57.40	75	»
22.05	20.95	57.00	75	$2CdSO_4 \cdot Tl_2SO_4 + Tl_2SO_4$
12.34	19.66	68.00	75	Tl_2SO_4
36.85	5.20	57.95	27	$CdSO_4 \cdot H_2O$
36.25	7.33	56.42	27	$CdSO_4 \cdot H_2O + 2CdSO_4 \cdot Tl_2SO_4$
34.51	8.07	57.42	27	$2CdSO_4 \cdot Tl_2SO_4$
27.65	10.88	61.47	27	»
21.24	14.64	64.12	27	»
11.84	23.97	64.19	27	$2CdSO_4 \cdot Tl_2SO_4 + Tl_2SO_4$

№ 5436 [2263]

CADMIUM SULFATE–THALLIUM CHLORIDE–WATER

$CdSO_4$—$TlCl$—H_2O

$t=25$

Concn. in aqueous soln., g/l	
Thallium Chloride	Cadmium Sulfate
4.933	6.255
6.081	16.41
7.399	32.82

№ 5437 [2039]

CADMIUM SULFATE–THALLIUM IODATE–WATER

$CdSO_4$—$TlIO_3$—H_2O

$t=25$

Concn. of Cadmium Sulfate in aqueous soln., g/l	Solubility of Thallium Iodate in aqueous soln. of Cadmium Sulfate, g/l
10.420	1.058

№ 5438 [1221]

CADMIUM CHLORIDE–CADMIUM BROMIDE–WATER

$CdCl_2—CdBr_2—H_2O$

Concn. in aqueous soln., wt%		d_4^t	t	Comp. of solid phase
Cadmium Bromide	Cadmium Chloride			
0.00	54.9	1.787	25	$CdCl_2 \cdot 5H_2O$
1.24	54.2	1.800	25	Solid soln. $\cdot 5/2H_2O$
4.78	51.7	1.843	25	»
14.07	44.4	1.903	25	»
18.49	40.6	1.922	25	»
26.5	32.7	1.938	25	Solid soln. $\cdot 4H_2O$
32.5	26.3	1.934	25	»
43.4	14.56	1.902	25	»
47.9	6.78	1.838	25	»
50.8	3.26	1.810	25	»
52.9	0.0	1.785	25	$CdBr_2 \cdot 4H_2O$
0.00	56.2	1.827	30	$CdCl_2 \cdot 5H_2O$
5.57	51.9	1.857	30	Solid soln. $\cdot 4H_2O$
6.42	52.7	1.915	30	»
11.74	47.6	1.939	30	»
15.70	44.2	1.966	30	»
25.9	35.2	1.992	30	»
30.4	30.3	1.994	30	»
36.0	24.2	1.981	30	»
41.0	18.51	1.988	30	»
43.4	16.09	1.981	30	»
46.2	13.36	1.979	30	»
49.2	10.20	1.975	30	»
51.9	6.65	1.945	30	»
53.3	4.63	1.927	30	»
56.2	0.0	1.873	30	$CdBr_2 \cdot 4H_2O$
0.0	57.3	1.858	35	$CdCl_2 \cdot H_2O$
2.51	55.4	1.877	35	Solid soln. $\cdot H_2O$
6.73	52.1	1.816	35	»
12.44	47.4	1.951	35	»
15.92	44.2	1.964	35	»
22.0	39.2	1.979	35	»
28.1	33.8	2.018	35	Solid soln. $\cdot H_2O$ + + Solid soln. $\cdot 4H_2O$
30.9	30.9	2.032	35	Solid soln. $\cdot 4H_2O$
34.5	27.5	2.036	35	»
41.5	20.9	2.042	35	»
46.2	15.88	2.031	35	»
46.3	15.22	2.031	35	»
48.8	12.75	2.016	35	»
48.9	12.24	2.022	35	»
51.6	9.39	2.017	35	»
54.0	6.66	2.013	35	»
55.9	4.60	2.002	35	»
57.1	3.27	1.992	35	»

Table 5438 (continuation)

Concn. in aqueous soln., wt%		d_4^t	t	Comp. of solid phase
Cadmium Bromide	Cadmium Chloride			
59.2	0.0	1.966	35	$CdBr_2 \cdot 4H_2O$
0.0	57.2	1.853	50	$CdCl_2 \cdot H_2O$
4.88	53.6	1.901	50	Solid soln. $\cdot H_2O$
9.64	48.6	1.936	50	»
14.25	46.1	1.968	50	»
22.4	38.9	2.004	50	»
25.2	36.5	2.029	50	»
31.0	30.4	2.048	50	»
34.7	27.5	—	50	Solid soln.
35.0	27.5	—	50	»
37.7	24.8	2.058	50	»
45.7	16.24	2.078	50	»
50.3	11.84	2.061	50	»
50.0	10.68	2.055	50	»
52.6	8.98	2.054	50	»
55.5	5.67	2.040	50	»
60.8	0.0	2.047	50	$CdBr_2$
0.0	58.0	1.908	75	$CdCl_2 \cdot H_2O$
1.64	56.9	1.926	75	Solid soln. $\cdot H_2O$
7.14	52.5	1.980	75	»
9.44	50.7	1.983	75	»
19.41	42.3	2.043	75	»
22.9	39.4	2.053	75	»
25.5	37.1	2.078	75	»
33.5	30.2	2.110	75	»
39.2	24.7	2.119	75	Solid soln. $\cdot H_2O$ + + Solid soln.
41.5	22.1	2.128	75	Solid soln.
41.9	21.6	2.134	75	»
44.6	19.13	2.134	75	»
48.6	14.80	2.133	75	»
51.3	11.97	2.126	75	»
52.6	10.50	2.127	75	»
53.6	8.18	2.081	75	»
58.0	3.68	2.077	75	»
60.2	1.28	2.052	75	»
61.2	0.0	2.047	75	$CdBr_2$
0.0	59.2	1.993	100	$CdCl_2 \cdot H_2O$
3.82	56.3	2.023	100	Solid soln. $\cdot H_2O$
14.72	47.9	2.095	100	»
23.4	40.3	2.151	100	»
30.9	34.0	2.185	100	»
37.3	28.2	2.217	100	»
43.1	24.4	2.214	100	»
47.0	18.85	2.237	100	Solid solution
48.4	16.94	2.230	100	»
54.8	8.75	2.177	100	»
57.2	5.26	2.150	100	»
61.4	0.0	2.110	100	$CdBr_2$

$CdCl_2—BaCl_2—H_2O$

Concn. in aqueous soln., wt%		d_4^t	t	Comp. of solid phase
Barium Chloride	Cadmium Chloride			
0.0	54.9	1.788	25	$2CdCl_2 \cdot 5H_2O$
3.00	52.9	1.815	25	»
3.38	52.8	1.820	25	$2CdCl_2 \cdot 5H_2O + 2CdCl_2 \cdot BaCl_2 \cdot 5H_2O$
3.95	52.5	1.814	25	$2CdCl_2 \cdot BaCl_2 \cdot 5H_2O$
7.85	43.2	1.737	25	»
9.89	39.4	1.706	25	»
14.25	33.1	1.640	25	»
18.40	27.7	1.606	25	»
18.89	27.0	1.602	25	$2CdCl_2 \cdot BaCl_2 \cdot 5H_2O + CdCl_2 \cdot BaCl_2 \cdot 4H_2O$
19.53	26.2	1.599	25	$CdCl_2 \cdot BaCl_2 \cdot 4H_2O$
20.0	24.8	1.583	25	»
22.0	21.3	1.550	25	»
23.5	18.3	1.528	25	»
24.3	17.27	1.518	25	»
26.6	13.52	1.494	25	»
26.7	13.39	1.495	25	$CdCl_2 \cdot BaCl_2 \cdot 4H_2O + BaCl_2 \cdot 2H_2O$
26.8	12.56	1.480	25	$BaCl_2 \cdot 2H_2O$
26.7	11.56	1.465	25	»
27.0	6.31	1.386	25	»
27.0	5.35	1.370	25	»
27.0	0.0	1.303	25	»
0.0	56.2	1.872	30	$2CdCl_2 \cdot 5H_2O$
3.22	54.2	1.859	30	$2CdCl_2 \cdot 5H_2O + 2CdCl_2 \cdot BaCl_2 \cdot 5H_2O$
4.36	51.2	1.814	30	$2CdCl_2 \cdot BaCl_2 \cdot 5H_2O$
6.01	47.8	1.770	30	»
12.62	36.6	1.665	30	»
16.26	31.4	1.635	30	»
20.0	26.9	1.620	30	$2CdCl_2 \cdot BaCl_2 \cdot 5H_2O + CdCl_2 \cdot BaCl_2 \cdot 4H_2O$
20.4	26.4	1.618	30	$CdCl_2 \cdot BaCl_2 \cdot 4H_2O$
21.6	24.1	1.596	30	»
22.6	20.0	1.576	30	»
26.6	15.46	1.527	30	»
27.4	14.10	1.520	30	$CdCl_2 \cdot BaCl_2 \cdot 4H_2O + BaCl_2 \cdot 2H_2O$
27.5	9.90	1.449	30	$BaCl_2 \cdot 2H_2O$
27.6	6.26	1.396	30	»
27.7	0.0	1.313	30	»
0.0	57.3	1.858	35	$CdCl_2 \cdot H_2O$
3.23	55.2	1.886	35	»
3.35	54.9	1.877	35	$CdCl_2 \cdot H_2O + 2CdCl_2 \cdot BaCl_2 \cdot 5H_2O$
5.54	52.3	1.808	35	$2CdCl_2 \cdot BaCl_2 \cdot 5H_2O$
10.89	40.1	1.707	35	»
16.87	31.7	1.652	35	»
21.5	26.3	1.636	35	$2CdCl_2 \cdot BaCl_2 \cdot 5H_2O + CdCl_2 \cdot BaCl_2 \cdot 4H_2O$
22.6	24.0	1.618	35	$CdCl_2 \cdot BaCl_2 \cdot 4H_2O$
24.2	21.0	1.586	35	»
16.8	16.6	1.556	35	»
27.9	15.15	1.550	35	$CdCl_2 \cdot BaCl_2 \cdot 4H_2O + BaCl_2 \cdot 2H_2O$

Table № 5439 (continuation)

Concn. in aqueous soln., wt%		d_4^t	t	Comp. of solid phase
Barium Chloride	Cadmium Chloride			
28.1	10.44	1.469	35	$BaCl_2 \cdot 2H_2O$
28.2	4.86	1.383	35	»
28.5	0.0	1.319	35	»
0.0	57.2	1.853	50	$CdCl_2 \cdot H_2O$
3.38	55.1	1.890	50	»
4.86	54.3	1.903	50	$CdCl_2 \cdot H_2O + 2CdCl_2 \cdot BaCl_2 \cdot 5H_2O$
5.53	52.9	1.883	50	$2CdCl_2 \cdot BaCl_2 \cdot 5H_2O$
12.8	40.0	1.750	50	»
16.95	34.5	1.719	50	»
23.2	27.2	1.699	50	»
24.5	26.1	1.699	50	$2CdCl_2 \cdot BaCl_2 \cdot 5H_2O + CdCl_2 \cdot BaCl_2 \cdot 4H_2O$
25.3	24.9	1.681	50	$CdCl_2 \cdot BaCl_2 \cdot 4H_2O$
26.8	22.4	1.670	50	»
27.7	20.7	1.660	50	»
29.6	17.91	1.639	50	$CdCl_2 \cdot BaCl_2 \cdot 4H_2O + BaCl_2 \cdot 2H_2O$
29.9	16.08	1.604	50	$BaCl_2 \cdot 2H_2O$
30.1	5.17	1.405	50	»
30.5	0.0	1.328	50	»
0.0	58.0	1.908	75	$CdCl_2 \cdot H_2O$
3.90	55.3	1.942	75	»
7.64	52.8	1.986	75	»
8.53	52.6	1.998	75	$CdCl_2 \cdot H_2O + 2CdCl_2 \cdot BaCl_2 \cdot 5H_2O$
9.09	51.6	1.994	75	$2CdCl_2 \cdot BaCl_2 \cdot 5H_2O$
15.73	41.1	1.880	75	»
24.4	30.6	1.836	75	»
29.2	27.1	1.846	75	»
29.7	27.1	1.850	75	$2CdCl_2 \cdot BaCl_2 \cdot 5H_2O + CdCl_2 \cdot BaCl_2 \cdot 4H_2O$
30.0	26.3	1.850	75	$CdCl_2 \cdot BaCl_2 \cdot 4H_2O$
31.5	23.9	1.833	75	»
32.2	22.7	1.822	75	$CdCl_2 \cdot BaCl_2 \cdot 4H_2O + BaCl_2 \cdot 2H_2O$
32.7	20.5	1.764	75	$BaCl_2 \cdot 2H_2O$
32.7	14.55	1.628	75	»
32.7	9.57	1.542	75	»
33.4	0.0	1.422	75	»
0.0	59.2	1.993	100	$CdCl_2 \cdot H_2O$
3.47	56.7	2.004	100	»
6.90	54.0	2.028	100	»
12.44	50.7	2.090	100	$CdCl_2 \cdot H_2O + 2CdCl_2 \cdot BaCl_2 \cdot 5H_2O$
15.45	46.7	2.064	100	$2CdCl_2 \cdot BaCl_2 \cdot 5H_2O$
23.8	36.5	1.989	100	»
30.5	30.8	2.021	100	»
34.1	28.9	2.078	100	$2CdCl_2 \cdot BaCl_2 \cdot 5H_2O + CdCl_2 \cdot BaCl_2 \cdot 3H_2O$
34.9	28.0	2.066	100	$CdCl_2 \cdot BaCl_2 \cdot 3H_2O$
35.0	28.0	2.065	100	»
35.0	28.9	2.065	100	$CdCl_2 \cdot BaCl_2 \cdot 3H_2O + BaCl_2 \cdot H_2O$
35.1	22.3	1.936	100	$BaCl_2 \cdot H_2O$
35.3	19.16	1.848	100	»
35.7	10.34	1.672	100	»
36.3	3.46	1.550	100	»
36.7	0.0	1.508	100	$BaCl_2 \cdot 2H_2O$

№ 5440 [1677, 2039]

CADMIUM CHLORIDE–LANTHANUM IODATE–WATER

$CdCl_2—La(IO_3)_3—H_2O$

$t=25$

Concn. of Cadmium Chloride in aqueous soln., g/l	Solubility of Lanthanum Iodate in aqueous soln. of Cadmium Chloride, g/l
9.1705	1.1209

№ 5441 **CADMIUM CHLORIDE–THALLIUM CHLORIDE–WATER** [2264]

$CdCl_2—TlCl—H_2O$

$t=25$

Concn. of Cadmium Chloride in aqueous soln., g/l	Solubility of Thallium Chloride in aqueous soln. of Cadmium Chloride, g/l	Concn. of Cadmium Chloride in aqueous soln., g/l	Solubility of Thallium Chloride in aqueous soln. of Cadmium Chloride, g/l
2.292	2.495	9.17	1.386
4.585	1.871	18.34	1.024

№ 5442 **CADMIUM CHLORIDE–THALLIUM CHLORIDE–WATER** [1197]

$CdCl_2—TlCl—H_2O$

$t=25$

Comp. of saturated soln., wt%			Comp. of solid phase
Thallium Chloride	Cadmium Chloride	Water	
0.374	0.0	99.626	TlCl
0.024	8.481	91.495	$TlCl \cdot CdCl_2$
0.013	16.15	83.837	»
0.015	19.09	80.895	»
0.016	27.31	72.674	»
0.021	34.61	65.369	»
0.019	43.22	56.761	»
0.021	48.36	51.619	»
0.021	49.80	50.179	»
0.023	55.54	44.437	»
0.0	56.80	43.20	$CdCl_2$

№ 5443 CADMIUM CHLORIDE–THALLIUM CHLORIDE–WATER [923]

$CdCl_2—TlCl—H_2O$

Comp. of saturated soln., wt%			t	Comp. of solid phase
Thallium Chloride	Cadmium Chloride	Water		
0.020	2.10	97.88	0	$TlCl+TlCl \cdot CdCl_2$
0.085	2.86	97.05	25	»
0.299	2.84	96.86	50	»
0.802	2.85	96.35	75	»
0.010	47.42	52.57	0	$TlCl \cdot CdCl_2+CdCl_2 \cdot 2.5H_2O$
0.024	54.68	45.30	25	»
0.051	57.83	42.12	50	$TlCl \cdot CdCl_2+CdCl_2 \cdot H_2O$
0.106	58.60	41.29	75	»

№ 5444 [2037]

CADMIUM CHLORIDE–THALLIUM IODATE–WATER

$CdCl_2—TlIO_3—H_2O$

$t=25$

Concn. of Cadmium Chloride in aqueous soln., g/l	Solubility of Thallium Iodate in aqueous soln. of Cadmium Chloride, g/l
9.170	1.024

№ 5445 CADMIUM CHLORIDE–LEAD (II) CHLORIDE–WATER [2262]

$CdCl_2—PbCl_2—H_2O$

$t=25$

Concn. in aqueous soln., g/l			
Cadmium Chloride	Lead Chloride	Cadmium Chloride	Lead Chloride
0	10.79	6	7.7
1	10.2	8	7.0
2	9.7	10	6.3
3	9.2	14	5.4
4	8.6	20	4.7

$CdBr_2—BaBr_2—H_2O$

Concn. in aqueous soln., wt%		d	t	Comp. of solid phase
Barium Bromide	Cadmium Bromide			
0.0	52.9	1.785	25	$CdBr_2 \cdot 4H_2O$
4.21	49.6	1.814	25	»
11.76	45.0	1.878	25	»
19.31	41.2	1.983	25	»
25.9	40.0	2.040	25	$CdBr_2 \cdot 4H_2O + CdBr_2$
26.7	39.6	2.165	25	$CdBr_2$
29.4	38.1	2.167	25	»
31.5	37.4	2.257	25	»
32.7	37.4	2.295	25	$CdBr_2 + CdBr_2 \cdot BaBr_2 \cdot 4H_2O$
34.9	36.4	2.312	25	$CdBr_2 \cdot BaBr_2 \cdot 4H_2O$
34.6	36.5	—	25	»
36.1	34.4	2.303	25	»
38.7	31.1	2.300	25	»
39.3	30.2	2.300	25	$CdBr_2 \cdot BaBr_2 \cdot 4H_2O + BaBr_2 \cdot 2H_2O$
38.2	30.9	2.259	25	$BaBr_2 \cdot 2H_2O$
40.6	24.6	2.115	25	»
44.1	14.15	1.917	25	»
46.8	7.46	1.803	25	»
50.0	0.0	1.734	25	»
0.0	56.2	1.873	30	$CdBr_2 \cdot 4H_2O$
9.53	49.3	1.937	30	»
18.52	44.3	2.056	30	»
20.7	43.4	2.095	30	$CdBr_2 \cdot 4H_2O + CdBr_2$
20.6	43.5	2.093	30	$CdBr_2$
24.7	40.6	2.132	30	»
29.7	38.0	2.205	30	»
31.8	37.3	2.253	30	»
35.0	36.6	2.353	30	$CdBr_2 + CdBr_2 \cdot BaBr_2 \cdot 4H_2O$
36.1	34.9	2.334	30	$CdBr_2 \cdot BaBr_2 \cdot 4H_2O$
39.1	31.5	2.311	30	»
39.3	30.7	2.293	30	$CdBr_2 \cdot BaBr_2 \cdot 4H_2O + BaBr_2 \cdot 2H_2O$
39.4	30.3	2.277	30	$BaBr_2 \cdot 2H_2O$
40.7	24.3	2.119	30	»
43.0	17.68	1.983	30	»
46.6	8.91	1.852	30	»
50.6	0.0	1.745	30	»
0.0	59.2	1.966	35	$CdBr_2 \cdot 4H_2O$
3.34	56.4	1.991	35	»
7.36	52.7	2.016	35	$CdBr_2$
17.46	46.6	2.038	35	»
15.67	47.3	2.060	35	»
34.9	40.9	2.127	35	»
25.6	40.0	—	35	»
31.9	37.1	2.260	35	»
34.8	36.5	2.344	35	»
35.5	36.4	2.379	35	»

Table № 5446 (continuation)

Concn. in aqueous soln., wt%		d	t	Comp. of solid phase
Barium Bromide	Cadmium Bromide			
35.4	36.3	2.366	35	$CdBr_2 + CdBr_2 \cdot BaBr_2 \cdot 4H_2O$
38.0	33.2	2.335	35	$CdBr_2 \cdot BaBr_2 \cdot 4H_2O$
39.5	31.0	2.321	35	$CdBr_2 \cdot BaBr_2 \cdot 4H_2O + BaBr_2 \cdot 2H_2O$
40.9	25.4	2.152	35	$BaBr_2 \cdot 2H_2O$
43.8	16.49	1.972	35	»
44.5	15.02	1.946	35	»
48.0	6.57	1.831	35	»
50.9	0.0	1.756	35	»
0.0	60.8	1.963	50	$CdBr_2$
9.87	51.8	2.017	50	»
19.11	44.6	2.086	50	»
24.6	40.8	2.142	50	»
31.6	37.6	2.289	50	»
37.1	36.6	2.470	50	$CdBr_2 + CdBr_2 \cdot BaBr_2 \cdot 4H_2O$
37.0	37.0	2.471	50	»
37.3	36.7	2.467	50	$CdBr_2 \cdot BaBr_3 \cdot 4H_2O$
37.5	36.1	2.470	50	»
38.0	35.6	2.453	50	»
38.8	34.3	2.434	50	»
39.9	33.0	2.421	50	$CdBr_2 \cdot BaBr_2 \cdot 4H_2O + BaBr_2 \cdot 2H_2O$
40.2	31.5	2.363	50	$BaBr_2 \cdot 2H_2O$
41.9	23.7	2.188	50	»
45.0	15.74	2.042	50	»
50.5	3.90	1.867	50	»
52.5	0.0	1.803	50	»
0.0	61.2	2.047	75	$CdBr_2$
9.59	52.4	2.066	75	»
18.46	45.6	2.127	75	»
26.7	40.5	2.227	75	»
33.9	37.9	2.442	75	»
38.9	37.8	2.650	75	$CdBr_2 + CdBr_2 \cdot BaBr_2 \cdot 4H_2O$
39.6	37.1	2.642	75	$CdBr_2 \cdot BaBr_2 \cdot 4H_2O$
40.0	36.7	2.639	75	»
40.4	35.9	2.637	75	$CdBr_2 \cdot BaBr_2 \cdot 4H_2O + BaBr_2 \cdot 2H_2O$
40.7	34.8	2.586	75	$BaBr_2 \cdot 2H_2O$
43.9	22.5	2.211	75	»
46.8	15.82	2.070	75	»
50.9	7.42	1.971	75	»
54.7	0.0	1.890	75	»
0.0	61.4	2.110	100	$CdBr_2$
5.66	56.7	2.128	100	»
11.32	51.9	2.153	100	»
28.7	40.3	2.326	100	»
40.1	38.8	2.860	100	$CdBr_2 + CdBr_2 \cdot BaBr_2 \cdot H_2O$
40.6	38.7	2.862	100	$CdBr_2 \cdot BaBr_2 \cdot H_2O$
41.0	37.2	2.790	100	$CdBr_2 \cdot BaBr_2 \cdot H_2O + BaBr_2 \cdot 2H_2O$
42.9	22.2	2.316	100	$BaBr_2 \cdot 2H_2O$
50.5	12.26	2.084	100	»
53.2	6.59	2.007	100	»
56.9	0.0	1.999	100	»

№ 5447 CADMIUM BROMIDE–LEAD (II) BROMIDE–WATER [2368]

$CdBr_2—PbBr_2—H_2O$

$t=25$

Concn. in aqueous soln., wt%			
Lead Bromide	Cadmium Bromide	Lead Bromide	Cadmium Bromide
0.036	0.713	6.82	0.248
0.073	0.698	10.60	0.235
0.182	0.667	15.46	0.245
0.364	0.622	17.11	0.256
0.725	0.535	26.76	0.318
1.74	3.81	38.12	0.408
3.44	2.85	60.11 *	0.694
4.58	0.269		

* Solid phase $PbBr_2 + CdBr_2 \cdot 4H_2O$.

№ 5448 CADMIUM IODIDE–MERCURY (II) IODIDE–WATER [1827]

$CdI_2—HgI_2—H_2O$

Concn. in aqueous soln., wt%		t	Concn. in aqueous soln., wt%		t
Cadmium Iodide	Mercury Iodide		Cadmium Iodide	Mercury Iodide	
48.4	0.0	50	12.8	5.5	75
44.8	3.4	50	6.3	3.5	75
43.5	3.5	50	55.0	0.0	100
43.5	3.3	50	53.3	4.3	100
43.2	4.1	50	52.2	3.7	100
38.0	5.5	50	53.1	3.9	100
34.1	6.2	50	52.5	4.2	100
32.5	6.6	50	51.3	4.7	100
28.0	6.7	50	49.8	5.4	100
24.4	6.6	50	46.0	7.0	100
19.9	6.6	50	46.0	7.0	100
13.7	5.8	50	40.6	8.3	100
8.4	3.7	50	40.3	8.2	100
49.4	3.7	50	35.3	9.2	100
49.8	3.7	75	34.0	9.5	100
49.6	3.3	75	26.3	9.3	100
49.0	3.3	75	22.5	9.0	100
44.3	4.9	75	12.6	7.0	100
38.9	6.8	75	10.3	5.8	100
33.0	7.6	75	6.5	3.6	100
24.0	7.6	75	0.0	0.2	100

№ 5449 INDIUM FLUORIDE – CESIUM FLUORIDE – WATER [249]

InF_3—CsF—H_2O

$t=25$

Comp. of saturated soln., wt%			Comp. of solid phase
Indium Fluoride	Cesium Fluoride	Water	
5.8	0.0	94.2	InF_3
6.6	0.9	92.5	$Cs[InF_4 \cdot (H_2O)_2]$
4.5	1.2	94.3	»
2.6	1.3	96.1	»
1.6	1.9	96.5	»
0.9	3.9	95.2	»
0.6	7.5	91.9	»
0.8	12.4	86.8	»
1.2	15.8	83.0	»
3.5	25.5	71.0	»
4.6	29.9	65.5	Cs_3InF_6
0.7	45.9	53.4	»
0.5	62.3	37.2	»
0.6	63.5	35.9	»

№ 5450 INDIUM SULFATE–CESIUM SULFATE–WATER [243]

$In_2(SO_4)_3$—Cs_2SO_4—H_2O

$t=20$

Comp. of saturated soln., wt%			Comp. of solid phase
Indium Sulfate	Cesium Sulfate	Water	
54.00	0.00	46.00	$In_2(SO_4)_3 \cdot 5H_2O$
51.26	0.01	48.73	»
33.12	0.12	66.76	$CsIn(SO_4)_2 \cdot 12H_2O$
28.60	0.18	71.22	»
19.86	0.19	79.95	»
5.18	0.36	94.46	»
4.86	0.60	94.54	»
3.66	0.71	95.63	»
3.01	1.10	95.89	»
2.24	1.19	96.57	»
1.16	1.96	96.88	»
0.74	2.15	97.11	»
0.65	3.72	95.63	»
0.45	6.97	92.58	»
0.38	10.25	89.37	»
0.35	21.54	78.11	»
0.36	28.13	71.51	»
0.65	30.85	68.50	$CsIn(SO_4)_2 \cdot 12H_2O + Cs_2In(OH) \cdot (SO_4)_2 \cdot 2H_2O$
0.59	32.01	67.40	$Cs_2In(OH)(SO_4)_2 \cdot 2H_2O$
0.58	38.09	61.33	»
0.61	39.86	59.53	»
0.65	47.73	51.62	»
0.68	58.30	41.02	»

Table № 5450 (continuation)

Comp. of saturated soln., wt%			Comp. of solid phase
Indium Sulfate	Cesium Sulfate	Water	
0.003	60.00	39.997	$Cs_2In(OH)_2(SO_4)_2 \cdot nH_2O$
0.002	62.00	37.998	»
0.0007	64.01	35.989	»
0.00	64.12	35.88	»

№ 5451 [1568]

INDIUM BROMIDE–THALLIUM BROMIDE–WATER

$InBr_3—TlBr—H_2O$

$t=20$

Concn. in aqueous soln., g/l		Comp. of solid phase
Indium Bromide	Thallium Bromide	
1760	4.2	$TlBr \cdot InBr_3 \cdot 3H_2O$
2430	5.2	»

№ 5452 [1568]

INDIUM IODIDE–BARIUM IODIDE–WATER

$InI_3—BaI_2—H_2O$

$t=20$

Concn. in aqueous soln., g/l		Comp. of solid phase
Indium Iodide	Barium Iodide	
2301	223	InI_3
1930	342	$BaI_2 \cdot 2InI_3 \cdot 10H_2O$
1887	404	»
1740	480	$BaI_2 \cdot InI_3 \cdot 5H_2O$
1000	812	»
1343	712	$BaI_2 \cdot H_2O$
1326	732	»

№ 5453 [1568]

INDIUM IODIDE–THALLIUM IODIDE–WATER

$InI_3—TlI—H_2O$

$t=20$

Concn. in aqueous soln., g/l		Comp. of solid phase
Indium Iodide	Thallium Iodide	
103	1.0	$TlI \cdot 2InI_3 \cdot 3H_2O$
280	5.1	»

№ 5454 IODINE PENTOXIDE–BARIUM IODATE–WATER [2408]

I_2O_5—$Ba(IO_3)_2$—H_2O

$t=25$

Comp. of saturated soln., wt%			d_4	Comp. of solid phase
Iodine Pentoxide	Barium Iodate	Water		
0.0	0.042	99.958	—	$Ba(IO_3)_2 \cdot H_2O$
53.09	0.000	46.91	1.805	»
53.29	0.000	46.71	1.804	$Ba(IO_3)_2 \cdot H_2O + Ba(IO_3)_2 \cdot I_2O_5$
58.92	0.000	41.08	1.970	$Ba(IO_3)_2 \cdot I_2O_5$
60.73	0.000	39.27	2.019	»
66.76	0.000	33.24	2.27	»
67.70	0.000	32.30	2.31	»
68.56	0.000	31.44	2.35	»
71.00	0.000	29.00	2.47	»
71.47	0.000	28.53	2.49	$Ba(IO_3)_2 \cdot I_2O_5 \cdot HIO_3$
71.47	0.000	28.53	2.49	HIO_3

№ 5455 CESIUM NITRATE–CERIUM NITRATE–WATER [298]

$CsNO_3$—$Ce(NO_3)_3$—H_2O

Comp. of saturated soln., wt%			t	Comp. of solid phase
Cesium Nitrate	Cerium Nitrate	Water		
11.8	0.0	88.2	10	$CsNO_3$
8.7	18.6	72.7	10	»
14.5	34.5	51.0	10	»
19.2	40.5	40.3	10	»
22.2	44.4	33.4	10	»
23.8	45.8	30.4	10	$CsNO_3 + Ce(NO_3)_3 \cdot 2CsNO_3 \cdot 4H_2O$
18.5	51.5	30.0	10	$Ce(NO_3)_3 \cdot 2CsNO_3 \cdot 4H_2O$
15.8	55.0	29.2	10	$Ce(NO_3)_3 \cdot 2CsNO_3 \cdot 4H_2O + Ce(NO_3)_3 \cdot 6H_2O$
14.2	55.4	30.4	10	$Ce(NO_3)_3 \cdot 6H_2O$
7.7	57.7	34.6	10	»
0.0	60.0	40.0	10	»
17.5	0.0	82.5	20	$CsNO_3$
13.3	17.7	69.0	20	»
18.1	33.0	48.9	20	»
22.2	39.0	38.8	20	»
25.2	42.7	32.1	20	»
27.0	44.8	28.2	20	$CsNO_3 + Ce(NO_3)_3 \cdot 2CsNO_3 \cdot 4H_2O$
26.6	45.5	27.9	20	$Ce(NO_3)_3 \cdot 2CsNO_3 \cdot 4H_2O$
21.8	50.3	27.9	20	»
19.0	54.3	26.7	20	»
15.2	58.5	26.3	20	$Ce(NO_3)_3 \cdot 2CsNO_3 \cdot 4H_2O + Ce(NO_3)_3 \cdot 6H_2O$
11.7	59.2	29.1	20	$Ce(NO_3)_3 \cdot 6H_2O$
6.7	60.4	32.9	20	»
0.0	61.5	38.5	20	»
25.5	0.0	74.5	30	$CsNO_3$
18.8	16.3	64.9	30	»

Comp. of saturated soln., wt%			t	Comp. of solid phase
Cesium Nitrate	Cerium Nitrate	Water		
21.2	31.3	47.4	30	$CsNO_3$
24.5	37.8	37.7	30	»
27.3	41.3	31.4	30	»
30.5	43.7	25.8	30	$CsNO_3+Ce(NO_3)_3\cdot 2CsNO_3\cdot 4H_2O$
29.0	44.9	26.4	30	$Ce(NO_3)_3\cdot 2CsNO_3\cdot 4H_2O$
24.5	49.7	25.8	30	»
21.2	53.5	25.3	30	»
14.3	62.3	23.7	30	$Ce(NO_3)_3\cdot 2CsNO_3\cdot 4H_2O+Ce(NO_3)_3\cdot 6H_2O$
9.3	62.8	27.9	30	$Ce(NO_3)_3\cdot 6H_2O$
5.3	63.0	31.7	30	»
0.0	63.0	37.0	30	»

№ 5456 CESIUM THIOCYANATE–THALLIUM THIOCYANATE–WATER [432]

$CsCNS—TlCNS—H_2O$

$t=25$

Concn. in aqueous soln., mol/l			
Thallium Thiocyanate	Cesium Thiocyanate	Thallium Thiocyanate	Cesium Thiocyanate
$(32\pm 2)\cdot 10^{-4}$	0.10	$(79\pm 1)\cdot 10^{-4}$	3.0
$(20\pm 1)\cdot 10^{-4}$	0.50	$(133\pm 4)\cdot 10^{-4}$	4.0
$(25\pm 1)\cdot 10^{-4}$	1.0	$(196\pm 4)\cdot 10^{-4}$	4.86
$(47\pm 2)\cdot 10^{-4}$	2.0	$(290\pm 10)\cdot 10^{-4}$	5.97

№ 5457 CESIUM NITRITE–BARIUM NITRITE–WATER [719]

$CsNO_2—Ba(NO_2)_2—H_2O$

$t=25$

Comp. of saturated soln., wt%			Comp. of solid phase
Cesium Nitrite	Barium Nitrite	Water	
84.33	0.00	15.67	$CsNO_2$
81.71	2.24	16.05	$CsNO_2+2CsNO_2\cdot Ba(NO_2)_2\cdot H_2O$
81.65	2.36	15.99	»
77.45	2.89	19.66	$2CsNO_2\cdot Ba(NO_2)_2\cdot H_2O$
69.33	5.73	24.94	»
51.69	15.74	32.57	»
44.90	24.28	30.82	»
38.83	30.57	30.60	$2CsNO_2\cdot Ba(NO_2)_2\cdot H_2O+$ $+CsNO_2\cdot 2Ba(NO_2)_2\cdot H_2O$
38.86	30.73	30.41	»

Table № 5457 (continuation)

Comp. of saturated soln., wt%			Comp. of solid phase
Cesium Nitrite	Barium Nitrite	Water	
37.72	30.95	31.33	$CsNO_2 \cdot 2Ba(NO_2)_2 \cdot H_2O$
34.03	33.47	32.50	»
30.81	35.64	33.55	$CaNO_2 \cdot 2Ba(NO_2)_2 \cdot H_2O + Ba(NO_2)_2 \cdot H_2O$
30.75	35.73	33.52	»
24.18	37.63	38.19	$Ba(NO_2)_2 \cdot H_2O$
13.59	40.24	46.17	»
8.76	42.15	49.09	»
0.00	45.30	54.70	»

№ 5458 [160]

CESIUM NITRITE–THALLIUM NITRITE–WATER

$CsNO_2—TlNO_2—H_2O$

$t=25$

Comp. of saturated soln., wt%			Comp. of solid phase
Thallium Nitrite	Cesium Nitrite	Water	
0.00	85.83	14.17	$CsNO_2$
1.15	85.08	13.77	Solid solns. ($CsNO_2 + TlNO_2$)
3.06	83.95	12.99	»
4.83	82.42	12.75	»
7.70	78.08	14.22	»
10.04	79.12	10.84	»
13.08	74.75	12.17	»
16.00	72.06	11.34	»
21.99	68.48	9.53	»
25.60	63.45	10.95	»
35.26	58.20	6.54	Eutonic from solid solns.
35.50	58.70	5.80	»
37.12	53.21	9.67	Solid solns. ($TlNO_2 + CsNO_2$)
38.90	52.10	9.00	»
39.99	51.40	8.61	»
42.78	48.87	8.35	Solid solns. $+ TlNO_2$
42.30	49.05	8.65	»
34.93	49.47	15.60	$TlNO_2$
34.00	46.87	19.13	»
29.23	45.85	24.92	»
26.73	42.38	30.89	»
25.07	38.29	36.64	»
24.61	34.81	40.58	»
24.49	26.03	49.48	»
28.34	12.44	59.22	»
31.21	4.45	64.34	»
32.25	0.00	67.75	»

№ 5459 [2125]

CESIUM NITRATE–LEAD (II) NITRATE–WATER

$CsNO_3—Pb(NO_3)_2—H_2O$

$t=25$

Comp. of saturated soln., wt%			Comp. of solid phase
Cesium Nitrate	Lead Nitrate	Water	
21.53	0.0	78.47	$CsNO_3$
19.88	4.37	75.75	»
18.21	10.79	71.00	»
17.65	18.43	63.92	»
17.08	29.30	53.62	»
15.23	34.65	50.12	»
14.09	38.13	47.78	»
12.84	42.10	45.06	$CsNO_3+Pb(NO_3)_2$
7.97	40.22	51.81	$Pb(NO_3)_2$
4.49	39.32	56.19	»
2.14	38.30	59.56	»
0.0	37.07	62.93	»

№ 5460 [1016]

CESIUM NITRATE–URANYL NITRATE–WATER

$CsNO_3—UO_2(NO_3)_2—H_2O$

Comp. of saturated soln., wt%			t	Comp. of solid phase
Uranyl Nitrate	Cesium Nitrate	Water		
0.00	0.86	99.14	0	$CsNO_3$
0.06	0.81	99.13	0	»
0.10	0.78	99.12	0	»
0.41	0.64	98.95	0	»
0.82	0.56	98.62	0	»
0.99	0.54	98.47	0	»
1.00	0.53	98.47	0	»
1.10	0.54	98.36	0	»
1.56	0.55	97.89	0	»
2.21	0.63	97.16	0	»
2.82	0.71	96.47	0	»
2.84	0.79	96.37	0	»
3.32	0.96	95.74	0	»
3.45	0.97	95.58	0	»
3.69	1.02	95.29	0	»
3.71	1.06	95.23	0	»
4.04	1.13	94.83	0	$CsNO_3+Cs_2UO_2(NO_3)_4$
4.00	1.02	94.98	0	$CsUO_2(NO_3)_4$
4.01	0.97	95.02	0	»
4.07	0.90	95.03	0	»
4.11	0.89	95.00	0	$CsUO_2(NO_3)_4+UO_2(NO_3)_2\cdot 6H_2O$

Table № 5460 (continuation)

Comp. of saturated soln., wt%			t	Comp. of solid phase
Uranyl Nitrate	Cesium Nitrate	Water		
4.11	0.80	95.09	0	$UO_2(NO_3)_2 \cdot 6H_2O$
4.13	0.68	95.19	0	»
4.20	0.33	95.47	0	»
4.29	0.00	95.71	0	»
0.00	2.46	97.54	25	$CsNO_3$
0.09	2.34	97.57	25	»
0.12	2.27	97.61	25	»
0.24	2.13	97.63	25	»
0.30	2.12	97.58	25	»
0.40	2.07	97.53	25	»
0.85	1.80	97.35	25	»
1.21	1.92	96.87	25	»
1.52	1.96	96.52	25	»
2.25	2.12	95.63	25	»
2.83	2.22	94.95	25	»
3.26	2.52	94.22	25	»
3.65	2.60	93.75	25	»
3.69	2.72	93.59	25	»
3.99	2.71	93.30	25	$CsNO_3+CsUO_2(NO_3)_4$
4.04	2.61	93.35	25	$CsUO_2(NO_3)_4$
4.06	2.32	93.62	25	»
4.39	2.09	93.52	25	»
4.34	2.00	93.66	25	»
4.40	1.97	93.63	25	»
4.38	1.86	93.76	25	»
4.52	1.92	93.56	25	»
4.71	1.79	93.50	25	»
4.77	1.75	93.48	25	»
4.81	1.70	93.49	25	»
5.00	1.76	93.24	25	$CsUO_2(NO_3)_4+CsUO_2(NO_3)_3$
4.96	1.46	93.58	25	$CsUO_2(NO_3)_3$
5.00	1.14	93.86	25	»
5.03	0.95	94.02	25	»
5.12	0.91	93.97	25	$CsUO_2(NO_3)_3+UO_2(NO_3)_2 \cdot 6H_2O$
5.18	0.57	94.25	25	$UO_2(NO_3)_2 \cdot 6H_2O$
5.30	0.30	94.40	25	»
5.49	0.00	94.51	25	»

№ 5461 CESIUM SULFATE–CERIUM SULFATE–WATER [2777]

$Cs_2SO_4—Ce_2(SO_4)_3—H_2O$

$t=25$

Concn. in aqueous Soln., wt%		Comp. of solid phase
Cesium Sulfate	Cerium Sulfate	
4.33	3.67	$Ce_2(SO_4)_3 \cdot Cs_2SO_4 \cdot 8H_2O$
2.29	4.20	»
1.46	5.38	»
0.45	8.06	$Ce_2(SO_4)_3 \cdot Cs_2SO_4 \cdot 8H_2O+Ce_2(SO_4)_3 \cdot 8H_2O$

№ 5462 CESIUM SULFATE–PRASEODYMIUM SULFATE–WATER [2388]

$Cs_2SO_4—Pr_2(SO_4)_3—H_2O$

$t=25$

Concn. in aqueous soln., wt%		Comp. of solid phase
Cesium Sulfate	Praseodymium Sulfate	
41.62	0.0	—
20.15	0.0	$Cs_2SO_4 \cdot Pr_2(SO_4)_3 \cdot 8H_2O$
16.06	0.0	»
7.21	0.0	»
1.92	0.47	»
0.45	3.70	»
0.40	6.10	»
0.35	7.00	»
0.31	8.16	—
0.33	8.21	—
0.15	8.39	—

№ 5463 CESIUM CHLORIDE–CERIUM CHLORIDE–WATER [94]

$CsCl—CeCl_3—H_2O$

$t=25$

Comp. of saturated soln., wt%			Comp. of solid phase
Cerium Chloride	Cesium Chloride	Water	
0.00	65.50	34.50	CsCl
2.98	61.09	35.93	»
5.23	58.79	35.98	»
7.13	57.38	35.49	»
8.96	55.87	35.17	»
12.20	53.01	34.79	»
13.59	52.07	34.34	»
15.22	51.06	33.72	»
17.43	49.50	33.07	»
20.08	48.32	31.60	$CsCl+CeCl_3 \cdot 3CsCl \cdot 5H_2O$
21.50	44.50	34.00	$CeCl_3 \cdot 3CsCl \cdot 5H_2O$
26.05	37.94	36.01	»
28.03	36.03	35.96	»
29.49	34.65	35.86	»
31.75	32.13	36.12	»
34.50	29.75	35.75	»
37.36	28.63	34.01	$CeCl_3 \cdot 3CsCl \cdot 5H_2O+CeCl_3 \cdot CsCl \cdot 7H_2O$
37.68	22.24	40.08	$CeCl_3 \cdot CsCl \cdot 7H_2O$
39.02	18.04	42.94	»
41.11	14.75	44.14	»
41.50	13.90	46.60	$CeCl_3 \cdot CsCl \cdot 7H_2O+CeCl_3 \cdot 7H_2O$
42.76	13.15	44.09	$CeCl_3 \cdot 7H_2O$
43.01	11.00	45.99	»
46.20	5.00	48.80	»
50.04	0.00	49.96	»

№ 5464 CESIUM CHLORIDE–MERCURY (II) CHLORIDE–WATER [1629]

$CsCl—HgCl_2—H_2O$

$t=25$

Comp. of saturated soln., wt%			Comp. of solid phase
Cesium Chloride	Mercury Chloride	Water	
65.61	0.00	34.39	CsCl
65.78	0.22	34.00	$CsCl+Cs_3HgCl_5$
62.36	0.32	37.32	Cs_3HgCl_5
57.01	0.64	42.35	»
52.35	1.23	46.42	»
51.08	1.44	47.48	$Cs_3HgCl_5+Cs_2HgCl_4$
49.30	1.49	49.21	Cs_2HgCl_4
45.95	1.69	52.36	»
45.23	1.73	53.04	$Cs_2HgCl_4+CsHgCl_3$
38.63	1.32	60.05	$CsHgCl_3$
17.03	0.51	82.46	»
1.53	0.42	98.05	»
0.61	2.64	96.75	$CsHgCl_3+CsHg_2Cl_5$
0.49	2.91	96.60	$CsHg_2Cl_5$
0.40	3.78	95.82	»
0.44	4.63	94.93	$CsHg_2Cl_5+CsHg_5Cl_{11}$
0.41	4.68	94.91	$CsHg_5Cl_{11}$
0.25	5.65	94.10	»
0.18	7.09	92.73	$CsHg_5Cl_{11}+HgCl_2$
0.00	6.90	93.10	$HgCl_2$

№ 5465 CESIUM CHLORIDE–THALLIUM TRICHLORIDE–WATER [2121]

$CsCl—TlCl_3—H_2O$

$t=15$

Comp. of saturated soln., wt%			Comp. of solid phase
Thallium Trichloride	Cesium Chloride	Water	
65.17	0.0	34.83	$TlCl_3 \cdot 4H_2O$
65.18	0.0	34.82	$TlCl_3 \cdot 4H_2O+2TlCl_3 \cdot 3CsCl$
61.18	0.0	38.82	$2TlCl_3 \cdot 3CsCl$
54.91	0.0	45.09	»
50.13	0.0	49.87	»
40.97	0.0	59.03	»
31.71	0.0	68.29	»
16.48	0.0	83.52	»
10.21	0.0	89.79	»
6.57	0.02	93.41	»
1.45	1.12	97.43	»
0.74	2.37	96.89	»
0.02	3.20	96.78	»
0.0	3.27	96.73	$2TlCl_3 \cdot 3CsCl+2TlCl_3 \cdot 2CsCl \cdot 2H_2O$
0.0	3.27	96.73	$2TlCl_3 \cdot 2CsCl \cdot 2H_2O$
0.0	8.31	91.69	»

Table № 5465 (continuation)

Comp. of saturated soln., wt%			Comp. of solid phase
Thallium Trichloride	Cesium Chloride	Water	
0.0	12.45	87.55	$2TlCl_3 \cdot 2CsCl \cdot 2H_2O$
0.0	20.02	79.98	»
0.0	26.93	73.07	»
0.0	27.17	72.83	$2TlCl_3 \cdot 2CsCl \cdot 2H_2O + TlCl_3 \cdot 3CsCl \cdot H_2O$
0.0	27.69	72.31	$TlCl_3 \cdot 3CsCl \cdot H_2O$
0.0	32.72	67.28	»
0.0	42.10	57.90	»
0.0	45.20	54.80	$TlCl_3 \cdot 3CsCl \cdot H_2O + TlCl_3 \cdot 3CsCl$
0.0	45.91	54.09	$TlCl_3 \cdot 3CsCl$
0.0	50.0	50.0	»
0.0	58.17	41.83	»
0.0	63.98	36.02	»
0.0	64.71	35.29	$TlCl_3 \cdot 3CsCl + CsCl$
0.0	64.75	35.25	CsCl

№ 5466 CESIUM BROMIDE–LEAD (II) BROMIDE–WATER [1633]

$CsBr—PbBr_2—H_2O$

$t=25$

Comp. of saturated soln., wt%			Comp. of solid phase
Cesium Bromide	Lead Bromide	Water	
0.24	0.00	99.43	$PbBr_2 + CsPb_2Br_5$
0.33	0.36	99.31	»
12.83	Traces	87.17	$CsPb_2Br_5$
17.24	»	82.76	»
17.68	»	82.32	$CsPb_2Br_5 + CsPbBr_3$
18.58	»	81.42	$CsPbBr_3$
33.65	»	66.35	»
36.7	»	63.3	$CsPbBr_3 + Cs_4PbBr_6$
46.4	»	53.6	Cs_4PbBr_6
51.15	»	48.85	»
54.4	»	45.6	$Cs_4PbBr_6 + CsBr$
55.23	0	44.73	CsBr

№ 5467 CESIUM IODIDE–THALLIUM TRIIODIDE–WATER [430]

$CsI—TlI_3—H_2O$

$t=25$

Concn. of Cesium Iodide in aqueous soln., mol/l	Solubility of Thallium Triiodide (mol/l) · 10^{3}	Concn. of Cesium Iodide in aqueous soln., mol/l	Solubility of Tallium Triiodide (mol/l) · 10^{3}
0.014	0.80	0.55	0.57
0.035	0.37	1.10	1.35
0.068	0.33	1.61	2.39
0.100	0.30	2.11	4.95
0.130	0.32	2.55	8.75
0.300	0.52		

№ 5468 BARIUM OXIDE–BARIUM CHLORIDE–WATER [2507]

$BaO—BaCl_2—H_2O$

$t=30$

Comp. of saturated soln., wt%			Comp. of solid phase
Barium Chloride	Barium Oxide	Water	
27.60	0.00	72.40	$BaCl_2 \cdot 2H_2O$
27.42	1.78	70.80	»
27.36	1.77	70.87	$BaCl_2 \cdot 2H_2O + BaClOH \cdot 2H_2O$
24.98	2.33	72.69	$BaClOH \cdot 2H_2O$
21.46	3.27	75.27	»
19.18	4.67	76.15	»
18.67	4.61	76.72	$BaClOH \cdot 2H_2O + BaO \cdot 9H_2O$
18.04	4.62	77.34	$BaO \cdot 9H_2O$
17.08	4.60	78.32	»
12.81	4.58	82.61	»
10.77	4.45	84.78	»
0.00	4.99	95.01	»

№ 5469 BARIUM OXIDE–BARIUM BROMIDE–WATER [2204]

$BaO—BaBr_2—H_2O$

$t=25$

Comp. of saturated soln., wt%			Comp. of solid phase
Barium Oxide	Barium Bromide	Water	
0.00	51.10	48.90	$BaBr_2 \cdot 2H_2O$
0.78	50.67	48.55	»
0.79	50.65	48.56	$BaBr_2 \cdot 2H_2O + BaBr_2 \cdot BaO \cdot 5H_2O$
0.91	49.30	49.79	$BaBr_2 \cdot BaO \cdot 5H_2O$
2.15	41.71	56.14	»
2.42	40.47	57.11	»
3.71	36.41	59.88	$BaBr_2 \cdot BaO \cdot 5H_2O + BaO \cdot 9H_2O$
3.46	27.91	68.63	$BaO \cdot 9H_2O$
3.48	25.03	71.49	»
3.41	18.42	78.17	»
3.53	13.45	83.02	»
4.05	0.0	95.95	»

№ 5470 BARIUM OXIDE–BARIUM IODIDE–WATER [2204]

$BaO—BaI_2—H_2O$

$t=15$

Comp. of saturated soln., wt%			Comp. of solid phase
Barium Oxide	Barium Iodide	Water	
0.00	68.59	31.41	$BaI_2 \cdot 2H_2O$
0.24	68.52	31.24	$BaI_2 \cdot 2H_2O + BaI_2 \cdot BaO \cdot 9H_2O$
0.36	62.76	36.88	$BaI_2 \cdot BaO \cdot 9H_2O$
0.40	60.33	39.27	»
0.64	56.77	42.59	»
1.04	53.68	45.28	»
1.25	52.39	46.36	»
1.67	49.85	48.48	»
3.14	44.99	51.87	$BaI_2 \cdot BaO \cdot 9H_2O + BaO \cdot 9H_2O$
3.08	42.28	54.64	$BaO \cdot 9H_2O$
3.00	32.92	64.08	»
3.01	28.43	68.56	»
3.25	18.38	78.37	»
4.05	0.00	95.95	»

№ 5471 BARIUM HYDROXIDE–BARIUM THIOCYANATE–WATER [1649]

$Ba(OH)_2—Ba(CNS)_2—H_2O$

$t=25$

Comp. of saturated soln., wt%			Comp. of solid phase
Barium Hydroxide	Barium Thiocyanate	Water	
0.00	62.63	37.37	$Ba(CNS)_2 \cdot 3H_2O$
2.05	61.97	35.98	»
3.17	61.67	35.16	$Ba(CNS)_2 \cdot 3H_2O + BaOHCNS$
7.75	57.34	37.91	BaOHCNS
6.16	54.49	39.35	»
8.47	50.94	40.59	$BaOHCNS + Ba(OH)_2 \cdot 8H_2O$
5.89	47.51	46.60	$Ba(OH)_2 \cdot 8H_2O$
3.44	26.54	70.02	»
3.58	15.83	80.59	»
4.49	0.00	95.51	»

№ 5472 BARIUM HYDROXIDE–BARIUM NITRATE–WATER [2305]

$Ba(OH)_2—Ba(NO_3)_2—H_2O$

$t=25$

Comp. of saturated soln., Wt%			d	Comp. of solid phase
Barium Nitrate	Barium Hydroxide	Water		
0.00	4.29	95.71	1.0512	$Ba(OH)_2 \cdot 8H_2O$
1.88	4.35	93.77	1.0651	»
3.47	4.48	92.05	1.0790	»
5.66	4.40	89.94	1.0975	»
7.55	4.72	87.73	1.1220	»
10.21	4.93	84.86	1.1371	»
11.48	5.02	83.50	1.1448	$Ba(OH)_2 \cdot 8H_2O+Ba(NO_3)_2$
11.04	3.22	85.74	1.1210	$Ba(NO_3)_2$
10.66	1.55	88.39	1.1002	»
10.30	0.00	89.70	1.0797	»

№ 5473 BARIUM HYDROXIDE–BARIUM METASILICATE–WATER [2020]

$Ba(OH)_2—BaSiO_3 \cdot 6H_2O—H_2O$

$t=20$

Concn. of Barium Hydroxide in aqueous soln., g/l	Solubility of Barium Metasilicate Hexahydrate in aqueous soln. of Barium Hydroxide g/l	Concn. of Barium Hydroxide in aqueous soln., g/l	Solubility of Barium Metasilicate Hexahydrate in aqueous soln. of Barium Hydroxide g/l
0.8569	0.6735	3.4276	0.1985
1.7138	0.3878	5.1414	0.1287

№ 5474 BARIUM HYDROXIDE–BARIUM HYDROGEN SULFIDE–WATER [2640]

$Ba(OH)_2—Ba(HS)_2—H_2O$

Comp. of saturated soln., wt%			t	Comp. of saturated soln., wt%			t
Barium Hydroxide	Barium Hydrogen Sulfide	Water		Barium Hydroxide	Barium Hydrogen Sulfide	Water	
1.49	0.92	97.59	0	3.00	4.87	92.13	20
0.88	5.45	93.67	0	2.85	6.20	90.95	20
0.69	13.60	85.80	0	1.03	15.60	83.37	20
0.60	19.60	79.80	0	1.08	17.90	81.02	20
0.65	26.30	73.05	0	1.00	29.00	70.00	20
0.69	22.50	76.81	0	1.95	17.10	80.95	20
0.65	27.80	71.55	0	1.10	26.20	72.70	20
0.70	28.60	70.70	0	1.00	29.00	70.00	20
0.60	29.00	70.40	0	0.90	30.00	69.10	20
0.96	32.60	66.44	0	1.18	32.80	66.02	20

Table № 5474 (continuation)

Comp. of saturated soln., wt%			t	Comp. of saturated soln., wt%			t
Barium Hydroxide	Barium Hydrogen Sulfide	Water		Barium Hydroxide	Barium Hydrogen Sulfide	Water	
4.65	4.50	90.85	40	2.07	37.20	60.73	60
3.80	9.55	86.65	40	22.30	1.81	75.89	80
1.90	15.30	82.80	40	19.00	5.55	75.45	80
1.50	20.40	78.10	40	18.50	9.60	71.90	80
1.50	27.50	71.00	40	9.00	12.40	78.60	80
1.65	17.60	80.75	40	6.00	21.00	73.00	80
1.44	20.70	77.86	40	4.60	25.00	70.40	80
1.39	25.10	73.51	40	4.00	30.00	66.00	80
1.32	30.60	68.08	40	4.20	31.40	64.40	80
1.29	35.20	63.51	40	3.70	34.95	61.35	80
12.90	1.15	85.95	60	2.05	39.00	58.95	80
11.00	7.82	81.18	60	41.90	1.32	56.78	100
8.00	10.03	81.97	60	30.50	7.10	62.40	100
5.38	12.25	82.37	60	29.50	7.60	62.90	100
3.00	19.90	77.10	60	27.30	11.90	60.80	100
2.56	27.00	70.44	60	24.40	13.00	62.60	110
2.52	28.80	68.68	60	5.80	35.20	59.00	100
2.13	33.00	64.87	60	5.00	41.80	53.20	100
2.10	33.50	64.40	60	1.96	43.70	54.34	100

№ 5475 [1649]

BARIUM HYDROXIDE–BARIUM CHLORIDE–WATER

$BaCl_2—Ba(OH)_2—H_2O$

$t=25$

Comp. of saturated soln., wt%			Comp. of solid phase
Barium Hydroxide	Barium Chloride	Water	
0.0	27.16	72.84	$BaCl_2 \cdot 2H_2O$
0.89	26.94	72.17	»
1.39	26.83	71.78	»
1.77	26.72	71.51	»
1.94	26.70	71.36	$BaCl_2 \cdot 2H_2O + BaOHCl \cdot 2H_2O$
2.65	23.57	73.78	$BaOHCl \cdot 2H_2O$
3.13	22.05	74.82	»
3.38	20.79	75.83	»
4.12	19.31	76.57	$BaOHCl \cdot 2H_2O + Ba(OH)_2 \cdot 8H_2O$
4.06	17.05	78.89	$Ba(OH)_2 \cdot 8H_2O$
4.05	15.33	80.62	»
4.03	12.62	83.35	»
4.04	10.10	85.86	»
4.10	6.788	89.112	»
4.224	3.649	92.127	»
4.489	0.0	95.511	»

№ 5476 **BARIUM HYDROXIDE–BARIUM CHLORATE–WATER** [1649]

$Ba(OH)_2—Ba(ClO_3)_2—H_2O$

$t=25$

Comp. of saturated soln., wt%			Comp. of solid phase
Barium Hydroxide	Barium Chlorate	Water	
0.00	27.58	72.42	$Ba(ClO_3)_2$
1.87	27.17	70.96	»
3.71	26.58	69.71	$Ba(ClO_3)_2+Ba(OH)_2 \cdot 8H_2O$
3.77	21.85	74.38	$Ba(OH)_2 \cdot 8H_2O$
3.85	15.98	80.17	»
4.02	8.79	87.19	»
4.49	0.00	95.51	»

№ 5477 **BARIUM HYDROXIDE–BARIUM IODIDE–WATER** [2205]

$Ba(OH)_2—BaI_2—H_2O$

$t=25$

Content in aqueous soln., wt%		Comp. of solid phase
Barium Iodide	Barium Oxide	
68.59	0	$BaI_2 \cdot 6H_2O$
68.52	0.24	$BaI_2 \cdot 6H_2O+BaI_2 \cdot Ba(OH)_2 \cdot 8H_2O$
62.76	0.36	$BaI_2 \cdot Ba(OH)_2 \cdot 8H_2O$
60.33	0.40	»
56.77	0.64	»
53.68	1.04	»
52.39	1.25	»
49.85	1.67	»
45.00	3.15	$BaI_2 \cdot Ba(OH)_2 \cdot 8H_2O+Ba(OH)_2 \cdot 8H_2O$
42.28	3.08	$Ba(OH)_2 \cdot 8H_2O$
32.92	3.00	»
28.43	3.01	»
18.38	3.25	»
0	4.05	»

№ 5478 **BARIUM NITRITE–THALLIUM NITRITE–WATER** [1370]

$Ba(NO_2)_2—TlNO_2—H_2O$

$t=25$

Comp. of saturated soln., wt%			Comp. of solid phase
Thallium Nitrite	Barium Nitrite	Water	
28.95	16.18	54.87	$TlNO_2$
31.82	26.74	41.44	$TlNO_2+TlNO_2 \cdot 2Ba(NO_2)_2$
29.62	27.36	43.02	$TlNO_2 \cdot 2Ba(NO_2)_2$
23.49	28.97	47.54	»
11.85	38.05	50.10	»
5.35	43.84	50.81	$TlNO_2 \cdot 2Ba(NO_2)_2+Ba(NO_2)_2 \cdot H_2O$
3.09	43.91	53.00	$Ba(NO_2)_2 \cdot H_2O$

№ 5479 BARIUM NITRITE–THALLIUM NITRITE–WATER [721]

$Ba(NO_2)_2—TlNO_2—H_2O$

Comp. of saturated soln., wt%			t	Comp. of solid phase
Thallium Nitrite	Barium Nitrite	Water		
32.27	0.00	67.73	25	$TlNO_2$
27.91	7.68	64.41	25	»
28.76	11.16	60.08	25	»
27.85	13.37	58.78	25	»
29.13	17.05	53.82	25	»
31.17	26.44	42.39	25	$TlNO_2+TlNO_2\cdot 2Ba(NO_2)_2$
31.18	26.39	42.43	25	»
31.16	26.45	42.39	25	»
25.74	28.44	45.82	25	$TlNO_2\cdot 2Ba(NO_2)_2$
24.20	29.03	46.77	25	»
22.46	29.92	47.62	25	»
20.52	31.16	48.32	25	»
16.09	33.57	50.34	25	»
13.46	35.00	51.54	25	»
9.80	36.63	53.57	25	»
6.96	39.82	53.22	25	»
5.60	43.54	50.86	25	$TlNO_2\cdot 2Ba(NO_2)_2+BaNO_2$
5.57	43.35	51.08	25	»
3.91	44.03	52.06	25	$Ba(NO_2)_2$
0.00	44.72	55.28	25	»
55.75	0.00	44.25	50	$TlNO_2$
52.70	3.17	44.13	50	»
52.49	9.20	38.31	50	»
51.52	17.52	30.96	50	»
53.93	22.78	23.29	50	$TlNO_2+TlNO_2\cdot 2Ba(NO_2)_2$
53.89	22.59	23.52	50	»
46.89	26.27	26.84	50	$TlNO_2\cdot 2Ba(NO_2)_2$
34.49	31.76	33.75	50	»
20.75	40.00	39.25	50	»
15.50	43.96	40.54	50	»
11.46	46.10	42.44	50	»
8.18	50.25	41.57	50	»
5.90	54.35	39.75	50	$TlNO_2\cdot 2Ba(NO_2)_2+Ba(NO_2)_2$
5.91	54.30	39.79	50	»
2.38	54.72	42.90	50	$Ba(NO_2)_2$
0.00	55.76	44.24	50	»

№ 5480 BARIUM NITRATE–BARIUM CHLORIDE–WATER [1577]

$Ba(NO_3)_2—BaCl_2—H_2O$

Comp. of saturated soln., wt%			t	Comp. of solid phase
Barium Chloride	Barium Nitrate	Water		
22.5	4.3	73.2	0	$BaCl_2\cdot 2H_2O+Ba(NO_3)_2$
24.5	6.0	69.5	20	»
26.5	7.5	66.0	40	»
28.5	9.5	62.0	60	»
31.0	14.0	55.0	100	»
32.0	20.0	48.0	140	»
33.0	26.0	41.0	180	»
32.0	32.0	36.0	210	»

№ 5481 BARIUM NITRATE–BARIUM CHLORIDE–WATER [1429]

$Ba(NO_3)_2—BaCl_2—H_2O$

$t=30$

Comp. of saturated soln., wt%			Comp. of solid phase
Barium Chloride	Barium Nitrate	Water	
6.06	9.55	84.39	$Ba(NO_3)_2$
13.75	8.20	78.05	»
16.14	7.92	75.94	»
22.70	7.94	69.36	»
26.11	7.88	66.01	$Ba(NO_3)_2+BaCl_2 \cdot 2H_2O$
26.64	5.37	67.99	$BaCl_2 \cdot 2H_2O$
26.91	4.13	68.96	»
27.38	1.58	71.04	»

№ 5482 BARIUM NITRATE–BARIUM CHLORIDE–WATER [1606]

$Ba(NO_3)_2—BaCl_2—H_2O$

$t=20$

Comp. of saturated soln., wt%			Comp. of solid phase
Barium Nitrate	Barium Chloride	Water	
0.00	26.32	73.68	$BaCl_2 \cdot 2H_2O$
1.31	26.07	72.62	»
6.93	25.09	67.98	$BaCl_2 \cdot 2H_2O+Ba(NO_3)_2$
6.54	20.81	72.65	$Ba(NO_3)_2$
6.72	12.31	80.97	»
7.17	7.51	85.32	»
8.41	0.00	91.59	»

№ 5483 BARIUM NITRATE–BARIUM CHLORIDE–WATER [920, 921]

$Ba(NO_3)_2—BaCl_2—H_2O$

Comp. of saturated soln., wt%			t	Comp. of solid phase
Barium Chloride	Barium Nitrate	Water		
0.00	8.50	91.50	20	$Ba(NO_3)_2$
4.64	7.50	87.86	20	»
9.33	6.70	83.97	20	»

Table № 5483 (continuation)

Comp. of saturated soln wt%			t	Comp. of solid phase
Barium Chloride	Barium Nitrate	Water		
14.03	6.50	79.47	20	»
18.76	6.20	75.04	20	»
20.64	6.20	73.16	20	»
21.55	6.30	72.15	20	»
24.60	6.08	69.32	20	$BaCl_2 \cdot 2H_2O + Ba(NO_3)_2$
26.40	2.20	71.40	20	$BaCl_2 \cdot 2H_2O$
27.00	1.09	71.91	20	»
27.50	0.00	72.50	20	»
0.00	12.50	87.50	40	$Ba(NO_3)_2$
4.47	10.70	84.83	40	»
9.01	9.70	81.29	40	$Ba(NO_3)_2$
13.65	9.00	77.35	40	»
18.28	8.60	73.12	40	»
21.20	8.20	70.60	40	»
21.14	8.10	70.76	40	»
22.95	8.20	68.85	40	»
26.00	7.40	66.60	40	$BaCl_2 \cdot 2H_2O + Ba(NO_3)_2$
26.50	5.88	67.62	40	$BaCl_2 \cdot 2H_2O$
30.50	0.00	69.50	40	»
0.00	16.17	83.83	60	»
4.30	14.00	81.70	60	»
8.75	12.50	78.75	60	»
13.25	11.70	75.05	60	»
17.80	11.00	71.20	60	»
20.63	10.30	69.07	60	»
22.65	10.20	67.15	60	»
28.20	7.18	65.52	60	$BaCl_2 \cdot 2H_2O$
29.12	5.67	65.21	60	»
33.50	0.00	66.50	60	»
0.00	4.75	95.25	−0.7	$Ice + Ba(NO_3)_2$
4.76	4.70	90.54	−2.1	»
9.50	4.50	86.00	−3.1	»
14.36	4.30	81.34	−4.0	»
19.10	4.50	71.40	−6.5	»
20.94	4.80	74.26	−7.8	»

Table № 5483 (continuation)

Comp. of saturated soln wt%			t	Comp. of solid phase
Barium Chloride	Barium Nitrate	Water		
21.85	5.00	73.15	—8.2	»
23.20	2.30	74.50	—8.3	»
23.00	1.15	75.85	—8.2	»
23.50	0.00	76.50	—7.6	Ice + $BaCl_2 \cdot 2H_2O$
22.50	5.10	72.40	—8.5	Ice + $BaCl_2 \cdot 2H_2O + Ba(NO_3)_2$
24.50	6.08	69.42	20	$BaCl_2 \cdot 2H_2O + Ba(NO_3)_2$
25.80	8.40	65.80	40	»
26.70	10.00	63.30	60	»
4.46	0.00	95.54	80	$BaCl_2 \cdot H_2O$
4.27	1.04	94.69	80	»
3.70	1.59	94.71	80	»
2.93	1.55	95.52	80	$Ba(NO_3)_2$
1.24	1.49	97.27	80	»
0.80	1.49	97.71	80	»
0.48	1.52	98.00	80	»
0.00	1.49	98.51	80	»
4.83	0.00	95.17	100	$BaCl_2$
5.01	1.71	93.28	100	»
4.83	2.07	93.10	100	»
3.70	1.96	94.34	100	$Ba(NO_3)_2$
1.71	2.04	96.25	100	»
1.13	2.00	96.87	100	»
0.65	2.09	97.26	100	»
0.00	2.36	97.64	100	»
5.14	0.00	94.86	120	$BaCl_2$
4.92	1.20	93.88	120	»
5.37	1.83	92.80	120	»
5.30	2.27	92.43	120	»
4.80	2.55	92.65	120	$Ba(NO_3)_2$
3.14	2.50	94.36	120	»
2.19	2.63	95.18	120	»
1.48	2.75	95.77	120	»
0.890	2.85	96.25	120	»
0.00	3.00	97.00	120	»

№ 5484 BARIUM NITRATE–BARIUM CHLORIDE–WATER [921]

$Ba(NO_3)_2$—$BaCl_2$—H_2O

Comp. of saturated soln., g-mol			t	Comp. of solid phase
Barium Nitrate	Barium Chloride	Water		
0.00	100.00	2143.0	80	$BaCl_2 \cdot H_2O$
16.60	83.40	1783.0	80	»
30.04	69.96	1789.0	80	$Ba(NO_3)_2$
34.70	65.30	2131.9	80	»
54.48	45.52	3558.8	80	»
65.02	34.98	4268.0	80	»
76.12	23.88	4898.0	80	»
100.00	0.00	6603.0	80	»
0.00	100.00	1970.7	100	$BaCl_2$
25.47	74.53	1388.1	100	»
30.04	69.96	1349.8	100	$Ba(NO_3)_2$
34.70	65.30	1665.7	100	»
54.48	45.52	2568.2	100	»
65.02	34.98	3095.0	100	»
76.12	23.88	3552.3	100	»
100.00	0.00	4130.0	100	»
0.00	100.00	1845.2	120	$BaCl_2$
16.60	83.40	1534.2	120	»
25.47	74.53	1288.5	120	»
30.04	69.96	1221.1	120	$Ba(NO_3)_2$
34.70	65.30	1260.2	120	»
44.35	55.65	1673.7	120	»
54.48	45.52	1975.0	120	»
65.02	34.98	2266.4	120	»
76.12	23.88	2565.7	120	»
100.00	0.00	3229.3	120	»
26.00	74.0	1600.0	80	$Ba(NO_3)_2+BaCl_2 \cdot H_2O$
28.50	71.50	1320.0	100	$Ba(NO_3)_2+BaCl_2$
30.00	70.00	1180.0	120	»

№ 5485 BARIUM NITRATE–BARIUM CHLORATE–WATER [2408]

$Ba(NO_3)_2$—$Ba(ClO_3)_2$—H_2O

Comp. of saturated soln., wt%			d_4^t	t	Comp. of solid phase
Barium Chlorate	Barium Nitrate	Water			
21.22	0.00	78.78	1.198	10	$Ba(ClO_3)_2 \cdot H_2O$
21.38	4.52	74.10	1.249	10	$Ba(ClO_3)_2 \cdot H_2O+Ba(ClO_3)_2 \cdot 6Ba(NO_3)_2 \cdot 12H_2O$
20.97	4.56	73.47	1.245	10	$Ba(ClO_3)_2 \cdot 6Ba(NO_3)_2 \cdot 12H_2O$
18.90	4.60	76.50	1.224	10	»
16.55	4.67	78.78	1.197	10	»
14.15	4.83	71.02	1.174	10	»
9.51	5.18	85.31	1.128	10	»
4.24	6.00	89.76	1.088	10	»
3.95	6.03	90.02	1.084	10	»
3.94	6.06	90.00	1.087	10	»

Table № 5485 (continuation)

Comp. of saturated soln., wt%			d_4^t	t	Comp. of solid phase
Barium Chlorate	Barium Nitrate	Water			
3.47	6.16	90.37	1.083	10	$Ba(ClO_3) \cdot 6Ba(NO_3)_2 \cdot 12H_2O+$ $+Ba(NO_3)_2$
0.00	6.361	93.639	1.051	10	$Ba(NO_3)_2$
27.42	0.00	72.58	1.263	25	$Ba(ClO_3)_2 \cdot H_2O$
26.82	7.94	65.24	1.362	25	$Ba(ClO_3)_2 \cdot H_2O+Ba(NO_3)_2$ (metastable)
26.67	7.28	36.05	1.347	25	$Ba(ClO_3)_2 \cdot H_2O+$ $+Ba(ClO_3)_2 \cdot 6Ba(NO_3)_2 \cdot 12H_2O$
26.12	7.25	66.63	1.344	25	$Ba(ClO_3)_2 \cdot 6Ba(NO_3)_2 \cdot 12H_2O$
25.91	7.23	66.86	1.337	25	»
25.39	7.28	67.33	1.333	25	»
25.32	7.28	67.40	1.332	25	»
24.27	7.36	68.37	1.318	25	»
22.48	7.53	69.99	1.301	25	»
20.24	7.81	71.95	1.277	25	»
20.20	7.78	72.02	1.275	25	»
18.84	7.84	73.32	1.258	25	»
18.29	7.96	73.75	1.254	25	»
17.32	8.09	74.59	1.243	25	»
15.08	8.39	76.53	1.222	25	$Ba(ClO_3)_2 \cdot 6Ba(NO_3)_2 \cdot 12H_2O+$ $+Ba(NO_3)_2$
0.00	9.246	90.754	1.079	25	$Ba(NO_3)_2$
34.90	0.00	65.10	1.347	45	$Ba(ClO_3)_2 \cdot H_2O$
32.48	10.16	57.36	1.465	45	$Ba(ClO_3)_2 \cdot H_2O+Ba(NO_3)_2$
0.00	13.60	86.40	1.110	45	$Ba(NO_3)_2$

№ 5486 [2408]

BARIUM NITRATE–BARIUM CHLORATE–BARIUM BROMIDE–WATER

$Ba(NO_3)_2—Ba(ClO_3)_2—BaBr_2—H_2O$

$t=10$

Comp. of saturated soln., wt%				d_4	Comp. of solid phase
Barium Chlorate	Barium Bromide	Barium Nitrate	Water		
4.50	44.68	3.90	46.92	1.779	$Ba(ClO_3)_2 \cdot H_2O+BaBr_2 \cdot 2H_2O+$ $+1:6:12$
2.0	24.3	4.2	69.5	—	$1:8:12+Ba(NO_3)_2+1:6:12$
1.60	46.58	3.93	47.89	1.762	$1:8:12+BaBr_2 \cdot 2H_2O+1:6:12$
4.40	44.71	3.99	46.90	1.779	$Ba(ClO_3)_2 \cdot H_2O+BaBr_2 \cdot 2H_2O+$ $+1:6:12$

Note. $1:6:12 = Ba(ClO_3)_2 \cdot 6Ba(NO_3)_2 \cdot 12H_2O$; $1:8:12 = BaBr_2 \cdot 8Ba(NO_3)_2 \cdot 12H_2O$.

№ 5487 BARIUM NITRATE–BARIUM BROMIDE–WATER [2408]

$Ba(NO_3)_2—BaBr_2—H_2O$

Comp. of saturated soln., wt%			d_4^t	t	Comp. of solid phase
Barium Bromide	Barium Nitrate	Water			
48.77	0.00	51.23	1.698	10	$BaBr_2 \cdot 2H_2O$
47.51	4.80	47.69	1.757	10	$BaBr_2 \cdot 2H_2O + Ba(NO_3)_2$ (metastable)
47.47	4.04	48.49	1.752	10	$BaBr_2 \cdot 2H_2O + BaBr_2 \cdot 8Ba(NO_3)_2 \cdot 12H_2O$
47.05	4.04	48.91	1.741	10	$BaBr_2 \cdot 8Ba(NO_3)_2 \cdot 12H_2O$
45.79	3.94	50.27	1.710	10	»
44.41	3.95	51.64	1.678	10	»
43.68	3.87	52.45	1.662	10	»
41.47	4.00	54.53	—	10	»
38.02	3.98	58.00	1.542	10	»
35.14	4.12	60.74	1.491	10	»
31.68	4.30	64.02	1.433	10	»
28.04	4.51	67.45	1.375	10	»
27.56	4.57	67.87	1.367	10	»
26.99	4.64	68.37	1.363	10	»
25.38	4.74	69.88	1.336	10	»
24.81	4.73	71.16	1.321	10	»
24.03	4.84	71.13	1.319	10	$BaBr_2 \cdot 8Ba(NO_3)_2 \cdot 12H_2O + Ba(NO_3)_2$
0.00	6.361	93.639	1.051	10	$Ba(NO_3)_2$
50.07	0.00	49.93	1.726	25	$BaBr_2 \cdot 2H_2O$
48.24	5.21	46.55	1.793	25	$BaBr_2 \cdot 2H_2O + BaBr_2 \cdot 8Ba(NO_3)_2 \cdot 12H_2O$
47.51	5.23	47.26	—	25	$BaBr_2 \cdot 8Ba(NO_3)_2 \cdot 12H_2O$
46.75	5.20	48.05	1.757	25	»

№ 5488 BARIUM NITRATE–BARIUM BROMATE–WATER [1777]

$Ba(NO_3)_2—Ba(BrO_3)_2—H_2O$

$t=25$

Content in aqueous soln., g/l		d	Content in aqueous soln., g/l		d
Barium Nitrate	Barium Bromate		Barium Nitrate	Barium Bromate	
0.00	7.93	1.0038	13.07	6.415	1.0132
3.27	7.22	1.0059	26.14	6.230	1.0233
6.53	6.83	1.0083			

№ 5489 BARIUM NITRATE–BARIUM BROMATE–WATER [2408]

$Ba(NO_3)_2—Ba(BrO_3)_2—H_2O$

$t=25$

Comp. of saturated soln., wt%			Comp. of solid phase
Barium Nitrate	Barium Bromate	Water	
0.00	0.788	99.212	$Ba(BrO_3)_2 \cdot H_2O$
12.71	0.395	86.89	»
24.15	0.320	75.53	»
36.19	0.257	63.55	»
44.21	0.229	55.57	»
49.92	0.220	49.86	$Ba(BrO_3)_2 \cdot H_2O + Ba(NO_3)_2$

№ 5490 **BARIUM NITRATE–BARIUM IODATE–WATER** **[1778]**

$Ba(NO_3)_2$—$Ba(IO_3)_2$—H_2O

$t=25$

Content in aqueous soln., g/l			
Barium Nitrate	Barium Iodate	Barium Nitrate	Barium Iodate
0.261	0.331	13.07	0.149
0.523	0.294	26.14	0.148
1.307	0.237	52.28	0.136
5.228	0.164		

№ 5491 **BARIUM NITRATE–BARIUM IODATE–WATER** **[2408]**

$Ba(NO_3)_2$—$Ba(IO_3)_2$—H_2O

$t=25$

Comp. of saturated soln., wt%			Comp. of solid phase
Barium Nitrate	Barium Iodate	Water	
0.00	0.042	99.958	$Ba(IO_3)_2 \cdot H_2O$
2.50	0.019	97.481	»
4.75	0.023	95.227	»
7.14	0.027	92.833	»
8.69	0.031	91.279	»
9.22	0.033	90.747	$Ba(IO_3)_2 \cdot H_2O + Ba(NO_3)_2$

№ 5492 **BARIUM NITRATE–THALLIUM BROMIDE–WATER** **[2581]**

$Ba(NO_3)_2$—$TlBr$—H_2O

$t=20$

Concn. in aqueous soln., g/l			
Thallium Bromide	Barium Nitrate	Thallium Bromide	Barium Nitrate
0.5083	1.307	0.5805	7.842
0.5311	2.614	0.6141	13.07
0.5592	5.228		

№ 5493 **[2702]**

BARIUM NITRATE–THALLIUM IODIDE–WATER

$Ba(NO_3)_2$—TlI—H_2O

$t=20$

Concn. in aqueous soln., g/l	
Thallium Iodide	Barium Nitrate
0.0697	1.305
0.0765	5.22
0.0828	13.05

№ 5494 BARIUM NITRATE–LEAD (II) NITRATE–WATER [1580]

$Ba(NO_3)_2—Pb(NO_3)_2—H_2O$

$t=25$

Content in aqueous soln., g/l		d	Content in aqueous soln., g/l		d
Barium Nitrate	Lead Nitrate		Barium Nitrate	Lead Nitrate	
102.2	0.0	1.079	66.0	177.3	1.198
54.9	17.63	1.088	57.5	247.7	1.252
86.5	49.80	1.108	25.9	334.3	1.294
79.7	68.1	1.119	28.0	429.7	1.376
77.0	97.2	1.140	0.0	553.8	1.459
69.8	130.7	1.163			

№ 5495 BARIUM NITRATE–LEAD (II) NITRATE–WATER [1728]

$Ba(NO_3)_2—Pb(NO_3)_2—H_2O$

Comp. of saturated soln., wt%			t
Lead Nitrate	Barium Nitrate	Water	
1.63	8.64	89.73	25
6.26	7.31	86.43	25
8.96	6.66	84.38	25
16.21	5.59	77.50	25
22.73	4.23	73.04	25
33.13	2.65	64.22	25
34.60	1.39	64.01	25
5.72	12.68	81.60	50
10.34	11.05	78.61	50
17.05	9.12	73.83	50
29.31	6.32	64.37	50
37.78	3.83	58.39	50
40.82	2.57	56.61	50

№ 5496 BARIUM NITRATE–LEAD (II) BROMIDE–WATER [2368]

$Ba(NO_3)_2—PbBr_2—H_2O$

$t=25$

Concn. in aqueous soln., wt%			
Lead Bromide	Barium Nitrate	Lead Bromide	Barium Nitrate
0.073	0.709	3.52	0.942
0.182	0.726	6.77	1.05
0.363	0.744	7.11	1.06
0.723	0.779	11.28	1.15
1.79	0.857	12.98 *	1.20
3.17	0.924		

* Solid phase $PbBr_2 + Ba(NO_3)_2$.

№ 5497 **BARIUM NITRATE–URANYL NITRATE–WATER** [1010]

$Ba(NO_3)_2—UO_2(NO_3)_2—H_2O$

Comp. of saturated soln., wt%			t	Comp. of solid phase
Uranyl Nitrate	Barium Nitrate	Water		
49.39	0.00	50.61	0.1	$UO_2(NO_3)_2 \cdot 6H_2O$
39.41	0.27	60.32	0.1	$Ba(NO_3)_2$
31.64	0.42	67.94	0.1	»
21.65	0.60	77.75	0.1	»
17.36	0.80	81.84	0.1	»
11.82	1.67	86.51	0.1	»
8.82	2.00	89.18	0.1	»
5.20	2.60	92.20	0.1	»
0.00	4.88	35.12	0.1	»
56.12	0.00	43.88	2.5	$UO_2(NO_3)_2 \cdot 6H_2O$
49.51	0.25	50.24	2.5	$Ba(NO_3)_2$
40.12	0.40	59.48	2.5	»
33.16	0.65	66.19	2.5	»
30.74	0.74	68.52	2.5	»
23.34	1.45	75.21	2.5	»
21.04	1.61	77.35	2.5	»
19.37	2.03	78.60	2.5	»
16.78	2.37	80.60	2.5	»
14.00	3.06	82.94	2.5	»
11.54	3.63	84.83	2.5	»
8.25	4.78	86.97	2.5	»
5.20	6.20	88.60	2.5	»
0.00	9.18	90.82	2.5	»
67.28	0.00	32.72	50	$UO_2(NO_3)_2 \cdot 6H_2O$
54.15	0.38	45.47	50	$Ba(NO_3)_2$
48.20	0.43	51.27	50	»
39.37	0.93	59.70	50	»
31.41	1.70	66.89	50	»
21.27	3.94	74.79	50	»
14.20	6.64	79.16	50	»
7.96	9.32	82.72	50	»
0.00	14.53	85.47	50	»

№ 5498 **BARIUM FLUROIDE–BARIUM CHLORIDE–WATER** [1336]

$BaF_2—BaCl_2—H_2O$

$t=25$

Content in aqueous soln., wt%		Comp. of solid phase
Barium Fluoride	Barium Chloride	
0.191	0	BaF_2
0.0905	0.995	»
0.051	1.07	$BaF_2+BaF_2 \cdot BaCl_2$
0.0051	1.07	»
	1.07—2.78	Solid soln. $BaF_2+BaF_2 \cdot BaCl_2$
	2.78—27.0	Solid soln. $BaF_2 \cdot BaCl_2+BaCl_2$
	27.0	Solid soln. $BaF_2 \cdot BaCl_2+BaCl_2+BaCl_2 \cdot 2H_2O$

№ 5499 BARIUM METASILICATE–BARIUM CHLORIDE–WATER [2020]

$BaSiO_3 \cdot 6H_2O—BaCl_2—H_2O$

$t=20$

Concn. of Barium Chloride in aqueous soln., g/l	Solubility of Barium Metasilicate Hexahydrate in aqueous soln. of Barium Chloride, g/l	Concn. of Barium Chloride in aqueous soln., g/l	Solubility of Barium Metasilicate Hexahydrate in aqueous soln. of Barium Chloride, g/l
1.0414	1.039	4.1654	0.8345
2.0827	0.9305	6.2481	0.804

№ 5500 [62]

BARIUM HYDROGEN SULFIDE–BARIUM CHLORIDE–WATER

$Ba(HS)_2—BaCl_2—H_2O$

Comp. of saturated soln., wt%		
Barium Hydrogen Sulfide	Barium Chloride	Water
22.3	14.0	63.7

№ 5501 BARIUM SULFATE–THORIUM CHLORIDE–WATER [1126]

$BaSO_4—ThCl_4—H_2O$

$t=20$

Content in aqueous soln., g/l			
Thorium Chloride	Barium Sulfate	Thorium Chloride	Barium Sulfate
32.57	3.57	3.81	1.28
17.45	2.40	1.94	0.58
9.39	1.54		

№ 5502 BARIUM CHLORIDE–BARIUM CHLORATE–WATER [1502]

$BaCl_2—Ba(ClO_3)_2—H_2O$

$t=20$

Comp. of saturated soln., wt%			Comp. of solid phase
Barium Chlorate	Barium Chloride	Water	
24.39	7.69	67.92	$Ba(ClO_3)_2 \cdot H_2O$

№ 5503 BARIUM CHLORIDE–BARIUM CHLORATE–WATER [1502]

$BaCl_2$—$Ba(ClO_3)_2$—H_2O

$t=20$

Comp. of saturated soln., wt%			Comp. of solid phase
Barium Chlorate	Barium Chloride	Water	
0.00	26.20	73.80	$BaCl_2 \cdot 2H_2O$
5.51	24.11	70.38	»
9.32	24.00	66.68	»
13.03	22.00	64.97	»
17.48	19.50	63.02	»
19.21	19.55	61.24	»
22.84	17.00	60.16	$BaCl_2 \cdot 2H_2O+Ba(ClO_3)_2 \cdot H_2O$
25.16	0.00	74.84	$Ba(ClO_3)_2 \cdot H_2O$

№ 5504 BARIUM CHLORIDE–BARIUM CHLORATE–WATER [2408]

$BaCl_2$—$Ba(ClO_3)_2$—H_2O

$t=25$

Comp. of saturated soln., wt%			d_4	Comp. of solid phase
Barium Chlorate	Barium Chloride	Water		
27.42	0.00	72.58	1.263	$Ba(ClO_3)_2 \cdot H_2O$
23.06	6.35	70.59	1.294	»
18.94	13.16	67.90	1.338	»
15.74	18.29	65.97	1.373	»
14.73	21.06	64.21	1.398	$Ba(ClO_3)_2 \cdot H_2O+BaCl_2 \cdot 2H_2O$
11.22	22.48	66.30	1.373	$BaCl_2 \cdot 2H_2O$
11.21	22.45	66.34	1.371	»
5.22	24.90	69.88	1.237	»

№ 5505 BARIUM CHLORIDE–BARIUM BROMIDE–WATER [1221]

$BaCl_2$—$BaBr_2$—H_2O

Comp. of saturated soln., wt%			t	d	Comp. of solid phase
Barium Chloride	Barium Bromide	Water			
27.0	0.0	73.0	25	1.323	$BaCl_2 \cdot 2H_2O$
22.8	6.9	70.3	25	1.338	$(BaCl_2, BaBr_2) \cdot 2H_2O$
22.4	8.16	69.44	25	1.351	»
21.7	9.37	68.93	25	1.361	»
17.86	17.03	65.11	25	1.400	»
19.95	26.8	53.25	25	1.498	»
10.44	31.9	57.66	25	1.560	»

Table № 5505 (continuation)

Comp. of saturated soln., wt%			t	d	Comp. of solid phase
Barium Chloride	Barium Bromide	Water			
9.55	34.2	56.25	25	1.588	$(BaCl_2, BaBr_2) \cdot 2H_2O$
9.12	35.4	55.48	25	1.614	»
8.35	36.3	55.34	25	1.606	»
6.17	41.3	52.57	25	—	»
4.28	44.2	51.52	25	1.696	»
3.31	46.6	50.09	25	1.731	»
2.7	47.4	49.0	25	1.730	$(BaBr_2, BaCl_2) \cdot 2H_2O$
0.0	50.0	50.0	25	1.734	$BaBr_2 \cdot 2H_2O$
30.5	0.0	69.5	50	1.328	$BaCl_2 \cdot 2H_2O$
25.7	7.13	67.17	50	1.363	$(BaCl_2, BaBr_2) \cdot 2H_2O$
21.0	15.08	63.92	50	1.412	»
18.21	20.1	61.69	50	1.488	»
15.22	25.8	58.98	50	1.525	»
10.37	35.8	53.83	50	1.641	»
9.11	38.5	52.39	50	1.678	»
3.65	48.9	47.45	50	1.801	»
1.34	51.9	46.76	50	1.797	$(BaBr_2, BaCl_2) \cdot 2H_2O$
0.0	52.5	47.5	50	1.803	$BaBr_2 \cdot 2H_2O$
33.4	0.0	66.6	75	1.422	$BaCl_2 \cdot 2H_2O$
31.5	2.89	65.61	75	1.437	$(BaCl_2, BaBr_2) \cdot 2H_2O$
29.3	6.34	64.36	75	1.457	»
26.5	11.09	62.41	75	1.489	»
25.0	13.46	61.54	75	1.505	»
21.7	19.03	59.27	75	1.552	»
19.47	22.6	57.93	75	1.562	»
17.70	26.2	56.1	75	1.611	»
16.30	28.7	55.0	75	1.635	»
15.24	30.7	54.06	75	1.661	»
12.47	36.1	51.43	75	1.718	»
11.15	38.8	50.05	75	1.757	»
9.58	42.3	48.12	75	1.772	»
7.26	45.3	47.44	75	1.818	»
6.46	46.8	46.74	75	1.843	»
6.16	47.9	45.94	75	—	»
5.00	49.1	45.9	75	1.855	»
4.36	50.8	44.84	75	1.894	$(BaBr_2, BaCl_2) \cdot 2H_2O + (BaCl_2, BaBr_2) \cdot 2H_2O$
3.95	51.3	44.75	75	—	»
1.61	53.5	44.89	75	1.889	$(BaBr_2, BaCl_2) \cdot 2H_2O$
0.0	54.7	45.3	75	1.890	$BaBr_2 \cdot 2H_2O$
36.7	0.0	63.3	100	1.508	$BaCl_2 \cdot 2H_2O$
35.3	2.44	62.26	100	1.530	»
31.2	8.55	60.25	100	1.561	»
28.2	13.30	58.5	100	1.593	»
22.7	21.7	55.6	100	1.657	»
17.69	30.4	51.91	100	1.752	»
11.74	40.7	47.56	100	1.860	»
7.47	48.8	43.73	100	1.968	»
6.46	50.3	43.24	100	1.977	$(BaCl_2, BaBr_2) \cdot 2H_2O$
2.26	55.3	42.44	100	2.016	$(BaBr_2, BaCl_2) \cdot 2H_2O$
0.0	56.9	43.1	100	1.998	$BaBr_2 \cdot 2H_2O$

№ 5506 BARIUM CHLORIDE–BARIUM BROMIDE–WATER [294]

$BaCl_2—BaBr_2—H_2O$

$t=25$

Comp. of saturated soln., wt%			d	Comp. of solid phase
Barium Chloride	Barium Bromide	Water		
0.00	50.09	49.91	1.7226	$BaBr_2 \cdot 2H_2O$
4.43	44.58	50.99	1.6948	BaCl+BaBr
8.07	37.02	54.91	1.6057	»
9.38	35.13	55.49	1.5860	»
14.41	25.13	60.46	1.4805	$BaCl_2 \cdot 2H_2O$
17.84	17.31	64.85	1.4061	»
18.08	16.85	65.07	1.4043	»
20.46	11.84	67.70	1.3678	»
21.02	11.32	67.66	1.3652	»
22.33	10.23	67.44	1.3563	»
27.53	0.00	72.47	1.2928	»

№ 5507 BARIUM CHLORIDE–BARIUM IODATE–WATER [2405]

$BaCl_2—Ba(IO_3)_2—H_2O$

$t=25$

Comp. of saturated soln., wt%			d,	Comp. of solid phase
Barium Iodate	Barium Chloride	Water		
0.0400	0.00	99.06	0.997	$Ba(IO_3)_2 \cdot H_2O$
0.0189	19.63	80.35	1.194	»
0.0187	20.50	79.48	1.206	»
0.0198	21.56	78.42	1.218	»
0.0231	25.29	74.69	1.268	»
0.0240	25.59	74.39	1.273	»
0.0267	27.09	72.88	1.286	$Ba(IO_3)_2 \cdot H_2O + BaCl_2 \cdot 2H_2O$
0.0201	22.06	77.92	1.223	$Ba(IO_3)_2 \cdot H_2O + Ba(IO_3)_2 \cdot BaCl_2 \cdot 2H_2O$
0.0189	22.72	77.26	1.232	$Ba(IO_3)_2 \cdot BaCl_2 \cdot 2H_2O$
0.0186	23.76	76.22	1.245	»
0.020	24.25	75.73	1.255	»
0.0179	24.73	72.2	1.258	»
0.0171	25.45	74.53	1.266	»
0.019	25.51	74.47	1.269	»
0.0165	26.01	73.97	1.274	»
0.0152	26.82	73.16	1.285	»
0.0153	27.11	72.87	1.289	$Ba(IO_3)_2 \cdot BaCl_2 \cdot 2H_2O + BaCl_2 \cdot 2H_2O$
0.00	27.11	72.89	1.289	$BaCl_2 \cdot 2H_2O$

№ 5508 BARIUM CHLORIDE–MERCURY (I) CHLORIDE–WATER [2412]

$BaCl_2—HgCl—H_2O$

$t=25$

Concn. in aqueous soln., g/l		d	Comp. of solid phase
Mercury Chloride	Barium Chloride		
0.044	104.15	1.088	HgCl+0.1 g Hg
0.088	156.22	1.134	»
0.107	208.30	1.174	»
0.231	312.54	1.263	»

№ 5509 BARIUM CHLORIDE–MERCURY (II) CHLORIDE–WATER [1631]

$BaCl_2—HgCl_2—H_2O$

Comp. of saturated soln., wt%			t	Comp. of solid phase
Barium Chloride	Mercury Chloride	Water		
23.58	50.54	25.88	10.4	$BaCl_2 \cdot 2H_2O + HgCl_2$
23.44	50.74	25.82	10.4	$BaCl_2 \cdot 3HgCl_2 \cdot 6H_2O$
22.58	51.23	26.19	10.4	»
22.48	51.41	26.11	10.4	
22.10	51.66	26.24	10.4	»
21.64	51.74	26.72	10.4	$BaCl_2 \cdot 2H_2O + HgCl_2$
23.02	54.83	22.15	25	»

№ 5510 BARIUM CHLORIDE–MERCURY (II) CHLORIDE–WATER [2508]

$BaCl_2—HgCl_2—H_2O$

Comp. of saturated soln., wt%			t	Comp. of solid phase
Mercury Chloride	Barium Chloride	Water		
0.00	23.70	76.30	0	$BaCl_2 \cdot 2H_2O$
14.25	24.00	61.75	0	»
36.20	24.89	38.91	0	»
46.08	24.05	29.87	0	$BaCl_2 \cdot 3HgCl_2 \cdot 6H_2O + BaCl_2 \cdot 2H_2O$
46.59	23.28	30.13	0	$BaCl_2 \cdot 3HgCl_2 \cdot 6H_2O$
47.78	21.05	31.17	0	»
48.47	20.67	30.87	0	»
44.33	18.50	37.17	0	$BaCl_2 \cdot 3HgCl_2 \cdot 6H_2O + HgCl_2$
29.00	11.59	59.71	0	$HgCl_2$
16.36	6.11	77.53	0	»
3.95	0.00	96.05	0	»

Table № 5510 (continuation)

Comp. of saturated soln., wt%			t	Comp. of solid phase
Mercury Chloride	Barium Chloride	Water		
0.00	27.77	72.23	30	$BaCl_2 \cdot 2H_2O$
2.90	27.56	69.54	30	»
12.98	26.99	60.03	30	»
34.57	26.69	38.74	30	»
46.50	25.22	28.28	30	»
55.22	23.17	21.61	30	$BaCl_2 \cdot 2H_2O + HgCl_2$
48.97	17.87	33.16	30	$HgCl_2$
41.30	14.26	44.44	30	»
27.62	8.41	63.97	30	»
14.19	2.65	83.16	30	»
7.67	0.00	92.33	30	»

№ 5511 BARIUM CHLORIDE–MERCURY (II) CHLORIDE–WATER [1815]

$BaCl_2—HgCl_2—H_2O$

$t=25$

Concn. in aqueous soln., g/l			
Mercury Chloride	Barium Chloride	Mercury Chloride	Barium Chloride
71.8	0	440	161.8
189.3	80.2	717.5	278.5
317	119.2	1453	631

№ 5512 [1162]

BARIUM CHLORIDE–MERCURY (II) CHLORIDE–WATER

$BaCl_2—HgCl_2—H_2O$

$t=25$

Concn. in aqueous soln wt%		Comp. of solid phase
Mercury Chloride	Barium Chloride	
6.90	0.00	$HgCl_2$
15.07	3.28	»
35.36	11.19	»
44.58	16.23	»
54.43	21.96	»
54.41	22.55	»
54.74	22.91	$HgCl_2 + BaCl_2 \cdot 2H_2O$
54.93	22.80	»
53.83	23.11	$BaCl_2 \cdot 2H_2O$
53.69	23.23	»
51.34	23.82	»
43.51	25.22	»
32.45	26.07	»
14.96	26.73	»
8.10	26.54	»

№ 5513 [2263]

BARIUM CHLORIDE–THALLIUM CHLORIDE–WATER

$BaCl_2$—$TlCl$—H_2O

$t=25$

Concn. in aqueous soln., g/l	
Thallium Chloride	Barium Chloride
2.052	5.895
0.773	30.59

№ 5514 [2264]

BARIUM CHLORIDE–THALLIUM CHLORIDE–WATER

$BaCl_2$—$TlCl$—H_2O

$t=25$

Concn. of Barium Chloride in aqueous soln., g/l	Solubility of Thallium Chloride in aqueous soln. of Barium Chloride, g/l
5.209	1.487
10.418	1.019

№ 5515 **BARIUM CHLORIDE–THALLIUM CHLORIDE–WATER** [1197]

$BaCl_2$—$TlCl$—H_2O

$t=25$

Comp. of saturated soln., wt%					
Thallium Chloride	Barium Chloride	Water	Thallium Chloride	Barium Chloride	Water
0.069	1.959	97.972	0.035	19.99	79.975
0.041	4.759	95.20	0.048	23.06	76.892
0.037	9.087	90.876	0.063	26.62	73.317
0.037	15.24	84.723			

№ 5516 **BARIUM CHLORIDE–LEAD (II) CHLORIDE–WATER** [1817]

$BaCl_2$—$PbCl_2$—H_2O

$t=25$

Concn. of Barium Chloride in aqueous soln., g/l	Solubility of Lead Chloride in aqueous soln. of Barium Chloride, g/l	Concn. of Barium Chloride in aqueous soln., g/l	Solubility of Lead Chloride in aqueous soln. of Barium Chloride, g/l
33.3	1.944	143.7	3.029
66.6	1.538	216.6	6.641

№ 5517 [1965]

BARIUM CHLORIDE–LEAD (II) CHLORIDE–WATER

$BaCl_2$—$PbCl_2$—H_2O

$t=25$

Concn. in aqueous soln., g/l		Comp. of solid phase
Lead Chloride	Barium Chloride	
3.23	10.41	$BaCl_2 \cdot PbCl_2$
2.78	20.82	»
2.73	52.06	»
3.03	99.96	»
3.23	113.50	»
3.61	140.6	»

Table № 5517 (continuation)

Concn. in aqueous soln., g/l		Comp. of solid phase
Lead Chloride	Barium Chloride	
3.92	160.2	$BaCl_2 \cdot PbCl_2$
4.55	183.4	»
4.90	210.1	»
5.09	218.4	»
5.60	239.2	»
6.06	254.8	»
6.39	266.2	»
7.14	287.0	»
7.78	301.6	»
8.23	309.9	»
9.17	324.5	»
9.94	333.8	»
10.23	339.0	»

№ 5518 BARIUM CHLORIDE–LEAD (II) CHLORIDE–WATER [187]

$BaCl_2—PbCl_2—H_2O$

Content in aqueous soln., g/l Water		t	Comp. of solid phase
Lead Chloride	Barium Chloride		
0.0	308.2	0	$BaCl_2 \cdot 2H_2O$
2.5	310.5	0	$BaCl_2 \cdot 2H_2O + PbCl_2$
0.627	124.2	0	$PbCl_2$
0.635	51.6	0	»
6.25	0.0	0	»
0.0	371.8	25	$BaCl_2 \cdot 2H_2O$
1.83	373.5	25	»
3.09	374.9	25	»
4.47	375.9	25	»
4.62	375.9	24	»
5.31	374.8	25	»
7.45	375.5	25	»
8.70	375.5	25	$BaCl_2 \cdot 2H_2O + PbCl_2$
6.59	338.8	25	$PbCl_2$
4.20	269.1	25	»
3.33	230.0	25	»
2.20	155.0	25	»
1.83	103.6	25	»
1.98	50.8	25	»
10.8	0.0	25	»

№ 5519 BARIUM CHLORIDE–LEAD (II) BROMIDE–WATER [1817]

$BaCl_2—PbBr_2—H_2O$

$t=25$

Concn. of Barium Chloride in aqueous soln. g/l	Solubility of Lead Bromide in aqueous soln. of Barium Chloride, g/l	Concn. of Barium Chloride in aqueous soln. g/l	Solubility of Lead Bromide in aqueous soln. of Barium Chloride, g/l
0.0	9.63	108.37	2.81
54.18	2.11	216.7	8.76

№ 5520 BARIUM CHLORATE–BARIUM BROMIDE–WATER [2408]

$Ba(ClO_3)_2—BaBr_2—H_2O$

Comp. of saturated soln., wt%			d_1^t	t	Comp. of solid phase
Barium Chlorate	Barium Bromide	Water			
21.22	0.00	78.78	1.198	10	$Ba(ClO_3)_2 \cdot H_2O$
4.32	46.02	49.66	1.724	10	$Ba(ClO_3)_2 \cdot H_2O + BaBr_2 \cdot 2H_2O$
0.00	48.77	51.23	1.698	10	$BaBr_2 \cdot 2H_2O$
27.42	0.00	72.58	1.263	25	$Ba(ClO_3)_2 \cdot H_2O$
5.68	46.53	47.79	1.755	25	$Ba(ClO_3)_2 \cdot H_2O + BaBr_2 \cdot 2H_2O$
0.00	50.07	49.93	1.726	25	$BaBr_2 \cdot 2H_2O$

№ 5521 BARIUM CHLORIDE–BARIUM BROMATE–WATER [2399]

$Ba(ClO_3)_2—Ba(BrO_3)_2—H_2O$

$t=25$

Concn. in aqueous soln., wt%		d_4	Comp. of solid phase
Barium Bromate	Barium Chlorate		
0.791	0.0	1.001	$Ba(BrO_3)_2 \cdot 2H_2O$
0.609	1.292	1.011	solid solution
0.553	2.304	1.016	»
0.491	4.850	1.043	»
0.446	7.019	1.060	»
0.423	9.37	1.985	»
0.402	10.50	1.095	»
0.347	14.41	1.131	»
0.347	14.56	1.132	»
0.310	16.83	1.151	»
0.282	18.07	1.165	»
0.249	19.53	1.178	»
0.235	20.50	1.186	»
0.207	21.72	1.202	»
0.176	22.91	1.209	»
0.145	23.85	1.217	»
0.112	24.87	1.227	»
0.078	25.87	1.242	»
0.057	26.51	1.241	»
0.0	27.54	1.249	»

№ 5522 BARIUM BROMIDE–BARIUM BROMATE–WATER [2408]

$BaBr_2—Ba(BrO_3)_2—H_2O$

$t=25$

Comp. of saturated soln., wt%			Comp. of solid phase
Barium Bromide	Barium Bromate	Water	
12.71	0.395	86.89	$Ba(BrO_3)_2 \cdot H_2O$
24.15	0.320	75.53	»
36.19	0.257	63.55	»
44.21	0.229	55.56	»
49.92	0.220	49.86	$Ba(BrO_3)_2 \cdot H_2O + BaBr_2 \cdot 2H_2O$

№ 5523 BARIUM BROMIDE–BARIUM IODIDE–WATER [1577]

$BaBr_2—BaI_2—H_2O$

Comp. of saturated soln., wt%			t	Comp. of solid phase
Barium Bromide	Barium Iodide	Water		
4.8	58.4	36.8	—16	$BaBr_2+BaI_2$
5.5	66.0	28.5	60	»
9.2	67.2	23.6	135	»
10.0	67.4	21.6	170	»
14.9	67.7	17.4	210	»

№ 5524 BARIUM BROMIDE–BARIUM IODIDE–WATER [1806]

$BaBr_2—BaI_2—H_2O$

Content in aqueous soln., wt%		t	Comp. of solid phase
Barium Bromide	Barium Iodide		
4.85	58.5	—16	$BaBr_2+BaI_2$
5.5	66.0	60	»
9.15	67.15	135	»
11.0	67.4	170	»
14.9	67.7	210	»

№ 5525 BARIUM BROMIDE–BARIUM IODATE–WATER [2408]

$BaBr_2—Ba(IO_3)_2—H_2O$

$t=25$

Comp. of saturated soln., wt%			Comp. of solid phase
Barium Iodate	Barium Bromide	Water	
0.042	0.000	99.958	$Ba(IO_3)_2 \cdot H_2O$
0.013	11.19	88.797	»
0.017	22.34	77.643	»
0.026	33.46	66.514	»
0.045	44.61	55.345	»
0.068	49.02	50.912	»
0.075	50.07	49.855	$Ba(IO_3)_2 \cdot H_2O + BaBr_2$

№ 5526 BARIUM BROMIDE–MERCURY (II) BROMIDE–WATER [1815]

$BaBr_2—HgBr_2—H_2O$

$t=25$

Concn. in aqueous soln., g/l			
Mercury Bromide	Barium Bromide	Mercury Bromide	Barium Bromide
6.12	0.0	273.3	172.2
133.2	81.3	532.5	325.3
194.3	117.6		

№ 5527 BARIUM BROMIDE–MERCURY (II) BROMIDE–WATER [1498]

$BaBr_2—HgBr_2—H_2O$

Comp. of saturated soln., wt%			t	Comp. of solid phase
Mercury Bromide	Barium Bromide	Water		
0.270	0.0	99.73	4.5	$HgBr_2$
14.3	8.0	77.7	4.5	»
30.1	16.9	53.0	4.5	»
45.3	24.2	30.5	4.5	»
47.6	25.5	25.9	4.5	»
52.4	27.6	20.0	4.5	»
51.7	28.3	20.0	4.5	Solid solution
51.2	28.7	20.1	4.5	»
51.0	29.4	19.6	4.5	»
50.0	30.4	19.6	4.5	»
50.0	30.1	19.9	4.5	»
49.7	31.7	18.6	4.5	»
49.7	30.8	19.5	4.5	»
49.7	31.6	18.7	4.5	»
49.0	32.1	18.9	4.5	$BaBr_2 \cdot 2H_2O$
35.9	35.4	28.7	4.5	»
21.9	39.7	38.4	4.5	»
10.0	45.0	45.0	4.5	»

Table № 5527 (continuation)

Comp. of saturated soln., wt%			t	Comp. of solid phase
Mercury Bromide	Barium Bromide	Water		
0.0	49.9	50.1	4.5	$BaBr_2 \cdot 2H_2O$
0.427	0.0	99.573	10.4	$HgBr_2$
20.0	10.1	69.9	10.4	»
35.4	19.0	45.6	10.4	»
48.8	26.1	25.1	10.4	»
53.2	28.7	18.1	10.4	»
53.0	29.7	17.3	10.4	Solid solution
53.0	29.6	17.4	10.4	»
52.0	30.5	17.5	10.4	»
52.0	30.0	18.0	10.4	»
51.9	30.8	17.3	10.4	»
51.8	30.6	17.6	10.4	»
51.8	31.0	17.2	10.4	$BaBr_2 \cdot 2H_2O$
32.3	35.8	31.9	10.4	»
11.6	44.6	43.8	10.4	»
0.0	50.2	49.8	10.4	»
0.609	0.0	99.391	25	$HgBr_2$
15.2	8.64	76.16	25	»
34.5	19.1	46.4	25	»
44.7	26.0	29.3	25	»
50.5	25.4	24.1	25	»
55.1	29.3	15.6	25	»
55.0	29.9	15.1	25	»
55.1	29.5	15.4	25	$BaBr_2 \cdot 2H_2O$
36.0	35.5	28.5	25	»
10.0	52.7	31.3	25	»
5.25	47.8	46.95	25	»
0.0	50.1	49.9	25	»

№ 5528 BARIUM BROMIDE–MERCURY (II) BROMIDE–WATER [2674]

$BaBr_2$—$HgBr_2$—H_2O

Comp. of saturated soln., wt%			t	Comp. of solid phase
Barium Bromide	Mercury Bromide	Water		
0.609	0.00	99.391	25	$HgBr_2$
15.20	8.64	76.16	25	»
34.50	19.10	46.50	25	»
44.70	26.00	29.30	25	»
50.50	25.40	24.10	25	»
55.10	29.30	15.60	25	»
55.00	29.90	15.10	25	»
55.10	29.50	15.40	25	$BaBr_2 \cdot 2H_2O$
36.00	35.50	28.50	25	»
16.00	42.70	41.30	25	»
5.25	47.80	46.95	25	»
0.00	50.10	49.90	25	»
0.427	0.00	99.573	10.4	$HgBr_2$
20.00	10.10	69.90	10.4	»
35.40	19.00	45.60	10.4	»

Table № 5528 (continuation)

Comp. of saturated soln., wt%			t	Comp. of solid phase
Barium Bromide	Mercury Bromide	Water		
48.80	26.10	25.10	10.4	$HgBr_2$
53.20	28.70	18.10	10.4	»
53.00	29.70	17.30	10.4	Intermediate solid phase of variable composition
53.00	29.60	17.40	10.4	
52.00	30.50	17.50	10.4	
52.00	30.00	18.00	10.4	
52.00	30.50	17.50	10.4	
51.90	30.80	17.70	10.4	
51.80	30.60	17.60	10.4	
51.80	31.00	17.20	10.4	$BaBr_2 \cdot 2H_2O$
32.30	35.80	31.90	10.4	»
11.60	44.60	43.80	10.4	»
0.00	50.20	49.80	10.4	»
0.270	0.00	99.73	4.5	$HgBr_2$
14.30	8.00	77.70	4.5	»
30.10	16.9	53.0	4.5	»
45.30	24.20	30.50	4.5	»
47.60	25.50	26.90	4.5	»
52.40	27.60	18.00	4.5	»
51.70	28.30	20.00	4.5	Solid solution $BaBr_2 \cdot 2HgBr_2 \cdot 7H_2O$ в $BaBr_2 \cdot 3HgBr_2 \cdot 8H_2O$
51.20	28.70	20.10	4.5	
51.00	29.40	19.60	4.5	
50.00	30.40	19.60	4.5	
50.00	30.10	19.90	4.5	
49.7	31.70	18.60	4.5	
49.7	30.80	19.50	4.5	
49.7	31.6	18.70	4.5	
49.00	32.10	18.90	4.5	$BaBr_2 \cdot 2H_2O$
35.90	35.40	28.70	4.5	»
21.90	39.70	38.40	4.5	»
10.00	45.00	45.00	4.5	»
0.00	49.90	50.10	4.5	»

№ 5529 [1817]

BARIUM BROMIDE–LEAD (II) CHLORIDE–WATER

$BaBr_2$—$PbCl_2$—H_2O

$t=25$

Concn. of Barium Bromide in aqueous soln., g/l	Solubility of Lead Chloride in aqueous soln. of Barium Bromide, g/l
63.89	1.677
135.22	3.048
271.9	10.730
401.2	35.238
545.3	82.323

№ 5530 [1817]

BARIUM BROMIDE–LEAD (II) BROMIDE–WATER

$BaBr_2$—$PbBr_2$—H_2O

$t=25$

Concn. of Barium Bromide in aqueous soln., g/l	Solubility of Lead Bromide in aqueous soln. of Barium Bromide, g/l
0.0	9.63
66.83	2.23
135.14	4.00
270.27	16.31
409.86	58.87
544.99	151.9

№ 5531 BARIUM IODIDE–MERCURY (II) IODIDE–WATER [1815]

$BaI_2—HgI_2—H_2O$

$t=25$

Concn. in aqueous soln., g/l			
Mercury Iodide	Barium Iodide	Mercury Iodide	Barium Iodide
26.75	38.73	408	382.6
337.5	292.7	664	590

№ 5532 BARIUM IODIDE–THALLIUM IODIDE–WATER [431]

$BaI_2—TlI—H_2O$

$t=25$

Content in aqueous soln.,			
Barium Iodide, g-eq/l	Thallium Iodide, [mol/l] · 10^5	Barium Iodide, g-eq/l	Thallium Iodide, [mol/l] · 10^5
1.91	0.76	2.60	4.52
1.35	1.01	2.96	6.48
0.79	1.61	3.27	8.97
2.22	2.76	3.76	15.5

№ 5533 [1779]
LANTHANUM NITRATE–LANTHANUM IODATE–WATER

$La(NO_3)_3—La(IO_3)_3—H_2O$

$t=25$

Concn. in aqueous soln., g/l		d_4
Lanthanum Iodate	Lanthanum Nitrate	
0.5595	0.217	0.997
0.5288	0.542	0.998
0.5194	1.083	0.998
0.5522	5.415	1.002
0.6214	10.83	1.007
0.7431	21.66	1.015

№ 5534 LANTHANUM NITRATE–LANTHANUM IODATE–WATER [1677, 2039]

$La(NO_3)_3—La(IO_3)_3—H_2O$

$t=25$

Concn. of Lanthanum Nitrate in aqueous soln., g/l	Solubility of Lanthanum Iodate in aqueous soln. of Lanthanum Nitrate, g/l	Concn. of Lanthanum Nitrate in aqueous soln., g/l	Solubility of Lanthanum Iodate in aqueous soln. of Lanthanum Nitrate, g/l
0.539	0.554	5.393	0.577
1.082	0.5376	10.819	0.6237

№ 5535 LANTHANUM NITRATE–NEODYMIUM NITRATE–WATER [1294]

$La(NO_3)_3$—$Nd(NO_3)_3$—H_2O

Comp. of saturated soln.			t	Comp. of solid phase
Neodymium Nitrate, g	Lanthanum Nitrate, g	Water, mol/100g salts		
0.0	100.0	1476	0	$La(NO_3)_3 \cdot 6H_2O$
15.5	84.5	1434	0	$(La, Nd)(NO_3)_3 \cdot 6H_2O$
35.0	65.0	1380	0	»
51.0	49.0	1321	0	»
57.2	42.8	1288	0	$(La, Nd)(NO_3)_3 \cdot 6H_2O +$ $+ (Nd, La)(NO_3)_3 \cdot 6H_2O$
57.3	42.7	1301	0	»
62.8	37.2	1310	0	$(Nd, La)(NO_3)_3 \cdot 6H_2O$
73.0	27.0	1343	0	»
83.7	16.3	1387	0	»
100.0	0.0	1446	0	$Nd(NO_3)_3 \cdot 6H_2O$
0.0	100.0	1255	20	$La(NO_3)_3 \cdot 6H_2O$
10.1	89.9	1227	20	$(La, Nd)(NO_3)_3 \cdot 6H_2O$
12.2	87.8	1214	20	»
27.8	72.2	1203	20	»
44.0	56.0	1158	20	»
53.4	46.6	1119	20	»
55.4	44.6	1110	20	»
57.5	42.5	1094	20	$(La, Nd)(NO_3)_3 \cdot 6H_2O +$ $+ (Nd, La)(NO_3)_3 \cdot 6H_2O$
57.9	42.1	1099	20	$(Nd, La)(NO_3)_3 \cdot 6H_2O$
61.0	39.0	1112	20	»
70.3	29.7	1162	20	»
89.2	10.8	1221	20	»
100.0	0.0	1259	20	$Na(NO_3)_3 \cdot 6H_2O$

№ 5536 LANTHANUM NITRATE–SAMARIUM NITRATE–WATER [1294]

$La(NO_3)_3$—$Sm(NO_3)_3$—H_2O

Comp. of saturated soln.,			t	Comp. of solid phase
Samarium Nitrate, g	Lanthanum Nitrate, g	Water mol/100g salts		
0.0	100.0	1255	20	$La(NO_3)_3 \cdot 6H_2O$
11.7	88.3	1243	20	$(La, Sm)(NO_3)_3 \cdot 6H_2O$
22.5	77.5	1217	20	»
33.3	66.7	1189	20	»
43.8	52.2	1157	20	»
48.2	51.8	1141	20	$(La, Sm)(NO_3)_3 \cdot 6H_2O +$ $+ (Sm, La)(NO_3)_3 \cdot 6H_2O$
51.8	48.2	1153	20	$(Sm, La)(NO_3)_3 \cdot 6H_2O$
61.6	38.4	1197	20	»
77.8	22.2	1268	20	»
100.0	0.0	1350	20	$Sm(NO_3)_3 \cdot 6H_2O$

Comp. of saturated soln.,			t	Comp. of solid phase
Samarium Nitrate, g	Lanthanum Nitrate, g	Water mol/100g salts		
0.0	100.0	917.2	50	$La(NO_3)_3 \cdot 6H_2O$
16.5	83.5	865.3	50	$(La, Sm)(NO_3)_3 \cdot 6H_2O$
32.4	67.6	810.8	50	»
40.7	59.3	781.9	50	$(La, Sm)(NO_3)_3 \cdot 6H_2O+$ $+(Sm, La)(NO_3)_3 \cdot 6H_2O$
45.3	54.7	800.6	50	$(Sm, La)(NO_3)_3 \cdot 6H_2O$
45.6	54.4	786.6	50	»
52.1	47.9	819.9	50	»
63.9	36.1	891.2	80	»
74.4	25.6	921.0	80	»
100.0	0.00	1020	50	$Sm(NO_3)_3 \cdot 6H_2O$
51.99	48.01	1327	0	$(La, Sm)(NO_3)_3 \cdot 6H_2O+$ $+(Sm, La)(NO_3)_3 \cdot 6H_2O$
44.31	55.69	956.5	35	»

№ 5537 LANTHANUM NITRATE–THALLIUM CHLORIDE–WATER [2367]

$La(NO_3)_3—TlCl—H_2O$

$t=25$

Concn. in aqueous soln., wt%			
Thallium Chloride	Lanthanum Nitrate	Thallium Chloride	Lanthanum Nitrate
0.415	0.168	0.501	1.333
0.423	0.284	0.565	2.570
0.462	0.650	0.604	5.979

№ 5538 [1677, 2039]
LANTHANUM SULFATE–LANTHANUM IODATE–WATER

$La_2(SO_4)_3— La(IO_3)_3—H_2O$

$t=25$

Concn. of Lanthanum Sulfate in aqueous soln., g/l	Solubility of Lanthanum Iodate in aqueous soln. of Lanthanum Sulfate, g/l
18.842	1.2145
9.393	1.0167

№ 5539 [2770]

LANTHANUM SULFATE–THALLIUM SULFATE–WATER

$La_2(SO_4)_3$—Tl_2SO_4—H_2O

$t=25$

Comp. of saturated soln., wt%			Comp. of solid phase
Lanthanum Sulfate	Thallium Sulfate	Water	
1.84	0.19	97.97	$La_2(SO_4)_3 \cdot 9H_2O + La_2(SO_4)_3 \cdot Tl_2SO_4 \cdot 2H_2O$
1.19	0.14	98.67	$La_2(SO_4)_3 \cdot Tl_2SO_4 \cdot 2H_2O$
0.29	0.49	99.22	»
0.19	0.84	98.97	»
0.15	1.27	98.58	»
0.11	1.75	98.14	$La_2(SO_4)_3 \cdot Tl_2SO_4 \cdot 2H_2O + La_2(SO_4)_3 \cdot 3Tl_2SO_4$
0.27	1.71	98.02	»
0.25	1.75	98.00	$La_2(SO_4)_3 \cdot 3Tl_2SO_4$
0.26	1.82	97.92	»
0.15	1.89	97.96	»
0.16	2.01	97.83	$La_2(SO_4)_3 \cdot 3Tl_2SO_4 + La_2(SO_4)_3 \cdot 4\frac{1}{2}Tl_2SO_4$
0.17	2.05	97.78	$La_2(SO_4)_3 \cdot 4^1/_2Tl_2SO_4$
0.12	2.37	97.51	»
—	3.19	96.81	»
—	4.07	95.93	»
—	5.25	94.75	»
—	5.16	94.84	$La_2(SO_4)_3 \cdot 4\frac{1}{2}Tl_2SO_4 + Tl_2SO_4$
—	5.28	94.72	»

№ 5540 [1306]

LANTHANUM SULFATE–THALLIUM CHLORIDE–WATER

$La_2(SO_4)_3$—$TlCl$—H_2O

Concn. of Lanthanum Sulfate in aqueous soln., g/l	Solubility of Thallium Chloride in aqueous soln. of Lanthanum Sulfate, g/l	*d*	*t*
5.658	1.789	1.0063	0
8.487	1.931	1.0088	0
11.317	1.986	1.0110	0
14.146	2.027	1.0135	0
5.658	4.320	1.0055	25
8.487	4.404	1.0079	25
11.317	4.589	1.0098	25
14.146	4.646	1.0122	25

№ 5541 [1677, 2039]

LANTHANUM CHLORIDE–LANTHANUM IODATE–WATER

$LaCl_3$—$La(IO_3)_3$—H_2O

$t=25$

Concn. of Lanthanum Chloride in aqueous soln., g/l	Solubility of Lanthanum Iodate in aqueous soln. of Lanthanum Chloride, g/l
8.098	0.6127

№ 5542 **CERIUM SULFATE–THALLIUM SULFATE–WATER** [2773]

$Ce_2(SO_4)_3$—Tl_2SO_4—H_2O

$t=25$

Concn. in aqueous soln., wt%		Comp. of solid phase
Thallium Sulfate	Cerium Sulfate	
1.01	1.02	$Ce_2(SO_4)_3 \cdot Tl_2SO_4 \cdot 4H_2O$
1.14	0.65	»
1.23	0.54	»
1.80	0.32	»
1.90	0.33	»
2.35	0.24	$Ce_2(SO_4)_3 \cdot Tl_2SO_4 \cdot H_2O + Ce_2(SO_4)_3 \cdot 3Tl_2SO_4 \cdot H_2O$
2.29	0.18	»
2.44	0.17	»
2.52	0.14	$Ce_2(SO_4)_3 \cdot 3Tl_2SO_4 \cdot H_2O$
2.44	0.16	$Ce_2(SO_4)_3 \cdot 3Tl_2SO_4 \cdot H_2O + Ce_2(SO_4)_3 \cdot 4\frac{1}{2}Tl_2SO_4$
2.55	0.12	»
3.46	0.05	$Ce_2(SO_4)_3 \cdot 4\frac{1}{2}Tl_2SO_4$
5.33	0.00	»
5.21	0.00	»
5.45	0.00	»

№ 5543 **CERIUM SULFATE–THORIUM SULFATE–WATER** [1061]

$Ce_2(SO_4)_3$—$Th(SO_4)_2$—H_2O

$t=25$

Comp. of saturated soln., wt%			Comp. of solid phase
Thorium Sulfate	Cerium Sulfate	Water	
1.82	0.00	98.18	$Th(SO_4)_2 \cdot 8H_2O$
2.66	2.81	94.53	»
3.08	4.38	92.54	»
3.36	5.39	91.25	»
3.96	8.05	87.99	»
4.22	9.40	86.38	»
4.24	9.45	86.31	»
4.23	9.62	86.15	$Th(SO_4)_2 \cdot 8H_2O + Ce_2(SO_4)_3 \cdot 8H_2O$

Comp. of saturated soln., wt%			Comp. of solid phase
Thorium Sulfate	Cerium Sulfate	Water	
4.26	9.54	86.20	$Th(SO_4)_2 \cdot 8H_2O + Ce_2(SO_4)_3 \cdot 8H_2O$
4.18	9.50	86.32	»
4.26	9.57	86.17	»
4.25	9.52	86.23	»
4.29	9.73	85.98	»
3.80	9.41	86.79	$Ce_2(SO_4)_3 \cdot 8H_2O$
3.12	9.15	87.73	»
1.82	8.57	89.61	»
1.08	8.37	90.55	»
0.73	8.00	91.27	»
0.00	7.62	92.38	»

№ 5544 PRASEODYMIUM NITRATE–NEODYMIUM NITRATE–WATER [1294]

$Pr(NO_3)_3—Nd(NO_3)_3—H_2O$

$t=20$

Comp. of saturated soln.			Comp. of solid phase
Neodymium Nitrate, g	Praseodymium Nitrate, g	Water mol/100g salts	
0.0	100.0	1149	$Pr(NO_3)_3 \cdot 6H_2O$
9.9	90.1	1155	$(Pr, Nd)(NO_3)_3 \cdot 6H_2O$
19.2	80.8	1152	»
34.9	65.1	1184	»
48.1	51.9	1187	»
63.8	36.2	1219	»
69.7	30.3	1221	»
76.9	23.1	1222	»
92.1	7.9	1254	»
100.0	0.0	1259	$Nd(NO_3)_3 \cdot 6H_2O$

№ 5545 PRASEODYMIUM NITRATE–SAMARIUM NITRATE–WATER [1294]

$Pr(NO_3)_3—Sm(NO_3)_3—H_2O$

$t=20$

Comp. of saturated soln.			Comp. of solid phase
Samarium Nitrate, g	Praseodymium Nitrate, g	Water mol/100g salts	
0.0	100.0	1259	$Nd(NO_3)_3 \cdot 6H_2O$
10.0	90.0	1254	$(Nd, Sm)(NO_3)_3 \cdot 6H_2O$
19.4	80.6	1271	»
47.3	52.7	1298	»
76.6	23.4	1326	»
88.5	11.5	1334	»
100.0	0.0	1350	$Sm(NO_3)_3 \cdot 6H_2O$

№ 5546 PRASEODYMIUM SULFATE–THALLIUM SULFATE–WATER [2778]

$Pr_2(SO_4)_3—Tl_2SO_4—H_2O$

$t=25$

Comp. of saturated soln., wt%			Comp. of solid phase
Thallium Sulfate	Praseodymium Sulfate	Water	
5.91	0.0	94.09	$Pr_2(SO_4)_3 \cdot 5Tl_2SO_4 + Tl_2SO_4$
5.83	0.0	94.17	$Pr_2(SO_4)_3 \cdot 5Tl_2SO_4$
5.27	0.0	94.73	»
4.98	0.0	95.02	$Pr_2(SO_4)_3 \cdot 4\frac{1}{2}Tl_2SO_4$
4.46	0.0	95.54	»
3.44	0.09	96.47	$Pr_2(SO_4)_3 \cdot 4\frac{1}{2}Tl_2SO_4 + Pr_2(SO_4)_3 \cdot 3Tl_2SO_4 \cdot H_2O$
3.41	0.11	96.48	$Pr_2(SO_4)_3 \cdot 3Tl_2SO_4 \cdot H_2O$
2.98	0.15	96.87	$Pr_2(SO_4)_3 \cdot 3Tl_2SO_4 \cdot H_2O + Pr_2(SO_4)_3 \cdot Tl_2SO_4 \cdot 4H_2O$
2.96	0.12	96.92	»
2.98	0.13	96.89	»
3.02	0.19	96.79	»
2.80	0.18	97.02	$Pr_2(SO_4)_3 \cdot Tl_2SO_4 \cdot 4H_2O$
0.73	0.52	98.75	»
0.13	6.18	93.69	»
0.03	8.43	91.54	$Pr_2(SO_4)_3 \cdot Tl_2SO_4 \cdot 4H_2O + Pr_2(SO_4)_3 \cdot 8H_2O$

№ 5547 NEODYMIUM SULFATE–THALLIUM SULFATE–WATER [2769]

$Nd_2(SO_4)_3—Tl_2SO_4—H_2O$

$t=25$

Comp. of saturated soln., wt%			Comp. of solid phase
Neodymium Sulfate	Thallium Sulfate	Water	
4.89	0.25	94.86	$Nd_2(SO_4)_3 \cdot 8H_2O + Nd_2(SO_4)_3 \cdot Tl_2SO_4 \cdot 3H_2O$
4.50	0.25	95.25	»
3.84	0.26	95.90	$Nd_2(SO_4)_3 \cdot Tl_2SO_4 \cdot 3H_2O$
3.81	0.27	95.92	»
3.80	0.27	95.93	»
3.78	0.28	95.94	$Nd_2(SO_4)_3 \cdot Tl_2SO_4 \cdot 8H_2O$
3.15	0.30	96.55	»
1.86	0.45	97.79	»
0.76	1.11	98.13	»
0.30	3.00	96.70	»
0.27	3.51	96.22	$Nd_2(SO_4)_3 \cdot Tl_2SO_4 \cdot 8H_2O + Nd_2(SO_4)_3 \cdot 4\frac{1}{2}Tl_2SO_4$
0.26	3.81	95.93	»
0.26	3.85	95.89	$Nd_2(SO_4)_3 \cdot 4\frac{1}{2}Tl_2SO_4$
0.24	4.10	95.66	»
0.19	4.70	95.11	»
0.16	4.85	94.99	$Nd_2(SO_4)_3 \cdot 4\frac{1}{2}Tl_2SO_4 + Tl_2SO_4$
0.11	4.90	94.99	»
0.09	5.02	94.89	»

№ 5548 [2652]

MERCURY (II) OXIDE (YELLOW)–MERCURY (II) CHLORIDE–WATER

$HgO—HgCl_2—H_2O$

$t=35$

Concn. in aqueous soln., wt%		Comp. of solid phase
Mercury Chloride	Mercury Oxide	
8.58	0.00	$HgCl_2$
8.72	0.14	$HgCl_2+HgCl_2\cdot 2HgO$
8.68	0.07	»
8.69	0.10	»
8.81	0.03	»
8.70	0.11	»
8.69	0.10	»
4.39	0.09	$HgCl_2\cdot 2HgO$
3.42	0.06	»
0.66	0.02	»
0.66	0.02	$HgCl_2\cdot 2H_2O+HgCl_2\cdot 4HgO$
0.61	0.02	$HgCl_2\cdot 4HgO$
0.23	0.03	»
0.20	0.06	»
0.13	0.04	$HgCl_2\cdot 4HgO+HgO$
0.07	0.04	»
0.05	0.03	»
0.05	0.04	»
0.04	0.04	HgO
0.01	0.03	»
0.00	0.001	»

№ 5549 [2221]

MERCURY (II) NITRATE–MERCURY (II) CHLORIDE–WATER

$Hg(NO_3)_2—HgCl_2—H_2O$

$t=25$

Concn. of Mercury Nitrate in aqueous soln., g/l	Solubility of Mercury Chloride in aqueous soln. of Mercury Nitrate, g/l
16.23	105

№ 5550 MERCURY (II) NITRATE–MERCURY (II) IODIDE–WATER [1051]

$Hg(NO_3)_2—HgI_2—H_2O$

$t=25$

Comp. of saturated soln., wt%			Comp. of saturated soln., wt%		
Mercury Nitrate	Mercury Iodide	Water	Mercury Nitrate	Mercury Iodide	Water
0.84	0.41	98.75	25.56	8.91	65.53
2.84	0.89	96.27	28.24	10.42	61.34
5.48	1.51	93.01	30.40	11.97	57.63
10.67	2.82	86.51	31.75	13.49	54.76
15.01	4.17	80.82	33.55	13.96	52.49
16.14	4.46	79.40	34.75	15.58	49.67
18.96	5.55	75.49	36.50	17.04	46.46
21.51	6.72	71.77	37.63	18.08	44.29
24.46	8.27	67.27			

№ 5551 MERCURY (II) NITRATE–URANYL NITRATE–WATER [1031]

$Hg(NO_3)_2—UO_2(NO_3)_2—H_2O$

Comp. of saturated soln., wt%			t	Comp. of solid phase
Uranyl Nitrate	Mercury Nitrate	Water		
1.9	69.6	28.5	0	$Hg(NO_3)_2 \cdot 0.5H_2O + UO_2(NO_3)_2 \cdot 6H_2O$
3.0	66.2	30.8	0	$UO_2(NO_3)_2 \cdot 6H_2O$
4.1	63.2	32.7	0	»
6.7	58.2	35.1	0	»
9.5	53.6	36.1	0	»
12.5	48.6	38.9	0	»
13.2	47.4	39.4	0	»
15.5	44.4	40.1	0	»
19.0	39.7	41.3	0	»
22.7	35.2	42.1	0	»
24.3	31.6	44.1	0	»
28.2	27.3	44.5	0	»
31.0	22.4	46.6	0	»
34.0	18.4	47.6	0	»
35.6	16.2	48.2	0	»
36.2	15.2	48.6	0	»
39.0	11.6	49.4	0	»
42.8	7.0	50.2	0	»
46.4	3.2	50.4	0	»
49.5	0.0	50.5	0	»
2.9	77.6	19.5	15	$Hg(NO_3)_2 \cdot 0.5H_2O + UO_2(NO_3)_2 \cdot 6H_2O$
11.6	55.3	33.1	15	$UO_2(NO_3)_2 \cdot 6H_2O$
13.8	52.5	33.7	15	»
15.3	49.8	34.9	15	»
19.5	44.4	36.1	15	»
22.0	41.2	36.8	15	»
26.4	34.0	39.6	15	»
30.8	27.8	41.4	15	»
36.5	20.7	42.8	15	»

Comp. of saturated soln., wt%			t	Comp. of solid phase
Uranyl Nitrate	Mercury Nitrate	Water		
39.0	16.0	45.0	15	$UO_2(NO_3)_2 \cdot 6H_2O$
44.3	7.7	48.0	15	»
49.0	3.3	47.7	15	»
52.9	0.0	47.1	15	»
4.8	77.6	17.6	15	$Hg(NO_3)_2 \cdot 0.5H_2O + UO_2(NO_3)_2 \cdot 6H_2O$
6.2	72.1	21.7	25	$UO_2(NO_3)_2 \cdot 6H_2O$
8.1	68.0	23.9	25	»
9.1	66.0	24.9	25	»
10.0	64.2	25.8	25	»
11.1	62.4	26.5	25	»
11.4	60.5	28.1	25	»
12.0	59.0	29.0	25	»
13.0	57.2	29.8	25	»
15.0	54.3	30.7	25	»
18.7	49.2	32.1	25	»
19.3	47.7	33.0	25	»
25.7	38.2	36.1	25	»
29.6	34.2	36.2	25	»
31.4	31.6	37.0	25	»
33.1	28.7	38.2	25	»
35.2	26.5	38.3	25	»
36.7	24.3	39.0	25	»
39.1	21.3	39.6	25	»
41.2	17.9	40.9	25	»
43.2	15.4	41.4	25	»
44.0	14.4	41.6	25	»
47.2	10.3	42.5	25	»
50.8	5.7	43.5	25	»
55.9	0.0	44.1	25	»

№ 5552 MERCURY (II) CHLORIDE–MERCURY (II) IODIDE–WATER [2616]

$HgCl_2$—HgI_2—H_2O

Comp. of saturated soln., wt%			t	Comp. of solid phase
Mercury Chloride	Mercury Iodide	Water		
7.55	0.00	92.45	30	Mixed crystals, white
7.67	0.41	91.92	30	»
7.75	0.21	92.04	30	»
7.77	0.43	91.8	30	White + red
7.72	0.54	91.74	30	Red
7.69	0.46	91.85	30	»
6.75	0.31	93.94	30	»
5.32	0.28	94.40	30	»
3.39	0.33	96.28	30	»
2.20	0.18	97.62	30	»
7.82	0.52	91.66	30	Yellow
5.36	0.37	94.27	30	»

Table № 5552 (continuation)

Comp. of saturated soln., wt%			t	Comp. of solid phase
Mercury Chloride	Mercury Iodide	Water		
19.12	0.00	80.88	70	Mixed crystals, white
19.5	0.44	80.06	70	»
20.5	1.10	78.4	70	Yellow
19.8	1.02	79.18	70	»
17.4	0.99	81.61	70	»
16.7	1.00	82.3	70	»
12.8	0.99	86.21	70	»
12.6	0.94	86.46	70	»
10.2	0.91	88.89	70	Yellow + red
9.9	0.84	89.26	70	»
7.35	0.75	91.9	70	Red
3.94	0.52	95.54	70	»

№ 5553 [1197]

MERCURY (II) CHLORIDE–THALLIUM CHLORIDE–WATER

$HgCl_2—TlCl—H_2O$

$t=25$

Concn. in aqueous soln., wt%		Comp. of solid phase
Mercury Chloride	Thallium Chloride	
0.037	0.288	TlCl
0.072	0.269	»
0.289	0.143	»
1.285	0.089	$TlCl \cdot HgCl_2$
1.535	0.076	»
2.099	0.063	»
3.610	0.057	»
4.478	0.054	»
5.965	0.032	»
6.732	0.0	$HgCl_2$

№ 5554

MERCURY (II) CHLORIDE–LEAD (II) CHLORIDE–WATER [1654]

$HgCl_2—PbCl_2—H_2O$

$t=20$

Concn. in aqueous soln., g/l		Concn. in aqueous soln., g/l	
Mercury Chloride	Lead Chloride	Mercury Chloride	Lead Chloride
0	9.71	8	11.6
1	9.8	10	12.2
2	10.0	14	13.2
3	10.3	20	14.8
4	10.5	40	19.0
6	11.0		

№ 5555 [1965]

MERCURY (II) CHLORIDE–LEAD (II) CHLORIDE–WATER

$HgCl_2—PbCl_2—H_2O$

$t=25$

Concn. in aqueous soln., g/l		Comp. of solid phase
Lead Chloride	Mercury Chloride	
12.30	8.16	$HgCl_2 \cdot PbCl_2$
13.66	13.60	»
14.93	20.40	»
19.24	68.0	»

№ 5556 [2265]

THALLIUM THIOCYANATE–THALLIUM BROMATE–WATER

$TlCNS—TlBrO_3—H_2O$

$t=39.75$

Concn. in aqueous soln., g/l	
Thallium Bromide	Thallium Thiocyanate
4.966	5.793

№ 5557 [2263]

THALLIUM THIOCYANATE–THALLIUM CHLORIDE–WATER

$TlCNS—TlCl—H_2O$

Concn. in aqueous soln., g/l		
Thallium Chloride	Thallium Thiocyanate	
2.849	2.802	25
4.326	5.632	39.75

№ 5558 [2271]

THALLIUM NITRATE–THALLIUM SULFATE–WATER

$TlNO_3—Tl_2SO_4—H_2O$

$t=25$

Concn. of Thallium Nitrate in aqueous soln., g/l	Solubility of Thallium Sulfate in aqueous soln. of Thallium Nitrate, g/l
26.51	42.17

№ 5559 [2264]

THALLIUM NITRATE–THALLIUM CHLORIDE–WATER

$TlNO_3—TlCl—H_2O$

$t=25$

Concn. of Thallium Nitrate in aqueous soln., g/l	Solubility of Thallium Chloride in aqueous soln. of Thallium Nitrate, g/l
6.659	2.118
13.319	1.502
26.639	1.015

№ 5560 [2268]

THALLIUM NITRATE–THALLIUM CHLORIDE–WATER

$TlNO_3—TlCl—H_2O$

$t=25$

Concn. in aqueous soln., g/l	
Thallium Chloride	Thallium Nitrate
1.987	7.518
1.368	14.89
0.795	39.05

№ 5561 [1306]

THALLIUM NITRATE–THALLIUM CHLORIDE–WATER

$TlNO_3—TlCl—H_2O$

Concn. of Thallium Nitrate in aqueous soln., g/l	Solubility of Thallium Chloride in aqueous soln. of Thallium Nitrate, g/l	*d*	*t*
13.319	1.475	1.0078	25
26.639	0.990	1.0202	25
53.278	0.719	1.0399	25
26.639	3.313	1.0131	50
53.278	2.423	1.0321	50

№ 5562 [2263]

THALLIUM NITRATE–THALLIUM BROMIDE–WATER

$TlNO_3—TlBr—H_2O$

$t=68.5$

Concn. in aqueous soln., g/l	
Thallium Bromide	Thallium Nitrate
2.47	0.0
1.16	4.34
0.82	7.82
0.420	25.40

№ 5563 [406]

THALLIUM NITRATE–THALLIUM HEXANITROCOBALTATE–WATER

$TlNO_3—Tl_3[Co(NO_2)_6]—H_2O$

$t=20$

Content in aqueous soln.	
Thallium Nitrate, mol/l	Thallium Hexanitrocobaltate, (mol/l) · 10^5
0.000	8.5
0.031	1.04
0.062	0.90
0.125	0.93
0.250	0.99

$TlNO_3—UO(NO_3)_2—H_2O$

Comp. of saturated soln., mol %			t	Comp. of solid phase
Uranyl Nitrate	Thallium Nitrate	Water		
0.00	0.27	99.73	0	$TlNO_3$
0.13	0.19	99.68	0	»
0.18	0.15	99.67	0	»
0.29	0.13	99.58	0	»
0.34	0.13	99.53	0	»
0.46	0.12	99.42	0	»
0.75	0.11	99.14	0	»
1.04	0.10	98.86	0	»
1.15	0.10	98.85	0	»
1.29	0.10	98.61	0	»
1.57	0.11	98.32	0	»
1.82	0.11	98.07	0	»
2.03	0.12	97.85	0	»
2.43	0.12	97.45	0	»
2.55	0.13	97.32	0	»
3.06	0.14	96.80	0	»
3.39	0.16	96.45	0	»
3.96	0.18	95.86	0	»
4.26	0.19	95.55	0	$TlNO_3+UO_2(NO_3)_2 \cdot 6H_2O$
4.27	0.10	95.63	0	$UO(NO_3)_2 \cdot 6H_2O$
4.29	0.00	95.71	0	»
0.00	0.79	99.21	25	$TlNO_3$
0.06	0.74	99.20	25	»
0.13	0.71	99.16	25	»
0.18	0.65	99.17	25	»
0.24	0.61	99.15	25	»
0.36	0.56	99.08	25	»
0.39	0.56	99.05	25	»
0.60	0.48	98.92	25	»
0.66	0.46	98.88	25	»
0.85	0.43	98.72	25	»
0.95	0.42	98.63	25	»
1.12	0.40	98.48	25	»
1.34	0.38	98.28	25	»
1.54	0.38	98.08	25	»
1.58	0.38	98.04	25	»
1.62	0.38	98.00	25	»
1.67	0.37	97.96	25	»
1.99	0.37	97.64	25	»
2.23	0.37	97.40	25	»
2.76	0.40	96.84	25	»
3.02	0.39	96.59	25	»
3.12	0.40	96.48	25	»
3.27	0.41	96.32	25	»
3.42	0.42	96.16	25	»
4.06	0.47	95.47	25	»
4.25	0.49	95.26	25	»
4.84	0.56	94.60	25	»
5.39	0.60	94.01	25	»

Comp. of saturated soln., mol %			t	Comp. of solid phase
Uranyl Nitrate	Thallium Nitrate	Water		
5.65	0.67	93.68	25	$TlNO_3+UO_2(NO_3)_2 \cdot 6H_2O$
5.60	0.47	93.93	25	$UO_2(NO_3)_2 \cdot 6H_2O$
5.49	0.00	94.51	25	»

№ 5565 [2263]

THALLIUM SULFATE–THALLIUM CHLORIDE–WATER

$Tl_2SO_4—TlCl—H_2O$

$t=25$

Concn. in aqueous soln., g/l	
Thallium Chloride	Thallium Sulfate
2.121	14.27
1.494	28.23

№ 5566 [1270]

THALLIUM SULFATE–THALLIUM CHLORIDE–WATER

$Tl_2SO_4—TlCl—H_2O$

$t=25$

Concn. of Thallium Sulfate in aqueous soln., g/l	Solubility of Thallium Chloride in aqueous soln. of Thallium Sulfate, g/l	d_4
5.045	2.480	1.0028
12.617	1.624	1.0090
25.231	1.122	1.0200

№ 5567 [1306]

THALLIUM SULFATE–THALLIUM CHLORIDE–WATER

$Tl_2SO_4—TlCl—H_2O$

Concn. of Thallium Sulfate in aqueous soln., g/l	Solubility of Thallium Chloride in aqueous soln. of Thallium Sulfate, g/l	d	t
12.619	1.631	1.0098	25
25.239	1.110	1.0186	25
12.619	4.944	1.0012	50
25.239	3.759	1.0140	50

№ 5568 [2270]

THALLIUM SULFATE–THALLIUM CHLORATE–WATER

$Tl_2SO_4—TlClO_3—H_2O$

$t=20$

Concn. in aqueous soln., g/l	
Thallium Chlorate	Thallium Sulfate
30.4	34.43

№ 5569 [2581]

THALLIUM SULFATE–THALLIUM BROMIDE–WATER

$Tl_2SO_4—TlBr—H_2O$

$t=20$

Concn. in aqueous soln., g/l	
Thallium Bromide	Thallium Sulfate
0.2732	0.5048
0.1376	1.5146
0.0660	4.0388

THALLIUM CHLORIDE–THALLIUM TRICHLORIDE–WATER
$TlCl—TlCl_3—H_2O$

Comp. of saturated soln., wt%			t	Comp. of solid phase
Thallium Trichloride	Thallium Chloride	Water		
0.00	0.05	99.95	35	TlCl (white)
0.17	0.37	99.46	35	$[TlCl_6]Tl_3$ (yellow)
0.81	0.26	98.93	35	»
2.12	0.24	97.64	35	»
2.78	0.23	96.99	35	»
3.44	0.22	96.34	35	»
4.24	0.22	95.54	35	»
8.01	0.21	91.78	35	»
13.08	0.23	86.69	35	»
21.90	0.19	77.91	35	»
28.47	0.22	71.31	35	»
42.47	0.44	57.09	35	»
46.69	0.54	52.77	35	»
53.93	0.70	45.37	35	»
58.60	0.92	40.48	35	»
60.48	1.04	38.48	35	»
60.93	1.11	37.96	35	»
61.34	1.08	37.58	35	$[TlCl_4]Tl$ (white)
61.78	1.05	37.17	35	»
62.72	1.01	36.27	35	»
65.85	0.88	33.27	35	»
0.00	0.07	99.93	45	TlCl (white)
0.21	0.49	99.30	45	$[TlCl_6]Tl$ (yellow)
1.83	0.40	97.77	45	»
4.70	0.34	94.96	45	»
16.73	0.35	82.92	45	»
28.24	0.35	71.41	45	»
39.42	0.48	60.10	45	»
47.26	0.62	52.12	45	»
53.56	0.90	45.54	45	»
58.41	1.26	40.33	45	»
60.07	1.45	38.48	45	»
61.50	1.63	36.87	45	»
62.09	1.60	36.31	45	$[TlCl_4]Tl$ (white)
65.23	1.45	33.32	45	»
66.76	1.37	31.87	45	»
0.00	0.09	99.91	55	TlCl (white)
0.29	9.67	99.04	55	$[TlCl_6]Tl_3$ (yellow)
1.30	0.58	98.12	55	»
4.24	0.56	95.20	55	»
20.86	0.42	78.72	55	»
28.68	0.49	70.83	55	$[TlCl_6]Tl$ (yellow)
46.55	1.04	52.41	55	»
53.93	1.34	44.73	55	»
59.65	2.07	38.28	55	»
59.80	2.12	38.08	55	»
60.69	2.30	37.01	55	»
61.15	2.43	36.42	55	»

Comp. of saturated soln., wt%			t	Comp. of solid phase
Thallium Trichloride	Thallium Chloride	Water		
62.41	2.36	35.23	55	$[TlCl_4]Tl$ (white)
63.31	2.28	34.41	55	»
65.35	2.14	32.51	55	»
66.83	2.02	31.15	55	»
61.77	3.59	34.64	70	$[TlCl_6]Tl_3$ (yellow)
62.93	3.55	33.52	70	$[TlCl_4]Tl$ (white)
60.93	5.18	33.89	80	$[TlCl_6]Tl_3$ (yellow)
66.01	4.69	29.30	80	»
67.89	4.41	27.70	80	$[TlCl_4]Tl$ (white)
70.53	4.93	24.54	90	$[TlCl_6]Tl_3$ (yellow)
70.81	4.84	24.35	90	$[TlCl_4]Tl$ (white)

№ 5571 [2264]

THALLIUM CHLORIDE–THALLIUM CHLORATE–WATER

$TlCl—TlClO_3—H_2O$

$t=25$

Concn. of Thallium Chlorate in aqueous soln., g/l	Solubility of Thallium Chloride in aqueous soln. of Thallium Chlorate, g/l
7.197	2.152
7.197	2.144

№ 5572 [2265]

THALLIUM CHLORIDE–THALLIUM BROMATE–WATER

$TlCl—TlBrO_3—H_2O$

$t=39.75$

Concn. in aqueous soln., g/l	
Thallium Chloride	Thallium Bromate
4.690	5.201

№ 5573 [1190]

THALLIUM BROMIDE–THALLIUM TRIBROMIDE–WATER

$TlBr—TlBr_3—H_2O$

Comp. of saturated soln., wt%			t	Comp. of solid phase
Thallium Tribromide	Thallium Bromide	Water		
0.0	0.07	99.93	30	TlBr (white)
1.13	0.20	98.67	30	»
1.53	0.26	98.21	30	»
1.94	0.35	97.71	30	$[TlBr_6]Tl_3$ (red)
2.24	0.31	97.45	30	»
2.76	0.37	96.87	30	»
3.36	0.37	96.27	30	»
3.71	0.43	95.86	30	»

Table № 5573 (continuation)

Comp. of saturated soln., wt%			t	Comp. of solid phase
Thallium Tribromide	Thallium Bromide	Water		
3.98	0.42	95.60	30	$[TlBr_4]Tl$ (yellow)
4.80	0.40	94.80	30	»
6.50	0.39	93.11	30	»
10.62	0.31	89.07	30	»
12.70	0.32	86.98	30	»
13.54	0.34	86.12	30	»
0.00	0.16	99.84	50	TlBr (yellow)
0.48	0.21	99.31	50	»
0.65	0.25	99.10	50	»
0.92	0.28	98.80	50	»
1.33	0.39	98.28	50	»
1.66	0.49	97.85	50	»
1.66	0.53	97.81	50	»
1.94	0.66	97.41	50	»
2.28	0.65	97.07	50	$[TlBr_6]Tl_3$ (red)
2.97	0.65	96.37	50	»
3.44	0.71	95.85	50	»
3.92	0.84	95.24	50	»
4.04	0.88	95.08	50	$Tl[Br_4]Tl$ (yellow)
4.51	0.85	94.64	50	»
4.98	0.84	94.18	50	»
6.17	0.82	93.01	50	»
7.78	0.80	91.43	50	»
9.23	0.79	89.98	50	»
11.46	0.75	87.79	50	»
13.35	0.81	85.84	50	»
5.53	2.57	91.90	80	TlBr (white)
5.69	2.72	91.59	80	$[TlBr_6]Tl_3$ (red)
6.78	2.93	90.29	80	»
7.21	2.87	89.92	80	$[TlBr_4]Tl$ (yellow)
7.35	3.34	89.31	90	TlBr (white)
7.91	3.98	88.11	90	$[TlBr_6]Tl_3$ (red)
8.33	4.36	87.31	90	$[TlBr_4]Tl$ (yellow)

№ 5574 LEAD (II) NITRATE–LEAD (II) SULFATE–WATER [1997]

$Pb(NO_3)_2—PbSO_4—H_2O$

$t=29$

Concn. of Lead Nitrate in aqueous soln., g/l	Solubility of Lead Sulfate in aqueous soln. of Lead Nitrate, g/l	Concn. of Lead Nitrate in aqueous soln., g/l	Solubility of Lead Sulfate in aqueous soln. of Lead Nitrate, g/l
0.0829	0.0224	0.2484	0.0115
0.0124	0.0180	0.4140	0.0071
0.1656	0.0133		

№ 5575 LEAD (II) NITRATE–LEAD (II) CHLORIDE–WATER [2262]

$Pb(NO_3)_2—PbCl_2—H_2O$

$t=25$

Concn. in aqueous soln., g/l		Concn. in aqueous soln., g/l	
Lead Nitrate	Lead Chloride	Lead Nitrate	Lead Chloride
0	10.79	8	11.00
1	10.8	10	11.05
2	10.85	14	11.15
3	10.87	20	11.20
4	10.90	40	11.70
6	10.95		

№ 5576 [1777]

LEAD (II) NITRATE–
LEAD (II) CHLORIDE–WATER

$Pb(NO_3)_2—PbCl_2—H_2O$

$t=25$

Concn. in aqueous soln., g/l		d
Lead Chloride	Lead Nitrate	
10.81	0	1.0069
10.67	3.31	1.0095
10.65	8.28	1.0139
10.84	16.56	1.0210
11.57	33.12	—

№ 5577 [1099]

LEAD (II) NITRATE–
LEAD (II) CHLORIDE–WATER

$Pb(NO_3)_2—PbCl_2—H_2O$

$t=25$

Comp. of solvent, wt%		Solubility of Lead Chloride wt%
Lead Nitrate	Water	
0.0	100	1.089
0.32	99.68	1.096
0.65	99.35	1.053
3.20	96.80	1.115
7.64	92.36	1.295

№ 5578 [2368]

LEAD (II) NITRATE–LEAD (II) BROMIDE–WATER

$Pb(NO_3)_2—PbBr_2—H_2O$

$t=25$

Concn. in aqueous soln., wt%	
Lead Bromide	Lead Nitrate
0.073	0.873
0.182	0.865
0.364	0.856
0.722	0.850
1.79	0.857
3.51	0.933
4.60	0.988
6.86	1.08
10.18	1.27
15.28	1.47
20.10	1.69
25.37	1.82
36.76	2.55
40.79 *	2.53

* Solid phase $PbBr_2 + Pb(NO_3)_2$.

№ 5579 [2165]

LEAD (II) NITRATE–LEAD (II) BROMATE–WATER

$Pb(NO_3)_2—Pb(BrO_3)_2—H_2O$

$t=25$

Concn. in aqueous soln., g/l		
Lead Bromate	Lead Nitrate	
15.91	0.0	1.0112
15.61	1.637	1.0104
15.43	3.284	1.0118
15.08	6.587	1.0132
14.64	16.49	1.0219
14.56	32.90	1.0356
15.11	65.47	1.0643
16.01	97.81	1.0936
16.93	128.49	1.1120
17.94	161.23	1.1480
18.99	192.11	1.1785
21.17	253.94	1.2290
23.36	313.46	1.2786
25.99	386.04	1.3402
28.63	457.01	1.4011

№ 5580 [55]

LEAD (II) NITRATE–LEAD (II) IODIDE–WATER

$Pb(NO_3)_2—PbI_2—H_2O$

$t=20$

Content of Lead Nitrate in soln., mol/l	Solubility of Lead Iodide, mol/l
0.20	$0.75 \cdot 10^{-3}$
0.60	$1.56 \cdot 10^{-3}$
1.00	$2.75 \cdot 10^{-3}$

№ 5581 LEAD (II) NITRATE–LEAD (II) IODIDE–WATER [891]

$Pb(NO_3)_2$—PbI_2—H_2O

t=18—20

Content of Lead Nitrate in aqueous soln., mol/l	Solubility of Lead Iodide, mol/l · 10^3	Content of Lead Nitrate in aqueous soln., mol/l	Solubility of Lead Iodide, mol/l · 10^3
0.1252	1.300	0.7449	2.826
0.1803	1.400	0.8486	3.503
0.2842	1.726	0.9453	3.761
0.3325	2.037	1.0227	4.057
0.4087	2.246	1.1003	4.211
0.6551	2.876	1.2855	4.770

№ 5582 LEAD (II) NITRATE–LEAD (II) IODATE–WATER [1778]

$Pb(NO_3)_2$—$Pb(IO_3)_2$—H_2O

t=25

Concn. in aqueous soln., g/l		Concn. in aqueous soln., g/l	
Lead Iodate	Lead Nitrate	Lead Iodate	Lead Nitrate
0.0242	0.0165	0.0045	16.561
0.0115	0.165	0.0078	82.805
0.0052	1.656	0.0418	496.83

№ 5583 [708]

LEAD (II) FLUORIDE–LEAD (II) BROMIDE–WATER

PbF_2—$PbBr_2$—H_2O

t=25

Content in aqueous soln., wt%		Comp. of solid phase
Lead Fluoride	Lead Bromide	
0.9699	0.00	$PbBr_2$
0.9689	0.00	»
0.9676	0.00	PbBrF
0.9652	0.00	»

Table № 5583 (continuation)

Content in aqueous soln., wt%		Comp. of solid phase
Lead Fluoride	Lead Bromide	
0.8238	0.00	PbBrF
0.6900	0.00	»
0.6437	0.00	»
0.5572	0.00	»
0.4923	0.00	»
0.5113	0.00	»
0.3917	0.00	»
0.3457	0.00	»
0.2820	Traces	»
0.2256	»	»
0.1456	$2.367 \cdot 10^{-5}$	»
$7.049 \cdot 10^{-2}$	$1.408 \cdot 10^{-3}$	»
$5.469 \cdot 10^{-2}$	$3.905 \cdot 10^{-3}$	»
$4.896 \cdot 10^{-2}$	$7.326 \cdot 10^{-3}$	»
$3.318 \cdot 10^{-2}$	$1.510 \cdot 10^{-2}$	»
$2.602 \cdot 10^{-2}$	$1.469 \cdot 10^{-2}$	»
$2.602 \cdot 10^{-2}$	$1.767 \cdot 10^{-2}$	»
$2.083 \cdot 10^{-2}$	$2.145 \cdot 10^{-2}$	»
$1.847 \cdot 10^{-2}$	$3.245 \cdot 10^{-2}$	»
$2.237 \cdot 10^{-2}$	$2.260 \cdot 10^{-2}$	»
$1.224 \cdot 10^{-2}$	$4.006 \cdot 10^{-2}$	»
$7.761 \cdot 10^{-3}$	$3.962 \cdot 10^{-2}$	»
$2.713 \cdot 10^{-3}$	$5.268 \cdot 10^{-2}$	PbF_2
Traces	$6.655 \cdot 10^{-2}$	»
0.00	$6.730 \cdot 10^{-2}$	»

№ 5584 [707]

LEAD (II) FLUORIDE–LEAD (II) IODIDE–WATER

PbF_2—PbI_2—H_2O

$t = 25$

Content in aqueous soln., wt%		Comp. of solid phase
Lead Fluoride	Lead Iodide	
0.0000	0.07238	PbI_2
0.01561	0.06987	PbFI
0.01802	0.06830	»

Table № 5584 (continuation)

Content in aqueous soln., wt%		Comp. of solid phase
Lead Fluoride	Lead Iodide	
0.01408	0.06682	PbFI
0.02818	0.03321	»
0.03330	0.03169	»
0.03626	0.03172	»
0.03750	0.02906	»
0.04080	0.02768	»
0.03993	0.02743	»
0.03922	0.02669	»
0.03931	0.02677	$4PbF_2 \cdot PbI_2$
0.04156	0.02599	»
0.04204	0.02496	»
0.04408	0.02418	»
0.04290	0.01746	»
0.04450	0.01032	»
0.04296	0.008728	»
0.05310	0.004332	$4PbF_2 \cdot PbI_2 + PbF_2$
0.06730	0.00	PbF_2

№ 5585 [870]

LEAD (II) FLUOROCHLORATE–URANYL NITRATE–WATER

$PbFCl—UO_2(NO_3)_2—H_2O$

$t=25$

Content of Uranyl Nitrate in aqueous soln., mol/1000g Water	Solubility of Lead Fluorochlorate in conversion to Cl^-, g-ion/l
0.005	0.005
0.025	0.013
0.040	0.021
0.050	0.021
0.080	0.030
0.123	0.040
0.25	0.060

№ 5586 [1126]

LEAD (II) SULFATE–THORIUM CHLORIDE–WATER

$PbSO_4—ThCl_4—H_2O$

$t=25$

Concn. in aqueous soln., g/l	
Lead Sulfate	Thorium Chloride
0.979	1.646
1.740	3.927
2.907	8.715
5.336	18.515
5.973	30.560

№ 5587 [2088]

LEAD (II) CHLORIDE–LEAD (II) IODIDE–WATER

$PbCl_2$—PbI_2—H_2O

$t=30$

Concn. in aqueous soln., g/l		Comp. of solid phase
Lead Iodide	Lead Chloride	
0.00	11.71	$PbCl_2$
0.081	11.71	»
0.120	11.71	»
0.168	11.71	»
0.235	11.71	$PbCl_2$+PbICl
0.241	11.40	PbICl
0.277	9.69	»
0.278	9.58	»
0.334	8.20	»
0.390	7.17	»
0.421	6.31	»
0.432	6.31	»
0.520	5.50	»
0.531	5.45	»
0.602	5.18	PbCl+PbI_2
0.606	4.17	PbI_2
0.620	3.52	»
0.630	2.70	»
0.647	2.09	»
0.687	1.44	»
0.712	1.00	»
0.788	0.46	»
0.910	0.00	»

Systematic Index

System name	Formulae of components	Table number
Sodium Chloride–Sodium Perchlorate–Ammonia	$NaCl—NaClO_4—NH_3$	28
Sodium Chloride–Sodium Bromide–Ammonia	$NaCl—NaBr—NH_3$	29
Potassium Chloride–Potassium Perchlorate–Ammonia	$KCl—KClO_4—NH_3$	30
Potassium Iodide–Potassium Iodate–Ammonia	$KI — KIO_3—NH_3$	31
Oxygen–Dinitrogen Tetroxide–Nitric Acid	$O_2—N_2O_4—HNO_3$	32
Hydrogen Cyanide–Silver Cyanide–Nitric Acid	$HCN—AgCN—HNO_3$	33
Sodium Fluoride–Potassium Hexafluorophosphate–Hydrogen Fluoride	$NaF—KPF_6—HF$	34
Sodium Fluoride–Thallium Hexafluoroantimonate–Hydrogen Fluoride	$NaF—TlSbF_6—HF$	35
Potassium Fluoride–Potassium Hexafluoroantimonate–Hydrogen Fluoride	$KF—KSbF_6—HF$	36
Potassium Fluoride–Silver Tetrafluoroborate–Hydrogen Fluoride	$KF—AgBF_4—HF$	37
Potassium Hexafluoroantimonate–Antimony Pentafluoride–Hydrogen Fluoride	$KSbF_6—SbF_5—HF$	38
Nitrogen Oxide–Sulfuric Acid–Nitrosylsulfuric Acid	$NO—H_2SO_4—HSO_4NO$	39
Hydrogen Cyanide–Silver Cyanide–Sulfuric Acid	$HCN—AgCN—H_2SO_4$	40
Chlorosulfuric Acid–Pyrosulfuric Acid–Sulfuric Acid	$HSO_3Cl—H_2S_2O_7—H_2SO_4$	41
Hydrogen–Carbon Dioxide–Water	$H_2—CO_2—H_2O$	42
Hydrogen–Nitrogen–Water	$H_2—N_2—H_2O$	43
Hydrogen–Nitric Acid–Water	$H_2—HNO_3—H_2O$	44
Hydrogen–Sulfuric Acid–Water	$H_2—H_2SO_4—H_2O$	45
Hydrogen–Hydrogen Chloride–Water	$H_2—HCl—H_2O$	46
Hydrogen–Lithium Chloride–Water	$H_2—LiCl—H_2O$	47
Hydrogen–Ammonium Nitrate–Water	$H_2—NH_4NO_3—H_2O$	48
Hydrogen–Sodium Hydroxide–Water	$H_2—NaOH—H_2O$	49
Hydrogen–Sodium Carbonate–Water	$H_2—Na_2CO_3—H_2O$	50
Hydrogen–Sodium Nitrate–Water	$H_2—NaNO_3—H_2O$	51, 52
Hydrogen–Sodium Sulfate–Water	$H_2—Na_2SO_4—H_2O$	53
Hydrogen–Sodium Chloride–Water	$H_2—NaCl—H_2O$	54
Hydrogen–Magnesium Sulfate–Water	$H_2—MgSO_4—H_2O$	55
Hydrogen–Potassium Hydroxide–Water	$H_2—KOH—H_2O$	56, 57
Hydrogen–Potassium Carbonate–Water	$H_2—K_2CO_3—H_2O$	58
Hydrogen–Potassium Nitrate–Water	$H_2—KNO_3—H_2O$	59, 60
Hydrogen–Potassium Chloride–Water	$H_2—KCl—H_2O$	61, 62

System name	Formulae of components	Table number
Hydrogen–Calcium Chloride–Water	H_2—$CaCl_2$—H_2O	63
Hydrogen–Barium Chloride–Water	H_2—$BaCl_2$—H_2O	64
Helium–Neon–Water	He—Ne—H_2O	65
Helium–Sodium Chloride–Water	He—NaCl—H_2O	67
Helium–Lithium Chloride–Water	He—LiCl—H_2O	68
Helium–Lithium Iodide–Water	He—LiI—H_2O	68
Helium–Sodium Nitrate–Water	He—$NaNO_3$—H_2O	68
Helium–Sodium Chloride–Water	He—NaCl—H_2O	68
Helium–Potassium Chloride–Water	He—KCl—H_2O	68
Helium–Seawater	He— — —H_2O	66
Nitrogen–Oxygen–Sulfuric Acid–Water	N_2—O_2—H_2SO_4—H_2O	69
Nitrogen–Chlorine–Water	N_2—Cl_2—H_2O	70
Nitrogen–Carbon Dioxide–Water	N_2—CO_2—H_2O	71
Nitrogen–Sulfuric Acid–Water	N_2—H_2SO_4—H_2O	72, 73
Nitrogen–Sodium Hydrogen Carbonate–Lead (II) Sulfide–Water	N_2—$NaHCO_3$—PbS—H_2O	74
Nitrogen–Sodium Chloride–Water	N_2—NaCl—H_2O	75
Nitrogen–Sodium Chloride–Lead (II) Sulfide–Water	N_2—NaCl—PbS—H_2O	76
Nitrogen–Barium Chloride–Water	N_2—$BaCl_2$—H_2O	77
Nitrogen–Lead (II) Sulfide–Water	N_2—PbS—H_2O	110
Air–Sulfuric Acid–Water	— —H_2SO_4—H_2O	78, 79
Air–Sodium Hydroxide–Water	— —NaOH—H_2O	80
Air–Sodium Chloride–Water	— —NaCl—H_2O	81
Oxygen–Nitric Acid–Water	O_2—HNO_3—H_2O	82, 83
Oxygen–Nitric Acid–Dinitrogen Tetroxide–Water	O_2—HNO_3—N_2O_4—H_2O	84
Oxygen–Sulfuric Acid–Water	O_2—H_2SO_4—H_2O	85—87
Oxygen–Hydrogen Chloride–Water	O_2—HCl—H_2O	88
Oxygen–Lithium Chloride–Water	O_2—LiCl—H_2O	89
Oxygen–Ammonium Chloride–Water	O_2—NH_4Cl—H_2O	90
Oxygen–Sodium Hydroxide–Water	O_2—NaOH—H_2O	91
Oxygen–Sodium Hydrogen Carbonate–Lead (II) Sulfide–Water	O_2—$NaHCO_3$—PbS—H_2O	77
Oxygen–Sodium Sulfate–Water	O_2—Na_2SO_4—H_2O	92
Oxygen–Sodium Chloride–Water	O_2—NaCl—H_2O	93, 94
Oxygen–Sodium Chloride–Lead (II) Sulfide–Water	O_2—NaCl—PbS—H_2O	76
Oxygen–Sodium Bromide–Water	O_2—NaBr—H_2O	95
Oxygen–Magnesium Chloride–Water	O_2—$MgCl_2$—H_2O	96
Oxygen–Potassium Hydroxide–Water	O_2—KOH—H_2O	97, 98
Oxygen–Potassium Cyanide–Water	O_2—KCN—H_2O	99
Oxygen–Potassium Nitrate–Water	O_2—KNO_3—H_2O	100

System name	Formulae of components	Table number
Oxygen–Potassium Sulfate–Water	O_2—K_2SO_4—H_2O	101, 102
Oxygen–Potassium Chloride–Water	O_2—KCl—H_2O	103
Oxygen–Potassium Bromide–Water	O_2—KBr—H_2O	104
Oxygen–Potassium Iodide–Water	O_2— KI —H_2O	105
Oxygen–Calcium Chloride–Water	O_2—$CaCl_2$—H_2O	106
Oxygen–Rubidium Chloride–Water	O_2—$RbCl$—H_2O	107
Oxygen–Cesium Chloride–Water	O_2—$CsCl$—H_2O	108
Oxygen–Barium Chloride–Water	O_2—$BaCl_2$—H_2O	109
Oxygen–Lead (II) Sulfide–Water	O_2—PbS—H_2O	110
Neon–Seawater	Ne— — —H_2O	66
Sulfur–Ammonium Hydroxide–Water	S—NH_4OH—H_2O	111
Sulfur–Sodium Oxide–Water	S—Na_2O—H_2O	112
Sulfur–Sodium Hydroxide–Water	S—$NaOH$—H_2O	113
Sulfur–Sodium Sulfide–Water	S—Na_2S—H_2O	114, 115
Sulfur–Magnesium Hydroxide–Water	S—$Mg(OH)_2$—H_2O	116
Sulfur–Potassium Hydroxide–Water	S—KOH—H_2O	117
Sulfur–Calcium Hydroxide–Water	S—$Ca(OH)_2$—H_2O	118
Sulfur–Barium Hydroxide–Water	S—$Ba(OH)_2$—H_2O	119
Chlorine–Argon–Water	Cl_2—Ar—H_2O	120
Chlorine–Sulfuric Acid–Water	Cl_2—H_2SO_4—H_2O	121
Chlorine–Hydrogen Chloride–Water	Cl_2—HCl—H_2O	122—124
Chlorine–Sodium Chloride–Water	Cl_2—$NaCl$—H_2O	125—128
Chlorine–Magnesium Chloride–Water	Cl_2—$MgCl_2$—H_2O	132
Chlorine–Potassium Chloride–Water	Cl_2—KCl—H_2O	129, 130
Chlorine–Calcium Chloride–Water	Cl_2—$CaCl_2$—H_2O	132
Chlorine–Manganese (II) Chloride–Water	Cl_2—$MnCl_2$—H_2O	132
Chlorine–Barium Chloride–Water	Cl_2—$BaCl_2$—H_2O	131
Argon–Perchloric Acid–Water	Ar—$HClO_4$—H_2O	133
Argon–Lithium Chloride–Water	Ar—$LiCl$—H_2O	133
Argon–Sodium Nitrate–Water	Ar—$NaNO_3$—H_2O	133
Argon–Sodium Chloride–Water	Ar—$NaCl$—H_2O	133
Argon–Magnesium Chloride–Water	Ar—$MgCl_2$—H_2O	133
Argon–Aluminum Chloride–Water	Ar—$AlCl_3$—H_2O	133
Argon–Potassium Chloride–Water	Ar—KCl—H_2O	133
Argon–Calcium Chloride–Water	Ar—$CaCl_2$—H_2O	133
Argon–Strontium Chloride–Water	Ar—$SrCl_2$—H_2O	133
Argon–Barium Chloride–Water	Ar—$BaCl_2$—H_2O	133
Argon–Seawater	Ar— — —H_2O	66
Copper–Sodium Hydrogen Carbonate–Sodium Carbonate–Water	Cu—$NaHCO_3$—Na_2CO_3—H_2O	134
Bromine–Sulfuric Acid–Water	Br_2—H_2SO_4—H_2O	151
Bromine–Hydrogen Chloride–Water	Br_2—HCl—H_2O	135
Bromine–Hydrogen Bromide–Water	Br_2—HBr—H_2O	136, 137
Bromine–Ammonium Nitrate–Water	Br_2—NH_4NO_3—H_2O	151
Bromine–Ammonium Sulfate–Water	Br_2—$(NH_4)_2SO_4$—H_2O	151
Bromine–Ammonium Chloride–Water	Br_2—NH_4Cl—H_2O	151
Bromine–Sodium Nitrate–Water	Br_2—$NaNO_3$—H_2O	138, 151
Brome–Sodium Sulfate–Water	Br_2—Na_2SO_4—H_2O	139, 151
Bromine–Sodium Chloride–Water	Br_2—$NaCl$—H_2O	140, 151
Bromine–Sodium Bromide–Water	Br_2—$NaBr$—H_2O	141

System name	Formulae of components	Table number
Bromine–Potassium Nitrate–Water	$Br_2—KNO_3—H_2O$	151
Bromine–Potassium Sulfate–Water	$Br_2—K_2SO_4—H_2O$	142, 151
Bromine–Potassium Chloride–Water	$Br_2—KCl—H_2O$	151
Bromine–Potassium Bromide–Water	$Br_2—KBr—H_2O$	140
Bromine–Strontium Bromide–Water	$Br_2—SrBr_2—H_2O$	148
Bromine–Cesium Bromide–Water	$Br_2—CsBr—H_2O$	149
Bromine–Mercury (II) Bromide–Water	$Br_2—HgBr_2—H_2O$	150
Krypton–Seawater	$Kr— — —H_2O$	66
Silver–Silver Iodide–Water	$Ag— AgI —H_2O$	152
Iodine–Boric Acid–Water	$I_2—H_3BO_3—H_2O$	154
Iodine–Nitric Acid–Water	$I_2—HNO_3—H_2O$	153
Iodine–Sulfuric Acid–Water	$I_2—H_2SO_4—H_2O$	155
Iodine–Hydrogen Chloride–Water	$I_2—HCl—H_2O$	156, 157
Iodine–Perchloric Acid–Water	$I_2—HClO_4—H_2O$	158
Iodine–Perchloric Acid–Sodium Perchlorate–Water	$I_2—HClO_4—NaClO_4—H_2O$	159
Iodine–Hydrogen Bromide–Water	$I_2—HBr—H_2O$	160, 161
Iodine–Hydrogen Iodide–Water	$I_2— HI —H_2O$	162–164
Iodine–Lithium Sulfate–Potassium Iodide–Water	$I_2—Li_2SO_4— KI —H_2O$	165, 190
Iodine–Lithium Nitrate–Potassium Iodide–Water	$I_2—LiNO_3—KI—H_2O$	165, 190
Iodine–Lithium Chloride–Water	$I_2—LiCl—H_2O$	166
Iodine–Ammonium Nitrate–Water	$I_2—NH_4NO_3—H_2O$	154
Iodine–Ammonium Sulfate–Water	$I_2—(NH_4)_2SO_4—H_2O$	154
Iodine–Ammonium Chloride–Water	$I_2—NH_4Cl—H_2O$	154
Iodine–Ammonium Bromide–Water	$I_2—NH_4Br—H_2O$	154
Iodine–Ammonium Iodide–Water	$I_2—NH_4I—H_2O$	167
Iodine–Sodium Nitrate–Water	$I_2—NaNO_3—H_2O$	154, 168
Iodine–Sodium Nitrate–Potassium Iodide–Water	$I_2—NaNO_3— KI —H_2O$	190
Iodine–Sodium Dihydrogen Orthophosphate–Water	$I_2—NaH_2PO_4—H_2O$	169
Iodine–Sodium Sulfate–Water	$I_2—Na_2SO_4—H_2O$	154, 170
Iodine–Sodium Sulfate–Sodium Iodide–Water	$I_2—Na_2SO_4— NaI —H_2O$	171
Iodine–Sodium Sulfate–Potassium Iodide–Water	$I_2—Na_2SO_4— KI —H_2O$	190
Iodine–Sodium Chloride–Water	$I_2—NaCl—H_2O$	154, 172, 173
Iodine–Sodium Chloride–Potassium Iodide–Water	$I_2—NaCl— KI —H_2O$	190
Iodine–Sodium Perchlorate–Water	$I_2—NaClO_4—H_2O$	174
Iodine–Sodium Bromide–Water	$I_2—NaBr—H_2O$	154, 175
Iodine–Sodium Iodide–Water	$I_2—NaI—H_2O$	176–179
Iodine–Magnesium Nitrate–Potassium Iodide–Water	$I_2—Mg(NO_3)_2— KI —H_2O$	190
Iodine–Magnesium Sulfate–Potassium Iodide–Water	$I_2—MgSO_4— KI —H_2O$	190
Iodine–Magnesium Chloride–Water	$I_2—MgCl_2—H_2O$	180, 181
Iodine–Magnesium Chloride–Potassium Iodide–Water	$I_2—MgCl_2— KI —H_2O$	190
Iodine–Potassium Nitrate–Water	$I_2—KNO_3—H_2O$	154
Iodine–Potassium Nitrate–Potassium Iodide–Water	$I_2—KNO_3— KI —H_2O$	190

System name	Formulae of components	Table number
Iodine–Potassium Sulfate–Water	I_2—K_2SO_4—H_2O	154
Iodine–Potassium Sulfate–Potassium Iodide–Water	I_2—K_2SO_4—KI—H_2O	190
Iodine–Potassium Chloride–Water	I_2—KCl—H_2O	154, 182, 183
Iodine–Potassium Chloride–Potassium Iodide–Water	I_2—KCl—KI—H_2O	190
Iodine–Potassium Bromide–Water	I_2—KBr—H_2O	154, 184
Iodine–Potassium Iodide–Water	I_2—KI—H_2O	185—193
Iodine–Potassium Iodide–Calcium Chloride–Water	I_2—KI—$CaCl_2$—H_2O	190
Iodine–Potassium Iodide–Barium Nitrate–Water	I_2—KI—$Ba(NO_3)_2$—H_2O	190
Iodine–Potassium Iodide–Barium Chloride–Water	I_2—KI—$BaCl_2$—H_2O	190
Iodine–Aluminum Iodide–Water	I_2—AlI_3—H_2O	210
Iodine–Hexamminecobalt Iodide–Water	I_2—$[Co(NH_3)_6]I_3$—H_2O	210
Iodine–Copper (I) Iodide–Water	I_2—CuI—H_2O	194
Iodine–Scandium Iodide–Water	I_2—ScI_3—H_2O	210
Iodine–Rubidium Iodide–Water	I_2—RbI—H_2O	195, 196
Iodine–Strontium Iodide–Water	I_2—$SrCl_2$—H_2O	197
Iodine–Yttrium Iodide–Water	I_2—YI_3—H_2O	210
Iodine–Cadmium Iodide–Water	I_2—CdI_2—H_2O	198
Iodine–Tellurium Tetraiodide–Water	I_2—TeI_4—H_2O	199
Iodine–Cesium Iodide–Water	I_2—CsI—H_2O	200—202
Iodine–Barium Chloride–Water	I_2—$BaCl_2$—H_2O	203, 204
Iodine–Barium Iodide–Water	I_2—BaI_2—H_2O	205, 206
Iodine–Lanthanum Iodide–Water	I_2—LaI_3—H_2O	207, 208
Iodine–Cerium Iodide–Water	I_2—CeI_3—H_2O	208, 210
Iodine–Praseodymium Iodide–Water	I_2—PrI_3—H_2O	210
Iodine–Neodymium Iodide–Water	I_2—NdI_3—H_2O	210
Iodine–Samarium Iodide–Water	I_2—SmI_3—H_2O	210
Iodine–Europium Iodide–Water	I_2—EuI_3—H_2O	210
Iodine–Gadolinium Iodide–Water	I_2—GdI_3—H_2O	210
Iodine–Terbium Iodide–Water	I_2—TbI_3—H_2O	210
Iodine–Dysprosium Iodide–Water	I_2—DyI_3—H_2O	210
Iodine–Holmium Iodide–Water	I_2—HoI_3—H_2O	210
Iodine–Erbium Iodide–Water	I_2—ErI_3—H_2O	210
Iodine–Thulium Iodide–Water	I_2—TmI_3—H_2O	210
Iodine–Ytterbium Iodide–Water	I_2—YbI_3—H_2O	210
Iodine–Lutetium Iodide–Water	I_2—LuI_3—H_2O	210
Iodine–Mercury (II) Chloride–Water	I_2—$HgCl_2$—H_2O	209
Xenon–Seawater	Xe— — —H_2O	66
Gold–Potassium Cyanide–Water	Au—KCN—H_2O	211
Heavy Water (Deuterium Oxide)–Sodium Chloride–Water	D_2O—NaCl—H_2O	212
Heavy Water (Deuterium Oxide)–Potassium Chlorate–Water	D_2O—$KClO_3$—H_2O	213
Heavy Water (Deuterium Oxide)–Potassium Dichromate–Water	D_2O—$K_2Cr_2O_7$—H_2O	214, 215
Heavy Water (Deuterium Oxide)–Potassium Permanganate–Water	D_2O—$KMnO_4$—H_2O	216

System name	Formulae of components	Table number
Heavy Water (Deuterium Oxide)–Potassium Silver Cyanide–Water	$D_2O—KAg(CN)_2—H_2O$	217
Heavy Water (Deuterium Oxide)–Silver Chlorate–Water	$D_2O—AgClO_3—H_2O$	218
Heavy Water (Deuterium Oxide)–Cadmium Sulfate–Water	$D_2O—CdSO_4—H_2O$	219
Heavy Water (Deuterium Oxide)–Cadmium Iodide–Water	$D_2O—CdI_2—H_2O$	220, 221
Heavy Water (Deuterium Oxide)–Cesium Nitrate–Water	$D_2O—CsNO_3—H_2O$	222
Heavy Water (Deuterium Oxide)–Mercury (II) Cyanide–Water	$D_2O—Hg(CN)_2—H_2O$	223
Heavy Water (Deuterium Oxide)–Thallium Nitrate–Water	$D_2O—TlNO_3—H_2O$	224
Heavy Water (Deuterium Oxide)–Thallium Perchlorate–Water	$D_2O—TlClO_4—H_2O$	225
Hydrogen Peroxide–Boric Acid–Water	$H_2O_2—H_3BO_3—H_2O$	226
Hydrogen Peroxide–Nitric Acid–Uranium Trioxide–Water	$H_2O_2—HNO_3—UO_3 \cdot 0.5H_2O—H_2O$	227
Hydrogen Peroxide–Nitric Acid–Neptunium Peroxide–Water	$H_2O_2—HNO_3—Np_3O_8—H_2O$	228
Hydrogen Peroxide–Sulfuric Acid–Uranium Peroxide–Water	$H_2O_2—H_2SO_4—UO_4 \cdot 2H_2O—H_2O$	229
Hydrogen Peroxide–Lithium Hydroxide–Water	$H_2O_2—LiOH—H_2O$	230, 231
Hydrogen Peroxide–Lithium Sulfate–Water	$H_2O_2—Li_2SO_4—H_2O$	232
Hydrogen Peroxide–Lithium Selenate–Water	$H_2O_2—Li_2SeO_4—H_2O$	233
Hydrogen Peroxide–Beryllium Hydroxide–Water	$H_2O_2—Be(OH)_2—H_2O$	234
Hydrogen Peroxide–Ammonium Sulfate–Water	$H_2O_2—(NH_4)_2SO_4—H_2O$	235
Hydrogen Peroxide–Ammonium Selenate–Water	$H_2O_2—(NH_4)_2SeO_4—H_2O$	236
Hydrogen Peroxide–Sodium Hydroxide–Water	$H_2O_2—NaOH—H_2O$	237
Hydrogen Peroxide–Sodium Carbonate–Water	$H_2O_2—Na_2CO_3—H_2O$	238
Hydrogen Peroxide–Sodium Nitrate–Water	$H_2O_2—NaNO_3—H_2O$	239
Hydrogen Peroxide–Sodium Fluoride–Water	$H_2O_2—NaF—H_2O$	240
Hydrogen Peroxide–Sodium Silicate–Water	$H_2O_2—Na_2SiO_3—H_2O$	241
Hydrogen Peroxide–Sodium Hydrogen Orthophosphate–Water	$H_2O_2—Na_2HPO_4—H_2O$	242, 243
Hydrogen Peroxide–Sodium Orthophosphate–Water	$H_2O_2—Na_3PO_4—H_2O$	244

System name	Formulae of components	Table number
Hydrogen Peroxide–Sodium Pyrophosphate–Water	$H_2O_2—Na_4P_2O_7—H_2O$	245, 246
Hydrogen Peroxide–Sodium Sulfate–Water	$H_2O_2—Na_2SO_4—H_2O$	247—249
Hydrogen Peroxide–Sodium Chloride–Water	$H_2O_2—NaCl—H_2O$	250
Hydrogen Peroxide–Sodium Selenate–Water	$H_2O_2—Na_2SeO_4—H_2O$	251
Hydrogen Peroxide–Magnesium Hydroxide–Water	$H_2O_2—Mg(OH)_2—H_2O$	252
Hydrogen Peroxide–Potassium Hydroxide–Potassium Chlorate–Water	$H_2O_2—KOH—KClO_3—H_2O$	253
Hydrogen Peroxide–Potassium Carbonate–Water	$H_2O_2—K_2CO_3—H_2O$	254
Hydrogen Peroxide–Potassium Nitrate–Water	$H_2O_2—KNO_3—H_2O$	255
Hydrogen Peroxide–Potassium Dihydrogen Orthophosphate–Water	$H_2O_2—KH_2PO_4—H_2O$	256
Hydrogen Peroxide–Potassium Sulfate–Water	$H_2O_2—K_2SO_4—H_2O$	257, 258
Hydrogen Peroxide–Potassium Chlorate–Water	$H_2O_2—KClO_3—H_2O$	260
Hydrogen Peroxide–Potassium Perchlorate–Water	$H_2O_2—KClO_4—H_2O$	261
Hydrogen Peroxide–Potassium Selenate–Water	$H_2O_2—K_2SeO_4—H_2O$	262
Hydrogen Peroxide–Calcium Hydroxide–Water	$H_2O_2—Ca(OH)_2—H_2O$	263—265
Hydrogen Peroxide–Copper (II) Hydroxide–Water	$H_2O_2—Cu(OH)_2—H_2O$	266
Hydrogen Peroxide–Zinc Hydroxide–Water	$H_2O_2—Zn(OH)_2—H_2O$	267
Hydrogen Peroxide–Rubidium Carbonate–Water	$H_2O_2—Rb_2CO_3—H_2O$	268
Hydrogen Peroxide–Strontium Hydroxide–Water	$H_2O_2—Sr(OH)_2—H_2O$	269
Hydrogen Peroxide–Cadmium Hydroxide–Water	$H_2O_2—Cd(OH)_2—H_2O$	270
Hydrogen Peroxide–Cesium Carbonate–Water	$H_2O_2—Cs_2CO_3—H_2O$	271
Hydrogen Peroxide–Barium Hydroxide–Water	$H_2O_2—Ba(OH)_2—H_2O$	272
Boric Acid–Nitric Acid–Water	$H_3BO_3—HNO_3—H_2O$	273—275, 281
Boric Acid–Metaphosphoric Acid–Water	$H_3BO_3—HPO_3—H_2O$	276
Boric Acid–Sulfuric Acid–Water	$H_3BO_3—H_2SO_4—H_2O$	277—282
Boric Acid–Sulfuric Acid–Magnesium Sulfate–Water	$H_3BO_3—H_2SO_4—MgSO_4—H_2O$	283—286
Boric Acid–Hydrogen Chloride–Water	$H_3BO_3—HCl—H_2O$	281, 287—292

System name	Formulae of components	Table number
Boric Acid–Hydrogen Chloride–Magnesium Chloride–Water	$H_3BO_3—HCl—MgCl_2—H_2O$	293
Boric Acid–Hydrogen Chloride–Magnesium Chloride–Calcium Chloride–Water	$H_3BO_3—HCl—MgCl_2—CaCl_2—H_2O$	294
Boric Acid–Hydrogen Chloride–Calcium Chloride–Water	$H_3BO_3—HCl—CaCl_2—H_2O$	295, 358
Boric Acid–Lithium Chloride–Water	$H_3BO_3—LiCl—H_2O$	296, 297, 361
Boric Acid–Ammonia–Water	$H_3BO_3—NH_3—H_2O$	298
Boric Acid–Ammonium Hydroxide–Water	$H_3BO_3—NH_4OH—H_2O$	299
Boric Acid–Ammonium Dihydrogen Orthophosphate–Water	$H_3BO_3—NH_4H_2PO_4—H_2O$	300
Boric Acid–Ammonium Hydrogen Orthophosphate–Water	$H_3BO_3—(NH_4)_2HPO_4—H_2O$	301
Boric Acid–Ammonium Sulfate–Water	$H_3BO_3—(NH_4)_2SO_4—H_2O$	302
Boric Acid–Ammonium Chloride–Water	$H_3BO_3—NH_4Cl—H_2O$	303
Boric Acid–Sodium Tetraborate–Water	$H_3BO_3—Na_2B_4O_7—H_2O$	368
Boric Acid–Sodium Tetraborate–Sodium Sulfate–Water	$H_3BO_3—Na_2B_4O_7—Na_2SO_4—H_2O$	368
Boric Acid–Sodium Tetraborate–Sodium Chloride–Water	$H_3BO_3—Na_2B_4O_7—NaCl—H_2O$	368
Boric Acid–Sodium Hydrogen Carbonate–Water	$H_3BO_3—NaHCO_3—H_2O$	304
Boric Acid–Sodium Carbonate–Water	$H_3BO_3—Na_2CO_3—H_2O$	305
Boric Acid–Sodium Nitrate–Water	$H_3BO_3—NaNO_3—H_2O$	306
Boric Acid–Sodium Fluoride–Water	$H_3BO_3—NaF—H_2O$	307
Boric Acid–Sodium Dihydrogen Orthophosphate–Water	$H_3BO_3—NaH_2PO_4—H_2O$	308
Boric Acid–Sodium Hydrogen Orthophosphate–Water	$H_3BO_3—Na_2HPO_4—H_2O$	309
Boric Acid–Sodium Sulfate–Water	$H_3BO_3—Na_2SO_4—H_2O$	310—316, 368
Boric Acid–Sodium Sulfate–Sodium Chloride–Water	$H_3BO_3—Na_2SO_4—NaCl—H_2O$	317
Boric Acid–Sodium Sulfate–Magnesium Sulfate–Water	$H_3BO_3—Na_2SO_4—MgSO_4—H_2O$	318—322
Boric Acid–Sodium Chloride–Water	$H_3BO_3—NaCl—H_2O$	323—327, 367, 368
Boric Acid–Magnesium Nitrate–Calcium Nitrate–Water	$H_3BO_3—Mg(NO_3)_2—Ca(NO_3)_2—H_2O$	328
Boric Acid–Magnesium Sulfate–Water	$H_3BO_3—MgSO_4—H_2O$	329—333
Boric Acid–Magnesium Sulfate–Potassium Sulfate–Water	$H_3BO_3—MgSO_4—K_2SO_4—H_2O$	334

System name	Formulae of components	Table number
Boric Acid–Magnesium Chloride–Water	$H_3BO_3—MgCl_2—H_2O$	335
Boric Acid–Aluminum Chloride–Water	$H_3BO_3—AlCl_3—H_2O$	336
Boric Acid–Potassium Pentaborate–Water	$H_3BO_3—K_2B_{10}O_{16}—H_2O$	368
Boric Acid–Potassium Pentaborate–Potassium Chloride–Water	$H_3BO_3—K_2B_{10}O_{16}—KCl—H_2O$	368
Boric Acid–Potassium Nitrate–Water	$H_3BO_3—KNO_3—H_2O$	337—339
Boric Acid–Potassium Fluoride–Water	$H_3BO_3—KF—H_2O$	340
Boric Acid–Potassium Dihydrogen Orthophosphate–Water	$H_3BO_3—KH_2PO_4—H_2O$	341, 342
Boric Acid–Potassium Sulfate–Water	$H_3BO_3—K_2SO_4—H_2O$	343—347
Boric Acid–Potassium Sulfate–Potassium Chloride–Water	$H_3BO_3—K_2SO_4—KCl—H_2O$	348
Boric Acid–Potassium Chloride–Water	$H_3BO_3—KCl—H_2O$	349—352, 367, 368
Boric Acid–Potassium Bromide–Water	$H_3BO_3—KBr—H_2O$	353
Boric Acid–Potassium Dihydrogen Orthoarsenate–Water	$H_3BO_3—KH_2AsO_4—H_2O$	354
Boric Acid–Potassium Iodide–Water	$H_3BO_3—KI—H_2O$	355, 367
Boric Acid–Calcium Dihydrogen Orthoborate–Water	$H_3BO_3—Ca(H_2BO_3)_2 \cdot \frac{1}{2} H_3BO_3—H_2O$	356
Boric Acid–Calcium Chloride–Water	$H_3BO_3—CaCl_2—H_2O$	357—359, 367
Boric Acid–Chromium Trioxide–Water	$H_3BO_3—CrO_3—H_2O$	360
Boric Acid–Iron (III) Chloride–Water	$H_3BO_3—FeCl_3—H_2O$	361
Boric Acid–Nickel (II) Sulfate–Water	$H_3BO_3—NiSO_4—H_2O$	362
Boric Acid–Copper (II) Sulfate–Water	$H_3BO_3—CuSO_4—H_2O$	363
Boric Acid–Zinc Sulfate–Water	$H_3BO_3—ZnSO_4—H_2O$	364
Boric Acid–Rubidium Chloride–Water	$H_3BO_3—RbCl—H_2O$	365, 367
Boric Acid–Cesium Chloride–Water	$H_3BO_3—CsCl—H_2O$	367
Boric Acid–Barium Oxide–Water	$H_3BO_3—BaO—H_2O$	366
Boric Acid–Barium Chloride–Water	$H_3BO_3—BaCl_2—H_2O$	367
Carbonic Acid–Sulfuric Acid–Sodium Carbonate–Sodium Sulfate–Water	$H_2CO_3—H_2SO_4—Na_2CO_3—Na_2SO_4—H_2O$	369

System name	Formulae of components	Table number
Carbonic Acid–Calcium Hydrogen Carbonate–Calcium Sulfate–Water	H_3CO_3—$Ca(HCO_3)_2$—$CaSO_4$—H_2O	370
Nitric Acid–Phosphoric Acid–Ammonia–Water	HNO_3—H_3PO_4—NH_3—H_2O	371
Nitric Acid–Phosphoric Acid–Ammonium Nitrate–Ammonium Phosphate–Water	HNO_3—H_3PO_4—NH_4NO_3—$(NH_4)_3PO_4$—H_2O	372, 373
Nitric Acid–Phosphoric Acid–Ammonium Nitrate–Ammonium Phosphate–Calcium Nitrate–Calcium Phosphate–Water	HNO_3—H_3PO_4—NH_4NO_3—$(NH_4)_3PO_4$—$Ca(NO_3)_2$—$Ca_3(PO_4)_2$—H_2O	374—379
Nitric Acid–Phosphoric Acid–Potassium Nitrate–Potassium Phosphate–Calcium Nitrate–Calcium Phosphate–Water	HNO_3—H_3PO_4—KNO_3—K_3PO_4—$Ca(NO_3)_2$—$Ca_3(PO_4)_2$—H_2O	380, 381
Nitric Acid–Phosphoric Acid–Calcium Nitrate–Calcium Phosphate–Water	HNO_3—H_3PO_4—$Ca(NO_3)_2$—$Ca_3(PO_4)_2$—H_2O	382—384
Nitric Acid–Sulfuric Acid–Lead (III) Sulfate–Water	HNO_3—H_2SO_4—$PbSO_4$—H_2O	385, 386
Nitric Acid–Hydrogen Chloride–Potassium Nitrate–Potassium Chloride–Water	HNO_3—HCl—KNO_3—KCl—H_2O	387
Nitric Acid–Antimonic Acid–Sodium Fluoride–Water	HNO_3—$Sb(OH)_3$—NaF—H_2O	388
Nitric Acid–Iodic Acid–Water	HNO_3—HIO_3—H_2O	389—391
Nitric Acid–Periodic Acid–Water	HNO_3—H_5IO_6—H_2O	392
Nitric Acid–Phosphorotungstic Acid–Water	HNO_3—$P_2O_5 \cdot 24WO_3 \cdot 51H_2O$—$H_2O$	393
Nitric Acid–Beryllium Nitrate–Water	HNO_3—$Be(NO_3)_2$—H_2O	394
Nitric Acid–Dinitrogen Oxide–Water	HNO_3—N_2O—H_2O	395, 396
Nitric Acid–Nitrogen Dioxide–Water	HNO_3—NO_2—H_2O	397
Nitric Acid–Ammonium Nitrate–Water	HNO_3—NH_4NO_3—H_2O	398—401
Nitric Acid–Ammonium Nitrate–Calcium Nitrate–Water	HNO_3—NH_4NO_3—$Ca(NO_3)_3$—H_2O	402
Nitric Acid–Ammonium Nitrate–Yttrium Nitrate–Water	HNO_3—NH_4NO_3—$Y(NO_3)_3$—H_2O	403
Nitric Acid–Ammonium Nitrate–Lanthanum Nitrate–Water	HNO_3—NH_4NO_3—$La(NO_3)_2$—H_2O	404
Nitric Acid–Ammonium Nitrate–Gadolinium Nitrate–Water	HNO_3—NH_4NO_3—$Gd(NO_3)_2$—H_2O	405
Nitric Acid–Ammonium Nitrate–Uranyl Nitrate–Water	HNO_3—NH_4NO_3—$UO_2(NO_3)_2$—H_2O	406
Nitric Acid–Ammonium Cerium (III) Nitrate–Water	HNO_3—$(NH_4)_2Ce(NO_3)_5$—H_2O	407

System name	Formulae of components	Table number
Nitric Acid–Ammonium Cerium (IV) Nitrate–Water	HNO_3—$(NH_4)_2Ce(NO_3)_6$—H_2O	408
Nitric Acid–Ammonium Uranyl Nitrate–Water	HNO_3—$NH_4NO_3 \cdot UO_2(NO_3)_2$—$H_2O$	409
Nitric Acid–Ammonium Sulfate–Lead (II) Sulfate–Water	HNO_3—$(NH_4)_2SO_4$—$PbSO_4$—H_2O	410
Nitric Acid–Sodium Oxide–Water	HNO_3—Na_2O—H_2O	411
Nitric Acid–Sodium Nitrate–Water	HNO_3—$NaNO_3$—H_2O	412—414
Nitric Acid–Sodium Nitrate–Potassium Nitrate–Water	HNO_3—$NaNO_3$—KNO_3—H_2O	415
Nitric Acid–Sodium Molybdate–Lead (II) Molybdate–Water	HNO_3—Na_2MoO_4—$PbMoO_4$—H_2O	416
Nitric Acid–Sodium Octauranate–Water	HNO_3—$Na_4UO_8 \cdot 9H_2O$—H_2O	417
Nitric Acid–Magnesium Nitrate–Water	HNO_3—$Mg(NO_3)_2$—H_2O	418
Nitric Acid–Magnesium Nitrate–Uranyl Nitrate–Water	HNO_3—$Mg(NO_3)_2$—$UO_2(NO_3)_2$—H_2O	419
Nitric Acid–Cerium Magnesium Nitrate–Water	HNO_3—$[Ce(NO_3)_6]Mg_3 \cdot 24H_2O$—$H_2O$	420
Nitric Acid–Magnesium Neodymium Nitrate–Water	HNO_3—$3Mg(NO_3)_2 \cdot 2Nd(NO_3)_3 \cdot 24H_2O$—$H_2O$	421
Nitric Acid–Magnesium Bismuth Nitrate–Water	HNO_3—$Bi_2Mg_3(NO_3)_2 \cdot 24H_2O$—$H_2O$	422
Nitric Acid–Aluminum Nitrate–Water	HNO_3—$Al(NO_3)_3$—H_2O	423—425
Nitric Acid–Aluminum Arsenate–Water	HNO_3—$AlAsO_4$—H_2O	426
Nitric Acid–Potassium Nitrate–Water	HNO_3—KNO_3—H_2O	427—431
Nitric Acid–Potassium Nitrate–Potassium Phosphoromolybdate–Water	HNO_3—KNO_3—$K_3PO_4 \cdot 12MoO_3$—H_2O	432
Nitric Acid–Potassium Uranyl Nitrate–Water	HNO_3—$UO_2(NO_3)_2 \cdot 2KNO_3$—$H_2O$	433, 434
Nitric Acid–Potassium Sulfate–Potassium Zirconium Sulfate–Water	HNO_3—K_2SO_4—$K_2Zr(SO_4)_3$—H_2O	435
Nitric Acid–Potassium Sulfate–Potassium Lanthanum Sulfate–Water	HNO_3—K_2SO_4—$K_5La(SO_4)_4$—H_2O	436
Nitric Acid–Potassium Sulfate–Potassium Cerium Sulfate–Water	HNO_3—K_2SO_4—$K_6Ce(SO_4)_3$—H_2O	437
Nitric Acid–Potassium Sulfate–Potassium Plutonium Sulfate–Water	HNO_3—K_2SO_4—$K_2Pu(SO_4)_3$—H_2O	438
Nitric Acid–Potassium Sulfate–Calcium Sulfate–Water	HNO_3—K_2SO_4—$CaSO_4$—H_2O	439
Nitric Acid–Potassium Sulfate–Lead (II) Sulfate–Water	HNO_3—K_2SO_4—$PbSO_4$—H_2O	440
Nitric Acid–Potassium Bromide–Bismuth Bromide Oxide–Water	HNO_3—KBr—$BiOBr$—H_2O	441
Nitric Acid–Potassium Bismuth Hexacyanoferrate–Water	HNO_3—$KBi[Fe(CN)_6]$—H_2O	442

System name	Formulae of components	Table number
Nitric Acid–Potassium Plutonium Sulfate–Water	HNO_3—$K_4Pu(SO_4)_4$—H_2O	443
Nitric Acid–Potassium Chloride–Bismuth Chloride Oxide–Water	HNO_3—KCl—$BiOCl$—H_2O	444
Nitric Acid–Calcium Nitrate–Water	HNO_3—$Ca(NO_3)_2$—H_2O	445—451
Nitric Acid–Calcium Nitrate–Strontium Nitrate–Water	HNO_3—$Ca(NO_3)_2$—$Sr(NO_3)_2$—H_2O	452, 453
Nitric Acid–Calcium Nitrate–Uranyl Nitrate–Water	HNO_3—$Ca(NO_3)_2$—$UO_2(NO_3)_2$—H_2O	454
Nitric Acid–Calcium Sulfate–Water	HNO_3—$CaSO_4$—H_2O	455
Nitric Acid–Scandium Oxide–Water	HNO_3—ScO_3—H_2O	456
Nitric Acid–Vanadium Pentoxide–Water	HNO_3—V_2O_5—H_2O	457
Nitric Acid–Chromium Trioxide–Water	HNO_3—CrO_3—H_2O	458
Nitric Acid–Chromium (III) Nitrate–Water	HNO_3—$Cr(NO_3)_3$—H_2O	458
Nitric Acid–Chromium (III) Arsenate–Water	HNO_3—$CrAsO_4$—H_2O	426
Nitric Acid–Manganese (II) Nitrate–Water	HNO_3—$Mn(NO_3)_2$—H_2O	460
Nitric Acid–Manganese (II) Hexacyanoferrate (II)–Water	HNO_3—$Mn_2[Fe(CN)_6]$—H_2O	461
Nitric Acid–Bismuth Manganese Nitrate–Water	HNO_3—$Bi_2Mn_3(NO_3)_{12}\cdot24H_2O$—$H_2O$	422
Nitric Acid–Cerium Manganese Nitrate–Water	HNO_3—$[Ce(NO_3)_6]Mn_3\cdot24H_2O$—$H_2O$	420
Nitric Acid–Iron (III) Oxide–Water	HNO_3—Fe_2O_3—H_2O	462
Nitric Acid–Iron (III) Nitrate–Water	HNO_3—$Fe(NO_3)_3$—H_2O	463, 464
Nitric Acid–Iron (III) Nitrate–Thorium Nitrate–Water	HNO_3—$Fe(NO_3)_3$—$Th(NO_3)_4$—H_2O	465
Nitric Acid–Iron (III) Arsenate–Water	HNO_3—$FeAsO_4$—H_2O	426
Nitric Acid–Iron (III) Selenite–Water	HNO_3—$Fe_2(SeO_3)_3$—H_2O	466
Nitric Acid–Cobalt (II) Nitrate–Water	HNO_3—$Co(NO_3)_2$—H_2O	467
Nitric Acid–Cobalt (II) Selenite–Water	HNO_3—$CoSeO_3$—H_2O	466
Nitric Acid–Bismuth Cobalt Nitrate–Water	HNO_3—$Bi_2Co_3(NO_3)_{12}\cdot24H_2O$—$H_2O$	422
Nitric Acid–Cobalt Hydrogen Phosphate–Water	HNO_3—$CoHPO_4$—H_2O	468
Nitric Acid–Cobalt Phosphate–Water	HNO_3—$Co_3(PO_4)_2$—H_2O	468
Nitric Acid–Cobalt (III) Arsenate–Water	HNO_3—$Co_3(AsO_4)_2$—H_2O	426

System name	Formulae of components	Table number
Nitric Acid–Cerium Cobalt Nitrate–Water	HNO_3—$[Ce(NO_3)_6]Co_3 \cdot 24H_2O$—$H_2O$	420
Nitric Acid–Nickel (II) Nitrate–Water	HNO_3—$Ni(NO_3)_2$—H_2O	469
Nitric Acid–Nickel (II) Phosphate–Water	HNO_3—$Ni_3(PO_4)_2$—H_2O	468
Nitric Acid–Nickel (II) Hexacyanoferrate (II)–Water	HNO_3—$Ni_2[Fe(CN)_6]$—H_2O	470
Nitric Acid–Nickel (II) Arsenate–Water	HNO_3—$Ni_3(AsO_4)_2$—H_2O	426
Nitric Acid–Nickel (II) Selenite–Water	HNO_3—$NiSeO_3$—H_2O	466
Nitric Acid–Cerium Nickel Nitrate–Water	HNO_3—$[Ce(NO_3)_6]Ni_3 \cdot 24H_2O$—$H_2O$	420
Nitric Acid–Bismuth Nickel Nitrate–Water	HNO_3—$Bi_2Ni_3(NO_3)_2 \cdot 24H_2O$—$H_2O$	422
Nitric Acid–Copper (II) Oxide–Water	HNO_3—CuO—H_2O	471
Nitric Acid–Copper (II) Nitrate–Water	HNO_3—$Cu(NO_3)_2$—H_2O	472
Nitric Acid–Copper (II) Nitrate–Thorium Nitrate–Water	HNO_3—$Cu(NO_3)_2$—$Th(NO_3)_4$—H_2O	473
Nitric Acid–Copper (II) Nitrate–Uranyl Nitrate–Water	HNO_3—$Cu(NO_3)_2$—$UO_2(NO_3)_2$—H_2O	474
Nitric Acid–Copper (II) Phosphate–Water	HNO_3—$Cu_3(PO_4)_2$—H_2O	468
Nitric Acid–Zinc Nitrate–Water	HNO_3—$Zn(NO_3)_2$—H_2O	475, 476
Nitric Acid–Zinc Nitrate–Uranyl Nitrate–Water	HNO_3—$Zn(NO_3)_2$—$UO_2(NO_3)_2$—H_2O	477
Nitric Acid–Zinc Hexacyanoferrate (II)–Water	HNO_3—$Zn_2[(CN)_6]$—H_2O	478
Nitric Acid–Cerium Zinc Nitrate–Water	HNO_3—$[Ce(NO_3)_6]Zn_3 \cdot 24H_2O$—$H_2O$	420
Nitric Acid–Bismuth Zinc Nitrate–Water	HNO_3—$Bi_2Zn_3(NO_3)_2 \cdot 24H_2O$—$H_2O$	422
Nitric Acid–Germanium Dioxide–Water	HNO_3—GeO_2—H_2O	479
Nitric Acid–Rubidium Nitrate–Yttrium Nitrate–Water	HNO_3—$RbNO_3$—$Y(NO_3)_3$—H_2O	480
Nitric Acid–Rubidium Nitrate–Praseodymium Nitrate–Water	HNO_3—$RbNO_3$—$Pr(NO_3)_3$—H_2O	481
Nitric Acid–Strontium Nitrate–Water	HNO_3—$Sr(NO_3)_2$—H_2O	482
Nitric Acid–Strontium Sulfate–Water	HNO_3—$SrSO_4$—H_2O	483
Nitric Acid–Strontium Chloride–Water	HNO_3—$SrCl_2$—H_2O	484
Nitric Acid–Zirconium Dioxide–Water	HNO_3—ZrO_2—H_2O	485
Nitric Acid–Niobium Pentoxide–Water	HNO_3—Nb_2O_5—H_2O	486
Nitric Acid–Molybdenum Oxide–Water	HNO_3—MoO_3—H_2O	487

System name	Formulae of components	Table number
Nitric Acid–Silver Nitrate–Water	HNO_3—$AgNO_3$—H_2O	488, 489
Nitric Acid–Silver Sulfate–Water	HNO_3—Ag_2SO_4—H_2O	490, 5403
Nitric Acid–Silver Chloride–Water	HNO_3—$AgCl$—H_2O	491—497
Nitric Acid–Silver Chromate–Water	HNO_3—Ag_2CrO_4—H_2O	498
Nitric Acid–Silver Dichromate–Water	HNO_3—$Ag_2Cr_2O_7$—H_2O	499
Nitric Acid–Silver Bromate–Water	HNO_3—$AgBrO_3$—H_2O	500
Nitric Acid–Silver Iodate–Water	HNO_3— $AgIO_3$—H_2O	501—503
Nitric Acid–Cadmium Nitrate–Water	HNO_3—$Cd(NO_3)_2$—H_2O	504, 505
Nitric Acid–Cadmium Phosphate–Water	HNO_3—$Cd_3(PO_4)_2$—H_2O	468
Nitric Acid–Indium Nitrate–Indium Periodate–Water	HNO_3—$In(NO_3)_3$— $InIO_5$—H_2O	506
Nitric Acid–Indium Iodate–Water	HNO_3—$In(IO_3)_3$—H_2O	507
Nitric Acid–Tellurium Dioxide–Water	HNO_3—TeO_2—H_2O	508
Nitric Acid–Cesium Nitrate–Water	HNO_3—$CsNO_3$—H_2O	509, 510
Nitric Acid–Cesium Fluoride–Uranyl Nitrate–Water	HNO_3—CsF —$UO_2(NO_3)_2$—H_2O	511
Nitric Acid–Cesium Silicomolybdate–Water	HNO_3—$1.54Cs_2O \cdot SiO_2 \cdot 11.5MoO_3 \cdot 6.8H_2O$—$H_2O$	512
Nitric Acid–Cesium Germanomolybdate–Water	HNO_3—$1.82Cs_2O \cdot GeO_2 \cdot 11.5MoO_3 \cdot 6H_2O$—$H_2O$	513
Nitric Acid–Barium Nitrate–Water	HNO_3—$Ba(NO_3)_2$—H_2O	514—517
Nitric Acid–Barium Sulfate–Water	HNO_3—$BaSO_4$—H_2O	518—520
Nitric Acid–Barium Iodate–Water	HNO_3— $Ba(IO_3)_2$—H_2O	521
Nitric Acid–Lanthanum Nitrate–Water	HNO_3—$La(NO_3)_3$—H_2O	522
Nitric Acid–Lanthanum Fluoride–Uranyl Nitrate–Water	HNO_3—LaF_3—$UO_2(NO_3)_2$—H_2O	511
Nitric Acid–Lanthanum Arsenate–Water	HNO_3—$LaAsO_4$—H_2O	523
Nitric Acid–Cerium Nitrate–Water	HNO_3—$Ce(NO_3)_3$—H_2O	524
Nitric Acid–Cerium Phosphate–Water	HNO_3—$Ce_3(PO_4)_2$—H_2O	525
Nitric Acid–Cerium Selenite–Water	HNO_3—$Ce_2(SeO_3)_3$—H_2O	466
Nitric Acid–Praseodymium Arsenate–Water	HNO_3—$PrAsO_4$—H_2O	523

System name	Formulae of components	Table number
Nitric Acid–Neodymium Nitrate–Water	HNO_3—$Nd(NO_3)_3$—H_2O	526
Nitric Acid–Neodymium Arsenate–Water	HNO_3—$NdAsO_4$—H_2O	523
Nitric Acid–Samarium Nitrate–Water	HNO_3—$Sm(NO_3)_3$—H_2O	527
Nitric Acid–Samarium Arsenate–Water	HNO_3—$SmAsO_4$—H_2O	523
Nitric Acid–Gadolinium Arsenate–Water	HNO_3—$GdAsO_4$—H_2O	523
Nitric Acid–Ytterbium Arsenate–Water	HNO_3—$YbAsO_4$—H_2O	523
Nitric Acid–Mercury (I) Selenite–Water	HNO_3—Hg_2SeO_3—H_2O	466
Nitric Acid–Mercury (II) Oxide–Water	HNO_3—HgO—H_2O	528
Nitric Acid–Mercury (I) Arsenate–Water	HNO_3—Hg_3AsO_4—H_2O	529
Nitric Acid–Mercury (II) Bromate–Water	HNO_3—$Hg(BrO_3)_2$—H_2O	530
Nitric Acid–Mercury (II) Bromate Hydroxide–Water	HNO_3—$HgOHBrO_3$—H_2O	531
Nitric Acid–Thallium (I) Nitrate–Water	HNO_3—$TlNO_3$—H_2O	532
Nitric Acid–Thallium (I) Chloride–Water	HNO_3—$TlCl$—H_2O	533
Nitric Acid–Lead (II) Nitrate–Water	HNO_3—$Pb(NO_3)_2$—H_2O	534
Nitric Acid–Lead (II) Nitrate–Lead (II) Sulfate–Water	HNO_3—$Pb(NO_3)_2$—$PbSO_4$—H_2O	535
Nitric Acid–Lead (II) Nitrate–Lead (II) Chromate–Water	HNO_3—$Pb(NO_3)_2$—$PbCrO_4$—H_2O	536
Nitric Acid–Lead (II) Nitrate–Lead (II) Molybdate–Water	HNO_3—$Pb(NO_3)_2$—$PbMoO_4$—H_2O	537
Nitric Acid–Lead (II) Sulfate–Water	HNO_3—$PbSO_4$—H_2O	538—543
Nitric Acid–Lead (II) Chromate–Water	HNO_3—$PbCrO_4$—H_2O	544
Nitric Acid–Lead (II) Arsenate–Water	HNO_3—$Pb_3(AsO_4)_2$—H_2O	426
Nitric Acid–Lead (II) Selenite–Water	HNO_3—$PbSeO_3$—H_2O	466
Nitric Acid–Lead (II) Molybdate–Water	HNO_3—$PbMoO_4$—H_2O	545
Nitric Acid–Bismuth Nitrate–Water	HNO_3—$Bi(NO_3)_3$—H_2O	546
Nitric Acid–Bismuth Nitrate Oxide–Water	HNO_3—$BiONO_3$—H_2O	547
Nitric Acid–Bismuth Hexacyanoferrate (II)–Water	HNO_3—$Bi[Fe(CN)_6]$—H_2O	548
Nitric Acid–Bismuth Chloride Oxide–Water	HNO_3—$BiOCl$—H_2O	549

System name	Formulae of components	Table number
Hydrogen Fluoride–Potassium Heptafluoroniobate–Potassium Heptafluorotantalate–Water	$HF—K_2NbF_7—K_2TaF_7—H_2O$	586
Hydrogen Fluoride–Potassium Hexafluorohafnate–Water	$HF—HfK_2F_6—H_2O$	587
Hydrogen Fluoride–Potassium Hexafluorotantalate–Water	$HF—KTaF_6—H_2O$	588–592
Hydrogen Fluoride–Calcium Fluoride–Water	$HF—CaF_2—H_2O$	593
Hydrogen Fluoride–Vanadium Dioxide–Water	$HF—VO_2—H_2O$	594
Hydrogen Fluoride–Vanadium Pentoxide–Water	$HF—V_2O_5—H_2O$	595
Hydrogen Fluoride–Chromium Trioxide–Water	$HF—CrO_3—H_2O$	596
Hydrogen Fluoride–Manganese (II) Fluoride–Water	$HF—MnF_2—H_2O$	597, 598
Hydrogen Fluoride–Iron (III) Fluoride–Water	$HF—FeF_3—H_2O$	599
Hydrogen Fluoride–Cobalt (II) Fluoride–Water	$HF—CoF_2—H_2O$	600
Hydrogen Fluoride–Nickel (II) Fluoride–Water	$HF—NiF_2—H_2O$	601
Hydrogen Fluoride–Copper (II) Oxide–Water	$HF—CuO—H_2O$	602
Hydrogen Fluoride–Copper (II) Fluoride–Water	$HF—CuF_2—H_2O$	603
Hydrogen Fluoride–Zinc Fluoride–Water	$HF—ZnF_2—H_2O$	604
Hydrogen Fluoride–Germanium Oxide–Water	$HF—GeO_2—H_2O$	605
Hydrogen Fluoride–Arsenic Trioxide–Water	$HF—As_2O_3—H_2O$	606
Hydrogen Fluoride–Rubidium Hexafluorogermanate–Water	$HF—Rb_2GeF_6—H_2O$	607
Hydrogen Fluoride–Zirconium Oxide–Water	$HF—ZrO_2—H_2O$	608, 609
Hydrogen Fluoride–Zirconium Tetrafluoride–Water	$HF—ZrF_4—H_2O$	610, 611
Hydrogen Fluoride–Niobium Fluoride–Water	$HF—NbF_5—H_2O$	612
Hydrogen Fluoride–Silver Fluoride–Water	$HF—AgF—H_2O$	613–615
Hydrogen Fluoride–Cadmium Fluoride–Water	$HF—CdF_2—H_2O$	616
Hydrogen Fluoride–Antimony Trifluoride–Water	$HF—SbF_3—H_2O$	617
Hydrogen Fluoride–Tellurium Dioxide–Water	$HF—TeO_2—H_2O$	618
Hydrogen Fluoride–Iodine Pentoxide–Water	$HF—I_2O_5—H_2O$	619
Hydrogen Fluoride–Cesium Hexafluorogermanate–Water	$HF—Cs_2GeF_6—H_2O$	620

System name	Formulae of components	Table number
Hydrogen Fluoride–Cerium Fluoride –Water	$HF—CeF_4—H_2O$	621
Hydrogen Fluoride–Hafnium Tetrafluoride–Water	$HF—HfF_4—H_2O$	622
Hydrogen Fluoride–Hafnium Tetrafluoride–Water	$HF—HfF_4—H_2O$	623
Hydrogen Fluoride–Tantalum Fluoride–Water	$HF—TaF_5—H_2O$	624
Hydrogen Fluoride–Tungsten Trioxide–Water	$HF—WO_3—H_2O$	625
Hydrogen Fluoride–Mercury (II) Oxide–Water	$HF—HgO—H_2O$	626
Hydrogen Fluoride–Thorium Fluoride–Water	$HF—ThF_4—H_2O$	627
Hydrogen Fluoride–Uranium Trioxide–Water	$HF—UO_3—H_2O$	628
Hydrogen Fluoride–Uranium Tetrafluoride–Water	$HF—UF_4—H_2O$	629
Hydrogen Fluoride–Plutonyl Fluoride–Water	$HF—PuO_2F_2—H_2O$	630
Silicic Acid–Carbon Dioxide–Water	$H_2SiO_3—CO_2—H_2O$	1737
Silicic Acid–Carbon Dioxide–Water	$H_2SiO_3—CO_2—H_2O$	631
Silicic Acid–Sodium Nitrate–Water	$H_2SiO_3—NaNO_3—H_2O$	632
Silicic Acid–Sodium Sulfate–Water	$H_2SiO_3—Na_2SO_4—H_2O$	632
Silicic Acid–Sodium Sulfate–Sodium Chloride–Magnesium Sulfate–Magnesium Chloride–Calcium Chloride–Water	$H_2SiO_3—Na_2SO_4—NaCl—MgSO_4—MgCl—CaCl_2—H_2O$	633
Silicic Acid–Sodium Chloride–Water	$H_2SiO_3—NaCl—H_2O$	632
Silicic Acid–Magnesium Sulfate–Water	$H_2SiO_3—MgSO_4—H_2O$	632
Silicic Acid–Magnesium Chloride–Water	$H_2SiO_3—MgCl_2—H_2O$	632
Hexafluorosilicic Acid– Nitrogen Pentoxide–Phosphorus Pentoxide–Calcium Oxide– Water	$H_2SiF_6—N_2O_5—P_2O_5—CaO—H_2O$	634
Hexafluorosilicic Acid–Nitrogen Pentoxide–Calcium Oxide–Water	$H_2SiF_6—N_2O_5—CaO—H_2O$	635
Hexafluorosilicic Acid–Magnesium Hexafluorosilicate–Water	$H_2SiF_6—MgSiF_6—H_2O$	636
Hexafluorosilicic Acid–Lead (II) Hexafluorosilicate–Water	$H_2SiF_6—PbSiF_6—H_2O$	637
Hexafluorosilicic Acid–Phosphoric Acid–Calcium Sulfate–Water	$H_2SiF_6—H_3PO_4—CaSO_4—H_2O$	638

System name	Formulae of components	Table number
Phosphoric Acid–Sulfuric Acid–Calcium Sulfate–Water	H_3PO_4—H_2SO_4—$CaSO_4$—H_2O	639
Phosphoric Acid–Dinitrogen Oxide–Water	H_3PO_4—N_2O—H_2O*	640, 641
Phosphoric Acid–Nitrogen Oxide–Water	H_3PO_4—NO—H_2O	1798
Phosphoric Acid–Ammonia–Water	H_3PO_4—NH_3—H_2O	642—645
Phosphoric Acid–Ammonia–Uranium Trioxide–Water	H_3PO_4—NH_3—UO_3—H_2O	646
Phosphoric Acid–Ammonium Dihydrogen Orthophosphate– Water	H_3PO_4—$NH_4H_2PO_4 \cdot H_3PO_4$—$H_2O$	647
Phosphoric Acid–Ammonium Dihydrogen Orthophosphate– Water	H_3PO_4—$NH_4H_2PO_4$—H_2O	648
Phosphoric Acid–Ammonium Phosphate–Water	H_3PO_4—$(NH_4)_3PO_4$—H_2O	649—651
Phosphoric Acid–Ammonium Dihydrogen Orthophosphate–Calcium Dihydrogen Orthophosphate–Water	H_3PO_4—$NH_4H_2PO_4$—$Ca(H_2PO_4)_2$—H_2O	652
Phosphoric Acid–Ammonium Phosphate–Calcium Phosphate– Water	H_3PO_4—$(NH_4)_3PO_4$—$Ca_3(PO_4)_2$—H_2O	653
Phosphoric Acid–Sodium Hydroxide–Water	H_3PO_4—$NaOH$—H_2O	654
Phosphoric Acid–Sodium Nitrate–Water	H_3PO_4—$NaNO_3$—H_2O	655
Phosphoric Acid–Sodium Hexafluorosilicate–Sodium Chloride–Water	H_3PO_4—Na_2SiF_6—$NaCl$—H_2O	656
Phosphoric Acid–Magnesium Hydrogen Orthophosphate–Magnesium Orthophosphate–Magnesium Chloride–Water	H_3PO_4—$MgHPO_4 \cdot 3H_2O$—$Mg_3(PO_4)_2 \cdot 8H_2O$—$MgCl_2$—H_2O	657a
Phosphoric Acid– Aluminum Phosphate–Water	H_3PO_4—$AlPO_4$—H_2O	657б
Phosphoric Acid–Potassium Hydroxide–Water	H_3PO_4—KOH—H_2O	658
Phosphoric Acid–Potassium Phosphate–Water	H_3PO_4—K_3PO_4—H_2O	659
Phosphoric Acid–Potassium Phosphate–Calcium Phosphate–Water	H_3PO_4—K_3PO_4—$Ca_3(PO_4)_2$—H_2O	660
Phosphoric Acid–Potassium Sulfate–Water	H_3PO_4—K_2SO_4—H_2O	661, 662
Phosphoric Acid–Potassium Sulfate–Calcium Sulfate–Water	H_3PO_4—K_2SO_4—$CaSO_4$—H_2O	663
Phosphoric Acid–Calcium Dihydrogen Orthophosphate–Water	H_3PO_4—$Ca(H_2PO_4)_2$—H_2O	664
Phosphoric Acid–Calcium Hydrogen Orthophosphate–Water	H_3PO_4—$CaHPO_4$—H_2O	665—671
Phosphoric Acid–Calcium Phosphate–Water	H_3PO_4—$Ca_3(PO_4)_2$—H_2O	672, 673
Phosphoric Acid–Calcium Sulfate–Water	H_3PO_4—$CaSO_4$—H_2O	674—676
Phosphoric Acid–Vanadium Pentoxide–Water	H_3PO_4—V_2O_5—H_2O	677

System name	Formulae of components	Table number
Phosphoric Acid–Manganese (II) Hydrogen Orthophosphate– Water	H_3PO_4—$MnHPO_4$—H_2O	678
Phosphoric Acid–Manganese (II) Phosphate–Water	H_3PO_4—$Mn_3(PO_4)_2$—H_2O	679
Phosphoric Acid–Iron (III) Phosphate –Water	H_3PO_4—$FePO_4$—H_2O	680
Phosphoric Acid–Molybdenum Oxide –Water	H_3PO_4—MoO_3—H_2O	681
Phosphoric Acid–Gallium Phosphate –Water	H_3PO_4—$GaPO_4$—H_2O	682
Phosphoric Acid–Arsenic Trioxide –Water	H_3PO_4—As_2O_3—H_2O	1270
Phosphoric Acid–Silver Phosphate– Water	H_3PO_4—Ag_3PO_4—H_2O	683
Phosphoric Acid–Lanthanum Phosphate–Water	H_3PO_4—$LaPO_4$—H_2O	684
Phosphoric Acid–Cerium Phosphate –Water	H_3PO_4—$CePO_4$—H_2O	685, 686
Hydrogen Sulfide–Sulfuric Acid– Water	H_2S—H_2SO_4—H_2O	687
Hydrogen Sulfide–Hydrogen Chloride–Water	H_2S—HCl—H_2O	687—689
Hydrogen Sulfide–Hydrogen Chloride–Iron (II) Chloride–Water	H_2S—HCl—$FeCl_2$—H_2O	690
Hydrogen Sulfide–Hydrogen Chloride–Zinc Chloride–Water	H_2S—HCl—$ZnCl_2$—H_2O	691
Hydrogen Sulfide–Perchloric Acid– Sodium Hydroxide–Sodium Perchlorate–Silver Sulfide–Water	H_2S—$HClO_4$—$NaOH$—$NaClO_4$—Ag_2S—H_2O	692
Hydrogen Sulfide–Hydrogen Iodide –Water	H_2S—HI—H_2O	693
Hydrogen Sulfide–Ammonia–Water	H_2S—NH_3—H_2O	694
Hydrogen Sulfide–Ammonium Nitrate–Water	H_2S—NH_4NO_3—H_2O	687
Hydrogen Sulfide–Ammonium Sulfate–Water	H_2S—$(NH_4)_2SO_4$—H_2O	687
Hydrogen Sulfide–Ammonium Chloride–Water	H_2S—NH_4Cl—H_2O	687
Hydrogen Sulfide–Ammonium Bromide–Water	H_2S—NH_4Br—H_2O	687
Hydrogen Sulfide–Sodium Hydroxide–Sodium Chloride–Silver Sulfide–Water	H_2S—$NaOH$—$NaCl$—Ag_2S—H_2O	695
Hydrogen Sulfide–Sodium Nitrate– Water	H_2S—$NaNO_3$—H_2O	687
Hydrogen Sulfide–Sodium Hydrogensulfide–Water	H_2S—$NaSH$—H_2O	696
Hydrogen Sulfide–Sodium Sulfate– Water	H_2S—Na_2SO_4—H_2O	687
Hydrogen Sulfide–Sodium Chloride –Water	H_2S—$NaCl$—H_2O	687
Hydrogen Sulfide–Sodium Bromide –Water	H_2S—$NaBr$—H_2O	687
Hydrogen Sulfide–Potassium Nitrate –Water	H_2S—KNO_3—H_2O	687

System name	Formulae of components	Table number
Hydrogen Sulfide–Potassium Sulfate –Water	$H_2S—K_2SO_4—H_2O$	687
Hydrogen Sulfide–Potassium Chloride–Water	$H_2S—KCl—H_2O$	687
Hydrogen Sulfide–Potassium Bromide–Water	$H_2S—KBr—H_2O$	687
Hydrogen Sulfide–Potassium Iodide –Water	$H_2S—KI—H_2O$	687
Hydrogen Sulfide–Calcium Hydrogensulfide–Water	$H_2S—Ca(HS)_2—H_2O$	697
Sulfurous Acid–Potassium Sulfite–Water	$H_2SO_3—K_2SO_3—H_2O$	698
Nitrosylsulfuric Acid–Sulfuric Acid –Water	$HO\cdot SO_2\cdot O\cdot NO—H_2SO_4—H_2O$	699
Sulfuric Acid–Sulfamic Acid–Water	$H_2SO_4—HSO_3NH_2—H_2O$	700, 701
Sulfuric Acid–Hydrogen Chloride–Water	$H_2SO_4—HCl—H_2O$	702—707
Sulfuric Acid–Hydrogen Chloride–Ammonium Sulfate– Ammonium Chloride–Water	$H_2SO_4—HCl—(NH_4)_2SO_4—NH_4Cl—H_2O$	708
Sulfuric Acid–Hydrochloric Acid–Sodium Sulfate–Sodium Chloride –Water	$H_2SO_4—HCl—Na_2SO_4—NaCl—H_2O$	709
Sulfuric Acid–Hydrogen Chloride–Aluminum Sulfate–Water	$H_2SO_4—HCl—Al_2(SO_4)_3—H_2O$	710
Sulfuric Acid–Hydrogen Chloride–Lead (II) Sulfate–Water	$H_2SO_4—HCl—PbSO_4—H_2O$	711
Sulfuric Acid–Perchloric Acid–Lead (II) Sulfate–Water	$H_2SO_4—HClO_4—PbSO_4—H_2O$	712
Sulfuric Acid– Phosphorotungstic Acid–Water	$H_2SO_4—P_2O_5\cdot 24WO_3\cdot 51H_2O—H_2O$	713
Sulfuric Acid–Arsine–Water	$H_2SO_4—AsH_3—H_2O$	714
Sulfuric Acid–Lithium Sulfate–Water	$H_2SO_4—Li_2SO_4—H_2O$	715—717
Sulfuric Acid–Lithium Sulfate–Beryllium Sulfate–Water	$H_2SO_4—Li_2SO_4—BeSO_4—H_2O$	718
Sulfuric Acid–Lithium Sulfate–Ammonium Sulfate–Water	$H_2SO_4—Li_2SO_4—(NH_4)_2SO_4—H_2O$	719, 720
Sulfuric Acid–Beryllium Sulfate–Water	$H_2SO_4—BeSO_4—H_2O$	721, 722
Sulfuric Acid–Carbon Dioxide–Water	$H_2SO_4—CO_2—H_2O$	723—729
Sulfuric Acid–Dinitrogen Oxide–Water	$H_2SO_4—N_2O—H_2O$	730—732
Sulfuric Acid–Nitrogen Oxide–Water	$H_2SO_4—NO—H_2O$	733—735
Sulfuric Acid–Hydrazine Sulfate–Water	$H_2SO_4—N_2H_6SO_4—H_2O$	736
Sulfuric Acid–Ammonium Sulfate–Water	$H_2SO_4—(NH_4)_2SO_4—H_2O$	737—746

System name	Formulae of components	Table number
Sulfuric Acid–Ammonium Sulfate–Ammonium Peroxodisulfate–Water	$H_2SO_4—(NH_4)_2SO_4—(NH_4)_2S_2O_8—H_2O$	747, 748
Sulfuric Acid–Ammonium Sulfate–Sodium Sulfate–Water	$H_2SO_4—(NH_4)_2SO_4—Na_2SO_4—H_2O$	749
Sulfuric Acid–Ammonium Sulfate–Titanium Dioxide–Water	$H_2SO_4—(NH_4)_2SO_4—TiO_2—H_2O$	750—752
Sulfuric Acid–Ammonium Aluminum Sulfate–Water	$H_2SO_4—(NH_4)Al(SO_4)_2\cdot 12H_2O—H_2O$	753
Sulfuric Acid–Ammonium Silicomolybdate–Water	$H_2SO_4—1.75(NH_4)_2OSiO_2\cdot 12MoO_3\cdot 5H_2O—H_2O$	754
Sulfuric Acid–Ammonium Germanomolybdate–Water	$H_2SO_4—1.85(NH_4)_2OGeO_2\cdot 12MoO_3\cdot 4.65H_2O—H_2O$	755
Sulfuric Acid–Sodium Sulfate–Water	$H_2SO_4—Na_2SO_4—H_2O$	756—765
Sulfuric Acid–Sodium Sulfate–Potassium Sulfate–Water	$H_2SO_4—Na_2SO_4—K_2SO_4—H_2O$	766
Sulfuric Acid–Sodium Chloride–Water	$H_2SO_4—NaCl—H_2O$	767
Sulfuric Acid–Magnesium Sulfate–Water	$H_2SO_4—MgSO_4—H_2O$	768—773
Sulfuric Acid–Aluminum Oxide–Water	$H_2SO_4—Al_2O_3—H_2O$	774
Sulfuric Acid–Aluminum Phosphate–Water	$H_2SO_4—AlPO_4—H_2O$	775
Sulfuric Acid–Aluminum Phosphate–Phosphorus Pentoxide–Water	$H_2SO_4—AlPO_4—P_2O_5—H_2O$	776
Sulfuric Acid–Aluminum Sulfate–Water	$H_2SO_4—Al_2(SO_4)_3—H_2O$	777—783
Sulfuric Acid–Aluminum Arsenate–Water	$H_2SO_4—AlAsO_4—H_2O$	426
Sulfuric Acid–Silicon Dioxide–Water	$H_2SO_4—SiO_2—H_2O$	784
Sulfuric Acid–Rubidium Silicomolybdate–Water	$H_2SO_4—1.7Rb_2O\cdot SiO_2\cdot 12.2MoO_3\cdot 8.1H_2O—H_2O$	785
Sulfuric Acid–Phosphorus Pentoxide–Potassium Sulfate–Water	$H_2SO_4—P_2O_5—K_2SO_4—H_2O$	786
Sulfuric Acid–Sulfur Dioxide–Water	$H_2SO_4—SO_2—H_2O$	787—792
Sulfuric Acid–Potassium Sulfate–Water	$H_2SO_4—K_2SO_4—H_2O$	793—798
Sulfuric Acid–Potassium Sulfate–Potassium Plutonium Sulfate–Water	$H_2SO_4—K_2SO_4—K_4Pu(SO_4)_4—H_2O$	799
Sulfuric Acid–Potassium Sulfate–Calcium Sulfate–Water	$H_2SO_4—K_2SO_4—CaSO_4—H_2O$	800
Sulfuric Acid–Potassium Sulfate–Mercury (I) Sulfate–Water	$H_2SO_4—K_2SO_4—Hg_2SO_4—H_2O$	801
Sulfuric Acid–Potassium Plutonium Sulfate–Water	$H_2SO_4—K_4Pu(SO_4)_4—H_2O$	802
Sulfuric Acid–Calcium Dihydrogen Orthophosphate–Water	$H_2SO_4—Ca(H_2PO_4)_2—H_2O$	803
Sulfuric Acid–Calcium Hydrogen Orthophosphate–Water	$H_2SO_4—CaHPO_4—H_2O$	803

System name	Formulae of components	Table number
Sulfuric Acid–Calcium Orthophosphate–Water	H_2SO_4—$Ca_3(PO_4)_2$—H_2O	803
Sulfuric Acid–Apatite–Water	H_2SO_4— —— —H_2O	803
Sulfuric Acid–Calcium Sulfate–Water	H_2SO_4—$CaSO_4$—H_2O	804—806
Sulfuric Acid–Scandium Sulfate–Water	H_2SO_4—$Sc_2(SO_4)_3$—H_2O	807
Sulfuric Acid–Titanium Dioxide–Water	H_2SO_4—TiO_2—H_2O	808
Sulfuric Acid–Titanium Sulfate–Water	H_2SO_4—$Ti_2(SO_4)_3$—H_2O	809
Sulfuric Acid–Vanadium Pentoxide–Water	H_2SO_4—V_2O_5—H_2O	810—813
Sulfuric Acid–Chromium Trioxide–Water	H_2SO_4—CrO_3—H_2O	814, 815
Sulfuric Acid–Chromium Sulfate–Water	H_2SO_4—$Cr_2(SO_4)_3$—H_2O	816
Sulfuric Acid–Chromium Arsenate–Water	H_2SO_4—$CrAsO_4$—H_2O	426
Sulfuric Acid–Manganese (II) Sulfate–Water	H_2SO_4—$MnSO_4$—H_2O	817, 818
Sulfuric Acid–Manganese (II) Arsenate–Water	H_2SO_4—$Mn_3(AsO_4)_2$—H_2O	819
Sulfuric Acid–Manganese (II) Selenate–Water	H_2SO_4—$MnSeO_3$—H_2O	466
Sulfuric Acid–Iron (III) Oxide–Water	H_2SO_4—Fe_2O_3—H_2O	820
Sulfuric Acid–Iron (II) Sulfate–Water	H_2SO_4—$FeSO_4$—H_2O	821—828
Sulfuric Acid–Iron (II) Sulfate–Copper (II) Sulfate–Water	H_2SO_4—$FeSO_4$—$CuSO_4$—H_2O	829, 830
Sulfuric Acid–Iron (II) Sulfate–Copper (II) Sulfate–Zinc Sulfate–Water	H_2SO_4—$FeSO_4$—$CuSO_4$—$ZnSO_4$—H_2O	831
Sulfuric Acid–Copper (II) Iron (II) Sulfate–Water	H_2SO_4—$(Cu, Fe)SO_4 \cdot 5H_2O$—H_2O	832
Sulfuric Acid–Iron (III) Sulfate–Water	H_2SO_4—$Fe_2(SO_4)_3$—H_2O	833
Sulfuric Acid–Iron (III) Arsenate–Water	H_2SO_4—$FeAsO_4$—H_2O	426
Sulfuric Acid–Iron (III) Selenate–Water	H_2SO_4—$Fe_2(SeO_3)_3$—H_2O	466
Sulfuric Acid–Hexamminecobalt Sulfate–Water	H_2SO_4—$[Co(NH_3)_6]_2(SO_4)_3$—H_2O	834
Sulfuric Acid–Cobalt (II) Sulfate–Water	H_2SO_4—$CoSO_4$—H_2O	835—837
Sulfuric Acid–Cobalt (II) Arsenate–Water	H_2SO_4—$Co_3(AsO_4)_2$—H_2O	426
Sulfuric Acid–Cobalt (II) Selenite–Water	H_2SO_4—$CoSeO_3$—H_2O	466
Sulfuric Acid–Nickel (II) Sulfide–Water	H_2SO_4—NiS—H_2O	838
Sulfuric Acid–Nickel (II) Arsenate–Water	H_2SO_4—$Ni_3(AsO_4)_2$—H_2O	426

System name	Formulae of components	Table number
Sulfuric Acid–Nickel (II) Selenite–Water	H_2SO_4—$NiSeO_3$—H_2O	466
Sulfuric Acid–Copper (II) Sulfate–Water	H_2SO_4—$CuSO_4$—H_2O	839—845
Sulfuric Acid–Copper Arsenate Sulfate Hydrate–Water	H_2SO_4—$3Cu_3(AsO_4)_2 \cdot 0.33CuSO_4 \cdot 13H_2O$—$H_2O$	846
Sulfuric Acid–Zinc Sulfide–Water	H_2SO_4—ZnS—H_2O	847
Sulfuric Acid–Zinc Sulfate–Water	H_2SO_4—$ZnSO_4$—H_2O	848—850
Sulfuric Acid–Germanium Dioxide–Water	H_2SO_4—GeO_2—H_2O	851, 852
Sulfuric Acid–Rubidium Germanomolybdate–Water	H_2SO_4—$2Rb_2O \cdot GeO_2 \cdot 12MoO_3 \cdot 9H_2O$—$H_2O$	853, 854
Sulfuric Acid–Arsenic Trioxide–Water	H_2SO_4—As_2O_3—H_2O	855, 1270
Sulfuric Acid–Strontium Sulfate–Water	H_2SO_4—$SrSO_4$—H_2O	856
Sulfuric Acid–Zirconium Dioxide–Water	H_2SO_4—ZrO_2—H_2O	857
Sulfuric Acid–Zirconium Sulfate–Water	H_2SO_4—$Zr(SO_4)_2$—H_2O	858
Sulfuric Acid–Niobium Pentoxide–Water	H_2SO_4—Nb_2O_5—H_2O	859, 860
Sulfuric Acid–Silver Sulfate–Water	H_2SO_4—Ag_2SO_4—H_2O	861—864, 5403
Sulfuric Acid–Silver Chloride–Water	H_2SO_4—$AgCl$—H_2O	865
Sulfuric Acid–Silver Bromate–Water	H_2SO_4—$AgBrO_3$—H_2O	866
Sulfuric Acid–Cadmium Sulfate–Water	H_2SO_4—$CdSO_4$—H_2O	867
Sulfuric Acid–Pyrophosphates of Indium–Water	H_2SO_4—In. . . —H_2O	868
Sulfuric Acid–Indium Sulfate–Water	H_2SO_4—$In_2(SO_4)_3$—H_2O	869, 870
Sulfuric Acid–Indium Selenate–Water	H_2SO_4—$In_2(SeO_3)_3$—H_2O	871
Sulfuric Acid–Iodine Pentoxide–Water	H_2SO_4—I_2O_5—H_2O	872
Sulfuric Acid–Cesium Silicomolybdate–Water	H_2SO_4—$1.54Cs_2O \cdot SiO_2 \cdot 11.5MoO_3 \cdot 6.8H_2O$—$H_2O$	873
Sulfuric Acid–Cesium Germanomolybdate–Water	H_2SO_4—$1.85Cs_2O \cdot GeO_2 \cdot 11.5MoO_3 \cdot 6H_2O$—$H_2O$	874
Sulfuric Acid–Cesium Hexachlororhenate–Water	H_2SO_4—$CsReCl_6$—H_2O	875
Sulfuric Acid–Barium Sulfate–Water	H_2SO_4—$BaSO_4$—H_2O	876—880
Sulfuric Acid–Lanthanum Sulfate–Water	H_2SO_4—$La_2(SO_4)_3$—H_2O	881
Sulfuric Acid–Lanthanum Arsenate–Water	H_2SO_4—$LaAsO_4$—H_2O	523

System name	Formulae of components	Table number
Sulfuric Acid–Cerium (IV) Phosphate –Water	H_2SO_4—$Ce_3(PO_4)_4$—H_2O	882
Sulfuric Acid–Cerium (IV) Pyrophosphate–Water	H_2SO_4—CeP_2O_7—H_2O	883
Sulfuric Acid–Cerium (III) Sulfate –Water	H_2SO_4—$Ce_2(SO_4)_3$—H_2O	884, 885
Sulfuric Acid–Cerium (III) Selenite –Water	H_2SO_4—$Ce_2(SeO_3)_3$—H_2O	466
Sulfuric Acid–Praseodymium Arsenate–Water	H_2SO_4—$PrAsO_4$—H_2O	523
Sulfuric Acid–Neodymium Sulfate –Water	H_2SO_4—$Nd_2(SO_4)_3$—H_2O	886
Sulfuric Acid–Neodymium Arsenate –Water	H_2SO_4—$NdAsO_4$—H_2O	523
Sulfuric Acid–Samarium Arsenate–Water	H_2SO_4—$SmAsO_4$—H_2O	523
Sulfuric Acid–Gadolinium Sulfate–Water	H_2SO_4—$Gd_2(SO_4)_3$—H_2O	887
Sulfuric Acid–Gadolinium Arsenate –Water	H_2SO_4—$GdAsO_4$—H_2O	523
Sulfuric Acid–Ytterbium Arsenate–Water	H_2SO_4—$YbAsO_4$—H_2O	523
Sulfuric Acid–Tantalum Pentoxide –Water	H_2SO_4—Ta_2O_5—H_2O	888
Sulfuric Acid–Mercury (I) Sulfate–Water	H_2SO_4—Hg_2SO_4—H_2O	889—892
Sulfuric Acid–Mercury (I) Arsenate –Water	H_2SO_4—Hg_3AsO_4—H_2O	893
Sulfuric Acid–Mercury (I) Selenite –Water	H_2SO_4—Hg_2SeO_3—H_2O	466
Sulfuric Acid–Thallium (III) Oxide –Water	H_2SO_4—Tl_2O_3—H_2O	894
Sulfuric Acid–Thallium (III) Hydroxide–Water	H_2SO_4—$Tl(OH)_3$—H_2O	895
Sulfuric Acid–Thallium (I) Sulfate –Water	H_2SO_4—Tl_2SO_4—H_2O	896, 897
Sulfuric Acid–Thallium (I) Thallium (III) Sulfate–Water	H_2SO_4—$Tl_2SO_4 \cdot Tl_2(SO_4)_3$—$H_2O$	898
Sulfuric Acid–Thallium (III) Selenite –Water	H_2SO_4—$Tl_2(SeO_3)_3$—H_2O	899
Sulfuric Acid–Lead Dioxide–Water	H_2SO_4—PbO_2—H_2O	900
Sulfuric Acid–Lead (II) Sulfate–Water	H_2SO_4—$PbSO_4$—H_2O	901—910
Sulfuric Acid–Bismuth Arsenate–Water	H_2SO_4—$BiAsO_4$—H_2O	426
Sulfuric Acid–Radium Sulfate–Water	H_2SO_4—$RaSO_4$—H_2O	911
Sulfuric Acid–Thorium Oxide– Water	H_2SO_4—ThO_2—H_2O	912
Sulfuric Acid–Thorium Hypophosphate–Water	H_2SO_4—ThP_2O_6—H_2O	913
Sulfuric Acid–Thorium Sulfate–Water	H_2SO_4—$Th(SO_4)_2$—H_2O	914—916

System name	Formulae of components	Table number
Sulfuric Acid–Uranium Tetrafluoride –Water	H_2SO_4—UF_4—H_2O	917
Sulfuric Acid–Uranium Sulfate–Water	H_2SO_4—$U(SO_4)_2\cdot 4H_2O$—H_2O	918, 919
Sulfuric Acid–Uranyl Sulfate–Water	H_2SO_4—UO_2SO_4—H_2O	920a
Sulfamic Acid–Ammonium Sulfito Amide–Water	HSO_3NH_2—$NH_4SO_3NH_2$—H_2O	920b, 920c
Hydrogen Chloride–Phosphorotungstic Acid–Water	HCl—$P_2O_5\cdot 24WO_3\cdot 51H_2O$—$H_2O$	921
Hydrogen Chloride–Hydrogen Bromide–Potassium Chloride–Potassium Bromide–Water	HCl—HBr—KCl—KBr—H_2O	922
Hydrogen Chloride–Molybdic Acid–Water	HCl—H_2MoO_4—H_2O	923
Hydrogen Chloride–Molybdic Acid – Tungstic Acid–Water	HCl—H_2MoO_4—H_2WO_4—H_2O	924
Hydrogen Chloride–Tungstic Acid	HCl—H_2WO_4—H_2O	925
Hydrogen Chloride–Lithium Chloride–Water	HCl—$LiCl$—H_2O	926—929
Hydrogen Chloride–Beryllium Hydroxide–Water	HCl—$Be(OH)_2$—H_2O	930
Hydrogen Chloride–Beryllium Chloride–Water	HCl—$BeCl_2$—H_2O	931, 932
Hydrogen Chloride–Carbon Dioxide –Water	HCl—CO_2—H_2O	933, 934
Hydrogen Chloride–Dinitrogen Oxide–Water	HCl—N_2O—H_2O	935, 936
Hydrogen Chloride–Hydrazine Sulfate–Water	HCl—$N_2H_6SO_4$—H_2O	937
Hydrogen Chloride–Ammonium Sulfate–Lead (II) Sulfate–Water	HCl—$(NH_4)_2SO_4$—$PbSO_4$—H_2O	938
Hydrogen Chloride–Ammonium Chloride–Water	HCl—NH_4Cl—H_2O	939—947
Hydrogen Chloride–Ammonium Chloride–Ammonium Hexachloroplatinate– Water	HCl—NH_4Cl—$(NH_4)_2PtCl_6$—H_2O	948
Hydrogen Chloride–Ammonium Chloride–Sodium Chloride–Water	HCl—NH_4Cl—$NaCl$—H_2O	949
Hydrogen Chloride–Ammonium Pentachlorotitanate–Water	HCl—$(NH_4)_2TiCl_5$—H_2O	950
Hydrogen Chloride–Ammonium Hexachlorotitanate–Water	HCl—$(NH_4)_2TiCl_6$—H_2O	951
Hydrogen Chloride–Ammonium Pentachloroxoniobate–Water	HCl—$(NH_4)_2NbOCl_5$—H_2O	952
Hydrogen Chloride–Ammonium Hexachloroplatinate–Water	HCl—$(NH_4)_2PtCl_6$—H_2O	948
Hydrogen Chloride–Sodium Hexafluorosilicate–Water	HCl—Na_2SiF_6—H_2O	953

System name	Formulae of components	Table number
Hydrogen Chloride–Sodium Hexafluorosilicate–Sodium Chloride–Water	$HCl—Na_2SiF_6—NaCl—H_2O$	954
Hydrogen Chloride–Sodium Chloride–Water	$HCl—NaCl—H_2O$	955—971
Hydrogen Chloride–Sodium Chloride–Magnesium Sulfate–Magnesium Chloride–Water	$HCl—NaCl—MgSO_4—MgCl_2—H_2O$	972
Hydrogen Chloride–Sodium Chloride–Magnesium Sulfate–Magnesium Chloride–Potassium Chloride–Calcium Sulfate–Water	$HCl—NaCl—MgSO_4—MgCl_2—KCl—CaSO_4—H_2O$	973
Hydrogen Chloride–Sodium Chloride–Magnesium Chloride–Potassium Chloride–Water	$HCl—NaCl—MgCl_2—KCl—H_2O$	974
Hydrogen Chloride–Sodium Chloride–Potassium Chloride–Water	$HCl—NaCl—KCl—H_2O$	975, 976
Hydrogen Chloride–Sodium Chloride–Aluminum Chloride–Potassium Chloride–Iron (III) Chloride–Water	$HCl—NaCl—AlCl_3—KCl—FeCl_3—H_2O$	977
Hydrogen Chloride–Sodium Chloride–Calcium Chloride–Water	$HCl—NaCl—CaCl_2—H_2O$	978
Hydrogen Chloride–Sodium Chloride–Cadmium Hexacyanoferrate (III)–Water	$HCl—NaCl—Cd_3[Fe(CN)_6]_2—H_2O$	979
Hydrogen Chloride–Sodium Chloride–Barium Chloride–Water	$HCl—NaCl—BaCl_2—H_2O$	980
Hydrogen Chloride–Magnesium Fluoride–Water	$HCl—MgF_2—H_2O$	981
Hydrogen Chloride–Magnesium Chloride–Water	$HCl—MgCl_2—H_2O$	982—988
Hydrogen Chloride–Aluminum Oxide–Water	$HCl—Al_2O_3—H_2O$	568
Hydrogen Chloride–Aluminum Chloride–Water	$HCl—AlCl_3—H_2O$	989—991
Hydrogen Chloride–Silicon Dioxide–Water	$HCl—SiO_2—H_2O$	992
Hydrogen Chloride–Potassium Sulfate–Calcium Sulfate–Water	$HCl—K_2SO_4—CaSO_4—H_2O$	993
Hydrogen Chloride–Potassium Chloride–Water	$HCl—KCl—H_2O$	994—1003
Hydrogen Chloride–Potassium Chloride–Cadmium Chloride– Water	$HCl—KCl—CdCl_2—H_2O$	4897
Hydrogen Chloride–Potassium Pentachloroaquotitanate–Water	$HCl—K_2TiCl_5—H_2O$	1004
Hydrogen Chloride–Potassium Hexachlorotitanate–Water	$HCl—K_2TiCl_6—H_2O$	1005
Hydrogen Chloride–Potassium Hexachlororhenate–Water	$HCl—K_2ReCl_6—H_2O$	1006
Hydrogen Chloride–Potassium Decachloroxodirhenate–Water	$HCl—K_4Re_2OCl_{10}—H_2O$	1007

System name	Formulae of components	Table number
Hydrogen Chloride–Calcium Oxide–Water	$HCl—CaO—H_2O$	1008
Hydrogen Chloride–Calcium Fluoride–Water	$HCl—CaF_2—H_2O$	1009
Hydrogen Chloride–Calcium Sulfate–Water	$HCl—CaSO_4—H_2O$	1010—1014
Hydrogen Chloride–Calcium Sulfate–Calcium Chloride–Water	$HCl—CaSO_4—CaCl_2—H_2O$	1015
Hydrogen Chloride–Calcium Chloride–Water	$HCl—CaCl_2—H_2O$	1016—1021
Hydrogen Chloride–Calcium Chloride–Barium Sulfate–Water	$HCl—CaCl_2—BaSO_4—H_2O$	1022
Hydrogen Chloride–Calcium Iodate–Water	$HCl—Ca(IO_3)_2—H_2O$	5190
Hydrogen Chloride–Titanium Chloride–Water	$HCl—TiCl_2—H_2O$	1023
Hydrogen Chloride–Rubidium Pentachloroaquotitanate–Water	$HCl—Rb_2TiCl_5—H_2O$	1024
Hydrogen Chloride–Rubidium Hexachlorotitanate–Water	$HCl—Rb_2TiCl_6—H_2O$	1025
Hydrogen Chloride–Vandadium Pentoxide–Water	$HCl—V_2O_5—H_2O$	1026, 1027
Hydrogen Chloride–Manganese (II) Hydroxide–Water	$HCl—Mn(OH)_2—H_2O$	1028
Hydrogen Chloride–Manganese (II) Hexacyanoferrate (II)–Water	$HCl—Mn_2[Fe(CN)_6]—H_2O$	1029
Hydrogen Chloride–Manganese (II) Arsenate–Water	$HCl—Mn_3(AsO_4)_2—H_2O$	1030
Hydrogen Chloride–Manganese (II) Selenite–Water	$HCl—MnSeO_3—H_2O$	466
Hydrogen Chloride–Iron (III) Oxide–Water	$HCl—Fe_2O_3—H_2O$	1031
Hydrogen Chloride–Iron (II) Chloride–Water	$HCl—FeCl_2—H_2O$	1032, 1033
Hydrogen Chloride–Iron (III) Chloride–Water	$HCl—FeCl_3—H_2O$	1034—1037
Hydrogen Chloride–Cobalt (II) Hydroxide–Water	$HCl—Co(OH)_2—H_2O$	2038
Hydrogen Chloride–Cobalt (II) Hydrogen Orthophosphate–Water	$HCl—CoHPO_4—H_2O$	468
Hydrogen Chloride–Cobalt (II) Orthophosphate–Water	$HCl—Co_3(PO_4)_2—H_2O$	468
Hydrogen Chloride–Cobalt (II) Chloride–Water	$HCl—CoCl_2—H_2O$	1039—1041
Hydrogen Chloride–Ytterbium Hexacyanocobaltate–Water	$HCl—Yb(CoC_6N_6)_2 \cdot 9H_2O—H_2O$	1042
Hydrogen Chloride–Nickel (II) Hydroxide–Water	$HCl—Ni(OH)_2—H_2O$	1043
Hydrogen Chloride–Nickel (II) Phosphate–Water	$HCl—Ni_3(PO_4)_2—H_2O$	468
Hydrogen Chloride–Nickel Hexacyanoferrate (II)–Water	$HCl—Ni_2[Fe(CN)_6]—H_2O$	1044

System name	Formulae of components	Table number
Hydrogen Chloride–Nickel (II) Chloride–Water	$HCl—NiCl_2—H_2O$	1045—1047
Hydrogen Chloride–Copper (II) Oxide–Water	$HCl—CuO—H_2O$	1048
Hydrogen Chloride–Copper (II) Phosphate–Water	$HCl—Cu_3(PO_4)_2—H_2O$	468
Hydrogen Chloride–Copper (II) Arsenate Sulfate Hydrate–Water	$HCl—3Cu_3(AsO_4)_2\cdot0.33CuSO_4\cdot13H_2O—H_2O$	1049
Hydrogen Chloride–Copper (I) Chloride–Water	$HCl—CuCl—H_2O$	1050—1052
Hydrogen Chloride–Copper (II) Chloride–Water	$HCl—CuCl_2—H_2O$	1053—1055
Hydrogen Chloride–Zinc Hexacyanoferrate (II)–Water	$HCl—Zn_2[Fe(CN)_6]—H_2O$	1056
Hydrogen Chloride–Zinc Chloride–Water	$HCl—ZnCl_2—H_2O$	1057
Hydrogen Chloride–Germanium Dioxide–Lanthanum Oxide–Water	$HCl—GeO_2—La_2O_3—H_2O$	1058
Hydrogen Chloride–Germanium Dioxide–Water	$HCl—GeO_2—H_2O$	1059, 1960
Hydrogen Chloride–Arsenic Trioxide–Water	$HCl—As_2O_3—H_2O$	1061—1064, 1270
Hydrogen Chloride–Rubidium Chloride–Water	$HCl—RbCl—H_2O$	1065
Hydrogen Chloride–Rubidium Pentachloroxoniobate–Water	$HCl—Rb_2NbOCl_5—H_2O$	1066
Hydrogen Chloride–Rubidium Dichloroiodate–Water	$HCl—RbCl_2I—H_2O$	1067
Hydrogen Chloride–Tellurium Rubidium Chloride–Water	$HCl—TeCl_2\cdot2RbCl_2—H_2O$	1068
Hydrogen Chloride–Strontium Oxide–Water	$HCl—SrO—H_2O$	1069
Hydrogen Chloride–Strontium Fluoride–Water	$HCl—SrF_2—H_2O$	1070
Hydrogen Chloride–Strontium Sulfate–Water	$HCl—SrSO_4—H_2O$	856, 1071, 1072
Hydrogen Chloride–Strontium Chloride–Water	$HCl—SrCl_2—H_2O$	1073—1075
Hydrogen Chloride–Yttrium Pyrophosphate–Water	$HCl—Y_4(P_2O_7)_3—H_2O$	1076
Hydrogen Chloride–Yttrium Chloride–Water	$HCl—YCl_2\cdot6H_2O—H_2O$	1077
Hydrogen Chloride–Zirconium Oxide–Water	$HCl—ZrO_2—H_2O$	1078
Hydrogen Chloride–Oxozirconium Dihydrogen Orthophosphate–Water	$HCl—ZrO(H_2PO_4)_2—H_2O$	1079
Hydrogen Chloride–Oxozirconium Chloride–Water	$HCl—ZrOCl_2\cdot8H_2O—H_2O$	1080, 1081
Hydrogen Chloride–Niobium Pentoxide–Water	$HCl—Nb_2O_5—H_2O$	1082, 1083
Hydrogen Chloride–Molybdenum Oxide–Water	$HCl—MoO_3—H_2O$	1084

System name	Formulae of components	Table number
Hydrogen Chloride–Silver Chloride–Water	$HCl—AgCl—H_2O$	1085—1090, 5412
Hydrogen Chloride–Cadmium Phosphate–Water	$HCl—Cd_3(PO_4)_2—H_2O$	468
Hydrogen Chloride–Cadmium Chloride–Water	$HCl—CdCl_2—H_2O$	1091
Hydrogen Chloride–Indium Chloride–Water	$HCl—InCl_3—H_2O$	1092
Hydrogen Chloride–Indium Selenite–Water	$HCl—In_2(SeO_3)_3—H_2O$	871
Hydrogen Chloride–Tin (II) Oxide–Water	$HCl—SnO—H_2O$	1093
Hydrogen Chloride–Antimony Trichloride–Water	$HCl—SbCl_3—H_2O$	1094
Hydrogen Chloride–Tellurium Cesium Chloride–Water	$HCl—TeCl_4 \cdot 2CsCl—H_2O$	1095
Hydrogen Chloride–Tellurium Oxide–Water	$HCl—TeO_2—H_2O$	1096, 1097
Hydrogen Chloride–Cesium Chloride–Water	$HCl—CsCl—H_2O$	1098
Hydrogen Chloride–Cesium Pentachloroaquotitanate–Water	$HCl—Cs_2TiCl_5—H_2O$	1099
Hydrogen Chloride–Cesium Pentachloroxoniabate–Water	$HCl—Cs_2NbOCl_5—H_2O$	1100
Hydrogen Chloride–Cesium Hexachlorotitanate–Water	$HCl—Cs_2TiCl_6—H_2O$	1101
Hydrogen Chloride–Cesium Antimony Chloride–Water	$HCl—3CsCl \cdot 2SbCl_3—H_2O$	1102
Hydrogen Chloride–Cesium Antimony Chloride–Water	$HCl—CsCl \cdot SbCl_3—H_2O$	1103
Hydrogen Chloride–Cesium Hexachlororhenate–Water	$HCl—CsReCl_6—H_2O$	1104
Hydrogen Chloride–Cesium Hexachlorouranate–Water	$HCl—CsUCl_6—H_2O$	1105
Hydrogen Chloride–Cesium Uranyl Chloride–Water	$HCl—Cs_2UO_2Cl_4—H_2O$	1106
Hydrogen Chloride–Cesium Hexachloroplutonate–Water	$HCl—CsPuCl_6—H_2O$	1107
Hydrogen Chloride–Cesium Dichloroiodate–Water	$HCl—CsCl_2I—H_2O$	1108
Hydrogen Chloride–Barium Oxide–Water	$HCl—BaO—H_2O$	1109
Hydrogen Chloride–Barium Fluoride–Water	$HCl—BaF_2—H_2O$	1110
Hydrogen Chloride–Barium Sulfate–Water	$HCl—BaSO_4—H_2O$	1111—1113
Hydrogen Chloride–Barium Chloride–Water	$HCl—BaCl_2—H_2O$	1114—1116
Hydrogen Chloride–Barium Iodate–Water	$HCl—Ba(IO_3)_2—H_2O$	1117
Hydrogen Chloride–Lanthanum Pyrophosphate–Water	$HCl—La_4(P_2O_7)_3—H_2O$	1118

System name	Formulae of components	Table number
Hydrogen Chloride–Lanthanum Arsenate–Water	$HCl—LaAsO_4—H_2O$	523
Hydrogen Chloride–Cerium (IV) Phosphate–Water	$HCl—Ce_3(PO_4)_2—H_2O$	1119
Hydrogen Chloride–Cerium Pyrophosphate–Water	$HCl—CeP_2O_7—H_2O$	1120, 1121
Hydrogen Chloride–Praseodymium Arsenate–Water	$HCl—PrAsO_4—H_2O$	523
Hydrogen Chloride–Neodymium Arsenate–Water	$HCl—NdAsO_4—H_2O$	523
Hydrogen Chloride–Samarium Arsenate–Water	$HCl—SmAsO_4—H_2O$	523
Hydrogen Chloride–Gadolinium Arsenate–Water	$HCl—GdAsO_4—H_2O$	523
Hydrogen Chloride–Ytterbium Pyrophosphate–Water	$HCl—YbP_2O_7—H_2O$	1122
Hydrogen Chloride–Ytterbium Chloride–Water	$HCl—YbCl_3 \cdot 6H_2O—H_2O$	1123
Hydrogen Chloride–Ytterbium Arsenate–Water	$HCl—IbAsO_4—H_2O$	523
Hydrogen Chloride–Hafrium Oxide Dihydrogen Orthophosphate –Water	$HCl—HfO(H_2PO_4)—H_2O$	1124
Hydrogen Chloride–Hafrium Chloride Oxide–Water	$HCl—HfOCl_2—H_2O$	1125, 1126
Hydrogen Chloride–Tantalum Pentoxide–Water	$HCl—Ta_2O_5—H_2O$	1127
Hydrogen Chloride–Tungsten Trioxide–Water	$HCl—WO_3—H_2O$	1128
Hydrogen Chloride–Mercury (II) Oxide (yellow)–Water	$HCl—HgO—H_2O$	1129
Hydrogen Chloride–Mercury (II) Chloride Hydroxide–Water	$HCl—Hg(OH)Cl—H_2O$	1130
Hydrogen Chloride–Mercury (I) Chloride–Water	$HCl—HgCl—H_2O$	1131
Hydrogen Chloride–Mercury (II) Chloride–Water	$HCl—HgCl_2—H_2O$	1132, 1133
Hydrogen Chloride–Thallium (I) Chloride–Water	$HCl—TlCl—H_2O$	1134, 1135
Hydrogen Chloride–Thallium (III) Selenite–Water	$HCl—Tl_2(SeO_3)_3—H_2O$	899
Hydrogen Chloride–Lead (II) Sulfate –Water	$HCl—PbSO_4—H_2O$	1136, 1137
Hydrogen Chloride–Lead (II) Chloride–Water	$HCl—PbCl_2—H_2O$	1138—1145
Hydrogen Chloride–Lead (II) Chromate–Water	$HCl—PbCrO_4—H_2O$	1146
Hydrogen Chloride–Lead (II) Selenite–Water	$HCl—PbSeO_3—H_2O$	466
Hydrogen Chloride–Bismuth Oxide–Water	$HCl—Bi_2O_3—H_2O$	1147—1150
Hydrogen Chloride–Bismuth Chloride–Water	$HCl—BiCl_2—H_2O$	1151

System name	Formulae of components	Table number
Hydrogen Chloride–Thorium Dioxide –Water	$HCl—ThO_2—H_2O$	1152
Hydrogen Chloride–Thorium Hypophosphate–Water	$HCl—ThP_2O_6 \cdot 11H_2O—H_2O$	1153
Hydrogen Chloride–Thorium Pyrophosphate–Water	$HCl—ThP_2O_7—H_2O$	1154
Hydrogen Chloride–Thorium Sulfate –Water	$HCl—Th(SO_4)_2—H_2O$	1155
Hydrogen Chloride–Uranium Tetrafluoride–Water	$HCl—UF_4—H_2O$	917
Hydrogen Chloride–Uranium Sulfate –Water	$HCl—U(SO_4)_2 \cdot 4H_2O—H_2O$	1156
Hydrogen Chloride–Beryllium Hydroxide–Water	$HClO_4—Be(OH)_2—H_2O$	1157
Perchloric Acid–Carbon Dioxide–Water	$HClO_4—CO_2—H_2O$	1158
Perchloric Acid–Sodium Perchlorate –Silver Sulfate–Water	$HClO_4—NaClO_4—Ag_2SO_4—H_2O$	1159
Perchloric Acid–Sodium Molybdate –Lead (II) Molybdate–Water	$HClO_4—Na_2MoO_4—PbMoO_4—H_2O$	1160
Perchloric Acid–Aluminum Hydroxide–Water	$HClO_4—Al(OH)_3—H_2O$	1161
Perchloric Acid–Calcium Perchlorate –Water	$HClO_4—Ca(ClO_4)_2—H_2O$	1162a
Perchloric Acid–Vanadium Pentoxide –Water	$HClO_4—V_2O_5—H_2O$	1162б
Perchloric Acid–Iron (III) Perchlorate–Water	$HClO_4—Fe(ClO_4)_3—H_2O$	1163
Perchloric Acid–Germanium Dioxide –Water	$HClO_4—GeO_2—H_2O$	1164
Perchloric Acid–Germanium Hydroxide–Water	$HClO_4—Ge(OH)_2—H_2O$	1165
Perchloric Acid–Arsenic Trioxide–Water	$HClO_4—As_2O_3—H_2O$	1270
Perchloric Acid–Silver Perchlorate–Water	$HClO_4—AgClO_4—H_2O$	1166
Perchloric Acid–Cadmium Perchlorate–Water	$HClO_4—Cd(ClO_4)_2—H_2O$	1167
Perchloric Acid–Indium Hydroxide –Water	$HClO_4—In(OH)_3—H_2O$	1168
Perchloric Acid–Tin (II) Oxide–Water	$HClO_4—SnO—H_2O$	1169
Perchloric Acid–Cerium (II) Perchlorate–Water	$HClO_4—Ce(ClO_4)_2—H_2O$	1170
Perchloric Acid–Mercury (I) Perchlorate–Water	$HClO_4—Hg_2(ClO_4)_2—H_2O$	1171
Perchloric Acid–Mercury (II) Bromate Hydroxide–Water	$HClO_4—HgOHBrO_3—H_2O$	1172
Perchloric Acid–Mercury (II) Bromate–Water	$HClO_4—Hg(BrO_3)_2—H_2O$	1173
Perchloric Acid–Thallium (III) Hydroxide–Water	$HClO_4—Tl(OH)_3—H_2O$	1174

System name	Formulae of components	Table number
Perchloric Acid–Lead (II) Sulfate –Water	$HClO_4$—$PbSO_4$—H_2O	1175
Perchloric Acid–Lead (II) Sulfate –Lead (II) Perchlorate–Water	$HClO_4$—$PbSO_4$—$Pb(ClO_4)_2$—H_2O	1176
Perchloric Acid–Lead (II) Chloride–Water	$HClO_4$—$PbCl_2$—H_2O	1177
Perchloric Acid–Lead (II) Perchlorate –Lead (II) Molybdate–Water	$HClO_4$—$Pb(ClO_4)_2$—$PbMoO_4$—H_2O	1178
Perchloric Acid–Lead (II) Chromate –Lead (I) Perchlorate–Water	$HClO_4$—$PbCrO_4$—$PbClO_4$—H_2O	1179
Perchloric Acid–Lead (II) Molybdate –Water	$HClO_4$—$PbMoO_4$—H_2O	1180
Perchloric Acid–Thorium Fluoride –Water	$HClO_4$—ThF_4—H_2O	1181
Perchloric Acid–Uranium Trioxide–Water	$HClO_4$—$UO_3 \cdot H_2O$—H_2O	1182
Perchloric Acid–Uranium Tetroxide Hydrate–Water	$HClO_4$—$UO_4 \cdot xH_2O$—H_2O	1183
Perchloric Acid–Uraninite–Water	$HClO_4$—$U_3O_8 \cdot xH_2O$—H_2O	1184
Perchloric Acid–Uranium Hydroxide –Water	$HClO_4$—$U(OH)_4$—H_2O	1185
Chromic Acid–Calcium Sulfate–Water	H_2CrO_4—$CaSO_4$—H_2O	1186
Arsine–Sodium Hydroxide–Water	AsH_3—$NaOH$—H_2O	1187
Arsine–Sodium Dithionite–Water	AsH_3—$Na_2S_2O_4$—H_2O	1188
Arsine–Sodium Chloride–Water	AsH_3—$NaCl$—H_2O	1189
Arsenious Acid–Ammonium Nitrate –Water	H_3AsO_3—NH_4NO_3—H_2O	1190
Arsenious Acid–Ammonium Sulfate –Water	H_3AsO_3—$(NH_4)_2SO_4$—[illegible]	1191
Arsenious Acid–Sodium Nitrate–Water	H_3AsO_3—$NaNO_3$—H_2O	1192
Arsenious Acid–Sodium Sulfate–Water	H_3AsO_3—Na_2SO_4—H_2O	1193
Arsenious Acid–Sodium Chloride–Water	H_3AsO_3—$NaCl$—H_2O	1194
Arsenious Acid–Magnesium Nitrate –Water	H_3AsO_3—$Mg(NO_3)_2$—H_2O	1195
Arsenious Acid–Magnesium Sulfate –Water	H_3AsO_3—$MgSO_4$—H_2O	1196
Arsenious Acid–Magnesium Bromide –Water	H_3AsO_3—$MgBr_2$—H_2O	1197
Arsenious Acid–Potassium Nitrate–Water	H_3AsO_3—KNO_3—H_2O	1198
Arsenious Acid–Potassium Fluoride –Water	H_3AsO_3—KF—H_2O	1199
Arsenious Acid–Potassium Sulfate –Water	H_3AsO_3—K_2SO_4—H_2O	1200

System name	Formulae of components	Table number
Arsenious Acid–Calcium Sulfate–Water	H_3AsO_4—$CaSO_4$—H_2O	1201
Arsenious Acid–Manganese (II) Arsenate–Water	H_3AsO_4—$Mn_3(AsO_4)_2$—H_2O	1202
Arsenious Acid–Iron (III) Oxide–Iron (III) Arsenate–Arsenic Pentoxide–Water	H_3AsO_4—Fe_2O_3—$FeAsO_4$—As_2O_5—H_2O	1203
Arsenious Acid–Arsenic Trioxide–Water	H_3AsO_4—As_2O_3—H_2O	1270
Hydrogen Selenide–Hydrogen Iodide–Water	H_2Se—HI—H_2O	1204
Selenic Acid–Ammonium Selenate–Water	H_2SeO_4—$(NH_4)_2SeO_4$—H_2O	1205
Selenic Acid–Magnesium Selenate–Water	H_2SeO_4—$MgSeO_4$—H_2O	1206
Selenic Acid–Chromium Trioxide–Water	H_2SeO_4—CrO_3—H_2O	1207
Selenic Acid–Nickel (II) Selenate–Water	H_2SeO_4—$NiSeO_4$—H_2O	1208
Selenic Acid–Copper (II) Selenate–Water	H_2SeO_4—$CuSeO_4$—H_2O	1209
Selenic Acid–Silver Selenate–Water	H_2SeO_4—Ag_2SeO_4—H_2O	1210
Selenic Acid–Lanthanum Selenate–Water	H_2SeO_4—$La_2(SeO_4)_3$—H_2O	1211
Selenic Acid–Neodymium Selenate–Water	H_2SeO_4—$Nd_2(SeO_4)_3$—H_2O	1212
Selenic Acid–Tantalum Pentoxide–Water	H_2SeO_4—Ta_2O_5—H_2O	1213
Selenic Acid–Uranium Sulfate–Water	H_2SeO_4—$U(SO_4)_2 \cdot 4H_2O$—H_2O	1214
Hydrogen Bromide–Lithium Bromide–Water	HBr—LiBr—H_2O	1215
Hydrogen Bromide–Ammonium Bromide–Water	HBr—NH_4Br—H_2O	1216
Hydrogen Bromide–Sodium Bromide–Water	HBr—NaBr—H_2O	1217—1220
Hydrogen Bromide–Magnesium Bromide–Water	HBr—$MgBr_2$—H_2O	1221
Hydrogen Bromide–Potassium Chloride–Water	HBr—KCl—H_2O	1222
Hydrogen Bromide–Potassium Bromide–Water	HBr—KBr—H_2O	1223, 1224
Hydrogen Bromide–Tellurium Potassium Bromide–Water	HBr—$TeBr_4 \cdot 2KBr$—H_2O	1225
Hydrogen Bromide–Calcium Bromide–Water	HBr—$CaBr_2$—H_2O	1226
Hydrogen Bromide–Copper (II) Bromide–Water	HBr—$CuBr_2$—H_2O	1227
Hydrogen Bromide–Germanium Dioxide–Water	HBr—GeO_2—H_2O	1228
Hydrogen Bromide–Rubidium Tellurium Bromide–Water	HBr—$TeBr_4 \cdot 2RbBr$—H_2O	1229

System name	Formulae of components	Table number
Hydrogen Bromide–Strontium Chloride–Water	$HBr—SrCl_2—H_2O$	1230
Hydrogen Bromide–Strontium Bromide–Water	$HBr—SrBr_2—H_2O$	1231
Hydrogen Bromide–Zirconium Oxide–Water	$HBr—ZrO_2—H_2O$	1232
Hydrogen Bromide–Silver Bromide–Water	$HBr—AgBr—H_2O$	1233, 5416
Hydrogen Bromide–Indium Bromide–Water	$HBr—InBr_3—H_2O$	1234
Hydrogen Bromide–Tellurium Dioxide–Water	$HBr—TeO_2—H_2O$	1235
Hydrogen Bromide–Tellurium Cesium Bromide–Water	$HBr—TeBr_4 \cdot 2CsBr—H_2O$	1236
Hydrogen Bromide–Barium Bromide–Water	$HBr—BaBr_2—H_2O$	1237—1239a
Hydrogen Bromide–Hafrium Dioxide–Water	$HBr—HfO_2—H_2O$	1239б
Hydrogen Bromide–Lead (II) Bromide–Water	$HBr—PbBr_2—H_2O$	1240, 1241
Antimonic Acid–Sodium Hydroxide–Sodium Fluoride–Water	$Sb(OH)_3—NaOH—NaF—H_2O$	1242
Hydrogen Iodide–Sodium Iodide–Water	$HI—NaI—H_2O$	1243, 1244
Hydrogen Iodide–Magnesium Iodide–Water	$HI—MgI_2—H_2O$	1945
Hydrogen Iodide–Potassium Iodide–Water	$HI—KI—H_2O$	1246, 1247
Hydrogen Iodide–Germanium Dioxide–Water	$HI—GeO_2—H_2O$	1248
Hydrogen Iodide–Strontium Chloride–Water	$HI—SrCl_2—H_2O$	1249
Hydrogen Iodide–Silver Iodide–Water	$HI—AgI—H_2O$	1250, 1251
Hydrogen Iodide–Indium Iodide–Water	$HI—InI_3—H_2O$	1252
Hydrogen Iodide–Tin (II) Iodide–Water	$HI—SnI_2—H_2O$	1253—1261
Iodic Acid–Ammonium Iodate–Water	$HIO_3—NH_4IO_3—H_2O$	1262
Iodic Acid–Sodium Iodate–Water	$HIO_3—NaIO_3—H_2O$	1263
Iodic Acid–Potassium Iodate–Water	$HIO_3—KIO_3—H_2O$	1264—1266
Iodic Acid–Dinitrogen Oxide–Water	$HIO_4—N_2O—H_2O$	1267
Iodic Acid–Sodium Periodate–Water	$HIO_4—NaIO_4—H_2O$	1268
Iodic Acid–Potassium Periodate–Water	$HIO_4—KIO_4—H_2O$	1269
Lithium Oxide–Boron Oxide–Water	$Li_2O—B_2O_3—H_2O$	1271—1273

System name	Formulae of components	Table number
Lithium Oxide–Phosphorus Pentoxide–Water	$Li_2O—P_2O_5—H_2O$	1274
Lithium Oxide–Arsenic Trioxide–Water	$Li_2O—As_2O_3—H_2O$	1275
Lithium Hydroxide–Lithium Carbonate–Water	$LiOH—Li_2CO_3—H_2O$	1276—1278
Lithium Hydroxide–Lithium Carbonate–Sodium Hydroxide–Sodium Carbonate–Water	$LiOH—Li_2CO_3—NaOH—Na_2CO_3—H_2O$	1279
Lithium Hydroxide–Lithium Sulfate–Water	$LiOH—Li_2SO_4—H_2O$	1280
Lithium Hydroxide–Lithium Tetrathioantimonate–Water	$LiOH—Li_3SbS_4—H_2O$	1281
Lithium Hydroxide–Ammonia–Water	$LiOH—NH_3—H_2O$	1282
Lithium Hydroxide–Aluminum Oxide–Water	$LiOH—Al_2O_3—H_2O$	1729
Lithium Hydroxide–Calcium Hydroxide–Water	$LiOH—Ca(OH)_2—H_2O$	1283
Lithium Hydroxide–Mercury (II) Oxide–Water	$LiOH—HgO—H_2O$	1284
Lithium Tetraborate–Lithium Chloride–Water	$Li_2B_4O_7—LiCl—H_2O$	1285
Lithium Tetraborate–Lithium Chloride–Potassium Tetraborate–Potassium Chloride–Water	$Li_2B_4O_7—LiCl—K_2B_4O_7—KCl—H_2O$	1286
Lithium Tetraborate–Potassium Tetraborate–Water	$Li_2B_4O_7—K_2B_4O_7—H_2O$	1287
Lithium Carbonate–Lithium Nitrate–Sodium Carbonate–Sodium Nitrate–Water	$Li_2CO_3—LiNO_3—Na_2CO_3—NaNO_3—H_2O$	1288
Lithium Carbonate–Lithium Sulfate–Water	$Li_2CO_3—Li_2SO_4—H_2O$	1289—1293
Lithium Carbonate–Lithium Sulfate–Sodium Carbonate–Sodium Sulfate–Water	$Li_2CO_3—Li_2SO_4—Na_2CO_3—Na_2SO_4—H_2O$	1294, 1295
Lithium Carbonate–Lithium Chloride–Water	$Li_2CO_3—LiCl—H_2O$	1296, 1297
Lithium Carbonate–Carbon Dioxide–Water	$Li_2CO_3—CO_2—H_2O$	1298
Lithium Carbonate–Ammonia–Water	$Li_2CO_3—NH_3—H_2O$	1299
Lithium Carbonate–Ammonium Sulfate–Water	$Li_2CO_3—(NH_4)_2SO_4—H_2O$	1300
Lithium Carbonate–Ammonium Chloride–Water	$Li_2CO_3—NH_4Cl—H_2O$	1301
Lithium Carbonate–Sodium Carbonate–Water	$Li_2CO_3—Na_2CO_3—H_2O$	1302—1306
Lithium Carbonate–Sodium Sulfate–Water	$Li_2CO_3—Na_2SO_4—H_2O$	1307
Lithium Carbonate–Sodium Chloride–Water	$Li_2CO_3—NaCl—H_2O$	1308
Lithium Carbonate–Potassium Carbonate–Water	$Li_2CO_3—K_2CO_3—H_2O$	1309

System name	Formulae of components	Table number
Lithium Carbonate–Potassium Nitrate–Water	$Li_2CO_3—KNO_3—H_2O$	1310
Lithium Carbonate–Potassium Sulfate–Water	$Li_2CO_3—K_2SO_4—H_2O$	1311, 1312
Lithium Carbonate–Potassium Chloride–Water	$Li_2CO_3—KCl—H_2O$	1313
Lithium Carbonate–Potassium Chlorate–Water	$Li_2CO_3—KClO_3—H_2O$	1314
Lithium Carbonate–Calcium Carbonate–Water	$Li_2CO_3—CaCO_3—H_2O$	1315
Lithium Thiocyanate–Thallium (I) Thiocyanate–Water	$LiCNS—TlCNS—H_2O$	1316
Lithium Tetracyanoplatinate (II)–Potassium Tetracyanoplatinate (II)–Water	$Li_2Pt(CN)_4—K_2Pt(CN)_4—H_2O$	1317
Lithium Nitrite–Thallium (I) Nitrite–Water	$LiNO_2—TlNO_2—H_2O$	1318
Lithium Nitrate–Lithium Sulfate–Water	$LiNO_3—Li_2SO_4—H_2O$	1319, 1320
Lithium Nitrate–Ammonium Nitrate–Water	$LiNO_3—NH_4NO_3—H_2O$	1321
Lithium Nitrate–Sodium Nitrate–Water	$LiNO_3—NaNO_3—H_2O$	1322
Lithium Nitrate–Silver Bromate–Water	$LiNO_3—AgBrO_3—H_2O$	1323
Lithium Nitrate–Thallium (I) Chloride–Water	$LiNO_3—TlCl—H_2O$	1324
Lithium Nitrate–Lead (II) Nitrate–Water	$LiNO_3—Pb(NO_3)_2—H_2O$	1325
Lithium Nitrate–Lead (II) Bromate–Water	$LiNO_3—Pb(BrO_3)_2—H_2O$	1326
Lithium Nitrate–Uranyl Nitrate–Water	$LiNO_3—UO_2(NO_3)_2—H_2O$	1327
Lithium Fluoride–Aluminum Fluoride–Water	$LiF—AlF_3—H_2O$	1328
Lithium Sulfate–Lithium Chloride–Water	$LiCl—Li_2SO_4—H_2O$	1329—1334
Lithium Sulfate–Lithium Chloride–Sodium Sulfate–Sodium Chloride–Water	$Li_2SO_4—LiCl—Na_2SO_4—NaCl—H_2O$	1335, 1336
Lithium Sulfate–Lithium Chloride–Sodium Sulfate–Sodium Chloride–Potassium Sulfate–Potassium Chloride–Water	$Li_2SO_4—LiCl—Na_2SO_4—NaCl—K_2SO_4—KCl—H_2O$	1337
Lithium Sulfate–Lithium Chloride–Potassium Sulfate–Potassium Chloride–Water	$Li_2SO_4—LiCl—K_2SO_4—KCl—H_2O$	1338, 1339
Lithium Sulfate–Berryllium Sulfate–Water	$Li_2SO_4—BeSO_4—H_2O$	1340
Lithium Sulfate–Dinitrogen Oxide–Water	$Li_2SO_4—N_2O—H_2O$	1341
Lithium Sulfate–Ammonia–Water	$Li_2SO_4—NH_3—H_2O$	1342
Lithium Sulfate–Ammonium Sulfate–Water	$Li_2SO_4—(NH_4)_2SO_4—H_2O$	1343—1347

System name	Formulae of components	Table number
Lithium Sulfate–Sodium Nitrate–Lead (II) Sulfate–Water	Li_2SO_4—$NaNO_3$—$PbSO_4$—H_2O	1348
Lithium Sulfate–Sodium Sulfate–Water	Li_2SO_4—Na_2SO_4—H_2O	1349—1356
Lithium Sulfate–Sodium Sulfate–Magnesium Sulfate–Water	Li_2SO_4—Na_2SO_4—$MgSO_4$—H_2O	1357, 1358
Lithium Sulfate–Sodium Sulfate–Potassium Sulfate–Water	Li_2SO_4—Na_2SO_4—K_2SO_4—H_2O	1359—1362
Lithium Sulfate–Magnesium Nitrate–Lead (II) Sulfate–Water	Li_2SO_4—$Mg(NO_3)_2$—$PbSO_4$—H_2O	1363
Lithium Sulfate–Magnesium Sulfate–Water	Li_2SO_4—$MgSO_4$—H_2O	1364—1367
Lithium Sulfate–Magnesium Sulfate–Potassium Sulfate–Water	Li_2SO_4—$MgSO_4$—K_2SO_4—H_2O	1368
Lithium Sulfate–Magnesium Sulfate-Rubidium Sulfate–Water	Li_2SO_4—$MgSO_4$—Rb_2SO_4—H_2O	1369
Lithium Sulfate–Aluminum Nitrate–Lead (II) Sulfate–Water	Li_2SO_4—$Al(NO_3)_3$—$PbSO_4$—H_2O	1370
Lithium Sulfate–Aluminum Sulfate–Water	Li_2SO_4—$Al_2(SO_4)_3$—H_2O	1371—1374
Lithium Sulfate–Potassium Sulfate–Water	Li_2SO_4—K_2SO_4—H_2O	1375—1380
Lithium Sulfate–Iron (II) Sulfate–Water	Li_2SO_4—$FeSO_4$—H_2O	1381, 1382
Lithium Sulfate–Cobalt (II) Sulfate–Water	Li_2SO_4—$CoSO_4$—H_2O	1383
Lithium Sulfate–Nickel (II) Sulfate–Water	Li_2SO_4—$NiSO_4$—H_2O	1384
Lithium Sulfate–Copper (II) Sulfate–Water	Li_2SO_4—$CuSO_4$—H_2O	1385, 1386
Lithium Sulfate–Zinc Sulfate–Water	Li_2SO_4—$ZnSO_4$—H_2O	1387, 1388
Lithium Sulfate–Rubidium Sulfate–Water	Li_2SO_4—Rb_2SO_4—H_2O	1389, 1390
Lithium Sulfate–Silver Sulfate–Water	Li_2SO_4—Ag_2SO_4—H_2O	1391
Lithium Sulfate–Silver Bromate–Water	Li_2SO_4—$AgBrO_3$—H_2O	1392
Lithium Sulfate–Cadmium Sulfate–Water	Li_2SO_4—$CdSO_4$—H_2O	1393
Lithium Sulfate–Cesium Sulfate–Water	Li_2SO_4—Cs_2SO_4—H_2O	1394
Lithium Sulfate–Lead (II) Sulfate–Thorium Nitrate–Water	Li_2SO_4—$PbSO_4$—$Th(NO_3)_4$—H_2O	1395
Lithium Sulfate–Bismuth Sulfate–Water	Li_2SO_4—$Bi_2(SO_4)_3$—H_2O	1396
Lithium Sulfate–Thorium Sulfate–Water	Li_2SO_4—$Th(SO_4)_2$—H_2O	1397
Lithium Chloride–Lithium Chlorate–Water	$LiCl$—$LiClO_3$—H_2O	1398
Lithium Chloride–Lithium Perchlorate–Water	$LiCl$—$LiClO_4$—H_2O	1399

System name	Formulae of components	Table number
Lithium Chloride–Beryllium Chloride–Water	$LiCl—BeCl_2—H_2O$	1400, 1401
Lithium Chloride–Carbon Dioxide–Water	$Li_2Cl—CO_2—H_2O$	1951, 1952
Lithium Chloride–Dinitrogen Oxide–Water	$LiCl—N_2O—H_2O$	1402, 1403
Lithium Chloride–Ammonia–Water	$LiCl—NH_3—H_2O$	1404—1406
Lithium Chloride–Ammonium Chloride–Water	$LiCl—NH_4Cl—H_2O$	1407, 1408
Lithium Chloride–Sodium Chloride–Water	$LiCl—NaCl—H_2O$	1409—1413
Lithium Chloride–Sodium Chloride–Magnesium Chloride–Water	$LiCl—NaCl—MgCl_2—H_2O$	1414
Lithium Chloride–Sodium Chloride–Potassium Chloride–Water	$LiCl—NaCl—KCl—H_2O$	1415—1417
Lithium Chloride–Magnesium Hydroxide–Water	$LiCl—Mg(OH)_2—H_2O$	1418
Lithium Chloride–Magnesium Chloride–Water	$LiCl—MgCl_2—H_2O$	1419
Lithium Chloride–Potassium Chloride–Water	$LiCl—KCl—H_2O$	1420—1425
Lithium Chloride–Potassium Chloride–Cadmium Chloride–Water	$LiCl—KCl—CdCl_2—H_2O$	4897
Lithium Chloride–Potassium Dichromate–Water	$LiCl—K_2Cr_2O_7—H_2O$	1426
Lithium Chloride–Potassium Permaganate–Water	$LiCl—KMnO_4—H_2O$	1427
Lithium Chloride–Calcium Hydroxide–Water	$LiCl—Ca(OH)_2—H_2O$	1428
Lithium Chloride–Calcium Fluoride–Water	$LiCl—CaF_2—H_2O$	1429
Lithium Chloride–Calcium Sulfate (Gypsum)–Water	$LiCl—CaSO_4\cdot 2H_2O—H_2O$	1430
Lithium Chloride–Calcium Chloride–Water	$LiCl—CaCl_2—H_2O$	1431—1433
Lithium Chloride–Calcium Molybdate–Calcium Tungstate–Water	$LiCl—CaMoO_4—CaWO_4—H_2O$	1434
Lithium Chloride–Calcium Iodate–Water	$LiCl—Ca(IO_3)_2—H_2O$	1435
Lithium Chloride–Calcium Tungstate–Water	$LiCl—CaWO_4—H_2O$	1436
Lithium Chloride–Manganese (II) Chloride–Water	$LiCl—MnCl_2—H_2O$	1437
Lithium Chloride–Cobalt (II) Chloride–Water	$LiCl—CoCl_2—H_2O$	1438, 1439
Lithium Chloride–Nickel (II) Chloride–Water	$LiCl—NiCl_2—H_2O$	1440
Lithium Chloride–Copper (II) Sulfate–Water	$LiCl—CuSO_4—H_2O$	1441
Lithium Chloride–Copper (II) Chloride–Water	$LiCl—CuCl_2—H_2O$	1442—1445
Lithium Chloride–Zinc Chloride–Water	$LiCl—ZnCl_2—H_2O$	1446, 1447

System name	Formulae of components	Table number
Lithium Chloride–Arsenic Trioxide –Water	$LiCl—As_2O_3—H_2O$	1448
Lithium Chloride–Rubidium Chloride–Water	$LiCl—RbCl—H_2O$	1449
Lithium Chloride–Strontium Carbonate–Water	$LiCl—SrCO_3—H_2O$	1450
Lithium Chloride–Strontium Sulfate –Water	$LiCl—SrSO_4—H_2O$	1451
Lithium Chloride–Strontium Chloride–Water	$LiCl—SrCl_2—H_2O$	1452—1454
Lithium Chloride–Silver Chloride–Water	$LiCl—AgCl—H_2O$	1455, 1456, 5412
Lithium Chloride–Cadrium Chloride –Water	$LiCl—CdCl_2—H_2O$	1457
Lithium Chloride–Indium Chloride–Water	$LiCl—InCl_3—H_2O$	1458
Lithium Chloride–Cesium Chloride –Water	$LiCl—CsCl—H_2O$	1459
Lithium Chloride–Barium Hydroxide–Water	$LiCl—Ba(OH)_2—H_2O$	1460
Lithium Chloride–Barium Carbonate–Water	$LiCl—BaCO_3—H_2O$	1461, 1462
Lithium Chloride–Barium Sulfate –Water	$LiCl—BaSO_4—H_2O$	1463
Lithium Chloride–Barium Chloride –Water	$LiCl—BaCl_2—H_2O$	1464, 1465
Lithium Chloride–Mercury (II) Oxide–Water	$LiCl—HgO—H_2O$	1466
Lithium Chloride–Mercury (II) Chloride–Water	$LiCl—HgCl_2—H_2O$	1467—1471
Lithium Chloride–Lead (II) Chloride –Water	$LiCl—PbCl_2—H_2O$	1472—1474
Lithium Selenate–Lead (II) Selenate –Water	$Li_2SeO_4—PbSeO_4—H_2O$	1475
Lithium Bromide–Lithium Iodide–Water	$LiBr— LiI —H_2O$	1476
Lithium Bromide–Ammonia–Water	$LiBr—NH_3—H_2O$	1477
Lithium Bromide–Magnesium Bromide–Water	$LiBr—MgBr_2—H_2O$	1478
Lithium Bromide–Calcium Bromide –Water	$LiBr—CaBr_2—H_2O$	1479, 1480
Lithium Bromide–Arsenic Trioxide–Water	$LiBr—As_2O_3—H_2O$	1481
Lithium Bromide–Indium Bromide –Water	$LiBr—InBr_3—H_2O$	1482
Lithium Bromide–Barium Bromide–Water	$LiBr—BaBr_2—H_2O$	1483, 1484
Lithium Bromide–Lead (II) Bromide–Water	$LiBr—PbBr_2—H_2O$	1485
Lithium Iodide–Ammonia–Water	$LiI —NH_3—H_2O$	1486
Lithium Iodide–Sodium Iodide–Water	$LiI —NaI —H_2O$	1487

System name	Formulae of components	Table number
Beryllium Sulfate–Iron (II) Sulfate –Water	$BeSO_4$—$FeSO_4$—H_2O	1522
Beryllium Sulfate–Copper (II) Sulfate –Water	$BeSO_4$—$CuSO_4$—H_2O	1523
Beryllium Sulfate–Zinc Sulfate– Water	$BeSO_4$—$ZnSO_4$—H_2O	1524, 1525
Beryllium Sulfate–Silver Sulfate– Water	$BeSO_4$—Ag_2SO_4—H_2O	1526
Beryllium Chloride–Ammonium Chloride–Water	$BeCl_2$—NH_4Cl—H_2O	1527
Beryllium Chloride–Sodium Chloride –Water	$BeCl_2$—$NaCl$—H_2O	1528
Beryllium Chloride–Magnesium Chloride–Water	$BeCl_2$—$MgCl_2$—H_2O	1529
Beryllium Chloride–Potassium Chloride–Water	$BeCl_2$—KCl—H_2O	1530
Beryllium Chloride–Calcium Chloride –Water	$BeCl_2$—$CaCl_2$—H_2O	1531
Beryllium Chloride–Copper (II) Chloride–Water	$BeCl_2$—$CuCl_2$—H_2O	1532
Beryllium Chloride–Zinc Chloride –Water	$BeCl_2$—$ZnCl_2$—H_2O	1533
Beryllium Chloride–Rubidium Chloride–Water	$BeCl_2$—$RbCl$—H_2O	1534
Beryllium Chloride–Strontium Chloride–Water	$BeCl_2$—$SrCl_2$—H_2O	1535
Beryllium Chloride–Cadmium Chloride–Water	$BeCl_2$—$CdCl_2$—H_2O	1536
Beryllium Chloride–Cesium Chloride –Water	$BeCl_2$—$CsCl$—H_2O	1537
Beryllium Chloride–Barium Chloride –Water	$BeCl_2$—$BaCl_2$—H_2O	1538, 1539
Beryllium Chloride–Mercury (II) Chloride–Water	$BeCl_2$—$HgCl_2$—H_2O	1540
Boron Oxide–Carbon Dioxide– Ammonia–Water	B_2O_3—CO_2—NH_3—H_2O	1541
Boron Oxide–Carbon Dioxide– Sodium Oxide–Water	B_2O_3—CO_2—Na_2O—H_2O	1542, 1543
Boron Oxide–Ammonium Oxide– Water	B_2O_3—$(NH_4)_2O$—H_2O	1544—1546
Boron Oxide–Sodium Oxide–Water	B_2O_3—Na_2O—H_2O	1547—1554
Boron Oxide–Sodium Oxide– Sodium Sulfate–Water	B_2O_3—Na_2O—Na_2SO_4—H_2O	1555
Boron Oxide–Sodium Oxide– Sodium Chloride–Water	B_2O_3—Na_2O—$NaCl$—H_2O	1556
Boron Oxide–Sodium Oxide– Calcium Oxide–Water	B_2O_3—Na_2O—CaO—H_2O	1557—1559
Boron Oxide–Sodium Chloride– Calcium Oxide–Water	B_2O_3—$NaCl$—CaO—H_2O	1560
Boron Oxide–Magnesium Oxide–Water	B_2O_3—MgO—H_2O	1561—1565
Boron Oxide–Magnesium Oxide– Magnesium Chloride–Water	B_2O_3—MgO—$MgCl_2$—H_2O	1566
Boron Oxide–Magnesium Oxide –Potassium Oxide–Water	B_2O_3—MgO—K_2O—H_2O	1567

System name	Formulae of components	Table number
Boron Oxide–Magnesium Oxide–Calcium Oxide–Water	B_2O_3—MgO—CaO—H_2O	1568
Boron Oxide–Phosphorus Pentoxide –Water	B_2O_3—P_2O_5—H_2O	1569
Boron Oxide–Sulfur Trioxide–Water	B_2O_3—SO_3—H_2O	1570
Boron Oxide–Potassium Oxide–Water	B_2O_3—K_2O—H_2O	1571—1573
Boron Oxide–Potassium Oxide–Potassium Chloride–Water	B_2O_3—K_2O—KCl—H_2O	1574
Boron Oxide–Potassium Metaborate –Water	B_2O_3—KBO_2—H_2O	1575
Boron Oxide–Calcium Oxide–Water	B_2O_3—CaO—H_2O	1576—1580
Boron Oxide–Chromium Trioxide–Water	B_2O_3—CrO_3—H_2O	1581
Boron Oxide–Silver Oxide–Water	B_2O_3—Ag_2O—H_2O	1582
Boron Oxide–Barium Oxide–Water	B_2O_3—BaO—H_2O	1583
Boron Oxide–Cesium Oxide–Water	B_2O_3—Cs_2O—H_2O	1584
Sodium Pentaborate–Sodium Chloride–Water	$Na_2O \cdot 5B_2O_3$—NaCl—H_2O	1585
Ulexite–Calcium Sulfate (Gypsum)–Water	$2CaO \cdot Na_2O \cdot 5B_2O_3 \cdot 16H_2O$—$CaSO_4 \cdot 2H_2O$—$H_2O$	1586
Kaliborite–Calcium Sulfate (Gypsum)–Water	$4MgO \cdot K_2O \cdot 11B_2O_3 \cdot 18H_2O$—$CaSO_4 \cdot 2H_2O$—$H_2O$	1587
Hydroboracite–Sodium Chloride (Halite)–Water	$CaO \cdot MgO \cdot 3B_2O_3 \cdot 6H_2O$—NaCl—$H_2O$	1588
Hydroboracite–Calcium Sulfate (Gypsum)–Water	$CaO \cdot MgO \cdot 3B_2O_3 \cdot 6H_2O$—$CaSO_4 \cdot 2H_2O$—$H_2O$	1589
Inyoite–Calcium Sulfate (Gypsum) –Water	$2CaO \cdot 3B_2O_3 \cdot 13H_2O$—$CaSO_4 \cdot 2H_2O$—$H_2O$	1590
Carbon Dioxide–Ammonia–Water	CO_2—NH_3—H_2O	1592—1596
Carbon Dioxide–Ammonia–Ammonium Sulfate–Water	CO_2—NH_3—$(NH_4)_2SO_4$—H_2O	1597
Carbon Dioxide–Ammonia–Ammonium Sulfate–Sodium Sulfate –Water	CO_2—NH_3—$(NH_4)_2SO_4$—Na_2SO_4—H_2O	1598, 1599
Carbon Dioxide–Ammonia–Ammonium Chloride–Sodium Chloride–Water	CO_2—NH_3—NH_4Cl—NaCl—H_2O	1600, 1601
Carbon Dioxide–Ammonia–Ammonium Chloride–Sodium Chloride–Potassium Sulfate–Potassium Chloride–Water	CO_2—NH_3—NH_4Cl—NaCl—K_2SO_4—KCl—H_2O	1602
Carbon Dioxide–Ammonia–Ammonium Chloride–Sodium Chloride–Potassium Chloride–Water	CO_2—NH_3—NH_4Cl—NaCl—KCl—H_2O	1603, 1604
Carbon Dioxide–Ammonia–Ammonium Chloride–Potassium Chloride–Water	CO_2—NH_3—NH_4Cl—KCl—H_2O	1605
Carbon Dioxide–Ammonia–Sodium Sulfate–Water	CO_2—NH_3—Na_2SO_4—H_2O	1606, 1607
Carbon Dioxide–Ammonia–Sodium Sulfate–Sodium Chloride–Water	CO_2—NH_3—Na_4SO_4—NaCl—H_2O	1608

System name	Formulae of components	Table number
Carbon Dioxide–Ammonia–Sodium Chloride–Potassium Chloride–Water	CO_2—NH_3—NaCl—KCl—H_2O	1609
Carbon Dioxide–Ammonia–Magnesium Oxide–Water	CO_2—NH_3—MgO—H_2O	1862
Carbon Dioxide–Ammonium Hydrogen Carbonate–Sodium Hydrogen Carbonate–Water	CO_2—NH_4HCO_3—$NaHCO_3$—H_2O	1610
Carbon Dioxide–Ammonium Sulfate–Water	CO_2—$(NH_4)_2SO_4$—H_2O	1611
Carbon Dioxide–Ammonium Sulfate–Sodium Sulfate–Water	CO_2—$(NH_4)_2SO_4$—Na_2SO_4—H_2O	1612
Carbon Dioxide–Ammonium Chloride–Water	CO_2—NH_4Cl—H_2O	1613—1615
Carbon Dioxide–Sodium Oxide–Gallium Oxide–Water	CO_2—Na_2O—Ga_2O_3—H_2O	1616
Carbon Dioxide–Sodium Hydrogen Carbonate–Lead (II) Sulfide–Water	CO_2—$NaHCO_3$—PbS—H_2O	74
Carbon Dioxide–Sodium Carbonate–Sodium Chloride–Magnesium Carbonate–Water	CO_2—Na_2CO_3—NaCl—$MgCO_3$—H_2O	1617
Carbon Dioxide–Sodium Carbonate Magnesium Carbonate–Water	CO_2—Na_2CO_3—$MgCO_3$—H_2O	1618
Carbon Dioxide–Sodium Carbonate–Calcium Carbobate–Water	CO_2—Na_2CO_3—$CaCO_3$—H_2O	1619—1621
Carbon Dioxide–Sodium Nitrate–Water	CO_2—$NaNO_3$—H_2O	1622
Carbon Dioxide–Sodium Phosphate–Water	CO_2—Na_3PO_4—H_2O	1737
Carbon Dioxide–Sodium Sulfate–Water	CO_2—Na_2SO_4—H_2O	1623—1625, 1737
Carbon Dioxide–Sodium Sulfate–Calcium Carbonate–Water	CO_2—Na_2SO_4—$CaCO_3$—H_2O	1626—1628
Carbon Dioxide–Sodium Chloride–Water	CO_2—NaCl—H_2O	1629—1637
Carbon Dioxide–Sodium Chloride–Magnesium Carbonate–Water	CO_2—NaCl—$MgCO_3$—H_2O	1638—1639
Carbon Dioxide–Sodium Chloride–Magnesium Carbonate–Calcium Carbonate–Water	CO_2—NaCl—$MgCO_3$—$CaCO_3$—H_2O	1640
Carbon Dioxide–Sodium Chloride–Magnesium Sulfate–Calcium Carbonate–Calcium Sulfate–Water	CO_2—NaCl—$MgSO_4$—$CaCO_3$—$CaSO_4$—H_2O	1641
Carbon Dioxide–Sodium Chloride–Calcium Carbonate–Water	CO_2—NaCl—$CaCO_3$—H_2O	1642—1651
Carbon Dioxide–Sodium Chloride–Calcium Carbonate–Calcium Sulfate–Water	CO_2—NaCl—$CaCO_3$—$CaSO_4$—H_2O	1652—1653
Carbon Dioxide–Sodium Chloride–Calcium Sulfate–Water	CO_2—NaCl—$CaSO_4$—H_2O	1655
Carbon Dioxide–Sodium Chloride–Lead (II) Sulfide–Water	CO_2—NaCl—PbS—H_2O	76

System name	Formulae of components	Table number
Carbon Dioxide–Sodium Bromide–Water	CO_2—$NaBr$—H_2O	1656
Carbon Dioxide–Magnesium Oxide–Water	CO_2—MgO—H_2O	1657
Carbon Dioxide–Magnesium Carbonate–Water	CO_2—$MgCO_3$—H_2O	1658—1667
Carbon Dioxide–Magnesium Carbonate–Potassium Hydrogen Carbonate–Water	CO_2—$MgCO_3$—$KHCO_3$—H_2O	1668, 1669
Carbon Dioxide–Magnesium Carbonate–Calcium Carbonate–Water	CO_2—$MgCO_3$—$CaCO_3$—H_2O	1670—1673
Carbon Dioxide–Magnesium Nitrate–Water	CO_2—$Mg(NO_3)_2$—H_2O	1674
Carbon Dioxide–Magnesium Sulfate–Water	CO_2—$MgSO_4$—H_2O	1675—1677
Carbon Dioxide–Magnesium Sulfate–Calcium Carbonate–Water	CO_2—$MgSO_4$—$CaCO_3$—H_2O	1678, 1679
Carbon Dioxide–Magnesium Chloride–Water	CO_2—$MgCl_2$—H_2O	1737
Carbon Dioxide–Magnesium Chloride–Calcium Carbonate–Water	CO_2—$MgCl_2$—$CaCO_3$—H_2O	1680, 1681
Carbon Dioxide–Potassium Nitrate–Water	CO_2—KNO_3—H_2O	1682
Carbon Dioxide–Potassium Hydrogen Sulfate–Water	CO_2—$KHSO_4$—H_2O	1683
Carbon Dioxide–Potassium Sulfate–Water	CO_2—K_2SO_4—H_2O	1684
Carbon Dioxide–Potassium Chloride–Water	CO_2—KCl—H_2O	1685—1692
Carbon Dioxide–Potassium Chloride–Calcium Carbonate–Water	CO_2—KCl—$CaCO_3$—H_2O	1693
Carbon Dioxide–Potassium Bromide–Water	CO_2—KBr—H_2O	1694—1696
Carbon Dioxide–Potassium Iodide–Water	CO_2—KI—H_2O	1697—1699
Carbon Dioxide–Aluminum Sulfate–Water	CO_2—$Al_2(SO_4)_3$—H_2O	1736
Carbon Dioxide–Aluminum Chloride–Water	CO_2—$AlCl_3$—H_2O	1736
Carbon Dioxide–Calcium Hydrogen Carbonate–Calcium Sulfate–Water	CO_2—$Ca(HCO_3)_2$—$CaSO_4$—H_2O	1700
Carbon Dioxide–Calcium Carbonate–Water	CO_2—$CaCO_3$—H_2O	1701—1717
Carbon Dioxide–Calcium Chloride–Water	CO_2—$CaCl_2$—H_2O	1718—1721, 1736
Carbon Dioxide–Iron (III) Hydroxide–Water	CO_2—$Fe(OH)_3$—H_2O	1737
Carbon Dioxide–Iron (II) Hydrogen Carbonate–Water	CO_2—$Fe(HCO_3)_2$—H_2O	1722
Carbon Dioxide–Cobalt (II) Carbonate–Water	CO_2—$CoCO_3$—H_2O	1723
Carbon Dioxide–Copper (I) Oxide–Water	CO_2—Cu_2O—H_2O	1724

System name	Formulae of components	Table number
Carbon Dioxide–Copper (II) Oxide–Water	CO_2—CuO—H_2O	1725
Carbon Dioxide–Copper (II) Carbonate Hydroxide–Water	CO_2—$CuCO_3 \cdot Cu(OH)_2$—H_2O	1726
Carbon Dioxide–Copper (II) Carbonate Hydroxide–Water	CO_2—$2CuCO_3 \cdot Cu(OH)_2$—H_2O	1727
Carbon Dioxide–Zinc Hydrogen Carbonate–Water	CO_2—$Zn(HCO_3)_2$—H_2O	1728
Carbon Dioxide–Zinc Chloride–Water	CO_2—$ZnCl_2$—H_2O	1736
Carbon Dioxide–Arsenic Trisulfide–Water	CO_2—As_2S_3—H_2O	1737
Carbon Dioxide–Rubidium Chloride–Water	CO_2—RbCl—H_2O	1729
Carbon Dioxide–Strontium Chloride–Water	CO_2—$SrCl_2$—H_2O	1730, 1731
Carbon Dioxide–Barium Hydrogen Carbonate–Barium Carbonate–Water	CO_2—$Ba(HCO_3)_2$—$BaCO_3$—H_2O	1732
Carbon Dioxide–Barium Chloride–Water	CO_2—$BaCl_2$—H_2O	1733
Carbon Dioxide–Lead (II) Carbonate–Water	CO_2—$PbCO_3$—H_2O	1734, 1735
Carbon Dioxide–Dolomitized Limestone–Water	CO_2— —— —H_2O	1738
Carbon Dioxide–Lead (II) Sulfide–Water	CO_2—PbS—H_2O	110
Carbon Dioxide–Seawater	CO_2— —— —H_2O	1739
Dinitrogen Oxide–Ammonium Nitrate–Water	N_2O—NH_4NO_3—H_2O	1740
Dinitrogen Oxide–Ammonium Sulfate–Water	N_2O—$(NH_4)_2SO_4$—H_2O	1741
Dinitrogen Oxide–Ammonium Chloride–Water	N_2O—NH_4Cl—H_2O	1742, 1743
Dinitrogen Oxide–Ammonium Bromide–Water	N_2O—NH_4Br—H_2O	1744
Dinitrogen Oxide–Sodium Hydroxide–Water	N_2O—NaOH—H_2O	1745
Dinitrogen Oxide–Sodium Nitrate–Water	N_2O—$NaNO_3$—H_2O	1746, 1747
Dinitrogen Oxide–Sodium Hydrogen Orthophosphate–Water	N_2O—Na_2HPO_4—H_2O	1748
Dinitrogen Oxide–Sodium Orthophosphate–Water	N_2O—Na_3PO_4—H_2O	1749
Dinitrogen Oxide–Sodium Sulfate–Water	N_2O—Na_2SO_4—H_2O	1750—1752
Dinitrogen Oxide–Sodium Chloride–Water	N_2O—NaCl—H_2O	1753—1756
Dinitrogen Oxide–Sodium Bromide–Water	N_2O—NaBr—H_2O	1757
Dinitrogen Oxide–Magnesium Nitrate–Water	N_2O—$Mg(NO_3)_2$—H_2O	1758, 1759
Dinitrogen Oxide–Magnesium Sulfate–Water	N_2O—$MgSO_4$—H_2O	1760—1762
Dinitrogen Oxide–Aluminum Nitrate–Water	N_2O—$Al(NO_3)_3$—H_2O	1763
Dinitrogen Oxide–Aluminum Sulfate–Water	N_2O—$Al_2(SO_4)_3$—H_2O	1764, 1765

System name	Formulae of components	Table number
Dinitrogen Oxide–Potassium Hydroxide–Water	$N_2O—KOH—H_2O$	1766, 1767
Dinitrogen Oxide–Potassium Nitrate –Water	$N_2O—KNO_3—H_2O$	1768—1770
Dinitrogen Oxide–Potassium Sulfate –Water	$N_2O—K_2SO_4—H_2O$	1771, 1772
Dinitrogen Oxide–Potassium Chloride–Water	$N_2O—KCl—H_2O$	1773—1776
Dinitrogen Oxide–Potassium Bromide–Water	$N_2O—KBr—H_2O$	1777, 1778
Dinitrogen Oxide–Potassium Iodide –Water	$N_2O—KI—H_2O$	1779
Dinitrogen Oxide–Potassium Periodate–Water	$N_2O—KIO_4—H_2O$	1780
Dinitrogen Oxide–Calcium Nitrate–Water	$N_2O—Ca(NO_3)_2—H_2O$	1781
Dinitrogen Oxide–Calcium Chloride –Water	$N_2O—CaCl_2—H_2O$	1782, 1783
Dinitrogen Oxide–Chromium (III) Sulfate–Water	$N_2O—Cr_2(SO_4)_3—H_2O$	1784
Dinitrogen Oxide–Manganese (II) Sulfate–Water	$N_2O—MnSO_4—H_2O$	1785
Dinitrogen Oxide–Iron (II) Sulfate –Water	$N_2O—FeSO_4—H_2O$	1786, 1787
Dinitrogen Oxide–Cobalt (II) Sulfate –Water	$N_2O—CoSO_4—H_2O$	1788
Dinitrogen Oxide–Nickel (II) Sulfate –Water	$N_2O—NiSO_4—H_2O$	1789
Dinitrogen Oxide–Copper (II) Nitrate–Water	$N_2O—Cu(NO_3)_2—H_2O$	1790
Dinitrogen Oxide–Zinc Nitrate–Water	$N_2O—Zn(NO_3)_2—H_2O$	1791
Dinitrogen Oxide–Zinc Sulfate–Water	$N_2O—ZnSO_4—H_2O$	1792
Dinitrogen Oxide–Rubidium Chloride–Water	$N_2O—RbCl—H_2O$	1793
Dinitrogen Oxide–Strontium Chloride–Water	$N_2O—SrCl_2—H_2O$	1794
Dinitrogen Oxide–Cadmium Nitrate –Water	$N_2O—Cd(NO_3)_2—H_2O$	1795
Dinitrogen Oxide–Cesium Chloride–Water	$N_2O—CsCl—H_2O$	1796
Dinitrogen Oxide–Barium Chloride–Water	$N_2O—BaCl_2—H_2O$	1797
Nitrogen Oxide–Ammonium Copper Chloride–Water	$NO—Cu_2(NH_3)_nCl_2—H_2O$	1798
Nitrogen Oxide–Sodium Hydroxide –Sodium Sulfate–Water	$NO—NaOH—Na_2SO_4—H_2O$	1798
Nitrogen Oxide–Manganese (II) Sulfate–Water	$NO—MnSO_4—H_2O$	1798
Nitrogen Oxide–Iron (II) Sulfate–Water	$NO—FeSO_4—H_2O$	1798
Nitrogen Oxide–Iron (II) Chloride –Water	$NO—FeCl_2—H_2O$	1798
Nitrogen Oxide–Cobalt (II) Sulfate–Water	$NO—CoSO_4—H_2O$	1798
Nitrogen Oxide–Nickel (II) Sulfate–Water	$NO—NiSO_4—H_2O$	1798
Nitrogen Oxide–Copper (II) Sulfate –Water	$NO—CuSO_4—H_2O$	1798
Nitrogen Oxide–Copper (II) Chloride –Water	$NO—CuCl_2—H_2O$	1798

System name	Formulae of components	Table number
Dinitrogen Pentoxide–Magnesium Oxide–Phosphorus Pentoxide–Water	N_2O_5—MgO—P_2O_5—H_2O	1799
Dinitrogen Pentoxide–Phosphorus Pentoxide–Iron (III) Oxide–Water	N_2O_5—P_2O_5—Fe_2O_3—H_2O	1800
Dinitrogen Pentoxide–Sulfur Trioxide–Uranium Trioxide–Water	N_2O_5—SO_3—UO_3—H_2O	1801
Dinitrogen Pentoxide–Zinc Oxide–Water	N_2O_5—ZnO—H_2O	1802
Dinitrogen Pentoxide–Mercury (I) Oxide–Water	N_2O_5—Hg_2O—H_2O	1803
Dinitrogen Pentoxide–Lead (II) Oxide–Water	N_2O_5—PbO—H_2O	1804
Ammonia–Ammonium Hydrogen Carbonate–Water	NH_3—NH_4HCO_3—H_2O	1805
Ammonia–Ammonium Carbonate–Water	NH_3—$(NH_4)_2CO_3$—H_2O	1806, 1807
Ammonia–Ammonium Carbonate–Ammonium Chloride–Water	NH_3—$(NH_4)_2CO_3$—NH_4Cl—H_2O	1808
Ammonia–Ammonium Carbonate–Copper (II) Hydroxide–Water	NH_3—$(NH_4)_2CO_3$—$Cu(OH)_2$—H_2O	1809
Ammonia–Ammonium Thiocyanate–Water	NH_3—NH_4CNS—H_2O	1810
Ammonia–Ammonium Thiocyanate–Copper (II) Hydroxide–Water	NH_3—NH_4CNS—$Cu(OH)_2$—H_2O	1809
Ammonia–Ammonium Nitrate–Water	NH_3—NH_4NO_3—H_2O	1811—1813
Ammonia–Ammonium Nitrate–Ammonium Chloride–Water	NH_3—NH_4NO_3—NH_4Cl—H_2O	1814
Ammonia–Ammonium Nitrate–Sodium Nitrate–Water	NH_3—NH_4NO_3—$NaNO_3$—H_2O	1815
Ammonia–Ammonium Nitrate–Calcium Nitrate–Water	NH_3—NH_4NO_3—$Ca(NO_3)_2$—H_2O	1816
Ammonia–Ammonium Nitrate–Nickel (II) Nitrate–Silver Iodate–Water	NH_3—NH_4NO_3—$Ni(NO_3)_2$—$AgIO_3$—H_2O	1817
Ammonia–Ammonium Nitrate–Copper (II) Nitrate–Water	NH_3—NH_4NO_3—$Cu(NO_3)_2$—H_2O	1818
Ammonia Ammonium Nitrate–Zinc Nitrate–Water	NH_3—NH_4NO_3—$Zn(NO_3)_2$—H_2O	1819
Ammonia–Ammonium Nitrate–Silver Iodate–Water	NH_3—NH_4NO_3—$AgIO_3$—H_2O	1820
Ammonia–Ammonium Sulfite–Sulfur Dioxide–Water	NH_3—$(NH_4)_2SO_3$—SO_2—H_2O	1821
Ammonia–Ammonium Sulfate–Water	NH_3—$(NH_4)_2SO_4$—H_2O	1822—1830
Ammonia–Ammonium Sulfate–Ammonium Chloride–Water	NH_3—$(NH_4)_2SO_4$—NH_4Cl—H_2O	1831, 1832
Ammonia–Ammonium Sulfate–Sodium Sulfate–Water	NH_3—$(NH_4)_2SO_4$—Na_2SO_4—H_2O	1833
Ammonia–Ammonium Sulfate–Potassium Sulfate–Water	NH_3—$(NH_4)_2SO_4$—K_2SO_4—H_2O	1834
Ammonia–Ammonium Sulfate–Copper (II) Hydroxide–Water	NH_3—$(NH_4)_2SO_4$—$Cu(OH)_2$—H_2O	1809
Ammonia–Ammonium Sulfate–Silver Sulfate–Water	NH_3—$(NH_4)_2SO_4$—Ag_2SO_4—H_2O	1835
Ammonia–Ammonium Sulfate–Cadmium Sulfate–Water	NH_3—$(NH_4)_2SO_4$—$CdSO_4$—H_2O	1836

System name	Formulae of components	Table number
Ammonia–Cadmium Ammonium Sulfate–Water	NH_3—$CdSO_4 \cdot (NH_4)_2SO_4$—$H_2O$	1837
Ammonia–Ammonium Chloride–Water	NH_3—NH_4Cl—H_2O	1838—1844
Ammonia–Ammonium Chloride–Sodium Carbonate–Water	NH_3—NH_4Cl—Na_2CO_3—H_2O	1845
Ammonia–Ammonium Chloride–Sodium Chloride–Water	NH_3—NH_4Cl—$NaCl$—H_2O	1846
Ammonia–Ammonium Chloride–Potassium Carbonate–Water	NH_3—NH_4Cl—K_2CO_3—H_2O	1847
Ammonia–Ammonium Chloride–Potassium Chloride–Water	NH_3—NH_4Cl—KCl—H_2O	1848
Ammonia–Ammonium Chloride–Copper (II) Chloride–Water	NH_3—NH_4Cl—$CuCl_2$—H_2O	1849
Ammonia–Ammonium Chloride–Zinc Chloride–Water	NH_3—NH_4Cl—$ZnCl_2$—H_2O	850
Ammonia–Ammonium Chloride–Molybdenum Oxide–Water	NH_3—NH_4Cl—MoO_3—H_2O	1851
Ammonia–Ammonium Chloride–Cadmium Chloride–Water	NH_3—NH_4Cl—$CdCl_2$—H_2O	1852
Ammonia–Ammonium Perchlorate–Water	NH_3—NH_4ClO_4—H_2O	1853
Ammonia–Ammonium Perchlorate–Ammonium Bromide–Silver Bromide–Water	NH_3—NH_4ClO_4—NH_4Br—$AgBr$—H_2O	1854
Ammonia–Ammonium Metavanadate–Water	NH_3—NH_4VO_3—H_2O	1855
Ammonia–Ammonium Bromide––Water	NH_3—NH_4Br—H_2O	1856
Ammonia–Ammonium Bromide–Copper (II) Bromide–Water	NH_3—NH_4Br—$CuBr_2$—H_2O	1857
Ammonia–Ammonium Bromide–Zinc Bromide–Water	NH_3—NH_4Br—$ZnBr_2$—H_2O	1858
Ammonia–Ammonium Iodide–Water	NH_4—NH_4I—H_2O	1859
Ammonia–Ammonium Iodide–Zinc Iodide–Water	NH_3—NH_4I—ZnI_2—H_2O	1860
Ammonia–Hexamminenickel Iodide–Water	NH_3—$[Ni(NH_3)_6]I_2$—H_2O	1861
Ammonia–Carbon Dioxide–Magnesium Oxide–Water	NH_3—CO_2—MgO—H_2O	1862
Ammonia–Sodium Hydroxide–Water	NH_3—$NaOH$—H_2O	1863—1865
Ammonia–Sodium Hydroxide–Sodium Hyponitrite–Silver Hyponitrite–Water	NH_3—$NaOH$—$Na_2N_2O_2$—$Ag_2N_2O_2$—H_2O	1866
Ammonia–Sodium Hydroxide–Silver Hyponitrite–Water	NH_3—$NaOH$—$Ag_2N_2O_2$—H_2O	1867
Ammonia–Sodium Carbonate–Water	NH_3—Na_2CO_3—H_2O	1868—1872, 1878
Ammonia–Sodium Nitrate–Water	NH_3—$NaNO_3$—H_2O	1873—1876
Ammonia–Sodium Sulfide–Water	NH_3—Na_2S—H_2O	1877
Ammonia–Sodium Sulfate–Water	NH_3—Na_2SO_4—H_2O	1879—1886
Ammonia–Sodium Sulfate–Sodium Chloride–Water	NH_3—Na_2SO_4—$NaCl$—H_2O	1887

System name	Formulae of components	Table number
Ammonia–Sodium Sulfate–Copper (II) Hydroxide–Water	NH_3—Na_2SO_4—$Cu(OH)_2$—H_2O	1809
Ammonia–Sodium Chloride–water	NH_3—$NaCl$—H_2O	1888—1895
Ammonia–Sodium Chloride–Potassium Chloride–Water	NH_3—$NaCl$—KCl—H_2O	1896
Ammonia–Sodium Chloride–Calcium Chloride–Water	NH_3—$NaCl$—$CaCl_2$—H_2O	1897
Ammonia–Sodium Bromide–Water	NH_3—$NaBr$—H_2O	1898, 1899
Ammonia–Sodium Iodide–Water	NH_3—NaI—H_2O	1900, 1901
Ammonia–Sodium Hexacyanoferrate (II)–Water	NH_3—$Na_4Fe(CN)_6$—H_2O	1902
Ammonia–Magnesium Oxide–Phosphorus Pentoxide–Water	NH_3—MgO—P_2O_5—H_2O	1903
Ammonia–Phosphorus Pentoxide–Water	NH_3—P_2O_5—H_2O	1904
Ammonia–Potassium Hydroxide–Water	NH_3—KOH—H_2O	1905—1908
Ammonia–Potassium Metaborate–Water	NH_3—KBO_2—H_2O	1909
Ammonia–Potassium Carbonate–Water	NH_3—K_2CO_3—H_2O	1910—1913
Ammonia–Potassium Cyanide–Water	NH_3—KCN—H_2O	1914
Ammonia–Potassium Hexacyanoferrate (II)–Water	NH_3—$K_4Fe(CN)_6$—H_2O	1915
Ammonia–Potassium Thiocyanate–Water	NH_3—$KCNS$—H_2O	1916
Ammonia–Potassium Nitrite–Water	NH_3—KNO_2—H_2O	1917
Ammonia–Potassium Nitrate–Water	NH_3—KNO_3—H_2O	1918—1920
Ammonia–Potassium Nitrate–Potassium Chloride–Water	NH_3—KNO_3—KCl—H_2O	1921
Ammonia–Potassium Nitrate–Silver Chloride–Water	NH_3—KNO_3—$AgCl$—H_2O	1922
Ammonia–Potassium Nitrate–Silver Iodate–Water	NH_3—KNO_3—$AgIO_3$—H_2O	1923
Ammonia–Potassium Fluoride–Water	NH_3—KF—H_2O	1924
Ammonia–Potassium Hydrogen Orthophosphate–Water	NH_3—K_2HPO_4—H_2O	1925
Ammonia–Potassium Sulfite–Water	NH_3—K_2SO_3—H_2O	1926
Ammonia–Potassium Sulfate–Water	NH_3—K_2SO_4—H_2O	1927—1933
Ammonia–Potassium Sulfate–Potassium Chloride–Water	NH_3—K_2SO_4—KCl—H_2O	1934
Ammonia–Potassium Chloride–Water	NH_3—KCl—H_2O	1935—1941
Ammonia–Potassium Chloride–Calcium Chloride–Water	NH_3—KCl—$CaCl_2$—H_2O	1942
Ammonia–Potassium Chloride–Silver Chloride–Water	NH_3—KCl—$AgCl$—H_2O	1943, 1944
Ammonia–Potassium Chlorate–Water	NH_3—$KClO_3$—H_2O	1945, 1946
Ammonia–Potassium Chromate–Water	NH_3—K_2CrO_4—H_2O	1947
Ammonia–Potassium Chromate–Thallium Chromate–Water	NH_3—K_2CrO_4—Tl_2CrO_4—H_2O	1948
Ammonia–Potassium Bromide–Water	NH_3—KBr—H_2O	1949—1951
Ammonia–Potassium Bromide–Silver Bromide–Water	NH_3—KBr—$AgBr$—H_2O	1952
Ammonia–Potassium Bromate–Water	NH_3—$KBrO_3$—H_2O	1953
Ammonia–Potassium Iodide–Water	NH_3—KI—H_2O	1954, 1955
Ammonia–Potassium Iodate–Water	NH_3—KIO_3—H_2O	1956
Ammonia–Calcium Oxide–Water	NH_3—CaO—H_2O	1957

System name	Formulae of components	Table number
Ammonia–Calcium Hydroxide–Water	NH_3—$Ca(OH)_2$—H_2O	1958, 1959
Ammonia–Calcium Nitrate–Water	NH_3—$Ca(NO_3)_2$—H_2O	1960
Ammonia–Calcium Iodate–Water	NH_3—$Ca(IO_3)_2$—H_2O	1961
Ammonia–Vanadium Pentoxide–Water	NH_3—V_2O_5—H_2O	1962
Ammonia–Iron (III) Pyrophosphate–Water	NH_3—$Fe_4(P_2O_7)_3$—H_2O	1963
Ammonia–Copper (I) Oxide–Water	NH_3—Cu_2O—H_2O	1964
Ammonia–Copper (II) Oxide–Water	NH_3—CuO—H_2O	1965
Ammonia–Copper (II) Hydroxide–	NH_3—$Cu(OH)_2$—H_2O	1966
Ammonia–Copper (II) Sulfate Hydroxide–Water	NH_3—$5Cu(OH)_2 \cdot CuSO_4$—H_2O	1967
Ammonia–Copper (II) Thiocyanate–Water	NH_3—$Cu(CNS)_2$—H_2O	1968
Ammonia–Copper (II) Nitrate–Water	NH_3—$Cu(NO_3)_2$—H_2O	1969
Ammonia–Copper (II) Nitrate–Silver Iodate–Water	NH_3—$Cu(NO_3)_2$—$AgIO_3$—H_2O	1970
Ammonia–Copper (II) Sulfate–Water	NH_3—$CuSO_4$—H_2O	1971, 1972
Ammonia–Copper (II) Chloride–Water	NH_3—$CuCl_2$—H_2O	1973, 1974
Ammonia–Copper (II) Bromide–Water	NH_3—$CuBr_2$—H_2O	1975
Ammonia–Zinc Nitrate–Water	NH_3—$Zn(NO_3)_2$—H_2O	1976, 1977
Ammonia–Zinc Sulfate–Water	NH_3—$ZnSO_4$—H_2O	1978
Ammonia–Zinc Chloride–Water	NH_3—$ZnCl_2$—H_2O	1979, 1980
Ammonia–Zinc Bromide–Water	NH_3—$ZnBr_2$—H_2O	1981
Ammonia–Zinc Iodide–Water	NH_3—ZnI_2—H_2O	1982
Ammonia–Gallium Oxide–Water	NH_3—Ga_2O_3—H_2O	1983
Ammonia–Arsenic Trioxide–Water	NH_3—As_2O_3—H_2O	1984, 1985
Ammonia–Rubidium Sulfate–Water	NH_3—Rb_2SO_4—H_2O	1986
Ammonia–Rubidium Chloride–Water	NH_3—$RbCl$—H_2O	1987
Ammonia–Rubidium Bromide–Water	NH_3—$RbBr$—H_2O	1988
Ammonia–Strontium Hydroxide–Water	NH_3—$Sr(OH)_2$—H_2O	1989
Ammonia–Molybdenum Trioxide–Water	NH_3—MoO_3—H_2O	1990, 1991
Ammonia–Silver Oxide–Water	NH_3—Ag_2O—H_2O	1992, 1993
Ammonia–Silver Cyanide–Water	NH_3—$AgCN$—H_2O	1994
Ammonia–Silver Thiocyanate–Water	NH_3—$AgCNS$—H_2O	1995
Ammonia–Silver Orthophosphate–Water	NH_3—Ag_3PO_4—H_2O	1996
Ammonia–Silver Chloride–Water	NH_3—$AgCl$—H_2O	1997—2002
Ammonia–Silver Chromate–Water	NH_3—Ag_2CrO_4—H_2O	2003
Ammonia–Silver Selenide–Water	NH_3—Ag_2Se—H_2O	2004
Ammonia–Silver Bromide–Water	NH_3—$AgBr$—H_2O	2005—2010
Ammonia–Silver Bromate–Water	NH_3—$AgBrO_3$—H_2O	2011
Ammonia–Silver Telluride–Water	NH_3—Ag_2Te—H_2O	2012
Ammonia–Silver Iodide–Water	NH_3—AgI—H_2O	2016
Ammonia–Silver Iodate–Water	NH_3—$AgIO_3$—H_2O	2017—2019
Ammonia–Silver Iodate–Cadmium Nitrate–Water	NH_3—$AgIO_3$—$Cd(NO_3)_2$—H_2O	2020
Ammonia–Cadmium Hydroxide–Water	NH_3—$Cd(OH)_2$—H_2O	2021
Ammonia–Cadmium Nitrate–Water	NH_3—$Cd(NO_3)_2$—H_2O	2022
Ammonia–Cadmium Sulfate–Water	NH_3—$CdSO_4$—H_2O	2023, 2024

System name	Formulae of components	Table number
Ammonia–Cadmium Chloride– Water	$NH_3—CdCl_2—H_2O$	2025
Ammonia–Cadmium Bromide– Water	$NH_3—CdBr_2—H_2O$	2026
Ammonia–Cadmium Iodide–Water	$NH_3—CdI_2—H_2O$	2027
Ammonia–Tin (IV) Sulfide–Water	$NH_3—SnS_2—H_2O$	2028
Ammonia–Antimony Sulfide– Water	$NH_3—Sb_2S_3—H_2O$	2029
Ammonia–Cesium Sulfate–Water	$NH_3—Cs_2SO_4—H_2O$	2030
Ammonia–Cesium Chloride–Water	$NH_3—CsCl—H_2O$	2031
Ammonia–Cesium Iodide–Water	$NH_3—CsI—H_2O$	2032
Ammonia–Barium Iodate–Water	$NH_3—BaIO_3—H_2O$	2033
Ammonia–Thallium (I) Chloride– Water	$NH_3—TlCl—H_2O$	2034
Ammonia–Mercury (II) Cyanide– Water	$NH_3—Hg(CN)_2—H_2O$	2035
Ammonia–Mercury (II) Amidochlorate –Water	$NH_3—HgNH_2Cl—H_2O$	2036
Ammonia–Mercury (II) Bromide– Water	$NH_3—HgBr_2—H_2O$	2037
Ammonium Oxide–Sulfur Dioxide– Iron (III) Oxide–Water	$(NH_4)_2O—SO_2—Fe_2O_3—H_2O$	2038
Ammonium Oxide–Sulfur Trioxide –Niobium Pentoxide–Water	$(NH_4)_2O—SO_3—Nb_2O_5—H_2O$	2039
Ammonium Oxide–Sulfur Trioxide –Tantalum Pentoxide–Water	$(NH_3)_2O—SO_3—Ta_2O_5—H_2O$	2040
Hydrazine–Sodium Hydroxide– Water	$N_2H_4—NaOH—H_2O$	2041
Hydrazine–Potassium Hydroxide– Water	$N_2H_4—KOH—H_2O$	2042
Hydrazine–Potassium Carbonate–Water	$N_2H_4—K_2CO_3—H_2O$	2043
Ammonium Hydroxide–Ammonium Carbonate–Ammonium Sulfate– Sodium Hydroxide–Sodium Carbonate–Sodium Sulfate– Water	$NH_4OH—(NH_4)_2CO_3—(NH_4)_2SO_4—NaOH—Na_2CO_3—Na_2SO_4—H_2O$	2044
Ammonium Hydroxide–Ammonium Hydrogen Carbonate–Ammonium Carbonate–Sodium Hydroxide– Sodium Hydrogen Carbonate– Sodium Carbonate–Water	$NH_4OH—NH_4HCO_3—(NH_4)_2CO_3—NaOH—NaHCO_3—Na_2CO_3—H_2O$	2045
Ammonium Hydroxide–Ammonium Carbonate–Nickel Carbonate Hydroxide–Water	$NH_4OH—(NH_4)_2CO_3—Ni(OH)_2\cdot 3NiCO_3—H_2O$	2046
Ammonium Hydroxide–Ammonium Carbonate–Copper (II) Oxide– Water	$NH_4OH—(NH_4)_2CO_3—CuO—H_2O$	2046
Ammonium Hydroxide–Ammonium Carbonate–Copper Carbonate Hydroxide (Malachite)–Water	$NH_4OH—(NH_4)_2CO_3—Cu_2(OH)_2CO_3—H_2O$	2046
Ammonium Hydroxide–Ammonium Nitrate–Aluminum Hydroxide– Water	$NH_4OH—NH_4NO_3—Al(OH)_3—H_2O$	2047
Ammonium Hydroxide–Ammonium Sulfate–Magnesium Sulfate– Water	$NH_4OH—(NH_4)_2SO_4—MgSO_4—H_2O$	2048
Ammonium Hydroxide–Ammonium Sulfate–Copper (II) Oxide–Water	$NH_4OH—(NH_4)_2SO_4—CuO—H_2O$	2049
Ammonium Hydroxide–Ammonium Sulfate–Cadmium Hydroxide– Cadmium Sulfate–Water	$NH_4OH—(NH_4)_2SO_4—Cd(OH)_2—CdSO_4—H_2O$	2050

System name	Formulae of components	Table number
Ammonium Hydroxide–Ammonium Chloride–Ammonium Metavanadate–Sodium Chloride–Sodium Metavanadate–Water	$NH_4OH—NH_4Cl—NH_4VO_3—NaCl—NaVO_3—H_2O$	2051
Ammonium Hydroxide–Ammonium Chloride–Ammonium Magnesium Arsenate–Water	$NH_4OH—NH_4Cl—NH_4MgAsO_4—H_2O$	2052
Ammonium Hydroxide–Ammonium Chloride–Magnesium Chloride–Water	$NH_4OH—NH_4Cl—MgCl_2—H_2O$	2053—2056
Ammonium Hydroxide–Ammonium Magnesium Arsenate–Water	$NH_4OH—NH_4MgAsO_4—H_2O$	2057
Ammonium Hydroxide–Aluminum Oxide–Water	$NH_4OH—Al_2O_3—H_2O$	2729
Ammonium Hydroxide–Aluminum Hydroxide–Water	$NH_4OH—Al(OH)_3—H_2O$	2058
Ammonium Hydroxide–Aluminum Hydroxide–Potassium Nitrate–Water	$NH_4OH—Al(OH)_3—KNO_3—H_2O$	2059
Ammonium Hydroxide–Zinc Hydroxide–Water	$NH_4OH—Zn(OH)_2—H_2O$	2060—2062
Ammonium Tetraborate–Ammonium Sulfate–Water	$(NH_4)_2B_4O_7 \cdot 4H_2O—(NH_4)_2SO_4—H_2O$	2063—2065
Ammonium Tetraborate–Ammonium Sulfate–Sodium Tetraborate–Sodium Sulfate–Water	$(NH_4)_2B_4O_7—(NH_4)_2SO_4—Na_2B_4O_7—Na_2SO_4—H_2O$	2066, 2067
Ammonium Tetraborate–Ammonium Sulfate–Sodium Sulfate– Water	$(NH_4)_2B_4O_7—(NH_4)_2SO_4—Na_2SO_4—H_2O$	2068
Ammonium Tetraborate–Ammonium Chloride–Water	$(NH_4)_2B_4O_7—NH_4Cl—H_2O$	2069
Ammonium Tetraborate–Sodium Tetraborate–Water	$(NH_4)_2B_4O_7—Na_2B_4O_7—H_2O$	2070—2073
Ammonium Tetraborate–Sodium Tetraborate–Sodium Sulfate– Water	$(NH_4)_2B_4O_7—Na_2B_4O_7—Na_2SO_4—H_2O$	2074
Ammonium Hydrogen Carbonate–Ammonium Nitrate–Water	$NH_4HCO_3—NH_4NO_3—H_2O$	2075—2079
Ammonium Hydrogen Carbonate–Ammonium Sulfate–Sodium Hydrogen Carbonate–Sodium Sulfate–Water	$NH_4HCO_3—(NH_4)_2SO_4—NaHCO_3—Na_2SO_4—H_2O$	2080—2082
Ammonium Hydrogen Carbonate–Ammonium Sulfate–Sodium Sulfate–Water	$NH_4HCO_3—(NH_4)_2SO_4—Na_2SO_4—H_2O$	2083
Ammonium Hydrogen Carbonate–Ammonium Sulfate–Magnesium Hydrogen Carbonate–Magnesium Sulfate–Water	$NH_4HCO_3—(NH_4)_2SO_4—Mg(HCO_3)_2—MgSO_4—H_2O$	2084
Ammonium Hydrogen Carbonate–Ammonium Chloride–Water	$NH_4HCO_3—NH_4Cl—H_2O$	2085—2088
Ammonium Hydrogen Carbonate–Ammonium Chloride–Sodium Hydrogen Carbonate–Water	$NH_4HCO_3—NH_4Cl—NaHCO_3—H_2O$	2089—2092
Ammonium Hydrogen Carbonate–Ammonium Chloride–Sodium Hydrogen Carbonate–Sodium Chloride–Water	$NH_4HCO_3—NH_4Cl—NaHCO_3—NaCl—H_2O$	2093, 2094

System name	Formulae of components	Table number
Ammonium Hydrogen Carbonate–Ammonium Chloride–Sodium Hydrogen Carbonate–Sodium Chloride–Potassium Chloride–Water	NH_4HCO_3—NH_4Cl—$NaHCO_3$—$NaCl$—KCl—H_2O	2095
Ammonium Hydrogen Carbonate–Ammonium Chloride–Sodium Chloride–Potassium Chloride–Water	NH_4HCO_3—NH_4Cl—$NaCl$—KCl—H_2O	2096, 2097
Ammonium Hydrogen Carbonate–Ammonium Chloride–Magnesium Hydrogen Carbonate–Magnesium Chloride–Water	NH_4HCO_3—NH_4Cl —$Mg(HCO_3)_2$—$MgCl_2$—H_2O	2098
Ammonium Hydrogen Carbonate–Ammonium Chloride–Potassium Chloride–Water	NH_4HCO_3—NH_4Cl—KCl—H_2O	2099
Ammonium Hydrogen Carbonate–Ammonium Heptoxodiuranate–Water	NH_4HCO_3—$(NH_4)_2U_2O_7$—H_2O	2100
Ammonium Hydrogen Carbonate–Sodium Hydrogen Carbonate–Water	NH_4HCO_3—$NaHCO_3$—H_2O	2101—2106
Ammonium Hydrogen Carbonate–Sodium Hydrogen Carbonate–Potassium Hydrogen Carbonate–Water	NH_4HCO_3—$NaHCO_3$—$KHCO_3$—H_2O	2107
Ammonia Hydrogen Carbonate–Sodium Chloride–Water	NH_4HCO_3—$NaCl$—H_2O	2108
Ammonium Hydrogen Carbonate–Potassium Hydrogen Carbonate–Water	NH_4HCO_3—$KHCO_3$—H_2O	2109
Ammonium Carbonate–Ammonium Plutonyl Carbonate–Water	$(NH_4)_2CO_3$—$(NH_4)_2CO_3 \cdot PuO_2CO_3$—$H_2O$	2110
Ammonium Carbonate–Ammonium Plutonyl Carbonate–Ammonium Nitrate–Water	$(NH_4)_2CO_3$—$(NH_4)_2CO_3 \cdot PuO_2CO_3$—$NH_4NO_3$—$H_2O$	2111
Ammonium Carbonate–Ammonium Sulfate–Sodium Sulfate–Water	$(NH_4)_2CO_3$—$(NH_4)_2SO_4$—Na_2SO_4—H_2O	2112
Ammonium Carbonate–Ammonium Chloride–Water	$(NH_4)_2CO_3$—NH_4Cl—H_2O	2113, 2114
Ammonium Carbonate–Ammonium Chloride–Sodium Carbonate–Sodium Chloride–Water	$(NH_4)_2CO_3$—NH_4Cl—Na_2CO_3—$NaCl$—H_2O	2115
Ammonium Carbonate–Ammonium Heptoxodiuranate–Water	$(NH_4)_2CO_3$—$(NH_4)_2U_2O_7$—H_2O	2116
Ammonium Carbonate–Sodium Carbonate–Water	$(NH_4)_2CO_3$—Na_2CO_3—H_2O	2117
Ammonium Carbonate–Sodium Sulfate–Water	$(NH_4)_2CO_3$—Na_2SO_4—H_2O	2118
Ammonium Carbonate–Magnesium Carbonate–Water	$(NH_4)_2CO_3$—$MgCO_3$—H_2O	2119
Ammonium Carbonate–Scandium Oxide–Water	$(NH_4)_2CO_3$—Sc_2O_3—H_2O	2110
Ammonium Carbonate–Silver Chloride–Water	$(NH_4)_2CO_3$—$AgCl$—H_2O	2569
Ammonium Carbonate–Silver Bromide–Water	NH_4CO_3—$AgBr$—H_2O	5420
Ammonium Plutonyl Carbonate–Ammonium Nitrate–Water	$(NH_4)_2CO_3 \cdot PuO_2CO_3$—$NH_4NO_3$—$H_2O$	2121

System name	Formulae of components	Table number
Ammonium Thiocyanate–Ammonium Nitrate–Water	$NH_4CNS—NH_4NO_3—H_2O$	2122
Ammonium Thiocyanate–Ammonium Sulfate–Water	$NH_4CNS—(NH_4)_2SO_4—H_2O$	2123
Ammonium Thiocyanate–Sodium Thiocyanate–Water	$NH_4CNS—NaCNS—H_2O$	2124
Ammonium Thiocyanate–Sodium Chloride–Water	$NH_4CNS—NaCl—H_2O$	2125
Ammonium Thiocyanate–Sulfur Dioxide–Water	$NH_4CNS—SO_2—H_2O$	2126
Ammonium Thiocyanate–Calcium Thiocyanate–Water	$NH_4CNS—Ca(CNS)_2—H_2O$	2127
Ammonium Thiocyanate–Calcium Fluoride–Iron (III) Nitrate– Water	$NH_4CNS—CaF_2—Fe(NO_3)_3—H_2O$	2128
Ammonium Thiocyanate–Silver Thiocyanate–Water	$NH_4CNS—AgCNS—H_2O$	2129
Ammonium Thiocyanate–Silver Chloride–Water	$NH_4CNS—AgCl—H_2O$	2569
Ammonium Thiocyanate–Silver Bromide–Water	$NH_4CNS—AgBr—H_2O$	5418, 5420
Ammonium Thiocyanate–Silver Iodide–Water	$NH_4CNS—AgI—H_2O$	5427
Ammonium Thiocyanate–Barium Thiocyanate–Water	$NH_4CNS—Ba(CNS)_2—H_2O$	2130
Ammonium Hexacyanoferrate (II) –Ammonium Chloride–Water	$(NH_4)_4[Fe(CN)_6]—NH_4Cl—H_2O$	2131
Ammonium Nitrate–Ammonium Dihydrogen Orthophosphate– Water	$NH_4NO_3—NH_4H_2PO_4—H_2O$	2132
Ammonium Nitrate–Ammonium Hydrogen Orthophosphate– Water	$NH_4NO_3—(NH_4)_2HPO_4—H_2O$	2133
Ammonium Nitrate–Ammonium Magnesium Orthophosphate– Water	$NH_4NO_3—NH_4MgPO_4—H_2O$	2134, 2135
Ammonium Nitrate–Ammonium Manganese Phosphate–Water	$NH_4NO_3—NH_4MnPO_4—H_2O$	2136
Ammonium Nitrate–Ammonium Sulfito Amide–Water	$NH_4NO_3—NH_4SO_3NH_2—H_2O$	2137
Ammonium Nitrate–Ammonium Hydrogen Sulfate–Water	$NH_4NO_3—NH_4HSO_4—H_2O$	2138
Ammonium Nitrate–Ammonium Sulfate–Water	$NH_4NO_3—(NH_4)_2SO_4—H_2O$	2139—2142
Ammonium Nitrate–Ammonium Sulfito Amide–Ammonium Sulfate–Water	$NH_4NO_3—NH_4SO_3NH_2—(NH_4)_2SO_4—H_2O$	2143
Ammonium Nitrate–Ammonium Chloride–Water	$NH_4NO_3—NH_4Cl—H_2O$	2144—2149
Ammonium Nitrate–Ammonium Chloride–Sodium Nitrate–Sodium Chloride–Water	$NH_4NO_3—NH_4Cl—NaNO_3—NaCl—H_2O$	2150
Ammonium Nitrate–Ammonium Chloride–Magnesium Hydroxide –Water	$NH_4NO_3—NH_4Cl—Mg(OH)_2—H_2O$	2151

System name	Formulae of components	Table number
Ammonium Nitrate–Ammonium Chloride–Potassium Nitrate–Potassium Chloride–Water	$NH_4NO_3—NH_4Cl—KNO_3—KCl—H_2O$	2152
Ammonium Nitrate–Diamminepalladium Chloride–Water	$NH_4NO_3—Pd(NH_3)_2Cl_2—H_2O$	2153
Ammonium Nitrate–Ammonium Perchlorate–Water	$NH_4NO_3—NH_4ClO_4—H_2O$	2154, 2155
Ammonium Nitrate–Ammonium Perchlorate–Sodium Nitrate–Sodium Perchlorate–Water	$NH_4NO_3—NH_4ClO_4—NaNO_3—NaClO_4—H_2O$	2156
Ammonium Nitrate–Ammonium Perchlorate–Potassium Nitrate–Potassium Perchlorate–Water	$NH_4NO_3—NH_4ClO_4—KNO_3—KClO_4—H_2O$	2157
Ammonium Nitrate–Ammonium Metavanadate–Water	$NH_4NO_3—NH_4VO_3—H_2O$	2158
Ammonium Nitrate–Ammonium Magnesium Arsenate–Water	$NH_4NO_3—NH_4MgAsO_4—H_2O$	2159
Ammonium Nitrate–Ammonium Bromide–Water	$NH_4NO_3—NH_4Br—H_2O$	2160
Ammonium Nitrate–Ammonium Molybdate–Water	$NH_4NO_3—(NH_4)_2MoO_4—H_2O$	2161
Ammonium Nitrate–Ammonium Iodide–Water	$NH_4NO_3—NH_4I—H_2O$	2162
Ammonium Nitrate–Sodium Nitrate–Water	$NH_4NO_3—NaNO_3—H_2O$	2163—2172
Ammonium Nitrate–Sodium Nitrate–Sodium Chloride–Water	$NH_4NO_3—NaNO_3—NaCl—H_2O$	2173
Ammonium Nitrate–Sulfur Dioxide–Water	$NH_4NO_3—SO_2—H_2O$	2174, 2175
Ammonium Nitrate–Potassium Nitrate–Water	$NH_4NO_3—KNO_3—H_2O$	2176—2180
Ammonium Nitrate–Potassium Dihydrogen Orthophosphate– Water	$NH_4NO_3—KH_2PO_4—H_2O$	2181
Ammonium Nitrate–Potassium Sulfate–Water	$NH_4NO_3—K_2SO_4—H_2O$	2182
Ammonium Nitrate–Calcium Carbonate–Water	$NH_4NO_3—CaCO_3—H_2O$	2183, 2184
Ammonium Nitrate–Calcium Nitrate–Water	$NH_4NO_3—Ca(NO_3)_2—H_2O$	2185—2188
Ammonium Nitrate–Calcium Hydrogen Orthophosphate–Calcium Sulfite–Water	$NH_4NO_3—CaHPO_4—CaSO_3—H_2O$	2189
Ammonium Nitrate–Calcium Hydrogen Orthophosphate–Calcium Sulfite–Calcium Sulfate–Water	$NH_4NO_3—CaHPO_4—CaSO_3—CaSO_4—H_2O$	2189
Ammonium Nitrate–Calcium Sulfite–Calcium Sulfate–Water	$NH_4NO_3—CaSO_3—CaSO_4—H_2O$	2189
Ammonium Nitrate–Calcium Sulfate–Water	$NH_4NO_3—CaSO_4—H_2O$	2189, 2190
Ammonium Nitrate–Copper (II) Nitrate–Water	$NH_4NO_3—Cu(NO_3)_2—H_2O$	2191

System name	Formulae of components	Table number
Ammonium Nitrate–Rubidium Nitrate–Water	$NH_4NO_3—RbNO_3—H_2O$	2192
Ammonium Nitrate–Strontium Sulfate–Water	$NH_4NO_3—SrSO_4—H_2O$	2193
Ammonium Nitrate–Yttrium Nitrate –Water	$NH_4NO_3—Y(NO_3)_3—H_2O$	2194
Ammonium Nitrate–Silver Nitrate–Water	$NH_4NO_3—AgNO_3—H_2O$	2195
Ammonium Nitrate–Silver Nitrate –Barium Nitrate–Water	$NH_4NO_3—AgNO_3—Ba(NO_3)_2—H_2O$	2196
Ammonium Nitrate–Silver Chloride –Water	$NH_4NO_3—AgCl—H_2O$	2197
Ammonium Nitrate–Silver Chromate–Water	$NH_4NO_3—Ag_2CrO_4—H_2O$	2198, 5413
Ammonium Nitrate–Barium Nitrate –Water	$NH_4NO_3—Ba(NO_3)_2—H_2O$	2199, 2200
Ammonium Nitrate–Barium Arsenate–Water	$NH_4NO_3—Ba_3(AsO_4)_2—H_2O$	2201
Ammonium Nitrate–Lanthanum Nitrate–Water	$NH_4NO_3—La(NO_3)_3—H_2O$	2202
Ammonium Nitrate–Thallium (I) Chloride–Water	$NH_4NO_3—TlCl—H_2O$	2203
Ammonium Nitrate–Lead (II) Nitrate–Water	$NH_4NO_3—Pb(NO_3)_2—H_2O$	2204, 2205
Ammonium Nitrate–Uranyl Nitrate –Water	$NH_4NO_3—UO_2(NO_3)_2—H_2O$	2206
Lanthanum Ammonium Nitrate–Lanthanum Iodate–Water	$NH_4NO_3 \cdot La(NO_3)_3—La(IO_3)_3—H_2O$	2207
Ammonium Fluoride–Ammonium Bromide–Water	$NH_4F—NH_4Br—H_2O$	2208
Ammonium Fluoride–Zirconium Ammonium Fluoride–Water	$NH_4F—Zr(NH_4)_3F_7—H_2O$	2209
Ammonium Fluoride–Beryllium Fluoride–Water	$NH_4F—BeF_2—H_2O$	2210
Ammonium Fluoride–Sodium Fluoride–Water	$NH_4F—NaF—H_2O$	2211
Ammonium Fluoride–Aluminum Fluoride–Water	$NH_4F—AlF_3—H_2O$	2212
Ammonium Fluoride–Potassium Fluoride–Water	$NH_4F—KF—H_2O$	2213
Ammonium Fluoride–Cobalt (II) Fluoride–Water	$NH_4F—CoF_2—H_2O$	2214
Ammonium Fluoride–Nickel (II) Fluoride–Water	$NH_4F—NiF_2—H_2O$	2215
Ammonium Fluoride–Copper (II) Fluoride–Water	$NH_4F—CuF_2—H_2O$	2216
Ammonium Fluoride–Zinc Fluoride –Water	$NH_4F—ZnF_2—H_2O$	2217
Ammonium Fluoride–Molybdenum Trioxide–Water	$NH_4F—MoO_3—H_2O$	2218
Ammonium Fluoride–Cadmium Fluoride–Water	$NH_4F—CdF_2—H_2O$	2219

System name	Formulae of components	Table number
Ammonium Fluoride–Tungsten Trioxide–Water	$NH_4F—WO_3—H_2O$	2220
Ammonium Fluoride–Thorium Nitrate–Water	$NH_4F—Th(NO_3)_4—H_2O$	2221
Ammonium Fluoride–Uranyl Fluoride–Water	$NH_4F—UO_2F_2—H_2O$	2222
Zirconium Ammonium Hexafluoride –Zirconium Ammonium Heptafluoride–Water	$Zr(NH_4)_2F_6—Zr(NH_4)_3F_7—H_2O$	2223
Ammonium Dihydrogen Orthophosphate–Ammonium Sulfate–Water	$NH_4H_2PO_4—(NH_4)_2SO_4—H_2O$	2224, 2225
Ammonium Dihydrogen Orthophosphate–Ammonium Chloride–Water	$NH_4H_2PO_4—NH_4Cl—H_2O$	2226, 2227
Ammonium Dihydrogen Orthophosphate–Ammonium Chloride–Ammonium Magnesium Orthophosphate–Magnesium Chloride–Water	$NH_4H_2PO_4—NH_4Cl—MgNH_4PO_4\cdot 6H_2O—MgCl_2—H_2O$	2228
Ammonium Dihydrogen Orthophosphate–Sodium Nitrate–Water	$NH_4H_2PO_4—NaNO_3—H_2O$	2229—2231
Ammonium Dihydrogen Orthophosphate–Sodium Dihydrogen Orthophosphate–Water	$NH_4H_2PO_4—NaH_2PO_4—H_2O$	2232
Ammonium Dihydrogen Orthophosphate–Potassium Nitrate–Water	$NH_4H_2PO_4—KNO_3—H_2O$	2233
Ammonium Dihydrogen Orthophosphate–Potassium Nitrate–Potassium Chloride–Water	$NH_4H_2PO_4—KNO_3—KCl—H_2O$	2234
Ammonium Dihydrogen Orthophosphate–Potassium Dihydrogen Orthophosphate–Water	$NH_4H_2PO_4—KH_2PO_4—H_2O$	2235—2237
Ammonium Dihydrogen Orthophosphate–Potassium Chloride–Water	$NH_4H_2PO_4—KCl—H_2O$	2238
Ammonium Dihydrogen Orthophosphate–Rubidium Dihydrogen Orthophosphate–Water	$NH_4H_2PO_4—RbH_2PO_4—H_2O$	2239
Ammonium Dihydrogen Orthophosphate–Cesium Dihydrogen Orthophosphate–Water	$NH_4H_2PO_4—CsH_2PO_4—H_2O$	2240
Ammonium Dihydrogen Orthophosphate–Thallium Dihydrogen Orthophosphate–Water	$NH_4H_2PO_4—TlH_2PO_4—H_2O$	2241
Ammonium Hydrogen Orthophosphate–Ammonium Sulfate–Water	$(NH_4)_2HPO_4—(NH_4)_2SO_4—H_2O$	2242
Ammonium Hydrogen Orthophosphate–Ammonium Chloride–Sodium Hydrogen Orthophosphate–Sodium Chloride–Water	$(NH_4)_2HPO_4—NH_4Cl—Na_2HPO_4—NaCl—H_2O$	2243
Ammonium Magnesium Orthophosphate–Sodium Sulfate–Water	$NH_4MgPO_4—Na_2SO_4—H_2O$	2244
Ammonium Magnesium Orthophosphate–Sodium Chloride–Water	$NH_4MgPO_4—NaCl—H_2O$	2245
Ammonium Sulfito Amide–Ammonium Sulfito Imide–Water	$NH_4SO_3NH_2—(NH_4SO_3)_2NH—H_2O$	2246
Ammonium Sulfito Amide–Ammonium Sulfito Imide–Ammonium Sulfate–Water	$NH_4SO_3NH_2—(NH_4SO_3)_2NH—(NH_4)_2SO_4—H_2O$	2247
Ammonium Sulfito Amide–Ammonium Sulfate–Water	$NH_4SO_3NH_2—(NH_4)_2SO_4—H_2O$	2248, 2249
Ammonium Sulfito Imide–Ammonium Sulfate–Water	$(NH_4SO_3)_2NH—(NH_4)_2SO_4—H_2O$	2250

Продолжение

Наименование системы	Химические формулы компонентов	Номера таблиц
Ammonium Sulfito Amide–Potassium Sulfito Amide–Water	$NH_4SO_3NH_2—KSO_3NH_2—H_2O$	2251
Аммония бисульфит—аммония сульфит—вода	$NH_4HSO_3—(NH_4)_2SO_3—H_2O$	2252
Ammonium Hydrogen Sulfite–Ammonium Sulfite–Ammonium Sulfate–Water	$NH_4HSO_3—(NH_4)_2SO_3—(NH_4)_2SO_4—H_2O$	2253, 2254
Ammonium Hydrogen Sulfite–Ammonium Sulfate–Water	$NH_4HSO_3—(NH_4)_2SO_4—H_2O$	2255
Ammonium Hydrogen Sulfite–Ammonium Chloride–Sodium Hydrogen Sulfite–Sodium Chloride–Water	$NH_4HSO_3—NH_4Cl—NaHSO_3—NaCl—H_2O$	2256
Ammonium Sulfite–Ammonium Sulfate–Water	$(NH_4)_2SO_3—(NH_4)_2SO_4—H_2O$	2257
Ammonium Sulfite–Ammonium Sulfate–Ammonium Chloride –Water	$(NH_4)_2SO_3—(NH_4)_2SO_4—NH_4Cl—H_2O$	2258
Ammonium Sulfite–Ammonium Chloride–Water	$(NH_4)_2SO_3—NH_4Cl—H_2O$	2259
Ammonium Sulfite–Ammonium Chloride–Sodium Sulfite–Sodium Chloride	$(NH_4)_2SO_3—NH_4Cl—Na_2SO_3—NaCl—H_2O$	2260, 2261
Ammonium Sulfite–Ammonium Heptoxodiuranate–Water	$(NH_4)_2SO_3—(NH_4)_2U_2O_7—H_2O$	2262
Ammonium Sulfite–Sodium Sulfite –Water	$(NH_4)_2SO_3—Na_2SO_3—H_2O$	2263, 2264
Ammonium Hydrogen Sulfate–Molybdenum Trioxide–Water	$NH_4HSO_4—MoO_3 \cdot 2H_2O—H_2O$	2265
Ammonium Sulfate–Ammonium Peroxodisulfate–Water	$(NH_4)_2SO_4—(NH_4)_2S_2O_8—H_2O$	2266
Ammonium Sulfate–Ammonium Chloride–Water	$(NH_4)_2SO_4—NH_4Cl—H_2O$	2267—2274
Ammonium Sulfate–Ammonium Chloride–Sodium Sulfate–Sodium Chloride–Water	$(NH_4)_2SO_4—NH_4Cl—Na_2SO_4—NaCl—H_2O$	2275
Ammonium Sulfate–Ammonium Chloride–Potassium Sulfate–Potassium Chloride–Water	$(NH_4)_2SO_4—NH_4Cl—K_2SO_4—KCl—H_2O$	2276
Ammonium Sulfate–Ammonium Perchlorate–Water	$(NH_4)_2SO_4—NH_4ClO_4—H_2O$	2277—2278
Ammonium Sulfate–Ammonium Chromate–Water	$(NH_4)_2SO_4—(NH_4)_2CrO_4—H_2O$	2279
Ammonium Sulfate–Ammonium Bromide–Water	$(NH_4)_2SO_4—NH_4Br—H_2O$	2280
Ammonium Sulfate–Ammonium Magnesium Orthophosphate– Water	$(NH_4)_2SO_4—NH_4MgPO_4—H_2O$	2281
Ammonium Sulfate–Ammonium Molybdate–Water	$(NH_4)_2SO_4—(NH_4)_3MoO_4—H_2O$	2282
Ammonium Sulfate–Ammonium Paratungstate–Water	$(NH_4)_2SO_4—5(NH_4)_2O \cdot 12WO_3 \cdot 9H_2O—H_2O$	2283
Ammonium Sulfate–Sodium Tetraborate–Sodium Sulfate– Water	$(NH_4)_2SO_4—Na_2B_4O_7—Na_2SO_4—H_2O$	2284, 2285
Ammonium Sulfate–Sodium Hydrogen Carbonate–Sodium Sulfate–Water	$(NH_4)_2SO_4—NaHCO_3—Na_2SO_4—H_2O$	2286, 2287

System name	Formulae of components	Table number
Ammonium Sulfate–Sodium Sulfate –Water	$(NH_4)_2SO_4—Na_2SO_4—H_2O$	2288—2301
Ammonium Sulfate–Sodium Sulfate –Potassium Sulfate–Water	$(NH_4)_2SO_4—Na_2SO_4—K_2SO_4—H_2O$	2302, 2303
Ammonium Sulfate–Magnesium Sulfate–Water	$(NH_4)_2SO_4—MgSO_4—H_2O$	2304—2308
Ammonium Sulfate–Magnesium Sulfate–Potassium Sulfate– Water	$(NH_4)_2SO_4—MgSO_4—K_2SO_4—H_2O$	2309
Ammonium Sulfate–Aluminum Sulfate–Water	$(NH_4)_2SO_4—Al_2(SO_4)_3—H_2O$	2310—2313
Ammonium Sulfate–Sulfur Dioxide –Water	$(NH_4)_2SO_4—SO_2—H_2O$	2314
Ammonium Sulfate–Sulfur Trioxide –Water	$(NH_4)_2SO_4—SO_3—H_2O$	2315
Ammonium Sulfate–Potassium Sulfate–Water	$(NH_4)_2SO_4—K_2SO_4—H_2O$	2316—2323
Ammonium Sulfate–Calcium Orthophosphate–Water	$(NH_4)_2SO_4—Ca_3(PO_4)_2—H_2O$	5113
Ammonium Sulfate–Calcium Sulfate –Water	$(NH_4)_2SO_4—CaSO_4—H_2O$	2324—2336
Ammonium Sulfate–Oxotitanium Sulfate–Water	$(NH_4)_2SO_4—TiOSO_4—H_2O$	2337
Ammonium Sulfate–Manganese (II) Sulfate–Water	$(NH_4)_2SO_4—MnSO_4—H_2O$	2338—2342
Ammonium Sulfate–Iron (II) Sulfate –Water	$(NH_4)_2SO_4—FeSO_4—H_2O$	2343, 2344
Ammonium Sulfate–Cobalt (II) Sulfate–Water	$(NH_4)_2SO_4—CoSO_4—H_2O$	2345, 2346
Ammonium Sulfate–Nickel (II) Sulfate–Water	$(NH_4)_2SO_4—NiSO_4—H_2O$	2347, 2348
Ammonium Sulfate–Copper (II) Sulfate–Water	$(NH_4)_2SO_4—CuSO_4—H_2O$	2349—2352
Ammonium Sulfate–Copper (II) Sulfate–Thallium (I) Sulfate–Water	$(NH_4)_2SO_4—CuSO_4—Tl_2SO_4—H_2O$	2353
Ammonium Sulfate–Zinc Sulfate–Water	$(NH_4)_2SO_4—ZnSO_4—H_2O$	2354, 2355
Ammonium Sulfate–Rubidium Sulfate–Water	$(NH_4)_2SO_4—Rb_2SO_4—H_2O$	2356
Ammonium Sulfate–Strontium Sulfate–Water	$(NH_4)_2SO_4—SrSO_4—H_2O$	2357, 2358
Ammonium Sulfate–Molybdenum Trioxide–Water	$(NH_4)_2SO_4—MoO_3\cdot 2H_2O—H_2O$	2359, 2360
Ammonium Sulfate–Silver Sulfate –Water	$(NH_4)_2SO_4—Ag_2SO_4—H_2O$	2360—2362
Ammonium Sulfate–Cadmium Suofate–Water	$(NH_4)_2SO_4—CdSO_4—H_2O$	2363
Ammonium Sulfate–Indium Sulfate –Water	$(NH_4)_2SO_4—In_2(SO_4)_3—H_2O$	2364
Ammonium Sulfate–Cesium Sulfate –Water	$(NH_4)_2SO_4—Cs_2SO_4—H_2O$	2365
Ammonium Sulfate–Lanthanum Sulfate–Water	$(NH_4)_2SO_4—La_2(SO_4)_3—H_2O$	2366—2368

System name	Formulae of components	Table number
Ammonium Sulfate–Cerium Sulfate –Water	$(NH_4)_2SO_4—Ce_2(SO_4)_3—H_2O$	2369—2371
Ammonium Sulfate–Praseodymium Sulfate–Water	$(NH_4)_2SO_4—Pr_2(SO_4)_3—H_2O$	2372
Ammonium Sulfate–Neodymium Sulfate–Water	$(NH_4)_2SO_4—Nd_2(SO_4)_3—H_2O$	2373
Ammonium Sulfate–Samarium Sulfate–Water	$(NH_4)_2SO_4—Sm_2(SO_4)_3—H_2O$	2374
Ammonium Sulfate–Thallium Iodide –Water	$(NH_4)_2SO_4—TlI—H_2O$	2375
Ammonium Sulfate–Lead (II) Sulfate–Water	$(NH_4)_2SO_4—PbSO_4—H_2O$	2376
Ammonium Sulfate–Bismuth Sulfate –Water	$(NH_4)_2SO_4—Bi_2(SO_4)_3—H_2O$	2377
Ammonium Sulfate–Thorium Sulfate–Water	$(NH_4)_2SO_4—Th(SO_4)_2—H_2O$	2378, 2379
Ammonium Sulfate–Uranyl Sulfite –Water	$(NH_4)_2SO_4—UO_2SO_3—H_2O$	2380
Ammonium Sulfate–Uranyl Sulfate –Water	$(NH_4)_2SO_4—UO_2SO_4—H_2O$	2381
Ammonium Sodium Sulfate–Sodium Hydrogen Carbonate–Sodium Sulfate–Water	$Na_2SO_4\cdot(NH_4)_2SO_4—NaHCO_3—Na_2SO_4—H_2O$	2382
Ammonium Magnesium Sulfate–Ammonium Magnesium Chromate–Water	$(NH_4)_2Mg(SO_4)_2—(NH_4)_2Mg(CrO_4)_2—H_2O$	2383
Ammonium Magnesium Sulfate–Ammonium Magnesium Selenate–Water	$(NH_4)_2Mg(SO_4)_2—(NH_4)_2Mg(SeO_4)_2—H_2O$	2384
Ammonium Magnesium Sulfate–Ammonium Manganese Sulfate– Water	$(NH_4)_2Mg(SO_4)_2—(NH_4)_2Mn(SO_4)_2—H_2O$	2385
Ammonium Magnesium Sulfate–Ammonium Manganese Sulfate–Ammonium Nickel Sulfate–Water	$MgSO_4\cdot(NH_4)_2SO_4—MnSO_4\cdot(NH_4)_2SO_4—NiSO_4\cdot(NH_4)_2SO_4—H_2O$	2386
Ammonium Magnesium Sulfate–Ammonium Iron Sulfate–Water	$(NH_4)_2Mg(SO_4)_2—(NH_4)_2Fe(SO_4)_2—H_2O$	2387
Ammonium Magnesium Sulfate–Ammonium Iron Sulfate–Water	$(NH_4)_2Mg(SO_4)_2—(NH_4)_2Fe(SO_4)_2—(NH_4)_2Ni(SO_4)_2—H_2O$	2388
Ammonium Magnesium Sulfate–Ammonium Nickel Sulfate–Water	$(NH_4)_2SO_4\cdot MgSO_4—(NH_4)_2SO_4\cdot NiSO_4—H_2O$	2389, 2390
Ammonium Magnesium Sulfate–Ammonium Copper Sulfate–Water	$(NH_4)_2MgSO_4—Cu(NH_4)_2SO_4—H_2O$	2391
Aluminum Ammonium Sulfate–Ammonium Chromium Sulfate–Water	$NH_4Al(SO_4)_2—NH_4Cr(SO_4)_2—H_2O$	2392
Aluminium Ammonium Sulfate–Ammonium Iron Sulfate–Water	$NH_4Al(SO_4)_2—NH_4Fe(SO_4)_2\cdot 12H_2O—H_2O$	2393, 2394
Aluminium Ammonium Sulfate–Aluminum Cesium Sulfate–Water	$(NH_4)_2Al_2(SO_4)_2—Cs_2Al_2(SO_4)_4—H_2O$	2395
Aluminum Ammonium Sulfate–Aluminum Thallium Sulfate–Water	$NH_4Al(SO_4)_2—TlAl(SO_4)_2—H_2O$	2396
Ammonium Chromium Sulfate–Ammonium Iron Sulfate–Water	$NH_4Cr(SO_4)_2—NH_4Fe(SO_4)_2—H_2O$	2397

System name	Formulae of components	Table number
Ammonium Chromium Sulfate–Chromium Potassium Sulfate–Water	$NH_4Cr(SO_4)_2$—$KCr(SO_4)_2$—H_2O	2398
Ammonium Manganese Sulfate–Ammonium Iron Sulfate–Water	$(NH_4)_2Mn(SO_4)_2$—$(NH_4)_2Fe(SO_4)_2$—H_2O	2399
Ammonium Manganese Sulfate–Ammonium Iron Sulfate–Ammonium Copper Sulfate–Water	$MnSO_4(NH_4)_2SO_4$—$FeSO_4(NH_4)_2SO_4$—$CuSO_4 \cdot (NH_4)_2SO_4$—$H_2O$	2400
Ammonium Manganese Sulfate–Ammonium Iron Sulfate–Ammonium Zinc Sulfate–Water	$MnSO_4 \cdot (NH_4)_2SO_4$—$FeSO_4 \cdot (NH_4)_2SO_4$—$ZnSO_4 \cdot (NH_4)_2SO_4$—$H_2O$	2401
Ammonium Manganese Sulfate–Ammonium Nickel Sulfate–Water	$(NH_4)_2SO_4 \cdot MnSO_4$—$(NH_4)_2SO_4 \cdot NiSO_4$—$H_2O$	2402
Ammonium Manganese Sulfate–Ammonium Copper Sulfate–Water	$(NH_4)_2SO_4 \cdot MnSO_4$—$CuSO_4 \cdot (NH_4)_2SO_4$—$H_2O$	2403, 2404
Ammonium Manganese Sulfate–Ammonium Zinc Sulfate–Water	$(NH_4)_2SO_4 \cdot MnSO_4$—$(NH_4)_2SO_4 \cdot ZnSO_4$—$H_2O$	2405
Ammonium Iron Sulfate–Ammonium Cobalt Sulfate–Water	$(NH_4)_2Fe(SO_4)_2$—$(NH_4)_2Co(SO_4)_2$—H_2O	2406
Ammonium Iron Sulfate–Ammonium Nickel Sulfate–Water	$(NH_4)_2Fe(SO_4)_2$—$(NH_4)_2Ni(SO_4)_2$—H_2O	2407, 2408
Ammonium Iron Sulfate–Ammonium Copper Sulfate–Water	$(NH_4)_2SO_4 \cdot FeSO_4$—$CuSO_4 \cdot (NH_4)_2SO_4$—$H_2O$	2409, 2410
Ammonium Iron Sulfate–Ammonium Zinc Sulfate–Water	$Fe(NH_4)_2(SO_4)_2$—$Zn(NH_4)_2(SO_4)_2$—H_2O	2411, 2412
Ammonium Iron Sulfate–Aluminum Potassium Sulfate–Water	$NH_4Fe(SO_4)_2$—$KAl(SO_4)_2$—H_2O	2413
Ammonium Cobalt Sulfate–Ammonium Nickel Sulfate–Water	$(NH_4)_2Co(SO_4)_2$—$(NH_4)_2Ni(SO_4)_2$—H_2O	2414
Ammonium Cobalt Sulfate–Ammonium Copper Sulfate–Water	$CoSO_4 \cdot (NH_4)_2SO_4$—$CuSO_4 \cdot (NH_4)_2SO_4$—$H_2O$	2415, 2416
Ammonium Nickel Sulfate–Ammonium Copper Sulfate–Water	$(NH_4)_2SO_4 \cdot NiSO_4$—$CuSO_4 \cdot (NH_4)_2SO_4$—$H_2O$	2417—2420
Ammonium Nickel Sulfate–Ammonium Zinc Sulfate–Water	$(NH_4)_2SO_4 \cdot NiSO_4$—$(NH_4)_2SO_4 \cdot ZnSO_4$—$H_2O$	2421
Ammonium Copper Sulfate–Ammonium Zinc Sulfate–Water	$CuSO_4 \cdot (NH_4)_2SO_4$—$(NH_4)_2SO_4 \cdot ZnSO_4$—$H_2O$	2422—2424
Ammonium Copper Sulfate–Copper Potassium Sulfate–Water	$CuSO_4 \cdot (NH_4)_2SO_4$—$CuSO_4 \cdot K_2SO_4$—$H_2O$	2425, 2526
Ammonium Copper Sulfate–Copper Thallium Sulfate–Water	$CuSO_4 \cdot (NH_4)_2SO_4 \cdot 6H_2O$—$CuSO_4 \cdot Tl_2SO_4 \cdot 6H_2O$—$H_2O$	2427
Ammonium Thiosulfate–Silver Chloride–Water	$(NH_4)_2S_2O_3$—$AgCl$—H_2O	2569
Ammonium Dithionate–Copper (II) Dithionate–Water	$(NH_4)_2S_2O_6$—CuS_2O_6—H_2O	2428
Ammonium Dithionate–Strontium Dithionate–Water	$(NH_4)_2S_2O_6$—SrS_2O_6—H_2O	2429
Ammonium Dithionate–Barium Dithionate–Water	$(NH_4)_2S_2O_6$—BaS_2O_6—H_2O	2430
Ammonium Persulfate–Potassium Dithionate–Water	$(NH_4)_2S_2O_8$—$K_2S_2O_8$—H_2O	2431

System name	Formulae of components	Table number
Ammonium Trithionate–Ammonium Tetrathionate–Ammonium Pentathionate–Water	$(NH_4)_2S_3O_6$—$(NH_4)_2S_4O_6$—$(NH_4)_2S_5O_6$—H_2O	2432
Ammonium Chloride–Ammonium Perchlorate–Water	NH_4Cl—NH_4ClO_4—H_2O	2434
Ammonium Chloride–Ammonium Metavanadate–Water	NH_4Cl—NH_4VO_3—H_2O	2435—2437
Ammonium Chloride–Ammonium Metavanadate–Sodium Chloride–Sodium Metavanadate–Water	NH_4Cl—NH_4VO_3—$NaCl$—$NaVO_3$—H_2O	2438
Ammonium Chloride–Ammonium Dichromate–Water	NH_4Cl—$(NH_4)_2Cr_2O_7$—H_2O	2439, 2440
Ammonium Chloride–Ammonium Dichromate–Potassium Chloride–Potassium Dichromate–Water	NH_4Cl—$(NH_4)_2Cr_2O_7$—KCl—$K_2Cr_2O_7$—H_2O	2441
Ammonium Chloride–Ammonium Paramolybdate–Water	NH_4Cl—$3(NH_4)_2O\cdot 7MoO_3\cdot 4H_2O$—$H_2O$	2442
Ammonium Chloride–Ammonium Bromide–Water	NH_4Cl—NH_4Br—H_2O	2443
Ammonium Chloride–Ammonium Bromide–Ammonium Iodide–Water	NH_4Cl—NH_4Br—NH_4I—H_2O	2444
Ammonium Chloride–Ammonium Molybdate–Water	NH_4Cl—$(NH_4)_2MoO_4$—H_2O	2445
Ammonium Chloride–Diamminepalladium Chloride–Water	NH_4Cl—$Pd(NH_3)_2Cl_2$—H_2O	2452
Ammonium Chloride–Ammonium Iodide–Water	NH_4Cl—NH_4I—H_2O	2446
Ammonium Chloride–Ammonium Paratungstate–Water	NH_4Cl—$5(NH_4)_2O\cdot 12WO_3$—H_2O	2447
Ammonium Chloride–Ammonium Magnesium Orthophosphate–Water	NH_4Cl—NH_4MgPO_4—H_2O	2448, 2449
Ammonium Chloride–Ammonium Magnesium Arsenate–Water	NH_4Cl—NH_4MgAsO_4—H_2O	2450
Ammonium Chloride–Ammonium Manganese Orthophosphate–Water	NH_4Cl—NH_4MnPO_4—H_2O	2451
Ammonium Chloride–Ammonium Hexachloroiridate–Water	NH_4Cl—$(NH_4)_2IrCl_6$—H_2O	2453
Ammonium Chloride–Ammonium Hexachloroplatinate–Water	NH_4Cl—$(NH_4)_2PtCl_6$—H_2O	948, 2454
Ammonium Chloride–Sodium Tetraborate–Sodium Chloride–Water	NH_4Cl—$Na_2B_4O_7$—$NaCl$—H_2O	2455
Ammonium Chloride–Sodium Hydrogen Carbonate–Sodium Chloride–Water	NH_4Cl—$NaHCO_3$—$NaCl$—H_2O	2456
Ammonium Chloride–Sodium Thiocyanate–Water	NH_4Cl—$NaCNS$—H_2O	2457
Ammonium Chloride–Sodium Chloride–Water	NH_4Cl—$NaCl$—H_2O	2458—2473

System name	Formulae of components	Table number
Ammonium Chloride–Sodium Chloride–Potassium Chloride– Water	$NH_4Cl—NaCl—KCl—H_2O$	2474, 2475
Ammonium Chloride–Sodium Chloride–Potassium Chloride– Water	$NH_4Cl—NaCl—KCl—H_2O$	2476
Ammonium Chloride–Sodium Metavanadate–Water	$NH_4Cl—NaVO_3—H_2O$	2477
Ammonium Chloride–Magnesium Chloride–Water	$NH_4Cl—MgCl_2—H_2O$	2478—2480
Ammonium Chloride–Magnesium Chloride–Potassium Chloride– Water	$NH_4Cl—MgCl_2—KCl—H_2O$	2481
Ammonium Chloride–Sulfur Dioxide –Water	$NH_4Cl—SO_2—H_2O$	2482
Ammonium Chloride–Potassium Nitrate–Water	$NH_4Cl—KNO_3—H_2O$	2483
Ammonium Chloride–Potassium Sulfate–Water	$NH_4Cl—K_2SO_4—H_2O$	2484
Ammonium Chloride–Potassium Chloride–Water	$NH_4Cl—KCl—H_2O$	2485—2494
Ammonium Chloride–Potassium Chromate–Water	$NH_4Cl—K_2CrO_4—H_2O$	2495
Ammonium Chloride–Potassium Dichromate–Water	$NH_4Cl—K_2Cr_2O_7—H_2O$	2496
Ammonium Chloride–Potassium Manganate–Water	$NH_4Cl—KMnO_4—H_2O$	2497
Ammonium Chloride–Calcium Oxide–Arsenic Pentoxide–Water	$NH_4Cl—CaO—As_2O_5—H_2O$	2499
Ammonium Chloride–Calcium Hydroxide–Water	$NH_4Cl—Ca(OH)_2—H_2O$	2500
Ammonium Chloride–Calcium Carbonate–Water	$NH_4Cl—CaCO_3—H_2O$	2501—2503
Ammonium Chloride–Calcium Orthophosphate–Water	$NH_4Cl—Ca_3(PO_4)_2—H_2O$	5113
Ammonium Chloride–Calcium Sulfate–Water	$NH_4Cl—CaSO_4—H_2O$	2504, 2505
Ammonium Chloride–Calcium Iodate–Water	$NH_4Cl—Ca(IO_3)_2—H_2O$	2506
Ammonium Chloride–Titanium (III) Chloride–Water	$NH_4Cl—TiCl_3—H_2O$	2507
Ammonium Chloride–Manganese (II) Chloride–Water	$NH_4Cl—MnCl_2—H_2O$	2508, 2509
Ammonium Chloride–Iron (II) Hydroxide–Water	$NH_4Cl—Fe(OH)_2—H_2O$	2510
Ammonium Chloride–Iron (II) Chloride–Water	$NH_4Cl—FeCl_2—H_2O$	2511, 2512
Ammonium Chloride Iron (III) Chloride–Water	$NH_4Cl—FeCl_3—H_2O$	2513—2515
Ammonium Chloride–Cobalt (II) Chloride–Water	$NH_4Cl—CoCl_2—H_2O$	2516—2519
Ammonium Chloride–Cobalt Mercury Thiocyanate–Water	$NH_4Cl—CoHg(CNS)_4—H_2O$	2520
Ammonium Chloride–Nickel (II) Chloride–Water	$NH_4Cl—NiCl_2—H_2O$	2521—2523

System name	Formulae of components	Table number
Ammonium Chloride–Copper (II) Chloride–Water	$NH_4Cl—CuCl_2—H_2O$	2524—2526
Ammonium Chloride–Zinc Hydroxide–Water	$NH_4Cl—Zn(OH)_2—H_2O$	2527
Ammonium Chloride–Zinc Chloride –Water	$NH_4Cl—ZnCl_2—H_2O$	2528
Ammonium Chloride–Mercury Zinc Thiocyanate–Water	$NH_4Cl—HgZn(CNS)_4—H_2O$	2529
Ammonium Chloride–Strontium Carbonate–Water	$NH_4Cl—SrCO_3—H_2O$	2530
Ammonium Chloride–Strontium Sulfate–Water	$NH_4Cl—SrSO_4—H_2O$	2531
Ammonium Chloride–Silver Chloride –Water	$NH_4Cl—AgCl—H_2O$	2532—2535, 5411, 5412
Ammonium Chloride–Cadmium Chloride–Water	$NH_4Cl—CdCl_2—H_2O$	2536
Ammonium Chloride–Barium Carbonate–Water	$NH_4Cl—BaCO_3—H_2O$	2537
Ammonium Chloride–Barium Nitrate–Water	$NH_4Cl—Ba(NO_3)_2—H_2O$	2538
Ammonium Chloride–Barium Chloride–Water	$NH_4Cl—BaCl_2—H_2O$	2539, 2540
Ammonium Chloride–Barium Hydrogen Arsenate–Water	$NH_4Cl—BaHAsO_4—H_2O$	2542
Ammonium Chloride–Barium Arsenate–Water	$NH_4Cl—Ba_3(AsO_4)_2—H_2O$	2541
Ammonium Chloride–Barium Iodate –Water	$NH_4Cl—Ba(IO_3)_2—H_2O$	2543
Ammonium Chloride–Mercury (II) Chloride–Water	$NH_4Cl—HgCl_2—H_2O$	2544
Ammonium Chloride–Thallium (I) Chloride–Water	$NH_4Cl—TlCl—H_2O$	2545
Ammonium Chloride–Lead (II) Chloride–Water	$NH_4Cl—PbCl_2—H_2O$	2546—2551
Ammonium Chloride–Cadmium Tetrammonium Chloride–Water	$NH_4Cl—4NH_4Cl\cdot CdCl_2—H_2O$	2552
Diamminepalladium Chloride–Sodium Chloride–Water	$Pd(NH_3)_2Cl_2—NaCl—H_2O$	2452
Ammonium Perchlorate–Sodium Perchlorate–Water	$NH_4ClO_4—NaClO_4—H_2O$	2553, 2554
Ammonium Perchlorate–Aluminum Perchlorate–Water	$NH_4ClO_4—Al(ClO_4)_3—H_2O$	2555
Ammonium Perchlorate–Potassium Perchlorate–Water	$NH_4ClO_4—KClO_4—H_2O$	2556
Ammonium Perchlorate–Copper (II) Perchlorate–Water	$NH_4ClO_4—Cu(ClO_4)_2—H_2O$	2557
Ammonium Metavanadate–Sodium Chloride–Water	$NH_4VO_3—NaCl—H_2O$	2558
Ammonium Metavanadate–Sodium Metavanadate–Water	$NH_4VO_3—NaVO_3—H_2O$	2559
Ammonium Metavanadate–Sodium Pyrovanadate–Water	$NH_4VO_3—Na_4V_2O_7—H_2O$	2560

System name	Formulae of components	Table number
Ammonium Chromate–Magnesium Chromate–Water	$(NH_4)_2CrO_4—MgCrO_4—H_2O$	2561
Ammonium Chromate–Potassium Chromate–Water	$(NH_4)_2CrO_4—K_2CrO_4—H_2O$	2562
Ammonium Dichromate–Sodium Didromate–Water	$(NH_4)_2Cr_2O_7—Na_2Cr_2O_7—H_2O$	2563
Ammonium Dichronate–Potassium Dichromate–Water	$(NH_4)_2Cr_2O_7—K_2Cr_2O_7—H_2O$	2564, 2565
Ammonium Selenate–Magnesium Selenate–Water	$(NH_4)_2SeO_4—MgSeO_4—H_2O$	2566
Ammonium Selenate–Calcium Selenate–Water	$(NH_4)_2SeO_4—CaSeO_4—H_2O$	2567
Ammonium Selenate–Copper (II) Selenate–Water	$(NH_4)_2SeO_4—CuSeO_4—H_2O$	2568
Ammonium Bromide–Ammonium Hexabromoplatinate–Water	$NH_4Br—(NH_4)_2PtBr_6—H_2O$	2570
Ammonium Bromide–Sodium Sulfide–Water	$NH_4Br—Na_2S—H_2O$	2571
Ammonium Bromide–Sodium Chloride–Water	$NH_4Br—NaCl—H_2O$	3724
Ammonium Bromide–Magnesium Bromide–Water	$NH_4Br—MgBr_2—H_2O$	2572
Ammonium Bromide–Sulfur Dioxide –Water	$NH_4Br—SO_2—H_2O$	2573
Ammonium Bromide–Potassium Bromide–Water	$NH_4Br—KBr—H_2O$	2574—2576
Ammonium Bromide–Iron (II) Bromide–Water	$NH_4Br—FeBr_2—H_2O$	2577
Ammonium Bromide–Cobalt (II) Bromide–Water	$NH_4Br—CoBr_2—H_2O$	2578
Ammonium Bromide–Copper (I) Iodide–Water	$NH_4Br—CuI—H_2O$	2579
Ammonium Bromide–Arsenic Trioxide–Water	$NH_4Br—As_2O_3—H_2O$	2580
Ammonium Bromide–Silver Bromide–Water	$NH_4Br—AgBr—H_2O$	2581
Ammonium Bromide–Lead (II) Bromide–Water	$NH_4Br—PbBr_2—H_2O$	2582
Ammonium Iodide–Sodium Chloride –Water	$NH_4I—NaCl—H_2O$	3724
Ammonium Iodide–Antimony Triiodide–Water	$NH_4I—SbI_3—H_2O$	2583, 2584
Ammonium Iodide–Lead (II) Iodide –Water	$NH_4I—PbI_2—H_2O$	2585
Ammonium Iodide–Thallium Triiodide–Water	$NH_4I—TlI_3—H_2O$	2586
Ammonium Magnesium Chromate–Ammonium Magnesium Selenate–Water	$(NH_4)_2H_3IO_6—Na_2H_3IO_6—H_2O$	2587
Ammonium Magnesium Chromate–Ammonium Magnesium Selenate –Water	$(NH_4)_2Mg(CrO_4)_2—(NH_4)_2Mg(SeO_4)_2—H_2O$	2588
Ammonium Magnesium Selenate–Magnesium Potassium Selenate–Water	$(NH_4)_2Mg(SeO_4)_2—K_2Mg(SeO_4)_2—H_2O$	2589

System name	Formulae of components	Table number
Sodium Oxide–Sodium Chloride–Barium Oxide–Barium Chloride–Water	$Na_2O—NaCl—BaO—BaCl_2—H_2O$	2590
Sodium Oxide–Sodium Bromide–Water	$Na_2O—NaBr—H_2O$	2591
Sodium Oxide–Aluminum Oxide–Water	$Na_2O—Al_2O_3—H_2O$	2592—2595
Sodium Oxide–Silicon Dioxide–Water	$Na_2O—SiO_2—H_2O$	2596—2600
Sodium Oxide–Phosphorus Tetroxide–Water	$Na_2O—P_2O_4—H_2O$	2601
Sodium Oxide–Phosphorus Pentoxide–Water	$Na_2O—P_2O_5—H_2O$	2602, 2603
Sodium Oxide [illegible]assium Oxide–Calcium Ox[illegible]–Calcium Sulfate–Water	$Na_2O—K_2O—CaO—CaSO_4—H_2O$	2604
Sodium Oxide–Sulfur Trioxide–Vanadium Pentoxide–Water	$Na_2O—SO_3—V_2O_5—H_2O$	2605
Sodium Oxide–Vandaium Pentoxide–Water	$Na_2O—V_2O_5—H_2O$	2606—2608
Sodium Oxide–Zinc Oxide–Water	$Na_2O—ZnO—H_2O$	2609, 2610
Sodium Oxide–Gallium Oxide–Water	$Na_2O—Ga_2O_3—H_2O$	2611, 2612
Sodium Oxide–Arsenic Trioxide–Water	$Na_2O—As_2O_3—H_2O$	2613, 2614
Sodium Oxide–Arsenic Pentoxide–Water	$Na_2O—As_2O_5—H_2O$	2615—2619a,
Sodium Oxide–Selenium Dioxide–Water	$Na_2O—SeO_2—H_2O$	2619б
Sodium Oxide–Molybdenum Oxide–Water	$Na_2O—MoO_3—H_2O$	2620
Sodium Oxide–Antimony Pentoxide–Water	$Na_2O—Sb_2O_5—H_2O$	2621
Sodium Oxide–Tin (II) Oxide–Water	$Na_2O—SnO_2—H_2O$	2622
Sodium Oxide–Tellurium Oxide–Water	$Na_2O—TeO_2—H_2O$	2623, 2624
Sodium Oxide–Lead (II) Oxide–Water	$Na_2O—PbO—H_2O$	2625
Sodium Hydroxide–Sodium Tetrahydroborate–Water	$NaOH—NaBH_4—H_2O$	2626, 2627
Sodium Hydroxide–Sodium Tetrahydroborate–Sodium Chloride–Water	$NaOH—NaBH_4—NaCl—H_2O$	2628
Sodium Hydroxide–Sodium Carbonate–Water	$NaOH—Na_2CO_3—H_2O$	2629—2646
Sodium Hydroxide–Sodium Carbonate–Sodium Sulfate–Water	$NaOH—Na_2CO_3—Na_2SO_4—H_2O$	2647—2651
Sodium Hydroxide–Sodium Carbonate–Sodium Sulfate–Sodium Chloride–Water	$NaOH—Na_2CO_3—Na_2SO_4—NaCl—H_2O$	2652—2655
Sodium Hydroxide–Sodium Carbonate–Sodium Sulfate–Calcium Hydroxide–Calcium Carbonate–Calcium Sulfate–Water	$NaOH—Na_2CO_3—Na_2SO_4—Ca(OH)_2—CaCO_3—CaSO_4—H_2O$	2656
Sodium Hydroxide–Sodium Carbonate–Sodium Chloride–Water	$NaOH—Na_2CO_3—NaCl—H_2O$	2657—2659

System name	Formulae of components	Table number
Sodium Hydroxide–Sodium Carbonate–Sodium Chloride–Potassium Hydroxide–Potassium Carbonate–Potassium Chloride–Water	$NaOH—Na_2CO_3—NaCl—KOH—K_2CO_3—KCl—H_2O$	2660
Sodium Hydroxide–Sodium Cyanide–Water	$NaOH—NaCN—H_2O$	2661
Sodium Hydroxide–Sodium Nitrate–Water	$NaOH—NaNO_3—H_2O$	2662
Sodium Hydroxide–Sodium Nitrate–Potassium Nitrate–Water	$NaOH—NaNO_3—KNO_3—H_2O$	2663
Sodium Hydroxide–Sodium Nitrate–Mercury (II) Oxide–Water	$NaOH—NaNO_3—HgO—H_2O$	2664
Sodium Hydroxide–Sodium Fluoride–Water	$NaOH—NaF—H_2O$	2665, 2666
Sodium Hydroxide–Sodium Aluminum Oxide–Sodium Sulfate–Water	$NaOH—NaAlO_2—Na_2SO —H_2O$	2667
Sodium Hydroxide–Sodium Aluminum Oxide–Sodium Sulfate–Potassium Hydroxide–Potassium Aluminum Oxide–Potassium Sulfate–Water	$NaOH—NaAlO_2—Na_2SO_4—KOH—KAlO_4—K_2SO_4—H_2O$	2668
Sodium Hydroxide–Sodium Aluminum Oxide–Potassium Hydroxide–Potassium Aluminum Oxide–Water	$NaOH—NaAlO_2—KOH—KAlO_2—H_2O$	2669
Sodium Hydroxide–Sodium Silicate–Water	$NaOH—Na_2SiO_3—H_2O$	2670
Sodium Hydroxide–Sodium Hydrogen Orthophosphate–Water	$NaOH—Na_2HPO_4—H_2O$	2671
Sodium Hydroxide–Sodium Orthophosphate–Water	$NaOH—Na_3PO_4—H_2O$	2672
Sodium Hydroxide–Sodium Sulfide–Bismuth Sulfide–Water	$NaOH—Na_2S—Bi_2S_3—H_2O$	2673
Sodium Hydroxide–Sodium Sulfite–Water	$NaOH—Na_2SO_3—H_2O$	2674
Sodium Hydroxide–Sodium Sulfate–Water	$NaOH—Na_2SO_4—H_2O$	2675—2689
Sodium Hydroxide–Sodium Sulfate–Sodium Chloride–Water	$NaOH—Na_2SO_4—NaCl—H_2O$	2690—2697
Sodium Hydroxide–Sodium Sulfate–Calcium Hydroxide–Calcium Sulfate–Water	$NaOH—Na_2SO_4—Ca(OH)_2—CaSO_4—H_2O$	2698—2701
Sodium Hydroxide–Sodium Sulfate–Calcium Carbonate–Calcium Sulfate–Water	$NaOH—Na_2SO_4—CaCO_3—CaSO_4—H_2O$	2702
Sodium Hydroxide–Sodium Sulfate–Mercury (II) Oxide–Water	$NaOH—Na_2SO_4—HgO—H_2O$	2703
Sodium Hydroxide–Sodium Chloride–Water	$NaOH—NaCl—H_2O$	2704—2712
Sodium Hydroxide–Sodium Chloride–Magnesium Oxide–Water	$NaOH—NaCl—MgO—H_2O$	2713

System name	Formulae of components	Table number
Sodium Hydroxide–Sodium Chloride –Calcium Oxide–Water	$NaOH—NaCl—CaO—H_2O$	2714
Sodium Hydroxide–Sodium Arsenate –Water	$NaOH—Na_3AsO_4—H_2O$	2715
Sodium Hydroxide–Sodium Arsenate –Sodium Stannate–Water	$NaOH—Na_3AsO_4—Na_2SnO_3—H_2O$	2716
Sodium Hydroxide–Sodium Bromide –Water	$NaOH—NaBr—H_2O$	2717, 2718
Sodium Hydroxide–Sodium Hexa-hydroxostannate (IV)–Water	$NaOH—Na_2[Sn(OH)_6]—H_2O$	2719
Sodium Hydroxide–Sodium Stannate –Sodium Antimonate–Water	$NaOH—Na_2SnO_3—NaSbO_3—H_2O$	2720
Sodium Hydroxide–Sodium Tetra-thioantimonate–Water	$NaOH—Na_3SbS_4—H_2O$	2721
Sodium Hydroxide–Sodium Iodide –Water	$NaOH—NaI—H_2O$	2722
Sodium Hydroxide–Sodium Period-ate–Water	$NaOH—NaIO_4—H_2O$	2723
Sodium Hydroxide–Sodium Octa-oxouranate–Water	$NaOH—Na_4UO_8\cdot 9H_2O—H_2O$	2724
Sodium Hydroxide–Magnesium Oxide–Silicon Dioxide–Water	$NaOH—MgO—SiO_2—H_2O$	2725
Sodium Hydroxide–Mangesium Chloride–Silicon Dioxide–Water	$NaOH—MgCl_2—SiO_2—H_2O$	2726
Sodium Hydroxide–Magnesium Chloride–Silicon Dioxide–Calcium Chloride–Water	$NaOH—MgCl_2—SiO_2—CaCl_2—H_2O$	2727
Sodium Hydroxide–Aluminum Oxide–Water	$NaOH—Al_2O_3—H_2O$	2728, 2729
Sodium Hydroxide–Aluminum Hydroxide–Water	$NaOH—Al(OH)_3—H_2O$	2730—2733
Sodium Hydroxide–Aluminum Orthophosphate–Water	$NaOH—AlPO_4—H_2O$	2734
Sodium Hydroxide–Aluminum Orthophosphate–Phosphorus Pentoxide–Water	$NaOH—AlPO_4—P_2O_5—H_2O$	2735
Sodium Hydroxide–Aluminum Sulfate–Lanthanum Sulfate– Water	$NaOH—Al_2(SO_4)_3—La_2(SO_4)_3—H_2O$	2736
Sodium Hydroxide–Silicon Dioxide –Water	$NaOH—SiO_2—H_2O$	2737
Sodium Hydroxide–Potassium Hy-droxide–Water	$NaOH—KOH—H_2O$	2738
Sodium Hydroxide–Potassium Nitrate–Mercury (II) Oxide–Water)	$NaOH—KNO_3—HgO—H_2O$	2739
Sodium Hydroxide–Potassium Iodide–Water	$NaOH—KI—H_2O$	2740

System name	Formulae of components	Table number
Sodium Hydroxide–Calcium Oxide–Water	$NaOH—CaO—H_2O$	2741, 2742
Sodium Hydroxide–Calcium Hydroxide–Water	$NaOH—Ca(OH)_2—H_2O$	2743—2747
Sodium Hydroxide–Calcium Carbonate–Water	$NaOH—CaCO_3—H_2O$	2748
Sodium Hydroxide–Calcium Dihydrogen Orthophosphate–Water	$NaOH—Ca(H_2PO_4)_2—H_2O$	2749
Sodium Hydroxide–Calcium Hydrogen Orthophosphate–Water	$NaOH—CaHPO_4—H_2O$	2749
Sodium Hydroxide–Calcium Orthophosphate–Water	$NaOH—Ca_3(PO_4)_2—H_2O$	2749
Sodium Hydroxide–Calcium Sulfate–Lanthanum Sulfate–Water	$NaOH—CaSO_4—La_2(SO_4)_3—H_2O$	2750
Sodium Hydroxide–Calcium Sulfate–Cerium Sulfate–Water	$NaOH—CaSO_4—Ce_2(SO_4)_3—H_2O$	2751
Sodium Hydroxide–Calcium Arsenate–Water	$NaOH—Ca_3(AsO_4)_2—H_2O$	2752
Sodium Hydroxide–Calcium Arsenate–Calcium Stannate–Water	$NaOH—Ca_3(AsO_4)_2—CaSnO_3—H_2O$	2753
Sodium Hydroxide–Calcium Iodate–Water	$NaOH—Ca(IO_3)_2—H_2O$	5190
Sodium Hydroxide–Scandium Hydroxide–Water	$NaOH—Sc(OH)_3—H_2O$	2755
Sodium Hydroxide–Titanium Oxide–Water	$NaOH—TiO_2—H_2O$	2756
Sodium Hydroxide–Chromium Trioxide–Water	$NaOH—Cr_2O_3—H_2O$	2757
Sodium Hydroxide–Manganese (II) Hydroxide–Water	$NaOH—Mn(OH)_2—H_2O$	2758, 2759
Sodium Hydroxide–Iron (III) Sulfate–Water	$NaOH—Fe_2(SO_4)_3—H_2O$	2760
Sodium Hydroxide–Cabalt (II) Hydroxide–Water	$NaOH—Co(OH)_2—H_2O$	2761, 2762
Sodium Hydroxide–Nickel (II) Hydroxide–Water	$NaOH—Ni(OH)_2—H_2O$	2763
Sodium Hydroxide–Copper (II) Hydroxide–Water	$NaOH—Cu(OH)_2—H_2O$	2764
Sodium Hydroxide–Zinc Oxide–Water	$NaOH—ZnO—H_2O$	2765
Sodium Hydroxide–Zinc Hydroxide–Water	$NaOH—Zn(OH)_2—H_2O$	2766, 2767
Sodium Hydroxide–Gallium Hydroxide–Water	$NaOH—Ga(OH)_3—H_2O$	2768
Sodium Hydroxide–Germanium Dioxide–Water	$NaOH—GeO_2—H_2O$	2769
Sodium Hydroxide–Germanium Hydroxide–Water	$NaOH—Ge(OH)_2—H_2O$	2770
Sodium Hydroxide–Arsenic Trioxide–Water	$NaOH—As_2O_3—H_2O$	2771
Sodium Hydroxide–Arsenic Pentoxide–Water	$NaOH—As_2O_5—H_2O$	2772, 2773

System name	Formulae of components	Table number
Sodium Hydroxide–Yttrium Hydroxide–Water	$NaOH—Y(OH)_3—H_2O$	2774
Sodium Hydroxide–Zirconium Hydroxide–Water	$NaOH—Zr(OH)_4—H_2O$	2775
Sodium Hydroxide–Silver Oxide –Water	$NaOH—Ag_2O—H_2O$	2776, 2777
Sodium Hydroxide–Cadmium Hydroxide–Water	$NaOH—Cd(OH)_2—H_2O$	2778, 2779
Sodium Hydroxide–Indium Hydroxide–Water	$NaOH—In(OH)_3—H_2O$	2780, 2781
Sodium Hydroxide–Tin (II) Oxide –Water	$NaOH—SnO—H_2O$	2782
Sodium Hydroxide–Tin (II) Hydroxide–Water	$NaOH—Sn(OH)_2—H_2O$	2783
Sodium Hydroxide–Tellurium Trioxide–Water	$NaOH—TeO_3—H_2O$	2784
Sodium Hydroxide–Barium Hydroxide–Water	$NaOH—Ba(OH)_2—H_2O$	2785—2787
Sodium Hydroxide–Barium Silicate –Water	$NaOH—BaSiO_3·6H_2O—H_2O$	2788
Sodium Hydroxide–Lanthanum Sulfate–Water	$NaOH—La_2(SO_4)_3—H_2O$	2789
Sodium Hydroxide–Lanthanum Sulfate–Cerium Sulfate–Water	$NaOH—La_2(SO_4)_3—Ce_2(SO_4)_3—H_2O$	2780
Sodium Hydroxide–Cerium Sulfate–Water	$NaOH—Ce_2(SO_4)_3—H_2O$	2781, 2782
Sodium Hydroxide–Hafnium Hydroxide–Water	$NaOH—Hf(OH)_4—H_2O$	2783
Sodium Hydroxide–Gold (I) Hydroxide–Water	$NaOH—AuOH—H_2O$	2784
Sodium Hydroxide–Mercury (II) Oxide–Water	$NaOH—HgO—H_2O$	2785—2789
Sodium Hydroxide–Thallium (III) Hydroxide–Water	$NaOH—Tl(OH)_3—H_2O$	2790
Sodium Hydroxide–Thallium (I) Iodate–Water	$NaOH—TlIO_3—H_2O$	2791
Sodium Hydroxide–Lead (II) Oxide –Water	$NaOH—PbO·xH_2O—H_2O$	2792—2794
Sodium Hydroxide–Lead (II) Hydroxide–Water	$NaOH—Pb(OH)_2—H_2O$	2795
Sodium Hydroxide–Lead (II) Sulfate –Water	$NaOH—PbSO_4—H_2O$	2796
Sodium Hydroxide–Bismuth Oxide –Water	$NaOH—Bi_2O_3—H_2O$	2797
Sodium Hydroxide–Bismuth Hydroxide–Water	$NaOH—Bi(OH)_3—H_2O$	2798
Sodium Hydroxide–Thorium Oxide –Water	$NaOH—ThO_2—H_2O$	2799
Sodium Hydroxide–Uranium Trioxide–Water	$NaOH—UO_3·H_2O—H_2O$	2800

System name	Formulae of components	Table number
Sodium Hydroxide–Uranium Tetroxide–Water	$NaOH—UO_4 \cdot xH_2O—H_2O$	2801
Sodium Hydroxide–Uranium Hydroxide–Water	$NaOH—U(OH)_4—H_2O$	2802, 2803
Sodium Orthophosphate Hydroxide –Sodium Carbonate–Water	$Na_3PO_4 \cdot {}^1/_7NaOH—Na_2CO_3—H_2O$	2804
Sodium Metaborate–Sodium Orthoborate–Water	$NaBO_2 \cdot 2H_2O—NaBO_3 \cdot H_2O—H_2O$	2805
Sodium Metaborate–Sodium Tetraborate–Water	$NaBO_2—Na_2B_4O_7—H_2O$	2806
Sodium Metaborate–Sodium Tetraborate–Sodium Hydrogen Carbonate–Sodium Carbonate–Sodium Chloride–Water	$Na_2B_2O_4—Na_2B_4O_7—NaHCO_3—Na_2CO_3—NaCl—H_2O$	2807
Sodium Metaborate–Sodium Tetraborate–Sodium Chloride–Water	$Na_2B_2O_4—Na_2B_4O_7—NaCl—H_2O$	2808, 2809
Sodium Metaborate–Sodium Carbonate–Sodium Chloride–Water	$NaBO_2—Na_2CO_3—NaCl—H_2O$	2810
Sodium Metaborate–Sodium Orthophosphate–Sodium Chloride–Water	$Na_2B_2O_4—Na_3PO_4—NaCl—H_2O$	2811
Sodium Metaborate–Sodium Chloride–Water	$NaBO_2—NaCl—H_2O$	2812
Sodium Tetraborate–Sodium Hydrogen Carbonate–Water	$Na_2B_4O_7—NaHCO_3—H_2O$	2813
Sodium Tetraborate–Sodium Hydrogen Carbonate–Sodium Carbonate–Water	$Na_2B_4O_7—NaHCO_3—Na_2CO_3—H_2O$	2814
Sodium Tetraborate–Sodium Hydrogen Carbonate–Sodium Sulfate–Water	$Na_2B_4O_7—NaHCO_3—Na_2SO_4—H_2O$	2815
Sodium Tetraborate–Sodium Hydrogen Carbonate–Sodium Sulfate–Sodium Chloride–Water	$Na_2B_4O_7—NaHCO_3—Na_2SO_4—NaCl—H_2O$	2816
Sodium Tetraborate–Sodium Hydrogen Carbonate–Sodium Chloride–Water	$Na_2B_4O_7—NaHCO_3—NaCl—H_2O$	2817, 2818
Sodium Tetraborate–Sodium Carbonate–Water	$Na_2B_4O_7—Na_2CO_3—H_2O$	2819—2821
Sodium Tetraborate–Sodium Fluoride–Water	$Na_2B_4O_7 \cdot 10H_2O—NaF—H_2O$	2822
Sodium Tetraborate–Sodium Hydrogen Orthophosphate–Sodium Chloride–Water	$Na_2B_4O_7—Na_2HPO_4—NaCl—H_2O$	2823
Sodium Tetraborate–Sodium Sulfate –Water	$Na_2B_4O_7—Na_2SO_4—H_2O$	2824—2828
Sodium Tetraborate–Sodium Sulfate –Sodium Chloride–Water	$Na_2B_4O_7—Na_2SO_4—NaCl—H_2O$	2929, 2830
Sodium Tetraborate–Sodium Sulfate –Sodium Chloride–Potassium Chloride–Water	$Na_2B_4O_7—Na_2SO_4—NaCl—KCl—H_2O$	2831
Sodium Tetraborate–Sodium Chloride–Water	$Na_2B_4O_7—NaCl—H_2O$	2832—2835
Sodium Tetraborate–Sodium Chloride–Magnesium Sulfate–Water	$Na_2B_4O_7—NaCl—MgSO_4—H_2O$	2836

System name	Formulae of components	Table number
Sodium Tetraborate–Sodium Chloride–Potassium Chloride –Water	$Na_2B_4O_7$—NaCl—KCl—H_2O	2837, 2838
Sodium Tetraborate–Potassium Tetraborate–Water	$Na_2B_4O_7$—$K_2B_4O_7$—H_2O	2839
Sodium Tetraborate–Calcium Sulfate–Water	$Na_2B_4O_7$—$CaSO_4$—H_2O	2840
Sodium Pentaborate–Sodium Chloride–Water	NaB_5O_8—NaCl—H_2O	2841—2842
Sodium Pentaborate–Sodium Chloride–Potassium Chloride–Water	NaB_5O_8—NaCl—KCl—H_2O	2843
Sodium Pentaborate–Potassium Pentaborate–Water	NaB_5O_8—KB_5O_8—H_2O	2844
Sodium Hexafluorotrioxotriborate–Sodium Iodide–Water	$Na_3B_3O_3F_6$—NaI—H_2O	2845
Sodium Hydrogen Carbonate–Sodium Carbonate–Water	$NaHCO_3$—Na_2CO_3—H_2O	2846—2856
Sodium Hydrogen Carbonate–Sodium Carbonate–Sodium Sulfate–Water	$NaHCO_3$—Na_2CO_3—Na_2SO_4—H_2O	2857—2862
Sodium Hydrogen Carbonate–Sodium Carbonate–Sodium Sulfate–Sodium Chloride–Water	$NaHCO_3$—Na_2CO_3—Na_2SO_4—NaCl—H_2O	2863—2866
Sodium Hydrogen Carbonate–Sodium Carbonate–Sodium Chloride–Water	$NaHCO_3$—Na_2CO_3—NaCl—H_2O	2867—2872
Sodium Hydrogen Carbonate–Sodium Carbonate–Potassium Hydrogen Carbonate–Potassium Carbonate–Water	$NaHCO_3$—Na_2CO_3—$KHCO_3$—K_2CO_3—H_2O	2873—2876
Sodium Hydrogen Carbonate–Sodium Carbonate–Potassium Carbonate–Water	$NaHCO_3$—Na_2CO_3—K_2CO_3—H_2O	2877
Sodium Hydrogen Carbonate–Sodium Nitrite–Water	$NaHCO_3$—$NaNO_2$—H_2O	2878
Sodium Hydrogen Carbonate–Sodium Nitrate–Water	$NaHCO_3$—$NaNO_3$—H_2O	2879
Sodium Hydrogen Carbonate–Sodium Metasilicate–Water	$NaHCO_3$—Na_2SiO_3—H_2O	2880
Sodium Hydrogen Carbonate–Sodium Sulfate–Water	$NaHCO_3$—Na_2SO_4—H_2O	2881—2891
Sodium Hydrogen Carbonate–Sodium Sulfate–Sodium Chloride–Water	$NaHCO_3$—Na_2SO_4—NaCl—H_2O	2892—2896
Sodium Hydrogen Carbonate–Sodium Sulfate–Magnesium Hydrogen Carbonate–Magnesium Sulfate–Water	$NaHCO_3$—Na_2SO_4—$Mg(HCO_3)_2$—$MgSO_4$—H_2O	2897
Sodium Hydrogen Carbonate–Sodium Sulfate–Potassium Hydrogen Carbonate–Potassium Sulfate–Water	$NaHCO_3$—Na_2SO_4—$KHCO_3$—K_2SO_4—H_2O	2898, 2899
Sodium Hydrogen Carbonate–Sodium Sulfate–Calcium Hydrogen Carbonate–Water	$NaHCO_3$—Na_2SO_4—$Ca(HCO_3)_2$—H_2O	2900
Sodium Hydrogen Carbonate–Sodium Sulfate–Calcium Hydrogen Carbonate–Calcium Carbonate–Water	$NaHCO_3$—Na_2SO_4—$Ca(HCO_3)_2$—$CaCO_3$—H_2O	2901

System name	Formulae of components	Table number
Sodium Hydrogen Carbonate–Sodium Sulfate–Calcium Hydrogen Carbonate–Calcium Sulfate–Water	$NaHCO_3$—Na_2SO_4—$Ca(HCO_3)_2$—$CaSO_4$—H_2O	2902
Sodium Hydrogen Carbonate–Sodium Chloride–Water	$NaHCO_3$—$NaCl$—H_2O	2903—2914
Sodium Hydrogen Carbonate–Sodium Chloride–Magnesium Hydrogen Carbonate–Magnesium Chloride–Water	$NaHCO_3$—$NaCl$—$Mg(HCO_3)_2$—$MgCl_2$—H_2O	2915, 2916
Sodium Hydrogen Carbonate–Sodium Chloride–Potassium Hydrogen Carbonate–Potassium Chloride–Water	$NaHCO_3$—$NaCl$—$KHCO_3$—KCl—H_2O	2917
Sodium Hydrogen Carbonate–Sodium Bromide–Water	$NaHCO_3$—$NaBr$—H_2O	2918
Sodium Hydrogen Carbonate–Sodium Bromide–Sodium Bromate–Water	$NaHCO_3$—$NaBr$—$NaBrO_3$—H_2O	2919
Sodium Hydrogen Carbonate–Sodium Bromate–Water	$NaHCO_3$—$NaBrO_3$—H_2O	2920
Sodium Hydrogen Carbonate–Potassium Hydrogen Carbonate–Water	$NaHCO_3$—$KHCO_3$—H_2O	2921—2934
Sodium Hydrogen Carbonate–Potassium Peroxodisulfate–Water	$NaHCO_3$—$K_2S_2O_8$—H_2O	2925
Sodium Hydrogen Carbonate–Titanium Oxide–Water	$NaHCO_3$—TiO_2—H_2O	2926
Sodium Carbonate–Sodium Nitrite–Water	Na_2CO_3—$NaNO_2$—H_2O	2927
Sodium Carbonate–Sodium Nitrate–Water	Na_2CO_3—$NaNO_3$—H_2O	2928, 2929
Sodium Carbonate–Sodium Fluoride–Water	Na_2CO_3—NaF—H_2O	2930
Sodium Carbonate–Sodium Fluoride–Sodium Chloride–Water	Na_2CO_3—NaF—$NaCl$—H_2O	2931
Sodium Carbonate–Sodium Metasilicate–Water	Na_2CO_3—Na_2SiO_3—H_2O	2933
Sodium Carbonate–Sodium Sulfate–Potassium Carbonate–Water	Na_2CO_3—Na_2SO_4—K_2CO_3—H_2O	2933
Sodium Carbonate–Sodium Metasilicate–Water	Na_2CO_3—Na_2SiO_3—H_2O	2934, 2935
Sodium Carbonate–Sodium Metasilicate–Potassium Carbonate–Potassium Metasilicate–Water	Na_2CO_3—Na_2SiO_3—K_2CO_3—K_2SiO_3—H_2O	2936
Sodium Carbonate–Sodium Orthophosphate–Water	Na_2CO_3—Na_3PO_4—H_2O	2937, 2938
Sodium Carbonate–Sodium Sulfide–Water	Na_2CO_3—Na_2S—H_2O	2939, 2940
Sodium Carbonate–Sodium Sulfate–Water	Na_2CO_3—Na_2SO_4—H_2O	2941—2955
Sodium Carbonate–Sodium Sulfate–Sodium Chloride–Water	Na_2CO_3—Na_2SO_4—$NaCl$—H_2O	2956—2962
Sodium Carbonate–Sodium Sulfate–Sodium Chloride–Potassium Carbonate–Potassium Chloride–Water	Na_2CO_3—Na_2SO_4—$NaCl$—K_2CO_3—KCl—H_2O	2963

System name	Formulae of components	Table number
Sodium Carbonate–Sodium Sulfate –Sodium Chloride–Calcium Hydroxide–Calcium Carbonate –Calcium Sulfate–Water	Na_2CO_3—Na_2SO_4—NaCl—$Ca(OH)_2$—$CaCO_3$—$CaSO_4$—H_2O	2964
Sodium Carbonate–Sodium Sulfate –Sodium Chloride–Calcium Carbonate–Water	Na_2CO_3—Na_2SO_4—NaCl—$CaCO_3$—H_2O	2965
Sodium Carbonate–Sodium Sulfate–Magnesium Carbonate– Magnesium Sulfate–Calcium Carbonate–Calcium Sulfate–Water	Na_2CO_3—Na_2SO_4—$MgCO_3$—$MgSO_4$—$CaCO_3$—$CaSO_4$—H_2O	2966
Sodium Carbonate–Sodium Sulfate –Potassium Carbonate–Potassium Sulfate–Water	Na_2CO_3—Na_2SO_4—K_2CO_3—K_2SO_4—H_2O	2967, 2968
Sodium Carbonate–Sodium Sulfate–Calcium Carbonate–Water	Na_2CO_3—Na_2SO_4—$CaCO_3$—H_2O	2969
Sodium Carbonate–Sodium Chloride–Water	Na_2CO_3—NaCl—H_2O	2970—2981
Sodium Carbonate–Sodium Chloride–Potassium Carbonate–Potassium Chloride–Water	Na_2CO_3—NaCl—K_2CO_3—KCl—H_2O	2982, 2983
Sodium Carbonate–Sodium Chloride–Calcium Carbonate–Water	Na_2CO_3—NaCl—$CaCO_3$—H_2O	2984
Sodium Carbonate–Sodium Chlorate –Water	Na_2CO_3—$NaClO_3$—H_2O	2985
Sodium Carbonate–Sodium Chlorate –Potassium Carbonate–Potassium Chlorate–Water	Na_2CO_3—$NaClO_3$—K_2CO_3—$KClO_3$—H_2O	2986
Sodium Carbonate–Sodium Arsenate –Copper (II) Arsenate–Water	Na_2CO_3—Na_3AsO_4—$Cu_3(AsO_4)_2$—H_2O	2987
Sodium Carbonate–Sodium Selenate –Water	Na_2CO_3—Na_2SeO_4—H_2O	2988
Sodium Carbonate–Sodium Bromide –Water	Na_2CO_3—NaBr—H_2O	2989, 2990
Sodium Carbonate–Sodium Bromide –Sodium Bromate–Water	Na_2CO_3—NaBr—$NaBrO_3$—H_2O	2991
Sodium Carbonate–Sodium Bromate –Water	Na_2CO_3—$NaBrO_3$—H_2O	2992
Sodium Carbonate–Sodium Molybdate–Water	Na_2CO_3—Na_2MoO_4—H_2O	2993
Sodium Carbonate–Sodium Hexahydroxostannate (IV)–Water	Na_2CO_3—$Na_2[Sn(OH)_6]$—H_2O	2994
Sodium Carbonate–Sodium Iodide–Water	Na_2CO_3—NaI—H_2O	2995
Sodium Carbonate–Sodium Iodate–Water	Na_2CO_3—$NaIO_3$—H_2O	2996
Sodium Carbonate–Sodium Tungstate–Water	Na_2CO_3—Na_2WO_4—H_2O	2997
Sodium Carbonate–Sodium Octaoxouranate–Water	Na_2CO_3—$Na_4UO_8 \cdot 9H_2O$—H_2O	2998

System name	Formulae of components	Table number
Sodium Carbonate–Magnesium Carbonate–Water	Na_2CO_3—$MgCO_3$—H_2O	2999—3002
Sodium Carbonate–Aluminum Oxide–Water	Na_2CO_3—Al_2O_3—H_2O	3655
Sodium Carbonate–Silicon Dioxide –Water	Na_2CO_3—SiO_2—H_2O	3003, 3004
Sodium Carbonate–Potassium Carbonate–Water	Na_2CO_3—K_2CO_3—H_2O	3005—3017
Sodium Carbonate–Potassium Peroxodisulfate–Water	Na_2CO_3—$K_2S_2O_8$—H_2O	3018
Sodium Carbonate–Potassium Chlorate–Water	Na_2CO_3—$KClO_3$—H_2O	3019
Sodium Carbonate–Calcium Carbonate–Water	Na_2CO_3—$CaCO_3$—H_2O	3020—3022
Sodium Carbonate–Strontium Sulfate–Water	Na_2CO_3—$SrSO_4$—H_2O	3023
Sodium Carbonate–Tellurium Trioxide–Water	Na_2CO_3—TeO_3—H_2O	3024
Sodium Carbonate–Lead (II) Sulfate –Water	Na_2CO_3—$PbSO_4$—H_2O	3025
Sodium Carbonate–Thorium Oxide–Water	Na_2CO_3—ThO_2—H_2O	3026
Sodium Carbonate–Thorium Carbonate–Water	Na_2CO_3—$Th(CO_3)_2$—H_2O	3027
Sodium Carbonate–Plutonium Hydroxide–Water	Na_2CO_3—$Pu(OH)_4$—H_2O	3028
Sodium Thiocyanate–Sodium Sulfate–Water	$NaCNS$—Na_2SO_4—H_2O	3029
Sodium Thiocyanate–Sodium Chloride–Water	$NaCNS$—$NaCl$—H_2O	3030
Sodium Thiocyanate–Sulfur Dioxide –Water	$NaCNS$—SO_2—H_2O	3031, 3032
Sodium Thiocyanate–Calcium Thiocyanate–Water	$NaCNS$—$Ca(CNS)_2$—H_2O	3033
Sodium Thiocyanate–Silver Thiocyanate–Water	$NaCNS$—$AgCNS$—H_2O	3034
Sodium Thiocyanate–Barium Thiocyanate–Water	$NaCNS$—$Ba(CNS)_2$—H_2O	3035
Sodium Thiocyanate–Thallium Thiocyanate–Water	$NaCNS$—$TlCNS$—H_2O	3036, 3037
Sodium Hexacyanoferrate (II)–Sodium Sulfate–Water	$Na_4Fe(CN)_6$—Na_2SO_4—H_2O	3038
Sodium Hexacyanoferrate (II)–Potassium Hexacyanoferrate (II) –Water	$Na_4Fe(CN)_6$—$K_4Fe(CN)_6$—H_2O	3039
Sodium Hexacyanoferrate (II)–Calcium Hexacyanoferrate (II)–Water	$Na_4Fe(CN)_6$—$Ca_2Fe(CN)_6$—H_2O	3040
Sodium Nitrite–Sodium Nitrate–Water	$NaNO_2$—$NaNO_3$—H_2O	3041, 3042
Sodium Nitrite–Sodium Nitrate–Sodium Chloride–Water	$NaNO_2$—$NaNO_3$—$NaCl$—H_2O	3043, 3044

System name	Formulae of components	Table number
Sodium Nitrite–Sodium Sulfate–Water	$NaNO_2—Na_2SO_4—H_2O$	3045
Sodium Nitrite–Sodium Chloride–Water	$NaNO_2—NaCl—H_2O$	3046, 3047
Sodium Nitrite–Potassium Nitrite–Water	$NaNO_2—KNO_2—H_2O$	3048, 3049
Sodium Nitrite–Calcium Hydroxide–Water	$NaNO_2—Ca(OH)_2—H_2O$	5074
Sodium Nitrite–Strontium Nitrite–Water	$NaNO_2—Sr(NO_2)_2—H_2O$	3050
Sodium Nitrite–Silver Nitrite–Water	$NaNO_2—AgNO_2—H_2O$	3051—3053
Sodium Nitrite–Barium Nitrite–Water	$NaNO_2—Ba(NO_2)_2—H_2O$	3054
Sodium Nitrate–Sodium Dihydrogen Orthophosphate–Water	$NaNO_3—NaH_2PO_4—H_2O$	3055
Sodium Nitrate–Sodium Hydrogen Orthophosphate–Water	$NaNO_3—Na_2HPO_4—H_2O$	3056, 3057
Sodium Nitrate–Sodium Sulfate–Water	$NaNO_3—Na_2SO_4—H_2O$	3058—3060
Sodium Nitrate–Sodium Sulfate–Sodium Chloride–Water	$NaNO_3—Na_2SO_4—NaCl—H_2O$	3061
Sodium Nitrate–Sodium Sulfate–Lead (II) Sulfate–Water	$NaNO_3—Na_2SO_4—PbSO_4—H_2O$	3062
Sodium Nitrate–Sodium Thiosulfate–Water	$NaNO_3—Na_2S_2O_3—H_2O$	3063
Sodium Nitrate–Sodium Chloride–Water	$NaNO_3—NaCl—H_2O$	3064—3075
Sodium Nitrate–Sodium Chloride–Potassium Nitrate–Water	$NaNO_3—NaCl—KNO_3—H_2O$	3076, 3077
Sodium Nitrate–Sodium Chloride–Potassium Nitrate–Potassium Chloride–Water	$NaNO_3—NaCl—KNO_3—KCl—H_2O$	3078—3080
Sodium Nitrate–Sodium Chloride–Potassium Perchlorate–Water	$NaNO_3—NaCl—KClO_4—H_2O$	3081
Sodium Nitrate–Sodium Chloride–Barium Nitrate–Barium Chloride–Water	$NaNO_3—NaCl—Ba(NO_3)_2—BaCl—H_2O$	3082, 3083
Sodium Nitrate–Sodium Chlorate–Water	$NaNO_3—NaClO_3—H_2O$	3084
Sodium Nitrate–Sodium Perchlorate–Water	$NaNO_3—NaClO_4—H_2O$	3085, 3086
Sodium Nitrate–Sodium Perchlorate–Potassium Nitrate–Potassium Perchlorate–Water	$NaNO_3—NaClO_4—KNO_3—KClO_4—H_2O$	3087, 3088
Sodium Nitrate–Sodium Chromate–Water	$NaNO_3—Na_2CrO_4—H_2O$	3089, 3090
Sodium Nitrate–Sodium Chromate–Calcium Nitrate–Calcium Chromate–Water	$NaNO_3—Na_2CrO_4—Ca(NO_3)_2—CaCrO_4—H_2O$	3091
Sodium Nitrate–Sodium Bromide–Water	$NaNO_3—NaBr—H_2O$	3092

System name	Formulae of components	Table number
Sodium Nitrate–Sodium Bromate–Water	$NaNO_3$—$NaBrO_3$—H_2O	3093
Sodium Nitrate–Sodium Molybdate–Water	$NaNO_3$—Na_2MoO_4—H_2O	3094, 3095
Sodium Nitrate–Sodium Hexahydroxostannate (IV)–Water	$NaNO_3$—$Na_2[Sn(OH)_6]$—H_2O	3096
Sodium Nitrate–Sodium Iodide–Water	$NaNO_3$—NaI—H_2O	3097
Sodium Nitrate–Sodium Iodate–Water	$NaNO_3$—$NaIO_3$—H_2O	3098
Sodium Nitrate–Sodium Tungstate–Water	$NaNO_3$—Na_2WO_4—H_2O	3099
Sodium Nitrate–Sodium Octaoxourante–Water	$NaNO_3$—$NaUO_3 \cdot 9H_2O$—H_2O	3100
Sodium Nitrate–Magnesium Nitrate–Water	$NaNO_3$—$Mg(NO_3)_2$—H_2O	3101, 3102
Sodium Nitrate–Aluminum Nitrate–Water	$NaNO_3$—$Al(NO_3)_3$—H_2O	3103
Sodium Nitrate–Potassium Hydroxide–Potassium Nitrate–Water	$NaNO_3$—KOH—KNO_3—H_2O	3104
Sodium Nitrate–Potassium Nitrate–Water	$NaNO_3$—KNO_3—H_2O	3105—3113
Sodium Nitrate–Potassium Nitrate–Potassium Chloride–Water	$NaNO_3$—KNO_3—KCl—H_2O	3114
Sodium Nitrate–Potassium Peroxodisulfate–Water	$NaNO_3$—$K_2S_2O_8$—H_2O	3115
Sodium Nitrate–Potassium Perchlorate–Water	$NaNO_3$—$KClO_4$—H_2O	3116, 3117
Sodium Nitrate–Potassium Bromide–Water	$NaNO_3$—KBr—H_2O	3118
Sodium Nitrate–Potassium Bromate–Silver Bromate–Water	$NaNO_3$—$KBrO_3$—$AgBrO_3$—H_2O	3119
Sodium Nitrate–Calcium Hydroxide–Water	$NaNO_3$—$Ca(OH)_2$—H_2O	5074
Sodium Nitrate–Calcium Carbonate–Water	$NaNO_3$—$CaCO_3$—H_2O	3739
Sodium Nitrate–Calcium Nitrate–Water	$NaNO_3$—$Ca(NO_3)_2$—H_2O	3120—3123
Sodium Nitrate–Calcium Orthophosphate–Water	$NaNO_3$—$Ca_3(PO_4)_2$—H_2O	5113
Sodium Nitrate–Calcium Sulfate–Water	$NaNO_3$—$CaSO_4$—H_2O	3124
Sodium Nitrate–Copper (II) Nitrate–Water	$NaNO_3$—$Cu(NO_3)_2$—H_2O	3125
Sodium Nitrate–Zinc Hydroxide–Water	$NaNO_3$—$Zn(OH)_2$—H_2O	3126
Sodium Nitrate–Strontium Sulfate–Water	$NaNO_3$—$SrSO_4$—H_2O	3127
Sodium Nitrate–Strontium Chloride–Water	$NaNO_3$—$SrCl_2$—H_2O	3128
Sodium Nitrate–Silver Nitrite–Water	$NaNO_3$—$AgNO_2$—H_2O	3129
Sodium Nitrate–Silver Orthophosphate–Water	$NaNO_3$—Ag_3PO_4—H_2O	3130

System name	Formulae of components	Table number
Sodium Nitrate–Silver Sulfate–Water	$NaNO_3$—Ag_2SO_4—H_2O	3131, 3132
Sodium Nitrate–Silver Chloride–Water	$NaNO_3$—$AgCl$—H_2O	3133—3135
Sodium Nitrate–Silver Chromate–Water	$NaNO_3$—Ag_2CrO_4—H_2O	3136, 5413
Sodium Nitrate–Silver Bromate–Water	$NaNO_3$—$AgBrO_3$—H_2O	3137, 3138
Sodium Nitrate–Cesium Nitrate–Water	$NaNO_3$—$CsNO_3$—H_2O	3139
Sodium Nitrate–Cesium Hexanitrocobaltate–Water	$NaNO_3$—$Cs_3[Co(NO_2)_6]$—H_2O	3140
Sodium Nitrate–Barium Nitrate–Water	$NaNO_3$—$Ba(NO_3)_2$—H_2O	3141, 3142
Sodium Nitrate–Lanthanum Iodate–Water	$NaNO_3$—$La(IO_3)_3$—H_2O	3143, 3144
Sodium Nitrate–Cerium Nitrate–Water	$NaNO_3$—$Ce(NO_3)_3$—H_2O	3145
Sodium Nitrate–Thallium Hexanitrocobaltate–Water	$NaNO_3$—$Tl_3[Co(NO_2)_6]$—H_2O	3146
Sodium Nitrate–Thallium Chloride–Water	$NaNO_3$—$TlCl$—H_2O	3147, 3148
Sodium Nitrate–Thallium Bromide–Water	$NaNO_3$—$TlBr$—H_2O	3149
Sodium Nitrate–Thallium Iodide–Water	$NaNO_3$—TlI—H_2O	3150
Sodium Nitrate–Thallium Iodate–Water	$NaNO_3$—$TlIO_3$—H_2O	3151
Sodium Nitrate–Lead (II) Nitrate–Water	$NaNO_3$—$Pb(NO_3)_2$—H_2O	3152, 3153
Sodium Nitrate–Lead (II) Sulfate–Water	$NaNO_3$—$PbSO_4$—H_2O	3154
Sodium Nitrate–Lead (II) Bromate–Water	$NaNO_3$—$Pb(BrO_3)_2$—H_2O	3155
Sodium Nitrate–Lead (II) Iodide–Water	$NaNO_3$—PbI_2—H_2O	3156
Sodium Nitrate–Lead (II) Iodate–Water	$NaNO_3$—$Pb(IO_3)_2$—H_2O	3157
Sodium Fluoride–Sodium Sulfate–Water	NaF—Na_2SO_4—H_2O	3158
Sodium Fluoride–Sodium Sulfate–Sodium Chloride–Water	NaF—Na_2SO_4—$NaCl$—H_2O	3159
Sodium Fluoride–Sodium Chloride–Water	NaF—$NaCl$—H_2O	3160—3162
Sodium Fluoride–Sodium Chloride–Potassium Chloride–Water	NaF—$NaCl$—KCl—H_2O	3163
Sodium Fluoride–Sodium Hexahydroxostannate (IV)–Water	NaF—$Na_2[Sn(OH)_6]$—H_2O	3164
Sodium Fluoride–Magnesium Fluoride–Water	NaF—MgF_2—H_2O	3165
Sodium Fluoride–Aluminum Fluoride–Water	NaF—AlF_3—H_2O	3166, 3167

System name	Formulae of components	Table number
Sodium Fluoride–Calcium Fluoride–Iron (III) Nitrate–Water	$NaF—CaF_2—Fe(NO_3)_3—H_2O$	3168
Sodium Fluoride–Calcium Fluoride–Thorium Nitrate–Water	$NaF—CaF_2—Th(NO_3)_4—H_2O$	3169
Sodium Fluoride–Chromium (III) Fluoride–Water	$NaF—CrF_3—H_2O$	3170
Sodium Fluoride–Iron (III) Fluoride–Water	$NaF—FeF_3—H_2O$	3171
Sodium Fluoride–Gallium Fluoride–Water	$NaF—GaF_3—H_2O$	3172
Sodium Fluoride–Strontium Fluoride–Water	$NaF—SrF_2—H_2O$	3173
Sodium Fluoride–Silver Fluoride–Water	$NaF—AgF—H_2O$	3174
Sodium Fluoride–Iodine Pentoxide–Water	$NaF—I_2O_5—H_2O$	3175
Sodium Fluoride–Barium Fluoride–Water	$NaF—BaF_2—H_2O$	3176
Sodium Fluoride–Cerium Fluoride–Water	$NaF—CeF_4—H_2O$	3177, 3178
Sodium Fluoride–Lead (II) Fluoride–Water	$NaF—PbF_2—H_2O$	3179
Sodium Fluoride–Thorium Nitrate–Water	$NaF—Th(NO_3)_4—H_2O$	3180
Sodium Hexafluoroaluminate–Aluminum Nitrate–Water	$Na_3AlF_6—Al(NO_3)_3—H_2O$	3181
Sodium Hexafluoroaluminate–Aluminum Sulfate–Water	$Na_3AlF_6—Al_2(SO_4)_3—H_2O$	3182
Sodium Hexafluoroaluminate–Aluminum Chloride–Water	$Na_3AlF_6—AlCl_3—H_2O$	3183
Sodium Hexafluoroaluminate–Iron (III) Nitrate–Water	$Na_3AlF_6—Fe(NO_3)_3—H_2O$	3184
Sodium Hexafluoroaluminate–Iron (III) Sulfate–Water	$Na_3AlF_6—Fe_2(SO_4)_3—H_2O$	3185
Sodium Hexafluoroaluminate–Iron (III) Chloride–Water	$Na_3AlF_6—FeCl_3—H_2O$	3186
Sodium Hedafluorosilicate–Sodium Sulfate–Water	$Na_2SiF_6—Na_2SO_4—H_2O$	3187
Sodium Metasilicate–Sodium Chloride–Water	$Na_2SiO_3—NaCl—H_2O$	3188, 3189
Sodium Metasilicate–Potassium Hydroxide–Water	$Na_2SiO_3—KOH—H_2O$	3190
Sodium Metasilicate–Potassium Metasilicate–Water	$Na_2SiO_3—K_2SiO_3—H_2O$	3191
Sodium Dihydrogen Orthosilicate–Barium Metasilicate–Water	$Na_2H_2SiO_4—BaSiO_3\cdot 6H_2O—H_2O$	3192
Sodium Phosphinate–Sodium Phosphonate–Water	$NaH_2PO_2—Na_2HPO_3—H_2O$	3193
Sodium Dihydrogen Orthophosphate–Sodium Sulfate–Water	$NaH_2PO_4—Na_2SO_4—H_2O$	3194
Sodium Dihydrogen Orthophosphate–Sodium Chloride–Potassium Dihydrogen Orthophosphate–Potassium Chloride–Water	$NaH_2PO_4—NaCl—KH_2PO_4—KCl—H_2O$	3195

System name	Formulae of components	Table number
Sodium Dihydrogen Orthophosphate –Potassium Dihydrogen Orthophosphate–Water	$NaH_2PO_4—KH_2PO_4—H_2O$	3196
Sodium Dihydrogen Orthophosphate –Arsenic Trioxide–Water	$NaH_2PO_4—As_2O_3—H_2O$	3197
Sodium Hydrogen Orthophosphate–Sodium Sulfate–Water	$Na_2HPO_4—Na_2SO_4—H_2O$	3198, 3199
Sodium Hydrogen Orthophosphate–Sodium Chloride–Water	$Na_2HPO_4—NaCl—H_2O$	3200
Sodium Hydrogen Orthophosphate–Potassium Hydrogen Orthophosphate–Water	$Na_2HPO_4—K_2HPO_4—H_2O$	3201
Sodium Hydrogen Orthophosphate–Potassium Peroxodisulfate–Water	$Na_2HPO_4—K_2S_2O_8—H_2O$	3202
Sodium Hydrogen Orthophosphate–Silver Orthophosphate–Water	$Na_2HPO_4—Ag_3PO_4—H_2O$	3203
Sodium Hydrogen Orthophosphate–Uranyl Nitrate–Water	$Na_2HPO_4—UO_2(NO_3)_2—H_2O$	3205
Sodium Orthophosphate–Sodium Sulfate–Water	$Na_3PO_4—Na_2SO_4—H_2O$	3206, 3207
Sodium Orthophosphate–Sodium Chloride–Water	$Na_3PO_4—NaCl—H_2O$	3208
Sodium Dihydrogen Pyrophosphate –Sodium Pyrophosphate–Water	$Na_2H_2P_2O_7—Na_4P_2O_7—H_2O$	3209
Sodium Pyrophosphate–Potassium Pyrophosphate–Water	$Na_4P_2O_7—K_4P_2O_7—H_2O$	3210
Sodium Pyrophosphate–Nickel (II) Pyrophosphate–Water	$Na_4P_2O_7—Ni_2P_2O_7—H_2O$	3211
Sodium Pyrophosphate–Copper (II) Pyrophosphate–Water	$Na_4P_2O_7—Cu_2P_2O_7—H_2O$	3212
Sodium Pyrophosphate–Silver Pyrophosphate–Water	$Na_4P_2O_7—Ag_4P_2O_7—H_2O$	3213, 3214
Sodium Pyrophosphate–Thorium Pyrophosphate–Water	$Na_4P_2O_7—ThP_2O_7—H_2O$	3215
Sodium Sulfide–Sodium Sulfate–Water	$Na_2S—Na_2SO_4—H_2O$	3216—3218
Sodium Sulfide–Aluminum Oxide–Water	$Na_2S—Al_2O_3—H_2O$	3219
Sodium Sulfide–Iron (II) Chloride–Water	$Na_2S—FeCl_2—H_2O$	3220
Sodium Sulfide–Copper (II) Sulfide –Water	$Na_2S—CuS—H_2O$	3221
Sodium Sulfide–Mercury (II) Sulfide –Water	$Na_2S—HgS—H_2O$	3222
Sodium Hydrogen Sulfite–Sulfur Dioxide–Water	$NaHSO_3—SO_2—H_2O$	3223
Sodium Sulfite–Sodium Sulfate–Water	$Na_2SO_3—Na_2SO_4—H_2O$	3224
Sodium Sulfite–Sodium Sulfate–Sodium Thiosulfate–Water	$Na_2SO_3—Na_2SO_4—Na_2S_2O_3—H_2O$	3225
Sodium Sulfite–Sodium Disulfite–Water	$Na_2SO_3—Na_2S_2O_5—H_2O$	3226, 3227
Sodium Sulfite–Sodium Chloride–Water	$Na_2SO_3—NaCl—H_2O$	3228

System name	Formulae of components	Table number
Sodium Sulfite–Sodium Hexa-hydroxostannate (IV)–Water	Na_2SO_3—$Na_2[Sn(OH)_6]$—H_2O	3229
Sodium Sulfite–Silver Chloride–Water	Na_2SO_3—$AgCl$—H_2O	3230
Sodium Sulfite–Silver Bromide–Water	Na_2SO_3—$AgBr$—H_2O	3231—3234, 5418
Sodium Sulfito Amide–Sodium Chloride–Water	$NaSO_3NH_2$—$NaCl$—H_2O	3235
Sodium Hydrogen Sulfate–Potassium Peroxodisulfate–Water	$NaHSO_4$—$K_2S_2O_8$—H_2O	3236
Sodium Hydrogen Sulfate–Vanadium Pentoxide–Water	$NaHSO_4$—V_2O_5—H_2O	3237
Sodium Hydrogen Sulfate–Thallium Sulfate–Water	$NaHSO_4$—Tl_2SO_4—H_2O	3238
Sodium Sulfate–Sodium Thiosulfate–Water	Na_2SO_4—$Na_2S_2O_3$—H_2O	3239
Sodium Sulfate–Sodium Pyrosulfate–Water	Na_2SO_4—$Na_2S_2O_5$—H_2O	3240
Sodium Sulfate–Sodium Chloride–Water	Na_2SO_4—$NaCl$—H_2O	3241—3286
Sodium Sulfate–Sodium Chloride–Sodium Dichromate–Water	Na_2SO_4—$NaCl$—$Na_2Cr_2O_7$—H_2O	3287
Sodium Sulfate–Sodium Chloride–Magnesium Sulfate–Water	Na_2SO_4—$NaCl$—$MgSO_4$—H_2O	3288—3289
Sodium Sulfate–Sodium Chloride–Magnesium Sulfate–Magnesium Chloride–Water	Na_2SO_4—$NaCl$—$MgSO_4$—$MgCl$—H_2O	3290—3311
Sodium Sulfate–Sodium Chloride–Magnesium Sulfate–Magnesium Chloride–Potassium Sulfate–Potassium Chloride–Water	Na_2SO_4—$NaCl$—$MgSO_4$—$MgCl$—K_2SO_4—KCl—H_2O	3312, 3313
Sodium Sulfate–Sodium Chloride–Magnesium Sulfate–Mangesium Chloride–Potassium Chloride–Water	Na_2SO_4—$NaCl$—$MgSO_4$—$MgCl_2$—KCl—H_2O	3314
Sodium Sulfate–Sodium Chloride–Magnesium Sulfate–Magnesium Chloride–Potassium Chloride–Calcium Sulfate–Water	Na_2SO_4—$NaCl$—$MgSO_4$—$MgCl_2$—KCl—$CaSO_4$—H_2O	3315—3317
Sodium Sulfate–Sodium Chloride–Potassium Sulfate–Potassium Chloride–Water	Na_2SO_4—$NaCl$—K_2SO_4—KCl—H_2O	3318—3324
Sodium Sulfate–Sodium Chloride–Potassium Chloride–Water	Na_2SO_4—$NaCl$—KCl—H_2O	3325—3327
Sodium Sulfate–Sodium Chloride–Calcium Oxide–Water	Na_2SO_4—$NaCl$—CaO—H_2O	3328
Sodium Sulfate–Sodium Chloride–Calcium Carbonate–Water	Na_2SO_4—$NaCl$—$CaCO_3$—H_2O	3329
Sodium Sulfate–Sodium Chloride–Calcium Carbonate–Calcium Sulfate–Water	Na_2SO_4—$NaCl$—$CaCO_3$—$CaSO_4$—H_2O	3330
Sodium Sulfate–Sodium Chloride–Calcium Sulfate–Water	Na_2SO_4—$NaCl$—$CaSO_4$—H_2O	3331—3336

System name	Formulae of components	Table number
Sodium Sulfate–Sodium Chloride–Calcium Sulfate–Calcium Chloride–Water	$Na_2SO_4—NaCl—CaSO_4—CaCl_2—H_2O$	3337, 3338
Sodium Sulfate–Sodium Chloride–Copper (II) Sulfate–Copper (II) Chloride–Water	$Na_2SO_4—NaCl—CuSO_4—CuCl_2—H_2O$	3339, 3340
Sodium Sulfate–Sodium Chloride–Lead (II) Chloride–Water	$Na_2SO_4—NaCl—PbCl_2—H_2O$	3341
Sodium Sulfate–Sodium Chlorate–Water	$Na_2SO_4—NaClO_3—H_2O$	3342
Sodium Sulfate–Sodium Perchlorate–Water	$Na_2SO_4—NaClO_4—H_2O$	3343
Sodium Sulfate–Sodium Chromate–Water	$Na_2SO_4—Na_2CrO_4—H_2O$	3344, 3345
Sodium Sulfate–Sodium Dichromate–Water	$Na_2SO_4—Na_2Cr_2O_7—H_2O$	3346, 3347
Sodium Sulfate–Sodium Arsenate–Water	$Na_2SO_4—Na_3AsO_4—H_2O$	3348
Sodium Sulfate–Sodium Bromide–Water	$Na_2SO_4—NaBr—H_2O$	3349—3351
Sodium Sulfate–Sodium Bromate–Water	$Na_2SO_4—NaBrO_3—H_2O$	3352, 3353
Sodium Sulfate–Sodium Molybdate–Water	$Na_2SO_4—Na_2MoO_4—H_2O$	3354—3356
Sodium Sulfate–Sodium Hexahydroxostannate (IV)–Water	$Na_2SO_4—Na_2[Sn(OH)_6]—H_2O$	3357
Sodium Sulfate–Sodium Iodide–Water	$Na_2SO_4—NaI—H_2O$	3358—3360
Sodium Sulfate–Sodium Iodate–Water	$Na_2SO_4—NaIO_3—H_2O$	3361
Sodium Sulfate–Magnesium Hydroxide–Water	$Na_2SO_4—Mg(OH)_2—H_2O$	3362, 3363
Sodium Sulfate–Magnesium Hydrogen Carbonate–Magnesium Carbonate–Water	$Na_2SO_4—Mg(HCO_3)_2—MgCO_3—H_2O$	3364
Sodium Sulfate–Magnesium Carbonate–Water	$Na_2SO_4—MgCO_3—H_2O$	3365, 3366
Sodium Sulfate–Magnesium Sulfate–Water	$Na_2SO_4—MgSO_4—H_2O$	3367—3395
Sodium Sulfate–Magnesium Sulfate–Potassium Sulfate–Water	$Na_2SO_4—MgSO_4—K_2SO_4—H_2O$	3396—3403
Sodium Sulfate–Magnesium Sulfate–Potassium Chloride–Water	$Na_2SO_4—MgSO_4—KCl—H_2O$	3404
Sodium Sulfate–Magnesium Sulfate–Calcium Sulfate–Water	$Na_2SO_4—MgSO_4—CaSO_4—H_2O$	3405
Sodium Sulfate–Magnesium Molybdate–Water	$Na_2SO_4—MgMoO_4—H_2O$	3406
Sodium Sulfate–Aluminum Sulfate–Water	$Na_2SO_4—Al_2(SO_4)_3—H_2O$	3407—3409
Sodium Sulfate–Aluminum Sulfate–Nickel (II) Sulfate–Water	$Na_2SO_4—Al_2(SO_4)_3—NiSO_4—H_2O$	3419, 3411
Sodium Sulfate–Sulfur Dioxide–Water	$Na_2SO_4—SO_2—H_2O$	3412—3414

System name	Formulae of components	Table number
Sodium Sulfate–Sulfur Trioxide–Water	Na_2SO_4—SO_3—H_2O	3415
Sodium Sulfate–Potassium Sulfate–Water	Na_2SO_4—K_2SO_4—H_2O	3416—3438
Sodium Sulfate–Potassium Peroxodisulfate–Water	Na_2SO_4—$K_2S_2O_8$—H_2O	3439
Sodium Sulfate–Potassium Perchlorate–Water	Na_2SO_4—$KClO_4$—H_2O	3440
Sodium Sulfate–Potassium Permanganate–Water	Na_2SO_4—$KMnO_4$—H_2O	3441
Sodium Sulfate–Potassium Iodide–Water	Na_2SO_4—KI—H_2O	3442
Sodium Sulfate–Calcium Hydroxide–Water	Na_2SO_4—$Ca(OH)_2$—H_2O	3443
Sodium Sulfate–Calcium Carbonate–Water	Na_2SO_4—$CaCO_3$—H_2O	3444—3449, 3736
Sodium Sulfate–Calcium Hydrogen Orthophosphate–Water	Na_2SO_4—$CaHPO_4 \cdot 2H_2O$—H_2O	5111
Sodium Sulfate–Calcium Orthophosphate–Water	Na_2SO_4—$Ca_3(PO_4)_2$—H_2O	5113
Sodium Sulfate–Calcium Sulfate–Water	Na_2SO_4—$CaSO_4$—H_2O	3450—3458
Sodium Sulfate–Manganese Tetrahydrogen Orthoborate–Water	Na_2SO_4—$MnH_4(BO_3)_2$—H_2O	3459
Sodium Sulfate–Manganese (II) Sulfate–Water	Na_2SO_4—$MnSO_4$—H_2O	3460—3463
Sodium Sulfate–Iron (II) Hydroxide–Water	Na_2SO_4—$Fe(OH)_2$—H_2O	3464
Sodium Sulfate–Iron (II) Sulfate–Water	Na_2SO_4—$FeSO_4$—H_2O	3465, 3466
Sodium Sulfate–Cobalt (II) Sulfate–Water	Na_2SO_4—$CoSO_4$—H_2O	3467—3471
Sodium Sulfate–Nickel (II) Sulfate–Water	Na_2SO_4—$NiSO_4$—H_2O	3472—3475
Sodium Sulfate–Copper (II) Sulfate–Water	Na_2SO_4—$CuSO_4$—H_2O	3476—3481
Sodium Sulfate–Zinc Hydroxide–Water	Na_2SO_4—$Zn(OH)_2$—H_2O	3482
Sodium Sulfate–Zinc Sulfate–Water	Na_2SO_4—$ZnSO_4$—H_2O	3483—3487
Sodium Sulfate–Strontium Sulfate–Water	Na_2SO_4—$SrSO_4$—H_2O	856
Sodium Sulfate–Yttrium Sulfate–Water	Na_2SO_4—$Y_2(SO_4)_3$—H_2O	3488
Sodium Sulfate–Silver Orthophosphate–Water	Na_2SO_4—Ag_3PO_4—H_2O	3489
Sodium Sulfate–Silver Sulfate–Water	Na_2SO_4—Ag_2SO_4—H_2O	3490—3493
Sodium Sulfate–Silver Bromide–Water	Na_2SO_4—$AgBr$—H_2O	5420
Sodium Sulfate–Silver Bromate–Water	Na_2SO_4—$AgBrO_3$—H_2O	3494, 5425

System name	Formulae of components	Table number
Sodium Sulfate –Cadmium Sulfate–Water	Na_2SO_4—$CdSO_4$—H_2O	3495, 3496
Sodium Sulfate–Cesium Sulfate–Water	Na_2SO_4—Cs_2SO_4—H_2O	3497
Sodium Sulfate–Cesium Hexanitrocobaltate–Water	Na_2SO_4—$Cs_3[Co(NO_2)_6]$—H_2O	3498
Sodium Sulfate–Lanthanum Sulfate–Water	Na_2SO_4—$La_2(SO_4)_3$—H_2O	3499—3501
Sodium Sulfa·e–Lanthanum Iodate–Water	Na_2SO_4—$La(IO_3)_3$—H_2O	3502
Sodium Sulfate–Cerium Sulfate–Water	Na_2SO_4—$Ce_2(SO_4)_3$—H_2O	3503
Sodium Sulfate–Praseodymium Sulfate–Water	Na_2SO_4—$Pr_2(SO_4)_3$—H_2O	3504
Sodium Sulfate–Neodymium Sulfate–Water	Na_2SO_4—$Nd_2(SO_4)_3$—H_2O	3505
Sodium Sulfate–Samarium Sulfate–Water	Na_2SO_4—$Sm_2(SO_4)_3$—H_2O	3506, 3507
Sodium Sulfate–Gadolinium Sulfate–Water	Na_2SO_4—$Gd_2(SO_4)_3$—H_2O	3508
Sodium Sulfate–Thallium Hexanitrocobaltate–Water	Na_2SO_4—$Tl_3[Co(NO_2)_6]$—H_2O	3509
Sodium Sulfate–Thallium Sulfate–Water	Na_2SO_4—Tl_2SO_4—H_2O	3510, 3511
Sodium Sulfate–Thallium Chloride–Water	Na_2SO_4—$TlCl$—H_2O	3512
Sodium Sulfate–Thallium Bromide–Water	Na_2SO_4—$TlBr$—H_2O	3513
Sodium Sulfate–Thallium Iodide–Water	Na_2SO_4—TlI—H_2O	3514
Sodium Sulfate–Thallium Iodate–Water	Na_2SO_4—$TlIO_3$—H_2O	3515
Sodium Sulfate–Lead (II) Sulfate–Water	Na_2SO_4—$PbSO_4$—H_2O	3516—3518
Sodium Sulfate–Thorium Sulfate–Water	Na_2SO_4—$Th(SO_4)_2$—H_2O	3519, 3520a
Sodium Calcium Sulfate (Glauberite)–Sea-water	$Na_2SO_4CaSO_4$— . . . —H_2O	3520b
Sodium Thiosulfate–Sodium Chloride–Water	$Na_2S_2O_3$—$NaCl$—H_2O	3521
Sodium Thiosulfate–Sodium Chloride–Magnesium Thiosulfate–Magnesium Chloride–Water	$Na_2S_2O_3$—$NaCl$—MgS_2O_3—$MgCl_2$—H_2O	3522
Sodium Thiosulfate–Sodium Chloride–Calcium Thiosulfate–Water	$Na_2S_2O_3$—$NaCl$—CaS_2O_3—H_2O	3523
Sodium Thiosulfate–Calcium Sulfate–Water	$Na_2S_2O_3$—$CaSO_4$—H_2O	3524
Sodium Thiosulfate–Calcium Thiosulfate–Water	$Na_2S_2O_3$—CaS_2O_3—H_2O	3525
Sodium Thiosulfate–Calcium Iodate–Water	$Na_2S_2O_3$—$Ca(IO_3)_2$—H_2O	3526
Sodium Thiosulfate–Silver Thiosulfate–Water	$Na_2S_2O_3$—$Ag_2S_2O_3$—H_2O	3527

System name	Formulae of components	Table number
Sodium Thiosulfate–Silver Bromide–Water	$Na_2S_2O_3$—AgBr—H_2O	3528, 3529, 5418—5420
Sodium Thiosulfate–Silver Iodide–Water	$Na_2S_2O_3$—AgI—H_2O	5427
Sodium Thiosulfate–Barium Thiosulfate–Water	$Na_2S_2O_3$—BaS_2O_3—H_2O	3530
Sodium Thiosulfate–Barium Iodate–Water	$Na_2S_2O_3$—$Ba(IO_3)_2$—H_2O	3531, 3532
Sodium Thiosulfate–Uranyl Thiosulfate–Water	$Na_2S_2O_3$—$UO_2S_2O_3$—H_2O	3533
Sodium Thiosulfate–Zinc Nitrate Hydroxide–Water	$Na_2S_2O_3\cdot 5H_2O$—$3Zn(OH)_2\cdot Zn(NO_3)_2\cdot xH_2O$—$H_2O$	3534
Sodium Dithionate–Strontium Dithionate–Water	$Na_2S_2O_6$—SrS_2O_6—H_2O	3535
Sodium Dithionate–Barium Dithionate–Water	$Na_2S_2O_6$—BaS_2O_6—H_2O	3536
Sodium Trithionate–Sodium Tetrathionate–Sodium Pentathionate–Water	$Na_2S_3O_6$—$Na_2S_4O_6$—$Na_2S_5O_5$—H_2O	3537
Sodium Tetrathionate–Sodium Pentathionate–Water	$Na_2S_4O_6$—$Na_2S_5O_6$—H_2O	3538, 3539
Sodium Chloride–Sodium Chlorite–Sodium Chlorate–Water	NaCl—$NaClO_2$—$NaClO_3$—H_2O	3540
Sodium Chloride–Sodium Chlorate–Water	NaCl—$NaClO_3$—H_2O	3541—3546
Sodium Chloride–Sodium Chlorate–Potassium Chloride–Potassium Chlorate–Water	NaCl—$NaClO_3$—KCl—$KClO_3$—H_2O	3547
Sodium Chloride–Sodium Perchlorate–Water	NaCl—$NaClO_4$—H_2O	3548—3551
Sodium Chloride–Sodium Metavanadate–Water	NaCl—$NaVO_3$—H_2O	3552
Sodium Chloride–Sodium Pyrovanadate–Water	NaCl—$Na_4V_2O_7$—H_2O	3553
Sodium Chloride–Sodium Dichromate–Water	NaCl—$Na_2Cr_2O_7$—H_2O	3554, 3555
Sodium Chloride–Sodium Bromide–Water	NaCl—NaBr—H_2O	3556—3561
Sodium Chloride–Sodium Bromide–Magnesium Chloride–Magnesium Bromide–Water	NaCl—NaBr—$MgCl_2$—$MgBr_2$—H_2O	3562
Sodium Chloride–Sodium Bromide–Potassium Chloride–Water	NaCl—NaBr—KCl—H_2O	3563
Sodium Chloride–Sodium Bromide–Potassium Bromide–Water	NaCl—NaBr—KBr—H_2O	3564
Sodium Chloride–Sodium Bromate Water	NaCl—$NaBrO_3$—H_2O	3565
Sodium Chloride–Sodium Molybdate–Water	NaCl—Na_2MoO_4—H_2O	3566
Sodium Chloride–Sodium Hexahydroxostannate (IV)–Water	NaCl—$Na_2[Sn(OH)_6]$—H_2O	3567

System name	Formulae of components	Table number
Sodium Chloride–Sodium Iodide–Water	$NaCl—NaI—H_2O$	3568—3572
Sodium Chloride–Sodium Iodide–Potassium Chloride–Potassium Iodide–Water	$NaCl—NaI—KCl—KI—H_2O$	3573, 3574
Sodium Chloride–Sodium Iodate–Water	$NaCl—NaIO_3—H_2O$	3575
Sodium Chloride–Magnesium Hydroxide–Water	$NaCl—Mg(OH)_2—H_2O$	3576
Sodium Chloride–Magnesium Hydrogen Carbonate–Water	$NaCl—Mg(HCO_3)_2—H_2O$	3577, 3578
Sodium Chloride–Magnesium Hydrogen Carbonate–Calcium Hydrogen Carbonate–Calcium Sulfate–Water	$NaCl—Mg(HCO_3)_2—Ca(HCO_3)_2—CaSO_4—H_2O$	3579
Sodium Chloride–Magesium Carbonate–Water	$NaCl—MgCO_3—H_2O$	3580—3582
Sodium Chloride–Magnesium Carbonate–Magnesium Sulfate–Calcium Carbonate–Calcium Sulfate–Water	$NaCl—MgCO_3—MgSO_4—CaCO_3—CaSO_4—H_2O$	3583
Sodium Chloride–Magnesium Sulfate–Water	$NaCl—MgSO_4—H_2O$	3584
Sodium Chloride–Magnesium Sulfate–Magnesium Chloride–Water	$NaCl—MgSO_4—MgCl_2—H_2O$	3585—3595
Sodium Chloride–Magnesium Sulfate–Magnesium Chloride–Potassium Sulfate–Water	$NaCl—MgSO_4—MgCl_2—K_2SO_4—H_2O$	3596
Sodium Chloride–Magnesium Sulfate–Magnesium Chloride–Potassium Sulfate–Potassium Chloride–Water	$NaCl—MgSO_4—MgCl_2—K_2SO_4—KCl—H_2O$	3597
Sodium Chloride–Magnesium Sulfate–Magnesium Chloride–Potassium Chloride–Water	$NaCl—MgSO_4—MgCl_2—KCl—H_2O$	3598—3614
Sodium Chloride–Magnesium Chloride–Potassium Chloride–Water	$NaCl—MgCl_2—KCl—H_2O$	3615
Sodium Chloride–Magnesium Chloride–Water	$NaCl—MgCl_2—H_2O$	3616—3636
Sodium Chloride–Magnesium Chloride–Potassium Chloride–Water	$NaCl—MgCl_2—KCl—H_2O$	3637—3649
Sodium Chloride–Magnesium Chloride–Potassium Chloride–Calcium Chloride–Water	$NaCl—MgCl_2—KCl—CaCl_2—H_2O$	3650, 3651
Sodium Chloride–Magnesium Chloride–Calcium Sulfate–Water	$NaCl—MgCl_2—CaSO_4—H_2O$	3652
Sodium Chloride–Magnesium Chloride–Calcium Chloride–Water	$NaCl—MgCl_2—CaCl_2—H_2O$	3653, 3654
Sodium Chloride–Aluminum Oxide–Water	$NaCl—Al_2O_3—H_2O$	3655
Sodium Chloride–Quartz–Water	$NaCl— . . . —H_2O$	3656

System name	Formulae of components	Table number
Sodium Chloride–Sulfur Dioxide–Water	$NaCl—SO_2—H_2O$	3657, 3658
Sodium Chloride–Potassium Hydrogen Carbonate–Potassium Chloride–Water	$NaCl—KHCO_3—KCl—H_2O$	3659
Sodium Chloride–Potassium Nitrate–Water	$NaCl—KNO_3—H_2O$	3660—3666
Sodium Chloride–Potassium Sulfate–Water	$NaCl—K_2SO_4—H_2O$	3667
Sodium Chloride–Potassium Sulfate–Potassium Chloride–Water	$NaCl—K_2SO_4—KCl—H_2O$	3668
Sodium Chloride–Potassium Sulfate–Calcium Sulfate–Water	$NaCl—K_2SO_4—CaSO_4—H_2O$	3669
Sodium Chloride–Potassium Chloride–Water	$NaCl—KCl—H_2O$	3670—3706
Sodium Chloride–Potassium Chloride–Potassium Bromide–Water	$NaCl—KCl—KBr—H_2O$	3707
Sodium Chloride–Potassium Chloride–Calcium Chloride–Water	$NaCl—KCl—CaCl_2—H_2O$	3708—3711
Sodium Chloride–Potassium Chloride–Calcium Molybdate–Water	$NaCl—KCl—CaMoO_4—H_2O$	3712
Sodium Chloride–Potassium Chloride–Calcium Tungstate–Water	$NaCl—KCl—CaWO_4—H_2O$	3713
Sodium Chloride–Potassium Chloride–Strontium Chloride–Water	$NaCl—KCl—SrCl_2—H_2O$	3714
Sodium Chloride–Potassium Chloride–Barium Chloride–Water	$NaCl—KCl—BaCl_2—H_2O$	3715
Sodium Chloride–Potassium Chlorate–Water	$NaCl—KClO_3—H_2O$	3716—3718
Sodium Chloride–Potassium Perchlorate–Water	$NaCl—KClO_4—H_2O$	3719, 3720
Sodium Chloride–Potassium Hexachloroplatinate–Water	$NaCl—K_2PtCl_6—H_2O$	3721
Sodium Chloride–Potassium Dichromate–Water	$NaCl—K_2Cr_2O_7—H_2O$	3722
Sodium Chloride–Potassium Permanganate–Water	$NaCl—KMnO_4—H_2O$	3723
Sodium Chloride–Potassium Bromide–Silver Bromide–Water	$NaCl—KBr—AgBr—H_2O$	5417
Sodium Chloride–Potassium Bromide–Water	$NaCl—KBr—H_2O$	3724
Sodium Chloride–Potassium Bromate–Water	$NaCl—KBrO_3—H_2O$	3725
Sodium Chloride–Potassium Iodide–Water	$NaCl—KI—H_2O$	3724
Sodium Chloride–Potassium Iodate–Water	$NaCl—KIO_3—H_2O$	3726
Sodium Chloride–Calcium Oxide–Water	$NaCl—CaO—H_2O$	3727

System name	Formulae of components	Table number
Sodium Chloride–Calcium Hydroxide–Water	$NaCl—Ca(OH)_2—H_2O$	3728—3730
Sodium Chloride–Calcium Hydrogen Carbonate–Water	$NaCl—Ca(HCO_3)_2—H_2O$	3731, 3732
Sodium Chloride–Calcium Hydrogen Carbonate–Calcium Sulfate–Water	$NaCl—Ca(HCO_3)_2—CaSO_4—H_2O$	3733
Sodium Chloride–Calcium Carbonate–Water	$NaCl—CaCO_3—H_2O$	3734—3741
Sodium Chloride–Calcium Nitrate–Water	$NaCl—Ca(NO_3)_2—H_2O$	3724
Sodium Chloride–Calcium Hydrogen Orthophosphate–Water	$NaCl—CaHPO_4 \cdot 2H_2O—H_2O$	5111
Sodium Chloride–Calcium Orthophosphate–Water	$NaCl—Ca_3(PO_4)_2—H_2O$	5113
Sodium Chloride–Calcium Sulfate–Water	$NaCl—CaSO_4—H_2O$	3742—3758
Sodium Chloride–Calcium Sulfate–Calcium Chloride–Water	$NaCl—CaSO_4—CaCl_2—H_2O$	3759—3761
Sodium Chloride–Calcium Thiosulfate–Calcium Chloride–Water	$NaCl—CaS_2O_3—CaCl_2—H_2O$	3762
Sodium Chloride–Calcium Chloride–Water	$NaCl—CaCl_2—H_2O$	3710, 3763—3777, 5181
Sodium Chloride–Calcium Chloride–Zinc Chloride–Water	$NaCl—CaCl_2—ZnCl_2—H_2O$	3777
Sodium Chloride–Calcium Iodate–Water	$NaCl—Ca(IO_3)_2—H_2O$	3778, 3779, 5190, 5191
Sodium Chloride–Manganese Tetrahydrogen Orthoborate–Water	$NaCl—MnH_4(BO_3)_2—H_2O$	3780
Sodium Chloride–Manganese (II) Chloride–Water	$NaCl—MnCl_2—H_2O$	3781
Sodium Chloride–Iron (II) Hydroxide–Water	$NaCl—Fe(OH)_2—H_2O$	3782
Sodium Chloride–Iron (III) Chloride–Water	$NaCl—FeCl_3—H_2O$	3783
Sodium Chloride–Cobalt (II) Chloride–Water	$NaCl—CoCl_2—H_2O$	3784—3787
Sodium Chloride–Cobalt (II) Iodate–Water	$NaCl—Co(IO_3)_2—H_2O$	3788
Sodium Chloride–Copper (II) Sulfate–Water	$NaCl—CuSO_4—H_2O$	3789
Sodium Chloride–Copper (I) Chloride–Water	$NaCl—CuCl—H_2O$	3790
Sodium Chloride–Copper (II) Chloride–Water	$NaCl—CuCl_2—H_2O$	3791—3797
Sodium Chloride–Zinc Hydroxide–Water	$NaCl—Zn(OH)_2—H_2O$	3798
Sodium Chloride–Zinc Carbonate–Water	$NaCl—ZnCO_3—H_2O$	3799
Sodium Chloride–Zinc Sulfate–Water	$NaCl—ZnSO_4—H_2O$	3724

System name	Formulae of components	Table number
Sodium Chloride–Zinc Bromide–Water	$NaCl—ZnBr_2—H_2O$	3724
Sodium Chloride–Strontium Carbonate–Water	$NaCl—SrCO_3—H_2O$	856, 3800—3802
Sodium Chloride–Strontium Chloride–Water	$NaCl—SrCl_2—H_2O$	3803
Sodium Chloride–Oxozirconium Chloride–Water	$NaCl—ZrOCl—H_2O$	3804
Sodium Perchlorate–Silver Sulfate–Water	$NaClO_4—Ag_2SO_4—H_2O$	3805
Sodium Chloride–Silver Chloride–Water	$NaCl—AgCl—H_2O$	3806—3814, 5410, 5412
Sodium Chloride–Silver Bromide–Water	$NaCl—AgBr—H_2O$	5417
Sodium Chloride–Silver Iodide–Water	$NaCl—AgI—H_2O$	3815
Sodium Chloride–Cadmium Sulfate–Water	$NaCl—CdSO_4—H_2O$	3724
Sodium Chloride–Cadmium Chloride–Water	$NaCl—CdCl_2—H_2O$	3816—3818
Sodium Chloride–Cesium Chloride–Water	$NaCl—CsCl—H_2O$	3 819
Sodium Chloride–Barium Hydroxide–Water	$NaCl—Ba(OH)_2—H_2O$	3820
Sodium Chloride–Barium Carbonate-Water	$NaCl—BaCO_3—H_2O$	3821, 3822
Sodium Chloride–Barium Nitrate–Water	$NaCl—Ba(NO_3)_2—H_2O$	3724
Sodium Chloride–Barium Metasilicate–Water	$NaCl—BaSiO_3 \cdot 6H_2O—H_2O$	3823
Sodium Chloride–Barium Sulfate–Water	$NaCl—BaSO_4—H_2O$	3824—3831
Sodium Chloride–Barium Chromate–Water	$NaCl—BaCrO_4—H_2O$	3832
Sodium Chloride–Barium Iodide–Water	$NaCl—BaI_2—H_2O$	3724
Sodium Chloride–Lanthanum Chloride–Water	$NaCl—LaCl_3—H_2O$	3833
Sodium Chloride–Lanthanum Iodate–Water	$NaCl—La(IO_3)_3—H_2O$	3834
Sodium Chloride–Cerium Chloride–Water	$NaCl—CeCl_3—H_2O$	3835
Sodium Chloride–Neodymium Chloride–Water	$NaCl—NdCl_3—H_2O$	3836
Sodium Chloride–Platinum Chloride–Water	$NaCl—PtCl_4—H_2O$	3837
Sodium Chloride–Mercury (II) Oxide–Water	$NaCl—HgO—H_2O$	3836
Sodium Chloride–Mercury (I) Chloride–Water	$NaCl—HgCl—H_2O$	3839
Sodium Chloride–Mercury (II) Chloride–Water	$NaCl—HgCl_2—H_2O$	3840—3842

System name	Formulae of components	Table number
Sodium Chloride–Thallium Hexanitrocobaltate–Water	$NaCl—Tl_3[Co(NO_2)_6]—H_2O$	3843
Sodium Chloride–Thallium (I) Chloride–Water	$NaCl—TlCl—H_2O$	3844
Sodium Chloride–Thallium (I) Iodate–Water	$NaCl—TlIO_3—H_2O$	3845, 3846
Sodium Chloride–Lead (II) Sulfate–Water	$NaCl—PbSO_4—H_2O$	3847, 3848
Sodium Chloride–Lead (II) Chloride–Water	$NaCl—PbCl_2—H_2O$	3849—3854
Sodium Chloride–Lead (II) Bromide–Water	$NaCl—PbBr_2—H_2O$	3855
Sodium Chloride–Lead (II) Iodide–Water	$NaCl—PbI_2—H_2O$	3856
Sodium Chloride–Lead (II) Iodate–Water	$NaCl—Pb(IO_3)_2—H_2O$	3857
Sodium Chlorate–Sodium Chromate–Water	$NaClO_3—Na_2CrO_4—H_2O$	3858
Sodium Chlorate–Sodium Bromide–Water	$NaClO_3—NaBr—H_2O$	3859
Sodium Chlorate–Sodium Bromate–Water	$NaClO_3—NaBrO_3—H_2O$	3860
Sodium Chlorate–Sodium Molybdate–Water	$NaClO_3—Na_2MoO_4—H_2O$	3861
Sodium Chlorate–Sodium Iodide–Water	$NaClO_3—NaI—H_2O$	3862
Sodium Chlorate–Sodium Iodate–Water	$NaClO_3—NaIO_3—H_2O$	3863
Sodium Chlorate–Potassium Chlorate–Water	$NaClO_3—KClO_3—H_2O$	3864—3868
Sodium Chlorate–Calcium Hydroxide–Water	$NaClO_3—Ca(OH)_2—H_2O$	5074
Sodium Chlorate–Calcium Carbonate–Water	$NaClO_3—CaCO_3—H_2O$	3739
Sodium Chlorate–Silver Chlorate–Water	$NaClO_3—AgClO_3—H_2O$	3869
Sodium Chlorate–Barium Chlorate–Water	$NaClO_3—Ba(ClO_3)_2—H_2O$	3870
Sodium Chlorate–Thallium (I) Chloride–Water	$NaClO_3—TlCl—H_2O$	3871
Sodium Perchlorate–Sodium Hexahydroxostannate–Water	$NaClO_4—Na_2[Sn(OH)_6]—H_2O$	3872
Sodium Perchlorate–Sodium Iodide–Silver Iodide–Water	$NaClO_4—NaI—AgI—H_2O$	3873
Sodium Perchlorate–Aluminum Perchlorate–Water	$NaClO_4—Al(ClO_4)_3—H_2O$	3874
Sodium Perchlorate–Potassium Perchlorate–Water	$NaClO_4—KClO_4—H_2O$	3875—3877a
Sodium Perchlorate–Calcium Hydroxide–Water	$NaClO_4—Ca(OH)_2—H_2O$	5074
Sodium Perchlorate–Silver Chloride–Water	$NaClO_4—AgCl—H_2O$	3877б

System name	Formulae of components	Table number
Sodium Perchlorate–Thorium Oxocarbonate Hydrate–Water	$NaClO_4—ThOCO_3\cdot 8H_2O—H_2O$	3878
Sodium Chromate–Sodium Molybdate–Water	$NaCrO_4—Na_2MoO_4—H_2O$	3879
Sodium Chromate–Sodium Hexahydroxostannate (IV)–Water	$Na_2CrO_4—Na_2[Sn(OH)_6]—H_2O$	3880
Sodium Chromate–Potassium Chromate–Water	$Na_2CrO_4—K_2CrO_4—H_2O$	3881
Sodium Chromate–Calcium Chromate–Water	$Na_2CrO_4—CaCrO_4—H_2O$	3882
Sodium Chromate–Rubidium Chromate–Water	$Na_2CrO_4—Rb_2CrO_4—H_2O$	3883
Sodium Chromate–Cesium Chromate–Water	$Na_2CrO_4—Cs_2CrO_4—H_2O$	3884
Sodium Dichromate–Potassium Dichromate–Water	$Na_2Cr_2O_7—K_2Cr_2O_7—H_2O$	3885
Sodium Dichromate–Cesium Dichromate–Water	$Na_2Cr_2O_7—Cs_2Cr_2O_7—H_2O$	3886
Sodium Dichydrogen Orthoarsenate–Arsenic Trioxide–Water	$NaH_2AsO_4—As_2O_3—H_2O$	3887
Sodium Selenite–Mercury (II) Selenite–Water	$Na_2SeO_3—HgSeO_3—H_2O$	3888
Sodium Selenate–Magnesium Selenate–Water	$Na_2SeO_4—MgSeO_4—H_2O$	3889
Sodium Selenate–Potassium Selenate–Water	$Na_2SeO_4—K_2SeO_4—H_2O$	3890
Sodium–Calcium Selenate–Water	$Na_2SeO_4—CaSeO_4—H_2O$	3891
Sodium Selenate–Lead (II) Selenate–Water	$Na_2SeO_4—PbSeO_4—H_2O$	3892
Sodium Bromide–Sodium Bromate–Water	$NaBr—NaBrO_3—H_2O$	3893, 3894
Sodium Bromide–Sodium Hexahydroxostannate (IV)–Water	$NaBr—Na_2[Sn(OH)_6]—H_2O$	3895
Sodium Bromide–Sodium Iodide–Water	$NaBr—NaI—H_2O$	3896—3899
Sodium Bromide–Sodium Iodide–Potassium Bromide–Water	$NaBr—NaI—KBr—H_2O$	3900, 3901
Sodium Bromide–Sodium Iodide–Potassium Bromide–Potassium Iodide–Water	$NaBr—NaI—KBr—KI—H_2O$	3902, 3903
Sodium Bromide–Sodium Iodide–Potassium Iodide–Water	$NaBr—NaI—KI—H_2O$	3904
Sodium Bromide–Sodium Iodate–Water	$NaBr—NaIO_3—H_2O$	3905
Sodium Bromide–Magnesium Bromide–Water	$NaBr—MgBr_2—H_2O$	3906
Sodium Bromide–Magnesium Bromide–Potassium Bromide–Water	$NaBr—MgBr_2—KBr—H_2O$	3907
Sodium Bromide–Sulfur Dioxide–Water	$NaBr—SO_2—H_2O$	3908, 3909
Sodium Bromide–Potassium Bromide–Water	$NaBr—KBr—H_2O$	3910—3913

System name	Formulae of components	Table number
Sodium Bromide–Potassium Iodide–Water	$NaBr—KI—H_2O$	3915
Sodium Bromide–Calcium Hydroxide–Water	$NaBr—Ca(OH)_2—H_2O$	3915, 5074
Sodium Bromide–Calcium Carbonate–Water	$NaBr—CaCO_3—H_2O$	3739
Sodium Bromide–Arsenic Trioxide–Water	$NaBr—As_2O_3—H_2O$	3916
Sodium Bromide–Silver Bromide–Water	$NaBr—AgBr—H_2O$	3917, 3918
Sodium Bromide–Cesium Bromide–Water	$NaBr—CsBr—H_2O$	3919
Sodium Bromide–Mercury (II) Bromide–Water	$NaBr—HgBr_2—H_2O$	3920, 3921
Sodium Bromide–Lead (II) Sulfate-Water	$NaBr—PbSO_4—H_2O$	3922
Sodium Bromide–Lead (II) Chloride–Water	$NaBr—PbCl_2—H_2O$	3923
Sodium Bromide–Lead (II) Bromide–Water	$NaBr—PbBr_2—H_2O$	3924, 3925
Sodium Bromate–Sodium Molybdate–Water	$NaBrO_3—Na_2MoO_4—H_2O$	3926
Sodium Bromate–Sodium Iodide–Water	$NaBrO_3—NaI—H_2O$	3927
Sodium Bromate–Potassium Bromate–Silver Bromate–Water	$NaBrO_3—KBrO_3—AgBrO_3—H_2O$	3928
Sodium Bromate–Silver Bromate–Water	$NaBrO_3—AgBrO_3—H_2O$	3929
Sodium Molybdate–Sodium Iodate–Water	$Na_2MoO_4—NaIO_3—H_2O$	3930
Sodium Molybdate–Magnesium Molybdate–Water	$Na_2MoO_4—MgMoO_4—H_2O$	3931
Sodium Molybdate–Molybdenum Oxide–Water	$Na_2MoO_4—MoO_3—H_2O$	3932
Sodium Molybdate–Lanthanum Molybdate–Water	$Na_2MoO_4—La_2(MoO_4)_3—H_2O$	3933
Sodium Molybdate–Silver Molybdate–Water	$Na_2MoO_4—Ag_2MoO_4—H_2O$	3934
Sodium Hexahydroxostannate–Sodium Iodide–Water	$Na_2[Sn(OH)_6]—NaI—H_2O$	3935
Sodium Hexahydroxostannate–Sodium Tungstate–Water	$Na_2[Sn(OH)_6]—Na_2WO_4—H_2O$	3936
Sodium Iodide–Sodium Iodate–Water	$NaI—NaIO_3—H_2O$	3937, 3938
Sodium Iodide–Potassium Bromide–Water	$NaI—KBr—H_2O$	3939
Sodium Iodide–Potassium Bromide–Potassium Iodide–Water	$NaI—KBr—KI—H_2O$	3940
Sodium Iodide–Potassium Iodide–Water	$NaI—KI—H_2O$	3941–3943
Sodium Iodide–Calcium Hydroxide–Water	$NaI—Ca(OH)_2—H_2O$	3944

System name	Formulae of components	Table number
Sodium Iodide–Silver Iodide–Water	$NaI—AgI—H_2O$	3945–3947
Sodium Iodide–Antimony Triiodide –Water	$NaI—SbI_3—H_2O$	3948, 3949
Sodium Iodide–Cesium Iodide– Water	$NaI—CsI—H_2O$	3950
Sodium Iodide–Mercury (II) Iodide– Water	$NaI—HgI_2—H_2O$	3951
Sodium Iodide–Lead (II) Iodide– Water	$NaI—PbI_2—H_2O$	3952
Sodium Iodide–Thallium (III) Iodide Water	$NaI—TlI_3—H_2O$	3953
Sodium Iodate–Magnesium Iodate–Water	$NaIO_3—Mg(IO_3)_2—H_2O$	3954
Sodium Iodate–Potassium Iodate–Water	$NaIO_3—KIO_3—H_2O$	3955
Sodium Iodate–Calcium Iodate– Water	$NaIO_3—Ca(IO_3)_2—H_2O$	3956
Sodium Iodate–Cobalt (II) Iodate– Water	$NaIO_3—Co(IO_3)_2—H_2O$	3957
Sodium Iodate–Lanthanum Iodate–Water	$NaIO_3—La(IO_3)_3—H_2O$	3958
Sodium Tungstate–Cerium Tungstate–Water	$Na_2WO_4—Ce_2(WO_4)_3—H_2O$	3959
Sodium Tungstate–Tungsten Tri- oxide–Water	$Na_2WO_4—WO_3—H_2O$	3960
Magnesium Oxide–Molybdenum Oxide–Water	$MgO—MoO_3—H_2O$	3961
Magnesium Oxide–Magnesium Chloride–Water	$MgO—MgCl_2—H_2O$	3962, 3963
Magnesium Oxide–Phosphorus Pent- oxide–Water	$MgO—P_2O_5—H_2O$	3964, 3965
Magnesium Oxide–Phosphorus Pent- oxide–Calcium Oxide–Water	$MgO—P_2O_5—CaO—H_2O$	3966, 3967
Magnesium Oxide–Arsenic Pent- odixe–Water	$MgO—As_2O_5—H_2O$	3968
Magnesium Hydroxide–Magnesium Sulfate–Water	$Mg(OH)_2—MgSO_4—H_2O$	3969
Magnesium Hydroxide–Magnesium Sulfate–Magnesium Chloride– Potassium Sulfate–Potassium Chloride–Water	$Mg(OH)_2—MgSO_4—MgCl_2—K_2SO_4—KCl—H_2O$	3970
Magnesium Hydroxide–Magnesium Chloride–Water	$Mg(OH)_2—MgCl_2—H_2O$	3971–3973
Magnesium Hydroxide–Potassium Chloride–Water	$Mg(OH)_2—KCl—H_2O$	3974, 3975
Magnesium Hydrogen Carbonate– Magnesium Sulfate–Potassium Hydrogen Carbonate–Potassium Sulfate–Water	$Mg(HCO_3)_2—MgSO_4—KHCO_3—K_2SO_4—H_2O$	3976
Magnesium Hydrogen Carbonate– Magnesium Sulfate–Calcium Hydrogen Carbonate–Calcium Sulfate–Water	$Mg(HCO_3)_2—MgSO_4—Ca(HCO_3)_2—CaSO_4—H_2O$	3977
Magnesium Hydrogen Carbonate– Magnesium Chloride–Potassium Hydrogen Carbonate–Potassium Chloride–Water	$Mg(HCO_3)_2—MgCl_2—KHCO_3—KCl—H_2O$	3978
Magnesium Hydrogen Carbonate– Calcium Hydrogen Carbonate– Water	$Mg(HCO_3)_2—Ca(HCO_3)_2—H_2O$	3979

System name	Formulae of components	Table number
Magnesium Carbonate–Magnesium Sulfate–Calcium Carbonate–Calcium Sulfate–Water	$MgCO_3$—$MgSO_4$—$CaCO_3$—$CaSO_4$—H_2O	3980–3982
Magnesium Carbonate–Magnesium Chloride–Water	$MgCO_3$—$MgCl_2$—H_2O	3983
Magnesium Carbonate–Potassium Hydrogen Carbonate–Water	$MgCO_3$—$KHCO_3$—H_2O	3984
Magnesium Carbonate–Calcium Carbonate–Water	$MgCO_3$—$CaCO_3$—H_2O	3985–3988
Magnesium Carbonate–Calcium Sulfate–Water	$MgCO_3$—$CaSO_4$—H_2O	3989
Magnesium Carbonate–Seawater	$MgCO_3 \cdot 3H_2O$— — —H_2O	3990
Magnesium Potassium Carbonate–Potassium Chloride–Water	$K_2Mg(CO_3)_2$—KCl—H_2O	3991
Magnesium Nitrate–Magnesium Sulfate–Water	$Mg(NO_3)_2$—$MgSO_4$—H_2O	3992, 3993
Magnesium Nitrate–Magnesium Chloride–Water	$Mg(NO_3)_2$—$MgCl_2$—H_2O	3994
Magnesium Nitrate–Magnesium Iodate–Water	$Mg(NO_3)_2$—$Mg(IO_3)_2$—H_2O	3995
Magnesium Nitrate–Magnesium Neodymium Nitrate–Water	$Mg(NO_3)_2 \cdot 6H_2O$—$3Mg(NO_3)_2 \cdot 2Nd(NO_3)_3 \cdot 2H_2O$—$H_2O$	3996
Magnesium Nitrate–Potassium Nitrate–Water	$Mg(NO_3)_2$—KNO_3—H_2O	3997
Magnesium Nitrate–Potassium Nitrate–Thallium Chromate– Water	$Mg(NO_3)_2$—KNO_3—Tl_2CrO_4—H_2O	3998
Magnesium Nitrate–Calcium Nitrate Water	$Mg(NO_3)_2$—$Ca(NO_3)_2$—H_2O	3999
Magnesium Nitrate–Calcium Sulfate Water	$Mg(NO_3)_2$—$CaSO_4$—H_2O	4000
Magnesium Nitrate–Manganese (Iı, Nitrate–Water	$Mg(NO_3)_2$—$Mn(NO_3)_2$—H_2O	4001
Magnesium Nitrate–Silver Sulfate–Water	$Mg(NO_3)_2$—Ag_2SO_4—H_2O	5404
Magnesium Nitrate–Silver Chromate–Water	$MgNO_3$—Ag_2CrO_4—H_2O	4002, 5413
Magnesium Nitrate–Silver Bromate–Water	$MgNO_3$—$AgBrO_3$—H_2O	4003, 4004, 5425
Magnesium Nitrate–Cesium Hexanitrocobaltate–Water	$Mg(NO_3)_2$—$Cs_3[Co(NO_2)_6]$—H_2O	4005
Magnesium Nitrate–Barium Bromate–Water	$Mg(NO_3)_2$—$Ba(BrO_3)_2$—H_2O	4006
Magnesium Nitrate–Lanthanum Nitrate–Water	$Mg(NO_3)_2$—$La(NO_3)_3$—H_2O	4007, 4008
Magnesium Nitrate–Lanthanum Iodate–Water	$Mg(NO_3)_2$—$La(IO_3)_3$—H_2O	4009
Magnesium Nitrate–Thallium Chromate–Water	$MgNO_3$—$TlCrO_4$—H_2O	4010
Magnesium Nitrate–Thallium Iodate–Water	$Mg(NO_3)_2$—$TlIO_3$—H_2O	4011
Magnesium Nitrate–Uranyl Nitrate–Water	$Mg(NO_3)_2$—$UO_2(NO_3)_2$—H_2O	4012

System name	Formulae of components	Table number
Magnesium Fluoride–Potassium Fluoride–Water	$MgF—KF_2—H_2O$	4013
Magnesium Fluoride–Silver Fluoride–Water	$MgF_2—AgF—H_2O$	4014
Magnesium Hydrogen Sulfite–Magnesium Sulfite–Magnesium Sulfate–Water	$Mg(HSO_3)_2—MgSO_3—MgSO_4—H_2O$	4015
Magnesium Hydrogen Sulfite–Magnesium Sulfite–Magnesium Sulfate–Sulfur Dioxide–Water	$Mg(HSO_3)_2—MgSO_3—MgSO_4—SO_2—H_2O$	4016
Magnesium Sulfate–Magnesium Chloride–Water	$MgSO_4—MgCl_2—H_2O$	4017—4035
Magnesium Sulfate–Magnesium Chloride–Potassium Sulfate–Potassium Chloride–Water	$MgSO_4—MgCl_2—K_2SO_4—KCl—H_2O$	4036—4041
Magnesium Sulfate–Magnesium Chloride–Potassium Chloride–Water	$MgSO_4—MgCl_2—KCl—H_2O$	4042—4050
Magnesium Sulfate–Magnesium Chloride–Calcium Sulfate–Calcium Chloride–Water	$MgSO_4—MgCl_2—CaSO_4—CaCl_2—H_2O$	4051
Magnesium Sulfate–Magnesium Molybdate–Water	$MgSO_4—MgMoO_4—H_2O$	4052
Magnesium Sulfate–Magnesium Bromide–Water	$MgSO_4—MgBr_2—H_2O$	4053
Magnesium Sulfate–Aluminum Sulfate–Water	$MgSO_4—Al_2(SO_4)_3—H_2O$	4054
Magnesium Sulfate–Potassium Sulfate–Water	$MgSO_4—K_2SO_4—H_2O$	4055—4075
Magnesium Sulfate–Potassium Sulfate–Calcium Sulfate–Water	$MgSO_4—K_2SO_4—CaSO_4—H_2O$	4076—4080
Magnesium Sulfate–Potassium Chloride–Water	$MgSO_4—KCl—H_2O$	4081
Magnesium Sulfate–Calcium Carbonate–Water	$MgSO_4—CaCO_3—H_2O$	3736, 4082, 4083
Magnesium Sulfate–Calcium Hydrogen Orthophosphate–Water	$MgSO_4—CaHPO_4 \cdot 2H_2O—H_2O$	5111
Magnesium Sulfate–Calcium Orthophosphate–Water	$MgSO_4 \cdot 7H_2O—Ca_3(PO_4)_2—H_2O$	5113
Magnesium Sulfate–Calcium Sulfate–Water	$MgSO_4—CaSO_4—H_2O$	4084—4090
Magnesium Sulfate–Calcium Chloride–Water	$MgSO_4—CaCl_2—H_2O$	5191
Magnesium Sulfate–Calcium Iodate–Water	$MgSO_4—Ca(IO_3)_2—H_2O$	4091
Magnesium Sulfate–Manganese (II) Sulfate–Water	$MgSO_4—MnSO_4—H_2O$	4092
Magnesium Sulfate–Nickel (II) Sulfate–Water	$MgSO_4—NiSO_4—H_2O$	4093
Magnesium Sulfate–Copper (II) Sulfate–Water	$MgSO_4—CuSO_4—H_2O$	4094, 4095
Magnesium Sulfate–Zinc Hydroxide–Water	$MgSO_4—Zn(OH)_2—H_2O$	4096
Magnesium Sulfate–Rubidium Sulfate–Water	$MgSO_4—Rb_2SO_4—H_2O$	4097

System name	Formulae of components	Table number
Magnesium Sulfate–Silver Sulfate–Water	$MgSO_4$—Ag_2SO_4—H_2O	4098, 4099, 5404
Magnesium Sulfate–Silver Chloride–Water	$MgSO_4$—AlCl—H_2O	4100
Magnesium Sulfate–Silver Bromate–Water	$MgSO_4$—$AgBrO_3$—H_2O	5425
Magnesium Sulfate–Silver Iodate Water	$MgSO_4$—$AgIO_3$—H_2O	4101
Magnesium Sulfate–Cadmium Sulfate–Water	$MgSO_4$—$CdSO_4$—H_2O	4102
Magnesium Sulfate–Cadmium Iodate–Water	$MgSO_4$—$Cd(IO_3)_2$—H_2O	4103
Magnesium Sulfate– Cesium Sulfate –Water	$MgSO_4$—Cs_2SO_4—H_2O	4104
Magnesium Sulfate–Lanthanum Iodate–Water	$MgSO_4$—$La(IO_3)_3$—H_2O	4105, 4106
Magnesium Sulfate–Cerium Iodate–Water	$MgSO_4$—$Ce(IO_3)_3$—H_2O	4107
Magnesium Sulfate–Thallium Sulfate–Water	$MgSO_4$—Tl_2SO_4—H_2O	4108
Magnesium Sulfate–Thallium Chloride–Water	$MgSO_4$—TlCl—H_2O	4109, 4110
Magnesium Sulfate–Thallium Bromide–Water	$MgSO_4$—TlBr—H_2O	4111
Magnesium Sulfate–Thallium Iodide –Water	$MgSO_4$—TlI—H_2O	4112
Magnesium Sulfate–Thallium Iodate –Water	$MgSO_4$—$TlIO_3$—H_2O	4113
Magnesium Potassium Sulfate, Magnesium Chloride–Calcium Orthophosphate–Water	$MgSO_2 \cdot K_2SO_4 \cdot MgCl_2$—$Ca_3(PO_4)_2$—$H_2O$	5113
Magnesium Dithionate–Barium Dithionate–Water	MgS_2O_6—BaS_2O_6—H_2O	4114
Magnesium Chloride–Magnesium Bromide–Water	$MgCl_2$—$MgBr_2$—H_2O	4115, 4116
Magnesium Chloride–Magnesium Bromide–Potassium Chloride–Potassium Bromide–Water	$MgCl_2$—$MgBr_2$—KCl—KBr—H_2O	4117
Magnesium Chloride–Magnesium Molybdate–Water	$MgCl_2$—$MgMoO_4$—H_2O	4118
Magnesium Chloride–Magnesium Iodide–Water	$MgCl_2$—MgI_2—H_2O	4119
Magnesium Chloride–Magnesium Potassium Orthophosphate Hydrate-Potassium Orthophosphate–Water	$MgCl_2$—$MgKPO_4 \cdot 6H_2O$—K_3PO_4—H_2O	4120
Magnesium Chloride–Potassium Nitrate–Water	$MgCl_2$—KNO_3—H_2O	4121
Magnesium Chloride–Potassium Chloride–Water	$MgCl_2$—KCl—H_2O	4122–4137
Magnesium Chloride–Potassium Chloride–Calcium Chloride–Water	$MgCl_2$—KCl—$CaCl_2$—H_2O	4138–4143

System name	Formulae of components	Table number
Magnesium Chloride–Potassium Chloride–Rubidium Chloride–Water	$MgCl_2$—KCl—$RbCl$—H_2O	4144—4146
Magnesium Chloride–Potassium Chromate–Water	$MgCl_2$—K_2CrO_4—H_2O	4147
Magnesium Chloride–Potassium Dichromate–Water	$MgCl_2$—$K_2Cr_2O_7$—H_2O	4148
Magnesium Chloride–Potassium Permanganate–Water	$MgCl_2$—$KMnO_4$—H_2O	4149
Magnesium Chloride–Calcium Oxide–Water	$MgCl_2$—CaO—H_2O	4150
Magnesium Chloride–Calcium Carbonate–Water	$MgCl_2$—$CaCO_3$—H_2O	3636, 4151
Magnesium Chloride–Calcium Orthophosphate–Water	$MgCl_2 \cdot 6H_2O$—$Ca_3(PO_4)_2$—H_2O	5113
Magnesium Chloride–Calcium Sulfate–Water	$MgCl_2$—$CaSO_4$—H_2O	1430, 4152, 4153
Magnesium Chloride–Calcium Sulfate–Calcium Chloride–Water	$MgCl_2$—$CaSO_4$—$CaCl_2$—H_2O	4154
Magnesium Chloride–Calcium Chloride–Water	$MgCl_2$—$CaCl_2$—H_2O	4155—4165
Magnesium Chloride–Calcium Iodate–Water	$MgCl_2$—$Ca(JO_3)_2$—H_2O	4166, 5190
Magnesium Chloride–Iron (II) Hydroxide–Water	$MgCl_2$—$Fe(OH)_2$—H_2O	4167
Magnesium Chloride–Iron (II) Chloride–Water	$MgCl_2$—$FeCl_2$—H_2O	4168
Magnesium Chloride–Cobalt (II) Chloride–Water	$MgCl_2$—$CoCl_2$—H_2O	4169, 4170
Magnesium Chloride–Zinc Hydroxide–Water	$MgCl_2$—$Zn(OH)_2$—H_2O	4171
Magnesium Chloride–Rubidium Chloride–Water	$MgCl_2$—$RbCl$—H_2O	4172—4174
Magnesium Chloride–Strontium Sulfate–Water	$MgCl_2$—$SrSO_4$—H_2O	4175
Magnesium Chloride–Strontium Chloride–Water	$MgCl_2$—$SrCl_2$—H_2O	4176
Magnesium Chloride–Silver Chloride–Water	$MgCl_2$—$AgCl$—H_2O	4177, 5410
Magnesium Chloride–Cadmium Chloride–Water	$MgCl_2$—$CdCl_2$—H_2O	4178
Magnesium Chloride–Indium Chloride–Water	$MgCl_2$—$InCl_3$—H_2O	4179
Magnesium Chloride–Cesium Chloride–Water	$MgCl_2$—$CsCl$—H_2O	4180
Magnesium Chloride–Barium Sulfate–Water	$MgCl_2$—$BaSO_4$—H_2O	4181
Magnesium Chloride–Barium Chloride–Water	$MgCl_2$—$BaCl_2$—H_2O	4182
Magnesium Chloride–Lanthanum Iodate–Water	$MgCl_2$—$La(IO_3)_3$—H_2O	4183, 4184
Magnesium Chloride–Cerium Iodate–Water	$MgCl_2$—$Ce(IO_3)_3$—H_2O	4185

System name	Formulae of components	Table number
Magnesium Chloride–Mercury (II) Chloride–Water	$MgCl_2—HgCl_2—H_2O$	4186, 4187
Magnesium Chloride–Thallium Chloride–Water	$MgCl_2—TlCl—H_2O$	4188, 4189
Magnesium Chloride–Thallium Iodate–Water	$MgCl_2—TlIO_3—H_2O$	4190
Magnesium Chloride–Lead (II) Chloride–Water	$MgCl_2—PbCl_2—H_2O$	4191, 4192
Magnesium Potassium Chloride–Calcium Orthophosphate–Water	$MgCl_2 \cdot KCl \cdot 6H_2O—Ca_3(PO_4)_2—H_2O$	5113
Magnesium Selenate–Potassium Selenate–Water	$MgSeO_4—K_2SeO_4—H_2O$	4070, 4193, 4194
Magnesium Bromide–Potassium Bromide–Water	$MgBr_2—KBr—H_2O$	4195, 4196
Magnesium Bromide–Rubidium Bromide–Water	$MgBr_2—RbBr—H_2O$	4197
Magnesium Bromide–Indium Bromide–Water	$MgBr_2—InBr_3—H_2O$	4198
Magnesium Iodide–Indium Iodide–Water	$MgI_2—InI_3—H_2O$	4199
Magnesium Iodide–Thallium Iodide–Water	$MgI_2—TlI—H_2O$	4200
Magnesium Iodate–Calcium Iodate–Water	$Mg(IO_3)_3—Ca(IO_3)_2—H_2O$	4201
Aluminum Oxide–Aluminum Sulfate–Water	$Al_2O_3—Al_2(SO_4)_3—H_2O$	4202
Aluminum Oxide–Phosphorus Pentoxide–Water	$Al_2O_3—P_2O_5—H_2O$	4203—4210
Aluminum Oxide–Potassium Fluoride–Water	$Al_2O_3—KF—H_2O$	3655
Aluminum Oxide–Potassium Chloride–Water	$Al_2O_3—KCl—H_2O$	3655
Aluminium Oxide–Potassium Hydroxide–Water	$Al_2O_3—KOH—H_2O$	2729
Aluminum Oxide–Potassium Carbonate–Water	$Al_2O_3—K_2CO_3—H_2O$	3655
Aluminum Oxide–Calcium Oxide–Water	$Al_2O_3—CaO—H_2O$	4211—4214
Aluminum Oxide–Arsenic Pentoxide–Water	$Al_2O_3—As_2O_5—H_2O$	4215
Aluminum Oxide–Cesium Hydroxide–Water	$Al_2O_3—CsOH—H_2O$	2729
Aluminum Oxide–Cesium Carbonate–Water	$Al_2O_3—Cs_2CO_3—H_2O$	3655
Aluminum Oxide–Barium Oxide–Water	$Al_2O_3—BaO—H_2O$	4216, 4217
Aluminum Oxide–Barium Hydroxide–Water	$Al_2O_3—BaOH—H_2O$	2729
Calcium Sulfoaluminate–Calcium Oxide–Water	$Al_2O_3 \cdot 3CaO \cdot 3CaSO_4 \cdot 32H_2O—CaO—H_2O$	4218

System name	Formulae of components	Table number
Aluminum Hydroxide–Aluminum Sulfate–Water	$Al(OH)_3—Al_2(SO_4)_3—H_2O$	4220
Aluminum Hydroxide–Potassium Hydroxide–Water	$Al(OH)_3—KOH—H_2O$	4219, 4221
Aluminum Thiocyanate–Silver Chloride–Water	$Al(CNS)_3—AgCl—H_2O$	4222
Aluminum Thiocyanate–Silver Iodide–Water	$Al(CNS)_3—AgI—H_2O$	5427
Aluminum Thiocyanate–Silver Bromide–Water	$Al(CNS)_3—AgBr—H_2O$	5418
Aluminum Nitrate–Aluminium Fluoride–Water	$Al(NO_3)_3—AlF_3—H_2O$	4223
Aluminum Nitrate–Potassium Nitrate–Water	$Al(NO_3)_3—KNO_3—H_2O$	4224, 4225
Aluminum Nitrate–Potassium Nitrate–Thallium Iodate–Water	$Al(NO_3)_2—KNO_3—TlIO_3—H_2O$	4226
Aluminum Nitrate–Calcium–Fluoride–Water	$Al(NO_3)_3—CaF_2—H_2O$	4227
Aluminum Nitrate–Iron (III) Nitrate–Water	$Al(NO_3)_3—Fe(NO_3)_3—H_2O$	4228, 4229
Aluminum Nitrate–Silver Sulfate–Water	$Al(NO_3)_3—Ag_2SO_4—H_2O$	4230
Aluminum Nitrate–Thallium Iodate–Water	$Al(NO_3)_3—TlIO_3—H_2O$	4231
Aluminum Nitrate Hydrate–Lead (II) Iodide–Water	$Al(NO_3)_3 \cdot 9H_2O—PbI_2—H_2O$	4232
Aluminum Nitrate–Thorium Nitrate–Water	$Al(NO_3)_3—Th(NO_3)_4—H_2O$	4233
Aluminum Fluoride–Aluminum Sulfate–Water	$AlF_3—Al_2(SO_4)_3—H_2O$	4234
Aluminum Fluoride–Aluminum Chloride–Water	$AlF_3—AlCl_3—H_2O$	4235
Aluminum Fluoride–Potassium Fluoride–Water	$AlF_3—KF—H_2O$	4236
Aluminum Sulfate–Potassium Sulfate–Water	$Al_2(SO_4)_3—K_2SO_4—H_2O$	4237—4241
Aluminum Sulfate–Potassium Sulfate–Cesium Sulfate–Water	$Al_2(SO_4)_3—K_2SO_4—Cs_2SO_4—H_2O$	4242, 4243
Aluminum Sulfate–Manganese (II) Sulfate–Water	$Al_2(SO_4)_3—MnSO_4—H_2O$	4244
Aluminum Sulfate–Iron (II) Sulfate–Water	$Al_2(SO_4)_3—FeSO_4—H_2O$	4245, 4246
Aluminum Sulfate–Iron (III) Sulfate–Water	$Al_2(SO_4)_3—Fe_2(SO_4)_3—H_2O$	4247, 4248
Aluminum Sulfate–Nickel (II) Sulfate–Water	$Al_2(SO_4)_3—NiSO_4—H_2O$	4249—4252
Aluminum Sulfate–Copper (II) Sulfate–Water	$Al_2(SO_4)_3—CuSO_4—H_2O$	4253—4255
Aluminum Sulfate–Silver Sulfate–Water	$Al_2(SO_4)_3—Ag_2SO_4—H_2O$	4256—4258
Aluminum Sulfate–Cesium Sulfate–Water	$Al_2(SO_4)_3—Cs_2SO_4—H_2O$	4259, 4260

System name	Formulae of components	Table number
Aluminum Sulfate–Thallium Sulfate–Water	$Al_2(SO_4)_3$—Tl_2SO_4—H_2O	4261
Aluminum Sulfate–Thallium Bromide–Water	$Al_2(SO_4)_3$—$TlBr$—H_2O	4262
Aluminum Chloride–Potassium Chloride–Water	$AlCl_3$—KCl—H_2O	4263
Aluminum Chloride–Iron (III) Chloride–Water	$AlCl_3$—$FeCl_3$—H_2O	4264
Aluminum Chloride–Barium Sulfate–Water	$AlCl_3$—$BaSO_4$—H_2O	4265
Aluminum Chloride–Mercury (II) Chloride–Water	$AlCl_3$—$HgCl_2$—H_2O	4266
Aluminum Perchlorate–Potassium Perchlorate–Water	$Al(ClO_4)_3$—$KClO_4$—H_2O	4267
Aluminum Potassium Sulfate–Aluminum Cesium Sulfate–Water	$K_2Al_2(SO_4)_4$—$Cs_2Al_2(SO_4)_4$—H_2O	4268
Aluminum Potassium Sulfate–Aluminum Thallium Sulfate– Water	$KAl(SO_4)_2$—$TlAl(SO_4)_2$—H_2O	4269, 4270
Phosphorus Pentoxide–Silicon Dioxide–Water	P_2O_5—SiO_2—H_2O	4271
Phosphorus Pentoxide–Sulfur Dioxide–Calcium Oxide–Water	P_2O_5—SO_2—CaO—H_2O	4272
Phosphorus Pentoxide–Sulfur Trioxide–Calcium Oxide–Water	P_2O_5—SO_3—CaO—H_2O	4273
Phosphorus Pentoxide–Potassium Oxide–Water	P_2O_5—K_2O—H_2O	4274, 4275
Phosphorus Pentoxide–Calcium Oxide–Water	P_2O_5—CaO—H_2O	4276–4281
Phosphorus Pentoxide–Calcium Sulfate–Water	P_2O_5—$CaSO_4$—H_2O	4282, 4283
Phosphorus Pentoxide–Vanadium Pentoxide–Water	P_2O_5—V_2O_5—H_2O	4284
Phosphorus Pentoxide–Chromium Trioxide–Water	P_2O_5—Cr_2O_3—H_2O	4285, 4286
Phosphorus Pentoxide–Manganese (II) Oxide–Water	P_2O_5—MnO—H_2O	4287, 4288
Phosphorus Pentoxide–Iron (II) Oxide–Water	P_2O_5—FeO—H_2O	4289
Phosphorus Pentoxide–Iron (III) Oxide–Water	P_2O_5—Fe_2O_3—H_2O	4290–4293
Phosphorus Pentoxide–Tetramminecobalt (III) Hydrogen Orthophosphate Oxide–Water	P_2O_5—$[Co(NH_3)_4HPO_4]_2O$—H_2O	4294
Phosphorus Pentoxide–Aquapentamminecobalt (III) Trioxide–Water	P_2O_5—$[Co(NH_3)_5 \cdot H_2O]_2O_3$—$H_2O$	4295
Phosphorus Pentoxide–Hexamminecobalt (III) Trioxide–Water	P_2O_5—$[Co(NH_3)_6]_2O_3$—H_2O	4296
Phosphorus Pentoxide–Copper (II) Oxide–Water	P_2O_5—CuO—H_2O	4297, 4298
Phosphorus Pentoxide–Zinc Oxide–Water	P_2O_5—ZnO—H_2O	4299, 4300
Phosphorus Pentoxide–Strontium Oxide–Water	P_2O_5—SrO—H_2O	4301

System name	Formulae of components	Table number
Phosphorus Pentoxide–Silver Oxide –Water	P_2O_5—Ag_2O—H_2O	4302
Phosphorus Pentoxide–Barium Oxide–Water	P_2O_5—BaO—H_2O	4303—4306
Phosphorus Pentoxide–Lanthanum Orthophosphate–Water	P_2O_5—$LaPO_4$—H_2O	4306
Phosphorus Pentoxide–Lead (II) Oxide–Water	P_2O_5—PbO—H_2O	4307
Phosphorus Pentoxide–Thorium Oxide–Water	P_2O_5—ThO_2—H_2O	4308
Sulfur Dioxide–Sulfur Trioxide–Water	SO_2—SO_3—H_2O	4309
Sulfur Dioxide–Potassium Thiocyanate–Water	SO_2—KCNS—H_2O	4310, 4311
Sulfur Dioxide–Potassium Nitrate–Water	SO_2—KNO_3—H_2O	4312, 4313
Sulfur Dioxide–Potassium Sulfate–Water	SO_2—K_2SO_4—H_2O	4314, 4315
Sulfur Dioxide–Potassium Chloride–Water	SO_2—KCl—H_2O	4316—4318
Sulfur Dioxide–Potassium Bromide–Water	SO_2—KBr—H_2O	4319
Sulfur Dioxide–Potassium Iodide–Water	SO_2— KI —H_2O	4320, 4321
Sulfur Dioxide–Calcium Oxide–Water	SO_2—CaO—H_2O	4322
Sulfur Dioxide–Iron (II) Oxide–Water	SO_2—FeO—H_2O	4323
Sulfur Dioxide–Zinc Oxide–Water	SO_2—ZnO—H_2O	4324
Sulfur Dioxide–Zinc Hydroxide–Water	SO_2—$Zn(OH)_2$—H_2O	4325
Sulfur Dioxide–Cadmium Sulfate–Water	SO_2—$CdSO_4$—H_2O	4326
Sulfur Dioxide–Cadmium Chloride–Water	SO_2—$CdCl_2$—H_2O	4327
Sulfur Dioxide–Cadmium Bromide–Water	SO_2—$CdBr_2$—H_2O	4328
Sulfur Dioxide–Lead (II) Oxide–Water	SO_2—PbO—H_2O	4329
Sulfur Trioxide–Potassium Sulfate–Water	SO_3—K_2SO_4—H_2O	4330
Sulfur Trioxide–Titanium Dioxide–Water	SO_3—TiO_2—H_2O	4331, 4332
Sulfur Trioxide–Vanadium Dioxide–Water	SO_3—VO_2—H_2O	4333
Sulfur Trioxide–Vanadium Pentoxide–Water	SO_3—V_2O_5—H_2O	4334
Sulfur Trioxide–Chromium Trioxide–Water	SO_3—CrO_3—H_2O	4335, 4336
Sulfur Trioxide–Iron (II) Oxide–Water	SO_3—FeO—H_2O	4337

System name	Formulae of components	Table number
Sulfur Trioxide–Iron (III) Oxide–Water	$SO_3—Fe_2O_3—H_2O$	4338—4343
Sulfur Trioxide–Nickel (II) Oxide–Water	$SO_3—NiO—H_2O$	4344
Sulfur Trioxide–Nickel (II) Oxide–Copper (II) Oxide–Uranium Trioxide–Water	$SO_3—NiO—CuO—UO_3—H_2O$	4345, 4346
Sulfur Trioxide–Nickel (II) Oxide–Uranium Trioxide–Water	$SO_3—NiO—UO_3—H_2O$	4347, 4348
Sulfur Trioxide–Copper (II) Oxide–Water	$SO_3—CuO—H_2O$	4349—4352
Sulfur Trioxide–Copper (II) Oxide–Uranium Trioxide–Water	$SO_3—CuO—UO_3—H_2O$	4353—4355
Sulfur Trioxide–Gallium Oxide–Water	$SO_3—Ga_2O_3—H_2O$	4356
Sulfur Trioxide–Niobium Pent-Oxide–Water	$SO_3—Nb_2O_5—H_2O$	4357, 4358
Sulfur Trioxide–Tin (II) Oxide–Water	$SO_3—SnO—H_2O$	4359
Sulfur Trioxide–Tantalum Pent-oxide–Water	$SO_3—Ta_2O_5—H_2O$	4360
Sulfur Trioxide–Mercury (II) Oxide–Water	$SO_3—HgO—H_2O$	4361
Sulfur Trioxide–Bismuth Oxide–Water	$SO_3—Bi_2O_3—H_2O$	4362
Sulfur Trioxide–Uranium Trioxide–Water	$SO_3—UO_3—H_2O$	4363—4365
Potassium Oxide–Potassium Sulfite–Water	$K_2O—K_2SO_3—H_2O$	4366
Potassium Oxide–Chromium Tri-oxide–Water	$K_2O—CrO_3—H_2O$	4367, 4368
Potassium Oxide–Arsenic Trioxide–Water	$K_2O—As_2O_3—H_2O$	4369
Potassium Oxide–Arsenic Pentoxide–Water	$K_2O—As_2O_5—H_2O$	4370
Potassium Hydroxide–Potassium Tetrahydroborate–Water	$KOH—KBH_4—H_2O$	4371
Potassium Hydroxide–Potassium Car-bonate–Water	$KOH—K_2CO_3—H_2O$	4372—4376
Potassium Hydroxide–Potassium Car-bonate–Potassium Permanganate/–Water	$KOH—K_2CO_3—KMnO_4—H_2O$	4377
Potassium Hydroxide–Potassium Carbonate–Potassium Manganate–Water	$KOH—K_2CO_3—K_2MnO_4—H_2O$	4378
Potassium Hydroxide–Potassium Cyanide–Water	$KOH—KCN—H_2O$	4379
Potassium Hydroxide–Potassium Hexacyanoferrate (III)–Water	$KOH—K_3Fe(CN)_6—H_2O$	4380
Potassium Hydroxide–Potassium Hexacyanoferrate (III)–Potas-sium Hexacyanoferrate (II)–Water	$KOH—K_3Fe(CN)_6—K_4Fe(CN)_6—H_2O$	4381
Potassium Hydroxide–Potassium Hexacyanoferrate (II)–Water	$KOH—K_4Fe(CN)_6—H_2O$	4382
Potassium Hydroxide–Potassium Nitrate–Water	$KOH—KNO_3—H_2O$	4383, 4383a

System name	Formulae of components	Table number
Potassium Hydroxide–Potassium Fluoride–Water	$KOH—KF—H_2O$	4384—4386
Potassium Hydroxide– Aluminum Potassium Oxide–Potassium Sulfate–Water	$KOH—KAlO_2—K_2SO_4—H_2O$	4387
Potassium Hydroxide–Potassium Sulfide–Bismuth Sulfide–Water	$KOH—K_2S—Bi_2S_3—H_2O$	4388
Potassium Hydroxide–Potassium Sulfate–Water	$KOH—K_2SO_4—H_2O$	4389—4391
Potassium Hydroxide–Potassium Sulfate–Calcium Hydroxide–Calcium Sulfate–Water	$KOH—K_2SO_4—Ca(OH)_2—CaSO_4—H_2O$	4392
Potassium Hydroxide–Potassium Sulfate–Calcium Sulfate–Water	$KOH—K_2SO_4—CaSO_4—H_2O$	4393
Potassium Hydroxide–Potassium Sulfate–Lead (II) Sulfate–Water	$KOH—K_2SO_4—PbSO_4—H_2O$	4394
Potassium Hydroxide–Potassium Chloride–Water	$KOH—KCl—H_2O$	4395—4401
Potassium Hydroxide–Potassium Chloride–Silver Chloride–Water	$KOH—KCl—AgCl—H_2O$	4402
Potassium Hydroxide–Potassium Chlorate–Water	$KOH—KClO_3—H_2O$	4403—4405
Potassium Hydroxide–Potassium Permanganate–Water	$KOH—KMnO_4—H_2O$	4406, 4407
Potassium Hydroxide–Potassium Permanganate–Potassium Manganate–Water	$KOH—KMnO_4—K_2MnO_4—H_2O$	4408
Potassium Hydroxide–Potassium Manganate–Water	$KOH—K_2MnO_4—H_2O$	4409, 4410
Potassium Hydroxide–Potassium Bromide–Water	$KOH—KBr—H_2O$	4411, 4412
Potassium Hydroxide–Potassium Tetrathioantimonate–Water	$KOH—K_3SbS_4—H_2O$	4413
Potassium Hydroxide–Potassium Iodide–Water	$KOH—KI—H_2O$	4414—4416
Potassium Hydroxide–Potassium Iodate–Water	$KOH—KIO_3—H_2O$	4417
Potassium Hydroxide–Potassium Periodate–Water	$KOH—KIO_4—H_2O$	4418
Potassium Hydroxide–Calcium Oxide–Water	$KOH—CaO—H_2O$	4419
Potassium Hydroxide–Calcium Sulfate–Water	$KOH—CaSO_4—H_2O$	4420
Potassium Hydroxide–Calcium Iodate–Water	$KOH—Ca(IO_3)_2—H_2O$	4421
Potassium Hydroxide–Titanium Oxide–Water	$KOH—TiO_2—H_2O$	4422
Potassium Hydroxide–Copper (II) Oxide–Water	$KOH—CuO—H_2O$	4423
Potassium Hydroxide–Zinc Hydroxide–Water	$KOH—Zn(OH)_2—H_2O$	4424
Potassium Hydroxide–Mercury (II) Oxide–Water	$KOH—HgO—H_2O$	4425

System name	Formulae of components	Table number
Potassium Carbonate–Potassium Chloride–Water	K_2CO_3—KCl—H_2O	4472—4476
Potassium Carbonate–Potassium Chlorate–Water	K_2CO_3—$KClO_3$—H_2O	4477, 4478
Potassium Carbonate–Potassium Permanganate–Water	K_2CO_3—$KMnO_4$—H_2O	4479
Potassium Carbonate–Calcium Carbonate–Water	K_2CO_3—$CaCO_3$—H_2O	4480
Potassium Carbonate–Titanium Oxide–Water	K_2CO_3—TiO_2—H_2O	4481
Potassium Carbonate–Zinc Hydroxide–Water	K_2CO_3—$Zn(OH)_2$—H_2O	4482
Potassium Carbonate–Thorium Oxide–Water	K_2CO_3—ThO_2—H_2O	4483
Potassium Carbonate–Thorium Hypophosphate–Water	K_2CO_3—$ThP_2O_6 \cdot 11H_2O$—H_2O	4484
Potassium Carbonate–Plutonium Hydroxide–Water	K_2CO_3—$Pu(OH)_4$—H_2O	4485
Potassium Cyanide–Nickel (II) Cyanide–Water	KCN—$Ni(CN)_2$—H_2O	4486
Potassium Cyanide–Copper (I) Cyanide–Water	KCN—$CuCN$—H_2O	4487
Potassium Cyanide–Zinc Cyanide–Water	KCN—$Zn(CN)_2$—H_2O	4488
Potassium Cyanide–Cadmium Cyanide–Water	KCN—$Cd(CN)_2$—H_2O	4489
Potassium Cyanide–Silver Cyanide–Water	KCN—$AgCN$—H_2O	4490, 4491
Potassium Cyanide–Silver Bromide–Water	KCN—$AgBr$—H_2O	5418, 5419
Potassium Cyanide–Silver Iodide–Water	KCN—AgI—H_2O	4492
Potassium Cyanide–Gold (I) Cyanide–Water	KCN—$AuCN$—H_2O	4493
Potassium Cyanide–Thallium Cyanide–Water	KCN—$TlCN$—H_2O	4494
Potassium Cyanide–Mercury (II) Cyanide–Water	KCN—$Hg(CN)_2$—H_2O	4495, 4496
Potassium Cyanide–Potassium Nitrite–Water	KCN—KNO_2—H_2O	4497
Potassium Thiocyanate–Potassium Nitrate–Water	$KCNS$—KNO_3—H_2O	4498—4500
Potassium Thiocyanate–Potassium Sulfate–Water	$KCNS$—K_2SO_4—H_2O	4501
Potassium Thiocyanate–Potassium Chloride–Water	$KCNS$—KCl—H_2O	4502
Potassium Thiocyanate–Potassium Iodide–Water	$KCNS$—KI—H_2O	4503
Potassium Thiocyanate–Potassium Bromide–Water	$KCNS$—KBr—H_2O	4504
Potassium Thiocyanate–Potassium Iodide–Copper (I) Thiocyanate–Water	$KCNS$—KI—$CuCNS$—H_2O	4505

System name	Formulae of components	Table number
Potassium Thiocyanate–Calcium Thiocyanate–Water	$KCNS—Ca(CNS)_2—H_2O$	4506
Potassium Thiocyanate–Copper (I) Thiocyanate–Water	$KCNS—CuCNS—H_2O$	4507
Potassium Thiocyanate–Zinc Nitrate Trihydroxide–Water	$KCNS—3Zn(OH)_2\cdot Zn(NO_3)_2\cdot xH_2O—H_2O$	4508
Potassium Thiocyanate–Silver Thiocyanate–Water	$KCNS—AgCNS—H_2O$	4509—4512
Potassium Thiocyanate–Silver Chloride–Water	$KCNS—AgCl—H_2O$	5410
Potassium Thiocyanate–Silver Bromide–Water	$KCNS—AgBr—H_2O$	4513, 5418
Potassium Thiocyanate–Silver Iodide–Water	$KCNS—AgI—H_2O$	4514, 5427
Potassium Thiocyanate–Barium Thiocyanate–Water	$KCNS—Ba(CNS)_2—H_2O$	4515
Potassium Thiocyanate–Mercury (II) Thiocyanate–Water	$KCNS—Hg(CNS)_2—H_2O$	4516
Potassium Thiocyanate–Thallium Thiocyanate–Water	$KCNS—TlCNS—H_2O$	4517—4519
Potassium Selenocyanate–Lead (II) Selenocyanate–Water	$KCNSe—Pb(CNSe)_2—H_2O$	4520
Potassium Hexacyanoferrate (III)–Potassium Hexacyanoferrate (II)–Water	$K_3[Fe(CN)_6]—K_4[Fe(CN)_6]—H_2O$	4521—4523
Potassium Hexacyanoferrate (II)–Potassium Sulfate–Water	$K_4[Fe(CN)_6]—K_2SO_4—H_2O$	4524—4526
Potassium Hexacyanoferrate (II)–Calcium Chloride–Water	$K_4Fe(CN)_6—CaCl_2—H_2O$	5191
Potassium Hexacyanoferrate (II)–Barium Chloride–Water	$K_4[Fe(CN)_6]—BaCl_2—H_2O$	4527
Potassium Nitrite–Potassium Nitrate–Water	$KNO_2—KNO_3—H_2O$	4528, 4529
Potassium Nitrite–Potassium Chloride–Water	$KNO_2—KCl—H_2O$	4530, 4531
Potassium Nitrite–Strontium Nitrite–Water	$KNO_2—Sr(NO_2)_2—H_2O$	4532
Potassium Nitrite–Silver Nitrite–Water	$KNO_2—AgNO_2—H_2O$	4532—4534
Potassium Nitrite–Barium Nitrite–Water	$KNO_2—Ba(NO_2)_2—H_2O$	4535
Potassium Nitrate–Potassium Fluoride–Lead (II) Fluoride–Water	$KNO_3—KF—PbF_2—H_2O$	4536
Potassium Nitrate–Potassium Dihydrogen Orthophosphate–Water	$KNO_3—KH_2PO_4—H_2O$	4537, 4538
Potassium Nitrate–Potassium Sulfate–Water	$KNO_3—K_2SO_4—H_2O$	4539—4544
Potassium Nitrate–Potassium Sulfate–Potassium Chloride–Water	$KNO_3—K_2SO_4—KCl—H_2O$	4545—4547
Potassium Nitrate–Potassium Sulfate–Potassium Chloride–Potassium Bromate–Water	$KNO_3—K_2SO_4—KCl—KBrO_3—H_2O$	4548
Potassium Nitrate–Potassium Sulfate–Potassium Bromate–Water	$KNO_3—K_2SO_4—KBrO_3—H_2O$	4549

System name	Formulae of components	Table number
Potassium Nitrate–Potassium Sulfate–Thallium Chromate–Water	KNO_3—K_2SO_4—Tl_2CrO_4—H_2O	4550
Potassium Nitrate–Potassium Sulfate–Thallium Iodate–Water	KNO_3—K_2SO_4—$TlIO_3$—H_2O	4551
Potassium Nitrate–Potassium Sulfate–Lead (II) Sulfate–Water	KNO_3—K_2SO_4—$PbSO_4$—H_2O	4552
Potassium Nitrate–Potassium Chloride–Water	KNO_3—KCl—H_2O	4553—4565
Potassium Nitrate–Potassium Chloride–Potassium Dichromate– Water	KNO_3—KCl—$K_2Cr_2O_7$—H_2O	4566
Potassium Nitrate–Potassium Chloride–Calcium Nitrate–Calcium Chloride–Water	KNO_3—KCl—$Ca(NO_3)_2$—$CaCl_2$—H_2O	4567
Potassium Nitrate–Potassium Chlorate–Water	KNO_3—$KClO_3$—H_2O	4568—4572
Potassium Nitrate–Potassium Perchlorate–Water	KNO_3—$KClO_4$—H_2O	4573, 4574
Potassium Nitrate–Potassium Chromate–Water	KNO_3—K_2CrO_4—H_2O	4575
Potassium Nitrate–Potassium Chromate–Yttrium Nitrate–Yttrium Chromate–Water	KNO_3—K_2CrO_4—$Y(NO_3)_3$—$Y_2(CrO_4)_3$—H_2O	4576
Potassium Nitrate–Potassium Chromate–Praseodymium Nitrate–Praseodymium Chromate–Water	KNO_3—K_2CrO_4—$Pr(NO_3)_3$—$Pr_2(CrO_4)_3$—H_2O	4577
Potassium Nitrate–Potassium Bromide–Water	KNO_3—KBr—H_2O	4578—4581
Potassium Nitrate–Potassium Bromate–Water	KNO_3—$KBrO_3$—H_2O	4582
Potassium Nitrate–Potassium Molybdate–Water	KNO_3—K_2MoO_4—H_2O	4583
Potassium Nitrate–Potassium Iodide–Water	KNO_3—KI—H_2O	4584, 4585
Potassium Nitrate–Potassium Iodide–Lead (II) Iodide–Water	KNO_3—KI—PbI_2—H_2O	4586
Potassium Nitrate–Potassium Iodate–Water	KNO_3—KIO_3—H_2O	4587
Potassium Nitrate–Calcium Nitrate–Water	KNO_3—$Ca(NO_3)_2$—H_2O	4588—4590
Potassium Nitrate–Calcium Sulfate–Water	KNO_3—$CaSO_4$—H_2O	4591, 4592
Potassium Nitrate–Calcium Chloride–Water	KNO_3—$CaCl_2$—H_2O	4593
Potassium Nitrate–Calcium Iodate–Water	KNO_3—$Ca(IO_3)_2$—H_2O	4594
Potassium Nitrate–Iron (III) Nitrate–Water	KNO_3—$Fe(NO_3)_3$—H_2O	4595
Potassium Nitrate–Zinc Hydroxide–Water	KNO_3—$Zn(OH)_2$—H_2O	4596
Potassium Nitrate–Strontium Nitrate–Water	KNO_3—$Sr(NO_3)_2$—H_2O	4597, 4598
Potassium Nitrate–Strontium Chloride–Water	KNO_3—$SrCl_2$—H_2O	4599
Potassium Nitrate–Silver Nitrate–Water	KNO_3—$AgNO_3$—H_2O	4600—4603

System name	Formulae of components	Table number
Potassium Nitrate–Silver Orthophosphate–Water	$KNO_3—Ag_3PO_4—H_2O$	4604
Potassium Nitrate–Silver Sulfate–Water	$KNO_3—Ag_2SO_4—H_2O$	4605, 5404
Potassium Nitrate–Silver Chloride–Water	$KNO_3—AgCl—H_2O$	4606, 4607
Potassium Nitrate–Silver Chromate–Water	$KNO_3—Ag_2CrO_4—H_2O$	4608, 5413
Potassium Nitrate–Silver Bromate–Water	$KNO_3—AgBrO_3—H_2O$	4609, 4610, 5425
Potassium Nitrate–Silver Iodate–Water	$KNO_3—AgIO_3—H_2O$	4611, 4612, 5429
Potassium Nitrate–Cadmium Iodate–Water	$KNO_3—Cd(IO_3)_2—H_2O$	4613
Potassium Nitrate–Antimony Trifluoride–Water	$KNO_3—SbF_3—H_2O$	4614
Potassium Nitrate–Barium Nitrate–Water	$KNO_3—Ba(NO_3)_2—H_2O$	4615—4617
Potassium Nitrate–Barium Bromate–Water	$KNO_3—Ba(BrO_3)_2—H_2O$	4618
Potassium Nitrate–Barium Iodate–Water	$KNO_3—Ba(IO_3)_2—H_2O$	4619—4621
Potassium Nitrate–Lanthanum Iodate–Water	$KNO_3—La(IO_3)_3—H_2O$	4622
Potassium Nitrate–Cerium Nitrate–Water	$KNO_3—Ce(NO_3)_3—H_2O$	4623
Potassium Nitrate–Cerium Iodate–Water	$KNO_3—Ce(IO_3)_2—H_2O$	4624
Potassium Nitrate–Thallium Nitrate–Water	$KNO_3—Tl(NO_3)_3—H_2O$	4625, 4626
Potassium Nitrate–Thallium Chloride–Water	$KNO_3—TlCl—H_2O$	4627—4629
Potassium Nitrate–Thallium Bromide–Water	$KNO_3—TlBr—H_2O$	4630
Potassium Nitrate–Thallium Iodate–Water	$KNO_3—TlIO_3—H_2O$	4231, 4631
Potassium Nitrate–Lead (II) Fluoride–Water	$KNO_3—PbF_2—H_2O$	4632
Potassium Nitrate–Lead (II) Nitrate–Water	$KNO_3—Pb(NO_3)_2—H_2O$	4633, 4634
Potassium Nitrate–Lead (II) Bromate–Water	$KNO_3—Pb(BrO_3)_2—H_2O$	4635
Potassium Nitrate–Lead (II) Iodide–Water	$KNO_3—PbI_2—H_2O$	4636, 4637
Potassium Nitrate–Lead (II) Iodate–Water	$KNO_3—Pb(IO_3)_2—H_2O$	4638, 4639
Potassium Nitrate–Uranyl Nitrate–Water	$KNO_3—UO_2(NO_3)_2—H_2O$	4640, 4641
Potassium Hydrogen Fluoride–Potassium Chloride–Water	$KHF_2—KCl—H_2O$	4642
Potassium Tetrafluoraborate–Potassium Fluoride–Water	$KBF_4—KF—H_2O$	4643
Potassium Fluoride–Potassium Chloride–Water	$KF—KCl—H_2O$	4644, 4645
Potassium Fluoride–Titanium Fluoride–Water	$KF—TiF_4—H_2O$	4646
Potassium Fluoride–Chromium Fluoride–Water	$KF—CrF_3—H_2O$	4647

System name	Formulae of components	Table number
Potassium Fluoride–Iron (III) Nitrate–Water	$KF—FeF_3—H_2O$	4648
Potassium Fluoride–Cobalt (II) Fluoride–Water	$KF—CoF_2—H_2O$	4649
Potassium Fluoride–Nickel (II) Fluoride–Water	$KF—NiF_2—H_2O$	4650
Potassium Fluoride–Copper (II) Fluoride–Water	$KF—CuF_2—H_2O$	4651
Potassium Fluoride–Zinc Fluoride–Water	$KF—ZnF_2—H_2O$	4652
Potassium Fluoride–Strontium Fluoride–Water	$KF—SrF_2—H_2O$	4653
Potassium Fluoride–Zirconium Fluoride–Water	$KF—ZrF_4—H_2O$	4654
Potassium Fluoride–Molybdenum Trioxide–Water	$KF—MoO_3—H_2O$	4655
Potassium Fluoride–Cadmium Fluoride–Water	$KF—CdF_2—H_2O$	4656
Potassium Fluoride–Iodine Pentoxide–Water	$KF—I_2O_5—H_2O$	4657
Potassium Fluoride–Barium Fluoride–Water	$KF—BaF_2—H_2O$	4658
Potassium Fluoride–Cerium (IV) Fluoride–Water	$KF—CeF_4—H_2O$	4659—4661
Potassium Fluoride–Lead (II) Fluoride–Water	$KF—PbF_2—H_2O$	4662, 4663
Potassium Fluoride–Thorium Nitrate–Water	$KF—Th(NO_3)_4—H_2O$	4664
Potassium Dihydrogen Orthophosphate–Potassium Chloride–Water	$KH_2PO_4—KCl—H_2O$	4665—4667
Potassium Dihydrogen Orthophosphate–Potassium Sulfate–Water	$KH_2PO_4—K_2SO_4—H_2O$	4668, 4669
Potassium Dihydrogen Orthophosphate–Arsenic Trioxide–Water	$KH_2PO_4—As_2O_3—H_2O$	4670
Potassium Hydrogen Orthophosphate–Silver Orthophosphate–Water	$K_2HPO_4—Ag_3PO_4—H_2O$	4671
Potassium Sulfito Amide–Potassium Sulfate–Water	$KSO_3NH_2—K_2SO_4—H_2O$	4672
Potassium Hydrogen Sulfate–Silver Sulfate–Water	$KHSO_4—Ag_2SO_4—H_2O$	5403
Potassium Sulfate–Potassium Peroxodisulfate–Water	$K_2SO_4—K_2S_2O_8—H_2O$	4673
Potassium Sulfate–Potassium Chloride–Water	$K_2SO_4—KCl—H_2O$	4674—4688
Potassium Sulfate–Potassium Chloride–Potassium Bromate–Water	$K_2SO_4—KCl—KBrO_3—H_2O$	4689
Potassium Sulfate–Potassium Chloride–Calcium Sulfate–Calcium Chloride–Water	$K_2SO_4—KCl—CaSO_4—CaCl_2—H_2O$	4690—4693
Potassium Sulfate–Potassium Chloride–Zinc Sulfate–Zinc Chloide–Water	$K_2SO_4—KCl—ZnSO_4—ZnCl_2—H_2O$	4694

System name	Formulae of components	Table number
Potassium Sulfate–Potassium Chlorate–Water	K_2SO_4—$KClO_3$—H_2O	4695—4697
Potassium Sulfate–Potassium Perchlorate–Water	K_2SO_4—$KClO_4$—H_2O	4698, 4699
Potassium Sulfate–Potassium Chromate–Water	K_2SO_4—K_2CrO_4—H_2O	4700—4703
Potassium Sulfate–Potassium Dichromate–Water	K_2SO_4—$K_2Cr_2O_7$—H_2O	4704
Potassium Sulfate–Potassium Permanganate–Water	K_2SO_4—$KMnO_4$—H_2O	4705
Potassium Sulfate–Potassium Bromide–Water	K_2SO_4—KBr—H_2O	4706, 4707
Potassium Sulfate–Potassium Bromide–Calcium Sulfate–Calcium Bromide–Water	K_2SO_4—KBr—$CaSO_4$—$CaBr_2$—H_2O	4708
Potassium Sulfate–Potassium Bromate–Water	K_2SO_4—$KBrO_3$—H_2O	4609, 4710
Potassium Sulfate–Potassium Molybdate–Water	K_2SO_4—K_2MoO_4—H_2O	4711—4713
Potassium Sulfate–Potassium Iodide–Water	K_2SO_4— KI —H_2O	4614, 4715
Potassium Sulfate–Potassium Iodide–Calcium Sulfate–Calcium Iodide–Water	K_2SO_4— KI —$CaSO_4$—CaI_2— H_2O	4716
Potassium Sulfate–Potassium Iodate–Water	K_2SO_4—KJO_3—H_2O	4717
Potassium Sulfate–Calcium Carbonate–Water	K_2SO_4—$CaCO_3$—H_2O	4718
Potassium Sulfate–Calcium Orthophosphate–Water	K_2SO_4—$Ca_3(PO_4)_2$—H_2O	5113
Potassium Sulfate–Calcium Sulfate–Water	K_2SO_4—$CaSO_4$—H_2O	4719—4731
Potassium Sulfate–Calcium Iodate–Water	K_2SO_4—$Ca(IO_3)_2$—H_2O	4732, 4733
Potassium Sulfate–Manganese (II) Sulfate–Water	K_2SO_4—$MnSO_4$—H_2O	4734, 4735
Potassium Sulfate–Iron (III) Sulfate–Water	K_2SO_4—$Fe_2(SO_4)_3$—H_2O	4736
Potassium Sulfate–Cobalt (II) Sulfate–Water	K_2SO_4—$CoSO_4$—H_2O	4737—4739
Potassium Sulfate–Cobalt (II) Chloride–Water	K_2SO_4—$CoCl_2$—H_2O	4740
Potassium Sulfate–Nickel (II) Sulfate–Water	K_2SO_4—$NiSO_4$—H_2O	4741, 4742
Potassium Sulfate–Copper (II) Sulfate–Water	K_2SO_4—$CuSO_4$—H_2O	4743, 4744
Potassium Sulfate–Zinc Hydroxide–Water	K_2SO_4—$Zn(OH)_2$—H_2O	4745
Potassium Sulfate–Zinc Sulfate–Water	K_2SO_4—$ZnSO_4$—H_2O	4746, 4747
Potassium Sulfate–Strontium Sulfate–Water	K_2SO_4—$SrSO_4$—H_2O	4748

System name	Formulae of components	Table number
Potassium Sulfate–Silver Orthophosphate–Water	$K_2SO_4—Ag_3PO_4—H_2O$	4749
Potassium Sulfate–Silver Sulfate–Water	$K_2SO_4—Ag_2SO_4—H_2O$	4750—4753, 5403, 5404
Potassium Sulfate–Silver Chloride–Water	$K_2SO_4—AgCl—H_2O$	4754
Potassium Sulfate–Silver Bromate–Water	$K_2SO_4—AgBrO_3—H_2O$	4755, 5425
Potassium Sulfate–Silver Iodate–Water	$K_2SO_4—AgIO_3—H_2O$	5429
Potassium Sulfate–Cadmium Sulfate–Water	$K_2SO_4—CdSO_4—H_2O$	4756, 4757
Potassium Sulfate–Indium Sulfate–Water	$K_2SO_4—In_2(SO_4)_3—H_2O$	4758
Potassium Sulfate–Antimony Trifluoride–Water	$K_2SO_4—SbF_3—H_2O$	4759
Potassium Sulfate–Cesium Sulfate–Water	$K_2SO_4—CsSO_4—H_2O$	4760, 4761
Potassium Sulfate–Lanthanum Sulfate–Water	$K_2SO_4—La_2(SO_4)_3—H_2O$	4762, 4763
Potassium Sulfate–Lanthanum Iodate–Water	$K_2SO_4—La(IO_3)_3—H_2O$	4764, 4765
Potassium Sulfate–Cerium Sulfate–Water	$K_2SO_4—Ce_2(SO_4)_3—H_2O$	4766—4769
Potassium Sulfate–Cerium Iodate–Water	$K_2SO_4—Ce(IO_3)_2—H_2O$	4770
Potassium Sulfate–Praseodymium Sulfate–Water	$K_2SO_4—PrSO_4—H_2O$	4771
Potassium Sulfate–Neodymium Sulfate–Water	$K_2SO_4—Nd_2(SO_4)_3—H_2O$	4772
Potassium Sulfate–Mercury (I) Sulfate–Water	$K_2SO_4—Hg_2SO_4—H_2O$	4773
Potassium Sulfate–Thallium Sulfate–Water	$K_2SO_4—Tl_2SO_4—H_2O$	4774, 4775
Potassium Sulfate–Thallium Chloride–Water	$K_2SO_4—TlCl—H_2O$	4776
Potassium Sulfate–Thallium Bromide–Water	$K_2SO_4—TlBr—H_2O$	4777
Potassium Sulfate–Thallium Iodate–Water	$K_2SO_4—TlIO_3—H_2O$	4778
Potassium Sulfate–Lead (II) Sulfate–Water	$K_2SO_4—PbSO_4—H_2O$	4779, 4780
Potassium Sulfate–Thorium Sulfate–Water	$K_2SO_4—Th(SO_4)_2—H_2O$	4781
Potassium Sulfate–Uranyl Sulfate–Water	$K_2SO_4—UO_2SO_4—H_2O$	4782
Potassium Dithionate–Barium Dithionate–Water	$K_2S_2O_6—BaS_2O_6—H_2O$	4783
Potassium Trithionate–Potassium Tetrathionate–Water	$K_2S_3O_6—K_2S_4O_6—H_2O$	4784
Potassium Trithionate–Potassium Pentathionate–Water	$K_2S_3O_6—K_2S_5O_6—H_2O$	4785

System name	Formulae of components	Table number
Potassium Tetrathionate–Potassium Pentathionate–Water	$K_2S_4O_6$—$K_2S_5O_6$—H_2O	4786
Potassium Chloride–Potassium Chlorate–Water	KCl—$KClO_3$—H_2O	4787—4792
Potassium Chloride–Potassium Perchlorate–Water	KCl—$KClO_4$—H_2O	4793—4799
Potassium Chloride–Potassium Hexachloroplatinate–Water	KCl—K_2PtCl_6—H_2O	4800—4802
Potassium Chloride–Potassium Metavanadate–Water	KCl—KVO_3—H_2O	4803
Potassium Chloride–Potassium Chromate–Water	KCl—K_2CrO_4—H_2O	4804
Potassium Chloride–Potassium Dichromate–Water	KCl—$K_2Cr_2O_7$—H_2O	4805—4807
Potassium Chloride–Potassium Permanganate–Water	KCl—$KMnO_4$—H_2O	4808, 4809
Potassium Chloride–Potassium Bromide–Water	KCl—KBr—H_2O	4810—4821
Potassium Chloride–Potassium Bromide–Potassium Iodide–Water	KCl—KBr— KI —H_2O	4822, 4823
Potassium Chloride–Potassium Bromate–Water	KCl—$KBrO_3$—H_2O	4824
Potassium Chloride–Potassium Molybdate–Water	KCl—K_2MoO_4—H_2O	4825
Potassium Chloride–Potassium Iodide–Water	KCl— KI —H_2O	4826—4832
Potassium Chloride–Potassium Perrhenate–Water	KCl—$KReO_4$—H_2O	4833—4835
Potassium Chloride–Calcium Oxide–Water	KCl—CaO—H_2O	4836
Potassium Chloride–Calcium Oxide–Arsenic Pentoxide–Water	KCl—CaO—As_2O_5—H_2O	4837
Potassium Chloride–Calcium Hydroxide–Water	KCl—$Ca(OH)_2$—H_2O	4838—4840
Potassium Chloride–Calcium Carbonate–Water	KCl—$CaCO_3$—H_2O	3739. 4841, 4842
Potassium Chloride–Calcium Nitrate–Water	KCl—$Ca(NO_3)_2$—H_2O	4843
Potassium Chloride–Calcium Hydrogen Orthophosphate–Water	KCl—$CaHPO_4 \cdot 2H_2O$—H_2O	5111
Potassium Chloride–Calcium Sulfate–Water	KCl—$CaSO_4$—H_2O	4844—4847
Potassium Chloride–Calcium Chloride–Water	KCl—$CaCl_2$—H_2O	4848—4860, 5182
Potassium Chloride–Calcium Chloride–Strontium Chloride–Water	KCl—$CaCl_2$—$SrCl_2$—H_2O	4861
Potassium Chloride–Calcium Chloride–Cadmium Chloride–Water	KCl—$CaCl_2$—$CdCl_2$—H_2O	4897

System name	Formulae of components	Table number
Potassium Chloride–Calcium Iodate–Water	$KCl—Ca(IO_3)_2—H_2O$	4862, 4863, 5181, 5191
Potassium Chloride–Manganese (II) Chloride–Water	$KCl—MnCl_2—H_2O$	4864, 4865
Potassium Chloride–Iron (III) Chloride–Water	$KCl—FeCl_3—H_2O$	4866—4868
Potassium Chloride–Cobalt (II) Chloride–Water	$KCl—CoCl_2—H_2O$	4869—4871
Potassium Chloride–Copper (II) Sulfate–Water	$KCl—CuSO_4—H_2O$	4872
Potassium Chloride–Copper (I) Chloride–Water	$KCl—CuCl—H_2O$	4873, 4874
Potassium Chloride–Copper (II) Chloride–Water	$KCl—CuCl_2—H_2O$	4875, 4876
Potassium Chloride–Copper (II) Iodate–Water	$KCl—Cu(IO_3)_2—H_2O$	4877
Potassium Chloride–Zinc Hydroxide –Water	$KCl—Zn(OH)_2—H_2O$	4878
Potassium Chloride–Zinc Chloride –Water	$KCl—ZnCl_2—H_2O$	4879
Potassium Chloride–Rubidium Chloride–Water	$KCl—RbCl—H_2O$	4880, 4881
Potassium Chloride–Strontium Carbonate–Water	$KCl—SrCO_3—H_2O$	4882
Potassium Chloride–Strontium Sulfate–Water	$KCl—SrSO_4—H_2O$	4883
Potassium Chloride–Strontium Chloride–Water	$KCl—SrCl_2—H_2O$	4884—4886
Potassium Chloride–Oxozirconium Chloride–Water	$KCl—ZrOCl_2—H_2O$	4887
Potassium Chloride–Silver Chloride–Water	$KCl—AgCl—H_2O$	4888—4893, 5410—5412
Potassium Chloride–Cadmium Chloride–Water	$KCl—CdCl_2—H_2O$	4894—4897
Potassium Chloride–Cadmium Iodate–Water	$KCl—Cd(IO_3)_2—H_2O$	4898
Potassium Chloride–Tin (II) Chloride–Water	$KCl—SnCl_2—H_2O$	4899
Potassium Chloride–Antimony Trifluoride–Water	$KCl—SbF_3—H_2O$	4900
Potassium Chloride–Barium Hydroxide–Water	$KCl—Ba(OH)_2—H_2O$	4901
Potassium Chloride–Barium Carbonate–Water	$KCl—BaCO_3—H_2O$	4902, 4903
Potassium Chloride–Barium Sulfate–Water	$KCl—BaSO_4—H_2O$	4904
Potassium Chloride–Barium Chloride–Water	$KCl—BaCl_2—H_2O$	4905, 4906
Potassium Chloride–Barium Chromate–Water	$KCl—BaCrO_4—H_2O$	4907
Potassium Chloride–Barium Hydrogen Orthoarsenate–Water	$KCl—BaHAsO_4—H_2O$	4908

System name	Formulae of components	Table number
Potassium Chloride–Barium Orthoarsenate–Water	$KCl—Ba_3(AsO_4)_2—H_2O$	4909
Potassium Chloride–Barium Bromate–Water	$KCl—Ba(BrO_3)_2—H_2O$	4910
Potassium Chloride–Barium Iodate–Water	$KCl—Ba(IO_3)_2—H_2O$	4911—4913
Potassium Chloride–Lanthanum Iodate–Water	$KCl—La(IO_3)_3—H_2O$	4914, 4915
Potassium Chloride–Cerium Chloride–Water	$KCl—CeCl_3—H_2O$	4916
Potassium Chloride–Mercury (II) Oxide–Water	$KCl—HgO—H_2O$	4917
Potassium Chloride–Mercury Amidochloride–Water	$KCl—HgNH_2Cl—H_2O$	4918
Potassium Chloride–Mercury (II) Chloride–Water	$KCl—HgCl_2—H_2O$	4919—4922
Potassium Chloride–Thallium Chloride–Water	$KCl—TlCl—H_2O$	4923, 4924
Potassium Chloride–Thallium Iodate–Water	$KCl—TlIO_3—H_2O$	4925
Potassium Chloride–Lead (II) Chloride–Water	$KCl—PbCl_2—H_2O$	4926—4932
Potassium Chloride–Lead (II) Bromide–Water	$KCl—PbBr_2—H_2O$	4933
Potassium Chloride–Lead (II) Iodate–Water	$KCl—Pb(IO_3)_2—H_2O$	4934
Potassium Chlorate–Potassium Bromide–Water	$KClO_3—KBr—H_2O$	4935, 4936
Potassium Chlorate–Potassium Bromate–Water	$KClO_3—KBrO_3—H_2O$	4937
Potassium Chlorate–Potassium Iodide–Water	$KClO_3—KI—H_2O$	4938, 4939
Potassium Chlorate–Potassium Iodate–Water	$KClO_3—KIO_3—H_2O$	4940
Potassium Chlorate–Calcium Chlorate–Water	$KClO_3—Ca(ClO_3)_2—H_2O$	4941
Potassium Chlorate–Rubidium Chlorate–Water	$KClO_3—RbClO_3—H_2O$	4942
Potassium Chlorate–Silver Iodate–Water	$KClO_3—AgIO_3—H_2O$	5429
Potassium Chlorate–Cesium Chlorate–Water	$KClO_3—CsClO_3—H_2O$	4943
Potassium Chlorate–Barium Iodate–Water	$KClO_3—Ba(IO_3)_2—H_2O$	4944
Potassium Chlorate–Thallium Chloride–Water	$KClO_3—TlCl—H_2O$	4945
Potassium Chlorate–Thallium Chlorate–Water	$KClO_3—TlClO_3—H_2O$	4946, 4947
Potassium Perchlorate–Potassium Permanganate–Water	$KClO_4—KMnO_4—H_2O$	4948, 4949
Potassium Perchlorate–Silver Bromate–Water	$KClO_4—AgBrO_3—H_2O$	5425
Potassium Perchlorate–Barium Nitrate–Water	$KClO_4—Ba(NO_3)_2—H_2O$	4950

System name	Formulae of components	Table number
Potassium Perchlorate–Barium Chloride–Water	$KClO_4$—$BaCl_2$—H_2O	4951
Potassium Perchlorate–Barium Iodate–Water	$KClO_4$—$Ba(IO_3)_2$—H_2O	4952
Potassium Chromate–Potassium Dichromate–Water	K_2CrO_4—$K_2Cr_2O_7$—H_2O	4953
Potassium Chromate–Potassium Bromide–Water	K_2CrO_4—KBr—H_2O	4954
Potassium Chromate–Potassium Molybdate–Water	K_2CrO_4—K_2MoO_4—H_2O	4955
Potassium Dichromate–Potassium Trichromate–Water	$K_2Cr_2O_7$—$K_2Cr_3O_{10}$—H_2O	4956
Potassium Dichromate–Potassium Bromide–Water	$K_2Cr_2O_7$—KBr—H_2O	4957
Potassium Dichromate–Calcium Chloride–Water	$K_2Cr_2O_7$—$CaCl_2$—H_2O	4958, 4959
Potassium Dichromate–Calcium Dichromate–Water	$K_2Cr_2O_7$—$CaCr_2O_7$—H_2O	4960
Potassium Dichromate–Strontium Chloride–Water	$K_2Cr_2O_7$—$SrCl_2$—H_2O	4961
Potassium Trichromate–Potassium Tetrachromate–Water	$K_2Cr_3O_{10}$—$K_2Cr_4O_{13}$—H_2O	4962
Potassium Tetrachromate–Chromium Trioxide–Water	$K_2Cr_4O_{13}$—CrO_3—H_2O	4963
Potassium Permanganate–Calcium Chloride–Water	$KMnO_4$—$CaCl_2$—H_2O	4964
Potassium Permanganate–Rubidium Permanganate–Water	$KMnO_4$—$RbMnO_4$—H_2O	4965
Potassium Permanganate–Strontium Chloride–Water	$KMnO_4$—$SrCl_2$—H_2O	4966
Potassium Dihydrogen Orthoarsenate–Arsenic Trioxide–Water	KH_2AsO_4—As_2O_3—H_2O	4967
Potassium Selenate–Calcium Selenate–Water	K_2SeO_4—$CaSeO_4$—H_2O	4968
Potassium Selenate–Lead (II) Selenate–Water	K_2SeO_4—$PbSeO_4$—H_2O	4969
Potassium Bromide–Potassium Bromate–Water	KBr—$KBrO_3$—H_2O	4970, 4971
Potassium Bromide–Potassium Iodide–Water	KBr—KI—H_2O	4972—4974
Potassium Bromide–Potassium Iodate–Water	KBr—KIO_3—H_2O	4975, 4976
Potassium Bromide–Calcium Hydroxide–Water	KBr—$Ca(OH)_2$—H_2O	4977
Potassium Bromide–Calcium Carbonate–Water	KBr—$CaCO_3$—H_2O	3739
Potassium Bromide–Calcium Sulfate–Water	KBr—$CaSO_4$—H_2O	4978
Potassium Bromide–Copper (I) Iodide–Water	KBr—CuI—H_2O	4979
Potassium Bromide–Arsenic Trioxide–Water	KBr—As_2O_3—H_2O	4980

System name	Formulae of components	Table number
Potassium Bromide–Rubidium Bromide–Water	$KBr—RbBr—H_2O$	4981
Potassium Bromide–Silver Bromide–Water	$KBr—AgBr—H_2O$	4982—4985, 5416, 5417
Potassium Bromide–Silver Iodide–Water	$KBr—AgI—H_2O$	4986
Potassium Bromide–Cadmium Bromide–Water	$KBr—CdBr_2—H_2O$	4987
Potassium Bromide–Antimony Trifluoride–Water	$KBr—SbF_3—H_2O$	4988
Potassium Bromide–Isomeric Diamminedibromoplatinates–Water	KBr—*cis-* and *trans-* $[Pt(NH_3)_2Br_2]—H_2O$	4989
Potassium Bromide–Mercury (II) Bromide–Water	$KBr—HgBr_2—H_2O$	4990, 4991
Potassium Bromide–Lead (II) Chloride–Water	$KBr—PbCl_2—H_2O$	4992
Potassium Bromide–Lead (II) Bromide–Water	$KBr—PbBr_2—H_2O$	4993—4995
Potassium Bromate–Potassium Iodide–Water	$KBrO_3—KI—H_2O$	4996
Potassium Bromate–Silver Bromate–Water	$KBrO_3—AgBrO_3—H_2O$	4997, 4998
Potassium Bromate–Barium Bromate–Water	$KBrO_3—Ba(BrO_3)_2—H_2O$	4999
Potassium Molybdate–Potassium Iodate–Water	$KMoO_4—KIO_3—H_2O$	5000
Potassium Iodide–Potassium Iodate–Water	$KI—KIO_3—H_2O$	5001, 5002
Potassium Iodide–Calcium Nitrate–Lead (II) Iodide–Water	$KI—Ca(NO_3)_2—PbI_2—H_2O$	5003
Potassium Iodide–Calcium Sulfate–Water	$KI—CaSO_4—H_2O$	5004
Potassium Iodide–Copper (I) Thiocyanate–Water	$KI—CuCNS—H_2O$	5005
Potassium Iodide–Copper (I) Iodide–Water	$KI—CuI—H_2O$	5006
Potassium Iodide–Arsenic Trioxide–Water	$KI—As_2O_3—H_2O$	5007
Potassium Iodide–Strontium Chloride–Water	$KI—SrCl_2—H_2O$	5008
Potassium Iodide–Palladium Iodide–Water	$KI—PdI_2—H_2O$	5009
Potassium Iodide–Silver Iodide–Water	$KI—AgI—H_2O$	5010—5013
Potassium Iodide–Cadmium Iodide–Water	$KI—CdI_2—H_2O$	5014
Potassium Iodide–Antimony (III) Iodide–Water	$KI—SbI_3—H_2O$	5015
Potassium Iodide–Mercury (II) Iodide–Water	$KI—HgI_2—H_2O$	5016—5019
Potassium Iodide–Thallium Iodide–Water	$KI—TlI—H_2O$	5020, 5021
Potassium Iodide–Lead (II) Iodide–Water	$KI—PbI_2—H_2O$	5022—5028

System name	Formulae of components	Table number
Potassium Iodate–Calcium Iodate–Water	KIO_3—$Ca(IO_3)_2$—H_2O	5190
Potassium Iodate–Silver Iodate–Water	KIO_3—$AgIO_3$—H_2O	5029
Potassium Iodate–Barium Iodate–Water	KIO_3—$Ba(IO_3)_2$—H_2O	5030
Potassium Iodate–Lanthanum Iodate–Water	KIO_3—$La(IO_3)_3$—H_2O	5031
Potassium Iodate–Lead (II) Iodate–Water	KIO_3—$Pb(IO_3)_2$—H_2O	5032
Iron (II) Potassium Sulfate–Potassium Zinc Sulfate–Water	$K_2Fe(SO_4)_2$—$K_2Zn(SO_4)_2$—H_2O	5033
Cobalt (II) Potassium Sulfate–Copper (II) Potassium Sulfate–Water	$CoK_2(SO_4)_2$—$CuK_2(SO_4)_2$—H_2O	5034, 5035
Nickel (II) Potassium Sulfate–Copper (II) Potassium Sulfate–Water	$K_2SO_4 \cdot NiSO_4$—$K_2SO_4 \cdot CuSO_4$—H_2O	5036, 5037
Nickel (II) Potassium Sulfate–Potassium Zinc Sulfate–Water	$K_2Ni(SO_4)_2$—$K_2Zn(SO_4)_2$—H_2O	5038
Copper (II) Potassium Sulfate–Potassium Zinc Sulfate–Water	$K_2SO_4 \cdot CuSO_4$—$K_2SO_4 \cdot ZnSO_4$—H_2O	5039, 5040
Copper (II) Potassium Sulfate–Copper (II) Thallium Sulfate–Water	$K_2SO_4 \cdot CuSO_4 \cdot 6H_2O$—$CuSO_4 \cdot Tl_2SO_4 \cdot 6H_2O$—$H_2O$	5041
Calcium Oxide–Calcium Nitrate–Water	CaO—$Ca(NO_3)_2$—H_2O	5042
Calcium Oxide–Calcium Sulfate–Water	CaO—$CaSO_4$—H_2O	5043
Calcium Oxide–Calcium Chloride–Water	CaO—$CaCl_2$—H_2O	5044—5048
Calcium Oxide–Calcium Hypochlorite–Water	CaO—$Ca(ClO)_2$—H_2O	5049
Calcium Oxide–Calcium Bromide–Water	CaO—$CaBr_2$—H_2O	5050, 5051
Calcium Oxide–Calcium Iodide–Water	CaO—CaI_2—H_2O	5052
Calcium Oxide–Arsenic Trioxide–Water	CaO—As_2O_3—H_2O	5053—5054
Calcium Oxide–Arsenic Pentoxide–Water	CaO—As_2O_5—H_2O	5055, 5056
Calcium Hydroxide–Calcium Carbonate–Water	$Ca(OH)_2$—$CaCO_3$—H_2O	5057
Calcium Hydroxide–Calcium Sulfate–Water	$Ca(OH)_2$—$CaSO_4$—H_2O	5058—5060
Calcium Hydroxide–Calcium Chloride–Water	$Ca(OH)_2$—$CaCl_2$—H_2O	5061—5067
Calcium Hydroxide–Calcium Chloride–Calcium Chlorate–Water	$Ca(OH)_2$—$CaCl_2$—$Ca(ClO_3)_2$—H_2O	5068
Calcium Hydroxide–Calcium Chlorate–Water	$Ca(OH)_2$—$Ca(ClO_3)_2$—H_2O	5069, 5070
Calcium Hydroxide–Calcium Iodate–Water	$Ca(OH)_2$—$Ca(IO_3)_2$—H_2O	5071, 5190
Calcium Hydroxide–Strontium Chloride–Water	$Ca(OH)_2$—$SrCl_2$—H_2O	5072
Calcium Hydroxide–Cesium Chloride–Water	$Ca(OH)_2$—$CsCl$—H_2O	5073
Calcium Hydroxide–Barium Chloride–Water	$Ca(OH)_2$—$BaCl_2$—H_2O	5074

System name	Formulae of components	Table number
Calcium Hydrogen Carbonate–Calcium Sulfate–Water	$Ca(HCO_3)_2$—$CaSO_4$—H_2O	5075, 5076
Calcium Carbonate–Calcium Sulfate–Water	$CaCO_3$—$CaSO_4$—H_2O	5077, 5078
Calcium Carbonate–Calcium Chloride–Water	$CaCO_3$—$CaCl_2$—H_2O	5079, 5080
Calcium Carbonate–Seawater	$CaCO_3$— —— —H_2O	5081–5084
Calcium Thiocyanate–Silver Thiocyanate–Wâter	$Ca(CNS)_2$—$AgCNS$—H_2O	5085
Calcium Thiocyanate–Silver Chloride–Water	$Ca(CNS)_2$—$AgCl$—H_2O	5410
Calcium Thiocyanate–Silver Iodide–Water	$Ca(CNS)_2$—AgI—H_2O	5427
Clacium Thiocyanate–Silver Bromide–Water	$Ca(CNS)_2$—$AgBr$—H_2O	5418
Calcium Nitrite–Calcium Nitrate–Water	$Ca(NO_2)_2$—$Ca(NO_3)_2$—H_2O	5086
Calcium Nitrite–Silver Nitrite–Water	$Ca(NO_2)_2$—$AgNO_2$—H_2O	5087
Calcium Nitrite–Thallium Nitrite–Water	$Ca(NO_2)_2$—$TlNO_2$—H_2O	5088
Calcium Nitrate–Calcium Orthophosphate–Water	$Ca(NO_3)_2$—$Ca_3(PO_4)_2$—H_2O	5089
Calcium Nitrate–Calcium Sulfate–Water	$Ca(NO_3)_2$—$CaSO_4$—H_2O	5090
Calcium Nitrate–Calcium Thiosulfate–Water	$Ca(NO_3)_2$—CaS_2O_3—H_2O	5091
Calcium Nitrate–Calcium Chloride–Water	$Ca(NO_3)_2$—$CaCl_2$—H_2O	5092, 5093
Calcium Nitrate–Calcium Chloride–Barium Nitrate–Water	$Ca(NO_3)_2$—$CaCl_2$—$Ba(NO_3)_2$—H_2O	5094
Calcium Nitrate–Calcium Chromate–Water	$Ca(NO_3)_2$—$CaCrO_4$—H_2O	5095
Calcium Nitrate–Strontium Nitrate–Water	$Ca(NO_3)_2$—$Sr(NO_3)_2$—H_2O	5096, 5097
Calcium Nitrate–Strontium Sulfate–Water	$Ca(NO_3)_2$—$SrSO_4$—H_2O	5098
Calcium Nitrate–Silver Bromate–Water	$Ca(NO_3)_2$—$AgBrO_3$—H_2O	4004, 5099
Calcium Nitrate–Barium Nitrate–Water	$Ca(NO_3)_2$—$Ba(NO_3)_2$—H_2O	5100
Calcium Nitrate–Barium Chloride–Water	$Ca(NO_3)_2$—$BaCl_2$—H_2O	5101
Calcium Nitrate–Lead (II) Bromate–Water	$Ca(NO_3)_2$—$Pb(BrO_3)_2$—H_2O	5102
Calcium Nitrate–Lead (II) Iodide–Water	$Ca(NO_3)_2$—PbI_2—H_2O	5103, 5104
Calcium Nitrate–Uranyl Nitrate–Water	$Ca(NO_3)_2$—$UO_2(NO_3)_2$—H_2O	5105
Calcium Fluoride–Calcium Chloride–Water	CaF_2—$CaCl_2$—H_2O	5106
Calcium Fluoride–Iron (III) Nitrate–Water	CaF_2—$Fe(NO_3)_3$—H_2O	5107

System name	Formulae of components	Table number
Calcium Fluoride–Iron (III) Chloride–Water	CaF_2—$FeCl_3$—H_2O	5108
Calcium Fluoride–Thorium Nitrate–Water	CaF_2—$Th(NO_3)_4$—H_2O	5109
Calcium Fluoride–Uranyl Nitrate–Water	CaF_2—$UO_2(NO_3)_2$—H_2O	5110
Calcium Orthophosphate–Calcium Sulfate–Water	$Ca_3(PO_4)_2$—$CaSO_4$—H_2O	5112
Calcium Sulfate–Calcium Chloride–Water	$CaSO_4$—$CaCl_2$—H_2O	5114—5121
Calcium Sulfate–Chromium Trioxide–Water	$CaSO_4$—CrO_3—H_2O	5122
Calcium Sulfate–Manganese (II) Sulfate–Water	$CaSO_4$—$MnSO_4$—H_2O	5123
Calcium Sulfate–Cobalt Hexammine-chloride–Water	$CaSO_4$—$Co(NH_3)_6Cl_3$—H_2O	5124
Calcium Sulfate–Nickel (II) Sulfate–Water	$CaSO_4$—$NiSO_4$—H_2O	5125
Calcium Sulfate–Copper (II) Sulfate–Water	$CaSO_4$—$CuSO_4$—H_2O	5126, 5127
Calcium Sulfate–Rubidium Sulfate–Water	$CaSO_4$—Rb_2SO_4—H_2O	5128
Calcium Sulfate–Silver Sulfate–Water	$CaSO_4$—Ag_2SO_4—H_2O	5129
Calcium Sulfate–Silver Chloride–Water	$CaSO_4$—$AgCl$—H_2O	5130
Calcium Sulfate–Cesium Sulfate–Water	$CaSO_4$—Cs_2SO_4—H_2O	5131
Calcium Sulfate (Gypsum)–Cesium Chloride–Water	$CaSO_4 \cdot 2H_2O$—$CsCl$—H_2O	1430
Calcium Sulfate (Gypsum)–Lanthanum Chloride–Water	$CaSO_4 \cdot 2H_2O$—$LaCl_3$—H_2O	5132
Calcium Sulfate–Cerium Sulfate–Water	$CaSO_4$—$Ce_2(SO_4)_3$—H_2O	5133
Calcium Sulfate (Gypsum)–Praseodymium Chloride–Water	$CaSO_4 \cdot 2H_2O$—$PrCl_3$—H_2O	5132
Calcium Sulfate (Gypsum)–Samarium Chloride–Water	$CaSO_4 \cdot 2H_2O$—$SmCl_3$—H_2O	5132
Calcium Sulfate–Seawater	$CaSO_4$— —— —H_2O	5134—5138
Calcium Chloride–Calcium Thiosulfate–Water	$CaCl_2$—CaS_2O_3—H_2O	5139
Calcium Chloride–Calcium Hypochlorite–Water	$CaCl_2$—$Ca(ClO)_2$—H_2O	5140
Calcium Chloride–Calcium Chlorate–Water	$CaCl_2$—$Ca(ClO_3)_2$—H_2O	5141—5143
Calcium Chloride–Calcium Dichromate–Water	$CaCl_2$—$CaCr_2O_7$—H_2O	5144
Calcium Chloride–Calcium Iodate–Water	$CaCl_2$—$Ca(IO_3)_2$—H_2O	5190
Calcium Chloride–Manganese Tetrahydrogen Orthoborate–Water	$CaCl_2$—$MnH_4(BO_3)_2$—H_2O	5145
Calcium Chloride–Iron (II) Hydroxide–Water	$CaCl_2$—$Fe(OH)_2$—H_2O	5146

System name	Formulae of components	Table number
Calcium Chloride–Cobalt (II) Chloride–Water	$CaCl_2—CoCl_2—H_2O$	5147
Calcium Chloride–Zinc Hydroxide–Water	$CaCl_2—Zn(OH)_2—H_2O$	5148
Calcium Chloride–Zinc Chloride–Water	$CaCl_2—ZnCl_2—H_2O$	5149, 5150
Calcium Chloride–Arsenic Trioxide–Water	$CaCl_2—As_2O_3—H_2O$	5151
Calcium Chloride–Strontium Sulfate–Water	$CaCl_2—SrSO_4—H_2O$	5152, 4153
Calcium Chloride–Strontium Chloride–Water	$CaCl_2—SrCl_2—H_2O$	5154
Calcium Chloride–Strontium Chloride–Barium Chloride–Water	$CaCl_2—SrCl_2—BaCl_2—H_2O$	5155
Calcium Chloride–Zirconium Chloride Oxide–Water	$CaCl_2—ZrOCl_2—H_2O$	5156
Calcium Chloride–Silver Chloride–Water	$CaCl_2—AgCl—H_2O$	5157, 5410—5412
Calcium Chloride–Cadmium Chloride–Water	$CaCl_2—CdCl_2—H_2O$	5158
Calcium Chloride–Indium Chloride–Water	$CaCl_2—InCl_3—H_2O$	5159
Calcium Chloride–Cesium Chloride–Water	$CaCl_2—CsCl—H_2O$	5160
Calcium Chloride–Barium Nitrate–Water	$CaCl_2—Ba(NO_3)_2—H_2O$	5161, 5162
Calcium Chloride–Barium Chloride–Water	$CaCl_2—BaCl_2—H_2O$	5163—5166
Calcium Chloride–Barium Iodate–Water	$CaCl_2—Ba(IO_3)_2—H_2O$	5167
Calcium Chloride–Lanthanum Chloride–Water	$CaCl_2—LaCl_3—H_2O$	5168
Calcium Chloride–Neodymium Chloride–Water	$CaCl_2—NdCl_3—H_2O$	5169
Calcium Chloride–Mercury (I) Chloride–Water	$CaCl_2—HgCl—H_2O$	5170
Calcium Chloride–Mercury (II) Chloride–Water	$CaCl_2—HgCl_2—H_2O$	5171—5173
Calcium Chloride–Thallium Chloride–Water	$CaCl_2—TlCl—H_2O$	5174, 5175
Calcium Chloride–Lead (II) Chloride–Water	$CaCl_2—PbCl_2—H_2O$	5176—5179
Calcium Chloride–Lead (II) Bromide–Water	$CaCl_2—PbBr_2—H_2O$	5180, 5181
Calcium Bromide–Arsenic Trioxide–Water	$CaBr_2—As_2O_3—H_2O$	5182, 5183
Calcium Bromide–Silver Bromide–Water	$CaBr_2—AgBr—H_2O$	5416
Calcium Bromide–Indium Bromide–Water	$CaBr_2—InBr_3—H_2O$	5184
Calcium Bromide–Mercury (II) Bromide–Water	$CaBr_2—HgBr_2—H_2O$	5185

System name	Formulae of components	Table number
Calcium Bromide–Lead (II) Chloride–Water	$CaBr_2—PbCl_2—H_2O$	5186
Calcium Bromide–Lead (II) Bromide–Water	$CaBr_2—PbBr_2—H_2O$	5187
Calcium Iodide–Indium Iodide–Water	$CaI_2—INI_3—H_2O$	5188
Calcium Iodide–Mercury (II) Iodide–Water	$CaI_2—HgI_2—H_2O$	5189
Chromium Trioxide–Copper (II) Oxide–Water	$CrO_3—CuO—H_2O$	5192
Chromium Trioxide–Zinc Oxide–Water	$CrO_3—ZnO—H_2O$	5193, 5194
Chromium Trioxide–Arsenic Pentoxide–Water	$CrO_3—As_2O_5—H_2O$	5195
Chromium Trioxide–Rubidium Oxide–Water	$CrO_3—Rb_2O—H_2O$	5196
Chromium Trioxide–Silver Oxide–Water	$CrO_3—Ag_2O—H_2O$	5197
Chromium Trioxide–Thorium Oxide–Water	$CrO_3—ThO_2—H_2O$	5198
Chromium Fluoride–Rubidium Fluoride–Water	$CrF_3—RbF—H_2O$	5199
Chromium Fluoride–Cesium Fluoride–Water	$CrF_3—CsF—H_2O$	5200
Manganese (II) Oxide–Arsenic Pentoxide–Water	$MnO—As_2O_5—H_2O$	5201
Manganese (II) Nitrate–Lanthanum Nitrate–Water	$Mn(NO_3)_2—La(NO_3)_3—H_2O$	5202
Manganese (II) Sulfate–Manganese (II) Chloride–Water	$MnSO_4—MnCl_2—H_2O$	5203
Manganese (II) Sulfate–Iron (II) Sulfate–Water	$MnSO_4—FeSO_4—H_2O$	5204
Manganese (II) Sulfate–Copper (II) Sulfate–Water	$MnSO_4—CuSO_4—H_2O$	5205
Manganese (II) Sulfate–Rubidium Sulfate–Water	$MnSO_4—Rb_2SO_4—H_2O$	5206
Manganese (II) Sulfate–Thallium Sulfate–Water	$MnSO_4—Tl_2SO_4—H_2O$	5207
Manganese (II) Sulfate–Thorium Sulfate–Water	$MnSO_4—Th(SO_4)_2—H_2O$	5208, 5209
Manganese (II) Chloride–Silver Chloride–Water	$MnCl_2—AgCl—H_2O$	5410
Manganese (II) Chloride–Cadmium Chloride–Water	$MnCl_2—CdCl_2—H_2O$	5210
Manganese (II) Chloride–Thallium Chloride–Water	$MnCl_2—TlCl—H_2O$	5211
Manganese (II) Chloride–Lead (II) Chloride–Water	$MnCl_2—PbCl_2—H_2O$	5212
Iron (III) Oxide–Iron (II) Sulfate–Water	$Fe_2O_3—FeSO_4—H_2O$	5213
Iron (III) Oxide–Arsenic Pentoxide–Water	$Fe_2O_3—As_2O_5—H_2O$	5214, 5215

System name	Formulae of components	Table number
Iron (II) Sulfate–Iron (III) Sulfate–Water	$FeSO_4—Fe_2(SO_4)_3—H_2O$	5216
Iron (II) Sulfate–Cobalt (II) Sulfate–Water	$FeSO_4—CoSO_4—H_2O$	5217
Iron (II) Sulfate–Nickel (II) Sulfate–Water	$FeSO_4—NiSO_4—H_2O$	5218
Iron (II) Sulfate–Copper (II) Sulfate–Water	$FeSO_4—CuSO_4—H_2O$	5219, 5220
Iron (II) Sulfate–Zinc Sulfate–Water	$FeSO_4\cdot 7H_2O—ZnSO_4\cdot 7H_2O—H_2O$	5221
Iron (II) Sulfate–Cadmium Sulfate–Water	$FeSO_4—CdSO_4—H_2O$	5222
Iron (II) Chloride–Iron (III) Chloride–Water	$FeCl_2—FeCl_3—H_2O$	5223
Iron (II) Chloride–Cobalt (II) Chloride–Water	$FeCl_2—CoCl_2—H_2O$	5224
Iron (II) Chloride–Nickel (II) Chloride–Water	$FeCl_2—NiCl_2—H_2O$	5225
Iron (II) Chloride–Copper (I) Chloride–Water	$FeCl_2—CuCl—H_2O$	5226
Iron (II) Chloride–Zinc Chloride–Water	$FeCl_2—ZnCl_2—H_2O$	5227
Iron (II) Chloride–Silver Chloride–Water	$FeCl_2—AgCl—H_2O$	5410
Iron (III) Chloride–Cobalt (II) Chloride–Water	$FeCl_3—CoCl_2—H_2O$	5228
Iron (III) Chloride–Nickel (II) Chloride–Water	$FeCl_3—NiCl_2—H_2O$	5229
Iron (III) Chloride–Silver Chloride–Water	$FeCl_3—AgCl—H_2O$	5410
Iron (III) Chloride–Cesium Chloride–Water	$FeCl_3—CsCl—H_2O$	5230
Iron (III) Chloride–Barium Sulfate–Water	$FeCl_3—BaSO_4—H_2O$	5231
Iron (II) Bromide–Iron (III) Bromide–Water	$FeBr_2—FeBr_3—H_2O$	5232
Cobalt (II) Oxide–Cobalt (II) Sulfate–Water	$CoO—CoSO_4—H_2O$	5233
Cobalt (II) Nitrate–Nickel (II) Nitrate–Water	$Co(NO_3)_2—Ni(NO_3)_2—H_2O$	5234, 5235
Cobalt (II) Nitrate–Copper (II) Nitrate–Water	$Co(NO_3)_2—Cu(NO_3)_2—H_2O$	5236, 5237
Cobalt (II) Sulfate–Cobalt (II) Chloride–Water	$CoSO_4—CoCl_2—H_2O$	5238, 5239
Cobalt (II) Sulfate–Cobalt (II) Iodate–Water	$CoSO_4—Co(IO_3)_2—H_2O$	5240
Cobalt (II) Sulfate–Nickel (II) Sulfate–Water	$CoSO_4—NiSO_4—H_2O$	5241
Cobalt (II) Sulfate–Copper (II) Sulfate–Water	$CoSO_4—CuSO_4—H_2O$	5242
Cobalt (II) Sulfate–Cadmium Sulfate–Water	$CoSO_4—CdSO_4—H_2O$	5243

System name	Formulae of components	Table number
Cobalt (II) Sulfate–Rubidium Sulfate–Water	$CoSO_4$—Rb_2SO_4—H_2O	5244
Cobalt (II) Sulfate–Thallium Sulfate–Water	$CoSO_4$—Tl_2SO_4—H_2O	5245
Cobalt (II) Chloride–Nickel (II) Chloride–Water	$CoCl_2$—$NiCl_2$—H_2O	5246
Cobalt (II) Chloride–Copper (II) Chloride–Water	$CoCl_2$—$CuCl_2$—H_2O	5247
Cobalt (II) Chloride–Zinc Chloride–Water	$CoCl_2$—$ZnCl_2$—H_2O	5248, 5249
Cobalt (II) Chloride–Lead (I) Chloride–Water	$CoCl_2$—$PbCl$—H_2O	5250
Cobalt (II) Chloride–Strontium Chloride–Water	$CoCl_2$—$SrCl_2$—H_2O	5251
Cobalt (II) Chloride–Cadmium Chloride–Water	$CoCl_2$—$CdCl_2$—H_2O	5252
Cobalt (II) Chloride–Cesium Chloride–Water	$CoCl_2$—$CsCl$—H_2O	5253
Cobalt (II) Chloride–Barium Chloride–Water	$CoCl_2$—$BaCl_2$—H_2O	5254
Cobalt (II) Chloride–Mercury (II) Chlorde–Water	$CoCl_2$—$HgCl_2$—H_2O	5255, 5256
Cobalt (II) Chloride–Thorium Chloride–Water	$CoCl_2$—$ThCl_4$—H_2O	5257
Nickel (II) Nitrate–Lanthanum Nitrate–Water	$Ni(NO_3)_2$—$La(NO_3)_3$—H_2O	5258
Nickel (II) Sulfate–Copper (II) Sulfate–Water	$NiSO_4$—$CuSO_4$—H_2O	5259, 5260
Nickel (II) Sulfate–Rubidium Sulfate–Water	$NiSO_4$—Rb_2SO_4—H_2O	5261
Nickel (II) Sulfate–Thallium Sulfate–Water	$NiSO_4$—Tl_2SO_4—H_2O	5262
Nickel (II) Chloride–Cadmium Chloride–Water	$NiCl_2$—$CdCl_2$—H_2O	5263
Copper (II) Oxide–Copper (II) Sulfate–Water	CuO—$CuSO_4$—H_2O	5264
Copper (II) Oxide–Copper (II) Sulfate–Zinc Oxide–Zinc Sulfate–Water	CuO—$CuSO_4$—ZnO—$ZnSO_4$—H_2O	5265
Copper (II) Oxide–Arsenic Pentoxide–Water	CuO—As_2O_5—H_2O	5266
Copper (II) Hydroxide–Copper (II) Selenate–Water	$Cu(OH)_2$—$CuSeO_4$—H_2O	5267
Copper (II) Nitrate–Copper (II) Sulfate–Water	$Cu(NO_3)_2$—$CuSO_4$—H_2O	5268
Copper (II) Nitrate–Lead (II) Nitrate–Water	$Cu(NO_3)_2$—$Pb(NO_3)_2$—H_2O	5269
Copper (II) Thiocyanate–Zinc Sulfate–Water	$Cu_2(CNS)_2$—$ZnSO_4$—H_2O	5270
Copper (II) Sulfate–Copper (II) Chloride–Water	$CuSO_4$—$CuCl_2$—H_2O	5271
Copper (II) Sulfate–Copper (II) Arsenate–Water	$CuSO_4$—$Cu_3(AsO_4)_2$—H_2O	5272, 5273
Copper (II) Sulfate–Zinc Sulfate–Water	$CuSO_4$—$ZnSO_4$—H_2O	5274
Copper (II) Sulfate–Rubidium Chloride–Water	$CuSO_4$—$RbCl$—H_2O	5275

System name	Formulae of components	Table number
Copper (II) Sulfate–Silver Sulfate–Water	$CuSO_4—Ag_2SO_4—H_2O$	5276
Copper (II) Sulfate–Cerium Sulfate–Water	$CuSO_4—Ce_2(SO_4)_3—H_2O$	5277
Copper (II) Sulfate–Thallium Sulfate–Water	$CuSO_4—Tl_2SO_4—H_2O$	5278
Copper (II) Sulfate–Copper (II) Chloride–Water	$CuSO_4—CuCl_2—H_2O$	5279, 5280
Copper (I) Chloride–Copper (II) Chloride–Water	$CuCl—CuCl_2—H_2O$	5281
Copper (II) Chloride–Strontium Chloride–Water	$CuCl_2—SrCl_2—H_2O$	5282
Copper (II) Chloride–Silver Chloride–Water	$CuCl_2—AgCl—H_2O$	5410
Copper (II) Chloride–Cadmium Chloride–Water	$CuCl_2—CdCl_2—H_2O$	5283
Copper (II) Chloride–Barium Chloride–Water	$CuCl_2—BaCl_2—H_2O$	5284
Copper (II) Chloride–Cerium Chloride–Water	$CuCl_2—CeCl_3—H_2O$	5285
Copper (II) Chloride–Mercury (II) Chloride–Water	$CuCl_2—HgCl_2—H_2O$	5286, 5287
Copper (II) Chloride–Thallium Chloride–Water	$CuCl_2—TlCl—H_2O$	5288
Copper (II) Chloride–Thorium Chloride–Water	$CuCl_2—ThCl_4—H_2O$	5289
Zinc Oxide–Zinc Sulfate–Water	$ZnO—ZnSO_4—H_2O$	5290
Zinc Oxide–Zinc Chloride–Water	$ZnO—ZnCl_2—H_2O$	5291
Zinc Oxide–Barium Oxide–Water	$ZnO—BaO—H_2O$	5292
Zinc Hydroxide–Barium Nitrate–Water	$Zn(OH)_2—Ba(NO_3)_2—H_2O$	5293
Zinc Hydroxide–Barium Chloride–Water	$Zn(OH)_2—BaCl_2—H_2O$	5294
Zinc Nitrate–Zinc Sulfate–Thallium Chromate–Water	$Zn(NO_3)_2—ZnSO_4—Tl_2CrO_4—H_2O$	5295
Zinc Nitrate–Silver Bromate–Water	$Zn(NO_3)_2—AgBrO_3—H_2O$	5296, 5297
Zinc Nitrate–Cadmium Nitrate–Water	$Zn(NO_3)_2—Cd(NO_3)_2—H_2O$	5298
Zinc Nitrate–Lanthanum Nitrate–Water	$Zn(NO_3)_2—La(NO_3)_3—H_2O$	5299
Zinc Nitrate–Thallium Chromate–Water	$Zn(NO_3)_2—TlCrO_4—H_2O$	4010
Zinc Nitrate–Thallium Iodate–Water	$Zn(NO_3)_2—TlIO_3—H_2O$	4231
Zinc Nitrate–Uranyl Nitrate–Water	$Zn(NO_3)_2—UO_2(NO_3)_2—H_2O$	5300
Zinc Sulfate–Silver Sulfate–Water	$ZnSO_4—Ag_2SO_4—H_2O$	5301
Zinc Sulfate–Silver Bromate–Water	$ZnSO_4—AgBrO_3—H_2O$	5302
Zinc Sulfate–Thallium Sulfate–Water	$ZnSO_4—Tl_2SO_4—H_2O$	5303

System name	Formulae of components	Table number
Zinc Sulfate–Thallium Chloride–Water	$ZnSO_4$—$TlCl$—H_2O	5304, 5305
Zinc Sulfate–Thallium Chromate–Water	$ZnSO_4$—$TlCrO_4$—H_2O	4010
Zinc Sulfate–Thallium Iodate–Water	$ZnSO_4$—$TlIO_3$—H_2O	5306
Zinc Chloride–Silver Chloride–Water	$ZnCl_2$—$AgCl$—H_2O	5410
Zinc Chloride–Silver Chloride–Water	$ZnCl_2$—$AgCl$—H_2O	5307
Zinc Chloride–Indium Chloride–Water	$ZnCl_2$—$InCl_3$—H_2O	5308
Zinc Chloride–Thallium Chloride–Water	$ZnCl_2$—$TlCl$—H_2O	5309
Zinc Chloride–Lead (II) Chloride–Water	$ZnCl_2$—$PbCl_2$—H_2O	5310, 5311
Zinc Bromide–Indium Bromide–Water	$ZnBr_2$—$InBr_3$—H_2O	5312
Zinc Iodide–Indium Iodide–Water	ZnI_2—InI_3—H_2O	5313
Arsenic Trioxide–Strontium Oxide–Water	As_2O_3—SrO—H_2O	5314, 5315
Arsenic Trioxide–Strontium Chloride–Water	As_2O_3—$SrCl_2$—H_2O	5316
Arsenic Trioxide–Strontium Bromide–Water	As_2O_3—$SrBr_2$—H_2O	5317
Arsenic Trioxide–Barium Oxide–Water	As_2O_3—BaO—H_2O	5318—5320
Arsenic Trioxide–Barium Chloride–Water	As_2O_3—$BaCl_2$—H_2O	5321
Arsenic Trioxide–Barium Bromide–Water	As_2O_3—$BaBr_2$—H_2O	5322
Arsenic Pentoxide–Strontium Oxide–Water	As_2O_5—SrO—H_2O	5323
Arsenic Pentoxide–Barium Oxide–Water	As_2O_5—BaO—H_2O	5324
Selenium Dioxide–Barium Selenite–Water	SeO_2—$BaSeO_3$—H_2O	5325
Selenium Chloride Oxide–Barium Chloride–Water	$SeOCl_2$—$BaCl_2$—H_2O	5326
Rubidium Carbonate–Dioxoamericium Carbonate–Water	Rb_2CO_3—$(AmO_2)_2CO_3$—H_2O	5327
Rubidium Nitrite–Strontium Nitrite–Water	$RbNO_2$—$Sr(NO_2)_2$—H_2O	5328
Rubidium Nitrite–Barium Nitrite–Water	$RbNO_2$—$Ba(NO_2)_2$—H_2O	5329
Rubidium Nitrite–Thallium Nitrite–Water	$RbNO_2$—$TlNO_2$—H_2O	5330
Rubidium Nitrate–Cerium Nitrate–Water	$RbNO_3$—$Ce(NO_3)_3$—H_2O	5331
Rubidium Hydrogen Sulfate–Rubidium Hydrogen Tellurate–Water	$RbHSO_4$—$RbHTeO_4$—H_2O	5332

System name	Formulae of components	Table number
Rubidium Sulfate–Silver Sulfate–Water	$Rb_2SO_4—Ag_2SO_4—H_2O$	5333
Rubidium Sulfate–Cadmium Sulfate–Water	$Rb_2SO_4—CdSO_4—H_2O$	5334
Rubidium Sulfate–Indium Sulfate–Water	$Rb_2SO_4—In_2(SO_4)_3—H_2O$	5335
Rubidium Sulfate–Praseodymium Sulfate–Water	$Rb_2SO_4—Pr_2(SO_4)_3—H_2O$	5336
Rubidium Chloride–Rubidium Bromide–Water	$RbCl—RbBr—H_2O$	5337
Rubidium Chloride–Silver Chloride–Water	$RbCl—AgCl—H_2O$	5411
Rubidium Chloride–Barium Hydroxide–Water	$RbCl—Ba(OH)_2—H_2O$	5338
Rubidium Chloride–Mercury (II) Chloride–Water	$RbCl_2—HgCl_2—H_2O$	5339
Rubidium Chlorate–Cesium Chlorate–Water	$RbClO_3—CsClO_3—H_2O$	5340
Rubidium Hydrogen Selenate–Rubidium Hydrogen Tellurate–Water	$RbHSeO_4—RbHTeO_4—H_2O$	5341
Rubidium Bromide–Rubidium Iodide–Water	$RbBr_2—RbI_2—H_2O$	5342
Rubidium Iodide–Cesium Iodide–Water	$RbI—CsI—H_2O$	5343
Rubidium Iodide–Mercury (II) Iodide–Water	$RbI—HgI_2—H_2O$	5344
Rubidium Iodide–Thallium Triiodide–Water	$RbI—TlI_3—H_2O$	5345
Strontium Oxide–Strontium Chloride–Water	$SrO—SrCl_2—H_2O$	5346
Strontium Oxide–Strontium Bromide–Water	$SrO—SrBr_2—H_2O$	5347
Strontium Oxide–Strontium Iodide–Water	$SrO—SrI_2—H_2O$	5348
Strontium Hydroxide–Strontium Nitrate–Water	$Sr(OH)_2—Sr(NO_3)_2—H_2O$	5349
Strontium Hydroxide–Strontium Hydrogen Sulfide–Water	$Sr(OH)_2—Sr(SH)_2—H_2O$	5350
Stontium Nitrite–Silver Nitrite–Water	$Sr(NO_2)_2—AgNO_2—H_2O$	5351
Strontium Nitrite–Cesium Nitrite–Water	$Sr(NO_2)_2—CsNO_2—H_2O$	5352
Stontium Nitrite–Thallium Nitrite–Water	$Sr(NO_2)_2—TlNO_2—H_2O$	5353
Strontium Nitrate–Strontium Chloride–Water	$Sr(NO_3)_2—SrCl_2—H_2O$	5354, 5355
Strontium Nitrate–Strontium Bromide–Water	$Sr(NO_3)_2—SrBr_2—H_2O$	5356
Strontium Nitrate–Silver Bromate–Water	$Sr(NO_3)_2—AgBrO_3—H_2O$	4004, 5357
Strontium Nitrate–Thallium Chromate–Water	$Sr(NO_3)_2—Tl_2CrO_4—H_2O$	4010

System name	Formulae of components	Table number
Strontium Nitrate–Lead (II) Nitrate–Water	$Sr(NO_3)_2—Pb(NO_3)_2—H_2O$	5358
Strontium Nitrate–Lead (II) Bromate–Water	$Sr(NO_3)_2—Pb(BrO_3)_2—H_2O$	5359
Strontium Nitrate–Uranyl Nitrate–Water	$Sr(NO_3)_2—UO_2(NO_3)_2—H_2O$	5360
Strontium Dithionate–Lead (II) Dithionate–Water	$SrS_2O_6—PbS_2O_6—H_2O$	5361
Strontium Chloride–Silver Chloride–Water	$SrCl_2—AgCl—H_2O$	5362, 5410 5411
Strontium Chloride–Indium Chloride–Water	$SrCl_2—InCl_3—H_2O$	5363
Strontium Chloride–Barium Sulfate–Water	$SrCl_2—BaSO_4—H_2O$	5364
Strontium Chloride–Mercury (II) Chloride–Water	$SrCl_2—HgCl_2—H_2O$	5365, 5366
Strontium Chloride–Thallium Chloride–Water	$SrCl_2—TlCl—H_2O$	5367
Strontium Chloride–Lead (II) Chloride–Water	$SrCl_2—PbCl_2—H_2O$	5368, 5369
Strontium Chloride–Lead (II) Bromide–Water	$SrCl_2—PbBr_2—H_2O$	5370
Strontium Bromide–Indium Bromide–Water	$SrBr_2—InBr_3—H_2O$	5371
Strontium Bromide–Mercury (II) Bromide–Water	$SrBr_2—HgBr_2—H_2O$	5372
Strontium Bromide–Lead (II) Chloride–Water	$SrBr_2—PbCl_2—H_2O$	5373
Strontium Bromide–Lead (II) Bromide–Water	$SrBr_2—PbBr_2—H_2O$	5374
Strontium Iodide–Indium Iodide–Water	$SrI_2—InI_3—H_2O$	5375
Strontium Iodide–Mercury (II) Iodide–Water	$SrI_2—HgI_2—H_2O$	5376
Yttrium Hydroxide–Yttrium Nitrate–Water	$Y(OH)_3—Y(NO_3)_3—H_2O$	5377
Silver Oxide–Iodine Heptoxide–Water	$Ag_2O—I_2O_7—H_2O$	5378
Silver Cyanide–Silver Nitrate–Water	$AgCN—AgNO_3—H_2O$	5379
Silver Thiocyanate–Barium Thiocyanate–Water	$AgCNS—Ba(CNS)_2—H_2O$	5380
Silver Nitrite–Silver Nitrate–Water	$AgNO_2—AgNO_3—H_2O$	5381, 5382
Silver Nitrite–Barium Nitrite–Water	$AgNO_2—Ba(NO_2)_2—H_2O$	5383, 5384
Silver Nitrate–Silver Sulfate–Water	$AgNO_3—Ag_2SO_4—H_2O$	5385, 5404
Silver Nitrate–Silver Chloride–Water	$AgNO_3—AgCl—H_2O$	5386—5388
Silver Nitrate–Silver Permanganate–Water	$AgNO_3—AgMnO_4—H_2O$	5389

System name	Formulae of components	Table number
Silver Nitrate–Silver Bromide–Water	$AgNO_3—AgBr—H_2O$	5390—5392
Silver Nitrate–Silver Bromate–Water	$AgNO_3—AgBrO_3—H_2O$	5393
Silver Nitrate–Silver Molybdate–Water	$AgNO_3—Ag_2MoO_4—H_2O$	5394
Silver Nitrate–Silver Iodide–Water	$AgNO_3—AgI—H_2O$	5395—5397
Silver Nitrate–Silver Iodate–Water	$AgNO_3—AgIO_3—H_2O$	5398
Silver Nitrate–Barium Nitrate–Water	$AgNO_3—Ba(NO_3)_2—H_2O$	5399
Silver Nitrate–Lead (II) Nitrate–Water	$AgNO_3—Pb(NO_3)_2—H_2O$	5400
Silver Sulfate–Cesium Sulfate–Water	$Ag_2SO_4—Cs_2SO_4—H_2O$	5401
Silver Sulfate–Lanthanum Sulfate–Water	$Ag_2SO_4—La_2(SO_4)_3—H_2O$	5402
Silver Chloride–Lanthanum Nitrate–Water	$AgCl—La(NO_3)_3—H_2O$	5405
Silver Chloride–Lanthanum Sulfate–Water	$AgCl—La_2(SO_4)_3—H_2O$	5406
Silver Chloride–Barium Thiocyanate–Water	$AgCl—Ba(CNS)_2—H_2O$	5410
Silver Chloride–Barium Nitrate–Water	$AgCl—Ba(NO_3)_2—H_2O$	5407
Silver Chloride–Barium Chloride–Water	$AgCl—BaCl_2—H_2O$	5408, 5410, 5411
Silver Chloride–Mercury (I) Nitrate–Water	$AgCl—HgNO_3—H_2O$	5409
Silver Perchlorate–Silver Permanganate–Water	$AgClO_4—AgMnO_4—H_2O$	5414
Silver Bromide–Mercury (I) Nitrate–Water	$AgBr—HgNO_3—H_2O$	5415
Silver Bromate–Cadmium Nitrate–Water	$AgBrO_3—Cd(NO_3)_2—H_2O$	5421, 5422
Silver Bromate–Cadmium Sulfate–Water	$AgBrO_3—CdSO_4—H_2O$	5423, 5425
Silver Bromate–Barium Nitrate–Water	$AgBrO_3—Ba(NO_3)_2—H_2O$	5424, 5425
Silver Bromate–Cerium Nitrate–Water	$AgBrO_3—Ce(NO_3)_3—H_2O$	5425
Silver Iodide–Barium Thiocyanate–Water	$AgI—Ba(CNS)_2—H_2O$	5427
Silver Iodide–Mercury (II) Nitrate–Water	$AgI—Hg(NO_3)_2—H_2O$	5426
Silver Iodate–Iodine Pentoxide–Water	$AgIO_3—I_2O_5—H_2O$	5428
Silver Iodate–Barium Nitrate–Water	$AgIO_3—Ba(NO_3)_2—H_2O$	5429
Cadmium Oxide–Cadmium Sulfate–Water	$CdO—CdSO_4—H_2O$	5430
Cadmium Hydroxide–Cadmium Sulfate–Water	$Cd(OH)_2—CdSO_4—H_2O$	5431

System name	Formulae of components	Table number
Cadmium Nitrate–Lead (II) Iodide–Water	$Cd(NO_3)_2—PbI_2—H_2O$	5432
Cadmium Nitrate–Uranyl Nitrate–Water	$Cd(NO_3)_2—UO_2(NO_3)_2—H_2O$	5433
Cadmium Sulfate–Lanthanum Iodate–Water	$CdSO_4—La(IO_3)_3—H_2O$	5434
Cadmium Sulfate–Thallium Sulfate–Water	$CdSO_4—Tl_2SO_4—H_2O$	5435
Cadmium Sulfate–Thallium Chloride–Water	$CdSO_4—TlCl—H_2O$	5436
Cadmium Sulfate–Thallium Iodate–Water	$CdSO_4—TlIO_3—H_2O$	5437
Cadmium Chloride–Cadmium Bromide–Water	$CdCl_2—CdBr_2—H_2O$	5438
Cadmium Chloride–Barium Chloride–Water	$CdCl_2—BaCl_2—H_2O$	5439
Cadmium Chloride–Lanthanum Iodate–Water	$CdCl_2—La(IO_3)_3—H_2O$	5440
Cadmium Chloride–Thallium Chloride–Water	$CdCl_2—TlCl—H_2O$	5441—5443
Cadmium Chloride–Thallium Iodate–Water	$CdCl_2—TlIO_3—H_2O$	5444
Cadmium Chloride–Lead (II) Chloride–Water	$CdCl_2—PbCl_2—H_2O$	5445
Cadmium Bromide–Barium Bromide Water	$CdBr_2—BaBr_2—H_2O$	5446
Cadmium Bromide–Lead (II) Bromide–Water	$CdBr_2—PbBr_2—H_2O$	5447
Cadmium Iodide–Mercury (II) Iodide–Water	$CdI_2—HgI_2—H_2O$	5448
Indium Fluoride–Cesium Fluoride–Water	$InF_3—CsF—H_2O$	5449
Indium Sulfate–Cesium Sulfate–Water	$In_2(SO_4)_3—Cs_2SO_4—H_2O$	5450
Indium Bromide–Thallium Bromide–Water	$InBr_3—TlBr—H_2O$	5451
Indium Iodide–Barium Iodide–Water	$InI_3—BaI_2—H_2O$	5452
Indium Iodide–Thallium Iodide–Water	$InI_3—TlI—H_2O$	5453
Iodine Pentoxide–Barium Iodate–Water	$I_2O_5—Ba(IO_3)_2—H_2O$	5454
Cesium Nitrate–Cerium Nitrate–Water	$CsNO_3—Ce(NO_3)_3—H_2O$	5455
Cesium Thiocyanate–Thallium Thiocyanate–Water	$CsCNS—TlCNS—H_2O$	5456
Cesium Nitrite–Barium Nitrite–Water	$CsNO_2—Ba(NO_2)_2—H_2O$	5457
Cesium Nitrite–Thallium Nitrite–Water	$CsNO_2—TlNO_2—H_2O$	5458
Cesium Nitrate–Lead (II) Nitrate–Water	$CsNO_3—Pb(NO_3)_2—H_2O$	5459
Cesium Nitrate–Uranyl Nitrate–Water	$CsNO_3—UO_2(NO_3)_2—H_2O$	5460
Cesium Sulfate–Cerium Sulfate–Water	$Cs_2SO_4—Ce_2(SO_4)_3—H_2O$	5461

System name	Formulae of components	Table number
Cesium Sulfate–Praseodymium Sulfate–Water	Cs_2SO_4—$Pr_2(SO_4)_3$—H_2O	5462
Cesium Chloride–Cerium Chloride–Water	$CsCl$—$CeCl_3$—H_2O	5463
Cesium Chloride–Mercury (II) Chloride–Water	$CsCl$—$HgCl_2$—H_2O	5464
Cesium Chloride–Thallium Trichloride–Water	$CsCl$—$TlCl_3$—H_2O	5465
Cesium Bromide–Lead (II) Bromide–Water	$CsBr$—$PbBr_2$—H_2O	5466
Cesium Iodide–Thallium Triiodide–Water	CsI—TlI_3—H_2O	5467
Barium Oxide–Barium Chloride–Water	BaO—$BaCl_2$—H_2O	5468
Barium Oxide–Barium Bromide–Water	BaO—$BaBr_2$—H_2O	5469
Barium Oxide–Barium Iodide–Water	BaO—BaI_2—H_2O	5470
Barium Hydroxide–Barium Thiocyanate–Water	$Ba(OH)_2$—$Ba(CNS)_2$—H_2O	5471
Barium Hydroxide–Barium Nitrate–Water	$Ba(OH)_2$—$Ba(NO_3)_2$—H_2O	5472
Barium Hydroxide–Barium Metasilicate–Water	$Ba(OH)_2$—$BaSiO_3 \cdot 6H_2O$—H_2O	5473
Barium Hydroxide–Barium Hydrogen Sulfide–Water	$Ba(OH)_2$—$Ba(HS)_2$—H_2O	5474
Barium Hydroxide–Barium Chloride–Water	$Ba(OH)_2$—$BaCl_2$—H_2O	5475
Barium Hydroxide–Barium Chlorate–Water	$Ba(OH)_2$—$Ba(ClO_3)_2$—H_2O	5476
Barium Hydroxide–Barium Iodide–Water	$Ba(OH)_2$—BaI_2—H_2O	5477
Barium Thiocyanate–Silver Bromide–Water	$Ba(CNS)_2$—$AgBr$—H_2O	5418
Barium Nitrite–Thallium Nitrite–Water	$Ba(NO_2)_2$—$TlNO_2$—H_2O	5478, 5479
Barium Nitrate–Barium Chloride–Water	$Ba(NO_3)_2$—$BaCl_2$—H_2O	5480—5484
Barium Nitrate–Barium Chlorate–Water	$Ba(NO_3)_2$—$Ba(ClO_3)_2$—H_2O	5485
Barium Nitrate–Barium Chlorate–Barium Bromide–Water	$Ba(NO_3)_2$—$Ba(ClO_3)_2$—$BaBr_2$—H_2O	5486
Barium Nitrate–Barium Bromide–Water	$Ba(NO_3)_2$—$BaBr_2$—H_2O	5487
Barium Nitrate–Barium Bromate–Water	$Ba(NO_3)_2$—$Ba(BrO_3)_2$—H_2O	5488, 5489
Barium Nitrate–Barium Iodate–Water	$Ba(NO_3)_2$—$Ba(IO_3)_2$—H_2O	5490, 5491
Barium Nitrate–Thallium Bromide–Water	$Ba(NO_3)_2$—$TlBr$—H_2O	5492
Barium Nitrate–Thallium Iodide–Water	$Ba(NO_3)_2$—TlI—H_2O	5493
Barium Nitrate–Lead (II) Nitrate–Water	$Ba(NO_3)_2$—$Pb(NO_3)_2$—H_2O	5494, 5495
Barium Nitrate–Lead (II) Bromide–Water	$Ba(NO_3)_2$—$PbBr_2$—H_2O	5496

System name	Formulae of components	Table number
Barium Nitrate–Uranyl Nitrate–Water	$Ba(NO_3)_2$—$UO_2(NO_3)_2$—H_2O	5497
Barium Fluoride–Barium Chloride–Water	BaF_2—$BaCl_2$—H_2O	5498
Barium Metasilicate–Barium Chloride–Water	$BaSiO_3 \cdot 6H_2O$—$BaCl_2$—H_2O	5499
Barium Hydrogen Sulfide–Barium Chloride–Water	$Ba(SH)_2$—$BaCl_2$—H_2O	5500
Barium Sulfate–Thorium Chloride–Water	$BaSO_4$—$ThCl_4$—H_2O	5501
Barium Chloride–Barium Chlorate–Water	$BaCl_2$—$Ba(ClO_3)_2$—H_2O	5502—5504
Barium Chloride–Barium Bromide–Water	$BaCl_2$—$BaBr_2$—H_2O	5505, 5506
Barium Chloride–Barium Iodate–Water	$BaCl_2$—$Ba(IO_3)_2$—H_2O	5507
Barium Chloride–Mercury (I) Chloride–Water	$BaCl_2$—$HgCl$—H_2O	5508
Barium Chloride–Mercury (II) Chloride–Water	$BaCl_2$—$HgCl_2$—H_2O	5509—5512
Barium Chloride–Thallium Chloride–Water	$BaCl_2$—$TlCl$—H_2O	5513—5515
Barium Chloride–Lead (II) Chloride–Water	$BaCl_2$—$PbCl_2$—H_2O	5516—5518
Barium Chloride–Lead (II) Bromide–Water	$BaCl_2$—$PbBr_2$—H_2O	5519
Barium Chlorate–Barium Bromide–Water	$Ba(ClO_3)_2$—$BaBr_2$—H_2O	5520
Barium Chlorate–Barium Bromate–Water	$Ba(ClO_3)_2$—$Ba(BrO_3)_2$—H_2O	5521
Barium Bromide–Barium Bromate–Water	$BaBr_2$—$Ba(BrO_3)_2$—H_2O	5522
Barium Bromide–Barium Iodide–Water	$BaBr_2$—BaI_2—H_2O	5523, 5524
Barium Bromide–Barium Iodate–Water	$BaBr_2$—$Ba(IO_3)_2$—H_2O	5525
Barium Bromide–Mercury (II) Bromide–Water	$BaBr_2$—$HgBr_2$—H_2O	5526—5528
Barium Bromide–Lead (II) Chloride–Water	$BaBr_2$—$PbCl_2$—H_2O	5529
Barium Bromide–Lead (II) Bromide–Water	$BaBr_2$—$PbBr_2$—H_2O	5530
Barium Iodide–Mercury (II) Iodide–Water	BaI_2—HgI_2—H_2O	5531
Barium Iodide–Thallium Iodide–Water	BaI_2—TlI—H_2O	5532
Lanthanum Nitrate–Lanthanum Iodate–Water	$La(NO_3)_3$—$La(IO_3)_3$—H_2O	5533, 5534
Lanthanum Nitrate–Neodymium Nitrate–Water	$La(NO_3)_3$—$Nd(NO_3)_3$—H_2O	5535
Lanthanum Nitrate–Samarium Nitrate–Water	$La(NO_3)_3$—$Sm(NO_3)_3$—H_2O	5536
Lanthanum Nitrate–Thallium Chloride–Water	$La(NO_3)_3$—$TlCl$—H_2O	5537

System name	Formulae of components	Table number
Lanthanum Sulfate–Lanthanum Iodate–Water	$La_2(SO_4)_3$—$La(IO_3)_3$—H_2O	5538
Lanthanum Sulfate–Thallium Sulfate–Water	$La_2(SO_4)_3$—Tl_2SO_4—H_2O	5539
Lanthanum Sulfate–Thallium Chloride–Water	$La_2(SO_4)_3$—$TlCl$—H_2O	5540
Lanthanum Chloride–Lanthanum Iodate–Water	$LaCl_3$—$La(IO_3)_3$—H_2O	5541
Cerium Sulfate–Thallium Sulfate–Water	$Ce_2(SO_4)_3$—Tl_2SO_4—H_2O	5542
Cerium Sulfate–Thorium Sulfate–Water	$Ce_2(SO_4)_3$—$Th(SO_4)_2$—H_2O	5543
Praseodymium Nitrate–Neodymium Nitrate–Water	$Pr(NO_3)_3$—$Nd(NO_3)_3$—H_2O	5544
Praseodymium Nitrate–Samarium Nitrate–Water	$Pr(NO_3)_3$—$Sm(NO_3)_3$—H_2O	5545
Thallium Sulfate–Praseodymium Sulfate–Water	Tl_2SO_4—$Pr_2(SO_4)_3$—H_2O	5546
Neodymium Sulfate–Thallium Sulfate–Water	$Nd_2(SO_4)_3$—Tl_2SO_4—H_2O	5547
Mercury (II) Oxide (Yellow)–Mercury (II) Chloride–Water	HgO—$HgCl_2$—H_2O	5548
Mercury (II) Nitrate–Mercury (II) Chloride–Water	$Hg(NO_3)_2$—$HgCl_2$—H_2O	5549
Mercury (II) Nitrate–Mercury (II) Iodide–Water	$Hg(NO_3)_2$—HgI_2—H_2O	5550
Mercury (II) Nitrate–Uranyl Nitrate–Water	$Hg(NO_3)_2$—$UO_2(NO_3)_2$—H_2O	5551
Mercury (II) Chloride–Mercury (II) Iodide–Water	$HgCl_2$—HgI_2—H_2O	5552
Mercury (II) Chloride–Thallium Chloride–Water	$HgCl_2$—$TlCl$—H_2O	5554
Mercury (II) Chloride–Lead (II) Chloride–Water	$HgCl_2$—$PbCl_2$—H_2O	5554, 5555
Thallium Thiocyanate–Thallium Bromate–Water	$TlCNS$—$TlBrO_3$—H_2O	5556
Thallium Thiocyanate–Thallium Chloride–Water	$TlCNS$—$TlCl$—H_2O	5557
Thallium Nitrate–Thallium Sulfate–Water	$TlNO_3$—Tl_2SO_4—H_2O	5558
Thallium Nitrate–Thallium Chloride–Water	$TlNO_3$—$TlCl$—H_2O	5559—5561
Thallium Nitrate–Thallium Bromide–Water	$TlNO_3$—$TlBr$—H_2O	5562
Thallium Nitrate–Thallium Hexanitrocobaltate–Water	$TlNO_3$—$Tl_3[Co(NO_2)_6]$—H_2O	5563
Thallium Nitrate–Uranyl Nitrate–Water	$TlNO_3$—$UO_2(NO_3)_2$—H_2O	5564
Thallium Sulfate–Thallium Chloride–Water	Tl_2SO_4—$TlCl$—H_2O	5565—5567
Thallium Sulfate–Thallium Chlorate–Water	Tl_2SO_4—$TlClO_3$—H_2O	5568
Thallium Sulfate–Thallium Bromide–Water	Tl_2SO_4—$TlBr$—H_2O	5569

System name	Formulae of components	Table number
Thallium Chloride–Thallium Tri-chloride–Water	$TlCl—TlCl_3—H_2O$	5570
Thallium Chloride–Thallium Chlor-ate–Water	$TlCl—TlClO_3—H_2O$	5571
Thallium Chloride–Thallium Bro-mate–Water	$TlCl—TlBrO_3—H_2O$	5572
Thallium Bromide–Thallium Tribro-mide–Water	$TlBr—TlBr_3—H_2O$	5573
Lead (II) Nitrate–Lead (II) Sulfate–Water	$Pb(NO_3)_2—PbSO_4—H_2O$	5574
Lead (II) Nitrate–Lead (II) Chloride–Water	$Pb(NO_3)_2—PbCl_2—H_2O$	5575—5577
Lead (II) Nitrate–Lead (II) Bro-mide–Water	$Pb(NO_3)_2—PbBr_2—H_2O$	5578
Lead (II) Nitrate–Lead (II) Bromate–Water	$Pb(NO_3)_2—Pb(BrO_3)_2—H_2O$	5579
Lead (II) Nitrate–Lead (II) Iodide–Water	$Pb(NO_3)_2—PbI_2—H_2O$	5580, 5581
Lead (II) Nitrate–Lead (II) Iodate–Water	$Pb(NO_3)_2—Pb(IO_3)_2—H_2O$	5582
Lead (II) Fluoride–Lead (II) Bro-mide–Water	$PbF_2—PbBr_2—H_2O$	5583
Lead (II) Fluroide–Lead (II) Iodide–Water	$PbF_2—PbI_2—H_2O$	5584
Lead (II) Fluorochlorate–Uranyl Nitrate–Water	$PbFCl—UO_2(NO_3)_2—H_2O$	5585
Lead (II) Sulfate–Thorium Chlor-ide–Water	$PbSO_4—ThCl_4—H_2O$	5586
Lead (II) Chloride–Lead (II) Iodide–Water	$PbCl_2—PbI_2—H_2O$	5587

REFERENCES

A few references in rare and inaccessible publications are quoted in reference books, from which the material and the relevant references have been taken. In such cases a reference is given below to one of the following publications:

[1] A reference book of solubilities of salt systems (Spravochnik po rastvorimosti solevykh sistem). Edited by V. M. Bukshtein, M. G. Valyashko and A. D. Pel'sh. Vols. I and II. Izd. Vses. Nauch.-Issled. Inst. Goz., Goskhimizdat., Moscow–Leningrad (1953–1954).

[2] Seidell A., Solubilities of inorganic and metal organic compounds, 3rd Ed., Van Nostrand, Princeton, 1940–41. Supplement: Solubilities of inorganic and organic compounds, Van Nostrand, Princeton, 1952.

Reference number	Author's name and reference	Table number
1.	ABASHKINA T. F., BRUTSKUS E. B., Tr. Nauch.-Issled. Inst. Udobr. Insektofungits. (postgrad. papers), **22** (1963)	1903
2.	ABDULAEV YA. A., Zh. Prikl. Khim. **14**, 302 (1941)	11
3.	ABDULAEV D. A., TALIPOV SH. T., BABAEV N., Tr. Sredneaziat. Gos. Univ. **55**, Khim. Nauki, No. 7, 125 (1954).	3174
4.	ABLYAZINA R. A.. Symposium "Some questions on chemical technology and physico-chemical analysis of inorganic systems" (Nekotorye voprosy khimicheskoi tekhnologii i fiziko-khimicheskogo analiza neorganicheskikh sistem), Tashkent, Akad. Nauk Uzb. SSR, **87** (1963)	3913
5.	AVIRMED A., RESHETNIKOVA L. P., NOVOSELOVA A. V., Vestn. Mosk. Gos. Univ., Ser. II, No. 2, 47 (1962)	718
6.	AGAEV A. I., Azerb. Khim. Zh., No. 5, 87 (1960)	3360
7.	AGAEV A. I., DOLININA P. M., Uch. Zap. Azerb. Gos. Univ., No. 1., 73 (1961)	3351
8.	AGAEV A. I., Uch. Zap. Azerb. Univ., No. 5, 77 (1961)	4053
9.	AGAEV A. I., GASANOV I. A., Uch. Zap. Azerb. Univ., Ser. Fiz.-Mat. i Khim. Nauk, No. 4, 123 (1962)	4832
10.	AGAEV A. I., MAMEDOV A. A., Uch. Zap. Azerb. Univ., Ser. Fiz.-Mat. i Khim. Nauk, No. 4, 149 (1962)	4707
11.	AGAEV A. I., GASANOV I. A., Uch. Zap. Azerb. Univ., Ser. Khim., No. 4, 11 (1963)	3572
12.	AKOPOV E. K., Zh. Neorgan. Khim, **7**, vyp. 2, 387 (1962)	1425
13.	AKOPOV E. K., Zh. Prikl. Khim., **36**, vyp. 9, 1916 (1963)	1413

Reference number	Author's name and reference	Table number
14.	AKYEV D. A., SHUKAROV B., NABIEV M. N., Dokl. Akad. Nauk Uzb. SSR, No. 3, 39 (1962)	1896
15.	ALEKSANDROV N. P., Zh. Prikl. Khim., **14**, vyp. 7–8, 947	1812, 1844, 1940
16.	ALEKSANDROV N. P., Zh. Prikl. Khim., **19**, vyp. 1, 63 (1946)	1814, 1920, 1921
17.	ALEKSANDROV E. V., Zh. Russ. Fiz.-Khim. Obshch., **53**, 326 (1921)	3019, 3716 4477
18.	ALENCHIKOVA I. F., LIPIS L. V., NIKOLAEV N. S., At. Energ. (USSR), **10**, No. 6, 592 (1961)	630
19.	ALYMOVA L. N., KORF D. M., LEBEDEVA N. D., Zh. Neorgan. Khim., **8**, 1024 (1963)	3193
20.	ANGELOV I. I., Tr. Vses. Nauch.-Issled. Inst. Khim. Reaktivov, vyp. 22, 18 (1958)	524
21.	ANGELOV I. I., POSLAVSKAYA K. D., Tr. Vses. Nauch.-Issled. Inst. Khim. Reaktivov, vyp. 22, 22 (1958)	407, 408 885
22.	ANDREEV V. V., Zh. Obshch. Khim., **24**, No. 10, 1730 (1954)	2436
23.	ANDREEV V. V., Zh. Prikl. Khim., **30**, vyp. 2, 300 (1957)	3552, 3553
24.	ANDREEV V. V., Zh. Neorgan. Khim., **4**, vyp. 10, 2384 (1959)	2477, 2558–2560
25.	ANDRONOVA N. P., Tr. Yaroslav. Sel'skokhoz. Inst., **4**, 291 (1957)	3435
26.	ANDRONOVA N. P., Uch. Zap. Yaroslav. Gos. Ped. Inst., Estestv., Geogr., vyp. 24, 53 (1958)	3391, 3401 3437, 4075
27.	ANDRONOVA N. P., LEPESHKOV I. N., Zh. Neorgan. Khim., **3**, No. 9, 2156 (1958)	3402, 3436
28.	ANIKIN I. N., SHUSHKANOV A. D., Kristallografiya, **8**, No. 1, 128 (1963)	1429, 5106
29.	ANOSOV V.YA., BYZOVA E.A., Izv. Sektora Fiz.-Khim. Analiza, Inst. Obshch. Neorgan. Khim., Akad. Nauk SSSR, **15**, 119 (1947)	4682
30.	ANTONOVA V. I., POLYNTSEVA N. K., Tr. Inst. Khim. Nauk, Akad. Nauk Kaz. SSR, 1, 71 (1957)	2813, 2814 2821
31.	ARTYKBAEV T., TSYGANOV G. A., Dokl. Akad. Nauk Uzb. SSR, No. 2, 48 (1962)	987, 1021
32.	ARKHIPOV M. I., PAKSHVER A. B., Zh. Prikl. Khim. **12**, 894 (1939)	1965, 1967 1971, 1973
33.	ARKHIPOV M. I., PAKSHVER A. B., PODBORNOVA N. I., Zh. Prikl. Khim., **23**, 181, 650 (1950)	1966, 2761, 2764
34.	ARKHIPOV M. I., Izv. Vyshikh Uchebn. Zavedenii, Khim. i Khim. Tekhnol. No. 2, 102 (1959)	1809
35.	ASLANOV S. M., BLIDIN V. P., Zh. Neorgan. Khim., **4**, vyp. 7, 1661 (1959)	1470
36.	ASLANOV S. M., Zh. Neorgan. Khim., **8**, 1492 (1963)	1471
37.	AKHUMOV E. I., VASIL'EV B. B., Zh. Obshch. Khim., **2**, 282 (1932)	3623, 3645, 3689, 4127
38.	AKHUMOV E. I., GOLOVKOV M. P., Zh. Obshch. Khim., **5**, vyp. 4, 504 (1935)	3627, 3692
39.	AKHUMOV E. I., VASIL'EV B. B., Solikamskie Karnallity, p. 105 (1935); Izv. Sektora Fiz.-Khim. Analiza, Inst. Obshch. Neorgan. Khim, Akad. Nauk SSSR, **9**, 313 (1936)	3645
40.	AKHUMOV E. I., GOLOVKOV M. P., Zh. Obshch. Khim., **6**, No. 4, 545 (1936)	2468, 2491

Reference number	Author's name and reference	Table number
41.	AKHUMOV E. I., EZEROVA E. A., Zh. Obshch. Khim., **6**, No. 10, 1418 (1936)	20, 2469
42.	AKHUMOV E. I., EZEROVA E. A., Zh. Prikl. Khim., **9**, 1173 (1936)	1870, 1878
43.	AKHUMOV E. I., DISTANOV G. K., Zh. Obshch. Khim., **8**, No. 9, 792 (1938); **9**, No. 6, 490 (1939)	1892, 1899, 1901
44.	AKHUMOV E. I., Zh. Obshch. Khim., **10**, 670 (1940); **18**, 545 (1948)	1876
45.	AKHUMOV E. I., SPIRO N. S., Zh. Prikl. Khim., **27**, No. 11, 1164 (1954)	968, 976, 986, 991
46.	AKHUMOV E. I., Zh. Prikl. Khim., **27**, No. 12, 1326 (1954)	431
47.	AKHUMOV E. I., PYLKOVA E. V., Zh. Neorgan. Khim., **5**, No. 8, 1819 (1960)	3705
48.	BABAEVA A. V., Tr. Vses. Nauch.-Issled. Inst. Khim. Reaktivov Osobo Chist. Khim. Veshchestv., No. 11, 125 (1931)	797
49.	BABAEVA A. V., ARCHAKO G. A., Zh. Obshch. Khim., **5**, 216 (1935)	1047
50.	BABAEVA A. V., KLYACHKO-GURVICH L. P., Zh. Obshch. Khim., **5**, 216 (1935)	589
51.	BABAYAN G. G., SAYAMYAN E. A., GYUNASHYAN A. P., Izv. Akad. Nauk Arm. SSR, Ser. Khim. Nauk, **6**, No. 3, 221 (1963)	4460
52.	BABAYAN G. G., OGANESYAN E. A., Izv. Akad. Nauk Arm. SSR, Ser. Khim. Nauk, **6**, No. 6, 539 (1963)	2738
53.	BABIEVSKAYA I. Z., PEREL'MAN F. M., Zh. Neorgan. Khim., **10**, No. 3, 681 (1965)	405
54.	BABKIN M. P., VOLOSKOVETS P. L., Tr. Donetsk. Ind. Inst., **39**, 23 (1959–1960)	2036, 4918
55.	BABKIN M. P., SECHAN R. D., Tr. Donetsk. Ind. Inst., 39, 27 (1959–1960)	3156, 4232 4637, 5104 5580
56.	BAMBUROV V. G., DEMENEV N. V., POLYAKOVA V. M., Izv. Sib. Otd. Akad. Nauk SSSR, No. 4, 73 (1962)	4646
57.	BAMBUROV V. G., DEMENEVA N. V., POLYAKOVA V. M., Izv. Sib. Otd. Akad. Nauk SSSR, No. 5, 70 (1962)	4654
58.	BATUNER L. M., Zh. Prikl. Khim., **11**, 391 (1938)	280, 284
59.	BATSANOVA L. P., NOVOSELOVA A. V., Zh. Obshch. Khim., **26**, No. 7, 1827 (1956)	1509
60.	BASHILOV I., VIL'NYANSKII YA., Z. angew. Chem., **41**, 57 (1928)	5163
61.	BASHKOV M., Tr. Vses. Nauch.-Issled. Inst. Galurgii , (1932), cited in [1]	4157
62.	BESUGLYI D. V., KUTSAKOV F. M., Zh. Prikl. Khim., **16**, 37 (1939)	3831, 5500
63.	BEKTUROV A. B., LITVINENKO V. I., Zh. Neorgan. Khim., **4**, No. 7, 1677 (1959)	313, 318
64.	BEKTUROV A. B., KONOBRITSKII E. G., Zh. Neorgan. Khim., **5**, No. 4, 945 (1960)	315, 319, 332, 3393
65.	BEKTUROV A. B., KONOBRITSKII E. G., Zh. Neorgan. Khim., **7**, No. 7, 1705 (1962)	321, 333, 3394
66.	BEKTUROV A. B., KADUSHKIPA L. A., KURMANGUSHINA L. K., Symposium "Chemistry and chemical technology" (Khimiya i khimicheskaya tekhnologiya", vol. I, Alma-Ata, 100 (1963)	677
67.	BELOKOSKOV V. I., Zh. Neorgan. Khim., **6**, No. 6, 1443 (1961)	4331

Reference number	Author's name and reference	Table number
68.	BELOKOSKOV V I., Zh. Neorgan. Khim., **7**, No. 2, 381 (1962)	4332
69.	BELONOGOV P. S., Zavod. Lab., **8**, 217 (1939)	948
70.	BELOPOL'SKII A. P., LEBEDEV B. A., TRIFONOVA M. KH., *et al.*, Zh. Prikl. Khim. 4, No. 5, 570 (1931)	1825, 1882 2105
71.	BELOPOL'SKII A. P., ALEXSANDROV N. P., *et al.*, Zh. Prikl. Khim., **5**, Nos. 3–4, 294 (1932)	2826, 2883
72.	BELOPOL'SKII A. P., ALEKSANDROV N. P., Zh. Prikl. Khim., **6**, No. 3; 393 (1933)	1833
73.	BELOPOL'SKII A. P., SHPUNT S. YA., SEREBRENNIKOVA M. T., Zh. Prikl. Khim., 7, vyp. 5, 677–683 (1934)	2079, 2082 2299
74.	BELOPOL'SKII A. P., URUSOV V. V., Zh. Prikl. Khim., **8**, vyp. 2, 194 (1935)	2112, 2118
75.	BELOPOL'SKII A. P., SHPUNT S. YA., SEREBRENNIKOVA M. T., Zh. Prikl. Khim., 8, 195 (1935)	2079, 2083 2299, 2886
76.	BELOPOL'SKII A. P., SHPUNT S. YA., ALEKSANDROV N. P., Kalii, No. 3, 26 (1936)	2300, 2302, 2320, 3429
77.	BELOPOL'SKII A. P., SEREBRENNIKOVA M. T., SHPUNT S. YA., Zh. Prikl. Khim., **10**, 403 (1937)	5089
78.	BELOPOL'SKII A. P., URUSOV V. V., Zh. Prikl. Khim., **10**, 1178 (1937)	446
79.	BELOPOL'SKII A. P., URUSOV V. V., Zh. Prikl. Khim., **14**, No. 13, 908 (1937)	1598, 1607 1606, 1884
80.	BELOPOL'SKII A. P., SHPUNT S. YA., *et al.*, Tr. Nauch.-Issled. Inst. Udobr. Insektofungits., vyp. 144, 31 (1940)	2802, 2083 2105, 2886
81.	BELOPOL'SKII A. P., ALEKSANDROV N. P., Tr. Nauch.-Issled. Inst. Udobr. Insektofungits., vyp. 144, 60 (1940)	1882, 1883
82.	BELOPOL'SKII A. P., SEREBRENNIKOVA M. T., BILEVICH A. V., Zh. Prikl. Khim., 13, No. 1, 3 (1940)	668, 4276
83.	BELOPOL'SKII A. P., URUSOV V. V., Tr. Nauch.-Issled. Inst. Udobr. Insektofungits., vyp. 144, 67–69 (1940)	1598, 1606, 1607
84.	BELOPOL'SKII A. P., SHPUNT S. YA., Zh. Prikl. Khim., **14**, No. 6, 716 (1941)	823
85.	BELOPOL'SKII A. P., SHPUNT S. YA., Zh. Prikl. Khim., **18**, Nos. 11–12, 624 (1945)	2300, 2303, 3429
86.	BELOPOL'SKII A. P., URUSOV V. V., Zh. Prikl. Khim., **21**, No. 8, 781 (1948)	824
87.	BELOPOL'SKII A. P., KOLYCHEVA V. N., SHPUNT C. YA., Zh. Priklad. Khim., **21**, No. 8, 794 (1948)	825
88.	BELOPOL'SKII A. P., SHPUNT S. YA., SHUL'GINA M. N., Zh. Priklad. Khim., **23**, No. 8, 826 (1950)	3964
89.	BELOPOL'SKII A. P., SHPUNT S. YA. SHUL'GINA M. N., Zh. Prikl. Khim., **26**, No. 3, 280 (1953)	3966
90.	BELOPOL'SKII A. P., SHPUNT S. YA., SHUL'GINA M. N., Zh. Prikl. Khim., **27**, No. 4, 393 (1954)	3967
91.	BEL'CHEV F. V., BERGMAN A. G., Zh. Prikl. Khim., **17**, Nos. 9–10, 520 (1944)	2225
92.	BEL'CHEV F. V. Tr. Beloruss. Sel'skokhoz. Akad., **19**, 145 (1953)	4669
93.	BELYAEV I. N., LYUBAS L. M., Zh. Neorgan. Khim., **10**, No. 4, 946 (1965)	3804, 4887

Reference number	Author's name and reference	Table number
94.	BELYAEV I. N., LE T'YUK, Zh. Neorgan. Khim., **10,** No. 5 , 1229 (1965)	5463
95.	BERG L. G., NIKOLAEV A., Izv. Akad. Nauk SSSR, Ser. Khim. Nauk, **848** (1937)	4475
96.	BERG L. G., Izv. Akad. Nauk SSSR, Ser. Khim. Nauk, **147,** 161 (1938)	4275
97.	BERG L. G., BORISOVA L. A., Izv. Kazan. Filiala Akad. Nauk SSSR, Ser. Khim. Nauk, No. 6, 210 (1961)	2966
98.	BERGMAN A. G., DOMBROVSKAYA N. S., S.T.E., **7,** 212–213 (1931)	3563
99.	BERGMAN A. G., POLYAKOVA L. B., Kalii, No. 7, 32 (1934)	2489
100.	BERGMAN A. G., BOCHKAREV P. F., Zh. Prikl. Khim., **10,** 1531 (1937)	2132
101.	BERGMAN A. G., KOLOSKOVA Z. A., DOMBROVSKAYA N. S., Za Nedra Volgo-Prikaspiya, vyp. 2, pp. 312–313 (1937)	3592, 3630, 4029
102.	BERGMAN A. G., Izv. Akad. Nauk SSSR, Ser. Khim. Nauk, **203** (1938)	2483
103.	BERGMAN A. G., NAGORNYI G. I., Izv. Akad. Nauk SSSR, Ser. Khim. Nauk, **217** (1938)	4121
104.	BERGMAN A. G., BOCHKAREV P. F., Izv. Akad. Nauk SSSR, Ser. Khim. Nauk, **237** (1938)	2181, 2233, 4538
105.	BERGMAN A. G., VLASOV N. P., Dokl. Akad. Nauk SSSR, **36,** 64 (1942)	4818
106.	BERGMAN A. G., SHELOKHOVICH M. L., Zh. Prikl. Khim., **15,** No. 4, 191 (1942)	2322
107.	BERGMAN A. G., VLASOV N. A., Izv. Sektora Fiz.-Khim. Analiza, Inst. Obshch. Neorgan. Khim., Akad. Nauk SSSR, **18,** 213–214 (1949)	3564
108.	BERGMAN A. G., LUZHNAYA N. P., Fiz.-Khim. Osn. Izucheniya i Ispol'zovaniya Solyanykh Mestorozhdenii Khlorid-Sul'fatnogo Tipa, Izd. Akad. Nauk SSSR, pp. 96–97 (1951)	3305, 3306
109.	BERGMAN A. G., SHAKHPARONOV M. I., Izv. Sektora Fiz.-Ihim. Analiza, Inst. Obshch. Neorgan. Khim., Akad. Nauk SSSR, **21,** 331 (1952)	2234
110.	BERGMAN A. G., KUZNETSOVA A. I., Zh. Neorgan. Khim., **4,** No. 1, 194 (1959)	4859
111.	BEREMZHANOV B. A., UVALIEVA A. A., BEKTUROV A. B., Izv. Akad Nauk Kaz. SSR, Ser. Khim., No. 8, 50 (1955)	2939, 2940
112.	BEREMZHANOV B. A., Izv. Akad. Nauk Kaz. SSR, Ser. Khim., vyp. 10, 104 (1956)	3217
113.	BEREMZHANOV B. A., PRONINA G. G., Uch. Zap. Kazan. Gos. Univ., **22,** 30 (1956)	3218
114.	BERNSHTEIN V. A., MATSENOK E. A., Zh. Prikl.-Khim, **34,** No. 5, 982 (1961)	2595
115.	BIRON E. V., MAL'CHEVSKII S. P., Zh. Russ. Fiz.-Khim Obshchest., Chast. Khim., **11,** No. 9, 1622 (1908)	3245, 3349 3358
116.	BLIDIN V. P., Zh. Obshch. Khim., **117,** 1590 (1947)	1478, 4197
117.	BLIDIN V. P., Dokl. Akad. Nauk SSSR, **84,** No. 5, 947 (1952)	1401, 1431, 1451, 1452, 1464

Reference number	Author's name and reference	Table number
118.	BLIDIN V. P., Sb. Statei Obshch. Khim., M.-L., Izd. Akad. Nauk SSSR, **1,** 86 (1953)	1480, 1484
119.	BLIDIN V. P., Dokl. Akad. Nauk SSSR, **88,** No. 3, 457 (1953)	1410, 1421
120.	BLIDIN V. P., Izv. Akad. Nauk SSSR, Otd. Khim. Nauk, No 5, 814 (1963)	1149, 1459 1479, 1483
121.	BLIDIN V. P., Izv. Akad. Nauk SSSR' Otd. Khim. Nauk, No. 3, 400 (1954)	1401, 1432 1446, 1452, 1453, 1457, 1465, 1469
122.	BLIDIN V. P., GORDIENKO V. I., Dokl. Akad. Nauk SSSR, **24,** No. 6, 1081 (1954)	1444, 1447
123.	BLIDIN V. P., Zh. Obshch. Khim., **26,** No. 5, 1281 (1956)	1528
124.	BLIDIN V. P., GORDIENKO V. I., SHALVEROVA O. K., Zh. Neorgan. Khim., **1,** No. 11, 2623 (1956)	1531, 1535. 1539
125.	BLIDIN V. P., Zh. Neorgan. Khim., **1,** No. 12, 2828 (1956)	1445, 5289
126.	BLIDIN V. P., Zh. Neorgan. Khim., **2,** No. 5, 1149 (1957)	1529, 1533, 1536, 1540
127.	BLIDIN V. P., ANDREEVA T. A., Zh. Neorgan. Khim., **3,** No. 10, 2414 (1958)	1382, 1383, 1384
128.	BLIDIN V. P., ASLANOV S. M., Zh. Neorgan. Khim., **4,** No. 3, 688 (1959)	1454
129.	BLYUMBERG YA. B., ZDANOVSKII A. B., Zh. Obshch. Khim. **9,** No. 9, 816 (1939)	708, 744, 944, 2272
130.	BOBROVA V. N., Radiokhimiya, **2,** No. 3, 364 (1960)	435, 438
131.	BOBROVA V. N., Radiokhimiya, **2,** No. 4, 500 (1960)	436, 437
132.	BOGDANOVSKAYA R. L. Uch. Zap. Perm. Gos. Univ., **13,** vyp. 3, 67 (1959)	4529
133.	BOGOYAVLENSKII P. S., Zh. Obshch. Khim., **19,** 1987 (1949)	4957
134.	BOGOYAVLENSKII P. S., Zh. Obshch. Khim., **22** (84), No. 4, 551 (1952)	4704
135.	BOGOYAVLENSKII P. S., SUKMANSKAYA G. V., MANANNIKOVA A. S., Sb. Nauch. Tr. L'vov. Gos. Vet.-Zootekhn. Inst., **6,** 366 (1953)	2913, 3281
136.	BOGOYAVLENSKII P. S., PIRO S. K., Sb. Nauch. Tr. L'vov. Gos. Vet.-Zootekhn. Inst., **6,** 381 (1953)	4498
137.	BOGOYAVLENSKII P. S., SUKMANSKAYA G. V., Zh. Obshch. Khim., **23,** No. 7, 1092 (1953)	4501
138.	BOGOYAVLENSKII P. S., MANANNIKOVA A. S., Sb. Nauch. Tr. L'vov. Gos. Vet.-Zootekhn. Inst., **6,** 376 (1954)	4452
139.	BOGOYAVLENSKII P. S., Dokl. Akad. Nauk SSSR, **101,** No. 5, 865 (1955)	4499, 4579
140.	BOGOYAVLENSKII P. S., SUKMANSKAYA G. V., Zh. Prikl. Khim., **28,** No. 2, 209 (1955)	3282
141.	BOGOYAVLENSKII P. S., MANANNIKOVA A. S., Zh. Prikl. Khim., **28,** No. 3, 327 (1955)	2914
142.	BOGOYAVLENSKII P. S., Zh. Fiz. Khim., **32,** 2037 (1958)	2122, 2160, 4500, 4580
143.	BOGOYAVLENSKII P. S., Zh. Strukt. Khim., **1,** No. 3, 305, (1960)	2162, 4584
144.	BOGOYAVLENSKII P. S., WAN LI-HUA, Zh. Strukt. Khim., **1,** 330 (1960)	4497, 4530

Reference number	Author's name and reference	Table number
145.	BOGOYAVLENSKII P. S., SUI KE-MIN', Zh. Strukt. Khim., **1,** No. 4, 425 (1960)	4502, 4503
146.	BOGOYAVLENSKII P. S., SUI KE-MIN', Zh. Neorgan. Khim., **6,** No. 2, 369 (1961)	4504
147.	BOGOYAVLENSKII P. S., MANANNIKOVA A. S., Zh. Neorgan. Khim., **6,** No. 4, 977 (1961)	2576
148.	BODALEVA N. V., LEPESHKOV I. N., Zh. Neorgan. Khim., **1,** No. 5, 995 (1956)	4074, 4079, 4087, 4729
149.	BODALEVA N. V., HU KE-YUAN', Zh. Neorgan. Khim., **4,** No. 12, 1816 (1959)	1354
150.	BOL'SHAKOV K. A., DERGUNOVA G. M., Zh. Neorgan. Khim., **5,** vyp 1, 209 (1960)	3139
151.	BOL'SHAKOV, P. E., Tr. Gos. Nauch.-Issled. Proekt. Inst. Azotn. Prom. Prod. Org. Sin., vyp. 4, 12 (1954)	14
152.	BORINA A. F., RAVICH M. I., Zh. Neorgan. Khim., **9,** No. 4, 975 (1964)	1434
153.	BOROVAYA F. E., RAVICH M. I., Zh. Neorgan. Khim., **4,** No. 10, 2335 (1959)	2696
154.	BORODULINA E. K., Zh. Neorgan. Khim., **4,** No. 10, 2390 (1959)	5068, 5070
155.	BORYACHEK A. F., DROZIN N. N., ZUBAKHINA Z. K., KUTSYNA M. I., Zh. Prikl. Khim., **28,** No. 1, 102 (1955)	2968
156.	BORYACHEK A. F., DROZIN N. N., ZUBAKHINA Z. K., KUTSYNA M. I., Zh. Neorgan. Khim., **2,** No. 7, 1655 (1957)	2862
157.	BOSIK I. I., VOROB'EVA O. I., NOVOSELOVA A. V., Zh.Neorgan. Khim., **5,** No. 5, 1157 (1960)	1340
158.	BRUILE E. S., DOMBROVSKAYA N. S., Zh. Neorgan. Khim., **4,** No. 9, 2091 (1959)	2794
159.	BRUTSKUS E. B., Symposium: "Investigations in applied chemistry" (Issledovaniya po prikladnoi khimii), Izd. Akad. Nauk SSSR, M.-L., 184 (1955)	4207
160.	BRYKOVA N. A., PROTSENKO P. I., Zh. Fiz. Khim., **39,** No. 3, 638 (1965)	5353, 5458
161.	BUKSHTEIN V. M., DYMSHITS R. A., Tr. Vses. Nauch.-Issled. Inst. Gal., (1952), cited in [1].	3289
162.	BUSLAEV YU. A., NIKOLAEV N. S., Zh. Neorgan. Khim., **4,** No. 2, 465 (1959)	624
163.	BUSLAEV YU. A., NIKOLAEV N. S., Dokl. Akad. Nauk SSSR, **135,** No. 6, 1385 (1960)	622, 623
164.	BUSLAEV YU. A., NIKOLAEV N. S., GUSTYAKOVA M. P., Izv. Sib. Otd. Akad. Nauk SSSR, No. 10, 57 (1960)	571
165.	BUSLAEV YU. A., NIKOLAEV N. S., TANANAEV I. V., Dokl. Akad. Nauk SSSR, **148,** No. 4, 832 (1963)	628
166.	BUSLAEV YU. A., GUSTYAKOVA M. P., Zh. Neorgan. Khim., **10,** No. 3, 622 (1965)	627
167.	BUSLAEV YU. A., BOCHKAREVA V. A., KREMER S. M., Zh. Neorgan. Khim., **10,** No. 3, 727 (1965)	4386
168.	BUTUZOV V. P., BRATOV L. V., Kristallografiya, **2,** No. 5, 670 (1957)	3004
169.	BYZOVA E. A., Uch. Zap. Leningrad. Gos. Pedagog. Inst. im. A. I. Gertsena, **6,** No. 3, 154 (1937)	4465, 4476
170.	BYZOVA E. A., Zh. Neorgan. Khim., **8,** 1951 (1963)	4471
171.	VAL'DMAN A., KLYACHKO-GURVICH L.; Zh. Obshch. Khim. **5,** 791 (1935)	467

Reference number	Author's name and reference	Table number
172.	VALYASHKO M. G., PETROVA V. M., Tr. Vses. Nauch.-Issled. Inst. Gal., vyp. 23, 162 (1952)	2965, 2969 2984
173.	VANUKOVA L. V., ISAEVA M. M., KABANOV B. N., Dokl. Akad. Nauk SSSR, **143,** No. 2, 377 (1962)	900
174.	VARASOVA E. N., MISHCHENKO K. P., FROST O. I., Zh. Obshch. Khim., **7,** No. 8, 1285 (1937)	5164
175.	VASILENKO N. A., Zh. Prikl. Khim., **21,** No. 9, 922 (1948)	2255
176.	VASILENKO N. A., Zh. Prikl. Khim., **22,** 338 (1949)	2252
177.	VASILENKO N. A., Zh. Prikl. Khim., **26,** No. 6, 650 (1953)	2253
178.	VASILENKO N. A., Nauch.-Tekh. Inform. Byull. Inst. Udobr. Insektofungits., Nos. 5–6, 105 (1967)	2254
179.	VASILENKO N. A., CHEPELEVETSKII M. L., Zh. Neorgan. Khim., **2,** No. 10, 2486 (1957)	4306
180.	VASIL'EV B. B., Tr. Gos. Inst. Prikl. Khim., vyp. 23, 111 (1935)	3012
181.	VASIL'EV G. I., KAZANTSEVA S. P., SEREBRENNIKOV V. V., Tr. Tomsk. Gos. Univ., **154,** Ser. Khim., 110 (1962)	523
182.	VISYAGIN N. I., Byull Inst. Galurgii, Nos. 10–11, 23–26 (1939)	3272, 3288, 3385
183.	VLASOV N. A., BERGMAN A. G., Dokl. Akad. Nauk SSSR, **38,** 211 (1943)	3560, 3911
184.	VLASOV N. A., BERGMAN A. G., Dokl. Akad. Nauk SSSR, **39,** No. 4, 148 (1943); Izv. Sektora Fiz.-Khim. Analiza Akad. Nauk. SSSR, **17,** 323 (1949)	3912
185.	VLASOV N. A., OSHENKO S. V., Izv. Fiz.-Khim. Nauch.-Issled. Inst. Irkutsk. Gos. Univ., **4,** No. 1, 62 (1959)	4860
186.	VOLKOV G. I., GRINEVICH G. I., Zh. Neorgan. Khim., **3,** No. 8, 1968 (1958)	1297
187.	VOLKOV P. A., Izv. Inst. Fiz.-Khim. Analiza, **3,** 714 (1927)	5518
188.	VOLKHONSKII E., Zh. Russ. Fiz.-Khim. Obshchest., Chast. Khim., **41,** 1766 (1909)	876
189.	VOL'NOV I. I., Izv. Sektora Fiz.-Khim. Analiza, Inst. Obshch. Neorgan. Khim., Akad. Nauk SSSR, **24,** 277 (1954)	5069
190.	VOL'NOV I. I., Izv. Sektora Fiz.-Khim. Analiza, Inst.. Obshch. Neorgan. Khim., Akad. Nauk SSSR, **26,** 211 (1955)	234
191.	VOL'NOV I. I., LATYSHEVA N. E., Zh. Prikl. Khim., **30,** No. 7, 978 (1957)	5066
192.	VOL'NOV I. I., KOLOTOVA E. I., Zh. Neorgan. Khim., **4,** No. 9, 2142 (1959)	249
193.	VOL'NOV I. I., SHATUNINA A. N., Izv. Akad. Nauk SSSR, Otd. Khim. Nauk, No. 2, 201 (1963)	241
194.	VOL'F F. F., YATLOV V. S., Zh. Prikl. Khim., **5,** vyp. 16, 839 (1928)	3684
195.	VOL'FKOVICH S. I., BELOPOL'SKII A. P., LEBEDEV B. A., Zh. Prikl. Khim., **4,** Nos. 2–3, 190; No. 5, 699 (1931); Tr. Nauch.-Issled. Inst. Udobr. Insektofungits., vyp. 144, 29 (1940)	2078, 2081
196.	VOROB'EVA O. I., OSANOVA L. P., Zh. Obshch. Khim., **23,** No. 8, 1288 (1953)	1521
197.	VOROB'EVA O. I., NOVOSELOVA A. V., ZHASMIN A. G., SIMANOV YU. P., Zh. Neorgan. Khim., **1,** No. 3, 516 (1956)	1506
198.	VOROB'EVA O. I., VLADIMIROVA Z. A., Zh. Neorgan. Khim., **2,** No. 9, 2221 (1957)	502

Reference number	Author's name and reference	Table number
199.	VOSKRESENSKAYA N. K., Zh. Prikl. Khim., **3**, vyp. 3, 331, 334 (1930)	3262, 3297, 4022
200.	VOSKRESENSKAYA N. K., Zh. Obshch. Khim., **4**, 153 (1934)	2479
201.	VOSKRESENSKAYA N. K., YANAT'EVA O. K., Izv. Sektora Fiz.-Khim. Analiza, Akad. Nauk SSSR, 9, 292 (1936)	1408
202.	VOSKRESENSKAYA N. K., YANAT'EVA O. K., Izv. Akad. Nauk SSSR, Ser. Khim., No. 1, 97 (1937)	1419
203.	VOSKRESENSKAYA N. K., BERUL' S. I., Izv. Sektora Fiz.-Khim. Analiza, Inst. Obshch. Neorgan. Khim., Akad. Nauk SSSR, **25**, 314 (1854)	3048
204.	GAVRISH M. L., GALINKER I. S., Zap. Khar'kov. Sel'skokhoz. Inst., **14** (51), 29 (1957)	3390
205.	GALKIN N. P., STEPANOV M. A., At. Energ., **8**, No. 3, 258 (1960)	2803
206.	GALKIN I. P., MAIOROV A. A., POLONNIKOVA G. A., SHCHERBAKOVA V. G., UTKINA L. V., At. Energ., **10**, No. 3, 233 (1961)	2100, 2116
207.	GALKIN N. P., POLONNIKOVA G. A., At. Energ., **10**, No. 3, 277 (1961)	2262
208.	GALUSHKINA R. A., BERGMAN A. G., Zh. Neorgan. Khim., **7**, No. 9, 2251 (1962)	4544
209.	GALUSHKINA R. A., BERGMAN A. G., Zh. Neorgan. Khim., **8**, 2576 (1963)	2182
210.	GEVORKYAN S. V., GUROVICH N. A., Izv. Akad. Nauk Arm. SSR, **10**, No. 6, 387 (1957)	2611
211.	GERASIMOV YA. I., Zh. Obshch. Khim., **4**, 723 (1934)	4970
212.	GITTERMAN K. E., VIL'NER M. M., Tr. Solyanoi Lab., Akad. Nauk SSSR, vyp. 15, 15–16 (1937), cited in [1]	5084
213.	GOLOVATYI R. N., Zh. Prikl. Khim., **13**, vyp. 4, 586 (1940)	439, 440, 4393, 4394, 4420, 4536, 4552, 4586, 4632, 4636, 4662, 4727, 4780, 5003, 5026, 5103
214.	GOLOSHCHAPOV M. V., Zh. Prikl. Khim. **26**, No. 3, 304 (1953)	328
215.	GOLOSHCHAPOV M. V., Izv. Voronezh. Gos. Pedagog. Inst., **16**, 19 (1955)	3999
216.	GOLOSHCHAPOV M. V., Zh. Neorgan. Khim., **1**, No. 11, 2633 (1956)	4288
217.	GOLOSHCHAPOV M. V., Zh. Prikl. Khim., **34**, No. 1, 138 (1961)	2108
218.	GOLUB A. M., KILIMNIK G. M., Zh. Prikl. Khim., **29**, No. 1, 17 (1956)	5166
219.	GOLUB A. M., Zh. Neorgan. Khim., **4**, No. 7, 1577 (1959)	4520
220.	GOLUTVIN YU. M., MALYSHEVA T. V., SKOROBOGATOVA V. I., Izv. Sib. Otd. Akad. Nauk SSSR, No. 8, 83 (1958)	694
221.	GONCHAROV P. I. Zh. Russ. Fiz.-Khim. Obshchest., Chast. Khim., **61**, No. 9, 1522 (1929)	927, 1001, 1018, 1420
222.	GORELOV I. P., SEREBRENNIKOVA V. V., Tr. Tomsk. Gos. Univ., **157**, 73 (1963)	210

Reference number	Author's name and reference	Table number
223.	GOROSHCHENKO YA. G., Zh. Neorgan. Khim., **1**, No. 5, 903 (1956)	2039, 2040 4357
224.	GOROSHCHENKO YA. G., SPASIBENKO T. P., Zh. Neorgan. Khim., **7**, No. 5, 1160 (1962)	1078
225.	GOROSHCHENKO YA. G., ANDREEVA M. I., Zh. Neorgan. Khim., **8**, 982 (1963)	4358
226.	GOROSHCHENKO YA. G., SIKORSKAYA E. K., Zh. Neorgan. Khim., **10**, No. 4, 950 (1965)	4337
227.	GOROSHCHENKO YA. G., ANDREEVA M. I., Zh. Neorgan. Khim., **10**, No. 4, 955 (1965)	4360
228.	GORSHTEIN G. I., SILANT'EVA N. I., Zh. Obshch. Khim., **23**, 1290 (1953)	5217, 5218
229.	GORSHTEIN G. I., SILANT'EVA N. I., Zh. Obshch. Khim., **24**, 29 (1954)	2406, 2407, 2414, 5234, 5235
230.	GORSHTEIN G. I., BASHKINA N. F., Tr. Vses. Nauch.-Issled. Inst. Khim. Reaktivov, vyp. 22, 8 (1958)	2420
231.	GORSHTEIN G. I., SILANT'EVA N. I., Tr. Vses. Nauch.-Issled. Inst. Khim. Reaktivov, vyp. 22, 12 (1958)	2416, 5221
232.	GRIGOROVICH Z. I., SAMOILOV O. YA., Radiokhimiya, **5**, No. 6, 686 (1963)	5132
233.	GRINBERG A. A., DOBROBORSKAYA A. I., Zh. Neorgan. Khim., **9**, No. 8, 2026 (1964)	4989
234.	GROMOV B. V., Zh. Prikl. Khim., **13**, No. 3, 345 (1940)	3273, 3341, 3853
235.	GROMOV B. V., Zh. Prikl. Khim., **21**, 261 (1948)	4202, 5213 5233, 5264 5265, 5430
236.	GROMOVA E. T., Zh. Neorgan. Khim., **5**, No. 11, 2575 (1960)	3338, 3458 5120
237.	GRUSHVITSKI V. E., FLERINSKAYA E. M., Tr. Vses. Nauch.-Issled. Inst. Gal., (1932), cited in [1]	2815, 2816 2818, 2828, 2830, 2835, 3265
238.	GRUSHVITSKI V. E., SHMIDT N. E., Tr. Vses. Nauch.-Issled. Inst. Gal., (1936), cited in [1]	358
239.	GRUSHVITSKI V. E., SHMIDT N. E., Bol'shaya Emba, Izd. Akad. Nauk SSSR, vol. I, pp. 673–678 (1937)	291, 293–295, 985, 1019
240.	GRUSHVITSKI V. E., GLYADINSKAYA E. P., CHELISHCHEVA A. G., Bol'shaya Emba, Izd. Akad. Nauk SSSR, vol. I, p. 685 (1937)	279, 283, 329, 769
241.	GUZAIROV R. S., LEITSIN V. A., GREKOV S. D., Zh. Neorgan. Khim., **9**, No. 1, 20 (1964)	868
242.	DEINEKA O. I., Tr. Vses. Nauch.-Issled. Inst. Gal., (1936), cited in [1]	1621, 2900
243.	DEICHMAN E. N., Zh. Neorgan. Khim., **6**, No. 8, 1832 (1961)	5450
244.	DEICHMAN E. N., RODICHEVA G. V., Zh. Neorgan. Khim., **6**, No. 9, 2183 (1961)	5335
245.	DEICHMAN E. N., RODICHEVA G. V., BRITSYNA E. A., Zh. Neorgan. Khim., **7**, No. 4, 879 (1962)	870
246.	DEICHMAN E. N., RODICHEVA G. V., Zh. Neorgan. Khim., **7**, No. 8, 1994 (1962)	4758

Reference number	Author's name and reference	Table number
247.	DEICHMAN E. N., TARARUSHKINA L. S., Zh. Neorgan. Khim., 7, No. 10, 2331 (1962)	2364
248.	DEICHMAN E. N., TANANAEV I. V., Zh. Anal. Khim., **17**, No. 1, 134 (1962)	511
249.	DEICHMAN E. N., KRYSINA L. S., Zh. Neorgan. Khim., **10**, No. 2, 476 (1965)	5449
250.	DENISOV P. V., BELOVA O. I., Tr. Inst. Khim., Kirg. Filial Akad. Nauk SSSR, No. 5, 85 (1953)	2880, 3189
251.	DENISOV P. V., BLESHINSKII S. V., Tr. Inst. Khim. Akad. Nauk Kirg. SSR, vyp. 5, 79 (1953)	2932, 2934
252.	DZHABAROV A. I., AGAEV A. I., Azerb. Khim. Zh., No. 4, 103 (1960)	3904
253.	DZHABAROV A. I., AGAEV A. I., Uch. Zap. Azerb. Gos. Univ., No. 2, 77 (1960)	3897, 3900, 3940
254.	DZHABAROV A. I., AGAEV A. I., Uch. Zap. Azerb. Gos. Univ., No. 2, 65 (1961)	3902
255.	DZHABAROV A. I., AGAEV A. I., Uch. Zap. Azerb. Gos. Univ., Ser. Fiz.-Mat. Khim. Nauk, Nos. 3–4, 65 (1961)	3903
256.	DZHABAROV A. I., AGAEV A. I., Uch. Zap. Azerb. Gos. Univ., Ser. Fiz.-Mat. Khim. Nauk, No. 1, 75 (1962)	4974
257.	DZHABAROV A. I., AGAEV A. I., Azerb. Khim. Zh., No. 6, 82 (1962)	3901
258.	DZHABAROV A. I., AGAEV A. I., Uch. Zap. Azerb. Gos. Univ., Ser. Fiz.-Mat. Khim. Nauk, No. 1, 75 (1962)	3898, 3914 3939, 3942
259.	DISTANOV G. K., Nauch. Zap. Uzhgorod. Gos. Univ., 7, 129 (1953)	1894, 1951
260.	DISTANOV G. K., Uch. Zap. Biol.-Pochv. Fak., Kirg. Gos. Univ., vyp. 7, 163 (1958)	855, 2001, 2010, 2016
261.	DMITRIEVSKII G. E., ALEKSANDROVA L. I., POZIGUN L. I., Nauch. Ezheg. Odess. Univ., vyp. 2, Khim. Fak., 12 (1962)	4896
262.	DOBRYNINA T. A., Izv. Akad. Nauk SSSR, Otd. Khim. Nauk, No. 6, 961–965 (1960)	231
263.	DOBRYNINA T. A., DZYATKEVICH B. S., Izv. Akad. Nauk SSSR, Ser. Khim., No. 5, 790 (1964)	268, 271
264.	DOLININA R. M., PEREL'MAN F. M., Uch. Zap. Azerb. Gos. Univ., Ser. Khim. Nauk, No. 2, 29 (1963)	1487
265.	DRABKINA L. E., Radiokhimiya, **2**, No. 3, 377 (1960)	2110, 2111, 2121
266.	DRUZHININ I. G., Izv. Akad. Nauk SSSR, Ser. Khim., 1141 (1938)	2950
267.	DRUZHININ I. G., YANKO A. P., Dokl. Akad. Nauk SSSR, **94**, No. 3, 481 (1954)	1331
268.	DRUZHININ I. G., YANKO A. P., Izv. Kirg. Filiala Akad. Nauk SSSR, No. 1 (11), 63 (1954)	1376
269.	DRUZHININ I. G., YANKO A. P., Tr. Inst. Khim., Akad. Nauk Kirg. SSR, No. 6, 3 (1955)	1332, 4686
270.	DRUZHININ I. G., SHEPELEV A. I., Tr. Inst. Khim., Akad. Nauk Kirg. SSR, No. 7, 3 (1956)	969, 978, 1020, 3775
271.	DRUZHININ I. G., SHEVCHUK V. G., Uch. Zap. Yaroslav. Gos. Pedagog. Inst., 31 (41), 325 (1957)	3487

Reference number	Author's name and reference	Table number
272.	DRUZHININ I. G., MAKIN A. V., Izv. Akad. Nauk Kirg. SSR, Ser. Estestv. Tekh. Nauk, **2**, No. 3, 19 (1960)	3199
273.	DRUZHININ I. G., KUZNETSOV V. G., IMANAKUNOV B., Izv. Akad. Nauk Kirg. SSR, Ser. Estestv. Tekh. Nauk, **2**, No. 3, 25 (1960)	3409, 3410, 3475, 4252
274.	DRUZHININ I. G., AKBAEV A., Izv. Akad. Nauk Kirg. SSR, Ser. Estestv. Tekh. Nauk, **2**, No. 11, 5 (1960)	846, 1049, 5272
275.	DRUZHININ I. G., KYDYNOV M., Izv. Akad. Nauk Kirg. SSR, Ser. Estestv. Tekh. Nauk, **2**, No. 11, 21 (1960)	5133, 5277
276.	DRUZHININ I. G., KALBAEV A., Izv. Akad. Nauk Kirg. SSSR, Ser. Estestv. Tekh. Nauk, **2**, No. 11, 165 (1960)	2113
277.	DRUZHININ I. G., AKBAEV A., Dokl. Akad. Nauk SSSR, **137**, No. 6, 1364 (1961)	2987
278.	DRUZHININ I. G., GOROKHOVA V. N., Izv. Vyssh. Ucheb. Zaved., Khim. Tekhnol., No. 5, 765 (1961)	3781, 5203
279.	DRUZHININ I. G., KOSYAKINA O. A., Zh. Neorgan. Khim., **6**, No. 7, 1703 (1961)	3481, 5203 5280
280.	DRUZHININ I. G., KOSYAKINA O. A., Zh. Neorgan. Khim., **6**, No. 9, 1711 (1961)	3340
281.	DRUZHININ I. G., IMANAKUNOV B., KUZNETSOV V. G., Zh. Neorgan. Khim., **6**, No. 11, 2583 (1961)	3411
282.	DRUZHININ I. G., AKBAEV A., Zh. Prikl. Khim., **37**, No. 6, 1194 (1963)	3348, 5273
283.	DUBNIKOVA K. T., SAMOILOV O. YA., Radiokhimiya, **5**, No. 6, 683 (1963)	1430
284.	DUBOVAYA V. K., NABIEV M. N., Uzb. Khim. Zh., No. 5, 6 (1959)	2189
285.	DUKEL'SKII M. P., Zh. Russ. Fiz.-Khim. Obshchest., **39**, 980 (1907)	1271
286.	ELISEEV A. G., Izv. Inst. Fiz.-Khim. Anal., Akad. Nauk SSSR, **3**, No. 1, 447 (1926)	1116
287.	ZHDANOV A. K., Zh. Obshch. Khim., **17**, No. 12, 2213 (1947)	2444, 2446
288.	ZHDANOV A. K., ADAMENKOVA V., Zh. Obshch. Khim., **18**, 251 (1948)	3570
289.	ZHDANOV A. K., Zh. Obshch. Khim., **18**, 554 (1948)	4820, 4823
290.	ZHDANOV A. K., KOVALENKO N., Zh. Obshch. Khim., **18**, 559 (1948)	4831
291.	ZHDANOV A. K., YAKOVLEV V. V., Uzb. Khim. Zh., No. 2, 5 (1958)	3156, 3848 3922
292.	ZHDANOV A. K., Uzb. Khim. Zh., No. 5, 39 (1959)	3561, 4821
293.	ZHDANOV A. K., Dokl. Akad. Nauk Uzb. SSR, No. 8, 40 (1959)	2208
294.	ZHDANOV A. K., Dokl. Akad. Nauk Uzb. SSR, No. 11, 45 (1959)	5506
295.	ZHEMCHUZHNYI S. F., RONKIN B. L., S.T.E., **7**, 215 (1931)	3588, 4023
296.	ZHUKOV A. I., SHUTOVA V. I., Tr. Ural. Nauch.-Issled. Khim. Inst., No. 1, 29 (1954)	3287
297.	ZHURAVLEV E. F., BYCHKOVA M. N., Zh. Neorgan. Khim., **4**, No. 10, 2367 (1959)	2123–2125, 2457, 3029, 3030
298.	ZHURAVLEV E. F., SHEVELEVA A. D., BOGDANOVSKAYA R. L., KUDRYASHOV S. F., SHUROV V. A., Zh. Neorgan. Khim., **8**, 1961 (1963)	3145, 4623, 5331, 5455

Reference number	Author's name and reference	Table number
299.	ZHURAVLEV E. F., KUDRYASHOV S. F., Zh. Neorgan. Khim., **9**, No. 8, 1996 (1964)	2440, 2441, 2494, 2565, 4807
300.	ZASLAVSKII A. I., ETTINGER I. L., (1933), cited in [1]	4225
301.	ZASLAVSKII A. I., Tr. Gos. Inst. Prikl. Khim., vyp. 23, 72 (1935)	4851
302.	ZASLAVSKII A. I., ETTINGER I. L., Zh. Obshch. Khim., **7**, 1947 (1937)	3111
303.	ZASLAVSKII A. I., SINANI S. S., SOKOLOVA L. A., Izv. Akad. Nauk SSSR, Ser. Khim., No. 1, 50–57 (1938)	3269, 3320
304.	ZASLAVSKII A. I., RAVDIN YA. A., Zh. Obshch. Khim., **9**, 1473 (1939)	4229
305.	ZASLAVSKII A. I., LUK'YANOVA E. I., Tr. Gos. Inst. Prikl. Khim., vyp. 34, 9 (1940)	3697, 3698
306.	ZAKHAROV-NARTSISSOV O. I., MIKHAILOV G. G., Izv. Vyssh. Ucheb. Zaved., Khim. Khim. Tekhnol., **3**, No. 1, 45 (1960)	3878
307.	ZAKHARCHENKO G. A., TSITSORIN V. A., Zh. Prikl. Khim., **22**, 703 (1949)	826
308.	ZVORYKIN A. YA., KUZNETSOV V. G., Izv. Akad. Nauk SSSR, Ser. Khim. Nauk, 195 (1938)	2236
309.	ZVORYKIN A. YA., PEREL'MAN F. M., Khim. Redk. Elem., vyp. 1, 52 (1954)	2282
310.	ZVORYKIN A. YA., PEREL'MAN F. M., Khim. Redk. Elem., vp. 2, 64 (1955)	2283
311.	ZVORYKIN A. YA., PEREL'MAN F. M., Khim. Redk. Elem., vyp. 3, 100 (1957)	2993, 3566
312.	ZVORYKIN A. YA., VETKINA L. S., Zh. Neorgan. Khim., **6**, No. 11, 2572 (1961)	2239
313.	ZVORYKIN A. YA., RATNIKOVA V. D., Zh. Neorgan. Khim., **8**, 1018 (1963)	2240
314.	ZDANOVSKII A. B., Zh. Prikl. Khim., **20**, No. 12, 1252 (1947)	3772
315.	ZEL'VENSKII YA. D., Zh. Prikl. Khim., **12**, 1311 (1939)	42, 71
316.	ZIL'BERMAN YA. I., IVANOV P. T., Zh. Prikl. Khim., **14**, Nos. 7–8, 939 (1941)	2256, 2260 2263
317.	ZIL'BERMAN YA. I., IVANOV P. T., Zh. Obshch. Khim., **16**, 1589 (1946)	3522, 3523, 3762, 5139
318.	ZINOV'EV A. A., SHIROKOVA N. A., Zh. Neorgan. Khim., **5**, No. 6, 1299 (1960)	1170
319.	ZINOV'EV N. P., RACHKOVA L. P., Sb. Stud. Nauch. Rab. Yaroslav, Gos. Pedagog. Inst., vyp. 1, 35 (1958)	4572
320.	ZOLOTAREV L. L., Izv. Vyssh. Ucheb. Zaved., Tsvet. Met., **2**, No. 2, 107 (1958)	2442
321.	IVANOV-EMIN B. N., NISEL'SON L. A., GREKSA YU., Zh. Neorgan. Khim., **5**, No. 9, 1998 (1960)	2780
322.	IVANOV-EMIN B. N., NISEL'SON L. A., IVOLGINA A. P., Zh. Neorgan. Khim., **5**, No. 12, 2841 (1960)	2755
323.	IVANOV-EMIN B. N., NISEL'SON L. A., IVOLGINA A. P., Zh. Neorgan. Khim., **6**, No. 6, 1483 (1961)	2774
324.	IVANOV-EMIN B. N., RYBINA V. I., KORNEV V. I., Zh. Neorgan. Khim., **10**, No. 4, 1005 (1965)	2790

Reference number	Author's name and reference	Table number
325.	IVANOVA I. N., OZEROVA M. I., EGOROVA E. I., Zh. Neorgan. Khim., **8**, 977 (1963)	2390
326.	IVANOVA I. N., OZEROVA M. I., Symposium: "Physico-chemical analysis" (Fiziko-khimicheskii analiz), Sib. Otd. Akad. Nauk SSSR, Novosibirsk, 119 (1963)	2408
327.	IVANOVA I. N., OZEROVA M. I., Zh. Neorgan. Khim., **9**, No. 8, 1988 (1964)	2388
328.	IL'INSKII V. P., Zh. Russ. Fiz.-Khim. Obshchest., **34**, 29, 729 (1924)	2985, 2986, 3008, 3864, 4478
329.	IL'INSKII V. P., KLEBANOV G. S., Zh. Prikl. Khim., **5**, Nos. 1–4, 10 (1928)	3587
330.	IL'INSKII V. P., SAGAIDACHNYI A. F., Zh. Russ. Fiz.-Khim. Obshchest., Chast. Khim., **61**, vyp. 10, 1957–1070 (1929)	3259, 3295, 3379
331.	IL'INSKII V. P., SAGAIDACHNYI A. F., Zh. Obshch. Khim., **1**, vyp. 5, 584 (1931)	3264, 3589, 3622
332.	IL'INSKII V. P., SAGAIDACHNYI A. F., MORDBERG L. K., Tr. Vses. Nauch.-Issled. Inst. Gal., (1932), cited in [1]	972, 973, 3299, 3381, 3624, 4024
333.	IL'INSKII V. P., VARYPAEV N. A., GITTERMAN K. E., SHMIDT N. E., Tr. Solyanoi Lab., Akad. Nauk SSSR, vyp. 7, 8 (1936)	3708, 3767
334.	IL'INSKII V. P., GERMAN N. A., Tr. Vses. Nauch.-Issled. Inst. Gal., (1936), cited in [1]	3520b, 3748, 3760
335.	IL'INSKII V. P., OSTROUMOVA L. M., Tr. Vses. Nuach.-Issled. Inst. Gal., (1936), cited in [1]	5083, 5136
336.	IL'INSKII V. P., UVERSKAYA A. T., Tr. Gos. Inst. Prikl. Khim., vyp. 41, 112 (1958)	5232
337.	IMANAKUNOV B., Tr. Molodykh Nauch. Rab. Akad. Nauk Kirg. SSR, 69 (1958)	3474, 4251
338.	INYUTIN F. D., TARABAEV S. I., Izv. Akad. Nauk Kaz. SSR, Ser. Gorn. Dela, Stroimater. Met., vyp. 6, 89 (1956)	2796, 3025
339.	ITKINA L. S., Zh. Prikl. Khim., **22**, No. 3, 279 (1949)	2636, 2653, 2659, 2681, 2692
340.	ITKINA L. S., Zh. Prikl. Khim., **26**, No. 5, 496 (1953)	2637, 2654
341.	ITKINA L. S., KOKHOVA V. F., Izv. Sektora Fiz.-Khim. Anal., Inst. Obshch. Neorgan. Khim. Akad. Nauk SSSR, **23**, 284 (1953)	2649, 2655, 2693
342.	ITKINA L. S., KOKHOVA V. F., Izv. Sektora Fiz.-Khim. Anal., Inst. Obshch. Neorgan. Khim. Akad. Nauk SSSR, **26**, 242 (1954)	4467
343.	ITKINA L. S., KOKHOVA V. F., Izv. Sektora Fiz.-Khim. Anal., Inst. Obshch. Neorgan. Khim. Akad. Nauk SSSR, **26**, 242 (1955)	2953
344.	ITKINA L. S., KOKHOVA V. F., Izv. Sektora Fiz.-Khim. Anal., Inst. Obshch. Neorgan. Khim. Akad. Nauk SSSR, **27**, 335 (1956)	3433
345.	ITKINA L. S., RASSONSKAYA I. S., CHAPLYGINA N. M., Zh. Neorgan. Khim., **3**, No. 7, 1675 (1958)	646
346.	ITKINA L. S., KOKHOVA V. F., Zh. Neorgan. Khim., **5**, No. 6, 1290 (1960)	2641, 2651, 2685
347.	ITKINA L. S., KOKHOVA V. F., Zh. Neorgan. Khim., **5**, No. 9, 2102 (1960)	2687, 2696
348.	ITKINA L. S., Zh. Neorgan. Khim., **7**, No. 1, 186 (1962)	2688, 2697

Reference number	Author's name and reference	Table number
349.	ITKINA L. S., CHAPLYGINA N. M., Zh. Neorgan. Khim., **7,** 2795 (1962)	1306
350.	ITKINA L. S., CHAPLYGINA N. M., Zh. Neorgan. Khim., **8,** 1481 (1963)	1279, 2644
351.	ISHKIN I. P., BARBO P. Z., Zh. Fiz. Khim., **14,** 1337 (1939)	12
352.	IORDANOV N. PAVLOVA M., God. Sofii. Univ., Fiz.-Mat. Fak., **52,** No. 3, 69, 1957–1958 (1959)	74, 76, 110
353.	KAZANTSEV A. A., Izv. Vses. Nauch.-Issled. Inst. Khim. Reaktivov Osobo Chist. Khim. Veschchestv. No. 2, 10 (1923)	399, 429, 463, 472, 489
354.	KAMAEVA I. G., SEREBRENNIKOV V. V., Tr. Tomsk. Gos. Univ., **157,** 131 (1963)	5156
355.	KAPUSTINSKII A. F., ANVAER B. I., Dokl. Akad. Nauk SSSR, **30,** 625 (1941)	689–691
356.	KARNAUKHOV A. S., DRUZHININ I. G., Dokl. Akad. Nauk SSSR, **84,** No. 5, 963 (1952)	2554
357.	KARNAUKHOV A. S., Izv. Sektora Fiz.-Chem. Anal., Inst. Obshch. Neorgan. Khim., Akad. Nauk SSSR, **25,** 334 (1954)	2154, 2156, 2170, 2476, 3085
358.	KARNAUKHOV A. S., KAROV Z. G., Uch. Zap. Kabardino-Balkar. Gos. Pedagog. Inst., vyp. 8, 61 (1955)	2278
359.	KARNAUKHOV A. S., Uch. Zap. Yaroslav. Gos. Pedagog. Inst., 31 (41), 255 (1956)	2179, 3086, 3088, 3113, 3876, 4574
360.	KARNAUKHOV A. S., Zh. Obshch. Khim., **26,** No. 4, 1027 (1956)	2171, 2192
361.	KARNAUKHOV A. S., Zh. Obshch. Khim., **26,** No. 5, 1281 (1956)	1530, 1532, 1534, 1537
362.	KARNAUKHOV A. S., MAKIN A. V., Zh. Neorgan. Khim., **2,** No. 4, 915 (1957)	3877a
363.	KARNAUKHOV A. S., Zh. Neorgan. Khim., **2,** No. 4, 915 (1957)	2155, 2157, 2180, 2556, 2574
364.	KARNAUKHOV A. S., Izv. Vyssh. Ucheb. Zaved., Khim. Khim. Tekhnol., No. 3, 34 (1958)	2434, 3550, 4797
365.	KARNAUKHOV A. S., MAKIN A. V., Uch. Zap. Yaroslav. Gos. Pedagog. Inst., **32** (42), 309 (1950)	2434, 4799
366.	KAROV Z. G., PEREL'MAN F. M., Zh. Neorgan. Khim., **4,** No. 4, 936 (1959)	4825
367.	KAROV Z. G., PEREL'MAN F. M., Zh. Neorgan. Khim., **5,** No. 3, 713 (1960)	1851, 1991, 2445
368.	KAROV Z. G., PEREL'MAN F. M., VASECHKO R. F., Uch. Zap. Kabardino-Balkar. Gos. Univ., vyp. 10, 237 (1961)	4712
369.	KAROV Z. G., PEREL'MAN F. M., Uch. Zap. Kabardino-Balkar. Gos. Univ., vyp. 12, 261 (1961)	3095
370.	KAROV Z. G., PEREL'MAN F. M., Zh. Neorgan. Khim., **7,** No. 10, 1452 (1962)	2161
371.	KAROV Z. G., PEREL'MAN F. M., ROGOZHIN G. N., Zh. Neorgan. Khim., **10,** No. 5, 1233 (1965)	3099
372.	KASYNOVA M. A., NABIEV M. N., Uzb. Khim. Zh., No. 1, 37 (1964)	1800

Reference number	Author's name and reference	Table number
373.	KINDYAKOV P. S., PESINA A. YA., Tr. Mosk. Inst. Tonkoi Khim. Tekhnol., vyp. 7, 10 (1958)	1292
374.	KINDYAKOV P. S., GASILINA N. K., Tr. Mosk. Inst. Tonkoi Khim. Tekhnol., vyp 7, 16 (1958)	1304
375.	KINDYAKOV P. S., KYSHTYMOVA E. A., RONOVA S. T., Tr. Mosk. Inst. Tonkoi Khim. Tekhnol., vyp. 7, 22 (1958)	1411
376.	KINDYAKOV P. S., KHOKHLOVA A. V., LAPCHINSKAYA L. L., Tr. Mosk. Inst. Tonkoi Khim. Tekhnol., vyp 7, 32 (1958)	1278, 1283, 1315, 5057
377.	KINDYAKOV P. S., KURTOVA L. V., Izv. Vyssh. Ucheb. Zaved., Khim. Khim. Tekhnol., **3**, No. 6, 967 (1960)	1305, 1355
378.	KINDYAKOV P. S., KURTOVA L. V., APRAKSINA G. Z., Zh. Neorgan. Khim., **6**, No. 12, 2763 (1961)	1294
379.	KIRAKOSYAN A. K., GALUSTYAN V. D., Zh. Neorgan. Khim., **3**, No. 8, 1925 (1958)	1818, 1969
380.	KIRAKOSYAN A. K., Zh. Neorgan. Khim., **4**, No. 12, 2777 (1959)	2024
381.	KIRAKOSYAN A. K., Zh. Neorgan. Khim., **5**, No. 1, 214 (1960)	2026
382.	KIRAKOSYAN A. K., Zh. Neorgan. Khim., **5**, No. 8, 1806 (1960)	2027
383.	KIRGINTSEV A. I., KASHINA N. I., VULIKH A. I., KOROTKEVICH B. I., Zh. Neorgan. Khim., **10**, No. 5, 1225 (1965)	4942, 4943, 5340
384.	KIREEV V. A., VAGRANSKAYA L. I., Zh. Obshch. Khim., **5**, 963 (1935)	2661
385.	KIRILLOV I. P., USTINOVA Z. A., Tr. Ivanov. Khim.-Tekhnol. Inst., vyp. 5, 59 (1956)	464
386.	KLEBANOV G. S., BASOVA E. P., Zh. Prikl. Khim., **12**, 1601 (1939)	2918–2920, 2990–2992, 3894
387.	KLEBANOV G. S., PINCHUK G. YA., Zh. Prikl. Khim., 37, No. 2, 289 (1964)	1244, 1245, 1247, 3943
388.	KLEVKE V. A., KIL'MAN YA. I., KANTOR A. S., Tr. Gos. Nauch.-Issled. Proekt. Inst. Azotn. Prom. Prod. Org. Sin., No. 9, 183 (1959)	1813, 1816
389.	KLEINER K. E., Zh. Prikl. Khim., **17**, Nos. 7–8, 409 (1944)	953, 954
390.	KLEINER K. E., GRIDCHINA G. I., Zh. Neorgan. Khim., **4**, No. 9, 2020 (1959)	388, 1242
391.	KLEMENT'EV V., Tr. Vses. Alyumin.-Magniev. Inst., **14**, 5 (1937)	3631
392.	KLYGIN A. E., KOLYADA N. S., Zh. Neorgan. Khim., **4**, No. 1, 239 (1959)	2380
393.	KLYGIN A. E., KOLYADA N. S., Zh. Neorgan. Khim., **5**, No. 5, 1170 (1960)	3533
394.	KLYACHKO-GURVICH L. L., GANDGANOV N. O., Zh. Obshch. Khim., **5**, 1781 (1935)	836
395.	KLYASHTORNYI M. I., OZHEREL'EV D. I., Zh. Prikl. Khim., **35**, No. 3, 679 (1962)	2642
396.	KNASTER M. B., APEL'BAUM L. A., Zh. Fiz. Khim., **38**, No. 1, 223 (1954)	57, 98
397.	KOLOSOV A. S., Tr. Khim.-Met.Inst., Zapad.-Sib. Filial, Akad. Nauk SSSR, No. 12, 29 (1958)	4089
398.	KOLOSOV A. S., Izv. Sib. Otd. Akad. Nauk SSSR, No. 3, 67 (1959)	3392, 3405, 4090
399.	KOLOSOV A. S., Izv. Sib. Otd. Akad. Nauk SSSR, No. 12, 64 (1962)	3335

Reference number	Author's name and reference	Table number
400.	KOMISSAROVA L. N., SHATSKII V. M., MOZHEICHENKO G. I., Zh. Neorgan. Khim., **10**, No. 4, 755 (1965)	807
401.	KONOBRITSKII E. G., BEKTUROV A. B., Izv. Akad. Nauk Kaz. SSR, Ser. Khim., No. 2 (16), 10 (1960)	314, 320, 331
402.	KONOBRITSKII E. G., Tr. Inst. Khim. Nauk, Akad. Nauk Kaz. SSR, **10**, 145 (1964)	316, 322, 3395
403.	KONOVALOV D. P., Zh. Russ. Fiz.-Chim. Obshchest., **31**, 985 (1899)	1928, 1945, 1958
404.	KONOPKINA Z. I., Zh. Prikl. Khim., **29**, No. 1, 22 (1956)	4960
405.	KONOPKINA Z. I., Zh. Prikl. Khim., **31**, No. 9, 1312 (1958)	4959, 5144
406.	KORENMAN I. M., SHEYANOVA F. R., VISHNEVSKAYA T. N., BRATANOV B. I., Zh. Neorgan. Khim., **3**, No. 5, 1188 (1958)	3140, 3146, 3498, 3509, 3843, 4005, 5563
407.	KORENMAN. I. M., GANINA V. G., PYATNITSKAYA N. M., PARSHUTKIN YU. M., Tr. Khim. Khim. Tekhnol., (Gor'kii), vyp. 1, 81 (1960)	3998, 4010, 4226, 4231, 4550, 4551, 5295
408.	KOROVIN S. S., IVANOVA R. V., SAAKOVA O. V., BOL'SHAKOV K. A., Zh. Prikl. Khim., **34**, No. 5, 1007 (1961)	767
409.	KOROLEV V. F., Tr. Solyanoi Lab. Akad. Nauk SSSR, vyp. 15, 28 (1937), cited in [1]	3268, 3337, 3769
410.	KORF D. M., SHABROVSKAYA L. P., Byull. V.I.G., Nos. 4–5, 45 (1940), cited in [1]	709, 764, 966
411.	KORF D. M., SHABROVSKAYA L. P., Zh. Obshch. Khim., **10**, No. 13, 1231 (1940)	709, 764, 966, 3274
412.	KORF D. M., BALYASNAYA A. M., Zh. Prikl. Khim., **14**, Nos. 4–5, 475 (1941)	2938
413.	KORF D. M., ALYMOVA L. N., Zh. Neorgan. Khim., **7**, No. 3, 697 (1962)	2988
414.	KORF D. M. SUGAKEVICH I. P., Zh. Neorgan. Khim., **7**, No. 3, 699 (1962)	4298
415.	KORF D. M., FOMINA E. A., Zh. Neorgan. Khim., **8**, 1022 (1963)	5123
416.	KOSYAKINA O. A., Uch. Zap. Yaroslav. Gos. Pedagog. Inst., **31** (41, 235 (1957)	3794
417.	KOSYAKINA O. A., Uch. Zap. Yaroslav. Gos. Pedagog. Inst., Estestv., Geograf., vyp. 24, 5 (1958)	3339, 3480, 3795, 5279
418.	KOSYAKINA O. A., Dokl. Nauch. Konf., Yaroslav. Gos. Pedagog. Inst., **1**, No. 4, 21 (1962)	3797
419.	KRASNOV K. S., Zh. Prikl. Khim., **28**, No. 12, 1277 (1955)	4279
420.	KRAT V. N., Tr. Gos. Inst. Prikl. Khim., vyp. 23, 118 (1935)	2922
421.	KRYLOV E. I., CHUKHLANTSEV V. G., CHUNIN V. S., Izv. Vyssh. Ucheb. Zaved., Tsvet. Met., No. 3, 97 (1958)	1213
422.	KRYUKOVA T. A., Zh. Fiz. Khim., **13**, 693 (1939)	909
423.	KUZNETSOV A. M., Zh. Prikl. Khim., **19**, No. 12, 1335 (1946)	1013, 1015, 5117
424.	KUZNETSOV V. G., Izv. Akad. Nauk SSSR, Ser. Khim. Nauk, 385 (1937)	4030
425.	KUZNETSOV D. I., KOZHUKHOVSKII A. A., Zh. Prikl. Khim., **9**, 185 (1936)	1057
426.	KUZNETSOV D. I., KOZHUKHOVSKII A. A., Zh. Prikl. Khim., **9**, 1178 (1936)	1091

Reference number	Author's name and reference	Table number
427.	KUZNETSOV D. I., KOZHUKHOVSKII A. A., Borovaya F.E., Zh. Prikl. Khim., **21**, No. 12, 1278 (1948)	2133, 2237
428.	KUZNETSOVA G. P., PLYUSHCHEV V. E., OBOZNENKO YU. V., Izv. Vyssh. Ucheb. Zaved., Khim. Khim. Tekhnol., 7, No. 3, 355 (1964)	1390
429.	KUZNETSOVA G. P., STEPIN B. D., Zh. Neorgan. Khim., **10**, No. 2, 472 (1965)	5342
430.	KUL'BA F. YA., MIRONOV V. E., Zh. Neorgan. Khim., **2**, vyp. 12, 2734, 2741 (1957)	1490, 2586, 3953, 5021, 5345, 5467
431.	KUL'BA F. YA. MIRONOV V. E., Zh. Neorgan. Khim., **3**, vyp. 11, 2480 (1958)	4200, 5020, 5532
432.	KUL'BA F. YA., MIRONOV V. E., Zh. Neorgan. Khim., **5**, vyp. 2, 287 (1960)	1316, 3037, 4519, 5456
433.	KUPERMAN M. E., BELOPOL'SKII A. P., Symposium: "Studies in applied chemistry" (Issledovaniya po prikladnoi khimii), Izd. Akad. Nauk SSSR, M.-L., 225 (1955)	5201
434.	KURBATOV D. I., DEMENEV N. V., Zh. Prikl. Khim., **29**, No. 6, 944 (1956)	486, 860, 1082
435.	KURNAKOV N. S., ZHEMCHUZHNII S. F., Izv. Inst. Fiz.-Khim. Anal., **1**, No. 7, 199 (1919)	3249
436.	KURNAKOV N. S., ZHEMCHUZNII S. F., Zh. Russ. Fiz.-Khim. Obshchest., **51**, 1 (1919)	3294, 3249, 3372, 3616, 4018
437.	KURNAKOV N. S., GENKE T. A., Dokl. Ross. Akad. Nauk, Ser. A, p. 167 (1924)	961, 2708
438.	KURNAKOV I. S., MUROMTSEV B. A., Dokl. Akad. Nauk SSSR, Ser. A, No. 17, 263 (1927)	997, 998, 4400
439.	KURNAKOV I. S., MAZEL' A. V., Izv. Inst. Fiz.-Khim. Anal., Akad. Nauk SSSR, **4**, No. 2, 315 (1930)	2947
440.	KURNAKOV N. S., MAKAROV S. Z., Izv. Inst. Fiz.-Khim. Anal., Akad. Nauk SSSR, **4**, No. 2, 329 (1930)	2946
441.	KURNAKOV N. S., OPYKHTINA M. A., Izv. Inst. Fiz.-Khim. Anal., Akad. Nauk SSSR, **4**, No. 2, 368, 373 (1930)	3298
442.	KURNAKOV N. S., OSOKOREVA N. A., *et al.*, Kalii, No. 2, 26 (1932)	3626, 3646, 3688, 4129
443.	KURNAKOV N. S., OSOKOREVA N. A., *et al.*, Tr. Gos. Inst. Prikl. Khim., vyp. 16, 38 (1932)	3626, 3646, 3688
444.	KURNAKOV N. S., KUZNETSOV V. G., Izv. Inst. Fiz.-Khim. Anal., Akad. Nauk SSSR, **7**, 186 (1935)	4027, 4028
445.	KURNAKOV N. S., SHOIKHET D. N., Izv. Inst. Fiz.-Khim. Anal., Akad. Nauk SSSR, **7**, 247 (1935)	943, 1841, 1842
446.	KURNAKOV N. S., OSOKAREVA N. A., *et al.*, Solikamskii Karnallity, p. 59 (1935)	3626, 3646, 3688, 4129
447.	KURNAKOV N. S., ZASLAVSKII A. I., LUK'YANOVA E. I., Gos. Inst. Prikl. Khim., (1936), cited in [1]	3629, 3649, 3694, 4132
448.	KURNAKOV N. S., LUZHNAYA N. P., KUZNETSOV V. G., Izv. Akad. Nauk SSSR, Ser. Khim. Nauk, 557 (1937)	2518
449.	KURNAKOV N. S., LUK'YANOVA E. I., Izv. Akad. Nauk SSSR, Ser. Khim., No. 1, 41–42 (1938)	4036
450.	KURNAKOV N. S., ZASLAVSKII A. I., LUK'YANOVA E. I., Izv. Akad. Nauk SSSR, Ser. Khim., No. 1, 54 (1938)	3695, 3699

Reference number	Author's name and reference	Table number
451.	KURNAKOV N. S., NIKOLAEV A. V., Izv. Akad. Nauk SSSR, Ser. Khim., No. 2, 404 (1938)	3653, 4161
452.	KURNAKOV N. S., NIKITINA E. A., Izv. Akad. Nauk SSSR, Ser. Khim., 433 (1938)	1035
453.	KURNAKOV N. S., RAVICH M. I., TROITSKAYA N. V., Izv. Sektora Fiz.-Khim. Anal., Akad. Nauk SSSR, **10**, 298 (1938)	1552, 1553
454.	KURNAKOV N. S., SHOIKHET D. N., Izv. Sektora Fiz.-Khim. Anal., Akad. Nauk SSSR, **10**, 310 (1938)	4037, 4068, 4681
455.	KURNAKOV N. S., NIKITINA E. A., Zh. Obshch. Khim., **10**, 577 (1940)	3837
456.	KURNAKOV N. S., FILIPPEO V. M., Izv. Sektora Fiz.-Khim. Anal., Akad. Nauk SSSR, **13**, 325 (1940)	798
457.	KURNAKOVA A. G., Izv. Ser. Fiz.-Khim. Anal., Akad. Nauk SSSR, **16**, 126 (1947)	1562, 1578
458.	KURNAKOVA A. G., NIKOLAEV A. V., Izv. Otd. Khim. Nauk, Akad. Nauk SSSR, No. 4, 380 (1939)	1557
459.	KURNAKOVA A. G., Izv. Ser. Fiz.-Khim. Anal., Akad. Nauk SSSR, **18**, 222–227 (1949)	1557, 1558
460.	KURNAKOVA A. G., NIKOLAEV A. V., Zh. Neorgan. Khim., **3**, No. 4, 1028 (1958)	406, 419, 454, 465, 473, 474, 477, 4233
461.	KURNAKOVA A. G., "The chemistry of borates" (Khimiya boratov.), Conf. material on questions in the chemistry of borates. Izd. Akad. Nauk Latv. SSR, Riga, p. 45 (1953)	1559, 1563, 1567, 1568, 1579
462.	KURTEVA O. I., BRUCHKUS E. B., Zh. Prikl. Khim., **34**, No. 8, 1714 (1961)	638, 639 786
463.	KURTEVA L. V., BOL'SHAKOVA L. P., PLYUSHCHEV V. E., Zh. Neorgan. Khim., **8**, 1993 (1963)	1322
464.	KYDYNOV M., DRUZHININ I. G., Izv. Akad. Nauk Kirg. SSSR, No. 4, 89 (1957)	3456
465.	KYDYNOV M., DRUZHININ I. G., Collected papers of a conference dedicated to the 100th anniversary of the birth of Academician Kurnakov, Frunze, 1963, p. 53	3336
466.	LAVUT E. A., VOROB'EVA O. I. Zh. Neorgan. Khim., **5**, No. 8, 1813 (1960)	2623
467.	LAVUT E. A., VOROB'EVA O. I., SHUL'GINA I. M., Zh. Neorgan. Khim., **6**, No. 12, 2759 (1961)	2624
468.	LADEINOVA L. V., LOZHKINA L. G., CHERNYSHEVA A. M., Izv. Akad. Nauk SSSR, Otd. Khim. Nauk, No. 1, 12 (1961)	270
469.	LAMP V. N., CHEPELEVETSKI M. L., Tr. Nauch.-Issled. Inst. Udobr. Insektofungits. (postgrad. papers), 9 (1963)	661
470.	LAMP V. N., CHEPELEVETSKI M. L., Zh. Prikl. Khim., **36**, No. 12, 2650 (1963)	662, 663, 4731
471.	LAMP V. N., CHEPELEVETSKI M. L., Zh. Prikl. Khim., **36**, No. 12, 2650 (1963)	663
472.	LAPAN A. A., TSEFT A. A., Tr. Irkutsk. Gornomet. Inst., vyp. 13, 168 (1958)	2046, 2049
473.	LAPITSKII A. V., POSPELOVA L. A., ARTAMONOVA E. P., Zh. Neorgan.-Khim., **1**, No. 4, 650 (1956)	859, 888, 1083, 1127
474.	LEVINA E. YU., Zh. Prikl. Khim., **11**, No. 5, 762 (1938)	3383

Reference number	Author's name and reference	Table number
475.	LEVINA E. YU., Zh. Prikl. Khim., **11**, No. 9, 1274 (1938)	3271, 3300
476.	LEVINA M. I., SUMBAROVSKAYA N. P., Zh. Fiz. Khim., **12**, Nos. 5–6, 653 (1938)	80
477.	LEIKINA B. N., NOVOSELOVA A. V., Zh. Obshch. Khim., **7**, 241 (1937)	931
478.	LEL'CHUK YU. L., SOSONKO S. M., Izv. Tomsk. Politekh. Inst., **71**, 52 (1952)	1323, 3138, 4610
479.	LEL'CHUK YU. L., Zh. Obshch. Khim., **25**, No. 7, 1273 (1955)	5296, 5297, 5421, 5422
480.	LEL'CHUK YU. L., SURNINA L. V., BARKHATOVA V. I., Zh. Obshch. Khim., **25**, No. 9, 1685 (1955)	3119, 3928, 3929, 4998
481.	LEL'CHUK YU. L., Izv. Tomsk. Politekh. Inst., **83**, 189 (1956)	4004, 5296, 5297, 5421, 5422
482.	LEL'CHUK Yu. L., Zh. Neorgan. Khim., **3**, No. 11, 2453 (1958)	866, 1392, 3494, 4755
483.	LEL'CHUK YU. L., SKRIPOVA L. L., CHASHCHINA O. V., Zh. Neorgan. Khim., **6**, No. 2, 474 (1961)	5302, 5423
484.	LEPESHKOV I. N., BODALEVA N. V., "Potassium salts of the Volga-Emba and Prikarpat'e" (Kaliinye soli Volga-Emba i Prikarpat'e), p. 103 (1946)	3317, 3612
485.	LEPESHKOV I. N., BODALEVA N. V., Izv. Sektora Fiz.-Khim. Anal., Akad. Nauk SSSR, **17**, 342 (1949)	4038, 4072, 4137, 4684
486.	LEPESHKOV I. N., NOVIKOVA L. V., Zh. Neorgan. Khim., **3**, No. 10, 2395 (1958)	4080, 4730
487.	LEPESHKOV I. N., BODALEVA N. V., KOTOVA L. G., Zh. Neorgan. Khim., **3**, No. 12, 2781 (1958)	1359
488.	LEPESHKOV I. N., FRADKINA KH. B., Zh. Neorgan. Khim., **4**, No. 12, 2803 (1959)	3333, 3334 3754
489.	LEPESHKOV I. N., ROMASHOVA N. N., Zh. Neorgan. Khim., **4**, No. 12, 2812 (1959)	1357, 1364
490.	LEPESHKOV I. N., ROMASHOVA N. N., Zh. Neorgan. Khim., **5**, No. 11, 2512 (1960)	1356, 1358, 1365
491.	LEPESHKOV I. N., FRADKINA KH. B., Zh. Neorgan. Khim., **6**, No. 1, 199 (1961)	1295
492.	LEPESHKOV I. N., BODALEVA N. V., KOTOVA L. T., Zh. Neorgan. Khim., **6**, No. 7, 1696 (1961)	1361, 1379
493.	LEPESHKOV I. N., ROMASHOVA N. N., Zh. Neorgan. Khim., **6**, No. 8, 1968 (1961)	1414
494.	LEPESHKOV I. N., BODALEVA N. V., KOTOVA L. T., Zh. Neorgan. Khim., **7**, No. 7, 1702 (1962)	1362
495.	LEPESHKOV I. N., FRADKINA KH. B., Zh. Neorgan. Khim., **8**, 447 (1963)	1293
496.	LEPESHKOV I. N., BODALEVA N. V., KOTOVA L. T., Zh. Neorgan. Khim., **8**, 2601 (1963)	1285–1287
497.	LILICH L. S., CHERNYKH L. V., SHALYGIN V. M., Zh. Neorgan. Khim., **8**, 2775 (1963)	1162a, 1167
498.	LIPATOVA N. P., MOROZOV I. S., Zh. Neorgan. Khim., **10**, No. 2, 429 (1965)	952, 1066 1100
499.	LIPIS L. V., POZHARSKII B. G., POZHARSKAYA M. E., Zh. Neorgan. Khim., **5**, No. 10, 2190 (1960)	443, 799, 802
500.	LOPATKINA G. A., Zh. Prikl. Khim., **32**, No. 12, 2677 (1959)	2930, 2931, 3162
501.	LOPATTO E. K., SAVINAEV A. M., Zh. Prikl. Khim., **7**, 881 (1934)	791

Reference number	Author's name and reference	Table number
502.	LUZHNAYA N. P., VERESHCHETINA I. A., Zh. Prikl. Khik., **19**, No. 7, 725 (1946)	3635, 3654
503.	LUZHNAYA N. P., Izv. Akad. Nauk SSSR, Otd. Khim. Nauk, No 1, 27 (1949)	2512
504.	LUZHNAYA N. P., Izv. Sektora Fiz.-Khim. Anal., Inst. Obshch. Neorgan. Khim., **20**, 212 (1950)	2536, 4868
505.	LUZHNAYA N. P., KOSYACHKOVA S. N., Izv. Sektora Fiz.-Khim. Anal., Inst. Obshch. Neorgan. Khim., Akad. Nauk SSSR, **25**, 345 (1954)	4453, 4466
506.	LUZHNAYA N. P., KOSYACHKOVA S. N., Izv. Sektora Fiz.-Khim. Anal., Inst. Obshch. Neorgan. Khim., Akad. Nauk SSSR, **26**, 259 (1955)	4447, 4448
507.	LUZHNAYA N. P., KOSYACHKOVA S. N., Izv. Sektora Fiz.-Khim. Anal., Inst. Obshch. Neorgan. Khim., Akad. Nauk SSSR, **27**, 358 (1956)	2888, 2898 2923
508.	LUZHNAYA N. P., KOVALEVA I. S., Zh. Neorgan. Khim., **6**, No. 6, 1440 (1961)	3027
509.	LUK'YANOVA E. I., Tr. Gos. Inst. Prikl. Khim., vyp. 18, 51 (1933)	3691
510.	LUK'YANOVA E. I., SHOIKHET D. N., Tr. Gos. Inst. Prikl. Khim., vyp. 34, 10 (1940)	3709, 3770 4853
511.	LUK'YANOVA E. I., Sb. Statei Obshch. Khim., Izd. Akad. Nauk SSSR, M.-L., **1**, 106 (1953)	765
512.	LUK'YANOVA E. I., SOKOL V.I., SOKOLOVA G. N., Zh. Neorgan. Khim., **1**, No. 2, 298 (1956)	4035, 4040, 4073, 4074
513.	LUK'YANOVA E. I., PODZOREI G. N., Zh. Neorgan. Khim., **3**, No. 12, 2786 (1958)	3667
514.	LUK'YANOVA E. I., PODZOREI G. N., Zh. Neorgan. Khim., **5**, No. 8, 1827 (1960)	4050
515.	LUK'YANOVA E. I., KOKHOVA V. F., Zh. Neorgan. Khim., **8**, 219 (1963)	2627, 2628
516.	LUK'YANYCHEV YU. A., NIKOLAEV N. S., At. Energ., **15**, No. 5, 423 (1963)	917
517.	LUCHINSKII G. P., LIKHACHEVA A. I., Zh. Fiz. Khim., **9**, No. 2, 199 (1937)	707
518.	LUCHINSKII G. P., Zh. Obshch. Khim., **10**, No. 14, 1266 (1940)	41
519.	LYALIKOV K. S., PISKUPOVA V. N., Zh. Fiz. Khim., **28**, 127 (1954)	1952, 2581, 4984
520.	MAZEL' V. A., Izv. Sektora Fiz.-Khim. Anal., Akad. Nauk SSSR, **13**, 323 (1940)	763, 2679
521.	MAZOVER N. D., Zh. Prikl. Khim., **26**, No. 6, 613 (1953)	830, 832
522.	MAZOVER N. D., Zh. Prikl. Khim., **26**, No. 7, 757 (1953)	831
523.	MAKAROV S. Z., Zh. Prikl. Khim., **3**, No. 3, 342–345, 348 (1930)	2864
524.	MAKAROV S. Z., YAKIMOV M. N., Zh. Russ. Fiz.-Khim. Obshchest., **62**, Nos. 8–9 (1930)	2858
525.	MAKAROV S. Z., VAKSBERG N. M., Zh. Russ. Fiz.-Khim. Obshchest., **62**, Nos. 8–9, 1866 (1930)	2884
526.	MAKAROV S. Z., S. T. E., **7**, 221 (1931)	2865, 2869, 2870, 2893, 2910, 2958, 2975, 2976
527.	MAKAROV S. Z., VALYASHKO M. G., Inst. Obshch. Neorgan. Khim., Akad. Nauk SSSR, 1031, cited in [1]	2894, 2911

Reference number	Author's name and reference	Table number
528.	MAKAROV S. Z., Zh. Obshch. Khim., **3**, No. 2, 246 (1933)	2884, 2975–2978
529.	MAKAROV S. Z., YAKIMOV M. N., Zh. Obshch. Khim., **3**, No. 8, 993 (1933)	2853, 2859, 2860, 2885, 2948
530.	MAKAROV S. Z., DRUZHININ I. G., Izv. Akad. Nauk SSSR, Ser. Khim. Nauk, 283 (1937)	4702
531.	MAKAROV S. Z., DRUZHININ I. G., Izv. Akad. Nauk SSSR, Ser. Khim. Nauk, 1291 (1937)	3344, 3430
532.	MAKAROV S. Z., BLIDIN V. P., Izv. Akad. Nauk SSSR, Ser. Khim. Nauk, 879 (1938)	2959, 2960
533.	MAKAROV S. Z., BLIDIN V. P., Izv. Novocherkassk. Industr. Inst., **6**, Ser. Khim., 77 (1940)	2960
534.	MAKAROV S. Z., SEDEL'NIKOV G. S., Izv. Otd. Khim. Nauk, Akad. Nauk SSSR, No. 2, 843 (1940); Zh. Prikl. Khim., **18**, Nos. 7–8, 432 (1945)	2861
535.	MAKAROV S. Z., REPA A. G., Izv. Akad. Nauk SSSR, Ser. Khim. Nauk, 451 (1938)	811
536.	MAKAROV S. Z., SEDEL'NIKOV G. S., Izv. Otd. Khim. Nauk, Akad. Nauk SSSR, No. 6, 845 (1940)	2871
537.	MAKAROV S. Z., REPA A. G., Izv. Akad. Nauk SSSR, Ser. Khim. Nauk, 349 (1940)	2605, 2607
538.	MAKAROV S. Z., SHUL'GINA M. P., Izv. Akad. Nauk SSSR, Ser. Khim. Nauk, 511 (1940)	3015
539.	MAKAROV S. Z., PEREL'MAN F. M., LEGKOVA T. K., Izv. Otd. Khim. Nauk, Akad. Nauk SSSR, No. 2, 234 (1941)	4146, 4174 4881
540.	MAKAROV S. Z., ITKINA L. S., Izv. Otd. Khim. Nauk, Akad. Nauk SSSR, No. 5, 285 (1944)	2652, 2658 2691
541.	MAKAROV S. Z., VOL'NOV I. I., Izv. Akad. Nauk SSSR, Otd. Khim. Nauk, No. 2, 201 (1951)	5143
542.	MAKAROV S. Z., CHAMOVA V. N., Izv. Akad. Nauk SSSR, Izv. Akad. Nauk SSSR, Otd. Khim. Nauk, No. 3, 255 (1951)	238
543.	MAKAROV S. Z., GRIGOR'EVA N. K., Izv. Akad. Nauk SSSR, Otd. Khim. Nauk, No. 3, 385 (1954)	263
544.	MAKAROV S. Z., VOL'NOV I. I., Izv. Akad. Nauk SSSR, Otd. Khim. Nauk, No. 5, 765 (1954)	252
545.	MAKAROV S. Z., VOL'NOV I. I., Izv. Sektora Fiz.-Khim. Anal., Inst. Obshch. Neorgan. Khim., Akad. Nauk SSSR, **25**, 320 (1954)	5065
546.	MAKAROV S. Z., GRIGOR'EVA N. K., Izv. Akad. Nauk SSSR, Otd. Khim. Nauk, No. 1, 17 (1955)	237
547.	MAKAROV S. Z., DOBRYNINA T. A., Izv. Akad. Nauk SSSR, Otd. Khim. Nauk, No. 3, 411 (1955)	230
548.	MAKAROV S. Z., KRASNIKOV S. N., Izv. Sektora Fiz.-Khim. Anal., Inst. Obshch. Neorgan. Khim., Akad. Nauk SSSR, **27**, 367 (1956)	2954, 2980 3283
549.	MAKAROV S. Z., DALEINOVA L. V., Izv. Akad. Nauk SSSR, Otd. Khim. Nauk, No. 1, 3 (1957)	267
550.	MAKAROV S. Z., CHAMOVA V. N., Izv. Akad. Nauk SSSR, Otd. Khim. Nauk, No. 9, 1025 (1958)	254
551.	MAKAROV S. Z., GRIGOR'EVA N. K., Izv. Akad. Nauk SSSR, Otd. Khim. Nauk, No. 11, 1289 (1958)	265
552.	MAKAROV S. Z., ARNOL'D G. I., Izv. Akad. Nauk SSSR, Otd. Khim. Nauk, No. 12, 1407 (1958)	269

Reference number	Author's name and reference	Table number
553.	MAKAROV S. Z., GRIGOR'EVA N. K., Izv. Akad. Nauk SSSR, Otd. Khim. Nauk, No. 1, 3 (1959)	272
554.	MAKAROV L. L., POPOV G. S., Dokl. Akad. Nauk SSSR, **129**, No. 4, 854 (1959)	4981
555.	MAKAROV S. Z., ARNOL'D T. I., STASEVICH N. N., SHORINA E. V., Izv. Akad. Nauk SSSR, Otd. Khim. Nauk, No. 11, 1914 (1960)	266
556.	MAKAROV L. L., PANKOV A. G., Zh. Fiz. Khim., **36**, No. 10, 2241 (1962)	5343
557.	MAKAROV L. L., VLASOV YU. G., IZOTOVA V. I., Zh. Fiz. Khim., **38**, No. 10, 2403 (1964)	5337
558.	MAKIN A. V., KARNAUKHOV A. S., Zh. Neorgan. Khim., **2**, No. 6, 1420 (1957)	3056
559.	MAKIN A. V., Zh. Neorgan. Khim., **2**, No. 12, 2794 (1957)	3200
560.	MAKIN A. V., Zh. Neorgan. Khim., **3**, No. 12, 2794 (1958)	3057
561.	MAKIN A. V., Zh. Neorgan. Khim., **4**, No. 5, 1190 (1959)	3060
562.	MAKIN A. V., Uch. Zap. Yaroslav. Gos. Pedagog. Inst., **32** (42), 291 (1959)	3198
563.	MAKSIMYCHEVA Z. T., DZHIYANBAEVA R. KH., Uzb. Khim. Zh., No. 2, 33 (1960)	4643
564.	MAKSIMYCHEVA Z. T., TOISKINA L., Uzb. Khim. Zh., No. 6, 16 (1964)	4440
565.	MANVELYAN M. G., BABAYAN G. G., SAYAMYAN E. V., VOSKANYAN S. S., Izv. Akad. Nauk Arm. SSR, Khim. Nauki, **12**, No. 2, 95 (1959)	3191
566.	MANVELYAN M. G., BABAYAN G. G., MAYAMYAN E. A., VOSKANYAN S. S., Izv. Akad. Nauk Arm. SSR, **13**, No. 1, 25 (1960)	3190
567.	MANVELYAN M. G., BABAYAN G. G., SAYAMYAN E. A., VOSKANYAN S. S., Izv. Akad. Nauk Arm. SSR, **14**, No. 4, 303 (1961)	2935
568.	MANVELYAN M. G., BABAYAN G. G., VOSKANYAN S. S., SAYAMYAN E. A., OGANESYAN E. B., Zh. Prikl. Khim., **36**, No. 11, 2402 (1963)	2936
569.	MANOM-BEZHAN B. A., Zh. Prikl. Khim., **29**, No. 8, 1147 (1956)	569
570.	MEERSON G. A., ZVEREV G. L., ZUBKOVA F. M., Tsvet. Met., **8**, 97 (1939)	585, 590
571.	MENKOVSKII M. A., PETROV N. A., LITVIN K. I., CHERNOVSKII D. S., Zh. Neorgan. Khim., **1**, No. 7, 1658 (1956)	137
572.	MILOSLAVSKII S. YA., Zh. Obshch. Khim., **1**, No. 7, 900 (1931)	1864
573.	MIRONOV N. N., TROFIMOVA L. M., Tr. Khim. Khim. Tekhnol., Gor'kii, No. 3, 551 (1961)	2750
574.	MIRONOV N. N., TROFIMOVA L. M., Tr. Khim. Khim. Tekhnol., Gor'kii, No. 3, 551 (1961)	2751, 2760
575.	MIRONOV K. E., SINITSINA E. D., Izv. Sib. Fil., Akad. Nauk SSSR, Ser. Khim. Nauk, vyp. 3, 3 (1963)	2736
576.	MITYUREVA T. T., Dokl. Akad. Nauk Ukr. SSR, No. 2, 166 (1959)	871
577.	MITYUREVA T. T., Dokl. Akad. Nauk Ukr. SSR, No. 2, 166 (1959)	899
578.	MIKHEEVA V. I., BREITSIS V. B., Zh. Neorgan. Khim., **5**, No. 11, 2553 (1960)	2626
579.	MIKHEEVA V. I., SELIVOKHINA M. S., Zh. Neorgan. Khim., **8**, 442 (1963)	4371

Reference number	Author's name and reference	Table number
580.	MOROZOV I. S., USTAVSHCHIKOVA G. V., Izv. Akad. Nauk SSSR, Ser. Khim. Nauk, 451, (1944)	946, 1052
581.	MOROZOV I. S., USTAVSHCHIKOVA G. V., Izv. Akad. Nauk SSSR, Ser. Khim. Nauk, 74 (1945	946, 1055, 2526
582.	MOROZOV I. S., Izv. Otd. Khim. Nauk, Akad. Nauk SSSR, No. 1, 39 (1946)	947
583.	MOROZOV I. S., TOPTYGINA G. M., Zh. Neorgan. Khim., **5**, No. 7, 1637, (1960)	2507
584.	MOROZOV I. S., TOPTYGINA G. M., Zh. Neorgan. Khim., **6**, No. 11, 2518 (1960)	951, 1005, 1025, 1101
585.	MOROZOV I. S., TOPTYGINA G. M., LIPATOVA N. P., Zh. Neorgan. Khim., **6**, No. 11, 2532 (1961)	950, 1004, 1034, 1099
586.	MOSKVIN A. I., GEL'MAN A. D., Zh. Neorgan. Khim., **3**, No. 4, 962 (1958)	3028, 4485
587.	MOTOV D. L., Zh. Neorgan. Khim., **2**, No. 11, 2661 (1957)	2337
588.	MOTOV D. L., Zh. Neorgan. Khim., **2**, No. 12, 1797 (1957)	750
589.	MOTOV D. L., Izv. Karel'. Kol'sk. Filialov. Akad. Nauk SSSR, No. 3, 137 (1958)	751
590.	MOTOV D. L., Sb. Tr. Khim. Tekhnol. Miner. Syr'ya Kol'sk. Polnostrova, vyp. 1, 101 (1959)	752
591.	MUKANOV I. P., BESKOV S. D., Zh. Prikl. Khim., **34**, No. 10, 2282 (1961)	391
592.	MUKIMOV S., GREBENSHCHIKOVA N., KAPKAEVA F., Byull. Akad. Nauk Usb. SSSR, **11**, 4 (1947)	4830
593.	MUROMTSEV B. A., S. T. E., 7, 214 (1931)	3907, 3910
594.	MUROMTSEV B. A., NAZAROVA L. A., Izv. Akad. Nauk SSSR, Ser. Khim. Nauk, 177 (1938)	644
595.	M KHITARYAN R. S., ZATIKYAN L. O., Izv. Akad. Nauk Arm. SSR, **12**, No. 3, 175 (1959)	1977
596.	MKHITARYAN R. S., ZATIKYAN L. O., Izv. Akad. Nauk Arm. SSSR., Ser. Khim., **6**, No. 4, 343 (1963)	1980
597.	NABIEV M. N., VISHNYAKOVA A. N., RACHIMDZHANOVA I., Uzb. Khim. Zh., No. 5, 3 (1961)	1799
598.	NAGORSKAYA N. D., NOVOSELOVA A. V., Zh. Obshch. Khim., **5**, 182 (1935)	2665, 2666
599.	NAIMUSHINA R. F., Khim. Khim. Tecknol., Vol. I, Alma-Ata, 92 (1963)	285, 772
600.	NAIMUSHINA R. F., Tr. Inst. Khim. Nauk, Akad. Nauk Kaz. SSR, **10**, 165 (1964)	286, 773
601.	NEIMAN M. V., MILLER V. B., FEDOSEEVA A. I., Dokl. Akad. Nauk. SSSR, **75**, 719 (1950)	5270
602.	NEKRASOV YU. D., MARKOV S. S., Sb. Tr. Gos. Inst. Prikl. Khim., vyp. 46, 309 (1960)	561
603.	NENNO E. S., DROZIN N. N., Zh. Prikl. Khim., **28**, No. 12, 1350 (1955)	1608, 1887
604.	NIKITINA E. A., Zh. Obshch. Khim., **3**, 513 (1933)	2168, 2169
605.	NIKITINA I. S., VELLER R. L., Sb. Nauch. Tr., Gos. Nauch.-Issled. Inst. Tsvet. Metal, No. 13, 129 (1957)	923–925
606.	NIKOLAEV A. V., CHELISHCHEVA A. G., Dokl. Akad. Nauk SSSR, **28**, 127 (1940)	1561, 1577

Reference number	Author's name and reference	Table number
607.	NIKOLAEV A. V., DURASQVA S. A., Izv. Sib. Otd., Akad. Nauk SSSR, Ser. Khim. Nauk, **3**, No. 1, 9 (1963)	229
608.	NIKOLAEV A. V., MIKHAILOVA M. P., Izv. Sib. Otd., Akad. Nauk SSSR, No. 3, 46 (1961)	1037
609.	NIKOLAEV A. V., KOLESNIKOV A. A., Symposium: "Physico-chemical analysis" (Fiziko-khimicheskii analiz), Sib. Otd., Akad. Nauk SSSR, Novosibirsk, p. 104 (1963)	404
610.	NIKOLAEV V. I., Zh. Russ. Fiz.-Khim. Obshchest., **60**, 893 (1928)	415, 2663, 3104, 3109, 4383
611.	NIKOLAEV V. I., RAVICH M. I., Zh. Obshchest. Khim., **1**, No. 7, 787 (1931)	1217, 1219, 2591, 2718
612.	NIKOLAEV V. I., GRITSEVICH E. V., STYAZHKINA E. G., RUDENKO E. I., Zh. Prikl. Khim., **5**, vyp. 6–7, 746 (1932)	3590
613.	NIKOLAEV V. I., VINTOVKINA N. P., Zh. Prikl. Khim., **7**, vyp. 1–2, 30 (1934)	3591
614.	NIKOLAEV V. I., GLINSKIKH S. A., Dokl. Akad. Nauk SSSR, **1**, No. 9, 571 (1934)	5062
615.	NIKOLAEV V. I., Izv. Inst. Fiz.-Khim. Anal., Akad. Nauk SSSR, **7**, 146 (1935)	4116, 4117, 4196, 4816
616.	NIKOLAEV V. I., BUROVAYA E. E., Izv. Sektora Fiz.-Khim. Anal., Akad. Nauk SSSR, **10**, 248 (1938)	3270, 3301, 3384, 3632
617.	NIKOLAEV V. I., KACHALOV A. E., Zh. Prikl. Khim., **15**, vyp. 1–2, 53 (1946)	3594
618.	NIKOLAEV N. S., IVANOV N. A., KOLTYNIN S. G. Zh. Prikl. Khim., **9**, 1183 (1936)	561, 569
619.	NIKOLAEV N. S., Khim. Redk. Elem. vyp. 1, 45 (1954)	625
620.	NIKOLAEV N. S., Izv. Sektora Fiz.-Khim. Anal., Inst. Obshch. Neorgan. Khim., Akad. Nauk SSSR, **25**, 375 (1954)	572
621.	NIKOLAEV N. S., BUSLAEV YU. A., Khim. Redk. Elem., No. 2, 57 (1955)	595
622.	NIKOLAEV N. S., BUSLAEV YU. A., Izv. Sektora Fiz.-Khim. Anal., Akad. Nauk SSSR, **26**, 270 (1955)	596
623.	NIKOLAEV N. S., BUSLAEV YU. A., Zh. Neorgan. Khim., **1**, No. 7, 1672 (1956)	619
624.	NIKOLAEV N. S., BUSLAEV YU. A., Zh. Neorgan. Khim., **4**, No. 1, 205 (1959)	612
625.	NIKOLAEV N. S., BUSLAEV YU. A., GUSTYAKOVA M. P., Zh. Neorgan. Khim., **7**, No. 7, 1686 (1962)	610
626.	NIKOLAEV N. S., LUK'YANYCHEV YU. A., At. Energ., **12**, No. 4, 334 (1962)	1181
627.	NIKOL'SKAYA YU. P., MOSHKINA I. A., Zh. Neorgan. Khim., **3**, No. 2, 498 (1958)	2897, 2902
628.	NIKOL'SKAYA YU. P., MOSHKINA I. A. Tr. Khim.-Met. Inst. Zapad.-Sib. Filial, Akad. Nauk SSSR, No. 12, 17 (1958)	2901, 3364
629.	NOVAKOVSKII M. S., MUSHKINA M. G., VOROB'EVA E. G., Uch. Zap. Khar'kov. Univ., 82, Tr. Khim. Fak., Nauch.-Issled. Inst. Khim., **16**, 107 (1957)	3534, 4508
630.	NOVIKOVA L. V., Zh. Neorgan. Khim., **2**, No. 3, 662 (1957)	4088
631.	NOVOSELOVA A. V., Zh. Obshch. Khim., **4**, 1206 (1934)	1510

Reference number	Author's name and reference	Table number
632.	NOVOSELOVA A. V., NAGORSKAYA N. D., Zh. Obshch. Khim., **5**, 1002 (1935)	394
633.	NOVOSELOVA A. V., BODALEVA N. V., GERSHTEIN M. M., Zh. Obshch. Khim., **8**, 732 (1938)	1523
634.	NOVOSELOVA A. V., Zh. Obshch. Khim., **9**, 1063 (1939)	2210
635.	NOVOSELOVA A. V., Zh. Obshch. Khim., **10**, 1547 (1940)	2212
636.	NOVOSELOVA A. V., DANILEVICH R., TIKHONOVA A., Zh. Obshch. Khim., **16**, 439 (1946)	1538
637.	NOVOSELOVA A. V., SOSNOVSKAYA I. G., Zh. Obshch. Khim., **21**, 813, 893 (1951)	1400
638.	NOVOSELOVA A. V., VOROB'EVA O. I., KNYAZEVA N. I., PASKUTSKAYA L. L., Zh. Obshch. Khim., **23**, No. 8, 1284 (1953)	1522
639.	NOVOSELOVA A. V., PASHINKIN A. S., SEMENENKO K. I., Vestn. Mosk. Gos. Univ., No. 3, 49 (1955)	1527
640.	NOVOSELOVA A. V., TAMM N. S., VOROB'EVA O. I., Khim. Redk. Elem., No. 2, 3 (1955)	1507
641.	NOVOSELOVA A. V., RESHETNIKOVA L. P., Khim. Redk. Elem., No. 3, 6 (1957)	1514
642.	NONOBASHVILI E. M., NADARSHEISHVILI SH. A., Tr. Inst. Khim., Akad. Nauk Gruz. SSSR, **13**, 129 (1957)	3220
643.	OBUKHOV A. P., MIKHAILOV M. N., Kalii, No. 3, 21 (1932)	984
644.	OBUKHOV A. P., LAVROV V. P., Zh. Prikl. Khim., **4**, Nos. 7–8, 966 (1931); Tr. Gos. Inst. Prikl. Khim., vyp. 25, 83 (1934)	983
645.	OBUKHOV A. P., GORDON E. YA. Zh. Prikl. Khim., **8**, 1143 (1935)	3001
646.	OBUKHOV A. P., MIKHAILOV M. N., Zh. Prikl. Khim., **8**, 1148 (1935)	3208
647.	OZEROV L. A., Tr. Voronezh. Gos. Univ., **38**, 24 (1953)	4865
648.	OZEROVA M. I., KHOMYAKOV K. G., Vestn. Mosk. Gos. Univ., Ser. Fiz.-Mat. Estestv. Nauk, No. 10, 97 (1955)	247
649.	OZEROVA M. I., BOITLER E. M., EGOROVA E. I., Zh. Neorgan. Khim., **6**, No. 4, 966 (1961)	2387
650.	OZEROVA M. I., SHCHEDRINA A. P., Zh. Neorgan. Khim., **8**, 2608 (1963)	2385
651.	OZEROVA M. I., KOCHANOVA N. N., EGOROVA E. I., Symposium: "Physico-chemical analysis" (Fiziko-khimicheskii analiz). Sib. Otd. Akad. Nauk SSSR, Novosibirsk, p. 122 (1963)	2399
652.	OL'SHANSKII YA. I., IVANENKO V. V., KHROMOV A. V., Dokl. Akad. Nauk SSSR, **124**, No. 2, 410 (1954)	695
653.	OPALOVSKII A. A., KUZNETSOVA Z. M., LUK'YANOVA A. A., Izv. Sib. Otd., Akad. Nauk SSSR, No. 6, 54 (1962)	3175, 4657
654.	OPALOVSKII A. A., KUZNETSOVA Z. M., BATSANOV S. S., Izv. Sib. Otd., Akad. Nauk SSSR, No. 9, 46 (1962)	2218
655.	OPALOVSKII A. A., BATSANOV S. S., KUZNETSOVA Z. M., Izv. Akad. Nauk SSSR, Ser. Khim. Nauk, **12**, 2110 (1963)	2220
656.	ORLOV I. E., Agressivnost' estestvennykh vod. O.N.T.I., p. 99 (1932)	1620, 1628, 1648

Reference number	Author's name and reference	Table number
657.	ORLOV N. A., Zh. Russ. Fiz.-Khim. Obshchest., Chast. Khim., **34**, 951 (1902)	3743, 5115
658.	PAVLOV B. A., BULOVICH N. KH., BERGMAN A. G., Dokl. Akad. Nauk SSSR, **39**, 265 (1943)	2148
659.	PALKIN A. P., Byull. Sredneaziat. Gos. Univ., **18**, 74 (1929)	3974
660.	PALKIN A. P., BRIKMAN N. M., Tr. Gos. Inst. Prikl. Khim., vyp. 16, 8–21 (1932)	4691
661.	PALKIN A. P. VARASOVA E. I., Zh. Prikl. Khim., **5**, Nos. 3–4, 319, 321 (1932)	3625, 3644, 4128
662.	PALKIN A. P., *et al.*, Solikamskie Karnallity, p. 70 et seq. (1935)	3628, 3647 3693, 4131
663.	PALKIN A. P., GOLOSHCHAPOV M. V., Tr. Voronezh. Gos. Univ., Khim. Otd., **11**, No. 1, 17–23 (1939)	311, 326
664.	PALKIN A. P., BOGOYAVLENSKII P. S., Zh. Prikl. Khim., **15**, No. 6, 734 (1941)	4566
665.	PALKIN A. P., PETROVA K. G., Nauch. Soobshch. Voronezh. Gos. Univ., vyp. 1, 36 (1941)	351
666.	PALKIN A. P., IL'IN P. V., Nauch. Soobshch. Voronezh. Gos. Univ., vyp. 1, 37 (1941)	359
667.	PALKINA N. A., Tr. Voronezh. Gos. Univ., **40**, 69 (1956)	3225
668.	PEL'SH A. D., Byull. Inst. Gallurgii, Nos. 6–7, 33 (1940)	3275
669.	PEL'SH A. D., Tr. Vses. Nauch.-Issled. Inst. Gal., (1949), cited in [1]	3388, 3389, 4034
670.	PEL'SH A. D., Tr. Vses.Nauch.-Issled. Inst. Gal., vyp. 21, 148 et seq. (1949)	3279, 3304
671.	PEL'SH A. D., Tr. Vses. Nauch.-Issled. Inst. Gal., vyp. 24, 3 (1952)	3307
672.	PEL'SH A. D., Tr. Vses. Nauch.-Issled. Inst. Gal., vyp. 27, 3 (1953)	3308
673.	PEREL'MAN F. I., ZVORYKIN A. YA., YAKUBOVSKAYA T. N., Zh. Neorgan. Khim., **3**, No. 6, 1364 (1958)	513, 754, 755, 785, 853, 854, 873, 874
674.	PEREL'MAN F. M., ZVORYKIN A. YA., DEMINA G. A., Zh. Neorgan. Khim., **5**, No. 4, 960 (1960)	2194
675.	PEREL'MAN F. M., KAROV Z. G., Zh. Neorgan. Khim., **6**, No. 6, 1431 (1961)	4713
676.	PEREL'MAN F. M., ZVORYKIN A. YA., DEMINA G. A., Zh. Neorgan. Khim., **7**, No. 3, 641 (1962)	481
677.	PEREL'MAN F. M., BABIEVSKAYA I. Z., Zh. Neorgan. Khim., **7**, No. 6, 1480 (1962)	403
678.	PEREL'MAN F. M., DOLININA R. M., Zh. Neorgan. Khim., **7**, No. 7, 1683 (1962)	1476
679.	PEREL'MAN F. M., DOLININA R. M., Zh. Neorgan. Khim., **7**, No. 10, 2460 (1962)	3899
680.	PEREL'MAN F. M., FEDOSEEVA E. I., Zh. Neorgan. Khim., **8**, 1256 (1963)	4576
681.	PEREL'MAN F. M., FEDOSEEVA E. I., Zh. Neorgan. Khim., **8**, 2604 (1963)	4577
682.	PEREL'MAN F. M., DEMINA G. A., Zh. Neorgan. Khim., **9**, No. 7, 1772 (1964)	480
683.	PEROVA A. P., Zh. Prikl. Khim., **11**, Nos. 7–8, 1045 (1938)	345
684.	PEROVA A. P., Sb. Statei Obshch. Khim., m.-L., Izd. Akad. Nauk SSSR, **1**, 91 (1953)	334

Reference number	Author's name and reference	Table number
685.	PEROVA A. P., Zh. Prikl. Khim., **27**, No. 12, 1278 (1954)	347
686.	PEROVA A. P., Zh. Prikl. Khim., **28**, No. 7, 747 (1955)	282
687.	PEROVA A. P., Soobshch. Nauch. Rab. Vses. Khim. Obshch. im. Mendeleeva, vyp. 2, 46 (1955)	4165
688.	PEROVA A. P., Zh. Neorgan. Khim., **1**, No. 2, 323 (1956)	4641
689.	PEROVA A. P., Zh. Neorgan. Khim., **2**, No. 12, 2789 (1957)	4143
690.	PEROVA A. P., Sb. Tr. Voronezh. Otd., Vses. Khim. Obshch. im. Mendeleeva, vyp. 1, 155 (1957)	4051
691.	PETROV N., RYBAKOV, Tr. Inst. Chist. Khim. Reaktivov, **16**, 113 (1939), cited in [1]	1224
692.	PETROV N., IL'IN A., Tr. Inst. Chist. Khim. Reaktivov, vyp. 16, 115 (1939), cited in [1]	922
693.	PINAEV V. A., Zh. Prikl. Khim., **36**, No. 10, 2116 (1963)	4016
694.	PINAEV G. F., Zh. Neorgan. Khim., **10**, No. 4, 965 (1965)	2051, 2438
695.	PINAEVSKAYA E. N., GOLUBCHENKO N. P., Zh. Prikl. Khim., **26**, No. 1, 102 (1953)	4642, 4645
696.	PINAEVSKAYA E. N., ANTOSHKINA N. L., GUSEVA G. G., Zh. Prikl. Khim., **34**, No. 8, 1722 (1961)	3090, 3091, 3123, 3882, 5095
697.	PIRYUTKO M. M., Izv. Akad. Nauk SSSR, Otd. Khim. Nauk, No. 3, 379 (1959)	631–633
698.	PLYUSHCHEV V. E., KUZNETSOVA G. P., Tr. Mosk. Inst. Tonkoi Khim. Tekhnol., vyp. 6, 15 (1956)	1065, 1098
699.	PLYUSHCHEV V. E., TULINOVA V. B., Zh. Neorgan. Khim., **2**, No. 2, 467 (1957)	1289
700.	PLYUSHCHEV V. E., TULINOVA V. B., KUZNETSOVA T. P., KOROVIN S. S., PETROVA R. G., Zh. Neorgan. Khim., **2**, No. 9, 2212 (1957)	5160
701.	PLYUSHCHEV V. E., TULINOVA V. B., KUZNETSOVA T. P., KOROVIN S. S., SHIPETINA N. S., Zh. Neorgan. Khim., **2**, No. 11, 2654 (1957)	3819
702.	PLYUSHCHEV V. E., TULINOVA V. B., Khim. Redk. Elem., vyp. 3, 3 (1957)	1290
703.	PLYUSHCHEV V. E., TULINOVA V. B., Nauch. Dokl. Vyssh. Shk., Khim. Khim. Tekhnol., No. 1, 9 (1958)	1291
704.	PLYUSHCHEV V. E., TULINOVA V. B., Zh. Neorgan. Khim., **4**, No. 5, 1184 (1959)	1333
705.	PLYUSHCHEV V. E., KUZNETSOVA G. P., STEPINA S. B., Zh. Neorgan. Khim., **4**, No. 6, 1449 (1959)	1423
706.	PLYUSHCHEV V. E., KURTOVA L. V., Zh. Neorgan. Khim., **10**, No. 6, 1471 (1965)	1288, 2929
707.	PODGORNOVA V. S., TALIPOV SH. T., Uzb. Khim. Zh., No. 6, 15 (1961)	5584
708.	PODGORNOVA V. S., TALIPOV SH. T., Uzb. Khim. Zh., No. 2, 12 (1962)	5583
709.	POZIN M. E., ZUBOV V. V., TERESHCHENKO L. YA., TARAT E. YA., PONOMAREV YU. L., Izv. Vyssh. Ucheb. Zaved., Khim. Khim. Tekhnol., **6**, No. 4, 610 (1963)	1798
710.	POLOSIN V. A., SHAKHPARONOV M. I., (1939), cited in [1]	4666
711.	POLOSIN V. A., Zh. Fiz., Khim., **20**, No. 12, 1471 (1946)	2227
712.	POLOSIN V. A., SHAKHPARONOV M. I., Zh. Fiz. Khim., **21**, No. 1, 121 (1947)	2238

Reference number	Author's name and reference	Table number
713.	PONIZOVSKII A. M., VLADIMIROVA N. M., Dokl. Akad. Nauk SSSR, **126**, No. 1, 97 (1959)	2916
714.	PONIZOVSKII A. M., VLADIMIROVA N. M., GORDON-YANOVSKII F. A., Zh. Neorgan. Khim., **5**, No. 11, 1587 (1960)	1667, 3578
715.	PONOMAREV V. D., RUBAN N. N., Izv. Akad. Nauk Kaz. SSR, Ser. Khim., No. 118, No. 6, 16 (1953)	2728, 3219
716.	PORTNOV M. A., RAVDIN YA. A., Zh. Obshch. Khim., **6**, 19, 2478 (1937)	27, 29
717.	PRIKHOD'KO G. B., Zh. Prikl. Khim., **5**, No. 1, 31 (1932)	2632, 2677
718.	PROTSENKO P. I., SHOKINA O. N., Zh. Neorgan. Khim., **7**, No. 6, 1425 (1962)	3049, 3054, 4535
719.	PROTSENKO P. I., ANDREEVA T. A., Zh. Neorgan. Khim., **9**, No. 6, 1435 (1964)	5329, 5457
720.	PROTSENKO P. I., BRYKOVA N. A., Ukr. Khim. Zh., **30**, No. 5, 448 (1964)	1318
721.	PROTSENKO P. I., BRYKOVA N. A., Zh. Neorgan. Khim., **9**, No. 4, 982 (1964)	5479
722.	PROTSENKO P. I., BRYKOVA N. A., Izv. Vyssh. Ucheb. Zaved., Khim. Khim. Tekhnol., **7**, No. 1, 3 (1964)	5330
723.	PROTSENKO P. I., BRYKOVA N. A., IVANOVA E. M., Zh. Neorgan. Khim., **10**, No. 6, 1477 (1965)	5458
724.	PROTSENKO P. I., MEDVEDEV B. S., Zh. Prikl. Khim., **38**, No. 3, 676 (1965)	5086
725.	PROTSENKO P. I., SHURDUMOV G. K., Zh. Fiz. Khim., **39**, No. 3, 613 (1965)	5252, 5328
726.	PUSHKINA G. YA., KOMISSAROVA L. N., Zh. Neorgan. Khim., **8**, 1499 (1963)	456
727.	PYLKOVA E. V., AKHUMOV E. I., Zh. Neorgan. Khim., **7**, No. 1, 200 (1962)	3286
728.	RAVICH M. I., Izv. Akad. Nauk SSSR, Ser. Khim. Nauk, 167 (1938)	4275
729.	RAVICH M. I., POPOVA Z. V., Izv. Akad. Nauk SSSR, Ser. Khim. Nauk, 268 (1942)	3201
730.	RAVICH M. I., Izv. Akad. Nauk SSSR, Ser. Khim. Nauk, 233 (1943)	3089
731.	RAVICH M. I., FROLOVA E. V., Izv. Akad. Nauk SSSR, Ser. Khim. Nauk, 92 (1944)	4575
732.	RAVICH M. I., BOROVAYA E. E., Izv. Sektora Fiz.-Khim. Anal., Akad. Nauk SSSR, **20**, 173 (1950)	3700, 3701
733.	RAVICH M. I., BOROVAYA F. E., KETKOVICH V. YA., Izv. Sektora Fiz.-Khim. Anal., Inst. Obshch. Neorgan. Khim., Akad. Nauk SSSR, **22**, 225 (1953)	4685
734.	RAVICH M. I., BOROVAYA F. E., KETKOVICH V. YA., Izv. Sektora Fiz.-Khim. Anal., Inst. Obshch. Neorgan. Khim., Akad. Nauk SSSR, **22**, 240 (1953)	3280
735.	RAVICH M. I., ITKINA L. S., KOKHOVA V. F., Izv. Sektora Fiz.-Khim. Anal., Inst. Obshch. Neorgan. Khim., **25**, 351 (1954)	3017
736.	RAVICH M. I., BOROVAYA F. E., Izv. Sektora Fiz.-Khim. Anal., Inst. Obshch. Neorgan. Khim., Akad. Nauk SSSR, **26**, 229 (1955)	2683
737.	RAVICH M. I., SHCHERBAKOVA L. G., Izv. Sektora Fiz.-Khim. Anal., Inst. Obshch. Neorgan. Khim., Akad. Nauk SSSR, **26**, 248 (1955)	2671

Reference number	Author's name and reference	Table number
738.	RAVICH M. I., ELENEVSKAYA V. M., Izv. Sektora Fiz.-Khim. Anal., Inst. Obshch. Neorgan. Khim., Akad. Nauk SSSR, **26**, 290 (1955)	2682
739.	RAVICH M. I., YASTREBOVA L. F., Zh. Neorgan. Khim., **3**, No. 12, 2771 (1958)	3206
740.	RAVICH M. I., YASTREBOVA L. F., Zh. Neorgan. Khim., **4**, No. 1, 169 (1959)	3207
741.	RAVICH M. I., BOROVAYA F. E., Zh. Neorgan. Khim., **4**, No. 9, 2100 (1959)	2694
742.	RAVICH M. I., YASTREBOVA L. F., Zh. Neorgan. Khim., **6**, No. 2, 431 (1961)	1436
743.	RAVICH M. I., VALYASHKO V. M., Zh. Neorgan. Khim., **10**, No. 1, 204 (1965)	3163
744.	RADISHCHEV V. P., Zh. Russ. Fiz.-Khim. Obshchest., **60**, 322 (1928)	3521
745.	RAKOVSKII A. V., POLYANSKII V. V., Tr. Inst. Chist. Khim. Reaktivov, 7 (1927)	3557
746.	RATNER A. P., GUREVICH A. M., POLOZHENSKAYA L. P., Zh. Neorgan. Khim., **2**, No. 10, 2316 (1957)	417, 2724, 2998, 3100
747.	RAKHIMOV KH. R., SADYKOVA G. YA., Dokl. Akad. Nauk SSR, No. 10, 37 (1954)	4527
748.	RAKHMATULLAEV K., TALIPOV SH. T., Dokl. Akad. Nauk Uzb. SSR, No. 11, 31 (1961)	4660
749.	RAKHMATULLAEV K., TALIPOV SH. T., YUSUPOVA R., Dokl. Akad. Nauk Uzb. SSR, No. 4, 46 (1962)	621
750.	RASHKOVSKAYA E. A., AVERBAKH R. A., DANILEVSKAYA M. F., NISENGOL'TS F. S., Ukr. Khim. Zh., **24**, No. 4, 510 (1958)	2875, 2890, 2899, 4449, 4454, 4468
751.	RASHKOVSKAYA E. A., LOMAKINA A. K., USENKO L. T., Ukr. Khim. Zh., **28**, No. 5, 574 (1962)	4581, 4585
752.	REPA A. G., LEGKOVA T. K., Vestn. Dal'nevost. Filial, Akad. Nauk SSSR, 32 (5), 63 (1938)	5064, 5076
753.	RESHETNIKOVA M. P., NOVOSELOVA A. V., KIRKINA D. F., Zh. Prikl. Khim., **3**, No. 2, 378 (1958)	1520
754.	RZA-ZADE P. F., RUSTAMOV P. G., Azerb. Khim. Zh., No. 6, 119 (1960)	3636, 3652
755.	RZA-ZADE P. F., RUSTAMOV P. G., Azerb. Khim. Zh., No. 4, 127 (1961)	4154
756.	RZA-ZADE P. F., RUSTAMOV P. G., Azerb. Khim. Zh., No. 1, 57 (1963)	3761, 5121
757.	RZA-ZADE P. F., GANF K. L., Azerb. Khim. Zh., No. 3, 119 (1963)	1560, 1580
758.	RZA-ZADE P. F., BAGIROV G., SEDEL'NIKOV G. S., Azerb. Khim. Zh., No. 2, 117 (1964)	1565
759.	RODE T., Izv. Sektora Fiz.-Khim. Anal., Akad. Nauk SSSR, **14**, 397 (1941)	3276, 3386, 3633, 4032
760.	RODE T., Izv. Sektora Fiz.-Khim. Anal., Akad. Nauk SSSR, **15**, 237 (1947)	3276, 3386, 3392, 4032
761.	RONKIN B. L., S.T.E., 7, 214 (1931)	3906
762.	RUSADZE A. V., SHPUNT S. YA., Zh. Neorgan. Khim., **4**, No. 1, 182 (1959)	450, 635
763.	RUSADZE A. V., SHPUNT S. YA., Zh. Neorgan. Khim., **5**, No. 11, 2539 (1960)	634
764.	RUSTAMOV P. G., Izv. Akad. Nauk Azerb. SSR, No. 5, 45 (1957)	3322

Reference number	Author's name and reference	Table number
765.	RUSTAMOV P. G., BERGMAN A. G., Zh. Neorgan. Khim., **3**, No. 9, 2184, 2192 (1958)	3323
766.	RUSTAMOV P. G., MARDAKHAEV B. N., Azerb. Khim. Zh., No. 4, 131 (1963)	2619b
767.	RYSS I. G., ZAYARNII A. E., ZELYANSKAYA A. I., Zh. Prikl. Khim., **14**, No. 1, 46 (1941)	1186, 5122
768.	RYSS I. G., SLUTSKAYA M. M., VITUKHNOVSKAYA B. S., Dokl. Akad. Nauk SSSR, 78, No. 2, 287 (1951)	2822
769.	RYSS I. G., SLUTSKAYA M. M., VITUKHNOVSKAYA B. S., Zh. Prikl. Khim., **25**, No. 2, 151 (1952)	307
770.	RYSS I. G., ZHUKOV A. I., Tr. Ural'. Nauch.-Issled. Khim. Inst., No. 1, 23 (1954)	3345
771.	RYSS I. G., VITUKHNOVSKAYA B. S., Zh. Obshch. Khim., **25**, No. 4, 643 (1955)	592
772.	RYSS I. G., NILUS E. L., Zh. Obshch. Khim., **25**, No. 6, 1076 (1955)	1014
773.	RYSS I. G., BOGDANOVA L. P., Zh. Neorgan. Khim., **3**, No. 7, 1669 (1958)	2845
774.	SAVCHENKO G. S., TANANAEV I. V., Zh. Prikl. Khim., **19**, No. 10, 1093 (1946)	582, 583, 586, 588, 591
775.	SAVCHENKO G. S., TANANAEV I. V., Zh. Prikl. Khim., **20**, 385 (1947)	578, 579, 584, 592
776.	SAGAIDACHNYI A. F., MORDBERG L. K., Tr. Vses. Nauch. -Issled. Inst. Gal. (1933–1934), cited in [1]	3266, 3454
777.	SAGAIDACHNYI A. F., SOLOV'EVA E. F., DUBROVO S. K., Tr. Vses. Nauch.-Issled. Inst. Gal., (1936), cited in [1]	2855
778.	SAMUSEVA R. G., EGOROVA R. S., PLYUSHCHEV V. E., Zh. Neorgan. Khim., 7, No. 7, 1667 (1962)	3919
779.	SAMUSEVA R. G., PLYUSHCHEV V. E., EGOROVA R. S., Zh. Neorgan. Khim., 7, No. 6, 1416 (1962)	3950
780.	SARSEMBAEV M. S., PONOMAREV V. D., Vestn. Akad. Nauk Kaz. SSR, No. 8, 69 (1960)	2784, 3024
781.	SAKHAROVA M. F., Uch. Zap. Yaroslav. Gos. Pedagog. Inst., 31, (41), 247, (1956)	4525
782.	SVESHNIKOVA V. N., Izv. Sektora Fiz.-Khim. Anal., Akad. Nauk SSSR, **17**, 348 (1949)	4692, 5118
783.	SVESHNIKOVA V. N., Izv. Otd. Khim. Nauk, Akad. Nauk SSSR, No. 3, 448 (1949)	4693
784.	SVESHNIKOVA V. N., Izv. Akad. Nauk, Otd. Khim. Nauk, No. 1, 45 (1952)	4693, 4728, 4858, 5119
785.	SVESHNIKOVA V. N., Dokl. Akad. Nauk SSSR, **85**, No. 2, 357 (1952)	1738
786.	SVESHNIKOVA V. N., DANILOVA E. P., Zh. Neorgan. Khim., 2, No. 4, 928 (1957)	4271
787.	SVESHNIKOVA V. N., Zh. Neorgan. Khim., **5**, No. 2, 477 (1960)	4208
788.	SVESHNIKOVA V. N., GINZBURG V. L., Zh. Neorgan. Khim., 7, No. 5, 1172 (1962)	685
789.	SEDEL'NIKOV G. S., Zh. Prikl. Khim., **17**, No. 6, 338 (1944)	2951, 2961, 2979, 3277
790.	SEDEL'NIKOV G. S., Zh. Prikl. Khim., **18**, Nos. 7–8, 430 (1945)	2866, 2872, 2887, 2895, 2912

Reference number	Author's name and reference	Table number
791.	SEDEL'NIKOV G. S., LAZAREVA A. I., Inst. Obshch. Neorgan. Khim., Akad. Nauk SSSR, 1948, cited in [1]	2856
792.	SEDEL'NIKOV G. S., STEPINA S. B., Tr. Mosk. Inst. Tonkoi Khim. Tekhnol., vyp. 7, 102 (1958)	452
793.	SEDEL'NIKOV G. S., TROFIMOVICH A. A., Zh. Neorgan. Khim., **4**, No. 6, 1443 (1959)	2876
794.	SELEZNEV K. A., Tr. Khim. Khim. Tekhnol., Gor'kii, vyp. 2, 265 (1958)	4603
795.	SELEZNEV K. A., SHATALINA G. A., Tr. Khim. Khim. Tekhnol., Gor'kii, vyp. 1, 30 (1962)	339
796.	SELIVANOVA A. S., Tr. Mosk. Inst. Tonkoi Khim. Tekhnol., vyp. 3, 23 (1952)	3665, 3724
797.	SELIVANOVA A. S., Tr. Mosk. Inst. Tonkoi Khim., Tekhnol., vyp. 20, 70 (1955)	1475, 3892, 4969
798.	SEMENCHENKO V. K., ZAVADA E. A., Zh. Obshch. Khim., **1**, vyp. 8–9, 1121 (1931)	949, 963, 2466, 3584
799.	SEMISHIN V. I., ABRAMOV I. I., VOROTNITSKAYA L. T., Izv. Vyssh. Ucheb. Zaved., Khim. Khim. Tekhnol., No. 6, 834 (1959)	4015
800.	SEREBRENNIKOV V. V., VOTINOVA V. P., Zh. Neorgan. Khim., **4**, No. 6, 1428 (1959)	207, 208
801.	SEREBRENNIKOVA M. T., Zh. Prikl. Khim., **12**, 577 (1939)	2617
802.	SEREBRENNIKOVA M. T., Zh. Prikl. Khim., **26**, No. 10, 1083 (1953)	459
803.	SECHENOV I. M., Zh. Russ. Fiz.-Khim. Obshchest., **18**, No. 5, 239 (1886)	1623, 1630
804.	SIGOV S. A., MIRZAEV F. M., Dokl. Akad. Nauk Uzb. SSR, No. 10, 29 (1954)	655
805.	SIGOV S. A., SADYKOVA G. YA., Uzb. Khim. Zh., No. 4, 18 (1959)	4293
806.	SIGOV S. A., SADYKOVA G. YA., Uzb. Khim. Zh., No. 1, 7 (1961)	4210
807.	SINANI S. S., KHUDYAKOV V. N., Tr. Gos. Inst. Prikl. Khim., vyp. 23, 102 (1935)	3013
808.	SLOVINSKAYA V. M., MUKIMOV S. M., Dokl. Akad. Nauk Uzb. SSR, No. 11, 35 (1956)	3573
809.	SLOVINSKAYA V. M., MUKIMOV S. M., Uzb. Khim. Zh., No. 2, 12 (1959)	3571, 3574, 3704
810.	SMUROV A. A., Zap. Vses. Mineral. Obshchest., Akad. Nauk SSSR, **67**, 465 (1938)	1723
811.	SOBOLEVA O. S., Nauch. Zap. L'vov. Univ., **34**, 56 (1955)	5241
812.	SOKOLOV V. A., Izv. Akad. Nauk SSSR, Ser. Khim. Nauk, 123 (1938)	2140
813.	SOKOLOVA E. I., CHIZHIKOV ZH. M., Zh. Neorgan. Khim., **2**, No. 7, 1662 (1957)	2625
814.	SOLOV'EVA E. F., Tr. Vses. Nauch.-Issled. Inst. Gal., (1946), cited in [1]	3595
815.	SOLOV'EVA E. F., Tr. Vses. Nauch.-Issled. Inst. Gal., **21**, 221–222 (1949)	4049
816.	SOLOV'EVA E. F., Tr. Vses. Nauch.-Issled. Inst. Gal., **27**, No. 48, 67 (1953)	4039
817.	SOLOV'EVA E. F., Tr, Vses. Nauch.-Issled. Inst. Gal., vyp. 31, 180 (1956)	3309
818.	SPERANSKAYA E. I., Izv. Sektora Fiz.-Khim. Anal., Inst. Obshch. Neorgan. Khim., Akad. Nauk SSSR, **24**, 212 (1954)	3083

Reference number	Author's name and reference	Table number
819.	SPIRYAGINA A. I., Tr. Vses.-Nauch.-Issled. Inst. Gal., vyp. 27, 77. (1953)	1586–1590
820.	SPITSYN V. I., MATYAZH I. YA., Zh. Neorgan. Khim., **4**, No. 4, 830 (1959)	393, 713, 921
821.	SPITSYN V. I., KOMISSAROVA L. N., SHATSKII V. M., PUSHKINA G. YA., Zh. Neorgan. Khim., **5**, No. 10, 2223 (1960)	2120
822.	STARKOVA Z. P., Zh. Obshch. Khim., **1**, No. 6, 749 (1931)	4446
823.	STEPINA S. B., SEDEL'NIKOV G. S., KOSTIKOVA R. V., Zh. Neorgan. Khim., 7, No. 3, 638 (1962)	451, 453 5097
824.	STRAKHOV N. M., ZARUBITSKAYA A. N., Tr. Inst. Geol. Nauk, Akad. Nauk SSSR, vyp. 124, Geolog. Ser. (No. 45), 18 (1951)	3449, 3741, 4082, 4151, 5078
825.	SUKMANSKAYA G. V., BOGOYAVLENSKII P. S., Zh. Prikl. Khim., **30**, No. 6, 948 (1957)	2889
826.	SUKMANSKAYA G. V., BOGOYAVLENSKII P. S., Zh. Neorgan. Khim., **5**, No. 4, 978 (1960)	2891, 2896, 3285
827.	SUNDUKOVA I. D., IL'IN K. G., VANINA G. F., Izv. Vyssh. Ucheb. Zaved., Khim. Khim. Tekhnol., 7, No. 3, 360 (1964)	1399
828.	TALIPOV SH. T., KHADEEV V. A., Dokl. Akad. Nauk Uzb. SSSR, No. 7, 17 (1950)	4013
829.	TALIPOV SH. T., KHADEEV V. A., Dokl. Akad. Nauk Uzb. SSR, No. 8, 16 (1950)	3165
830.	TALIPOV SH. T., KHADEEV V. A., Zh. Obshch. Khim., **20** (82), No. 5, 774 (1950)	3173, 3176, 4653, 4658
831.	TALIPOV SH. T., ANTIPOV V. E., Tr. Inst. Khim., Akad. Nauk Uzb. SSR,, 3, 206 (1952)	1328, 3170
832.	TALIPOV SH. T., OBEL'CHENKO P. F., Dokl. Akad. Nauk Uzb. SSR, No. 1, 24 (1952)	3131, 3492
833.	TALIPOV SH. T., OBEL'CHENKO P. F., Dokl. Akad. Nauk Uzb. SSR, No. 3, 20 (1952)	3130, 4604, 4671, 4749
834.	TALIPOV SH. T., ABDULLAEV M. A., KIRSANOVA M. V., Tr. Sredneaziat. Gos. Univ., vyp. 40, Khim. Nauki, kniga 5, 43 (1953)	4014
835.	TALIPOV SH. T., FEDOROVA T. I., Tr. Sredneaziat. Gos. Univ., vyp. 40, Khim. Nauki, kniga 5, 47 (1953)	4647
836.	TALIPOV SH. T., OBEL'CHENKO P. F., Dokl. Akad. Nauk Uzb. SSR, No. 5, 42 (1953)	3203
837.	TALIPOV SH. T., OBEL'CHENKO P. F., Dokl. Akad. Nauk Uzb. SSR, No. 7, 29 (1953)	3489
838.	TALIPOV SH. T., OBEL'CHENKO P. F., Tr. Sredneaziat. Gos. Univ., 55, Khim. Nauki, No. 7, 3 (1954)	3132, 3204, 3493, 4604, 4671, 4749
839.	TALIPOV SH. T., KUTUMOVA O. F., Dokl. Akad. Nauk Uzb. SSR, No. 6, 15 (1956)	3179, 4663
840.	TALIPOV SH. T., KRUKOVSKAYA E. L., Tr. Sredneaziat. Gos. Univ., vyp. 84, 3 (1958)	5199, 5200
841.	TALIPOV SH. T., PODGORNOVA V. S., ZININA G. N., Uzb. Khim. Zh., No. 4, 11 (1961)	1504
842.	TALIPOV SH. T., RAKHMATULLAEV K., Uzb. Khim. Zh., No. 6, 8 (1961)	4659

Reference number	Author's name and reference	Table number
843.	TALIPOV SH. T., RAKHMATULLAEV K., TASHPULATOV YU., Symposium: "Some questions on chemical technology and physico-chemical analysis of inorganic systems" (Nekotorye voprosy khimicheskoi tekhnologii i fiziko-khimicheskogo analiza neorganicheskikh sistem), Tashkent, Akad. Nauk Uzb. SSR, 96 et seq. (1963)	3177, 4661
844.	TALIPOV SH. T., RAKHMATULLAEV K., TASHPULATOV YU., Dokl. Akad. Nauk Uzb. SSR, No. 7, 34 (1964)	3178
845.	TAMM N. S., NOVOSELOVA A. V., Zh. Neorgan. Khim., **2**, No. 6, 1428 (1957)	1508
846.	TANANAEV I. V., CHRELASHVILI S., Zh. Obshch. Khim., **6**, 696 (1936)	981, 1009, 1070
847.	TANANAEV I. V., Zh. Prikl. Khim., **11**, 214 (1938)	576
848.	TANANAEV I., CHRELASHVILI S., Zh. Obshch. Khim., **8**, 1120 (1938)	570, 1110
849.	TANANAEV I. V., Izv. Sektora Fiz.-Khim. Anal., Akad. Nauk SSSR, **14**, 364 (1941)	577, 4385
850.	TANANAEV I. V., Zh. Obshch. Khim., **11**, 267 (1941)	565
851.	TANANAEV I. V., LEL'CHUK YU. L., Dokl. Akad. Nauk SSSR, **11**, 114 (1943)	3166
852.	TANANAEV I. V., DEICHMAN E. N., Zavod. Lab., **11**, 763 (1945)	3171
853.	TANANAEV I. V., DEICHMAN E. N., Zh. Prikl. Khim., **19**, No. 9, 1018 (1946)	599
854.	TANANAEV I. V., DEICHMAN E. N., Izv. Akad. Nauk SSSR, Ser. Khim. Nauk, No. 4, 373 (1946)	4648
855.	TANANAEV I. V., Zh. Anal. Khim., **3**, 276 (1948)	5009
856.	TANANAEV I. V., LEL'CHUK YU. L., PETROVITSKAYA V. KH., Zh. Obshch. Khim., **19**, 1207 (1949)	4003, 5099, 5357, 5424
857.	TANANAEV I. V., NEKHAMKINA M. A., Izv. Sektora Fiz.-Khim. Anal., Inst. Obshch. Neorgan. Khim., Akad. Nauk SSSR, **20**, 227 (1950)	4236
858.	TANANAEV I. V., DEICHMAN E. N., Izv. Akad. Nauk SSSR, Otd. Khim. Nauk, 26 (1951)	1501
859.	TANANAEV A. V., MIZETSKAYA I. B., Zh. Anal. Khim., **6**, No. 6, 337 (1951)	1348, 1363, 1370
860.	TANANAEV I. V., MIZETSKAYA I. B., VINOGRADOVA A. D., Zh. Anal. Khim., **7**, No. 1, 14 (1952)	1395, 4227
861.	TANANAEV I. V., Khim. Redk. Elem., vyp. 2, 33 (1954)	560
862.	TANANAEV I. V., BAUSOVA N. V., Khim. Redk. Elem., Inst. Obshch. Neorgan. Khim., Akad. Nauk SSSR, No. 2, 21 (1955)	3172
863.	TANANAEV I. V., NIKOLAEV N. S., BUSLAEV YU. A., Zh. Noergan. Khim., **1**, No. 2, 274 (1956)	609
864.	TANANAEV I. V., SAVCHENKO G. S., Zh. Neorgan. Khim., **2**, No. 10, 2449 (1957)	606
865.	TANANAEV I. V., AVDUEVSKAYA K. A., Zh. Neorgan. Khim., **3**, No. 9, 2165 (1958)	605
866.	TANANAEV I. V., LU CHAO-DA, Zh. Neorgan. Khim., **4**, No. 9, 2122 (1959)	2221, 3169, 3180, 5109
867.	TANANAEV I. V., LU CHAO-DA, Zh. Neorgan. Khim., **4**, No. 12, 2796 (1959)	4664
868.	TANANAEV I. V., VINOGRADOVA A. D., Zh. Neorgan. Khim., **5**, No. 2, 321 (1960)	1502, 1503, 2128, 3168, 5107
869.	TANANAEV I. V., VINOGRADOVA A. D., Zh. Neorgan. Khim., **5**, No. 2, 321 (1960)	5108

Reference number	Author's name and reference	Table number
870.	TANANAEV I. V., DEICHMAN E. N., Radiokhimiya, No. 6, 712 (1961)	5110, 5585
871.	TANANAEV I. V., SAVCHENKO G. S., At. Energ., **12**, No. 5, 397 (1962)	559, 629
872.	TANANAEV I. V., CHUDINOVA N. N., Zh. Neorg. Khim., 7, No. 10, 2286 (1962)	682
873.	TANANAEV I. V., NIKOLAEV N. S., BUSLAEV YU. A., Symposium: "Physico-chemical analysis" (Fiziko-khimicheskii analiz), Sib. Otd., Akad. Nauk SSSR, Novosibirsk, p. 85 (1963)	594, 611
874.	TANANAEV I. V., RODICHEVA G. V., At. Energ., **14**, No. 4, 395 (1963)	3205
875.	TANANAEV I. V., VASIL'EVA V. P., Zh. Neorgan. Khim., **9**, No. 1, 213 (1964)	684
876.	TAPEROVA A. A., Zh. Prikl. Khim., **13**, No. 5, 643 (1940)	4282
877.	TAPEROVA A. A., SHUL'GINA M. N., Zh. Prikl. Khim., **18**, Nos. 9–10, 521 (1945)	4283
878.	TAPEROVA A. A., ISAEVA T. T., Zh. Prikl. Khim., **22**, 343 (1949)	678
879.	TIKHOMIROV V., Zh. Russ. Fiz.-Khim. Obshchest., **39**, 731 (1909)	4919
880.	TISHCHENKO V. E., KRAT V. N., RAINES M. M., Tr. Gos. Inst. Prikl. Khim., vyp. 29, 173 (1936)	3014
881.	TOPTYGINA G. M., MOROZOV I. S., Zh. Neorgan. Khim., **6**, No. 7, 1686 (1961)	1023
882.	TRANDAFELOV D., GEORGIEV G., God. Sofii. Univ., Fiz.-Mat. Fak., **51**, No. 3, 71, 1956–1957 (1958)	2280
883.	TRANDAFELOV D., POPYANKOV B., Khim. Ind. (Sofia), No. 5, 142 (1961)	5067
884.	TRET'YAKOV YU. D., BOGDASAR'YAN A. KH., Zh. Neorgan. Khim., **6**, No. 9, 2203 (1961)	2394
885.	TRET'YAKOV YU. D., SIMAKOVA L. K., Zh. Neorgan. Khim., **6**, No. 9, 2203 (1961)	2400, 2404, 2410
886.	TRET'YAKOV YU. D., BAGDASAR'YAN A. KH., Zh. Neorgan. Khim., 7, No. 7, 1709 (1962)	2401, 2405, 2412
887.	TRET'YAKOV YU. D., BAGDASAR'YAN A. KH., Zh. Neorgan. Khim., 7, No. 7, 1717 (1962)	2386, 2389, 2402
888.	TULINOVA V. B., MORZHANOVA L. G., PLYUSHCHEV V. E., Zh. Neorgan. Khim., **4**, No. 5, 1170 (1959)	1280
889.	TULINOVA V. B., PLYUSHCHEV V. E., TERNOVSKAYA I. V., Zh. Neorgan. Khim., **5**, No. 3, 695 (1960)	3501
890.	TULINOVA V. B., STATSENKO A. A., PLYUSHCHEV V. E., Zh. Neorgan. Khim., **5**, No. 11, 2530 (1960)	4391
891.	TUR'YAN YA. I., Zh. Neorgan. Khim., **6**, No. 1, 162 (1961)	5028, 5432, 5581
892.	TYCHINSKAYA I. I., OPALOVSKII A. A., NIKOLAEV N. S., Zh. Neorgan. Khim., **9**, No. 7, **1696 (1964)**	607, 620
893.	UGAI YA. A., Zh. Prikl. Khim., **23**, No. 5, 484 (1950)	342
894.	UGAI YA., SIMKINA T., Zh. Obshch. Khim., **20** (82), No. 10, 1768 (1950)	338
895.	UDINTSEVA V. S., POPOVICH I. A., Zh. Prikl. Khim., **4**, Nos. 2–3, 221 (1931)	2631

Reference number	Author's name and reference	Table number
896.	UKRAINTSEVA E. A., Izv. Sib. Otd., Akad. Nauk SSSR, Ser. Khim. Nauk, vyp. 1, 14 (1963)	243, 244, 246
897.	UL'YANOV A. I., KAZAKOVA T. I., Izv. Akad. Nauk SSSR, Ser. Khim., 7, 1157 (1963)	686
898.	URAZOV G. G., LIFATOVA Z. I., Zh. Prikl. Khim., **17**, No. 1, 16 (1944)	1303, 1309
899.	URAZOV G. G., KIRAKOSYAN A. K., Izv. Sektora Fiz.-Khim. Anal., Inst. Obshch. Neorgan. Khim., Akad. Nauk SSSR, **22**, 255 (1953)	1837, 2023
900.	URAZOV G. G., KIRAKOSYAN A. K., Izv. Sektora Fiz.-Khim. Anal., Inst. Obshch. Neorgan. Khim., Akad. Nauk SSSR, **23**, 241 (1953)	1836, 2050
901.	URAZOV G. G., KIRAKOSYAN A. K., Izv. Sektora Fiz.-Khim. Anal., Inst. Obshch. Neorgan. Khim., Akad. Nauk SSSR, **25**, 268 (1954)	5431
902.	URAZOV G. G., SEDEL'NIKOV G. S., TITOVA M. YA., Izv. Sektora Fiz.-Khim. Anal., Inst. Obshch. Neorgan. Khim., Akad. Nauk SSSR, **25**, 361 (1954)	2044, 2045
903.	URAZOV G. G., KIRAKOSYAN A.K., LYASHCHENKO M.N., Izv. Sektora Fiz.-Khim. Anal., Inst. Obshch. Neorgan. Khim., Akad. Nauk SSSR, **27**, 315 (1956)	2022
904.	URAZOV G. G., SHEVTSOVA Z. N., Zh. Neorgan. Khim., **2**, No. 3, 656 (1957)	2202, 4008, 5258, 5299
905.	URAZOV G. G., KIRAKOSYAN A. K., GALUSTYAN V. D., Zh. Neorgan. Khim., **2**, No. 5, 1094 (1957)	1849, 1857, 1974, 1975
906.	URAZOV G. G., KIRAKOSYAN A. K., Zh. Neorgan. Khim., **2**, No. 6, 1332 (1957)	1852, 2025
907.	URAZOV G. G., BASHILOVA N. I., Zh. Neorgan. Khim., **2**, No. 8, 1922 (1957)	897
908.	URAZOV G. G., TULINOVA V. B., PLYUSHCHEV V. E., CHUIKINA N. I., Khim. Redk. Elem., vyp. 3, 14 (1957)	2368
909.	URAZOV G. G., KIRAKOSYAN A. K., MKHITARYAN R. S., Zh. Neorgan. Khim., **2**, No. 2, 464 et seq. (1958)	1819, 1850, 1858, 1860, 1976, 1979, 1981, 1982
910.	URAZOV G. G., KINDYAKOV P. S., Tr. Mosk. Inst. Tonkoi Khim. Tekhnol., vyp. 7, 81 (1958)	2312
911.	URAZOV G. G., KINDYAKOV P. S., KHOKHLOVA A. V., Tr. Mosk. Inst. Tonkoi Khim. Tekhnol., vyp. 7, 90 (1958)	1517
912.	URAZOV G. G., SNAIDERMAN L. I., PRYANISHNIKOVA T. A., Tr. Mosk. Inst. Tonkoi Khim. Tekhnol., vyp. 7, 123 (1958)	3511
913.	URAZOV G. G., KINDYAKOV P. S., SNAIDERMAN L. I., Tr. Mosk. Inst. Tonkoi Khim. Tekhnol., vyp. 7, 140 (1958)	1150
914.	URAZOV G. G., KINDYAKOV P. S., CHUKAN A. K., Tr. Mosk. Inst. Tonkoi Khim. Tecknol., vyp. 7, 144 (1958)	4362
915.	URAZOV G. G., KIRAKOSYAN A. K., GALSTYAN V. D., Izv. Akad. Nauk Arm. SSR, Khim. Nauki, **11**, No. 4, 249 (1958)	1972
916.	URAZOV G. G., LIPSHITS B. M., LOVCHIKOV V. S., Zh. Neorgan. Khim., **4**, No. 2, 439 (1959)	2621
917.	URAZOV G. G., LIPSHITS B. M., LOVCHIKOV V. S., Zh. Neorgan. Khim., **4**, No. 10, 2380 (1959)	2622

Reference number	Author's name and reference	Table number
918.	URAZOV G. G., LIPSHITS B. M., LOVCHIKOV V. S., Zh. Neorgan. Khim., **5**, No. 4, 950 (1960)	2619a, 2715
919.	URAZOV G. G., LIPSHITS B. M., LOVCHIKOV V. S., Zh. Neorgan. Khim., **5**, No. 11, 2509 (1960)	2716, 2720, 2752, 2753
920.	USPENSKAYA L. N., BERGMAN A. G., Zh. Obshch. Khim., **25**, No. 7, 1277 (1955)	5094, 5100, 5161, 5165, 5483
921.	USPENSKAYA L. N., GLUSHKOVA N. P., BERGMAN A. G., Zh. Obshch. Khim., **25**, No. 9, 1658 (1955)	5101, 5162, 5483, 5484
922.	USPENSKAYA L. N., BERGMAN A. G., Zh. Obshch. Khim., **25**, No. 11, 2028 (1955)	3072, 3077, 3114, 3666, 4565
923.	FEDOROV P. I. Acta Chim. Sinica, **23**, No. 6, 469 (1957)	5443
924.	FEDOROV P. I., POSTNIKOVA S. V., STIFIENKO L. A., Zh. Neorgan. Khim., **7**, No. 7, 1673 (1962)	4835
925.	FEDOROV P. I., AKULKINA L. M., RAZGON E. S., Zh. Neorgan. Khim., **8**, 260 (1963)	2437
926.	FEDOROV P. I., SLOTVINSKII-SIDAK N. P., TSIMBALIST ZH. N., Neorgan. Khim., **8**, 2594 (1963)	813
927.	FEDOROV P. I., EVRASOVA I. I., Sb. Stud. Rab. Yaroslav. Gos. Pedagog. Inst., vyp. 1, Estestv. Geogr., 29 (1958)	4798
928.	FEIGEL'SON I. B., LARINA A. P., Byull. Inst. Galurgii, Nos. 4–5, 45 (1939)	3593
929.	FILINOV F. M., TEKSTER E. N., KOLPAKOVA A. A., PANTELEEVA E. P., Zh. Neorgan. Khim., **5**, No. 5, 1147 (1960)	3215
930.	FILIPPEO V. M., Izv. Sektora Fiz.-Khim. Anal., Akad. Nauk SSSR, **13**, 325 (1940)	4390
931.	FRIDMAN YA. D., SARBAEV D. S., Zh. Neorgan. Khim., **4**, No. 8, 1849 (1959)	4505, 4507, 5005, 5006
932.	KHAIDUKOV N. I., LINETSKAYA Z. G., Kalii, No. 8, 33 (1935)	3648
933.	KHITROVA N. N., Zh. Prikl. Khim., **27**, No. 12, 1283 (1954)	3071
934.	KHITROVA N. N., Izv. Sektora Fiz.-Khim. Anal., Inst. Obshch. Neorgan. Khim., Akad. Nauk SSSR, **27**, 344 (1956)	3079
935.	KHLOPIN V., Z. anorg. Chem., **143**, 108 (1925)	515
936.	KHRIPIN L. A., Zh. Neorgan. Khim., **5**, No. 1, 181 (1960)	4240, 4242, 4259, 4760
937.	KHRIPIN L. A., LEPESHKOV M. N., Zh. Neorgan. Khim., **5**, No. 2, 481 (1960)	2395, 4241, 4243, 4260, 4268, 4761
938.	KHRIPIN L. A., Zh. Neorgan. Khim., **10**, No. 1, 220 (1965)	1394
939.	HU KE-YUAN', Zh. Neorg. Khim., **4**, No. 8, 1910 (1959)	1353
940.	HU KE-YUAN', Zh. Neorg. Khim., **5**, No. 1, 191 (1960)	1336, 1412
941.	TSEITLIN A. N., Zh. Prikl. Khim., **19**, 820 (1946)	39

Reference number	Author's name and reference	Table number
942.	CHAMOVA V. P., SERGEEVA V. P., Zh. Neorgan. Khim., **2**, No. 8, 1938 (1957)	264
943.	CHEPELEVETSKII M. L., BOL'TS TS. S., Zh. Prikl. Khim., **10**, 1183 (1937)	656
944.	CHEPELEVETSKII M. L., BOL'TS TS. S., VASILENKO N. A., RUBINOVA S. S., Symposium: "Studies in applied chemistry" (Issledovaniya po prikladnoi khimii", Akad. Nauk SSSR, M.-L., 175 (1955)	670, 4280
945.	CHERNOVA K. S., KORZH E. V., Zh. Obshch. Khim., **16**, 171 (1946)	2242
946.	CHUN HUAN-BAN, FEDOROV P. I., Zh. Neorgan. Khim., **6**, No. 4, 971 (1961)	3520a
947.	CHIZHIKOV D. M., SHAKOV A. S., Zh. Prikl. Khim., **9**, 1387 (1936)	5149, 5179, 5311
948.	CHUKHLANTSEV V. G., Zh. Anal. Khim., **11**, No. 5, 529 (1956)	426, 529, 819, 893, 1030
949.	CHUKHLANTSEV V. G., Zh. Anal. Khim., **12**, No. 3, 296 (1957)	466
950.	CHUKHLANTSEV V. G., ALYAMOVSKAYA K. V., Izv. Vyssh. Ucheb. Zaved., Khim. Khim. Tekhnol., No. 5, 706 (1961)	468
951.	SHAPIRO K. YA., YURKEVICH YU. N., KULAKOVA V. V., Zh. Neorgan. Khim., **10**, No. 2, 555 (1965)	2447
952.	SHAKHNO I. V., PLYUSHCHEV V. E., TITULINA E. M., SAMUSEVA R. G., Zh. Neorgan. Khim., **8**, 1467 (1963)	3884
953.	SHAKHNO I. V., PLYUSHCHEV V. E., AVZHIEVA E. M., Zh. Neorgan. Khim., **10**, No. 2, 552 (1965)	3883
954.	SHAKHNO I. V., PLYUSHCHEV V. E., Zh. Neorgan. Khim., **10**, No. 5, 1237 (1965)	3886
955.	SHVARTSMAN B. KH., VOLKOVA N. S., Zh. Prikl. Khim., **34**, No. 5, 1155 (1961)	2600
956.	SHVARTSMAN B. KH., VOLKOVA N. S., Zh. Prikl. Khim., **34**, No. 7, 1490 (1961)	2667, 4387
957.	SHVARTSMAN B. KH., VOLKOVA N. S., Zh. Prikl. Khim., **34**, No. 12, 2634 (1961)	2668, 2669
958.	SHEVELEVA A. D., KUDRYASHOV S. F., Uch. Zap. Perm. Gos. Univ., **13**, vyp. 3, 63 (1959)	2564
959.	SHEVELEVA A. D., BOGDANOVSKAYA R. L., ZHURAVLEV E. F., Zh. Neorgan. Khim., **9**, No. 6, 1435 (1964)	3835, 4916, 5285
960.	SHEVCHUK V. G., Soobshch. Nauch. Rab. Chlenov Vses. Khim. Obshchest., vyp. 2, 42 (1954)	2301
961.	SHEVCHUK V. G., Tr. Yaroslav. Sel'skokhoz. Inst., **2**, vyp. 2, 74 (1955)	3486
962.	SHEVCHUK V. G., Izv. Vyssh. Ucheb. Zaved., Khim. Khim. Tekhnol., No. 2, 25 (1958)	2355
963.	SHEVCHUK V. G., Uch. Zap. Yaroslav. Gos. Pedagog. Inst., **32** (42), 341 (1959)	2434, 3551
964.	SHEVCHUK V. G., Zh. Neorgan. Khim., **6**, No. 8, 1957 (1961)	1366, 4097
965.	SHEVCHUK V. G., VAISFEL'D M. I., TRET'YAK S. S., Zh. Neorgan. Khim., **7**, No. 8, 1992 (1962)	1387, 1525
966.	SHEVCHUK V. G., Zh. Neorgan. Khim., **7**, No. 10, 2464 (1962)	1389
967.	SHEVCHUK V. G., Zh. Neorgan. Khim., **8**, 1965 (1963)	1369
968.	SHEVCHUK V. G., KOST' L. L., Zh. Neorgan. Khim., **9**, No. 2 432 (1964)	2308, 4104

Reference number	Author's name and reference	Table number
969.	SHEVCHUK V. G., KOST' L. L., Zh. Noergan. Khim., **9**, No. 5, 1242 (1964)	1368, 1380
970.	SHEKA I. A., PEVZNER TS. V., Zh. Neorgan. Khim., **5**, No. 10, 2311 (1960)	2775
971.	SHELUD'KO M. K., KULISH N. F., Tr. Dnepropetrovsk. Khim.-Tekhnol. Inst., vyp. 5, 201 (1956)	2149, 2150, 2172, 2473, 3073
972.	SHEPELEV A. I., Uch. Zap., Mord. Gos. Univ., vyp. XVI, Ser. Khim. Nauk, 57 (1962)	1433
973.	SHISHKIN N. V., POKHVALENSKII E. D., Zh. Obshch. Khim., **8**, 1125 (1938)	743
974.	SHISHKIN N. V., KROGIUS E. A., POKHVALENSKII E. D., Uch. Zap. Saratov. Gos. Univ., vyp. Khim., **43**, 3 (1956)	745, 2038
975.	SHISHKIN N. V., LIPOVSKAYA B. A., Uch. Zap. Saratov. Gos. Univ., vyp. Khim., **43**, 19 (1956)	2393, 4703
976.	SHLEZINGER N. A., ZORKIN F. P., "Problems of Lake El'ton" (Problemy ozera El'ton), vyp. 1, p. 115 et seq., Stalingrad (1936)	3559, 3562
977.	SHLEZINGER N. A., ZORKIN F. P., NOVOZHENOVA L. V., Zh. Prikl. Khim., **11**, 1259 (1938)	4817
978.	SHLEZINGER N. A., FEIGEL'SON I. B., SPIRYAGINA A. I., Dokl. Akad. Nauk SSSR, **20**, Nos. 2–3, 142 (1938)	2836
979.	SHLEZINGER N. A., ZORKIN F. P., PETUKHOVA E. V., Dokl. Akad. Nauk SSSR, **27**, No. 5, 466 (1940)	3609, 4046
980.	SHMIDT E., Zh. Obshch. Khim., **9**, 13, 1220 (1939)	1939
981.	SHOIKHET D. N., Izv. Sektora Fiz.-Khim. Anal., Akad. Nauk SSSR, **10**, 331 (1938)	4031
982.	SHOKIN I. N., KIM SEN SIK, Tr. Mos. Khim.-Tekhnol. Inst., No. 25, 12 (1961)	1600
983.	SHOKIN I. N., YAKHONTOVA E. A., TIMOFEEVA N. N., Tr. Mosk. Khim.-Tekhnol. Inst. im. Mendeleeva, Khim. Tekhnol. Neorgan. Veshchestv, **35**, 24 (1961)	1601, 1604, 1605, 1609
984.	SHITTSER V. YA., Zh. Russ. Fiz.-Khim. Obshch., **49**, 357 (1917)	411, 552, 911, 912, 984, 1152, 2799, 3026, 4483, 4484
985.	SHPUNT S. YA., Zh. Prikl. Khim., **13**, No. 1, 19 (1940)	2232, 3055
986.	SHPUNT S. YA., Zh. Prikl. Khim., **19**, No. 3, 293 (1946)	3055
987.	SHPUNT S. YA., Zh. Prikl. Khim., **20**, No. 7, 686 (1947)	2229
988.	SHPUNT S. YA., BELOPOL'SKII A. P., SHUL'GINA M. N., Zh. Prikl. Khim., **24**, No. 4, 404 (1951)	3965
989.	SHPUNT S. YA., Zh. Prikl. Khim., **30**, No. 7, 986 (1957)	2230
990.	SHPUNT S. YA., Zh. Prikl. Khim., **30**, No. 8, 1149 (1957)	2231
991.	SHTERNINA E. B., FROLOVA E. V., Dokl. Akad. Nauk SSSR, **47**, No. 1, 35 (1945)	1649, 1653, 1655
992.	SHTERNINA E. B., FROLOVA E. V., Dokl. Akad. Nauk SSSR, **50**, 264 (1945)	1649
993.	SHTERNINA E. B., Dokl. Akad. Nauk SSSR, **40**, 247 (1948)	3750
994.	SHTERNINA E. B., FROLOVA E. V., Izv. Sektora Fiz.-Khim. Anal., Akad. Nauk SSSR, **17**, 354 (1949)	3751
995.	SHTERNINA E. B., Izv. Sektora Fiz.-Khim. Anal., Akad. Nauk SSSR, **19**, 348 (1949)	1650
996.	SHTERNINA E. V., Zh. Neorgan. Khim., **2**, No. 4, 933 (1957)	3732, 3776

Reference number	Author's name and reference	Table number
997.	SHTERNINA E. B., FROLOVA E. V., Zh. Neorgan. Khim., **2**, No. 7, 1649 (1957)	1654
998.	SHUL'GINA M. P., KHARCHUK O. S., YANAT'EVA O. K., Izv. Sektora Fiz.-Khim. Anal., Inst. Obshch. Neorgan. Khim., Akad. Nauk SSSR, **26**, 198 (1955)	4687
999.	SHUL'TS M. M., MAKAROV L. L. MARINICHEV A. N., SU YU-CHEN, Zh. Fiz. Khim., **37**, No. 6, 1219 (1963)	2523
1000.	SHURDUMOV G. K., PROTSENKO P. I., Zh. Neorgan. Khim., **9**, No. 5, 1237 (1964)	3050, 4531
1001.	SHEVTSOVA Z. N., ZELOVA V. S., USHAKOVA L. I., Nauch. Dokl. Vyssh. Shk., Khim. Khim. Tekhnol., No. 3, 417 (1958)	3833, 3836, 5168, 5169
1002.	SHCHEDRINA A. P., OZEROVA M. I., KHOMYAKOV K. G., Zh. Neorgan. Khim., **9**, No. 3, 724 (1964)	4168
1003.	SHCHEDRINA A. P., OZEROVA M. I., Zh. Neorgan. Khim., **9**, No. 6, 1505 (1964)	5224
1004.	SHCHEDRINA A. R., OZEROVA M. I., Zh. Neorgan. Khim., **10**, No. 6, 1504 (1965)	5227
1005.	SHCHIGOL' M. B., Zh. Obshch. Khim., **22**, No. 5, 721 (1952)	5029
1006.	SHCHUKAREV S. A., KUDRYASHOVA N. V., Gidrokhim. Mater., Izd. Akad. Nauk SSSR, **11**, 194 (1939); **17**, 70 (1950)	3752
1007.	SHCHUKAREV S. A., YAKIMOV M. P., MISHIN V. YA., Zh. Neorgan. Khim., **3**, No. 7, 1661 (1958)	503
1008.	EPSHTEIN D. A., KOSTERINA Z. N., Zh. Obshch. Khim., **9**, 623 (1939)	387
1009.	YAVORSKII S. I., Rab. Khim. Chlenov Krym. Otd. Vses. Khim. Obshchest. im. Mendeleeva, 29 (1961)	3938, 5002
1010.	YAKIMOV M. A., NOSOVA N. F., GRISHIN V. A., Zh. Neorgan. Khim., **3**, No. 2, 504 (1958)	5105, 5360, 5497
1011.	YAKIMOV M. A., NOSOVA N. F., Zh. Neorgan. Khim., **5**, No. 3, 720 (1960)	4012
1012.	YAKIMOV M. A., NOSOVA N. F., Zh. Neorgan. Khim., **6**, No. 1, 208 (1961)	1327, 1505
1013.	YAKIMOV M. A., NOSOVA N. F., Vestn. Leningrad. Gos. Univ., No. 4, Ser. Fiz.-Khim., vyp. 1, 106 (1962)	5300, 5433, 5551
1014.	YAKIMOV M. A., MISHIN V. YA., Zh. Neorgan. Khim., **8**, 227 (1963)	510
1015.	YAKIMOV M. A., MISHIN V. YA., GOLOVA L. F., Zh. Neorgan. Khim., **8**, 1470 (1963)	532
1016.	YAKIMOV M. A., NOSOVA N. F., DEGTEREV A. YA., YU TSIEN-TSI, Radiokhimiya, **5**, No. 1, 73 (1963)	5460, 5564
1017.	YAKIMOV M. A., MISHIN V. YA., Radiokhimiya, **6**, No. 4, 454 (1964)	557
1018.	YAKOVKIN G. A., Tr. Gos. Inst. Prikl. Khim., vyp. 8, 6 (1927)	1881
1019.	YAKOVKIN G. A., Tr. Gos. Inst. Prikl. Khim., vyp. 14, 4 (1930)	1869
1020.	YANAT'EVA O. K., Zh. Prikl. Khim., **19**, No. 7, 709 (1946)	3771, 4162
1021.	YANAT'EVA O. K., Zh. Obshch. Khim., **17**, 1039 (1947)	1415, 3710, 3715

Reference number	Author's name and reference	Table number
1022.	YANAT'EVA O. K., Zh. Prikl. Khim., **21**, No. 1, 26 (1948)	3278, 3303 4033
1023.	YANAT'EVA O. K., Dokl. Akad. Nauk SSSR, **67**, No. 3, 479 (1949)	1672, 3583
1024.	YANAT'EVA O. K., Izv. Sektora Fiz.-Khim. Anal., Akad. Nauk SSSR, **17**, 373 et seq. (1949)	1672, 3312, 3321, 3399, 3432
1025.	YANAT'EVA O. K., Izv. Sektora Fiz.-Khim. Anal., Akad. Nauk SSSR, **20**, 259 (1950)	1641, 3977
1026.	YANAT'EVA O. K., Izv. Sektora Fiz.-Khim. Anal., Inst. Obshch. Neorgan. Khim., Akad. Nauk SSSR, **20**, 252 (1952)	3986
1027.	YANAT'EVA O. K., Izv. Sektora Fiz.-Khim. Anal., Inst. Obshch. Neorgan. Khim., Akad. Nauk SSSR, **26**, 266 (1955)	3981, 3986
1028.	YANAT'EVA O. K., Zh. Obshch. Khim., **25**, No. 2, 234 (1955)	1673
1029.	YANAT'EVA O. K., DANILOVA E. P., Zh. Prikl. Khim., **29**, No. 8, 1152 (1956)	3579, 3979
1030.	YANAT'EVA O. K., ORLOVA V. G., Zh. Neorgan. Khim., **1**, No. 5, 988 (1956)	3400, 3434
1031.	YANAT'EVA O. K., Zh. Neorgan. Khim., **2**, No. 9, 2183 (1957)	4083
1032.	YANAT'EVA O. K., Dokl. Akad. Nauk SSSR, **112**, No. 1056 (1957)	3982
1033.	YANAT'EVA O. K., ORLOVA V. G., Zh. Neorgan. Khim., **3**, No. 10, 2408 (1958)	3403, 3404, 4081
1034.	YANAT'EVA O. K., ORLOVA V. G., Zh. Neorgan. Khim., **4**, No. 8, 1903 (1959)	3313, 3324, 3438
1035.	YANAT'EVA O. K., Zh. Neorgan. Khim., **5**, No. 11, 2582 (1960)	3989
1036.	YANAT'EVA O. K., RAPOPORT G. S., RASSONSKAYA I. S., USTINOVA M. B., Zh. Prikl. Khim., **34**, No. 10, 2347 (1961)	3988
1037.	YANAT'EVA O. K., RASSONSKAYA I. S., Zh. Neorgan. Khim., **6**, No. 6, 1424 (1961)	3987
1038.	YANKO A. P., DRUZHININ I. G., Zh. Obshch. Khim., **25**, No. 1, 17 (1955)	1377
1039.	YANKO A. P., MOSHKINA G. M., Izv. Fiz.-Khim. Nauch.-Issled. Inst. Irkutsk. Gos. Univ., **4**, No. 1, 137 (1959)	1334
1040.	YANKO A. P., Symposium: "Physico-chemical analysis" (Fiziko-khimicheskii analiz), Otd. Akad. Nauk SSSR, Novosibirsk 141 (1963)	1339
1041.	YARLYKOV M. M., Zh. Prikl. Khim., **7**, 902 (1934)	2467, 2490
1042.	YARYM-AGAEV N. L., RUDIN V. YA., TSEITLENOK T. A., Zh. Neorgan. Khim., **10**, No. 4, 976 (1965)	3080
1043.	YASTREBOVA L. F., BORINA A. F., RAVICH M. I., Zh. Neorgan. Khim., **8**, 211 (1963)	3712, 3713
1044.	YATLOV V. S., Zh. Prikl. Khim., **7**, Nos. 11–12, 754 (1930)	1603, 2474, 2476
1045.	YATLOV V. S., Zh. Prikl. Khim., **9**, No. 3, 34–36 (1932)	1602
1046.	YATLOV V. S., BELOVA N. S., Tr. Ural'. Nauch.-Issled. Khim. Inst., vyp. 2, 145 (1933), cited in [1]	1602, 1603, 2095
1047.	YATLOV V. S., PINAEVSKAYA E. N., Zh. Obshch. Khim., **8**, 1665 (1938)	636, 637
1048.	YATLOV V. S., PINAEVSKAYA E. N., Zh. Obshch. Khim., **19**, 24 (1949)	3167
1049.	YATSENKO S. P., DEMENEV N. V., Zh. Neorgan. Khim., **4**, 1437	1616, 2612
1050.	YATSENKO S. P., Zh. Neorgan. Khim., **6**, No. 8, 1922 (1961)	4356
1051.	YATSIMIRSKII K. B., Dokl. Akad. Nauk SSSR, **77**, 819 (1951); YATSIMIRSKII K. B., SHUTOV A. A., Zh. Fiz. Khim., **26**, 842 (1952)	5550

Reference number	Author's name and reference	Table number
1052.	YATSIMIRSKII K. B., GRAFOVA Z. M., Zh. Obshch. Khim., **22**, 10, 1726 (1952)	1861
1053.	YATSIMIRSKII K. B., Sb. Statei Obshch. Khim., Izd. Akad. Nauk SSSR, M.-L., **1**, 97 (1953)	441, 444, 549, 550
1054.	YATSIMIRSKII K. B., VASIL'EV V. P., Zh. Anal. Khim., **11**, No. 5, 529 (1956)	3211, 3212
1055.	BOGOYAVLENSKII P. S., NI SHI-SHEN, Sci. Sinica, **8**, No. 6, 580 (1959)	2924
1056.	GAO CHI, SUN SHI-SHUN, KAN CHUN-SHUN, KE YUI-TSUN, LUN LIAN-YUN, TSAI SHU-FAN, CHEN SHU-LAN, Acta Sci. Natur. Univ. Szechuan., No. 2, 73 (1959)	1417, 5155
1057.	KUBO, TAMIGUTI, TSZUAUMI, J. Chem. Soc. Jap., Ind. Chem. Sect., **63**, No. 10, 1679, A93 (1960)	710
1058.	NIKOLAEV N. S., OPALOVSKII A. A., MALYGINA N. D., Bul. Inst. politenic din, Iasi, **5**, Nos. 1–2, 189 (1959)	4655
1059.	NI SHI-SHEN, BOGOYAVLENSKII P. S., Acta Chim. Sinica, **26**, No. 2, 61 (1960)	2917, 4456
1060.	CHANG TSIN-LIEN, WANG KI-SHANG, Acta Chim. Sinica, **24**, No. 3, 274 (1958)	221
1061.	CHUN HUAN-BAN, SUN I-CHEN, FEDOROV P. I., Acta Chim. Sinica, **22**, No. 5, 414 (1956)	5543
1062.	ABEGG R., RIESENFELD H., Z. physik. Chem., **40**, 84 (1902)	1282, 1404, 1477, 1486, 1863, 1877, 1888, 1898, 1900, 1905, 1909, 1910, 1914 1916–1918, 1924–1926, 1929, 1935, 1946, 1947, 1949, 1953, 1954, 1956,
1063.	ADDELSTONE J. A., MARSH L. R., HALL G. C., J. Phys. Chem., **44**, 256 (1940)	4257
1064.	ADGE G., BARKHOLT H., Z. angew. Chem., **39**, 851 (1926)	5219
1065.	ADGE G., BARKHOLT H., Z. angew. Chem., **40**, 378 (1927)	843
1066.	ADGE G., SCHIMMEL, Z. angew. Chem., **41**, 340 (1928)	848
1067.	AKAHANE M., JANO H., J. Chem. Soc. Japan, Ind. Chem. Sec., **62**, No. 12, 1819 (1959)	2725–2727
1068.	AKERLOF G., THOMAS H. C., J. Am. Chem. Soc., **56**, 593 (1934)	862, 5276, 5301
1069.	AKERLOF G., TURCK H., J. Am. Chem. Soc., **56**, 1876 (1934)	964

Reference number	Author's name and reference	Table number
1070	AKERLOF G., TURCK H. E., J. Am. Chem. Soc. **56**, 593 (1935)	239
1071	AKERLOF G., J. Am. Chem. Soc., **57**, 1196 (1935)	133
1072	AKERLOF G., TURCK H. E., J. Am. Chem. Soc., **57**, 1746 (1935)	240, 250, 255, 257, 259, 261
1073	AKERLOF G., SHORT O., J. Am. Chem. Soc., **59**, 1912 (1937)	946, 2271, 4401
1074	ALLMAND A. J., BURRAGE L. J., Trans. Faraday Soc., **29**, 679 (1933)	4932
1075	AMADORI M., PAMPANINI G., Atti accad. Lincei, **20**, 11, 475 (1911)	4814, 4827, 4973
1076	AMADORI M., Atti acad. Lincei, **32**, 2, 467 (667 (1912)	4701, 4711, 4955
1077	AMADORI M., Atti del reale istituto Veneto, **79**, 2 sem. 296 (1919–1920), cited in [1]	3421, 3678
1078	ANDERSON E., NESTELL R. J., Ind. Eng. Chem., **12**, 243 (1920)	4721
1079	ANDO K., Mem. Coll. Sci. Kyoto Imp. Univ., (1), 8, 283 (1925)	2177, 2178
1080	ANHORN V. J., HUNT H., J. Phys. Chem., **45**, 351 (1941)	31, 2498 3726
1081	ANSELME A., Bull Soc. Chim., (3), **29**, 938 (1903)	2743
1082	ANTROPOFF A., MARCAU A., Sommer W., Z. Elektrochem., **30**, 457 (1924)	2709, 4398
1083	ANTROPOFF A., SOMMER W., Z. Phys. Chem., **123**, 185 (1926)	2710
1084	APFEL O., Dissertation, Technische Hochschule, Darmstadt (1911), cited in [2]	2224, 3194, 4461, 4537, 4668
1085	APPLEBEY M. P., LANE K. W., J. Chem. Soc., **113**, 614 (1918)	134, 2848, 4441
1086	APPLEBEY M. P., WILKES S. H., J. Chem. Soc., **121**, 337 (1922)	4339
1087	APPLEBEY M. P., LEISHMAN M. A., J. Chem. Soc. (Lond.), 1603 (1932)	1911
1088	APPLEBEY M. P., WINDRIDGE E. D., J. Chem. Soc., 1608 (1932)	1978
1089	ARAKI S., Mem. Coll. Sci. Kyoto Imp. Univ., (A), 8, 213 (1925)	2279, 2562
1090	ARAVAMUDAN G., Current Sci., **26**, No. 2, 49 (1957)	1320
1091	ARAVAMUDAN G., Canad. J. Chem., **40**, No. 5, 1035 (1962)	1367
1092	ARAVAMUDAN G., Canad. J. Chem., **41**, No. 1, 203 (1963)	1388
1093	ARCHIBALD E. J., WILCOX W. G., BUCKLEY B. G., J. Am. Chem. Soc., **30**, 747 (1908)	3721, 4800
1094	ARCHIBALD E. H., KERN J. W., Trans. Roy. Soc. (Canada), (3), **11**, (sec. III), 716 (1917)	2047, 2058, 2059, 2453, 2454, 2570, 4801
1095	ARCHIBALD E. H., GALE W. A., J. Am. Chem. Soc., **46**, 1760 (1924)	3372
1096	ARII (1927), cited in [2]	3226
1097	ARMSTRONG H. E., EYRE J. V., HUSSEY A. V., PADDISON W. P., Proc. Roy. Soc., London, A, 79, 569 (1907)	940, 958, 995
1098	ARMSTRONG H. E., Eyre J. V., Proc. Roy. Soc., (A), **84**, 123 (1910–1911)	941, 958, 995, 4557
1099	ARMSTRONG H. E., EYRE J. V., Proc. Roy. Soc., **88**, 234 (1913)	5577
1100	ARNTSON R. H., DICKSON F. W., TUNELL G., Science, **128**, No. 3326, 716 (1958)	115
1101	ARNSTON R. H., DICKSON F. W., TUNELL G., Amer. J. Sci., **258**, No. 8, 574 (1960)	112

Reference number	Author's name and reference	Table number
1102	ARRHENIUS S., Z. Phys. Chem., **11**, 397 (1893)	4568
1103	ARTH, CRÉTIEN, Bull. Soc. Chim., (3), **35–36**, 780 (1906)	3759
1104	ASKENASY P., NESSLER F., Z. anorg. Chem., **189**, 305 (1930)	2226, 2235, 4665
1105	ASKEW H. O., Trans. New-Zealand Inst., **54**, 791 (1923)	3739
1106	ASSARSSON G., Sveriges Geologiska Undersökning, Stockholm, Ser. C., No. 501 (1949) [Årsbok, 42, No. 10 (1948)], cited in [1]	3773, 4856
1107	ASSARSSON G. O., J. Am. Chem. Soc., **72**, 1437 (1950)	3702, 3711, 3774, 4164, 4857
1108	ASSARSSON G. O., J. Phys. Chem., **57**, No. 2, 207 (1953)	3803, 4886
1109	ASSARSSON G. O., Balder A., J. Phys. Chem., **57**, No. 7, 717 (1953)	5154
1110	ASSARSSON G. O., BALDER A., J. Phys. Chem., **58**, No. 3, 253 (1954)	4861
1111	ASSARSSON G. O., BALDER A., J. Phys. Chem., **58**, No. 5, 416 (1954)	4176
1112	ASSARSSON G. O., J. Phys. Chem., **59**, No. 2, 207 (1955)	3714
1113	ASSARSSON G. O., BALDER A., J. Phys. Chem., **59**, No. 7, 631 (1955)	4142
1114	ASSARSSON G. O., J. Phys. Chem., **60**, No. 10, 1436 (1956)	4182
1115	ATANASIN J. A., BABOR M., Bull. Sect. sci. Acad. roumaine, 20, No. 1–3, 27 (1938)	525, 882 1119
1116	AUERBACH G., Elektrochem., **10**, 163 (1904)	1669, 3989
1117	AUGER V., Compt. rend., **177**, 1302 (1923)	2756, 2926, 4422, 4457, 4481
1118	AUTENRIETEN H., BRAUN G., Kali und Steinsalz, **3**, No. 1, 15 (1960)	3310
1119	AUTENRIETEN H., BRAUN G., Kali und Steinsalz, **3**, No. 3, 85 (1960)	3311
1120	AYRES F. D., J. Phys. Chem., **49**, 4, 366 (1945)	1865
1121	BAAT W. C., de, Chem. Weekblad., **15**, 463 (1918)	2196, 2200, 5399
1122	BAAT W. C., de, Rec. trav. chim., **45**, 237 (1926)	2428–2430, 3535, 3536, 4114, 4783
1123	BÄCKSTRÖM H. L., Z. physik. Chem., **97**, 213 (1921); Medd-Vetenskapsakad. Nobel. Inst., **4**, No. 11 (1921)	1710
1124	BAHL R. K., SINGH S., J. Ind. Chem. Soc., **18**, 307 (1941)	2141
1125	BAIN H. W., J. Am. Chem. Soc., **49**, 2734 (1927)	3009
1126	BALCAR F. R., STEGMAN G., J. Phys. Chem., **32**, 1411 (1928)	5501, 5586
1127	BALEJ J., REGNER A., Coll. Chechosl. Chem. Comm., **25**, No. 6, 1955–1957 (1960)	4673
1128	BALEJ J., REGNER A., Coll. Czechosl. Chem. Comm., **27**, No. 9, 2208 (1962)	2431
1129	BALEJ J., REGNER A., Coll. Czechosl. Chem. Comm., **28**, No. 5, 1266 (1963)	2266
1130	BANCROFT W. D., GOULD L. P., J. Phys. Chem., **38**, 197 (1934)	3032, 3414, 3658, 3909, 4311, 4313, 4315, 4317, 4319, 4321,

Reference number	Author's name and reference	Table number
1131	BANER, JUE, WILLS, J. Am. Chem. Soc., **72**, 5369 (1950)	2599
1132	BANTHISCH, J. prakt. Chem. (2), **29**, 54 (1884)	455, 483, 518, 1111
1133	BÄR O., Centralbl. Mineral., A, 53 (1932)	3985
1134	BARBAUDY J., Rec. trav. chim., 42, 639 (1923)	4559, 4588, 4593, 4843, 4848, 5092
1135	BARLOW, J. Am. Chem. Soc., **28**, 1446 (1906)	2533, 3807
1136	BARNEY J. E., ARGERSINGER W. J., REYNOLDS C. A., J. Am. Chem. Soc., 73, 3785 (1951)	4893
1137	BARNS R. L., LAUDISE R. A., SHIELDS R. M., J. Phys. Chem., **67**, No. 4, 835 (1963)	568, 2729 3655
1138	BARONI A., Gazz. Chim. ital., **72**, 19 (1942)	5267
1139	BARRE M., Compt. rend., **148**, 1605 (1909)	2329, 2330, 4720
1140	BARRE M., Compt. rend., **149**, 222, ff. (1909)	2357, 4748
1141	BARRE M., Compt. rend. **150**, 1321, 1699 (1910)	2366, 2369, 3499, 3519, 3563, 4766
1142	BARRE M., Compt. rend., **151**, 871 (1910)	2369, 3503, 4766
1143	BARRE M., Ann. chim. phys., (8), **24**, 149, 202, 210 (1911)	889, 890, 2326, 2357, 2360, 2366, 2378, 3452, 3490, 3499, 4720, 4748, 4751, 4762, 4781
1144	BARRE M., Bull. Soc. chim., (4), 11, 646 (1912)	915, 1397
1145	BARRE M., Compt. rend., **154**, 280 (1912)	4480
1146	BARTON C. J., HEBERT G. M., MARSHALL W. L., J. Inorg. Nucl. Chem., **21**, No. 1–2, 141 (1961)	1801
1147	BASINSKA H., ORYLSKI Z., Chem. Anal., **6**, No. 3, 307 (1961)	548
1148	BASINSKA H., ORYLSKI Z., PERKOWSKI T., Chem. Anal., **7**, No. 5, 911 (1962)	442
1149	BASINSKY A., KONECZNA M., Roczn. chem., **30**, No. 4, 1077 (1956)	3214
1150	BASINSKY A., PASGRETA M., Roczn. chem., **34**, No. 1, 42 (1960)	461, 1029
1151	BASINSKY A., ROZWADOWSKY M., Roczn. chem., **34**, No. 1, 47 (1960)	470, 1044
1152	BASINSKY A., MUCHA J., Roczn. chem., **34**, No. 8, 11 (1960)	478, 1056
1153	BASKERVILLE W. H., CAMERON F. K., J. Phys. Chem., **39**, 769 (1935)	4341
1154	BASSETT H., Z. Anorg. Chem., **59**, 11 (1908)	666, 667
1155	BASSETT H., TAYLOR H. S., J. Chem. Soc., **101**, 576 (1912)	445
1156	BASSETT H., TAYLOR H. S., J. Chem. Soc., **105**, 1926 (1914)	5042
1157	BASSETT H. J., Chem. Soc. (London), **111**, 620 (1917)	667
1158	BASSETT H., CORBET A. S., J. Chem. Soc., **125**, 1660 (1924)	4379, 4487, 4490, 4491, 4493, 4494
1159	BASSETT H., CROUCHER H. H., J. Chem. Soc., 8, 1809 (1930)	4170, 5249, 5256
1160	BASSETT H., HARRY H., J. Chem. Soc. (London), 1784 (1930)	1041, 4169, 5248, 5255

Reference number	Author's name and reference	Table number
1161	BASSETT H., SANDERSON J., J. Chem. Soc., **10**, 1855 (1932)	1438
1162	BASSETT H., BARTON G. W., FOSTER A. R., PATEMAN C. R. J., J. Chem. Soc., 2, 163 (1933)	4187, 5172, 5287, 5366, 5512
1163	BASSETT H., LEMAN J. T., J. Chem. Soc., **10**, 1423 (1933)	3527
1164	BASSETT H., SANDERSON J., J. Chem. Soc. (London), 1116 (1934)	5243
1165	BASSETT H., GORDON H. F., HENSHALL J. H., J. Chem. Soc., **56**, 971 (1937)	5147, 5251, 5257
1166	BASSETT H., HENSHALL J. H., SERGEANT G. A., SHIPLEY H. S., J. Chem. Soc., 12, 657 (1939)	3817, 5247, 5252, 5263
1167	BASSETT H., GOODWIN T. H., J. Chem. Soc., 9, 2244 (1949)	782
1168	BASSETT H., WATT W., J. Chem. Soc., 5, 1409 (1950)	4054
1169	BASSETT H., HENSHALL J. H., J. Chem. Soc., 17, 1971 (1950)	3471
1170	BASSETT H., STRAIN R. N. C., J. Chem. Soc., **5**, 1795 (1952)	4178, 5158, 5210, 5283
1171	BASSETT H., J. Chem. Soc., **9**, 2950 (1958)	4281
1172	BAUBIGNY H., Bull. Soc. chim., (4) **3**, 772 (1908); Compt. rend., **146**, 1263 (1908)	2015
1173	BAXTER W. P., J. Am. Chem. Soc., **48**, 615 (1926)	4101, 5429
1174	BAYLISS N. S., COLE A. R. H., EWERS W. E., JONES N. S., J. Am. Chem. Soc., **69**, 2033 (1947)	3387, 3398, 3431, 4071
1175	BECK K., STEGMÜLLER Ph., Arb. Kais. Gesundheitsamt., **34**, 447 (1910), cited in [2]	539, 544, 1136, 1146, 3847
1176	BECK K., STEGMÜLLER Ph., Z. Elektrochem., **17**, 843 (1911)	539, 544, 1146
1177	BECKURTS, Arch. Pharm., **18**, 429 (1881)	1658
1178	BELL J., J. Chem. Soc., **21**, 350 (1868)	1139
1179	BELL J., TABER W., J. Phys. Chem., **10**, 119 (1906)	2324
1180	BELL J. M., TABER W. C., J. Phys. Chem., **11**, 637 (1907)	5126
1181	BELL J. M., TABER W. C., J. Phys. Chem., **12**, 174 (1908)	840
1182	BELL J. M., BUCKLEY M. L., J. Am. Chem. Soc., **34**, 10 (1912)	141, 175, 184
1183	BELL R. P., WAIND G. M., J. Chem. Soc., 1979 (1950)	2754, 3779
1184	BELL R. P., GEORGE J. H. B., Trans. Faraday Soc., **49**, 619 (1953)	4421, 4733
1185	BENDER P., J. Am. Chem. Soc., **67**, 1771 (1945)	1102
1186	BENDER P., STREHLOW R. A., J. Am. Chem. Soc., **70**, 1995 (1948)	1067
1187	BENDER P., BUNDE R. E., J. Am. Chem. Soc., **72**, 2273 (1950)	1103
1188	BENNETT S. L., BIERMANN W. J., Canad. J. Chem., **40**, No. 1, 73 (1962)	1494, 1495, 1516
1189	BENRATH A., ESPENSCHIED H., Z. anorg. Chem., **121**, 361 (1922)	898
1190	BENRATH A., Z. anorg. Chem., **135**, 248 (1924)	5570, 5573
1191	BENRATH A., Z. anorg. Chem., **136**, 358 (1924)	5570, 5573
1192	BENRATH A., Z. anorg. Chem., **163**, 396 (1927)	2517, 3785
1193	BENRATH A., Z. anorg. Chem., **170**, 257 (1928)	3101, 3102, 3373, 3992
1194	BENRATH A., SCHLOEMER A., CLERMONT J., KOJITSCH S., Z. anorg. Chem., **170**, 277 (1928)	3376
1195	BENRATH A., BENRATH H., Z. anorg. Chem., **170**, 283 (1928), Caliche, **11**, 117 (1929)	3376, 3377
1196	BENRATH A., BEN W., Z. anorg. Chem., **170**, 270 (1928); Caliche, **11**, 110 (1929)	3375

Reference number	Author's name and reference	Table number
1197	BENRATH A., AMMER G., Z. anorg. Chem., **177**, 129 (1929)	4189, 5175, 5309, 5367, 5442, 5515, 5553
1198	BENRATH A., BENRATH H., Z. anorg. Chem., **179**, 369 (1929)	3462, 3466, 3468, 3473, 3479, 3485, 3495
1199	BENRATH A., Z. anorg. Chem., **183**, 296 (1929)	3463
1200	BENRATH A., BENRATH H., Z. anorg. Chem., **184**, 359 (1929)	3997
1201	BENRATH A., Caliche, **11**, 99 (1929)	3101, 3102, 3392
1202	BENRATH A., WAZELLE, BENRATH H., Z. anorg. Chem., **188**, 74 (1930)	4061
1203	BENRATH A., BENRATH H., Z. anorg. Chem., **189**, 72 (1930)	3993, 3997
1204	BENRATH A., Z. anorg. Chem., **189**, 86 (1930)	3485, 4735, 4747
1205	BENRATH A., Z. anorg. Chem., **195**, 247 (1931)	2341, 2342, 5206, 5207
1206	BENRATH A., SICHELSCHMIDT A., Z. anorg. Chem., **197**, 113 (1931)	3997, 4064
1207	BENRATH A., Z. anorg. Chem., **202**, 161 (1931)	3469, 3473, 4108, 5303
1208	BENRATH A., THONESSEN C., Z. anorg. Chem., **203**, 405 (1931)	4757, 5334, 5435
1209	BENRATH H., Z. anorg. Chem., **205**, 417 (1932)	1440, 1443
1210	BENRATH A., Z. anorg. Chem., **208**, 169 (1932)	2345, 2347, 4738, 4742, 5244, 5261, 5262
1211	BENRATH A., THIEMANN W., Z. anorg. Chem., **208**, 177 (1932)	2305, 2363, 3496, 5245
1212	BENRATH A., BLANKENSTEIN A., Z. anorg. Chem., **216**, 41 (1934)	4092
1213	BENRATH H., Z. anorg. Chem., **220**, 145 (1934)	1437
1214	BENRATH A., SCHIFFERS B., Z. anorg. Chem., **240**, 1, 67 (1939)	2578
1215	BENRATH A., NEUMANN E., Z. anorg. Chem., **240**, 1, 81 (1939)	3787
1216	BENRATH H., Z. anorg. Chem., **240**, 1, 90 (1939)	1439
1217	BENRATH A., NEUMANN E., Z. anorg. Chem., **242**, 72 (1939)	4093
1218	BENRATH A., NEUMANN E., Z. anorg. Chem., **243**, 2, 177 (1939)	2519
1219	BENRATH A., RITTER G., J. prakt. Chem., **154**, 40 (1939)	4739, 4740, 4871, 5239
1220	BENRATH A., BRAUN A., Z. anorg. Chem., **244**, 348 (1940)	4791, 4795
1221	BENRATH A., LECHNER K., Z. anorg. Chem., **244**, 359 (1940)	5438, 5439, 5446, 5505
1222	BENRATH A., BRAUN A., Z. anorg. Chem., **244, 4, 350 (1940)**	4563
1223	BENRATH A., Z. anorg. Chem., **252**, 86 (1943)	3059, 3070, 3112, 4542, 4564, 4796
1224	BERECZ E., ANDRAS L., GEIGER J., Magyar Kem. folyóirat, **66**, No. 3, 85 (1960)	970
1225	BERECZ E., HORANYI G., Acta chim. Hung. Tomus, **29**, No. 1–2, 75 (1961)	929
1226	BERJU, KOSMINKO, Landw. vers. Sta, **60**, 422 (1904), cited in [2]	2183
1227	BERTISCHOWRA B., Roczn. chem., **6**, 705 (1926)	2403, 2415

Reference number	Author's name and reference	Table number
1228	BESSLEY G. E., Proc. of the Symposium on the chemistry of cements, Stockholm, 178 (1938)	4211
1229	BEYER G. L., RIEMAN W., J. Am. Chem. Soc., **65**, 971 (1943)	3832, 4907
1230	BHAGWAT W. F., DHAR N. R., J. Ind. Chem. Soc., **6**, 807 (1929)	340, 360
1231	BILLITER J., Monatsh. Chem., **41**, 287 (1920)	3542
1232	BILTZ W., MARCUS E., Z. anorg. Chem., **71**, 167 (1911)	2478, 2481, 2486
1233	BINDER O., Ann. chim. (11), **5**, 337 (1936)	4350
1234	BISSELL D. W., JAMES C., J. Am. Chem. Soc., **38**, 873 (1916)	887, 3508
1235	BITTRICH H. J., LEIBNITZ E., J. prakt. Chem., **4**, No. 7, 311 (1959)	2685, 3457, 5060
1236	BITTRICH H. J., LEIBNITZ E., J. prakt. Chem., **14**, No. 4–6, 218 (1961)	2700
1237	BITTRICH H. J., LEIBNITZ E., J. prakt. Chem., **17**, No. 1–2, 92 (1962)	2701
1238	BLAREZ Ch., Compt. rend., **112**, 434, 939, 1213 (1891)	4675, 4706, 4714
1239	BLAREZ Ch., (1911), cited in [2]	4404, 4569, 4695, 4788, 4935, 4938
1240	BLASDALE W. C., Inl. Eng. Chem., **10**, 344 (1918)	3248, 3319, 3420, 3677, 4677
1241	BLASDALE W., Ind. Eng. **12**, 165 (1920)	3250, 3291
1242	BLASDALE W., J. Am. Chem. Soc., **45**, 2938 (1923)	2967, 2982, 3007, 3677, 4473
1243	BLASDALE W. C., ROBSON H. L., J. Am. Chem. Soc., **50**, 35 (1928)	3374
1244	BLASDALE W., SLANSKY C., J. Am. Chem. Soc., **61**, 919 (1939)	1554
1245	BNAWWAN W. V., DHAR N. R., J. Ind. Chem. Soc., **6**, 807 (1929)	1199
1246	BOCK E., Canad. J. Chem., **39**, No. 6, 1746 (1961)	3758
1247	BOCK E., Canad. J. Chem., **39**, No. 9, 1746 (1961)	3755
1248	BODLÄNDER G., Z. physik. Chem., **7**, 317, 361 (1891)	4554
1249	BODLÄNDER G., Z. physik. Chem., **9**, 734 (1892)	2006
1250	BODLÄNDER G., BREULL P., Z. angew. Chem., **16**, 389 (1901)	2904
1251	BODLÄNDER G., FITTING R., Z. physik. Chem., **39**, 597 (1901–1902)	1943, 2008
1252	BODLÄNDER G., Z. angew Chem., **18**, 1138 (1905)	2744
1253	BOEKE H., Z. Kryst. und Miner., **45**, 363 (1908)	3556, 3568, 3707, 4115, 4119, 4195, 4813
1254	BOERICKE, (1905), cited in [2]	143
1255	BOGDAN P., Ann. Sci. Univ. Jassy, **2**, 47 (1902–1903)	306, 310, 323, 337, 343, 349
1256	BOHR C., Wied. Ann. Physik., **68**, 504 (1899)	1631
1257	BOHR C., Z. physik. Chem., **71**, 47 (1910)	69, 73, 87
1258	BOSDORFF W., Z. anorg. Chem., **41**, 180 (1904)	2021, 2061
1259	BOVALINI E., FABRIS E., Gazz. chim. ital., **63**, 338 (1933)	4524
1260	BOVALINI E., FABRIS E., Gazz. chim. ital., **65**, 617 (1935)	2319

Reference number	Author's name and reference	Table number
1261	BOZORTH R. M., J. Am. Chem. Soc., **45**, 2653 (1923)	3116, 3440, 3719, 3875, 4573, 4699, 4794, 4950, 4951
1262	BOZZA G., Gior. chim. Ind. applicata, **16**, 109 (1934)	4065
1263	BRAIHANTI A., CHIERICI J., Gazz. chim. ital., **88**, No. 10, 793 (1958)	3036, 4518
1264	BRAM L., Z. physik. Chem., **33**, 735 (1900)	5354
1265	BRANTLEY J. C., IMIZENGA J. R., J. Am. Chem. Soc., **74**, 6101 (1952)	883
1266	BRAUER G., HAAK H., Z. anorg. Chem., **261**, No. 3–4, 183 (1950)	1058
1267	BRAUER G., MÜLLER H., Z. anorg. Chem., **287**, No. 1–2, 74 (1956)	479, 852, 1060, 1164, 1228, 1248
1268	BRAUN L., Z. physik. Chem., **33**, 732 (1900)	64, 75, 77
1269	BRAY W. C., MACKAY G. M. Z., J. Am. Chem. Soc., **32**, 914, 1207 (1910)	186
1270	BRAY W. C., WINNINGHOFF, J. Am. Chem. Soc., **33**, 1663 (1911)	4628, 4776, 5566
1271	BRESTSZNAJDER S., KOTOWSKA W., Przem. chem. **11**, No. 6, 285 (1955)	753
1272	BRICKWEDDE L. H., J. Res. Nat. Bur. Standards, **36**, 377 (1946)	219
1273	BRIGGS T. R., GREENWALD J. A., J. Phys. Chem., **34**, 1951 (1930)	201
1274	BRIGGS T. R., BALLARD K. H., ALRICH F. R., WIRSWO J. P., J. Phys. Chem., **44**, 325 (1940)	167
1275	BRIGGS T. R., CLACK K. D. G., BALLARD K. H., SASSAMAN W. A., J. Phys. Chem., **44**, 350 (1940)	193
1276	BRIGGS T. R., GEIGLE W. F., EATON J. L., J. Phys. Chem., **45**, 595 (1941)	179
1277	BRIGGS T. R., CONRAD C. C., GREGG C. C., REED W. H., J. Phys. Chem., **45**, 614 (1941)	196
1278	BRIGGS T. R., HUBARD S. S., J. Phys. Chem., **45**, 806 (1941)	202
1279	BRINKLEY S. R., J. Am. Chem. Soc., **44**, 1210 (1922)	2035
1280	BRITTON H. T. S., ALLMAND A. J., J. Chem. Soc., **119**, 1463 (1921)	1518
1281	BRITTON H. T. S., J. Chem. Soc., **119**, 1967 (1921)	721
1282	BRITTON H. T. S., J. Chem. Soc., **121**, 982 (1922)	4238
1283	BRITTON H. T. S., J. Chem. Soc., **121**, 2612 (1922)	1511
1284	BRITTON H. T. S., J. Chem. Soc., **123**, 1429 (1923)	5198
1285	BRITTON H. T. S., J. Chem. Soc., **127**, 2796 (1925)	1048
1286	BRITTON H. T. S., WELFORD G., J. Chem. Soc., 895 (1940)	457, 812, 1026, 1027, 1162b, 3237
1287	BRÖNSTED J. N., 7-th Int. Congress Applied Chem., 10, 110 (1909)	2547
1288	BRÖNSTED J. N., Z. physik. Chem., **77**, 132 (1911)	2548
1289	BRÖNSTED J. N. Z. physik. Chem. **80**, 208, 214 (1912)	4873, 4928
1290	BRÖNSTED J. N., J. Am. Chem. Soc., **42**, 1448 (1920)	4383, 4397, 4405, 4412, 4415, 4417
1291	BROSHEER J. C., ANDERSON J. F., J. Am. Chem. Soc., 68, 902 (1946)	645
1292	BROSHEER J. C., LENFESTY F. A., ANDERSON J. F., J. Am. Chem. Soc., **76**, 5951 (1954)	657b, 4204, 4205

Reference number	Author's name and reference	Table number
1293	BRUNISHOLZ G., BODMER M., Helv. chim. acta, **46**, No. 7, 2577 (1963)	3195, 3196, 3706, 4667
1294	BRUNISHOLZ G., QUINCHE, ABDEL MAGID KALO, Helv. chim. acta, **47**, No. 1, 14 (1964)	5535, 5536, 5544, 5545
1295	BRUZAV, Bull. Soc. chim. France, **15**, 1177 (1948)	2241
1296	BUETTNER J., JACHE A. W., J. Inorg. Chem., **19**, No. 3–4, 377 (1961)	615
1297	BUETTNER J. P., JAHE A. W., J. Inorg. Chem., **2**, No. 1, 19 (1963)	562
1298	BULLOUGH W., GANNING T. A., STRAWBRIDGE M. J., J. Appl. Chem., **2**, No. 12, 703 (1952)	827
1299	BUREAU J., Ann. Chim., (11), 8, 97 (1937)	2878, 2927, 4528
1300	BURRAGE L. J., J. Chem. Soc., **129**, 1703, 1896 (1926)	3856, 4931, 4994, 5024
1301	BURRAGE L. J., Trans. Faraday Soc., **28**, 529 (1932)	4932
1302	BURY C. R., DAVIES E. R. H., J. Chem. Soc., 2008 (1932)	3963
1303	BURY C. R., DAVIES D. G., J. Chem. Soc., **9**, 2419 (1932)	5208
1304	BURY C. R., DAVIES E. R. H., J. Chem. Soc., **6**, 703 (1933)	4150, 4159, 5048
1305	BURY C. R., REDD R., J. Chem. Soc., 1161 (1933)	3022
1306	BUTTLER J. A. V., HISCOCKS E. S., J. Chem. Soc., **129**, 2554 (1926)	4629, 4924, 5304, 5540, 5561, 5567
1307	BUTTLER F. G., TAYLOR F. W., J. Chem. Soc., No. 6, 2103 (1958)	4213
1308	BYE J., Bull. Soc. chim. France, (5), **10**, 239 (1943); Ann. Chim., (11), **20**, 515 (1945)	2620
1309	CABOT G. L., J. Soc. Chem. Ind., **16**, 417 (1897)	3727, 4836
1310	CADBURY W. E., J. Am. Chem. Soc., **67**, 262 (1945)	3879
1311	CAGLIOTT V., MALOSSI L., Atti accad. Lincei, (6), **10**, 97 (1929)	2377
1312	CALCAGNI G., Gazz. chim. ital., **50**, 11, 331 (1920)	111, 113, 116–119
1313	CALVERT H. T., Z. physik. Chem., **38**, 521 (1901)	253, 260, 4403
1314	CALVO C., SIMONS E., J. Am. Chem. Soc., **74**, 5, 1202 (1952)	2323, 2356, 2365
1315	CAMERON F., J. Phys. Chem., **5**, 559 (1901); Bull. U.S. Dept. Agricult., No. 18, (1901)	3742
1316	CAMERON F., SEIDELL A., J. Phys. Chem., 5, 643 (1901)	1652, 3450 3733, 5114
1317	CAMERON F., SEIDELL A., SMITH, J. Phys. Chem., **5**, 645 (1901)	4000, 4152
1318	CAMERON F. K., SEIDELL A., J. Phys. Chem., **6**, 50 (1902)	1626, 1643, 3444, 3734
1319	CAMERON F. K., BREAZEALE J. F., J. Phys. Chem., **7**, 574 (1903)	804
1320	CAMERON F., SEIDELL A., J. Phys. Chem., 7, 582 (1903)	2999, 3365, 3577, 3580
1321	CAMERON F. K., BREAZEALE J. F., J. Phys. Chem., **8**, 335 (1904)	3451, 4591, 4719
1322	CAMERON F. K., BROWN B., J. Phys. Chem., **9**, 210 (1905)	2190, 2504
1323	CAMERON F. K., SEIDELL A., J. Am. Chem. Soc., **27**, 1508 (1905)	665
1324	CAMERON F. K., BELL J. M., J. Am. Chem. Soc., 28, 1220, 1222 (1906)	5043
1325	CAMERON F. K., Bell J. M., J. Phys. Chem., **10**, 210 (1906)	4084
1326	CAMERON F., BELL J., Bull. U.S. Dept. Agricult., No. 41 (1907)	5058

Reference number	Author's name and reference	Table number
1327	CAMERON F. K., BELL J. M., J. Phys. Chem., **11**, 363 (1907)	680
1328	CAMERON F. K., BELL J. M., ROBINSON W. O., J. Phys. Chem., **11**, 396 (1907)	1644, 3244, 3328–3331, 3445, 3734, 3763
1329	CAMERON F. K., ROBINSON W. O., J. Phys. Chem., **11**, 577, 641, 691 (1907)	820, 1031 1693, 4718, 4841
1330	CAMERON F. K., ROBINSON W. O., J. Phys. Chem., **13**, 157, 251 (1909)	462
1331	CAMERON F. K., CROCKFORD H. D., J. Phys. Chem., **33**, 709 (1929)	829, 844, 5220
1332	CAMERON F. K., J. Phys. Chem., **34**, 692 (1930)	822
1333	CAMERON F. K., J. Phys. Chem., **40**, 689 (1936)	2344, 4736, 5216
1334	CAMPBELL A. N., YANICK N. S., Trans. Faraday Soc., **28**, 657 (1932)	5125
1335	CAMPBELL A. N., DOWNS K. W., SAMIS C. S., J. Am. Chem. Soc., **56**, 2507 (1934)	4026, 4066, 4130, 4680
1336	CAMPBELL A. N., CAMPBELL A. J., Trans. Faraday, Soc., **35**, 243 (1939)	3161, 4644, 5498
1337	CAMPBELL A. N., J. Am. Chem. Soc., **64**, 2680 (1942)	1321
1338	CAMPBELL A. N., J. Am. Chem. Soc., **65**, 2268 (1943)	1330
1339	CAMPBELL A. N., SHEMILT L. W., Trans. Roy. Soc. Can., **40**, Sect. III, 17 (1946)	1
1340	CAMPBELL A. N., LEMAIRE H. P., Canad. J. Research (sec. B), **25**, 228 (1947)	4197, 5197
1341	CAMPBELL A. N., COUTTS J. W., Ind. Eng. Chem., **40**, 7, 1295 (1948)	4273, 5112
1342	CAMPBELL A. N., SUKAVA A. J., KOOP J., J. Am. Chem. Soc., **73**, 6, 2833 (1951)	722
1343	CAMPBELL A. N., McGULLOCH W. J. G., KARTZMARK E. M., Canad. J. Chem., **32**, No. 7, 696 (1954)	1347
1344	CAMPBELL A. N., KARTZMARK E. M., Canad. J. Chem., **34**, No. 5, 672 (1956)	1416, 1422
1345	CAMPBELL A. N., GRIFFITHS J. E., Canad. J. Chem., **34**, No. 11, 1647 (1956)	1398
1346	CAMPBELL A. N., KARTZMARK E. M., Canad. J. Chem., **36**, No. 1, 171 (1958)	1352, 1378
1347	CAMPBELL A. N., KARTZMARK E. M., LOVERING E. G., Canad. J. Chem., **36**, No. 11, 1511 (1958)	1335, 1338
1348	CAMPBELL A. N., KARTZMARK E. M., Canad. J. Chem., **37**, No. 9, 1409 (1959)	1360
1349	CAMPBELL J. A., MARSH F. L., J. Phys. Chem., **63**, No. 2, 316 (1959)	2572
1350	CAMPBELL A. N., KARTZMARK E. M., HUTTON H. M., Canad. J. Chem., **29**, No. 7, 1462 (1961)	1337
1351	CANNON P., J. Inorg. Nucl. Chem., **11**, No. 2, 124 (1959)	487, 681, 1084
1352	CANTONI H., GOGUELIA G., Bull. Soc. chim. (3), 33, 24 (1905)	2501, 2530, 3735, 3821, 4902
1353	CANTONI H., PASSAMONIK, Ann. chim. anal. appl., **10**, 258 (1905)	3799
1354	CARBONELL L., Bull. Soc. chim. France, No. 11–12, 1990 (1959)	4376

Reference number	Author's name and reference	Table number
1355	CARLSON E. T., CHAVONAS T. J., WELLS L. S., J. Res. Nat. Bur. Standards, **45**, 381 (1950)	4217
1356	CARLSON E. T., J. Res. Nat. Bur. Standards, **61**, No. 1, 1 (1958)	4214
1357	CAROBBI G., Gazz. chim. ital., **58**, 45 (1928)	3933, 3959
1358	CARPENI G., Bull. Soc. chim. France, 1327 (1955)	1575
1359	CARPENTER, J. Soc. Chem. Ind., 5, 286 (1886)	5413
1360	CARTER S. R., HARTSHORNE N. H., J. Chem. Soc., **123**, 2223 (1923)	4290
1361	CARTER J. S., J. Chem. Soc., **127**, 2861 (1925)	153, 155, 168, 169, 170, 172
1362	CARTER J. S., J. Chem. Soc., 2227 (1928)	171, 177
1363	CARTER J. S., HOSKINS C. R., J. Chem. Soc., 580 (1929)	157, 163, 181, 183, 204
1364	CARTER S. R., HARTSHORNE N. H., J. Chem. Soc., **129**, 363 (1926)	4289
1365	CARTER S. R., MEGSON M. J. L., J. Chem. Soc., 3, 2954 (1928)	1227
1366	CASPARI W. A., J. Chem. Soc., **125**, 2381 (1924)	2942
1367	CASTAGNOW R., LARCEBAY S., Bull. Soc. pharm., Bordeaux, **89**, 187 (1951)	880
1368	CAUSSE, Compt. rend., 114, 414 (1892)	665
1369	CAVALCA L., NARDELLI M., Gazz. chim. ital., **82**, No. 6, 394 (1952)	1350
1370	CAVALCA L., NARDELLI M., BRAIBANTI A., Gazz. chim. ital., **83**, 476 (1953)	3053, 5478
1371	CAVAZZI A., Gazz. chim. ital., **46**, 11, 122 (1916)	1709
1372	CAVE G. C. B., HUME D. N., J. Am. Chem. Soc., **75**, 2893 (1953)	4512
1373	CAVEN R. M., MITCHELL T. C., J. Chem. Soc., **125**, 1428 (1924)	2351
1374	CAVEN R. M., MITCHELL T. C., J. Chem. Soc., **127**, 527 (1925)	4244, 4253
1375	CAVEN R. M., MITCHELL T. C., J. Chem. Soc., **127**, 2549 (1925)	4249, 4256
1376	CAVEN R. M., JOHNSTON W., J. Chem. Soc., **129**, 2628 (1926)	4741, 4746
1377	CAVEN R. M., JOHNSTON W., J. Chem. Soc., 3, 2358 (1927)	2340, 3478, 4734
1378	CAVEN R. M., JOHNSTON W., J. Chem. Soc., 3, 2506 (1928)	3461, 3484, 4737
1379	CAVEN R. M., J. Chem. Soc., 2417 (1932)	5209
1380	CHANG T. L., CHU T. C., Z. physik. Chem., A184, 411 (1939)	212
1381	CHANG TSIN-LIEN, CHANG JUNG-SEN, Sci. Sinica, **4**, No. 4, 555 (1955)	215
1382	CHEESMAN G. H., DUNCAN D. R., HARRIS J. W., J. Chem. Soc., 837 (1940)	178
1383	CHLOPIN V., Z. anorg. Chem., **143**, 108 (1925)	1237
1384	CHLOPIN V., NIKITIN B., Z. anorg. Chem., **166**, 311 (1927)	1238
1385	CHLOUPEK J. B., DANES V. Z., Coll. Czechosl. Chem. Comm., **4**, 8 (1932)	2361, 4605, 4752
1386	CHLOUPEK J. B., DANES V. Z., DANESOVA B. A., Coll. Czechosl. Chem. Comm., **4**, 473 (1932)	4107, 4185, 4624, 4770
1387	CHLOUPEK J. B., DANES V. Z., DANESOVA B. A., Chem. listy, **26**, 531 (1932)	4107, 4185, 4624, 4770
1388	CHLOUPEK J. B., DANES V. Z., DANESOVA B. A., Coll. Czechosl. Chem. Comm., **5**, 339 (1933)	4091, 4166, 4594, 4732
1389	CHOROWER Ch., Z. angew. Chem., **33**, 201–202 (1920)	2660
1390	CHRETIEN A., Caliche, 7, 439; 8, 355, 390 (1926)	3061, 3258
1391	CHRETIEN A., Caliche, 9, 248 (1927)	3061
1392	CHRETIEN A., Ann. chim., (10), **12**, 9 (1929)	3058, 3258

Reference number	Author's name and reference	Table number
1393	CHRETIEN A., WEIL R., Bull. Soc. chim., (5), **2**, 1577 (1935)	1876
1394	CHRISTOFF A., Z. physik. Chem., **53**, 325 (1905), cited in [2]	725, 1541, 1542, 1592, 1611, 1632, 1676, 1683, 1684, 1688, 1696, 1699
1395	CHRISTOFF A., Z. physik. Chem., **55**, 622 (1906)	72, 78, 86, 726
1396	CHRISTY S. B., Z. Elektrochem., **7**, 205 (1901)	211
1397	CLARK G., J. Am. Chem. Soc., **41**, 1489 (1919)	557, 563, 575
1398	CLARK L., HUNTER E., J. Chem. Soc., 386 (1935)	2656, 2702, 2964, 5077, 5080
1399	CLARKE L., PARTRIDGE E., Ind. Eng. Chem., **26**, 898 (1934)	4724
1400	CLENDINNEN F. W. J., RIVETT A. C. D., J. Chem. Soc., **119**, 1329 (1921)	2509
1401	CLENDINNEN F. W. J., J. Chem. Soc., **121**, 801 (1922)	2511, 2516, 2522
1402	CLENDINNEN F. W. J., J. Chem. Soc., **123**, 1338 (1923)	2515
1403	CLIFFORD A. F., KONGPRICHA S., J. Inorg. Nucl. Chem., **29**, No. 1–2, 148 (1961)	34–38
1404	CLOEZ C., Bull. Soc. chim., (3), **29**, 167 (1903)	3745
1405	COCHERET D. H., Dissertation, Leiden (1911); "Tables Annuelles", 2, 439, 444 (1911), cited in [2]	2971, 2989, 2995
1406	COHN E., Z. physik. Chem., **18**, 61 (1895)	5419
1407	COLANI A., Bull. Soc. chim., **39**, 1243 (1926)	554
1408	COLANI A., Compt. rend., **185**, 1475 (1927)	2206
1409	COLANI A., Bull. Soc. chim., (4), **43**, 194, 754 (1928)	2206, 2381, 4640, 4782
1410	COLLINS S. C., CAMERON F. K., J. Phys. Chem., **32**, 1705 (1928)	1406
1411	CONINCK O., de, Ann. chim. phys., (7), **28**, 7 (1903)	558, 918, 1156, 1214
1412	CONLEY J., GABRIEL A., PATRIDGE E., J. Phys. Chem., **42**, 596 (1938)	4078
1413	CONRAD F. H., BENSCHLEIN W. L., J. Am. Chem. Soc., **56**, 2554 (1934)	4322
1414	CONSTABLE F. H., ISNEL A. H., Rev. fac. sci. Univ. Istanbul, **18A**, 34 (1953)	275, 281, 292, 298, 300–305, 308, 309, 312, 327, 336, 346, 352, 356, 361–364
1415	CONTET M., Compt. rend., **217**, 277 (1943)	3921
1416	COPELAND L. C., SHORT O. A., J. Am. Chem. Soc., **62**, 3285 (1940)	850, 5290
1417	CORBET A. S., J. Chem. Soc., **129**, 3190 (1926)	4486, 4488, 4489, 4496
1418	CORNEC E., Ann. chim. phys., (8), **30**, 132 (1913)	299, 1572
1419	CORNEC E., DUCKELEY J., Bull. Soc. chim., (4), **25**, 215 (1919); Compt. rend., **184**, 1555 (1927)	3548, 3549
1420	CORNEC E., CHRETIEN A., Caliche, **6**, 358 (1924)	3067

Reference number	Author's name and reference	Table number
1421	CORNEC E., KLUG P., Bull. Soc. chim., (4), **41**, 1016 (1927)	3558, 4815
1422	CORNEC E., KROMBACH H., Ann. chim., (10), **12**, 203 (1929)	3076, 3110, 3424, 3426
1423	CORNEC E., KROMBACH H., SPAK A., Caliche, **11**, 224 (1929)	3423
1424	CORBEC E., NEUMEISTER A., Revista Caliche (Chile), **10**, 492 (1929), cited in [2]	3081, 3117, 3720
1425	CORNEC E., SPACK A., Bull. Soc. chim. (France), (4), **49**, 582 (1931)	3098
1426	CORNEC E., KROMBACH H., Compt. rend., **194**, 714 (1932); Ann. chim., (10), **18**, 5 (1932)	3690
1427	CORNEC E., LEHNE M., Compt. rend., **23**, 1817 (1939)	2307, 4069, 4070
1428	COPPADORO A., Gazz. chim. ital., **39**, 625 (1909)	702, 703
1429	COPPADORO A., Atti accad. Lincei, (5), **21**, 11, 842 (1912); cited in [2]	5481
1430	COPPADORO A., Gazz. chim. ital., **42**, 1 420 (1912); **43**, 1, 138 (1913)	3141
1431	CRAIG D. N., VINAL G. W., VINAL F. E., J. Res. U.S. Bur. Standards, **17**, 709 (1936)	892
1432	CRAIG D. N., VINAL G. W., J. Res. Nat. Bur. Standards, **22**, 55 (1939)	908
1433	CRAVEN R. M., GARDNER W. K., J. Chem. Soc. (London), 943 (1933)	2346, 2354, 3470
1434	CRAVEN R. M., BRYCE G., J. Chem. Soc. (London), 514 (1934)	2555, 2557, 3874, 4267
1435	CREIGHTON H. J. M., WARD W. H., J. Am. Chem. Soc., **37**, 2333 (1915)	4533, 5382
1436	CROCKFORD H. D., WARRICK L. E., J. Phys. Chem., **34**, 1064 (1930)	845
1437	CROCKFORD H. D., WEBSTER M. M., J. Phys. Chem., **34**, 2375 (1930)	1386
1438	CROCKFORD H. D., BRAWLEY D. J., J. Phys. Chem., **36**, 1594 (1932)	5242
1439	CROCKFORD H. D., ADDLESTANE J. A., J. Phys. Chem., **40**, 303 (1936)	907
1440	CUPERY M. E., Ind. Eng. Chem., **30**, 627 (1938)	700
1441	CUPR V., Rec. trav. chim., **44**, 478 (1925)	704
1442	CUPR V., Publ. fac. sci. Univ. Masaryk, **63**, 16 (1926)	705, 706
1443	CUPR V., Publ. fac. sci. Univ. Masaryk, **68**, 8 (1926)	789
1444	CUPR V., Rec. trav. chim., **47**, 55 (1928)	706, 789
1445	CUVELIER B. V. J., Z. anal. Chem., **102**, 16 (1935)	2529
1446	CUVELIER B. V. J., Z. anorg. Chem., **226**, 197 (1936)	2520
1447	DALTON R. H., POMEROY R., WEYMOUTH L. E., J. Am. Chem. Soc. **46**, 60 (1924)	5425
1448	DANIEL S. S., SALMON J. E., J. Chem. Soc., **1**, 87 (1961)	4294–4296
1449	DANK A. D., Chem. Weekblad, **5**, 529, 629, 767 (1908)	1281, 2721
1450	D'ANS J., Z. anorg. Chem., **61**, 91 (1909)	756
1451	D'ANS J., Z. anorg. Chem., **62**, 140 (1909)	2326–2328
1452	D'ANS J., SCHREINER O., Z. anorg. Chem., **62**, 152 (1909)	800, 993, 3669, 4690, 4708, 4716, 5128, 5131
1453	D'ANS J., Z. anorg. Chem., **63**, 225 (1909)	794
1454	D'ANS J., FRITSCHE O., Z. anorg. Chem., **65**, 231 (1909)	896

Reference number	Author's name and reference	Table number
1455	D'ANS J., Z. anorg. Chem., **65**, 229 (1910)	737
1456	D'ANS J., SCHREINER O., Z. physik. Chem., **50**, 101 (1910)	654, 658
1457	D'ANS J., SCHREINER O., Z. anorg. Chem., **67**, 437 (1910)	1822, 2675, 4389
1458	D'ANS J., Z. anorg. Chem., **80**, 235 (1913)	739, 796, 2315, 3415, 4330
1459	D'ANS J., BERTSCH A., GESSNER A., Kali, No. 9, 153 (1915)	3638
1460	D'ANS J., BERTSCH A., GESSNER A., Kali, No. 11, 165 (1915)	3325
1461	D'ANS J., BERTSCH A., GESSNER A., Kali, No. 12, 181 (1915)	3368, 3585
1462	D'ANS J., BERTSCH A., GESSNER A., Kali, No. 13, 193 (1915)	4044, 4056
1463	D'ANS J., BERTSCH A., GESSNER A., Kali, No. 14, 218, 220 (1915)	3396
1464	D'ANS J., BERTSCH A., GESSNER A., Kali, No. 15, 233; No. 16, 245–246 (1915)	3599
1465	D'ANS J., BERTSCH A., GESSNER A., Kali, No. 17, 262 (1915)	3316, 4690
1466	D'ANS J., DAWIHL W., Z. anorg. Chem., **178**, 252 (1929)	4308
1467	D'ANS J., BUSCH F., Z. anorg. Chem., **232**, 359 (1937)	4145, 4173, 4180, 4880
1468	D'ANS J., HOFER P., Z. angew. Chem., **50**, 101, (1937)	676
1469	D'ANS J., KATZ W., Kali, No. 35, 41 (1941)	3969, 3973
1470	D'ANS J., Sypiena G., Kali, No. 36, 90 (1942)	3634, 4134
1471	D'ANS J., Kali, No. 38, 89, 184 (1944)	3611, 4047
1472	D'ANS J., Kanakowsky Th., Kali, No. 4, 72 (1944)	3610, 4048
1473	D'ANS J., KNUTTER R., Z. angew. Chem., **65**, 578 (1953)	4278
1474	D'ANS J., BEGRENDT K. H., Kali und Steinsalz, **2**, No. 4, 121 (1957)	1564, 1566
1475	D'ANS J., KAUFMANN E., Z. anorg. Chem., **290**, 5–6, 345 (1957)	4694, 4879
1476	D'ANS J., FREUNOL H. E., Kali und Steinsalz, **3**, No. 1, 31 (1960)	771
1477	D'ANSELME A., Bull. Soc. chim., (3), **29**, 373 (1903)	3744
1478	D'ANSELME, Bull. Soc. chim., (3), **29**, 938 (1903)	2741
1479	DAVE P. C., KRISHNASWAMI K. R., J. Indian Inst. Sci., (A), **16**, 153 (1933)	495, 2197, 3135, 4607
1480	DAVIES C. W., WYATT P. A. H., Trans. Faraday Soc., **45**, 770 (1949)	3526, 3531, 3532
1481	DAVIES C. W., HOYLE B. E., J. Chem. Soc., 233 (1951)	5071
1482	DAWKINS A. E., J. Chem. Soc., **121**, 776 (1922)	2941
1483	DAWSON H., J. Chem. Soc., Trans., **113**, 679 (1918)	749, 2290
1484	DEACON G. E. R., J. Chem. Soc., **130**, 2063 (1927)	1474, 3852
1485	DEDE L., WALTER Th., Z. anorg. Chem., **163**, 185 (1927)	5412, 5416
1486	DELAPINE M., Bull. Soc. chim. belges, **62**, 91 (1953)	1983
1487	DEMASSIEUX N., Compt. rend., **156**, 892 (1913)	2546
1488	DEMASSIEUX N., (1922), cited in [2]	3849
1489	DEMASSIEUX N., Ann. chim., (9), **20**, 233 (1923)	1472, 4929, 5023
1490	DEMASSIEUX N., Compt. rend., **177**, 51 (1923)	1472, 2549, 2585, 4929
1491	DEMASSIEUX N., ROGER L., Compt. rend., **204**, 1818 (1937)	1491
1492	DENHAM H. G., DICK D. A., J. Chem., Soc., 1753 (1931)	1802
1493	DENHAM H. G., KIDSON J. O., J. Chem. Soc., 1757 (1931)	1804
1494	DENHAM H. G., FIFE C. V., J. Chem. Soc., **10**, 1416 (1933)	1803
1495	DENHAM H. G., KING W. E., J. Chem. Soc., 1251 (1935)	4359
1496	DERR P. F., STOCKDALE R. M., VOSHBURGH M. C., J. Am. Chem. Soc., **63**, 2670 (1941)	1820, 1922, 1923, 1944, 1970, 2018, 2020

Reference number	Author's name and reference	Table number
1497	DERR P. F., VOSHBURGH W. C., J. Am. Chem. Soc., **65**, 2408 (1943)	1817, 2033, 2543
1498	DERVAULLE M. L., HEEMS J., Compt. rend., **235**, 161 (1952)	5527
1499	DEY A. K., J. Inorg. Nucl. Chem., **6**, No. 1, 71, (1958)	3529
1500	DIACON, Jahrsber. Chem., 61, cited in [2]	4094
1501	DI CAPUA C., SCALETTI U., Gazz. chim. ital., **57**, 394 (1927)	3543, 3682
1502	DI CAPUA C., BERTONI A., Gazz. chim. ital., **58**, 249 (1928)	3830, 3865, 3870, 5502, 5503
1503	DI CAPUA C., Gazz. chim. ital., **59**, 164 (1929)	4001, 4007, 5202
1504	DICKSON F. W., TUNELL G., Science, **119**, No. 3093, 467 (1954)	3222
1505	DIETRICH, Kaminer Handbuch der Balneologie, vol. 1, p. 205 (1916), cited in [2]	3746, 4085
1506	DIETRICH H. G., JOHNSTON J., J. Am. Chem. Soc., **49**, 1419 (1927)	2062, 2767
1507	DITTE A., Compt. rend., **92**, 242, 719 (1881)	1045, 1139, 1240, 5022
1508	DITTE A., Compt. rend., **123**, 1282 (1896)	574
1509	DITTE A., Compt. rend., **124**, 30 (1897)	2717, 4411
1510	DITTE A., Ann. chim. phys., (7), **10**, 564 (1897)	1243, 1246, 4384, 4395, 4414
1511	DITTE A., Ann. chim. phys., (7), **14**, 294 (1898)	4844, 4978, 5004
1512	DOBBINS J. T., BYRD R. M., J. Phys. Chem., **35**, 3673 (1931)	3407
1513	DOBBINS J. T., ADDESTON J. A., J. Phys. Chem., **39**, 637 (1935)	3407
1514	DOBBINS J. T., THOMAS L. C., J. Phys. Chem., **41**, 655 (1937)	779
1515	DOLEZALEK F., FINCKLI K., Z. anorg. Chem., **51**, 320 (1906)	903
1516	DOHLEN W. C., SIMONS E. L., J. Am. Chem. Soc., **73**, 1, 461 (1951)	5333, 5401
1517	DOLIQUE R., PAUC M., Trav. Soc. pharm. (Montpellier), **6**, 86 (1946–1947)	2273, 2493, 4683
1518	DOMINIK W., Przem. chem., Warszawa, **5**, 185 (1921)	759
1519	DOMINIK W., Przem. chem., **6**, 317 (1922)	3038
1520	DOMKE R., NEUMANN B., Z. Elektrochem., **34**, 136 (1928)	2091
1521	DONK A. D., Chem. Weekblad, **13**, 92 (1916)	905
1522	DONK M. G., Chem. Weekblad, **5**, 529, 629, 767 (1908)	4413
1523	DONNAN F. G., THOMAS J. S., J. Chem. Soc. (London), **99**, 1788 (1911)	1964
1524	DORP G. C. A., von, Z. physik. Chem., **73**, 284 (1910)	715, 738
1525	DORP G. C. A., von, Chem. Weekblad, 8, 269 (1911)	738
1526	DORP G. C. A., von, Z. physik. Chem., **86**, 112 (1914)	719
1527	DORP G. C. A., Rec. trav. chim., **42**, 769 (1923)	720
1528	DRATOVSKY M., TURNWALD J. Coll. Czechosl. Chem. Comm., **25**, No. 10, 2503 (1960)	2587
1529	DRIOT, Compt. rend., **150**, 1426 (1910)	5291
1530	DRUCKER K., Z. anorg. Chem., **28**, 362 (1901)	861, 889, 4750, 4773
1531	DUBRISSAY R., Compt. rend., **153**, 1077 (1911)	546
1532	DUBRISSAY R., FRANCOIS R., Compt. rend., **102**, 742 (1931)	4842
1533	DUC-MAUGÉ C., Bull. Soc. chim. France, No. 2, 227 (1957)	4370
1534	DUKELSKI M. P., Z. anorg. Chem., **50**, 42 (1906)	1547, 1571
1535	DUKELSKI M. P., Z. anorg. Chem., **53**, 327; **54**, 45 (1907)	1271
1536	DUKES E. K., BURNEY G. A., J. Inorg. Nucl. Chem., **24**, 899 (1962)	228
1537	DUNN, Chem. News, **45**, 272 (1882)	787

Reference number	Author's name and reference	Table number
1538	DUNNINGHAM A. C., J. Chem. Soc., **105**, 368, 733, 2630 (1914)	5017
1539	EBERLY N. E., GROSS C. V., CROWELL W. S., J. Am. Chem. Soc., **42**, 1433 (1920)	4299
1540	ECKSTEIN H., LISSNER A., Chem. Tech., **6**, 18, (1954)	5260
1541	EDDY R. D., MENZIES A. W. C., J. Phys. Chem., **44**, 207 (1940)	220
1542	EHLERT H., HEMPEL W., Z. Elektrochem., **18**, 727 (1912)	1645, 1679, 1680, 3736, 5113
1543	EHRET Wm. F., J. Am. Chem. Soc., **54**, 3126 (1932)	4634, 5093, 5142, 5355
1544	EHRET W. F., FRERE F. J., J. Am. Chem. Soc., **67**, 68 (1945)	4223, 4234, 4235
1545	EICHELBERGER W. C., Ind. Eng. Chem., **47**, No. 10, 2223 (1955)	81
1546	EKERLOF, (1935), cited in [2]	68
1547	ELLINGBOE J. L., J. Chem. Engng Data, **9**, No. 1, 91 (1964)	3347
1548	ELLIOTT G. A., KLEIST L. L., WILKINS F. J., WEBB H. W., J. Chem. Soc., **129**, 1231 (1926)	699
1549	ELLIS A. J., Am. J. Sci., **257**, No. 5, 354 (1958)	1716
1550	ELLIS A. J., GOLDING R. M., Am. J. Sci., **261**, No. 1, 47 (1963)	1637
1551	ELLIS A. J., Am. J. Sci., **261**, No. 3, 259 (1963)	1651
1552	ELMORE K. L., FARR T. D., Ind. Eng. Chem., **32**, 4, 580 (1940)	669, 4277
1553	EL WAKKAD S. E. S., RIZK H. A. M., J. Chem. Soc., 3248 (1956)	1128
1554	ENDE C. L., von, Z. anorg. Chem., **26**, 148 (1901)	1141, 4927
1555	ENGEL R., VILLE J., Compt. rend., **93**, 340 (1881)	1659
1556	ENGEL R., Bull. Soc. chim., **44**, 359 (1885)	1668
1557	ENGEL R., Compt. rend., **102**, 114 (1886)	926, 2349
1558	ENGEL R., Compt. rend., **104**, 433, 507, 913 (1887)	412, 427, 839, 867, 982, 1016
1559	ENGEL R., Ann. chim. phys., (6), **13**, 348 (1888)	939, 955, 994, 1073, 1114, 1659
1560	ENGEL R., Ann. chim. phys., (6), **17**, 351 (1889)	1039, 1050, 1053, 1132, 1140
1561	ENGEL R., Bull. Soc. chim., (3), **6**, **17** (1891)	1838, 2662
1562	ENGEL R., Compt. rend., **121**, 529 (1895)	1050
1563	ENKLAAR J. E., Rec. trav. chim., **20**, 183 (1901)	956
1564	ENSSLIN F., DREYER H., Z. anorg. Chem., **249**, 119 (1942)	1092, 1252
1565	ENSSLIN F., Z. anorg. Chem., **250**, 199 (1942–1943)	507
1566	ENSSLIN F., LESSMANN O., ZIEMECK B., Z. anorg. Chem., **254**, 83 (1947)	869
1567	ENSSLIN F., DREYER H., Z. anorg. Chem., **254**, 92 (1942)	1234
1568	ENSSLIN F., ZIEMECK B., SCHAEPDRYVER L., Z. anorg. Chem., 254, 293 (1947)	506, 1458, 1482, 1488, 4179, 4198, 4199, 5159, 5184, 5188, 5308, 5312, 5313, 5363, 5371, 5375, 5451–5453

Reference number	Author's name and reference	Table number
1569	EPIK P. A., Z. anal. Chem., **89**, 17 (1932)	2028, 2029
1570	ERBER W., Z. anorg. Chem., **248**, 32 (1941)	1233, 1250, 1251
1571	ERBER W., SCHÜHLY A., J. prakt. Chem., **158**, 176 (1941)	1089
1572	ERDÖS E., Chem. listy, **51**, No. 9, 1626 (1957)	3703
1573	ERDÖS E., SIMKOVA H., Chem. listy, **51**, No. 12, 2200 (1957)	3046
1574	ERDÖS E., SIMKOVA H., Coll. Czechosl. Chem. Comm., **24**, No. 2, 503 (1959)	3047
1575	ERDÖS E., JURU Z., Coll. Czechosl. Chem. Coom., **25**, No. 7, 1720 (1960)	4543, 4547–4549, 4688, 4689, 4710
1576	ERVIN G. J., GIORGI A. L., McCARTHY C. E., J. Am. Chem. Soc., **66**, 384 (1944)	3016
1577	ETARD, Ann. chim. phys., (7), **3**, 283, (1894)	3660, 4553, 4600, 4822, 4826, 4828, 5480, 5523
1578	ETARD, Ann. chim. phys., (7), **3**, 275 (1897)	4810, 4972
1579	EULER H., Ber., **36**, 2879, 3400 (1903)	2060
1580	EULER H., Z. physik. Chem., **49**, 313 (1904)	3673, 4539 4615, 5129, 5494
1581	EVERSOLE W. G., McLACHLAN R. W., J. Am. Chem. Soc., **54**, 864 (1932)	494
1582	EWING W. W., RICHARD A. J., TAYLOR W. J. J., WINKLER D. M., J. Am. Chem. Soc., **55**, 4830 (1933)	476
1583	EWING W. W., KLINGER E., J. Am. Chem. Soc., **55**, 4825 (1933)	418
1584	EWING W. W., GLICK C. F., J. Am. Chem. Soc., **62**, 2174 (1940)	460
1585	FAIBY C. F., J. Am. Chem. Soc., **55**, 3112 (1933)	2791, 3846
1586	FAIRHALL L. T., J. Am. Chem. Soc., **46**, 1593 (1924)	4307
1587	FALINSKI M., Ann. chim. phys., **16**, 237 (1941)	485, 857
1588	FARR T. D., TARBUTTON G., LEWIS H. T., J. Phys. Chem., **66**, No. 2, 318 (1962)	573
1589	FARROW M., J. Chem. Soc. (London), 1153 (1927)	3040
1590	FAUST O., ESSELMAN P., Z. anorg. Chem., **157**, 290 (1926)	758
1591	FEDOTIEFF P. P., REICH, Monatsh. Chem., **12**, 464 (1891)	2903
1592	FEDOTIEFF P. P., Z. physik. Chem., **49**, 168 (1904)	2085, 2089, 2103, 2460
1593	FEDOTIEFF P. P., Z. anorg. Chem., **69**, 26 (1910–1911)	194
1594	FEDOTIEFF P. P., Z. anorg. Chem., **73**, 178 (1911–1912)	5269
1595	FEDOTIEFF P. P., KOLTUNOFF J., Z. anorg. Chem., **85**, 251 (1914)	1815, 1873, 2075, 2164
1596	FEDOTIEFF P. P., KOLOSOFF A., Z. anorg. Chem., **130**, 39 (1923)	1879, 2077, 2080, 2087, 2103, 2293, 2882
1597	FEDOTIEFF P. P., Z. anorg. Chem., **173**, 81 (1928)	1051, 3793, 5281
1598	FEDOTIEFF P. P., KOLTUNOFF J., J. anorg. Chem., **199**, 268 (1931)	2879
1599	FEIT W., KUBIERSCHKY K., Chem. Ztg., **16**, No. 21, 336 (1892)	4144, 4172
1600	FEIT W., PRSIBYLLA K., Kali, No. 18, 394 (1909)	3615

Reference number	Author's name and reference	Table number
1601	FERRIS, L. M., J. Chem. Engng Data, **5**, No. 3, 241, (1960)	534, 2222
1602	FINDLAY A., HARBY W. H., Kolloid Z., **3**, 169 (1908).	1737
1603	FINDLAY A., CREIGHTON H., J. Chem. Soc., **97**, 538, 557 (1910)	1689
1604	FINDLAY A., SHEN B., J. Chem. Soc., **101**, 1461 (1912)	1615, 1690
1605	FINDLAY A., MORGAN J., MORRIS J. P., J. Chem. Soc., **105**, 779 (1914)	4598, 4616
1606	FINDLAY A., CRUICKSHANK J., J. Chem. Soc., **129**, 316 (1926)	3068, 3082, 3142, 3829, 5482
1607	FLATT R., BURKHARDT G., Helv. chim. acta, **27**, 1605 (1944)	2443, 2492, 2575, 4819
1608	FLATT R., FRITZ P., Helv. chim. acta, **33**, No. 7, 2045 (1950)	400, 447, 2186
1609	FLATT R., FRITZ P., Helv. chim. acta, **34**, No. 1, 231 (1951)	402
1610	FLATT R., BRUNISHOLZ G., CHAPUIS-GOTTREUX S., Helv. chim. acta, **34**, No. 2, 683 (1951)	649, 672
1611	FLATT R., BRUNICHOLZ G., Helv. chim. acta, **34**, No. 2, 692 (1951)	683, 4302
1612	FLATT R., BRUNISHOLZ G., CHAPUIS-GOTTREUX S., Helv. chim. acta, **34**, No. 3, 884 (1951)	653
1613	FLATT R., BRUNISHOLZ G., CLERC P., Helv. chim. acta, **35**, No. 1, 336 (1952)	374
1614	FLATT R., BRUNISHOLZ G., LAUBER E. Helv. chim. acta, **36**, No. 7, 1980 (1953)	375, 376
1615	FLATT R., BRUNISHOLZ G., JAUNIN M., Helv. chim. acta, **37**, No. 1, 282 (1954)	377
1616	FLATT R., WILHELM J., BRUNISHOLZ G., FELL G., Helv. chim. acta, **37**, No. 2, 607 (1954)	382
1617	FLATT R., BRUNISHOLZ G., FELL G., Helv. chim. acta, **37**, No. 7, 2363 (1954)	383, 448
1618	FLATT R., BRUNISHOLZ G., BLUMER O., Helv. chim. acta, **38**, No. 3, 753 (1955)	372, 650
1619	FLATT R., BRUNISHOLZ G., ROD P., Helv. chim. acta, **38**, No. 3, 769 (1955)	371
1620	FLATT R., BRUNISHOLZ G., DENEREAZ A., Helv. chim. acta, **39**, No. 2, 473 (1956)	373, 384, 401, 449, 651, 673
1621	FLATT R., BRUNISHOLZ G., BOURGEOIS J., Helv. chim. acta, **39**, No. 3, 841 (1956)	380, 381, 659, 660
1622	FLATT R., BRUNISHOLZ G., FELL M., Helv. chim. acta, **39**, No. 4, 1130 (1956)	378
1623	FLATT R., BRUNISHOLZ G. HOTZ R., Helv. chim. acta, **39**, No. 5, 1406 (1956)	379
1624	FLATT R., BRUNISHOLZ G., MUHLETHALER R., Helv. chim. acta, **44**, No. 6, 1582 (1961)	2188
1625	FLATT R. BRUNISHOLZ G., DAGON R., Helv. chim. acta, **44**, No. 7, 2173 (1961)	648, 652, 664
1626	FLECK J., Bull. Soc. chim., (5), **4**, 558 (1937)	4790
1627	FOCK A., Z. kryst. Min., **28**, 365, 397 (1897)	2417, 2422, 2425, 2316, 2574, 4269, 4601, 4625, 4700, 4811, 5259, 5358, 5361, 5494

Reference number	Author's name and reference	Table number
1628	FOOTE H. W., Am. Chem. J., 29, 210 (1903)	200
1629	FOOTE H. W., Am. Chem. J., **30**, 341 (1903)	4509, 5464
1630	FOOTE H. W., Z. physik. Chem., **46**, 81 (1903)	4509
1631	FOOTE H. W., BRISTOL, Am. Chem. J., **32**, 248 (1904)	4905, 5509
1632	FOOTE H. W., LEVY, Am. Chem. J., **35**, 239 (1906)	3841, 5339
1633	FOOTE H. W., LEVY, Am. Chem. J., **37**, 119 (1907)	2546, 4920, 5466
1634	FOOTE H. W., CHALKER W. C., Am. Chem. J., **39**, 564, 567 (1908)	185, 195
1635	FOOTE H. W., J. Am. Chem. Soc., **33**, 465 (1911)	3418, 3497
1636	FOOTE H. W., J. Am. Chem. Soc., **34**, 880 (1912)	2521
1637	FOOTE H. W., SAXON B., J. Am. Chem. Soc., **36**, 1695 (1914)	2508
1638	FOOTE H. W., J. Am. Chem. Soc., 37, 290, 1200 (1915)	840
1639	FOOTE H. W., Ind. Eng. Chem., **11**, 629 (1919)	758, 841
1640	FOOTE H. W., BRINKLEY S. R., J. Am. Chem. Soc., **43**, 1018 (1921)	16
1641	FOOTE H. W., J. Am. Chem. Soc., **43**, 1031 (1921)	1810
1642	FOOTE H. W., J. Am. Chem. Soc., **45**, 663 (1923)	1040, 1046, 1054
1643	FOOTE H. W., Am. J. Sci., (5), **13**, 158 (1927)	3786, 4870, 5250, 5253
1644	FOOTE H. W., VANCE J. E., Am. J. Sci., (5), **17**, 425 (1929)	3098, 3575
1645	FOOTE H. W., SCHAIRER J. F., J. Am. Chem. Soc., **52**, 4202 (1930)	3158–3160, 3263
1646	FOOTE H. W., VANCE J. E., Am. J. Sci., (5), **19**, 203 (1930)	3361
1647	FOOTE H. W., VANCE J. E., Am. J. Sci., (5), **25**, 499 (1933)	2996
1648	FOOTE H. W., BRADLEY W. M., J. Am. Chem. Soc., **58**, 930 (1936)	1990
1649	FOOTE H. W., HICKEY F. C., J. Am. Chem. Soc., **56**, 648 (1937)	5471, 5475, 5476
1650	FORBES G. S., J. Am. Chem. Soc., **33**, 1937 (1911)	1086, 2534, 3808, 4889, 5157, 5307, 5362, 5408
1651	FORBES G. S., COLE H. J., J. Am. Chem. Soc., **43**, 2492 (1921)	1087, 3809, 5411
1652	FORCRAND De, FONZES-DIACON, Ann. chim. phys., (7), **26**, 253 (1902)	1204
1653	FORET J., Compt. rend., **217**, 264 (1943)	3749
1654	FORMANEK, Chem. Centralbl., 18, 270 (1887)	5554
1655	FORREST R. H., HARRY H. S., J. Am. Chem. Soc., **79**, 2999 (1957)	15
1656	FOX R. K., CHAS J. J., Z. physik. Chem., **41**, 458 (1902)	2126, 2174, 2175, 2314, 2482, 2573, 3031, 3412, 3657, 3908, 4310, 4312, 4314, 4316, 4320, 4326–4328
1657	FOX R. K., SWINEHART D. F., GARRETT A. B., J. Am. Chem. Soc., **63**, 1779 (1941)	1028, 2758
1658	FRANCOIS M., Compt. rend., **200**, 393 (1935)	5015
1659	FRANCOIS M., Compt. rend., **201**, 215, 1489 (1935)	2583, 3948
1660	FRANCOIS M., DELWAULLE M. L., Bull. Soc. chim., (5), 3, 687 (1936)	1489, 2584, 3949, 5015
1661	FRATINI N., Ann. chim. applicata, **39**, 616 (1949)	2742, 4419

Reference number	Author's name and reference	Table number
1662	FRAPS G. S., Am. Chem. J., **27**, 290 (1901)	4181, 4265, 5231
1663	FREAR G. L., JOHNSTON J., J. Am. Chem. Soc., **51**, 2082 (1929)	1700, 1713, 3731, 5075
1664	FRED A., SCHIMMEL, J. Am. Chem. Soc., **74**, 18, 4689 (1952)	1033, 5223
1665	FREDHOLM H., Z. anorg. Chem., **218**, 237, (1934)	2055
1666	FREETH F. A., Phil. Trans. Roy. Soc. (London), Ser. A., **223**, 35 (1922)	2629, 2657, 2707, 2849, 2867, 2906, 2972
1667	FREETH F. A., Dissertation, Leiden, (1924), cited in [2]	2277, 2297, 2553, 3342
1668	FREETH F. A., Rec. trav. chim., **43**, 475 (1924)	2277, 2294, 2553, 3343
1669	FRERE F. J., J. Am. Chem. Soc., **58**, 1965 (1936)	3181–3186
1670	FRICKE R., WINDHAUSEN O., Z. anorg. Chem., **132**, 226 (1924)	2757
1671	FRICKE R., BLENCKE W., Z. anorg. Chem., **143**, 193 (1925)	2768
1672	FRICKE R., Z. anorg. Chem., **172**, 234 (1928)	2765
1673	FRICKE R., HUMME H., Z. anorg. Chem., **178**, 400 (1929)	1499
1674	FRICKE R., JUCTIGIS P., Z. anorg. Chem., **191**, 129 (1930)	2731, 4221
1675	FRICKE R., BRUMMER F., Z. anorg. Chem., **213**, 319 (1933)	516
1676	FRICKE R., BRUMMER F., Z. anorg. Chem., **223**, 397 (1935)	965
1677	FRIEDMAN H. B., LA MER V. K., J. Am. Chem. Soc. **53**, 103 (1931)	3144, 3502, 3834, 4009, 4105, 4183, 4622, 4764, 4914, 5434, 5440, 5534, 5538, 5541
1678	FRIEDMAN J. J., J. Am. Chem. Soc., **70**, 2649 (1948)	3003
1679	FRIEND J. A. N., J. Chem. Soc., 1633, 1903 (1930)	421, 886, 3996
1680	FRIEND J. A. N., J. Chem. Soc., 1802, 2225 (1931)	1212
1681	FRIEND J. A. N., J. Chem. Soc. (London), 1597 (1932)	1211
1682	FRISAK A., Metal. u. Erz., **19**, 200 (1922)	906
1683	FROEHLICH W., Mitt. Kali Forsch. Anstalt, **68**, 37 (1929)	3605, 3685, 4021
1684	FROWEIN F., MÜHLENDAHL E., von, Z. anorg. Chem., **39**, 1491, (1926)	3620, 3642, 4126
1685	FROWEIN F., Z. anorg. Chem., **169**, 336 (1928)	3122, 4589
1686	FUJIMURA T., Mem. Coll. Sci. Kyoto, **1**, 63 (1914)	4899
1687	FUNAKI K., Bull. Tokyo Inst. Technol., Ser. B, No. 1, 165 (1950)	2311, 4248
1688	FURUKAWA H., KING G. B., J. Phys. Chem., **48**, 174 (1944)	1206
1689	GABER K. H., LEIDER H., J. Am. Chem. Soc., **77**, 1448 (1955)	1182
1690	GABER K. H., LEIDER H., J. Am. Chem. Soc., **77**, 1448 (1955)	2800
1691	GALL J. F., CHURCH G. L., BROWN R. L., J. Phys. Chem., **47**, 645 (1943)	747
1692	GALLIN G., Ann. chim., (12), **1**, 308 (1946)	2048, 2056
1693	GALLO G., Ann. chim. applicata, **25**, 628 (1935)	2193, 2358, 2531, 2746, 3127, 3802, 5153
1694	GARRAN R. R., J. Chem. Soc., **129**, 848 (1926)	3239

Reference number	Author's name and reference	Table number
1695	GARRETT A. B., HIRSCHLER A. E., J. Am. Chem. Soc., **60**, 299 (1938)	1284, 2664, 2703, 2739, 4425
1696	GARRETT A. B., VELLENGA S., FONTENA C. N., J. Am. Chem. Soc., **61**, 367 (1939)	2793
1697	GARRETT A. B., HOWELL W. W., J. Am. Chem. Soc., **61**, 1730 (1939)	528
1698	GARRETT A. B., HOLMES O., LAUBE A., J. Am. Chem. Soc., **62**, 2024 (1940)	1062, 2771
1699	GARRETT A. B., HEINS R. E., J. Am. Chem. Soc., **63**, 562 (1941)	1093, 2782
1700	GARRETT A. B., LEMLEY J. J. Am. Chem. Soc., **64**, 2380 (1942)	1129, 1130
1701	GARRICK F. J., WILSON C. L., J. Chem. Soc. (London), 835 (1932)	1995
1702	GAUDECHON H., Compt. rend. **150**, 467 (1910)	1037
1703	GAUNT J., BASTIEN J. J., ADELMAN M., Canad. J. Chem., **41**, No. 2, 527 (1963)	556
1704	GAYER K. H., GARRETT A. B., J. Am. Chem. Soc., **71**, 2973 (1949)	1043, 2763
1705	GAYER K. H., GARRETT A. B., J. Am. Chem. Soc., **72**, 3921 (1950)	1038, 2761
1706	GAYER K. H., LEIDER H., Canad. J. Chem., **35**, No. 1, 5 (1957)	1185, 2802
1707	GAYER K. H., THOMPSON L. C., ZAJICEK O. T., Canad. J. Chem., **36**, No. 9, 1268 (1958)	2733
1708	GAYER K. H., THOMPSON L. C., Canad. J. Chem., **36**, No. 12, 1649 (1958)	1183, 2801
1709	GAYER K. H., HAAS R. M., THOMPSON L. C., J. Inorg. Nucl. Chem., **26**, No. 8, 1401 (1964)	1184
1710	GAYER K. H., ZAJICEK O. T., J. Inorg. Nucl. Chem., **26**, No. 12, 2133 (1964)	1165, 2770
1711	GEFFECKEN G., Z. physik. Chem., **49**, 268 (1904)	44, 45, 46, 49, 56, 82, 85, 88, 91, 93, 97, 101, 395, 724, 731, 933, 935, 1324, 1403, 1687, 1695, 1698, 1729, 1742, 1767, 1774, 1777, 1779, 1793, 1796, 2203, 3118, 3147, 3188, 3725, 3871, 4945
1712	GEFFECKEN G., Z. anorg. Chem., **43**, 197 (1905)	1300, 1301, 1307, 1308, 1310, 1311, 1313, 1314
1713	GEIGER A., Inaug. Dissert., Berlin, (1904), cited in [1]	4055, 4076
1714	GELBACH R. W., KING G. B., J. Am. Chem. Soc., **64**, 1054 (1942)	1210
1715	GERASSIMOW J. I., Z. anorg. Chem., **187**, 321 (1930)	2439, 2465, 2563, 3555
1716	GERLACH H., Kali, No. 1, 8 (1926)	3681
1717	GHANI M. O., ALLEM S. A., Indian J. Agr. Sci., **13**, 142 (1943)	775, 776, 803, 2734, 2735
1718	GHIRON D., MANGIKI L., Gazz. chim. ital., **65**, 1244 (1935)	1270

Reference number	Author's name and reference	Table number
1719	GHIRON D., MONTICELLI M., Gazz. chim. ital., 68, 555 (1938)	3197, 3887, 4670, 4967
1720	GILBERT E. C., J. Phys. Chem., **33**, 1235 (1929)	3137
1721	GILBERT L. F., BUCKLEY H., MASSON J., J. Chem. Soc., **121**, 1934 (1922)	4335
1722	GILBERT L. F., J. Chem. Soc., **127**, 1541 (1925)	1581
1723	GILBERT R. A., GARRETT A. B., J. Am. Chem. Soc., **78**, 5501 (1956)	930, 1157, 1500
1724	GILL H. W., J. Chem., Met. Soc. S. Africa, **14**, 290 (1914)	176
1725	GIRARD H., Bull. Soc. chim., (2), **43**, 552 (1885)	1927
1726	GLASSTONE S., J. Chem. Soc., **121**, 58 (1922)	2792, 4427
1727	GLASSTONE S., SAUNDERS R. N., J. Chem. Soc., **123**, 2134 (1923)	5633
1728	GLASSTONE S., RIGGS E. J., J. Chem. Soc., **127**, 2846 (1925)	4617, 5495
1729	GLOSS G., Dissert., Über Magnesiumcarbonate, Naumburg, (1937–1938), cited in [1]	2084, 2098, 2306, 2480, 2915, 3002, 3976, 3978, 4451
1730	GLOWCZYNSKI Z., Kolloidchem. Beihefte, **6**, 147 (1914), cited in [2]	491, 2535, 4890
1731	GOLDSCHMIDT, LARSON, Z. physik. Chem., **71**, 437 (1910)	696
1732	GOODWIN H. M., HORSCH W. G., Chem. Met. Eng., **21**, 181 (1919)	842
1733	GOODWIN W. L., Ber., **15**, 3039 (1882)	129
1734	GORDON V., Z. physik. Chem., **18**, 1 (1895)	47, 50, 51, 53, 55, 58, 59, 61, 63, 1341, 1402, 1750, 1753, 1760, 1771, 1773, 1782, 1794
1735	GORMAN M., LEIGHTON P. A., J. Am. Chem. Soc., **64**, 719 (1942)	1169
1736	GOTHE F., Chem. Ztg., **39**, No. 48, 306 (1915)	3000, 3366, 3446, 3581, 3737, 3983, 5079
1737	GOUDRIAAN F., Proc. Acad. Sci. (Amsterdam), **22**, 179 (1919); Rec. trav. chim., **39**, 505 (1920)	2609
1738	GOUDRIAAN F., Rec. trav. chim., **41**, 22 (1922)	2592
1739	GRACE N. S., J. Chem. Soc., 594 (1931)	191
1740	GREEN S. J., FRATTALI F. J., J. Am. Chem. Soc., **68**, 1789 (1946)	2635, 2648, 2680, 2952
1741	GREENE C. H., J. Am. Chem. Soc., **56**, 1186 (1937)	517
1742	GREENBERG A. L., WALDEN G. H., jr., J. Chem. Phys., **8**, 645 (1940)	4949
1743	GRÄGER M., Z. anorg. Chem., 70, 135 (1911)	5193
1744	GROSCHUFF E., Z. anorg. Chem., **58**, 102 (1908)	389
1745	GROSS P., KLINGHOFFER St. S., Monatsheft. Chem., **55**, 338 (1930)	1435, 2506, 3778, 4862
1746	GRUBE G., Z. Elektrochem., **20**, 342 (1914)	4380–4382, 4521
1747	GRUBE G., STAESCHE M., Z. physik. Chem., **130**, 572 (1927)	679
1748	GRUBE, HELLER, HERRMANN, (1936), cited in [2]	1202
1749	GUERIN H., Bul. Soc. chim. (France), (5), **5**, 1472 (1938)	5324
1750	GUERIN H., Ann. chim. (Ser. 11), **16**, 101 (1941)	5056

Reference number	Author's name and reference	Table number
1751	GUERIN H., Compt. rend., **213**, 1012 (1941)	2201, 2499, 2541, 2542, 4837, 4908, 4909
1752	GUERIN H., ARTUR A., Bull. Soc. chim. France, 562 (1952)	4304
1753	GUERIN H., MARTIN R., Compt. rend., **234**, 1777 (1952)	4203
1754	GUERIN H., KOZICKI H., Compt. rend., **235**, 52 (1952)	4297
1755	GUILLIER R., Caliche, **9**, 292 (1927)	3061
1756	GUNTZ A., GUNTZ A. A., jr., Ann. chim., **2**, 101 (1914)	613
1757	GUYER A., BIELER A., SCHMID G., Helv. chim. acta, **17**, 1530 (1934)	26, 1874, 1919, 1937
1758	GUYER A., BIELER A., ORELLI E., Helv. chim. acta, **23**, 28 (1940)	1807, 1829, 1872, 1885, 1912, 1932
1759	GUYER A., PIECHOWICZ T., Helv. chim. acta, **27**, 28 (1944)	1805
1760	GUYER A., PIECHOWICZ T., Helv. chim. acta, **27**, 864 (1944)	1596
1761	GUYER A., PIECHOWICZ T., Helv. chim. acta, **28**, 402–404 (1945)	1597
1762	GYANI P. P., J. Phys. and Colloid. Chem., **55**, 1111 (1951)	5378
1763	HABER F., OORDT G., van, Z. anorg. Chem., **38**, 387 (1904)	1498, 2730
1764	HABER-CHUWIS Q., Rocsz. chem., **6**, 700 (1926)	2409, 5033
1765	HAEHNEL P., J. prakt. Chem., (2), **107**, 165 (1924)	1664, 1712, 1732, 1735
1766	HAEHNEL O., J. prakt. Chem., (2), **108**, 61 (1924)	1664, 1712, 1732, 1735
1767	HAEHNEL O., J. prakt. Chem., (2), **148**, 295 (1937)	1298
1768	HAENDLER H., JACHE, J. Am. Chem. Soc., **72**, 4137 (1950)	2213
1769	HAENDLER H., GLOW A., J. Am. Chem. Soc., **74**, 1843 (1952)	2211
1770	HALPERN E., Roszn. chem., **6**, 661 (1926)	2423, 5036
1771	HALLA F., Z. physik. Chem., **175**, 63, 396 (1935)	1671
1772	HALLA F., Z. anorg. Chem., **226**, 139 (1926)	3991
1773	HAMID M. A., J. Chem. Soc., **129**, 199 (1926)	3108, 3422, 4540
1774	HAMID M. A., DAS, RAS, J. Ind. Chem. Soc., **7**, 881 (1930)	4590
1775	HAMMICK D. L., CURRIE J. A., J. Chem. Soc., **127**, 1623 (1925)	2674
1776	HARA R., TANAKA Y., NAKAMURA K., The Technology reports Tôhuku Imp. Univ. Sendai, Japan, **11**, No. 2, 102–103 (1934)	5125, 5138
1777	HARKINS W. D., J. Am. Chem. Soc., **33**, 1807 (1911)	4006, 4618, 4999, 5404, 5488, 5576
1778	HARKINS W. D., WINNINGHOFF W. J., J. Am. Chem. Soc., **33**, 1827 (1911)	4619, 4638, 5030, 5032, 5490, 5582
1779	HARKINS W. D., PEARCE W. T., J. Am. Chem. Soc., **38**, 2694, 2717 (1916)	2207, 3039, 3143, 3958, 5031, 5356, 5533
1780	HARKINS W. D., PAINE H. M., J. Am. Chem. Soc., **38**, 2709 (1916)	484, 1074, 1230, 1249, 3128, 4599, 4885, 5008, 5282, 5354
1781	HARKINS W. D., PAINE H. M., J. Am. Chem. Soc., **41**, 1155 (1919)	4086, 4592, 5127
1782	HARMAN R. W., J. Phys. Chem., **32**, 511 (1927)	2596
1783	HARNED H. S., DAVIS R., jr., J. Am. Chem. Soc., **65**, 2030 (1943)	1636
1784	HARRIS J. W. H., J. Chem. Soc. (London), 582, 1694, 2709 (1932)	146, 149

Reference number	Author's name and reference	Table number
1785	HARRIS S. E., CHRISTIANSEN W. G., J. Am. Pharmac. Assoc., 563 (1934)	4829
1786	HARTLEY, ROMAGE, J. Chem. Soc., **63**, 137 (1893)	3459, 3780, 5145
1787	HARTSHORNE N. H., J. Chem. Soc. (London), 1759 (1927)	5214
1788	HASTINGS, MURRAY, SENDROY, J. Biol. Chem., **71**, 746 (1927)	1646
1789	HATTA S., The Technology reports Tôhoku Imp. Univ. Sendai, Japan, **10**, No. 4, 147 (1932)	1691
1790	HAUSER O., Z. anorg. Chem., **54**, 196 (1907)	858
1791	HAUSEN W. C., PRESSER E. E., Ind. Eng. Chem., 1280 (1947)	2604
1792	HAYEK E., Z. anorg. Chem., **219**, 297 (1934)	3972, 5063
1793	HEIDENHAIN H., Ind. Eng. Chem., **10**, 426 (1918)	432
1794	HEIN Fr., DANIEL W., Z. anorg. Chem., **234**, 155 (1937)	5389, 5414
1795	HELLWIG, Z. anorg. Chem., **25**, 166 (1900)	1994, 4511, 5011, 5379, 5386, 5390, 5396
1796	HELVENSTON E. P., CUEVAS E. A., J. Chem. Eng. Data, **9**, No. 3, 321 (1964)	3777
1797	HEMPEL W., TEDESCO H., Z. anorg. Chem., **24**, 2462 (1911)	1889
1798	HENDRICKS S. B., J. Phys. Chem., **30**, 248 (1926)	5324
1799	HENRY J. L., KING G. B., J. Am. Chem. Soc., **71**, 4, 1143 (1949)	781
1800	HERING E., These, Univ. Strasbourg (1926), cited in [2]	3087, 4541, 4545, 4560, 4678
1801	HERING E., Revista Caliche, Chile (1926, 1927) cited in [2]	4541
1802	HERING H., Ann. chim., (11), **5**, 483 (1936)	4895, 4987, 5014
1803	HERING H., These, Paris (1936), cited in [2]	4895, 4987, 5014
1804	HERING H. Bull. Soc. chim. (France), **12**, 557 (1945); Bull. Soc. chim. (France), 278 (1946)	3818
1805	HERRMAN E., STIPETIC J., Z. anorg. Chem., **262**, 269 (1950)	2732
1806	HERZ W., Ann. chim. phys., (7) **3**, 287 (1894)	5524
1807	HERZ W., Z. anorg. Chem., **33**, 355 (1903)	273, 277, 287
1808	HERZ W., Z. anorg. Chem., **34**, 205 (1903)	287
1809	HERZ W., MUHS G., Z. anorg. Chem., **38**, 140 (1904)	2054, 2151
1810	HERZ W. Dissertation (Berlin), (1905), cited in [2]	4601
1811	HERZ W. Z., anorg. Chem., **66**, 359 (1910)	365
1812	HERZ W., Z. anorg. Chem., **68**, 69, 165 (1910)	296, 325, 350, 1441, 1460, 3023, 3789, 3820, 4872, 4901, 5275
1813	HERZ W. Z. anorg. Chem., 68, 165 (1910)	5338
1814	HERZ W., Z. anorg. Chem., 73, 277 (1911–1912)	757, 795, 925, 926, 960, 996, 1222

Reference number	Author's name and reference	Table number
1815	HERZ W., PAUL W., Z. anorg. Chem., **82**, 431 (1913)	1467, 3842, 3920, 3951, 4186, 4921, 4990, 5016, 5171, 5185, 5189, 5365, 5372, 5376, 5511, 5526, 5531
1816	HERZ W., PAUL W., Z. anorg. Chem., **85**, 214 (1914)	150, 209
1817	HERZ W., HELLEBRANDT M., Z. anorg. Chem., **130**, 188 (1923)	3850, 3855, 3923, 3924. 4933, 4992, 5180, 5181, 5186, 5187, 5368, 5370, 5373, 5374, 5516, 5519, 5529, 5530
1818	HERZ W., HIEBENTHAL F., Z. anorg. Chem., **177**, 368 (1929)	166, 173, 180, 182, 197, 203, 1426, 1427, 1466, 2495 –2497, 3722, 3723, 3838, 4147 –4149, 4804, 4806, 4809, 4917, 4954, 4958, 4961, 4964, 4966
1819	HEVESY G., von, KIMURA K., J. Am. Chem. Soc., **47**, 2540 (1922)	1079
1820	HEVESY V., von, CHRISTIANSEN J. A., BERGLUND V., Z. anorg. Chem., **144**, 69 (1925)	581, 580, 587, 2209, 2223
1821	HEVESY G., von, DANSKE K., Vidensky Selskah Mat. fys. Med., **6**, 7, 1 (1925), cited in [2]	581, 587, 1080, 1125, 2209, 2223
1822	HEVESY G., von, KIMURA K., Z. angew. Chem., **38**, 775 (1925)	1079, 1124
1823	HEVESY G., von, KIMURA K., J. Am. Chem. Soc., **47**, 2540 (1925)	1124
1824	HEVESY G., von, WAGNER O. H., Z. anorg. Chem., **191**, 194 (1930)	608, 1232, 1239b, 1241
1825	HEYROVSKY J., BEREZICKY S., Coll. Czechosl. Chem. Soc., **1**, 19 (1929)	1072, 1451, 1461, 1463, 1561
'6	HICKS W. B., J. Am. Chem. Soc., **37**, 844 (1915)	975
'	HIETANEN S., SILLEN L. G., Svensk. kem. Tidskr., **12**, 228 (1947)	5448
	HILL A. E., SIMMONS J. D., J. Am. Chem. Soc., **31**, 821 (1909)	490, 502, 533
	HILL A. E., SIMMONS J. D., Z. physik. Chem., **67**, 594 (1909)	490, 502, 533

Reference number	Author's name and reference	Table number
1830	HILL A. E., MILLER F. W., jr., J. Am. Chem. Soc., **49**, 669 (1927)	3010
1831	HILL A. H., HILL D., J. Am. Chem. Soc., **49**, 968 (1927)	4442
1832	HILL A. E., BACON L. R., J. Am. Chem. Soc., **49**, 2487 (1927)	2850
1833	HILL A. E., J. Am. Chem. Soc., **50**, 2678 (1928)	1268, 1269, 2723, 4418
1834	HILL A., SMITH S., J. Am. Chem. Soc., **51**, 1630 (1929)	2873
1835	HILL A. E., MOSKOWITZ S., J. Am. Chem. Soc., **51**, 2396 (1929)	4463
1836	HILL A., J. Am. Chem. Soc., **52**, 3815 (1930)	2874, 4445
1837	HILL A. E., MOSKOWITZ S., J. Am. Chem. Soc., **53**, 941 (1931)	3995
1838	HILL A. E., DONAVAN J. E., J. Am. Chem. Soc., **49**, 2487 (1931)	3098
1839	HILL A. E., RICCI J. E., J. Am. Chem. Soc., **53**, 4305 (1931)	3954, 3955, 4717, 4828
1840	HILL A. E., BROWN S. F., J. Am. Chem. Soc., **53**, 4316 (1931)	3956, 4587
1841	HILL A. E., WILLSON H. S., BISHOP J. A., J. Am. Chem. Soc., **55**, 520 (1933)	3941
1842	HILL A. E., J. Am. Chem. Soc., **56**, 1071 (1934)	4725
1843	HILL A. E., YANICK N. S., J. Am. Chem. Soc., **57**, 645 (1935)	2333–2336
1844	HILL A. E., KAPLAN N., J. Am. Chem. Soc., **58**, 1644 (1936)	2205
1845	HILL A. E., KOUCKS C. M., J. Am. Chem. Soc., **59**, 2094 (1937)	1827, 1832, 1834, 1843, 1848, 1931, 1934, 1938, 2276, 2321
1846	HILL A. E., J. Am. Chem. Soc., **59**, 2243 (1937)	4726
1847	HILL A. E., KAPLAN N., J. Am. Chem. Soc., **60**, 550 (1938)	2310
1848	HILL A. E., TAYLOR W. J., J. Am. Chem. Soc., **60**, 1099 (1938)	2348, 2354, 2419, 2421, 2424
1849	HILL A. E., WILLS J. H., J. Am. Chem. Soc., **60**, 1647 (1938)	3455
1850	HILL A. E., SMITH N. O., RICCI J. G., J. Am. Chem. Soc., **62**, 858 (1940)	2396, 2398, 2413, 4261, 4270
1851	HILL A. E., DURHAM G. S., RICCI J. E., J. Am. Chem. Soc., **62**, 1031 (1940)	2391, 2426, 5035, 5037, 5038, 5040
1852	HILL A. E., SOTH G. C., RICCI J. E., J. Am. Chem. Soc., **62**, 2717 (1940)	2383, 2384, 2561, 2588, 2589, 4194
1853	HILL L. M., J. Chem. Soc., 476 (1945)	1543, 2820
1854	HINRICHSEN F. W., SACHSEL E., Z. physik. Chem., **50**, 81 (1904–1905)	3783, 4866, 5230
1855	HOGG A. R., J. Chem. Soc., **129**, 855 (1926)	3216
1856	HOITSEMA C., Z. physik. Chem., **17**, 651 (1895)	4361
1857	HOLLAND A., Ann. chim. anal., **2**, 243 (1897)	3221
1858	HOLLUTA J., MOUTNER S., Z. physik. Chem., **127**, 455 (1927)	3683, 4561
1859	HOLLUTA J., PETER F., Z. physik. Chem., **127**, 455 (1929)	3717, 3866, 4571, 4696, 4789
1860	HOLT L. E., PIERCE J. A., KAJDI C. N., J. Colloid. Sci., **9**, 409 (1954)	4305
1861	HÖLZL F., Z. Elektrochem., **43**, 202 (1937)	698, 4366
1862	HOMEYER, TITSERT, Pharm. Ztg., **33**, 738 (1888)	3840
1863	HOOKER A. H., Chem. Met. Eng., **23**, 967 (1920)	2706

Reference number	Author's name and reference	Table number
1864	HORAN H. A., SKARULIS J. A., J. Am. Chem. Soc., **61**, 2689 (1939)	1373
1865	HORAN H. A., DUANE J. J., J. Am. Chem. Soc., **63**, 3533 (1941)	1374
1866	HORN D. W., Am. Chem. J., **37**, 471 (1907)	1968
1867	HOSTÁLEK Z., Chem. listy, **50**, No. 5, 216 (1956)	2638
1868	HOSTÁLEK Z., KASPAROVÁ J., Chem. listy, **50**, No. 6, 979 (1956)	4373
1869	HOSTÁLEK Z., Chem. listy, **50**, No. 6, 981 (1956)	1276
1870	HOSTÁLEK Z., Coll. Czechosl. Chem. Comm., **22**, No. 2, 532 (1957)	2639
1871	HOSTÁLEK Z., Coll. Czechosl. Chem. Comm., **22**, No. 2. 618 (1957)	1277
1872	HOSTÁLEK Z., KASPAROVÁ J., Coll. Czechosl. Chem. Comm., **22**, No. 2, 621 (1957)	4374
1873	HUDGINS C. M., J. Chem. Eng. Data, **9**, No. 3, 434 (1964)	5150
1874	HUDSON J. C., J. Chem. Soc., **127**, 1332 (1925)	3413
1875	HUDSON J. C., J. Chem. Soc., **127**, 1332 (1934)	4318
1876	HUNT, Am. Journ. Sci., (2), **40**, 154 (1870)	3791
1877	HUNT H., J. Am. Chem. Soc., 54, 3509 (1932)	1811
1878	HUNT H., BONCYK L., J. Am. Chem. Soc., **55**, 3528 (1933)	17, 18, 25
1879	HUYBRECHTS M., RAMELOT H., Bull. Soc. chim. belges, **35**, 239 (1926)	385, 410, 540, 711, 938, 1137, 2376, 4779
1880	HUYBRECHTS M., LANGERAN N., de, Bull. Soc. chim. belges, **39**, 43 (1930)	3516
1881	IGELSRUND I., THOMPSON T. G., J. Am. Chem. Soc., **58**, 318 (1936)	3768, 4133, 4139, 4160, 4852
1882	INAMURA, Mem. Coll. Sci. Kyoto Imp. Univ., **3**, 105 (1919), cited in [2]	423
1883	INGHAM J. W., J. Chem. Soc. (London), **131**, 1917 (1928)	962, 999
1884	INGHAM J. W., J. Chem. Soc., **2**, 2059 (1929)	942
1885	INGUZZI C. M., Atti Soc. Toscana Sci. nat. Merv., 55, Ser. A, 26 (1948)	4095
1886	INOUYE R., Mem. Coll. Sci. Kyoto Imp. Univ., (A), **8**, 287 (1925)	4540
1887	IOKOTA N., Chem. Engng (Japan), **22**, No. 8, 476 (1958)	128
1888	IRVING H., CHERRY G. W., J. Chem. Soc., **25**, (1941)	2131
1889	ISAAC F., J. Chem. Soc., 93, 398, 927 (1908)	3153
1890	ISCHIKAWA F., MUROOKA T., Bull. Inst. Phys. Chem. Res. (Tokyo), 8, 77 (1929)	2257
1891	ITO J., J. Chem. Soc. Japan, Ind. Chem. Sec., **63**, No. 10, 1685, A93 (1960)	701, 746
1892	ITO J., Rep. gov. Chem. Ind. Res. Inst. Tokyo, **56**, No. 7, 266 (1961)	701, 746, 920b
1893	ITO J., KOBAYASHI E., J. Chem. Soc. Japan, Ind. Chem. Sec., **63**, No. 11, 1909 (1960)	2137, 2142, 2143, 2246, 2247, 2250
1894	ITO J., KOBAYASHI E., Tokyo Kogyo Shikensho Hokoku Repts. Govt. Chem. Ind. Res. Inst., Tokyo, **57**, No. 8, 397 (1962)	2246, 2247, 2249, 2250
1895	ITO J., Tokyo Kogyo Shikensho Hokoku Repts. Govt. Chem. Ind. Res. Inst., Tokyo, 57, No. 8, 403 (1962)	2137, 2142, 2143
1896	JACKMAN D. N., BROWNE A., J. Chem. Soc., **121**, 694 (1922)	3101, 3102
1897	JACKOVKIN A. A., Z. Phys. Chem., **20**, 38 (1896)	138, 139, 142

Reference number	Author's name and reference	Table number
1898	JACOBS W., Chem. Weekblad, **14**, 208 (1917)	1147
1899	JAEGER A., Z. anorg. Chem., **27**, 25 (1901)	593, 626
1900	JÄGER L., Chem. prumysl, **7**, No. 10, 544 (1957)	4325
1901	JÄGER L., Chem. prumysl, 8, No. 3, 136 (1948)	1312
1902	JÄGER L., Chem. listy, **52**, No. 4, 734 (1958)	542
1903	JÄGER L., Coll. Czechosl. Chem. Comm., **24**, No. 5, 1103 (1959)	543
1904	JÄGER L., REJLEK M., KLIMECEK R., MACHALA J., Chem. prumysl, 9 (34), No. 7, 361 (1959)	3227
1905	JÄGER L., REJLEK M., KLIMECEK R., MACHALA R., Chem. prumysl, **10**, No. 10, 518 (1960)	3240
1906	JAMES C., PRATT L. A., J. Am. Chem. Soc., **32**, 873 (1910)	5377
1907	JAMES C., HOLDEN H. C., J. Am. Chem. Soc., **35**, 559 (1913)	3488
1908	JAMES C., WILLAND P. S., J. Am. Chem. Soc., **38**, 1499 (1916)	1042
1909	JAMES S. M., JACHE W. W., J. Am. Chem. Soc., **81**, 1821 (1959)	566
1910	JAMESON R. F., SALMON J. E., J. Chem. Soc., **11**, 4014 (1954)	4206, 4292
1911	JAMESON R. F., SALMON J. E., J. Chem. Soc., **2**, 361 (1955)	4285
1912	JANATJEWA O. K., ORLOWA W. T., Freib. Forschungs. **A123**, 119 (1959)	4041
1913	JANDER G., GRÜTTNER B., Ber., **81**, 107 (1947)	33, 40
1914	JÄNECKE E., ERDMANN, Jahresber. Hallesch. Verbandes, **2**, 257 (1920), cited in [1]	4045, 4058
1915	JÄNECKE E., Z. physik. Chem., **127**, 71 (1927)	647, 1904, 4274
1916	JÄNECKE E., Z. angew. Chem., **41**, 916 (1928)	2146, 2488
1917	JÄNECKE E., Z. Elektrochem., **35**, 332, 716 (1929)	1595, 2318
1918	JÄNECKE E., Z. angew. Chem., **42**, 1169 (1929)	2318
1919	JÄNECKE E., Z. Elektrochem., **39**, 682 (1933)	1907
1920	JÄNECKE E., Z. Elektrochem., **40**, 738 (1934)	3382
1921	JÄNECKE E., Kali, No. 29, 145 (1935)	3382, 3397, 3428, 4067
1922	JÄNECKE E., Z. physik. Chem., **177**, 7, (1936)	643
1923	JANTSCH G., Z. anorg. Chem., **76**, 321 (1912)	420, 422
1924	JAQUES S., Helv. chim. acta, **29**, 1041 (1946)	1090
1925	JARRY R., Ann. chim. phys., (7), **17**, 342 (1899)	1007
1926	JEANNEL G., Compt. rend., 103,381 (1886)	994
1927	JEHN D. B., HUDDLESTON L. J., Chem. Soc., **125**, 1451 (1924)	564
1928	JELLINEK K., KÜHN W., Z. physik. Chem., **105**, 340, (1923)	1148
1929	JONES E. V., MARSCHALL W. L., J. Inorg. Nucl. Chem., **23**, No. 3–4, 287 (1961)	3, 5, 8, 4346, 4348, 4354, 4364
1930	JONES F. E., Trans. Faraday Soc., **35**, 1484 (1939)	4218, 5059
1931	JONES F. E., J. Phys. Chem., **48**, 311 (1944)	2699, 4214, 4392
1932	JONES G., HARTMANN M. L., J. Am. Chem. Soc., **37**, 241 (1915)	188
1933	JONES G., HARTMANN M. L., Trans. Am. Electrochem. Soc., **30**, 295 (1916)	144
1934	JOHNSON C. R., LAW G. W., J. Am. Chem. Soc., **55**, 2262 (1933)	496
1935	JOHNSON W. C., WHEATLEY A. C., Z. anorg. Chem., **216**, 273 (1934)	2571
1936	JOHNSON H. L., CUTA F., GARRETT A. B., J. Am. Chem. Soc., **55**, 2311 (1933)	2777
1937	JOHNSTONE A. F., LEPPLA P. W., J. Am. Chem. Soc., **56**, 2233 (1934)	790
1938	JOHNSTON J., J. Am. Chem. Soc., **37**, 2011 (1915)	1661
1939	JOHNSTON J., J. Am. Chem. Soc., **37**, 2001 (1915)	1708

Reference number	Author's name and reference	Table number
1940	JOHNSTON J., GROVE C., J. Am. Chem. Soc., **53**, 3984 (1931)	1428, 3729, 3915, 3944, 4840, 4977, 5072–5074
1941	JONTE J. H., MARTIN D. S., J. Am. Chem. Soc., **74**, 2052 (1952)	3813
1942	JOSEPH J. Chem. Soc., **117**, 377 (1920)	145
1943	JOZEFOWICZ E., WITEKOWA S., ZUBRANSKA W., Roczn. chem., **24**, 64 (1950)	1190–1198, 1200
1944	JOSHI N. K., DEVADATTA S. C., Proc. Ind. Acad. Sci., **7A**, No. 2, 137 (1938)	770
1945	JOUNG R. S., GOLLEDGE A., Ind. Eng. Chem., **39**, 1299 (1947)	2004
1946	JOUNG S. W., J. Am. Chem. Soc., **19**, 851 (1897)	1253–1261
1947	JUHN E. L., STAUDE H., Z. Naturforsch., 6a 385 (1951)	3234
1948	JUNG F., Biochem. Z., **302**, 294 (1939)	714, 1187–1189
1949	KANAKOWSKY Th., D'ANS, Tables, p. 128 (1933), cited in [2]	4025
1950	KARAOGLANOW Z., Z. anal. Chem., **56**, 228 (1917)	518, 1022, 1112, 4904, 5364
1951	KARSTEN, Ann. der Chem. u. Pharm. Suppl., Bb. 3, 170 (1864–1865), cited in [2]	2459, 2538, 3825, 3662
1952	KARSTEN, NICOL S. M. G., Phil. Mag. (London), (5), **31**, 369 (1891)	3066
1953	KASARNOWSKY J., Z. physik. Chem., **109**, 289 (1924)	1096
1954	KATZIN L. J., GEBERT E., J. Am. Chem. Soc., 77, 5814 (1955)	158, 159, 174
1955	KAYSER E., Kali, No. 1, 2 (1923)	3680
1956	KAYSER E., Kali, No. 17, 7 (1923)	3640
1957	KEEFER R. M., RIEBER A. G., BISSON C. S., J. Am. Chem. Soc., **62**, 2951 (1940)	4863, 4910, 4913
1958	KEEFER R. M., REIBER H. G., J. Am. Chem. Soc., **63**, 3504 (1941)	4630
1959	KEEFER R. M., J. Am. Chem. Soc., **70**, 476 (1948)	4877
1960	KEITEL H., Mitt. KFA, **32**, 100 (1922)	3602, 3639, 3679
1961	KEITEL H., Kali, **17**, 248, 261 (1923)	3619, 4124
1962	KELMERS A. D., J. Inorg. Nucl. Chem., **17**, No. 1–2, 168 (1961)	1962
1963	KENDALL J., DAVIDSON A. W., J. Am. Chem. Soc., **43**, 979 (1921)	688
1964	KENDALL J., ANDREWS J. C., J. Am. Chem. Soc., **43**, 1545 (1921)	274, 288
1965	KENDALL J., SLOAN C. H., J. Am. Chem. Soc., **47**, 2306 (1925)	1143, 1455, 1473, 2550, 3810, 3851, 4177, 4192, 4930, 5178, 5369, 5517, 5555
1966	KENTTÄMAA J., Suomen Kem., **30**, No. 1, 189 (1957)	1159, 3877b
1967	KERNOT G., D'AGOSTINO E., PELLEGRINO M., Gzzz. chim. ital., **38**, 1, 532 (1908)	2537, 4839
1968	KEYES D. B., JAMES C., J. Am. Chem. Soc., **36**, 634 (1914)	2734, 3506
1969	KIEHL S. J., MANFREDO E. J., J. Am. Chem. Soc., **56**, 2118 (1937)	2606
1970	KILDE G., Z. anorg. Chem., **218**, 113 (1934)	5190
1971	KING E. L., J. Am. Chem. Soc., **71**, 319 (1949)	979

Reference number	Author's name and reference	Table number
1972	KING E. L., KRALL H. J., PANDOW M. L., J. Am. Chem. Soc., **74**, 3492 (1952)	3873
1973	KING G. B., J. Phys. Chem., **41**, 797 (1937)	1205
1974	KING G. B., BECKMAN W. A., J. Am. Chem. Soc., **67**, 957 (1945)	2568
1975	KIRSCHMAN H. D., POMEROY R., J. Am. Chem. Soc., **65**, 1695 (1943)	4416
1976	KIRSCHMAN H. D., POMEROY R., J. Am. Chem. Soc., **66**, 1793 (1944)	2740
1977	KISS A., URMANCZY A., Z. anorg. Chem. **202**, 172 (1931)	190
1978	KITAHARA S., Rev. Phys. Chem. Japan, **30**, No. 2, 115 (1960)	3656
1979	KLEIN O., Z. anorg. Chem., **74**, 158 (1912)	4424
1980	KLEMENC A., SPIESS T., Monatsh., **77**, 216 (1947)	397
1981	KLINE W. D., J. Am. Chem. Soc., **51**, 2095 (1929)	1666, 1714
1982	KOBE K. A., WILLIAMS J. S., Ind. Eng. Chem., Anal. Ed., **7**, No. 1, 38 (1935)	728, 1634, 1736
1983	KOBE K. A., LEIPPER A., Ind. Eng. Chem., **32**, 198 (1940)	2602, 2804, 2937
1984	KOBE K. A., STEWART P. B., J. Am. Chem. Soc., **64**, 1301 (1942)	5096
1985	KOBE K. A., FREDRICKSON R. E., Chem. End. Data Senes, **1**, 13 (1956)	828
1986	KOELICHEN K., Mitt. KFA, **1**, 15 (1919)	3600
1987	KOELICHEN K., PRZIBULLA C., Mitt. KFA, 1, **126** (1920)	3601
1988	KOELICHEN K., ALTHAMMER W., Mitt. KFA, **3**, No. 43, 73–78 (1923); Roczn. chem. (poln.), **6**, 713 (1926)	3586
1989	KOERKER F. W., COLDERWOOD H. N., J. Phys. Chem., **42**, 1151 (1938)	4343
1990	KOHN M., O'BRIEN, J. Soc. Chem. Ind., **17**, 100 (1898)	125, 132
1991	KOHN M., Z. anorg. Chem., **63**, 337 (1909)	2579, 4979
1992	KOHN M. KLEIN A., Z. anorg. Chem., **77**, 254 (1912)	4979
1993	KOHN M., Anal. chim. acta, **9**, 229 (1953)	3213
1994	KOLB, Bull. soc. ind. Mulhouse, 222 (1872), cited in [2]	787
1995	KOLTHOFF J. M., VOGELENZANG E. H., Z. anal. Chem., **58**, 20, 51 (1919)	520, 1113
1996	KOLTHOFF J., Rec. trav. chim., **45**, 609 (1926)	290, 297, 353, 355, 357
1997	KOLTHOFF J. M., ROSENBLUM C., J. Am. Chem. Soc., **55**, 2656 (1933)	3517, 5574
1998	KOLTHOFF J. M., STENGER V. A., J. Phys. Chem., **38**, 639 (1934)	1853, 1957, 1959, 1961
1999	KOLTHOFF J. M., LINGANE, J. Phys. Chem., **42**, 133 (1938)	4611
2000	KOLTHOFF J. M., PERLICH R. W., WEIBLEN D., J. Phys. Chem., **46**, 561 (1942)	910, 3062, 3518
2001	KOOI J., WEISSKOPF E., GRUEN D. M., J. Inorg. Nucl. Chem., **13**, No. 3–4, 210 (1960)	1105–1107
2002	KÖNIG H., Z. Naturforsch., **18a**, No. 3, 363 (1963)	66
2003	KOPPEL, J., Z. physik. Chem., **42**, 8 (1901–1902)	3476
2004	KOPPEL, WETZEL, Z. physik. Chem., **52**, 395 (1905)	3467, 3472
2005	KOPPEL, Z. physik. Chem., **52**, 405 (1905)	3465
2006	KOPPEL-GUMPERY, Z. physik. Chem., **52**, 413 (1905)	3483
2007	KOPPEL J., BLUMENTHAL R., Z. anorg. Chem., **53**, 228 (1907)	4368, 4953, 4956, 4962, 4963

Reference number	Author's name and reference	Table number
2008	KOPPEL J., HOLTKAMP H., Z. anorg. Chem., **67**, 274 (1910)	551, 914, 1155
2009	KNOPP, Z. physik. Chem., **48**, 103 (1904)	48, 52, 60, 62, 1746, 1768
2010	KNOX J., J. Chem. Soc. (London), **95**, 1760 (1909)	2673, 2798, 4388
2011	KRATOHVIL J., TEŞAK B., Arhiv Kem., **26**, 243 (1954)	3917
2012	KREMANN R., HÜTTINGER K., Jahrber. Ged. Reichsanstalt (Wien), **53**, 637 (1908), cited in [2]	4220
2013	KREMANN K., ZITEK A., Monatsh. Chem., **30**, 311 (1909)	2928, 3005, 3106, 4459
2014	KREMANN R., NOSS F., Monatsh. Chem., **33**, 1205 (1912)	3790, 5226
2015	KREMANN R., RODEMUND H., Z. anorg. Chem., **86**, 373 (1914)	3120, 5091
2016	KREMANN R., RODEMUND H., Monatsh. Chem., **35**, 1065 (1914)	3063, 3120, 3525, 5091
2017	KRITSCHEWSKY J., GOLDMANN E., Z. anorg. Chem., **214**, 307 (1933)	4545, 4546
2018	KRITSCHEWSKY J., ISCHKOWITZ R. K., Z. anorg. Chem., **215**, 103 (1933)	4567
2019	KRITSCHEWSKY J., GOLDMANN E., Z. anorg. Chem., **218**, 253 (1934)	2152
2020	KRÜGER G., THILO E., Z. anorg. Chem., **308**, 1–6, 246 (1961)	2788, 3192, 3823, 5473, 5499
2021	KRYM V., Chem. Zbb., **11**, 681 (1909)	3945
2022	KUBELKA V., Chem. zvesti, **13**, No. 7–8, 430 (1959)	3223
2023	KUKI T., HARA R., J. Soc. Chem. Ind. Japan, **45**, 119B (1940)	1893
2024	KUMPF, Wied. Ann. Bibl., **6**, 276 (1882), cited in [2]	125
2025	KURNAKOV N. S., ZEMCUZNYJ S. F., Z. anorg. Chem., **140**, 149 (1924)	3249, 3294, 3372, 3616, 4018
2026	KURTENACKER A., FLUSS W., Z. anorg. Chem., **210**, 125 (1933)	4784–4786
2027	KURTENACKER A., FINGER W., HEY F., Z. anorg. Chem., **211**, 83, 281 (1933)	597, 600, 601, 603, 604, 616, 2214–2217, 2219, 4649–4651, 4656
2028	KURTENACKER A., LASZLO G., Z. anorg. Chem., **237**, 359 (1938)	2432, 3537–3539
2029	KÜSTER F. W., HEBERLEIN E., Z. anorg. Chem., **43**, 56 (1905)	114
2030	LABASH J. A., LUSBY G. R., Canad. J. Chem., **33**, No. 5, 774 (1955)	2258, 2259, 2261, 2264, 2472, 3228
2031	LACHAUD M., LEPIERRE C., Bull. Soc. chim., (4), **35**, 321 (1891)	4428
2032	LADEBURG A., Ber., **35**, 1256 (1902)	2014
2033	LAFONTAINE G. H., Compt. rend., **180**, 2046 (1925)	1862, 2119
2034	LAGARDE Z., Compt. rend., **216**, 156 (1943)	1296, 1302
2035	LAMB A. B., PHILLIPS A. W., J. Am. Chem. Soc., **45**, 108 (1923)	872
2036	LAMBERGER J., PARIS R., Compt. rend., **227**, 637 (1948)	2185

Reference number	Author's name and reference	Table number
2037	LAMBERGER J., PARIS R., Bull. Soc. chim. France, No. 5–6, 546 (1950)	2187
2038	LAMBERT R. H., J. Phys. Chem., **30**, 973 (1926)	4982
2039	LA MER V. K., GOLDMAN F. H., J. Am. Chem. Soc., **51**, 2632 (1929)	3144, 3151, 3502, 3515, 3834, 3845, 4009, 4011, 4105, 4113, 4183, 4190, 4622, 4631, 4764, 4778, 4914, 4925, 5305, 5306, 5434, 5437, 5440, 5444, 5534, 5538, 5541
2040	LA MER V. K., GOLDMAN F. H., J. Am. Chem. Soc., **52**, 2791 (1930)	3157, 3857, 4639, 4934
2041	LANDSBERG R., ÖRGEL P., Chem. Techn., **13**, N 11, 665 (1961)	4377, 4378, 4407, 4408, 4410
2042	LANDFORD O. E., KIEHL S. J., J. Am. Chem. Soc., **62**, 1660 (1940)	4334
2043	LANDFORD O. E., KIEHL S. J., J. Am. Chem. Soc., **63**, 667 (1941)	5027
2044	LANG A. A., SUKAVA A. J., Canad. J. Chem., **36**, No. 7, 1064 (1958)	4375
2045	LANGAUER D., Roczn. chem., **11**, 485 (1931)	3597
2046	LANGAUER D., Roczn. chem., **12**, 265 (1932)	3606
2047	LANGAUER D., Roczn. chem., **13**, 204 (1933)	3607
2048	LANGAUER D., Z. anorg. Chem., **213**, 180 (1933)	5238
2049	LANGAUER D., OLPINSKI W., Roczn. chem., **14**, 249 (1934)	3608
2050	LANGE H., STACKELBERG M., von, Z. anorg. Chem., **256**, 273 (1948)	2598
2051	LARSON A. T., BLOCK C. A., J. Am. Chem. Soc., **47**, 1015 (1925)	13
2052	LASZCYNSKI S., von, Ber., **27**, 2285 (1894)	5395
2053	LAUFFENBURGER R., BRODSKY M., Compt. rend., **206**, 1383 (1938)	2243
2054	LAWRENCE R. W., KING G. B., J. Am. Chem., Soc., **60**, 1987 (1938)	2566
2055	LEATHER J. W., SEN J. N., Mem. Dept. Agr. (India), Chem. Ser., **1**, 117 (1909), cited in [2]	1704
2056	LEATHER J. W., MUKERJI J. M., Mem. Dept. Agr. (India), Chem. Ser., **3**, 177 (1913), cited in [2]	3064, 3107, 3664, 3671, 4558
2057	LE BLANC M., NOYES A. A., Z. physik. Chem., **6**, 386 (1890)	3152, 4597
2058	LE BLANC M., NOVOTNY K., Z. anorg. Chem., **51**, 181 (1906)	2748
2059	LEDEN J., PERSSON G., Acta Chem. Scand., **15**, No. 3, 607 (1961)	1854, 2002
2060	LEE F. H., LEE K. H., J. Chinese Chim. Soc., **4**, 126 (1936)	161
2061	LEE W. B., EGERTON A. C., J. Chem. Soc., **123**, 706 (1923)	4125, 4156, 4849
2062	LEICK J., Z. anal. chem., **87**, 417 (1932)	3447, 3740
2063	LEIMBACH G., Kali, No. 1, 12 (1926)	3643
2064	LEIMBACH G., Kali, No. 15, 236; No. 17, 266 (1926)	3604
2065	LEIMBACH G., Caliche, **11**, 469 (1929)	3686
2066	LEIMBACH G., PFEIFFENBERG A., Caliche, **11**, 64 (1929)	3260, 3296, 3621
2067	LENHER V., MERRILL H. B., J. Am. Chem. Soc., **39**, 2630 (1917)	784, 992

Reference number	Author's name and reference	Table number
2068	LEO R., Monatsh., **43**, 567 (1922)	3530
2069	LEPIERRE C., LACHAUD M., Compt. rend., **113**, 196 (1891)	4426
2070	LEVI G. R., AGUZZI A., Gazz. chim. ital., **68**, 179 (1938)	341, 354
2071	LEVI M., GILBERT F., J. Chem. Soc. (London), **130**, 2117 (1927)	276, 278, 1569, 1570
2072	LEVI S. M., Z. physik. Chem., **106**, 93 (1923)	4060
2073	LEVI S., MOSER A., Z. physik. Chem., **108**, 421 (1924)	2295
2074	LEWIN S. Z., VANCE J. E., NELSON L. B., J. Am. Chem. Soc., **75**, No. 11, 2768 (1953)	1145, 1177, 3854
2075	LEY H., SCHOEFER K., Ber., **39**, 1263 (1906)	3051, 3129
2076	LI N. C. C., LO Y. T., J. Am. Chem. Soc., **63**, 394 (1941)	503, 4612
2077	LIESER K. H., Z. anorg. Chem., **292**, 1–3, 103 (1957)	3814, 3918, 3947
2078	LIESER K. H., Z. anorg. Chem., **304**, 5–6, 299 (1960)	5388, 5392, 5397
2079	LIETZKE M. H., STOUGHTON R. W., J. Am. Chem. Soc., **79**, 2067 (1957)	4230, 4258, 5385, 5402
2080	LIETZKE M. H., STOUGHTON R. W., J. Phys. Chem., **63**, No. 7, 1188 (1959)	864
2081	LIETZKE M. H., STOUGHTON R. W., J. Phys. Chem., **63**, No. 11, 1984 (1959)	4098
2082	LIGHTFOOT W. J., PRUTTON C. F., J. Am. Chem. Soc., **68**, 1001 (1946)	4135, 4163, 4854
2083	LIGHTFOOT W. J., PRUTTON C. F., J. Am. Chem. Soc., **69**, 2098 (1947)	4136, 4163, 4855
2084	LIGHTFOOT W. J., PRUTTON C. F., J. Am. Chem. Soc., **70**, 4112 (1948)	4140
2085	LIGHTFOOT W. J., PRUTTON C. F., J. Am. Chem. Soc., **71**, 1233 (1949)	4141
2086	LIND S. C., UNDERWOOD J. E., WHITTEMORE C. F., J. Am. Chem. Soc., **40**, 465 (1918)	911
2087	LINDERSTROM-LANG K., Compt. rend. trav. Lab. Carlsberg, **15**, No. 4, 22 (1924)	289, 335, 367
2088	LINDSAY A., McRAE SOWERBY, J. Chem. Soc., No. 1, 1337 (1927)	5587
2089	LINDSTRAND F., Z. anorg. Chem., **230**, 187 (1936)	1163
2090	LINHART G. A., J. Am. Chem. Soc., **40**, 158 (1918)	189
2091	LINKE W. F., J. Phys. Chem., **60**, No. 1, 91 (1956)	1036
2092	LINKE W. F., COOPER J. A., J. Phys. Chem., **60**, No. 12, 1662 (1956)	3356
2093	LOCUTY P., LAFFITTE P., Compt. rend., **196**, 1311 (1934); Compt. rend., **199**, 950 (1934)	741
2094	LOCUTY P., Rev. ind. min., No. 348, 284 (1935)	742
2095	LONG J. Anal. Chem., **2**, 243 (1888)	2034
2096	LONGI, Gazz. chim. ital., **13**, 87 (1883)	500, 501, 1997, 2011, 2013, 2017
2097	LOVEN J. M., Z. anorg. Chem., **11**, 408 (1896)	2053
2098	LÖWENHERZ R., Z. physik. Chem., **13**, 479 (1894)	4017
2099	LÖWENHERZ R., Z. physik. Chem., **23**, 96 (1897)	4052
2100	LUCCHESI P. J., WHITNEY W. D., J. Appl. Chem., **12**, No. 6, 277 (1962)	856
2101	LUGG J. W. H., Trans. Faraday Soc., **27**, 297 (1931)	5111
2102	LUKASZEWSKI G. M., REDFERN J. P., SALMON J. E., J. Chem. Soc., No. 1, 39 (1961)	5195
2103	LUNGE G., Ber., **14**, 2188 (1881)	730, 1745, 1766

Reference number	Author's name and reference	Table number
2152	MASSON J., GUÉRIN H., Bull. Soc. chim. France, No. 3, 400 (1958)	5053, 5054, 5314, 5320
2153	MASSON J., Ann. Chim., 3, No. 3–4, 253 (1958)	5315, 5318, 5319
2154	MASSOL, MALDES, Compt. rend., **133**, 187 (1901)	3476
2155	MATHUR P. P. D., DHAR N. F., Z. anorg. Chem., **199**, 387 (1931)	2776
2156	MATIGNON C., MEYER F., Compt. rend., **165**, 788 (1917); **166**, 116 (1918); **166**, 687 (1918); Ann. chim., **9**, 265 (1918)	2289
2157	MATTRAT P., GUERIN H., Bull. Soc. chim. France, No. 4, 601 (1960)	3968
2158	MAYEDA T. J. Soc. Chem. Ind., Japan, **23**, 573 (1920)	3251, 3292, 3369, 3617, 4019
2159	MAYEDA T., J. Soc. Chem. Ind., Japan, **23**, 954 (1920)	3292, 3369, 4019
2160	MAYEDA T., YAMANE S., Scientific papers Inst. phys. and chem. Research, Tokyo, **4**, 101 (1926)	3971
2161	MAZZETTI C., Gazz. chim. ital., **56**, 601 (1926)	3784, 4869, 5254
2162	MAZZETTI C., Ann. chim. applicata, **19**, 273 (1929)	4850, 4941, 5141
2163	McCOY H.W., SMITH H. J., J. Am. Chem. Soc., **35**, 468 (1911)	1705
2164	McDOUGALL G., DAVIES C. W., J. Chem. Soc., (London), 1416 (1935)	4620, 4911, 4944, 4952, 5167
2165	MACDOUGALL F. H., HOFFMAN E. J., J. Phys. Chem., **40**, 317 (1936)	1326, 3155, 5102, 5359, 5579
2166	McDOWELL L. A., JOHNSTON H. L., J. Am. Chem. Soc., **58**, 2009 (1936)	4423
2167	McKERROW A. K., OCCLESSHAW V. J., DRABBLE F., J. Chem. Soc., 1–5, I (1946)	2127, 3033, 4506, 5085
2168	McLAUCHLAN W. H., J. Phys. Chem., **44**, 600 (1903)	151, 154, 687
2169	MEBONE W. M., DOBBINS J. T., CAMERON F. K., J. Phys. Chem., **33**, 961 (1929)	4272
2170	MECKE P., Zement, **24**, No. 48, 765 (1935)	806, 1012, 2698, 2747, 3443, 3730
2171	MEERBURG P. A., Z. anorg. Chem., **33**, 299 (1903)	1094
2172	MEERBURG P. A., Z. anorg. Chem., **37**, 203 (1903)	2528
2173	MEERBURG P. A., Z. anorg. Chem., **45**, 1, 324 (1905)	1262–1264, 2524
2174	MEERBURG P. A., Z. anorg. Chem., **59**, 136 (1908)	2544
2175	MEES C. E. K., PIPER C. W., Photogr. Journ., **33**, 227; **36**, 234; **52**, 221 (1912), cited in [2]	3232, 5420
2176	MEHL see: DOELTER, Handbuch der Mineralchemie, 1, 958 (1912)	2847
2177	MELLOR J. W., J. Chem. Soc., **79**, 216 (1901)	122
2178	MENKE J. B., Z. anorg. Chem., **77**, 283 (1812)	199
2179	MENZEL H., Z. anorg. Chem., **164**, 1, 34, 42; **166**, 67, 181 (1927)	226
2180	MENZEL H., GABLER C., Z. anorg. Chem., **177**, 187 (1929)	242, 245, 256
2181	MENZEL H., HAGEN W., Z. anorg. Chem., **233**, 49, 209 (1937)	2616
2182	MENZEL H., MÜLLER G., Z. anorg. Chem., **272**, 1–4, 84 (1953)	2608
2183	MENZIES J. A., OWEN L. W., J. Chem. Soc., 2064 (1958)	1063

Reference number	Author's name and reference	Table number
2184	MERCIER J. M., Compt. rend., **204**, 500 (1937)	2577
2185	METTOLA A., Chimia (Milan), 4, 422 (1949)	2019
2186	MEYER J., Z. Elektrochem., **15**, 266 (1909)	1855, 2158, 2435
2187	MEYER J., STATECZNY V., Z. anorg. Chem., **122**, 1 (1922)	814, 1207
2188	MEYER J., FRIEDRICH W., Z. physik. Chem., **101**, 499 (1922)	878
2189	MEYER J., Rec. trav. chim., **42**, 614 (1923)	894
2190	MEYER J., WILK H., Z. anorg. Chem., **132**, 239 (1924)	894
2191	MEYER J., AULICH N., Z. anorg. Chem., **172**, 321 (1928)	3889–3891, 4193, 4968
2192	MEYER J., AULICH M., Z. anorg. Chem., **194**, 278 (1930)	810
2193	MEYER R. J., NOCHOD H., Ann. chim. phys., **440**, 186 (1924)	919
2194	MEYER T. A., PRUTTON C. G., LIGHTFOOT W. J. J. Am. Chem. Soc., **71**, 4, 1237 (1949)	3651
2195	MEYERHOFFER W., Z. physik. Chem., **5**, 102 (1890)	4875
2196	MEYERHOFFER W., SAUNDERS A., Z. physik. Chem., **28**, 466 (1899)	3241, 3318, 3416, 3672
2197	MEYERHOFFER W., Z. physik. Chem., **53**, 518 (1905)	4462
2198	MICHELS, PRZIBYLLA, Die Kalirohsalze, (Leipzig), 104 (1916), cited in [2]	3615, 3676
2199	MILBAUER J., Bull. Soc. chim. (France), (5), **3**, 221 (1936)	792
2200	MILES F. D., FENTON J., J. Chem. Soc., **117**, 59 (920)	788
2201	MILES F. D., CARSON T., J. Chem., Soc., 786 (1946)	4309
2202	MILES, J. Am. Chem. Soc., **69**, 1716 (1947)	977
2203	MILLER J. P., Am. J. Sci., **250**, 161 (1952)	1715
2204	MILLIKAN J., Z. physik. Chem., **92**, 59 (1917)	5046, 5050, 5052, 5346, 5347, 5348, 5469, 5470
2205	MILLIKAN J., Z. physik. Chem., **91**, 496 (1918)	1008, 1017, 1069, 1075, 1109, 5050, 5477
2206	MILLIGAN, (1922), cited in [2]	424
2207	MILLS R. V., WELLS R. C., Bull. U. S. Geol. Survey, **693**, 72 (1918)	3765
2208	MIRSCH E., SEROWY F., Freiberger Forschungshefte, No. 118, 1 (1961)	974
2209	MITCHELL A. E., J. Chem. Soc., 123, 1887 (1923)	1711, 1663, 1670
2210	MOELLER T., SCHWEITZER G. K., Anal. Chem., **20**, No. 12, 1201 (1948)	1076, 1118, 1120, 1121, 1122, 1154
2211	MOHR, Z. physik. Chem., **27**, 197 (1898)	2514
2212	MOLES E., PEREZ-VITORIO A., An. Soc. esp. fis. quim., **30**, 200 (1932)	390
2213	MONDAIN-MONVAL P., Compt. rend., **174**, 1014 (1922)	2113, 2115, 2117, 2462, 2973
2214	MONDAIN-MONVAL P., Compt. rend., **175**, 163 (1922)	2115, 2462, 2973
2215	MONDAIN-MONVAL P., (1925), cited in [2]	2167
2216	MONK H., MORTIFEE A. V., J. Appl. Chem., **10**, 11, 456 (1960)	5292
2217	MONTEMARTINI G., LOSANA L., L'Industria Chimica, **4**, 107, 199, 291 (1928)	716, 760, 766, 768, 817, 835, 849

Reference number	Author's name and reference	Table number
2218	MOREY G., BURLEW J., Am. J. Sci., (5), **35**, 212 (1938)	2634
2219	MOREY G. W., BURLEW J. S., J. Phys. Chem., **68**, No. 7, 1706 (1964)	2645
2220	MORRISON T. J., JOHNSTONE N. B., J. Chem. Soc., 3655 (1955)	67
2221	MORSE H., Z. physik. Chem., **41**, 708 (1902)	5409, 5415, 5426, 5549
2222	MOSER L., Z. anorg. Chem., **61**, 384 (1909)	2798, 4429
2223	MOSER L., BEHR M., Z. anorg. Chem., **134**, 49 (1924)	838, 847
2224	MOSER L., BRUKL A., Monatsh. Chem., **47**, 711 (1926)	1948
2225	MUELLER P., ABEGG R., Z. physik. Chem., **57**, 526 (1907)	324
2226	MULDER G. J., Scheikundige Verhandelingen Onderzoekingen, vol. 3, pt. 2. Rotterdam (1864), cited in [2]	2176, 2199, 3133
2227	MULDER G. J., Jahresber. Fortschr. Chem., 67 (1886)	3442, 4884
2228	MULLEN R. T., CROWE W. H., J. Chem. Soc., **2**, 1764 (1927)	1203
2229	MÜLLER E., Z. physik. Chem., 110, 363 (1924)	2757
2230	MÜLLER J., Z. anorg. Chem., **96**, 29 (1916)	2601
2231	MUMFORD S. A., GILBERT L. F., J. Chem. Soc., **123**, 471 (1923)	458
2232	MUNTER P. A., BROUN R. L., J. Am. Chem. Soc., **65**, 2456 (1943)	3867
2233	MUTHMANN, KUNTZE, Z. kryst. Min., **23**, 368 (1894)	4948, 4965
2234	NACKEN R., Sitzber. k. preus. Akad. Wiss., 1016 (1910)	3417
2235	NAIDICH S., RICCI J. E., J. Am. Chem. Soc., **61**, 3268 (1939)	521, 1117, 4621, 4912
2236	NAKAGAWA S., ARAI Y., IKENO S., J. Soc. Chem. Ind. Japan, **45**, 445 (1942)	2114
2237	NAKAYAMA M., Bull. Agric. Chem. Soc. Japan, **24**, No. 4, 362 (1960)	3970
2238	NALLET A., PÂRIS R. A., Bull. Soc. chim. France, No. 3, 488 (1956)	3544, 3546, 3547, 3718, 3868, 4792
2239	NARDELLI M., CAVALCA L., BRAIBANTI A., Gazz. chim. ital., **82**, 413 (1952)	4534
2240	NARDELLI M., BRAIBANTI A., Gazz. chim. ital., **83**, No. 1, 28 (1953)	5400
2241	NARDELLI M., CAVALCA L., BRAIBANTI A., Gazz. chim. ital., **85**, No. 1–2, 161 (1955)	5384
2242	NAUDE S. M., Z. physik. Chem., **125**, 98 (1927)	5018
2243	NAUMANN A., RUCKER A., Ber., **38**, 2293 (1905)	5381
2244	NAUMANN D., Z. anorg. Chem., **309**, 1–2, 46 (1961)	1081, 1126
2245	NEAL J. L., McCRISKY C. R., J. Am. Chem. Soc., **60**, 911 (1938)	5315
2246	NEALE S. M., STRINGFELLOW S. A., Trans. Faraday Soc., **28**, 765 (1932)	2786
2247	NELSON O. A., J. Am. Chem. Soc., **63**, 1870 (1941)	2614
2248	NEUMANN B., DOMKE R., Z. Elektrochem., **34**, 136 (1928)	2088, 2092, 2094, 2104, 2456, 2464, 2908
2249	NEUMAN E. W., J. Am. Chem. Soc., **54**, 2195 (1932)	493, 865, 3134, 4100, 4606, 4754, 5130, 5405–5407
2250	NEWSBERY E., Trans. Electrochem. Soc., **69**, 611 (1936)	1171
2251	NICOL W. W. J., Phil. Mag., **31**, 369 (1891)	3065, 3105, 3670, 3671, 4553

Reference number	Author's name and reference	Table number
2252	NICHOLS E. L., HOWES H. L., Pub. Carnegie Inst., **298**, 207 (1919)	409, 428, 433, 434, 555
2253	NIGON J. P., PENNEMAN R. A., STARITSKY E., KEENAN T. K., ASPREY L. B., J. Phys. Chem., **58**, 403 (1954)	5327
2254	NIKOLAJEW V. I., Z. anorg. Chem., **181**, 249 (1929)	3069, 4562
2255	NISHIZAWA K. J. Chem. Ind. Tokyo, **23**, **25**, 1015 (1920)	2076, 2101, 2286, 2291, 2382, 2881
2256	NISHIZAWA K., J. Chem. Ind. Tokyo, **23**, 25, 1025 (1925)	2287
2257	NISHIZAWA K., J. Chem. Ind. Tokyo, **23**, 25, 1015 (1926)	2883
2258	NIXON A. C., SMITH R. E., J. Phys. Chem., **60**, 10, 1422 (1956)	5051
2259	NODDAK J., NODDAK W., Z. anorg. Chem., **215**, 167 (1933)	875, 1006, 1007, 1104
2260	NOONAN E. C., J. Am. Chem. Soc., **70**, 2915 (1948)	213, 214, 216–218, 222–225
2261	NORTON F., JOHNSTON J., Am. J. Sci., (5), **12**, No. 72, 478 (1926)	3257, 3350
2262	NOYES A. A., cited in [2]	1138, 4191, 4926, 4997, 5176, 5212, 5310, 5393, 5445, 5575
2263	NOYES A. A., Z. physik. Chem., **6**, 248 (1890)	1134, 4517, 4627, 5436, 5513, 5557, 5562, 5565
2264	NOYES A. A., Z. physik. Chem., 9, 606 (1892)	1135, 1140, 2545, 3844, 4188, 4923, 5174, 5211, 5288, 5441, 5514, 5559, 5571
2265	NOYES A. A., ABBOTT C. G., Z. physik. Chem., **16**, 130 (1895)	5556, 5572
2266	NOYES A., CHAPIN E., Z. physik. Chem., **28**, 520 (1899)	2500
2267	NOYES A. A., KOHR, J. Am. Chem. Soc., **24**, 1144 (1902)	4402
2268	NOYES A. A., Z. physik. Chem., **6**, 248 (1904)	5560
2269	NOYES A. A., BOGGS C. R., J. Am. Chem., Soc., **33**, 1650 (1911)	4698, 4793
2270	NOYES A. A., FARRELL F.S., J. Am. Chem. Soc., **33**, 1654 (1911)	5568
2271	NOYES A. A., STEWART M. A., J. Am. Chem. Soc., **33**. 1658 (1911)	3238, 3510, 5558
2272	NOYES A. A., HELL F. W., J. Am. Chem. Soc., **39**, 2529 (1917)	1151
2273	O'BRIEN, KENNY, FUXA, J. Chem. Education, **17**, 576 (1940)	1220
2274	OCCLESHAW V. J., J. Chem. Soc., **127**, 2598 (1925)	4245, 4254
2275	OCCLESHAW V. J., J. Chem. Soc. (London), **55**, 1436 (1931)	3035, 4515
2276	OCCLESHAW V. J., J. Chem. Soc. (London), No. 9, 2402 (1932)	2129, 3034, 4510
2277	OCCLESHAW V. J., J. Chem. Soc. (London), No. 12, 1892 (1934)	2130, 5380
2278	O'CONNOR E. A., J. Chem. Soc., **130**, 2700 (1927)	5047, 5049, 5140
2279	OELKE W. C., WAGNER C., Proc. Iowa Acad. Sci., **46**, 187 (1939)	4103, 4898
2280	OEY T. S., KOOPMAN D. E., J. Phys. Chem., **62**, No. 6, 755 (1958)	3545

Reference number	Author's name and reference	Table number
2281	OEY T. S., GUNNINGHAM G. L., KOOPMAN D. E., J. Chem. Engng Data, **5**, No. 3, 248 (1960)	3540
2282	OGLESBY N. E., J. Am. Chem. Soc., **51**, 2352 (1929)	2921
2283	OKADA, Mem. Coll. Sci. Kyoto Imp. Univ., **1**, No. 2, 96 (1914)	3419
2284	OKUNO T., MIYAZAKI K., J. Soc. Chem. Ind. Japan, **35**, 97B (1932)	761
2285	OKUNO T., J. Soc. Chem. Ind. Japan, **38**, 311B (1935)	762
2286	OLIVERI-MANDALA E., ANGENICA A., Gazz. chim. ital., **50**, 1, 273 (1920)	156, 160, 162
2287	OLIVERI-MANDALA E., Gazz. chim. ital., **50**, 11, 89 (1920)	123, 126, 135, 136, 140
2288	OLMER L. J., Bull. Soc. chim., (4), **35**, 333 (1924)	1993
2289	ORELLI E., Dokt. Dissert., Zürich (1940), cited in [2]	1806, 1808, 1828, 1845, 1847, 1871, 1886, 1913, 1933
2290	ORLICK, SIMONS, J. Am. Chem. Soc., **73**, 4325 (1951)	1526
2291	OSAKA J., Mem. Coll. Sci. Eng. (Kyoto), **3**, 58 (1910–1911)	3005, 3006
2292	OSAKA J., YAGINUMA T., Z. Physik. Chem., **130**, 480 (1927)	5225
2293	OSAKA J., YAGINUMA T., Bull. Chem. Soc. (Japan), **3**, 4 (1928)	5228, 5229, 5246
2294	OSTERSETZER D., Roczn. Chem., **6**, 679 (1926)	2411, 2418
2295	OSTWALD W., BANTHISCH W., J. prakt. Chem., (2), **29**, 54 (1884)	1010, 1071, 1072
2296	OSWALD M., Compt. rend., 155, 1504 (1912)	3041, 3052
2297	OSWALD M., 8-th Int. Congr. Appl. Chem., **2**, 205 (1912)	3041, 3052
2298	OSWALD M., Ann. Chim., **1**, 57 (1914)	3041, 3045, 3052, 4532, 5087, 5351, 5383
2299	PAEPE D., de, Bull. Soc. chim. belges, **25**, 174 (1911)	2846
2300	PAJETTA R., Gazz. chim. ital., **36**, 11, 67, 155, 300 (1906)	2925, 3018, 3115, 3202, 3236, 3439
2301	PARIS R., MONDAIN-MONVAL P., Bull. Soc. chim. (France), (5), **5**, 1142 (1938)	4455
2302	PARKER E. G., J. Phys. Chem., **18**, 653 (1914)	642, 658
2303	PARKER T. W., ROBINSON P. L., J. Chem. Soc., 1314 (1931)	1097, 1235
2304	PARKS H., PATTERSON M., VOSBURGH W. C., J. Am. Chem. Soc., **56**, 2141 (1937)	1393
2305	PARSONS C. L., CARSON H. P., J. Am. Chem. Soc., **32**, 1383 (1910)	5472
2306	PARSONS C. L., PERKINS C. L., J. Am. Chem. Soc., **32**, 1387 (1910)	5349
2307	PARSONS C. L., WHITTEMORE C. F., J. Am. Chem. Soc., **33**, 1933 (1911)	187
2308	PASKAL P., Ann. chim. phys., (8), **16**, 374 (1909)	1963
2309	PASKAL P., ERO, Bull. Soc. chim., (4), **25**, 35 (1919)	758
2310	PATRIDGE E., J. Am. Chem. Soc., **57**, 1541, 1543 (1935)	2678
2311	PEARCE J. N., O'LEARY J. K., Proc. Iowa Acad. Sci., **30**, 379 (1923)	148
2312	PEARCE J. N., EVERSOLE W. G., J. Phys. Chem., **28**, 245 (1924)	205
2313	PEARCE J. N., OELKE W. C., J. Phys. Chem., **42**, 95 (1938)	4106, 4184, 4765, 4915

Reference number	Author's name and reference	Table number
2314	PEDERSEN K. J., Kgl. Danske Videnskab., Selskab, Math.-fys. Medd., **18**, 12, 1 (1941)	4201
2315	PELLING A. J., ROBERTSON J. B., S. African J. Sci., **20**, 236 (1913), cited in [2]	3121, 3764, 3766
2316	PELLING A. J., J. South African Chem. Inst., **8**, No. 1, 3 (1925)	3254
2317	PELLING J., ROBERTSON J., J. Chem. met. mining Soc. South Africa, p. 196 (1926)	3766
2318	PELLING J., ROBERTSON J., Caliche, **8**, 532 (1927)	3766
2319	PELLINI G., Atti accad. Lincei, (5), **17**, 11, 78 (1909)	5332, 5341
2320	PENG, CHUNG-MING, Bull. Soc. chim. (France), (5), **2**, 985 (1935)	1585, 2842
2321	PENNEMAN R. A., ANDRIETH L. F., J. Am. Chem. Soc., **71**, No. 5, 1644 (1949)	2041–2043
2322	PEPPLER R. B., WELLS L. S., J. Res. Nat. Bur. Standards, **52**, No. 2, 75 (1954)	4212
2323	PERNOT M., Compt. rend., (1926, 1927, 1931), cited in [2]	5019
2324	PERNOT M., Compt. rend., **195**, 238 (1932)	4991
2325	PERNOT M., Compt. rend., **210**, 603 (1940)	5344
2326	PFEIFFENBERGER A., LEIMBACH G., Caliche, **9**, 485 (1928)	3061
2327	PILLOTON R. L., CRAWLEY G. E., J. Electrochem. Soc., **107**, No. 2, 122 (1960)	2997
2328	PINKUS A., JACOBI, Bull. Soc. chim. phys., **24**, 83, 137 (1927)	735
2329	PINKUS A., BERKOLAIKO W., J. Chim. Phys., **27**, 364 (1930)	3811, 4891
2330	PINKUS A., HAUGEN M., Bull. Soc. chim. belges, **45**, 693 (1936)	3811, 4891
2331	PINKUS A., TIMMERMANS A. M., Bull. Soc. chim. belges, **46**, 46 (1937)	1088, 1456, 3812, 4892
2332	PINKUS A., FREDERIC M. S., SHEEPMANS R., Bull. Soc. chim. belges, **47**, 304 (1938)	5387
2333	PINKUS A., SHEPMANS R., Bull. Soc. chim. belges, 337 (1939)	497
2334	PISCHINGER E., SZUFARSKI Z., Chem. Stosow., **2**, No. 2, 247 (1958)	2640, 2650, 2684
2335	PLAINOWNA A., Roczn. chem., **6**, 690 (1926)	5034
2336	PLATER, (1928), cited in [2]	2778
2337	PLEISSNER M., Arb. Kais. Gesundheitsamt., **26**, 384 (1907)	1142, 1734
2338	PLAIN, (1926), cited in [2]	5039
2339	POHL, Sitzber. Akad. Wiss. Wien, **41**, 267 (1860)	1997, 2005
2340	POLITZER F., Z. anorg. Chem., **64**, 121 (1909)	693
2341	POLYSOROPOULOS C. N., JANNAKOPOULOS T., J. Inorg. Nucl. Chem., **19**, No. 1–2, 107 (1961)	1866, 1867
2342	POMEROY R., KIRSCHMAN H. D., J. Am. Chem. Soc., **66**, 178 (1944)	2722
2343	PONI, TERREY, J. Chem. Soc., No. 9, 3056 (1955)	232, 233, 235, 236, 248, 251, 258, 262
2344	POPOFF S., NEUMAN E. W., J. Phys. Chem., **34**, 1853 (1930)	493, 865, 3134, 4100, 4606, 4754, 5130, 5405–5407
2345	POSNJAK E., MERWIN H. E., J. Am. Chem. Soc., **44**, 1965 (1922)	4340
2346	POSNJAK E., TUNELL G., Am. J. Sci., **18**, **1**, (1929)	4349
2347	POTDAR B., Caliche, 7, 391 (1925)	3255
2348	POURADIER J., VENET A. M., CHATEAU H., J. Chem. Phys., **51**, 375 (1954)	4985

Reference number	Author's name and reference	Table number
2349	POWELL C. F., CAMPBELL J. E., J. Am. Chem. Soc., **69**, 1227 (1947)	164
2350	PRATOLONGO, (1924), cited in [2]	3021
2351	PRECHT H., WITTGEN B., Ber., **14**, 1667 (1881)	3637, 3826, 4122
2352	PRECHT H., WITTGEN B., Ber., **15**, 1666 (1882)	3596, 3668
2353	PRECHT H., WITTGEN B., HOFF, MEYERHOFFER W., cited in [2]	4674
2354	PRIDEAUX E. B. R., MILBOTT J. O., J. Chem. Soc., 2703 (1928)	618
2355	PRUTTON C. F., TOWER P. F., J. Am. Chem. Soc., **54**, 3040 (1932)	4158
2356	PRUTTON C. G., BROSHEER J. C., MARAN S. H., J. Am. Chem. Soc., **57**, 1656 (1935)	2147
2357	PRUTTON C. F., SAVAGE R. L., J. Chem. Soc., **61**, 1553 (1945)	1721
2358	PUGH W., J. Chem. Soc., 1537 (1929)	851, 1059, 2769
2359	PUSHIN N. A., Bull. Soc. chim. Belgrade, **14**, 33 (1949)	5298
2360	QUADBURY W. E., J. Phys. Chem., **59**, No. 3, 257 (1955)	3355
2361	QUERIN H., MATTRAT P., Bull. Soc. Chim. France	2618
2362	QUILL L. L., ROBEY R. F., J. Am. Chem. Soc., **59**, 1071, 2591 (1937)	522,527
2363	RABE W. O., Z. anorg. Chem., **31**, 156 (1902)	4626, 4774, 4947
2364	RAFFO M., POSSI G., Gazz. chim. ital., **45**, 1, 45 (1915)	5098
2365	RAKESTRAW N. W., EMMEL V. M., J. Phys. Chem., **42**, 9, 1211 (1938)	70, 120
2366	RAKOWSKI A. W., TARASSENKOW D. N., Z. anorg. Chem., **174**, 91 (1928)	4336
2367	RANDALL, CHANG, (1928), cited in [2]	4109, 5537
2368	RANDALL M. VIETTI W. V. A., J. Am. Chem. Soc., **50**, 1526 (1928)	4995, 5447, 5496, 5578
2369	RANDALL M., HALFORD J. O., J. Am. Chem. Soc., **52**, 178 (1930)	2000, 4492, 4513, 4514
2370	RAOULT, Z. anorg. Chem., **141**, 38, 45 (1924)	1906, 1960
2371	RAY R. Ch., CHATTERLI K. K., J. Chem. Soc., **1**, 385 (1932)	4437
2372	RAY R. C., MITRA H. C., Trans. Faraday Soc., **30**, 1161 (1934)	4438
2373	RAY R. C., MITRA H. C., Trans. Faraday Soc., **31**, 1312 (1935)	4439
2374	REBURN W. T., GALE W. A., J. Phys. Chem., **59**, No. 1, 19 (1955)	1272
2375	REDFERN J. P., SALMON J. E., J. Chem. Soc., No. 1, 291 (1961)	4286
2376	REES A. G., HUDDLESTON L. J., J. Chem. Soc., 1648 (1931)	3187
2377	REICH K., Monatsh. Chem., **12**, 464 (1891)	2970
2378	REICHSTEIN S., EWENTOW L., KASARNOWSKY J., Z. anorg. Chem., **216**, 1 (1933)	192
2379	REIFF F., TOUSSAINT S. M., Z. anorg. Chem., **241**, 4, 372 (1939)	2719, 2994, 3096, 3164, 3229, 3357, 3567, 3872, 3880, 3895, 3935, 3936
2380	REINDERS W., Proc. k. Akad. Wet. (Amst.), **16**, 1065 (1914, 1915)	3078
2381	REINDERS W., Z. anorg. Chem., **93**, 204 (1915)	3685
2382	REINDERS W., KIES H. L., Rec. trav. chim., **59**, 785 (1940)	808
2383	REINDERS W., NICOLAI H., Rec. trav. chim., **66**, 477 (1947)	1846, 2106, 2471

Reference number	Author's name and reference	Table number
2384	REINHARDT R. A., J. Inorg. Nucl. Chem., **1**, No. 4, 839 (1962)	2153, 2452
2385	REGNER A., BALEJ J., Chem. listy, **51**, 2, 368 (1957)	4522
2386	REGNER A., BALEJ J., Coll. Czechosl. Chem. Comm., **22**, No. 5, 1683 (1957)	4523
2387	RENGADE E., Rev. gen. sci., **28**, 489 (1917); Chim. et Industr., **7**, 1090 (1922)	2173
2388	RESTIANO S., Atti accad. Lincei, (6), **20**, 192 (1934)	3507, 5336, 5462
2389	RESTIANO S., Atti del X convresso internazionale di chimica, **2**, 761 (1938)	2470
2390	RICCI J. E., J. Am. Chem. Soc., **56**, 290 (1934)	3093, 3352, 3565, 3893, 3905, 3927, 3937, 4582, 4709, 4824, 4971, 4975, 4976, 4996
2391	RICCI J. E., J. Am. Chem. Soc., **57**, 805 (1935)	3353
2392	RICCI J. E., JANICK N. S., J. Am. Chem. Soc., **58**, 313 (1936)	3569
2393	RICCI J. E., J. Am. Chem. Soc., **58**, 1077 (1936)	3359, 4715
2394	RICCI J. E., JANICK N. S., J. Am. Chem. Soc., **59**, 491 (1937)	3342, 4697, 5001
2395	RICCI J. E., J. Am. Chem. Soc., **59**, 866, 1763 (1937)	4936, 4939
2396	RICCI J. E., BUDISH J., BORODULIA N., J. Am. Chem. Soc., **59**, 868 (1937)	3092, 3097
2397	RICCI J. E., J. Am. Chem. Soc., **60**, 2040 (1938)	3863, 4940
2398	RICCI J. E., WELTMAN C., J. Am. Chem. Soc., **64**, 2746 (1942)	3858
2399	RICCI J. E., SMILEY S. H., J. Am. Chem. Soc., **66**, 1011 (1944)	5521
2400	RICCI J. E., J. Am. Chem. Soc., **66**, 1015 (1944)	3084, 3859, 3862
2401	RICCI J. E., DOPPELT L., J. Am. Chem. Soc., **66**, 1985 (1944)	3094, 3932
2402	RICCI J. E., SELIKSON, J. Am. Chem. Soc., **69**, 995 (1947)	920b
2403	RICCI J. E., LINKE W. F., J. Am. Chem. Soc., **69**, 1080 (1947)	3861, 3926, 3930
2404	RICCI J. E., OFFENBACH J. A., J. Am. Chem. Soc. **73**, 4, 1597 (1951)	3869
2405	RICCI J. E., LINKE W. F., J. Am. Chem. Soc., **73**, 3607 (1951)	3354, 3406, 3931, 3934, 4052, 4099, 4118, 5394, 5507
2406	RICCI J. E., AMRON J., J. Am. Chem. Soc., **73**, 3613 (1951)	5398, 5428
2407	RICCI J. E., FISCHER J., J. Am. Chem. Soc., **74**, 6, 1443 (1952)	2352, 2353, 2457, 4743, 4744, 4775, 5041, 5278
2408	RICCI J. E., FREEDMAN A. J., J. Am. Chem. Soc., **74**, 1765 (1952)	5454, 5485–5487, 5489, 5492, 5504, 5520, 5522, 5525
2409	RICCI J. E., SELIKSON B., J. Am. Chem. Soc., **74**, 8, 1956 (1952)	2248, 2251, 3235, 4672
2410	RICCI J. E., LOPREST F. J., J. Am. Chem. Soc., **75**, 5, 1224 (1953)	4583, 5000

Reference number	Author's name and reference	Table number
2411	RICHARDS T. W., FABER H. B., Am. Chem. J. **21**, 167 (1899)	3528
2412	RICHARDS T. W., ARCHIBALD E. H., Z. physik. Chem., **40**, 385 (1902)	1131, 3839, 5170, 5508
2413	RIESENFELD E. H., FELD H., Z. anorg. Chem., **116**, 213 (1921)	697
2414	RIMBACH E., Ber., **35**, 1300 (1902)	2433, 2552, 4897
2415	RINDELL A., Z. Physik. Chem., **70**, 452 (1910)	2184, 2502
2416	RITZEL A., Z. kryst. Min., **49**, 152 (1911)	3663
2417	RIVETT A. C., J. Chem. Soc., **121–122**, 383 (1922)	2270, 2275, 2292, 2463, 3253
2418	RIVETT A. C. D., LEWIS N. B., Rec. trav. chim., **42**, 954 (1963)	3224
2419	RIVETT A. C. D., CLENDINNEN F. W. J., J. Chem. Soc., **123**, 1634 (1923)	2525
2420	RIVETT A. C. D., PACKER J., J. Chem. Soc., (London), **130**, 1342 (1927)	206
2421	ROBERTSON J. B., J. Soc. Chem. Ind., **43**, 334 (1924)	3554, 3885, 4805
2422	ROBINSON F. W., WAGGAMAN W. H., J. Phys. Chem., **13**,673 (1909)	3962
2423	RODWELL, J. Chem. Soc., **15**, 59 (1862)	538, 902
2424	ROGER L., Ann. chim., (11), **19**, **362** (1944)	1485, 1492, 3925, 3952
2425	ROGERS A. F., Am. J. Sci., (5), **11**, No. 66, 474 (1926)	2944, 2954, 2956, 3256
2426	ROHMER R., Ann. chim., (11), 11, 611; Compt. rend., **209**, 315 (1939)	837
2427	ROHRER C.S., LANFORD O. E., KIEHL S. J., J. Am. Chem. Soc., **64**, 2810 (1942)	4333
2428	ROHRER, FRONING, J. Am. Chem. Soc., **72**, 4656 (1950)	1208
2429	ROLLET A., ANDRÈS L., Compt. rend., **191**, 377 (1930)	1584
2430	ROLLET A. P., Compt. rend., **191**, 488 (1930)	1582
2431	ROLLET A. P., LAUFFENBURGER R., Bull. Soc. chim. France, (5), **1**, 146 (1934)	1274
2432	ROLLET A. P., BOUAZIZ R., C. r. Acad. Sci., **240**, No. 11, 1227 (1955)	1273
2433	ROOZEBOOM H., Z. physik. Chem., **2**, 454 (1888)	3367
2434	ROOZEBOOM H., Z. physik. Chem., **10**, 148 (1892)	2513
2435	ROOZEBOOM H., SCHREINEMAKERS F. A. H., Z. physik. Chem., **15**, 633 (1894)	1034
2436	ROOZEBOOM H. W. B., cited in [2]	4946
2437	ROSENHEIM A., DAVIDSOHN J., Z. anorg. Chem., **37**, 325 (1903)	2265, 2359
2438	ROSENHEIM A., GRUNBAUM, Z. anorg. Chem., **61**, 187 (1909)	617, 4614, 4759, 4900, 4988
2439	ROSENHEIM A., PRITZE M., Z. anorg. Chem., **63**, 275 (1909)	3888
2440	ROSENHEIM A., THOR S., Z. anorg. Chem., **69**, 263 (1910–1911)	2615
2441	ROSENHEIM A., LEYSER F., Z. anorg. Chem., **119**, 12 (1921)	1549
2442	ROSENHEIM A., THOR S., Z. anorg. Chem., **167**, 1 (1927)	2772, 2773
2443	ROSENHEIM A., ZICKERMANN J., Z. anorg. Chem., **208**, 95 (1932)	2379
2444	ROSTANIO S., di, Atti accad. Lincei, (6), **20**, 192 (1934)	4771
2445	ROTH, Z. physik. Chem., **24**, 123 (1897)	640, 1754
2446	ROTHMUND Y., Z. physik. Chem., **69**, 531 (1909)	1299, 1989
2447	RUBENBAUER, Z. anorg. Chem., **30**, 334 (1902)	1497, 2766, 2783, 2795

Reference number	Author's name and reference	Table number
2448	RÜDORFF, Ber., **6**, 482 (1873)	3066, 4553
2449	RÜDORFF, Ber., **18**, 1160 (1885)	3827
2450	RÜDORFF F., Sitzungsber. preuss. Akad., **364** (1885), cited in [1]	1343, 1407
2451	RÜDORFF, cited in [2]	2144, 2163, 2267, 2458, 2539
2452	RÜDORFF, Karsten, cited in [2]	2484, 2485
2453	RUSSELL R. U., SALMON J. E., TIETZE H. R., J. Chem. Soc., **8**, 3212 (1961)	4284
2454	RYLANDS J. R., JENKINSON J. R., Engineering, **156**, 517 (1943)	774, 780
2455	SACKUR O., TAEGENER W., Z. Elektrochem., **18**, 722 (1912)	4406, 4409, 4479, 4808
2456	SAEGUSA F., J. Chem. Soc. Japan, Pure Chem. Sect., **71**, 223 (1950)	4613
2457	SAKAR, BARAT, (1930), cited in [2]	834
2458	SALAUN A., Bull. Soc. chim. France, No. 6, 1106 (1960)	1895, 1897, 1941, 1942
2459	SALMON J. E., TERREY H., J. Chem. Soc., **10**, 2813 (1950)	4287, 4300
2460	SALMON J. E., J. Chem. Soc., **6**, 2321 (1952)	4291
2461	SALSTROM E. J., SMITH G., J. Phys. Chem., **34**, 2241 (1930)	3427
2462	SANDERS J. P., DOBBINS J. T., J. Phys. Chem., **35**, 3086 (1931)	1372, 1374
2463	SANFOURCHE A., KRAPIVINE A., Bull. Soc. chim. France, (4), **53**, 970 (1933)	675
2464	SARASWAT H. C., Proc. Ind. Acad. Sci., **30A**, 329 (1949)	1996
2465	SASLAWSKY A. I., ETTINGER Y. L., Z. anorg. Chem., **223**, 277 (1935)	3103, 3111
2466	SASLAWSKY A. I., ETTINGER J. L., ESEROWA E. A., Z. anorg. Chem., **225**, 305 (1935)	414, 425, 3103
2467	SBORGI U., Atti accad. Lincei, (5), **21**, 11, 855 (1912)	1544
2468	SBORGI U., Atti accad. Lincei, (5), **22**, 1, 90 (1913)	366, 1544, 1576
2469	SBORGI U., Atti accad. Lincei, (5), **23**, 1, 719 (1914)	1583
2470	SBORGI U., MENACEI F., Atti accad. Lincei, (5), **24**, 1, 443, ff. (1915)	1544
2471	SBORGI U., MECACCI F., Atti accad. Lincei, (5), **25**, 11, 328 (1916)	1544, 1548
2472	SBORGI U., FERRI L., Mem. accad. Lincei, (5), **13**, 569 (1921)	1545
2473	SBORGI U., FRANCO C., Gazz. chim. ital., **51**, 1 (1921)	2069, 2070, 2455, 2461, 2833
2474	SBORGI U., MEZETTI, Atti accad. Lincei, (5), **30**, 11, 189 (1921)	1544
2475	SBORGI U., FERRI L., Atti accad. Lincei, (5), **31**, 1, 325 (1922)	1546
2476	SBORGI U., FRANCO C., Giorn. chim. ind. applicata, 4 (June), (1922)	2461
2477	SBORGI U., GALLICHI E., Gazz. chim. ital., **54**, 255 (1924)	2063, 2071
2478	SBORGI U., BOVALINI E., Gazz. chim. ital., **54**, 919 (1924)	2064
2479	SBORGI U., BOVALINI E., MEDICI M., Gazz. chim. ital., **54**, 934 (1924)	2065, 2066, 2068, 2073, 2074, 2284, 2285, 2298, 2825, 2826
2480	SBORGI U., BURICHETTI E., Gazz. chim. ital., **54**, 948 (1924)	2067, 2072, 2296, 2824
2481	SBORGI U., Atti Soc. Toscana, **37**, 183 (1926)	2331, 2332, 2505, 2840, 3747, 4722, 4845

Reference number	Author's name and reference	Table number
2482	SBORGI U., AMELOTTI L., Gazz. chim. ital., **60**, 468 (1930)	1551
2483	SBORGI U., Gazz. chim. ital., **62**, 3 (1932)	1551
2484	SCHAFER H., Z. anorg. Chem., **258**, 69 (1949)	1032
2485	SCHIMMEL F. A., J. Chem. Engng Data, 5, No. 4, 519 (1960)	928
2486	SCHIERHOLZ, Sitzber. k. Akad. Wiss., (Wien), **101**, 26, 4 (1890), cited in [2]	2532, 3805, 3806, 3815, 4888, 4986, 5010, 5417
2487	SCHLOESING A., Compt. rend., **171**, 979 (1920)	2165
2488	SCHMIDT E., Über Die Unsetzung von Alkalichloriden mit Calciumnitrat in flüssigem Ammoniak, 40 (1934), cited in [1]	1891
2489	SCHMIDT E., cited in [2]	1875
2490	SCHOLDER R., PATSCH R., Z. anorg. Chem., **222**, 135 (1935)	2787
2491	SCHOLDER R., HENDRICH G., Z. anorg. Chem., **241**, 1, 80 (1939)	2610
2492	SCHOLDER R., STAUFENBIEL E., Z. anorg. Chem., **247**, 259 (1941)	2779
2493	SCHOLDER R., KOLB A., Z. anorg. Chem., **264**, 2–4, 209 (1951)	2759
2494	SCHOLLE S., UHLIR Z., BENES J., Chem. listy, **52**, 6, 1191 (1958)	3896
2495	SCHREINEMAKERS F. A. H., COCHERET D. H., FILIPPO H., WAAL A. J. C., de, Z. physik. Chem., **59**, 645 (1905, 1907)	1344
2496	SCHREINEMAKERS F. A. H., Z. physik. Chem., **55**, 83 (1906)	4367
2497	SCHREINEMAKERS F. A. H., WAAL A. J. C., de, Chem. Weekblad, **3**, 539 (1906)	1371
2498	SCHREINEMAKERS F. A. H., FILIPPO A., (1906), cited in [2]	5196
2499	SCHREINEMAKERS F. A. H., Chem. Weekblad, 5, 847 (1907)	1385, 2540
2500	SCHREINEMAKERS F. A. H., BAAT W. C., de, Z. physik. Chem., **65**, 555 (1908–1909)	3792, 4602, 5284
2501	SCHREINEMAKERS F. A. H., BAAT W. C., de, Z. physik. Chem., **65**, 586; **67**, 551 (1909)	3246, 3828
2502	SCHREINEMAKERS F. A. H., Z. physik. Chem., **66**, 687 (1909)	1385, 2340
2503	SCHREINEMAKERS F. A. H., Z. physik. Chem., **68**, 83 (1909–1910)	957, 980, 1115, 2590, 2705, 2785, 5468
2504	SCHREINEMAKERS F. A. H., Z. physik. Chem., **69**, 562 (1909)	2269
2505	SCHREINEMAKERS F. A. H., Chem. Weekblad, **6**, 131, 140 (1909)	1385, 2338, 2339
2506	SCHREINEMAKERS F. A. H., Arch. Néer. sc. ex. nat., (2), **15**, 81, 117 (1910)	1381, 2268, 2350, 2705, 5271, 5510
2507	SCHREINEMAKERS F. A. H., BAAT W. C., de, Chem. Weekblad, 7, 259 (1910)	2195, 2540, 5468
2508	SCHREINEMAKERS F. A. H., Z. physik. Chem., **69**, 557 (1910)	1381, 2268, 2350, 5022, 5271, 5510
2509	SCHREINEMAKERS F. A. H., Z. physik. Chem., **71**, 109 (1910)	2343
2510	SCHREINEMAKERS F. A. H., BAAT W., de, Arch. Néer, sc. ex. nat., (2), **15**, 83 (1911)	3247
2511	SCHREINEMAKERS F. A. H., BAAT W., de, Proc. Ac. Sci. Amst., **13**, 1163 (1911)	3247
2512	SCHREINEMAKERS F. A. H., Arch. Néer. sci. ex. nat., (2), **15**, 490 (1911)	1345, 3247
2513	SCHREINEMAKERS F. A. H., FIGEE T., Chem. Weekblad, **8**, 683 (1911)	5045

Reference number	Author's name and reference	Table number
2514	SCHREINEMAKERS F. A. H., THONUS J. C., Proc. k. Akad. Wet. (Amst.), **15**, 472 (1912)	5286
2515	SCHREINEMAKERS F. A. H., PROVIJE D. J., Proc. k. Akad. Wet., **15**, 1326 (1913)	3460
2516	SCHREINEMAKERS F. A. H., BAAT W. C., de, Proc. sect. Sci., Amsterdam, **7**, 782 (1914), cited in [1]	4906
2517	SCHREINEMAKERS F. A. H., BAAT W. C., de, Proc. k. Akad. Wet. (Amst.), **17**, 533, 781 (1914)	4676
2518	SCHREINEMAKERS F., BAAT W., de,Verlag. k. Akad. Wet. (Amst.), **23**, 1097, May (1915)	1984
2519	SCHREINEMAKERS F. A. H., BAAT W. C., de, Chem. Weekblad, 14, 141 (1917)	1448, 1481, 2580, 2613, 3916, 4608, 4980, 5007, 5151, 5183, 5316, 5317, 5321, 5322
2520	SCHREINEMAKERS F. A. H., NOORDUYN A. C., Chem. Weekblad, **15**, 118 (1918)	1442
2521	SCHREINEMAKERS F. A. H., KAYSER G. M. A., Chem. Weekblad, **15**, 120 (1918)	1329
2522	SCHREINEMAKERS F. A. H., BAAT W. C., de, MISS W. C., Rec. trav. chim., **39**, 423 (1920)	1275, 1985, 4369
2523	SCHREINEMAKERS F. A. H., BERKHOFF V., POSTHUMUS K., Rev. trav. chim., **43**, 508 (1924)	2191
2524	SCHRÖEDER W., Z. anorg. Chem., **177**, 74 (1928); Caliche, **11**, 155 (1929)	3378
2525	SCHRÖEDER W., Z. anorg. Chem., 184, 81 (1929); Caliche, **12**, 560 (1930)	3380
2526	SCHRÖEDER W., Z. anorg. Chem., **185**, 153 (1930)	3102
2527	SCHRÖEDER W., SCHACKMANN H., Z. anorg. Chem., **220**, 389 (1934)	4769
2528	SCHRÖEDER W., GABRIEL A., PATRIDGE E., J. Am. Chem. Soc., **57**, 1545 (1935)	2690, 3267
2529	SCHRÖEDER W., Z. anorg. Chem., **228**, 129 (1936)	1524
2530	SCHRÖEDER W. C., BERK A. A., GABRIEL A., J. Am. Chem. Soc., **58**, 843 (1936)	2633, 2647, 2949
2531	SCHRÖEDER W. C., BERK A. A., GABRIEL A., J. Am. Chem. Soc., **59**, 1783 (1937)	2672
2532	SCHRÖEDER W., KEHNOR E., FRINGS K., POELVOODRE H., von, Z. anorg. Chem., **238**, 209, 305 (1938)	2371
2533	SCHRÖEDER W., Z. anorg. Chem., **239**, 39, 406 (1938)	1515, 1519
2534	SCHRÖEDER W., HOMPESCH H., MIRBACH P., Z. anorg. Chem., **239**, 225 (1938)	1515
2535	SCHRÖEDER W., Z. anorg. Chem., **240**, 1, 58 (1939)	1512
2536	SCHRÖEDER W., Z. anorg. Chem., **241**, 2–3, 183 (1939)	1513
2537	SCHULZ K., TEZAK B., Arhiv Kem., **23**, 200 (1951)	3946, 5013
2538	SCHUMB W. C., RITTNER E. S., J. Am. Chem. Soc., **65**, 1055 (1943)	2797
2539	SCHÜTZE H., PIECHOWICZ T., WAHL B., Helv. chim. acta, **26**, 237 (1943)	2475
2540	SCHÜTZE H., PIECHOWICZ T., PUSTELNIK W., Helv. chim. acta, **26**, 238 (1943)	2096, 2097, 2099, 2107, 2109, 3659
2541	SCHÜTZE H., PIECHOWICZ T., Helv. chim. acta, **26**, 244 (1943)	1830

Reference number	Author's name and reference	Table number
2542	SCOTT A. F., DURHAM E. J., J. Phys. Chem., **34**, 531 (1424 (1930)	1215, 1216, 1218, 1221, 1223, 1226, 1231, 1239
2543	SCOTT A. F., JOHNSON C. R.. J. Am. Chem. Soc., **52**, 3586 (1930)	492
2544	SECHENOV J. M., Bull. St. Petersb. Acad., **22**, 103 (1877)	723
2545	SECHENOV J. M., Mém. de l'Acad. de St. Petersb., (7), **34**, No. 3, 2–3 (1886), cited in [2]	1623, 1630
2546	SECHENOV J. M., Mém. de l'Acad. de St. Petersb., (7), **35**, No. 7, 24 (1887), cited in [2]	1614, 1675, 1719, 1731, 1733
2547	SECHENOV I. M., Nouv. mém. soc. natur. de Moscou, **15**, No. 6, 267–268 (1889), cited in [2]	1591, 1624, 1630, 1656, 1675, 1686, 1694, 1697, 1729
2548	SECHENOV I. M., Ann. chim. phys., (6), **25**, 252, (1892), cited in [2]	723, 1591, 1614, 1624, 1630, 1656, 1675, 1686, 1694
2549	SEGNIT E. R., HOLLAND H. D., BISCARDI C. J., Geochimica et Cosmochimica acta, **26**, 1301 (1962)	1717
2550	SEHNAL J., Compt. rend., **148**, 1394 (1909)	904
2551	SEIDELL A., Am. Chem. J., **27**, 52 (1902)	3242
2552	SEIDELL A., SMITH, J. Phys. Chem., 8, 493 (1904)	4591, 4719, 5090
2553	SEIDELL A., SMITH, CAMERON, BREAZEALE, cited in [2]	3124
2554	SEIDELL A., cited in [2]	901, 1108, 2288, 3881
2555	SEIDELL W., FISCHER W., Z. anorg. Chem., **247**, 367 (1941)	932, 945, 967, 990, 1002
2556	SELICHIRO KAWAKAMI, J. Chem. Soc., Japan, Chem. sec., **60**, No. 10, 1258 (1957)	3346
2557	SELVA, (1935), cited in [2]	3209
2558	SEROKY, Kali, No. 17, 347 (1923)	3326, 3603, 3641
2559	SEROWY F., KASSNER B., Chem. Technik, **9**, No. 1, 27 (1957)	3614
2560	SESHADRI H., LOBO J., J. Scient. Industr. Res., **BC16**, No. 12, 8531 (1957)	2955, 2962, 2981, 3284
2561	SEYLER C. A., LLOYD P. V., J. Chem. Soc., **95–96**, 1349 (1909)	370, 1619, 1627, 1678, 1703
2562	SHASSEVENT L., Ann. chim. (10), **6**, 334 (1926)	4846
2563	SHERRILL M. S., Z. physik. Chem., **43**, 705 (1903)	4495, 5016
2564	SHERRILL M. S., EATON F. M., J. Am. Chem. Soc., **29**, 1643 (1907)	2003
2565	SHERRILL M. S., RUSS D. E., J. Am. Chem. Soc., **29**, 1657 (1907)	498, 499
2566	SHERRILL M. S., IZARD E. F., J. Am. Chem. Soc., **50**, 1665 (1928)	124
2567	SHERRILL M. S., IZARD E. F., J. Am. Chem. Soc., **53**, 1667 (1931)	121, 124, 127, 130, 131
2568	SHERRILL M. S., HAAS A. J. J., J. Am. Chem. Soc., **58**, 952 (1936)	895, 1174
2569	SIEVERTS A., MÜLLER E. A., Caliche, **12**, 91 (1930)	3101

Reference number	Author's name and reference	Table number
2570	SIEVERTS A., MÜLLER E. L., Z. anorg. Chem., **189**, 241 (1930)	3101, 3102, 3994
2571	SIEVERTS A., MÜLLER E. L., Z. anorg. Chem., **200**, 305 (1931)	3102
2572	SIEVERTS A., PETZOLD W., Z. anorg. Chem., **214**, 396 (1933)	482, 505
2573	SIEVERTS A., SCHREINER L., Z. anorg. Chem., **219**, 105 (1934)	469
2574	SIDGWICK N. V., LEWIS N. B., J. Chem. Soc., **129**, 1287 (1926)	1493, 1496
2575	SIMKOVÁ H., ERDÖS E., Chem. listy, **52**, **4**, 568 (1958)	3042, 3043, 3074
2576	SIMKOVÁ H., ERDÖS E., Coll. Czechosl. Chem. Comm., **24**, No. 3, 694 (1959)	3044, 3075
2577	SIMONS E. L., RICCI J. E., J. Am. Chem. Soc., **68**, 1413 (1946)	863
2578	SIMONS E. L., RICCI J. E., J. Am. Chem. Soc., **68**, 2194 (1946)	1391, 2362, 3491, 4753
2579	SIMONS E. L., Inorg. Chem., **3**, No. 8, 1079 (1964)	3960
2580	SIREAR S. C., PRASED B., J. Ind. Chem. Soc., **33**, 361 (1956)	471
2581	SIRUCEK J., VICTORIN O., Coll. Czechosl. Chem. Comm., **11**, 403 (1939)	3149, 3513, 4111, 4262, 4777, 5492, 5569
2582	SIRUCEK J., VICTORIN O., Coll. Czechosl. Chem. Comm., **12**, 630 (1947)	2198, 3136, 4002, 4608
2583	SKARULIS J. A., HARAN H. A., MALEENY, J. Am. Chem. Soc., **76**, 5, 1540 (1954)	3408
2584	SKARULIS J. A., HARAN H. A., J. Am. Chem. Soc., **77**, 3489 (1955)	1351
2585	SMITH C. M., J. Am. Chem. Soc., **42**, 259 (1920)	5055
2586	SMITH G. F., J. Am. Chem. Soc., **45**, 2074 (1923)	547
2587	SMITH G. F., J. Am. Chem. Soc., **46**, 1581 (1924)	530, 531, 1172, 1173
2588	SMITH G. F., RING F., J. Am. Chem. Soc., **56**, 1889 (1937)	1166
2589	SMITH H. A., BALL T. R., J. Am. Chem. Soc., **39**, 217 (1917)	3417
2590	SMITH H. J., J. Am. Chem. Soc., **40**, 879 (1918)	1722, 1728
2591	SMITH S. B., J. Am. Chem. Soc., **69**, 2285 (1947)	1266
2592	SMITH N. O., J. Am. Chem. Soc., **64**, 41 (1942)	4246, 4250, 4255
2593	SMITH N. O., LENNOX C. S., J. Am. Chem. Soc., **70**, 1799 (1948)	2392, 2397
2594	SMITH W. T., J. Am. Chem. Soc. **68**, 394 (1946)	4833, 4834
2595	SMITS A., ELGERSMA J., HARDENBERG M. E., Rec. trav. chim., **43**, 671 (1924)	1409
2596	SNYDER, Ber. **11**, 936 (1878)	2527, 3126, 3482, 3798, 4096, 4171, 4482, 4596, 4745, 4878, 5148, 5293, 5294
2597	SOCH C. A., J. Phys. Chem., **2**, 43 (1898)	3661, 4555
2598	SOMMER, WEISE, Z. anorg. Chem., **94**, 51 (1916)	736, 937
2599	SORUM C. H., WOLF H. A., J. Chem. Education, **27**, 614 (1950)	1064
2600	SPIELREIN C., Compt. rend., **155**, 348 (1912)	1349
2601	SPIELREIN C., Compt. rend., **157**, 48 (1913)	1346, 1349, 1375
2602	SPRAGUE R. W., Industr. Engng Chem., **47**, No. 11, 2396 (1955)	32, 83, 84
2603	SPRANER J. W., PEARCE D. W., J. Phys. Chem., **44**, 909 (1940)	2594, 2597

Reference number	Author's name and reference	Table number
2604	STAMM H., Inaug. Dissert., Halle, (1926), cited in [2]	1342, 1405, 1823, 2831, 1840, 1856, 1859, 1868, 1880, 1890, 1930, 1936, 1950, 1955, 1986, 1987, 1988, 2030, 2031, 2032, 2271
2605	STARRS B. A., CLARK L., J. Phys. Chem., **34**, 1058 (1930)	4062
2606	STARRS B., STORCH H., J. Phys. Chem., **34**, 2371 (1930)	4063
2607	STORTENBECKER W., Z. physik. Chem., **22**, 62 (1897)	5274
2608	STORTENBECKER W., Z. physik. Chem., **34**, 109 (1900)	5205, 5222
2609	STORTENBECKER W., Rec. trav. chim., **21**, 407 (1902)	793
2610	STRAUB F., Univ. Ill. Bull., Eng. Expt. Sta., cited in [1]	2630, 2676
2611	STRAUB F., Ind. Eng. Chem., **24**, No. 8, 917 (1932)	3448, 3453
2612	STRAUB J., Z. physik. Chem., **77**, 332 (1911)	1999
2613	STRÖMHOLM D., Z. anorg. Chem., **57**, 72 (1908)	1839
2614	STUBICAN V., TEZAH B., Arhiv Kem., **22**, 75, 83 (1950)	4511
2615	SUDHAUS K., Neues Jahrb. Min. Geol., (Beil. Be.), **37**, 1, (1914), cited in [2]	3816, 4894
2616	SUGDEN R., J. Chem. Soc., 488 (1929)	5552
2617	SULLIVAN E. C., J. Am. Chem. Soc., **27**, 529 (1905)	2324, 2325
2618	SÜSS J., Z. kryst. Min., **51**, 262 (1913)	4864
2619	SWAN C. M., "Chemistry Thesis", Mass. Inst. Technology (1899), cited in [2]	5403
2620	SWENSON T., RICCI J. E., J. Amer. Chem. Soc., **61**, 1974 (1939)	3860, 4937
2621	SZUFARSKI Z., PISHINGER E., Przem. Chem., **41**, No. 12, 695 (1962)	2643, 2689, 2712
2622	TABER W. C., J. Phys. Chem., **10**, 593 (1906); Bull., **33**, Bureau of Soils, U. S. Dept. Agr., (1906), cited in [2]	674
2623	TAKAHASHI G., Bull. Imp. Hyg. Lab. (Tokyo), **29**, 165 (1927), cited in [2]	1657, 1665, 4443
2624	TAKAHASHI T., SASAKI K., J. Chem. Soc., Japan, Ind. chem. Sect., **53**, 383 (1950)	5215
2625	TAKEGAMI S., J. Chem. Soc. Tokyo, **41**, 831 (1920)	3252, 3293
2626	TAKEGAMI S., J. Chem. Soc., Tokyo, **42**, 441 (1921)	3371
2627	TAKEGAMI S., Mem. Coll. Sci. Kyoto, Imp. Univ., **4**, 317 (1921), cited in [2]	3253, 3293, 3370, 3618, 4020
2628	TAKEGAMI S., Mem. Coll. Sci. Kyoto. Imp. Univ., **5**, 191 (1922)	3371
2629	TANAKA Y., NAKAMURA K., HARA R., J. Soc. Chem. Ind., Japan, Suppl. **34**, 285B (1931); The Technology reports Tôhoku Imp. Univ. Sendai, Japan, **11**, No. 2, 101 (1934)	5134, 5137
2630	TARTAR H. V., LORCH J. R., J. Am. Chem. Soc., **51**, 1091 (1929)	4301, 4303
2631	TARTAR H. V., RICE M. R., SWEO B. J., J. Am. Chem. Soc., **53**, 3949 (1931)	5323
2632	TAYLOR D., J. Chem. Soc., **6**, 2370 (1952)	818
2633	TAYLOR D., BASSETT H., J. Chem. Soc., **11**, 1431 (1952)	783
2634	TAYLOR D., J. Chem. Soc., **8**, 2503 (1953)	816

Reference number	Author's name and reference	Table number
2635	TAYLOR A. W., FRAZIER A. W., GURNEY E. L., Trans. Faraday Soc., **59**, No. 7, 1580 (1962)	2228, 4120
2636	TAYLOR A. W., FRAZIER A. W., GURNEY E. L., SWITH J. P., Trans. Faraday Soc., **59**, No. 7, 1585 (1963)	657
2637	TEEPLE J. E., The Industrial Development of Searles Lake Brines (1929)	317, 344, 348, 368, 1550, 1555, 1556, 1573, 1574, 2806–2812, 2817, 2823, 2827, 2834, 2838, 2839, 2841, 2843, 2844, 2852, 2857, 2863, 2868, 2877, 2892, 2909, 2933, 2945, 2957, 2963, 2974, 2983, 3011, 3261, 3327, 3425, 3687, 4430, 4431, 4435, 4436, 4444, 4464, 4470, 4474, 4679
2638	TELLER G., BYE J., Bull. Soc. Chim. France, No. 10 1531 (1959)	3961
2639	TEMPLETON C. C., J. Chem. Eng. Data, **5**, No. 4, 514 (1960)	3824
2640	TERRES E., BRÜCKNER K., Z. Elektrochem., **26**, 1 (1920)	5350, 5474
2641	TERRES E., WEISER H., Z. Elektrochem., **27**, 177 (1921)	1593
2642	TERRES H., JOLLY V. G., J. Chem. Soc., **123**, 2217 (1923)	1317
2643	TERRES E., HAHN E., Das Gas- u. Wasserfach., **70**, 309 (1927)	1821
2644	TERRES E., SCHMIDT W., Das Gas- und Wasserfach., **70**, 765 (1927)	740, 1824
2645	TERRES E., BEHRENS H., Z. physik. Chem., **130**, 695 (1928)	1594
2646	TERRES E., RÜHL G., Z. angew. Chem., **47**, 331 (1934)	4323, 4324, 4329
2647	TETTAMANZI A., Gazz. chim. ital., **63**, 575 (1933)	1902, 1915
2648	THOMAS H. C., J. Am. Chem. Soc., **61**, 921 (1939)	1133, 1468, 4266, 5173
2649	THOMAS H. J., JACHE A. W., J. Inorg. Nucl. Chem., **13**, No. 1–2, 54 (1960)	614
2650	THOMPSON L. C. A., PACER R., J. Inorg. Nucl. Chem., **25**, No. 8, 1041 (1963)	1168, 2781
2651	TILMANS J., HEUBLEIN O., Gesundheitsingenieur, No. 34, 674 (1912), cited in [1]	1706
2652	TODA, Mem. Coll. Sci. Kyoto, Imp. Univ., **3**, 305 (1921)	4570, 5548
2653	TODA S., J. Chem. Soc. Japan, **43**, 320 (1922)	4570
2654	TOLLERT H., Z. physik. Chem., **6**, 3–4, 257 (1956)	3613
2655	TOPORESCU E., Compt. rend., **174**, 870, ff. (1922)	2085, 2090, 2093, 2102, 2461, 2907

Reference number	Author's name and reference	Table number
2656	TOPORESCU E., Compt. rend., **175**, 269 (1922)	2086, 2093, 2102, 2461, 2907
2657	TOPORESCU E., (1940), cited in [2]	152
2658	TOUREN C., Compt. rend., **130**, 259 (1900)	4556, 4578, 4812
2659	TOUREN C., Compt. rend., 131, 259 (1900)	4450, 4458
2660	TOWENEUX, Ann. chim., (9), **11**, 225 (1919)	4922
2661	TOWER, Z. anorg. Chem., **50**, 382 (1906)	79, 733
2662	TOWNLEY R. W., WHITNEY W. B., FELSING W. A., J. Am. Chem. Soc., **56**, 631 (1937)	1450, 1462, 3800, 3822, 4882, 4903
2663	TRANQUARD A., Bull. Soc. chim. France, No. 2, 264 (1964)	2646
2664	TRAVERS, NOUVEL, Compt. rend., **188**, 501 (1929)	3362
2665	TREADWELL F., REUTER M., Z. anorg. Chem., **17**, 193 (1898)	1642, 1660, 1701, 1702
2666	TREADWELL W. D., HEPENSTRICK H., Helv. Chim. Acta, **32**, No. 6, 1872 (1949)	692
2667	TRENNER N. R., TAYLOR H. A., J. Phys. Chem., **35**, 1336 (1931)	879
2668	TRIMBLE H. M., J. Am. Chem. Soc., **44**, 451 (1922)	3441, 4705
2669	TRIMBLE H. M., J. Am. Chem. Soc., **58**, 1868 (1936)	3788, 3957, 5240
2670	TRONSTAD L., VEIMO R., Water and Water Eng., **42**, 225 (1940)	1724–1727
2671	TRUJILLO R., MACHADO E. F., An. Real. soc. espanola, fis. y quim., **B51**, No. 12, 257 (1955)	4803
2672	TUNELL G., POSNJAK E., J. Phys. Chem., **35**, 929 (1931)	4341, 4349
2673	TUTTLE O. F., FRIEDMAN J. J., J. Am. Chem. Soc., **70**, 919 (1948)	2737
2674	TYRREN H. J. V., RICHARDS J., J. Chem. Soc., **12**, 3812 (1953)	5528
2675	TZ'AI LIU-SHENG, VEN WEN-HSING. J. Chinese Chem. Soc., 4, 178 (1936)	4239
2676	UMEDA J., J. Chem. Soc. Japan, **66**, 49 (1945)	2819
2677	UTIDA S., J. Soc. Chem. Ind. Japan, Suppl., Band **45**, 294B (1942)	369, 1599, 1610, 1612
2678	UYEDA K., Mem. Coll. Sci. and Eng., Kyoto Imp. Univ., **2**, 249–250 (1909–1910), cited in [2]	3674
2679	UYEDA K., Z. anorg. Chem., 93, 202 (1909–1910)	3078
2680	UYEDA K., Original commun. 8 Intern. Congress appl. chemistry, sect. **10b**, 22, 237 (1912), cited in [2]	2487
2681	VALENTA, Monatsh. Chem., **15**, 250 (1894)	1997, 2569, 4222, 5410, 5418, 5427
2682	VALENTINER S., Z. physik. Chem., **25**, 269 (1927)	65
2683	VALETON J. J., FRÖMEL W., Z. anorg. Chem., **137**, 91 (1924)	4874
2684	VAN DAMM W., DONK A. D., Chem. Weekblad, 8, 848 (1911)	5012
2685	VAN GELDER D. V., Req. trav. chim., **77**, No. 8, 739 (1958)	2805
2686	VAN KLOOSTER H. S., J. Phys. Chem., **21**, 513 (1917)	4057, 4077
2687	VAN KLOOSTER H. S., BALON P. A., J. Am. Chem. Soc., **56**, 591 (1934)	5025
2688	VAN-NAME R. S., BROWN W. G., Am. J. Sci., (4), **44**, 105 (1917)	198
2689	VAN'T HOFF J., MEYERHOFFER W., Sitzungsber. preuss. Akad., No. 6–7, 1030 (1897)	4042

Reference number	Author's name and reference	Table number
2690	VAN'T HOFF J. H., MEYERHOFFER W., Z. physik. Chem., **27**, 75 (1898)	4123
2691	VAN'T HOFF J. H., MEYERHOFFER W., Sitzungsber. preuss. Akad., 37 (1898)	3290, 3598
2692	VAN'T HOFF J. H., MEYERHOFFER W., Sitzungsber. preuss. Akad., 376 (1899); 422 (1901)	3290, 3598
2693	VAN'T HOFF J. H., Z. anorg. Chem., **47**, 247 (1905)	3314, 3315, 3650
2694	VAN'T HOFF J. H., BLASDALE, Sitzungsber. preuss. Akad., 252 (1905)	2829, 2832, 2837, 3315, 4138
2695	VAN'T HOFF J. H., Zur Bild. ozean. Salzablag, v. 2, (1909), cited in [2]	2829, 2831, 2832, 2837, 4138
2696	VAN'T HOFF J. H., BLASDALE, Untersuch. Bild. ozean. Salzablag., (1912), cited in [2]	2837, 3315
2697	VAN'T HOFF J. H., KENRICK F. B., "Ozeanischen Salzablagerungen" (1912), cited in [2]	3598, 4155
2698	VAN'T HOFF J. H., MEYERHOFFER W., Z. physik. Chem., **30**, 64 (1912)	4123
2699	VELDHULZEN H., van, These, Utrecht (1929), cited in [2]	805, 1011, 3524, 4153, 4526, 4723, 4847, 5116, 5124
2700	VERICAD RAGA J. Bta, An. Real. Soc. espanola fis. y quim., **B54**, No. 3, 243 (1958)	815
2701	VESTERBERG A., 8th Inter. Congr. Appl. Chem., **2**, 238, 255 (1912)	2670, 3188
2702	VICTORIN O., SIRUCEK J., Coll. Czechosl. Chem. Comm., **11**, 474 (1939)	2375, 3148, 3150, 3512, 3514, 4110, 4112, 5493,
2703	VIRECK A., Chem. Zentralbl., **7**, 402 (1862)	3801, 4175, 4883, 5152
2704	VOGEL, HAHN, Wyand. Silver Smelting Works (1877), cited in [2]	3806
2705	VÖLZ-FLADRICH H., Freib. Forschungsh, C, No. 134, 75 (1962)	971, 988, 1003
2706	VOSBURGH W. C., LACKEY O. N., J. Am. Chem. Soc., **52**, 1407 (1930)	891
2707	VOSBURGH W. C., DIBELER V. H., PARKS H. C., MILLER W. J., J. Am. Chem. Soc., **62**, 1471 (1940)	4102
2708	VOSBURGH W. C., McCLURE R. S., J. Am. Chem. Soc., **65**, 1060 (1943)	1820, 1835
2709	VOSBURGH W. C., COGWELL S. A., J. Am. Chem. Soc., **65**, 2412 (1943)	4609
2710	VOUK V. B., KRATOHVIL J., TESAK B., Arhiv Kem., **25**, 219 (1953)	4983, 5391
2711	WAAL A. J. C., de, Dissertation, Leyden, "Tables annuelles" (1910), cited in [2]	4372, 4472
2712	WALDECK W. F., LYNN G., HILL A. E., J. Am. Chem. Soc., **56**, 43 (1934)	2854
2713	WARIS G., Quart. J. Indian. Chem. Soc., **1**, 307 (1925)	1149
2714	WARYNSKI T., KOUROPATURINSKA S., J. Chim. Phys., **14**, 328 (1916)	2503, 3738

Reference number	Author's name and reference	Table number
2715	WASILEWSKI L., KOBYLCZYK A., Przem. chem., **11**, No. 5, 232 (1955)	748
2716	WATSON L. R., METCALF J. S., J. Chem. Engng Data, **6**, No. 3, 331 (1962)	3210
2717	WATTENBERG H., Deutsche Atlantische Expedition "Meteor", 1925–1927, t. 8, 208 (1933)	5081
2718	WATTERBERG H., Fortschr. Min., **20**, 11, 178 (1936)	1739, 5082
2719	WATTENBERG H., TIMMERMANN E., Kieler Meeresforschungen, **11**, No. 1, 86 (1937)	1681, 3990
2720	WEGSCHELDER R., WALTER H., Monatsh. Chemie, **28**, 633 (1907)	3020
2721	WEGSCHELDER R., MEHL J., Monatsh. Chemie, **49**, 289 (1928)	2851
2722	WEIMARN P. P., von, Z. physik.Chem., **76**, 218 (1941)	877
2723	WEITZ, HEUBAUM, Ber., **66**, 790 (1933)	1908
2724	WELLS K. C., J. Wasch-Acad. Sci., Baltimore, **5**, 619 (1915), cited in [1]	1707
2725	WELLS R. C., J. Am. Chem. Soc., **37**, 1705 (1915)	1639, 1662
2726	WELTON R. C., KING G. B., J. Am. Chem. Soc., **61**, 1251 (1939)	2567
2727	WENDROW B., KOBE K. A., Ind. Eng. Chem., **44**, 6, 1439 (1952)	2603
2728	WENGER P., Dissertation, Geneve (1911), cited in [2]	2052, 2057, 2134, 2136, 2159, 2448, 2450, 2451
2729	WESTON A., J. Chem. Soc., **121**, 1223 (1922)	2304, 2317, 4059
2730	WESTON A., J. Chem. Soc., **121–122**, 1233 (1925)	2309
2731	WHEELER H. L., Z. anorg. Chem., **3**, 432 (1893)	1068, 1095, 1225, 1229, 1236
2732	WHITBY G. S., Z. anorg. Chem., **67**, 107 (1910)	1085
2733	WHITBY L., Trans. Faraday Soc., **31**, 647 (1935)	1418, 3363, 3576, 3975
2734	WHITE A. Mc. L., J. Am. Chem. Soc., **55**, 3182 (1933)	5204
2735	WHITMAN W. G., RUSSEL R. B., DAVIS G. H. B., J. Am. Chem. Soc., **47**, 70 (1925)	2510, 3464, 3782, 4167, 5146
2736	WHITNEY W. R., MELCHER A. C., J. Am. Chem. Soc., **25**, 78 (1903)	1992, 1998, 2009
2737	WIBAUT J. P., Chem. Weekblad, **6**, 401, (1909)	2268
2738	WIEBE R., GADDY V. L., J. Am. Chem. Soc., **57**, 847, 1487 (1935)	43
2739	WILCOX, BAILEY, (1927), cited in [2]	5236
2740	WILLARD H. H., KASSNER J. L. (1926), cited in [2]	536, 1179
2741	WILLARD H. H., KASSNER J. L., J. Am. Chem. Soc., **52**, 2381 (1930)	386, 416, 535, 537, 541, 545, 712, 1160, 1175, 1176, 1178, 1180
2742	WILLARD, Inorganic Synthesis, **1**, p. 173, Mc-Graw Hill, London, (1939)	392
2743	WILLIAMS M. D., FOGG H. G., JAMES C., J. Am. Chem. Soc., **47**, 297 (1925)	1077, 1123

Reference number	Author's name and reference	Table number
2744	WILKERSON L., BATHHURST N. O., PARTON H. N., Trans. Faraday Soc., **33**, 623 (1937)	1144, 2551, 2582
2745	WILSMORE, Z. physik. Chem., **35**, 305 (1900)	889, 4773
2746	WILSON E. O., YÜ CH'IH–CH'IU, Ind. Eng. Chem., **26**, 1101 (1934)	1617, 1618, 3582
2747	WINIFRED K. W., BAYLCY C. R., J. Chem. Soc., **1**, 151 (1927)	5237
2748	WINTELER F., Z. Elektrochem., **7**, 360 (1900)	2704, 3541, 4396, 4787
2749	WIRTH F., Z. anorg. Chem., **76**, 174 (1912)	777, 778 881, 884, 916
2750	WIRTH F., Z. anorg. Chem., **79**, 357 (1912–1913)	821, 833
2751	WIRTH F., BAKKE B., Z. anorg. Chem., **87**, 29, 47 (1914)	4247, 4338
2752	WISE C R., J. Am. Chem. Soc., **45**, 1233 (1923)	5326
2753	WISE W. C. A., DAVIES C. W., J. Chem. Soc. (London), 273 (1938)	5191
2754	WÖHLER L., SCHÄFFER W., Z. anorg. Chem., **149**, 391 (1925)	2138
2755	WOLF, KRAUSE, Archiv für Wärmewirtschaft, **8**, 218 (1927), cited in [1]	727, 934, 1633, 1720
2756	WOLINGER E., Dissert. Beiträge zur Kenntnis der ozean. Salzablagerungen, 23–26 (1913), cited in [2]	4043
2757	WOOD J. K., J. Chem. Soc. (London), **93**, 412 (1908)	1061
2758	WOODWARD A. E., ALLEN E. R., ANDERSON R. H., J. Phys. Chem., **60**, No. 7, 939 (1956)	5194
2759	WORLEY F. P., J. Chem. Soc. (London), **87**, 1107 (1905)	144
2760	WRIGHT, THOMSON, Phil. Mag., (5), **17**, 288; **19**, 1 (1884–1885), cited in [2]	4773
2761	WRIGHT A., ASKEW H., Trans. and Proc. New Zeal., Inst., **60**, 267 (1929)	2745, 5061
2762	WURMSER M., Compt. rend., **174**, 1466 (1922)	2145, 2166, 2762
2763	YIN CHING–CHIN, SHEN YUAN-LUNG, SHEN JIEN-KING, CHEN HSIANG-CHUAN, LIU TUNG, Acta Chim. Sinica, **30**, No. 6, 572 (1964)	4432–4434, 4469
2764	YOUNG R. S., GOLLEDGE A., Ind. Eng. Chem., **39**, 1299 (1947)	2012
2765	YUCATIS P., Z. anorg. Chem., **220**, 257 (1934)	2593, 4219
2766	ZAHORSKY, Z. anorg. Chem., **3**, 41 (1893)	5044
2767	ZAMBONINI F., CARIBBI G., Atti accad. Lincei, (5), **33**, 11, 301 (1924)	4763
2768	ZAMBONINI F., CAGLIOTTI V., Atti accad. Lincei, (5), **33**, 308 (1924)	4772
2769	ZAMBONINI F., CAGLIOTTI V., Atti accad. Lincei (6), **2**, 153 (1925)	5547
2770	ZAMBONINI F., CAROBBI G., Atti accad. Lincei, (6), **1**, 278 (1925)	5539
2771	ZAMBONINI F., CAROBBI G., Atti accad. Lincei, (6), **2**, 300, 374 (1925)	3500, 3505
2772	ZAMBONINI F., RESTAINO S., Atti accad. Lincei, (6), **3**, 178 (1926)	2370
2773	ZAMBONINI F., RESTAINO S., Atti accad. Lincei, (6), **5**, 175 (1926)	4767, 4768, 5542
2774	ZAMBONINI F., STOLFI A., Atti accad. Lincei, (6), **4**, 424 (1926)	2367
2775	ZAMBONINI F., STOLFI A., Atti accad. Lincei, (6), **5**, 832 (1927)	2373

Reference number	Author's name and reference	Table number
2776	ZAMBONINI F., RESTAINO S., Atti accad. Lincei, (6), **9**, 131 (1928)	2372
2777	ZAMBONINI F., RESTAINO S., Atti accad. Lincei, (6), **9**, 131; **10**, 774 (1929)	5461
2778	ZAMBONINI F., RESTAINO S., Atti accad. Lincei, (6), **13**, 650 (1931)	5546
2779	ZAMBONINI F., RESTAINO S., Atti accad. Lincei, (6), **14**, 69 (1931)	3504
2780	ZATLOUKAL J., JÄGER L., MACHALA J., JERMER Z., Chem. prumysl., **9 (34)**, No. 6, 304 (1959)	1424
2781	ZERNIKE J., NAWAB A., AZIZ A., Rec. trav. chim., **70**, 784 (1951)	147
2782	ZIELER H., Z. anorg. Chem., **162**, 190 (1927)	1201